This book series is published on behalf of the American Physiological Society (APS) by Springer. Access to APS books published with Springer is free to APS members.

APS publishes three book series in partnership with Springer: *Physiology in Health and Disease* (formerly *Clinical Physiology*), *Methods in Physiology*, and *Perspectives in Physiology* (formerly *People and Ideas*), as well as general titles.

Physiology in Health and Disease

Published on behalf of the American Physiological Society
by Springer

Ronald F. Tuma • Jerome W. Breslin
Shayn Peirce-Cottler • Walter Lee Murfee
Editors

Handbook of Microcirculation

Third Edition

Springer

american physiological society

Editors
Ronald F. Tuma
Emeritus Professor of Physiology
and Neurosurgery, Lewis Katz
School of Medicine
Temple University
Philadelphia, PA, USA

Jerome W. Breslin
Department of Molecular Pharmacology
and Physiology, Morsani College of Medicine
University of South Florida
Tampa, FL, USA

Shayn Peirce-Cottler
Department of Biomedical Engineering
University of Virginia
Charlottesville, VA, USA

Walter Lee Murfee
Department of Biomedical Engineering
University of Florida
Gainesville, FL, USA

ISSN 2625-252X ISSN 2625-2538 (electronic)
Physiology in Health and Disease
ISBN 978-3-031-99608-5 ISBN 978-3-031-99609-2 (eBook)
https://doi.org/10.1007/978-3-031-99609-2

Jointly published with The American Physiological Society

This Springer imprint is published by the registered company Springer Nature Switzerland AG
The registered company address is: Gewerbestrasse 11, 6330 Cham, Switzerland

Introduction: Updating Our Understanding to Motivate Future Opportunities

The existence of the microcirculation was in a sense postulated in 1628 by William Harvey when explaining that the circulation of blood required the flow of blood through the "porosities of the flesh" with the first observation of microvessels being reported in 1661 by Marcello Malpigi when describing the capillaries in the lung of frogs. Antoni van Leewenhoek followed up these observations in 1688 by describing capillary perfusion in the fin of a glass eel. Van Leewenhoek's work is even more remarkable when appreciating that his quantitate blood velocity measurement obtained a value of 2 mm/s, a value well within the range of velocities measured by modern instrumentation. This remains an amazing and a humble reminder of what can be learned through curiosity and scientific ingenuity. During the eighteenth and nineteenth century, numerous investigators provided descriptions of microvessel structure, and interest in microcirculatory physiology was greatly enhanced by the publications of August Krogh, for which he was rewarded the Nobel Prize in 1920. Importantly Krogh explored the relationship between the regulation of capillary perfusion and tissue need for the delivery of oxygen. The pioneering efforts of these early investigators led to a progressive increase in the study of the microcirculation over the years, with thousands of papers focusing of the microcirculation now published each year.

Today we appreciate that the microcirculation plays a central role in all major cardiovascular functions. In addition to serving as the primary site of exchange across the vasculature, phenomena regulated at the level of the microcirculation include control of total peripheral resistance and therefore blood pressure, regulation of blood flow, and inflammatory responses. Alterations in microvascular function are a major contributor to a host of if not all diseases. Understanding the function of the microcirculation continues to be an essential element of cardiovascular physiology and serves to link fields including immunology, engineering, development, and stem cell biology. This third edition of the handbook updates the continued emergent impact of microvascular research for the future. Individual chapters overview microvascular structure and function across cell, vessel, and network scales; highlight the importance of organ specificity; and identify the intersection of the microcirculation

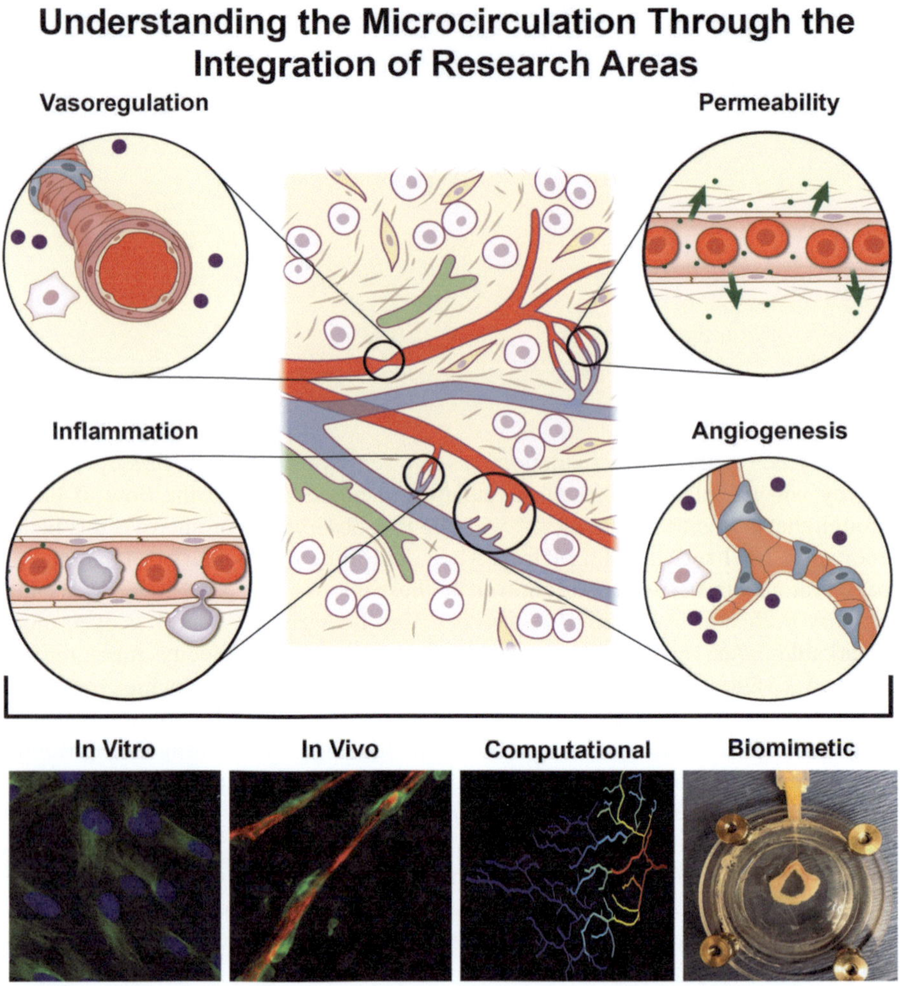

Fig. 1 Understanding the microcirculation requires integration of multiple systems, sub-processes, and microvascular research areas that incorporate interactions of common cell types. For example, angiogenesis involves the sprouting of new capillaries from existing ones and is achieved via the interactions of endothelial cells, pericytes, and interstitial cell populations that secrete and respond to local growth factors. The differentiation of pericytes into smooth muscle cells and associated vessel maturation leads to functional vasoregulation. Other cell dynamics related to these microvascular processes include permeability and white blood cell trafficking associated with inflammation. The motivation to understand the dynamic structure and function relationships between the various cell types across spatiotemporal scales and both physiological and pathological scenarios has led to the application of in vitro, in vivo, computational, and biomimetic modeling approaches

with lymphatic and nervous systems. A focus on cellular dynamics during pathophysiology and integrative approaches that involve new technologies further serve to emphasize the interdisciplinary nature of microvascular research.

With the goal to provide most comprehensive reference about the microcirculation covering emerging topics that link microvascular research to basic science and clinical impacts, the current work reflects the perspectives of four presidents of The Microcirculatory Society (MCS) as Editors, whose combined expertise spans computational modeling to vascular biology, and international microvascular research leaders. The compilation of chapters offers an up-to-date definitive reference about the microcirculation.

In conclusion the editors would like to express their sincere gratitude to each of their research advisors and mentors: Dr. Thomas Skalak and Dr. Brian R. Duling (Shayne Peirce Cottler), Dr. Walter N. Durán and Dr. Sarah Y. Yuan (Jerry Breslin), Dr. Thomas Skalak and Dr. Geert Schmid-Schoenbein (Walter Lee Murfee), and Dr. Mary P Wiedeman and Dr. Karl E. Arfors (Ron Tuma). Their guidance and inspiration created a long-lasting passion for microvascular research in each of us. We would like to also express our gratefulness to all authors, who contributed to the first two handbook volumes, and to all of our collaborators, colleagues, and MCS members, who have pushed us along (Fig. 1).

Philadelphia, PA, USA — Ronald F. Tuma
Tampa, FL, USA — Jerome W. Breslin
Charlottesville, VA, USA — Shayn Peirce-Cottler
Gainesville, FL, USA — Walter Lee Murfee

Contents

Chapter 1
Blood Flow in Microvascular Networks

Axel R. Pries and Timothy W. Secomb

Abstract A microvascular network is a system of conduits that distributes blood throughout tissues. Flow of blood is driven by arterial pressure. Blood is a concentrated suspension of red blood cells (RBCs), and its apparent viscosity changes depending on hematocrit, tube diameter, and flow rate. Red blood cells do not divide in proportion to flow in diverging bifurcations, leading to wide variations in microvascular hematocrits. Microvessels are lined by a relatively thick endothelial surface layer (ESL) that substantially increases resistance to blood flow and affects the interaction of blood cells with the vessel wall. The stochastic nature of growth and remodeling of microvascular networks leads to substantial heterogeneity in their topological and morphological structures. To achieve adequate transport functions, including oxygen delivery to tissues, this heterogeneity must be compensated for by active biological processes that control network and vessel geometry, including structural adaptation and control of vascular tone. In the resulting network structures, most parameters describing microvessel geometry and flow show wide dispersions and significant correlations. Consequently, estimation of system functions based on average or typical parameter values may result in incorrect or misleading conclusions. Understanding these phenomena requires analysis of their consequences at the network level. The study of blood flow in microvascular networks represents a key step in the process of translating advances in molecular and cellular biology into improved understanding of cardiovascular function in health and disease.

Supported by Deutsche Forschungsgemeinschaft (Ga 225/5-3, Pr 271/1-1, 1-2 & 5-1) and by NIH Grants HL034555, HL070657, and HL133362.

A. R. Pries
Charité – Universitätsmedizin Berlin, Institute of Physiology, Berlin, Germany

German Center for Cardiovascular Research (DZHK), partner site Berlin, Berlin, Germany

T. W. Secomb (✉)
Department of Physiology, University of Arizona, Tucson, AZ, USA
e-mail: secomb@arizona.edu

R. F. Tuma et al. (eds.), *Handbook of Microcirculation*, Physiology in Health and Disease, https://doi.org/10.1007/978-3-031-99609-2_1

Keywords Capillaries · Endothelium · Flow regulation · Hematocrit · Heterogeneity · Microcirculation · Oxygen transport · Red blood cells · Vascular remodeling

1.1 Introduction

Terminal vascular beds bring flowing blood into close proximity with parenchymal cells. The primary function of the circulatory system is to transport materials to and from all parts of the body. Transport within the blood occurs by convection, which is effective for moving substances over large distances. Exchange of materials between blood and parenchymal cells mainly occurs by passive diffusion, which is effective only over relatively short distances. Therefore, to provide adequate transport, the blood vessels must form a dense network within tissues. This constraint is particularly stringent with regard to oxygen diffusion. The maximum diffusion distance of oxygen (L) may be simply estimated by considering one-dimensional diffusion into an oxygen-consuming tissue (Secomb et al. 1993):

$$L = (2D\alpha P_0/M)^{1/2} \tag{1.1}$$

where D and α are the diffusivity and solubility of oxygen in the tissue, respectively, P_0 is the partial pressure of oxygen in the tissue immediately adjacent to the vessel, and M is the rate of oxygen consumption in the tissue, assumed to be uniform. Diffusion distances estimated using this formula range from about 20 μm in tissues with very high oxygen demand, such as maximally working skeletal muscle, to about 200 μm in vascularized tissues with low demand and 2–3 mm in nonvascularized cartilage (Lee et al. 2011). The architecture of the microvascular network, including vessel-to-vessel spacing and vascular density, is therefore a crucial determinant of the maximum rate of oxygen consumption in a tissue (Krogh 1919a; McGuire and Secomb 2001; Popel 1989; Roy and Popel 1996).

Blood is a suspension containing a high volume fraction of red blood cells (RBCs or erythrocytes) and other suspended elements. RBCs increase the oxygen-carrying capacity of blood, which would otherwise be relatively small because the solubility of oxygen in water is low. With a volume fraction (hematocrit) typically in the range 40–45%, RBCs exert a large influence on the flow properties of blood. Other suspended elements, including white blood cells (leukocytes) and platelets, play critical biological roles but are normally present at much lower volume fractions and therefore have less effect on flow properties.

The diameters of blood vessels forming the microcirculation range from a few hundred microns down to a few microns, while the diameter of a freely suspended human RBC is about 8 μm. So, when considering blood flow in microvessels, the particulate nature of blood and the finite dimensions of RBCs should be taken into account. This contrasts with the situation in larger blood vessels, where blood can be generally considered as a continuous material (a continuum) whose physical behavior is adequately described in terms of bulk properties such as viscosity. The particulate nature of blood results in uneven distribution of RBCs and plasma at diverging branch points (bifurcations) in the network such that the downstream branches may receive different hematocrits. Microvessels are lined by endothelial cells that present a specialized surface to flowing blood. Adjacent to the endothelial cells toward the vessel lumen is the relatively thick (approximately 0.5–1 μm) endothelial surface layer (ESL) of macromolecules that has significant impact on blood flow.

In this chapter, factors affecting blood flow in single microvessels and bifurcations are discussed in Sect. 1.2 and the phenomena associated with the complex and heterogeneous structures of microvascular networks are addressed in Sect. 1.3.

1.2 Flow in Single Microvessels and Bifurcations

1.2.1 Flow Resistance, Poiseuille's Law, and Apparent Viscosity

Flow in blood vessels is primarily driven by the hydrostatic pressure generated by the beating heart. The relationship between the rate of flow and the driving pressure, whether of a single vascular segment or of the entire peripheral circulation, is conveniently expressed in terms of the flow resistance R:

$$R = \frac{\Delta P}{Q} \tag{1.2}$$

where ΔP is the pressure difference driving the flow and Q is the volume flow rate. The flow resistance determines the pressure required to sustain a given flow rate and the rate of work that must be done, which is given by

$$W = \Delta P \cdot Q = R \cdot Q^2 \tag{1.3}$$

Within a network of vessels, the distribution of flow and pressure drop is determined by the distribution of flow resistances, similar to how electrical resistances affect current flow. In a set of parallel flow pathways subjected to a given pressure drop, for example, the flow in each pathway is inversely proportional to the flow resistance of that pathway. In a flow pathway consisting of several elements in series, all carrying the same flow, the pressure drop across each element is linearly proportional to the flow resistance of that element. Knowledge of the factors determining the flow resistance of vessel segments is therefore important to understand the distribution of blood flow in the circulatory system.

In the nineteenth century, Jean-Louis-Marie Poiseuille performed direct measurements of arterial and venous blood pressure (Poiseuille 1828, 1830) and found high pressure in arterial and low pressure in venous vessels irrespective of the size of the vessel. From these results, he concluded that the pressure drop in the circulation must mainly occur in the peripheral vascular beds. Direct pressure measurements in the small vessels of the microcirculation were not feasible at that time or indeed for another 100 years (Landis 1930). Instead, Poiseuille investigated the passage of simple fluids through long, narrow bore tubes (Poiseuille 1840; Pappenheimer 1984) to indirectly investigate the source of flow resistance in the circulation.

The main findings of Poiseuille's meticulous work can be represented by the following equation, which was actually derived later but is generally known as Poiseuille's law:

$$Q = \frac{\pi}{128} \frac{\Delta P D^4}{L\eta} \tag{1.4}$$

This equation gives the flow (Q) through a cylindrical tube as a function of the driving pressure difference along the tube (ΔP), the tube diameter (D), and length (L). Dynamic viscosity (η) is the material property of the fluid that describes its internal resistance to shearing motions, in which different parts of the fluid move with different velocities. Flow resistance is then given by:

$$R = \frac{128}{\pi} \frac{L\eta}{D^4} \tag{1.5}$$

Flow resistance is therefore inversely proportional to the fourth power of vessel diameter. This sensitive dependence on diameter has important consequences for the distribution and control of blood flow in the circulatory system.

The derivation of Poiseuille's law depends on a number of simplifying assumptions. The tube is assumed to be uniform with a circular cross section. The flow is assumed to be steady (unvarying in time) and laminar (free of turbulence), and the fluid is assumed to be Newtonian, with a fixed viscosity. In practice, these conditions are rarely if ever satisfied in the circulatory system. Nonetheless, Poiseuille's law is often a useful approximation and represents a logical starting point for any analysis of blood flow in microvessels. The Reynolds number for blood flow in microvessels is generally in the range of 10^{-3} to about 1. Therefore, the flow is laminar. The effects of fluid inertia may be neglected in many cases, particularly in smaller microvessels.

The dependence of the flow rate on tube radius according to Eq. (1.4) may be understood as follows: The fluid velocity in Poiseuille flow shows a parabolic variation with radial position, being maximal at the center of the tube and dropping to zero at the wall (i.e., the velocity profile is parabolic). The coefficient of this quadratic depends only on the pressure gradient and viscosity. Consequently, the center-line velocity increases as the square of the radius. The volume flow rate is obtained by integrating the velocity over the cross-sectional area, which itself increases in proportion to radius squared. Overall, flow increases as the fourth power of radius if the pressure gradient is held constant. For formal derivations of Poiseuille's law, see Ethier and Simmons (2007) and Secomb (2016).

The pressure force driving a fluid along a cylindrical tube is balanced by the shear stress generated by the velocity gradient of the fluid at the tube wall. This balance of forces leads to the following equation for wall shear stress:

$$\tau = \frac{D\Delta P}{4L} \tag{1.6}$$

independent of the fluid properties. Combining this with Poiseuille's law gives

$$\tau = \frac{32\eta Q}{D^3} \tag{1.7}$$

For fluids such as water consisting of a single physical phase, viscosity η can be treated as a constant at a given temperature, independent of the rate of flow or deformation. Such fluids are referred to as Newtonian. However, normal blood contains a high concentration of macromolecules and suspended RBCs, which strongly affect its flow properties. The resulting phenomena fall under two main categories: non-Newtonian effects and non-continuum effects.

Non-Newtonian effects refer to the dependence of viscosity on the rate of fluid deformation. Such effects can be observed in the bulk behavior of blood, when its viscosity is measured in a flow system that is large compared to the dimensions of individual RBCs. Measurements made in rotational viscometers show that the bulk viscosity of blood decreases with increasing shear rate (Chien et al. 1966). In such devices, blood is subjected to a nearly uniform shearing flow. Shear rate is defined as the gradient of velocity within a fluid, i.e., the difference in velocity between the two surfaces containing the fluid divided by the distance between them. Two main causes for non-Newtonian properties have been identified (Chien et al. 1966; Chien 1975). At very low shear rates, aggregation of RBCs causes increased viscous resistance to flow. As the shear rate increases, these aggregates are broken down and viscosity decreases. In the higher range of shear rates, RBCs are increasingly deformed by flow forces, allowing them to interfere less with the flow and resulting in a further decrease in viscosity with increasing shear rate. These studies have also shown a strong nonlinear dependence of viscosity on hematocrit, with viscosity increasing as hematocrit is increased (Chien et al. 1966).

When blood flows through microvessels, the dimensions of blood cells are comparable to the vessel diameters (Figs. 1.1, 1.2, and 1.3), and non-continuum effects consequently become significant. Cell deformability and aggregation do not necessarily have the same effects that they have in bulk shear flow, as discussed below.

Fluid mechanical interactions between RBCs and vessel walls can have a major effect on flow properties, including resistance to blood flow in narrow tubes. For the purpose of describing such effects, it is useful to define the apparent or effective

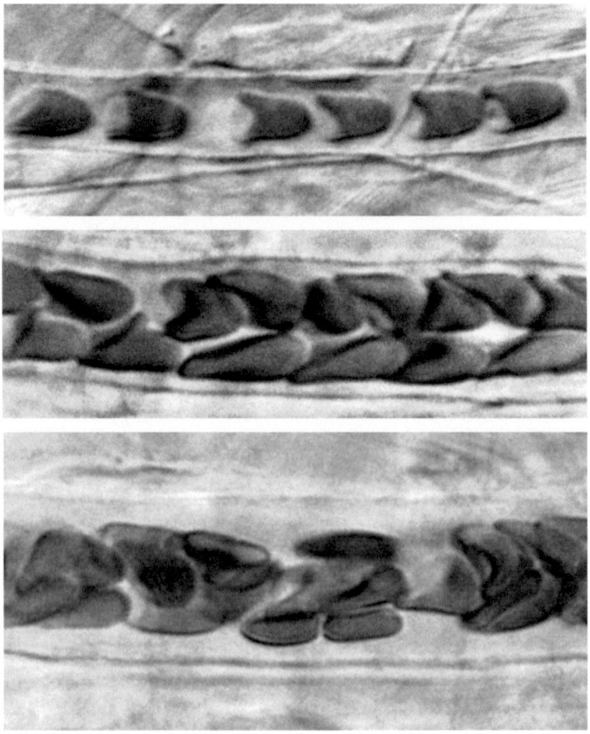

Fig. 1.1 Blood flow through microvessels in the rat mesentery with 7-, 12-, and 16-μm inner diameters (top to bottom). Flow is from left to right

viscosity of blood, i.e., the viscosity of a Newtonian fluid that would give the same volume flow rate for a given tube geometry and driving pressure. According to Eq. (1.3), the apparent viscosity is given by:

$$\eta_{\text{app}} = \frac{\pi}{128} \frac{\Delta P D^4}{LQ} \tag{1.8}$$

The apparent viscosity of blood in microvessels depends not only on the material properties of blood but also on other factors such as the dimensions of the flow channel and flow velocity. Relative apparent viscosity is defined as the apparent viscosity of a fluid divided by its suspending medium's viscosity. Non-continuum effects on blood flow are discussed in the following sections.

1.2.2 Cell-Free Layer

An important consequence of the particulate, non-continuum nature of blood is the formation of cell-free or cell-depleted regions near the vessel walls. It is observed that flowing RBCs are excluded from or infrequently present in a region adjacent to

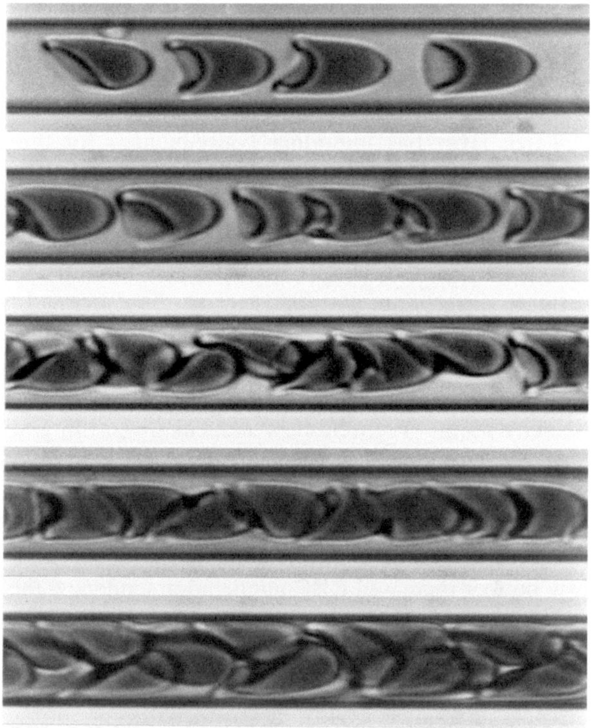

Fig. 1.2 Human erythrocytes during flow through a glass tube with an inner diameter of 7 μm. The discharge hematocrits are 0.2, 0.34, 0.41, 0.52, and 0.65 (top to bottom). Flow is from left to right

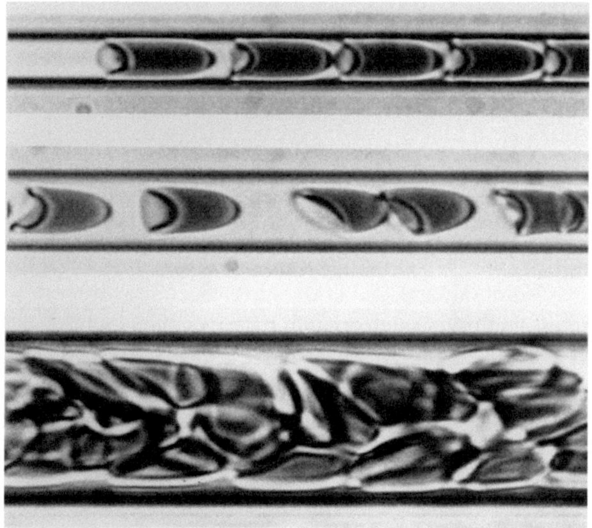

Fig. 1.3 Human erythrocytes during flow through glass tubes with 4-, 7-, and 17-μm inner diameters (top to bottom). Flow is from left to right

the endothelium, whose width varies but is typically at least 1 μm. Here, the term "cell-free layer" is used to describe this region, even if some RBCs are present there. Observations of the cell-free layer and speculations concerning its causes date back to the nineteenth century and even earlier (Goldsmith et al. 1989). The presence of a cell-free layer has important consequences for the flow properties of blood in microvessels (Kim et al. 2009), which are discussed in the following sections.

Several mechanisms contribute to the formation of the cell-free layer (Secomb 2017), as schematically illustrated in Fig. 1.4. (a) Because of the finite size of RBCs, the center of the mass of each cell cannot approach the wall closer than a certain distance. This minimum distance can be estimated as about 1 μm, based on the unstressed shape of an RBC, which is a disk about 2-μm-thick. (b) RBCs and other deformable particles tend to migrate away from solid boundaries when suspended in a shear flow (Goldsmith 1971; Grandchamp et al. 2013). (c) The curvature of the velocity profile in a tube flow induces inward migration, independent of the influence of solid boundaries (Kaoui et al. 2008). (d) Living microvessels are lined by a macromolecular layer that impedes plasma flow and largely excludes flowing RBCs (Pries et al. 2000; Weinbaum et al. 2007). This phenomenon, which tends to reduce the thickness of the layer, is discussed in Sect. 1.2.6. (e) Under conditions of very low flow or increased RBC aggregation tendency, the formation of a central aggregated core region in the vessel can substantially increase the width of the cell-depleted region (Reinke et al. 1987). (f) In shear flow of concentrated suspensions, particle–particle interactions induce random lateral particle motions, whose net effect is to produce motion down the concentration gradient, toward the wall, in a phenomenon known as shear-induced dispersion (Leighton and Acrivos 1987; Pranay et al. 2012), which leads to a decrease of the cell-free layer's width.

Numerous experimental studies on the cell-free layer, both in vitro and in vivo, have been carried out (Kim et al. 2009; Ng et al. 2016; Maeda et al. 1996). Figure 1.5 shows the distributions of hematocrit observed in the arterioles of the rat mesentery (Pries et al. 1983), with corresponding data for flow velocity and estimated local blood viscosity (Long et al. 2004). The absence of RBCs near the wall results in a region of reduced viscosity and a relatively high velocity gradient (i.e., shear rate). The shape of the velocity profile is consequently blunted compared to the parabolic profile that is characteristic of Poiseuille flow for homogeneous Newtonian fluids.

Computational methods have been developed for simulating the motion of multiple flowing deformable particles such as RBCs (Dupin et al. 2007; McWhirter et al. 2009; Fedosov et al. 2010; Freund 2014). These simulations have shown that formation of a cell-free layer and other flow phenomena in microvessels can be understood in fluid mechanical terms, based on the known mechanical properties of individual RBCs (Secomb 2017).

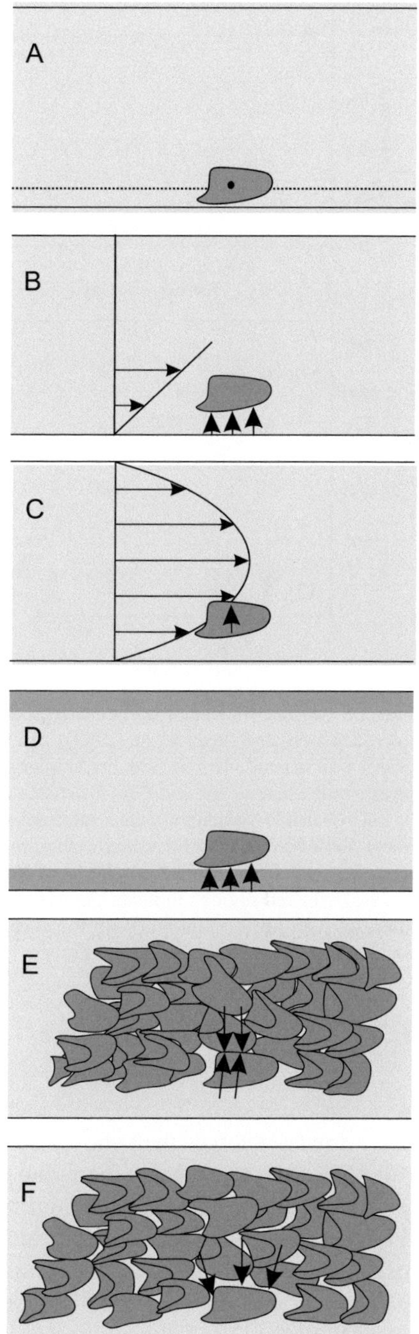

Fig. 1.4 Schematic illustration of the mechanisms determining formation of a cell-depleted layer. (**a**). Exclusion due to the finite size of RBCs. (**b**). Deformable particles tend to migrate away from solid boundaries in shear flow. (**c**). Curvature of the velocity profile. (**d**). Endothelial surface layer. (**e**). Aggregation in very slow flow or increased aggregation tendency. (**f**). Shear-induced dispersion. See text for explanation. *RBCs* red blood cells

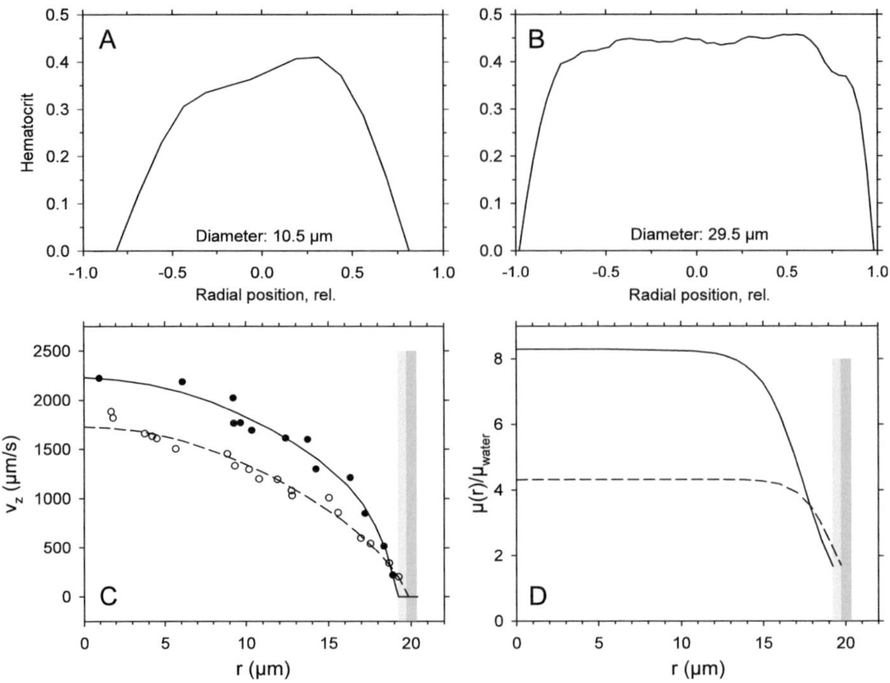

Fig. 1.5 (**a** and **b**) Radial hematocrit profiles in arterioles of the rat mesentery determined by microdensitometry (Pries et al. 1983). The absolute dimension of the cell-depleted layer at the vessel wall is similar for the smaller and larger arterioles. However, the impact of the cell-depleted layer on tube hematocrit and flow resistance is larger in smaller vessels. (**c** and **d**). Profiles obtained by microparticle velocimetry in a rat cremaster venule (diameter ~ 40.5 μm) before (continuous lines) and after (dashed lines) hemodilution (Long et al. 2004). (**c**). Flow velocity. (**d**). Relative viscosity, which correlates with the local hematocrit at a given radial position. The vertical dark- and light-shaded regions indicate the width of the endothelial surface layer before (0.82 μm) and after (0.62 μm) hemodilution. With the exception of this layer, the velocity profiles are similar for microvessels and similar-sized glass tubes

1.2.3 The Fåhraeus Effect

As a consequence of its proximity to the vessel wall, the plasma in the cell-free layer has a lower velocity than the blood in the cell-rich central region of the vessel (Fig. 1.5). More generally, the radial velocity profile exhibits a positive covariance with the radial profile of RBC concentration. The average velocity v_c of RBCs is, therefore, higher than the average velocity v_b of the blood. Fåhraeus (Fahraeus 1928, 1929) realized that this disparity in velocities leads to a change in the volume concentration of RBCs within the tube (tube hematocrit, H_T), relative to the volume concentration in the fluid entering or leaving the tube (discharge hematocrit, H_D). If the mean RBC velocity is higher than the mean flow velocity of the suspension, RBCs pass through the tube faster and their transit time is correspondingly reduced, leading to a reduction in tube hematocrit relative to the discharge hematocrit. This phenomenon is known as the Fåhraeus effect.

This effect may be mathematically represented as follows. The discharge hematocrit is given by the ratio of the volume flow rate of RBCs (Q_c) to the flow rate of blood (Q_b):

$$H_D = \frac{Q_c}{Q_b} = \frac{H_T v_c A}{v_b A} \tag{1.9}$$

where A is the cross-sectional area of the tube. Therefore,

$$\frac{H_T}{H_D} = \frac{v_b}{v_c} \tag{1.10}$$

If the distributions $v(r)$ and $H(r)$ of velocity and hematocrit are given as the functions of radial position r, then the mean blood flow velocity and mean RBC velocity are

$$v_b = \int_0^a v(r)\, r\, dr \ \Big/ \int_0^a r\, dr \tag{1.11}$$

$$v_c = \int_0^a v(r) H(r)\, r\, dr \ \Big/ \int_0^a H(r)\, r\, dr \tag{1.12}$$

where a is the tube radius. It follows using Eq. (1.10) that

$$\frac{H_T}{H_D} = \frac{\int_0^a v(r)\, r\, dr \cdot \int_0^a H(r)\, r\, dr}{\int_0^a v(r) H(r)\, r\, dr \cdot \int_0^a r\, dr} \tag{1.13}$$

This relationship shows that H_T is lower than H_D to the extent that $H(r)$ and $v(r)$ exhibit a positive covariance, considered as functions of position within the vessel's cross section.

Based on experimental data (Fig. 1.6) (Albrecht et al. 1979), a parametric description of the Fåhraeus effect in relation to the discharge hematocrit and the tube or vessel diameter was proposed (Pries et al. 1990):

$$\frac{H_T}{H_D} = H_D + (1 - H_D)\left(1 + 1.7e^{-0.415D} - 0.6e^{-0.011D}\right) \tag{1.14}$$

The numerical values given in this equation and the following parametric descriptions of rheological phenomena are obtained with (or scaled to) human blood with a mean corpuscular (RBC) volume (MCV) of 92 fl (MCVh). Scaling to other species with nonnucleated, circular RBCs (e.g., rats and mice) with a different MCV

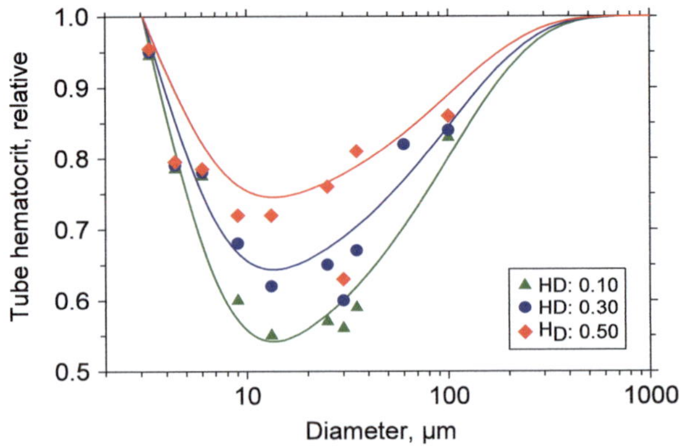

Fig. 1.6 Relation between tube hematocrit (H_T) and discharge hematocrit (H_D) as determined in glass tubes of different luminal diameters (Albrecht et al. 1979). Data for three different levels of H_D are given together with the respective parametric fits according to Eq. (1.14)

(MCVs) may be performed by multiplying the vessel diameters used in the respective equations with the factor $(\text{MCVh/MCVs})^{1/3}$ (e.g., for rat blood with an MCV of 55 fl, the factor would be 1.187) (Pries et al. 1990, 1998a).

1.2.4 The Fåhraeus–Lindqvist Effect

The effects of the cell-free layer and other flow properties of blood on the flow resistance of a microvessel are conveniently discussed in terms of effective or apparent viscosity, as defined in Eq. (1.8). Starting with the works of Martini et al. (1930) and Fåhraeus and Lindqvist (1931), many studies have been conducted on the apparent viscosity of blood and suspensions of blood cells flowing through narrow glass tubes (Reinke et al. 1987; Azelvandre and Oiknine 1976; Barbee and Cokelet 1971; Barbee 1973; Bayliss 1959; Braasch and Jenett 1968; Braasch 1984; Gerbstädt et al. 1966; Gupta and Seshadri 1977; Halikas and Sheppard 1969; Haynes and Burton 1959; Haynes 1960; McKay and Meiselman 1988, 1989; Reinke et al. 1986; Stadler and Linderkamp 1989; Stadler et al. 1990; Voss 1983). The resulting data were assembled in a comprehensive form by Pries et al. (1992a), as shown in Fig. 1.7. From these experimental data, an empirical description of relative apparent viscosity in vitro for suspensions of human erythrocytes as a function of tube diameter and hematocrit was derived according to the equations,

$$\eta_{\text{vitro}} = 1 + (\eta_{0.45\cdot} - 1)\frac{(1 - H_D)^C - 1}{(1 - 0.45)^C - 1} \tag{1.15}$$

$$\eta_{0.45} = 220\exp(-1.3D) + 3.2 - 2.44\exp\left(-0.06D^{0.645}\right) \tag{1.16}$$

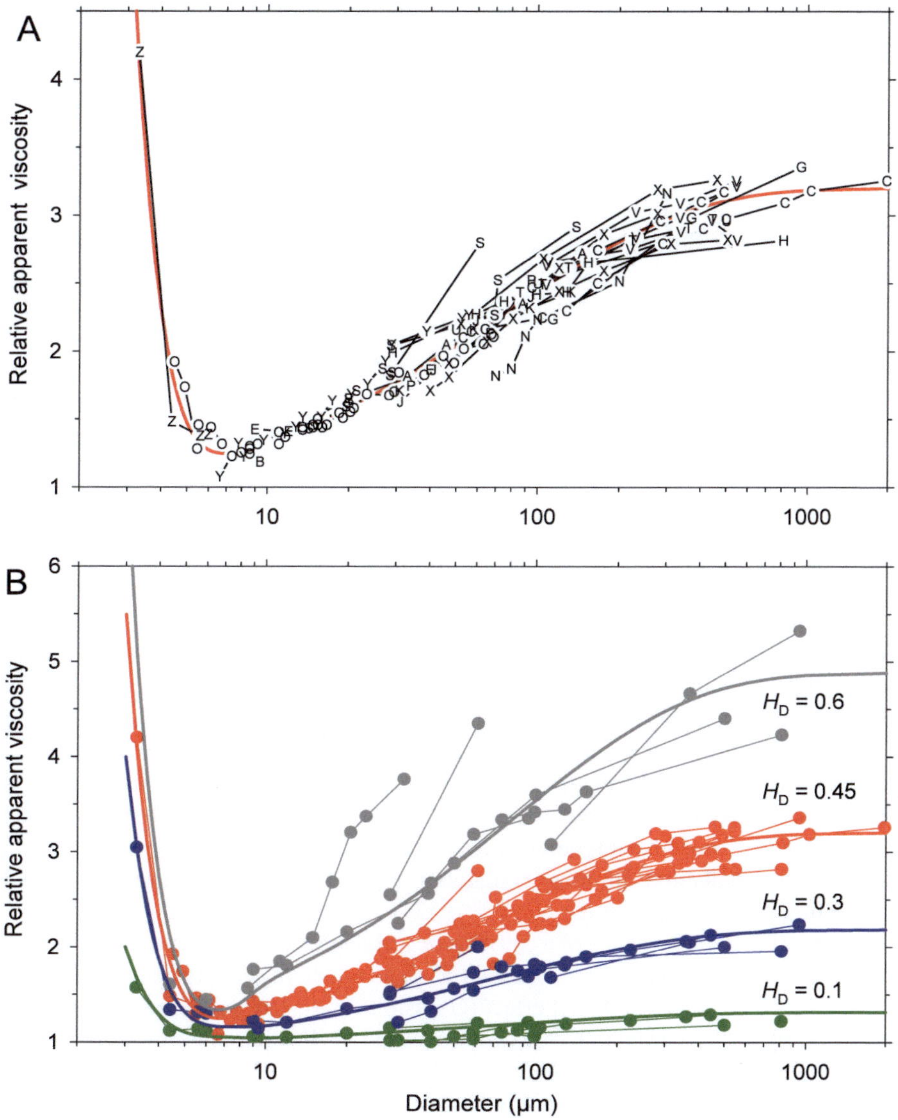

Fig. 1.7 (**a**). Relative apparent viscosity of blood in long, straight glass tubes at a discharge hematocrit of 0.45. Experimental data for human blood (Reinke et al. 1987; Martini et al. 1930; Fahraeus and Lindqvist 1931; Azelvandre and Oiknine 1976; Braasch and Jenett 1968; Gerbstädt et al. 1966; Gupta and Seshadri 1977; Halikas and Sheppard 1969; Haynes and Burton 1959; McKay and Meiselman 1988; Reinke et al. 1986; Stadler et al. 1990; Voss 1983; Pries et al. 1992a; Barbee and Cokelet 1990; Bayliss 1952; Neuhaus et al. 1992) are given together with the results of the parametric fit defined in Eqs. (1.15, 1.16, and 1.17) (scaling to other species can be attempted by scaling the diameter with the cubic root of the ratio of the respective mean RBC volumes). (**b**). Experimental data and approximations for different levels of discharge hematocrit. (Modified from Pries et al. 1992a)

where $\eta_{0.45}$ is the relative apparent blood viscosity for a fixed discharge hematocrit of 0.45, D is the luminal tube diameter in μm, and C describes the dependence of the viscosity on hematocrit:

$$C = \left(0.8 + e^{-0.075D}\right)\left(-1 + \frac{1}{1 + 10^{-11}D^{12}}\right) + \frac{1}{1 + 10^{-11}D^{12}} \tag{1.17}$$

A consistent finding is that apparent viscosity declines substantially with decreasing diameter below about 300 μm. This phenomenon is generally known as the Fåhraeus–Lindqvist effect. Some data are available for tubes with diameters below 7 μm. These results show that apparent viscosity reaches a minimum for diameters of about 5–7 μm and increases with further decreases in diameter (Pries et al. 1992a). Similar trends are observed when suspensions with hematocrit levels above or below the normal range are used. As would be expected, apparent viscosity increases with increasing hematocrit at any given diameter. For diameters in the range 5–7 μm, however, relative apparent viscosity is below 1.5 even at elevated hematocrit levels, showing that the presence of suspended RBCs has remarkably little effect on flow resistance in glass tubes with diameters in this range: the presence of RBCs at a hematocrit of 0.45 increases viscosity by only about 50%.

The mechanisms underlying the Fåhraeus–Lindqvist effect have been much studied and discussed. Barbee and Cokelet (1971) showed that the apparent viscosity of blood flowing in a 29-μm tube could be predicted from macroscopic rheological data, when the reduction in tube hematocrit associated with the Fåhraeus effect is taken into account. This led to the suggestion that the Fåhraeus–Lindqvist effect is a consequence of the Fåhraeus effect. However, this explanation is incomplete, as shown by the fact that apparent viscosity exhibits a minimum at diameters of about 5–7 μm, with somewhat higher values for lower hematocrits, and vice versa. In contrast, the reduction in tube hematocrit is maximal for diameters of about 10–15 μm.

It is clear that the presence of a cell-free or cell-depleted layer near the vessel wall, discussed earlier in connection with the Fåhraeus effect, is also the main underlying cause of the Fåhraeus–Lindqvist effect (Goldsmith et al. 1989). A simple two-phase model of blood flow (Secomb 1995) can be used to illustrate the impact of such a layer on apparent viscosity. In this model, a cylindrical central "core" region containing RBCs (which is neutrally buoyant relative to the plasma), assumed to have viscosity η_c, is surrounded by a layer of plasma with viscosity η_p, where $\eta_c > \eta_p$. An analysis similar to that used to derive Poiseuille's law shows that the apparent viscosity is

$$\eta_{\text{app}} = \frac{\eta_p}{1 - \lambda^4\left(1 - \eta_p/\eta_c\right)} = \frac{\eta_c}{1 + \left(1 - \lambda^4\right)\left(\eta_c/\eta_p - 1\right)} \tag{1.18}$$

where λ is the ratio of the radius of the core region to the tube radius, i.e., $\lambda = 1 - h/r$, where h is the width of the cell-free plasma layer and r is the tube radius. As

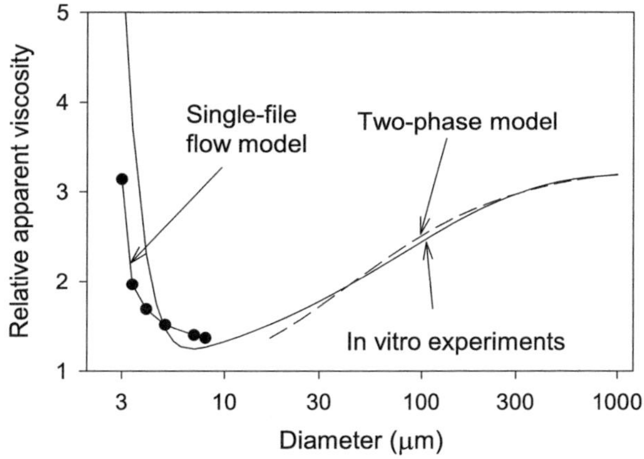

Fig. 1.8 Dependence of relative apparent viscosity on tube diameter. Comparison of experimental data, represented by the parametric description of Eqs. (1.15, 1.16, and 1.17), and predictions of theoretical models. (Modified from Secomb 1995)

h increases, λ decreases from 1. Because of its fourth-power dependence on λ, the factor $1 - \lambda^4$ increases rapidly and apparent viscosity declines from the core value. A relatively narrow plasma layer can have a substantial impact on apparent viscosity because it decreases the local viscosity in the region near the wall where the shear rate is highest and, therefore, where viscous energy dissipation would otherwise be concentrated.

The width h of the cell-free layer in glass tubes is largely determined by hydrodynamic interactions between the RBCs and the tube wall, as discussed in Sect. 1.2.2. Therefore, it might be expected to be approximately independent of tube diameter. If a fixed value of h is assumed, then λ decreases as r increases and a decrease in apparent viscosity with decreasing tube diameter is predicted according to Eq. (1.14). The resulting dependence of relative apparent viscosity on tube diameter is shown in Fig. 1.8. In this case, it is assumed that $\eta_c/\eta_p = 3.3$ and $h = 1.8$ μm. These values were chosen (Secomb 1995) to fit the empirical curve describing the experimental data (Pries et al. 1992a). Good agreement is seen for diameters down to about 30 μm. Detailed numerical simulations of the motion of multiple RBCs have led to predictions consistent with this empirical equation. Fedosov et al. (2010) predicted the probability distribution function of the cell-free layer width, with an average of close to 1.8 μm for a tube diameter of 40 μm and a hematocrit of 45%. The corresponding predictions of apparent viscosity were also in agreement with the experimental results in glass tubes (Fedosov et al. 2010, 2014).

At smaller diameters, however, the prediction of the two-phase model with h independent of diameter does not agree with the experimental data. This may be because the increasing curvature of the wall results in a decrease in the thickness of the cell-free layer or because non-continuum effects become increasingly important. In the range of diameters below 8 μm, RBCs often flow in single file and continuum models are clearly not appropriate. Detailed theoretical analyses of the single-file motion of RBCs (Secomb et al. 1986; Secomb 1987), based on the cells' mechanical properties, provided predictions that were in reasonable agreement with the observed behavior.

1.2.5 Phase Separation in Bifurcations

Blood flowing through a microvascular network encounters numerous successive branch points. The branches are almost always bifurcations, at which blood flow in a parent vessel is divided between two daughter vessels (diverging bifurcation) or flow in two daughter vessels combines to enter a parent vessel (converging bifurcation). When blood flows through a diverging branch point in the microcirculation, the RBCs and plasma may be distributed non-proportionally between the daughter vessels as a consequence of the particulate, non-continuum nature of blood (Fig. 1.9). As a result, discharge hematocrits in the daughter vessels may differ from each other and also from those in the parent vessel. In general, the branch receiving a higher flow rate tends to receive a higher hematocrit, a phenomenon sometimes called the Zweifach–Fung effect (Svanes and Zweifach 1968; Fung 1973; Doyeux et al. 2011). This phenomenon has been extensively investigated in vivo (Svanes and Zweifach 1968; Fung 1973; Johnson 1971; Pries et al. 1989), in artificial bifurcations (Bugliarello and Hsiao 1964; Levine and Goldsmith 1977; Yen and Fung 1978; Pries et al. 1981; Chien et al. 1985; Fenton et al. 1985; Carr and Wickham 1990; Mantegazza et al. 2020; Merlo et al. 2022) and using theoretical simulations (Doyeux et al. 2011; Perkkiö et al. 1986; Audet and Olbricht 1987; Barber et al. 2008; Obrist et al. 2010; Barber et al. 2011; Balogh and Bagchi 2018; Köry et al. 2022). This effect produces a large dispersion in vessel hematocrit within any given microvascular network, as discussed further below.

A major factor in the partial separation of the plasma and RBC phases is the presence of a cell-free or cell-depleted layer at the vessel wall, as already discussed.

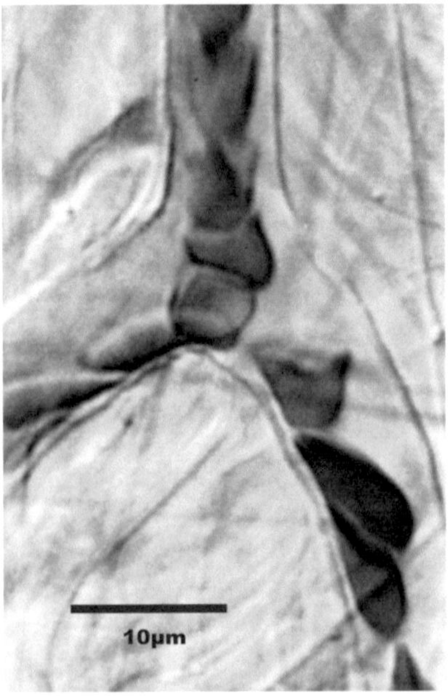

Fig. 1.9 Microvascular bifurcation in the rat mesentery (flow from top to bottom)

If one of the daughter vessels receives a small fraction of the total small flow rate, the fluid entering it is mainly drawn from this peripheral layer in the parent vessel, a phenomenon known as "plasma skimming." The consequence is that the discharge hematocrit in the low-flow branch is typically less than that in the high-flow branch. This effect depends on the nonuniform distribution of hematocrit in the upstream vessel. If it is assumed that the streamlines of plasma flow and the trajectories of RBC motion coincide, then a separating surface can be defined in the parent vessel, such that the plasma and RBCs lying on one side of the surface enter one daughter branch and those on the other side enter the other branch. The discharge hematocrits in the daughter branches are then defined by the fractional contributions of RBCs to the volume flow rates on each side of the separating surface.

With decreasing size of the vessels, the likelihood of significant deviations between RBC trajectories and the streamlines of plasma flow arises because the size of the cells becomes increasingly significant relative to the flow domain. Such deviations can also contribute to phase separation in bifurcations. For instance, cells may have difficulty entering a very small side branch even if they are initially on a streamline that enters the branch. This effect is known as RBC screening (Pries et al. 1981). Thus, plasma skimming results from the upstream establishment of a cell-free layer close to the endothelial surface, while RBC screening is caused by the flow forces at the bifurcation proper.

For theoretical simulations of hematocrit distribution in microvascular networks, a parametric representation of phase separation is needed, which can be applied over a range of vessel diameters and parent vessel hematocrits. The fractional flow rate of RBCs into one daughter branch, FQ_E, can be expressed as a function of the fractional blood flow rate into that branch, FQ_B. In the absence of phase separation, this relationship would be the identity $FQ_E = FQ_B$. With phase separation, a sigmoidal relationship is typically observed, which can be described by an equation of the form:

$$\begin{cases} FQ_E = 0 & \text{if } FQ_B \leq X_0 \\ \operatorname{logit}(FQ_E) = A + B\operatorname{logit}\dfrac{FQ_B - X_0}{1 - 2X_0} & \text{if } X_0 < FQ_B < 1 - X_0 \\ FQ_E = 1 & \text{if } FQ_B \geq 1 - X_0 \end{cases} \tag{1.19}$$

where logit $x = \ln[x/(1 - x)]$. The parameters A, B, and X_0 defining the phase separation characteristics of the bifurcation were obtained by fitting experimental data obtained in the rat mesentery (Pries et al. 1989) (Fig. 1.10). The version given here has small modifications to render predictions more robust for extreme combinations of input hematocrit and diameter distribution (Pries and Secomb 2005) and is supported by a reanalysis of the underlying experimental data (Rasmussen et al. 2017).

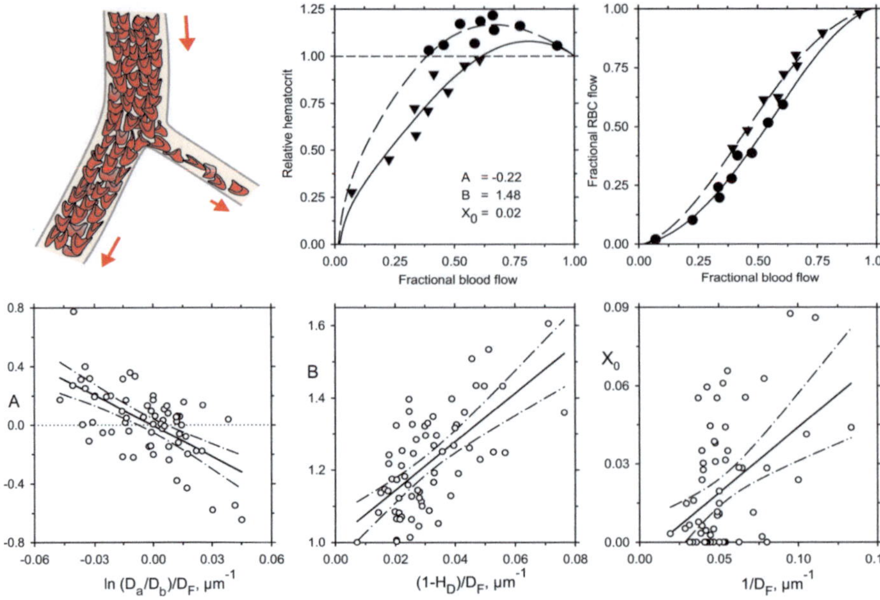

Fig. 1.10 Parameters of RBC distribution at microvascular bifurcations. Upper panels: Schematic drawing of a microvascular bifurcation (left). Data for a single bifurcation: Hematocrit relative to that in the feeding vessel (middle) and fractional erythrocyte flow (right) in the daughter branches versus the fractional blood flow in the respective branch. Experimental data, the fits obtained with Eq. (1.19), and the respective parameters are given. Lower panels: Parameters of the logit fit (*A*, *B*, X_0) obtained for 65 bifurcations in the mesentery plotted versus the relevant combinations of independent variables. The respective linear regression lines are shown with their 95% confidence intervals. (Adapted from Pries et al. 1989. *RBCs*, red blood cells)

$$A = -13.29 \frac{D_\alpha^2/D_\beta^2 - 1}{D_\alpha^2/D_\beta^2 + 1}(1 - H_D)/D_F \tag{1.20}$$

$$B = 1 + 6.98(1 - H_D)/D_F \tag{1.21}$$

$$X_0 = 0.964(1 - H_D)/D_F \tag{1.22}$$

Here, D_α, D_β, and D_F are the diameters of the daughter branches and the mother vessel and H_D is the discharge hematocrit in the mother vessel.

Development of advanced computational methods for motion of multiple RBCs in complex three-dimensional geometries (Freund 2014) has allowed simulations of blood flow in networks with multiple bifurcations. Such simulations have shown significant effects of upstream flow conditions on the partition of hematocrit in diverging bifurcations (Merlo et al. 2022; Balogh and Bagchi 2018; Köry et al. 2022). Side branches at upstream bifurcations lead to a shift of streamlines closer to the wall at the corresponding side of the main downstream vessel, which may cause a reduction of plasma skimming at subsequent bifurcations.

1.2.6 Aggregation and Sedimentation

Under normal conditions, RBCs of humans and some other mammalian species tend to adhere to each other, forming aggregates, often termed "rouleaux." This effect is countered by shear stresses induced by blood flow, which tend to pull aggregates apart. The extent of aggregation, therefore, depends on the balance between flow forces and the aggregation tendency of RBCs (Chien et al. 1984). In arteries and arterioles, shear stresses are usually sufficiently high that aggregation is inhibited. The narrow diameters of capillaries typically do not allow formation of aggregates. Aggregation in the microcirculation is most often observed in postcapillary venules, in which levels of shear stress are relatively low. Conditions of reduced flow increase the likelihood of aggregation, as does enhanced aggregation tendency, resulting, for instance, from changes in blood plasma protein composition.

Two theories have been proposed for the biophysical basis of aggregation: bridging and depletion. According to the first theory (Chien and Jan 1973), aggregation occurs when macromolecules are adsorbed onto the surfaces of adjacent cells, forming a bridge between them. The depletion theory (Bäumler et al. 1996; Neu and Meiselman 2002) proposes that protein concentration is reduced near the surface of the cell, as compared with the suspending medium. As a consequence, an osmotic gradient is set up that tends to withdraw water from the space between two cells, creating a force that tends to hold them together. Several types of evidence, including quantitative studies of aggregation induced by artificial polymers (dextrans and polyethylene glycol) support the depletion theory (Neu and Meiselman 2002; Baskurt and Meiselman 2013).

When blood in bulk is subjected to shear flow, as, for example, in a large-scale cone–plate rheometer, aggregation leads to increased apparent viscosity. As mentioned earlier, aggregation is one of the main factors contributing to the increase in apparent viscosity of blood as shear rate is decreased. However, the effects of aggregation in microvessels are more complex because they can lead to redistribution of RBCs within the vessel cross section and to consequent changes in the width of the cell-depleted layer adjacent to the cell wall.

Experiments performed using vertical glass tubes to limit the impact of aggregation with diameters in the range of 30–132 μm (Reinke et al. 1987) have shown that apparent viscosity decreased with decreasing flow rate over a certain range of shear rates. This effect was reduced when aggregation was inhibited by suspending the RBCs in serum or saline but was accentuated when a dextran molecular weight of 250 kD was added to increase aggregation. Decreases in apparent viscosity are associated with increases in the width of the cell-free layer at the wall. These experiments show that aggregate formation in narrow tubes can draw RBCs away from the walls and thereby decrease flow resistance. However, this effect is dependent on the orientation of the tube (Fig. 1.11). At low flow rates in horizontal tubes, the aggregated core, which is slightly denser than the surrounding plasma, eventually approaches the lower part of the tube because of gravity. In this case, the layer of RBCs may move slower than the overlying layer of plasma, resulting in inversion of

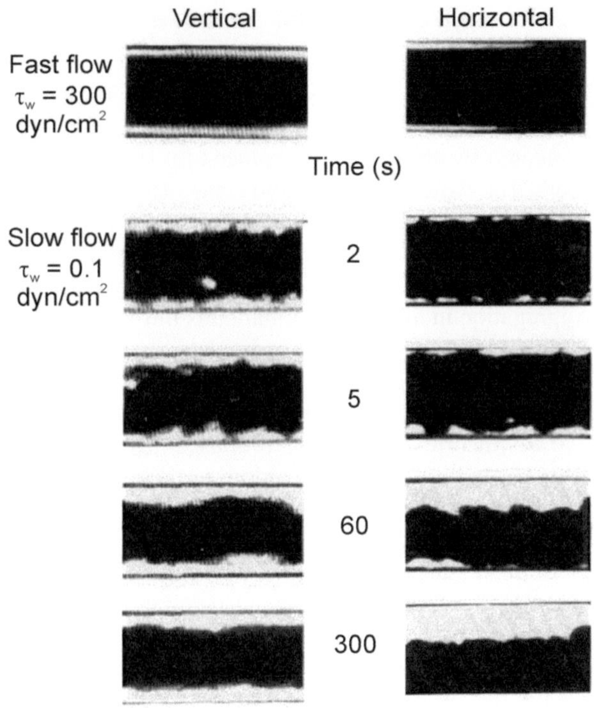

Fig. 1.11 Images of blood flowing in vertical (left) and horizontal (right) glass tubes of diameter 60 μm. Following a sudden reduction of flow rate, a central core of aggregated RBCs is formed, surrounded by enlarged plasma spaces near the tube walls. In the horizontal tube, this aggregated core eventually approaches the lower side of the tube as a result of sedimentation. (Reproduced by permission from Alonso et al. 1995. *RBCs*, red blood cells)

the Fåhraeus effect (Schmid-Schönbein 1988) and a marked increase in apparent viscosity (Alonso et al. 1995).

The significance of aggregation and sedimentation for flow resistance in the intact circulation is complicated by the range of diameters, flow geometries, and flow rates present in the microcirculation and cannot be directly predicted from the results described in the previous paragraph. Available evidence suggests that aggregation can lead to increased flow resistance in the venous circulation. Experiments in skeletal muscle (Cabel et al. 1997) have shown an increase in venous resistance when flow rate was reduced. This effect was reduced or absent when non-aggregating RBC suspensions were used to perfuse the muscle. Observations of velocity profiles in venules with diameters 45–75 μm (Bishop et al. 2001) showed significant blunting, relative to a parabolic profile, at low flow rates in the physiological range. Again, this blunting was not seen with non-aggregating RBC suspensions. A blunted velocity profile is consistent with formation of aggregates in the central region of the flow, increasing blood viscosity in that region.

1.2.7 Endothelial Surface Layer

The interior surfaces of blood vessels are lined by endothelial cells. By scaling anatomical data on the vascular system in the canine intestine (Mall 1888) from a 13-kg canine to a 70-kg human, the total area of the endothelial surface can be estimated as about 350 m^2 (Pries et al. 2000). The endothelium thus presents a large surface area for exchange of materials and is critically involved in many other functions of the microcirculation. Like other cells, endothelial cells are coated with a glycocalyx, a layer of membrane-bound macromolecules, which is typically shown by electron microscopy to be several tens of nanometers thick. A large body of evidence implies the existence of a much thicker structure on endothelial cell surfaces adjacent to flowing blood, referred to as the endothelial surface layer (ESL) here (Pries et al. 2000). The thickness is estimated to range from about 0.5 μm to >1 μm (Pries et al. 2000; Weinbaum et al. 2007; Vink and Duling 1996; Pries and Kuebler 2006; Reitsma et al. 2007; Chappell et al. 2009), which is much greater than the ~50-nm thickness of the endothelial glycocalyx. Whereas the glycocalyx is directly anchored to the endothelial plasma membrane, the overlying ESL consists of adsorbed, probably plasma-borne molecules (Pries et al. 1998b). To avoid confusion, it is advisable to use the above terminology, which distinguishes between the glycocalyx proper and ESL. The presence of ESL has a substantial effect on hematocrit and flow resistance in microvessels and may influence a number of other processes, including inflammatory responses and blood coagulation (Pries et al. 2000).

Initial evidence for the presence of an endothelial surface layer came from observations (Klitzman and Duling 1979) of tube hematocrit levels in capillaries that were much lower than expected based on systemic hematocrit and estimates of the Fåhraeus effect derived from observations in glass tubes. Such observations could be explained if a slow-moving or stationary plasma layer about 1-μm-thick was present on the interior surface of the capillaries. With such a layer, the mean flow velocity is substantially reduced for a given RBC velocity, leading to an enhanced Fåhraeus effect according to Eq. (1.10). Further support for this concept was derived from the observation that infusion of heparinase, an enzyme that cleaves sugar side chains from glycoproteins, led to an increase in tube hematocrit (Desjardins and Duling 1990). Such a change would be consistent with a decrease in the thickness of the layer resulting from the enzyme treatment. Vink and Duling (1996) showed an increase in the region within a capillary accessible to RBCs and macromolecules by about 0.8 μm, following light-induced excitation of dyes within the capillary. These changes were attributed to disruption of the endothelial surface layer by this light-dye treatment. Experimental studies have used a variety of techniques to probe the transport characteristics of the layer (Vink and Duling 1996; Henry and Duling 1999; Henry et al. 1997; Osterloh et al. 2002).

In subsequent work, microparticle image velocimetry (μPIV) was used to analyze the rheological characteristics of blood flow in microvessels (Long et al. 2004; Smith et al. 2003). This method is based on the determination of the location and speed of fluorescent microbeads using double exposure strobe illumination. From such images, radial velocity profiles are established (Fig. 1.5). Deviations of the observed profile from the paraboloid shape expected for tube flow of Newtonian fluids are used to assess radial distributions in blood viscosity, shear stress, and shear rate. With additional knowledge of the in vitro rheological properties of blood from the investigated species, axial pressure gradient, apparent viscosity, and endothelial cell surface layer thickness can be predicted for vessels recorded in vivo.

The approach is restricted to conditions in which the rheological properties of the blood can be adequately represented and recorded by imaging the microbeads. This is the case in venules with diameters from about 20 to 60 μm. In pericapillary and capillary vessels, the assumption of a continuous distribution of hematocrit and viscosity no longer holds, while, in larger vessels, the reliable visualization of particles in the central flow regions is compromised. For arterioles, an additional triggering of the recordings with respect to the heart cycle is required to eliminate the effects of time-dependent variations in flow velocity. Within these limits, microparticle image velocimetry is a promising approach for the in vivo investigation of microrheology by analyzing the movement of microscopic particles in intravital microscopic images. This approach has been used to investigate the hydrodynamically effective thickness of ESL in medium-sized microvessels. For venules with diameters ranging from 30 to about 50 μm, the thickness of ESL varies between about 0.5 and 0.8 μm (Long et al. 2004; Smith et al. 2003).

Using a double tracer infusion (indocyanine green for plasma, stained autologous RBCs), Rehm et al. (2001) estimated a whole-body volume of about 720 ml of ESL in humans. Assuming a whole-body endothelial surface area in the range of 350 m^2 (Pries et al. 2000), the average ESL thickness can be calculated to be in the order of 2.1 μm. Nieuwdorp et al. (2006) used a similar approach and reported substantially higher values for the whole-body ESL volume ranging from 1.5 to 1.7 l in healthy subjects. It may seem difficult to accommodate an average ESL thickness of ~4.7 μm, especially in microcirculatory vessels representing the largest component of the whole-body endothelial surface. However, the strong decline of the measured values in clinical conditions, e.g., diabetes (Nieuwdorp et al. 2006), indicates a significant degradation of ESL with possible pathophysiological implications.

An indirect approach can also be used to assess the thickness of ESL as a function of vessel diameter. For microvascular networks of the rat mesentery, measured flow velocities were compared to predictions obtained from a hemodynamic simulation (Pries and Secomb 2005). In these simulations, the blood viscosity relation derived from in vitro measurements (Eqs. 1.15, 1.16, and 1.17)) was used for the central part of the vessel and different diameter-dependent thickness profiles of ESL were tested. ESL was assumed to be impermeable to flow. The minimal deviation between measured and simulated results was obtained assuming a layer thickness of about 0.8–1 μm for vessel diameters ranging from 10 to 40 μm and a strong decline for smaller diameters. In addition, a further hematocrit-dependent impact on flow resistance, exhibiting a maximum for vessel diameters around 10 μm, had to be assumed (Fig. 1.12). This effect may be related to flow conditions in a transitional diameter range. In vessels with diameters up to about 8 μm, a single-file flow regime

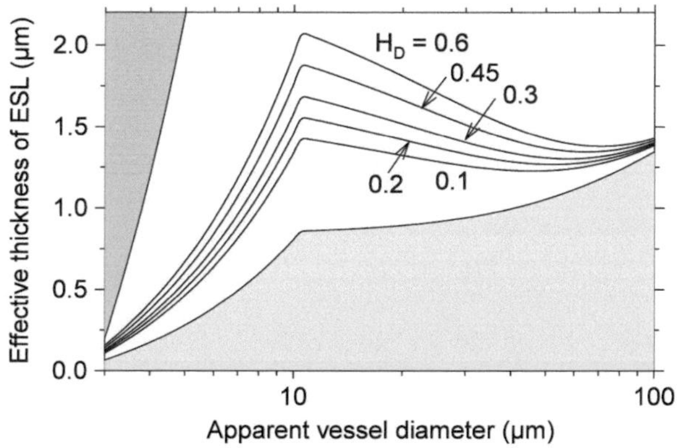

Fig. 1.12 Effective thickness of the endothelial surface layer (ESL), as a function of microvessel luminal diameter. Estimates were obtained by minimizing the differences between observed and predicted flow velocities in microvascular networks of the rat mesentery (Pries and Secomb 2005). Values in the shaded area to the upper left are excluded by the constraint that RBCs must be able to pass through the vessel when deformed to a minimal diameter of 2.8 µm. The lower shaded area indicates the estimated physical thickness of the layer. The optimization indicated that the layer exerts an additional hematocrit-dependent increase in flow resistance with a peak of about 10 µm for vessel diameters. For different levels of discharge hematocrit (H_D), this increase is given as an equivalent increase in the layer thickness (lines). *RBCs* red blood cells

prevails, with RBCs traveling one after another, and interactions with the vessel wall or ESL are limited and stable (Fig. 1.1, upper panel). Above this diameter range, multi-file flow patterns develop, in which cell-to-cell interactions cause increased interactions between cells and ESL. If two RBCs travel side by side (Fig. 1.1, middle panel), they tend to interfere and may push each other into ESL, generating additional flow resistance.

Apart from its effects on apparent viscosity and tube hematocrit, the endothelial surface layer can greatly influence interactions between endothelial cells and flowing blood. It limits the access of macromolecules to the endothelial surface in a manner dependent on molecular weight (Vink and Duling 1996). Constituting the interface between the flowing blood and the stationary endothelium, the endothelial surface layer must transmit the shear stress to the endothelial plasma membrane (Damiano 1998; Secomb et al. 2001a) and is thus important for the mechanotransduction of shear stress sensitivity. The mechanical interactions between flowing RBCs and the endothelial surface layer in capillaries have been examined using theoretical models (Damiano 1998; Secomb et al. 1998). It has been shown that RBCs are partly excluded from the layer by hydrodynamic forces generated within the layer by the motion of the cells (Feng and Weinbaum 2000). This effect increases with increasing cell velocity (Secomb et al. 2001b). Based on hemodynamic simulations, the endothelial surface layer may protect RBCs from damage during multiple passages through the irregular pathways presented by the microcirculation due to a reduction of the peak deformation forces (Secomb et al. 2002). Recent observations of ESL in murine kidney capillaries (Drost et al. 2023) have shown ESL structures completely

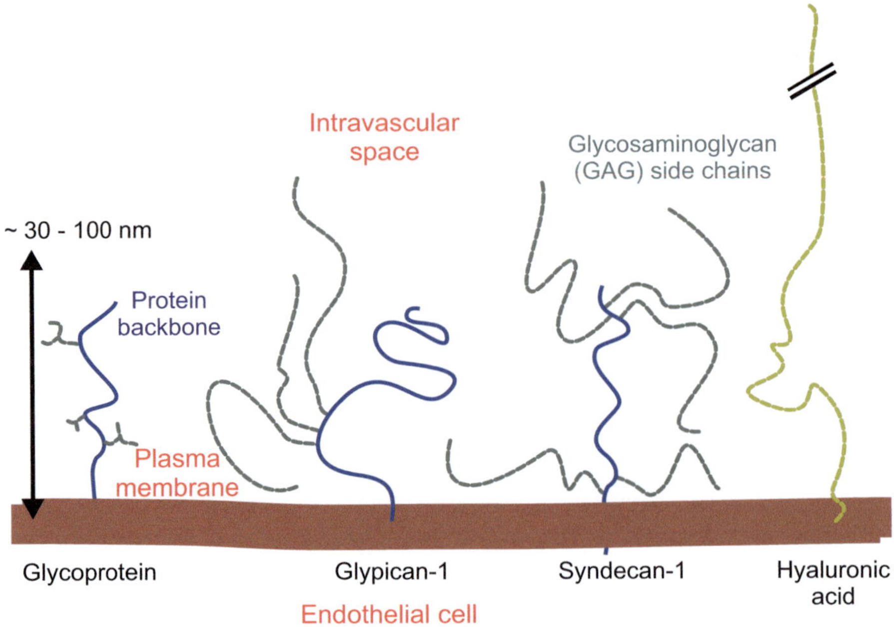

Fig. 1.13 Schematic representation of endothelial glycocalyx components. See text for explanation

filling the vessel. The layer is compressed by passing RBCs and leukocytes but appears to be reestablished after the cells have passed by. Theoretical approaches have been used to examine the mechanics of leukocyte interaction with the endothelial surface layer (Zhao et al. 2001) and the mechanical and transport properties of the layer (Weinbaum et al. 2003). Other phenomena influenced by the endothelial surface layer may include control of coagulation and hemostasis, angiogenesis, and ischemia–reperfusion (Weinbaum et al. 2007; Vink and Duling 1996; Pries and Kuebler 2006; Reitsma et al. 2007).

The exact composition of the endothelial surface layer remains unknown. In electron microscopy, the endothelial glycocalyx proper exhibits a width of <100 nm (Baldwin and Winlove 1984; Haldenby et al. 1994; Luft 1966). It consists of a layer of molecules with carbohydrate side chains directly bound to the endothelial plasma membrane (Bennett 1963), including glycoproteins, proteoglycans belonging to the syndecan and glypican families, and hyaluronan (hyaluronic acid) (Fig. 1.13). The carbohydrate side chains of glycoproteins (selectins, integrins, members of the immunoglobulin superfamily) are short and branched, while proteoglycans are characterized by long, unbranched side chains. Hyaluronan may be produced by endothelial cells (Suzuki et al. 2003) or adsorbed from the plasma (Saegusa et al. 2002) onto endothelial surface receptors (McCourt et al. 1999; Nandi et al. 2000). A computer-based analysis of electron micrographs suggests that the glycocalyx consists of a fibrous meshwork with a characteristic spacing of about 20 nm (Arkill et al. 2012; Squire et al. 2001).

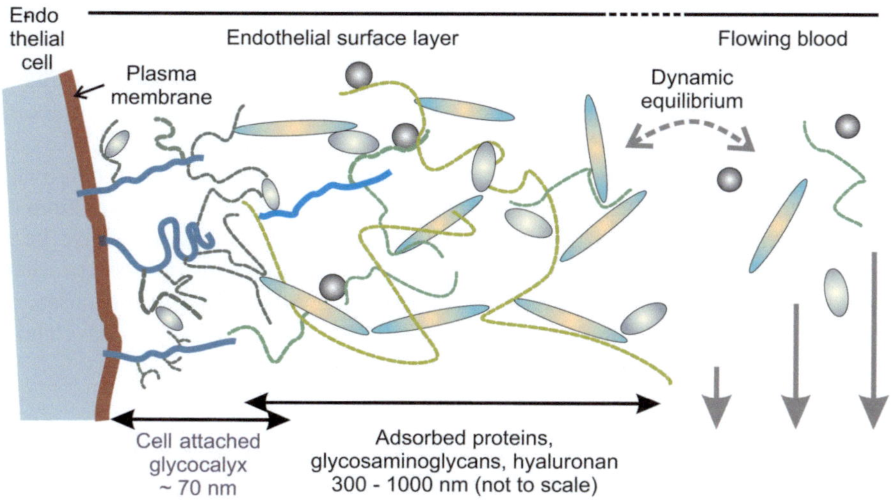

Fig. 1.14 Hypothesized composition of the endothelial surface layer. See text for explanation. Molecules and components of the layer are not drawn to scale

However, the much thicker endothelial surface layer contains additional structural components. Proposed sources include hyaluronan secreted by endothelial cells or derived from the blood (Henry and Duling 1998, 1999) and the adsorption of proteins and other macromolecules from the plasma onto the membrane-bound glycocalyx proper (Pries et al. 2000; Platts and Duling 2004). According to the latter proposal (Fig. 1.14), the main part of the endothelial surface layer consists of a complex array of soluble plasma components, possibly including a variety of proteins, solubilized glycosaminoglycans, and hyaluronan. This layer is in a dynamic equilibrium with the flowing plasma and stabilized by osmotic forces due to a low solid fraction in the range of ~0.001 (Secomb et al. 1998). The surface layer may be degraded by mechanisms targeting the glycocalyx proper (e.g., enzymes, inflammatory mediators) or by changing the plasma composition (e.g., by infusion of artificial plasma replacement fluids).

This proposal is supported by a meta-analysis of experimental hemodilution studies (Pries et al. 1998b). The available studies were separated into two groups according to whether plasma or artificial fluid (saline with or without added colloids) was used for dilution. The reduction of flow resistance for a given change in hematocrit in the "artificial fluid" group was twice that of the "plasma" group. The stronger resistance decrease was interpreted to indicate the removal of a thick (~1.2 μm) layer on the endothelial surface in the "artificial fluid" group. At present, while it is not known which plasma components are necessary to maintain the endothelial surface layer, albumin alone seems to be insufficient. Since the endothelial surface layer protects against edema (van den Berg et al. 2003; Rehm et al. 2004), maintaining its integrity during infusion therapy may be of significant clinical importance. Conversely, transfusion of whole blood appears to not reduce ESL thickness (Fuchs et al. 2021).

The role of ESL in protecting the endothelial surface from pathogenic insults is increasingly recognized. This has motivated the development of methods to assess ESL dimensions in human subjects (Nieuwdorp et al. 2008; Rovas et al. 2018). In this approach, orthogonal polarization spectral imaging of the sublingual microvasculature is used to measure the distance between the endothelium and the column of RBCs traversing capillaries. The resulting estimates of ESL thickness have been shown to correlate with whole-body ESL volume, determined as discussed earlier. Alterations of ESL thickness have been detected using this technique in a variety of clinical conditions, including resuscitated sepsis (Rovas et al. 2019), dialysis (Vlahu et al. 2012), and cardiac surgery (Koning et al. 2016).

1.2.8 Effective Blood Viscosity In Vivo

Flow resistance in individual microvessels is extremely difficult to measure directly in vivo because intravascular pressure must be simultaneously measured at both upstream and downstream locations in the vessels. The few available data (Lipowsky et al. 1980; Lipowsky and Zweifach 1977) indicate substantially higher apparent viscosity values than expected based on measurements in glass tubes of corresponding diameters. In a study of blood flow in mesenteric microvascular networks, Pries et al. (1994) compared the measured distributions of flow velocity and hematocrit with predictions of a theoretical model of network hemodynamics. Substantial discrepancies were found between the observations and predicted data based on measurements of apparent viscosity of the blood in glass tubes. However, satisfactory agreement between the observed and predicted behaviors was only reached when a stationary layer of plasma, about 1.1-μm-wide, adjacent to the vessel walls, was included in the model. The resulting predicted variation of apparent viscosity with diameter and hematocrit is shown in Fig. 1.15, along with experimental data of Lipowsky et al. (1978, 1980). These data were compiled in a parametric form

$$\eta_{\mathrm{vivo}} = \left[1 + \left(\eta^*_{0.45\cdot} - 1\right)\frac{(1 - H_D)^C - 1}{(1 - 0.45)^C - 1}\left(\frac{D}{D - 1.1}\right)^2\right]\left(\frac{D}{D - 1.1}\right)^2 \tag{1.23}$$

with C according to Eq. (1.17) and

$$\eta^*_{0.45} = 6\exp(-0.085D) + 3.2 - 2.44\exp\left(-0.06D^{0.645}\right) \tag{1.24}$$

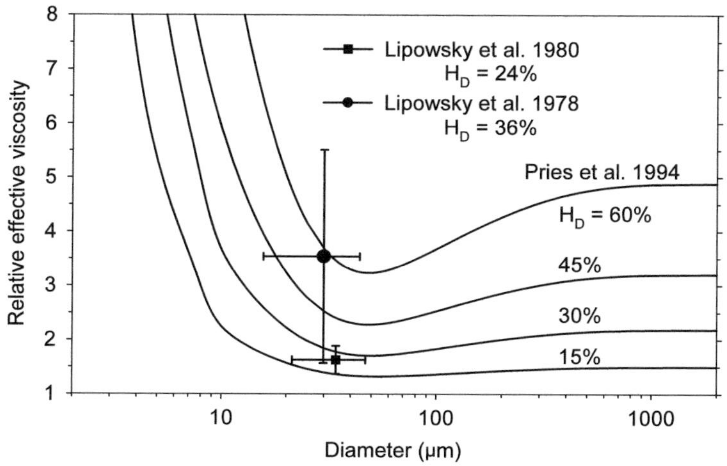

Fig. 1.15 Apparent viscosity in vivo derived from an analysis of blood flow in microvascular networks (Eqs. 1.23 and 1.24) compared with experimental measurements (Lipowsky et al. 1978, 1980). (Modified from Pries et al. 1996b)

According to this "in vivo viscosity law" (Pries et al. 1994), apparent viscosity is substantially higher in microvessels than in corresponding glass tubes, as shown by a comparison of Fig. 1.15 with Fig. 1.7. High values of effective blood viscosity were also found by microparticle image velocimetry (μPIV) in venules with diameters above 20 μm. For example, Long et al. report a relative viscosity of 3.08 for a venule in the rat cremaster with a diameter of 40.2 μm (Long et al. 2004).

Although other factors, such as the irregularity of vessel walls (Kiani et al. 1993; Pries et al. 1997a) and the presence of white blood cells, may contribute to this difference, available estimates suggest that their effects are relatively small under the conditions of the experiments just described (Pries et al. 1997a), leaving ESL as the major factor causing increased blood flow in vivo. Evidence in support of this conclusion was obtained by the observation of reduced flow resistance following infusion of heparinase into terminal vascular beds (Pries et al. 1997b).

Consequently, flow resistance in microvessels may be estimated from the free luminal diameter (anatomical inner vessel diameter as determined by standard intravital microscopy—twice the endothelial surface layer thickness) and the apparent viscosity relations derived from measurements in vitro (Eqs. 1.15, 1.16, and 1.17). This approach was adopted by Pries and Secomb (2005), using the effective thickness of the endothelial layer as shown in Fig. 1.12, which resulted in improved agreement between predicted velocity distributions in mesenteric networks and the values directly measured during intravital microscopy. Thus, accounting for the endothelial surface layer seems to reconcile flow resistance measurements in vivo with blood rheology assessment in small bore tubes in vitro.

1.3 Flow in Microvascular Networks

1.3.1 General Features

The number of microvessels in the human body is in the range 10^9–10^{10}, with a total length of order 10^7 m (based on the data of Mall (1888)). These numerous vessels form interconnected, complex networks, which exhibit substantial organ differences, reflecting their diverse structural and functional features (Fig. 1.16). As commonly described, the arteries supply blood through numerous diverging branches to feed the arterioles and capillaries. A series of converging branches then carries the blood through venules to the veins and back to the heart. Vessel diameters and segment lengths generally decrease from proximal to distal branches, corresponding to the subdivision of flow into multiple parallel segments. These features imply a significant degree of organization and structural adaptation in vascular networks.

Nonetheless, a considerable amount of heterogeneity and apparent randomness is immediately evident in observations of microcirculatory beds. Vessels do not form geometrically precise arrays. Their spacing is nonuniform, and their pathways are often tortuous. These features reflect the fact that the growth and proliferation of microvessels are stochastic rather than deterministic. Vessels feeding and draining a given region are frequently closely juxtaposed. This results in considerable heterogeneity in the length of flow pathways through the network, in which longer pathways connect the more distal branches of the feeding arteriole and the draining venule.

These features of microvascular networks can be visualized particularly clearly in the mesentery (Fig. 1.17). This tissue forms a thin sheet and permits the mapping and observation of complete microvascular networks including hundreds of vessel segments. The data on the network structure presented in this section are largely based on observations of the rat mesentery (Pries et al. 1995a). However, comparable heterogeneity of all parameters considered has been observed in studies of other tissues and is a hallmark of the microcirculation (Zuurbier et al. 1999; Vogel et al. 1997; Zheng et al. 1996; Duling 1994; Sarelius 1990; Kuhnle et al. 1993; Gonzalez and Bassingthwaighte 1990; Sweeney and Sarelius 1990; Duling and Damon 1987; Klitzman and Johnson 1982; Hertz and Paulson 1980; Schmid-Schonbein and Zweifach 1991). Table 1.1 presents estimates of the degree of heterogeneity in the structural and functional parameters within mesenteric networks. Here, the classification of segments as arterioles, capillaries, and venules is based on network topology and flow rather than on morphometric criteria. Vessels connecting diverging bifurcations are classified as arterioles, those connecting diverging with converging bifurcations are classified as capillaries, and those connecting converging bifurcations are classified as venules.

In this scheme, vessels that form part of a flow pathway from a converging bifurcation to a diverging bifurcation are termed "mesh" segments. If no mesh segments are present, then the structure can be strictly represented as two trees, arteriolar and venular, with capillaries forming connections between the terminal

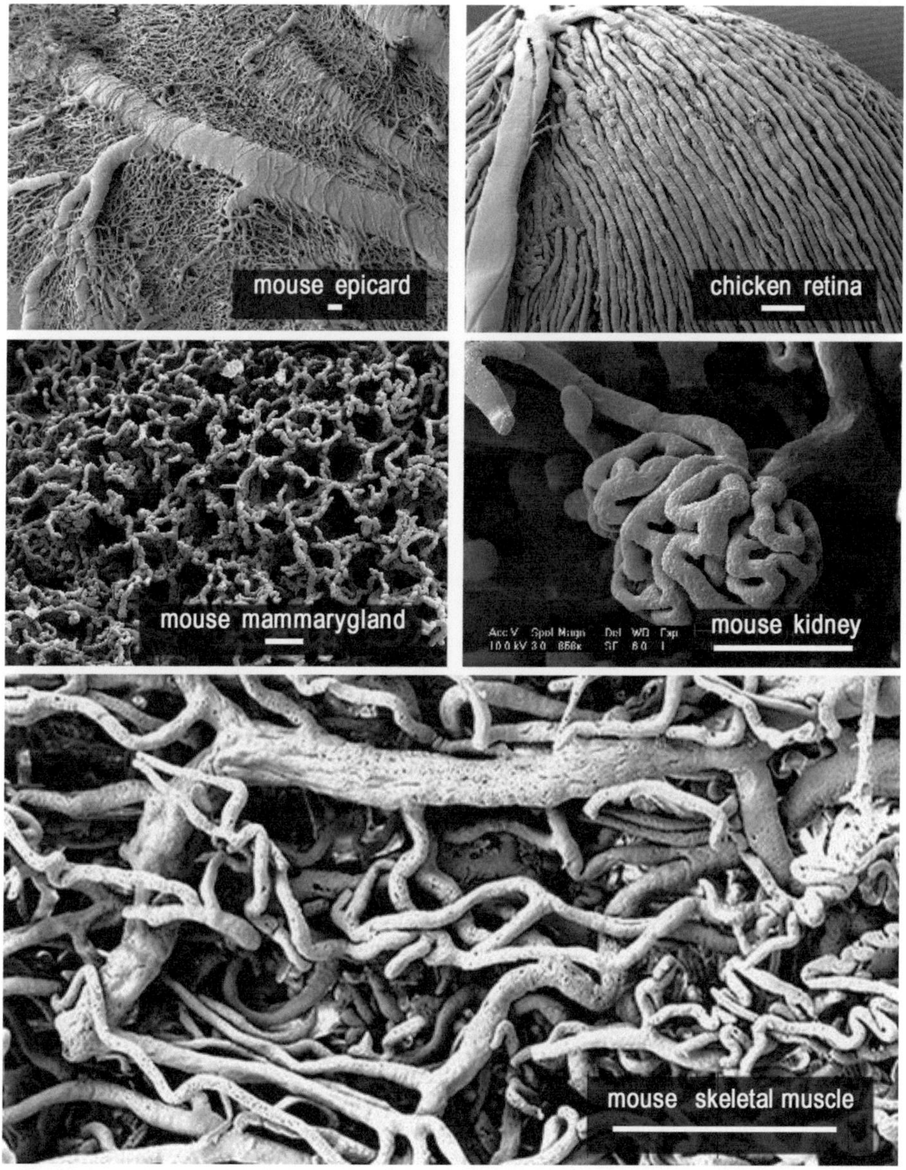

Fig. 1.16 Microvascular networks in different tissues as visualized by scanning electron microscopy. Scale bars indicate a length of 40 μm. These show a sample of the wide variety of microvascular patterns found in the body. Different network structures are seen in the heart muscle, skin, brain, lung, liver, spleen, lymph node, bone marrow, etc. (Micrographs courtesy of Valentin Djonov)

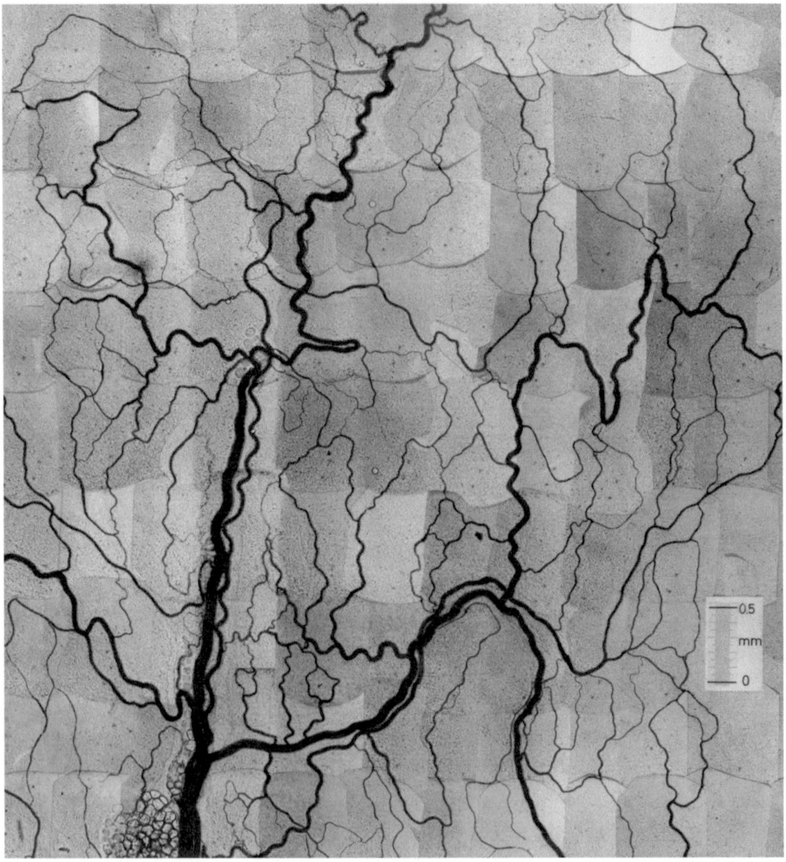

Fig. 1.17 The microvascular network in the rat mesentery with a total area of 31 mm^2. The main feeding arteriole and draining venule enter at the left part of the lower boundary. (Adapted from Pries et al. 1986)

branches of the two trees. In the mesenteric networks studied, mesh segments were relatively rare, being found in only 5% of the arterio-venous flow pathways (Pries et al. 1986). Thus, despite the mesh-like appearance of mesenteric networks (Fig. 1.17), they can be actually represented to a good approximation by a two-tree structure. However, this is not the case for microvascular networks that contain arcades, multiple capillaries in series (rete mirabile), or anastomosing networks. For such conditions, additional classification schemes have been developed (Schmid-Schonbein and Zweifach 1991; Chen and Battarbee 1987; Hill et al. 1992; Price et al. 2002; Torres Filho et al. 1994; Engelson et al. 1985). Irrespective of the classification scheme used, the coefficient of variation (standard deviation divided by average) is near or even above unity for many parameters, indicative of strongly heterogeneous (mostly right-skewed) distributions.

Table 1.1 Morphological and functional heterogeneity in microvessel networks of the rat mesentery

	Arterioles		Capillaries		Venules	
	Mean	CV	Mean	CV	Mean	CV
Generation number	10.2	0.43	11.9	0.37	13.9	0.44
Diameter[a] (μm)	13.2	0.41	8.72	0.28	20.6	0.53
Length[a] (μm)	337	0.83	424	0.65	334	0.82
Tube hematocrit[a]	0.29	0.41	0.23	0.60	0.31	0.41
Velocity[b] (mm/s)	2.03	0.84	0.85	0.99	1.07	0.65
Volume flow[c] (nl/min)	28.0	2.23	3.11	1.76	40.9	1.86
Pressure[d] (mmHg)	41.4	0.37	28.3	0.33	23.2	0.24
Shear stress[e] (dyn/cm^2)	64.9	0.90	47.5	1.19	20.6	1.17
Oxygen partial pressure[e] (mmHg)	88.70	0.06	74.90	0.25	76.20	0.10

CV coefficient of variation. The arteriolar and venular generation numbers are defined as the number of branch points between the vessel and feeding arteriole or draining venule in the tissue section investigated. For capillaries, the generation number is given with respect to the position in the arteriolar vessel tree

[a]Based on measurements in 3129 segments of 7 networks

[b]Based on measurements in 1321 segments of 3 networks

[c]Combined from measurements in three networks and model simulations in four networks

[d]Model simulations in seven networks

[e]Model simulation in a single network with 913 segments. Data compiled from Pries et al. (1995a, 2001a)

The data given in Table 1.1 are based on mesenteric networks, in which vessels do not exhibit vascular tone and the veins drain into the liver. These special conditions have to be considered when translating to other networks such as skeletal muscle or the brain for which complete datasets like the ones used here are not available. Network structures of microvascular networks and functional parameters exhibit fractal and chaotic properties (Baish and Jain 2000; Griffith 1996). For example, observed heterogeneity of perfusion in the heart exhibits fractal characteristics with respect to the size of the perfusion area (Bassingthwaighte et al. 1989; Deussen 1998), is stable over time (King et al. 1985), and is linked to a similar pattern of heterogeneity in tissue metabolism (Schwanke et al. 2000; Decking et al. 2001). Thus, the structure of microvascular networks (and vascular beds at large) is orderly enough to maintain satisfactory supply of the tissue. But it is also too irregular and heterogeneous to allow understanding of its functional properties without taking into account its specific architectural and topological features.

1.3.2 Topology

The substantial variation of topological, morphological, hemodynamic, and functional parameters exhibited by microvascular networks renders understanding and describing their properties and behavior on a vessel-to-vessel basis extremely difficult. Consequently, a number of ordering schemes have been introduced to group vessels into "classes" (Pries et al. 1986; Engelson et al. 1985; Fenton and Zweifach 1981; Ley et al. 1986; Wiedeman 1968; Fick 1888) with the hope that these vessels share similar properties (Pries et al. 1986).

In the most straightforward ordering schemes, microvascular networks are represented as various classes of vessels connected in series (e.g., large and small arterioles, capillaries, large and small venules) (Fick 1888). The vessels within each class are arranged in parallel and assumed to exhibit identical properties. Thus, all flow pathways from the arterial to the venous side are equivalent. In this "typical vessel" approach, the functional characteristics of complete vascular beds are then derived from the properties of the individual vessel classes and the number of vessels within each class.

A prominent example of this approach is the concept of "typical capillaries" used in the analysis of diffusive oxygen delivery to tissues, according to Krogh (1919b). These are assigned properties averaged from measurements on a number of real capillaries, e.g., in intravital microscopy. Especially in situations of hypoxia or maximal oxygen consumption, where extraction is high and individual capillaries are functionally isolated, such models provide helpful insights into oxygen delivery (McGuire and Secomb 2003, 2004). However, many functional properties of microvascular networks can only be understood if their irregular architecture and topological characteristics are taken into account.

Network topology can be described using the mathematical terminology of graph theory, in which vertices correspond to vascular branch points and edges correspond to segments. Graphs with no circular connections are called trees (Cayley 1859) and correspond to the arterial and venous portions of vascular networks, which do not exhibit arcading structures, to the bronchial tree (Weibel 1979; Mauroy et al. 2004) or to neuronal dendritic trees (Berry et al. 1986; Berry and Bradley 1976). Within such trees, the vessel segments can be rigorously attributed to different vessel classes and the results may be used to characterize the network topology by the respective summary statistics. The most prominent approaches for corresponding nomenclatures are the *centripetal Horton–Strahler ordering scheme* (Fenton and Zweifach 1981; Toroczkai 2002; Jiang et al. 1994), including its modification by Horsfield (1990) and Jiang et al. (1994), and the *centrifugal generation scheme* (Ley et al. 1986; Weibel 1979; Horsfield 1990; Gaehtgens et al. 1986).

The Horton–Strahler approach aims at grouping vessels with similar characteristics under one order (Fig. 1.18). The resulting data may then be used to determine the ratios (R) of numbers (R_b, branching ratio), diameters, or length between successive orders. The different R values were found to be reasonably well-conserved over the successive order levels within individual trees ("Horton's law") and have thus been used to characterize their topological and morphological properties (Fenton and Zweifach 1981; Schröder et al. 1990). However, it proved difficult to relate the determined parameters to the functional properties of the networks investigated. This may be due to the fact that the Horton–Strahler approach leads to substantial loss of topological information (Jarvis and Werritty 1975). Some of these deficiencies have been addressed by including morphological information in the ordering process

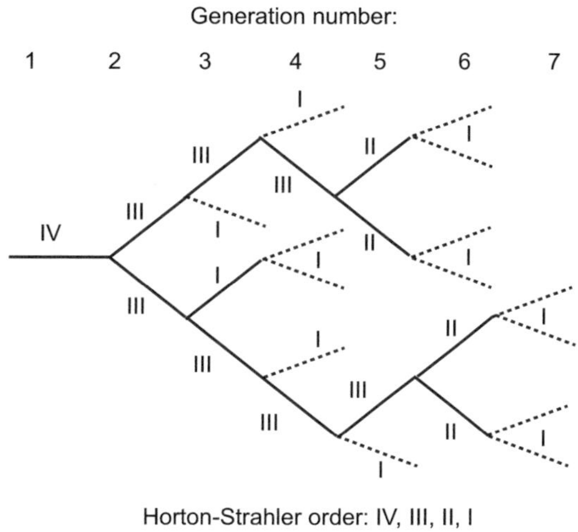

Fig. 1.18 Topological description of trees by the Horton–Strahler and the generation nomenclature. The Horton–Strahler approach starts at the capillary (or terminal) level and proceeds centripetally. The order is increased if two segments of equal order join at a bifurcation. The centrifugal generation scheme starts from the most central vessel considered and proceeds to the capillary level, increasing the generation number by one at every branch point

(Jiang et al. 1994), bridging the gap to nonmathematical nomenclatures using a combination of morphological and topological information available in intravital microscopy to classify microvessels (Wiedeman 1968).

The centrifugal generation scheme can be used to quantify the degree of asymmetry within vascular networks using distributions of generation numbers. In this scheme, the numbering starts from the most proximal vessel and increases by one at each bifurcation (Fig. 1.18). In a completely symmetric tree, in which the two subtrees at any branch point are identical, all capillaries would have the same generation. With increasing asymmetry, the distribution of capillary generation broadens and the mean capillary generation level increases. The mean generation number of capillaries for a vessel tree contains almost three times more information as compared to the branching ratio R_b (Jarvis and Werritty 1975) and is thus a suitable parameter to describe asymmetry of vascular networks (Ley et al. 1986; Wahl et al. 1984, 2004). Table 1.1 shows data derived from microvascular networks in the rat mesentery, for which the mean venular generation number (13.9) is higher than the mean arteriolar generation number (10.2), indicating that the venular tree has a more asymmetric branching pattern than the arteriolar tree.

In order to supply all tissue cells in a given organ volume, the exchange vessels of a microvascular network must form a space-filling mesh. The vessels supplying and draining the network are embedded in the tissue or attached to its surface. These conditions imply large heterogeneity in path length and number of vessel segments between the main vessels and the terminal branches (Pries and Secomb 2009). When the functional properties of microvascular networks are analyzed, incorrect estimates of derived quantities may be obtained if values of segment parameters such as diameter, flow rate, or RBC flux are assigned based on any of the classification schemes described above (typical vessel, Horton–Strahler, generation number). Such errors can arise if significant correlations between heterogeneously distributed parameters are not represented in the analysis (Pries et al. 1995a, b; Popel 1979;

Table 1.2 Correlation coefficients of linear regression analysis between the morphological, topological, and hemodynamic parameters for capillaries

Parameter A	Parameter B	r	SD_r
Pressure gradient	Velocity	0.67	0.17
Generation (art.)	Pressure	−0.53	0.39
Diameter	Discharge hematocrit	0.49	0.18
Pressure	Pressure gradient	0.49	0.45
Generation (art.)	Pressure gradient	−0.45	0.36
Generation (art.)	Velocity	−0.42	0.33
Pressure	Velocity	0.37	0.13

Average r values (± SD) are given for seven microvascular networks. Modified from Pries et al. (1995a)

Iversen and Nicolaysen 1995; Vicaut 1986; Duling et al. 1982; Honig et al. 1977). This problem can be addressed using a "network-oriented" approach, taking into account the heterogeneity and estimating the derived parameters for each segment based on directly measured parameters rather than on average values. For example, if the diameter, flow velocity, and discharge hematocrit for all capillaries of a given microvascular bed are known, it is possible to calculate a correct value of total flow of RBCs and plasma. This value may differ substantially from estimates based on arithmetic means of capillary diameter, flow velocity, and hematocrit (Duling et al. 1982) as used in the "typical vessel" approach, if the individual parameters (diameter, velocity, hematocrit) are significantly correlated between capillaries.

The "network-oriented" approach requires parallel estimation of the pertinent parameters (diameter, length, flow velocity, hematocrit, pressure, etc.) in a large number of microvessel segments. Major experimental difficulties can be partially overcome by a combination of direct experimental measurements with mathematical simulations (Pries et al. 1990, 1995a; Kiani et al. 1994; Popel 1980), as described below. Such simulations can, for example, be used to obtain estimates of pressure and flow in each segment of microvascular networks. Complete databases on microvascular network structure and hemodynamics permit an improved understanding of the functional characteristics of complete networks. As shown in Table 1.2, significant correlations exist between many pertinent topological, morphological, and functional parameters. This includes, in addition to the obvious relationships between velocity and pressure gradient and between pressure and arterial generation, a strong relation between discharge hematocrit and capillary diameter.

1.3.3 Simulation of Blood Flow in Microvascular Networks

The empirical equations presented in Sects. 1.2.4, 1.2.5, and 1.2.8 for apparent viscosity and partition of hematocrit in diverging bifurcations provide a basis for simulating blood flow in microvascular networks. The basic method was described by Lipowsky and Zweifach (1974). The structure of the network is specified as a set of segments with specified lengths and diameters, which are connected at a set of nodes, identified by node numbers. The conductance G_{ij} of a segment connecting nodes i and j is computed as the inverse of the flow resistance specified by Eq. (1.4), where the viscosity is typically estimated using the equations in Sect. 1.2.8. The condition that the sum of all flows entering internal node i is zero is represented by

$$\sum_j G_{ij}\left(P_j - P_i\right) = 0 \tag{1.25}$$

where the sum is over all nodes j directly connected to node i. If known pressures are imposed at all the boundary nodes, this results in a system of linear equations for pressures P_i, which can be solved by standard methods, and the segment flows can then be deduced. The same method can be used with slight modifications if flows instead of pressures are specified at some boundary nodes, provided that at least one pressure boundary condition is specified.

This basic method assumes that the hematocrit is known in each segment so that the segment conductance values can be computed. However, the partition of hematocrit in diverging bifurcations is dependent on the flow rates entering the downstream branches, as discussed in Sect. 1.2.5, resulting in a nonlinear interaction between segment flow rates and hematocrits. The simplest method to simulate this effect is to use an iterative approach. In the initial step, a fixed discharge hematocrit is assumed and flows are computed as described above. Segment hematocrits are assigned using Eq. (1.19) and the computed flows, and segment conductance values are recomputed. The procedure is then repeated until some convergence criterion is satisfied. In practice, convergence to a steady-state solution of these equations cannot always be achieved. Certain network configurations give rise to non-decaying oscillations in segment flow rates and hematocrits (Carr et al. 2005; Geddes et al. 2007; Karst et al. 2015). In realistic networks, this behavior is typically found in a small but significant fraction of segments. Rigorous analysis of such cases requires consideration of hematocrit as a function of both position and time in each segment or analysis at the level of individual RBCs (Balogh and Bagchi 2019).

Assigning boundary conditions in simulations of network flow can be problematical. Advanced techniques for mapping microvascular network structures in three dimensions, for instance, in the brain and the myocardium, have made extensive structural datasets available (Vigneshwaran et al. 2019; Cassot et al. 2006). However, measurements of pressure or flow rate are generally not available for all the numerous vessels that cross the boundaries of the observed regions. An approach for this problem was developed (Fry et al. 2012), based on the distributions of wall shear stress obtained in microvascular networks of the rat mesentery for which extensive data on flow rates were available (Pries et al. 1995b). In this approach, the sum of squared deviations of wall shear stresses from expected values is minimized, subject to the constraints implied by conservation of flow according to Eq. (1.25).

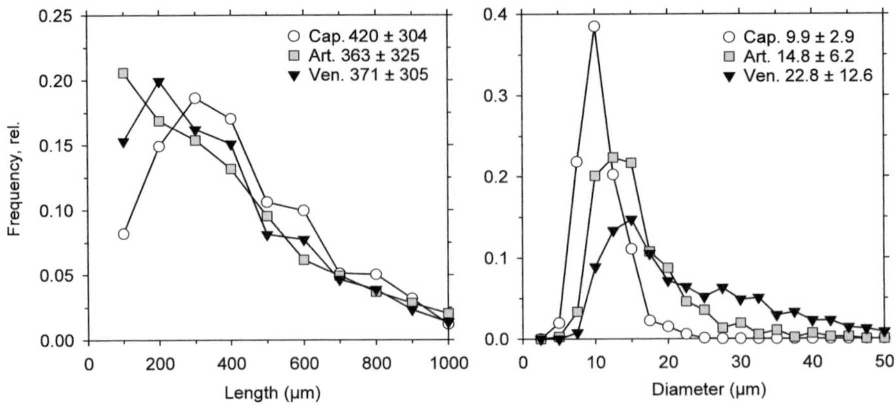

Fig. 1.19 Distributions of length and diameter for the arteriolar, capillary, and venular segments in microvascular networks in the rat mesentery (total 2720 segments). For each vessel category, the mean value and the standard deviation are given

1.3.4 Segment Lengths and Diameters

For six microvascular networks in the rat mesentery with a total of 3115 vessels, Fig. 1.19 shows the distributions of segment lengths and vessel diameters (Pries et al. 1995a). Segments are classified as arterioles, capillaries, or venules by topological criteria, as already described. Average segment lengths are around 400 μm with similar distributions for arterioles, capillaries, and venules. The diameter distributions for capillaries, arterioles, and venules exhibit pronounced differences with respect to shape and mean value. Again, there is substantial heterogeneity with strong overlap between classes. As would be expected, capillary segments, being the most distal branches, are generally smaller in diameter. A significant asymmetry is seen between the arteriolar and vascular tress, with venules larger in diameter than arterioles. This difference has the important consequence that flow resistance is higher in the arteriolar tree than in the venular tree, and, therefore, capillary pressure is much less than the mean of arteriolar and venous pressures. As an additional source of heterogeneity, individual microvascular segments exhibit significant variations in diameter along their lengths (Pries et al. 1997a).

In Fig. 1.20 (left two panels), the variations of diameter and length with the generation number are shown for arterioles, capillaries, and venules. Here, although capillaries are defined by topological criteria, their morphological properties differ significantly from those of the arterial and venous segments. All three vessel classes exhibit a decline in length with increasing generation but capillary segments are somewhat longer at all generation levels. While arterial and venous diameters decline with generation level, the diameter of capillaries shows a small increase. This feature is related to the structural adaptation of vessel diameters necessity to avoid shunting of blood flow through short, proximal pathways through the vascular network, as discussed below.

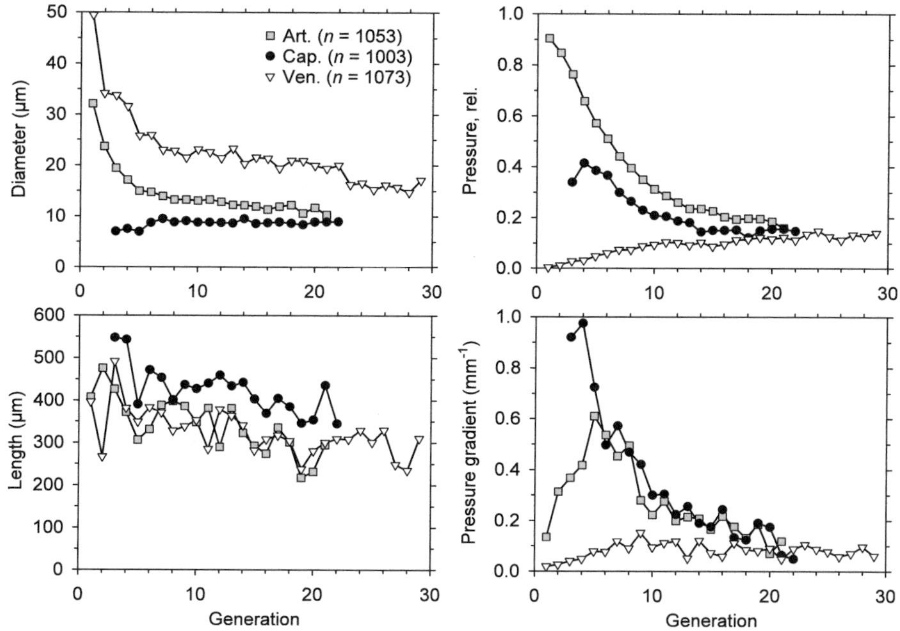

Fig. 1.20 Variation of diameter, length, relative intravascular pressure, and longitudinal (relative) pressure gradient of arterial, capillary, and venular vessel segments with vessel generation. The relative pressure scale was set to 1 for the pressure in the main feeding arterioles and to 0 in the main draining venules. For each parameter and class, mean values were determined for all segments fed by the main arteriolar inputs in each of the seven microvascular networks in the rat mesentery. These data were than averaged across networks. (Modified from Pries et al. 1995a)

1.3.5 Flow and Transit Time

In a completely symmetric network, the flow rate and transit time would be identical on every flow pathway. In the case of a network consisting of two interconnected tree structures, each pathway is identified with a specific terminal (capillary) segment. In real networks, dispersion in pathway lengths is inevitable because the feeding and draining vessels are embedded within the region being supplied and are necessarily close to some parts of the region and distant from others and because of the stochastic processes involved in the generation of vessel networks. This heterogeneity in network structure gives rise to dispersion in the distribution of pathway flow rates.

The relationship between structural (topological and geometrical) and hemodynamic heterogeneity in mesenteric microvessel networks has been examined using experimental data and mathematical simulations (Pries et al. 1996a). A broad distribution of pathway lengths is observed (Fig. 1.21). For a tissue area of 20–50 mm^2, the average pathway length is about 10 mm. Shorter flow pathways through a given vascular bed exhibit higher pressure drop per length. Therefore,

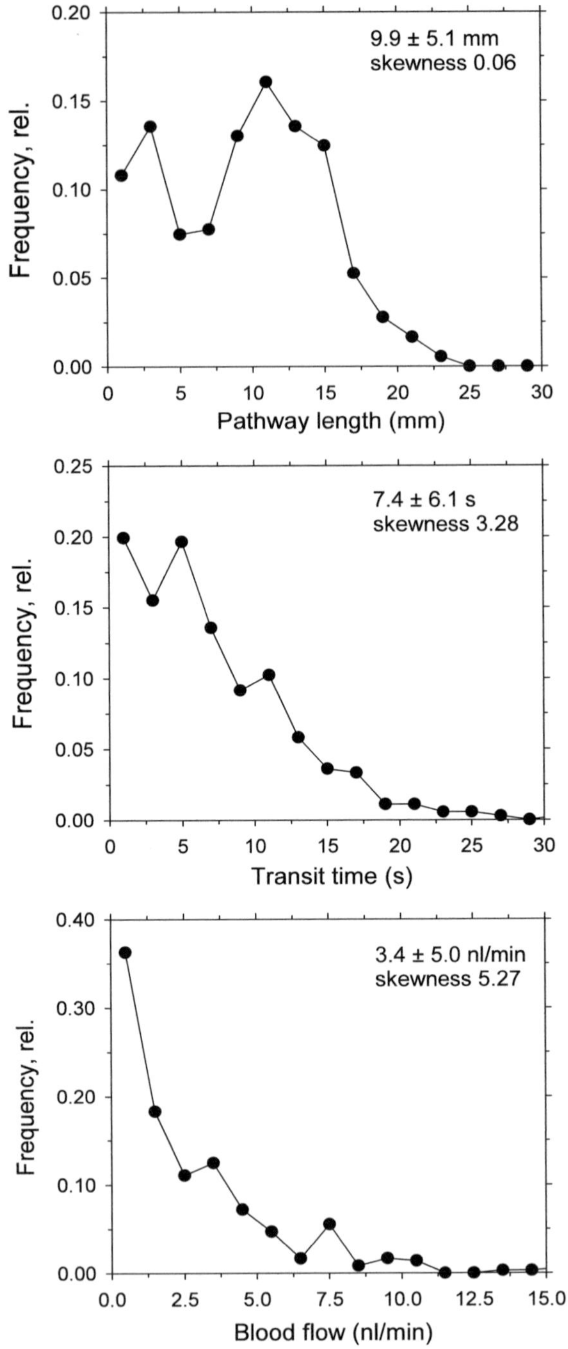

Fig. 1.21 Distributions of parameters for 361 arterio-venous flow pathways through a microvascular network with 913 segments. Data are based on experimental measurements combined with mathematical simulations of blood flow. The skewness given in addition to the mean values and standard deviations characterizes the asymmetry of the distribution

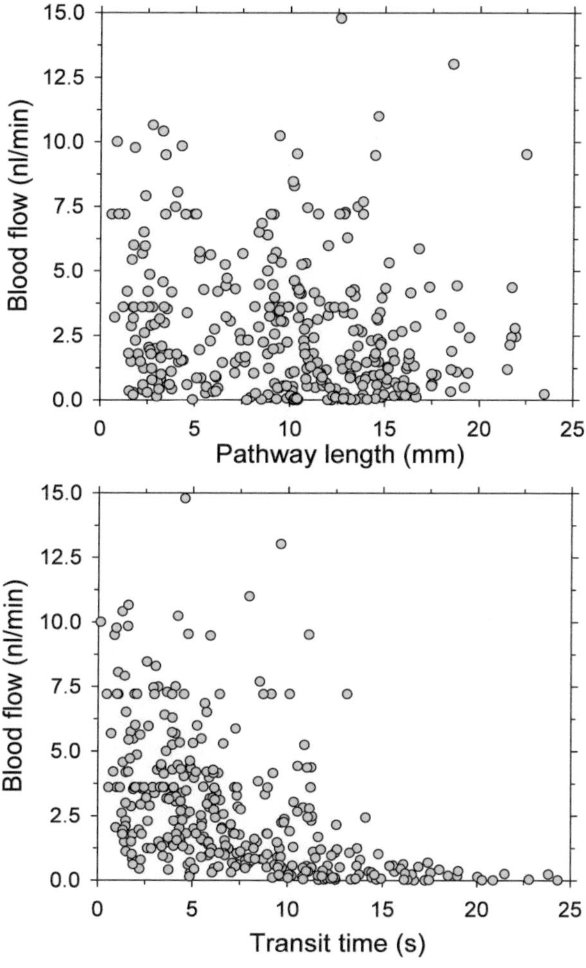

Fig. 1.22 Correlations between blood flow through 361 individual pathways through a microvascular network in the rat mesentery (913 segments) and pathway length (upper panel) and transit time (lower panel)

higher flow velocities would be expected on short pathways, leading to much shorter transit times. This effect is partially counteracted by a reduction of vessel diameters on short flow pathways resulting from structural adaptation, as discussed below in Sect. 1.3.8. As a result, blood flow does not show a strong inverse correlation with pathway length as might be expected, although it does show the expected inverse correlation with transit time (Fig. 1.22). These results indicate that faster (but not necessarily shorter) flow pathways transport the majority of blood flow. The presence of wide dispersion among flow pathways with respect to both transit times and blood flow rates leads to strongly left-skewed distributions of these parameters (Fig. 1.21). An analysis of the sources of heterogeneity in transit times (Pries et al. 1996a) showed that both topological heterogeneity (variation in capillary generation number) and geometric heterogeneity (variation in segment lengths and diameters) contribute substantially to transit time variability.

The inverse correlation between volume flow rate and transit time per pathway is also evidenced by the fact that the volume-weighted overall transit time for blood in the example shown is 4.3 s and is thus 58% of the mean transit time averaged over all flow pathways through the network (7.4 s). Such apparent discrepancies between the global parameters for microcirculatory networks and extrapolations from vessel or pathway averages (Duling et al. 1982) generally arise if strong correlations are present between the pertinent parameters. It is therefore important to perform an adequate analysis of the functional properties of microvascular networks to take such effects into account, by attributing the measured parameters to individual segments instead of using class averages.

1.3.6 Pressures and Wall Stresses

Experimental observations and theoretical simulations of blood flow in networks can also be used to estimate hemodynamic parameters including intravascular pressure and wall shear stress in every segment (Pries et al. 1995a). Results derived from such an approach applied to the rat mesentery are shown in Fig. 1.23. The frequency distribution of vessel segments as a function of intravascular pressure indicates that the majority of segments, and therefore of the vascular exchange surface, is concentrated at around 20 mmHg, close to the equilibrium of Starling forces governing fluid exchange with the interstitium. This is a functionally relevant consequence of the smaller diameter of arterioles relative to the corresponding venules.

A further consequence of this fundamental arterio-venous asymmetry in network structure is that wall shear stress is substantially reduced in venules relative to arterioles, as shown when shear stress is plotted as a function of diameter. The systematic variations of wall shear stress with diameter show that the concept of constant wall shear stress throughout the vasculature in order to minimize energy consumption (Murray's law) (Murray 1926; Griffith and Edwards 1990; Sherman 1981; Zhou et al. 1999) does not provide an adequate description of microvascular networks.

When wall shear stress is plotted as a function of pressure, a somewhat different picture emerges. Wall shear stress exhibits a near-linear decline with decreasing pressure by nearly one order of magnitude (Pries et al. 1995b), and the average wall shear stress at any given level of pressure is approximately the same for all three classes of vessels (arterioles, capillaries, and venules), as defined topologically. This behavior could not be predicted based on purely hemodynamic considerations. Pries et al. (1995b, 1998a) proposed that the observed correlation reflects the outcome of structural adaptation processes, in which diameters of all segments, regardless of classification, exhibit common structural responses to the levels of pressure and wall shear stress that they experience, as discussed in Sect. 1.3.8.

Figure 1.23 (upper two panels) shows that a large part of the overall pressure drop occurs in a relatively small number of vessel segments at high pressure and shear stress. These vessels are largely responsible for the regulation of blood flow to the

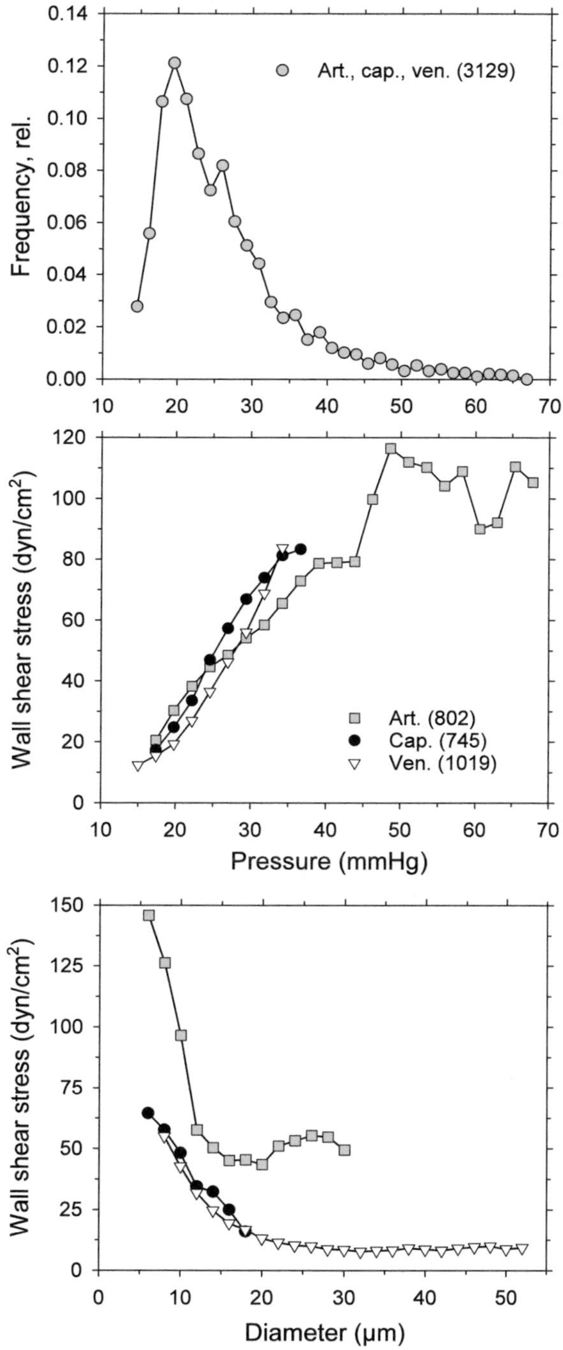

Fig. 1.23 Frequency distribution of intravascular pressure levels for all vessel segments of seven microvascular networks (top, modified from Pries et al. 1995a) and variation of wall shear stress with intravascular pressure (middle) and with vessel diameter (bottom) for arterial, capillary, and venular vessel segments of six microvascular networks. (Modified from Pries et al. 1995a)

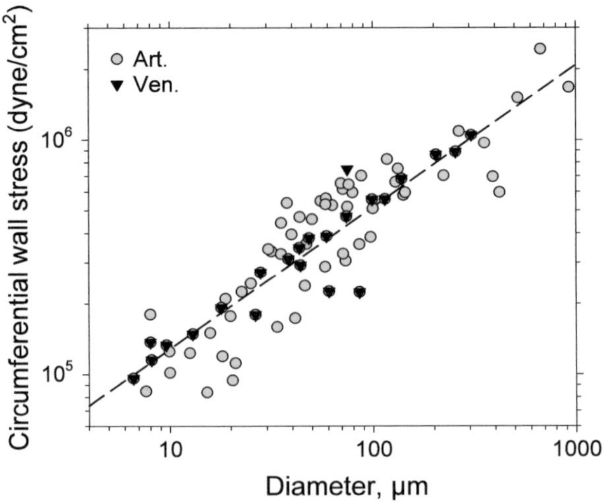

Fig. 1.24 Relationship between the log of circumferential wall stress and the log of vessel diameter. Data from experimental studies (Murray 1926; Bohlen and Niggl 1979; Gore 1974; Harper and Bohlen 1984; Hashimoto and Prewitt 1986; Lang and Johns 1987; Lee et al. 1983; Miller et al. 1987; Rakusan and Wicker 1990; Rhodin 1967, 1968; Tomanek et al. 1986) have been analyzed using the relation between microvascular pressure and diameter derived with a mathematical flow model (Pries et al. 1995a). The r^2 value of the linear regression to all data points shown is 0.85. (Modified from Pries et al. 1999, 2001b)

large number of vessels at low pressure and low shear stress vessels in the functional exchange domain. Here, this domain is defined on the basis of pressure levels and the total exchange surface (as represented by the number of segments) and includes vessels of the arteriolar, capillary, and venular types, as defined topologically. The apparent differences between the middle and lower panels of Fig. 1.23, which are based on the same network data, emphasize the complex relations of different parameters in heterogeneous microvascular networks, which often preclude the use of a "typical vessel" approach.

In addition to wall shear stress, vessel walls are subjected to circumferential stress resulting from intravascular pressure. As shown in Fig. 1.24, circumferential wall stress increases consistently with vessel diameter irrespective of the vessel category. This behavior is indicative of vascular responses to the mechanical stresses acting within the vessel wall (Secomb 1995; Pries et al. 1999, 2001a).

Microvascular networks show substantial heterogeneity in most structural and hemodynamic parameters, as has already been emphasized. For example, the variability of wall shear stress and circumferential wall stress, as estimated in rat mesenteric networks, is shown in Fig. 1.25. This heterogeneity is the inevitable result of the wide dispersion in pathway lengths due to the pairing of feeding and draining vessels and of the nondeterministic nature of microvessel growth. However, the emergence of the unexpected correlations shown in Figs. 1.23 and 1.24 shows

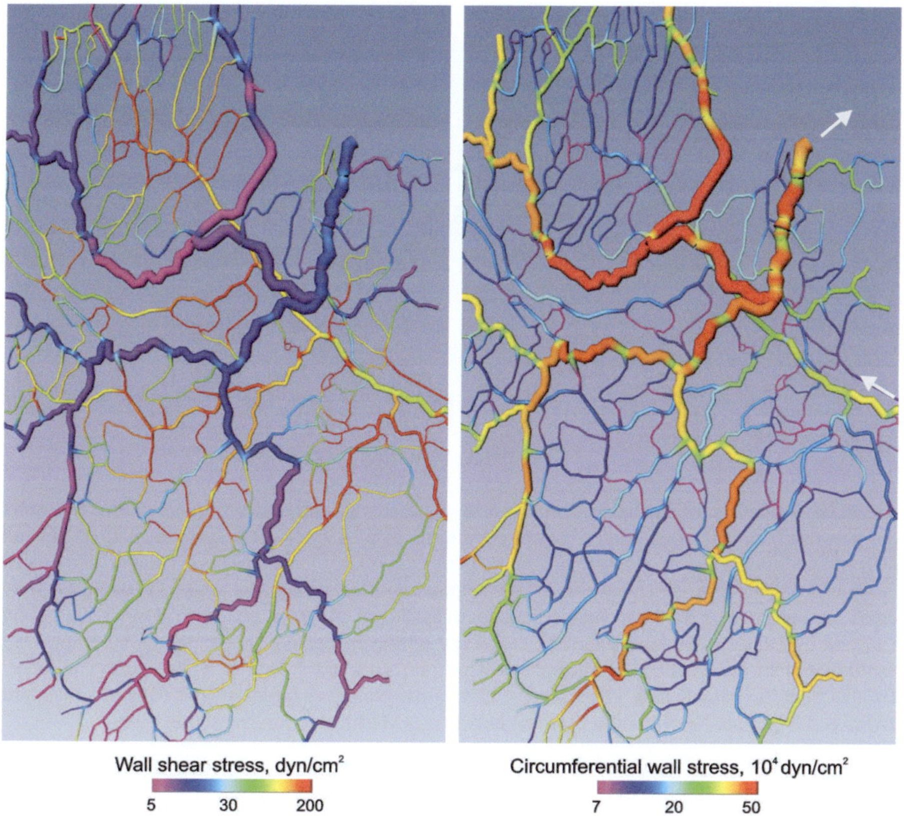

Fig. 1.25 The mesenteric microvascular network with an area of about 35 mm^2 and 546 vessel segments, color-coded for wall shear stress (left) and circumferential wall stress (right). The main inflow arteriole and outflow venule are marked. Circumferential stress is higher in larger vessels, while shear stress is high in the arteriolar and low venous portions of the network

that structural adaptation processes can generate a degree of order in network structure and hemodynamics, as discussed further in Sect. 1.3.8.

1.3.7 The Network Fåhraeus Effect

In microvascular networks, correlations between flow velocity and hematocrit among the vessel segments influence the hematocrit level in a way that is similar to the mechanisms underlying the Fåhraeus effect in single tubes. The resulting reduction in average (area-weighted) discharge hematocrit relative to the hematocrit of the blood flowing into the vascular network was named the "network Fåhraeus effect" (Pries et al. 1986).

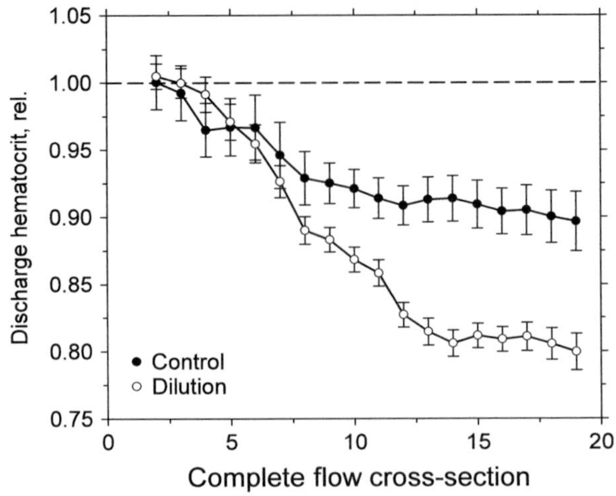

Fig. 1.26 Reduction of discharge hematocrit in the microvascular network in the rat mesentery by the network Fåhraeus effect under control conditions ($n = 5$, systemic hematocrit 0.48–0.51) and after hemodilution ($n = 6$, systemic hematocrit 0.29–0.35). The mean discharge hematocrit (weighted by the cross-sectional area of the segments) in all vessel segments of a complete flow cross section divided by the systemic hematocrit is shown. A given complete flow cross section consists of all arteriolar and capillary segments of the respective generation number plus all capillaries with lower generation numbers. The final flow cross section includes all capillaries. By this definition, each flow cross section carries the complete inflow of the network. Standard errors correspond to the variation between networks. (Modified from Pries et al. 1986)

The network Fåhraeus effect is generated by a positive covariance of flow velocity and hematocrit in the microvessels constituting a complete flow cross section through a microvascular network (a complete flow cross section is defined as a group of vessel segments, which together carry the complete inflow or outflow of the network considered). The correlation between flow velocity and hematocrit, in turn, depends on the extent of phase separation at the diverging arteriolar bifurcations. At a bifurcation, the daughter branch with the higher volume flow generally receives blood with an increased hematocrit, and vice versa, as already discussed. Thus, the phase separation effect generates a positive covariance between hematocrit and flow velocity in the daughter branches of a bifurcation and also among vessels constituting the consecutive flow cross sections of a network (Fig. 1.26). The network Fåhraeus effect, therefore, increases with the number of branch points passed and shows a maximum at the capillary level (Pries et al. 1986). Under control conditions, the network Fåhraeus effect leads to a reduction in capillary discharge hematocrit by about 10%. A much stronger reduction of about 35% is caused by the vessel Fåhraeus effect. Especially in capillary-sized vessels, ESL may be the dominant factor in causing the observed extremely low levels of tube hematocrit (Klitzman and Duling 1979; Duling and Desjardins 1987).

After hemodilution, phase separation is more pronounced and hematocrit reduction increases to about 20% (Pries et al. 1992b). However, this is accompanied by a

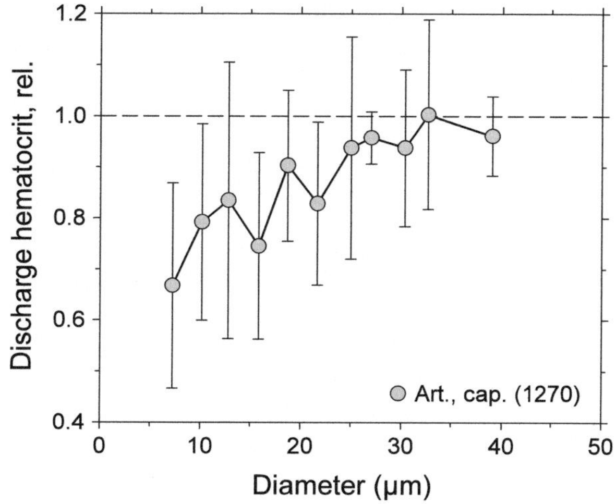

Fig. 1.27 Variation of discharge hematocrit averaged for arteriolar and capillary segments for a microvascular network with 913 segments. The hematocrit was normalized with respect to the discharge hematocrit in the main arteriole feeding the network

redistribution of RBC flow to distal network regions, which exhibit the lowest perfusion values during control. The network Fåhraeus effect is reflected by a decrease of averaged discharge hematocrit with decreasing vessel diameter (Fig. 1.27).

The Fåhraeus effect in single vessels, the network Fåhraeus effect, and the effect due to different average pathways of RBCs and plasma through networks ("pathway effect") together lead to a reduction of the mean hematocrit in all vessels of a given vascular bed (H_{tissue}) relative to the hematocrit of the blood flowing into the vascular bed (H_{inflow}) (Pries and Gaehtgens 1989). The reduction of H_{tissue} relative to H_{inflow} corresponds to a decreased residence time of RBCs within the tissue relative to the average residence time of blood and thus to the relation of the mean transit time of RBCs through a vascular bed (t_{cells}) to that of the blood (t_{blood}) (Pries and Gaehtgens 1989):

$$\frac{H_{\text{tissue}}}{H_{\text{inflow}}} = \frac{t_{\text{cells}}}{t_{\text{blood}}} \tag{1.26}$$

1.3.8 Biological Control of Heterogeneity

As already discussed, microvascular networks exhibit a high degree of heterogeneity in both structural and hemodynamic variables. The causes of heterogeneity can be classified as extrinsic or intrinsic (Roy and Secomb 2022). Spatially dependent functional demands, such as local variations in oxygen demand in the systemic circulation (Delashaw and Duling 1991; Lo et al. 2003) or in the ventilation/perfusion ratio in the lungs (Cano et al. 2015), are extrinsic drivers of heterogeneity

(Decking 2002). The stochastic growth of blood vessel networks and the variability in flow pathways due to geometric constraints, as already discussed, are intrinsic sources of heterogeneity.

In general, a high degree of heterogeneity would be expected to be detrimental to transport functions including oxygen delivery (Duling 1994; Hoffman 1995). For example, consider a set of geometrically equivalent vessels connected in parallel, with a fixed total flow rate. If the heterogeneity of the individual flow rates in these vessels is increased, then oxygen delivery may be reduced because some low-flow vessels are depleted of oxygen, while high-flow vessels retain a high oxygen saturation (Roy and Secomb 2021; Lücker et al. 2018; Walley 1996; Jespersen and Ostergaard 2012). In more realistic network structures, however, heterogeneous flow distributions may be necessary to compensate for the intrinsic geometrical heterogeneity. In particular, higher flow rates may be needed in longer flow pathways to avoid depletion of intravascular oxygen. According to this reasoning, active biological control mechanisms play an essential role in ensuring adequate tissue perfusion, not only by responding to changes in external demands (Balaban and Arai 2001) but also by overcoming the potentially deleterious effects of intrinsic heterogeneity (Pries and Secomb 2009). These mechanisms act both on short time scales, by local flow regulation, and on longer time scales, by structural adaptation of microvascular networks (Pries et al. 1995a, 2001a).

Acute flow regulation is primarily achieved by contraction and relaxation of vascular smooth muscle in the walls or arterioles. According to the inverse fourth-power dependence of flow resistance on diameter (Eq. (1.5)), changes in vessel diameters permit the variation of flow resistance over a wide range. Because most of the flow resistance in the circulatory system resides in the arterioles, changes in arteriolar flow resistance can result in substantial changes in the overall flow resistance and hence in the blood flow rate. Moreover, the fact that arterioles form a branching network with many fine ramifications makes possible the local control of blood flow with high spatial resolution. The ability of smooth muscle in arteriole walls to contract or dilate over time scales of seconds or longer, and to remain chronically constricted, provides the basis for dynamic control of flow according to time-varying needs.

Control of blood flow over a wide dynamic range, as occurs, for example, in skeletal muscle, requires coordinated contraction of the segments forming each flow pathway through a microvascular bed. Such coordination implies the existence of mechanisms for information transfer along vascular pathways, including the upstream direction (Segal et al. 1989; Secomb and Pries 2002). The responsiveness of vessel walls to shear stresses provides one such mechanism. Dilation of distal vessels causes increased flow, and the resulting increase in shear stress can cause dilation of upstream vessels. Furthermore, the vessel walls themselves act as information conduits. Conducted responses, involving changes in cell membrane potential, are communicated from cell to cell through gap junctions. By this mechanism, an increase in demand in a peripheral part of the network can be transmitted as a signal stimulating dilation in proximal feeding vessels (Segal 2000; Segal and Jacobs 2001; Bagher and Segal 2011).

The ability of microvascular networks to deliver blood in accordance to tissue needs is critically dependent on their structure, both in terms of the topology (number and connectivity of segments of various types) and geometry (particularly the length and diameter of each segment). In view of the large number of segments ($>10^9$) and the relatively limited number of genes ($\sim 4 \times 10^4$) in the human genome, it is evident that a predetermined individual control of these elements according to a genetic "blueprint" is not feasible. Furthermore, while short-term changes in tissue needs can be met by regulatory responses, as just discussed, long-term variations in tissue demands during growth, maturation, and other normal and abnormal physiological processes require the ability of the vessels to adapt their structures dynamically. Thus, cellular reactions to available stimuli, involving gene regulation, must generate, maintain, and adapt these complex networks. This process, which includes angiogenesis, remodeling, and pruning, has been termed "angioadaptation" (Zakrzewicz et al. 2002; Pries and Secomb 2014). Elucidating the mechanisms of angioadaptation can potentially contribute not only to understanding vascular physiology but also to developing methods for vascularizing regenerating tissues (Yin et al. 2021) and tissue-engineered grafts (Tronolone and Jain 2021).

Structural control of vessel diameters involves responses to wall shear stress (Rodbard 1975), blood pressure (Mulvany 2012), and metabolic states such as hypoxia or ischemia (Bailey et al. 2008). A theoretical model for structural remodeling in mesenteric tissues showed that conducted responses are needed to achieve a balance between flows in long and short pathways through the microvascular networks (Pries et al. 1998a). Attenuation of conducted responses was shown to lead to network properties characteristic of tumor microcirculation (Pries et al. 2010). Given that connections between endothelial cells via gap junctions are essential for the propagation of conducted responses, it appears likely that such responses are impaired in the inflammatory environment of tumors, with destabilized connection between endothelial cells. However, direct experimental evidence in support of this concept remains sparse.

Theoretical models have been used to explore the ability of the combined processes of angiogenesis remodeling and pruning to generate microvascular networks with structural and functional properties consistent with the observed networks (Secomb et al. 2013; Alberding and Secomb 2021). In these dynamic simulations, tissue oxygen levels are computed as a function of position for a given network structure and flow pattern. A growth factor generated in hypoxic regions is assumed to stimulate enlargement of existing vessels and generate new vessels by stochastic sprouting and growth. The simulations include remodeling of vessel diameters according to levels of shear stress, pressure, and conducted responses to metabolic conditions. A shrinking tendency is imposed so that vessels lacking a growth stimulus, either from wall shear stress or from growth factor, shrink below a diameter of 3 μm and are effectively pruned, since RBCs can no longer traverse them. The simulations show that this set of mechanisms is sufficient to generate stable functional networks with properties consistent with observations. In reality, multiple growth factors participate in angiogenesis and vascular remodeling, and further work is needed to integrate their respective roles into quantitative models for angioadaptation.

According to the preceding discussion, control of heterogeneity in microvascular networks is necessary for the adequate and efficient function of the circulatory system, particularly with regard to oxygen delivery. Maldistribution of flow at the local level can result in tissue hypoxia, even when overall oxygen supply to the tissue is adequate (Pries et al. 2010). In critically ill patients, for example, organ damage and failure due to hypoxia can occur even when cardiac output and arterial hemoglobin saturation are near normal levels (Roy and Secomb 2021). Such behavior can be logically attributed to dysfunction of the normal control mechanisms of flow regulation and/or structural adaptation. The term "endothelial dysfunction" is often used to describe pathologies attributed to the microcirculation, for example, in the myocardium (Bockus and Kim 2022) and in the brain (Quick et al. 2021). However, the endothelium has many functions, and their relative importance in causing tissue hypoxia and damage is not well-understood. Upstream conducted responses, primarily propagated via gap junctions connecting endothelial cells, play a critical role in both acute flow regulation and structural adaptation of the microcirculation, as already discussed. These signals are particularly vulnerable to endothelial dysfunction, since even a local loss of normal endothelial cell function or coupling is sufficient to interrupt their transmission (Segal and Jacobs 2001). The growing recognition of the effects of microcirculatory dysfunction in a range of diseases (Bockus and Kim 2022; Quick et al. 2021; Miranda et al. 2016) implies a need for more research on the role of conducted responses in the control of structure and blood flow in microvascular networks.

References

Alberding JP, Secomb TW (2021) Simulation of angiogenesis in three dimensions: application to cerebral cortex. PLoS Comput Biol 17(6):e1009164. https://doi.org/10.1371/journal.pcbi.1009164

Albrecht KH, Gaehtgens P, Pries A, Heuser M (1979) The Fahraeus effect in narrow capillaries (i.d. 3.3 to 11.0 micron). Microvasc Res 18(1):33–47

Alonso C, Pries AR, Kiesslich O, Lerche D, Gaehtgens P (1995) Transient rheological behavior of blood in low-shear tube flow: velocity profiles and effective viscosity. Am J Phys 268(1 Pt 2): H25–H32. https://doi.org/10.1152/ajpheart.1995.268.1.H25

Arkill KP, Neal CR, Mantell JM, Michel CC, Qvortrup K, Rostgaard J et al (2012) 3D reconstruction of the glycocalyx structure in mammalian capillaries using electron tomography. Microcirculation 19(4):343–351. https://doi.org/10.1111/j.1549-8719.2012.00168.x

Audet DM, Olbricht WL (1987) The motion of model cells at capillary bifurcations. Microvasc Res 33(3):377–396. https://doi.org/10.1016/0026-2862(87)90029-x

Azelvandre F, Oiknine C (1976) The Fahraeus and Fahraeus-Lindqvist effects: experimental testing of theoretical models. Biorheology 13(6):325–335

Bagher P, Segal SS (2011) Regulation of blood flow in the microcirculation: role of conducted vasodilation. Acta Physiol (Oxf) 202(3):271–284. https://doi.org/10.1111/j.1748-1716.2010.02244.x

Bailey AM et al (2008) Arteriolar remodeling following ischemic injury extends from capillary to large arteriole in the microcirculation. Microcirculation 15(5):389–404. https://doi.org/10.1080/10739680701708436

Baish JW, Jain RK (2000) Fractals and cancer. Cancer Res 60(14):3683–3688

Balaban RS, Arai A (2001) Function, metabolic, and flow heterogeneity of the heart: the view is getting better. Circ Res 88(3):265–267. https://doi.org/10.1161/01.res.88.3.265

Baldwin AL, Winlove CP (1984) Effects of perfusate composition on binding of ruthenium red and gold colloid to glycocalyx of rabbit aortic endothelium. J Histochem Cytochem 32(3):259–266. https://doi.org/10.1177/32.3.6198357

Balogh P, Bagchi P (2018) Analysis of red blood cell partitioning at bifurcations in simulated microvascular networks. Phys Fluids 30(5). https://doi.org/10.1063/1.5024783

Balogh P, Bagchi P (2019) Three-dimensional distribution of wall shear stress and its gradient in red cell-resolved computational modeling of blood flow in in vivo-like microvascular networks. Physiol Rep 7(9):e14067. https://doi.org/10.14814/phy2.14067

Barbee JH (1973) The effect of temperature on the relative viscosity of human blood. Biorheology 10(1):1–5. https://doi.org/10.3233/bir-1973-10101

Barbee JH, Cokelet GR (1971) The Fahraeus effect. Microvasc Res 3(1):6–16. https://doi.org/10.1016/0026-2862(71)90002-1

Barbee J, Cokelet G (1990) In-vitro capillary flow data.

Barber JO, Alberding JP, Restrepo JM, Secomb TW (2008) Simulated two-dimensional red blood cell motion, deformation, and partitioning in microvessel bifurcations. Ann Biomed Eng 36(10): 1690–1698

Barber JO, Restrepo JM, Secomb TW (2011) Simulated red blood cell motion in microvessel bifurcations: effects of cell-cell interactions on cell partitioning. Cardiovasc Eng Technol 2(4): 349–360

Baskurt OK, Meiselman HJ (2013) Erythrocyte aggregation: basic aspects and clinical importance. Clin Hemorheol Microcirc 53(1–2):23–37. https://doi.org/10.3233/ch-2012-1573

Bassingthwaighte JB, King RB, Roger SA (1989) Fractal nature of regional myocardial blood flow heterogeneity. Circ Res 65(3):578–590

Bäumler H, Donath E, Krabi A, Knippel W, Budde A, Kiesewetter H (1996) Electrophoresis of human red blood cells and platelets. Evidence for depletion of dextran. Biorheology 33(4–5): 333–351. https://doi.org/10.1016/0006-355x(96)00026-1

Bayliss L (1952) Rheology of blood and lymph. In: Frey-Wissling A (ed) Deformation and flow in biological systems. North-Holland, Amsterdam, pp 355–418

Bayliss LE (1959) The axial drift of the red cells when blood flows in a narrow tube. J Physiol 149(3):593–613. https://doi.org/10.1113/jphysiol.1959.sp006363

Bennett H (1963) Morphological aspects of extracellular polysaccharides. J Histochem Cytochem 11:14–23

Berry M, Bradley PM (1976) The application of network analysis to the study of branching patterns of large dendritic fields. Brain Res 109(1):111–132. https://doi.org/10.1016/0006-8993(76)90383-8

Berry M, Sadler M, Flinn R (1986) Vertex analysis of neural tree structures containing trichotomous nodes. J Neurosci Methods 18(1–2):167–177. https://doi.org/10.1016/0165-0270(86)90118-4

Bishop JJ, Nance PR, Popel AS, Intaglietta M, Johnson PC (2001) Effect of erythrocyte aggregation on velocity profiles in venules. Am J Physiol Heart Circ Physiol 280(1):H222–H236. https://doi.org/10.1152/ajpheart.2001.280.1.H222

Bockus L, Kim F (2022) Coronary endothelial dysfunction: from pathogenesis to clinical implications. Open heart 9(2). https://doi.org/10.1136/openhrt-2022-002200

Bohlen HG, Niggl BA (1979) Adult microvascular disturbances as a result of juvenile onset diabetes in Db/Db mice. Blood Vessels 16(5):269–276. https://doi.org/10.1159/000158215

Braasch D (1984) The missing negative effect of red cell aggregation upon blood flow in small capillaries at low shear forces. Biorheology Suppl 1:227–230. https://doi.org/10.3233/bir-1984-23s139

Braasch D, Jenett W (1968) Erythrocyte flexibility, hemoconcentration and blood flow resistance in glass capillaries with diameters between 6 and 50 microns. Pflugers Arch 302(3):245–254. https://doi.org/10.1007/bf00586729

Bugliarello G, Hsiao GC (1964) Phase separation in suspensions flowing through bifurcations: a simplified hemodynamic model. Science 143(3605):469–471. https://doi.org/10.1126/science.143.3605.469

Cabel M, Meiselman HJ, Popel AS, Johnson PC (1997) Contribution of red blood cell aggregation to venous vascular resistance in skeletal muscle. Am J Phys 272(2 Pt 2):H1020–H1032. https://doi.org/10.1152/ajpheart.1997.272.2.H1020

Cano I, Roca J, Wagner PD (2015) Effects of lung ventilation-perfusion and muscle metabolism-perfusion heterogeneities on maximal O2 transport and utilization. J Physiol 593(8):1841–1856. https://doi.org/10.1113/jphysiol.2014.286492

Carr RT, Wickham LL (1990) Plasma skimming in serial microvascular bifurcations. Microvasc Res 40(2):179–190. https://doi.org/10.1016/0026-2862(90)90017-1

Carr RT, Geddes JB, Wu F (2005) Oscillations in a simple microvascular network. Ann Biomed Eng 33(6):764–771

Cassot F, Lauwers F, Fouard C, Prohaska S, Lauwers-Cances V (2006) A novel three-dimensional computer-assisted method for a quantitative study of microvascular networks of the human cerebral cortex. Microcirculation 13(1):1–18

Cayley A (1859) On the analytical forms called trees, part II. Philos Mag 18:374–378

Chappell D, Jacob M, Paul O, Rehm M, Welsch U, Stoeckelhuber M et al (2009) The glycocalyx of the human umbilical vein endothelial cell: an impressive structure ex vivo but not in culture. Circ Res 104(11):1313–1317. https://doi.org/10.1161/circresaha.108.187831

Chen II, Battarbee HD (1987) The quantitative estimation of microvessels in microvascular networks. J Theor Biol 127(3):315–320. https://doi.org/10.1016/s0022-5193(87)80109-1

Chien S (1975) Biophysical behavior of red blood cells in suspension. In: Surgenor DM (ed) The red blood cell, vol II. Academic Press, New York, pp 1031–1133

Chien S, Jan K (1973) Ultrastructural basis of the mechanism of rouleaux formation. Microvasc Res 5(2):155–166. https://doi.org/10.1016/0026-2862(73)90068-x

Chien S, Usami S, Taylor HM, Lundberg JL, Gregersen MI (1966) Effects of hematocrit and plasma proteins on human blood rheology at low shear rates. J Appl Physiol 21(1):81–87

Chien S, Usami S, Skalak R (1984) Blood flow in small rubes. In: Renkin M, Michel C, Geiger S (eds) Handbook of physiology. The cardiovascular system IV part 1. Microcirculation. Section 2. American Physiological Society, Bethesda, pp 217–249

Chien S, Tvetenstrand CD, Epstein MA, Schmid-Schönbein GW (1985) Model studies on distributions of blood cells at microvascular bifurcations. Am J Phys 248(4 Pt 2):H568–H576. https://doi.org/10.1152/ajpheart.1985.248.4.H568

Damiano ER (1998) The effect of the endothelial-cell glycocalyx on the motion of red blood cells through capillaries. Microvasc Res 55(1):77–91

Decking UK (2002) Spatial heterogeneity in the heart: recent insights and open questions. News Physiol Sci 17:246–250. https://doi.org/10.1152/nips.01393.2002

Decking UK, Skwirba S, Zimmermann MF, Preckel B, Thämer V, Deussen A et al (2001) Spatial heterogeneity of energy turnover in the heart. Pflugers Arch 441(5):663–673. https://doi.org/10.1007/s004240000464

Delashaw JB, Duling BR (1991) Heterogeneity in conducted arteriolar vasomotor response is agonist dependent. Am J Phys 260(4 Pt 2):H1276–H1282. https://doi.org/10.1152/ajpheart.1991.260.4.H1276

Desjardins C, Duling BR (1990) Heparinase treatment suggests a role for the endothelial cell glycocalyx in regulation of capillary hematocrit. Am J Phys 258(3 Pt 2):H647–HH54

Deussen A (1998) Blood flow heterogeneity in the heart. Basic Res Cardiol 93(6):430–438. https://doi.org/10.1007/s003950050112

Doyeux V, Podgorski T, Peponas S, Ismail M, Coupier G (2011) Spheres in the vicinity of a bifurcation: elucidating the Zweifach-Fung effect. J Fluid Mech 674:359–388

Drost CC, Unger A, Linke WA, Vink H, Kümpers P (2023) The dynamics of the endothelial glycocalyx: a rare snapshot by electron microscopy. Angiogenesis 26:1–5. https://doi.org/10.1007/s10456-023-09880-x

Duling BR (1994) Is red cell flow heterogeneity a critical variable in the regulation and limitation of oxygen transport to tissue? Adv Exp Med Biol 361:237–247

Duling BR, Damon DH (1987) An examination of the measurement of flow heterogeneity in striated muscle. Circ Res 60(1):1–13

Duling B, Desjardins C (1987) Capillary hematocrit - what does it mean? News Physiol Sci 2:66–69

Duling BR, Sarelius IH, Jackson WF (1982) A comparison of microvascular estimates of capillary blood flow with direct measurements of total striated muscle flow. Int J Microcirc Clin Exp 1(4): 409–424

Dupin MM, Halliday I, Care CM, Alboul L, Munn LL (2007) Modeling the flow of dense suspensions of deformable particles in three dimensions. Phys Rev E 75(6)

Engelson ET, Skalak TC, Schmid-Schonbein GW (1985) The microvasculature in skeletal muscle. I. Arteriolar network in rat spinotrapezius muscle. Microvasc Res 30(1):29–44

Ethier CR, Simmons CA (2007) Introductory biomechanics. Cambridge University Press

Fahraeus R (1928) Die Strömungsverhältnisse und die Verteilung der Blutzellen im Gefäβsystem. Zur Frage der Bedeutung der intravasculären Erythrocytenaggregation. Klin Wochenschr 7(3): 100–106

Fåhraeus R (1929) The suspension stability of the blood. Physiol Rev 9:241–274

Fahraeus R, Lindqvist T (1931) The viscosity of the blood in narrow capillary tubes. Am J Phys 96: 562–568

Fedosov DA, Caswell B, Popel AS, Karniadakis GE (2010) Blood flow and cell-free layer in microvessels. Microcirculation 17(8):615–628

Fedosov DA, Dao M, Karniadakis GE, Suresh S (2014) Computational biorheology of human blood flow in health and disease. Ann Biomed Eng 42(2):368–387. https://doi.org/10.1007/s10439-013-0922-3

Feng J, Weinbaum S (2000) Lubrication theory in highly compressible porous media: the mechanics of skiing, from red cells to humans. J Fluid Mech 422:281–317

Fenton BM, Zweifach BW (1981) Microcirculatory model relating geometrical variation to changes in pressure and flow rate. Ann Biomed Eng 9:303–321

Fenton BM, Carr RT, Cokelet GR (1985) Nonuniform red cell distribution in 20 to 100 micrometers bifurcations. Microvasc Res 29(1):103–126. https://doi.org/10.1016/0026-2862(85)90010-x

Fick A (1888) Uber den Druck in den Blutkapillaren. Pflugers Arch 42:482–438

Freund JB (2014) Numerical simulation of flowing blood cells. Annu Rev Fluid Mech 46:67–95

Fry BC, Lee J, Smith NP, Secomb TW (2012) Estimation of blood flow rates in large microvascular networks. Microcirculation 19(6):530–538

Fuchs A, Groß S, Neumann T, Illerhaus A, Vink H, Klasen G et al (2021) Immediate effects of whole blood donation on the endothelial surface layer and glycocalyx shedding. Blood transfusion = Trasfusione del sangue 19(3):190–196. https://doi.org/10.2450/2020.0025-20

Fung YC (1973) Stochastic flow in capillary blood vessels. Microvasc Res 5(1):34–48

Gaehtgens P, Ley K, Pries A (1986) Topological approach to the analysis of microvessel structure and hemarocrit distribution. In: Popel A, Johnson P (eds) Microvascular networks: experimental and theoretical studies. Karger, Basel, pp 52–60

Geddes JB, Carr RT, Karst NJ, Wu F (2007) The onset of oscillations in microvascular blood flow. SIAM J Appl Dyn Syst 6(4):694–727

Gerbstädt H, Vogtmann C, Rüth P, Schöntube E (1966) The apparent viscosity of blood in glass capillaries of the smallest diameter. Naturwissenschaften 53(20):526. https://doi.org/10.1007/bf00600644

Goldsmith HL (1971) Red cell motions and wall interactions in tube flow. Fed Proc 30(5): 1578–1590

Goldsmith HL, Cokelet GR, Gaehtgens P (1989) Robin Fahraeus: evolution of his concepts in cardiovascular physiology. Am J Phys 257(3 Pt 2):H1005–H1H15

Gonzalez F, Bassingthwaighte JB (1990) Heterogeneities in regional volumes of distribution and flows in rabbit heart. Am J Phys 258(4 Pt 2):H1012–H1H24

Gore RW (1974) Pressures in cat mesenteric arterioles and capillaries during changes in systemic arterial blood pressure. Circ Res 34(4):581–591

Grandchamp X, Coupier G, Srivastav A, Minetti C, Podgorski T (2013) Lift and down-gradient shear-induced diffusion in red blood cell suspensions. Phys Rev Lett 110(10)

Griffith TM (1996) Temporal chaos in the microcirculation. Cardiovasc Res 31(3):342–358

Griffith TM, Edwards DH (1990) Basal EDRF activity helps to keep the geometrical configuration of arterial bifurcations close to the Murray optimum. J Theor Biol 146(4):545–573. https://doi.org/10.1016/s0022-5193(05)80378-9

Gupta BB, Seshadri V (1977) Flow of red blood cell suspensions through narrow tubes. Biorheology 14(2–3):133–143. https://doi.org/10.3233/bir-1977-142-310

Haldenby KA, Chappell DC, Winlove CP, Parker KH, Firth JA (1994) Focal and regional variations in the composition of the glycocalyx of large vessel endothelium. J Vasc Res 31(1):2–9. https://doi.org/10.1159/000159025

Halikas G, Sheppard CW (1969) The viscosity of water and of blood in small diameter capillary tubes. Anomalous viscosity of blood. Biorheology 6(2):137–142. https://doi.org/10.3233/bir-1969-6208

Harper SL, Bohlen HG (1984) Microvascular adaptation in the cerebral cortex of adult spontaneously hypertensive rats. Hypertension 6(3):408–419

Hashimoto H, Prewitt RL (1986) Arteriolar dimensions from unanesthetized rabbits. Jpn Circ J 50(5):449–454. https://doi.org/10.1253/jcj.50.449

Haynes RH (1960) Physical basis of the dependence of blood viscosity on tube radius. Am J Phys 198:1193–1200. https://doi.org/10.1152/ajplegacy.1960.198.6.1193

Haynes RH, Burton AC (1959) Role of the non-Newtonian behavior of blood in hemodynamics. Am J Phys 197:943–950. https://doi.org/10.1152/ajplegacy.1959.197.5.943

Henry C, Duling B (1998) Hyaluronidase treatment suggests a role for cell surface hyaluronan in determining vascular permeability. Abstract FASEB J 12:139

Henry CB, Duling BR (1999) Permeation of the luminal capillary glycocalyx is determined by hyaluronan. Am J Phys 277(2):H508–H514. https://doi.org/10.1152/ajpheart.1999.277.2.H508

Henry CB, Durán WN, DeFouw DO (1997) Permselectivity of angiogenic microvessels following alteration of the endothelial fiber matrix by oligosaccharides. Microvasc Res 53(2):150–155. https://doi.org/10.1006/mvre.1996.1996

Hertz MM, Paulson OB (1980) Heterogeneity of cerebral capillary flow in man and its consequences for estimation of blood-brain barrier permeability. J Clin Invest 65(5):1145–1151

Hill MA, Trippe KM, Li QX, Meininger GA (1992) Arteriolar arcades and pressure distribution in cremaster muscle microcirculation. Microvasc Res 44(1):117–124. https://doi.org/10.1016/0026-2862(92)90106-y

Hoffman JI (1995) Heterogeneity of myocardial blood flow. Basic Res Cardiol 90(2):103–111. https://doi.org/10.1007/bf00789440

Honig CR, Feldstein ML, Frierson JL (1977) Capillary lengths, anastomoses, and estimated capillary transit times in skeletal muscle. Am J Phys 233(1):H122–H129. https://doi.org/10.1152/ajpheart.1977.233.1.H122

Horsfield K (1990) Diameters, generations, and orders of branches in the bronchial tree. J Appl Physiol 68(2):457–461. https://doi.org/10.1152/jappl.1990.68.2.457

Iversen PO, Nicolaysen G (1995) High correlation of fractals for regional blood flows among resting and exercising skeletal muscles. Am J Phys 269(1 Pt 2):H7–H13. https://doi.org/10.1152/ajpheart.1995.269.1.H7

Jarvis RS, Werritty A (1975) Some comments on testing random topology stream network models. Water Resour Res 11(2):309–318. https://doi.org/10.1029/WR011i002p00309

Jespersen SN, Ostergaard L (2012) The roles of cerebral blood flow, capillary transit time heterogeneity, and oxygen tension in brain oxygenation and metabolism. J Cereb Blood Flow Metab 32(2):264–277

Jiang ZL, Kassab GS, Fung YC (1994) Diameter-defined Strahler system and connectivity matrix of the pulmonary arterial tree. Journal of applied physiology 76(2):882–892. https://doi.org/10.1152/jappl.1994.76.2.882

Johnson PC (1971) Red cell separation in the mesenteric capillary network. Am J Phys 221(1): 99–104. https://doi.org/10.1152/ajplegacy.1971.221.1.99

Kaoui B, Ristow GH, Cantat I, Misbah C, Zimmermann W (2008) Lateral migration of a two-dimensional vesicle in unbounded Poiseuille flow. Phys Rev E 77(2)

Karst NJ, Storey BD, Geddes JB (2015) Oscillations and multiple equilibria in microvascular blood flow. Bull Math Biol 77(7):1377–1400. https://doi.org/10.1007/s11538-015-0089-1

Kiani MF, Cokelet GR, Sarelius IH (1993) Effect of diameter variability along a microvessel segment on pressure drop. Microvasc Res 45(3):219–232

Kiani MF, Pries AR, Hsu LL, Sarelius IH, Cokelet GR (1994) Fluctuations in microvascular blood flow parameters caused by hemodynamic mechanisms. Am J Phys 266(5 Pt 2):H1822–H1828. https://doi.org/10.1152/ajpheart.1994.266.5.H1822

Kim S, Ong PK, Yalcin O, Intaglietta M, Johnson PC (2009) The cell-free layer in microvascular blood flow. Biorheology 46(3):181–189. https://doi.org/10.3233/bir-2009-0530

King RB, Bassingthwaighte JB, Hales JR, Rowell LB (1985) Stability of heterogeneity of myocardial blood flow in normal awake baboons. Circ Res 57(2):285–295. https://doi.org/10.1161/01.res.57.2.285

Klitzman B, Duling BR (1979) Microvascular hematocrit and red cell flow in resting and contracting striated muscle. Am J Phys 237(4):H481–HH90

Klitzman B, Johnson PC (1982) Capillary network geometry and red cell distribution in hamster cremaster muscle. Am J Phys 242(2):H211–H2H9

Koning NJ, Vonk AB, Vink H, Boer C (2016) Side-by-side alterations in Glycocalyx thickness and perfused microvascular density during acute microcirculatory alterations in cardiac surgery. Microcirculation 23(1):69–74. https://doi.org/10.1111/micc.12260

Köry J, Maini PK, Pitt-Francis JM, Byrne HM (2022) Dependence of cell-free-layer width on rheological parameters: combining empirical data on flow separation at microvascular bifurcations with geometrical considerations. Phys Rev E 105(1–1):014414. https://doi.org/10.1103/PhysRevE.105.014414

Krogh A (1919a) The number and the distribution of capillaries in muscle with the calculation of the oxygen pressure necessary for supplying the tissue. J Physiol Lond 52:409–515

Krogh A (1919b) The supply of oxygen to the tissues and the regulation of the capillary circulation. J Physiol 52(6):457–474

Kuhnle GE, Groh J, Leipfinger FH, Kuebler WM, Goetz AE (1993) Quantitative analysis of network architecture, and microhemodynamics in arteriolar vessel trees of the ventilated rabbit lung. Int J Microcirc Clin Exp 12(3):313–324

Landis EM (1930) Micro-injection studies of capillary blood pressure in human skin. Heart 15:209–228

Lang DJ, Johns BL (1987) Rat venule mechanical characteristics during venous pressure elevation. Am J Phys 252(4 Pt 2):H704–H713. https://doi.org/10.1152/ajpheart.1987.252.4.H704

Lee RM, Garfield RE, Forrest JB, Daniel EE (1983) Morphometric study of structural changes in the mesenteric blood vessels of spontaneously hypertensive rats. Blood Vessels 20(2):57–71

Lee JI, Sato M, Ushida K, Mochida J (2011) Measurement of diffusion in articular cartilage using fluorescence correlation spectroscopy. BMC Biotechnol 11:19. https://doi.org/10.1186/1472-6750-11-19

Leighton D, Acrivos A (1987) The shear-induced migration of particles in concentrated suspensions. J Fluid Mech 181:415–439

Levine R, Goldsmith HL (1977) Particle behavior in flow through small bifurcations. Microvasc Res 14(3):319–344. https://doi.org/10.1016/0026-2862(77)90029-2

Ley K, Pries AR, Gaehtgens P (1986) Topological structure of rat mesenteric microvessel networks. Microvasc Res 32:315–332

Lipowsky HH, Zweifach BW (1974) Network analysis of microcirculation of cat mesentery. Microvasc Res 7(1):73–83
Lipowsky HH, Zweifach BW (1977) Methods for the simultaneous measurement of pressure differentials and flow in single unbranched vessels of the microcirculation for rheological studies. Microvasc Res 14(3):345–361
Lipowsky HH, Kovalcheck S, Zweifach BW (1978) The distribution of blood rheological parameters in the mirovasculature of the cat mesentery. Circ Res 43(5):738–749
Lipowsky HH, Usami S, Chien S (1980) In vivo measurements of "apparent viscosity" and microvessel hematocrit in the mesentery of the cat. Microvasc Res 19(3):297–319
Lo A, Fuglevand AJ, Secomb TW (2003) Oxygen delivery to skeletal muscle fibers: effects of microvascular unit structure and control mechanisms. Am J Physiol Heart Circ Physiol 285(3): H955–HH63
Long DS, Smith ML, Pries AR, Ley K, Damiano ER (2004) Microviscometry reveals reduced blood viscosity and altered shear rate and shear stress profiles in microvessels after hemodilution. Proc Natl Acad Sci USA 101(27):10060–10065
Lücker A, Secomb TW, Weber B, Jenny P (2018) The relation between capillary transit times and hemoglobin saturation heterogeneity. Part 1: theoretical models. Front Physiol 9:420. https://doi.org/10.3389/fphys.2018.00420
Luft JH (1966) Fine structures of capillary and endocapillary layer as revealed by ruthenium red. Fed Proc 25(6):1773–1783
Maeda N, Suzuki Y, Tanaka S, Tateishi N (1996) Erythrocyte flow and elasticity of microvessels evaluated by marginal cell-free layer and flow resistance. Am J Phys Heart Circ Phys 271(6): H2454–H2H61
Mall FP (1888) Die Blut und Lymphwege in Dunndarm des Hundes. Abhandlungen der Mathematisch-Physischen Classe der Königlich Sachsischen Gessellschaft der Wissenschaften 14:151–200
Mantegazza A, Clavica F, Obrist D (2020) In vitro investigations of red blood cell phase separation in a complex microchannel network. Biomicrofluidics 14(1):014101. https://doi.org/10.1063/1.5127840
Martini P, Pierach A, Schreyer E (1930) Die Strömung des Blutes in engen Gefäβen. Eine Abweichung vom Poiseuille'schen Gesetz. Dt Arch. Klin Med 169:212–222
Mauroy B, Filoche M, Weibel ER, Sapoval B (2004) An optimal bronchial tree may be dangerous. Nature 427(6975):633–636. https://doi.org/10.1038/nature02287
McCourt PA, Smedsrød BH, Melkko J, Johansson S (1999) Characterization of a hyaluronan receptor on rat sinusoidal liver endothelial cells and its functional relationship to scavenger receptors. Hepatology 30(5):1276–1286. https://doi.org/10.1002/hep.510300521
McGuire BJ, Secomb TW (2001) A theoretical model for oxygen transport in skeletal muscle under conditions of high oxygen demand. J Appl Physiol 91(5):2255–2265
McGuire BJ, Secomb TW (2003) Estimation of capillary density in human skeletal muscle based on maximal oxygen consumption rates. Am J Physiol Heart Circ Physiol 285(6):H2382–H2H91
McGuire BJ, Secomb TW (2004) Theoretical predictions of maximal oxygen consumption in hypoxia: effects of transport limitations. Respir Physiol Neurobiol 143(1):87–97
McKay CB, Meiselman HJ (1988) Osmolality-mediated Fahraeus and Fahraeus-Lindqvist effects for human RBC suspensions. Am J Phys 254(2 Pt 2):H238–H249. https://doi.org/10.1152/ajpheart.1988.254.2.H238
McKay CB, Meiselman HJ (1989) Osmolality- and hematocrit-mediated flow behavior of RBC suspensions in 33 micrometer ID tubes. Biorheology 26(4):863–874. https://doi.org/10.3233/bir-1989-26416
McWhirter JL, Noguchi H, Gompper G (2009) Flow-induced clustering and alignment of vesicles and red blood cells in microcapillaries. Proc Natl Acad Sci USA 106(15):6039–6043
Merlo A, Berg M, Duru P, Risso F, Davit Y, Lorthois S (2022) A few upstream bifurcations drive the spatial distribution of red blood cells in model microfluidic networks. Soft Matter 18(7): 1463–1478. https://doi.org/10.1039/d1sm01141c

Miller BG, Connors BA, Bohlen HG, Evan AP (1987) Cell and wall morphology of intestinal arterioles from 4- to 6- and 17- to 19-week-old Wistar-Kyoto and spontaneously hypertensive rats. Hypertension 9(1):59–68. https://doi.org/10.1161/01.hyp.9.1.59

Miranda M, Balarini M, Caixeta D, Bouskela E (2016) Microcirculatory dysfunction in sepsis: pathophysiology, clinical monitoring, and potential therapies. Am J Physiol Heart Circ Physiol 311(1):H24–H35. https://doi.org/10.1152/ajpheart.00034.2016

Mulvany MJ (2012) Small artery remodelling in hypertension. Basic Clin Pharmacol Toxicol 110(1):49–55

Murray CD (1926) The physiological principle of minimum work. I.The vascular system and the cost of blood volume. Proc Natl Acad Sci USA 12:207–214

Nandi A, Estess P, Siegelman MH (2000) Hyaluronan anchoring and regulation on the surface of vascular endothelial cells is mediated through the functionally active form of CD44. J Biol Chem 275(20):14939–14948. https://doi.org/10.1074/jbc.275.20.14939

Neu B, Meiselman HJ (2002) Depletion-mediated red blood cell aggregation in polymer solutions. Biophys J 83(5):2482–2490. https://doi.org/10.1016/s0006-3495(02)75259-4

Neuhaus D, Fedde MR, Gaehtgens P (1992) Changes in haemorheology in the racing greyhound as related to oxygen delivery. Eur J Appl Physiol Occup Physiol 65(3):278–285. https://doi.org/10.1007/bf00705094

Ng YC, Fisher LK, Salim V, Kim S, Namgung B (2016) Visualization and quantification of the cell-free layer in arterioles of the rat Cremaster muscle. J Vis Exp 116. https://doi.org/10.3791/54550

Nieuwdorp M, Mooij HL, Kroon J, Atasever B, Spaan JA, Ince C et al (2006) Endothelial glycocalyx damage coincides with microalbuminuria in type 1 diabetes. Diabetes 55(4): 1127–1132. https://doi.org/10.2337/diabetes.55.04.06.db05-1619

Nieuwdorp M, Meuwese MC, Mooij HL, Ince C, Broekhuizen LN, Kastelein JJ et al (2008) Measuring endothelial glycocalyx dimensions in humans: a potential novel tool to monitor vascular vulnerability. Journal of applied physiology (Bethesda, Md : 1985) 104(3):845–852. https://doi.org/10.1152/japplphysiol.00440.2007

Obrist D, Weber B, Buck A, Jenny P (2010) Red blood cell distribution in simplified capillary networks. Philos Transact A Math Phys Eng Sci 2010(368):2897–2918. https://doi.org/10.1098/rsta.2010.0045

Osterloh K, Ewert U, Pries AR (2002) Interaction of albumin with the endothelial cell surface. Am J Physiol Heart Circ Physiol 283(1):H398–H405. https://doi.org/10.1152/ajpheart.00558.2001

Pappenheimer JR (1984) Contributions to microvascular research of Jean Léonard Marie Poiseuille. In: Renkin EM, Michel CC (eds) Handbook of physiology: the cardiovascular system. American Physiological Society, Bethesda, pp 1–10

Perkkiö J, Hokkanen J, Keskinen R (1986) Theoretical model of phase separation of erythrocytes, platelets, and plasma at branches. Med Phys 13(6):882–886. https://doi.org/10.1118/1.595813

Platts SH, Duling BR (2004) Adenosine A3 receptor activation modulates the capillary endothelial glycocalyx. Circ Res 94(1):77–82. https://doi.org/10.1161/01.res.0000108262.35847.60

Poiseuille JLM (1828) Recherches sur la force du coeur aortique. Didot le jeune, Paris

Poiseuille JLM (1830) Recherches sur les causes du mouvement du sang dans les veins. J Phvsiol Exp Pathol 10:277–295

Poiseuille JLM (1840) Recherches expérimentelles sur le mouvement des liquids dans les tubes de très petits diamètres. Comptes Rendus 11:961–967

Popel AS (1979) Effect of heterogeneity of capillary flow on the capillary hematocrit. American Society of Mechanical Engineers, Applied Mechanics Division 32:83–84

Popel AS (1980) A model of pressure and flow distribution in branching networks. J Appl Mech 47(2):247–253. https://doi.org/10.1115/1.3153650

Popel AS (1989) Theory of oxygen transport to tissue. Crit Rev Biomed Eng 17(3):257–321

Pranay P, Henriquez-Rivera RG, Graham MD (2012) Depletion layer formation in suspensions of elastic capsules in Newtonian and viscoelastic fluids. Phys Fluids 24(6)

Price RJ, Less JR, Van Gieson EJ, Skalak TC (2002) Hemodynamic stresses and structural remodeling of anastomosing arteriolar networks: design principles of collateral arterioles. Microcirculation 9(2):111–124. https://doi.org/10.1038/sj/mn/7800127
Pries A, Gaehtgens P (1989) Dispersion of blood cell flow in microvascular networks. In: Lee J, Skalak T (eds) Microvascular mechanics, hemodynamics of systemic and pulmonary microcirculation. Springer, New York, pp 39–49
Pries AR, Kuebler WM (2006) Normal endothelium. In: Handbook of experimental pharmacology, pp 1–40. https://doi.org/10.1007/3-540-32967-6_1
Pries AR, Secomb TW (2005) Microvascular blood viscosity in vivo and the endothelial surface layer. Am J Physiol Heart Circ Physiol 289(6):H2657–H2H64
Pries AR, Secomb TW (2009) Origins of heterogeneity in tissue perfusion and metabolism. Cardiovasc Res 81(2):328–335
Pries AR, Secomb TW (2014) Making microvascular networks work: angiogenesis, remodeling, and pruning. Physiology 29(6):446–455. https://doi.org/10.1152/physiol.00012.2014
Pries AR, Albrecht KH, Gaehtgens P (1981) Model studies on phase separation at a capillary orifice. Biorheology 18(3–6):355–367. https://doi.org/10.3233/bir-1981-183-605
Pries AR, Kanzow G, Gaehtgens P (1983) Microphotometric determination of hematocrit in small vessels. Am J Phys 245(1):H167–H177. https://doi.org/10.1152/ajpheart.1983.245.1.H167
Pries AR, Ley K, Gaehtgens P (1986) Generalization of the Fahraeus principle for microvessel networks. Am J Phys 251(6 Pt 2):H1324–H1332
Pries AR, Ley K, Claassen M, Gaehtgens P (1989) Red cell distribution at microvascular bifurcations. Microvasc Res 38(1):81–101
Pries AR, Secomb TW, Gaehtgens P, Gross JF (1990) Blood flow in microvascular networks. Experiments and simulation Circ Res 67(4):826–834
Pries AR, Neuhaus D, Gaehtgens P (1992a) Blood viscosity in tube flow: dependence on diameter and hematocrit. Am J Phys 263:H1770–H17H8
Pries AR, Fritzsche A, Ley K, Gaehtgens P (1992b) Redistribution of red blood cell flow in microcirculatory networks by hemodilution. Circ Res 70(6):1113–1121. https://doi.org/10.1161/01.res.70.6.1113
Pries AR, Secomb TW, Gessner T, Sperandio MB, Gross JF, Gaehtgens P (1994) Resistance to blood flow in microvessels in vivo. Circ Res 75(5):904–915
Pries AR, Secomb TW, Gaehtgens P (1995a) Structure and hemodynamics of microvascular networks: heterogeneity and correlations. Am J Phys 269(5 Pt 2):H1713–H1H22
Pries AR, Secomb TW, Gaehtgens P (1995b) Design principles of vascular beds. Circ Res 77(5):1017–1023
Pries AR, Secomb TW, Gaehtgens P (1996a) Relationship between structural and hemodynamic heterogeneity in microvascular networks. Am J Phys 270(2 Pt 2):H545–HH53
Pries AR, Secomb TW, Gaehtgens P (1996b) Biophysical aspects of blood flow in the microvasculature. Cardiovasc Res 32(4):654–667
Pries AR, Schonfeld D, Gaehtgens P, Kiani MF, Cokelet GR (1997a) Diameter variability and microvascular flow resistance. Am J Phys 272(6 Pt 2):H2716–H2H25
Pries AR, Secomb TW, Jacobs H, Sperandio M, Osterloh K, Gaehtgens P (1997b) Microvascular blood flow resistance: role of endothelial surface layer. Am J Phys 273(5 Pt 2):H2272–H22H9
Pries AR, Secomb TW, Gaehtgens P (1998a) Structural adaptation and stability of microvascular networks: theory and simulations. Am J Phys 275(2 Pt 2):H349–HH60
Pries AR, Secomb TW, Sperandio M, Gaehtgens P (1998b) Blood flow resistance during hemodilution: effect of plasma composition. Cardiovasc Res 37(1):225–235
Pries AR, Secomb TW, Gaehtgens P (1999) Structural autoregulation of terminal vascular beds: vascular adaptation and development of hypertension. Hypertension 33(1):153–161
Pries AR, Secomb TW, Gaehtgens P (2000) The endothelial surface layer. Pflugers Arch 440(5):653–666
Pries AR, Reglin B, Secomb TW (2001a) Structural adaptation of microvascular networks: functional roles of adaptive responses. Am J Physiol Heart Circ Physiol 281(3):H1015–H1H25

Pries AR, Reglin B, Secomb TW (2001b) Structural adaptation of vascular networks: role of the pressure response. Hypertension 38(6):1476–1479. https://doi.org/10.1161/hy1201.100592

Pries AR, Hopfner M, le Noble F, Dewhirst MW, Secomb TW (2010) The shunt problem: control of functional shunting in normal and tumour vasculature. Nat Rev Cancer 10(8):587–593

Quick S, Moss J, Rajani RM, Williams A (2021) A vessel for change: endothelial dysfunction in cerebral small vessel disease. Trends Neurosci 44(4):289–305. https://doi.org/10.1016/j.tins.2020.11.003

Rakusan K, Wicker P (1990) Morphometry of the small arteries and arterioles in the rat heart: effects of chronic hypertension and exercise. Cardiovasc Res 24(4):278–284

Rasmussen PM, Smith AF, Sakadzic S, Boas DA, Pries AR, Secomb TW et al (2017) Model-based inference from microvascular measurements: combining experimental measurements and model predictions using a Bayesian probabilistic approach. Microcirculation 24(4). https://doi.org/10.1111/micc.12343

Rehm M, Haller M, Orth V, Kreimeier U, Jacob M, Dressel H et al (2001) Changes in blood volume and hematocrit during acute preoperative volume loading with 5% albumin or 6% hetastarch solutions in patients before radical hysterectomy. Anesthesiology 95(4):849–856. https://doi.org/10.1097/00000542-200110000-00011

Rehm M, Zahler S, Lötsch M, Welsch U, Conzen P, Jacob M et al (2004) Endothelial glycocalyx as an additional barrier determining extravasation of 6% hydroxyethyl starch or 5% albumin solutions in the coronary vascular bed. Anesthesiology 100(5):1211–1223. https://doi.org/10.1097/00000542-200405000-00025

Reinke W, Johnson PC, Gaehtgens P (1986) Effect of shear rate variation on apparent viscosity of human blood in tubes of 29 to 94 microns diameter. Circ Res 59(2):124–132. https://doi.org/10.1161/01.res.59.2.124

Reinke W, Gaehtgens P, Johnson PC (1987) Blood viscosity in small tubes: effect of shear rate, aggregation, and sedimentation. Am J Phys 253(3 Pt 2):H540–H5H7

Reitsma S, Slaaf DW, Vink H, van Zandvoort MA, Oude Egbrink MG (2007) The endothelial glycocalyx: composition, functions, and visualization. Pflugers Arch 454(3):345–359

Rhodin JA (1967) The ultrastructure of mammalian arterioles and precapillary sphincters. J Ultrastruct Res 18(1):181–223. https://doi.org/10.1016/s0022-5320(67)80239-9

Rhodin JA (1968) Ultrastructure of mammalian venous capillaries, venules, and small collecting veins. J Ultrastruct Res 25(5):452–500. https://doi.org/10.1016/s0022-5320(68)80098-x

Rodbard S (1975) Vascular caliber. Cardiology 60(1):4–49

Rovas A, Lukasz AH, Vink H, Urban M, Sackarnd J, Pavenstädt H et al (2018) Bedside analysis of the sublingual microvascular glycocalyx in the emergency room and intensive care unit - the GlycoNurse study. Scand J Trauma Resusc Emerg Med 26(1):16. https://doi.org/10.1186/s13049-018-0483-4

Rovas A, Seidel LM, Vink H, Pohlkötter T, Pavenstädt H, Ertmer C et al (2019) Association of sublingual microcirculation parameters and endothelial glycocalyx dimensions in resuscitated sepsis. Crit Care 23(1):260. https://doi.org/10.1186/s13054-019-2542-2

Roy TK, Popel AS (1996) Theoretical predictions of end-capillary PO2 in muscles of athletic and nonathletic animals at VO2max. Am J Phys 271(2 Pt 2):H721–HH37

Roy TK, Secomb TW (2021) Effects of impaired microvascular flow regulation on metabolism-perfusion matching and organ function. Microcirculation 28(3):e12673. https://doi.org/10.1111/micc.12673

Roy TK, Secomb TW (2022) Functional implications of microvascular heterogeneity for oxygen uptake and utilization. Physiol Rep 10(10):e15303. https://doi.org/10.14814/phy2.15303

Saegusa S, Isaji S, Kawarada Y (2002) Changes in serum hyaluronic acid levels and expression of CD44 and CD44 mRNA in hepatic sinusoidal endothelial cells after major hepatectomy in cirrhotic rats. World J Surg 26(6):694–699. https://doi.org/10.1007/s00268-001-0292-0

Sarelius IH (1990) An analysis of microcirculatory flow heterogeneity using measurements of transit time. Microvasc Res 40(1):88–98

Schmid-Schönbein H (1988) Fahraeus-effect-reversal (FER) in compaction stasis (CS): microrheological and haemodynamic consequences of intravascular sedimentation of red cell aggregates. Biorheology 25(1–2):355–366. https://doi.org/10.3233/bir-1988-251-245

Schmid-Schonbein G, Zweifach B (1991) Biophysical framework of the microcirculation as an organized functional unit in skeletal muscle: topology pressure, tone, resistance. In: Mulvany M, Aalkjaer C, Heagerty A, Nyborg N, Strandgaard S (eds) Resistance arteries, structure and function. Elsevier Science Publishers B.V.. Excerpta Medica, Amsterdam, New York, Oxford, pp 65–68

Schröder S, Brab M, Schmid-Schönbein GW, Reim M, Schmid-Schönbein H (1990) Microvascular network topology of the human retinal vessels. Fortschritte der Ophthalmologie : Zeitschrift der Deutschen Ophthalmologischen Gesellschaft 87(1):52–58

Schwanke U, Deussen A, Heusch G, Schipke JD (2000) Heterogeneity of local myocardial flow and oxidative metabolism. Am J Physiol Heart Circ Physiol 279(3):H1029–H1035. https://doi.org/10.1152/ajpheart.2000.279.3.H1029

Secomb TW (1987) Flow-dependent rheological properties of blood in capillaries. Microvasc Res 34(1):46–58

Secomb TW (1995) Mechanics of blood flow in the microcirculation. Symp Soc Exp Biol 49:305–321

Secomb TW (2016) Hemodynamics. Compr Physiol 6(2):975–1003. https://doi.org/10.1002/cphy.c150038

Secomb TW (2017) Blood flow in the microcirculation. Annu Rev Fluid Mech 49:443–461. https://doi.org/10.1146/annurev-fluid-010816-060302

Secomb TW, Pries AR (2002) Information transfer in microvascular networks. Microcirculation 9(5):377–387

Secomb TW, Skalak R, Ozkaya N, Gross JF (1986) Flow of axisymmetric red blood cells in narrow capillaries. J Fluid Mech 163:405–423

Secomb TW, Hsu R, Dewhirst MW, Klitzman B, Gross JF (1993) Analysis of oxygen transport to tumor tissue by microvascular networks. Int J Radiat Oncol Biol Phys 25(3):481–489

Secomb TW, Hsu R, Pries AR (1998) A model for red blood cell motion in glycocalyx-lined capillaries. Am J Phys 274(3 Pt 2):H1016–H1H22

Secomb TW, Hsu R, Pries AR (2001a) Effect of the endothelial surface layer on transmission of fluid shear stress to endothelial cells. Biorheology 38(2–3):143–150

Secomb TW, Hsu R, Pries AR (2001b) Motion of red blood cells in a capillary with an endothelial surface layer: effect of flow velocity. Am J Physiol Heart Circ Physiol 281(2):H629–HH36

Secomb TW, Hsu R, Pries AR (2002) Blood flow and red blood cell deformation in nonuniform capillaries: effects of the endothelial surface layer. Microcirculation 9(3):189–196

Secomb TW, Alberding JP, Hsu R, Dewhirst MW, Pries AR (2013) Angiogenesis: an adaptive dynamic biological patterning problem. PLoS Comput Biol 9(3):e1002983. https://doi.org/10.1371/journal.pcbi.1002983

Segal SS (2000) Integration of blood flow control to skeletal muscle: key role of feed arteries. Acta Physiol Scand 168(4):511–518

Segal SS, Jacobs TL (2001) Role for endothelial cell conduction in ascending vasodilatation and exercise hyperaemia in hamster skeletal muscle. J Physiol 536(Pt 3):937–946

Segal SS, Damon DN, Duling BR (1989) Propagation of vasomotor responses coordinates arteriolar resistances. Am J Phys 256(3 Pt 2):H832–H8H7

Sherman TF (1981) On connecting large vessels to small. The meaning of Murray's law. J Gen Physiol 78(4):431–453. https://doi.org/10.1085/jgp.78.4.431

Smith ML, Long DS, Damiano ER, Ley K (2003) Near-wall micro-PIV reveals a hydrodynamically relevant endothelial surface layer in venules in vivo. Biophys J 85(1):637–645

Squire JM, Chew M, Nneji G, Neal C, Barry J, Michel C (2001) Quasi-periodic substructure in the microvessel endothelial glycocalyx: a possible explanation for molecular filtering? J Struct Biol 136(3):239–255. https://doi.org/10.1006/jsbi.2002.4441

Stadler A, Linderkamp O (1989) Flow behavior of neonatal and adult erythrocytes in narrow capillaries. Microvasc Res 37(3):267–279. https://doi.org/10.1016/0026-2862(89)90045-9

Stadler AA, Zilow EP, Linderkamp O (1990) Blood viscosity and optimal hematocrit in narrow tubes. Biorheology 27(5):779–788. https://doi.org/10.3233/bir-1990-27513

Suzuki K, Yamamoto T, Usui T, Suzuki K, Heldin P, Yamashita H (2003) Expression of hyaluronan synthase in intraocular proliferative diseases: regulation of expression in human vascular endothelial cells by transforming growth factor-beta. Jpn J Ophthalmol 47(6):557–564. https://doi.org/10.1016/j.jjo.2003.09.001

Svanes K, Zweifach BW (1968) Variations in small blood vessel hematocrits produced in hypothermic rats by micro-occlusion. Microvasc Res 1(2):210–220. https://doi.org/10.1016/0026-2862(68)90019-8

Sweeney TE, Sarelius IH (1990) Spatial heterogeneity in striated muscle arteriolar tone, cell flow, and capillarity. Am J Phys 259(1 Pt 2):H124–HH36

Tomanek RJ, Palmer PJ, Peiffer GL, Schreiber KL, Eastham CL, Marcus ML (1986) Morphometry of canine coronary arteries, arterioles, and capillaries during hypertension and left ventricular hypertrophy. Circ Res 58(1):38–46. https://doi.org/10.1161/01.res.58.1.38

Toroczkai Z (2002) Topological classification of binary trees using the Horton-Strahler index. Phys Rev E 65(1 Pt 2):016130. https://doi.org/10.1103/PhysRevE.65.016130

Torres Filho IP, Cyrino FZ, Popel AS, Bouskela E, Johnson PC (1994) Morphometric analysis of the anastomosing arteriolar network in cat sartorius muscle. Int J Microcirc Clin Exp 14(1–2): 3–13. https://doi.org/10.1159/000178200

Tronolone JJ, Jain A (2021) Engineering new microvascular networks on-chip: ingredients, assembly, and best practices. Adv Funct Mater 31(14). https://doi.org/10.1002/adfm.202007199

van den Berg BM, Vink H, Spaan JA (2003) The endothelial glycocalyx protects against myocardial edema. Circ Res 92(6):592–594. https://doi.org/10.1161/01.res.0000065917.53950.75

Vicaut E (1986) Statistical estimation of parameters in microcirculation. Microvasc Res 32(2): 244–247. https://doi.org/10.1016/0026-2862(86)90058-0

Vigneshwaran V, Sands GB, LeGrice IJ, Smaill BH, Smith NP (2019) Reconstruction of coronary circulation networks: a review of methods. Microcirculation 26:e12542. https://doi.org/10.1111/micc.12542

Vink H, Duling BR (1996) Identification of distinct luminal domains for macromolecules, erythrocytes, and leukocytes within mammalian capillaries. Circ Res 79(3):581–589

Vlahu CA, Lemkes BA, Struijk DG, Koopman MG, Krediet RT, Vink H (2012) Damage of the endothelial glycocalyx in dialysis patients. J Am Soc Nephrol 23(11):1900–1908. https://doi.org/10.1681/asn.2011121181

Vogel J, Waschke KF, Kuschinsky W (1997) Flow-independent heterogeneity of brain capillary plasma perfusion after blood exchange with a Newtonian fluid. Am J Phys 272(4 Pt 2):H1833–H18H7

Voss R (1983) Enwicklung einer kapillarviskosimetrischen Methode zur Untersuchung der Schubspannungsahhangigkeit des Fahraeus-Lindqvist-Effektes. Dissertation, University of Cologne

Wahl EM, Daniels FH, Leonard EF, Levinthal C, Cortell S (1984) A graph theory model of the glomerular capillary network and its development. Microvasc Res 27(1):96–109. https://doi.org/10.1016/0026-2862(84)90044-x

Wahl EM, Quintas LV, Lurie LL, Gargano ML (2004) A graph theory analysis of renal glomerular microvascular networks. Microvasc Res 67(3):223–230. https://doi.org/10.1016/j.mvr.2003.11.005

Walley KR (1996) Heterogeneity of oxygen delivery impairs oxygen extraction by peripheral tissues: theory. J Appl Physiol 81(2):885–894

Weibel ER (1979) Stereological methods. Academic Press, New York

Weinbaum S, Zhang X, Han Y, Vink H, Cowin SC (2003) Mechanotransduction and flow across the endothelial glycocalyx. Proc Natl Acad Sci USA 100(13):7988–7995

Weinbaum S, Tarbell JM, Damiano ER (2007) The structure and function of the endothelial glycocalyx layer. Annu Rev Biomed Eng 9:121–167

Wiedeman MP (1968) Blood flow through terminal arterial vessels after denervation of the bat wing. Circ Res 22(1):83–89. https://doi.org/10.1161/01.res.22.1.83

Yen RT, Fung YC (1978) Effect of velocity of distribution on red cell distribution in capillary blood vessels. Am J Phys 235(2):H251–H257. https://doi.org/10.1152/ajpheart.1978.235.2.H251

Yin H, Arpino JM, Lee JJ, Pickering JG (2021) Regenerated microvascular networks in ischemic skeletal muscle. Front Physiol 12:662073. https://doi.org/10.3389/fphys.2021.662073

Zakrzewicz A, Secomb TW, Pries AR (2002) Angioadaptation: keeping the vascular system in shape. News Physiol Sci 17:197–201

Zhao Y, Chien S, Weinbaum S (2001) Dynamic contact forces on leukocyte microvilli and their penetration of the endothelial glycocalyx. Biophys J 80(3):1124–1140. https://doi.org/10.1016/s0006-3495(01)76090-0

Zheng L, Golub AS, Pittman RN (1996) Determination of PO2 and its heterogeneity in single capillaries. Am J Phys 271(1 Pt 2):H365–HH72

Zhou Y, Kassab GS, Molloi S (1999) On the design of the coronary arterial tree: a generalization of Murray's law. Phys Med Biol 44(12):2929–2945

Zuurbier CJ, Van Iterson M, Ince C (1999) Functional heterogeneity of oxygen supply-consumption ratio in the heart. Cardiovasc Res 44(3):488–497

Chapter 2
Microcirculatory Exchange Physiology and Molecular Mechanisms Controlling Microvascular Permeability

Jerome W. Breslin, Zeinab Y. Motawe, and Mengmeng Chang

Abstract Exchange between the blood and tissues is the primary function of the microcirculation. The small size and large collective surface area of capillaries facilitate delivery of oxygen, water, and nutrients to the surrounding tissues. However, under inflammatory conditions, the semipermeable barrier of capillaries and postcapillary venules becomes compromised, allowing for excessive leakage of fluids and solutes, which can result in edema. This chapter focuses on the biophysics of fluid and solute exchange across microvascular barriers and the cellular and molecular mechanisms that control microvascular permeability.

Keywords Microvascular permeability · Capillary permeability · Microvascular filtration · Capillary filtration · Edema

2.1 Introduction

Exchange of oxygen, water, ions, nutrients, and waste metabolites between the blood and surrounding tissues is a critical function of the microcirculation. The microvascular endothelium of capillaries and postcapillary venules forms a semipermeable barrier that allows selective transport of fluids and solutes. Normal extravasation of plasma proteins serves to transport certain water-insoluble substances, such as nonesterified fatty acids and hormones, into the interstitium and plays important roles in fluid balance, inflammatory responses, and lymph formation. However, abnormally excessive microvascular leakage can lead to edema and tissue dysfunction. Mild increases in microvascular permeability are associated with diabetes, hypertension, rheumatoid arthritis, and heart disease (Michel and Curry 1999; Yuan et al. 2000, 2007; Jaap and Tooke 1995; Daniels 1993; Yamaji et al. 1993; Mazzoni and Schmid-Schonbein 1996; McDonagh and Hokama 2000). Serious, life-threatening microvascular hyperpermeability is involved in acute respiratory distress

J. W. Breslin (✉) · Z. Y. Motawe · M. Chang
Department of Molecular Pharmacology and Physiology, Morsani College of Medicine, University of South Florida, Tampa, FL, USA
e-mail: breslin@usf.edu

R. F. Tuma et al. (eds.), *Handbook of Microcirculation*, Physiology in Health and Disease, https://doi.org/10.1007/978-3-031-99609-2_2

syndrome (Groeneveld 2002) and multiple organ failure caused by sepsis (Groeneveld et al. 1986; Lush and Kvietys 2000) and thermal injury (Arturson 2000; Lund et al. 1992; Gibran and Heimbach 2000; Cioffi 2001). A major research focus in the past several decades has been to develop a better understanding of the mechanisms that regulate the vascular endothelial barrier, particularly at the cellular level.

Increased permeability of the endothelium is the primary cause of excessive microvascular leakage (Michel and Curry 1999; Renkin 1985). It is important to note, however, that extravasation of fluids and solutes can also be influenced by changes in blood flow, capillary pressure, and efficiency of lymphatic drainage and by the homeostasis of interstitial fluid (Bates and Harper 2003; Curry 1984). Numerous studies have described the biophysics of fluid and solute transport across the endothelium [reviewed in Curry (1984), Michel (1984), Crone and Levitt (1984), and Taylor and Granger (1984)]. In this updated handbook chapter, a brief review of fluid and solute transport and conditions that lead to increased extravasation will be provided, followed by a discussion of the mechanisms that control microvascular permeability.

2.2 Physical Principles Controlling Fluid and Solute Transport

2.2.1 General Characteristics of Fluid and Solute Flux

"Flux," "permeability," and "clearance" are all terms that are commonly used to describe the diffusive transport of water and solutes across biological membranes. While these terms are sometimes used loosely to describe extravasation, they all have specific definitions. Chemical flux is defined as the rate of movement of molecules across a unit area. In the microvasculature, we are concerned with the net fluxes of fluids and solutes across the microvascular endothelium, which separates the blood and the interstitial space.

Permeability is defined as the rate at which a substance can cross the microvascular endothelium, with units of distance/time. The influence of permeability of a biological membrane on flux across that membrane is apparent when examining Fick's first law of diffusion:

$$J_s = P_s S \left(C_{\text{plasma}} - C_{\text{ISF}} \right) \tag{2.1}$$

Fick's first law of diffusion describes that the net diffusive flux of a solute (J_s) across a membrane is the product of the permeability of the membrane to the solute (P_s), the area of the membrane (S), and the concentration gradient of the solute across the membrane. For microvascular transport, the concentration gradient is written as the plasma concentration of the solute (C_{pl}) minus its concentration in the interstitial fluid (C_{ISF}). From Fick's first law of diffusion, it is apparent that an increase in J_s can result by increasing P_s, S, or (C_{pl}–C_{ISF}). While permeability is often cited as the underlying cause for an increase in solute flux, an increase in the number of microvessels perfused will increase the area of exchange and will also cause an increase in flux.

Solutes can also cross the microvascular barrier by convection. When water is filtered from the plasma to the interstitial space, solutes dissolved in the filtrate are carried along. The flux of a solute due to convection, when the concentration gradient of the solute across the endothelium is zero, can be expressed as:

$$J_s = J_v(1 - \sigma)C_{pl} \tag{2.2}$$

where J_v is the volume flux (filtration rate) of water and σ is the reflection coefficient of the endothelial barrier to the solute (sometimes referred to as the ultrafiltration coefficient of the solvent drag coefficient). σ has a value ranging from 0, which represents a solute permeability equal to that of water, to 1, which represents an impermeable solute. It should be noted that as J_v increases, J_s also increases and that J_s is lower for solutes with higher reflection coefficients (as σ approaches 1).

To determine solute flux, first, we need to determine the filtration rate of water. Starling recognized that under normal steady-state conditions, the balance between hydrostatic and osmotic pressures prevents the shifting of fluid into or out of the microvasculature (Starling 1896). Thus, the net flux of fluid across the microvascular barrier is due to an imbalance between hydrostatic and osmotic pressures. At filtration equilibrium, Starling's hypothesis is expressed as:

$$P_c - P_t = \Pi_{pl} - \Pi_t \tag{2.3}$$

where P_c and P_t are the capillary and tissue hydrostatic pressures, respectively, $\prod_{pl}$ is the colloid osmotic pressure of the plasma, and $\prod_t$ is the colloid osmotic pressure of the tissues.

Landis (1927) introduced a dynamic form of Starling's equation to describe the filtration rate of water (F) across capillaries:

$$F = K_{fc}\left[(P_c - P_t) - \left(\Pi_{pl} - \Pi_t\right)\right] \tag{2.4}$$

This equation introduces the capillary filtration coefficient, K_{fc}, which is an indicator of the ease of fluid transfer through a membrane. It is expressed as flow rate divided by driving force (e.g., volume/time/pressure). Kedem and Katchalsky (1958) subsequently added an osmotic reflection coefficient (σ). In their equation, flow became volume flux (J_v) and K_{fc} is expressed as the product of the hydraulic conductivity (L_p) and the surface area for exchange (S):

$$J_v = L_p S\left[(P_c - P_t) - \sigma\left(\Pi_{pl} - \Pi_t\right)\right] \tag{2.5}$$

For simplicity, we can assume that this reflection coefficient is the same as in Eq. 2.2. Because S is expressed in units of area, this makes the units of L_p distance/time/pressure.

Kedem and Katchalsky (1958) also derived an equation to compute solute flux, considering both osmotic pressure and convection:

$$J_s = \omega\Delta\Pi + (1-\sigma)c_{avg}J_v \tag{2.6}$$

In this equation, ω is a permeation coefficient, $\Delta\Pi$ is the osmotic pressure gradient across the microvascular endothelium, σ is the reflection coefficient, and c_{avg} is the mean solute concentration.

Solute flux can also be shown as a function of volume flux with the following relation based on the Patlak equation (Patlak et al. 1963):

$$J_s = J_v(1-\sigma)\left(\frac{C_e - C_T e^{-Pe}}{1 - e^{-Pe}}\right) \tag{2.7}$$

where σ is the solute reflection coefficient, C_c and C_T are the solute concentrations in the blood and tissue, respectively, e is the base of natural logarithms, and *Pe* is a modified Péclet number. *Pe* is a dimensionless value that represents the ratio of imposed velocity to diffusion velocity of a solute and is calculated as:

$$Pe = \frac{J_v(1-\sigma)}{P_s S} \tag{2.8}$$

In Eq. 2.7, the convective component is $J_v(1-\sigma)$, while the remainder of the right side of the equation is related to the ratio of diffusive to convective transport, a function of Pe. It should be noted that as Pe approaches zero and J_v becomes very small, Eq. 2.7 reduces to Fick's first law of diffusion (Eq. 2.1).

Michel incorporated the relationship between solute flux and volume flux, as described in Eq. 2.7, into the formula for J_v. His derivation (Michel 1984) yielded the equation:

$$J_v = L_p S\left[(P_c - P_t) - \sigma^2\Pi_{pl}\left(\frac{1 - e^{-Pe}}{1 - \sigma e^{-Pe}}\right)\right] \tag{2.9}$$

This formula, unlike Eq. 2.5, describes how the relationship between $\prod_t$ and J_v affects the steady-state filtration rate for a given capillary pressure.

This brief review of the theoretical parameters that influence instantaneous changes in volume and solute flux serves as a background for understanding the strengths and weaknesses of different models used to study microvascular permeability. One additional parameter that has been used in studies of microvascular transport is solute clearance. Clearance of a substance from the blood into a tissue can be defined as the solute flux divided by the arterial concentration of the solute (C_a):

$$Cl_s = \frac{J_s}{C_a} \tag{2.10}$$

In some experimental models, clearance may serve as a convenient approximation of the product of diffusional permeability for the solute and exchange surface area (P_sS). However, the following assumptions must be made: (1) J_s must be unidirectional, as may be the case with a tracer molecule infused into the bloodstream. (2) P_s should be low, such that the tracer concentration in microvascular beds is the same as C_a. For large molecules with high reflection coefficients, this assumption is reasonable but not for small hydrophilic molecules. (3) Transport should be dominated by diffusion and not convection. The issues of fluid movement and associated solutes have been reviewed in detail by Curry and colleagues (Michel and Curry 1999; Adamson et al. 2004).

2.2.2 Exchange Microvessels

Capillaries and postcapillary venules function as the main sites of blood–tissue exchange. Several characteristics of these microvessels make them favorable for fluid and solute exchange. One factor is that the flow in capillaries and venules is relatively slow, providing favorable conditions for passive diffusion between the blood and tissues. A key factor is that capillaries present a vast surface area for exchange. In addition, capillaries and postcapillary venules are more permeable to solutes than arterioles. For example, the permeability coefficient for albumin is generally an order of magnitude higher in venules than in arterioles (Table 2.1). Moreover, electrical resistance across the endothelial barrier is higher in arterioles than in venules (Crone and Christensen 1981; Olesen et al. 1984). It is also worth noting that some capillaries are fenestrated for specialized functions, such as glomerular filtration in the kidney and nutrient uptake in the intestinal mucosa.

An important characteristic of capillaries and postcapillary venules is their ability to alter their permeability to fluids and solutes in response to physiological and inflammatory stimuli. Early studies revealed that the initial increase in microvascular permeability caused by histamine and serotonin exclusively affected venules (Majno et al. 1961). Longer, sustained increases in permeability, on the other hand, involve capillaries and venules (Cotran and Majno 1964). Subsequently, with imaging studies utilizing fluorescent tracers, it has been frequently observed that rapid increases in permeability in response to different inflammatory stimuli occur primarily in postcapillary venules (Gawlowski et al. 1982; Gawlowski and Duran 1986; Fox et al. 1980; Horan et al. 1986; Svensjo 1990; Svensjo and Joyner 1984; Wu and Baldwin 1992; Lindbom 2003; Boric et al. 1987; Dillon and Duran 1988).

Table 2.1 Estimated solute permeability coefficients of isolated porcine coronary arterioles and venules

Vessel type	Solute	P_s ($\times 10^{-6}$ cm/s)	Vessel diameter (μm)	References
Arteriole	FITC–BSA	1.7 ± 0.1*	57 (avg.)	Huang et al. (1998)
Arteriole	TRITC–PSA	0.71 ± 0.14*	53 ± 6*	Huxley and Williams (1996)
Arteriole	TRITC–PSA	0.40 ± 0.22*	35 ± 5*	Huxley and Williams (2000)
Arteriole	FITC–BSA or TRITC–BSA	0.64 (average)	29 ± 5	Bingaman et al. (2003)
Venule	FITC–BSA	3.22 ± 0.23*†	20–50	Yuan et al. (1992, 1993a, b, 1995, 1997, 1998, 2000, 2002), Wu et al. (1996, 1999, 2001), Huang and Yuan (1997)
Venule	TRITC–PSA	1.48 ± 0.62*	58 ± 6*	Rumbaut and Huxley (2002)
Venule	FITC–BSA	3.98 (average)	67 ± 14	Bingaman et al. (2003)
Venule	TRITC–BSA	2.98 (average)	67 ± 14	Bingaman et al. (2003)

*Mean ± S.E.M. †Mean ± S.E.M. calculated from the means reported in the studies cited in the same row

FITC fluorescein isothiocyanate, *BSA* bovine serum albumin, *TRITC* tetramethylrhodamine isothiocyanate, *PSA Pisum sativum* agglutinin

2.2.3 *Edema*

The swelling of tissues with excess interstitial fluid, termed "edema," arises due to increased microvascular extravasation of plasma fluid that is not sufficiently removed by lymphatic clearance. Edema, or swelling of tissues, is a serious clinical problem because it increases the diffusion distance for oxygen from capillaries to cells in the surrounding tissues, allowing for potential formation of oxygen radicals that can damage tissues (Scallan et al. 2010; Lampejo et al. 2022). A variety of clinical conditions feature edema, with differential causes such as increased L_p or enhanced permeability, elevated P_c, low $\prod_{pl}$, or combinations of these factors as the primary cause (see Eq. 2.5). In addition, insufficient function of lymphatic vessels to remove excess interstitial fluid can also be a primary cause of edema, which is reviewed elsewhere (Breslin et al. 2019; Breslin 2023). Conditions featuring elevated P_c include deep vein thrombosis, venous valve dysfunction, cardiogenic pulmonary edema, and pregnancy-induced lower limb edema, which all feature elevated venous hydrostatic pressure (P_v), leading to increased P_c and causing microvascular extravasation of fluid (Seem and Stranden 1986; Szabo et al. 1980; McDonagh 1993; Vreim et al. 1976; Fein et al. 1979; Jacobs et al. 1982; Bamigboye and Hofmeyr 2006). Low $\prod_{pl}$ can result from decreased production of albumin in the liver due to malnutrition or loss of plasma proteins via the kidney (proteinuria) due to renal failure or nephritis (Zweifach 1972; Goetsch et al. 2021; Jaryal et al. 2015). In liver diseases such as hepatitis, cirrhosis, or liver cancer, the damaged liver tissue can result in combined high splanchnic P_c and low $\prod_{pl}$, due to portal hypertension and impaired synthesis of albumin, respectively, resulting in formation of ascites and swelling of the abdominal cavity (Breslin 2023). Rapid infusion of crystalloid fluids like normal saline or lactated Ringer's solution can also lead to high P_c and low $\prod_{pl}$ due to the increased fluid volume and dilution of plasma (Spinella et al. 2016; Peters and Hogan 1980; Torres et al. 2017).

Increased L_p and permeability to plasma proteins are a hallmark of inflammation, whether localized such as with an insect bite or widespread as with systemic inflammatory response syndrome due to traumatic injury. A variety of stimuli including inflammatory mediators and activated neutrophils can directly stimulate the endothelium to alter its barrier function, increasing permeability to fluids and solutes. Depending upon the source of inflammation, the increase in permeability may be rapid, such as with the cytokine storm that accompanies traumatic injury, or gradual as diseases such as diabetes mellitus develop over time. In the case of trauma or sepsis, an extreme inflammatory response that includes mediators such as histamine, bradykinin, cytokines, chemokines, and leukocyte-derived factors can all stimulate endothelial cells to increase microvascular permeability. The elevated permeability and resulting edema can lead to hypoxia and tissue dysfunction and, in the case of the lungs, may cause alveolar flooding and acute respiratory distress (Kumar et al. 2009; Durán et al. 2008; Dudek and Garcia 2001; Ware and Matthay 2000). In more slowly developing pathologies such as diabetes mellitus, the increase in oxidative stress, formation of advanced glycation end products (AGEs), and activation of protein kinase C (PKC) family members in endothelial cells all contribute to chronic microvascular hyperpermeability, which, in turn, results in swelling of the lower extremities (Yuan et al. 2007). It is also worth noting that many injuries and illnesses that feature an increase in microvascular permeability may also have additional circulatory problems that compound the problem (Kumar et al. 2009; Durán et al. 2008; Yuan 2000).

2.2.4 Experimental Models for Studying Microvascular Transport

2.2.4.1 Background

Many different experimental models have been developed to study microvascular transport in vivo, in situ, and in vitro. In part, the emergence of several models can be attributed to the diversity of research goals, such as (1) determining the permeability to distinct types of solutes, (2) understanding the structural components of the endothelium that regulate solute flux, (3) testing whether a substance increases endothelial permeability, (4) assessing the ability of pharmacological agents to inhibit enhanced permeability, (5) ascertaining whether altered hemodynamic parameters affect permeability, (6) determining how disease or injury affects permeability, (7) evaluating the endothelial barrier during development, and (8) determining the specific roles of signaling molecules in regulating permeability within the endothelium. Each method has its own strengths and weaknesses, and may or may not be suitable for a particular study, based on several factors, as discussed below.

2.2.4.2 Dye Accumulation Assays

A simple approach to evaluate the movement of solutes from the blood to the tissues is to intravenously inject an animal with a tracer dye, such as Evans blue, or a fluorophore-conjugated protein or tracer, and measure dye accumulation in a particular organ. Evans blue dye quantitatively binds to albumin (Allen and Dralpvats 1948; Rawson 1943), making it a useful marker for macromolecule extravasation. The Miles assay (Miles and Miles 1952) has been used to measure dye accumulation in the skin as a way to assess permeability changes caused by different topically applied agents. The advantages of the Miles assay are that it is a simple method to evaluate extravasation of albumin, each animal can serve as its own control, and it has the ability to test multiple topical agents. One disadvantage, however, is the requirement of accumulation of relatively large amounts of dye in the tissues to observe increased extravasation visually or using spectrophotometric methods (Saria and Lundberg 1983), and it is strongly influenced by changes in blood flow.

Accumulation of dye-labeled albumin has also been examined in other organs and has yielded results similar to those with injected radiolabeled albumin (Udaka et al. 1970). After extracting the dye from the target organ, it can be quantified by either spectrophotometry (Sirois and Edelman 1997; Evans et al. 1987; Sheppard et al. 1986) or fluorometry (Saria and Lundberg 1983). The fluorescence measurement method is more sensitive. The dye can also be observed in sectioned tissues using fluorescence microscopy (Saria and Lundberg 1983) or whole organs using a gel imager equipped for fluorescence (Villalba et al. 2020).

While these assays detect extravasation of macromolecules, they are not appropriate for the determination of solute permeability coefficients. Factors that increase capillary filtration also affect the degree of extravasation. These factors include blood flow (which influences the surface area for exchange), perfusion pressure, L_p, the plasma to tissue solute concentration gradient, and the colloid osmotic pressure gradient (Michel and Curry 1999; Curry 1984). Assays of dye accumulation cannot account for the relative contributions of these parameters. Moreover, several agents that cause extravasation do so in part due to their ability to increase blood flow and thus the surface area for exchange. One example is vascular endothelial growth factor (VEGF)-165, which reportedly causes marked extravasation when tested with the Miles assay (Dvorak et al. 1999). However, when applied to isolated microvessels, VEGF elicits an extremely brief, eightfold increase in capillary L_p (Bates and Curry 1996) and a transient, threefold increase in venular permeability to albumin (Wu et al. 1996) that cannot fully account for the large degree of extravasation observed with the Miles assay. On the other hand, VEGF-induced dilation of arterioles, which increases blood flow and capillary perfusion, plays an important role in microvascular leakage observed with the Miles assay (Bates et al. 1999). Thus, while they provide a quick answer regarding the overall phenomenon, these assays are not useful for studying the mechanistic aspects of solute permeability regulation.

2.2.4.3 Analysis of Solute Accumulation in Lymph

This approach requires several assumptions (Taylor et al. 1985). First, the sampled lymph should reflect the interstitial fluid, i.e., the sampled lymph should be exclusively from the organ of interest prior to reaching a lymph node. Moreover, for this assumption to be true, there should be no exchange of fluids and solutes across the collecting lymphatic wall and no gradients of protein concentrations in the interstitial spaces. The second assumption is that the microvascular endothelium is the major barrier to protein extravasation. Third, charge effects of the microvascular barrier are incorporated into the calculated permeability measurements (P_sS) or reflection coefficients (σ).

Studies utilizing lymph flow and solute concentration have proven extremely useful for the determination of the average σ and P_sS for different solutes in various organs and has aided in understanding increased extravasation of fluids and proteins in response to various stimuli. However, there are several limitations to this method. First, there is no way to tightly control arteriolar diameter and thus the blood flow and surface area of exchange; therefore, P_s cannot be directly determined. Second, several assumptions need to be satisfied to study the signal transduction pathways involved in permeability regulation, particularly the assumption that experimental interventions do not alter the surface area for exchange. Some agents and conditions have also been shown to affect lymphatic wall permeability (Scallan et al. 2013, 2015; Ivanov et al. 2016; Jannaway and Scallan 2021), so this factor would also need to be ruled out. Other concerns arise when targeting a particular signaling molecule because pharmacological agents that block certain signaling pathways are not selective for microvascular endothelial cells. This method is also limited by the assumption that the sampled lymph reflects the composition of the interstitial fluid.

2.2.4.4 Intravital Microscopy Techniques

Several tissues can serve as a "natural window" for microscopic observation of the microcirculation (with varying degrees of surgical preparation), such as the bat wing, rodent ear, hamster cheek pouch, rodent mesentery, and various muscle preparations. Intravital microscopy allows the observation and video recording of blood flow, vessel diameter changes, intravascular behavior of blood cells, and movement of labeled macromolecules. Early studies of microvascular transport utilized

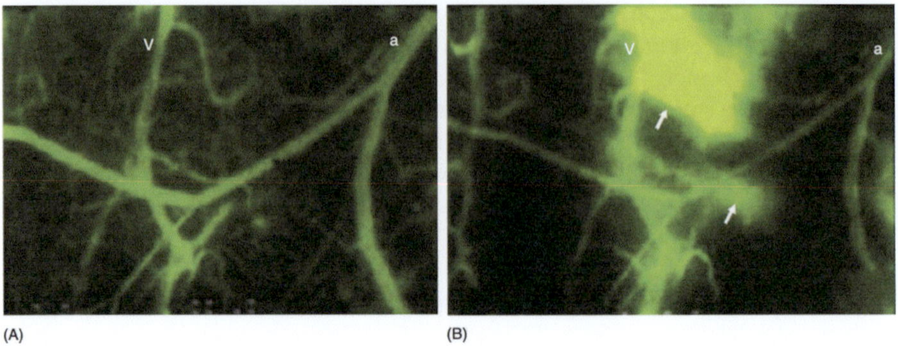

Fig. 2.1 Microvascular hyperpermeability response to platelet-activating factor (PAF; 1.0×10^{-7} M) in the hamster cheek pouch. The details of the microvascular network are observed by fluorescence intravital microscopy using a macromolecular fluorochrome (i.e., FITC-dextran 150 kDa). The branching pattern of the arterioles (a) and tributaries to venules (v) can be followed easily. The control panel (**a**) shows the absence of FITC-dextran 150 in the interstitium. Topical application of PAF (**b**) induced arteriolar constriction and permeation of the fluorescent macromolecular tracer into the interstitium across postcapillary venules (arrows). (Modified from Durán and Dillon 1990. With permission. *FITC* fluorescein isothiocyanate)

intravascularly injected dyes that bind to plasma proteins (Landis 1927). However, this approach was limited by the requirement of high concentration of the labeling dye, the low contrast images obtained, and uncertainty regarding the degree of binding between the dyes and plasma proteins (Levick and Michel 1973).

The development of fluorescence microscopy, photometric imaging techniques, and fluorescein isothiocyanate (FITC)-labeled dextrans and proteins for intravascular injection provided a much more detailed view of the microcirculation, including microvascular leakage sites, upon stimulation with an inflammatory mediator (Fig. 2.1). Using the hamster cheek pouch model as an example, we will first discuss methods for assessing changes in microvascular permeability. Then, we will discuss the differences between the hamster cheek pouch and other "natural windows" from various animals and tissues.

The cheek pouch of the hamster is an invagination of the oral mucosa that extends under the subcutaneous tissue down to the shoulder region. The blood supply for the pouches is mainly from the external branches of the carotid arteries, although some blood is also supplied from the retractor muscle. For the hamster cheek pouch preparation, a hamster is anesthetized and either a jugular vein or a femoral vein is cannulated for administration of supplemental anesthesia, fluorescent tracer, and, in some cases, test agents. The hamster cheek pouch to be studied is placed on a stage with the skin pinned back and the areolar connective tissue carefully removed to expose the microvasculature. The cheek pouch is enclosed in a chamber and continuously suffused with a bicarbonate buffer (Dillon and Duran 1988; Mayhan and Joyner 1984). This preparation remains stable generally for 5–6 h, as determined by arteriolar reactivity to acetylcholine (ACh) (Duling 1973).

A fluorescent tracer, e.g., FITC-dextran, FITC-albumin, or another fluorophore-conjugated molecule of choice, is intravenously injected as an initial bolus and then continuously infused to establish a steady-state plasma concentration (Ramirez et al. 1995; Kim et al. 1993). The time for stabilization of tracer gradients between the blood and interstitial compartments depends on the tracer used. For FITC-dextrans, 30 min has been reported as sufficient (Svensjo 1990). The experiment consists of assessing temporal changes in extravasation by one of three methods: (1) measuring clearance of the tracer in the suffusate over time, (2) counting the number of visible leakage sites in a predetermined viewing window at planned time points (Gawlowski et al. 1982; Svensjo et al. 1978), or (3) measuring the integrated optical intensity (IOI) of the tracer in a predetermined area of the tissue at planned time points (Bekker et al. 1989). The latter approach allows for calculation of permeability and the relative contributions of diffusion and convection to transport macromolecules (Kim et al. 1993; Armenante et al. 1991).

In vivo methods generally present some difficulty when studying endothelial cell signal transduction pathways, due to the presence of several cell types in the tissue and the usual lack of selectivity encountered with pharmacological inhibitors. On the other hand, when compared to cultured endothelial cell monolayer permeability methods (discussed in a subsequent section), in vivo methods have significant physiological relevance. It is worth noting that test substances can be topically applied within the chamber to stimulate localized changes in macromolecular extravasation and minimize systemic cardiovascular effects. Furthermore, although arteriolar diameter cannot be tightly controlled during tissue preparation, it can be monitored to determine whether changes in blood flow occur during the experiment. Experimental strategies can also be employed to avoid potential problems with changes in blood flow, such as the utilization of agents that either increase macromolecule extravasation without increasing arteriolar diameter (Dillon and Duran 1988; Hatakeyama et al. 2006) or change diameter without enhancing permeability (Gawlowski and Duran 1986; Ramirez et al. 1995).

Clearance of a Macromolecular Tracer

The ability to accurately measure clearance of macromolecules in tissues such as the hamster cheek pouch and rat cremaster muscle arose from the development of specialized chambers that allow observation by intravital microscopy yet keep suffusion flow contained (Boric et al. 1987; Mayhan and Joyner 1984). These chambers have an outflow port in which the suffusate may be collected and the concentration of a fluorescent tracer measured. To calculate clearance, the flux term in Eq. 2.10 is expanded to yield the following relation:

$$Cl_s = \frac{(C_s V/t)}{C_a} \tag{2.11}$$

where C_s is the solute concentration in the suffusate and V/t represents the volume of the suffusate collected divided by the time at collection. The total output rate of the tracer into the suffusate is a direct function of the amount of tracer in the interstitial compartment. Likewise, the concentration within the interstitium is a function of the diffusive and convective factors that govern exudation of the tracer from the blood to the tissue. Thus, Cl_s is proportional to P_sS for the tissue under investigation.

The determination of Cl_s is an extremely convenient method for the evaluation of the mean macromolecular extravasation when using intravital preparations. There is no bias in choosing specific leakage sites within the tissue as with imaging methods because the suffusate is collected from the whole tissue. On the other hand, a disadvantage of collecting fluorescence from the suffusion fluid is the loss of spatial resolution. Problems may also arise if the amount of tracer trapped within the interstitium is large with respect to the total tracer output. In this situation, upon stimulation with an inflammatory mediator, the rise in tracer concentration in the suffusate may be delayed with respect to the rise in fluorescence in the tissue. Increasing the rate of suffusion may alleviate this, as higher flow dilutes the concentration of the tracer (Mayhan and Joyner 1984). An additional factor to consider is photobleaching of the tracer, which can potentially occur when fluorescent microscopy is employed during the experiment (Gawlowski and Duran 1986). Extended periods of epifluorescence of FITC-dextran tracers can also cause platelet aggregation and affect hemodynamics (Bekker et al. 1986). However, these potential problems can be minimized by limiting the exposure times when acquiring images (Boric et al. 1987; Bekker et al. 1986). Methods like stroboscopic epifluorescence or two-photon microscopy limit exposure to the excitation light source and are less phototoxic than laser confocal microscopy.

Localized Leakage Sites

The use of a fluorescent tracer in combination with intravital microscopy also allows for the qualitative assessment of macromolecular permeability by visual detection of extravascular fluorescence emitted by the tracer. Several studies have reported that extravascular fluorescence originates at distinct leakage sites, sometimes called "hot spots," which can be counted for semiquantitative evaluation of permeability (Gawlowski et al. 1982; Svensjo et al. 1982; Bekker et al. 1988). A dose–response relationship exists between the number of leakage sites and clearance of the tracer after topical application of inflammatory mediators (Gawlowski and Duran 1986; Mayhan and Joyner 1984). Moreover, there exists a linear relationship such that as the number of leaky sites increases, clearance of the tracer also increases (Mayhan and Joyner 1984). The appearance of leaky sites, and the peak values observed at those sites, may precede the observed overall changes in FITC-dextran clearance (Mayhan and Joyner 1984).

There are, however, a few potential problems when determining the number of leakage sites to quantify changes in permeability; these include the investigator's subjectivity in counting and the possibility of diffuse-type leakage that is below the level of detection. Moreover, the definition of a leakage site is ambiguous because the actual number of intercellular junctions open is unknown. There is also the possibility of overlap between two or more leakage sites in close proximity, merging and appearing as a single stie, particularly when permeability is dramatically elevated throughout the tissue (Gawlowski and Duran 1986; Boric et al. 1987). Furthermore, technical parameters, such as the rate of suffusion, can alter the number of leaky sites observed. High suffusion flow rates dilute the amount of tracer per volume of the suffusate and can essentially wash out some areas where the tracer accumulates, i.e., leakage sites (Mayhan and Joyner 1984). Despite these issues, determining the number of leakage sites can be an extremely useful way to understand temporal changes in macromolecule permeability.

IOI Measurement

Extravasation of a fluorescent tracer can be determined by quantifying its intensity or brightness in the tissue. The development of procedures to digitize fluorescein angiograms combined with intravital fluorescence microscopy spawned methods to determine the apparent interstitial diffusion coefficients in microvascular beds (Bekker et al. 1989). These methods involve sampling and quantifying the brightness of each pixel in an image to determine fluorochrome concentration. Transport parameters are then determined by comparing curves based upon experimentally derived concentration profiles and theoretically predicted concentration distributions. Permeability coefficients to fluorescent tracers have also been determined using changes in integral light intensity in an area adjacent to the microvascular wall as the independent variable (Gerlowski and Jain 1986; Huxley et al. 1987; Ley and Arfors 1982, 1986). Ley and Arfors measured the increase in perivascular tracer concentration over time following microinjection of a second- or third-order arteriole in the hamster cheek pouch. Using a model for one-dimensional diffusion in a composite medium, they determined the apparent permeability coefficient for FITC-dextran-3 kDa in arterioles and postcapillary venules as 0.97 ± 0.31 and $2.64 \pm 0.75 \times 10^{-5}$ cm/s, respectively (Ley and Arfors 1986).

Durán and colleagues developed and validated a variation of the IOI approach to estimate permeability coefficients. A field of view that included a single postcapillary venule (~20-μm diameter) and the surrounding tissue was chosen for study. The window size ranged from 60–120-μm long to 100–180-μm wide, depending on the microvascular geometry. Leakage of FITC-dextran was monitored and video recorded, and frame-by-frame images were digitized into *x*,*y* arrays for analysis. The light intensity in each pixel can be quantitatively represented by a grayscale number (Bekker et al. 1989; Ritter et al. 1985). Integrated optical intensity, a measure of the total gray value for an image or selected area, is calculated as:

$$\text{IOI} = \sum_{x=1}^{n_c} \sum_{y=1}^{n_r} D(x,y) \tag{2.12}$$

where $D(x,y)$ is the intensity (gray level) of a pixel at positions x and y, and n_c and n_r are the total number of columns and rows, respectively, of pixels chosen from the selected frame. Plasma concentrations of FITC-dextran can be maintained at fairly constant levels by continuous intravenous infusion. Thus, the degree of light intensity emitted intravascularly is nearly constant and represents the baseline IOI, which is subtracted from IOI values from subsequent time points (Bekker et al. 1989).

IOI measurements can be used to determine the amount of tracer molecules that have accumulated over time, $W(\tau)$, which, in turn, can be used to determine effective permeability (P) with a nonlinear regression algorithm using a one-dimensional, two-compartmental diffusion model and a step change in macromolecular concentration at the boundary (Bekker et al. 1989):

$$W'(\tau) = \frac{2SD_2KC_0}{K+1} \int_0^{\tau} \left\{ \sum_{n=0}^{\infty} \frac{\alpha^n K}{\sqrt{\pi D_1 t}} \exp\left[- \frac{(2n+1)l}{2\sqrt{D_1 t}} \right]^2 \right\} dt \tag{2.13}$$

An explanation of the variables in Eq. 2.13 is provided in Table 2.2. P is calculated as $P = D_1 l$. The values of P obtained with this analysis were higher than those obtained by whole organ techniques (Table 2.3). This can be explained by (1) the fact that whole organ models do not account for the role that the interstitial space plays in reducing the rate of transfer of macromolecules from the blood to the lymph (Fox and Wayland 1979; Nugent and Jain 1984) and (2) that the P values derived from whole organ studies represent the average measurements of the organ's microvasculature, including both leaky and non-leaky sites, whereas the P values calculated from IOI measurements represent a specific microvascular leakage site.

Table 2.2 Determination of effective permeability coefficients from IOI measurements (Bekker et al. 1989)

Equation 2.13	$W'(\tau) = \frac{2SD_2KC_0}{K+1} \int_0^{\tau} \left\{ \sum_{n=0}^{\infty} \frac{\alpha^n K}{\sqrt{\pi D_1 t}} \exp\left[- \frac{(2n+1)l}{2\sqrt{D_1 t}} \right]^2 \right\} dt$
IOI, integrated optical intensity D_1	Unknown variable; effective diffusion coefficient for solute inside the vessel wall
$W(\tau)$	Total amount of tracer accumulated in the interstitial space in time τ calculated from IOI values, C_0, and the effective interstitial volume
S	Surface area of the leaking vessel
D_2	Effective diffusion coefficient for solute in the interstitial space; from Nugent and Jain (1984)
l	Thickness of the vessel wall
C_0	Tracer concentration in arterial blood
K	$K = (D_1/D_2)^{1/2}$
α	$\alpha = (1—K)/(1 + K)$

Table 2.3 Effective permeability of macromolecules in subcutaneous tissue

Solute	Method	Tissue	$P \times 10^{-8}$ (cm/s)	References
Dextran (dx) 10	Steady-state lymph data	Dog paw	43.71 ± 15.43	Garlick and Renkin (1970)
Dx 20			6.34 ± 0.11	
Dx 40			2.23	
Dx 80			1.71	
Dx 110			1.03 ± 0.03	
Dx 150			0.26	
Dx 150	Intravital microscopy	Rabbit ear	7.26 ± 3.29	Gerlowski and Jain (1986)
Dx 70	Transient analysis of the superfusate	Hamster cheek pouch	4.1	Baxter et al. (1987)
Dx 20	Intravital microscopy	Hamster cheek pouch	42.8 ± 14.7	Bekker et al. (1989)
Dx 40			31.2 ± 9.5	
Dx 70			16.6 ± 5.1	

Adapted from Bekker et al. (1989)

It is worth noting that the calculation of *P* values should take into account the relative contributions of the convective and diffusive fluxes through the vascular wall and interstitial compartment. IOI measurements can be used to ascertain interstitial concentrations of FITC-dextrans, using an in vivo calibration procedure. With these data and mathematical modeling, diffusion was shown to predominate over convection in macromolecule transport in the hamster cheek pouch (Kim et al. 1990, 1993; Armenante et al. 1991).

In some experimental models, particularly in those related to hemorrhagic shock or burn injury, it can be difficult to have consistent intravascular concentrations of tracers due to the massive systemic microvascular leakage in test groups compared to controls. In these models, investigators have adopted the method of collecting data from regions of interest in the interior of a postcapillary venule (I_i) and the regions surrounding the same venule (I_o) and, then, using I_o divided by I_i as an index of microvascular permeability (Alves et al. 2019a; Breslin et al. 2008; Reynoso et al. 2007; Tharakan et al. 2010). Other general issues also worth considering with IOI models are the quenching of fluorescence with high fluorochrome density in the intravascular space, interactions between the tracers, blood, and the extracellular matrix of endothelial cells, and the fact that variations in tissue thickness may affect the observed fluorescence intensity of the tracer.

Comparison of Different Preparations

The hamster cheek pouch, rodent mesentery, and rodent cremaster muscle are commonly used for intravital microscopy. The choice of a particular model will

Table 2.4 Comparison of three different intravital microscopy permeability models

Model	Surgery	Methods	Anatomy	Relevance
Hamster cheek pouch	Minimally invasive: Requires pinning back of the skin and careful dissection of connective tissue above microcirculation. Stable 5–6 h	Clearance, leakage sites, and IOI all OK	Thin muscle layer with no lymph nodes (Svensjo 1990)	Can be used as a model of ischemia–reperfusion (I/R) injury. No transgenic or knockout animals available
Rodent cremaster	Moderately invasive: Requires splaying of cremaster muscle. Stable 5–6 h. Mouse models may be less stable than rat models	Clearance and IOI both OK. Leakage sites merge 5 minutes after bradykinin treatment (Boric et al. 1987)	Thin muscle layer	Can be used as a model of ischemia–reperfusion (I/R) injury. Transgenic and knockout mice available for study
Rodent mesentery	Major abdominal surgery: The mesentery is exteriorized. Necessary to not let blood or fluids from incision come in contact with the mesentery. Stable 5–6 h. Mouse models may be less stable than rat models	Clearance, leakage sites, and IOI all OK	Translucent to transparent tissue	Can be used as a model for septic shock-induced microvascular leakage. Transgenic and knockout mice available for study. Mouse mesentery may have less microcirculation in translucent areas compared to rat mesentery

IOI integrated optical intensity

depend on the objective of the study. For example, if the investigator's interest is cardiac ischemia–reperfusion (I/R) injury, either the hamster cheek pouch or cremaster muscle model would be more relevant than the mesentery model (Suval et al. 1987a, b; Noel et al. 1995; Yasuhara et al. 1991). On the other hand, the mesentery may be more suitable for studies of septic shock (Suzuki et al. 2000). Differences between these models, and their unique features, are summarized in Table 2.4.

2.2.4.5 Studies in Isolated Microvessels

One limitation of in vivo models is that the investigator cannot tightly control changes in arteriolar diameter and blood flow that may alter extravasation of macromolecules. Studies utilizing cannulated or single perfused microvessels address this issue. In addition, isolated microvessel models have the advantages of the ability to directly measure the surface area for exchange, intravascular and interstitial pressures, and solute concentrations. The disadvantages of isolated microvessel models include direct interference with the vessels being studied, which may itself increase permeability, and the limited number of convenient and practical vessel types (e.g., superficial coronary venules) for study (Michel and Curry 1999; Yuan 2000).

Modified Landis Technique to Measure L_p

The determination of capillary L_p is described in detail elsewhere (Michel 1984), and, therefore, here, we will only provide a summary. With this technique, a long (>450 μm), straight capillary is cannulated with a single perfusion micropipette. This micropipette is filled with Ringer's solution containing red blood cells (RBCs) (1% suspension) as flow markers and is connected to a water manometer to control perfusion pressure. At a short distance from the pipette, the vessel is occluded with a glass rod. J_v is then measured at different capillary pressures (Fig. 2.2), using the relation:

$$J_v/S = \left(\frac{\Delta x}{\Delta t}\right)\left(\frac{1}{x_0}\right)\left(\frac{r}{2}\right) \tag{2.14}$$

where Δx is the distance a marker red blood cell travels during a given time (Δt), x_0 is the initial distance between the red blood cell and the occluding rod, and r is the radius of the capillary. L_p is the slope of the line $(\Delta J_v/S)/\Delta P$. The major assumptions are (1) the vessel is cylindrical, and, therefore, surface to volume is equal to r/2, (2) the pressure in the pipette equals the pressure in the capillary upon occlusion,

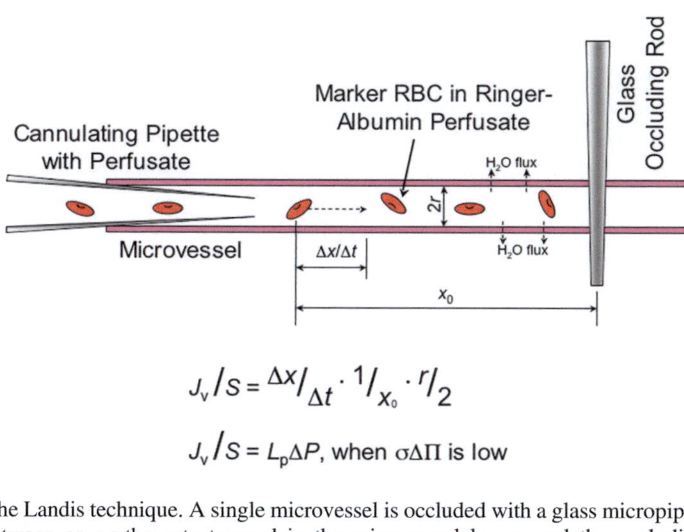

Fig. 2.2 The Landis technique. A single microvessel is occluded with a glass micropipette, and the distance between an erythrocyte trapped in the microvessel lumen and the occluding probe is measured as a function of time. The microvessel diameter is also measured. Volume flux (J_v/S) is estimated using the initial erythrocyte velocity. Hydraulic conductivity (L_p) is determined as the slope of J_v/S as a function of mean occluded intraluminal pressure (P)

(3) the movement of the marker red blood cell reflects filtration and not compliance, and (4) solute concentrations across the microvascular wall do not change during the measurement period (a few seconds or less).

Determination of P_s to Albumin in Isolated Venules

Yuan et al. developed an isolated coronary venule model to determine apparent permeability to albumin (Yuan et al. 1992). This procedure differs from all the previously discussed models, such that the venule is excised from the tissue and cannulated on both ends with glass micropipettes (Fig. 2.3). One micropipette is connected to an inflow reservoir and one to an outflow reservoir, with both containing an albumin-physiological salt solution (APSS). A third, smaller glass micropipette is inserted into the inflow pipette and is connected to a reservoir containing a fluorescent tracer (e.g., FITC-albumin) solution in APSS. The reservoirs are mounted on wall manometers to allow controlling pressure and intraluminal flow by changing the heights of the reservoirs. The venule is viewed with a fluorescent microscope equipped with a camera, recording equipment, and a computer for analysis. The advantage to this system is that myogenic, neurogenic, and hemodynamic factors, as well as the composition of the bath solution, can be tightly controlled. Moreover, the absence of any influence by nonvascular cells simplifies studies regarding venular signal transduction pathways (Yuan 2000).

The calculation of P_s is based on quantitative fluorescence microscopy analysis of single perfused capillaries in the rat mesentery (Huxley et al. 1987). The venular lumen is rapidly filled with the fluorescent tracer by switching flow from the outer inflow pipette to the inner inflow pipette (Fig. 2.3a). This causes a step increase in the luminal concentration of the tracer (Fig. 2.3b), which is directly related to the measured step increase in fluorescence intensity within the lumen (ΔI_{f0}) divided by the luminal volume (V_c). Following the initial step increase, there is also an additional rate of increase in fluorescence intensity $(dI_f/dt)_0$, due to transmural leakage of the solute. In terms of Fick's first law of diffusion (Eq. 2.1),

$$J_s = \left(\frac{dI_f}{dt}\right)_0 \tag{2.15}$$

and

$$\left(C_{\text{plasma}} - C_{ISF}\right) = \frac{\Delta I_{f_0}}{V_c} \tag{2.16}$$

assuming that the fluorescent tracer crosses the venular wall exclusively by diffusion. Assuming the venule has a cylindrical geometry, V_c/S is equal to $r/2$. Thus, Fick's first law can be written as:

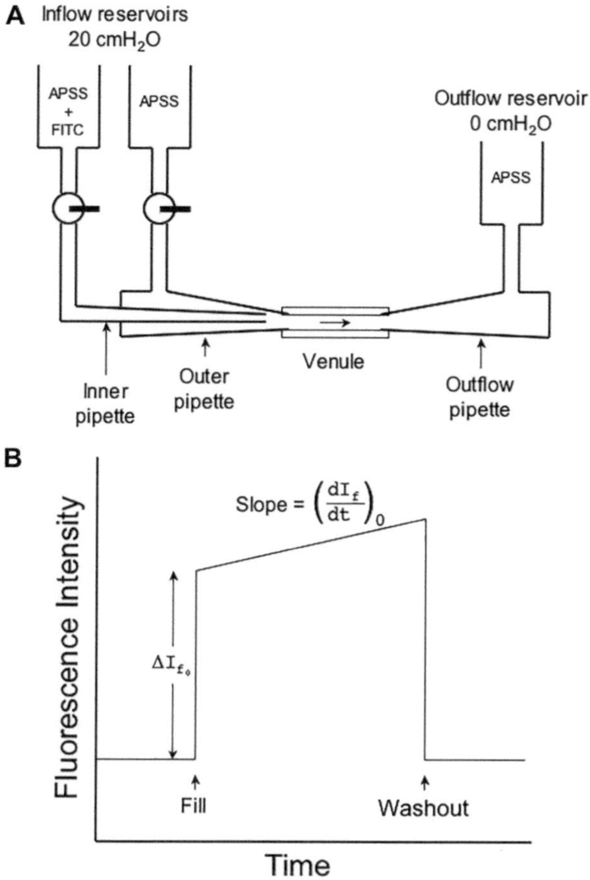

Fig. 2.3 The isolated venule technique. (**a**) A schematic of the experimental setup. A venule is dissected, excised, and mounted on two glass micropipettes and bathed with an albumin-physiological saline solution at 37 °C. Pressure and flow are controlled by altering the heights of the inflow and outflow reservoirs, which are mounted on manometers. (**b**) To assess permeability, the fluorescence intensity of a window containing the venule and the nearby extraluminal area is measured. The inflow is quickly switched from normal APSS to APSS containing a fluorochrome-labeled tracer (e.g., FITC-albumin), causing a step increase in fluorescence intensity (ΔI_{f0}) that is proportional to the number of solute tracer molecules that have entered the lumen. This is followed by a gradual increase in intensity $(dI_f/d_t)_0$. The rate of $(dI_f/d_t)_0$ is proportional to solute transport across the microvascular wall. The intensity returns to the basal level when the tracer is washed out from the lumen. (Adapted from Yuan et al. 1992 and Huxley et al. 1987. *APSS* albumin-physiological salt solution, *FITC* fluorescein isothiocyanate)

$$\left(\frac{\Delta I_f}{dt}\right)_0 = P_s\left(\frac{2}{r}\right)(\Delta I_f) \tag{2.17}$$

or

$$P_s = \left(\frac{1}{\Delta I_f}\right)\left(\frac{\Delta I_f}{dt}\right)_0\left(\frac{r}{2}\right) \tag{2.18}$$

to obtain the apparent permeability for the solute. It is worth noting that the calculated P_s overestimates the true diffusive permeability because macromolecular extravasation is also coupled to water filtration. However, measures can be taken to stabilize the transmural pressure and account for pressure-dependent contributions to extravasation (Yuan et al. 1992).

The isolated venule model developed by Yuan et al. (1992) differs from other isolated microvessel models in which the cannulated vessel remains in the tissue (Huxley et al. 1987) or is excised along with a significant amount of non-microvessel tissue (Huxley and Williams 1996). While there is the concern of trauma to the adventitial layer, ultrastructural studies have shown that isolated coronary venules have an intact endothelial layer and that the venules display solute permeability coefficients within the physiological range. Moreover, the isolated venules respond to shear and inflammatory stimuli in a time- and dose-dependent manner and the responses are reversible (Yuan et al. 1993a). Thus, the advantage of Yuan's isolated venule model is that only venular endothelial cells and pericytes are present within the system and the surrounding environment can be tightly controlled with many extrinsic factors eliminated (Yuan et al. 1992). This has been particularly advantageous for studying signal transduction in the endothelium and allows experimental manipulations such as delivery of cell membrane-impermeable peptides or proteins into the venule by protein transfection methods. This technical approach enables direct study of the involvement of various intracellular signaling cascades in the regulation of microvascular function (Tinsley et al. 2001, 2002; Yuan et al. 2002).

2.2.4.6 Endothelial Cell Monolayer Models

The development of endothelial cell culture methods in the 1970s (Jaffe et al. 1973) propagated the use of in vitro endothelial cell monolayer models of solute permeability. A major advantage of these models is their ability to examine the barrier function of endothelial cells in the absence of other cell types. This feature allows for experimental manipulations that are not readily available in vivo, such as transfection of plasmid DNA, suppression of gene expression by antisense oligonucleotides, silencing RNA (siRNA) or short hairpin RNA (shRNA), or transfection of exogenous proteins into the endothelial layer. The investigator also has tight control over the hydrostatic pressure, shear rate, and chemical composition of the abluminal and luminal fluids as well as direct access to them. This type of model also reduces the number of animals needed for experimentation, which is particularly attractive to the pharmaceutical industry (Gumbleton and Audus 2001). Because of these features, endothelial cell monolayer permeability models have proven quite useful in the understanding of endothelial cell signal transduction pathways that influence microvascular permeability. However, the features that contribute to the simplicity of cell monolayer permeability models also raise questions about their relevance to barrier function in vivo.

Typically, cells are grown on a porous membrane such as polycarbonate. Commercially available membrane inserts (such as Transwell) have an upper and lower chamber separated by a membrane. Optimally, the pore size in the membrane is small enough to allow passive diffusion of macromolecules but prevent endothelial cells from migrating to the opposite side. The membrane is usually coated with matrix proteins such as fibronectin, gelatin, or collagen or with a mixture such as Matrigel, to facilitate attachment of endothelial cells. It is worth noting that matrix proteins influence the phenotype of endothelial cells, which needs to be considered when attempting to reproduce previous findings. Cells are seeded onto the membrane (in the upper chamber) and grown to form a confluent monolayer. Because the cells are difficult to visualize on polycarbonate, confluency can be confirmed by labeling with trypan blue at the end of an experiment or in a small sample of replicate wells (Albelda et al. 1988). Alternatively, cells can be grown on transparent polyester membranes, which allow visualization of the cells before, during, or after an experiment.

To determine permeability, a tracer is added to the chamber containing the cell monolayer (luminal chamber) and the flux of the tracer to the opposite chamber (abluminal chamber) is measured (Fig. 2.4). Experiments are typically performed using a serum-free medium at appropriate volumes in each chamber to prevent any hydrostatic pressure gradients. They are also generally performed under no-flow conditions or by continuous stirring to prevent the formation of unstirred layers. After flux values for the tracer are obtained, its permeability coefficient is calculated using Fick's first law of diffusion (Eq. 2.1).

One major concern about endothelial monolayer permeability models is the origin and lineage of the endothelial cells. Several studies have utilized endothelial cells obtained from large vessels rather than microvascular endothelial cells for a few reasons: (1) Endothelial cells from large vessels are relatively easy to obtain, whereas microvascular endothelial cells usually have small yields upon harvesting. (2) The fact that microvascular endothelial cells are harvested in low yields means they must be used at higher passages for experiments. (3) Microvascular endothelial cells often require additional serum and/or growth factors present in the medium during growth, which may confound data obtained during experiments. Nevertheless, it is generally believed that microvascular endothelial cells will better represent the microcirculation (Breslin and Yuan 2024).

Although microvascular endothelial cells have become more readily available in recent years, there is an additional issue regarding subpopulations of arteriolar, capillary, venular, and lymphatic lineages, which are expected to have varying degrees of barrier function based on their behavior in vivo. To address this concern, Schelling et al. isolated bovine coronary venular endothelial cells (CVECs) so that they could be studied directly (Schelling et al. 1988). Subsequently, several reports generated by Yuan et al. have shown a notable similarity between the permeability changes caused by activated neutrophils in isolated venules and in CVEC monolayers (Yuan et al. 2002; Tinsley et al. 1999; Breslin and Yuan 2004).

An additional concern arises when examining the morphological characteristics of an endothelial cell monolayer grown in vitro. Albelda et al. reported that gaps of

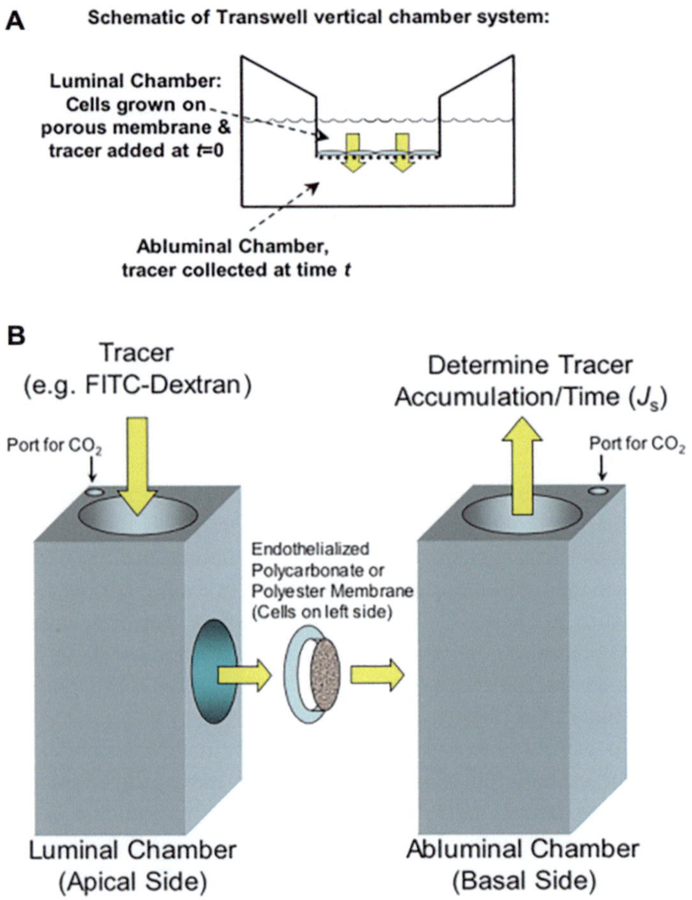

Fig. 2.4 Cultured endothelial cell permeability models. An endothelial cell monolayer is grown on a porous membrane that is placed between two chambers. (**a**) A vertical chamber system is shown, in which the upper chamber represents the "luminal" side and the lower chamber represents the "abluminal" side of the endothelial monolayer. It should be noted that the fluid levels of the upper and lower chambers are kept at the same height to minimize any hydrostatic differences between the two chambers. (**b**) A horizontal setup

0.5–2.0 μm were visible between 5% and 10% of cultured endothelial cells (Albelda et al. 1988), which are not seen in microvessels in vivo. This serves to explain why permeability coefficients from cultured endothelial cell monolayer models are typically 10–100 times higher than those obtained in vivo (Albelda et al. 1988). As a result, the magnitude of agent-induced permeability changes using cultured endothelial cells may possibly be diminished or even absent. Investigators often utilize exclusion criteria that reject the use of monolayers that have relatively high baseline permeability coefficients. There are also continuing studies with different culture conditions and endothelial cell types in the pursuit of establishing a tight monolayer with no or very few gaps (Gumbleton and Audus 2001; Tilling et al. 1998). Newer, microfluidic chambers now allow for growing endothelial cells in small channels with small pores that allow for the study of permeability under flow conditions more relevant to those seen in vivo. In some cases, coculture is also used to better model scenarios such as the neurovascular unit in brain capillaries (Kadry and Cucullo 2024).

An additional method for evaluating endothelial barrier function is the measurement of transendothelial electrical resistance (TER). Studies in the late 1980s utilized endothelial monolayers grown on filters and placed between two chambers containing a medium. The reported range of resistance values was 6.1–69 $\Omega{\cdot}cm^2$ for non-brain endothelial cells (Albelda et al. 1988; Shasby and Shasby 1986; Navab et al. 1986; Territo et al. 1984; Furie et al. 1984) and 160–800 $\Omega{\cdot}cm^2$ for bovine brain microvascular endothelial cells (Tilling et al. 1998; Raub et al. 1992; Abbruscato and Davis 1999). The range of TER values for non-brain endothelial cells is somewhat comparable to the reported resistance in vivo of frog skeletal muscle capillaries (23 $\Omega{\cdot}cm^2$) and skin arterioles and venules (70 and 24 $\Omega{\cdot}cm^2$, respectively) (Olesen et al. 1984; Olesen and Crone 1983). Brain microvessels in vivo, on the other hand, have a much higher electrical resistance (1870 $\Omega{\cdot}cm^2$) and thus a tighter and more restrictive barrier than cultured brain endothelial cell monolayers (Crone and Olesen 1982). TER has proven to be a useful tool for evaluating endothelial barrier function, particularly for ions and small molecules. However, when studying permeability to macromolecules, TER data must be interpreted cautiously.

Expanding upon the concept of using TER as an indicator of cell monolayer barrier function, Giaever and Keese developed a technique named electric cell-substrate impedance sensing (ECIS) (Giaever and Keese 1991, 1993). Briefly, cells are grown on a small gold electrode, and the remainder of the circuit consists of a medium, used as the electrolyte, and a large counter electrode. Generally, a 1-V, 4000-Hz AC signal is supplied through a 1-MΩ resistor to approximate a constant current source. Both in-phase voltage (proportional to resistance) and the out-of-phase voltage (proportional to capacitive resistance) are measured. With ECIS, investigators can evaluate TER of an endothelial cell monolayer grown on a small gold electrode. Additionally, using a mathematical model with some simple assumptions about the geometry of the cells in the monolayer, ECIS can be employed to resolve the relative contributions of intercellular adhesion and focal adhesion. While this technique is an extremely useful tool to study endothelial cell micromotion, again, interpretations must be made cautiously when it is used exclusively to evaluate barrier function. TER is not a direct measurement of flux of solutes and cannot provide the reflection or permeability coefficient for a given solute. However, when ECIS measurements and monolayer permeability experiments are integrated together, TER can be useful for examining the association between temporal changes in barrier function and changes in cell–cell and cell–matrix adhesion (Giaever and Keese 1993; Moy et al. 1996, 2000, 2002, 2004); Tiruppathi et al. 1992.

2.2.4.7 Selection of a Macromolecule Tracer

Ideally, a tracer for the study of permeability would be identical to the macromolecules found in the blood, in vivo. However, chemical modification to macromolecules is needed for their detection. Depending on the method used for detection, macromolecules can be labeled in different ways, such as Evans blue labeling of serum albumin for visual detection, labeling with fluorochromes for fluorescent imaging, and radioactive labeling for quantification.

When selecting a tracer for study, the labeled molecule should be chosen based on the experimental methods that will be employed. For example, today, fluorochrome-labeled tracers are used in most imaging studies because their detection level is superior to that of dyes such as Evans blue. The choice of the macromolecule tracer may also be based on the objectives of the study. For example, if the investigator wishes to study the permeability of charge-independent macromolecules, dextrans would be a better choice than albumin. On the other hand, several investigators choose albumin as the macromolecule to study due to its role in osmotic gradients across the endothelial wall, as a carrier protein, and in gp60 receptor binding (Curry 1984; Lum and Malik 1994).

There are several other criteria that should also be considered when selecting a tracer: (1) The tracer should be evenly labeled. (2) The labeled tracer needs to be chemically stable, i.e., the label should not dissociate from the macromolecule. (3) The tracer should represent the transport characteristics of native macromolecules in the blood. (4) Additionally, the tracer should ideally cause no toxic effects or alter the function of the endothelial barrier.

Even labeling and the chemical stability of the tracer are important for making an accurate assessment of permeability to macromolecules. One weakness of using intravenously injected Evans blue dye as a tracer is that even labeling, although assumed, is not certain (Michel and Curry 1999). Even in the case of commercially prepared FITC-albumin, uneven labeling occurs due to the high molar ratio of FITC to albumin. An excessive amount of FITC is needed because it fades rapidly due to photobleaching. However, considering that albumin has the natural function to act as a carrier of small molecules, a high FITC:albumin ratio increases the probability of much weaker, nonspecific FITC binding that can vary from molecule to molecule, particularly in albumin that has been poorly purified. Excessive amounts of FITC bound to albumin actually cause quenching of the emitted fluorescence (McDonagh and Williams 1984) and a nonlinearly emitted fluorescence to protein concentration ratio. It is also worth noting that noncovalently bound FITC may dissociate from albumin and pass more rapidly through the endothelial barrier, causing an overestimation of P_s.

Another important consideration with respect to the chemical stability of fluorochrome-labeled macromolecules is potential phototoxicity. All fluorochromes are toxic through photobleaching and free radical generation (Martin and Logsdon 1987), which may cause cellular damage. Prolonged light excitation of FITC-dextran or FITC-albumin causes changes in microvascular function in vivo, including increased functional capillary diameter, increased microvascular leakage, and decreased lymphatic pump activity (Vink and Duling 1996; Reed and Miller 1988; Gawlowski et al. 1989; Zhang et al. 1997). These phototoxic effects can be minimized, however, by limiting exposure to light excitation. Exposure times should be kept as short as possible.

It is implicitly assumed that the macromolecule tracer crosses the endothelium in a similar manner as native serum proteins. However, a recent study has shown variation in the permeability coefficients of albumin when conjugated to different fluorochromes (Bingaman et al. 2003). Moreover, a different study has reported that inhibition of nitric oxide synthase (NOS) caused increased microvascular leakage when FITC-albumin served as a tracer, decreased leakage with dichlorotriazinyl aminofluorescein (DTAF)-albumin as the tracer, and no change in leakage when Texas Red-albumin or Oregon Green-albumin were used (Rumbaut et al. 1999). These differences may be due to the varying levels of phototoxicity associated with different fluorochromes (Rumbaut and Sial 1999). Various labeling methods, particularly conjugation of FITC to albumin at a high FITC:albumin molar ratio, cause significant changes in the physiochemical characteristics of albumin (McDonagh and Williams 1984; Bingaman et al. 2003) that could possibly affect its flux across the endothelial wall.

2.2.5 Paracellular and Transcellular Transport

Based on the morphology of microvascular endothelial cells (as illustrated in Fig. 2.5), fluids and solutes have at least three routes of passage from the blood into the interstitial space. In a continuous, non-fenestrated endothelium, these routes of passage are (1) transcellular, through the endothelial cell membranes and cytoplasm, (2) transcellular, via vesicles that traverse the endothelium, and (3) paracellular, at junctions between endothelial cells. These routes are schematically illustrated in Fig. 2.6. With the exception of the microvasculature of the blood–brain barrier (BBB), fluids and solutes cross the endothelium predominantly via the latter two routes. Over the years, much evidence supporting both transcellular and paracellular transport has been gathered, although there are differing opinions regarding the relative importance of these two routes.

2.2.5.1 Theoretical Models of the Endothelial Wall

Linking physiological data regarding microvascular permeability with ultrastructural descriptions of the microvascular wall has been one of the great challenges in the field of microvascular research. Before morphological images of the capillary wall became available, functional studies of the transport of water and hydrophilic solutes across the endothelium led to the formation of the "classical" pore theory (Pappenheimer et al. 1951) and a subsequent two-pore model (Taylor and Granger

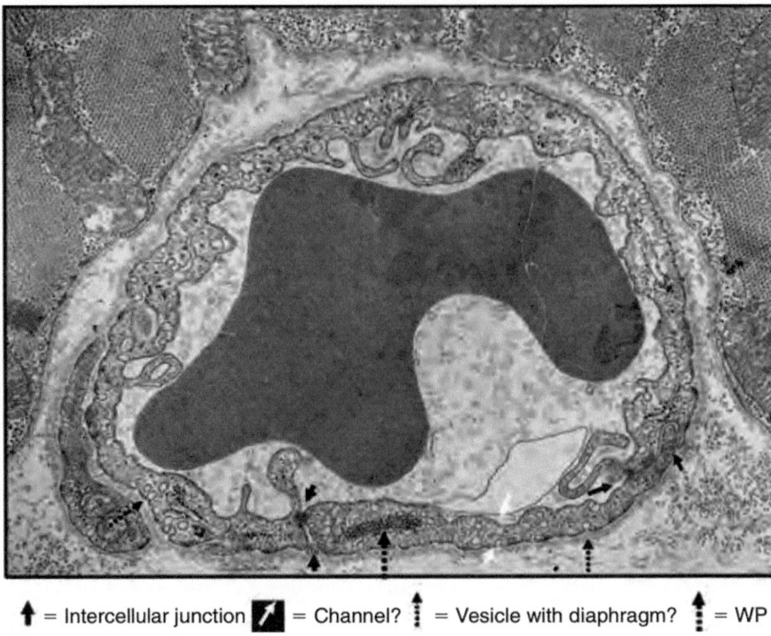

Fig. 2.5 Ultrastructure of canine cardiac postcapillary venule. The electron micrograph demonstrates the complex morphology of endothelial cells. Possible transport pathways for solutes are indicated by different arrows. Caveolae and vesicles are abundant throughout the endothelial cell. WP, Weibel–Palade body. Durán, WN & Berendsen P (unpublished)

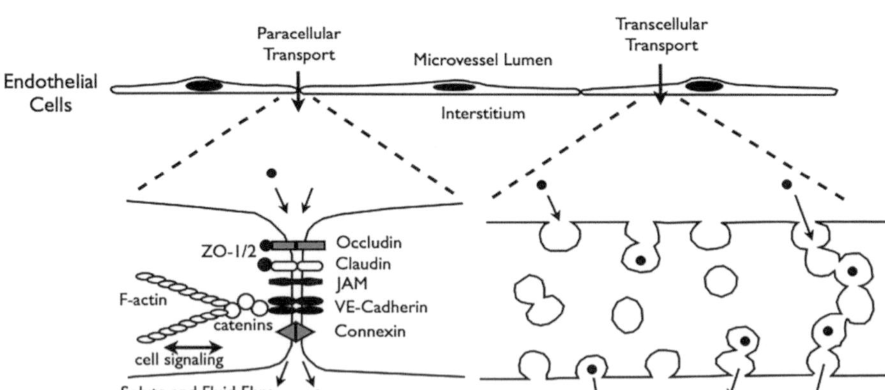

Fig. 2.6 Routes of fluid and solute transport across the endothelium. Two pathways, one paracellular and the other transcellular, have been proposed. In the paracellular pathway, fluids and solutes pass between endothelial cells, with the major barrier being the tightness of junctions between the endothelial cells. Several molecules believed to regulate junctional integrity are shown. In the transcellular pathway, active vesicular transport, or the formation of channels of VVOs, is proposed as a route for fluid and solute passage. *VVO* vesicular-vacuolar organelle

1984). In this theory, the vascular wall is described as having cylindrical pores or rectangular slits, which determine endothelial resistance to water and solute flow. A system of three parallel porous pathways, including an exclusive pathway for water, a population of small (4–5 nm radius) pores, and a much smaller population of large (20–30 nm radius) pores, was used to explain the permeability of microvessels (Taylor and Granger 1984; Renkin 1994; Rippe and Haraldsson 1994).

The pore theory is an idealized model of the endothelial wall and thus has limitations. It is not a good predictor of macromolecule transport when it is not coupled to water flow (Renkin et al. 1988). In addition, in a continuous endothelium, actual pores with the geometric properties described in the model have not been found. Subsequently, and on the basis of a model originally advanced by Chambers and Zweifach (1947), the fiber matrix that lines the gap between endothelial cells was proposed to act as the molecular sieve represented in the idealized pore models (Curry and Michel 1980). In this model, the permeability of the endothelium is not necessarily determined by the width of fenestrae or intercellular clefts between endothelial cells but rather by the inter-fiber spacing of the fiber matrix at the entrance of fenestrae or clefts. Details regarding this model, including structure function analyses, are reviewed elsewhere (Michel and Curry 1999).

2.2.5.2 Vesicular Transport

The discovery of plasmalemmal vesicles in the vascular endothelium led to the hypotheses of their involvement in water in solute transport across the endothelium. Initially, a transport mechanism involving shuttling of vesicles from the luminal membrane to the basal membrane was proposed and termed "transcytosis" (Palade 1960). Studies utilizing electron opaque tracers support that vesicles are involved in microvascular transport (Jennings and Florey 1967; Bruns and Palade 1968; Simionescu et al. 1975; Johansson 1979; Clough and Michel 1981). In addition, two pharmacological inhibitors of vesicular transport, filipin and *N*-ethylmaleimide (NEM), decrease albumin extravasation in isolated, perfused lung preparations (Predescu et al. 1994; Schnitzer et al. 1994).

There are several lines of evidence, however, suggesting that transcytosis does not play a significant role in transendothelial transport. First, the degree of observed transvascular exchange would require impossibly high rates of vesicular shuttling (Simionescu and Simionescu 1984). Second, three-dimensional reconstructions of ultrathin sections of frog mesenteric capillaries show that most vesicles are grouped together and appear to be attached to either the luminal or abluminal membrane (Bundgaard et al. 1979; Frokjaer-Jensen 1980). The relative paucity of free vesicles in the endothelial cytoplasm contradicts the notion of transcytosis in microvascular transport (Crone 1984). Third, in isolated perfused rat hindquarters, macromolecular transport is coupled to convection (Rippe and Haraldsson 1994), indicating that it does not occur via transcytosis. It is worth noting that a low coupling of albumin extravasation to fluid filtration has also been reported (Renkin et al. 1988; Reed 1988), suggesting the possibility that this argument may not necessarily apply in all cases. Fourth, the decrease in albumin extravasation caused by lowering temperature is much more subtle than would be predicted by a transcytotic pathway for microvascular exchange in isolated perfused rat lung or hindquarter preparations. Instead, the decrease in macromolecule transport is proportional to the predicted increase in viscosity caused by such a temperature drop, suggesting that a mechanism like transcytosis, which presumably would require energy, does not have a significant role in extravasation of macromolecules (Rippe and Haraldsson 1994; Carlsson et al. 2001; Rippe and Taylor 2001). Fifth, albumin extravasation is proportional to the mean hydrostatic pressure within the microcirculation (Rippe and Haraldsson 1994). Sixth, albumin transport in perfused rat hindquarter preparation is reportedly the same before and after fixation with glutaraldehyde (Haraldsson and Johansson 1985). Seventh, inhibitors of vesicular transport, NEM and filipin, actually increase permeability in the isolated lung perfused under isogravimetric conditions (neither gaining or losing weight), bringing the specificity of these compounds into question (Carlsson et al. 2001; Rippe and Taylor 2001).

While transcytosis is generally dismissed as a mechanism of endothelial transport for most solutes, an alternative possibility is that fusion of vesicles allows the mixture of their contents. Evidence supporting this concept comes from the gradient labeling of ferritin in vesicles after perfusion of single microvessels with known concentrations of this macromolecule (Clough 1991). An extension of the vesicle fusion concept is the hypothesis that fused vesicles may form tortuous channels from the lumen to the abluminal side of endothelial cells. Such channels, also referred to as vesicular-vacuolar organelles (VVOs), have been observed in the endothelium of venules and also in cultured endothelial cells (Dvorak et al. 1999; Vasile et al. 1999; Feng et al. 2000; Dvorak and Feng 2001). The evidence against transcytosis does not provide such a solid argument against the VVO hypothesis. With regard to the functional experiments comparing living versus glutaraldehyde fixed tissues, there is the possibility that VVO channels may remain intact after fixation, serving as a significant route for transport. With regard to macromolecule extravasation being proportional to intraluminal pressure, there is evidence that labeling of vesicles with ferritin is elevated by increased microvascular pressure in frog mesenteric capillaries (Michel and Curry 1999). Irrespective of whether VVOs play a role in macromolecule transport coupled to fluid filtration, their relative contribution o to normal microvascular permeability remains unclear. However, there is some evidence that increased microvascular permeability in tumors or in experimental metabolic syndrome might be due to elevated endothelial VVO formation (Taskaeva and Bgatova 2021; Liu et al. 2023).

There are also functional studies supporting microvascular transport of certain macromolecules via vesicles. Endothelial cells express an albumin-binding surface protein, gp60, also known as albondin, which is believed to play a role in vesicle formation. Activation of gp60 increases albumin flux across endothelial cell monolayers without changing L_p, suggesting directed vesicular transport of albumin (Minshall et al. 2000; Vogel et al. 2001). Vesicles may also be involved in the transport of other substances that bind to receptors on the endothelial wall, such as insulin (Holmang et al. 1992) or low-density lipoprotein (LDL) (Vasile et al. 1983).

In the case of LDL transport in single perfused frog mesenteric microvessels, lowering the temperature from 18–21 to 4–6 °C caused a reduction in LDL permeability that was much greater than estimated by the effects of increased fluid viscosity. Moreover, increasing microvascular pressure, which usually increases LDL clearance at 18–21 °C, did not do so at 4–6 °C. These data suggest that LDL may pass through the microvascular wall by lateral diffusion on endothelial cell plasma membranes, which may include diffusion through VVO channels and would predictably be much slower at lower temperatures (Rutledge 1992).

2.2.5.3 Paracellular Transport

It is widely accepted that water and solutes can cross the microvascular endothelium via the intercellular junctions between endothelial cells. Chambers and Zweifach hypothesized that the active role of endothelial cells in barrier function was to regulate the composition of "intercellular cement" located between them (Chambers and Zweifach 1947). The modern version of this hypothesis is the fiber-matrix theory (Michel and Curry 1999; Curry and Michel 1980). Cells are generally coated with a fibrous, gel-like meshwork that contains proteins and carbohydrates known as the glycocalyx. The endothelial glycocalyx found on the apical surface and in the junctional clefts serves as a diffusive barrier to plasma solutes in a size- and charge-selective manner and limits adhesion of leukocytes and platelets to endothelial cells (Curry and Michel 1980); Vink and Duling 2000; Henry and Duling 1999; Lipowsky 2012. Inflammatory conditions often involve degradation of the endothelial glycocalyx, allowing for increased leukocyte–endothelium adhesion and elevated microvascular permeability (Alves et al. 2019a; Schmidt et al. 2012; van den Berg et al. 2003; Mulivor and Lipowsky 2002; Torres Filho et al. 2016). Glycocalyx shedding from the endothelium has been documented with trauma, hemorrhagic shock and resuscitation (HSR), ischemia–reperfusion, hyperglycemia, sepsis, and tumor necrosis factor-α (TNF-α)-induced inflammation (Schmidt et al. 2012; Henry and Duling 2000; Zuurbier et al. 2005; Kozar et al. 2011; Hofmann-Kiefer et al. 2009; Ostrowski and Johansson 2012; Rehm et al. 2007). In contrast, reductions of microvascular permeability in response to sphingosine-1-phosphate (S1P) show improved endothelial glycocalyx stability (Alves et al. 2019a; Zeng et al. 2014; Zhang et al. 2016a; Diebel et al. 2019). Synthetic pathways and enzymatic degradation mechanisms (e.g., those by heparinase, hyaluronidase, and matrix metalloproteinases) are believed to determine the function and structure of the endothelial glycocalyx (Gouverneur et al. 2006; Mulivor and Lipowsky 2009). Disease conditions such as sepsis may also result in changes in glycocalyx composition, leading to altered immune and inflammatory responses (Oshima et al. 2019). The impact of the microvascular glycocalyx on health and disease is reviewed more extensively elsewhere in the book by Oshima and Schmidt (2025).

Increases in microvascular permeability in response to inflammatory mediators have been attributed to increased paracellular transport of water and solutes. Electron microscopy studies by Majno and Palade have suggested that histamine-induced increases in permeability are due to increased flux through gaps formed between endothelial cells (Majno and Palade 1961). Subsequently, several investigators have suggested that inflammatory mediator-induced gap formation between endothelial cells is caused by two concomitant actions: (1) decreased adhesiveness between endothelial cells and (2) contraction or retraction of endothelial cells related to the actin cytoskeleton (Dudek and Garcia 2001; Moy et al. 1994, 1996; Garcia et al. 1995; Yuan 2003). This can be viewed as a tensegrity model, in which actin–myosin interactions produce centripetal tension that is counterbalanced by adhesion-dependent centrifugal forces and noncontractile elements of the cytoskeleton, helping maintain cell shape in individual endothelial cells when the barrier is intact. When the junctions are compromised, the potential energy stored by the cytoskeleton is released, causing cell retraction. There is also evidence of increased tension produced by the cytoskeleton under these conditions, which might cause active cell contraction.

Evidence supporting contraction of endothelial cells comes from cell culture studies, which directly demonstrate inflammatory mediator-induced increases in isometric tension within endothelial cell monolayers (Moy et al. 1996; Kolodney and Wysolmerski 1992; Hixenbaugh et al. 1997). The force generation appears to be mediated by actin–myosin-mediated contraction (Dudek and Garcia 2001; Yuan 2000). The actin cytoskeleton also plays an integral role in helping maintain endothelial shape and promoting adhesion between endothelial cells. One mechanism characterized by actin-driven localized cell protrusions (local lamellipodia) is key for maintenance of cell–cell adhesions (Abu Taha et al. 2014). These mechanisms are discussed in further detail in the section on signal transduction pathways.

The role of intercellular adhesion proteins in microvascular permeability changes has been extensively studied. Endothelial cells express proteins that are typically found in tight junctions, adherens junctions, and gap junctions. In the epithelium, tight and adherens junctions occupy spatially distinct zones; however, in the endothelium, these zones are much closer to each other and hard to distinguish. Endothelial tight junctions typically contain transmembrane proteins such as occludins, claudins, and junctional adhesion molecule (JAM) family members as well as the cytoplasmic peripheral membrane proteins zona occludens (ZO)-1, ZO-2, and ZO-3 (Harhaj and Antonetti 2004). Vascular endothelial (VE)-cadherin, α-catenin, β-catenin, γ-catenin, and p120-catenin are the components of endothelial adherens junctions (Bazzoni and Dejana 2004). Connexin-37 (Cx37), Cx40, and Cx43 are also expressed in endothelial cells and give rise to connexons of various compositions and specificities (Beyer et al. 2000).

Claudins are tight junction proteins that, in some cases, create charge-specific paracellular channels, allowing passage of ions between cells at junctions (Harhaj and Antonetti 2004). Claudin-5 is the most well-studied member of this family due to its high expression in the brain microcirculation and its role in blood–brain barrier (BBB) function (Sirotkin et al. 1997; Hawkins and Davis 2005; Daneman and Prat 2015). Deletion of the claudin-5 gene in mice causes increased permeability of the blood–brain barrier to molecules smaller than 800 Da (Nitta et al. 2003). Moreover, animal models of neuroinflammation and neurogeneration and cultured brain endothelial cells exposed to inflammatory mediators also feature downregulation of claudin-5 in association with increased BBB permeability (Beard Jr. et al. 2011, 2014; McColl et al. 2008).

Occludin is an integral membrane protein with a cytoplasmic N-terminal region, four transmembrane regions, two extracellular loops, one intracellular loop, and a cytoplasmic C-terminal region (Harhaj and Antonetti 2004). Occludin expression correlates with endothelial barrier function, however, their role appears to be to strenghthen already existing tight juntions, as occludin knockout mice form normal tight junction strands, while overexpression of occluding enhances barrier properties (Sohet and Daneman 2013). In vitro, arterial endothelial cells form a tighter solute barrier and express 18-fold higher occludin protein levels than venous endothelial cells (Kevil et al. 1998a). The BBB expresses high levels of occludins, compared to a relatively less restrictive endothelium of nonneuronal tissue (Hirase et al. 1997). Under inflammatory conditions, occludin expression is decreased at cell–cell junctions serving as sites of neutrophil transmigration through the endothelium (DeMaio et al. 2001). Occludin expression also appears to be reduced in experimentally induced diabetic retinopathy (Antonetti et al. 1998).

JAMs are found in endothelial tight junctions and have roles in junctional formation, cell polarity, and transmigration of leukocytes across the endothelium (Martin-Padura et al. 1998; Aurrand-Lions et al. 2001; Ebnet et al. 2003; Weber et al. 2007). JAM-A knockout mice display impaired neutrophil transmigration in response to inflammatory mediators (Woodfin et al. 2007, 2009). Blockade of JAM-C with blocking antibodies caused increased reverse transmigration of monocytes, demonstrating the role of JAM-C in unidirectional extravasation of monocytes (Bradfield et al. 2007). Moreover, in an ischemia–reperfusion mouse model, endothelial junctional JAM-C was found to be reduced and leukocyte reverse transendothelial migration events (abluminal to luminal migration) were elevated, supporting the role of JAM-C in promoting luminal to abluminal migration (Woodfin et al. 2011).

The ZO family are cytoplasmic peripheral membrane proteins that bind to occludin, JAM, α-catenin, and the gap junction protein connexin-43. ZO proteins link occludins and claudins to the actin cytoskeleton (Hawkins and Davis 2005). Phosphorylation levels of ZO proteins appear to be correlated with the permeability of epithelial barriers (Harhaj and Antonetti 2004), and there is evidence that ZO-1 tyrosine phosphorylation is associated with VEGF-induced microvascular hyperpermeability in the rat retina (Antonetti et al. 1999).

VE-cadherin is a single-pass transmembrane molecule with an extracellular binding domain, a membrane domain, and an intracellular domain that is linked to the cytoskeleton via catenins. VE-cadherin molecules from adjacent endothelial cells bind homotypically to form adherens junctions (Bazzoni and Dejana 2004). Evidence supporting a role for VE-cadherin in the regulation of barrier function is mainly derived from studies showing that disrupting the homotypic binding of VE-cadherin molecules to adjacent endothelial cells increases permeability (Corada et al. 1999, 2001. Intuitively, proteins coming into contact with the intracellular, cytoplasmic domain likely control homotypic binding of VE-cadherin. Guo et al.

tested this concept by transfecting a recombinant VE-cadherin cytoplasmic domain (rVE-cad-CPD) protein construct into endothelial cells and isolated coronary venules. Their results show that rVE-cad-CPD competitively binds to β-catenin, displacing native VE-cadherin and altering its organization at cell–cell junctions in association with increased permeability (Guo et al. 2004). A more detailed description of VE-cadherin is provided in the section on molecular mechanisms.

Relatively little is known about the role of function gap junctions in regulating microvascular permeability. Knockout of multiple connexins by interbreeding $Cx37^{-/-}$ and $Cx40^{-/-}$ mice to generate $Cx37^{-/-}Cx40^{-/-}$ mice causes prenatal vascular abnormalities and perinatal death, indicating an important role for gap junction communication between endothelial cells in normal endothelial functions (Simon and McWhorter 2002, 2003). Upon stimulation with VEGF or bradykinin, gap junction communication between endothelial cells is reduced (Thuringer 2004; Suarez and Ballmer-Hofer 2001). In endothelial-specific Cx43 knockout mice, increases in microvascular leakage stimulated by thrombin was reduced (Parthasarathi et al. 2006). In addition, in mice exposed to endotoxin in the lungs, Cx43 in lung microvessels increased concomitantly with microvascular leakage and loss of VE-cadherin expression (Kandasamy et al. 2015). Notably, knockdown of endothelial Cx43 with intravenous Cx43 shRNA increased VE-cadherin expression, suggesting that Cx43 may have a role in controlling VE-cadherin expression (Kandasamy et al. 2015). While these studies suggest a role for gap junctions in endothelial barrier regulation, much remains to be investigated.

2.3 Regulatory Mechanisms

2.3.1 Agonists and Receptors: Localization and Function

Agents that rapidly increase microvascular permeability initiate signaling events by binding to their distinct receptors on endothelial cells, including to seven-transmembrane domain receptors (G protein-coupled receptors or GPCRs, tyrosine kinase receptors, pattern recognition receptors, cytokine receptors, and others). Platelet-activating factor (PAF), bradykinin, and histamine bind to GPCRs (Snyder 1990; de Weerd and Leeb-Lundberg 1997), while VEGF binds to single-pass membrane tyrosine kinase receptors, VEGFR-1 (Flt-1) and VEGFR-2 (Flk-1/KDR) (Millauer et al. 1993); Quinn et al. 1993. Receptor occupancy on GPCRs initiates sequestering of G proteins that activate phospholipase C (PLC)-β (PLCβ) directly and PLCγ via protein tyrosine kinases, adenylyl cyclase, and phosphatidylinositol 3-kinase (PI3K) (Snyder 1990; Izumi and Shimizu 1995; Kuruvilla et al. 1994). Binding of VEGF to its receptors activates the receptor tyrosine kinases, causing autophosphorylation (Kroll and Waltenberger 1997). Receptor autophosphorylation, in turn, promotes association with, and phosphorylation of, proteins that contain Src-homology-2 (SH2) domains, such as PLC, PI3K, the adaptor protein Nck, and Ras GTPase-activating protein (RasGAP) (Kroll and

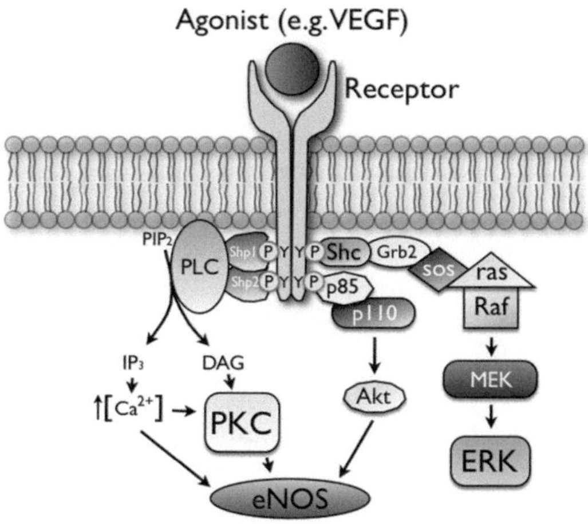

Fig. 2.7 Signal transduction pathways in microvascular permeability. Receptors are activated upon binding to their respective agonists and interact with downstream signaling pathways. The diagram reflects a typical set of cascades activated by permeability agonists; VEGF is used as an example. *VEGF* vascular endothelial growth factor

Waltenberger 1997; Guo et al. 1995; Mukhopadhyay et al. 1998; Wu et al. 1999). Tyrosine phosphorylation activates these signaling mediators, leading to initiation of their respective downstream signaling cascades (Fig. 2.7).

Agents that increase microvascular permeability often bind to multiple receptors. As stated earlier, VEGF binds to at least two different receptors on endothelial cells, although it has become apparent that VEGFR-2 mediates permeability changes while VEGFR-1 does not (Bates and Harper 2003). VEGF-C, on binding to VEGFR-2 and VEGFR-3, also appears to increase microvascular permeability through VEGFR-2 while influencing changes in the lymph flow through VEGFR-3 (Hillman et al. 2001; Breslin et al. 2007a, 2007b). Pharmacological evidence supports that one group of PAF receptors (PAFRs) promotes constriction of arterioles, while the other promotes elevated venular permeability (Tomeo and Duran 1991). Histamine binds to at least four receptors (H1–H4), increasing extravasation of water and solutes by concomitant vasodilation and increased microvascular permeability. Blockade of histamine receptors has been widely used to minimize allergic reactions (McLeod et al. 1999, 2005). In addition, some receptors may be activated by multiple endogenous agonists. VEGFR-2, mentioned above, was noted to be activated by both VEGF-A and VEGF-C. However, other receptors, such as the pattern recognition receptor RAGE (receptor for AGEs), can be activated by a variety of agonist types such as AGEs, HMGB1, and s100/calgranulin family members to elicit increases in endothelial permeability (Zhang et al. 2015; Guo et al. 2009; Wolfson et al. 2011; Weng et al. 2019; Zou et al. 2022).

Localization of receptors at specific sites on endothelial cells may also play an important role in their function. After stimulation of endothelial cells with VEGF, VEGFR-2 reportedly coimmunoprecipitates with VE-cadherin, indicating that these proteins either bind directly to or exist within a complex (Lampugnani et al. 2002). It is conceivable that permeability changes may be mediated by a subpopulation of receptors localized closer to sites regulating solute and fluid flux, but this notion remains to be tested experimentally.

2.3.2 *Signal Transduction Pathways*

When agonists occupy their receptors on endothelial cells, various signal transduction pathways are activated. These pathways involve several different kinds of mediators and enzymes. For example, intracellular calcium, lipid mediators, protein kinases, cyclic AMP (cAMP) and cGMP, and nitric oxide are all involved in the signaling pathways that control the endothelial barrier. In this section, we highlight evidence supporting the roles of different signaling pathways in microvascular permeability regulation.

2.3.2.1 Phosphorylation and Regulation of Permeability

Prior to the availability of more specific agents, nonspecific tyrosine kinase inhibitors such as genistein and tyrphostin-25 were used to study tyrosine phosphorylation. When applied to the hamster cheek pouch, these agents inhibit PAF-induced increases in extravasation of FITC-Dx 150, suggesting that tyrosine phosphorylation is involved in microvascular permeability regulation (Kim and Duran 1995). The importance of several specific serine/threonine kinases, such as protein kinase C (PKC), as well as cAMP- and cGMP-dependent protein kinases (protein kinase A (PKA) and protein kinase G (PKG), respectively), also became apparent using selective activators and inhibitors (Yuan 2003). Specific protein kinases and their influence on the endothelial barrier are discussed in this section.

PKC

Several edemagenic agents stimulate PKC activation (Yuan 2003). PKC is a family of serine/threonine kinases, with at least 10 distinct isoforms. These isoforms are organized into three classifications (conventional, novel, and atypical) based upon what stimuli can cause their activation. Conventional PKCs (α, β_I, β_{II}, and γ isoforms) require Ca^{2+} and diacylglycerol (DAG) for activation. Novel PKC isoforms (δ, ε, η, and θ) require DAG but not Ca^{2+} for activation. Atypical PKC isoforms (ζ, ι, and λ) do not require DAG or Ca^{2+} for activation (Newton 1997).

Exogenous phorbol esters, such as phorbol 12-myristate 13-acetate (PMA) or phorbol ester dibutyrate (PDBu), have been widely used for experimentally stimulating PKCs. A notable alternative is the application of exogenous DAG or DAG analogues. Both DAG and phorbol esters cause conventional and novel PKCs, as well as PKCμ, to translocate from the cytoplasm to the cell membrane and get activated (Newton 1997; Tinsley et al. 2004a). Phorbol esters and DAG analogues increase fluid and solute flux across endothelial cell monolayers, in perfused rat lungs, single perfused mesenteric microvessels, isolated coronary venules, and the hamster cheek pouch (Johnson et al. 1990; Pocock et al. 2004; Huang and Yuan 1997; Ramirez et al. 1996; Lynch et al. 1990; Kobayashi et al. 1994), providing evidence that PKC activation promotes increased microvascular permeability.

Direct pharmacological blockade of PKC activity attenuates increases in permeability produced by PAF, bradykinin, and VEGF in vivo (Kobayashi et al. 1994; Aramoto et al. 2004; Murray et al. 1991). PKC blockade also inhibits VEGF-induced hyperpermeability in isolated coronary venules (Wu et al. 1999) as well as H_2O_2- and neutrophil-stimulated extravasation in isolated perfused rat lungs (Tanita et al. 2000; Johnson et al. 1989). In endothelial cell monolayer permeability models, PKC blockade inhibits hyperpermeability responses to VEGF, thrombin, H_2O_2, and plasma from rats sustaining thermal injury (Lynch et al. 1990; Breslin et al. 2003; Tinsley et al. 2005; Siflinger-Birnboim et al. 1992). These data provide further evidence of the importance of PKC as a signaling element in agonist-induced microvascular hyperpermeability.

Studies of the upstream signaling pathways that mediate PKC activation also provide insights into the role of PKC in permeability changes. Upstream of PKC is phospholipase Cγ (PLCγ), which is activated upon receptor occupancy and catalyzes the formation of inositol triphosphate (IP_3) and diacylglycerol (DAG), with the latter serving as an activator of conventional and novel PKCs (Xia et al. 1996). Pharmacological blockade of PLCγ inhibits permeability increases elicited by PAF and VEGF (Wu et al. 1999; Kim et al. 2000), suggesting its role in permeability regulation.

IP_3, the other product of PLCγ activity, mobilizes Ca^{2+} from intracellular stores in the endoplasmic reticulum, resulting in elevated $[Ca^{2+}]$, and promotes Ca^{2+}-activated processes, including stimulation of conventional PKCs. It is worth noting that elevated $[Ca^{2+}]_i$ can also induce additional Ca^{2+} release from internal stores. Both VEGF and thrombin cause transient increases in endothelial cell $[Ca^{2+}]_i$ (Wu et al. 1999; Bates and Curry 1997; Sandoval et al. 2001a, b). Likewise, histamine induces increases in $[Ca^{2+}]_i$ in isolated coronary microvessels (Huang and Yuan 1997). Treatment of single perfused microvessels with a Ca^{2+} ionophore augments Ca^{2+} influx into the endothelium and increases L_p (He et al. 1990). The rate of increase in L_p is proportional to that in $[Ca^2+]_i$ (Glass et al. 2005). In the case of VEGF, the rise in $[Ca^2+]_i$ that is required for increased L_p is due to the influx of extracellular Ca^{2+}, probably through the transient receptor potential (TRP) channel TRPC-6, and is not dependent upon intracellular Ca^{2+} stores for all agonists (Pocock et al. 2000, 2004). These data support the involvement of a PLC-DAG/Ca^{2+}-PKC pathway in signaling, leading to enhanced microvascular permeability.

Advances in pharmacology and molecular biology allowed for a wave of studies in which investigators examined PKC isoforms, using selective inhibitors, antisense oligonucleotides, synthetic inhibitory peptides, siRNA, and transgenic and knockout animals. Some studies have indicated the involvement of PKC-α in endothelial barrier dysfunction caused by ischemia, thrombin, or TNF-α (Sandoval et al. 2001b; Mehta et al. 2001; Ferro et al. 2000). Involvement of PKCδ and PKCμ has also been recently reported in rat pulmonary microvascular endothelial cell monolayer hyperpermeability (Tinsley et al. 2004a). Currently, the most clinically relevant studies are those of PKCβ isoforms. Overexpression of PKCβ enhances phorbol ester-mediated barrier dysfunction (Nagpala et al. 1996). PKCβ also reportedly mediates H_2O_2-induced hyperpermeability in endothelial cell monolayers

(Siflinger-Birnboim et al. 1992). Oral administration of ruboxistaurin (LY333531) or LY290181, selective inhibitors of PKCβ (Jirousek et al. 1996), prevents microvascular leakage in the kidney and retina in diabetic animals (Aiello et al. 1997) (Ishii et al. 1996; Koya et al. 2000). Likewise, elevated baseline permeability in isolated coronary venules is coupled to increased PKC enzymatic activity and increased $PKC\beta_{II}$ expression in cardiac and vascular tissues during the early stages of experimentally induced diabetes in pigs (Yuan et al. 2000; Guo et al. 2003). The selective PKCβ inhibitor hispidin can ameliorate the elevated coronary venular permeability; however, the PKCα/γ inhibitor HBDDE does not (Yuan et al. 2000). The selective PKCβ inhibitor ruboxistaurin was initially found to prevent and/or reverse diabetic microvascular complications such as retinopathy (Comer and Ciulla 2004), nephropathy (Tuttle and Anderson 2003), and neuropathy (Kim et al. 2003) but did not advance into medical use following clinical trials. PKCθ has also been suggested to mediate interleukin (IL)-1β-induced disruption of brain endothelial barrier function through phosphorylation of ZO-1 (Rigor et al. 2012).

PKC-mediated hyperpermeability appears to involve multiple downstream pathways and effectors. One way that PKC increases permeability is through stimulation of NO synthesis, as blockade of endothelial NO synthesis inhibits phorbol ester-induced hyperpermeability in the hamster cheek pouch and in isolated porcine coronary venules (Huang and Yuan 1997; Ramirez et al. 1996). PKC also activates the cytoskeleton, specifically promoting myosin light chain (MLC) phosphorylation, actin polymerization, and phosphorylation of caldesmon and vimentin (Moy et al. 2004; Garcia et al. 1995; Stasek Jr. et al. 1992). PKC-mediated changes in the cytoskeleton appear to be mediated in part by the Rho pathway (Mehta et al. 2001). Another mechanism associated with PKC-induced hyperpermeability is reorganization of endothelial cell focal adhesions and intercellular junctions (Aramoto et al. 2004; Sandoval et al. 2001b; Rigor et al. 2012; Alexander et al. 1998; Abedi and Zachary 1997). Interestingly, VEGF-induced macromolecule hyperpermeability involves PKC-dependent activation of the Raf-MEK-ERK-1/2 mitogen-activated protein (MAP) kinase (MAPK) pathway (Breslin et al. 2003). VE-cadherin appears to be a potential target of this VEGF-induced interaction between PKC and MAPK (Aramoto et al. 2004).

PKA

Cyclic AMP (cAMP) and cAMP-dependent protein kinase (protein kinase A; PKA) appear to be negative regulators of microvascular permeability, inasmuch as they enhance the barrier properties of the microvascular wall. Increases in microvascular permeability correlate with decreases in intracellular [cAMP] and reduced PKA activity (Moore et al. 1998). In agreement with these observations, administration of PKA inhibitors increases microvascular permeability (He et al. 2000). Moreover, administration of forskolin or rolipram increases [cAMP], decreases permeability, and increases the mean number of tight junctions between endothelial cells (Adamson et al. 1998). Mechanisms involving interactions with [cGMP] via phosphodiesterases (He et al. 2000), regulation of mitogen-activated protein (MAP) kinase (MAPK) pathways (Cobb 1999; Hood and Granger 1998; Parenti et al. 1998), and PKA-mediated phosphorylation of vasodilator-stimulated phosphoprotein (VASP) contribute to enhancing endothelial barrier function (Bogatcheva et al. 2009). Additionally, a mechanism involving Epac (exchange protein directly activated by cAMP) has been extensively studied (Kooistra et al. 2005; Fukuhara et al. 2005; Cullere et al. 2005) and is covered later in this chapter.

PKG

cGMP-induced protein kinase, or PKG, seems to counteract PKA-mediated barrier enhancement upon stimulation of the endothelium with inflammatory mediators. Selective inhibition of NO synthase, guanylyl cyclase, or PKG attenuates microvascular hyperpermeability caused by increased shear stress, histamine, and VEGF in isolated venules (Wu et al. 1996; Yuan et al. 1992, 1993b). Blockade of PKG activity also inhibits microvascular leakage in the developing chick chorioallantoic membrane (DeFouw and DeFouw 2001) as well as bradykinin- and dimaprit-induced increases in cerebral venular capillaries (Sarker and Fraser 2002). Similarly, pharmacological blockade of guanylyl cyclase inhibits histamine-induced hyperpermeability in the hamster cheek pouch (Mayhan 1994). Agents that increase intracellular [cAMP] also increase L_p in single perfused microvessels (Meyer Jr. and Huxley 1992). Moreover, direct activation of PKG with cGMP analogues has been shown to increase the permeability of endothelial monolayers and increase L_p in single perfused microvessels (Varma et al. 2002; He et al. 1998).

There is limited information regarding the downstream pathways that mediate cGMP-PKG-induced increases in permeability. PKG can activate the ERK-1/2 MAP kinase pathway (Hood and Granger 1998; Parenti et al. 1998), while pharmacological blockade of the ERK-1/2 MAP kinase pathway inhibits cGMP-induced hyperpermeability (Varma et al. 2002). Another possible mechanism for PKG-induced increases in endothelial permeability is interaction with the PKA pathway. This interaction may occur at multiple levels, including competing for common downstream structural targets, opposing PKA's influence on leukocyte adhesion and platelet aggregation, or by feedback regulation through phosphodiesterases (Yuan 2003). These mechanisms were inferred in studies investigating rapid increases in endothelial permeability (within minutes). Other works have focused on more chronic increases in permeability that occur after hours or days of exposure to stimuli. For example, treatment of a brain microvascular endothelial cell–astrocyte coculture model with 8-Br-cGMP for 24 h results in a decrease in barrier function associated with a decrease in ZO-1 mRNA and protein expression, suggesting a transcriptional mechanism (Bohara et al. 2014). One study has suggested that PKG-mediated activation of nuclear factor-κB (NF-κB) p65 may result in downregulation of occludin and claudin-5, plus the upregulation of caveolin-1 and caveolin-2, to increase permeability of brain endothelium (Yang et al. 2014).

PI3K, Phosphoinositide-Dependent Kinase (PDK), and Akt

The regulatory p85 subunit of PI3K is constitutively associated with VEGF receptor-2, and, upon activation by VEGF, PI3K activity increases (Thakker et al. 1999). PI3K catalyzes the formation of phosphatidylinositol (3,4,5)-trisphosphate (PIP_3) from phosphatidylinositol (4,5)-bisphosphate (PIP_2). An increase in membrane PIP_3 promotes localization of phosphoinositide-dependent kinase (PDK) and Akt (also called protein kinase B; PKB) to the plasma membrane. Binding of Akt's pleckstrin homology (PH) domain to PIP_3 causes a conformational change that permits PDK to phosphorylate threonine-308, located in Akt's activation loop. In addition, undetermined enzymes phosphorylate Akt on serine-473 and tyrosine-474, located near the C-terminus. Phosphorylation of these three sites increases Akt activity (Schlessinger 2000; Downward 1998; Hemmings 1997; Conus et al. 2002).

Pharmacological inhibition of PI3K activity attenuates VEGF-induced hyperpermeability in endothelial cell monolayers and in the hamster cheek pouch (Aramoto et al. 2004; Lal et al. 2001). In addition, transgenic mice overexpressing caveolin-1 have reduced Akt-dependent phosphorylation of endothelial NOS (eNOS) in association with reduced VEGF-induced microvascular extravasation, compared to wild-type mice (Bauer et al. 2005). It is worth noting that these changes in extravasation may be due in part to VEGF-induced changes in arteriolar diameter and blood flow. However, it was shown in the hamster cheek pouch that while VEGF-induced arteriolar dilation is transient, increases in macromolecule extravasation persist and that PI3K/Akt-dependent vasodilation and extravasation appear to involve distinct mechanisms (Aramoto et al. 2004).

MAP Kinases

MAP kinases (MAPKs) are involved in several aspects of cellular activity. MAPK pathways are typically composed of three protein kinases in series. Usually there is a Ras-activated MAP kinase kinase kinase (MAPKKK), which phosphorylates and activates a MAP kinase kinase (MAPKK). MAPKK, in turn, phosphorylates MAPK, which mediates downstream signaling events. An interesting feature of MAPKs is that their activity is synergistically increased by the dual phosphorylation on specific Thr-X-Tyr activation sites. For example, ERK-1/2 activation strongly correlates with its phosphorylation at Thr-183 and Tyr-185. The dual phosphorylation of this particular site causes a conformational change that exposes the catalytic domain of ERK-1/2 (Cobb 1999).

The ERK-1/2 pathway consists of Raf-1, MEK-1/2, and ERK-1/2. Raf-1 can be activated by Ras-GTP, PKC, and PKG and inactivated by PKA (English et al. 1999; Chang and Karin 2001; Garrington and Johnson 1999; Chuang et al. 1994; Mischak et al. 1996; Wu et al. 1993). At least two other Raf isoforms exist (A-Raf and B-Raf); however, their expression occurs predominantly in urogenital tissues and the brain (Hagemann and Rapp 1999).

Several studies utilizing pharmacological inhibitors have suggested a role for the Raf-1/MEK-1/2-ERK-1/2 pathway in inflammatory mediator-induced hyperpermeability (Aramoto et al. 2004; Breslin et al. 2003; Varma et al. 2002; Lal et al. 2001; Kevil et al. 1998b, 2000; Becker et al. 2001; Yu et al. 2005) and in the regulation of baseline permeability (Varma et al. 2002). Moreover, reduction of ERK-1/2 expression with antisense oligonucleotides attenuates phorbol ester- and VEGF-induced decreases in electrical resistance across endothelial cell monolayers, supporting a role for ERK-1/2 in endothelial barrier regulation (Breslin et al. 2003; Verin et al. 2000). In isolated porcine coronary venules, transfection of active MEK-1 protein increases permeability to albumin and the pharmacological inhibitors of MEK-1/2 dose dependently inhibit increases in permeability caused by VEGF, histamine, NO donors, and the cGMP analogue 8-Br-cGMP (Wu et al. 2005). Interestingly, in the same study, transfection of active ERK-1 did not enhance venular permeability and inhibition of ERK-1/2 using a specific peptide did not reduce VEGF or histamine-induced hyperpermeability, suggesting that MEK-1/2 may regulate permeability independently of ERK-1/2 (Wu et al. 2005). In vivo (hamster cheek pouch) inhibition of ERK-1/2 reduces VEGF-induced hyperpermeability, without affecting VEGF-induced vasodilation (Aramoto et al. 2004).

It is also worth noting that other studies suggest that the p38 MAPK pathway is also involved in endothelial permeability regulation. Pharmacological inhibition of p38 MAPK attenuates hyperpermeability induced by VEGF, histamine, H_2O_2, pertussis toxin, and TNF in endothelial cell monolayers (Lal et al. 2001; Niwa et al. 2001; Garcia et al. 2002; Nwariaku et al. 2002a; Issbrucker et al. 2003; Adderley et al. 2015a, b). The same inhibitor, however, failed to alter baseline permeability or affect cGMP-induced increases in permeability (Varma et al. 2002). Interestingly, the ability of atrial naturetic peptide (ANP) to suppress inflammatory mediator-induced increases in endothelial permeability appears to be due to its induction of the MAPK phosphatase MKP-1, which dephosphorylates and inhibits p38 MAPK (Kiemer et al. 2002).

Src

The Src family of non-receptor protein kinases, with members c-Src, Lyn, Fyn, Lck, Hck, Fgr, Blk, and Yes, phosphorylates substrate proteins and modifies their activity through its interactions with their SH2 and SH3 domains. The best-characterized member is c-Src (also referred to as Src, Src kinase, and pp60c-Src). Src's activity is increased by its phosphorylation at Tyr-416, located in the catalytic domain, and downregulated by phosphorylation at Tyr-527, near the carboxyl terminus (Schlessinger 2000; Shi et al. 2000; Thomas and Brugge 1997). Src acts upon several signaling pathways and cellular structures involved in the regulation of microvascular permeability, including MAPK pathways, MLC kinase (MLCK), β-catenin, and focal adhesions (Yuan 2003).

In endothelial cell monolayers, pharmacological blockade of Src inhibits permeability increases caused by TNF and H_2O_2 (Nwariaku et al. 2002b; Kevil et al. 2001). C5a-activated polymorphonuclear leukocytes (PMNs) stimulate phosphorylation of Src at Tyr-416 and dephosphorylation at Tyr-527 in endothelial cells, suggesting that PMNs promote endothelial Src activity. Additionally, inhibition of Src by either the pharmacological agent protein phosphatase 2 (PP2) or transfection of a specific inhibiting peptide diminishes PMN-induced hyperpermeability in isolated coronary venules (Tinsley et al. 2002). Mice lacking the pp60c-src gene are resistant to VEGF-induced increases in extravasation and have decreased infarct volume, following experimentally induced stroke. Similar results are observed with pharmacological inhibition of Src in wild-type mice (Paul et al. 2001; Eliceiri et al. 1999). Pharmacological inhibition of Src has also been shown to attenuate increases in microvascular permeability caused by AGEs, HMGB1, TNF-α, and fibrinogen-γ-C- terminal fragments (Zhang et al. 2015; Weng et al. 2019; Guo et al. 2020; Adam et al. 2016).

Src influences multiple cellular structures involved in endothelial barrier regulation. Activation of gp60 receptors by albumin causes Src and Fyn phosphorylation, and inhibition of Src or dominant-negative Src expression prevents caveolae-mediated transcytosis of albumin (Minshall et al. 2000; Tiruppathi et al. 1997). Src-mediated phosphorylation of dynamin-2 reportedly mediates this process (Shajahan et al. 2004). Src also interacts with endothelial cell MLCK, phosphorylating it at Tyr-464 and Tyr-471, thus promoting MLC phosphorylation, actin–myosin-induced contraction, and increased endothelial permeability (Garcia et al. 1999; Birukov et al. 2001; Dudek et al. 2002; Mucha et al. 2003). β-Catenin, part of the adherens junction protein complex that maintains endothelial barrier function, is also phosphorylated by Src (Gumbiner 1995). Blockade of Src activity reduces VEGF-induced dissociation of the Flk/β-catenin/VE-cadherin complex in association with a reduction in VEGF-induced myocardial edema and tissue injury. In addition, Src promotes the coupling of focal adhesion kinase (FAK) to integrin $\alpha_v\beta_5$ (Eliceiri et al. 2002). Because of its involvement in several signaling pathways and in regulating different cellular structures, Src activation may serve as a coordinating signal that synchronizes the concomitant cytoskeletal contraction, junctional reorganization, and focal adhesion redistribution that shape endothelial barrier properties (Yuan 2003).

Phosphatases

In the cell, phosphorylation of proteins is a dynamic process that involves both kinases and phosphatases. Relatively little is known about the roles of specific protein phosphatases in endothelial barrier regulation due mainly to the fact that there are very few available inhibitors for specific phosphatases. Inhibition of tyrosine or serine/threonine phosphatase activity causes endothelial barrier dysfunction in cell monolayers (Garcia et al. 2000; Wang et al. 2001), increases permeability of isolated coronary venules (Yuan et al. 1998), and diminishes formation of the selective endothelial barrier in the developing chick chorioallantoic membrane (Cruz et al. 2000).

A few recent studies have begun to decipher the roles of specific protein phosphatases in endothelial barrier regulation. For example, the myosin-associated protein phosphatase, MYPT, has an important role in the regulation of actin–myosin-mediated contraction and may promote enhanced endothelial barrier integrity (Verin et al. 1995; Birukova et al. 2004a). Protein phosphatase 2A (PP2A) may also enhance endothelial barrier function by stabilization of microtubules (Tar et al. 2004). In addition, PP2B may promote restoration of endothelial barrier integrity following PKC-mediated increases in permeability (Lum et al. 2001).

2.3.2.2 Rho Family GTPases

Signaling by Rho GTPases has been implicated in endothelial permeability changes induced by several vasoactive agonists. The Rho GTPase family belongs to the Ras superfamily and has 20 members. Of these, the most well-characterized in endothelial cells are Rho (A, B, C), Rac (Michel and Curry 1999; Yuan et al. 2000; Jaap and Tooke 1995), and Cdc42 (Wojciak-Stothard and Ridley 2002). Rho appears to promote disruption of the endothelial barrier, whereas Rac maintains the endothelial barrier function (Wojciak-Stothard et al. 2001) and Cdc42 promotes restoration of barrier function following an agonist-induced increase in permeability (Kouklis et al. 2004). An atypical member, Rnd3, was found to protect the endothelial barrier by inhibiting RhoA activation (Breslin et al. 2016).

Rho and Rac

Several studies utilizing endothelial cell monolayers suggest that RhoA and its downstream effector Rho kinase (Rho-associated coiled-coil forming kinase; ROCK) have a key role in endothelial permeability. Rho and ROCK promote actin polymerization and stress actin fiber formation as well as FAK activation and focal adhesion formation (Hall 1998; Amano et al. 1997; Needham and Rozengurt 1998; Flinn and Ridley 1996; Sinnett-Smith et al. 2001). Inhibition of Rho with *Clostridium botulinum* exoenzyme C3 transferase enhances cultured endothelial cell monolayer barrier function (Carbajal and Schaeffer Jr. 1999) and reportedly inhibits increases in permeability elicited by thrombin, histamine, transforming growth factor-β (TGF-β), and TNF (Wojciak-Stothard et al. 2001; Nwariaku et al. 2003; Clements et al. 2005). On the other hand, transfection of a constitutively active ROCK protein (lacking the regulatory domain) increases permeability in isolated coronary venules (Fig. 2.9) (Breslin et al. 2006). Other studies report that pharmacological blockade of ROCK attenuates thrombin-induced increases in endothelial cell F-actin content, tyrosine phosphorylation of FAK, and endothelial monolayer permeability (Birukova et al. 2004a; Carbajal et al. 2000; van Nieuw Amerongen et al. 2000). Inhibition of the Rho/ROCK pathway by transfection of a dominant-negative RhoA or pharmacological inhibition of ROCK also reduces thrombin- and histamine-induced increases in endothelial monolayer permeability (Wojciak-

Stothard et al. 2001). Blockade of ROCK activity with Y-27632 also decreases baseline L_p in isolated rat mesenteric venules (Adamson et al. 2002). Moreover, Rho/ROCK inhibition attenuates histamine-induced airway microvascular leakage in vivo (Tokuyama et al. 2002), neutrophil-induced hyperpermeability (Breslin and Yuan 2004; Breslin et al. 2006), VEGF-induced hyperpermeability (Sun et al. 2006), and microvascular leakage caused by burn injury (Zheng et al. 2003; Tinsley et al. 2004b). However, a universal role of the Rho/ROCK pathway in agonist-induced increases in microvascular permeability remains controversial, with other studies reporting that inhibition of this pathway does not affect histamine, TNF-α, or alcohol-induced hyperpermeability in endothelial monolayers (Adderley et al. 2015b); Petrache et al. 2001; Hirase et al. 2001; Doggett and Breslin 2014 or bradykinin- and platelet-activating factor-induced increases in L_p in isolated microvessels (Adamson et al. 2002). In one study utilizing lymphatic endothelial cell monolayers, inhibition of ROCK decreased baseline endothelial barrier function, suggesting that the RhoA/ROCK pathway may also be important for helping maintain normal endothelial integrity with this endothelial cell type (Breslin 2011). Other reports have suggested that activation of RhoA in the central region of the cell promotes barrier disruption, while activation of RhoA at the cell periphery helps maintain endothelial barrier function (Szulcek et al. 2013; Zhang et al. 2016b).

In endothelial cells, Rac (Michel and Curry 1999; Yuan et al. 2000; Jaap and Tooke 1995) promotes similar but less dramatic actin polymerization like Rho and also induces formation of lamellipodia (Wojciak-Stothard et al. 1998). However, Rac appears to oppose Rho in the regulation of endothelial permeability, promoting enhanced barrier function. Rac activation is linked to sphingosine-1-phosphate (S1P)-induced enhancement of barrier function (Garcia et al. 2001; Breslin et al. 2015). Decreased Rac1 activation is linked to thrombin-induced endothelial barrier dysfunction (Breslin et al. 2015, 2016). Toxin B, which inhibits Rac as well as Rho and Cdc42, causes elevated permeability in endothelial cell monolayers and increased L_p in isolated microvessels (Adamson et al. 2002; Hippenstiel et al. 1997). Considering that in this model, inhibition of Rho enhances endothelial barrier function (Carbajal and Schaeffer Jr. 1999), and that transfection of dominant-negative Cdc42 had no effect on baseline permeability (Wojciak-Stothard et al. 2001), by deduction, this increase in L_p must be due to inhibition of Rac. The demonstration that permeability is elevated when endothelial cell monolayers are transfected with a dominant-negative Rac supports this notion (Wojciak-Stothard et al. 2001). Moreover, *Clostridium sordellii* lethal toxin, which can selectively inhibit Rac with little effect on Rho or Cdc42, increased L_p in rat mesenteric microvessels (Waschke et al. 2004). Rac1 activity has been shown to correlate with the activity of local lamellipodia at endothelial cell–cell junctions, suggesting Rac1's role in the maintenance of the endothelial barrier to promote actin-mediated cellular protrusions that aid in junction formation and maintenance (Abu Taha et al. 2014; Breslin et al. 2015).

Cdc42

Transfection of a dominant-negative Cdc42 into endothelial cells inhibits thrombin-induced actin polymerization and stress fiber formation but does not affect baseline endothelial permeability or inhibit histamine- or thrombin-induced hyperpermeability (Wojciak-Stothard et al. 2001). These initial results suggested that Cdc42 might not have an important role in endothelial barrier regulation. However, one recent report, which examined temporal changes in endothelial barrier function, has suggested the involvement of Cdc42 in the restoration of normal endothelial barrier function, following thrombin-induced barrier disruption (Kouklis et al. 2004). Further investigations will be needed to provide insights into the role of Cdc42 as well as other Rho family small GTPases not yet studied, in the regulation of microvascular barrier function.

2.3.2.3 Endothelial Cell Cytoskeleton

The cytoskeleton is involved in all aspects of endothelial cell shape and movement. The endothelial cell cytoskeleton is composed of three different networks, namely, actin microfilaments, intermediate filaments composed of vimentin, and microtubules. Actin has been extensively studied and will be the main focus of this section. Less studies have examined the roles of vimentin and microtubules in endothelial barrier function; however, a few recent developments will also be summarized.

Overview of Actin-Mediated Mechanisms

Endothelial cells contain several different filamentous (F) actin-containing structures, and continual maintenance of these structures is critical for normal integrity of the endothelium (Dudek and Garcia 2001; Shen et al. 2009; Schnittler et al. 2014). These structures include the cortical actin network that supports the plasma membrane, larger cortical actin bundles near cell–cell junctions, ventral actin stress fibers, and transverse arc stress fibers (Alves et al. 2018). The role of the actin cytoskeleton in paracellular routes for solutes to cross the endothelium has been extensively studied, and the current model is a tensegrity model in which the equilibrium of adhesive forces at cell–cell and cell–matrix junctions plus the structural integrity of cortical actin balance against the centripetal tension generated by the contractile components of the cytoskeleton (Moy et al. 1994; Garcia et al. 1995). Paracellular permeability to fluids and solutes is increased when these forces become imbalanced in a way that compromises the integrity of intercellular junctions, thus allowing for more diffusion of fluids and solutes across the endothelial layer (Baldwin and Thurston 2001); McDonald et al. 1999; Schnittler et al. 1990.

Endothelial Cortical Actin Structures

Cortical actin bundles are typically observed lining the circumferential rims of endothelial cells in microvessels (Thurston and Baldwin 1994). These cortical actin rings are disrupted at microvascular leakage sites in postcapillary venules (Thurston et al. 1995; Baldwin and Thurston 1995). Stabilization of actin at endothelial cell–cell junctions by phalloidin treatment causes reduced permeability of endothelial cell monolayers and attenuates increases in permeability caused by inflammatory mediators (Alexander et al. 1988). Phalloidin treatment also causes a decrease in ischemia–reperfusion injury-induced microvascular leakage in vivo (Korthuis et al. 1991). These observations, plus others linking strengthening of the cortical actin network to improved endothelial barrier function (Garcia et al. 2001; Dudek et al. 2010), indicate the importance of actin dynamics in the control of endothelial barrier function.

New technologies allowing visualization of the actin cytoskeleton in live cultured endothelial cells expressing green fluorescent protein (GFP) fused to β-actin (GFP-actin) have revealed that the actin cytoskeleton is highly dynamic, featuring continuous formation and regression of local lamellipodia at junctions between cells (Breslin et al. 2015; Doggett and Breslin 2011). These intermittent lamellipodia facilitate formation of new adhesion sites for VE-cadherin on adjacent cells (Abu Taha et al. 2014; Abu Taha and Schnittler 2014). The local lamellipodia are formed by expanding branched actin networks, and inhibitors of the actin-related protein 2/3 complex (ARP2/3) prevent their formation and block VE-cadherin dynamics, resulting in the formation of intercellular gaps (Abu Taha and Schnittler 2014). Treatment of endothelial cells with either thrombin or histamine causes a rapid decrease in local lamellipodia activity, thus allowing for the opening of junctions (Abu Taha et al. 2014; Adderley et al. 2015a; Breslin et al. 2015, 2016; Schnittler et al. 2014; Alves et al. 2018). However, it is also worth noting that these lamellipodia specifically appear at junctions lacking VE-cadherin (Abu Taha et al. 2014) and at sites where clefts appear between cells (Breslin et al. 2015) to strengthen the junctions in these areas. In addition, when considering the degree of overlap at endothelial cell–cell junctions when local lamellipodia are present, the presence of lamellipodia may also increase the diffusion distance for solutes as they pass through junctions (Fig. 2.8).

Several signals are involved in the control of local lamellipodia. As noted above, proteins of the ARP2/3 complex are required, indicating the involvement of actively growing branched actin networks in local lamellipodia formation (Abu Taha and Schnittler 2014). The formation of local lamellipodia is nearly abolished when endothelial cells are treated with the myosin II inhibitor blebbistatin, which also causes increased permeability of both cultured endothelial cell monolayers and isolated rat mesenteric venules (Breslin et al. 2015). Rac1 activity has been observed to correlate with the lamellipodia protrusion frequency and barrier function of endothelial monolayers (Breslin et al. 2015). Moreover, overexpression of Rac1 elevated, while Rac1 inhibition lowered, both lamellipodia protrusion frequency and endothelial barrier function (Breslin et al. 2015). RhoA is suspected to also

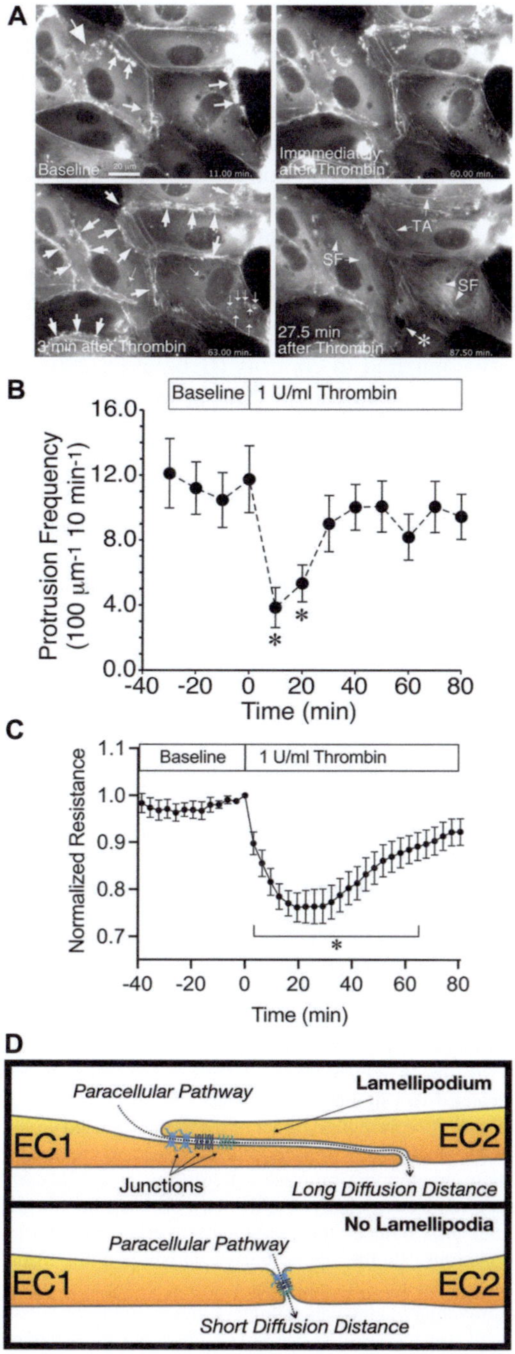

Fig. 2.8 Local lamellipodia and endothelial barrier function. (**a**) Frequent protrusion and withdrawal of local lamellipodia were observed at the junctions between HUVECs transfected with GFP-actin. In the top left is an image taken during baseline before addition of thrombin. The small arrows denote local lamellipodia protruding at the edges of cells, while the large arrowhead shows

contribute to lamellipodia formation, but its role remains somewhat elusive. On one hand, treatment of cultured endothelial cells with exogenous Rnd3, an atypical member of the Rho family that lacks GTPase activity and counteracts RhoA, rescues the drop in lamellipodia protrusion frequency and increase in permeability caused by thrombin (Breslin et al. 2016). However, RhoA also undergoes rapid activation near cell–cell junctions in response to S1P, correlating with phosphorylation of MLC at the cell periphery (Zhang et al. 2016b). VASP (vasodilator-stimulated phosphoprotein) and Ena-VASP-like (EVL) are members of the Ena/VASP family and mediate dynamic actin rearrangements in junctions, focal adhesions, and lamellipodia (Krause et al. 2003; Lambrechts et al. 2000). Notably, global VASP knockout mice show elevated bradykinin- and lipopolysaccharide (LPS)-induced microvascular hyperpermeability compared to controls (Benz et al. 2008; Henes et al. 2009). EVL appears to interact with cortactin, profilin-2, and modulate actin polymerization and the dynamics of lamellipodia (Mascarenhas et al. 2022). EVL overexpression causes amplified S1P-induced endothelial barrier enhancement together with formation of larger focal adhesions in response to S1P. However, with thrombin challenge, EVL overexpression also reduces the number of focal adhesions (Mascarenhas et al. 2021). Lastly, the absence of VE-cadherin at the cell periphery has also been shown to be a stimulus for local lamellipodia formation at sites where the endothelial barrier is compromised (Abu Taha et al. 2014). These advances suggest a complex mechanism that orchestrates local lamellipodia formation, and additional research is needed to understand their role in controlling microvascular permeability, particularly in vivo.

Fig. 2.8 (continued) an "actin cloud" that was a dynamic actin ring structure that expanded concentrically and, when found near the cell edge, would give rise to a lamellipodium. About 3 min after the addition of thrombin (bottom left image), a coordinated increase in GFP-actin was observed at several junctions (thin arrowheads) and around vesicles (very small arrows), followed by a decrease in local lamellipodia formation. Later (bottom right image, 27.5 min after thrombin addition), de novo formation of ventral stress fibers (SFs) was apparent. There was also an opening of a tricellular junction (denoted by *) that was closed soon after by the lamellipodia from at least two cells. Scale bar = 20 μm. (**b**) Quantitation of the local lamellipodia over time revealed that thrombin (1 U/ml) caused a significant decrease in lamellipodia protrusion frequency (*$P < 0.05$ vs. baseline). (**c**) A representative plot of thrombin-induced endothelial barrier function using ECIS shows that the decrease in transendothelial resistance temporally correlates with the time points during which lamellipodia protrusion frequency is significantly reduced in panel B (*$P < 0.05$ vs. baseline). (**d**) Conceptually, the decrease in lamellipodia protrusions may lead to opening of junctions but also reduces the amount of overlap of opposing cell membranes at endothelial cell–cell junctions. In the upper box, a lamellipodium from one endothelial cell (EC) travels over (or under) the adjacent endothelial cell, causing a localized increase in diffusion distance for any solutes traveling via the paracellular pathway. In the absence of lamellipodia (lower box), the point of cell–cell contact has little overlap of adjacent cell membranes, resulting in a very short diffusion distance. (Images and data from panels A–C are from Breslin et al. 2015 with permission. *HUVECs* human umbilical vein endothelial cell, *GFP* green fluorescent protein, *ECIS* electric cell-substrate impedance sensing)

Contractile Actin Bundles in Endothelial Cells

Several reports utilizing endothelial cell monolayers suggest that actin-mediated cellular contraction promotes increased paracellular leakage via the formation of intercellular gaps (Yuan 2000; Lum and Malik 1994; Moy et al. 1994; Wojciak-Stothard and Ridley 2002). In support of this notion, certain edemagenic agents, such as thrombin and phorbol esters, directly increase endothelial cell monolayer isometric tension in an actin-dependent manner within the time frame during which permeability is elevated (Moy et al. 1996, 2004; Kolodney and Wysolmerski 1992) Fig. 2.9.

Actin-mediated contraction is powered by interactions between F-actin and the myosin ATPase. Myosin ATPase activity is controlled by the phosphorylation state of myosin regulatory light chains (MLC) at Thr-18 and Ser-19, which, in turn, is determined by MLC kinase (MLCK) and MLC phosphatase (MLCP) activities (Fig. 2.10). The dual phosphorylation of MLC allows for self-assembly of myosin chains into small bipolar filaments that interact with and contract actin microfilaments. Increased MLC phosphorylation correlates with increased isometric forces and elevated permeability, suggesting roles for MLCK and MLC phosphatases in agonist-induced hyperpermeability (Verin et al. 1995; Moy et al. 1993; Sheldon et al. 1993; Shi et al. 1998). MLCK inhibition attenuates hyperpermeability caused by phorbol esters, NO donors, cGMP analogues, activated neutrophils, and burn injury (Yuan et al. 1997, 2002; Tinsley et al. 2004b). In addition, mice lacking the long isoform of MLCK, MLCK210, have an attenuated hyperpermeability response to traumatic burn injury (Reynoso et al. 2007). Transfection of active MLCK protein

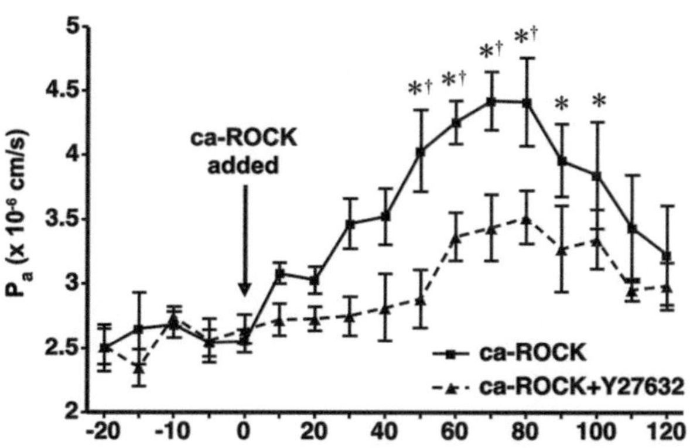

Fig. 2.9 Transfection of ca-ROCK protein increases coronary venular permeability. Isolated porcine coronary venules were transfected with 3 μg/ml ca-ROCK protein for 120 min in the absence (solid line, squares, $n = 6$) or presence (dashed line, triangles, $n = 4$) of the ROCK inhibitor Y-27632 (5×10^{-6} M). Y-27632 was added to the bath 20 min prior to the addition of ca-ROCK. Transfection was initiated at 0 min and continued throughout the time course. $^*P < 0.05$ vs. baseline permeability (0-minute time point). $^\dagger P < 0.05$ vs. ca-ROCK transfection when Y-27632 is present. (From Breslin et al. 2006, with permission. *ROCK* Rho-associated coiled-coil forming kinase)

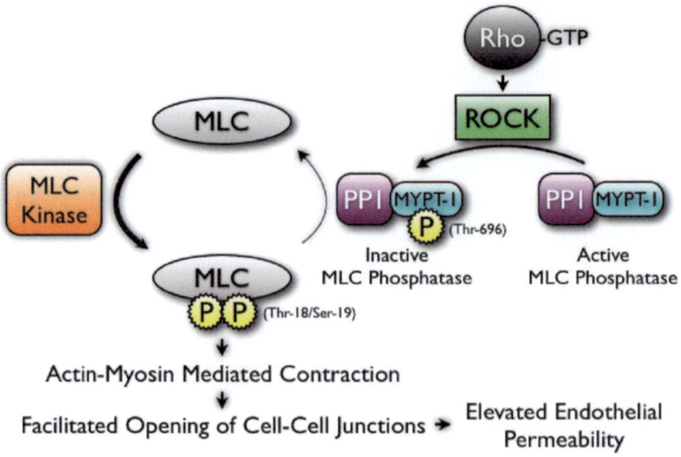

Fig. 2.10 MLC phosphorylation as a mechanism of endothelial barrier regulation. Inflammatory stimuli cause elevated phosphorylation of MLC at Thr-18/Ser-19 in endothelial cells, which, in turn, elevates actin–myosin-mediated contraction. This leads to increased cellular centrifugal force that promotes opening of endothelial cell–cell junctions, causing increased endothelial permeability. MLC phosphorylation increases when MLCK activity is elevated or MLCP activity is reduced. The Rho/ROCK pathway can inactivate MLCP through phosphorylation of the MLCP targeting subunit MYPT-1 at Thr-696. *MLC* myosin light chain, *MLCK* MLC kinase, *MLCP* MLC phosphatase, *ROCK* Rho-associated coiled-coil forming kinase

into endothelial cell monolayers concomitantly increases MLC phosphorylation and permeability to albumin (Tinsley et al. 2000).

Changes in endothelial permeability caused by Rho small GTPases are likely linked to their involvement in actin cytoskeleton organization and the determination of cell shape. RhoA and ROCK promote the formation of actin stress fibers (Amano et al. 1997), and ROCK activation leads to elevated phosphorylation of MLC at Thr-18 and Ser-19, promoting actomyosin contraction (Amano et al. 2000). ROCK phosphorylates MYPT-1, the MLC phosphatase regulatory subunit, at Thr-686 and Thr-850, decreasing MLC phosphatase activity and leading to a net increase in MLC phosphorylation (Birukova et al. 2004a). Transfection of constitutively active ROCK protein causes an increase in endothelial cell monolayer isometric tension that temporally correlates with barrier dysfunction (Fig. 2.11) (Breslin et al. 2006). Blockade of the RhoA/ROCK pathway inhibits actin polymerization and stress fiber formation, and inflammatory stimuli cause an increase in permeability (Breslin and Yuan 2004; Birukova et al. 2004a; Wojciak-Stothard et al. 2001; Carbajal and Schaeffer Jr. 1999). In contrast to RhoA inhibition, expression of dominant-negative Rac1 in endothelial cells causes increased actin stress fiber formation in addition to endothelial monolayer hyperpermeability (Wojciak-Stothard et al. 2001). The finding that Rho and Rac cause opposite responses in both the actin cytoskeleton and in barrier function further indicates the importance of actin in endothelial permeability regulation.

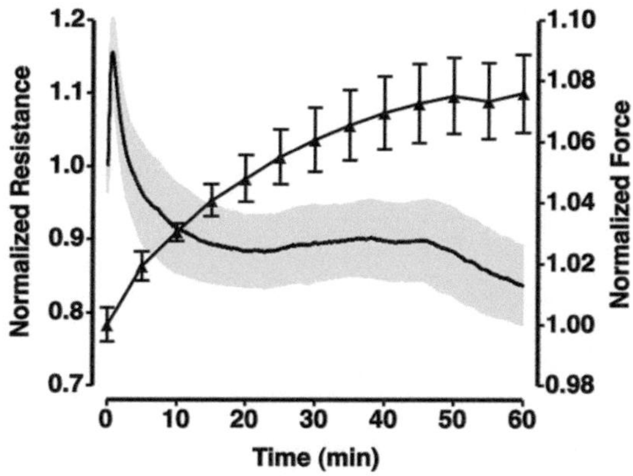

Fig. 2.11 ROCK-mediated tension and barrier dysfunction in coronary venular endothelial cell monolayers. Transfection of ca-ROCK (3 μg/ml) into cultured coronary venular endothelial cell (CVEC) monolayers increases isometric tension and causes a drop in TER. A time-course overlay of ca-ROCK-stimulated changes in TER and tension shows that these two events occur simultaneously. (From Breslin et al. 2006, with permission. *TER* transendothelial electrical resistance, *ROCK* Rho-associated coiled-coil forming kinase)

Multiple cellular mechanisms collectively determine permeability of the endothelium. Therefore, it is of interest to know the relative contribution of actin-mediated contraction to the endothelial barrier changes caused by different agonists. For example, phorbol esters stimulate temporal changes in TER that correlate with the time course of changes in actin-mediated isometric tension (Moy et al. 2004). This association indicates the likelihood that centripetal tension has a relatively important role in phorbol ester-induced hyperpermeability. Temporal changes in TER caused by thrombin, however, are not in phase with the observed time course of changes in isometric tension, with the peak change in TER preceding the peak rise in tension (Moy et al. 2002, 2004). Pharmacological blockade of MLCK, which inhibits thrombin-induced tension, also does not affect the peak change in TER but rather accelerates the reestablishment of baseline endothelial barrier function (Moy et al. 2004). In this case, tension development in endothelial cells appears to extend the time course of barrier dysfunction. Histamine-induced changes in TER, on the other hand, are not associated with any detectible change in endothelial isometric tension and is not affected by MLCK inhibition (Moy et al. 2002). This finding indicates that actin-mediated tension plays little or no role in the observed changes in permeability, which are instead attributed to changes in cell–cell and cell–matrix adhesion (Moy et al. 2000).

Although several studies with endothelial cells report that actin stress fibers are associated with increased endothelial permeability, it is not entirely clear whether these subcellular structures mediate in vivo permeability changes. In vivo, stress fibers are apparent in endothelial cells of large vessels and arterioles, but capillary and venular endothelial cells have few if any (Thurston and Baldwin 1994).

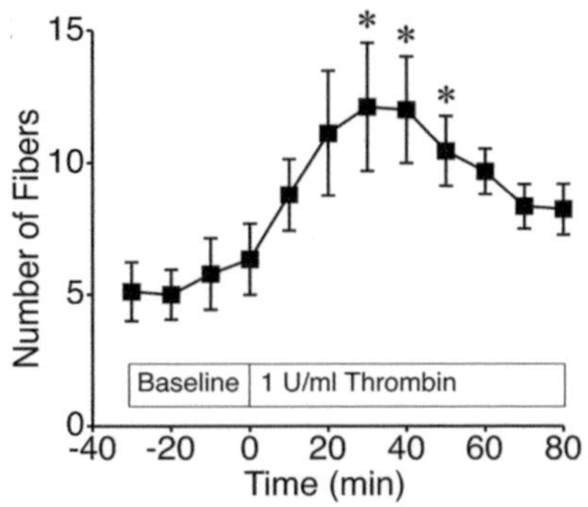

Fig. 2.12 In HUVECs expressing GFP-actin, the number of actin stress fibers significantly increased 30 min after 1 U/ml thrombin was added. This time course of the increase in stress fibers appears to correlate with the drop in endothelial barrier function, with the significant ($P < 0.05$) and maximal increase in actin stress fibers occurring at the point when endothelial barrier function is starting to recover toward baseline in Fig. 2.8c. (These data are from Breslin et al. 2015, shown here with permission. *HUVECs*, human umbilical vein endothelial cell, *GFP* green fluorescent protein)

Microvascular leakage in vivo is associated with disruption of the F-actin lining the rims of endothelial cells (Thurston et al. 1995; Baldwin and Thurston 1995; Wong et al. 1999).

A better understanding of the precise changes in actin organization that correlate with increased microvascular permeability was obtained in human umbilical vein endothelial cells (HUVECs) expressing GFP-actin (Doggett and Breslin 2011). In addition to revealing a role of local lamellipodia in endothelial barrier function, this model revealed a few new insights into actin stress fiber formation in endothelial cells. First, two types of stress fibers were typically observed in cultured endothelial cells: (1) ventral stress fibers connected to either focal adhesions or cell–cell junctions and (2) transverse arcs, which formed at the cortical actin ring and travelled laterally toward the cell center (Adderley et al. 2015a; Breslin et al. 2015). Second, thrombin-induced actin stress fiber formation (assembly of both ventral stress fibers and transverse arcs) occurred largely after the barrier function was compromised in cultured endothelial cells, with the peak in the number of stress fibers occurring during the beginning of recovery to normal barrier function (Fig. 2.12). Histamine caused increased formation of transverse arcs and accelerated their lateral movement toward the cell center, but this was after the histamine-induced barrier disruption returned to the baseline (Adderley et al. 2015a). While new stress fiber formation occurred too late to be involved in the disruption of the endothelial barrier, it was observed that when cell–cell adhesions became disrupted, the preexisting actin stress fibers connected to these disrupted adhesions rapidly contracted (Breslin et al. 2015). This finding suggests that during periods of normal endothelial barrier function, the actin fibers within the endothelial cells are under tension so that when a focal or cell–cell adhesion to which they are connected becomes compromised, they recoil (Alves et al. 2018). Lastly, the formation of ventral stress fibers appears to occur as a reaction to the disruption of barrier function and may serve as a mechanism to help stabilize cell shape and facilitate closure of gaps that form between cells (Breslin et al. 2015).

Intermediate Filaments and Microtubules

While studied less than actin, there is evidence of the involvement of tubulin and vimentin in microvascular permeability changes. With regard to intermediate filaments, one study reports that phosphorylation of vimentin increases concurrently with thrombin- and phorbol ester-induced endothelial barrier dysfunction (Stasek Jr. et al. 1992). Another report suggests that histamine causes a rapid dissociation between adherens junction proteins and vimentin (Shasby et al. 2002).

Microtubules have an important role in maintaining endothelial barrier function (Birukova et al. 2004b). Intravenous administration of anticancer drugs that disrupt the microtubule network can cause pulmonary edema (Cattan and Oberg 1999). In addition, thrombin causes microtubule disassembly in cultured endothelial cell monolayers and stabilization of the microtubule network with Taxol inhibits thrombin-induced barrier dysfunction in vitro (Birukova et al. 2004c). It was found that thrombin causes disassembly of peripheral endothelial microtubules within 5 minutes of treatment, prior to the formation of actin stress fibers (Alieva et al. 2013). Additionally, it appears that the stability of the microtubule network also influences Rho-mediated changes in the endothelial barrier through a direct interaction with the guanine exchange factor GEF-H1 (Birukova et al. 2006). In cyclic stretch-induced endothelial barrier dysfunction, knockdown of GEF-H1 prevented both microtubule destabilization and endothelial barrier dysfunction (Birukova et al. 2010). Inhibitory phosphorylation of GEF-H1 by the downstream effector of Rac1, PAK1 kinase, was shown to mediate atrial natriuretic peptide-induced protection against thrombin-induced microtubule disassembly (Tian et al. 2014a). Additional work found that GEF-H1 interacts with the adaptor protein cingulin that is involved in the organization of tight junctions and that this association serves as a mechanism for RhoA inactivation and the restoration of endothelial barrier function after challenge with agents such as thrombin (Tian et al. 2016). Other studies have shown microtubule-dependent endothelial barrier enhancement in response to hepatocyte growth factor, with a key role for the Rac1-specific guanine exchange factor Asef (Higginbotham et al. 2014; Tian et al. 2014b, 2015). These reports indicate an important role of microtubules both in injury-induced endothelial hyperpermeability and in potentially improving endothelial barrier function.

2.3.2.4 Intercellular Adhesion Proteins

VE-Cadherin and Catenins

Cadherins are a family of transmembrane adhesion proteins that mediate intercellular adhesion by homotypic binding. VE-cadherin is expressed in endothelial cells and is necessary for the organization of vascular structures and for maintaining the integrity of the vascular endothelial wall (Corada et al. 2001; Dejana et al. 2000; Vittet et al. 1997). Deletion of the VE-cadherin gene from murine embryonic stem cells impairs the formation of rudimentary vascular structures in 11-day-old embryos (Vittet et al. 1997). VE-cadherin is also a cell adhesion receptor and regulates cell growth (Caveda et al. 1996). Moreover, endothelial cells lacking the VE-cadherin gene adopt a fibroblast-like shape when grown in culture, as opposed to the usual well-spread, cobblestone endothelial morphology (Lampugnani et al. 2002). VE-cadherin may also act as part of a mechanoreceptor complex that transduces shear stress signals (Shay-Salit et al. 2002).

VE-cadherin is a single-pass transmembrane protein, with the amino terminal in the extracellular domain and the carboxy terminal in the intracellular domain. The amino terminal is the site of homotypic adhesion, whereas the carboxy terminal regulates cell growth (Caveda et al. 1996) and associates with several cytoplasmic proteins, including β-catenin, p120-catenin, and γ-catenin (also named plakoglobin). β-Catenin links the cadherin complex to α-catenin, which binds to the actin cytoskeleton (Fig. 2.13). In cultured endothelial cells, the degree of association between VE-cadherin, β-catenin, and p120-catenin remains relatively constant between non-confluent and confluent cells, whereas γ-catenin tends to associate with this complex only when cells become confluent and form mature junctions. This conditional association between γ-catenin and the VE-cadherin complex may be an early signaling event for contact-mediated inhibition of cell proliferation (Lampugnani et al. 1995).

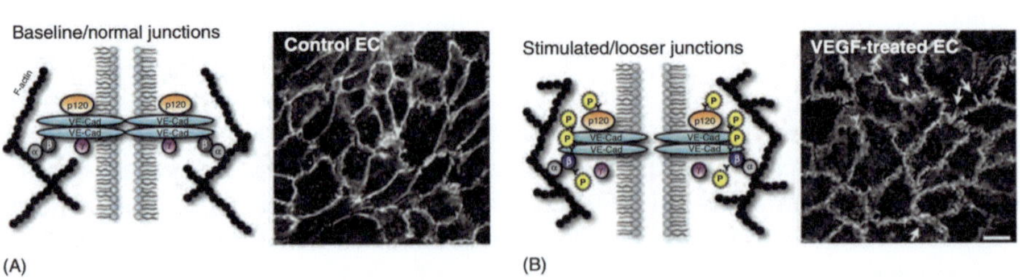

Fig. 2.13 Adherens junctions in endothelial cells. (**a**) Adhesion between endothelial cells is achieved by homotypic binding of VE-cadherin molecules at endothelial cell–cell junctions. Several catenins (p120, α, β, γ) are associated with VE-cadherin and regulate its homotypic binding as well as linkage to the cytoskeleton. (**b**) Under stimulated conditions, such as during VEGF treatment, VE-cadherin, β-catenin, and p120-catenin are phosphorylated on tyrosine residues and a reorganization of endothelial cell–cell junctions occurs, characterized by the appearance of several VE-cadherin-rich finger-like projections (arrows). Scale bar = 20 μm. (Images from Aramoto et al. 2004, with permission. *VEGF* vascular endothelial growth factor)

Cell confluence also correlates with tyrosine phosphorylation on VE-cadherin, β-catenin, and p120-catenin. While these proteins are phosphorylated on tyrosine in non-confluent cells, tyrosine phosphorylation is markedly reduced when the cells become more tightly confluent (Lampugnani et al. 1997). VEGF and histamine-stimulated endothelial permeability increases have been associated with elevated tyrosine phosphorylation of VE-cadherin, β-catenin, and p120-catenin (Shasby et al. 2002); Esser et al. 1998; Andriopoulou et al. 1999; Cohen et al. 1999 and with changes in the organization of VE-cadherin at intercellular junctions (Fig. 2.13) (Aramoto et al. 2004; Kevil et al. 1998b; Esser et al. 1998; Alexander et al. 2000. Activated neutrophils also increase tyrosine phosphorylation of VE-cadherin and β-catenin, cause a reorganization of interendothelial junctions, and increase permeability (Tinsley et al. 1998, 2002).

Several protein tyrosine kinases (including Src kinase and growth factor receptors) and tyrosine phosphatases associate with cadherin complexes (Tsukita et al. 1991; Downing and Reynolds 1991; Ukropec et al. 2000; Brady-Kalnay et al. 1998; Hoschuetzky et al. 1994). Upon VEGF stimulation, the kinase domain receptor (KDR, also known as VEGF receptor 2) transiently associates with VE-cadherin (Carmeliet et al. 1999). Phosphorylation of VE-cadherin on tyrosine occurs concomitantly with its dissociation from KDR. In addition, dephosphorylation of the adaptor protein Shc occurs when VE-cadherin dissociates from KDR (Zanetti et al. 2002). These interactions between VE-cadherin and KDR suggest that VE-cadherin may regulate VEGF-induced activation of signal transduction pathways.

Until recently, the intracellular signals modulated by VE-cadherin expression had been unknown. Lampugnani et al. approached this problem by comparing VE-cadherin$^{-/-}$ endothelial cells transfected with VE-cadherin to control VE-cadherin$^{-/-}$ endothelial cells. VE-cadherin expression increases the level of active Rac and also increases membrane localization of Tiam-1 (T-cell lymphoma invasion and metastasis protein), an activator of Rac. VE-cadherin expression also decreases the level of active Rho (Lampugnani et al. 2002). The decrease in Rho activity may be caused by p120-catenin (Anastasiadis and Reynolds 2001) or by Rac activation (Zondag et al. 2000; Sander et al. 1999). Rho and Rac have previously been reported to modulate cadherin-dependent adhesion (Braga et al. 1997, 2000; Kaibuchi et al. 1999; Takaishi et al. 1997); however, the observations of Lampugnani et al. (2002) indicate that the binding state of VE-cadherin also modulates the activity of Rho family GTPases (Lampugnani et al. 2002).

Binding between β-catenin and VE-cadherin, which links adherens junctions to the actin cytoskeleton, appears to be important for maintaining the homotypic binding between VE-cadherin molecules of adjacent endothelial cells. Introduction of a mutant VE-cadherin lacking the extracellular domain required for cell–cell adhesion partially dissociated native VE-cadherin from β-catenin and increased permeability in both endothelial cell monolayers and isolated coronary venules (Guo et al. 2004).

Few studies have examined those signal transduction pathways that influence agonist-induced changes in VE-cadherin organization. Pharmacological blockade of PKC inhibits VEGF-induced tyrosine phosphorylation and reorganization of β-catenin (Cohen et al. 1999). Inhibition of PKC also attenuates reorganization of VE-cadherin at intercellular junctions in response to VEGF, thrombin, thapsigargin, or plasma from animals suffering from burn injury (Aramoto et al. 2004; Tinsley et al. 2005; Sandoval et al. 2001a, b). Src kinases localize to adherens junctions (Tsukita et al. 1991), and phosphorylate p120-catenin may reduce cadherin adhesiveness (Mariner et al. 2001). VEGF-induced elevation of macromolecule extravasation, assessed by the Miles assay, occurs in wild-type mice but is absent in $Src^{-/-}$ mice (Eliceiri et al. 1999). Pharmacological blockade of Src kinase inhibits neutrophil-induced hyperpermeability in coronary venules and endothelial monolayers as well as tyrosine phosphorylation of β-catenin (Tinsley et al. 2002).

Occludin, ZO-1, and Claudin-5

The microvascular endothelium of the central nervous system expresses higher levels of occludin than in other microvascular beds (Hirase et al. 1997). Thus, occludin is believed to have an important role in maintaining the highly selective barrier between the bloodstream and brain tissue (blood–brain barrier or BBB), but our current knowledge of occludin's role in permeability regulation is limited. Studies indicate that occludin and ZO-1 localization in endothelial monolayer cell–cell junctions is disrupted after stimulation with agents that increase P_s, such as VEGF and H_2O_2 (Kevil et al. 1998b, 2000; Murphy and Duffy 2003). VEGF also increases occludin phosphorylation and tyrosine phosphorylation of ZO-1 in the same time frame of permeability increases (Antonetti et al. 1999). Serine phosphorylation of occludin in endothelial cells after treatment with H_2O_2 has also been reported (Kevil et al. 2000). Additional work will be required to extend our knowledge of the function of these tight junction proteins in the regulation of microvascular permeability.

Claudin-5 is a tight junction protein expressed in endothelial cells that has a critical role in maintaining BBB integrity. Disruption of claudin-5 has been linked to multiple neurological disorders including Alzheimer's disease, depression, schizophrenia, multiple sclerosis, and in the disease progression of HIV encephalitis (Greene et al. 2019; Alvarez et al. 2011; Persidsky et al. 2006). Claudin-5 is maintained at junctions by Akt2 signaling that represses the transcription factor FOXO1. When Akt2 is inactivated by IL-1β, FOXO1 accumulates in the nucleus and claudin-5-dependent barrier integrity of brain endothelial cells is reduced (Beard Jr. et al. 2014, 2020). Translocation of β-catenin to the nucleus and the importance of non-muscle myosin light chain kinase have also been identified in the IL-1β-induced repression of claudin-5 (Beard Jr. et al. 2014). Disruption of claudin-5 at junctions in association with increased BBB leakage or brain microvascular endothelial cell monolayer permeability has also been shown in response to *Pseudomonas aeruginosa* infection, extracellular histones, or alcohol (Villalba et al. 2020, 2023; Alves et al. 2019b). Notably, S1P-induced protection against alcohol-induced endothelial barrier dysfunction in vitro involves maintenance of claudin-5 at junctions (Alves et al. 2019b).

2.3.2.5 Focal Adhesion Proteins

FAK

Several agents that increase microvascular permeability also activate FAK (Abedi and Zachary 1997; Yuan et al. 1998; Leeb-Lundberg et al. 1994). Moreover, transfection of isolated coronary venules with FAK-related non-kinase (FRNK) inhibits increases in venular permeability elicited by VEGF or activated neutrophils (Wu et al. 2003; Guo et al. 2005). FAK is a tyrosine kinase that is also activated upon binding of integrins to extracellular matrix proteins (Guan and Shalloway 1992; Hanks et al. 1992; Huang et al. 1993; Abbi and Guan 2002) and plays an important role in promoting adhesion-dependent cell survival (Frisch et al. 1996). FAK directly phosphorylates paxillin, a focal adhesion protein believed to act as a scaffold for other signaling and cytoskeleton-associated proteins (Abbi and Guan 2002). Currently, indirect evidence indicates that the RhoA/ROCK pathway lies upstream of FAK and paxillin in the regulation of endothelial permeability. Exoenzyme C3, which inactivates RhoA signaling, causes a concurrent decrease in endothelial cell monolayer baseline permeability and tyrosine phosphorylation on paxillin (Carbajal and Schaeffer Jr. 1999). In addition, thrombin-induced hyperpermeability is associated with increased tyrosine phosphorylation of FAK and paxillin, both of which are inhibited by the ROCK inhibitor Y-27632 or transfection of dominant-negative RhoA (Carbajal et al. 2000; van Nieuw Amerongen et al. 2004).

2.3.2.6 Endothelial Nitric Oxide Synthase (eNOS)

Bird's Eye View of eNOS Regulation

Nitric oxide is an important regulator of vascular homeostasis. While endothelial cells express at least three NO synthase (NOS) isoforms, namely, endothelial NOS (eNOS), inducible NOS (iNOS), and neuronal NOS (nNOS) (Govers and Rabelink 2001), the prime source of NO in the cardiovascular system is eNOS. This enzyme is tightly regulated with respect to activity and localization. Myristoylation and dual palmitoylation at its extreme N-terminus target eNOS to the cytoplasmic face of the Golgi complex and to the plasma membrane (Liu et al. 1997), where eNOS is believed to be fully capable of activation (Sessa et al. 1995; Shaul et al. 1996). In addition, eNOS can be normally found in the Golgi complex, where it is also capable of producing NO (Zhang et al. 2006). Misrouting of acylation-deficient eNOS mutants impairs NO production (Liu et al. 1996; Sakoda et al. 1995), indicating that correct subcellular targeting is critical for stimulus-dependent activation of the enzyme (Sessa et al. 1995).

In the plasma membrane, most eNOS is localized to the caveolae as shown by a variety of microscopic and biochemical techniques (Shaul et al. 1996; Garcia-Cardena et al. 1996a; Pobiner et al. 1991). eNOS binds directly to caveolin-1 (cav-1) through a consensus site, and this interaction inhibits eNOS activity (Fleming and Busse 1999). When intracellular calcium ($[Ca^{2+}]_i$) becomes elevated and subsequent binding of calcium to calmodulin occurs, the calcium–calmodulin complexes interact with eNOS, facilitating dissociation of eNOS from Cav-1 (Michel et al. 1997a, b; Garcia-Cardena et al. 1997; Feron et al. 1998) and causing a rise in NO production (Kanai et al. 1995); Blatter et al. 1995.

After activation, eNOS shuttles between caveolae and other subcellular compartments such as cytosol, non-caveolar plasma membrane portions, Golgi apparatus, and/or perinuclear structures, depending on the cell type and mode of activation (Michel et al. 1997a; Wang et al. 1997; Venema et al. 1996; Prabhakar et al. 1998; Goetz et al. 1999; Feng et al. 1999). Whether agonist-induced eNOS internalization is dependent on membrane endocytosis or on the dissociation of eNOS from the plasma membrane, followed by internalization through interactions with the cytoskeleton, is unknown.

Transient phosphorylation and depalmitoylation further contribute to activity modulation and subcellular targeting of eNOS. While a specific enzyme capable of catalyzing eNOS depalmitoylation in vitro has been identified (Yeh et al. 1999), it is unknown whether removing the palmitate from eNOS causes its translocation to the cytoplasm or whether the enzyme is involved in this process. Translocation does not alter the activation of the enzyme as it is established that eNOS remains functional, demonstrated by its ability to produce NO at all cellular locations (Zhang et al. 2006).

The translocation of eNOS may be regulated, in part, by protein–protein interactions. NOSIP (eNOS interacting protein) (Dedio et al. 2001) and NOSTRIN (eNOS traffic inducer) (Zimmermann et al. 2002) are two proteins that interact with eNOS in this process. In cells overexpressing NOSIP, translocation of eNOS appears to be shifted to intracellular sites that colocalize with Golgi and/or cytoskeletal marker proteins. Overexpression of NOSTRIN in CHO (Chinese Hamster Ovarian) cells mobilizes eNOS to cytosolic vesicle-like structures, where it colocalizes with NOSTRIN. Redistribution of eNOS in NOSTRIN expressing CHO-eNOS cells was paralleled by a decrease in NO release, down to 33–38% of control cells (which are considered to represent 100% NO release). This marked inhibition could be brought about by (a) direct inhibition of eNOS enzyme activity, (b) indirect inhibition of eNOS via modulation of regulatory mechanisms such as phosphorylation or protein, or (c) intracellular redistribution of eNOS (Zimmermann et al. 2002).

Relevance of eNOS-Derived NO in the Regulation of Microvascular Transport

Signaling interactions between the vascular wall and blood cells provide a unique way of communicating, coordinating, and integrating an appropriate physiological response to the changing tissue environment in vivo. Inflammatory processes are characterized by an increase in microvascular permeability (hyperpermeability) to macromolecules. Under inflammatory conditions, microvascular permeability to macromolecules is controlled mainly at the postcapillary venules. Relatively few studies have investigated the permeability-related signaling mechanisms in the complex cellular environment of the in vivo microcirculation.

NO was initially recognized as an important signaling regulator of vasodilatory function (Ignarro 1990; Furchgott and Zawadzki 1980; Furchgott 1984), but it also has a significant role in the control of microvascular permeability. The potential involvement of NO in permeability was initially advanced by speculation by Christian Crone in his Malpighi lecture (Crone 1987), and the first experimental evidence for it in relation to transport of macromolecules was probably published in 1992 (Yuan et al. 1992). The literature highlights controversial reports on the impact of NO on the control of microvascular permeability. Evidence in tissues, isolated venules, and cultured endothelial cell monolayers indicates that elevated eNOS activity increases microvascular permeability to macromolecules in response to inflammatory agents (Wu et al. 1996; Ramirez et al. 1995, 1996; Yuan et al. 1993b; Lal et al. 2001; Ialenti et al. 1992). Other reports, in animal models of vascular disease, have indicated that elevated NOS activity prevents increases in permeability (Kurose et al. 1993, 1994, 1995). These controversial results may be due to species differences or different mechanisms operating under experimental conditions. In either view, the compelling evidence for endogenous NO involvement as a positive or negative modulator of permeability was mainly based on either the specificity of L-arginine analogues to block NOS or the ability of NO donors to mimic endogenous NO actions. Furthermore, the published studies relied on the ability of pharmacological inhibitors to decrease NO production, but neither this capacity nor the ability of inflammatory agonists to increase NO production was determined directly in the cited studies.

This gap in knowledge was approached by investigations in which NO production was determined in vivo in response to challenges using PAF (platelet-activating factor) as a pro-inflammatory agent (Figs. 2.14 and 2.15). Topical application of PAF to the rat mesentery, hamster cheek pouch, and mouse cremaster induces production of NO and an increase in microvascular permeability (Durán et al. 2000; Zhu et al. 2005). Importantly, the time course in these series of studies suggests that NO production precedes the onset of hyperpermeability.

Studies performed in eNOS knockout mice more unequivocally established the role of eNOS activity in the control of microvascular permeability in both normal conditions and ischemia–reperfusion. These studies are (and should be) interpreted with the appropriate caveats associated with the adaptive changes occurring in response to deletion of important genes in genetically engineered mice. The hyperpermeability responses to PAF (Fig. 2.15) and VEGF are greatly reduced in eNOS$^{-/-}$ mice, while baseline permeability remains similar to eNOS wild-type mice (Hatakeyama et al. 2006; Fukumura et al. 2001). Because the responses to PAF in iNOS$^{-/-}$ mice were statistically the same as those to control mice, these data demonstrate that eNOS is the crucial isoform needed for increasing permeability in response to a pro-inflammatory challenge (Hatakeyama et al. 2006; Fukumura et al. 2001). This finding is further supported by evidence showing that specific inhibition of eNOS with a caveolin-1 scaffolding domain peptide (an endogenous cellular

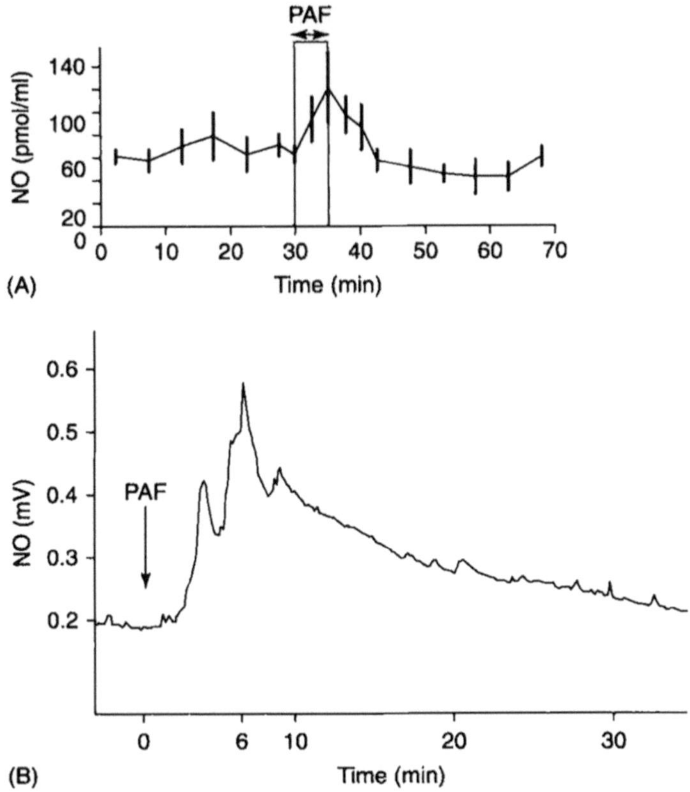

Fig. 2.14 PAF-induced NO production. Top panel: NO concentration was measured in the suffusate of the hamster cheek pouch using a nitric oxide analyzer (Durán et al. 2000). Bottom panel: Perivenular NO concentration was determined using a NO-sensitive electrode in the mouse cremaster muscle. NO concentration rose quickly upon topical administration of PAF in both tissues. *PAF* platelet-activating factor, *NO* nitric oxide

regulator of eNOS) reduced the ability of PAF to increase microvascular permeability (Hatakeyama et al. 2006). It is also important to note that because PAF is a vasoconstrictor, the observed increases in microvascular leakage found in this model reflect an increase in diffusive permeability and not higher convection due to increased capillary blood flow. Thus, it is fair to conclude that at this time the preponderance of the evidence favors the concept that activation of eNOS leads to hyperpermeability (Durán et al. 2006).

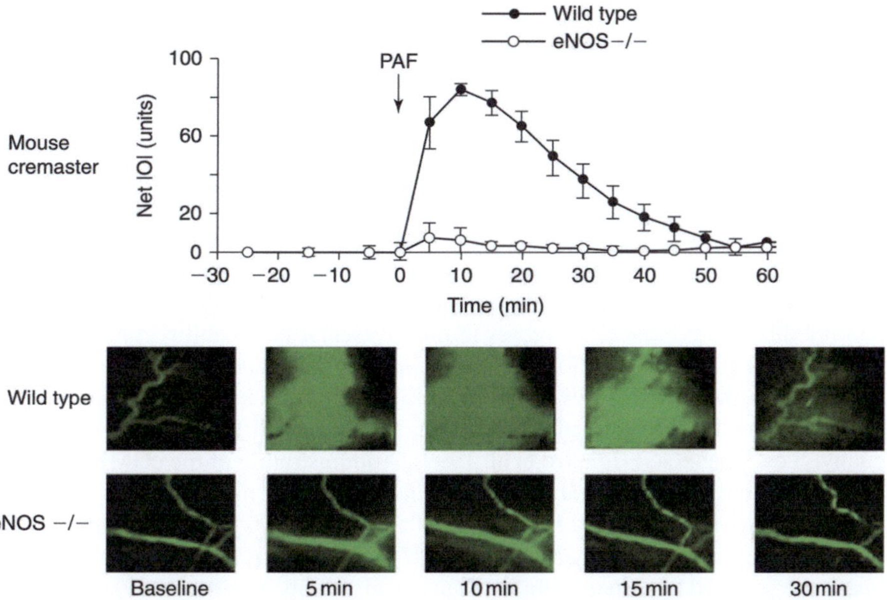

Fig. 2.15 Obligatory role of eNOS-mediated NO production in PAF-induced hyperpermeability in vivo. PAF-induced hyperpermeability is markedly reduced in $eNOS^{-/-}$ mice compared to wild-type control mice. Typical intravital fluorescence microscopic images of the respective cremaster muscles are shown below the IOI graph. (Adapted from Hatakeyama et al. 2006 and Durán et al. 2006. *eNOS* endothelial nitric oxide synthase, *PAF* platelet-activating factor, *NO* nitric oxide, *IOI* integrated optical intensity)

Biochemical Signaling Pathways Associated with eNOS Regulation of Microvascular Permeability

Agents that cause hyperpermeability increase the production of nitric oxide (NO) in the endothelium (Durán et al. 2000; Hood et al. 1998; Arnal et al. 1999; Ju et al. 1998; Garcia-Cardena et al. 1998; He et al. 1999). The activation of eNOS is a complex process involving—among other factors—dissociation from caveolin-1, association with heat shock protein (HSP)-90, Ca^{2+}–calmodulin binding, phosphorylation on serine-1177, and dephosphorylation of threonine-495 (Michel et al. 1997b; Garcia-Cardena et al. 1998; Fleming et al. 2001; Dimmeler et al. 1999; Fulton et al. 1999; Luo et al. 2000; Gratton et al. 2000; Garcia-Cardena et al. 1996b).

Figure 2.7 depicts a view of the signaling cascade associated with agonist-stimulated regulation of microvascular permeability via eNOS. When endothelial cells are stimulated by VEGF, insulin-like growth factor, or shear stress, Akt phosphorylates eNOS on serine-1177, causing increased eNOS activity (Dimmeler et al. 1999; Fulton et al. 1999; Michell et al. 1999). Stimulation with bradykinin causes calmodulin-dependent kinase II to phosphorylate on serine-1177 of eNOS

(Fleming et al. 2001). Activation of eNOS may also involve dephosphorylation of threonine-495 by the protein phosphatase PP1, a process likely promoted by protein kinase A (PKA) (Michell et al. 2001).

There are controversial reports on the role of PKC in the regulation of eNOS. PKC increases phosphorylation of eNOS on threonine-495 in endothelial cells, resulting in decreased eNOS activity (Fleming et al. 2001). It has also been reported that inhibition of PKC promotes increased release of nitrogen oxides in bovine aortic endothelial cells (Wedgwood et al. 2001; Marsen et al. 1999; Shen et al. 1999; Ohara et al. 1995). In contrast, activation of PKC in vivo promotes hyperpermeability (Kobayashi et al. 1994) and pharmacological blockade of eNOS inhibits PMA- and phorbol dibutyrate-induced increases in microvascular permeability, providing further evidence that PKC promotes the activation of eNOS and an increase in permeability to macromolecules in vivo (Wu et al. 1999; Ramirez et al. 1996).

NO is a well-described activator of cyclic GMP (cGMP) production by soluble guanylyl cyclase (sGC) (Ignarro et al. 1999). Administration of the cGMP analogue 8-Br-cGMP increases permeability across endothelial cell monolayers as well as in isolated microvessels (He et al. 1998, 2000; (Varma et al. 2002; Yonemaru et al. 1992). Decreasing cellular cGMP activity by inhibiting sGC attenuates VEGF- and histamine-induced hyperpermeability (Yuan et al. 1993b; Mayhan 1999). Moreover, sGC blockade attenuates phorbol ester- and sodium nitroprusside-mediated hyperpermeability, suggesting that cGMP activity is downstream from PKC and NOS (Yuan et al. 1993b). In addition, pharmacological blockade of cGMP-dependent protein kinase (protein kinase G; PKG) abolishes histamine-induced hyperpermeability. This occurs without affecting histamine-induced transient increases in $[Ca^{2+}]_i$, indicating that cGMP is a downstream mediator with respect to changes in $[Ca^{2+}]_i$ (Yuan et al. 1993b). Concordantly, direct administration of cGMP analogues to frog or rat mesenteric microvessels increased permeability without changing $[Ca^{2+}]_i$ (He et al. 1998, 2000).

The literature also includes different results regarding activation of eNOS by "functional" stimuli. Different rates of shear stress can cause opposite responses in eNOS phosphorylation (Corson et al. 1996). Shear stress at 25 dyne/cm^2 (a rate found in arteries) elicits eNOS phosphorylation in bovine aortic endothelial cells (BAEC), while shear stress at 5 dyne/cm^2 (a rate observed in microvessels) does not cause phosphorylation in arterial endothelial cells (Corson et al. 1996).

eNOS Translocation, S-Nitrosylation, and Microvascular Permeability

Functional changes in eNOS activity may be associated with molecular movement of eNOS (translocation) from the membrane to the cytosolic compartment (Goetz et al. 1999; Gratton et al. 2000; Michel and Feron 1997; Michel et al. 1993). Both membrane-bound and cytosolic eNOS release NO, but the relative importance of each pool for basal or stimulated release of NO has not been determined. It has been suggested, however, that "biologically active eNOS resides in different subcellular compartments and that each pool can be differentially regulated and responsive to different forms of stimulation" (Garcia-Cardena et al. 1996a). Thus, it is plausible that the cellular location of eNOS may represent a mechanism that determines the rate of NO production under different physiological or experimental conditions. A priori, it appears that a diffusible gas would not need translocation of its releasing enzyme for appropriate activity, but it has become more evident that the precise location of eNOS at subcellular compartments is necessary to achieve the specific, functionally efficacious concentration of NO (Sánchez et al. 2006).

Translocation of eNOS in cultured endothelial cells and in cells transfected with eNOS after agonist stimulation has been reported using fractionation or extraction techniques and confocal microscopy (Wang et al. 1997; Prabhakar et al. 1998; Goetz et al. 1999; Sánchez et al. 2006). eNOS translocation from the plasma membrane could be a mechanism to decrease the inhibitory association with Cav-1 in order to activate the enzyme. Agonists such as bradykinin and VEGF induce cytoskeletal reorganization during eNOS translocation (Thuringer 2004). Whether or not translocation of eNOS is a relevant step in microvascular function had not been investigated until it became possible to correlate in vitro and in vivo experiments.

One initial approach was to utilize agonists with a relatively defined functional impact (Sánchez et al. 2006). In particular, the ability of PAF to induce in vivo hyperpermeability without causing vasodilation in the hamster cheek pouch served as a starting point to dissect the possible functional role of translocation in the microcirculation. In contrast, ACh—a molecule that induces vasodilation without increasing microvascular permeability to macromolecules in the hamster cheek pouch—served as an indicator of "pure" vasodilating function. The first study attempting this correlation used both endothelial and ECV-304 cells transfected with eNOS-GFP. ECV-304 cells were initially characterized as a HUVEC-derived cell line lacking eNOS but were later found to be identical to a urinary bladder cell line due to cross contamination of cultures (Dirks et al. 1999). Still, the cells were a useful tool to study the localization and function of transfected eNOS (Sánchez et al. 2006; Sowa et al. 1999). PAF and ACh displaced eNOS from caveolae ascertained by a discontinuous sucrose gradient (Sánchez et al. 2006). Importantly, while eNOS distributes preferentially in the plasma membrane and Golgi under control conditions, treatment with PAF or ACh promotes translocation of eNOS to different subcellular locations in HUVECs, human dermal microvascular endothelial cells, and ECV-304 cells expressing eNOS-GFP. The translocation of eNOS was demonstrated by sucrose gradient centrifugation to identify lipid rafts and by fluorescence microscopy (Fig. 2.16). Specifically, ACh causes diminished staining in the plasma membrane and an increase in the perinuclear region of the cells. Stimulation with 10^{-7} M PAF results in reduced staining in the plasma membrane, but the label appeared more diffuse in the cytosol (Sánchez et al. 2006). Distribution of caveolin is not significantly influenced by either PAF or ACh. In addition, VEGF, a potent hyperpermeability-enhancing agent, induces translocation of eNOS to the cytosol in ECs, as evidenced through confocal imaging (Erwin et al. 2005). Importantly, the cited confocal images for PAF and VEGF report experimental results from independent laboratories.

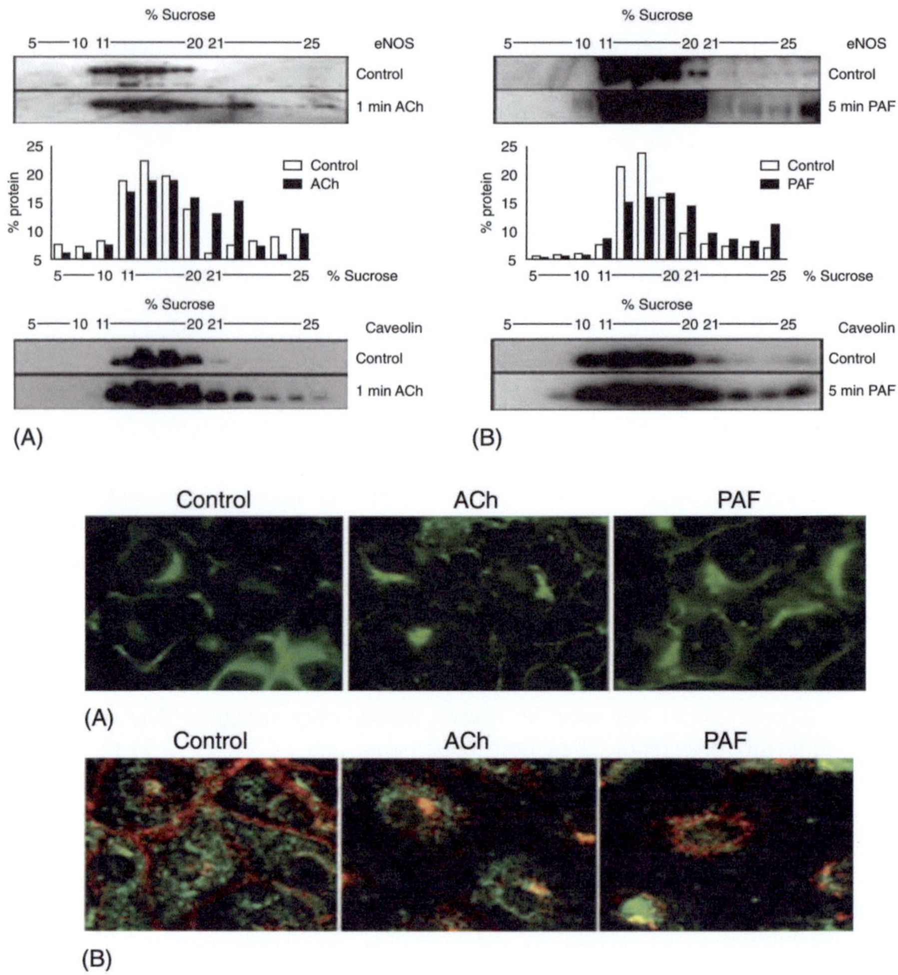

Fig. 2.16 Translocation of eNOS. Top panel (**a**) Isolation of lipid raft domains was done in controls and ACh- or PAF-treated cells. Fractions were probed against caveolin-1 and eNOS. (**a**) ACh treatment. (**b**) PAF treatment. The blots represent three independent experiments. Bottom panel: The influence of PAF and ACh on the distribution of eNOS-GFP was assessed by fluorescence of GFP. The images are representative of three independent experiments. (**a**) ECV-304–eNOS–GFP. (**b**) Human dermal microvascular cells stained with anti-eNOS (red) and anti-caveolin-1 (green) antibodies. (From Sánchez et al. 2006. *eNOS* endothelial nitric oxide synthase, *ACh* acetylcholine, *PAF* platelet-activating factor, *GFP* green fluorescent protein)

Blockade of caveolae scission with transfection of a dominant-negative dynamin-2 mutant or a mutant caveolin-1 prevents PAF-induced phosphorylation of eNOS at Ser-1177 and PAF-induced hyperpermeability, indicating that eNOS is internalized to its function-related intracellular targets via caveolae, as revealed by subsequent studies (Sanchez et al. 2008, 2009). In ECV-304 cells, which, when transfected with eNOS, display PAF-induced increases in monolayer permeability, studies of eNOS mutants that restrict the role of eNOS in different subcellular compartments have proven useful. When these cells express wild-type eNOS or an eNOS mutant that is restricted to the Golgi body, PAF can elicit increases in permeability. However, in ECV-304 cells expressing an eNOS mutant that is restricted to the plasma membrane, PAF fails to elicit increased permeability (Sánchez et al. 2011). Figure 2.17 shows preferential translocation of eNOS and signaling factors as well as the proteins involved in eNOS regulation.

The preceding results correlate well with the known PAF-induced increase in trans-monolayer transport of FITC-dextran 70, eNOS phosphorylation at Ser-1179, and dephosphorylation at Thr-495 in ECV–eNOS–GFP cells (Fig. 2.18). ACh also induces phosphorylation of eNOS at Ser-1177 in ECV-304 cells expressing eNOS-GFP and dephosphorylation at Thr-495 but causes no changes in the trans-monolayer transport of FITC-dextran 70. Because ACh and PAF lead to different physiological outcomes in vivo, these results suggest a functional significance for the molecular movement of eNOS. While further investigations are appropriate, it seems that the ACh-induced eNOS translocation from cell membrane to the Golgi is a preferential movement associated with a signal element important for vasodilation. Because translocation occurred after ACh applications as short as 30 s, these data indicate the translocation signals for the onset of vasodilation, an interpretation that is in agreement with in vivo data in the hamster cheek pouch (Dillon and Duran 1988; Figueroa et al. 2002). Because PAF induced an increase in permeability to FITC-dextran 70 in ECV-304 cell monolayers expressing eNOS-GFP (and cell constriction could not be documented), the PAF-induced translocation to the cytosol has been initially interpreted as an event or a step in the signaling mechanism encoding for endothelium-mediated microvascular hyperpermeability (Sánchez et al. 2006). The results obtained in cultured cells correlate very well and strongly support data obtained in vivo in the hamster cheek pouch (Sánchez et al. 2006; Figueroa et al. 2002). These results, which combine physiological and molecular biology approaches, contribute to advance our understanding of the regulatory mechanisms by integrating the molecular basis of the regulation of microvascular transport in a systematic manner.

Downstream of NO, there are two main signaling pathways. First, there is cGMP-PKG signaling, which was discussed in a previous section. Second, there is S-nitrosylation (nitrosation), the modification of protein cysteine thiol by NO to form S-nitrosothiols (Durán et al. 2013; Sánchez et al. 2013). Junctional proteins, including β-catenin, p120-catenin, and VE-cadherin, have been shown to be S-nitrosylation targets in endothelial cells, in connection with increased endothelial permeability (Marin et al. 2012; Thibeault et al. 2010; Guequén et al. 2016). For example, VEGF stimulates S-nitrosylation of β-catenin on Cys619, located at its interaction site with VE-cadherin. Inhibition of S-nitrosylation of β-catenin on Cys619 prevents VEGF-induced dissociation of the VE-cadherin–β-catenin complex, disassembly of adherens junctions, and increases in endothelial permeability (Thibeault et al. 2010). A similar mechanism was also found with PAF as the stimulus, with S-nitrosylation of VE-cadherin as a key step facilitating its

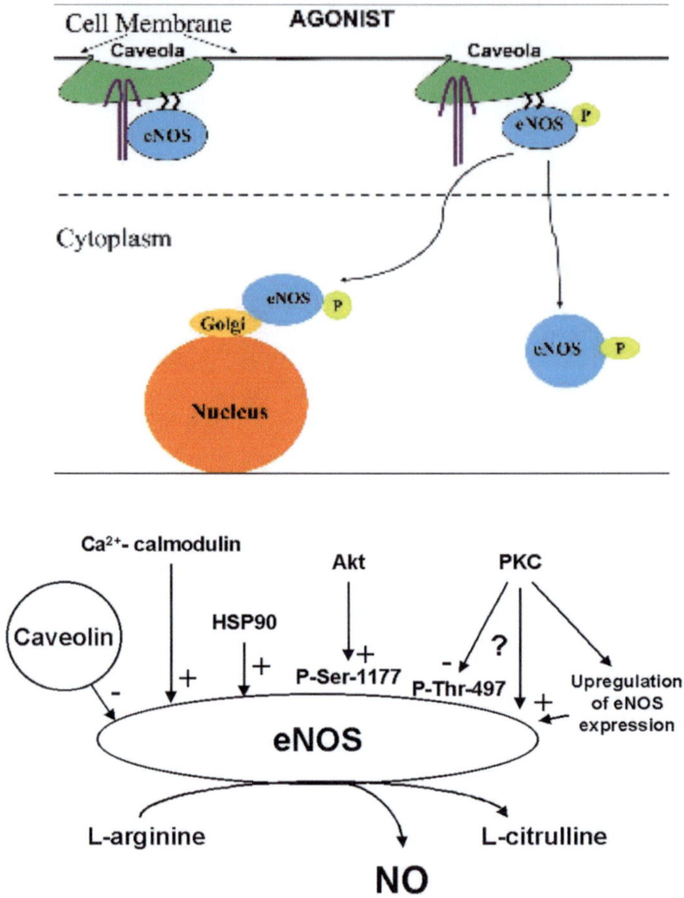

Fig. 2.17 Diagram of the eNOS translocation process and eNOS activation. (**a**) A simplified diagram of the eNOS translocation hypothesis. eNOS is anchored in caveolae through myristoylation and palmitoylation and is associated with caveolin-1. Upon stimulation with agonists, eNOS becomes phosphorylated, dissociates from caveolin-1, and translocates from the caveolae to either the Golgi apparatus or the cytosol. It is postulated that preferential translocation to different compartments is related to vascular function and may allow eNOS to interact with specific function-related target molecules. Even though the drawing depicts eNOS as being "free" in the cytosol, our recent unpublished data indicate that eNOS may be internalized to its intracellular targets via caveolae. (**b**) Several signals can regulate eNOS activity. While association with caveolin-1 and phosphorylation of Thr-497 decrease enzyme activity (NO production), interaction with $Ca2^{+}$-calmodulin and HSP90, phosphorylation at Ser-1177 by Akt, and PKC activation can all increase activity. *eNOS* endothelial nitric oxide synthase, *NO* nitric oxide, *PKC* protein kinase C

phosphorylation on tyrosine residues and its internalization (Guequén et al. 2016). PAF-induced S-nitrosylation of β-catenin and p120-catenin correlates with diminished abundance of these proteins at adherens junctions in both cultured cell models and in vivo. Moreover, using ECV-304 cells expressing either membrane-bound or

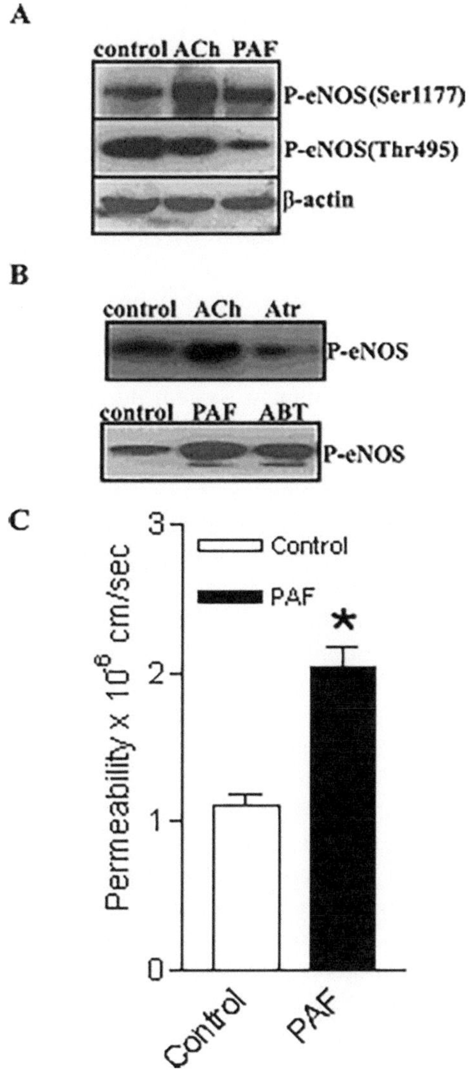

Fig. 2.18 PAF-induced eNOS phosphorylation and hyperpermeability in ECV-304–eNOS–GFP cells. (**a**) ACh and PAF activate eNOS by Ser-1177 phosphorylation and Thr-495 dephosphorylation. (**b**) Inhibition of muscarinic ACh and PAF receptors blocks Ser-1177 phosphorylation, and β-actin serves as a (loading) control in experiments shown in panels A and B. (**c**) PAF increases permeability in ECV–eNOS–GFP cells. Monolayers of ECV–eNOS–GFP cells were treated with PAF, and permeability to FITC-Dx 70 was measured. Data are expressed as permeability coefficients (mean ± S.E.M) for control and PAF-stimulated monolayers. $^{*}p < 0.05$, $n = 6$. *PAF* platelet-activating factor, *eNOS* endothelial nitric oxide synthase, *ACh* acetylcholine, *GFP* green fluorescent protein

cytosolic eNOS, it was found that the S-nitrosylation of β-catenin and p120-catenin is mediated by cytosolic eNOS (Guequén et al. 2016). PAF also causes S-nitrosylation of VASP on Cys64 (Zamorano et al. 2017). VASP is an important protein for maintaining endothelial barrier function (Benz et al. 2008; Henes et al. 2009; Schlegel et al. 2008; Schlegel and Waschke 2009). ECV-304 cells expressing a specific VASP mutant for Cys64 fail to display PAF-induced S-nitrosylation of VASP and undergo PAF-induced hyperpermeability (Zamorano et al. 2017). It is worth noting that PAF-induced increases in permeability are attenuated by pharmacological inhibition of NO synthesis or S-nitrosylation but not inhibition of the cGMP/PKG pathway (Guequén et al. 2016). S-nitrosylation by eNOS-derived NO also has an important role in regulating microvascular hyperpermeability in the tumor microenvironment (Zamorano et al. 2019; Guequen et al. 2019) and in the mechanisms for opening junctions for leukocytes or metastatic cells to cross the endothelium (Aguilar et al. 2020, 2021; Koning et al. 2023).

2.3.2.7 Venular Wall Reactivity in Inflammation and Immuno-Microvascular Permeability

The microvascular endothelium, in addition to its role in blood–tissue exchange, participates in immunity and host defense. Vascular endothelial cells come into direct contact with pathogens and toxins that enter the blood and interact with cells of the immune system during sepsis, inflammation, and allergic responses. Moreover, endothelial cells also act as antigen-presenting cells and are involved in T-cell homing (Pober et al. 2001).

Polymorphonuclear leukocytes (PMNs), predominantly neutrophils, play an important role in inflammatory responses by adhering to and migrating across the endothelium into surrounding tissues at the sites of injury. Often, this process is associated with microvascular hyperpermeability (Beynon et al. 1994; Burns et al. 1997; Yuan et al. 1995; Kubes et al. 1991). Eosinophils, which play an important role in allergic reactions, also modify the endothelial barrier. Eosinophil granule proteins and eosinophil-derived major basic protein increase microvascular leakage in the hamster cheek pouch (Minnicozzi et al. 1995; Minnicozzi et al. 1994).

Experimentally, C5a-activated PMNs increase permeability in both cell monolayer and isolated venule models (Tinsley et al. 1999, 2002; Yuan et al. 2002; Breslin and Yuan 2004). These permeability increases are associated with increased dual phosphorylation of myosin regulatory light chains (MLCs), actin–myosin-mediated cellular contraction, and tyrosine phosphorylation of focal adhesion kinase and the adherens junction protein β-catenin (Tinsley et al. 1999, 2002; Yuan et al. 2002; Breslin et al. 2006; Guo et al. 2005; Garcia et al. 1998).

PMN-induced increases in microvascular permeability can occur via both adhesion-dependent mechanisms and chemical mediators released from PMNs. The latter was initially evidenced by findings that 1) the supernatant derived from activated neutrophil suspensions activate endothelial monolayer permeability and 2) the permeability of isolated venules is increased when activated neutrophils are placed in the suffusion bath rather than in the lumen of the vessel (Tinsley et al. 2002; Yuan et al. 2002; Breslin and Yuan 2004). A different study found that fMLP (*N*-formyl-methionyl-leucyl-phenylalanine)-stimulated neutrophils increase endothelial $[Ca^{2+}]_i$ and permeability of isolated microvessels in an adhesion-independent manner, through the release of oxygen radicals (Zhu et al. 2005). In addition, it is now well-known that activated neutrophils can release their cytoplasmic contents as neutrophil extracellular traps (NETs), in a process known as NETosis that not only kills bacteria but is also involved in the systemic inflammatory response during infection or trauma (Brinkmann et al. 2004); Xu et al. 2009; Abrams et al. 2013. Among the contents of NETs, extracellular histones have been shown to significantly increase microvascular leakage (Villalba et al. 2020; Meegan et al. 2018). C-type lectin domain family 2 member D (Clec2d) has been recently identified as an endothelial cell receptor that mediates histone-induced increases in permeability (Yang et al. 2024).

Priming of the endothelium by an inflammatory mediator can enhance the hyperpermeability response upon subsequent exposure. For example, a subthreshold dose of PAF or of histamine does not initiate a hyperpermeability response in vivo; however, when one is applied 5 minutes after the application of the other agent, there is an explosive increase in microvascular permeability to macromolecules (Tomeo et al. 1991). A similar situation, with a much longer interval, applies to thrombin in its ability to increase L_p in individually perfused mesenteric microvessels (Curry et al. 2003).

Pattern recognition receptors, such as toll-like receptors (TLRs), which are well-known for their ability to sense bacterial toxins, may also have a role in priming of the endothelium during inflammatory responses. Lipopolysaccharide-induced activation of toll-like receptor 4 (TLR4) on endothelial cells causes modification of the expression of endothelial cell membrane proteins, such as G protein-coupled receptor kinases, to enhance PMN migration across the endothelium (Fan and Malik 2003). In addition, TLR4 appears to be an important sensor during sterile inflammatory responses, and there is evidence now of multiple endogenous agonists (Mollen et al. 2006). During severe burn injury, TLR4$^{-/-}$ mice had a significantly lower systemic hyperpermeability response than their wild-type counterparts, indicating the importance of TLR4 in mediating systemic inflammatory responses (Breslin et al. 2007c).

In addition, there is evidence that extracellular vesicles (EVs) can carry signals that can alter endothelial permeability (Chatterjee et al. 2020a). In one study in endothelial-derived EVs were produced in a mouse septic injury model, the EVs were found capable of activating cultured endothelial cells, causing cytoskeletal and junctional changes concomitant with barrier dysfunction (Chatterjee et al. 2020b). Moreover, EVs were also found to activate neutrophils, leading to NETosis, and to carry c-Src as cargo to transfer signals, thus leading to increased permeability (Chatterjee et al. 2020b). Palmitoylation, mediated by DHHC palmitoyl acyltransferases, is an important step for endothelial EV formation and neutrophil-induced endothelial hyperpermeability (Beard Jr. et al. 2016). EVs from mice with a loss of function in the palmitoyl acyltransferase DHHC21 display a reduced ability to activate neutrophils, and the mice have reduced neutrophil infiltration into the lungs in a septic shock model (Yang et al. 2022).

2.3.2.8 Termination of Microvascular Hyperpermeability and Signals that Enhance Microvascular Barrier Function

As stated above, inflammatory processes are characterized by an increase in microvascular hyperpermeability to macromolecules. On the other hand, maintenance of barrier properties (reduced or low permeability) seems to be regulated through factors controlling proteins that form intercellular adhesions. Cell–cell adhesion, in turn, is regulated through feedback signaling between small GTP-binding proteins and junctional proteins (Anastasiadis and Reynolds 2001). The notion that microvascular permeability is regulated by counterbalancing signaling mechanisms that maintain homeostasis by enhancing permeability or barrier properties is one that is simple and attractive.

Most of the research on inflammation has focused on prevention of hyperpermeability and on modulating or controlling pro-inflammatory agents that increase basal permeability. Much less attention has been given to elements that would normally contribute to maintain the integrity of the microvascular wall and that would enhance its barrier properties. In this regard, the search for the activation–inactivation mechanisms of eNOS has two main fronts: first, investigations of phosphorylation (activation) and de-phosphorylation (inactivation) and, the second more recent one, the determination of S-nitrosylation as an inactivation mechanism and de-nitrosylation as an activation stimulus. The location of eNOS also seems to be of significance with regard to its potential for S-nitrosylation. If so, S-nitrosylation of eNOS at a given location may determine (as phosphorylation does) eNOS–NO-linked signaling pathways. The mechanisms of phosphorylation and S-nitrosylation appear to be interrelated. These intricate connections have been reviewed recently (Dudzinski et al. 2006). While protein S-nitrosylation can be demonstrated in purified enzymes and in vitro, the functional in vivo significance and consequences of S-nitrosylation has not been clearly demonstrated. Particularly, the physiological role of S-nitrosylation in the microvascular endothelium, if any, is unknown. Thus, the question of how these complex interactions lead to activation and inactivation of functional outcomes at the tissue level remains.

Regulation of Inactivation (Reversal) of Hyperpermeability

Endothelial cells are the main controllers of microvascular permeability. This endothelial function is exquisitely modulated by endogenous and exogenous factors (endocrine, paracrine, and intracrine regulation). Deviations from the normal balance, owing to a deficiency or excess of regulatory factors, may lead to pathological states. Thus, a better understanding of the endothelial mechanisms involved in the initiation and termination of hyperpermeability in response to inflammatory agents may provide opportunities for successful clinical interventions.

cAMP and Cell Adhesion

Advances in cell research have provided a solid framework to consider cAMP as an important second messenger in regulating and promoting cell adhesion and, *by extension*, the integrity of the microvascular barrier. cAMP is a pleiotropic signaling second messenger. The classical views have been to consider that most, if not all, actions of cAMP proceed by activation of PKA and are terminated by phosphodiesterase 4-catalyzed degradation. These views have significantly changed since the relatively recent discovery of Epac (exchange protein directly activated by cAMP) (de Rooij et al. 1998), a guanine nucleotide exchange factor (GEF) that acts as an effector of cAMP. Importantly, Epac seems to respond with specificity to activation by cAMP. In fact, while cGMP binds to Epac, it does not activate it (Christensen et al. 2003). The interactions between cAMP effectors and the enzymes that regulate cAMP concentration within signaling complexes are the basis for the differential signaling through G protein-coupled receptors. The level of complexity of regulatory interactions is high and requires that cAMP signaling is compartmentalized in the cell. It is, thus, safe to conclude that cAMP dynamics and the activation of cAMP effectors are tightly controlled in terms of space (localization) and time (sequence) in vivo (DiPilato et al. 2004).

Epac/Rap1 and Cell Junctions

Epac, as a cAMP-stimulated GEF, serves to activate the small GTPase Rap1, its immediate effector. Rap1, as indicated in Fig. 2.19, may inhibit the pathway leading to ERK activation. If this is correct, then Rap1 would contribute to inhibiting increases in permeability and to stabilizing barrier properties by preventing the phosphorylation of ERK-1/2. However, this possibility may be incorrect (or cell type-dependent) inasmuch as 8CPT-2Me-cAMP (8-(4-chloro-phenylthio)-2'-*O*-methyl-adenosine-3′, 5′-cyclic monophosphate), a selective agonist of Epac/Rap1, did not influence ERK activity in cell lines (Enserink et al. 2002), and available evidence indicates that cAMP modulation of ERK may proceed via PKA. In the context of microvascular permeability, "other targets" of Rap1, as shown in Fig. 2.19, include reorganization of the actin cytoskeleton and stabilization of cell junctions via adhesion molecules (Kooistra et al. 2005). In human umbilical vein

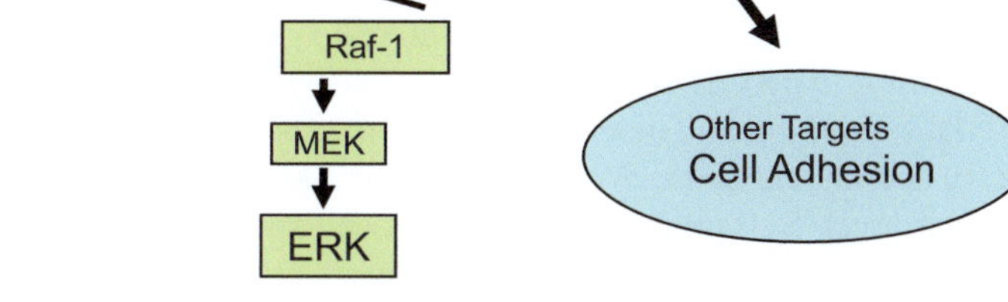

Fig. 2.19 Functions of Rap1. The diagram shows known and putative effectors and targets of Rap1. The arrows indicate stimulation, and the blunt ends denote inhibition. (Adapted from Zwartkruis and Bos 1999)

endothelial cells (HUVEC), passage 2–4 and cultured for about 72 hours, stimulation of Epac leads to accelerated formation and tightening of cell junctions through VE-cadherin (Kramer et al. 1982). Stimulation of Epac/Rap1 with 8CPT-2Me-cAMP enhances the endothelial barrier properties of HUVECs (Fukuhara et al. 2005; Cullere et al. 2005). There is a claim that 8CPT-2Me-cAMP reduced permeability in the mouse skin obtained with the Miles assay (Fukuhara et al. 2005). However, this finding is not so compelling considering that the same compound failed to reduced baseline Lp in single perfused microvessels (Adamson et al. 2008). Still, application of 8CPT-2Me-cAMP attenuates PAF-induced increases in Lp (Adamson et al. 2008), and the reports on HUVECs indicate that stimulation of Epac/Rap1 enhances the barrier properties of endothelial monolayers (Kooistra et al. 2005; Fukuhara et al. 2005; Cullere et al. 2005). These data are encouraging as they provide a literature basis for exploring the possible role of cAMP-activated Epac in restoring the barrier (Kramer et al. 1982) properties of the microvascular wall.

Reduction in baseline permeability, which has been demonstrated in isolated venules (Yuan et al. 1997) and in cultured cells (Yuan 2000; Fukuhara et al. 2005; Varma et al. 2002), is rarely observed in vivo (Kramer et al. 1982). The potential reasons for this difference include the different environments, interactions among different cells, etc., which occur in vivo relative to the more controlled conditions of isolated microvessels and tissue culture. Studies in this field will contribute to fill significant gaps in our knowledge of regulation of microvascular permeability and will have clinical significance inasmuch as they may lead to development of new pharmacological agents and treatment modalities to assist physicians and vascular surgeons in treating conditions, such as inflammation and ischemia–reperfusion injury, in which the microvascular barrier requires enhancement.

Sphingosine-1-Phosphate (S1P) and Endothelial Permeability

Sphingosine-1-phosphate (S1P) is a bioactive lipid produced in the cell membrane by sphingosine kinase-1 (SphK1) and Sphk2-mediated phosphorylation of sphingosine (Strub et al. 2010). The importance of S1P as a vital lipid mediator that regulates a large amount of physiological and pathophysiological processes has been highlighted in several reports describing its promotion of cell growth and survival, migration and differentiation, platelet aggregation, inflammation responses, and angiogenesis (Spiegel and Milstien 2003; Hla 2003). S1P is stored and continuously released by red blood cells, platelets, and other cell types (Singleton et al. 2005; Curry and Adamson 2013). It is carried in plasma by albumin and high-density lipoprotein (HDL), working as a modulator of microvascular permeability (Hla et al. 2008; McVerry and Garcia 2005; Wang and Dudek 2009).

At the cellular level, S1P promotes endothelial cell spreading and contributes to enhanced endothelial barrier function in vitro (Garcia et al. 2001; Lee et al. 1999; Rosen et al. 2007; Mehta and Malik 2006) and in vivo (Sanna et al. 2006; Li et al. 2008). S1P can regulate cell behavior via five different G protein-coupled receptors known as S1P receptors 1–5 (S1P1–S1P5) (Spiegel and Milstien 2003; Hla 2003;

Rosen et al. 2007). In endothelial cells, S1P binds to the S1P1 receptor to stimulate Rac1 activation via G_i-dependent recruitment of PI3K and Tiam1. Activation of Rac1 then induces the localization of focal adhesion-associated proteins such as paxillin and focal adhesion kinase (FAK) to focal adhesions, promotes reorganization of the cytoskeleton, and enhances assembly of adherens junctions and tight junctions, all of which together enhance endothelial barrier function (Garcia et al. 2001; Shikata et al. 2003). Cytoskeletal reorganization features the formation of local lamellipodia that protrude at cell–cell junctions, helping promote junctional integrity and maximizing diffusion distance for paracellular flux (Breslin et al. 2015). Interestingly, activation of RhoA at the cell periphery, in addition to Rac1, also contributes to S1P-induced endothelial barrier enhancement (Zhang et al. 2016b). This peripheral RhoA activation is in contrast to thrombin-induced RhoA activation within the central region of the cell associated with stress fiber formation and increased permeability sites but similar to brief peripheral RhoA activation localized to endothelial junctional gaps that coincides with their closure (Szulcek et al. 2013). There is also evidence that S1P increases the intracellular free Ca^{2+} concentration via a G_i–PLC pathway and transactivation of S1P1 signaling by other barrier-enhancing agents (Yoon et al. 2008).

Endothelial junctional stability may require continuous activation of signaling pathways by S1P. In transgenic mice in which S1P was not detectable, edema was present as evidenced by albumin accumulation in the lungs and increased interstitial thickness in the skin (Camerer et al. 2009). To answer the question of whether S1P1 agonists could decrease inflammation, S1P1 agonists, including a novel, engineered S1P chaperone, ApoM-Fc, have been tested both in vivo and in vitro. The data support that S1P1 agonists prevent barrier dysfunction in response to activated neutrophil-induced injury in vitro and diminish neutrophil transmigration and red blood cell (RBC) extravasation in vivo (Burg et al. 2018). Additionally, there are studies suggesting that S1P maintains normal vascular permeability by preserving endothelial surface glycocalyx in both cultured endothelial monolayer and intact microvessels, possibly via inhibiting syndecan-1 shedding (Zeng et al. 2014; Zhang et al. 2016a).

The potential therapeutic use of S1P has been explored in rat models of hemorrhagic shock and resuscitation (Alves et al. 2019a; Doggett et al. 2017). Increased microvascular permeability to plasma proteins is a hallmark of traumatic injuries and sepsis, and failure to control the excessive extravasation of fluids and proteins results in edema, impairing normal tissue homeostasis, and, in severe cases, leading to multiple organ failure and death. Hemorrhagic shock involves significant loss of intravascular volume that decreases tissue perfusion to most tissues that are not essential to maintain blood pressure. When fluids are administered intravenously to elevate blood pressure and improve perfusion, the events that lead to microvascular leakage are complex. First, the reperfusion of ischemic tissues causes ischemia–reperfusion injury and the release of several inflammatory mediators that stimulate endothelial hyperpermeability. Second, resuscitation fluids, often balanced salt solutions (crystalloids), dilute the plasma protein levels in the blood, reducing the plasma oncotic pressure that helps retain water in the blood. It is also worth noting that overall injury severity is often worsened by additional factors such as alcohol intoxication (Vonghia et al. 2008; Jurkovich et al. 1992; Hadfield et al. 2001; Carden and Granger 2000). A series of recent findings have highlighted significant insights into hemorrhagic shock and resuscitation (HSR)-induced microvascular permeability and the potential use of S1P as a therapeutic agent. First, alcohol intoxication exacerbates HSR-induced hypotension and microvascular leakage (Doggett et al. 2018) likely due to alcohol-induced disruption of VE-cadherin binding at cell–cell junctions (Doggett and Breslin 2014). Second, S1P could effectively ameliorate microvascular leakage following HSR (Fig. 2.20) or HSR combined with alcohol intoxication (Alves et al. 2019a; Doggett et al. 2017). With further investigations of the underlying mechanisms of S1P-induced protection of barrier function during HSR, it was found that S1P preserves the paracellular junctions, prevents loss of the glycocalyx surface layer, and protects endothelial cells from damage induced by mitochondrial dysfunction (Fig. 2.20) (Alves et al. 2019a). Furthermore, in other studies, S1P was found to protect against brain microvascular endothelial junctional protein disorganization and barrier dysfunction caused by alcohol (Alves et al. 2019b). With regard to the increased permeability scenario, reports from several laboratories demonstrate that the activation of signaling pathways by S1P in endothelial cells attenuate increased permeability (Garcia et al. 1995, 2001; Adamson et al. 2003; van Nieuw Amerongen and van Hinsbergh 2002; McVerry and Garcia 2005; Minnear et al. 2005; Lee et al. 2009), and S1P or its analogues have protective effect in other models, such as in murine lung injury (Peng et al. 2004; Mathew et al. 2011).

There is also increasing evidence that the principal action of S1P to attenuate microvascular hyperpermeability involves regulation of adhesion between endothelial cells, leading to junction complex stabilization (Kooistra et al. 2005; Adamson et al. 2008; Mehta and Malik 2006; Bogatcheva and Verin 2008). What we want to highlight is a concept that mechanisms that strengthen endothelial cell–cell adhesion are a common and essential feature of the action of agents that protect the endothelial barrier after exposure to injury and inflammatory conditions (Cullere et al. 2005; Waschke et al. 2004; McVerry and Garcia 2005). This is especially highlighted in a recent work showing that S1P helps preserve VE-cadherin architecture at junctions in postcapillary venules following HSR (Alves et al. 2019a) and in a work with single perfused microvessels in which S1P prevents changes in the distribution of VE-cadherin, occludin, and cortactin in intact microvessels caused by inflammatory mediator exposure (Adamson et al. 2010). These results demonstrate that S1P inhibits acute permeability increases in association with enhanced stabilization of peripheral endothelial adhesion proteins.

Sigma Receptors and Barrier Function of Endothelial Cells

Sigma-1 receptor agonists have been long-studied as therapeutic targets for many conditions, including neurological disorders, stroke, brain injuries, and myocardial infarction (Motawe et al. 2020a). Recent studies have shown the effectiveness of

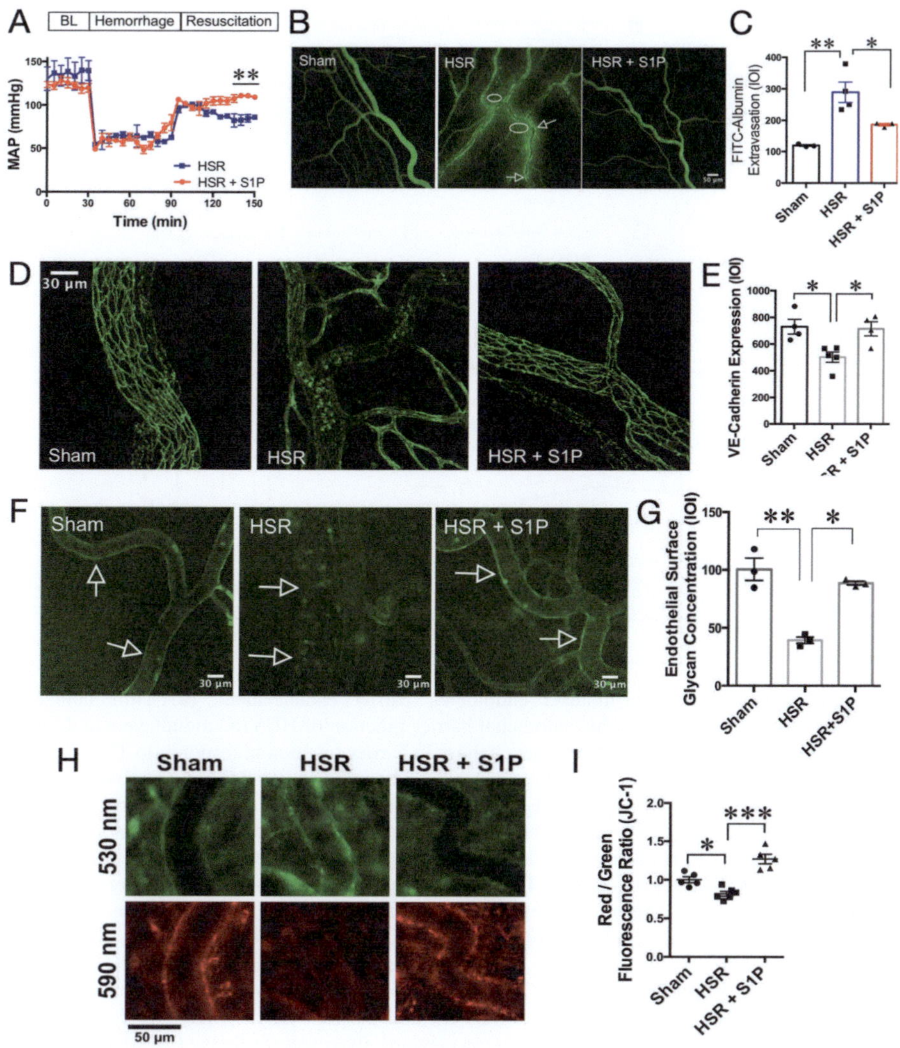

Fig. 2.20 S1P and hemorrhagic shock/resuscitation (HSR)-induced microvascular hyperpermeability. The ability of S1P to ameliorate microvascular leakage caused by hemorrhagic shock and crystalloid fluid resuscitation was investigated in a rat fixed-pressure hemorrhagic shock model. It was found that adding 0.1 mg/kg S1P to the resuscitation fluid regimen improved mean arterial blood pressure during resuscitation (**a**) and extravasation of FITC-albumin from postcapillary venules in the mesentery (**b** and **c**). Immunofluorescence labeling of VE-cadherin in postcapillary venules (images in panel **d**) revealed that HSR causes severe disruption of VE-cadherin labeling at endothelial cell–cell junctions, which is not apparent when S1P is included in resuscitation fluids (HSR + S1P). (**e**) Accordingly, the intensity of VE-cadherin labeling is decreased by HSR, but this is prevented when S1P is present. (**f** and **g**). FITC-lectin of the mesenteric microcirculation reveals the level of glycocalyx present. HSR reduces the FITC-lectin intensity on the microvasculature, and inclusion of S1P in resuscitation fluids prevents this decrease. (**h** and **i**). Intravital imaging with JC-1 labeling indicates the integrity of mitochondria. JC-1 fluorescence is green and diffuse when it is in the cytoplasm but turns red when it enters the mitochondria; the punctate labeling reveals individual mitochondria. HSR reduces the ratio of red:

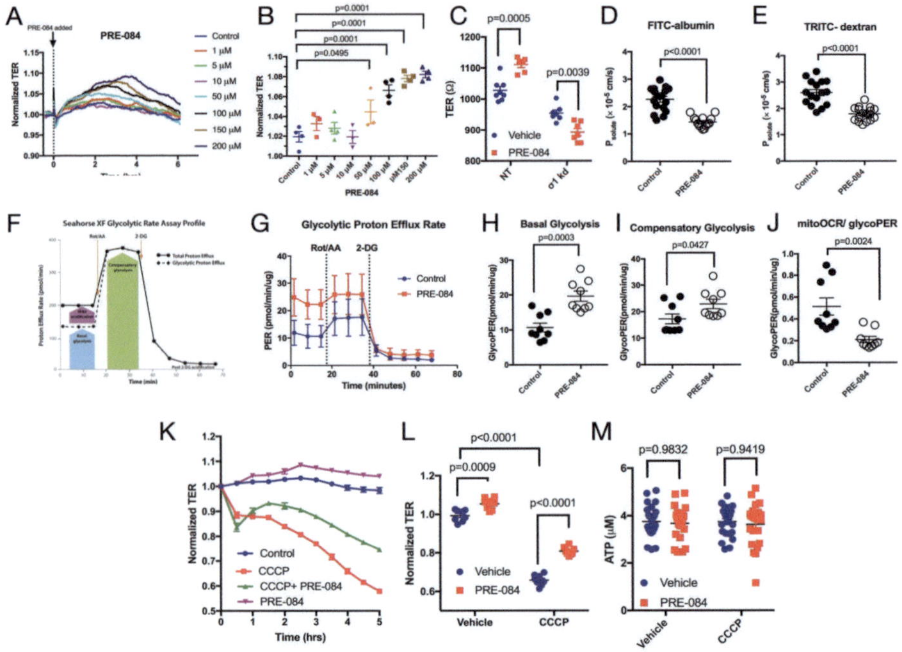

Fig. 2.21 Sigma-1 receptor-induced endothelial barrier enhancement. (**a**) Time course of PRE-084-induced increases in the endothelial barrier function of HUVEC monolayers at different concentrations of PRE-084. (**b**) The mean changes in barrier function in response to TER, 3 h after addition of PRE-084, normalized to the time point just prior to its addition. (**c**) Knockdown of sigma-1 with siRNA abolishes the PRE-084-induced increase in the endothelial barrier function. (**d** and **e**) PRE-084 (100 μM) also significantly reduces permeability of HUVEC monolayers to FITC-albumin and TRITC-dextran-4.4 kDa. (**f**) The glycolytic rate assay profile for the Seahorse instrument (© Agilent Technologies, Inc). (Reproduced with permission, courtesy of Agilent Technologies, Inc). (**g**) The glycolytic rate assay determines the glycolytic proton efflux rate (glycoPER) before and after addition of a rotenone/antimycin-A cocktail (Rot/AA), followed by the addition of 2-deoxyglucose (2-DG). The differences for the control and PRE-084 groups are shown. The oxygen consumption rate (OCR) is also measured (data not shown). From these data, basal glycolysis (**h**), compensatory glycolysis (**i**), and the ratio of the mitochondrial OCR and glycoPER (**j**) are calculated. These data show an elevated glycolysis rate in the PRE-084 group. (**k**). Time course of changes in endothelial barrier function of HUVEC monolayers treated with the mitochondrial uncoupler CCCP in the absence or presence of PRE-084. (**j**) PRE-084 partially rescues the drop in barrier function caused by CCCP, and this is independent of total ATP modulation (**m**). P-values for significant differences are shown where appropriate. (The data are from Motawe et al. 2020b and reproduced with permission)

Fig. 2.20 (continued) green labeling in postcapillary venules, suggesting that the mitochondrial wall is compromised by HSR. Addition of S1P to resuscitation fluids prevents the drop in the JC-1 red:green ratio, suggesting that S1P protects endothelial mitochondria in postcapillary venules. $*P < 0.05$, ** or $***P < 0.001$ for the comparisons indicated. (From Alves et al. 2019a with permission. *S1P* sphingosine-1-phosphate, *FITC* fluorescein isothiocyanate) sigma-1 receptor agonists on modulating endothelial barrier function. Experimental results in endothelial cell models have shown that the selective sigma-1 receptor agonist PRE-084 could enhance barrier function (Motawe et al. 2020b). Activation of sigma-1 with PRE-084 increases TER of HUVEC in a concentration-dependent manner, yet the ability of PRE-084 to enhance barrier function is inhibited when the cells are treated with siRNA directed against sigma-1 (Fig. 2.21). PRE-084 also enhances glycolysis in HUVECs. When the cells are challenged with a mitochondrial oxidative phosphorylation uncoupler (CCCP), the barrier function is disrupted. However, CCCP-induced endothelial barrier disruption is partially blocked when PRE-084 is present (Fig. 2.21), suggesting that this sigma-1 receptor agonist's barrier-enhancing properties are due to its ability to protect mitochondrial function (Motawe et al. 2020b). Sigma-1 receptor agonists have also been shown to contribute to lymphatic barrier function in addition to counteracting inflammation-induced lymphatic barrier dysfunction, via a similar mechanism (Motawe et al. 2021).

2.4 Concluding Remarks

Microvascular permeability remains a significant clinical problem for which there are currently limited therapeutic strategies. While it can be difficult to predict what the future holds for research on microvascular transport, one could anticipate that discovery of additional agents that reduce microvascular permeability and a better understanding of the endothelial mechanisms that reduce permeability will help drive the field toward finding specific therapeutics that address endothelial dysfunction. This ongoing challenge warrants additional work focusing on the fundamental balance between active mechanisms to maintain the barrier and those that compromise the barrier.

References

Abbi S, Guan JL (2002) Focal adhesion kinase: protein interactions and cellular functions. Histol Histopathol 17:1163–1171

Abbruscato TJ, Davis TP (1999) Combination of hypoxia/aglycemia compromises in vitro blood-brain barrier integrity. J Pharmacol Exp Ther 289:668–675

Abedi H, Zachary I (1997) Vascular endothelial growth factor stimulates tyrosine phosphorylation and recruitment to new focal adhesions of focal adhesion kinase and paxillin in endothelial cells. J Biol Chem 272:15442–15451

Abrams ST, Zhang N, Manson J, Liu T, Dart C, Baluwa F, Wang SS, Brohi K, Kipar A, Yu W, Wang G, Toh CH (2013) Circulating histones are mediators of trauma-associated lung injury. Am J Respir Crit Care Med 187:160–169

Abu Taha A, Schnittler HJ (2014) Dynamics between actin and the VE-cadherin/catenin complex: novel aspects of the ARP2/3 complex in regulation of endothelial junctions. Cell Adhes Migr 8: 125–135

Abu Taha A, Taha M, Seebach J, Schnittler HJ (2014) ARP2/3-mediated junction-associated lamellipodia control VE-cadherin-based cell junction dynamics and maintain monolayer integrity. Mol Biol Cell 25:245–256

Adam AP, Lowery AM, Martino N, Alsaffar H, Vincent PA (2016) Src family kinases modulate the loss of endothelial barrier function in response to TNF-alpha: crosstalk with p38 signaling. PLoS One 11:e0161975

Adamson RH, Liu B, Fry GN, Rubin LL, Curry FE (1998) Microvascular permeability and number of tight junctions are modulated by cAMP. Am J Phys 274:H1885–H1894

Adamson RH, Curry FE, Adamson G, Liu B, Jiang Y, Aktories K, Barth H, Daigeler A, Golenhofen N, Ness W, Drenckhahn D (2002) Rho and rho kinase modulation of barrier properties: cultured endothelial cells and intact microvessels of rats and mice. J Physiol 539: 295–308

Adamson RH, Zeng M, Adamson GN, Lenz JF, Curry FE (2003) PAF- and bradykinin-induced hyperpermeability of rat venules is independent of actin-myosin contraction. Am J Physiol Heart Circ Physiol 285:H406–H417

Adamson RH, Lenz JF, Zhang X, Adamson GN, Weinbaum S, Curry FE (2004) Oncotic pressures opposing filtration across non-fenestrated rat microvessels. J Physiol 557:889–907

Adamson RH, Ly JC, Sarai RK, Lenz JF, Altangerel A, Drenckhahn D, Curry FE (2008) Epac/Rap1 pathway regulates microvascular hyperpermeability induced by PAF in rat mesentery. Am J Physiol Heart Circ Physiol 294:H1188–H1196

Adamson RH, Sarai RK, Altangerel A, Thirkill TL, Clark JF, Curry FR (2010) Sphingosine-1-phosphate modulation of basal permeability and acute inflammatory responses in rat venular microvessels. Cardiovasc Res 88:344–351

Adderley SP, Lawrence C, Madonia E, Olubadewo JO, Breslin JW (2015a) Histamine activates p38 MAP kinase and alters local lamellipodia dynamics, reducing endothelial barrier integrity and eliciting central movement of actin fibers. Am J Physiol Cell Physiol 309:C51–C59

Adderley SP, Zhang XE, Breslin JW (2015b) Involvement of the H1 histamine receptor, p38 MAP kinase, myosin light chains kinase, and rho/ROCK in histamine-induced endothelial barrier dysfunction. Microcirculation 22:237–248

Aguilar G, Koning T, Ehrenfeld P, Sanchez FA (2020) Role of NO and S-nitrosylation in the expression of endothelial adhesion proteins that regulate leukocyte and tumor cell adhesion. Front Physiol 11:595526

Aguilar G, Cordova F, Koning T, Sarmiento J, Boric MP, Birukov K, Cancino J, Varas-Godoy M, Soza A, Alves NG, Mujica PE, Duran WN, Ehrenfeld P, Sanchez FA (2021) TNF-alpha-activated eNOS signaling increases leukocyte adhesion through the S-nitrosylation pathway. Am J Physiol Heart Circ Physiol 321:H1083–H1095

Aiello LP, Bursell SE, Clermont A, Duh E, Ishii H, Takagi C, Mori F, Ciulla TA, Ways K, Jirousek M, Smith LE, King GL (1997) Vascular endothelial growth factor-induced retinal permeability is mediated by protein kinase C in vivo and suppressed by an orally effective beta-isoform-selective inhibitor. Diabetes 46:1473–1480

Albelda SM, Sampson PM, Haselton FR, McNiff JM, Mueller SN, Williams SK, Fishman AP, Levine EM (1988) Permeability characteristics of cultured endothelial cell monolayers. J Appl Physiol 64:308–322

Alexander JS, Hechtman HB, Shepro D (1988) Phalloidin enhances endothelial barrier function and reduces inflammatory permeability in vitro. Microvasc Res 35:308–315

Alexander JS, Jackson SA, Chaney E, Kevil CG, Haselton FR (1998) The role of cadherin endocytosis in endothelial barrier regulation: involvement of protein kinase C and actin-cadherin interactions. Inflammation 22:419–433

Alexander JS, Alexander BC, Eppihimer LA, Goodyear N, Haque R, Davis CP, Kalogeris TJ, Carden DL, Zhu YN, Kevil CG (2000) Inflammatory mediators induce sequestration of VE-cadherin in cultured human endothelial cells. Inflammation 24:99–113

Alieva IB, Zemskov EA, Smurova KM, Kaverina IN, Verin AD (2013) The leading role of microtubules in endothelial barrier dysfunction: disassembly of peripheral microtubules leaves behind the cytoskeletal reorganization. J Cell Biochem 114:2258–2272

Allen TH, Dralpvats PD (1948) Spectrophotometric measurement of traces of the dye T-1824 by extraction with cellophane from both blood serum and urine of normal dogs. Am J Phys 154:27–38

Alvarez JI, Cayrol R, Prat A (2011) Disruption of central nervous system barriers in multiple sclerosis. Biochim Biophys Acta 1812:252–264

Alves NG, Motawe ZY, Yuan SY, Breslin JW (2018) Endothelial protrusions in junctional integrity and barrier function. Curr Top Membr 82:93–140

Alves NG, Trujillo AN, Breslin JW, Yuan SY (2019a) Sphingosine-1-phosphate reduces hemorrhagic shock and resuscitation-induced microvascular leakage by protecting endothelial mitochondrial integrity. Shock 52:423–433

Alves NG, Yuan SY, Breslin JW (2019b) Sphingosine-1-phosphate protects against brain microvascular endothelial junctional protein disorganization and barrier dysfunction caused by alcohol. Microcirculation 26:e12506

Amano M, Chihara K, Kimura K, Fukata Y, Nakamura N, Matsuura Y, Kaibuchi K (1997) Formation of actin stress fibers and focal adhesions enhanced by rho-kinase. Science 275: 1308–1311

Amano M, Fukata Y, Kaibuchi K (2000) Regulation and functions of rho-associated kinase. Exp Cell Res 261:44–51

Anastasiadis PZ, Reynolds AB (2001) Regulation of rho GTPases by p120-catenin. Curr Opin Cell Biol 13:604–610

Andriopoulou P, Navarro P, Zanetti A, Lampugnani MG, Dejana E (1999) Histamine induces tyrosine phosphorylation of endothelial cell-to-cell adherens junctions. Arterioscler Thromb Vasc Biol 19:2286–2297

Antonetti DA, Barber AJ, Khin S, Lieth E, Tarbell JM, Gardner TW (1998) Vascular permeability in experimental diabetes is associated with reduced endothelial occludin content: vascular endothelial growth factor decreases occludin in retinal endothelial cells. Diabetes 47:1953–1959

Antonetti DA, Barber AJ, Hollinger LA, Wolpert EB, Gardner TW (1999) Vascular endothelial growth factor induces rapid phosphorylation of tight junction proteins occludin and zonula occluden 1. A potential mechanism for vascular permeability in diabetic retinopathy and tumors. J Biol Chem 274:23463–23467

Aramoto H, Breslin JW, Pappas PJ, Hobson RW 2nd, Durán WN (2004) Vascular endothelial growth factor stimulates differential signaling pathways in in vivo microcirculation. Am J Physiol Heart Circ Physiol 287:H1590–H1598

Armenante PM, Kim D, Durán WN (1991) Experimental determination of the linear correlation between in vivo TV fluorescence intensity and vascular and tissue FITC-DX concentrations. Microvasc Res 42:198–208

Arnal JF, Dinh-Xuan AT, Pueyo M, Darblade B, Rami J (1999) Endothelium-derived nitric oxide and vascular physiology and pathology. Cell Mol Life Sci 55:1078–1087

Arturson G (2000) Forty years in burns research - the postburn inflammatory response. Burns 26: 599–604

Aurrand-Lions M, Duncan L, Ballestrem C, Imhof BA (2001) JAM-2, a novel immunoglobulin superfamily molecule, expressed by endothelial and lymphatic cells. J Biol Chem 276:2733–2741

Baldwin AL, Thurston G (1995) Changes in endothelial actin cytoskeleton in venules with time after histamine treatment. Am J Phys 269:H1528–H1537

Baldwin AL, Thurston G (2001) Mechanics of endothelial cell architecture and vascular permeability. Crit Rev Biomed Eng 29:247–278

Bamigboye AA, Hofmeyr GJ (2006) Interventions for leg edema and varicosities in pregnancy. What evidence? Eur J Obstet Gynecol Reprod Biol 129:3–8
Bates DO, Curry FE (1996) Vascular endothelial growth factor increases hydraulic conductivity of isolated perfused microvessels. Am J Phys 271:H2520–H2528
Bates DO, Curry FE (1997) Vascular endothelial growth factor increases microvascular permeability via a ca(2+)-dependent pathway. Am J Phys 273:H687–H694
Bates DO, Harper SJ (2003) Regulation of vascular permeability by vascular endothelial growth factors. Vasc Pharmacol 39:225–237
Bates DO, Lodwick D, Williams B (1999) Vascular endothelial growth factor and microvascular permeability. Microcirculation 6:83–96
Bauer PM, Yu J, Chen Y, Hickey R, Bernatchez PN, Looft-Wilson R, Huang Y, Giordano F, Stan RV, Sessa WC (2005) Endothelial-specific expression of caveolin-1 impairs microvascular permeability and angiogenesis. Proc Natl Acad Sci USA 102:204–209
Baxter LT, Jain RK, Svensjo E (1987) Vascular permeability and interstitial diffusion of macromolecules in the hamster cheek pouch: effects of vasoactive drugs. Microvasc Res 34:336–348
Bazzoni G, Dejana E (2004) Endothelial cell-to-cell junctions: molecular organization and role in vascular homeostasis. Physiol Rev 84:869–901
Beard RS Jr, Reynolds JJ, Bearden SE (2011) Hyperhomocysteinemia increases permeability of the blood-brain barrier by NMDA receptor-dependent regulation of adherens and tight junctions. Blood 118:2007–2014
Beard RS Jr, Haines RJ, Wu KY, Reynolds JJ, Davis SM, Elliott JE, Malinin NL, Chatterjee V, Cha BJ, Wu MH, Yuan SY (2014) Non-muscle Mlck is required for beta-catenin- and FoxO1-dependent downregulation of Cldn5 in IL-1beta-mediated barrier dysfunction in brain endothelial cells. J Cell Sci 127:1840–1853
Beard RS Jr, Yang X, Meegan JE, Overstreet JW, Yang CG, Elliott JA, Reynolds JJ, Cha BJ, Pivetti CD, Mitchell DA, Wu MH, Deschenes RJ, Yuan SY (2016) Palmitoyl acyltransferase DHHC21 mediates endothelial dysfunction in systemic inflammatory response syndrome. Nat Commun 7: 12823
Beard RS Jr, Hoettels BA, Meegan JE, Wertz TS, Cha BJ, Yang X, Oxford JT, Wu MH, Yuan SY (2020) AKT2 maintains brain endothelial claudin-5 expression and selective activation of IR/AKT2/FOXO1-signaling reverses barrier dysfunction. J Cereb Blood Flow Metab 40:374–391
Becker PM, Verin AD, Booth MA, Liu F, Birukova A, Garcia JG (2001) Differential regulation of diverse physiological responses to VEGF in pulmonary endothelial cells. Am J Physiol Lung Cell Mol Physiol 281:L1500–L1511
Bekker AY, Ritter AB, Duran WN (1986) Reduction of pressure in postcapillary venules induced by EPI-fluorescent illumination of FITC-dextrans. Microcirc Endothel Lymphat 3:411–423
Bekker AY, Dillon PK, Paul J, Ritter AB, Duran WN (1988) Dose-response relationships between platelet activating factor and permeability—surface area product of FITC-dextran 150 in the hamster cheek pouch. Microcirc Endothel Lymphat 4:433–446
Bekker AY, Ritter AB, Duran WN (1989) Analysis of microvascular permeability to macromolecules by video-image digital processing. Microvasc Res 38:200–216
Benz PM, Blume C, Moebius J, Oschatz C, Schuh K, Sickmann A, Walter U, Feller SM, Renne T (2008) Cytoskeleton assembly at endothelial cell-cell contacts is regulated by alphaII-spectrin-VASP complexes. J Cell Biol 180:205–219
Beyer EC, Gemel J, Seul KH, Larson DM, Banach K, Brink PR (2000) Modulation of intercellular communication by differential regulation and heteromeric mixing of co-expressed connexins. Braz J Med Biol Res 33:391–397
Beynon HL, Davies KA, Haskard DO, Walport MJ (1994) Erythrocyte complement receptor type 1 and interactions between immune complexes, neutrophils, and endothelium. J Immunol 153: 3160–3167
Bingaman S, Huxley VH, Rumbaut RE (2003) Fluorescent dyes modify properties of proteins used in microvascular research. Microcirculation 10:221–231

Birukov KG, Csortos C, Marzilli L, Dudek S, Ma SF, Bresnick AR, Verin AD, Cotter RJ, Garcia JG (2001) Differential regulation of alternatively spliced endothelial cell myosin light chain kinase isoforms by p60(Src). J Biol Chem 276:8567–8573

Birukova AA, Smurova K, Birukov KG, Kaibuchi K, Garcia JG, Verin AD (2004a) Role of rho GTPases in thrombin-induced lung vascular endothelial cells barrier dysfunction. Microvasc Res 67:64–77

Birukova AA, Smurova K, Birukov KG, Usatyuk P, Liu F, Kaibuchi K, Ricks-Cord A, Natarajan V, Alieva I, Garcia JG, Verin AD (2004b) Microtubule disassembly induces cytoskeletal remodeling and lung vascular barrier dysfunction: role of rho-dependent mechanisms. J Cell Physiol 201:55–70

Birukova AA, Birukov KG, Smurova K, Adyshev D, Kaibuchi K, Alieva I, Garcia JG, Verin AD (2004c) Novel role of microtubules in thrombin-induced endothelial barrier dysfunction. FASEB J 18:1879–1890

Birukova AA, Adyshev D, Gorshkov B, Bokoch GM, Birukov KG, Verin AD (2006) GEF-H1 is involved in agonist-induced human pulmonary endothelial barrier dysfunction. Am J Physiol Lung Cell Mol Physiol 290:L540–L548

Birukova AA, Fu P, Xing J, Yakubov B, Cokic I, Birukov KG (2010) Mechanotransduction by GEF-H1 as a novel mechanism of ventilator-induced vascular endothelial permeability. Am J Physiol Lung Cell Mol Physiol 298:L837–L848

Blatter LA, Taha Z, Mesaros S, Shacklock PS, Wier WG, Malinski T (1995) Simultaneous measurements of Ca2+ and nitric oxide in bradykinin-stimulated vascular endothelial cells. Circ Res 76:922–924

Bogatcheva NV, Verin AD (2008) The role of cytoskeleton in the regulation of vascular endothelial barrier function. Microvasc Res 76:202–207

Bogatcheva NV, Zemskova MA, Kovalenkov Y, Poirier C, Verin AD (2009) Molecular mechanisms mediating protective effect of cAMP on lipopolysaccharide (LPS)-induced human lung microvascular endothelial cells (HLMVEC) hyperpermeability. J Cell Physiol 221:750–759

Bohara M, Kambe Y, Nagayama T, Tokimura H, Arita K, Miyata A (2014) C-type natriuretic peptide modulates permeability of the blood-brain barrier. J Cereb Blood Flow Metab 34:589–596

Boric MP, Roblero JS, Duran WN (1987) Quantitation of bradykinin-induced microvascular leakage of FITC-dextran in rat cremaster muscle. Microvasc Res 33:397–412

Bradfield PF, Scheiermann C, Nourshargh S, Ody C, Luscinskas FW, Rainger GE, Nash GB, Miljkovic-Licina M, Aurrand-Lions M, Imhof BA (2007) JAM-C regulates unidirectional monocyte transendothelial migration in inflammation. Blood 110:2545–2555

Brady-Kalnay SM, Mourton T, Nixon JP, Pietz GE, Kinch M, Chen H, Brackenbury R, Rimm DL, Del Vecchio RL, Tonks NK (1998) Dynamic interaction of PTPmu with multiple cadherins in vivo. J Cell Biol 141:287–296

Braga VM, Machesky LM, Hall A, Hotchin NA (1997) The small GTPases rho and Rac are required for the establishment of cadherin-dependent cell-cell contacts. J Cell Biol 137:1421–1431

Braga VM, Betson M, Li X, Lamarche-Vane N (2000) Activation of the small GTPase Rac is sufficient to disrupt cadherin-dependent cell-cell adhesion in normal human keratinocytes. Mol Biol Cell 11:3703–3721

Breslin JW (2011) ROCK and cAMP promote lymphatic endothelial cell barrier integrity and modulate histamine and thrombin-induced barrier dysfunction. Lymphat Res Biol 9:3–11

Breslin JW (2023) Edema and lymphatic clearance: molecular mechanisms and ongoing challenges. Clin Sci (Lond) 137:1451–1476

Breslin JW, Yuan SY (2004) Involvement of RhoA and rho kinase in neutrophil-stimulated endothelial hyperpermeability. Am J Physiol Heart Circ Physiol 286:H1057–H1062

Breslin JW, Yuan SY (2024) Determination of solute permeability of microvascular endothelial cell monolayers in vitro. Methods Mol Biol 2711:1–12

Breslin JW, Pappas PJ, Cerveira JJ, Hobson RW 2nd, Durán WN (2003) VEGF increases endothelial permeability by separate signaling pathways involving ERK-1/2 and nitric oxide. Am J Physiol Heart Circ Physiol 284:H92–H100

Breslin JW, Sun H, Xu W, Rodarte C, Moy AB, Wu MH, Yuan SY (2006) Involvement of ROCK-mediated endothelial tension development in neutrophil-stimulated microvascular leakage. Am J Physiol Heart Circ Physiol 290:H741–H750

Breslin JW, Yuan SY, Wu MH (2007a) VEGF-C alters barrier function of cultured lymphatic endothelial cells through a vEGFR-3 dependent mechanism. Lymphat Res Biol 5:105

Breslin JW, Gaudreault N, Watson KD, Reynoso R, Yuan SY, Wu MH (2007b) Vascular endothelial growth factor-C stimulates the lymphatic pump by a VEGF receptor-3-dependent mechanism. Am J Physiol Heart Circ Physiol 293:H709–H718

Breslin JW, Wu MH, Guo M, Reynoso R, Yuan SY (2007c) Toll-like receptor-4 contributes to microvascular inflammation and barrier dysfunction in thermal injury. Shock 29:349

Breslin JW, Wu MH, Guo M, Reynoso R, Yuan SY (2008) Toll-like receptor 4 contributes to microvascular inflammation and barrier dysfunction in thermal injury. Shock 29:349–355

Breslin JW, Zhang XE, Worthylake RA, Souza-Smith FM (2015) Involvement of local lamellipodia in endothelial barrier function. PLoS One 10:e0117970

Breslin JW, Daines DA, Doggett TM, Kurtz KH, Souza-Smith FM, Zhang XE, Wu MH, Yuan SY (2016) Rnd3 as a novel target to ameliorate microvascular leakage. J Am Heart Assoc 5: e003336

Breslin JW, Yang Y, Scallan JP, Sweat RS, Adderley SP, Murfee WL (2019) Lymphatic vessel network structure and physiology. Compr Physiol 9:207–299

Brinkmann V, Reichard U, Goosmann C, Fauler B, Uhlemann Y, Weiss DS, Weinrauch Y, Zychlinsky A (2004) Neutrophil extracellular traps kill bacteria. Science 303:1532–1535

Bruns RR, Palade GE (1968) Studies on blood capillaries. II. Transport of ferritin molecules across the wall of muscle capillaries. J Cell Biol 37:277–299

Bundgaard M, Frokjaer-Jensen J, Crone C (1979) Endothelial plasmalemmal vesicles as elements in a system of branching invaginations from the cell surface. Proc Natl Acad Sci USA 76:6439–6442

Burg N, Swendeman S, Worgall S, Hla T, Salmon JE (2018) Sphingosine 1-phosphate receptor 1 signaling maintains endothelial cell barrier function and protects against immune complex-induced vascular injury. Arthritis Rheumatol 70:1879–1889

Burns AR, Walker DC, Brown ES, Thurmon LT, Bowden RA, Keese CR, Simon SI, Entman ML, Smith CW (1997) Neutrophil transendothelial migration is independent of tight junctions and occurs preferentially at tricellular corners. J Immunol 159:2893–2903

Camerer E, Regard JB, Cornelissen I, Srinivasan Y, Duong DN, Palmer D, Pham TH, Wong JS, Pappu R, Coughlin SR (2009) Sphingosine-1-phosphate in the plasma compartment regulates basal and inflammation-induced vascular leak in mice. J Clin Invest 119:1871–1879

Carbajal JM, Schaeffer RC Jr (1999) RhoA inactivation enhances endothelial barrier function. Am J Phys 277:C955–C964

Carbajal JM, Gratrix ML, Yu CH, Schaeffer RC Jr (2000) ROCK mediates thrombin's endothelial barrier dysfunction. Am J Physiol Cell Physiol 279:C195–C204

Carden DL, Granger DN (2000) Pathophysiology of ischaemia-reperfusion injury. J Pathol 190: 255–266

Carlsson O, Rosengren BI, Rippe B (2001) Transcytosis inhibitor N-ethylmaleimide increases microvascular permeability in rat muscle. Am J Physiol Heart Circ Physiol 281:H1728–H1733

Carmeliet P, Lampugnani MG, Moons L, Breviario F, Compernolle V, Bono F, Balconi G, Spagnuolo R, Oostuyse B, Dewerchin M, Zanetti A, Angellilo A, Mattot V, Nuyens D, Lutgens E, Clotman F, de Ruiter MC, Gittenberger-de Groot A, Poelmann R, Lupu F, Herbert JM, Collen D, Dejana E (1999) Targeted deficiency or cytosolic truncation of the VE-cadherin gene in mice impairs VEGF-mediated endothelial survival and angiogenesis. Cell 98:147–157

Cattan CE, Oberg KC (1999) Vinorelbine tartrate-induced pulmonary edema confirmed on rechallenge. Pharmacotherapy 19:992–994

Caveda L, Martin-Padura I, Navarro P, Breviario F, Corada M, Gulino D, Lampugnani MG, Dejana E (1996) Inhibition of cultured cell growth by vascular endothelial cadherin (cadherin-5/VE-cadherin). J Clin Invest 98:886–893

Chambers R, Zweifach BW (1947) Intercellular cement and capillary permeability. Physiol Rev 27: 436–463

Chang L, Karin M (2001) Mammalian MAP kinase signalling cascades. Nature 410:37–40

Chatterjee V, Yang X, Ma Y, Wu MH, Yuan SY (2020a) Extracellular vesicles: new players in regulating vascular barrier function. Am J Physiol Heart Circ Physiol 319:H1181–H1196

Chatterjee V, Yang X, Ma Y, Cha B, Meegan JE, Wu M, Yuan SY (2020b) Endothelial microvesicles carrying Src-rich cargo impair adherens junction integrity and cytoskeleton homeostasis. Cardiovasc Res 116:1525–1538

Christensen AE, Selheim F, de Rooij J, Dremier S, Schwede F, Dao KK, Martinez A, Maenhaut C, Bos JL, Genieser HG, Doskeland SO (2003) cAMP analog mapping of Epac1 and cAMP kinase. Discriminating analogs demonstrate that Epac and cAMP kinase act synergistically to promote PC-12 cell neurite extension. J Biol Chem 278:35394–35402

Chuang E, Barnard D, Hettich L, Zhang XF, Avruch J, Marshall MS (1994) Critical binding and regulatory interactions between Ras and Raf occur through a small, stable N-terminal domain of Raf and specific Ras effector residues. Mol Cell Biol 14:5318–5325

Cioffi WG (2001) What's new in burns and metabolism. J Am Coll Surg 192:241–254

Clements RT, Minnear FL, Singer HA, Keller RS, Vincent PA (2005) RhoA and rho-kinase dependent and independent signals mediate TGF-{beta}-induced pulmonary endothelial cytoskeletal reorganization and permeability. Am J Physiol Lung Cell Mol Physiol 288:294–306

Clough G (1991) Relationship between microvascular permeability and ultrastructure. Prog Biophys Mol Biol 55:47–69

Clough G, Michel CC (1981) The role of vesicles in the transport of ferritin through frog endothelium. J Physiol 315:127–142

Cobb MH (1999) MAP kinase pathways. Prog Biophys Mol Biol 71:479–500

Cohen AW, Carbajal JM, Schaeffer RC Jr (1999) VEGF stimulates tyrosine phosphorylation of beta-catenin and small-pore endothelial barrier dysfunction. Am J Phys 277:H2038–H2049

Comer GM, Ciulla TA (2004) Pharmacotherapy for diabetic retinopathy. Curr Opin Ophthalmol 15: 508–518

Conus NM, Hannan KM, Cristiano BE, Hemmings BA, Pearson RB (2002) Direct identification of tyrosine 474 as a regulatory phosphorylation site for the Akt protein kinase. J Biol Chem 277: 38021–38028

Corada M, Mariotti M, Thurston G, Smith K, Kunkel R, Brockhaus M, Lampugnani MG, Martin-Padura I, Stoppacciaro A, Ruco L, McDonald DM, Ward PA, Dejana E (1999) Vascular endothelial-cadherin is an important determinant of microvascular integrity in vivo. Proc Natl Acad Sci USA 96:9815–9820

Corada M, Liao F, Lindgren M, Lampugnani MG, Breviario F, Frank R, Muller WA, Hicklin DJ, Bohlen P, Dejana E (2001) Monoclonal antibodies directed to different regions of vascular endothelial cadherin extracellular domain affect adhesion and clustering of the protein and modulate endothelial permeability. Blood 97:1679–1684

Corson MA, James NL, Latta SE, Nerem RM, Berk BC, Harrison DG (1996) Phosphorylation of endothelial nitric oxide synthase in response to fluid shear stress. Circ Res 79:984–991

Cotran RS, Majno G (1964) The delayed and prolonged vascular leakage in inflammation. I. Topography of the leaking vessels after thermal injury. Am J Pathol 45:261–281

Crone C (1984) Recent advances in physiology. Chruchill Livingstone, London

Crone C (1987) The Malpighi lecture. From 'Porositates carnis' to cellular microcirculation. Int J Microcirc Clin Exp 6:101–122

Crone C, Christensen O (1981) Electrical resistance of a capillary endothelium. J Gen Physiol 77: 349–371

Crone C, Levitt DG (1984) Capillary permeability to small solutes. In: Handbook of physiology: the cardiovascular system microcirculation. American Physiological Society, Bethesda, pp 411–466
Crone C, Olesen SP (1982) Electrical resistance of brain microvascular endothelium. Brain Res 241:49–55
Cruz A, DeFouw LM, DeFouw DO (2000) Restrictive endothelial barrier function during normal angiogenesis in vivo: partial dependence on tyrosine dephosphorylation of beta-catenin. Microvasc Res 59:195–203
Cullere X, Shaw SK, Andersson L, Hirahashi J, Luscinskas FW, Mayadas TN (2005) Regulation of vascular endothelial barrier function by Epac, a cAMP-activated exchange factor for rap GTPase. Blood 105:1950–1955
Curry FE (1984) Mechanics and thermodynamics of transcapillary exchange. In: Renkin EM, Michel CC (eds) Handbook of physiology: the cardiovascular system microcirculation. American Physiological Society, Bethesda, pp 411–466
Curry FR, Adamson RH (2013) Tonic regulation of vascular permeability. Acta Physiol (Oxf) 207: 628–649
Curry FE, Michel CC (1980) A fiber matrix model of capillary permeability. Microvasc Res 20:96–99
Curry FE, Zeng M, Adamson RH (2003) Thrombin increases permeability only in venules exposed to inflammatory conditions. Am J Physiol Heart Circ Physiol 285:H2446–H2453
Daneman R, Prat A (2015) The blood-brain barrier. Cold Spring Harb Perspect Biol 7:a020412
Daniels BS (1993) Regional versus systemic modulation of the microcirculation in diabetes mellitus. J Lab Clin Med 121:375–376
de Rooij J, Zwartkruis FJ, Verheijen MH, Cool RH, Nijman SM, Wittinghofer A, Bos JL (1998) Epac is a Rap1 guanine-nucleotide-exchange factor directly activated by cyclic AMP. Nature 396:474–477
de Weerd WF, Leeb-Lundberg LM (1997) Bradykinin sequesters B2 bradykinin receptors and the receptor-coupled Galpha subunits Galphaq and Galphai in caveolae in DDT1 MF-2 smooth muscle cells. J Biol Chem 272:17858–17866
Dedio J, Konig P, Wohlfart P, Schroeder C, Kummer W, Muller-Esterl W (2001) NOSIP, a novel modulator of endothelial nitric oxide synthase activity. FASEB J 15:79–89
DeFouw LM, DeFouw DO (2001) Differential phosphodiesterase activity contributes to restrictive endothelial barrier function during angiogenesis. Microvasc Res 62:263–270
Dejana E, Lampugnani MG, Martinez-Estrada O, Bazzoni G (2000) The molecular organization of endothelial junctions and their functional role in vascular morphogenesis and permeability. Int J Dev Biol 44:743–748
DeMaio L, Chang YS, Gardner TW, Tarbell JM, Antonetti DA (2001) Shear stress regulates occludin content and phosphorylation. Am J Physiol Heart Circ Physiol 281:H105–H113
Diebel ME, Diebel LN, Liberati DM (2019) Protective effects of plasma products on the endothelial-glycocalyx barrier following trauma-hemorrhagic shock: is sphingosine-1 phosphate responsible? J Trauma Acute Care Surg 87:1061–1069
Dillon PK, Duran WN (1988) Effect of platelet-activating factor on microvascular permselectivity: dose-response relations and pathways of action in the hamster cheek pouch microcirculation. Circ Res 62:732–740
Dimmeler S, Fleming I, Fisslthaler B, Hermann C, Busse R, Zeiher AM (1999) Activation of nitric oxide synthase in endothelial cells by Akt-dependent phosphorylation. Nature 399:601–605
DiPilato LM, Cheng X, Zhang J (2004) Fluorescent indicators of cAMP and Epac activation reveal differential dynamics of cAMP signaling within discrete subcellular compartments. Proc Natl Acad Sci USA 101:16513–16518
Dirks WG, MacLeod RA, Drexler HG (1999) ECV304 (endothelial) is really T24 (bladder carcinoma): cell line cross- contamination at source. In Vitro Cell Dev Biol Anim 35:558–559
Doggett TM, Breslin JW (2011) Study of the actin cytoskeleton in live endothelial cells expressing GFP-actin. J Vis Exp 57:3187

Doggett TM, Breslin JW (2014) Acute alcohol intoxication-induced microvascular leakage. Alcohol Clin Exp Res 38:2414–2426

Doggett TM, Alves NG, Yuan SY, Breslin JW (2017) Sphingosine-1-phosphate treatment can ameliorate microvascular leakage caused by combined alcohol intoxication and hemorrhagic shock. Sci Rep 7:4078

Doggett TM, Tur JJ, Alves NG, Yuan SY, Tipparaju SM, Breslin JW (2018) Assessment of cardiovascular function and microvascular permeability in a conscious rat model of alcohol intoxication combined with hemorrhagic shock and resuscitation. Methods Mol Biol 1717:61–81

Downing JR, Reynolds AB (1991) PDGF, CSF-1, and EGF induce tyrosine phosphorylation of p120, a pp60src transformation-associated substrate. Oncogene 6:607–613

Downward J (1998) Mechanisms and consequences of activation of protein kinase B/Akt. Curr Opin Cell Biol 10:262–267

Dudek SM, Garcia JG (2001) Cytoskeletal regulation of pulmonary vascular permeability. J Appl Physiol 91:1487–1500

Dudek SM, Birukov KG, Zhan X, Garcia JG (2002) Novel interaction of cortactin with endothelial cell myosin light chain kinase. Biochem Biophys Res Commun 298:511–519

Dudek SM, Chiang ET, Camp SM, Guo Y, Zhao J, Brown ME, Singleton PA, Wang L, Desai A, Arce FT, Lal R, Van Eyk JE, Imam SZ, Garcia JG (2010) Abl tyrosine kinase phosphorylates nonmuscle myosin light chain kinase to regulate endothelial barrier function. Mol Biol Cell 21: 4042–4056

Dudzinski DM, Igarashi J, Greif D, Michel T (2006) The regulation and pharmacology of endothelial nitric oxide synthase. Annu Rev Pharmacol Toxicol 46:235–276

Duling BR (1973) The preparation and use of the hamster cheek pouch for studies of the microcirculation. Microvasc Res 5:423–429

Durán WN, Dillon PK (1990) Acute microcirculatory effects of platelet-activating factor. J Lipid Mediat 2(Suppl):S215–S227

Durán WN, Seyama A, Yoshimura K, Gonzalez DR, Jara PI, Figueroa XF, Boric MP (2000) Stimulation of NO production and of eNOS phosphorylation in the microcirculation in vivo. Microvasc Res 60:104–111

Durán WN, Hatakeyama T, Sánchez FA (2006) Endothelial nitric oxide synthase-derived NO signals regulate microvascular permeability. Physiol News 65:21–23

Durán WN, Sánchez FA, Breslin JW (2008) Microcirculatory exchange function. In: Tuma RF, Durán WN, Ley K (eds) Handbook of physiology: microcirculation. Academic Press, San Diego, pp 81–124

Durán WN, Beuve AV, Sánchez FA (2013) Nitric oxide, S-nitrosation, and endothelial permeability. IUBMB Life 65:819–826

Dvorak AM, Feng D (2001) The vesiculo-vacuolar organelle (VVO). A new endothelial cell permeability organelle. J Histochem Cytochem 49:419–432

Dvorak HF, Nagy JA, Feng D, Brown LF, Dvorak AM (1999) Vascular permeability factor/vascular endothelial growth factor and the significance of microvascular hyperpermeability in angiogenesis. Curr Top Microbiol Immunol 237:97–132

Ebnet K, Aurrand-Lions M, Kuhn A, Kiefer F, Butz S, Zander K et al (2003) The junctional adhesion molecule (JAM) family members JAM-2 and JAM-3 associate with the cell polarity protein PAR-3: a possible role for JAMs in endothelial cell polarity. J Cell Sci 116:3879–3891

Eliceiri BP, Paul R, Schwartzberg PL, Hood JD, Leng J, Cheresh DA (1999) Selective requirement for Src kinases during VEGF-induced angiogenesis and vascular permeability. Mol Cell 4:915–924

Eliceiri BP, Puente XS, Hood JD, Stupack DG, Schlaepfer DD, Huang XZ, Sheppard D, Cheresh DA (2002) Src-mediated coupling of focal adhesion kinase to integrin alpha(v)beta5 in vascular endothelial growth factor signaling. J Cell Biol 157:149–160

English J, Pearson G, Wilsbacher J, Swantek J, Karandikar M, Xu S, Cobb MH (1999) New insights into the control of MAP kinase pathways. Exp Cell Res 253:255–270

Enserink JM, Christensen AE, de Rooij J, van Triest M, Schwede F, Genieser HG, Doskeland SO, Blank JL, Bos JL (2002) A novel Epac-specific cAMP analogue demonstrates independent regulation of Rap1 and ERK. Nat Cell Biol 4:901–906
Erwin PA, Lin AJ, Golan DE, Michel T (2005) Receptor-regulated dynamic S-nitrosylation of endothelial nitric-oxide synthase in vascular endothelial cells. J Biol Chem 280:19888–19894
Esser S, Lampugnani MG, Corada M, Dejana E, Risau W (1998) Vascular endothelial growth factor induces VE-cadherin tyrosine phosphorylation in endothelial cells. J Cell Sci 111(Pt 13): 1853–1865
Evans TW, Chung KF, Rogers DF, Barnes PJ (1987) Effect of platelet-activating factor on airway vascular permeability: possible mechanisms. J Appl Physiol 63:479–484
Fan J, Malik AB (2003) Toll-like receptor-4 (TLR4) signaling augments chemokine-induced neutrophil migration by modulating cell surface expression of chemokine receptors. Nat Med 9:315–321
Fein A, Grossman RF, Jones JG, Overland E, Pitts L, Murray JF, Staub NC (1979) The value of edema fluid protein measurement in patients with pulmonary edema. Am J Med 67:32–38
Feng Y, Venema VJ, Venema RC, Tsai N, Caldwell RB (1999) VEGF induces nuclear translocation of Flk-1/KDR, endothelial nitric oxide synthase, and caveolin-1 in vascular endothelial cells. Biochem Biophys Res Commun 256:192–197
Feng D, Nagy JA, Brekken RA, Pettersson A, Manseau EJ, Pyne K, Mulligan R, Thorpe PE, Dvorak HF, Dvorak AM (2000) Ultrastructural localization of the vascular permeability factor/ vascular endothelial growth factor (VPF/VEGF) receptor-2 (FLK-1, KDR) in normal mouse kidney and in the hyperpermeable vessels induced by VPF/VEGF-expressing tumors and adenoviral vectors. J Histochem Cytochem 48:545–556
Feron O, Saldana F, Michel JB, Michel T (1998) The endothelial nitric-oxide synthase-caveolin regulatory cycle. J Biol Chem 273:3125–3128
Ferro T, Neumann P, Gertzberg N, Clements R, Johnson A (2000) Protein kinase C-alpha mediates endothelial barrier dysfunction induced by TNF-alpha. Am J Physiol Lung Cell Mol Physiol 278:L1107–L1117
Figueroa XF, Gonzalez DR, Martinez AD, Duran WN, Boric MP (2002) ACh-induced endothelial NO synthase translocation, NO release and vasodilatation in the hamster microcirculation in vivo. J Physiol 544:883–896
Fleming I, Busse R (1999) Signal transduction of eNOS activation. Cardiovasc Res 43:532–541
Fleming I, Fisslthaler B, Dimmeler S, Kemp BE, Busse R (2001) Phosphorylation of Thr(495) regulates ca(2+)/calmodulin-dependent endothelial nitric oxide synthase activity. Circ Res 88: E68–E75
Flinn HM, Ridley AJ (1996) Rho stimulates tyrosine phosphorylation of focal adhesion kinase, p130 and paxillin. J Cell Sci 109(Pt 5):1133–1141
Fox JR, Wayland H (1979) Interstitial diffusion of macromolecules in the rat mesentery. Microvasc Res 18:255–276
Fox J, Galey F, Wayland H (1980) Action of histamine on the mesenteric microvasculature. Microvasc Res 19:108–126
Frisch SM, Vuori K, Ruoslahti E, Chan-Hui PY (1996) Control of adhesion-dependent cell survival by focal adhesion kinase. J Cell Biol 134:793–799
Frokjaer-Jensen J (1980) Three-dimensional organization of plasmalemmal vesicles in endothelial cells. An analysis by serial sectioning of frog mesenteric capillaries. J Ultrastruct Res 73:9–20
Fukuhara S, Sakurai A, Sano H, Yamagishi A, Somekawa S, Takakura N, Saito Y, Kangawa K, Mochizuki N (2005) Cyclic AMP potentiates vascular endothelial cadherin-mediated cell-cell contact to enhance endothelial barrier function through an Epac-Rap1 signaling pathway. Mol Cell Biol 25:136–146
Fukumura D, Gohongi T, Kadambi A, Izumi Y, Ang J, Yun CO, Buerk DG, Huang PL, Jain RK (2001) Predominant role of endothelial nitric oxide synthase in vascular endothelial growth factor-induced angiogenesis and vascular permeability. Proc Natl Acad Sci USA 98:2604–2609

Fulton D, Gratton JP, McCabe TJ, Fontana J, Fujio Y, Walsh K, Franke TF, Papapetropoulos A, Sessa WC (1999) Regulation of endothelium-derived nitric oxide production by the protein kinase Akt. Nature 399:597–601

Furchgott RF (1984) The role of endothelium in the responses of vascular smooth muscle to drugs. Annu Rev Pharmacol Toxicol 24:175–197

Furchgott RF, Zawadzki JV (1980) The obligatory role of endothelial cells in the relaxation of arterial smooth muscle by acetylcholine. Nature 288:373–376

Furie MB, Cramer EB, Naprstek BL, Silverstein SC (1984) Cultured endothelial cell monolayers that restrict the transendothelial passage of macromolecules and electrical current. J Cell Biol 98:1033–1041

Garcia JG, Davis HW, Patterson CE (1995) Regulation of endothelial cell gap formation and barrier dysfunction: role of myosin light chain phosphorylation. J Cell Physiol 163:510–522

Garcia JG, Verin AD, Herenyiova M, English D (1998) Adherent neutrophils activate endothelial myosin light chain kinase: role in transendothelial migration. J Appl Physiol 84:1817–1821

Garcia JG, Verin AD, Schaphorst K, Siddiqui R, Patterson CE, Csortos C, Natarajan V (1999) Regulation of endothelial cell myosin light chain kinase by rho, cortactin, and p60(src). Am J Phys 276:L989–L998

Garcia JG, Schaphorst KL, Verin AD, Vepa S, Patterson CE, Natarajan V (2000) Diperoxovanadate alters endothelial cell focal contacts and barrier function: role of tyrosine phosphorylation. J Appl Physiol 89:2333–2343

Garcia JG, Liu F, Verin AD, Birukova A, Dechert MA, Gerthoffer WT, Bamberg JR, English D (2001) Sphingosine 1-phosphate promotes endothelial cell barrier integrity by Edg-dependent cytoskeletal rearrangement. J Clin Invest 108:689–701

Garcia JG, Wang P, Schaphorst KL, Becker PM, Borbiev T, Liu F, Birukova A, Jacobs K, Bogatcheva N, Verin AD (2002) Critical involvement of p38 MAP kinase in pertussis toxin-induced cytoskeletal reorganization and lung permeability. FASEB J 16:1064–1076

Garcia-Cardena G, Oh P, Liu J, Schnitzer JE, Sessa WC (1996a) Targeting of nitric oxide synthase to endothelial cell caveolae via palmitoylation: implications for nitric oxide signaling. Proc Natl Acad Sci USA 93:6448–6453

Garcia-Cardena G, Fan R, Stern DF, Liu J, Sessa WC (1996b) Endothelial nitric oxide synthase is regulated by tyrosine phosphorylation and interacts with caveolin-1. J Biol Chem 271:27237–27240

Garcia-Cardena G, Martasek P, Masters BS, Skidd PM, Couet J, Li S, Lisanti MP, Sessa WC (1997) Dissecting the interaction between nitric oxide synthase (NOS) and caveolin. Functional significance of the nos caveolin binding domain in vivo. J Biol Chem 272:25437–25440

Garcia-Cardena G, Fan R, Shah V, Sorrentino R, Cirino G, Papapetropoulos A, Sessa WC (1998) Dynamic activation of endothelial nitric oxide synthase by Hsp90. Nature 392:821–824

Garlick DG, Renkin EM (1970) Transport of large molecules from plasma to interstitial fluid and lymph in dogs. Am J Phys 219:1595–1605

Garrington TP, Johnson GL (1999) Organization and regulation of mitogen-activated protein kinase signaling pathways. Curr Opin Cell Biol 11:211–218

Gawlowski DM, Duran WN (1986) Dose-related effects of adenosine and bradykinin on microvascular permselectivity to macromolecules in the hamster cheek pouch. Circ Res 58:348–355

Gawlowski DM, Ritter AB, Duran WN (1982) Reproducibility of microvascular permeability responses to successive topical applications of bradykinin in the hamster cheek pouch. Microvasc Res 24:354–363

Gawlowski DM, Harding NR, Granger HJ (1989) Leukocyte phagocytosis and alterations in microvascular integrity elicited by FITC-dextran 150 and epi-illumination in the microcirculation of the hamster cheek pouch. Microvasc Res 37:1–15

Gerlowski LE, Jain RK (1986) Microvascular permeability of normal and neoplastic tissues. Microvasc Res 31:288–305

Giaever I, Keese CR (1991) Micromotion of mammalian cells measured electrically. Proc Natl Acad Sci USA 88:7896–7900
Giaever I, Keese CR (1993) A morphological biosensor for mammalian cells. Nature 366:591–592
Gibran NS, Heimbach DM (2000) Current status of burn wound pathophysiology. Clin Plast Surg 27:11–22
Glass CA, Pocock TM, Curry FE, Bates DO (2005) Cytosolic Ca2+ concentration and the rate of increase of the cytosolic Ca2+ concentration in the regulation of vascular permeability in Rana in vivo. J Physiol 564:817
Goetsch MR, Plott C, Woller JA 3rd, Fine DM, Arend LJ, Locke CF, Walker SL, Gelber AC (2021) Facial swelling and pancytopenia: first features and clues to the etiology of acute kidney injury. Am J Med 134:1238–1241
Goetz RM, Thatte HS, Prabhakar P, Cho MR, Michel T, Golan DE (1999) Estradiol induces the calcium-dependent translocation of endothelial nitric oxide synthase. Proc Natl Acad Sci USA 96:2788–2793
Gouverneur M, Spaan JA, Pannekoek H, Fontijn RD, Vink H (2006) Fluid shear stress stimulates incorporation of hyaluronan into endothelial cell glycocalyx. Am J Physiol Heart Circ Physiol 290:H458–H452
Govers R, Rabelink TJ (2001) Cellular regulation of endothelial nitric oxide synthase. Am J Physiol Renal Physiol 280:F193–F206
Gratton JP, Fontana J, O'Connor DS, Garcia-Cardena G, McCabe TJ, Sessa WC (2000) Reconstitution of an endothelial nitric-oxide synthase (eNOS), hsp90, and caveolin-1 complex in vitro. Evidence that hsp90 facilitates calmodulin stimulated displacement of eNOS from caveolin-1. J Biol Chem 275:22268–22272
Greene C, Hanley N, Campbell M (2019) Claudin-5: gatekeeper of neurological function. Fluids Barriers CNS 16:3
Groeneveld AB (2002) Vascular pharmacology of acute lung injury and acute respiratory distress syndrome. Vasc Pharmacol 39:247–256
Groeneveld AB, van Lambalgen TA, Thijs LG (1986) Microvascular permeability in endotoxin and bacterial shock. Acute Care 12:195–218
Guan JL, Shalloway D (1992) Regulation of focal adhesion-associated protein tyrosine kinase by both cellular adhesion and oncogenic transformation. Nature 358:690–692
Guequén A, Carrasco R, Zamorano P, Rebolledo L, Burboa P, Sarmiento J, Boric MP, Korayem A, Durán WN, Sánchez FA (2016) S-nitrosylation regulates VE-cadherin phosphorylation and internalization in microvascular permeability. Am J Physiol Heart Circ Physiol 310:H1039–H1044
Guequen A, Zamorano P, Cordova F, Koning T, Torres A, Ehrenfeld P, Boric MP, Salazar-Onfray F, Gavard J, Duran WN, Quezada C, Sarmiento J, Sanchez FA (2019) Interleukin-8 secreted by glioblastoma cells induces microvascular Hyperpermeability through NO signaling involving S-Nitrosylation of VE-cadherin and p120 in endothelial cells. Front Physiol 10: 988
Gumbiner BM (1995) Signal transduction of beta-catenin. Curr Opin Cell Biol 7:634–640
Gumbleton M, Audus KL (2001) Progress and limitations in the use of in vitro cell cultures to serve as a permeability screen for the blood-brain barrier. J Pharm Sci 90:1681–1698
Guo D, Jia Q, Song HY, Warren RS, Donner DB (1995) Vascular endothelial cell growth factor promotes tyrosine phosphorylation of mediators of signal transduction that contain SH2 domains. Association with endothelial cell proliferation. J Biol Chem 270:6729–6733
Guo M, Wu MH, Korompai F, Yuan SY (2003) Upregulation of PKC genes and isozymes in cardiovascular tissues during early stages of experimental diabetes. Physiol Genomics 12:139–146
Guo M, Wu MH, Granger HJ, Yuan SY (2004) Transference of recombinant VE-cadherin cytoplasmic domain alters endothelial junctional integrity and porcine microvascular permeability. J Physiol 554:78–88

Guo M, Wu MH, Granger HJ, Yuan SY (2005) Focal adhesion kinase in neutrophil-induced microvascular hyperpermeability. Microcirculation 12:223–232

Guo X, Wang L, Chen B, Li Q, Wang J, Zhao M, Wu W, Zhu P, Huang X, Huang Q (2009) ERM protein moesin is phosphorylated by advanced glycation end products and modulates endothelial permeability. Am J Physiol Heart Circ Physiol 297:H238–H246

Guo X, Eitnier RA, Beard RS Jr, Meegan JE, Yang X, Aponte AM, Wang F, Nelson PR, Wu MH (2020) Focal adhesion kinase and Src mediate microvascular hyperpermeability caused by fibrinogen- gammaC- terminal fragments. PLoS One 15:e0231739

Hadfield RJ, Mercer M, Parr MJ (2001) Alcohol and drug abuse in trauma. Resuscitation 48:25–36

Hagemann C, Rapp UR (1999) Isotype-specific functions of Raf kinases. Exp Cell Res 253:34–46

Hall A (1998) Rho GTPases and the actin cytoskeleton. Science 279:509–514

Hanks SK, Calalb MB, Harper MC, Patel SK (1992) Focal adhesion protein-tyrosine kinase phosphorylated in response to cell attachment to fibronectin. Proc Natl Acad Sci USA 89: 8487–8491

Haraldsson B, Johansson BR (1985) Changes in transcapillary exchange induced by perfusion fixation with glutaraldehyde, followed by measurements of capillary filtration coefficient, diffusion capacity and albumin clearance. Acta Physiol Scand 124:99–106

Harhaj NS, Antonetti DA (2004) Regulation of tight junctions and loss of barrier function in pathophysiology. Int J Biochem Cell Biol 36:1206–1237

Hatakeyama T, Pappas PJ, Hobson RW 2nd, Boric MP, Sessa WC, Duran WN (2006) Endothelial nitric oxide synthase regulates microvascular hyperpermeability in vivo. J Physiol 574:275–281

Hawkins BT, Davis TP (2005) The blood-brain barrier/neurovascular unit in health and disease. Pharmacol Rev 57:173–185

He P, Pagakis SN, Curry FE (1990) Measurement of cytoplasmic calcium in single microvessels with increased permeability. Am J Phys 258:H1366–H1374

He P, Zeng M, Curry FE (1998) cGMP modulates basal and activated microvessel permeability independently of [Ca2+]i. Am J Phys 274:H1865–H1874

He H, Venema VJ, Gu X, Venema RC, Marrero MB, Caldwell RB (1999) Vascular endothelial growth factor signals endothelial cell production of nitric oxide and prostacyclin through flk-1/KDR activation of c-Src. J Biol Chem 274:25130–25135

He P, Zeng M, Curry FE (2000) Dominant role of cAMP in regulation of microvessel permeability. Am J Physiol Heart Circ Physiol 278:H1124–H1133

Hemmings BA (1997) Akt signaling: linking membrane events to life and death decisions. Science 275:628–630

Henes J, Schmit MA, Morote-Garcia JC, Mirakaj V, Kohler D, Glover L, Eldh T, Walter U, Karhausen J, Colgan SP, Rosenberger P (2009) Inflammation-associated repression of vasodilator-stimulated phosphoprotein (VASP) reduces alveolar-capillary barrier function during acute lung injury. FASEB J 23:4244–4255

Henry CB, Duling BR (1999) Permeation of the luminal capillary glycocalyx is determined by hyaluronan. Am J Phys 277:H508–H514

Henry CB, Duling BR (2000) TNF-alpha increases entry of macromolecules into luminal endothelial cell glycocalyx. Am J Physiol Heart Circ Physiol 279:H2815–H2823

Higginbotham K, Tian Y, Gawlak G, Moldobaeva N, Shah A, Birukova AA (2014) Hepatocyte growth factor triggers distinct mechanisms of Asef and Tiam1 activation to induce endothelial barrier enhancement. Cell Signal 26:2306–2316

Hillman NJ, Whittles CE, Pocock TM, Williams B, Bates DO (2001) Differential effects of vascular endothelial growth factor-C and placental growth factor-1 on the hydraulic conductivity of frog mesenteric capillaries. J Vasc Res 38:176–186

Hippenstiel S, Tannert-Otto S, Vollrath N, Krull M, Just I, Aktories K, von Eichel-Streiber C, Suttorp N (1997) Glucosylation of small GTP-binding rho proteins disrupts endothelial barrier function. Am J Phys 272:L38–L43

Hirase T, Staddon JM, Saitou M, Ando-Akatsuka Y, Itoh M, Furuse M, Fujimoto K, Tsukita S, Rubin LL (1997) Occludin as a possible determinant of tight junction permeability in endothelial cells. J Cell Sci 110(Pt 14):1603–1613

Hirase T, Kawashima S, Wong EY, Ueyama T, Rikitake Y, Tsukita S, Yokoyama M, Staddon JM (2001) Regulation of tight junction permeability and occludin phosphorylation by Rhoa-p160ROCK-dependent and -independent mechanisms. J Biol Chem 276:10423–10431

Hixenbaugh EA, Goeckeler ZM, Papaiya NN, Wysolmerski RB, Silverstein SC, Huang AJ (1997) Stimulated neutrophils induce myosin light chain phosphorylation and isometric tension in endothelial cells. Am J Phys 273:H981–H988

Hla T (2003) Signaling and biological actions of sphingosine 1-phosphate. Pharmacol Res 47:401–407

Hla T, Venkataraman K, Michaud J (2008) The vascular S1P gradient-cellular sources and biological significance. Biochim Biophys Acta 1781:477–482

Hofmann-Kiefer KF, Kemming GI, Chappell D, Flondor M, Kisch-Wedel H, Hauser A, Pallivathukal S, Conzen P, Rehm M (2009) Serum heparan sulfate levels are elevated in endotoxemia. Eur J Med Res 14:526–531

Holmang A, Bjorntorp P, Rippe B (1992) Tissue uptake of insulin and inulin in red and white skeletal muscle in vivo. Am J Phys 263:H1170–H1176

Hood J, Granger HJ (1998) Protein kinase G mediates vascular endothelial growth factor-induced Raf-1 activation and proliferation in human endothelial cells. J Biol Chem 273:23504–23508

Hood JD, Meininger CJ, Ziche M, Granger HJ (1998) VEGF upregulates ecNOS message, protein, and NO production in human endothelial cells. Am J Phys 274:H1054–H1058

Horan KL, Adamski SW, Ayele W, Langone JJ, Grega GJ (1986) Evidence that prolonged histamine suffusions produce transient increases in vascular permeability subsequent to the formation of venular macromolecular leakage sites. Proof of the Majno-Palade hypothesis. Am J Pathol 123:570–576

Hoschuetzky H, Aberle H, Kemler R (1994) Beta-catenin mediates the interaction of the cadherin-catenin complex with epidermal growth factor receptor. J Cell Biol 127:1375–1380

Huang Q, Yuan Y (1997) Interaction of PKC and NOS in signal transduction of microvascular hyperpermeability. Am J Phys 273:H2442–H2451

Huang MM, Lipfert L, Cunningham M, Brugge JS, Ginsberg MH, Shattil SJ (1993) Adhesive ligand binding to integrin alpha IIb beta 3 stimulates tyrosine phosphorylation of novel protein substrates before phosphorylation of pp125FAK. J Cell Biol 122:473–483

Huang Q, Wu M, Meininger C, Kelly K, Yuan Y (1998) Neutrophil-dependent augmentation of PAF-induced vasoconstriction and albumin flux in coronary arterioles. Am J Phys 275:H1138–H1147

Huxley VH, Williams DA (1996) Basal and adenosine-mediated protein flux from isolated coronary arterioles. Am J Phys 271:H1099–H1108

Huxley VH, Williams DA (2000) Role of a glycocalyx on coronary arteriole permeability to proteins: evidence from enzyme treatments. Am J Physiol Heart Circ Physiol 278:H1177–H1185

Huxley VH, Curry FE, Adamson RH (1987) Quantitative fluorescence microscopy on single capillaries: alpha-lactalbumin transport. Am J Phys 252:H188–H197

Ialenti A, Ianaro A, Moncada S, Di Rosa M (1992) Modulation of acute inflammation by endogenous nitric oxide. Eur J Pharmacol 211:177–182

Ignarro LJ (1990) Biosynthesis and metabolism of endothelium-derived nitric oxide. Annu Rev Pharmacol Toxicol 30:535–560

Ignarro LJ, Cirino G, Casini A, Napoli C (1999) Nitric oxide as a signaling molecule in the vascular system: an overview. J Cardiovasc Pharmacol 34:879–886

Ishii H, Jirousek MR, Koya D, Takagi C, Xia P, Clermont A, Bursell SE, Kern TS, Ballas LM, Heath WF, Stramm LE, Feener EP, King GL (1996) Amelioration of vascular dysfunctions in diabetic rats by an oral PKC beta inhibitor. Science 272:728–731

Issbrucker K, Marti HH, Hippenstiel S, Springmann G, Voswinckel R, Gaumann A, Breier G, Drexler HC, Suttorp N, Clauss M (2003) p38 MAP kinase – a molecular switch between VEGF-induced angiogenesis and vascular hyperpermeability. FASEB J 17:262–264

Ivanov S, Scallan JP, Kim KW, Werth K, Johnson MW, Saunders BT, Wang PL, Kuan EL, Straub AC, Ouhachi M, Weinstein EG, Williams JW, Briseno C, Colonna M, Isakson BE, Gautier EL, Forster R, Davis MJ, Zinselmeyer BH, Randolph GJ (2016) CCR7 and IRF4-dependent dendritic cells regulate lymphatic collecting vessel permeability. J Clin Invest 126:1581–1591

Izumi T, Shimizu T (1995) Platelet-activating factor receptor: gene expression and signal transduction. Biochim Biophys Acta 1259:317–333

Jaap AJ, Tooke JE (1995) Pathophysiology of microvascular disease in non-insulin-dependent diabetes. Clin Sci (Lond) 89:3–12

Jacobs MK, McCance KL, Stewart ML (1982) External pneumatic intermittent compression for treatment of dependent pregnancy edema. Nurs Res 31:159–162

Jaffe EA, Nachman RL, Becker CG, Minick CR (1973) Culture of human endothelial cells derived from umbilical veins. Identification by morphologic and immunologic criteria. J Clin Invest 52: 2745–2756

Jannaway M, Scallan JP (2021) VE-cadherin and vesicles differentially regulate lymphatic vascular permeability to solutes of various sizes. Front Physiol 12:687563

Jaryal A, Kumar V, Sharma V (2015) Renal disease in patients infected with hepatitis B virus. Trop Gastroenterol 36:220–228

Jennings MA, Florey L (1967) An investigation of some properties of endothelium related to capillary permeability. Proc R Soc Lond B Biol Sci 167:39–63

Jirousek MR, Gillig JR, Gonzalez CM, Heath WF, McDonald JH 3rd, Neel DA, Rito CJ, Singh U, Stramm LE, Melikian-Badalian A, Baevsky M, Ballas LM, Hall SE, Winneroski LL, Faul MM (1996) (S)-13-[(dimethylamino)methyl]-10,11,14,15-tetrahydro-4,9:16, 21-dimetheno-1H, 13H-dibenzo[e,k]pyrrolo[3,4-h][1,4,13]oxadiazacyclohexadecene-1,3(2H)-d ione (LY333531) and related analogues: isozyme selective inhibitors of protein kinase C beta. J Med Chem 39: 2664–2671

Johansson BR (1979) Size and distribution of endothelial plasmalemmal vesicles in consecutive segments of the microvasculature in cat skeletal muscle. Microvasc Res 17:107–117

Johnson A, Phillips P, Hocking D, Tsan MF, Ferro T (1989) Protein kinase inhibitor prevents pulmonary edema in response to H2O2. Am J Phys 256:H1012–H1022

Johnson A, Hocking DC, Ferro TJ (1990) Mechanisms of pulmonary edema induced by a diacylglycerol second messenger. Am J Phys 258:H85–H91

Ju H, Venema VJ, Marrero MB, Venema RC (1998) Inhibitory interactions of the bradykinin B2 receptor with endothelial nitric-oxide synthase. J Biol Chem 273:24025–24029

Jurkovich GJ, Rivara FP, Gurney JG, Seguin D, Fligner CL, Copass M (1992) Effects of alcohol intoxication on the initial assessment of trauma patients. Ann Emerg Med 21:704–708

Kadry H, Cucullo L (2024) Evaluation of barrier integrity using a two-layered microfluidic device mimicking the blood-brain barrier. Methods Mol Biol 2711:77–88

Kaibuchi K, Kuroda S, Fukata M, Nakagawa M (1999) Regulation of cadherin-mediated cell-cell adhesion by the rho family GTPases. Curr Opin Cell Biol 11:591–596

Kanai AJ, Strauss HC, Truskey GA, Crews AL, Grunfeld S, Malinski T (1995) Shear stress induces ATP-independent transient nitric oxide release from vascular endothelial cells, measured directly with a porphyrinic microsensor. Circ Res 77:284–293

Kandasamy K, Escue R, Manna J, Adebiyi A, Parthasarathi K (2015) Changes in endothelial connexin 43 expression inversely correlate with microvessel permeability and VE-cadherin expression in endotoxin-challenged lungs. Am J Physiol Lung Cell Mol Physiol 309:L584–L592

Kedem O, Katchalsky A (1958) Thermodynamic analysis of the permeability of biological membranes to non-electrolytes. Biochim Biophys Acta 27:229–246

Kevil CG, Okayama N, Trocha SD, Kalogeris TJ, Coe LL, Specian RD, Davis CP, Alexander JS (1998a) Expression of zonula occludens and adherens junctional proteins in human venous and arterial endothelial cells: role of occludin in endothelial solute barriers. Microcirculation 5:197–210

Kevil CG, Payne DK, Mire E, Alexander JS (1998b) Vascular permeability factor/vascular endothelial cell growth factor-mediated permeability occurs through disorganization of endothelial junctional proteins. J Biol Chem 273:15099–15103

Kevil CG, Oshima T, Alexander B, Coe LL, Alexander JS (2000) H(2)O(2)-mediated permeability: role of MAPK and occludin. Am J Physiol Cell Physiol 279:C21–C30

Kevil CG, Okayama N, Alexander JS (2001) H(2)O(2)-mediated permeability II: importance of tyrosine phosphatase and kinase activity. Am J Physiol Cell Physiol 281:C1940–C1947

Kiemer AK, Weber NC, Furst R, Bildner N, Kulhanek-Heinze S, Vollmar AM (2002) Inhibition of p38 MAPK activation via induction of MKP-1: atrial natriuretic peptide reduces TNF-alpha-induced actin polymerization and endothelial permeability. Circ Res 90:874–881

Kim D, Duran WN (1995) Platelet-activating factor stimulates protein tyrosine kinase in hamster cheek pouch microcirculation. Am J Phys 268:H399–H403

Kim D, Armenante PM, Duran WN (1990) Mathematical modeling of mass transfer in microvascular wall and interstitial space. Microvasc Res 40:358–378

Kim D, Armenante PM, Duran WN (1993) Transient analysis of macromolecular transport across microvascular wall and into interstitium. Am J Phys 265:H993–H999

Kim DD, Ramirez MM, Duran WN (2000) Platelet-activating factor modulates microvascular dynamics through phospholipase C in the hamster cheek pouch. Microvasc Res 59:7–13

Kim H, Sasaki T, Maeda K, Koya D, Kashiwagi A, Yasuda H (2003) Protein kinase Cbeta selective inhibitor LY333531 attenuates diabetic hyperalgesia through ameliorating cGMP level of dorsal root ganglion neurons. Diabetes 52:2102–2109

Kobayashi I, Kim D, Hobson RW 2nd, Duran WN (1994) Platelet-activating factor modulates microvascular transport by stimulation of protein kinase C. Am J Phys 266:H1214–H1220

Kolodney MS, Wysolmerski RB (1992) Isometric contraction by fibroblasts and endothelial cells in tissue culture: a quantitative study. J Cell Biol 117:73–82

Koning T, Cordova F, Aguilar G, Sarmiento J, Mardones GA, Boric M, Varas-Godoy M, Lladser A, Duran WN, Ehrenfeld P, Sanchez FA (2023) S-Nitrosylation in endothelial cells contributes to tumor cell adhesion and extravasation during breast cancer metastasis. Biol Res 56:51

Kooistra MR, Corada M, Dejana E, Bos JL (2005) Epac1 regulates integrity of endothelial cell junctions through VE-cadherin. FEBS Lett 579:4966–4972

Korthuis RJ, Carden DL, Kvietys PR, Shepro D, Fuseler J (1991) Phalloidin attenuates postischemic neutrophil infiltration and increased microvascular permeability. J Appl Physiol 71:1261–1269

Kouklis P, Konstantoulaki M, Vogel S, Broman M, Malik AB (2004) Cdc42 regulates the restoration of endothelial barrier function. Circ Res 94:159–166

Koya D, Haneda M, Nakagawa H, Isshiki K, Sato H, Maeda S, Sugimoto T, Yasuda H, Kashiwagi A, Ways DK, King GL, Kikkawa R (2000) Amelioration of accelerated diabetic mesangial expansion by treatment with a PKC beta inhibitor in diabetic db/db mice, a rodent model for type 2 diabetes. FASEB J 14:439–447

Kozar RA, Peng Z, Zhang R, Holcomb JB, Pati S, Park P, Ko TC, Paredes A (2011) Plasma restoration of endothelial glycocalyx in a rodent model of hemorrhagic shock. Anesth Analg 112:1289–1295

Kramer GC, Harms BA, Bodai BI, Demling RH, Renkin EM (1982) Mechanisms for redistribution of plasma protein following acute protein depletion. Am J Phys 243:H803–H809

Krause M, Dent EW, Bear JE, Loureiro JJ, Gertler FB (2003) Ena/VASP proteins: regulators of the actin cytoskeleton and cell migration. Annu Rev Cell Dev Biol 19:541–564

Kroll J, Waltenberger J (1997) The vascular endothelial growth factor receptor KDR activates multiple signal transduction pathways in porcine aortic endothelial cells. J Biol Chem 272: 32521–32527

Kubes P, Grisham MB, Barrowman JA, Gaginella T, Granger DN (1991) Leukocyte-induced vascular protein leakage in cat mesentery. Am J Phys 261:H1872–H1879

Kumar P, Shen Q, Pivetti CD, Lee ES, Wu MH, Yuan SY (2009) Molecular mechanisms of endothelial hyperpermeability: implications in inflammation. Expert Rev Mol Med 11:e19

Kurose I, Kubes P, Wolf R, Anderson DC, Paulson J, Miyasaka M, Granger DN (1993) Inhibition of nitric oxide production. Mechanisms of vascular albumin leakage. Circ Res 73:164–171

Kurose I, Wolf R, Grisham MB, Granger DN (1994) Modulation of ischemia/reperfusion-induced microvascular dysfunction by nitric oxide. Circ Res 74:376–382

Kurose I, Wolf R, Grisham MB, Aw TY, Specian RD, Granger DN (1995) Microvascular responses to inhibition of nitric oxide production. Role of active oxidants. Circ Res 76:30–39

Kuruvilla A, Pielop C, Shearer WT (1994) Platelet-activating factor induces the tyrosine phosphorylation and activation of phospholipase C-gamma 1, Fyn and Lyn kinases, and phosphatidylinositol 3-kinase in a human B cell line. J Immunol 153:5433–5442

Lal BK, Varma S, Pappas PJ, Hobson RW 2nd, Duran WN (2001) VEGF increases permeability of the endothelial cell monolayer by activation of PKB/akt, endothelial nitric-oxide synthase, and MAP kinase pathways. Microvasc Res 62:252–262

Lambrechts A, Kwiatkowski AV, Lanier LM, Bear JE, Vandekerckhove J, Ampe C, Gertler FB (2000) cAMP-dependent protein kinase phosphorylation of EVL, a Mena/VASP relative, regulates its interaction with actin and SH3 domains. J Biol Chem 275:36143–36151

Lampejo AO, Jo M, Murfee WL, Breslin JW (2022) The microvascular-lymphatic Interface and tissue homeostasis: critical questions that challenge current understanding. J Vasc Res:1–16

Lampugnani MG, Corada M, Caveda L, Breviario F, Ayalon O, Geiger B, Dejana E (1995) The molecular organization of endothelial cell to cell junctions: differential association of plakoglobin, beta-catenin, and alpha-catenin with vascular endothelial cadherin (VE-cadherin). J Cell Biol 129:203–217

Lampugnani MG, Corada M, Andriopoulou P, Esser S, Risau W, Dejana E (1997) Cell confluence regulates tyrosine phosphorylation of adherens junction components in endothelial cells. J Cell Sci 110(Pt 17):2065–2077

Lampugnani MG, Zanetti A, Breviario F, Balconi G, Orsenigo F, Corada M, Spagnuolo R, Betson M, Braga V, Dejana E (2002) VE-cadherin regulates endothelial actin activating Rac and increasing membrane association of Tiam. Mol Biol Cell 13:1175–1189

Landis EM (1927) Microinjection studies of capillary permeability. II. The relation between capillary pressure and the rate at which fluid passes through the walls of single capillaries. Am J Phys 82:217–238

Lee MJ, Thangada S, Claffey KP, Ancellin N, Liu CH, Kluk M, Volpi M, Sha'afi RI, Hla T (1999) Vascular endothelial cell adherens junction assembly and morphogenesis induced by sphingosine-1-phosphate. Cell 99:301–312

Lee JF, Gordon S, Estrada R, Wang L, Siow DL, Wattenberg BW, Lominadze D, Lee MJ (2009) Balance of S1P1 and S1P2 signaling regulates peripheral microvascular permeability in rat cremaster muscle vasculature. Am J Physiol Heart Circ Physiol 296:H33–H42

Leeb-Lundberg LM, Song XH, Mathis SA (1994) Focal adhesion-associated proteins p125FAK and paxillin are substrates for bradykinin-stimulated tyrosine phosphorylation in Swiss 3T3 cells. J Biol Chem 269:24328–24334

Levick JR, Michel CC (1973) The permeability of individually perfused frog mesenteric capillaries to T1824 and T1824-albumin as evidence for a large pore system. Q J Exp Physiol Cogn Med Sci 58:67–85

Ley K, Arfors KE (1982) Changes in macromolecular permeability by intravascular generation of oxygen-derived free radicals. Microvasc Res 24:25–33

Ley K, Arfors KE (1986) Segmental differences of microvascular permeability for FITC-dextrans measured in the hamster cheek pouch. Microvasc Res 31:84–99

Li X, Stankovic M, Bonder CS, Hahn CN, Parsons M, Pitson SM, Xia P, Proia RL, Vadas MA, Gamble JR (2008) Basal and angiopoietin-1-mediated endothelial permeability is regulated by sphingosine kinase-1. Blood 111:3489–3497

Lindbom L (2003) Regulation of vascular permeability by neutrophils in acute inflammation. Chem Immunol Allergy 83:146–166
Lipowsky HH (2012) The endothelial glycocalyx as a barrier to leukocyte adhesion and its mediation by extracellular proteases. Ann Biomed Eng 40:840–848
Liu J, Garcia-Cardena G, Sessa WC (1996) Palmitoylation of endothelial nitric oxide synthase is necessary for optimal stimulated release of nitric oxide: implications for caveolae localization. Biochemistry 35:13277–13281
Liu J, Hughes TE, Sessa WC (1997) The first 35 amino acids and fatty acylation sites determine the molecular targeting of endothelial nitric oxide synthase into the Golgi region of cells: a green fluorescent protein study. J Cell Biol 137:1525–1535
Liu IF, Lin TC, Wang SC, Yen CH, Li CY, Kuo HF, Hsieh CC, Chang CY, Chang CR, Chen YH, Liu YR, Lee TY, Huang CY, Hsu CH, Lin SJ, Liu PL (2023) Long-term administration of Western diet induced metabolic syndrome in mice and causes cardiac microvascular dysfunction, cardiomyocyte mitochondrial damage, and cardiac remodeling involving caveolae and caveolin-1 expression. Biol Direct 18:9
Lum H, Malik AB (1994) Regulation of vascular endothelial barrier function. Am J Phys 267: L223–L241
Lum H, Podolski JL, Gurnack ME, Schulz IT, Huang F, Holian O (2001) Protein phosphatase 2B inhibitor potentiates endothelial PKC activity and barrier dysfunction. Am J Physiol Lung Cell Mol Physiol 281:L546–L555
Lund T, Onarheim H, Reed RK (1992) Pathogenesis of edema formation in burn injuries. World J Surg 16:2–9
Luo Z, Fujio Y, Kureishi Y, Rudic RD, Daumerie G, Fulton D, Sessa WC, Walsh K (2000) Acute modulation of endothelial Akt/PKB activity alters nitric oxide-dependent vasomotor activity in vivo. J Clin Invest 106:493–499
Lush CW, Kvietys PR (2000) Microvascular dysfunction in sepsis. Microcirculation 7:83–101
Lynch JJ, Ferro TJ, Blumenstock FA, Brockenauer AM, Malik AB (1990) Increased endothelial albumin permeability mediated by protein kinase C activation. J Clin Invest 85:1991–1998
Majno G, Palade G (1961) Studies on inflammation I. The effect of histamine and serotonin on vascular permeability: an electron microscopic study. J Biophys Cytol 11:597–605
Majno G, Palade GE, Schoefl GI, Studies on inflammation. II. (1961) The site of action of histamine and serotonin along the vascular tree: a topographic study. J Biophys Biochem Cytol 11:607–626
Marin N, Zamorano P, Carrasco R, Mujica P, Gonzalez FG, Quezada C, Meininger CJ, Boric MP, Duran WN, Sanchez FA (2012) S-Nitrosation of beta-catenin and p120 catenin: a novel regulatory mechanism in endothelial hyperpermeability. Circ Res 111:553–563
Mariner DJ, Anastasiadis P, Keilhack H, Bohmer FD, Wang J, Reynolds AB (2001) Identification of Src phosphorylation sites in the catenin p120ctn. J Biol Chem 276:28006–28013
Marsen TA, Egink G, Suckau G, Baldamus CA (1999) Tyrosine-kinase-dependent regulation of the nitric oxide synthase gene by endothelin-1 in human endothelial cells. Pflugers Arch 438:538–544
Martin JP, Logsdon N (1987) Oxygen radicals mediate cell inactivation by acridine dyes, fluorescein, and lucifer yellow CH. Photochem Photobiol 46:45–53
Martin-Padura I, Lostaglio S, Schneemann M, Williams L, Romano M, Fruscella P, Panzeri C, Stoppacciaro A, Ruco L, Villa A, Simmons D, Dejana E (1998) Junctional adhesion molecule, a novel member of the immunoglobulin superfamily that distributes at intercellular junctions and modulates monocyte transmigration. J Cell Biol 142:117–127
Mascarenhas JB, Gaber AA, Larrinaga TM, Mayfield R, Novak S, Camp SM, Gregorio C, Jacobson JR, Cress AE, Dudek SM, Garcia JGN (2021) EVL is a novel focal adhesion protein involved in the regulation of cytoskeletal dynamics and vascular permeability. Pulm Circ 11: 20458940211049002

Mascarenhas JB, Song JH, Gaber AA, Jacobson JR, Cress AE, Camp SM, Dudek SM, Garcia JGN (2022) An actin-, Cortactin- and Ena-VASP-linked complex contributes to endothelial cell focal adhesion and vascular barrier regulation. Cell Physiol Biochem 56:329–339

Mathew B, Jacobson JR, Berdyshev E, Huang Y, Sun X, Zhao Y, Gerhold LM, Siegler J, Evenoski C, Wang T, Zhou T, Zaidi R, Moreno-Vinasco L, Bittman R, Chen CT, LaRiviere PJ, Sammani S, Lussier YA, Dudek SM, Natarajan V, Weichselbaum RR, Garcia JG (2011) Role of sphingolipids in murine radiation-induced lung injury: protection by sphingosine 1-phosphate analogs. FASEB J 25:3388–3400

Mayhan WG (1994) Nitric oxide accounts for histamine-induced increases in macromolecular extravasation. Am J Phys 266:H2369–H2373

Mayhan WG (1999) VEGF increases permeability of the blood-brain barrier via a nitric oxide synthase/cGMP-dependent pathway. Am J Phys 276:C1148–C1153

Mayhan WG, Joyner WL (1984) The effect of altering the external calcium concentration and a calcium channel blocker, verapamil, on microvascular leaky sites and dextran clearance in the hamster cheek pouch. Microvasc Res 28:159–179

Mazzoni MC, Schmid-Schonbein GW (1996) Mechanisms and consequences of cell activation in the microcirculation. Cardiovasc Res 32:709–719

McColl BW, Rothwell NJ, Allan SM (2008) Systemic inflammation alters the kinetics of cerebrovascular tight junction disruption after experimental stroke in mice. J Neurosci 28:9451–9462

McDonagh PF (1993) The microvascular pathophysiology of chronic venous insufficiency. Yale J Biol Med 66:27–36

McDonagh PF, Hokama JY (2000) Microvascular perfusion and transport in the diabetic heart. Microcirculation 7:163–181

McDonagh PF, Williams SK (1984) The preparation and use of fluorescent-protein conjugates for microvascular research. Microvasc Res 27:14–27

McDonald DM, Thurston G, Baluk P (1999) Endothelial gaps as sites for plasma leakage in inflammation. Microcirculation 6:7–22

McLeod RL, Mingo GG, Herczku C, DeGennaro-Culver F, Kreutner W, Egan RW, Hey JA (1999) Combined histamine H1 and H3 receptor blockade produces nasal decongestion in an experimental model of nasal congestion. Am J Rhinol 13:391–399

McLeod RL, Mingo GG, Kreutner W, Hey JA (2005) Effect of combined histamine H1 and H3 receptor blockade on cutaneous microvascular permeability elicited by compound 48/80. Life Sci 76:1787–1794

McVerry BJ, Garcia JGN (2005) In vitro and in vivo modulation of vascular barrier integrity by sphingosine 1-phosphate: mechanistic insights. Cell Signal 17:131–139

Meegan JE, Yang X, Beard RS Jr, Jannaway M, Chatterjee V, Taylor-Clark TE, Yuan SY (2018) Citrullinated histone 3 causes endothelial barrier dysfunction. Biochem Biophys Res Commun 503:1498–1502

Mehta D, Malik AB (2006) Signaling mechanisms regulating endothelial permeability. Physiol Rev 86:279–367

Mehta D, Rahman A, Malik AB (2001) Protein kinase C-alpha signals rho-guanine nucleotide dissociation inhibitor phosphorylation and rho activation and regulates the endothelial cell barrier function. J Biol Chem 276:22614–22620

Meyer DJ Jr, Huxley VH (1992) Capillary hydraulic conductivity is elevated by cGMP-dependent vasodilators. Circ Res 70:382–391

Michel CC (1984) Fluid movements through capillary walls. In: Handbook of physiology: the cardiovascular system microcirculation. American Physiological Society, Bethesda, pp 375–409

Michel CC, Curry FE (1999) Microvascular permeability. Physiol Rev 79:703–761

Michel T, Feron O (1997) Nitric oxide synthases: which, where, how, and why? J Clin Invest 100: 2146–2152

Michel T, Li GK, Busconi L (1993) Phosphorylation and subcellular translocation of endothelial nitric oxide synthase. Proc Natl Acad Sci USA 90:6252–6256
Michel JB, Feron O, Sase K, Prabhakar P, Michel T (1997a) Caveolin versus calmodulin. Counterbalancing allosteric modulators of endothelial nitric oxide synthase. J Biol Chem 272: 25907–25912
Michel JB, Feron O, Sacks D, Michel T (1997b) Reciprocal regulation of endothelial nitric-oxide synthase by Ca2+-calmodulin and caveolin. J Biol Chem 272:15583–15586
Michell BJ, Griffiths JE, Mitchelhill KI, Rodriguez-Crespo I, Tiganis T, Bozinovski S, de Montellano PR, Kemp BE, Pearson RB (1999) The Akt kinase signals directly to endothelial nitric oxide synthase. Curr Biol 9:845–848
Michell BJ, Chen Z, Tiganis T, Stapleton D, Katsis F, Power DA, Sim AT, Kemp BE (2001) Coordinated control of endothelial nitric-oxide synthase phosphorylation by protein kinase C and the cAMP-dependent protein kinase. J Biol Chem 276:17625–17628
Miles AA, Miles EM (1952) Vascular reactions to histamine, histamine-liberator and leukotaxine in the skin of Guinea-pigs. J Physiol 118:228–257
Millauer B, Wizigmann-Voos S, Schnurch H, Martinez R, Moller NP, Risau W, Ullrich A (1993) High affinity VEGF binding and developmental expression suggest Flk-1 as a major regulator of vasculogenesis and angiogenesis. Cell 72:835–846
Minnear FL, Zhu L, He P (2005) Sphingosine 1-phosphate prevents platelet-activating factor-induced increase in hydraulic conductivity in rat mesenteric venules: pertussis toxin sensitive. Am J Physiol Heart Circ Physiol 289:H840–H844
Minnicozzi M, Duran WN, Gleich GJ, Egan RW (1994) Eosinophil granule proteins increase microvascular macromolecular transport in the hamster cheek pouch. J Immunol 153:2664–2670
Minnicozzi M, Ramirez MM, Egan RW, Gleich GJ, Kobayashi I, Kim D, Duran WN (1995) Polyarginine and eosinophil-derived major basic protein increase microvascular permeability independently of histamine or nitric oxide release. Microvasc Res 50:56–70
Minshall RD, Tiruppathi C, Vogel SM, Niles WD, Gilchrist A, Hamm HE, Malik AB (2000) Endothelial cell-surface gp60 activates vesicle formation and trafficking via G(i)-coupled Src kinase signaling pathway. J Cell Biol 150:1057–1070
Mischak H, Seitz T, Janosch P, Eulitz M, Steen H, Schellerer M, Philipp A, Kolch W (1996) Negative regulation of Raf-1 by phosphorylation of serine 621. Mol Cell Biol 16:5409–5418
Mollen KP, Anand RJ, Tsung A, Prince JM, Levy RM, Billiar TR (2006) Emerging paradigm: toll-like receptor 4-sentinel for the detection of tissue damage. Shock 26:430–437
Moore TM, Chetham PM, Kelly JJ, Stevens T (1998) Signal transduction and regulation of lung endothelial cell permeability. Interaction between calcium and cAMP. Am J Phys 275:L203–L222
Motawe ZY, Abdelmaboud SS, Cuevas J, Breslin JW (2020a) PRE-084 as a tool to uncover potential therapeutic applications for selective sigma-1 receptor activation. Int J Biochem Cell Biol 126:105803
Motawe ZY, Farsaei F, Abdelmaboud SS, Cuevas J, Breslin JW (2020b) Sigma-1 receptor activation-induced glycolytic ATP production and endothelial barrier enhancement. Microcirculation e12620
Motawe ZY, Abdelmaboud SS, Breslin JW (2021) Involvement of sigma Receptor-1 in lymphatic endothelial barrier integrity and bioenergetic regulation. Lymphat Res Biol 19:231–239
Moy AB, Shasby SS, Scott BD, Shasby DM (1993) The effect of histamine and cyclic adenosine monophosphate on myosin light chain phosphorylation in human umbilical vein endothelial cells. J Clin Invest 92:1198–1206
Moy AB, Sheldon R, Lindsley K, Shasby S, Shasby DM (1994) Centripetal tension and endothelial retraction. Chest 105:107S–108S
Moy AB, Van Engelenhoven J, Bodmer J, Kamath J, Keese C, Giaever I, Shasby S, Shasby DM (1996) Histamine and thrombin modulate endothelial focal adhesion through centripetal and centrifugal forces. J Clin Invest 97:1020–1027

Moy AB, Winter M, Kamath A, Blackwell K, Reyes G, Giaever I, Keese C, Shasby DM (2000) Histamine alters endothelial barrier function at cell-cell and cell-matrix sites. Am J Physiol Lung Cell Mol Physiol 278:L888–L898

Moy AB, Blackwell K, Kamath A (2002) Differential effects of histamine and thrombin on endothelial barrier function through actin-myosin tension. Am J Physiol Heart Circ Physiol 282:H21–H29

Moy AB, Blackwell K, Wang N, Haxhinasto K, Kasiske MK, Bodmer J, Reyes G, English A (2004) Phorbol ester-mediated pulmonary artery endothelial barrier dysfunction through regulation of actin cytoskeletal mechanics. Am J Physiol Lung Cell Mol Physiol 287:L153–L167

Mucha DR, Myers CL, Schaeffer RC Jr (2003) Endothelial contraction and monolayer hyperpermeability are regulated by Src kinase. Am J Physiol Heart Circ Physiol 284:H994–H1002

Mukhopadhyay D, Nagy JA, Manseau EJ, Dvorak HF (1998) Vascular permeability factor/vascular endothelial growth factor-mediated signaling in mouse mesentery vascular endothelium. Cancer Res 58:1278–1284

Mulivor AW, Lipowsky HH (2002) Role of glycocalyx in leukocyte-endothelial cell adhesion. Am J Physiol Heart Circ Physiol 283:H1282–H1291

Mulivor AW, Lipowsky HH (2009) Inhibition of glycan shedding and leukocyte-endothelial adhesion in postcapillary venules by suppression of matrixmetalloprotease activity with doxycycline. Microcirculation 16:657–666

Murphy JT, Duffy S (2003) ZO-1 redistribution and F-actin stress fiber formation in pulmonary endothelial cells after thermal injury. J Trauma 54:81–89

Murray MA, Heistad DD, Mayhan WG (1991) Role of protein kinase C in bradykinin-induced increases in microvascular permeability. Circ Res 68:1340–1348

Nagpala PG, Malik AB, Vuong PT, Lum H (1996) Protein kinase C beta 1 overexpression augments phorbol ester-induced increase in endothelial permeability. J Cell Physiol 166:249–255

Navab M, Hough GP, Berliner JA, Frank JA, Fogelman AM, Haberland ME, Edwards PA (1986) Rabbit beta-migrating very low density lipoprotein increases endothelial macromolecular transport without altering electrical resistance. J Clin Invest 78:389–397

Needham LK, Rozengurt E (1998) Galpha12 and Galpha13 stimulate rho-dependent tyrosine phosphorylation of focal adhesion kinase, paxillin, and p130 Crk-associated substrate. J Biol Chem 273:14626–14632

Newton AC (1997) Regulation of protein kinase C. Curr Opin Cell Biol 9:161–167

Nitta T, Hata M, Gotoh S, Seo Y, Sasaki H, Hashimoto N, Furuse M, Tsukita S (2003) Size-selective loosening of the blood-brain barrier in claudin-5-deficient mice. J Cell Biol 161:653–660

Niwa K, Inanami O, Ohta T, Ito S, Karino T, Kuwabara M (2001) p38 MAPK and Ca2+ contribute to hydrogen peroxide-induced increase of permeability in vascular endothelial cells but ERK does not. Free Radic Res 35:519–527

Noel AA, Fallek SR, Hobson RW 2nd, Duran WN (1995) Inhibition of nitric oxide synthase attenuates primed microvascular permeability in the in vivo microcirculation. J Vasc Surg 22: 661–669

Nugent LJ, Jain RK (1984) Plasma pharmacokinetics and interstitial diffusion of macromolecules in a capillary bed. Am J Phys 246:H129–H137

Nwariaku FE, Chang J, Zhu X, Liu Z, Duffy SL, Halaihel NH, Terada L, Turnage RH (2002a) The role of p38 map kinase in tumor necrosis factor-induced redistribution of vascular endothelial cadherin and increased endothelial permeability. Shock 18:82–85

Nwariaku FE, Liu Z, Zhu X, Turnage RH, Sarosi GA, Terada LS (2002b) Tyrosine phosphorylation of vascular endothelial cadherin and the regulation of microvascular permeability. Surgery 132: 180–185

Nwariaku FE, Rothenbach P, Liu Z, Zhu X, Turnage RH, Terada LS (2003) Rho inhibition decreases TNF-induced endothelial MAPK activation and monolayer permeability. J Appl Physiol 95:1889–1895
Ohara Y, Sayegh HS, Yamin JJ, Harrison DG (1995) Regulation of endothelial constitutive nitric oxide synthase by protein kinase C. Hypertension 25:415–420
Olesen SP, Crone C (1983) Electrical resistance of muscle capillary endothelium. Biophys J 42:31–41
Olesen SP, de Saint-Aubain ML, Bundgaard M (1984) Permeabilities of single arterioles and venules in the frog skin: a functional and morphological study. Microvasc Res 28:1–22
Oshima K, Schmidt EP (2025) The endothelial Glycocalyx in health and disease. In: Handbook of microcirculation
Oshima K, Han X, Ouyang Y, El Masri R, Yang Y, Haeger SM, McMurtry SA, Lane TC, Davizon-Castillo P, Zhang F, Yue X, Vives RR, Linhardt RJ, Schmidt EP (2019) Loss of endothelial sulfatase-1 after experimental sepsis attenuates subsequent pulmonary inflammatory responses. Am J Physiol Lung Cell Mol Physiol 317:L667–L677
Ostrowski SR, Johansson PI (2012) Endothelial glycocalyx degradation induces endogenous heparinization in patients with severe injury and early traumatic coagulopathy. J Trauma Acute Care Surg 73:60–66
Palade GE (1960) Transport in quanta across the endothelium of blood capillaries. Anat Rec 116: 254
Pappenheimer JR, Renkin EM, Borrero LM (1951) Filtration, diffusion and molecular sieving through peripheral capillary membranes; a contribution to the pore theory of capillary permeability. Am J Phys 167:13–46
Parenti A, Morbidelli L, Cui XL, Douglas JG, Hood JD, Granger HJ, Ledda F, Ziche M (1998) Nitric oxide is an upstream signal of vascular endothelial growth factor-induced extracellular signal-regulated kinase1/2 activation in postcapillary endothelium. J Biol Chem 273:4220–4226
Parthasarathi K, Ichimura H, Monma E, Lindert J, Quadri S, Issekutz A, Bhattacharya J (2006) Connexin 43 mediates spread of Ca2+-dependent proinflammatory responses in lung capillaries. J Clin Invest 116:2193–2200
Patlak CS, Goldstein DA, Hoffman JF (1963) The flow of solute and solvent across a two-membrane system. J Theor Biol 5:426–442
Paul R, Zhang ZG, Eliceiri BP, Jiang Q, Boccia AD, Zhang RL, Chopp M, Cheresh DA (2001) Src deficiency or blockade of Src activity in mice provides cerebral protection following stroke. Nat Med 7:222–227
Peng X, Hassoun PM, Sammani S, McVerry BJ, Burne MJ, Rabb H, Pearse D, Tuder RM, Garcia JG (2004) Protective effects of sphingosine 1-phosphate in murine endotoxin-induced inflammatory lung injury. Am J Respir Crit Care Med 169:1245–1251
Persidsky Y, Heilman D, Haorah J, Zelivyanskaya M, Persidsky R, Weber GA, Shimokawa H, Kaibuchi K, Ikezu T (2006) Rho-mediated regulation of tight junctions during monocyte migration across the blood-brain barrier in HIV-1 encephalitis (HIVE). Blood 107:4770–4780
Peters RM, Hogan JS (1980) Mechanism of death in massive fluid infusion. J Trauma 20:452–459
Petrache I, Verin AD, Crow MT, Birukova A, Liu F, Garcia JG (2001) Differential effect of MLC kinase in TNF-alpha-induced endothelial cell apoptosis and barrier dysfunction. Am J Physiol Lung Cell Mol Physiol 280:L1168–L1178
Pober JS, Kluger MS, Schechner JS (2001) Human endothelial cell presentation of antigen and the homing of memory/effector T cells to skin. Ann N Y Acad Sci 941:12–25
Pobiner BF, Northup JK, Bauer PH, Fraser ED, Garrison JC (1991) Inhibitory GTP-binding regulatory protein Gi3 can couple angiotensin II receptors to inhibition of adenylyl cyclase in hepatocytes. Mol Pharmacol 40:156–167
Pocock TM, Williams B, Curry FE, Bates DO (2000) VEGF and ATP act by different mechanisms to increase microvascular permeability and endothelial [ca(2+)](i). Am J Physiol Heart Circ Physiol 279:H1625–H1634

Pocock TM, Foster RR, Bates DO (2004) Evidence of a role for TRPC channels in VEGF-mediated increased vascular permeability in vivo. Am J Physiol Heart Circ Physiol 286:H1015–H1026
Prabhakar P, Thatte HS, Goetz RM, Cho MR, Golan DE, Michel T (1998) Receptor-regulated translocation of endothelial nitric-oxide synthase. J Biol Chem 273:27383–27388
Predescu D, Horvat R, Predescu S, Palade GE (1994) Transcytosis in the continuous endothelium of the myocardial microvasculature is inhibited by N-ethylmaleimide. Proc Natl Acad Sci USA 91: 3014–3018
Quinn TP, Peters KG, De Vries C, Ferrara N, Williams LT (1993) Fetal liver kinase 1 is a receptor for vascular endothelial growth factor and is selectively expressed in vascular endothelium. Proc Natl Acad Sci USA 90:7533–7537
Ramirez MM, Quardt SM, Kim D, Oshiro H, Minnicozzi M, Duran WN (1995) Platelet activating factor modulates microvascular permeability through nitric oxide synthesis. Microvasc Res 50: 223–234
Ramirez MM, Kim DD, Duran WN (1996) Protein kinase C modulates microvascular permeability through nitric oxide synthase. Am J Phys 271:H1702–H1705
Raub TJ, Kuentzel SL, Sawada GA (1992) Permeability of bovine brain microvessel endothelial cells in vitro: barrier tightening by a factor released from astroglioma cells. Exp Cell Res 199: 330–340
Rawson RA (1943) The binding of T-1824 and structurally related diazo dyes by plasma proteins. Am J Phys 138:708–717
Reed RK (1988) Transcapillary albumin extravasation in rat skin and skeletal muscle: effect of increased venous pressure. Acta Physiol Scand 134:375–382
Reed MW, Miller FN (1988) Importance of light dose in fluorescent microscopy. Microvasc Res 36:104–107
Rehm M, Bruegger D, Christ F, Conzen P, Thiel M, Jacob M, Chappell D, Stoeckelhuber M, Welsch U, Reichart B, Peter K, Becker BF (2007) Shedding of the endothelial glycocalyx in patients undergoing major vascular surgery with global and regional ischemia. Circulation 116: 1896–1906
Renkin EM (1985) Capillary transport of macromolecules: pores and other endothelial pathways. J Appl Physiol 58:315–325
Renkin EM (1994) Cellular aspects of transvascular exchange: a 40-year perspective. Microcirculation 1:157–167
Renkin EM, Gustafson-Sgro M, Sibley L (1988) Coupling of albumin flux to volume flow in skin and muscles of anesthetized rats. Am J Phys 255:H458–H466
Reynoso R, Perrin RM, Breslin JW, Dalnes DA, Watson KD, Watterson DM, Wu MH, Yuan S (2007) A role for long chain myosin light chain kinase (Mlck-210) in microvascular hyperpermeability during severe burns. Shock 28:589–595
Rigor RR, Beard RS Jr, Litovka OP, Yuan SY (2012) Interleukin-1beta-induced barrier dysfunction is signaled through PKC-theta in human brain microvascular endothelium. Am J Physiol Cell Physiol 302:C1513–C1522
Rippe B, Haraldsson B (1994) Transport of macromolecules across microvascular walls: the two-pore theory. Physiol Rev 74:163–219
Rippe B, Taylor A (2001) NEM and filipin increase albumin transport in lung microvessels. Am J Physiol Heart Circ Physiol 280:H34–H41
Ritter AB, Braun W, Stein A, Duran W (1985) Visualization of the coronary microcirculation using digital image processing. Comput Biol Med 15:361–374
Rosen H, Sanna MG, Cahalan SM, Gonzalez-Cabrera PJ (2007) Tipping the gatekeeper: S1P regulation of endothelial barrier function. Trends Immunol 28:102–107
Rumbaut RE, Huxley VH (2002) Similar permeability responses to nitric oxide synthase inhibitors of venules from three animal species. Microvasc Res 64:21–31
Rumbaut RE, Sial AJ (1999) Differential phototoxicity of fluorescent dye-labeled albumin conjugates. Microcirculation 6:205–213

Rumbaut RE, Harris NR, Sial AJ, Huxley VH, Granger DN (1999) Leakage responses to L-NAME differ with the fluorescent dye used to label albumin. Am J Phys 276:H333–H339
Rutledge JC (1992) Temperature and hydrostatic pressure-dependent pathways of low-density lipoprotein transport across microvascular barrier. Am J Phys 262:H234–H245
Sakoda T, Hirata K, Kuroda R, Miki N, Suematsu M, Kawashima S, Yokoyama M (1995) Myristoylation of endothelial cell nitric oxide synthase is important for extracellular release of nitric oxide. Mol Cell Biochem 152:143–148
Sánchez FA, Savalia NB, Duran RG, Lal BK, Boric MP, Durán WN (2006) Functional significance of differential eNOS translocation. Am J Physiol Heart Circ Physiol 291:H1058–H1064
Sanchez FA, Kim DD, Duran RG, Meininger CJ, Duran WN (2008) Internalization of eNOS via caveolae regulates PAF-induced inflammatory hyperpermeability to macromolecules. Am J Physiol Heart Circ Physiol 295:H1642–H1648
Sánchez FA, Rana R, Kim DD, Iwahashi T, Zheng R, Lal BK, Gordon DM, Meininger CJ, Durán WN (2009) Internalization of eNOS and NO delivery to subcellular targets determine agonist-induced hyperpermeability. Proc Natl Acad Sci USA 106:6849–6853
Sánchez FA, Rana R, Gonzalez FG, Iwahashi T, Duran RG, Fulton DJ, Beuve AV, Kim DD, Durán WN (2011) Functional significance of cytosolic endothelial nitric-oxide synthase (eNOS): regulation of hyperpermeability. J Biol Chem 286:30409–30414
Sánchez FA, Ehrenfeld IP, Durán WN (2013) S-nitrosation of proteins: an emergent regulatory mechanism in microvascular permeability and vascular function. Tissue Barr 1:e23896
Sander EE, ten Klooster JP, van Delft S, van der Kammen RA, Collard JG (1999) Rac downregulates rho activity: reciprocal balance between both GTPases determines cellular morphology and migratory behavior. J Cell Biol 147:1009–1022
Sandoval R, Malik AB, Naqvi T, Mehta D, Tiruppathi C (2001a) Requirement for Ca2+ signaling in the mechanism of thrombin-induced increase in endothelial permeability. Am J Physiol Lung Cell Mol Physiol 280:L239–L247
Sandoval R, Malik AB, Minshall RD, Kouklis P, Ellis CA, Tiruppathi C (2001b) Ca(2+) signalling and PKCalpha activate increased endothelial permeability by disassembly of VE-cadherin junctions. J Physiol 533:433–445
Sanna MG, Wang SK, Gonzalez-Cabrera PJ, Don A, Marsolais D, Matheu MP, Wei SH, Parker I, Jo E, Cheng WC, Cahalan MD, Wong CH, Rosen H (2006) Enhancement of capillary leakage and restoration of lymphocyte egress by a chiral S1P1 antagonist in vivo. Nat Chem Biol 2:434–441
Saria A, Lundberg JM (1983) Evans blue fluorescence: quantitative and morphological evaluation of vascular permeability in animal tissues. J Neurosci Methods 8:41–49
Sarker MH, Fraser PA (2002) The role of guanylyl cyclases in the permeability response to inflammatory mediators in pial venular capillaries in the rat. J Physiol 540:209–218
Scallan J, Huxley VH, Korthuis RJ (2010) Chapter 4: pathophysiology of edema formation. In: Scallan J, Huxley VH, Korthuis RJ (eds) Capillary fluid exchange: regulation, functions, and pathology. Morgan & Claypool Life Sciences, San Rafael
Scallan JP, Davis MJ, Huxley VH (2013) Permeability and contractile responses of collecting lymphatic vessels elicited by atrial and brain natriuretic peptides. J Physiol 591:5071–5081
Scallan JP, Hill MA, Davis MJ (2015) Lymphatic vascular integrity is disrupted in type 2 diabetes due to impaired nitric oxide signalling. Cardiovasc Res 107:89–97
Schelling ME, Meininger CJ, Hawker JR Jr, Granger HJ (1988) Venular endothelial cells from bovine heart. Am J Phys 254:H1211–H1217
Schlegel N, Waschke J (2009) Impaired cAMP and Rac 1 signaling contribute to TNF-alpha-induced endothelial barrier breakdown in microvascular endothelium. Microcirculation 16:521–533
Schlegel N, Burger S, Golenhofen N, Walter U, Drenckhahn D, Waschke J (2008) The role of VASP in regulation of cAMP- and Rac 1-mediated endothelial barrier stabilization. Am J Physiol Cell Physiol 294:C178–C188

Schlessinger J (2000) Cell signaling by receptor tyrosine kinases. Cell 103:211–225

Schmidt EP, Yang Y, Janssen WJ, Gandjeva A, Perez MJ, Barthel L, Zemans RL, Bowman JC, Koyanagi DE, Yunt ZX, Smith LP, Cheng SS, Overdier KH, Thompson KR, Geraci MW, Douglas IS, Pearse DB, Tuder RM (2012) The pulmonary endothelial glycocalyx regulates neutrophil adhesion and lung injury during experimental sepsis. Nat Med 18:1217–1223

Schnittler HJ, Wilke A, Gress T, Suttorp N, Drenckhahn D (1990) Role of actin and myosin in the control of paracellular permeability in pig, rat and human vascular endothelium. J Physiol 431: 379–401

Schnittler H, Taha M, Schnittler MO, Taha AA, Lindemann N, Seebach J (2014) Actin filament dynamics and endothelial cell junctions: the Ying and Yang between stabilization and motion. Cell Tissue Res 355:529–543

Schnitzer JE, Oh P, Pinney E, Allard J (1994) Filipin-sensitive caveolae-mediated transport in endothelium: reduced transcytosis, scavenger endocytosis, and capillary permeability of select macromolecules. J Cell Biol 127:1217–1232

Seem E, Stranden E (1986) Transcapillary forces in subcutaneous tissue of lower limbs with deep venous thrombosis. Scand J Clin Lab Invest 46:417–422

Sessa WC, Garcia-Cardena G, Liu J, Keh A, Pollock JS, Bradley J, Thiru S, Braverman IM, Desai KM (1995) The Golgi association of endothelial nitric oxide synthase is necessary for the efficient synthesis of nitric oxide. J Biol Chem 270:17641–17644

Shajahan AN, Timblin BK, Sandoval R, Tiruppathi C, Malik AB, Minshall RD (2004) Role of Src-induced dynamin-2 phosphorylation in caveolae-mediated endocytosis in endothelial cells. J Biol Chem 279:20392–20400

Shasby DM, Shasby SS (1986) Effects of calcium on transendothelial albumin transfer and electrical resistance. J Appl Physiol 60:71–79

Shasby DM, Ries DR, Shasby SS, Winter MC (2002) Histamine stimulates phosphorylation of adherens junction proteins and alters their link to vimentin. Am J Physiol Lung Cell Mol Physiol 282:L1330–L1338

Shaul PW, Smart EJ, Robinson LJ, German Z, Yuhanna IS, Ying Y, Anderson RG, Michel T (1996) Acylation targets emdothelial nitric-oxide synthase to plasmalemmal caveolae. J Biol Chem 271:6518–6522

Shay-Salit A, Shushy M, Wolfovitz E, Yahav H, Breviario F, Dejana E, Resnick N (2002) VEGF receptor 2 and the adherens junction as a mechanical transducer in vascular endothelial cells. Proc Natl Acad Sci USA 99:9462–9467

Sheldon R, Moy A, Lindsley K, Shasby S, Shasby DM (1993) Role of myosin light-chain phosphorylation in endothelial cell retraction. Am J Phys 265:L606–L612

Shen BQ, Lee DY, Zioncheck TF (1999) Vascular endothelial growth factor governs endothelial nitric-oxide synthase expression via a KDR/Flk-1 receptor and a protein kinase C signaling pathway. J Biol Chem 274:33057–33063

Shen Q, Wu MH, Yuan SY (2009) Endothelial contractile cytoskeleton and microvascular permeability. Cell Health Cytoskeleton 1:43–50

Sheppard D, Scypinski L, Horn J, Gordon T, Thompson JE (1986) Granulocyte-mediated airway edema in Guinea pigs. J Appl Physiol 60:1213–1220

Shi S, Verin AD, Schaphorst KL, Gilbert-McClain LI, Patterson CE, Irwin RP, Natarajan V, Garcia JG (1998) Role of tyrosine phosphorylation in thrombin-induced endothelial cell contraction and barrier function. Endothelium 6:153–171

Shi S, Garcia JG, Roy S, Parinandi NL, Natarajan V (2000) Involvement of c-Src in diperoxovanadate-induced endothelial cell barrier dysfunction. Am J Physiol Lung Cell Mol Physiol 279:L441–L451

Shikata Y, Birukov KG, Garcia JG (2003) S1P induces FA remodeling in human pulmonary endothelial cells: role of Rac, GIT1, FAK, and paxillin. J Appl Physiol 94:1193–1203

Siflinger-Birnboim A, Goligorsky MS, Del Vecchio PJ, Malik AB (1992) Activation of protein kinase C pathway contributes to hydrogen peroxide-induced increase in endothelial permeability. Lab Investig 67:24–30

Simionescu M, Simionescu N (1984) Ultrastructure of the microvascular wall: functional correlations. In: Renkin EM, Michel CC (eds) Handbook of physiology: the cardiovascular system microcirculation. American Physiological Society, Bethesda, pp 41–101
Simionescu N, Siminoescu M, Palade GE (1975) Permeability of muscle capillaries to small heme-peptides. Evidence for the existence of patent transendothelial channels. J Cell Biol 64:586–607
Simon AM, McWhorter AR (2002) Vascular abnormalities in mice lacking the endothelial gap junction proteins connexin37 and connexin40. Dev Biol 251:206–220
Simon AM, McWhorter AR (2003) Role of connexin37 and connexin40 in vascular development. Cell Commun Adhes 10:379–385
Singleton PA, Dudek SM, Chiang ET, Garcia JG (2005) Regulation of sphingosine 1-phosphate-induced endothelial cytoskeletal rearrangement and barrier enhancement by S1P1 receptor, PI3 kinase, Tiam1/Rac1, and alpha-actinin. FASEB J 19:1646–1656
Sinnett-Smith J, Lunn JA, Leopoldt D, Rozengurt E (2001) Y-27632, an inhibitor of rho-associated kinases, prevents tyrosine phosphorylation of focal adhesion kinase and paxillin induced by bombesin: dissociation from tyrosine phosphorylation of p130(CAS). Exp Cell Res 266:292–302
Sirois MG, Edelman ER (1997) VEGF effect on vascular permeability is mediated by synthesis of platelet-activating factor. Am J Phys 272:H2746–H2756
Sirotkin H, Morrow B, Saint-Jore B, Puech A, Das Gupta R, Patanjali SR, Skoultchi A, Weissman SM, Kucherlapati R (1997) Identification, characterization, and precise mapping of a human gene encoding a novel membrane-spanning protein from the 22q11 region deleted in velo-cardio-facial syndrome. Genomics 42:245–251
Snyder F (1990) Platelet-activating factor and related acetylated lipids as potent biologically active cellular mediators. Am J Phys 259:C697–C708
Sohet F, Daneman R (2013) Genetic mouse models to study blood-brain barrier development and function. Fluids Barriers CNS 10:3
Sowa G, Liu J, Papapetropoulos A, Rex-Haffner M, Hughes TE, Sessa WC (1999) Trafficking of endothelial nitric-oxide synthase in living cells. Quantitative evidence supporting the role of palmitoylation as a kinetic trapping mechanism limiting membrane diffusion. J Biol Chem 274: 22524–22531
Spiegel S, Milstien S (2003) Sphingosine-1-phosphate: an enigmatic signalling lipid. Nat Rev Mol Cell Biol 4:397–407
Spinella PC, Perkins JG, Cap AP (2016) Lessons learned for the resuscitation of traumatic hemorrhagic shock. US Army Med Dep J:37–42
Starling EH (1896) On the absorption of fluids from the connective tissue spaces. J Physiol 19:312–326
Stasek JE Jr, Patterson CE, Garcia JG (1992) Protein kinase C phosphorylates caldesmon77 and vimentin and enhances albumin permeability across cultured bovine pulmonary artery endothelial cell monolayers. J Cell Physiol 153:62–75
Strub GM, Maceyka M, Hait NC, Milstien S, Spiegel S (2010) Extracellular and intracellular actions of sphingosine-1-phosphate. Adv Exp Med Biol 688:141–155
Suarez S, Ballmer-Hofer K (2001) VEGF transiently disrupts gap junctional communication in endothelial cells. J Cell Sci 114:1229–1235
Sun H, Breslin JW, Zhu J, Yuan SY, Wu MH (2006) Rho and ROCK signaling in VEGF-induced microvascular endothelial hyperpermeability. Microcirculation 13:237–247
Suval WD, Hobson RW 2nd, Boric MP, Ritter AB, Duran WN (1987a) Assessment of ischemia reperfusion injury in skeletal muscle by macromolecular clearance. J Surg Res 42:550–559
Suval WD, Duran WN, Boric MP, Hobson RW 3rd, Berendsen PB, Ritter AB (1987b) Microvascular transport and endothelial cell alterations preceding skeletal muscle damage in ischemia and reperfusion injury. Am J Surg 154:211–218
Suzuki Y, Deitch EA, Mishima S, Duran WN, Xu DZ (2000) Endotoxin-induced mesenteric microvascular changes involve iNOS-derived nitric oxide: results from a study using iNOS knock out mice. Shock 13:397–403

Svensjo E (1990) The hamster cheek pouch as a model in microcirculation research. Eur Respir J Suppl 12:595s–600s

Svensjo E, Joyner WL (1984) The effects of intermittent and continuous stimulation of microvessels in the cheek pouch of hamsters with histamine and bradykinin on the development of venular leaky sites. Microcirc Endothel Lymphat 1:381–396

Svensjo E, Arfors KE, Arturson G, Rutili G (1978) The hamster cheek pouch preparation as a model for studies of macromolecular permeability of the microvasculature. Ups J Med Sci 83:71–79

Svensjo E, Adamski SW, Su K, Grega GJ (1982) Quantitative physiological and morphological aspects of microvascular permeability changes induced by histamine and inhibited by terbutaline. Acta Physiol Scand 116:265–273

Szabo G, Posch E, Magyar Z (1980) Interstitial fluid, lymph and oedema formation. Acta Physiol Acad Sci Hung 56:367–378

Szulcek R, Beckers CM, Hodzic J, de Wit J, Chen Z, Grob T, Musters RJ, Minshall RD, van Hinsbergh VW, van Nieuw Amerongen GP (2013) Localized RhoA GTPase activity regulates dynamics of endothelial monolayer integrity. Cardiovasc Res 99:471–482

Takaishi K, Sasaki T, Kotani H, Nishioka H, Takai Y (1997) Regulation of cell-cell adhesion by rac and rho small G proteins in MDCK cells. J Cell Biol 139:1047–1059

Tanita T, Song C, Kubo H, Ono S, Fujimura S (2000) Endothelial signal transduction system enhances neutrophil-induced pulmonary vascular permeability. Eur Respir J 15:452–458

Tar K, Birukova AA, Csortos C, Bako E, Garcia JG, Verin AD (2004) Phosphatase 2A is involved in endothelial cell microtubule remodeling and barrier regulation. J Cell Biochem 92:534–546

Taskaeva I, Bgatova N (2021) Microvasculature in hepatocellular carcinoma: An ultrastructural study. Microvasc Res 133:104094

Taylor AE, Granger DN (1984) Exchange of macrfomolecules across the microcirculation. In: Renkin EM, Michel CC (eds) Handbook of physiology: the cardiovascular system microcirculation. American Physiological Society, Bethesda, pp 467–520

Taylor A, Townsley MI, Korthuis RJ (1985) Macromolecule transport across the pulmonary microvessel walls. Exp Lung Res 8:97–123

Territo M, Berliner JA, Fogelman AM (1984) Effect of monocyte migration on low density lipoprotein transport across aortic endothelial cell monolayers. J Clin Invest 74:2279–2284

Thakker GD, Hajjar DP, Muller WA, Rosengart TK (1999) The role of phosphatidylinositol 3-kinase in vascular endothelial growth factor signaling. J Biol Chem 274:10002–10007

Tharakan B, Whaley JG, Hunter FA, Smythe WR, Childs EW (2010) (−)-Deprenyl inhibits vascular hyperpermeability after hemorrhagic shock. Shock 33:56–63

Thibeault S, Rautureau Y, Oubaha M, Faubert D, Wilkes BC, Delisle C, Gratton JP (2010) S-nitrosylation of beta-catenin by eNOS-derived NO promotes VEGF-induced endothelial cell permeability. Mol Cell 39:468–476

Thomas SM, Brugge JS (1997) Cellular functions regulated by Src family kinases. Annu Rev Cell Dev Biol 13:513–609

Thuringer D (2004) The vascular endothelial growth factor-induced disruption of gap junctions is relayed by an autocrine communication via ATP release in coronary capillary endothelium. Ann N Y Acad Sci 1030:14–27

Thurston G, Baldwin AL (1994) Endothelial actin cytoskeleton in rat mesentery microvasculature. Am J Phys 266:H1896–H1909

Thurston G, Baldwin AL, Wilson LM (1995) Changes in endothelial actin cytoskeleton at leakage sites in the rat mesenteric microvasculature. Am J Phys 268:H316–H329

Tian X, Tian Y, Gawlak G, Sarich N, Wu T, Birukova AA (2014a) Control of vascular permeability by atrial natriuretic peptide via a GEF-H1-dependent mechanism. J Biol Chem 289:5168–5183

Tian X, Tian Y, Moldobaeva N, Sarich N, Birukova AA (2014b) Microtubule dynamics control HGF-induced lung endothelial barrier enhancement. PLoS One 9:e105912

Tian X, Tian Y, Gawlak G, Meng F, Kawasaki Y, Akiyama T, Birukova AA (2015) Asef controls vascular endothelial permeability and barrier recovery in the lung. Mol Biol Cell 26:636–650

Tian Y, Gawlak G, Tian X, Shah AS, Sarich N, Citi S, Birukova AA (2016) Role of Cingulin in agonist-induced vascular endothelial permeability. J Biol Chem 291:23681–23692

Tilling T, Korte D, Hoheisel D, Galla HJ (1998) Basement membrane proteins influence brain capillary endothelial barrier function in vitro. J Neurochem 71:1151–1157

Tinsley JH, Hawker J, Yuan Y (1998) Efficient protein transfection of cultured coronary venular endothelial cells. Am J Phys 275:H1873–H1878

Tinsley JH, Wu MH, Ma W, Taulman AC, Yuan SY (1999) Activated neutrophils induce hyperpermeability and phosphorylation of adherens junction proteins in coronary venular endothelial cells. J Biol Chem 274:24930–24934

Tinsley JH, De Lanerolle P, Wilson E, Ma W, Yuan SY (2000) Myosin light chain kinase transference induces myosin light chain activation and endothelial hyperpermeability. Am J Physiol Cell Physiol 279:C1285–C1289

Tinsley JH, Zawieja DC, Wu MH, Ustinova EE, Xu W, Yuan SY (2001) Protein transfection of intact microvessels specifically modulates vasoreactivity and permeability. J Vasc Res 38:444–452

Tinsley JH, Ustinova EE, Xu W, Yuan SY (2002) Src-dependent, neutrophil-mediated vascular hyperpermeability and beta -catenin modification. Am J Physiol Cell Physiol 283:C1745–C1751

Tinsley JH, Teasdale NR, Yuan SY (2004a) Involvement of PKCdelta and PKD in pulmonary microvascular endothelial cell hyperpermeability. Am J Physiol Cell Physiol 286:C105–C111

Tinsley JH, Teasdale NR, Yuan SY (2004b) Myosin light chain phosphorylation and pulmonary endothelial cell hyperpermeability in burns. Am J Physiol Lung Cell Mol Physiol 286:L841–L847

Tinsley JH, Breslin JW, Teasdale NR, Yuan SY (2005) PKC-dependent, burn-induced adherens junction reorganization and barrier dysfunction in pulmonary microvascular endothelial cells. Am J Phys

Tiruppathi C, Malik AB, Del Vecchio PJ, Keese CR, Giaever I (1992) Electrical method for detection of endothelial cell shape change in real time: assessment of endothelial barrier function. Proc Natl Acad Sci USA 89:7919–7923

Tiruppathi C, Song W, Bergenfeldt M, Sass P, Malik AB (1997) Gp60 activation mediates albumin transcytosis in endothelial cells by tyrosine kinase-dependent pathway. J Biol Chem 272: 25968–25975

Tokuyama K, Nishimura H, Iizuka K, Kato M, Arakawa H, Saga R, Mochizuki H, Morikawa A (2002) Effects of Y-27632, a rho/rho kinase inhibitor, on leukotriene D(4)- and histamine-induced airflow obstruction and airway microvascular leakage in Guinea pigs in vivo. Pharmacology 64:189–195

Tomeo AC, Duran WN (1991) Resistance and exchange microvessels are modulated by different PAF receptors. Am J Phys 261:H1648–H1652

Tomeo AC, Egan RW, Duran WN (1991) Priming interactions between platelet activating factor and histamine in the in vivo microcirculation. FASEB J 5:2850–2855

Torres Filho IP, Torres LN, Salgado C, Dubick MA (2016) Plasma syndecan-1 and heparan sulfate correlate with microvascular glycocalyx degradation in hemorrhaged rats after different resuscitation fluids. Am J Physiol Heart Circ Physiol 310:H1468–H1478

Torres LN, Chung KK, Salgado CL, Dubick MA, Torres Filho IP (2017) Low-volume resuscitation with normal saline is associated with microvascular endothelial dysfunction after hemorrhage in rats, compared to colloids and balanced crystalloids. Crit Care 21:160

Tsukita S, Oishi K, Akiyama T, Yamanashi Y, Yamamoto T (1991) Specific proto-oncogenic tyrosine kinases of src family are enriched in cell-to-cell adherens junctions where the level of tyrosine phosphorylation is elevated. J Cell Biol 113:867–879

Tuttle KR, Anderson PW (2003) A novel potential therapy for diabetic nephropathy and vascular complications: protein kinase C beta inhibition. Am J Kidney Dis 42:456–465

Udaka K, Takeuchi Y, Movat HZ (1970) Simple method for quantitation of enhanced vascular permeability. Proc Soc Exp Biol Med 133:1384–1387

Ukropec JA, Hollinger MK, Salva SM, Woolkalis MJ (2000) SHP2 association with VE-cadherin complexes in human endothelial cells is regulated by thrombin. J Biol Chem 275:5983–5986

van den Berg BM, Vink H, Spaan JA (2003) The endothelial glycocalyx protects against myocardial edema. Circ Res 92:592–594

van Nieuw Amerongen GP, van Hinsbergh VW (2002) Targets for pharmacological intervention of endothelial hyperpermeability and barrier function. Vasc Pharmacol 39:257–272

van Nieuw Amerongen GP, van Delft S, Vermeer MA, Collard JG, van Hinsbergh VW (2000) Activation of RhoA by thrombin in endothelial hyperpermeability: role of rho kinase and protein tyrosine kinases. Circ Res 87:335–340

van Nieuw Amerongen GP, Natarajan K, Yin G, Hoefen RJ, Osawa M, Haendeler J, Ridley AJ, Fujiwara K, van Hinsbergh VW, Berk BC (2004) GIT1 mediates thrombin signaling in endothelial cells: role in turnover of RhoA-type focal adhesions. Circ Res 94:1041–1049

Varma S, Breslin JW, Lal BK, Pappas PJ, Hobson RW 2nd, Duran WN (2002) p42/44MAPK regulates baseline permeability and cGMP-induced hyperpermeability in endothelial cells. Microvasc Res 63:172–178

Vasile E, Simionescu M, Simionescu N (1983) Visualization of the binding, endocytosis, and transcytosis of low-density lipoprotein in the arterial endothelium in situ. J Cell Biol 96:1677–1689

Vasile E, Qu H, Dvorak HF, Dvorak AM (1999) Caveolae and vesiculo-vacuolar organelles in bovine capillary endothelial cells cultured with VPF/VEGF on floating Matrigel-collagen gels. J Histochem Cytochem 47:159–167

Venema VJ, Marrero MB, Venema RC (1996) Bradykinin-stimulated protein tyrosine phosphorylation promotes endothelial nitric oxide synthase translocation to the cytoskeleton. Biochem Biophys Res Commun 226:703–710

Verin AD, Patterson CE, Day MA, Garcia JG (1995) Regulation of endothelial cell gap formation and barrier function by myosin-associated phosphatase activities. Am J Phys 269:L99–L108

Verin AD, Liu F, Bogatcheva N, Borbiev T, Hershenson MB, Wang P, Garcia JG (2000) Role of ras-dependent ERK activation in phorbol ester-induced endothelial cell barrier dysfunction. Am J Physiol Lung Cell Mol Physiol 279:L360–L370

Villalba N, Baby S, Cha BJ, Yuan SY (2020) Site-specific opening of the blood-brain barrier by extracellular histones. J Neuroinflammation 17:281

Villalba N, Ma Y, Gahan SA, Joly-Amado A, Spence S, Yang X, Nash KR, Yuan SY (2023) Lung infection by Pseudomonas aeruginosa induces neuroinflammation and blood-brain barrier dysfunction in mice. J Neuroinflammation 20:127

Vink H, Duling BR (1996) Identification of distinct luminal domains for macromolecules, erythrocytes, and leukocytes within mammalian capillaries. Circ Res 79:581–589

Vink H, Duling BR (2000) Capillary endothelial surface layer selectively reduces plasma solute distribution volume. Am J Physiol Heart Circ Physiol 278:H285–H289

Vittet D, Buchou T, Schweitzer A, Dejana E, Huber P (1997) Targeted null-mutation in the vascular endothelial-cadherin gene impairs the organization of vascular-like structures in embryoid bodies. Proc Natl Acad Sci USA 94:6273–6278

Vogel SM, Minshall RD, Pilipovic M, Tiruppathi C, Malik AB (2001) Albumin uptake and transcytosis in endothelial cells in vivo induced by albumin-binding protein. Am J Physiol Lung Cell Mol Physiol 281:L1512–L1522

Vonghia L, Leggio L, Ferrulli A, Bertini M, Gasbarrini G, Addolorato G (2008) Acute alcohol intoxication. Eur J Intern Med 19:561–567

Vreim CE, Snashall PD, Staub NC (1976) Protein composition of lung fluids in anesthetized dogs with acute cardiogenic edema. Am J Phys 231:1466–1469

Wang L, Dudek SM (2009) Regulation of vascular permeability by sphingosine 1-phosphate. Microvasc Res 77:39–45

Wang Q, Patton WF, Hechtman HB, Shepro D (1997) A novel anti-inflammatory peptide inhibits endothelial cell cytoskeletal rearrangement, nitric oxide synthase translocation, and paracellular permeability increases. J Cell Physiol 172:171–182

Wang P, Verin AD, Birukova A, Gilbert-McClain LI, Jacobs K, Garcia JG (2001) Mechanisms of sodium fluoride-induced endothelial cell barrier dysfunction: role of MLC phosphorylation. Am J Physiol Lung Cell Mol Physiol 281:L1472–L1483

Ware LB, Matthay MA (2000) The acute respiratory distress syndrome. N Engl J Med 342:1334–1349

Waschke J, Baumgartner W, Adamson RH, Zeng M, Aktories K, Barth H, Wilde C, Curry FE, Drenckhahn D (2004) Requirement of Rac activity for maintenance of capillary endothelial barrier properties. Am J Physiol Heart Circ Physiol 286:H394–H401

Weber C, Fraemohs L, Dejana E (2007) The role of junctional adhesion molecules in vascular inflammation. Nat Rev Immunol 7:467–477

Wedgwood S, Bekker JM, Black SM (2001) Shear stress regulation of endothelial NOS in fetal pulmonary arterial endothelial cells involves PKC. Am J Physiol Lung Cell Mol Physiol 281: L490–L498

Weng J, Yu L, Chen Z, Su H, Yu S, Zhang Y, Lei X, Chen L, Cui Y, Huang Q, Jiang Y, Guo X (2019) Beta-catenin phosphorylation at Y654 and Y142 is crucial for high mobility group box-1 protein-induced pulmonary vascular hyperpermeability. J Mol Cell Cardiol 127:174–184

Wojciak-Stothard B, Ridley AJ (2002) Rho GTPases and the regulation of endothelial permeability. Vasc Pharmacol 39:187–199

Wojciak-Stothard B, Entwistle A, Garg R, Ridley AJ (1998) Regulation of TNF-alpha-induced reorganization of the actin cytoskeleton and cell-cell junctions by rho, Rac, and Cdc42 in human endothelial cells. J Cell Physiol 176:150–165

Wojciak-Stothard B, Potempa S, Eichholtz T, Ridley AJ (2001) Rho and Rac but not Cdc42 regulate endothelial cell permeability. J Cell Sci 114:1343–1355

Wolfson RK, Chiang ET, Garcia JG (2011) HMGB1 induces human lung endothelial cell cytoskeletal rearrangement and barrier disruption. Microvasc Res 81:189–197

Wong RK, Baldwin AL, Heimark RL (1999) Cadherin-5 redistribution at sites of TNF-alpha and IFN-gamma-induced permeability in mesenteric venules. Am J Phys 276:H736–H748

Woodfin A, Reichel CA, Khandoga A, Corada M, Voisin MB, Scheiermann C, Haskard DO, Dejana E, Krombach F, Nourshargh S (2007) JAM-A mediates neutrophil transmigration in a stimulus-specific manner in vivo: evidence for sequential roles for JAM-A and PECAM-1 in neutrophil transmigration. Blood 110:1848–1856

Woodfin A, Voisin MB, Imhof BA, Dejana E, Engelhardt B, Nourshargh S (2009) Endothelial cell activation leads to neutrophil transmigration as supported by the sequential roles of ICAM-2, JAM-A, and PECAM-1. Blood 113:6246–6257

Woodfin A, Voisin MB, Beyrau M, Colom B, Caille D, Diapouli FM, Nash GB, Chavakis T, Albelda SM, Rainger GE, Meda P, Imhof BA, Nourshargh S (2011) The junctional adhesion molecule JAM-C regulates polarized transendothelial migration of neutrophils in vivo. Nat Immunol 12:761–769

Wu NZ, Baldwin AL (1992) Transient venular permeability increase and endothelial gap formation induced by histamine. Am J Phys 262:H1238–H1247

Wu J, Dent P, Jelinek T, Wolfman A, Weber MJ, Sturgill TW (1993) Inhibition of the EGF-activated MAP kinase signaling pathway by adenosine 3′,5′-monophosphate. Science 262:1065–1069

Wu HM, Huang Q, Yuan Y, Granger HJ (1996) VEGF induces NO-dependent hyperpermeability in coronary venules. Am J Phys 271:H2735–H2739

Wu HM, Yuan Y, Zawieja DC, Tinsley J, Granger HJ (1999) Role of phospholipase C, protein kinase C, and calcium in VEGF-induced venular hyperpermeability. Am J Phys 276:H535–H542

Wu MH, Ustinova E, Granger HJ (2001) Integrin binding to fibronectin and vitronectin maintains the barrier function of isolated porcine coronary venules. J Physiol 532:785–791

Wu MH, Guo M, Yuan SY, Granger HJ (2003) Focal adhesion kinase mediates porcine venular hyperpermeability elicited by vascular endothelial growth factor. J Physiol 552:691–699

Wu MH, Yuan SY, Granger HJ (2005) The protein kinase MEK1/2 mediate vascular endothelial growth factor- and histamine-induced hyperpermeability in porcine coronary venules. J Physiol 563:95–104

Xia P, Aiello LP, Ishii H, Jiang ZY, Park DJ, Robinson GS, Takagi H, Newsome WP, Jirousek MR, King GL (1996) Characterization of vascular endothelial growth factor's effect on the activation of protein kinase C, its isoforms, and endothelial cell growth. J Clin Invest 98:2018–2026

Xu J, Zhang X, Pelayo R, Monestier M, Ammollo CT, Semeraro F, Taylor FB, Esmon NL, Lupu F, Esmon CT (2009) Extracellular histones are major mediators of death in sepsis. Nat Med 15: 1318–1321

Yamaji T, Fukuhara T, Kinoshita M (1993) Increased capillary permeability to albumin in diabetic rat myocardium. Circ Res 72:947–957

Yang ZH, Liu LB, Zhao LN, Liu YH, Xue YX (2014) Permeability of the blood-tumor barrier is enhanced by combining vascular endothelial growth factor with papaverine. J Neurosci Res 92: 703–713

Yang X, Zheng E, Chatterjee V, Ma Y, Reynolds A, Villalba N, Wu MH, Yuan SY (2022) Protein palmitoylation regulates extracellular vesicle production and function in sepsis. J Extracell Biol 1

Yang X, Zheng E, Sun X, Reynolds A, Gonzalez M, Hernandez Villamil J, Pando BD, Smith DJ, Yuan SY, Wu MH (2024) C-type lectin-2d receptor contributes to histone-induced vascular barrier dysfunction during burn injury. Shock 61:592–600

Yasuhara H, Hobson RW 2nd, Dillon PK, Duran WN (1991) A new model for studying ischemia-reperfusion injury in hamster cheek pouch. Am J Phys 261:H1626–H1629

Yeh DC, Duncan JA, Yamashita S, Michel T (1999) Depalmitoylation of endothelial nitric-oxide synthase by acyl-protein thioesterase 1 is potentiated by ca(2+)-calmodulin. J Biol Chem 274: 33148–33154

Yonemaru M, Ishii K, Murad F, Raffin TA (1992) Atriopeptin-induced increases in endothelial cell permeability are associated with elevated cGMP levels. Am J Phys 263:L363–L369

Yoon CM, Hong BS, Moon HG, Lim S, Suh P-G, Kim Y-K, Chae C-B, Gho YS (2008) Sphingosine-1-phosphate promotes lymphangiogenesis by stimulating S1P1/Gi/PLC/Ca2+ signaling pathways. Blood 112:1129–1138

Yu P, Hatakeyama T, Aramoto H, Miyata T, Shigematsu H, Nagawa H, Hobson RW, Duran WN (2005) Mitogen-activated protein kinases regulate platelet-activating factor-induced hyperpermeability. Microcirculation 12:637–643

Yuan SY (2000) Signal transduction pathways in enhanced microvascular permeability. Microcirculation 7:395–403

Yuan SY (2003) Protein kinase signaling in the modulation of microvascular permeability. Vasc Pharmacol 39:213–223

Yuan Y, Granger HJ, Zawieja DC, Chilian WM (1992) Flow modulates coronary venular permeability by a nitric oxide-related mechanism. Am J Phys 263:H641–H646

Yuan Y, Chilian WM, Granger HJ, Zawieja DC (1993a) Permeability to albumin in isolated coronary venules. Am J Phys 265:H543–H552

Yuan Y, Granger HJ, Zawieja DC, DeFily DV, Chilian WM (1993b) Histamine increases venular permeability via a phospholipase C-NO synthase-guanylate cyclase cascade. Am J Phys 264: H1734–H1739

Yuan Y, Mier RA, Chilian WM, Zawieja DC, Granger HJ (1995) Interaction of neutrophils and endothelium in isolated coronary venules and arterioles. Am J Phys 268:H490–H498

Yuan Y, Huang Q, Wu HM (1997) Myosin light chain phosphorylation: modulation of basal and agonist-stimulated venular permeability. Am J Phys 272:H1437–H1443

Yuan Y, Meng FY, Huang Q, Hawker J, Wu HM (1998) Tyrosine phosphorylation of paxillin/pp125FAK and microvascular endothelial barrier function. Am J Phys 275:H84–H93
Yuan SY, Ustinova EE, Wu MH, Tinsley JH, Xu W, Korompai FL, Taulman AC (2000) Protein kinase C activation contributes to microvascular barrier dysfunction in the heart at early stages of diabetes. Circ Res 87:412–417
Yuan SY, Wu MH, Ustinova EE, Guo M, Tinsley JH, De Lanerolle P, Xu W (2002) Myosin light chain phosphorylation in neutrophil-stimulated coronary microvascular leakage. Circ Res 90: 1214–1221
Yuan SY, Breslin JW, Perrin R, Gaudreault N, Guo M, Kargozaran H, Wu MH (2007) Microvascular permeability in diabetes and insulin resistance. Microcirculation 14:363–373
Zamorano P, Marin N, Cordova F, Aguilar A, Meininger C, Boric MP, Golenhofen N, Contreras JE, Sarmiento J, Duran WN, Sanchez FA (2017) S-nitrosylation of VASP at cysteine 64 mediates the inflammation-stimulated increase in microvascular permeability. Am J Physiol Heart Circ Physiol 313:H66–H71
Zamorano P, Koning T, Oyanadel C, Mardones GA, Ehrenfeld P, Boric MP, Gonzalez A, Soza A, Sanchez FA (2019) Galectin-8 induces endothelial hyperpermeability through the eNOS pathway involving S-nitrosylation-mediated adherens junction disassembly. Carcinogenesis 40: 313–323
Zanetti A, Lampugnani MG, Balconi G, Breviario F, Corada M, Lanfrancone L, Dejana E (2002) Vascular endothelial growth factor induces SHC association with vascular endothelial cadherin: a potential feedback mechanism to control vascular endothelial growth factor receptor-2 signaling. Arterioscler Thromb Vasc Biol 22:617–622
Zeng Y, Adamson RH, Curry FR, Tarbell JM (2014) Sphingosine-1-phosphate protects endothelial glycocalyx by inhibiting syndecan-1 shedding. Am J Physiol Heart Circ Physiol 306:H363–H372
Zhang JL, Yokoyama S, Ohhashi T (1997) Inhibitory effects of fluorescein isothiocyanate photoactivation on lymphatic pump activity. Microvasc Res 54:99–107
Zhang Q, Church JE, Jagnandan D, Catravas JD, Sessa WC, Fulton D (2006) Functional relevance of Golgi- and plasma membrane-localized endothelial NO synthase in reconstituted endothelial cells. Arterioscler Thromb Vasc Biol 26:1015–1021
Zhang W, Xu Q, Wu J, Zhou X, Weng J, Xu J, Wang W, Huang Q, Guo X (2015) Role of Src in vascular Hyperpermeability induced by advanced glycation end products. Sci Rep 5:14090
Zhang L, Zeng M, Fan J, Tarbell JM, Curry FR, Fu BM (2016a) Sphingosine-1-phosphate maintains Normal vascular permeability by preserving endothelial surface Glycocalyx in intact microvessels. Microcirculation 23:301–310
Zhang XE, Adderley SP, Breslin JW (2016b) Activation of RhoA, but not Rac1, mediates early stages of S1P-induced endothelial barrier enhancement. PLoS One 11:e0155490
Zheng HZ, Zhao KS, Zhou BY, Huang QB (2003) Role of rho kinase and actin filament in the increased vascular permeability of skin venules in rats after scalding. Burns 29:820–827
Zhu L, Castranova V, He P (2005) fMLP-stimulated neutrophils increase endothelial [Ca2+]i and microvessel permeability in the absence of adhesion: role of reactive oxygen species. Am J Physiol Heart Circ Physiol 288:H1331–H1338
Zimmermann K, Opitz N, Dedio J, Renne C, Muller-Esterl W, Oess S (2002) NOSTRIN: a protein modulating nitric oxide release and subcellular distribution of endothelial nitric oxide synthase. Proc Natl Acad Sci USA 99:17167–17172
Zondag GC, Evers EE, ten Klooster JP, Janssen L, van der Kammen RA, Collard JG (2000) Oncogenic Ras downregulates Rac activity, which leads to increased rho activity and epithelial-mesenchymal transition. J Cell Biol 149:775–782
Zou Z, Li L, Li Q, Zhao P, Zhang K, Liu C, Cai D, Maegele M, Gu Z, Huang Q (2022) The role of S100B/RAGE-enhanced ADAM17 activation in endothelial glycocalyx shedding after traumatic brain injury. J Neuroinflammation 19:46

Zuurbier CJ, Demirci C, Koeman A, Vink H, Ince C (2005) Short-term hyperglycemia increases endothelial glycocalyx permeability and acutely decreases lineal density of capillaries with flowing red blood cells. J Appl Physiol 99:1471–1476
Zwartkruis FJ, Bos JL (1999) Ras and Rap1: two highly related small GTPases with distinct function. Exp Cell Res 253:157–165
Zweifach BW (1972) Capillary filtration and mechanisms of edema formation. Pflugers Arch 336: S81–S95

Chapter 3
Angiogenesis and Microvascular Remodeling

John C. Chappell

Abstract Angiogenesis and microvascular remodeling are essential phenomena in the development and maintenance of blood vessel networks that sustain the health and homeostasis of all tissues and organs within the human body. Here, we provide an overview and discuss the release of pro-angiogenic cues and the different modes of angiogenic remodeling with a specific focus on sprouting and intussusceptive angiogenesis (IA), comparing and contrasting the distinct stages and mechanisms underlying each. Vessel maturation processes -- such as extracellular matrix (ECM) deposition into the vascular basement membrane and pericyte investment -- are presented, along with the mechanisms of coordinated pruning and regression associated with a destabilized microvasculature. Therapeutic targeting of angiogenesis and microvascular remodeling in various disease states such as cancer and diabetes is reviewed. Current in vitro, ex vivo, in vivo, and in silico approaches to modeling angiogenesis are discussed alongside key limitations that need to be considered when interpreting data from each. We conclude this chapter with a brief summary and offer forward-looking thoughts on the opportunities and challenges ahead for the field of angiogenesis and microvascular remodeling.

Keywords Angiogenesis · Microvascular Remodeling · Intussusceptive Angiogenesis · Endothelial Cell · Pericyte · Extracellular Matrix · Capillary Pruning · Tumor Vascularization · Angiogenesis Models

J. C. Chappell (✉)
Fralin Biomedical Research Institute (FBRI) at Virginia Tech-Carilion (VTC), Roanoke, VA, USA

FBRI Center for Vascular and Heart Research, Roanoke, VA, USA

Department of Biomedical Engineering and Mechanics and School of Biomedical Engineering and Sciences, Virginia Tech, Blacksburg, VA, USA

Virginia Tech Carilion School of Medicine, Roanoke, VA, USA
e-mail: JChappell@vtc.vt.edu

R. F. Tuma et al. (eds.), *Handbook of Microcirculation*, Physiology in Health and Disease, https://doi.org/10.1007/978-3-031-99609-2_3

3.1 Introduction

Controlled movement of fluid within a tube is one of the oldest and most basic phenomena conserved across living organisms (Dodd et al. 2017; Papineau et al. 2022). Numerous plant species, if not all, depend on a circulatory system for proper nutrient transport and waste clearance, as seen in the intricate and stereotyped patterns within a single leaf (Radhakrishnan et al. 2021; Byatt and Martin 2023). Animals of all sizes also require a series of interconnected channels that support the flow of nutrients and waste, sustaining the health and homeostasis of surrounding cells and tissues (Rye et al. 2016). In this way, the evolutionary adaptation of biological tube formation has emerged as and remains an essential feature of complex living organisms.

"Vessels" or "containers" of fluid are captured by the ancient Greek words *angeîon* or *ángos*, giving rise to the prefix "angio-." Angiogenesis, therefore, is the creation, origin, or development (i.e., -genesis) of new vessels from existing conduits, such as the vasculature containing blood, specifically in mammals. This term contrasts with "vasculogenesis," which refers to the de novo formation of vascular structures from newly differentiated endothelial cells derived from the primitive mesoderm or other stem cell sources (Risau et al. 1988; Cleaver et al. 1997; Boscolo and Bischoff 2009). Moreover, although they share a subset of common cellular and molecular mechanisms, angiogenesis is also distinct from "lymphangiogenesis," which refers to the formation of new lymphatic vessels from preexisting ones. These terms have emerged as the most appropriate descriptors of the structures and processes underlying microvessel formation (Carmeliet and Jain 2011), as scientists over the years have explored the vascular system and identified the mechanisms by which it forms.

Leonardo da Vinci (1452–1519) proposed that the blood vasculature "sprouts" from the heart, much as a tree trunk and its roots and limbs might emerge from a seed (Risau 1997; Sterpetti 2019). While sprouting angiogenesis is indeed a mechanism by which vascular networks grow, we now know that a primitive yet intact circulatory system must emerge before the heart begins to beat. Scientists such as William Harvey (1578–1657) and Marcello Malpighi (1628–1694) built upon da Vinci's early descriptions to refine our understanding of how blood circulates within this system and becomes oxygenated (Ribatti 2009; Sterpetti 2019). However, John Hunter (1728–1793), a British surgeon, is credited with first using the term "angiogenesis" in 1787, and Judah Folkman (1933–2008) describes this and other seminal contributions to the field, including his own, in his "History of Angiogenesis" chapter (Folkman 2008). We continue to advance the paths forged by these and many other pioneers such as Florence Sabin (1871–1953) (Sabin 1909), August Krogh (1874–1949) (Poole et al. 2021), and Olga Hudlická (1926–2014) (Egginton and Brown 2014) to further understand the cellular, molecular, and genetic mechanisms that facilitate angiogenesis and microvascular remodeling in health and disease (Eelen et al. 2020).

A PubMed search of "angiogenesis" reveals that articles published on this topic have flourished in recent history with 314 in 1992, 5068 in 2007, and 9425 in 2022. As such, this chapter will highlight specific concepts and select studies that have informed our understanding of angiogenesis and microvessel remodeling while encouraging the reader to delve deeper into excellent contemporary reviews (Mentzer and Konerding 2014; Korn and Augustin 2015; Potente and Makinen 2017; Eelen et al. 2020; Coelho-Santos and Shih 2020; Dudley and Griffioen 2023; Walchli et al. 2023) and the supporting primary literature that is regrettably not covered here due to space constraints.

3.2 Modes of Angiogenesis: A Brief Conceptual Overview and Comparison

As discussed above, early observations in studying the mammalian vascular system suggested parallels with plant biology such that structures essential for supporting organism health "sprouted" to form new conduits for transporting nutrients and waste (Radhakrishnan et al. 2021; Byatt and Martin 2023). With the advent of higher-powered imaging modalities such as electron and confocal microscopy, scientists have been able to appreciate cellular features within various experimental models that indeed resembled the emergence of a new structure from an existing one, i.e., a sprout (Ausprunk and Folkman 1977; Glaser et al. 1980; Hansen-Smith et al. 1996; Gerhardt et al. 2003). Endothelial cells composed of an existing vessel were found to be migrating outward from a parent vessel (Gerhardt et al. 2003), typically emitting thin cellular extensions, or filopodia ("filo-" = thread or string; "-podia" = condition of the foot). They utilize filopodia to detect and traverse along gradients of growth factors such as vascular growth factor-A (VEGF-A) (Ruhrberg et al. 2002), physically engaging the surrounding extracellular matrix (ECM) (Anderson et al. 2004). The lumen of the nascent vessel forms via cell and cord hollowing as sprouting proceeds and is completed following anastomotic connection with the target vasculature (Kamei et al. 2006; Iruela-Arispe and Davis 2009; Strilic et al. 2010; Xu and Cleaver 2011).

In contrast to the sprouting angiogenesis described above, splitting or intussusceptive ("intus-" = within; "suscept" = to take, receive) angiogenesis (IA) involves the formation of new capillaries by dividing an existing lumen into two separate channels (Hansen-Smith et al. 1996; Burri and Djonov 2002). This mode of microvessel remodeling is more challenging to identify and describe, owing in large part to the nature of how it occurs. Specifically, an endothelial or tissue micro-pillar extends into the lumen of an existing vessel and connects with the opposing apical surface (Egginton et al. 2001). This pillar expands in size to divide the single channel into two conduits, thereby creating a new vessel branch while maintaining the lumen of both branches (Mentzer and Konerding 2014). Since this process occurs within the lumen of an existing vessel, visualizing this type of angiogenesis in fixed specimens is extremely difficult. Additionally, identifying a tissue pillar extending into the capillary lumen is fairly challenging without the use of higher-powered imaging modalities (e.g., scanning electron microscopy) (Djonov et al. 2003; Du Cheyne et al. 2021). Furthermore, recent evidence has suggested that this mode of microvascular remodeling may require specific hemodynamic conditions in addition to exposure to pro-angiogenic cues (Karthik et al. 2018).

Studies within the last 10 years have suggested that additional modes of angiogenesis may occur in certain scenarios. Vascular "island" incorporation, for example, was observed in adult rat mesenteric tissues wherein endothelial cell segments that had disconnected from neighboring vascular networks were able to grow and reintegrate into existing vessels (Kelly-Goss et al. 2013). This mode may occur following a phase of coordinated or pathology-induced vascular regression in which these cell segments are stimulated to proliferate and reintegrate into the existing microvasculature and contribute to network perfusion (Kelly-Goss et al. 2012). This mode also mirrors the developmental scenario of primitive "blood islands" wherein endothelial cells form vessel-like structures that contain hematopoietic precursor cells (Doetschman et al. 1985; Risau et al. 1988; Shalaby et al. 1995; Bautch et al. 1996; Drake and Fleming 2000). Additional work will be needed to establish the involvement of coordinated endothelial cell migration and perhaps endothelial sprouting from these isolated vessel regions. Furthermore, it is intriguing to speculate that in conditions such as ischemic stroke or myocardial infarction, capillary segments downstream of an occluded vessel might transform into these "islands" that are targets for rapid therapeutic revascularization of damaged tissue (Kelly-Goss et al. 2014). A conceptually similar mode of angiogenesis has been proposed to involve the coalescence of adjacent vessel segments such that two branches fuse together to form one larger channel (Drake and Little 1995; Blum et al. 2008; Nitzsche et al. 2022). In a sense, this lumen fusion process is perhaps the reverse of intussusceptive angiogenesis, and it could be viewed as a mode of microvessel remodeling that expands the lumen diameter or caliber of a vascular branch rather than generating a new branch from existing vessels.

Each type of microvascular remodeling may contribute to capillary network formation within all tissues and organs, but one mode may predominate based on the location or context (Zhan et al. 2020). Thus, the mode of angiogenesis may be somewhat tissue-dependent, as the microvasculature within certain tissues such as the lungs and gastrointestinal tract seems to harbor more intussusceptive angiogenic growth as compared to other organs like the retina, brain, and heart (De Spiegelaere et al. 2012; Karthik et al. 2018; Cahill et al. 2018). Developing neurological tissues such as the brain and retina appear to exhibit a larger degree of sprouting angiogenesis as compared to intussusceptive capillary expansion. However, this may depend on the specific timeframe in which these tissues are examined. A study of the zebrafish caudal vein plexus by Karthik et al. (2018) suggests that intussusceptive growth may depend on specific hemodynamic conditions following a robust sprouting phase. During conditions such as cerebral ischemia, however, both modes appear to be relevant in restoring tissue perfusion concurrently within discrete regions affected by the sudden loss of blood flow (Zhang et al. 2002). Furthermore, the type of angiogenesis assay that has emerged may also contribute to an imbalance in the perceived frequency of each type of microvascular remodeling. For instance, in vitro angiogenesis assays have largely focused on sprouting endothelial cells (Feraud et al. 2001; Nakatsu and Hughes 2008), so the creation of new branches by vessel splitting may be underappreciated. It may also be that many of these culture-based assays simply lack intussusceptive angiogenesis due to the lack of relevant hemodynamics (Lucitti et al. 2007) and the variability in lumen formation within in vitro assays.

3.2.1 Angiogenic Cue Release

Cells and tissues requiring new vessel growth release molecular cues to stimulate the cellular processes that drive angiogenesis (Carmeliet and Jain 2011; Lugano et al. 2020). One highly relevant scenario in which this occurs is hypoxia, that is, when tissue oxygenation falls below the level needed to sustain metabolic demand, a state that appears to be tissue- or organ-dependent (Liao and Johnson 2007; Krock et al. 2011). Under normoxic conditions, hypoxia-inducible factors (HIFs) are continually degraded by (i) poly-ubiquitination via prolyl hydroxylase (PHD) activity and (ii) subsequent proteasomal degradation mediated by von Hippel–Lindau (VHL) proteins (Marxsen et al. 2004). During hypoxia, HIFs are spared from enzymatic destruction and begin to accumulate, becoming phosphorylated at increasing levels (Ziello et al. 2007). Phosphorylated HIFs facilitate and increase their association with other regulatory partners to promote expression of genes containing hypoxia-responsive elements (Wigerup et al. 2016). Vascular endothelial growth factor-A (VEGF-A) is one of the most well-known and potent pro-angiogenic cues produced by HIF-mediated transcriptional upregulation (Ferrara 2004; Shibuya 2011). Other pro-angiogenic cues synthesized and secreted under hypoxic conditions include platelet-derived growth factor-B (PDGF-B), angiopoietin-1 and -2 (Ang1/2), basic fibroblast growth factor (bFGF/FGF2), and stromal-derived factor 1-α (SDF1α) (Rodriguez et al. 2021), though antiangiogenic cues can also be upregulated by HIF signaling activity (Krock et al. 2011). Local concentrations of these factors increase, and they begin to diffuse outward from their cellular sources, creating gradients that ultimately engage with vascular cells of neighboring microvessels and initiate endothelial cell sprouting (Ruhrberg et al. 2002; Helm et al. 2005; Hashambhoy et al. 2011) or vessel splitting (Gianni-Barrera et al. 2014).

3.2.2 *Sprouting Angiogenesis*

3.2.2.1 Sprout Initiation

Hypoxia stimulates production of not only pro-angiogenic ligands but also their cognate receptors (Rodriguez et al. 2021). In particular, VEGF receptor-2 (VEGFR2 in humans or Flk-1/KDR in mice) expression on endothelial cells facilitates VEGF-A signaling to initiate activation of these cells to commence sprouting (Feraud et al. 2001; Jakobsson et al. 2010). VEGF receptor-1 (VEGFR1 in humans or Flt-1 in mice) is also produced under low oxygen conditions (Gerber et al. 1997), but this VEGF-A receptor exhibits a ten-fold higher affinity for ligand binding relative to VEGFR2 and acts primarily as a negative regulator via a "decoy" function (Meyer et al. 2006; Chappell et al. 2019; Moore et al. 2020). In this way, endothelial cells receiving an oncoming gradient of VEGF-A exhibit differential and heterogeneous responses based on their level of activation (Blancas et al. 2013; Krueger et al. 2011).

This phenotypic heterogeneity within the endothelium is also facilitated by VEGF-A pathway cross talk with Notch–Delta signaling, which has been extensively studied (Lawson et al. 2002; Gale et al. 2004; Noguera-Troise et al. 2006; Ridgway et al. 2006; Lobov et al. 2007; Leslie et al. 2007; Suchting et al. 2007; Hellstrom et al. 2007; Siekmann and Lawson 2007; Harrington et al. 2008; Siekmann et al. 2008; Bentley et al. 2008; Gerhardt 2008; Jakobsson et al. 2009; Cao et al. 2009; Funahashi et al. 2010; Jakobsson et al. 2010; Chappell et al. 2013; Ubezio et al. 2016). Current working models for these interactions suggest that, as an endothelial cell begins to emerge as the "tip" cell leading the new vessel sprout, it engages with neighboring cells via surface presentation of Delta-like 4 (Dll4) ligands (Blanco and Gerhardt 2013). Adjacent endothelial cells receive these signals via their own Notch1 receptors, which are enzymatically cleaved to release an intracellular domain (Takeshita et al. 2007). This fragment joins a complex to regulate downstream effector genes such as those in the Hey and Hes family of transcriptional regulators as well as VEGF-A receptor genes (Roca and Adams 2007; Suchting et al. 2007; Siekmann et al. 2008). Within the dynamics of this sprout initiation, it appears that endothelial cells compete for the lead position via a complex interplay between these and other signaling axes such as the bone morphogenetic protein (BMP) pathway (Larrivee et al. 2012; Aspalter et al. 2015; Mouillesseaux et al. 2016).

These signaling cues activate the endothelium to commence structural remodeling, which requires endothelial cells to alter their local microenvironment to facilitate their outward migration. For established vessels induced to remodel, the surrounding extracellular matrix (ECM) comprising the vascular basement membrane, for instance, must be degraded to allow for their "escape" from the capillary wall (del Toro et al. 2010). Enzymes such as matrix metalloproteinase-9 (MMP9), among others (Rundhaug 2005), have been implicated in activated endothelial cells reshaping their local ECM environment to begin sprouting outward (Funahashi et al. 2011). This remodeling can be seen in the targeted degradation of type IV collagens, laminins, and vitronectin, among other ECM constituents. Endothelial cells must also rearrange their junctions with neighboring endothelial cells (Szymborska and Gerhardt 2018), specifically vascular endothelial cadherin (VE-cadherin, gene *Cdh5*) (Bentley et al. 2014), platelet endothelial cell adhesion molecule-1 (PECAM-1/CD31) (Cao et al. 2002), or claudin-5 (Escudero-Esparza et al. 2012), to allow their movement from the vessel wall into the parenchyma. Physical associations with perivascular cells such as pericytes must also be altered, diminished, or abandoned altogether to gain the plasticity necessary for outward migration (Gerhardt and Betsholtz 2003; Bergers and Song 2005; Bodnar et al. 2013; Birbrair et al. 2014; Stapor et al. 2014).

On encountering an incoming gradient of chemoattractants and/or growth factors, endothelial cells undergo cytoskeletal reorganization (Bayless and Johnson 2011) and emit actin-based filopodia (Gerhardt et al. 2003; Vaeyens et al. 2020), which, although not required for migration, likely enhance sprouting efficiency (Phng et al. 2013). Microtubules not associated with centrosomes are also remodeled to facilitate persistent endothelial cell sprouting in three-dimensional (3D) culture and zebrafish angiogenesis models (Martin et al. 2018). Interestingly, an excess number of centrosomes can disrupt microtubule organization and lead to defective migration and proliferation during angiogenic sprouting (Taylor et al. 2010; Kushner et al. 2014, 2016). In this way, endothelial cells undergoing sprouting angiogenesis experience a shift in their polarity from a luminal–abluminal orientation (also termed "apical-basal" or "apical-basolateral") to a front–rear polarity, which must be reversed during vessel lumen formation (Lee and Bautch 2011; Pelton et al. 2014). Rearrangement of intracellular organelles also includes a repositioning of the Golgi apparatus (Martin et al. 2018), which is typically positioned upstream relative to the fluid flow through a capillary (Lee et al. 2022). Uncovering these subcellular processes continues to shed light on the complex interplay between molecular and cellular mechanisms that facilitate the initiation of endothelial cell sprouting and angiogenic sprouting.

3.2.2.2 Sprout Elongation and Guidance

As a new vessel sprout begins its outward trajectory from a parent vessel, additional processes must facilitate the extension of an emerging capillary. Endothelial cells at the leading front of the new vessel sprout engage in collective migration (Arima et al. 2011) while simultaneously competing to lead the sprout forward (Bentley et al. 2014), a process conserved in other tube formation scenarios such as the *Drosophila* trachea (Ghabrial and Krasnow 2006). Much work has been performed to elucidate the role that the Notch pathway plays in this "tip"–"stalk" cell competition dynamic (Hellstrom et al. 2007; Bentley et al. 2009; Jakobsson et al. 2010; del Toro et al. 2010; Blancas et al. 2013). Cross talk exists with other signaling axes such as the VEGF-A and bone morphogenetic protein (BMP) pathways (Itoh et al. 2004; Larrivee et al. 2012; Aspalter et al. 2015; Mouillesseaux et al. 2016), to determine which cell will maintain the lead position and whether a new sprout may emerge from the one that is elongating. The intersection of these signaling cascades also dictates the level of endothelial cell division that must occur to support the lengthening of the new vessel sprout (Ubezio et al. 2016). Mechanical forces may also be an important factor in how the tip and stalk cells relate to one another and perhaps facilitate cell division by maintaining a biophysical force on the trailing stalk cells (Kilarski et al. 2009; Barrasa-Ramos et al. 2022). These forces are transmitted through the junctions that are maintained between the cells that comprise the new vessel sprout. Shear stress and intraluminal pressure also appear to influence the rate of sprout elongation as well as the location of sprout initiation (Ghaffari et al. 2015).

Endothelial "tip" cells are tasked with finding their way along the chemotactic gradient as the sprout extends away from the parent vessel. Current models suggest that soluble Flt-1 (sFlt-1) may reduce VEGF-A ligand availability in regions adjacent to the parent vessel to prevent the sprout from collapsing back into the originating capillary (Chappell et al. 2009). As the elongating sprout advances further away from the parent vessel, other guidance mechanisms appear to either promote or restrict the migration of the endothelial tip cell into specific regions (Gerhardt et al. 2004; Carmeliet and Tessier-Lavigne 2005; Gerhardt 2008; Adams and Eichmann 2010; Zygmunt et al. 2011). For example, during zebrafish development, the somite boundaries are maintained by specific repulsion cues such as UNC5b (Lu et al. 2004) and Sema3a (Torres-Vazquez et al. 2004), among others, which limit or prevent the in-growth of new vessels. These types of cues may be more essential in the early stages of development when vascular patterning of major vessels must be precisely stereotyped. Formation of a vascular plexus that is more freeform may not require these types of guidance cues but may be more dependent on other vessel patterning mechanisms. For instance, the spatial distribution of VEGF-A isoforms can directly influence the architecture and segment length of capillary networks (Ruhrberg et al. 2002).

In addition to soluble cues and cell–cell signaling, the extracellular matrix (ECM) provides an element of control over angiogenic sprout guidance and morphology (Davis and Senger 2005). Endothelial "tip" cells that engage with ECM fibers extend filopodia containing integrins such as integrin-αVβ3 (Cheresh 1987) to anchor and facilitate directional migration (Avraamides et al. 2008). Endothelial cells within an elongating sprout also appear to engage with the adjacent ECM such that, in a synthetic biomaterials environment, sprout stalk diameter corresponds to the level of physical compaction and degradation that occurs in the surrounding matrix (Wang et al. 2021). In this way, basic ECM properties such as fiber density, stiffness/deformability, and orientation can alter angiogenic sprouting in a biophysical manner (Rauff et al. 2019) and by influencing the distribution of soluble pro- and antiangiogenic cues.

3.2.2.3 Anastomotic Connection

Once an angiogenic sprout has successfully extended toward a target, the constituent endothelial cells must engage with the vessel to forge a new connection. This process likely involves many of the molecular mechanisms that set the stage for angiogenic remodeling (Blum et al. 2008; Lenard et al. 2013). For example, VEGF-A signaling regulation may predispose a vessel region for connection with an oncoming sprout led by an endothelial tip cell (Nesmith et al. 2017). Such a region must harbor a sufficient level of plasticity to engage with a sprout and form a new connection, likely induced by pro-angiogenic factors secreted by hypoxic or activated cells (Moreira-Soares et al. 2018). In developmental contexts, two sprouts may extend such that their respective "tip" cells interact directly via filopodial contact points (Lenard et al. 2013). Computational and experimental models have suggested that Notch signaling and local tissue mechanics may provide important cues for establishing anastomotic connections (Outtz et al. 2011; Vega et al. 2020). The location for these interactions may also be orchestrated in part by macrophages that are recruited during injury (Martin and Gurevich 2021) and those that are tissue-resident such as in the murine embryonic hindbrain (Fantin et al. 2010), which have been shown to act as "guideposts" for endothelial sprouts.

As endothelial cells engage with one another, they must begin forming junctions to stabilize the vessel wall. Concurrently, and perhaps even as the sprout is extending, the vessel lumen must begin to open for carrying blood (Betz et al. 2016). The proposed mechanisms for lumen formation include rapid deposition of negatively charged sialic acids that repel opposing endothelial cell surfaces, "pushing" the lumen open, which is then sustained by cytoskeletal rearrangements (Strilic et al. 2009; Strilic et al. 2010). Observations from zebrafish and cell-based models suggest that intracellular vesicles may be shuttled to the interface between two endothelial cells and that they accumulate to create a void that grows into the lumen space (Kamei et al. 2006). Hemodynamic forces have also been implicated in lumen formation, specifically in causing deformation on the apical side of the sprout described as "inverse blebbing" (Gebala et al. 2016). These mechanisms may work in parallel or discretely depending on the vessel size or region.

3.2.3 Intussusceptive Angiogenesis

3.2.3.1 Transluminal Endothelial Extensions

Capillary regions that become activated to initiate intussusceptive angiogenesis (IA) appear to require less initial remodeling compared to angiogenic sprouting. The surrounding ECM and pericyte investment seem to remain largely intact as IA commences. Endothelial cells begin this process by extending a thin tissue pillar into the capillary lumen toward the opposite wall (Paku et al. 2011). Unlike the actin-based filopodia emitted by sprouting endothelial cells, it is unclear whether the intraluminal projections of the endothelial cell membrane contain actin filaments during IA onset (Konerding et al. 2010). These extensions do not appear to contain contractile elements and therefore have been classified more as "pseudopodia" (Hansen-Smith et al. 1996). Interestingly, nascent tissue pillar extensions lack an obvious gradient of growth factors or tissue morphogens for their patterning, and they must have sufficient structural rigidity to withstand the force of intraluminal blood flow. Transluminal endothelial cell projections engage with the opposite microvessel wall and create a tissue pillar as IA proceeds. This “bridge” across the capillary lumen may be initially comprised only of endothelial cells, but associated myofibroblasts and/or pericytes and their microfilaments have been detected within these structures (Djonov et al. 2000). The involvement of other cell types in this process may, however, be tissue-dependent. In this context, in a porcine model of kidney glomeruli vascular development, perivascular cells and pericytes did not reveal any association with IA formation (Logothetidou et al. 2018).

3.2.3.2 Tissue Pillar Expansion

To continue the process of splitting a single microvessel into two new channels, the initial transluminal tissue pillar must expand and separate these new vessel segments. Very little, if any, endothelial cell proliferation appears to occur at sites of tissue pillar expansion (Djonov et al. 2003), in contrast to sprouting angiogenesis. Nevertheless, these “bridges” can expand in size, presumably with endothelial cell and/or perivascular cell cytoplasm. Recent advances in live imaging of the developing zebrafish caudal vein plexus, however, have suggested that these structures are more dynamic, with a subset of pillars being unstable and hence removed (Ross et al. 2023). Tissue pillars that are reinforced may collectively expand when those in close proximity actually fuse together (Mentzer and Konerding 2014; Ross et al. 2023), creating a larger region of separation between the two new microvessels.

Because the luminal spaces of both daughter capillaries are maintained throughout IA, distinct mechanisms to open the lumen are not required as in sprouting angiogenesis. It is intriguing to consider this aspect from an evolutionary perspective and the advantage conveyed by this mode of angiogenesis compared to others. For example, IA allows sustained blood flow, and thus metabolic and waste exchange, throughout the remodeling process, which may be critical for specific organs such as the liver, kidney (Logothetidou et al. 2018), or placenta (Cahill et al. 2018) that are largely dedicated to facilitating sustained exchange. Splitting angiogenesis in conjunction with capillary enlargement may also have a distinct impact on overall peripheral resistance through capillary networks (Wirth et al. 2023) as compared to sprouting-mediated microvascular remodeling. Future studies employing high-resolution imaging approaches (Du Cheyne et al. 2021) coupled with computational fluid dynamics (Motlana and Ngoepe 2023) will likely provide further insights into the complex flow profiles that may be generated during IA as well as those discussed above.

3.2.4 Maturation of New Vessel Segments

After a new vessel segment is formed by the various modes of angiogenesis described, a series of events must occur to transition this nascent capillary from immature to stable. For instance, as the lumen space opens and blood components begin to circulate, the ultrathin layer of negatively charged proteoglycans, glycoproteins, and glycolipids known as the glycocalyx must also be synthesized and localized to the apical or lumen-facing surface of the endothelium. The presence of blood flow itself is one of the major regulators of glycocalyx synthesis (Maroski et al. 2011; Mensah et al. 2020), suggesting that IA may require less new glycocalyx synthesis, if any, due to the nature of this type of microvascular remodeling. On the abluminal or basal side of the maturing vessel wall, ECM also experiences a transition from a more primitive, provisional composition into a mature configuration containing type IV collagens and laminins, among other key proteins (Marchand et al. 2019). This vascular basement membrane must gain a level of structural integrity to support the new capillary segment in counterbalancing intraluminal pressure changes (Davis and Kemp 2023). The capillary ECM will also shift to a more quiescent state with regard to the pro- and antiangiogenic factors that are retained within this meshwork of unique proteins (Mongiat et al. 2016), thus minimizing vessel instability while also being primed for a future remodeling event.

An additional component of the microvessel wall that participates in angiogenic remodeling is the pericyte. These specialized mural cells extend along the basal/basolateral side of capillaries, secreting ECM to (i) support the basement membrane and (ii) physically separate themselves from the endothelium, adjacent pericytes, and surrounding parenchymal cells (Stratman et al. 2009; Abdelazim et al. 2022). Pericytes collaborate with endothelial cells during angiogenesis (Bergers and Song 2005; Birbrair et al. 2014; Eilken et al. 2017; Chiaverina et al. 2019; Payne et al. 2021), though they do not appear to be required for sprouting (Nakatsu and Hughes 2008). Sprouting endothelial cells respond to growth factor stimulation in part by secreting platelet-derived growth factor-BB (PDGF-BB) (Lindblom et al. 2003; Hellstrom et al. 2007), which is retained in the adjacent ECM and provides a chemotactic cue for pericyte recruitment (Abramsson et al. 2007). Endothelial cells undergoing IA likely secrete factors to promote pericyte association with both daughter microvessels, though these molecular details remain to be established. Pericytes may be recruited to newly formed vessels from distant locations, but the predominant source of these stabilizing cells is likely from the microvasculature adjacent to regions undergoing remodeling (Armulik et al. 2011; Berthiaume et al. 2018; Payne et al. 2020). Therefore, pericyte recruitment presumably occurs concurrently with angiogenic sprouting (Chiaverina et al. 2019) or splitting (Diaz-Flores et al. 2022) to ensure new microvessel segments are stabilized and begin maturing as soon as an anastomotic connection forms.

3.2.5 Capillary Pruning and Rarefaction

Under physiological conditions, angiogenic growth and microvascular remodeling occur to match the metabolic and/or functional demands of a given tissue or organ. In certain scenarios, such as tissue and organ development, capillaries may be pruned or systematically removed if they are no longer needed for metabolic exchange for various reasons (Korn and Augustin 2015). This phenomenon parallels plant biology as seen in the process of abscission or the selective loss of a branch or appendage that is no longer contributing to organismal health and growth (Olsson and Butenko 2018). For both plants and animals, an ideal balance must be achieved between metabolic delivery and demand. In the blood vasculature of vertebrates, this optimal point is governed in large part by network perfusion and oxygen delivery (Santamaria et al. 2023). Within murine retina and zebrafish microvessels, capillary segments receiving less perfusion undergo a programmed regression through coordinated rearrangement of endothelial cells (Lobov et al. 2011; Chen et al. 2012; Kochhan et al. 2013; Franco et al. 2015; Lenard et al. 2015). This cellular mechanism of vessel pruning has also been observed in a recent study using longitudinal imaging of murine skin neonatal endothelial cells (Kam et al. 2023). In addition to perfusion dynamics, pericytes have also been implicated in the coordinated regression of capillaries, specifically through the expression of the endosialin (CD248) receptor, which induces endothelial cell apoptosis and detachment (Simonavicius et al. 2012). It is intriguing to speculate that pericytes and vascular smooth muscle cells may also influence capillary pruning by restricting blood flow into select microvessels or networks, resulting in their orchestrated removal from the microvasculature.

While capillary pruning occurs to tune microvessel density to the specific metabolic demands within a given tissue, capillary rarefaction is viewed more as a pathological reduction in microvascular density. Clinically relevant scenarios involving microvessel loss include the aging brain, which has been suggested to be at an increased risk of prolonged capillary obstruction and associated perfusion impairment that leads to rarefaction (Schager and Brown 2020). A wide range of additional mechanisms have been proposed to reduce capillary density, including chronic inflammation, pericyte loss, and an imbalance between pro- and antiangiogenic factors, among others (Afsar et al. 2018). In tissues such as the kidney, this can lead to unresolved hypoxia, deficient hemodynamic responsiveness, and, ultimately, disease manifestation and hypertension onset and progression (Afsar et al. 2018). Hypertension itself can also cause capillary rarefaction in many organs and tissues, including the eye (Frost et al. 2021). Long COVID has also been recently linked to persistent capillary loss as measured by sublingual (i.e., under the tongue) video microscopy (Osiaevi et al. 2023). Premature infants exposed to elevated oxygen can also experience a reduction in capillary density within the eye, setting the stage for aberrant vessel overgrowth once they return to ambient oxygen levels—a condition known as retinopathy of prematurity (Hartnett 2015).

3.2.6 *Therapeutic Targeting of Angiogenesis and Microvascular Remodeling*

A tremendous amount of research effort has been aimed at understanding the basic mechanisms underlying angiogenesis so that this knowledge may be applied to clinically relevant therapies for disease conditions such as those described above. However, the clinical scenario receiving a fairly substantial amount of attention is tumor angiogenesis (Hanahan and Weinberg 2011). Solid cancers often exhibit elevated metabolic demand due to the rapid proliferation of transformed cells, which, in turn, leads to the hypoxia-driven secretion of pro-angiogenic factors (Harris 2002; Liao and Johnson 2007; Wigerup et al. 2016). A broad range of pharmacological agents has been developed to limit and reverse vessel formation in tumors, with anti-VEGF-A therapies (bevacizumab/Avastin) being one of the most widely explored (Apte et al. 2019). While improved clinical outcomes can be achieved with the discerning use of antiangiogenic therapies, some cancers become refractory to these treatments (Bergers and Hanahan 2008; Ribatti 2016; Ma et al. 2018; Lugano et al. 2020) and develop adaptations to subvert them (Rahbari et al. 2016). Certain drugs like those targeting VEGF-A can also fuel patient noncompliance due to severe side effects such as hypertension, proteinuria, and gastrointestinal symptoms, among many others (Kamba and McDonald 2007). Therefore, while antiangiogenic therapies are effective in certain tumor treatment strategies, they have not achieved the level of efficacy that was initially envisioned when first proposed: to starve and destroy the tumor, reduce metastatic burden, and move patients toward cancer-free survival and complete remission (Goel et al. 2011). Clinical data from combination treatment using chemotherapeutics alongside anti-VEGF-A therapy, however, demonstrated improved survival outcomes. These observations inspired the notion of "tumor vessel normalization," whereby an abnormal tumor vasculature is rehabilitated by antiangiogenic agents (Tong et al. 2004; Jain 2005; Ferrara 2005; Heath and Bicknell 2009; Greenberg and Cheresh 2009; Falcon et al. 2009; Goel et al. 2011). Transitioning cancer vessels to a more stable and functional state may reduce hypoxia and improve the tumor microenvironment with respect to metabolism, immune cell activity, and radiotherapy efficacy, among other benefits (Goel et al. 2011). Therefore, next-generation agents targeting tumor vasculature are being developed with this perspective, aiming to enhance the therapeutic effects of existing chemotherapeutics and emerging immunotherapy agents (Ramjiawan et al. 2017; Schmittnaegel et al. 2017; Allen et al. 2017; Lee et al. 2020).

Another pathological condition envisioned to benefit from targeting blood vessel growth is diabetic retinopathy. Microvascular damage caused by chronic hyperglycemia leads to sustained inflammation and hypoxia in various tissues, including the eye (Antonetti et al. 2012). Such conditions in the retina induce secretion of pro-angiogenic growth factors including VEGF-A to stimulate neovascularization (Penn et al. 2008). Unfortunately, these new vessels are defective due to the excess growth factors, inflammatory cytokines, and glucose (Wei et al. 2022), exacerbating an already unfavorable environment for eye health and homeostasis. Anti-VEGF-A therapies have therefore been developed to reduce the abundance of this highly potent angiogenic stimulus, with intraocular injections of agents such as ranibizumab (Lucentis®) and aflibercept (Eylea®) becoming first-line therapies for diabetic retinopathy as well as other causes of macular edema (Simo and Hernandez 2008). These VEGF-A inhibitors appear to be well-tolerated, though concerns have emerged regarding effects on blood pressure regulation, proteinuria, impaired wound healing, and nonvascular retinal cells that also utilize VEGF-A receptors for survival (Froger et al. 2020). Alternatives to VEGF-A-targeted therapies are also being explored to potentially avoid these unintended consequences. For instance, a recent study from the lab of Dr. Courtney Griffin has identified endothelial transcription factors that could be blocked to induce the regression of low-flow vessels, independent of anti-VEGF-A treatment (Schafer et al. 2020). Similar avenues of research exploring transcription-modulating drug targets within microvascular remodeling mechanisms represent an important direction for blood vessel-focused therapeutic strategies, given the known and emerging limitations of antibody-based pharmaceuticals.

In contrast to strategies aimed at reducing microvessel density, the goal of therapeutic neovascularization is to stimulate the formation of a functional and stable microvasculature within ischemic tissues while minimizing side effects (Cao et al. 2005). Growth factors, alone or in combination, have been shown to improve vascular network remodeling and function in preclinical disease models of tissue ischemia (Luttun et al. 2002; Cao et al. 2009; Chen et al. 2023). Due to the multistage nature of vessel growth, as discussed above, efforts are ongoing to develop biomaterials capable of distributing pro-angiogenic cues (Marsico et al. 2021) with various release profiles to enhance each stage when appropriate (Brey et al. 2005; Matsui and Tabata 2012). These types of constructs are also being developed free from growth factor incorporation (Serbo and Gerecht 2013; Pal et al. 2018) or in conjunction with seeding cells that may promote new vessel formation and stability (Lee et al. 2015). As with all clinical interventions, though, therapeutic neovascularization strategies will need to be optimized for maximum benefit without inducing or exacerbating pathological angiogenesis within the local microenvironment of the treatment area (Ozawa et al. 2004) as well as systemically.

3.2.7 Advancing Modeling Approaches

To uncover the basic mechanisms underlying microvascular remodeling and determine the potential of the therapeutic targets discussed above, a wide spectrum of experimental and computational models have been developed. In vitro models involve culturing, often in nonphysiological conditions, primary endothelial cells alone or in combination with other vascular cells (e.g., pericytes) (Eglinger et al. 2017) and/or nonvascular cells depending on the tissue type being modeled (Goodwin 2007). Human umbilical vein endothelial cells (HUVECs) have been a favored cell line for many years due to the relative ease of their collection (Adair and Montani 2010), though recent efforts have been made to use endothelial cells sourced from the tissue or organ of interest to address concerns about phenotypic heterogeneity. Embryonic and induced pluripotent stem cells (iPSCs) have also been utilized as differentiated embryoid (Risau et al. 1988; Feraud et al. 2001; Jakobsson et al. 2007; Walpole et al. 2015) or organoid models (Wimmer et al. 2019; Ryan and Cleaver 2022). A variety of two-dimensional (2D) and 3D configurations have been explored depending on the specific research question, such as coating a bead with endothelial cells, placing it in a select matrix, and stimulating angiogenesis with exogenous recombinant or cell-derived growth factors (Nakatsu and Hughes 2008). In vitro angiogenesis assays can replicate many aspects of new vessel formation including endothelial sprouting and lumen formation, but they lack blood flow and other key biomechanical, chemical, and cellular inputs such as immune cells. Recent microfluidics approaches have been addressing some of these limitations (Song and Munn 2011; Murfee and Peirce 2017; Akbari et al. 2017; Zhao and Chappell 2019), though microvessel-sized structures appear difficult to maintain over the course of being cultured. These limitations do not preclude their use in exploring the basic aspects of angiogenesis and microvascular remodeling, but they should be openly considered when offering interpretations of data from such models.

To complement cell-based approaches to studying angiogenesis, ex vivo models such as the aortic ring assay have been developed in an effort to capture more of the biological complexity inherent to a given vascular bed (Goodwin 2007; Chiaverina et al. 2019). Although these assays also lack blood flow due to their removal from the host circulation, culturing tissues allows researchers to study the endogenous cellular constituents of capillary networks in their natural configuration (Azimi et al. 2017; Suarez-Martinez et al. 2018a; Payne et al. 2021). The culturing conditions do not adequately match the physiological environment from which the tissue was collected, e.g., exposure to normal oxygen tensions (Colom et al. 2014), but these assays can offer an orthogonal view of microvascular remodeling processes (Sawamiphak et al. 2010; Rezzola et al. 2013; Lu et al. 2023). Tissue explant cultures have been adapted to incorporate some aspects of fluid flow (Motherwell et al. 2019), though challenges remain in long-term culturing and the perfusion of a blood-like substance. While no experimental model of angiogenesis is free from key limitation, ex vivo tissue explant culture assays provide a complementary approach to better understand the molecules and mechanisms that regulate angiogenic and microvascular remodeling.

Because angiogenesis research is largely focused on the translational application of new insights into clinical solutions, in vivo models remain essential to testing new hypotheses in the complex, intact vascular networks within a living organism (Staton et al. 2009; Adair and Montani 2010). Murine models, particularly the developing postnatal retina, have received significant attention in angiogenesis research (Stahl et al. 2010), but other species have been invaluable in understanding basic microvascular remodeling mechanisms including zebrafish-based (Chavez et al. 2016) and avian-based systems (Makanya et al. 2015), among others. Recent advances in transcriptional profiling have ushered in a greater appreciation for the heterogeneity in vascular cells throughout the circulatory system and within each tissue and organ (Kalucka et al. 2020; Dumas et al. 2020; Trimm and Red-Horse 2023). Therefore, observations of angiogenesis within a given tissue bed are unlikely to translate to all scenarios of microvascular remodeling. Furthermore, specific experimental interventions (Suarez-Martinez et al. 2018b) and approaches used for long-term imaging (Biel et al. 2014) may artificially stimulate the immune system and induce systemic hormone release that may introduce confounding variables when observing angiogenic remodeling. Nevertheless, in vivo angiogenesis assays remain a critical component of the research spectrum in uncovering new mechanisms that drive physiological and pathological microvascular remodeling that may be targeted for therapeutic benefit.

As the complexity of the biological process underlying angiogenic remodeling continues to emerge, it is becoming increasingly critical to simultaneously build tools that aid in the interpretation and synthesis of the data being generated. Computational models represent one such approach to creating a framework for integrating existing information while identifying key gaps in knowledge and generating new hypotheses (Peirce 2008). Sprouting angiogenesis, in particular, has been the subject of various computational approaches (Stefanini et al. 2009; Peirce et al. 2012; Bentley et al. 2014; Walpole et al. 2015; Walpole et al. 2017; Kuhn and Checa 2019; Vega et al. 2020), which have yielded new insights into the cellular and molecular cross talk involved. Numerical modeling offers the additional benefit of being able to integrate information across different scales (Walpole et al. 2013; Logsdon et al. 2014), such as in exploring how intraluminal and interstitial fluid flow impact microvascular remodeling (Welter and Rieger 2013; Ghaffari et al. 2015; Motlana and Ngoepe 2023) and associated oxygen delivery (Secomb et al. 2013; Al-Shammari et al. 2014). Computer simulations can also illuminate the interaction between the local microenvironment of a particular vascular bed and the systemic circulation, specifically when considering the distribution of pro- and antiangiogenic cues within and across each tissue compartment (Mac Gabhann and Popel 2007, 2008; Stefanini et al. 2009; Vempati et al. 2010). These approaches also offer the advantage of testing certain hypotheses in a computational environment when experimental approaches may not be feasible for technical reasons. Computational models may also inform the design of certain experimental approaches (e.g., physiological range of concentrations) as well as the interpretation of the data generated from the angiogenesis assays described above (Cardenas et al. 2020). Lastly, as the breadth of available clinical data increases, computational approaches will yield biologically informed models to predict the efficacy of specific pro- or antiangiogenic interventions for a specific individual (Collin et al. 2022) and across subsets of patient populations (Mousavi et al. 2022).

3.2.8 Perspectives

Over the last three decades, tremendous progress has been made in our understanding of angiogenesis and microvascular remodeling, and, with recent technological advances in molecular biology tools and imaging modalities, more discoveries are certainly on the horizon. RNA sequencing (RNA-Seq) approaches including spatial transcriptomics and serial block face-scanning electron microscopy (SBF-SEM), among others, will likely usher in new insights into the cellular and molecular mechanisms relevant to physiological and pathological microvascular remodeling. Endothelial cell and pericyte heterogeneity across tissues and in different disease states is being appreciated to a greater extent and will inspire the design of therapeutic strategies that account for this cellular diversity. These tools and new perspectives will also be useful in determining the impact of SARS-CoV-2 infection on capillaries and how long COVID syndrome may involve the microcirculation, especially as we move into an era with this new pathogen circulating throughout the human population. Additionally, as plans continue to unfold for human exploration and life beyond our planet, it will be critical to understand how prolonged existence in space affects the microvasculature and associated organ function. Along with these open questions, key limitations remain in translating new findings into clinical therapies to improve patient health and enhance quality of life. Effectively modulating tumor vasculature to increase chemo−/immunotherapy efficacy remains an important goal in the battle against cancer and metastatic burden. Furthermore, in diseases of excess and aberrant microvascular remodeling such as proliferative diabetic retinopathy and wet age-related macular degeneration, novel therapeutic targets such as those related to vessel pruning may be important to develop alongside new delivery platforms. Cutting-edge approaches to stimulate new capillary growth will also be needed in scenarios of tissue ischemia such as peripheral artery disease, myocardial infarction, and stroke, perhaps in conjunction with next-generation tissue engineering and reparative medicine strategies. These may include human iPSC-based systems derived from advances in organoid models, which must also continue to advance, along with other cell-based assays, to incorporate fluid flow and relevant hemodynamics. In overcoming these and other challenges, we will continue building new paths of discovery and our knowledge of angiogenesis and microvascular remodeling for the advancement of human health and longevity.

References

Abdelazim H, Payne LB, Nolan K, Paralkar K, Bradley V, Kanodia R, Gude R, Ward R, Monavarfeshani A, Fox MA, Chappell JC (2022) Pericyte heterogeneity identified by 3D ultrastructural analysis of the microvessel wall. Front Physiol 13:1016382. https://doi.org/10.3389/fphys.2022.1016382

Abramsson A, Kurup S, Busse M, Yamada S, Lindblom P, Schallmeiner E, Stenzel D, Sauvaget D, Ledin J, Ringvall M, Landegren U, Kjellen L, Bondjers G, Li JP, Lindahl U, Spillmann D,

Betsholtz C, Gerhardt H (2007) Defective N-sulfation of heparan sulfate proteoglycans limits PDGF-BB binding and pericyte recruitment in vascular development. Genes Dev 21(3): 316–331

Adair TH, Montani JP (2010) Angiogenesis. Integrated systems physiology: from molecule to function to disease. San Rafael. https://doi.org/10.4199/C00017ED1V01Y201009ISP009

Adams RH, Eichmann A (2010) Axon guidance molecules in vascular patterning. Cold Spring Harb Perspect Biol 2(5):a001875

Afsar B, Afsar RE, Dagel T, Kaya E, Erus S, Ortiz A, Covic A, Kanbay M (2018) Capillary rarefaction from the kidney point of view. Clin Kidney J 11(3):295–301. https://doi.org/10.1093/ckj/sfx133

Akbari E, Spychalski GB, Song JW (2017) Microfluidic approaches to the study of angiogenesis and the microcirculation. Microcirculation 24(5). https://doi.org/10.1111/micc.12363

Allen E, Jabouille A, Rivera LB, Lodewijckx I, Missiaen R, Steri V, Feyen K, Tawney J, Hanahan D, Michael IP, Bergers G (2017) Combined antiangiogenic and anti-PD-L1 therapy stimulates tumor immunity through HEV formation. Sci Transl Med 9(385). https://doi.org/10.1126/scitranslmed.aak9679

Al-Shammari AA, Gaffney EA, Egginton S (2014) Modelling capillary oxygen supply capacity in mixed muscles: capillary domains revisited. J Theor Biol 356:47–61. https://doi.org/10.1016/j.jtbi.2014.04.016

Anderson CR, Ponce AM, Price RJ (2004) Immunohistochemical identification of an extracellular matrix scaffold that microguides capillary sprouting in vivo. J Histochem Cytochem 52(8): 1063–1072. https://doi.org/10.1369/jhc.4A6250.2004

Antonetti DA, Klein R, Gardner TW (2012) Diabetic retinopathy. N Engl J Med 366(13): 1227–1239. https://doi.org/10.1056/NEJMra1005073

Apte RS, Chen DS, Ferrara N (2019) VEGF in Signaling and Disease: Beyond Discovery and Development. Cell 176(6):1248–1264. https://doi.org/10.1016/j.cell.2019.01.021

Arima S, Nishiyama K, Ko T, Arima Y, Hakozaki Y, Sugihara K, Koseki H, Uchijima Y, Kurihara Y, Kurihara H (2011) Angiogenic morphogenesis driven by dynamic and heterogeneous collective endothelial cell movement. Development 138(21):4763–4776

Armulik A, Genove G, Betsholtz C (2011) Pericytes: developmental, physiological, and pathological perspectives, problems, and promises. Dev Cell 21(2):193–215

Aspalter IM, Gordon E, Dubrac A, Ragab A, Narloch J, Vizan P, Geudens I, Collins RT, Franco CA, Abrahams CL, Thurston G, Fruttiger M, Rosewell I, Eichmann A, Gerhardt H (2015) Alk1 and Alk5 inhibition by Nrp1 controls vascular sprouting downstream of Notch. Nat Commun 6: 7264. https://doi.org/10.1038/ncomms8264

Ausprunk DH, Folkman J (1977) Migration and proliferation of endothelial cells in preformed and newly formed blood vessels during tumor angiogenesis. Microvasc Res 14(1):53–65. https://doi.org/10.1016/0026-2862(77)90141-8

Avraamides CJ, Garmy-Susini B, Varner JA (2008) Integrins in angiogenesis and lymphangiogenesis. Nat Rev Cancer 8(8):604–617. https://doi.org/10.1038/nrc2353

Azimi MS, Motherwell JM, Murfee WL (2017) An Ex vivo method for time-lapse imaging of cultured rat mesenteric microvascular networks. J Vis Exp 120. https://doi.org/10.3791/55183

Barrasa-Ramos S, Dessalles CA, Hautefeuille M, Barakat AI (2022) Mechanical regulation of the early stages of angiogenesis. J R Soc Interface 19(197):20220360. https://doi.org/10.1098/rsif.2022.0360

Bautch VL, Stanford WL, Rapoport R, Russell S, Byrum RS, Futch TA (1996) Blood Island formation in attached cultures of murine embryonic stem cells. Dev Dyn 205(1):1–12

Bayless KJ, Johnson GA (2011) Role of the cytoskeleton in formation and maintenance of angiogenic sprouts. J Vasc Res 48(5):369–385. https://doi.org/10.1159/000324751

Bentley K, Gerhardt H, Bates PA (2008) Agent-based simulation of notch-mediated tip cell selection in angiogenic sprout initialisation. J Theor Biol 250(1):25–36

Bentley K, Mariggi G, Gerhardt H, Bates PA (2009) Tipping the balance: robustness of tip cell selection, migration and fusion in angiogenesis. PLoS Comput Biol 5(10):e1000549. https://doi.org/10.1371/journal.pcbi.1000549

Bentley K, Franco CA, Philippides A, Blanco R, Dierkes M, Gebala V, Stanchi F, Jones M, Aspalter IM, Cagna G, Westrom S, Claesson-Welsh L, Vestweber D, Gerhardt H (2014) The role of differential VE-cadherin dynamics in cell rearrangement during angiogenesis. Nat Cell Biol 16(4):309–321. https://doi.org/10.1038/ncb2926

Bergers G, Hanahan D (2008) Modes of resistance to anti-angiogenic therapy. Nat Rev Cancer 8(8): 592–603. https://doi.org/10.1038/nrc2442

Bergers G, Song S (2005) The role of pericytes in blood-vessel formation and maintenance. Neuro-Oncology 7(4):452–464. https://doi.org/10.1215/S1152851705000232

Berthiaume AA, Grant RI, McDowell KP, Underly RG, Hartmann DA, Levy M, Bhat NR, Shih AY (2018) Dynamic Remodeling of Pericytes In Vivo Maintains Capillary Coverage in the Adult Mouse Brain. Cell Rep 22(1):8–16. https://doi.org/10.1016/j.celrep.2017.12.016

Betz C, Lenard A, Belting HG, Affolter M (2016) Cell behaviors and dynamics during angiogenesis. Development 143(13):2249–2260. https://doi.org/10.1242/dev.135616

Biel NM, Lee JA, Sorg BS, Siemann DW (2014) Limitations of the dorsal skinfold window chamber model in evaluating anti-angiogenic therapy during early phase of angiogenesis. Vasc Cell 6:17. https://doi.org/10.1186/2045-824X-6-17

Birbrair A, Zhang T, Wang ZM, Messi ML, Olson JD, Mintz A, Delbono O (2014) Type-2 pericytes participate in normal and tumoral angiogenesis. Am J Physiol Cell Physiol 307(1): C25–C38. https://doi.org/10.1152/ajpcell.00084.2014

Blancas AA, Wong LE, Glaser DE, McCloskey KE (2013) Specialized tip/stalk-like and phalanx-like endothelial cells from embryonic stem cells. Stem Cells Dev 22(9):1398–1407. https://doi.org/10.1089/scd.2012.0376

Blanco R, Gerhardt H (2013) VEGF and Notch in tip and stalk cell selection. Cold Spring Harb Perspect Med 3(1):a006569. https://doi.org/10.1101/cshperspect.a006569

Blum Y, Belting HG, Ellertsdottir E, Herwig L, Luders F, Affolter M (2008) Complex cell rearrangements during intersegmental vessel sprouting and vessel fusion in the zebrafish embryo. Dev Biol 316(2):312–322

Bodnar RJ, Rodgers ME, Chen WC, Wells A (2013) Pericyte regulation of vascular remodeling through the CXC receptor 3. Arterioscler Thromb Vasc Biol 33(12):2818–2829. https://doi.org/10.1161/ATVBAHA.113.302012

Boscolo E, Bischoff J (2009) Vasculogenesis in infantile hemangioma. Angiogenesis 12(2): 197–207. https://doi.org/10.1007/s10456-009-9148-2

Brey EM, Uriel S, Greisler HP, McIntire LV (2005) Therapeutic neovascularization: contributions from bioengineering. Tissue Eng 11(3–4):567–584. https://doi.org/10.1089/ten.2005.11.567

Burri PH, Djonov V (2002) Intussusceptive angiogenesis – the alternative to capillary sprouting. Mol Asp Med 23(6S):S1–S27. https://doi.org/10.1016/s0098-2997(02)00096-1

Byatt TC, Martin P (2023) Parallel repair mechanisms in plants and animals. Dis Model Mech 16(1). https://doi.org/10.1242/dmm.049801

Cahill LS, Rennie MY, Hoggarth J, Yu LX, Rahman A, Kingdom JC, Seed M, Macgowan CK, Sled JG (2018) Feto- and utero-placental vascular adaptations to chronic maternal hypoxia in the mouse. J Physiol 596(15):3285–3297. https://doi.org/10.1113/JP274845

Cao G, O'Brien CD, Zhou Z, Sanders SM, Greenbaum JN, Makrigiannakis A, DeLisser HM (2002) Involvement of human PECAM-1 in angiogenesis and in vitro endothelial cell migration. Am J Physiol Cell Physiol 282(5):C1181–C1190. https://doi.org/10.1152/ajpcell.00524.2001

Cao Y, Hong A, Schulten H, Post MJ (2005) Update on therapeutic neovascularization. Cardiovasc Res 65(3):639–648. https://doi.org/10.1016/j.cardiores.2004.11.020

Cao L, Arany PR, Wang YS, Mooney DJ (2009) Promoting angiogenesis via manipulation of VEGF responsiveness with notch signaling. Biomaterials 30(25):4085–4093

Cardenas R, Martinez-Seoane J, Amero C (2020) Combining experimental data and computational methods for the non-computer specialist. Molecules 25(20). https://doi.org/10.3390/molecules25204783

Carmeliet P, Jain RK (2011) Molecular mechanisms and clinical applications of angiogenesis. Nature 473(7347):298–307

Carmeliet P, Tessier-Lavigne M (2005) Common mechanisms of nerve and blood vessel wiring. Nature 436(7048):193–200

Chappell JC, Taylor SM, Ferrara N, Bautch VL (2009) Local guidance of emerging vessel sprouts requires soluble Flt-1. Dev Cell 17(3):377–386. https://doi.org/10.1016/j.devcel.2009.07.011

Chappell JC, Mouillesseaux KP, Bautch VL (2013) Flt-1 (Vascular endothelial growth factor receptor-1) is essential for the vascular endothelial growth factor-notch feedback loop during angiogenesis. Arterioscler Thromb Vasc Biol 33(8):1952–1959

Chappell JC, Darden J, Payne LB, Fink K, Bautch VL (2019) Blood vessel patterning on retinal astrocytes requires endothelial flt-1 (VEGFR-1). J Dev Biol 7(3). https://doi.org/10.3390/jdb7030018

Chavez MN, Aedo G, Fierro FA, Allende ML, Egana JT (2016) Zebrafish as an Emerging Model Organism to Study Angiogenesis in Development and Regeneration. Front Physiol 7:56. https://doi.org/10.3389/fphys.2016.00056

Chen Q, Jiang L, Li C, Hu D, Bu JW, Cai D, Du JL (2012) Haemodynamics-driven developmental pruning of brain vasculature in zebrafish. PLoS Biol 10(8):e1001374. https://doi.org/10.1371/journal.pbio.1001374

Chen X, Yu W, Zhang J, Fan X, Liu X, Liu Q, Pan S, Dixon RAF, Li P, Yu P, Shi A (2023) Therapeutic angiogenesis and tissue revascularization in ischemic vascular disease. J Biol Eng 17(1):13. https://doi.org/10.1186/s13036-023-00330-2

Cheresh DA (1987) Human endothelial cells synthesize and express an Arg-Gly-Asp-directed adhesion receptor involved in attachment to fibrinogen and von Willebrand factor. Proc Natl Acad Sci USA 84(18):6471–6475. https://doi.org/10.1073/pnas.84.18.6471

Chiaverina G, di Blasio L, Monica V, Accardo M, Palmiero M, Peracino B, Vara-Messler M, Puliafito A, Primo L (2019) Dynamic interplay between pericytes and endothelial cells during sprouting angiogenesis. Cells 8(9). https://doi.org/10.3390/cells8091109

Cleaver O, Tonissen KF, Saha MS, Krieg PA (1997) Neovascularization of the Xenopus embryo. Dev Dyn 210(1):66–77

Coelho-Santos V, Shih AY (2020) Postnatal development of cerebrovascular structure and the neurogliovascular unit. Wiley Interdiscip Rev Dev Biol 9(2):e363. https://doi.org/10.1002/wdev.363

Collin CB, Gebhardt T, Golebiewski M, Karaderi T, Hillemanns M, Khan FM, Salehzadeh-Yazdi-A, Kirschner M, Krobitsch S, Eu-Stands Pm C, Kuepfer L (2022) Computational models for clinical applications in personalized medicine-guidelines and recommendations for data integration and model validation. J Pers Med 12(2). https://doi.org/10.3390/jpm12020166

Colom A, Galgoczy R, Almendros I, Xaubet A, Farre R, Alcaraz J (2014) Oxygen diffusion and consumption in extracellular matrix gels: implications for designing three-dimensional cultures. J Biomed Mater Res A 102(8):2776–2784. https://doi.org/10.1002/jbm.a.34946

Davis GE, Kemp SS (2023) Extracellular matrix regulation of vascular morphogenesis, maturation, and stabilization. Cold Spring Harb Perspect Med 13(4). https://doi.org/10.1101/cshperspect.a041156

Davis GE, Senger DR (2005) Endothelial extracellular matrix: biosynthesis, remodeling, and functions during vascular morphogenesis and neovessel stabilization. Circ Res 97(11): 1093–1107. https://doi.org/10.1161/01.RES.0000191547.64391.e3

De Spiegelaere W, Casteleyn C, Van den Broeck W, Plendl J, Bahramsoltani M, Simoens P, Djonov V, Cornillie P (2012) Intussusceptive angiogenesis: a biologically relevant form of angiogenesis. J Vasc Res 49(5):390–404. https://doi.org/10.1159/000338278

del Toro R, Prahst C, Mathivet T, Siegfried G, Kaminker JS, Larrivee B, Breant C, Duarte A, Takakura N, Fukamizu A, Penninger J, Eichmann A (2010) Identification and functional analysis of endothelial tip cell-enriched genes. Blood 116(19):4025–4033

Diaz-Flores L, Gutierrez R, Garcia MP, Gonzalez-Gomez M, Diaz-Flores L Jr, Carrasco JL, Madrid JF, Rodriguez Bello A (2022) Comparison of the behavior of perivascular cells (Pericytes and CD34+ stromal cell/Telocytes) in sprouting and intussusceptive angiogenesis. Int J Mol Sci 23(16). https://doi.org/10.3390/ijms23169010

Djonov V, Schmid M, Tschanz SA, Burri PH (2000) Intussusceptive angiogenesis: its role in embryonic vascular network formation. Circ Res 86(3):286–292

Djonov V, Baum O, Burri PH (2003) Vascular remodeling by intussusceptive angiogenesis. Cell Tissue Res 314(1):107–117. https://doi.org/10.1007/s00441-003-0784-3

Dodd MS, Papineau D, Grenne T, Slack JF, Rittner M, Pirajno F, O'Neil J, Little CT (2017) Evidence for early life in Earth's oldest hydrothermal vent precipitates. Nature 543(7643): 60–64. https://doi.org/10.1038/nature21377

Doetschman TC, Eistetter H, Katz M, Schmidt W, Kemler R (1985) The in vitro development of blastocyst-derived embryonic stem cell lines: formation of visceral yolk sac, blood islands and myocardium. J Embryol Exp Morphol 87:27–45

Drake CJ, Fleming PA (2000) Vasculogenesis in the day 6.5 to 9.5 mouse embryo. Blood 95(5): 1671–1679

Drake CJ, Little CD (1995) Exogenous vascular endothelial growth factor induces malformed and hyperfused vessels during embryonic neovascularization. Proc Natl Acad Sci USA 92(17): 7657–7661

Du Cheyne C, Smeets M, De Spiegelaere W (2021) Techniques used to assess intussusceptive angiogenesis: A systematic review. Dev Dyn 250(12):1704–1716. https://doi.org/10.1002/dvdy.382

Dudley AC, Griffioen AW (2023) Pathological angiogenesis: mechanisms and therapeutic strategies. Angiogenesis 26(3):313–347. https://doi.org/10.1007/s10456-023-09876-7

Dumas SJ, Meta E, Borri M, Goveia J, Rohlenova K, Conchinha NV, Falkenberg K, Teuwen LA, de Rooij L, Kalucka J, Chen R, Khan S, Taverna F, Lu W, Parys M, De Legher C, Vinckier S, Karakach TK, Schoonjans L, Lin L, Bolund L, Dewerchin M, Eelen G, Rabelink TJ, Li X, Luo Y, Carmeliet P (2020) Single-Cell RNA Sequencing Reveals Renal Endothelium Heterogeneity and Metabolic Adaptation to Water Deprivation. J Am Soc Nephrol 31(1):118–138. https://doi.org/10.1681/ASN.2019080832

Eelen G, Treps L, Li X, Carmeliet P (2020) Basic and Therapeutic Aspects of Angiogenesis Updated. Circ Res 127(2):310–329. https://doi.org/10.1161/CIRCRESAHA.120.316851

Egginton S, Brown MD (2014) In Memoriam: Olga Hudlicka (11.07.26-03.05.14). Microcirculation 21. https://doi.org/10.1111/micc.12151

Egginton S, Zhou AL, Brown MD, Hudlicka O (2001) Unorthodox angiogenesis in skeletal muscle. Cardiovasc Res 49(3):634–646. https://doi.org/10.1016/s0008-6363(00)00282-0

Eglinger J, Karsjens H, Lammert E (2017) Quantitative assessment of angiogenesis and pericyte coverage in human cell-derived vascular sprouts. Inflammation and Regeneration 37(1):2. https://doi.org/10.1186/s41232-016-0033-2

Eilken HM, Dieguez-Hurtado R, Schmidt I, Nakayama M, Jeong HW, Arf H, Adams S, Ferrara N, Adams RH (2017) Pericytes regulate VEGF-induced endothelial sprouting through VEGFR1. Nat Commun 8(1):1574. https://doi.org/10.1038/s41467-017-01738-3

Escudero-Esparza A, Jiang WG, Martin TA (2012) Claudin-5 participates in the regulation of endothelial cell motility. Mol Cell Biochem 362(1–2):71–85. https://doi.org/10.1007/s11010-011-1129-2

Falcon BL, Hashizume H, Koumoutsakos P, Chou J, Bready JV, Coxon A, Oliner JD, McDonald DM (2009) Contrasting actions of selective inhibitors of angiopoietin-1 and angiopoietin-2 on the normalization of tumor blood vessels. Am J Pathol 175(5):2159–2170. https://doi.org/10.2353/ajpath.2009.090391

Fantin A, Vieira JM, Gestri G, Denti L, Schwarz Q, Prykhozhij S, Peri F, Wilson SW, Ruhrberg C (2010) Tissue macrophages act as cellular chaperones for vascular anastomosis downstream of VEGF-mediated endothelial tip cell induction. Blood 116(5):829–840

Feraud O, Cao Y, Vittet D (2001) Embryonic stem cell-derived embryoid bodies development in collagen gels recapitulates sprouting angiogenesis. Lab Investig 81(12):1669–1681

Ferrara N (2004) Vascular endothelial growth factor: basic science and clinical progress. Endocr Rev 25(4):581–611. https://doi.org/10.1210/er.2003-0027

Ferrara N (2005) VEGF as a therapeutic target in cancer. Oncology 69(Suppl 3):11–16

Folkman J (2008) History of Angiogenesis. In: Figg WD, Folkman J (eds) Angiogenesis: An Integrative Approach From Science to Medicine. Springer US, Boston, pp 1–14. https://doi.org/10.1007/978-0-387-71518-6_1

Franco CA, Jones ML, Bernabeu MO, Geudens I, Mathivet T, Rosa A, Lopes FM, Lima AP, Ragab A, Collins RT, Phng LK, Coveney PV, Gerhardt H (2015) Dynamic endothelial cell rearrangements drive developmental vessel regression. PLoS Biol 13(4):e1002125. https://doi.org/10.1371/journal.pbio.1002125

Froger N, Matonti F, Roubeix C, Forster V, Ivkovic I, Brunel N, Baudouin C, Sahel JA, Picaud S (2020) VEGF is an autocrine/paracrine neuroprotective factor for injured retinal ganglion neurons. Sci Rep 10(1):12409. https://doi.org/10.1038/s41598-020-68488-z

Frost S, Nolde JM, Chan J, Joyson A, Gregory C, Carnagarin R, Herat LY, Matthews VB, Robinson L, Vignarajan J, Prentice D, Kanagasingam Y, Schlaich MP (2021) Retinal capillary rarefaction is associated with arterial and kidney damage in hypertension. Sci Rep 11(1):1001. https://doi.org/10.1038/s41598-020-79594-3

Funahashi Y, Shawber CJ, Vorontchikhina M, Sharma A, Outtz HH, Kitajewski J (2010) Notch regulates the angiogenic response via induction of VEGFR-1. J Angiogenes Res 2(1):3

Funahashi Y, Shawber CJ, Sharma A, Kanamaru E, Choi YK, Kitajewski J (2011) Notch modulates VEGF action in endothelial cells by inducing Matrix Metalloprotease activity. Vasc Cell 3(1):2. https://doi.org/10.1186/2045-824X-3-2

Gale NW, Dominguez MG, Noguera I, Pan L, Hughes V, Valenzuela DM, Murphy AJ, Adams NC, Lin HC, Holash J, Thurston G, Yancopoulos GD (2004) Haploinsufficiency of delta-like 4 ligand results in embryonic lethality due to major defects in arterial and vascular development. Proc Natl Acad Sci USA 101(45):15949–15954

Gebala V, Collins R, Geudens I, Phng LK, Gerhardt H (2016) Blood flow drives lumen formation by inverse membrane blebbing during angiogenesis in vivo. Nat Cell Biol 18(4):443–450. https://doi.org/10.1038/ncb3320

Gerber HP, Condorelli F, Park J, Ferrara N (1997) Differential transcriptional regulation of the two vascular endothelial growth factor receptor genes. Flt-1, but not Flk-1/KDR, is up-regulated by hypoxia. J Biol Chem 272(38):23659–23667. https://doi.org/10.1074/jbc.272.38.23659

Gerhardt H (2008) VEGF and endothelial guidance in angiogenic sprouting. Organogenesis 4(4): 241–246

Gerhardt H, Betsholtz C (2003) Endothelial-pericyte interactions in angiogenesis. Cell Tissue Res 314(1):15–23. https://doi.org/10.1007/s00441-003-0745-x

Gerhardt H, Golding M, Fruttiger M, Ruhrberg C, Lundkvist A, Abramsson A, Jeltsch M, Mitchell C, Alitalo K, Shima D, Betsholtz C (2003) VEGF guides angiogenic sprouting utilizing endothelial tip cell filopodia. J Cell Biol 161(6):1163–1177

Gerhardt H, Ruhrberg C, Abramsson A, Fujisawa H, Shima D, Betsholtz C (2004) Neuropilin-1 is required for endothelial tip cell guidance in the developing central nervous system. Dev Dyn 231(3):503–509. https://doi.org/10.1002/dvdy.20148

Ghabrial AS, Krasnow MA (2006) Social interactions among epithelial cells during tracheal branching morphogenesis. Nature 441(7094):746–749

Ghaffari S, Leask RL, Jones EA (2015) Flow dynamics control the location of sprouting and direct elongation during developmental angiogenesis. Development 142(23):4151–4157. https://doi.org/10.1242/dev.128058

Gianni-Barrera R, Bartolomeo M, Vollmar B, Djonov V, Banfi A (2014) Split for the cure: VEGF, PDGF-BB and intussusception in therapeutic angiogenesis. Biochem Soc Trans 42(6): 1637–1642. https://doi.org/10.1042/BST20140234

Glaser BM, D'Amore PA, Seppa H, Seppa S, Schiffmann E (1980) Adult tissues contain chemoattractants for vascular endothelial cells. Nature 288(5790):483–484. https://doi.org/10.1038/288483a0

Goel S, Duda DG, Xu L, Munn LL, Boucher Y, Fukumura D, Jain RK (2011) Normalization of the vasculature for treatment of cancer and other diseases. Physiol Rev 91(3):1071–1121. https://doi.org/10.1152/physrev.00038.2010

Goodwin AM (2007) In vitro assays of angiogenesis for assessment of angiogenic and anti-angiogenic agents. Microvasc Res 74(2–3):172–183. https://doi.org/10.1016/j.mvr.2007.05.006

Greenberg JI, Cheresh DA (2009) VEGF as an inhibitor of tumor vessel maturation: implications for cancer therapy. Expert Opin Biol Ther 9(11):1347–1356. https://doi.org/10.1517/14712590903208883

Hanahan D, Weinberg RA (2011) Hallmarks of cancer: the next generation. Cell 144(5):646–674. https://doi.org/10.1016/j.cell.2011.02.013

Hansen-Smith FM, Hudlicka O, Egginton S (1996) In vivo angiogenesis in adult rat skeletal muscle: early changes in capillary network architecture and ultrastructure. Cell Tissue Res 286(1):123–136. https://doi.org/10.1007/s004410050681

Harrington LS, Sainson RC, Williams CK, Taylor JM, Shi W, Li JL, Harris AL (2008) Regulation of multiple angiogenic pathways by Dll4 and Notch in human umbilical vein endothelial cells. Microvasc Res 75(2):144–154

Harris AL (2002) Hypoxia – a key regulatory factor in tumour growth. Nat Rev Cancer 2(1):38–47. https://doi.org/10.1038/nrc704

Hartnett ME (2015) Pathophysiology and mechanisms of severe retinopathy of prematurity. Ophthalmology 122(1):200–210. https://doi.org/10.1016/j.ophtha.2014.07.050

Hashambhoy YL, Chappell JC, Peirce SM, Bautch VL, Mac Gabhann F (2011) Computational modeling of interacting VEGF and soluble VEGF receptor concentration gradients. Front Physiol 2:62. https://doi.org/10.3389/fphys.2011.00062

Heath VL, Bicknell R (2009) Anticancer strategies involving the vasculature. Nat Rev Clin Oncol 6(7):395–404

Hellstrom M, Phng LK, Hofmann JJ, Wallgard E, Coultas L, Lindblom P, Alva J, Nilsson AK, Karlsson L, Gaiano N, Yoon K, Rossant J, Iruela-Arispe ML, Kalen M, Gerhardt H, Betsholtz C (2007) Dll4 signalling through Notch1 regulates formation of tip cells during angiogenesis. Nature 445(7129):776–780

Helm CL, Fleury ME, Zisch AH, Boschetti F, Swartz MA (2005) Synergy between interstitial flow and VEGF directs capillary morphogenesis in vitro through a gradient amplification mechanism. Proc Natl Acad Sci USA 102(44):15779–15784

Iruela-Arispe ML, Davis GE (2009) Cellular and molecular mechanisms of vascular lumen formation. Dev Cell 16(2):222–231

Itoh F, Itoh S, Goumans MJ, Valdimarsdottir G, Iso T, Dotto GP, Hamamori Y, Kedes L, Kato M, ten Dijke PP (2004) Synergy and antagonism between Notch and BMP receptor signaling pathways in endothelial cells. EMBO J 23(3):541–551. https://doi.org/10.1038/sj.emboj.7600065

Jain RK (2005) Normalization of tumor vasculature: an emerging concept in antiangiogenic therapy. Science 307(5706):58–62

Jakobsson L, Domogatskaya A, Tryggvason K, Edgar D, Claesson-Welsh L (2007) Laminin deposition is dispensable for vasculogenesis but regulates blood vessel diameter independent of flow. FASEB J 22:1530

Jakobsson L, Bentley K, Gerhardt H (2009) VEGFRs and Notch: a dynamic collaboration in vascular patterning. Biochem Soc Trans 37(Pt 6):1233–1236

Jakobsson L, Franco CA, Bentley K, Collins RT, Ponsioen B, Aspalter IM, Rosewell I, Busse M, Thurston G, Medvinsky A, Schulte-Merker S, Gerhardt H (2010) Endothelial cells dynamically compete for the tip cell position during angiogenic sprouting. Nat Cell Biol 12(10):943–953

Kalucka J, de Rooij L, Goveia J, Rohlenova K, Dumas SJ, Meta E, Conchinha NV, Taverna F, Teuwen LA, Veys K, Garcia-Caballero M, Khan S, Geldhof V, Sokol L, Chen R, Treps L, Borri M, de Zeeuw P, Dubois C, Karakach TK, Falkenberg KD, Parys M, Yin X, Vinckier S, Du Y, Fenton RA, Schoonjans L, Dewerchin M, Eelen G, Thienpont B, Lin L, Bolund L, Li X, Luo Y, Carmeliet P (2020) Single-Cell Transcriptome Atlas of Murine Endothelial Cells. Cell 180(4):764–779. https://doi.org/10.1016/j.cell.2020.01.015

Kam CY, Singh ID, Gonzalez DG, Matte-Martone C, Sola P, Solanas G, Bonjoch J, Marsh E, Hirschi KK, Greco V (2023) Mechanisms of skin vascular maturation and maintenance captured by longitudinal imaging of live mice. Cell 186(11):2345–2360. https://doi.org/10.1016/j.cell.2023.04.017

Kamba T, McDonald DM (2007) Mechanisms of adverse effects of anti-VEGF therapy for cancer. Br J Cancer 96(12):1788–1795. https://doi.org/10.1038/sj.bjc.6603813

Kamei M, Saunders WB, Bayless KJ, Dye L, Davis GE, Weinstein BM (2006) Endothelial tubes assemble from intracellular vacuoles in vivo. Nature 442(7101):453–456

Karthik S, Djukic T, Kim JD, Zuber B, Makanya A, Odriozola A, Hlushchuk R, Filipovic N, Jin SW, Djonov V (2018) Synergistic interaction of sprouting and intussusceptive angiogenesis during zebrafish caudal vein plexus development. Sci Rep 8(1):9840. https://doi.org/10.1038/s41598-018-27791-6

Kelly-Goss MR, Winterer ER, Stapor PC, Yang M, Sweat RS, Stallcup WB, Schmid-Schonbein GW, Murfee WL (2012) Cell proliferation along vascular islands during microvascular network growth. BMC Physiol 12:7. https://doi.org/10.1186/1472-6793-12-7

Kelly-Goss MR, Sweat RS, Azimi MS, Murfee WL (2013) Vascular islands during microvascular regression and regrowth in adult networks. Front Physiol 4:108. https://doi.org/10.3389/fphys.2013.00108

Kelly-Goss MR, Sweat RS, Stapor PC, Peirce SM, Murfee WL (2014) Targeting pericytes for angiogenic therapies. Microcirculation 21(4):345–357. https://doi.org/10.1111/micc.12107

Kilarski WW, Samolov B, Petersson L, Kvanta A, Gerwins P (2009) Biomechanical regulation of blood vessel growth during tissue vascularization. Nat Med 15(6):657–664

Kochhan E, Lenard A, Ellertsdottir E, Herwig L, Affolter M, Belting HG, Siekmann AF (2013) Blood flow changes coincide with cellular rearrangements during blood vessel pruning in zebrafish embryos. PLoS One 8(10):e75060. https://doi.org/10.1371/journal.pone.0075060

Konerding MA, Turhan A, Ravnic DJ, Lin M, Fuchs C, Secomb TW, Tsuda A, Mentzer SJ (2010) Inflammation-induced intussusceptive angiogenesis in murine colitis. Anat Rec (Hoboken) 293(5):849–857. https://doi.org/10.1002/ar.21110

Korn C, Augustin HG (2015) Mechanisms of Vessel Pruning and Regression. Dev Cell 34(1):5–17. https://doi.org/10.1016/j.devcel.2015.06.004

Krock BL, Skuli N, Simon MC (2011) Hypoxia-induced angiogenesis: good and evil. Genes Cancer 2(12):1117–1133. https://doi.org/10.1177/1947601911423654

Krueger J, Liu D, Scholz K, Zimmer A, Shi Y, Klein C, Siekmann A, Schulte-Merker S, Cudmore M, Ahmed A, le Noble F (2011) Flt1 acts as a negative regulator of tip cell formation and branching morphogenesis in the zebrafish embryo. Development 138(10):2111–2120

Kuhn C, Checa S (2019) Computational Modeling to Quantify the Contributions of VEGFR1, VEGFR2, and Lateral Inhibition in Sprouting Angiogenesis. Front Physiol 10:288. https://doi.org/10.3389/fphys.2019.00288

Kushner EJ, Ferro LS, Liu JY, Durrant JR, Rogers SL, Dudley AC, Bautch VL (2014) Excess centrosomes disrupt endothelial cell migration via centrosome scattering. J Cell Biol 206(2):257–272. https://doi.org/10.1083/jcb.201311013

Kushner EJ, Ferro LS, Yu Z, Bautch VL (2016) Excess centrosomes perturb dynamic endothelial cell repolarization during blood vessel formation. Mol Biol Cell 27(12):1911–1920. https://doi.org/10.1091/mbc.E15-09-0645

Larrivee B, Prahst C, Gordon E, del Toro R, Mathivet T, Duarte A, Simons M, Eichmann A (2012) ALK1 signaling inhibits angiogenesis by cooperating with the Notch pathway. Dev Cell 22(3): 489–500

Lawson ND, Vogel AM, Weinstein BM (2002) sonic hedgehog and vascular endothelial growth factor act upstream of the Notch pathway during arterial endothelial differentiation. Dev Cell 3(1):127–136

Lee CY, Bautch VL (2011) Ups and downs of guided vessel sprouting: the role of polarity. Physiology (Bethesda) 26(5):326–333. https://doi.org/10.1152/physiol.00018.2011

Lee S, Valmikinathan CM, Byun J, Kim S, Lee G, Mokarram N, Pai SB, Um E, Bellamkonda RV, Yoon YS (2015) Enhanced therapeutic neovascularization by CD31-expressing cells and embryonic stem cell-derived endothelial cells engineered with chitosan hydrogel containing VEGF-releasing microtubes. Biomaterials 63:158–167. https://doi.org/10.1016/j.biomaterials.2015.06.009

Lee WS, Yang H, Chon HJ, Kim C (2020) Combination of anti-angiogenic therapy and immune checkpoint blockade normalizes vascular-immune crosstalk to potentiate cancer immunity. Exp Mol Med 52(9):1475–1485. https://doi.org/10.1038/s12276-020-00500-y

Lee HW, Shin JH, Simons M (2022) Flow goes forward and cells step backward: endothelial migration. Exp Mol Med 54(6):711–719. https://doi.org/10.1038/s12276-022-00785-1

Lenard A, Ellertsdottir E, Herwig L, Krudewig A, Sauteur L, Belting HG, Affolter M (2013) In vivo analysis reveals a highly stereotypic morphogenetic pathway of vascular anastomosis. Dev Cell 25(5):492–506

Lenard A, Daetwyler S, Betz C, Ellertsdottir E, Belting HG, Huisken J, Affolter M (2015) Endothelial cell self-fusion during vascular pruning. PLoS Biol 13(4):e1002126. https://doi.org/10.1371/journal.pbio.1002126

Leslie JD, Ariza-McNaughton L, Bermange AL, McAdow R, Johnson SL, Lewis J (2007) Endothelial signalling by the Notch ligand Delta-like 4 restricts angiogenesis. Development 134(5):839–844

Liao D, Johnson RS (2007) Hypoxia: a key regulator of angiogenesis in cancer. Cancer Metastasis Rev 26(2):281–290. https://doi.org/10.1007/s10555-007-9066-y

Lindblom P, Gerhardt H, Liebner S, Abramsson A, Enge M, Hellstrom M, Backstrom G, Fredriksson S, Landegren U, Nystrom HC, Bergstrom G, Dejana E, Ostman A, Lindahl P, Betsholtz C (2003) Endothelial PDGF-B retention is required for proper investment of pericytes in the microvessel wall. Genes Dev 17(15):1835–1840. https://doi.org/10.1101/gad.266803

Lobov IB, Renard RA, Papadopoulos N, Gale NW, Thurston G, Yancopoulos GD, Wiegand SJ (2007) Delta-like ligand 4 (Dll4) is induced by VEGF as a negative regulator of angiogenic sprouting. Proc Natl Acad Sci USA 104(9):3219–3224

Lobov IB, Cheung E, Wudali R, Cao J, Halasz G, Wei Y, Economides A, Lin HC, Papadopoulos N, Yancopoulos GD, Wiegand SJ (2011) The Dll4/Notch pathway controls postangiogenic blood vessel remodeling and regression by modulating vasoconstriction and blood flow. Blood 117(24):6728–6737

Logothetidou A, De Spiegelaere W, Vandecasteele T, Tschulenk W, Walter I, Van den Broeck W, Cornillie P (2018) Intussusceptive Pillar Formation in Developing Porcine Glomeruli. J Vasc Res 55(5):278–286. https://doi.org/10.1159/000490905

Logsdon EA, Finley SD, Popel AS, Mac Gabhann F (2014) A systems biology view of blood vessel growth and remodelling. J Cell Mol Med 18(8):1491–1508. https://doi.org/10.1111/jcmm.12164

Lu X, Le Noble F, Yuan L, Jiang Q, De Lafarge B, Sugiyama D, Breant C, Claes F, De Smet F, Thomas JL, Autiero M, Carmeliet P, Tessier-Lavigne M, Eichmann A (2004) The netrin receptor UNC5B mediates guidance events controlling morphogenesis of the vascular system. Nature 432(7014):179–186

Lu P, Wu B, Wang Y, Zhang J, Zhou B (2023) A sandwiched ventricular explant assay to model mouse coronary angiogenesis ex vivo. STAR Protoc 4(4):102619. https://doi.org/10.1016/j.xpro.2023.102619

Lucitti JL, Jones EA, Huang C, Chen J, Fraser SE, Dickinson ME (2007) Vascular remodeling of the mouse yolk sac requires hemodynamic force. Development 134(18):3317–3326

Lugano R, Ramachandran M, Dimberg A (2020) Tumor angiogenesis: causes, consequences, challenges and opportunities. Cell Mol Life Sci 77(9):1745–1770. https://doi.org/10.1007/s00018-019-03351-7

Luttun A, Tjwa M, Moons L, Wu Y, Angelillo-Scherrer A, Liao F, Nagy JA, Hooper A, Priller J, De Klerck B, Compernolle V, Daci E, Bohlen P, Dewerchin M, Herbert JM, Fava R, Matthys P, Carmeliet G, Collen D, Dvorak HF, Hicklin DJ, Carmeliet P (2002) Revascularization of ischemic tissues by PlGF treatment, and inhibition of tumor angiogenesis, arthritis and atherosclerosis by anti-Flt1. Nat Med 8(8):831–840. https://doi.org/10.1038/nm731

Ma S, Pradeep S, Hu W, Zhang D, Coleman R, Sood A (2018) The role of tumor microenvironment in resistance to anti-angiogenic therapy. F1000Res 7:326. https://doi.org/10.12688/f1000research.11771.1

Mac Gabhann F, Popel AS (2007) Interactions of VEGF isoforms with VEGFR-1, VEGFR-2, and neuropilin in vivo: a computational model of human skeletal muscle. Am J Physiol Heart Circ Physiol 292(1):H459–H474

Mac Gabhann F, Popel AS (2008) Systems biology of vascular endothelial growth factors. Microcirculation 15(8):715–738

Makanya AN, Styp-Rekowska B, Dimova I, Djonov V (2015) Avian area vasculosa and CAM as rapid in vivo pro-angiogenic and antiangiogenic models. Methods Mol Biol 1214:185–196. https://doi.org/10.1007/978-1-4939-1462-3_11

Marchand M, Monnot C, Muller L, Germain S (2019) Extracellular matrix scaffolding in angiogenesis and capillary homeostasis. Semin Cell Dev Biol 89:147–156. https://doi.org/10.1016/j.semcdb.2018.08.007

Maroski J, Vorderwulbecke BJ, Fiedorowicz K, Da Silva-Azevedo L, Siegel G, Marki A, Pries AR, Zakrzewicz A (2011) Shear stress increases endothelial hyaluronan synthase 2 and hyaluronan synthesis especially in regard to an atheroprotective flow profile. Exp Physiol 96(9):977–986. https://doi.org/10.1113/expphysiol.2010.056051

Marsico G, Martin-Saldana S, Pandit A (2021) Therapeutic Biomaterial Approaches to Alleviate Chronic Limb Threatening Ischemia. Adv Sci (Weinh) 8(7):2003119. https://doi.org/10.1002/advs.202003119

Martin P, Gurevich DB (2021) Macrophage regulation of angiogenesis in health and disease. Semin Cell Dev Biol 119:101–110. https://doi.org/10.1016/j.semcdb.2021.06.010

Martin M, Veloso A, Wu J, Katrukha EA, Akhmanova A (2018) Control of endothelial cell polarity and sprouting angiogenesis by non-centrosomal microtubules. elife 7. https://doi.org/10.7554/eLife.33864

Marxsen JH, Stengel P, Doege K, Heikkinen P, Jokilehto T, Wagner T, Jelkmann W, Jaakkola P, Metzen E (2004) Hypoxia-inducible factor-1 (HIF-1) promotes its degradation by induction of HIF-alpha-prolyl-4-hydroxylases. Biochem J 381(Pt 3):761–767. https://doi.org/10.1042/BJ20040620

Matsui M, Tabata Y (2012) Enhanced angiogenesis by multiple release of platelet-rich plasma contents and basic fibroblast growth factor from gelatin hydrogels. Acta Biomater 8(5):1792–1801. https://doi.org/10.1016/j.actbio.2012.01.016

Mensah SA, Nersesyan AA, Harding IC, Lee CI, Tan X, Banerjee S, Niedre M, Torchilin VP, Ebong EE (2020) Flow-regulated endothelial glycocalyx determines metastatic cancer cell activity. FASEB J 34(5):6166–6184. https://doi.org/10.1096/fj.201901920R

Mentzer SJ, Konerding MA (2014) Intussusceptive angiogenesis: expansion and remodeling of microvascular networks. Angiogenesis 17(3):499–509. https://doi.org/10.1007/s10456-014-9428-3

Meyer RD, Mohammadi M, Rahimi N (2006) A single amino acid substitution in the activation loop defines the decoy characteristic of VEGFR-1/FLT-1. J Biol Chem 281(2):867–875

Mongiat M, Andreuzzi E, Tarticchio G, Paulitti A (2016) Extracellular matrix, a hard player in angiogenesis. Int J Mol Sci 17(11). https://doi.org/10.3390/ijms17111822

Moore KH, Chapman H, George EM (2020) Unfractionated heparin displaces sFlt-1 from the placental extracellular matrix. Biol Sex Differ 11(1):34. https://doi.org/10.1186/s13293-020-00311-w

Moreira-Soares M, Coimbra R, Rebelo L, Carvalho J et al (2018) Angiogenic Factors produced by Hypoxic Cells are a leading driver of Anastomoses in Sprouting Angiogenesis-a computational study. Sci Rep 8(1):8726. https://doi.org/10.1038/s41598-018-27034-8

Motherwell JM, Rozenblum M, Katakam PVG, Murfee WL (2019) Bioreactor System to Perfuse Mesentery Microvascular Networks and Study Flow Effects During Angiogenesis. Tissue Eng Part C Methods 25(8):447–458. https://doi.org/10.1089/ten.TEC.2019.0119

Motlana MK, Ngoepe MN (2023) Computational fluid dynamics (CFD) model for analysing the role of shear stress in angiogenesis in rheumatoid arthritis. Int J Mol Sci 24(9). https://doi.org/10.3390/ijms24097886

Mouillesseaux KP, Wiley DS, Saunders LM, Wylie LA, Kushner EJ, Chong DC, Citrin KM, Barber AT, Park Y, Kim JD, Samsa LA, Kim J, Liu J, Jin SW, Bautch VL (2016) Notch regulates BMP responsiveness and lateral branching in vessel networks via SMAD6. Nat Commun 7:13247. https://doi.org/10.1038/ncomms13247

Mousavi M, Manshadi MD, Soltani M, Kashkooli FM, Rahmim A, Mosavi A, Kvasnica M, Atkinson PM, Kovacs L, Koltay A, Kiss N, Adeli H (2022) Modeling the efficacy of different anti-angiogenic drugs on treatment of solid tumors using 3D computational modeling and machine learning. Comput Biol Med 146:105511. https://doi.org/10.1016/j.compbiomed.2022.105511

Murfee WL, Peirce SM (2017) Microfluidics technologies and approaches for studying the microcirculation. Microcirculation 24(5). https://doi.org/10.1111/micc.12377

Nakatsu MN, Hughes CC (2008) An optimized three-dimensional in vitro model for the analysis of angiogenesis. Methods Enzymol 443:65–82

Nesmith JE, Chappell JC, Cluceru JG, Bautch VL (2017) Blood vessel anastomosis is spatially regulated by Flt1 during angiogenesis. Development 144(5):889–896. https://doi.org/10.1242/dev.145672

Nitzsche B, Rong WW, Goede A, Hoffmann B, Scarpa F, Kuebler WM, Secomb TW, Pries AR (2022) Coalescent angiogenesis-evidence for a novel concept of vascular network maturation. Angiogenesis 25(1):35–45. https://doi.org/10.1007/s10456-021-09824-3

Noguera-Troise I, Daly C, Papadopoulos NJ, Coetzee S, Boland P, Gale NW, Lin HC, Yancopoulos GD, Thurston G (2006) Blockade of Dll4 inhibits tumour growth by promoting non-productive angiogenesis. Nature 444(7122):1032–1037

Olsson V, Butenko MA (2018) Abscission in plants. Curr Biol 28(8):R338–R339. https://doi.org/10.1016/j.cub.2018.02.069

Osiaevi I, Schulze A, Evers G, Harmening K, Vink H, Kumpers P, Mohr M, Rovas A (2023) Persistent capillary rarefication in long COVID syndrome. Angiogenesis 26(1):53–61. https://doi.org/10.1007/s10456-022-09850-9

Outtz HH, Tattersall IW, Kofler NM, Steinbach N, Kitajewski J (2011) Notch1 controls macrophage recruitment and Notch signaling is activated at sites of endothelial cell anastomosis during retinal angiogenesis in mice. Blood 118(12):3436–3439. https://doi.org/10.1182/blood-2010-12-327015

Ozawa CR, Banfi A, Glazer NL, Thurston G, Springer ML, Kraft PE, McDonald DM, Blau HM (2004) Microenvironmental VEGF concentration, not total dose, determines a threshold between normal and aberrant angiogenesis. J Clin Invest 113(4):516–527. https://doi.org/10.1172/JCI18420

Paku S, Dezso K, Bugyik E, Tovari J, Timar J, Nagy P, Laszlo V, Klepetko W, Dome B (2011) A new mechanism for pillar formation during tumor-induced intussusceptive angiogenesis: inverse sprouting. Am J Pathol 179(3):1573–1585. https://doi.org/10.1016/j.ajpath.2011.05.033

Pal A, Vernon BL, Nikkhah M (2018) Therapeutic neovascularization promoted by injectable hydrogels. Bioact Mater 3(4):389–400. https://doi.org/10.1016/j.bioactmat.2018.05.002

Papineau D, She Z, Dodd MS, Iacoviello F, Slack JF, Hauri E, Shearing P, Little CTS (2022) Metabolically diverse primordial microbial communities in Earth's oldest seafloor-hydrothermal jasper. Sci Adv 8(15):eabm2296. https://doi.org/10.1126/sciadv.abm2296

Payne LB, Hoque M, Houk C, Darden J, Chappell JC (2020) Pericytes in Vascular Development. Curr Tissue Microenviron Rep 1(3):143–154. https://doi.org/10.1007/s43152-020-00014-9

Payne LB, Darden J, Suarez-Martinez AD, Zhao H, Hendricks A, Hartland C, Chong D, Kushner EJ, Murfee WL, Chappell JC (2021) Pericyte migration and proliferation are tightly synchronized to endothelial cell sprouting dynamics. Integr Biol (Camb) 13(2):31–43. https://doi.org/10.1093/intbio/zyaa027

Peirce SM (2008) Computational and mathematical modeling of angiogenesis. Microcirculation 15(8):739–751

Peirce SM, Mac Gabhann F, Bautch VL (2012) Integration of experimental and computational approaches to sprouting angiogenesis. Curr Opin Hematol 19(3):184–191. https://doi.org/10.1097/MOH.0b013e3283523ea6

Pelton JC, Wright CE, Leitges M, Bautch VL (2014) Multiple endothelial cells constitute the tip of developing blood vessels and polarize to promote lumen formation. Development 141(21): 4121–4126. https://doi.org/10.1242/dev.110296

Penn JS, Madan A, Caldwell RB, Bartoli M, Caldwell RW, Hartnett ME (2008) Vascular endothelial growth factor in eye disease. Prog Retin Eye Res 27(4):331–371. https://doi.org/10.1016/j.preteyeres.2008.05.001

Phng LK, Stanchi F, Gerhardt H (2013) Filopodia are dispensable for endothelial tip cell guidance. Development 140(19):4031–4040. https://doi.org/10.1242/dev.097352

Poole DC, Kano Y, Koga S, Musch TI (2021) August Krogh: Muscle capillary function and oxygen delivery. Comp Biochem Physiol A Mol Integr Physiol 253:110852. https://doi.org/10.1016/j.cbpa.2020.110852

Potente M, Makinen T (2017) Vascular heterogeneity and specialization in development and disease. Nat Rev Mol Cell Biol 18(8):477–494. https://doi.org/10.1038/nrm.2017.36

Radhakrishnan D, Shanmukhan AP, Kareem A, Mathew MM, Varaparambathu V, Aiyaz M, Radha RK, Mekala KR, Shaji A, Prasad K (2021) Age, Wound Size, and Position of Injury - Dependent Vascular Regeneration Assay in Growing Leaves. Bio Protoc 11(9):e4010. https://doi.org/10.21769/BioProtoc.4010

Rahbari NN, Kedrin D, Incio J, Liu H, Ho WW, Nia HT, Edrich CM, Jung K, Daubriac J, Chen I, Heishi T, Martin JD, Huang Y, Maimon N, Reissfelder C, Weitz J, Boucher Y, Clark JW, Grodzinsky AJ, Duda DG, Jain RK, Fukumura D (2016) Anti-VEGF therapy induces ECM remodeling and mechanical barriers to therapy in colorectal cancer liver metastases. Sci Transl Med 8(360):360ra135. https://doi.org/10.1126/scitranslmed.aaf5219

Ramjiawan RR, Griffioen AW, Duda DG (2017) Anti-angiogenesis for cancer revisited: Is there a role for combinations with immunotherapy? Angiogenesis 20(2):185–204. https://doi.org/10.1007/s10456-017-9552-y

Rauff A, LaBelle SA, Strobel HA, Hoying JB, Weiss JA (2019) Imaging the Dynamic Interaction Between Sprouting Microvessels and the Extracellular Matrix. Front Physiol 10:1011. https://doi.org/10.3389/fphys.2019.01011

Rezzola S, Belleri M, Ribatti D, Costagliola C, Presta M, Semeraro F (2013) A novel ex vivo murine retina angiogenesis (EMRA) assay. Exp Eye Res 112:51–56. https://doi.org/10.1016/j.exer.2013.04.014

Ribatti D (2009) William Harvey and the discovery of the circulation of the blood. J Angiogenes Res 1:3. https://doi.org/10.1186/2040-2384-1-3

Ribatti D (2016) Tumor refractoriness to anti-VEGF therapy. Oncotarget 7. https://doi.org/10.18632/oncotarget.8694

Ridgway J, Zhang G, Wu Y, Stawicki S, Liang WC, Chanthery Y, Kowalski J, Watts RJ, Callahan C, Kasman I, Singh M, Chien M, Tan C, Hongo JA, de Sauvage F, Plowman G, Yan M (2006) Inhibition of Dll4 signalling inhibits tumour growth by deregulating angiogenesis. Nature 444(7122):1083–1087

Risau W (1997) Mechanisms of angiogenesis. Nature 386(6626):671–674

Risau W, Sariola H, Zerwes HG, Sasse J, Ekblom P, Kemler R, Doetschman T (1988) Vasculogenesis and angiogenesis in embryonic-stem-cell-derived embryoid bodies. Development 102(3):471–478

Roca C, Adams RH (2007) Regulation of vascular morphogenesis by Notch signaling. Genes Dev 21(20):2511–2524

Rodriguez D, Watts D, Gaete D, Sormendi S, Wielockx B (2021) Hypoxia pathway proteins and their impact on the blood vasculature. Int J Mol Sci 22(17). https://doi.org/10.3390/ijms22179191

Ross H, Aaldijk D, Vladymyrov M, Odriozola A, Djonov V (2023) Transluminal pillars-their origin and role in the remodelling of the zebrafish caudal vein plexus. Int J Mol Sci 24(23). https://doi.org/10.3390/ijms242316703

Ruhrberg C, Gerhardt H, Golding M, Watson R, Ioannidou S, Fujisawa H, Betsholtz C, Shima DT (2002) Spatially restricted patterning cues provided by heparin-binding VEGF-A control blood vessel branching morphogenesis. Genes Dev 16(20):2684–2698

Rundhaug JE (2005) Matrix metalloproteinases and angiogenesis. J Cell Mol Med 9(2):267–285. https://doi.org/10.1111/j.1582-4934.2005.tb00355.x

Ryan AR, Cleaver O (2022) Plumbing our organs: Lessons from vascular development to instruct lab generated tissues. Curr Top Dev Biol 148:165–194. https://doi.org/10.1016/bs.ctdb.2022.02.013

Rye C, Wise R, Jurukovski V, DeSaix J, Choi J (2016) In: Avissar Y (ed) Biology. OpenStax, Houston

Sabin FR (1909) The Lymphatic System in Human Embryos, with a Consideration of the Morphology of the System As a Whole. American Journal of Anatomy 9:43–91

Santamaria R, Cruz-Caballero J, Gkontra P, Jimenez-Montiel A, Clemente C, Lopez JA, Villalba-Orero M, Vazquez J, Hutloff A, Lara-Pezzi E, Arroyo AG (2023) Capillary pruning couples tissue perfusion and oxygenation with cardiomyocyte maturation in the postnatal mouse heart. Front Cell Dev Biol 11:1256127. https://doi.org/10.3389/fcell.2023.1256127

Sawamiphak S, Ritter M, Acker-Palmer A (2010) Preparation of retinal explant cultures to study ex vivo tip endothelial cell responses. Nat Protoc 5(10):1659–1665

Schafer CM, Gurley JM, Kurylowicz K, Lin PK, Chen W, Elliott MH, Davis GE, Bhatti F, Griffin CT (2020) An inhibitor of endothelial ETS transcription factors promotes physiologic and therapeutic vessel regression. Proc Natl Acad Sci USA 117(42):26494–26502. https://doi.org/10.1073/pnas.2015980117

Schager B, Brown CE (2020) Susceptibility to capillary plugging can predict brain region specific vessel loss with aging. J Cereb Blood Flow Metab 40(12):2475–2490. https://doi.org/10.1177/0271678X19895245

Schmittnaegel M, Rigamonti N, Kadioglu E, Cassara A, Wyser Rmili C, Kiialainen A, Kienast Y, Mueller HJ, Ooi CH, Laoui D, De Palma M (2017) Dual angiopoietin-2 and VEGFA inhibition elicits antitumor immunity that is enhanced by PD-1 checkpoint blockade. Sci Transl Med 9(385). https://doi.org/10.1126/scitranslmed.aak9670

Secomb TW, Alberding JP, Hsu R, Dewhirst MW, Pries AR (2013) Angiogenesis: an adaptive dynamic biological patterning problem. PLoS Comput Biol 9(3):e1002983. https://doi.org/10.1371/journal.pcbi.1002983

Serbo JV, Gerecht S (2013) Vascular tissue engineering: biodegradable scaffold platforms to promote angiogenesis. Stem Cell Res Ther 4(1):8. https://doi.org/10.1186/scrt156

Shalaby F, Rossant J, Yamaguchi TP, Gertsenstein M, Wu XF, Breitman ML, Schuh AC (1995) Failure of blood-Island formation and vasculogenesis in Flk-1-deficient mice. Nature 376(6535):62–66. https://doi.org/10.1038/376062a0

Shibuya M (2011) Vascular Endothelial Growth Factor (VEGF) and Its Receptor (VEGFR) Signaling in Angiogenesis: A Crucial Target for Anti- and Pro-Angiogenic Therapies. Genes Cancer 2(12):1097–1105. https://doi.org/10.1177/1947601911423031

Siekmann AF, Lawson ND (2007) Notch signalling limits angiogenic cell behaviour in developing zebrafish arteries. Nature 445(7129):781–784

Siekmann AF, Covassin L, Lawson ND (2008) Modulation of VEGF signalling output by the Notch pathway. BioEssays 30(4):303–313. https://doi.org/10.1002/bies.20736

Simo R, Hernandez C (2008) Intravitreous anti-VEGF for diabetic retinopathy: hopes and fears for a new therapeutic strategy. Diabetologia 51(9):1574–1580. https://doi.org/10.1007/s00125-008-0989-9

Simonavicius N, Ashenden M, van Weverwijk A, Lax S, Huso DL, Buckley CD, Huijbers IJ, Yarwood H, Isacke CM (2012) Pericytes promote selective vessel regression to regulate vascular patterning. Blood 120(7):1516–1527. https://doi.org/10.1182/blood-2011-01-332338

Song JW, Munn LL (2011) Fluid forces control endothelial sprouting. Proc Natl Acad Sci USA 108(37):15342–15347. https://doi.org/10.1073/pnas.1105316108

Stahl A, Connor KM, Sapieha P, Chen J, Dennison RJ, Krah NM, Seaward MR, Willett KL, Aderman CM, Guerin KI, Hua J, Lofqvist C, Hellstrom A, Smith LE (2010) The mouse retina as an angiogenesis model. Invest Ophthalmol Vis Sci 51(6):2813–2826. https://doi.org/10.1167/iovs.10-5176

Stapor PC, Sweat RS, Dashti DC, Betancourt AM, Murfee WL (2014) Pericyte dynamics during angiogenesis: new insights from new identities. J Vasc Res 51(3):163–174. https://doi.org/10.1159/000362276

Staton CA, Reed MW, Brown NJ (2009) A critical analysis of current in vitro and in vivo angiogenesis assays. Int J Exp Pathol 90(3):195–221. https://doi.org/10.1111/j.1365-2613.2008.00633.x

Stefanini MO, Wu FT, Mac Gabhann F, Popel AS (2009) The presence of VEGF receptors on the luminal surface of endothelial cells affects VEGF distribution and VEGF signaling. PLoS Comput Biol 5(12):e1000622. https://doi.org/10.1371/journal.pcbi.1000622

Sterpetti AV (2019) Cardiovascular Research by Leonardo da Vinci (1452-1519). Circ Res 124(2):189–191. https://doi.org/10.1161/CIRCRESAHA.118.314253

Stratman AN, Malotte KM, Mahan RD, Davis MJ, Davis GE (2009) Pericyte recruitment during vasculogenic tube assembly stimulates endothelial basement membrane matrix formation. Blood 114(24):5091–5101

Strilic B, Kucera T, Eglinger J, Hughes MR, McNagny KM, Tsukita S, Dejana E, Ferrara N, Lammert E (2009) The molecular basis of vascular lumen formation in the developing mouse aorta. Dev Cell 17(4):505–515

Strilic B, Eglinger J, Krieg M, Zeeb M, Axnick J, Babal P, Muller DJ, Lammert E (2010) Electrostatic cell-surface repulsion initiates lumen formation in developing blood vessels. Curr Biol 20(22):2003–2009

Suarez-Martinez AD, Bierschenk S, Huang K, Kaplan D, Bayer CL, Meadows SM, Sperandio M, Murfee WL (2018a) A Novel ex vivo Mouse Mesometrium Culture Model for Investigating Angiogenesis in Microvascular Networks. J Vasc Res 55(3):125–135. https://doi.org/10.1159/000489102

Suarez-Martinez AD, Peirce SM, Isakson BE, Nice M, Wang J, Lounsbury KM, Scallan JP, Murfee WL (2018b) Induction of microvascular network growth in the mouse mesentery. Microcirculation 25(8):e12502. https://doi.org/10.1111/micc.12502

Suchting S, Freitas C, le Noble F, Benedito R, Breant C, Duarte A, Eichmann A (2007) The Notch ligand Delta-like 4 negatively regulates endothelial tip cell formation and vessel branching. Proc Natl Acad Sci USA 104(9):3225–3230

Szymborska A, Gerhardt H (2018) Hold me, but not too tight-endothelial cell-cell junctions in angiogenesis. Cold Spring Harb Perspect Biol 10(8). https://doi.org/10.1101/cshperspect.a029223

Takeshita K, Satoh M, Ii M, Silver M, Limbourg FP, Mukai Y, Rikitake Y, Radtke F, Gridley T, Losordo DW, Liao JK (2007) Critical role of endothelial Notch1 signaling in postnatal angiogenesis. Circ Res 100(1):70–78

Taylor SM, Nevis KR, Park HL, Rogers GC, Rogers SL, Cook JG, Bautch VL (2010) Angiogenic factor signaling regulates centrosome duplication in endothelial cells of developing blood vessels. Blood 116(16):3108–3117. https://doi.org/10.1182/blood-2010-01-266197

Tong RT, Boucher Y, Kozin SV, Winkler F, Hicklin DJ, Jain RK (2004) Vascular normalization by vascular endothelial growth factor receptor 2 blockade induces a pressure gradient across the vasculature and improves drug penetration in tumors. Cancer Res 64(11):3731–3736. https://doi.org/10.1158/0008-5472.CAN-04-0074

Torres-Vazquez J, Gitler AD, Fraser SD, Berk JD, Van NP, Fishman MC, Childs S, Epstein JA, Weinstein BM (2004) Semaphorin-plexin signaling guides patterning of the developing vasculature. Dev Cell 7(1):117–123. https://doi.org/10.1016/j.devcel.2004.06.008

Trimm E, Red-Horse K (2023) Vascular endothelial cell development and diversity. Nat Rev Cardiol 20(3):197–210. https://doi.org/10.1038/s41569-022-00770-1

Ubezio B, Blanco RA, Geudens I, Stanchi F, Mathivet T, Jones ML, Ragab A, Bentley K, Gerhardt H (2016) Synchronization of endothelial Dll4-Notch dynamics switch blood vessels from branching to expansion. elife 5. https://doi.org/10.7554/eLife.12167

Vaeyens MM, Jorge-Penas A, Barrasa-Fano J, Shapeti A, Roeffaers M, Van Oosterwyck H (2020) Actomyosin-dependent invasion of endothelial sprouts in collagen. Cytoskeleton (Hoboken) 77(7):261–276. https://doi.org/10.1002/cm.21624

Vega R, Carretero M, Travasso RDM, Bonilla LL (2020) Notch signaling and taxis mechanisms regulate early stage angiogenesis: A mathematical and computational model. PLoS Comput Biol 16(1):e1006919. https://doi.org/10.1371/journal.pcbi.1006919

Vempati P, Mac Gabhann F, Popel AS (2010) Quantifying the proteolytic release of extracellular matrix-sequestered VEGF with a computational model. PLoS One 5(7):e11860. https://doi.org/10.1371/journal.pone.0011860

Walchli T, Bisschop J, Carmeliet P, Zadeh G, Monnier PP, De Bock K, Radovanovic I (2023) Shaping the brain vasculature in development and disease in the single-cell era. Nat Rev Neurosci 24(5):271–298. https://doi.org/10.1038/s41583-023-00684-y

Walpole J, Papin JA, Peirce SM (2013) Multiscale computational models of complex biological systems. Annu Rev Biomed Eng 15:137–154. https://doi.org/10.1146/annurev-bioeng-071811-150104

Walpole J, Chappell JC, Cluceru JG, Mac Gabhann F, Bautch VL, Peirce SM (2015) Agent-based model of angiogenesis simulates capillary sprout initiation in multicellular networks. Integr Biol (Camb) 7(9):987–997. https://doi.org/10.1039/c5ib00024f

Walpole J, Gabhann FM, Peirce SM, Chappell JC (2017) Agent-based computational model of retinal angiogenesis simulates microvascular network morphology as a function of pericyte coverage. Microcirculation 24. https://doi.org/10.1111/micc.12393

Wang WY, Jarman EH, Lin D, Baker BM (2021) Dynamic endothelial stalk cell-matrix interactions regulate angiogenic sprout diameter. Front Bioeng Biotechnol 9:620128. https://doi.org/10.3389/fbioe.2021.620128

Wei L, Sun X, Fan C, Li R, Zhou S, Yu H (2022) The pathophysiological mechanisms underlying diabetic retinopathy. Front Cell Dev Biol 10:963615. https://doi.org/10.3389/fcell.2022.963615

Welter M, Rieger H (2013) Interstitial fluid flow and drug delivery in vascularized tumors: a computational model. PLoS One 8(8):e70395. https://doi.org/10.1371/journal.pone.0070395

Wigerup C, Pahlman S, Bexell D (2016) Therapeutic targeting of hypoxia and hypoxia-inducible factors in cancer. Pharmacol Ther 164:152–169. https://doi.org/10.1016/j.pharmthera.2016.04.009

Wimmer RA, Leopoldi A, Aichinger M, Wick N, Hantusch B, Novatchkova M, Taubenschmid J, Hammerle M, Esk C, Bagley JA, Lindenhofer D, Chen G, Boehm M, Agu CA, Yang F, Fu B, Zuber J, Knoblich JA, Kerjaschki D, Penninger JM (2019) Human blood vessel organoids as a model of diabetic vasculopathy. Nature 565(7740):505–510. https://doi.org/10.1038/s41586-018-0858-8

Wirth G, Juusola G, Tarvainen S, Laakkonen JP, Korpisalo P, Yla-Herttuala S (2023) Capillary dynamics regulate post-ischemic muscle damage and regeneration in experimental Hindlimb ischemia. Cells 12(16). https://doi.org/10.3390/cells12162060

Xu K, Cleaver O (2011) Tubulogenesis during blood vessel formation. Semin Cell Dev Biol 22(9): 993–1004. https://doi.org/10.1016/j.semcdb.2011.05.001

Zhan K, Bai L, Hu Q (2020) Selective induction of sprouting and intussusception is associated with the concentration distributions of oxygen and hypoxia-induced VEGF. Microvasc Res 132: 104041. https://doi.org/10.1016/j.mvr.2020.104041

Zhang ZG, Zhang L, Tsang W, Soltanian-Zadeh H, Morris D, Zhang R, Goussev A, Powers C, Yeich T, Chopp M (2002) Correlation of VEGF and angiopoietin expression with disruption of blood-brain barrier and angiogenesis after focal cerebral ischemia. J Cereb Blood Flow Metab 22(4):379–392. https://doi.org/10.1097/00004647-200204000-00002

Zhao H, Chappell JC (2019) Microvascular bioengineering: a focus on pericytes. J Biol Eng 13:26. https://doi.org/10.1186/s13036-019-0158-3

Ziello JE, Jovin IS, Huang Y (2007) Hypoxia-Inducible Factor (HIF)-1 regulatory pathway and its potential for therapeutic intervention in malignancy and ischemia. Yale J Biol Med 80(2):51–60

Zygmunt T, Gay CM, Blondelle J, Singh MK, Flaherty KM, Means PC, Herwig L, Krudewig A, Belting HG, Affolter M, Epstein JA, Torres-Vazquez J (2011) Semaphorin-PlexinD1 signaling limits angiogenic potential via the VEGF decoy receptor sFlt1. Dev Cell 21(2):301–314. https://doi.org/10.1016/j.devcel.2011.06.033

Chapter 4
The Endothelial Glycocalyx in Health and Human Disease

Kaori Oshima and Eric P. Schmidt

Abstract The extracellular matrix of blood vessels is well-known to contribute to multiple facets of vascular homeostasis and disease (Xu, Shi, Biochim Biophys Acta 1842:2106–2119, 2014; Lin, Davis, Arterioscler Thromb Vasc Biol 43:1599–1616, 2023). Over the past three decades, however, there has been increasing recognition that the vascular matrix is not limited to the medial and adventitial spaces of blood vessels. Indeed, the endovascular surface of the vascular intima also features a luminal matrix of substantial importance to health and disease: the endothelial glycocalyx. In this review, we will discuss what is known about the structure of the endothelial glycocalyx and its substantial heterogeneity as well as the key physiological functions of this endothelial surface layer. We will also review how the endothelial glycocalyx is damaged during acute and chronic vascular diseases, with a focus on what has been learned from human studies, including an increasing literature of randomized controlled studies testing the efficacy of glycocalyx-protective therapeutics.

Keywords Glycocalyx · Glycosaminoglycans · Heparan sulfate · Hyaluronic acid · Chondroitin Sulfate · Endothelium · Critical illness · Proteoglycans · Sepsis · Human

4.1 Part One: Composition of the Endothelial Glycocalyx

As the name suggests, the glyco (sugar) calyx (husk) is a carbohydrate-rich layer composed of various proteoglycans, glycoproteins, glycolipids, and plasma proteins. In this review, we will primarily focus on the endothelial glycocalyx (Fig. 4.1), acknowledging that multiple cell types (e.g., epithelial cells) also have cell surface glycocalyces (Haeger et al. 2016; Rizzo et al. 2022).

The major constituent of the endothelial glycocalyx is proteoglycans. Proteoglycans consist of a "core" protein (often, but not always, transmembrane- or

---

K. Oshima · E. P. Schmidt (✉)
Division of Pulmonary and Critical Care Medicine, Department of Medicine, Massachusetts General Hospital, Boston, MA, USA
e-mail: eschmidt1@mgh.harvard.edu

R. F. Tuma et al. (eds.), *Handbook of Microcirculation*, Physiology in Health and Disease, https://doi.org/10.1007/978-3-031-99609-2_4

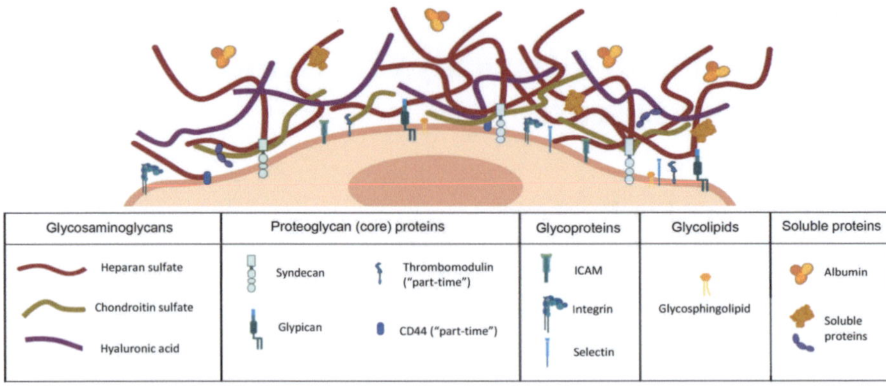

Fig. 4.1 Structural components of the endothelial glycocalyx. The major components of the endothelial glycocalyx include glycosaminoglycans depicted here in solid lines: heparan sulfate (red), chondroitin sulfate (taupe), and hyaluronic acid(purple), which are attached to or associated with proteoglycan "core" proteins anchored to the cell membrane. Glycoproteins, glycolipids, and soluble proteins also contribute to this substantial endovascular layer

membrane-anchored) with covalently attached glycosaminoglycan (GAG) chains that are much larger than the branched *N*- or *O*-glycans that decorate glycoproteins (Varki et al. 2022). Glycosaminoglycans (GAGs) are linear polysaccharides with repeating disaccharide units, each consisting of an amino sugar and a uronic acid or a galactose (Varki et al. 2022). GAGs include heparan sulfate, chondroitin sulfate, dermatan sulfate, keratan sulfate, and hyaluronic acid, with each GAG classified based on its constituent sugars. Of those, heparan sulfate, consisting of 50–150 units of repeating disaccharides of glucosamine and hexuronic acid, is believed to account for 50–90% of the endothelial glycocalyx GAGs (Weinbaum et al. 2007; Ihrcke et al. 1993). Because of its abundance and size, heparan sulfate has been the major focus of studies of the endothelial glycocalyx. Chondroitin sulfate and dermatan sulfate are similar in structure; the disaccharide unit of chondroitin sulfate is comprised of galactosamine and glucuronic acid, while that of dermatan sulfate is comprised of galactosamine and iduronic acid (Mizumoto and Yamada 2022). Heparan sulfate and chondroitin/dermatan sulfate are tethered to the cell surface by "core" proteoglycan proteins and sulfated at specific sites along the polysaccharide chains, which imparts a negative charge. This negative charge allows for the electrostatic binding to positively charged cognate ligands, allowing GAGs to exert biological functions as discussed below. Hyaluronic acid (also known as hyaluronan) is a GAG comprised of repeating disaccharide units of glucosamine and glucuronic acid (Varki et al. 2022). Unlike other GAGs, hyaluronic acid is not covalently bound to proteins on the cell surface. Instead, it is associated with cell surface receptors, such as CD44 (Thorne et al. 2004). Hyaluronic acid also does not have any sulfation along the polysaccharide chain (Varki et al. 2022). The distribution of keratan sulfate within the endothelial glycocalyx is poorly understood (Reitsma et al. 2007) (Funderburgh 2000).

The major “core” proteoglycans within the endothelial glycocalyx are syndecans and glypicans. Other, “part-time” proteoglycans such as thrombomodulin and CD44 may also carry GAGs (Varki et al. 2022). Syndecans are transmembrane proteins that carry both heparan sulfate and chondroitin sulfate polysaccharide chains (Varki et al. 2022). There are four syndecan subtypes, with syndecan-1 often used as a biomarker of endothelial injury and (as discussed later) associated with worse outcomes in various diseases such as trauma (Ostrowski et al. 2012), sepsis, and acute respiratory distress syndrome (ARDS) (Schmidt et al. 2016; Murphy et al. 2017). Glypicans are glycosylphosphatidylinositol-anchored proteins that carry only heparan sulfate (Varki et al. 2022). While there are six subtypes of glypicans, glypican-1 is the only one found on endothelial cells (Rosenberg et al. 1997). Thrombomodulin is a transmembrane protein that carries one chondroitin or dermatan sulfate chain (Varki et al. 2022). It is widely recognized as an anticoagulant that binds to thrombin and propagates protein C-mediated anticoagulation; its anticoagulant property is significantly increased with the presence of chondroitin sulfate (Martin et al. 2013). Finally, CD44 is a transmembrane protein that is widely recognized as a hyaluronic acid receptor (Thorne et al. 2004), and it can also carry 0–2 chains of chondroitin, dermatan, or heparan sulfate (Varki et al. 2022).

Glycoproteins are proteins with relatively small and branched *N*- or *O*-glycosylation (1–20 residues) side chains. *N*-glycans are glycans attached to proteins at asparagine residues of the minimal sequence of Asn-X-Ser/Thr (X cannot be Pro) (Varki et al. 2022). About 70% of proteins contain the Asn-X-Ser/Thr sequence, and, of those, about 70% carry *N*-glycans (Varki et al. 2022). At the end of complex *N*-glycan synthesis, “capping” of carbohydrate branches often involves addition of the sialic acid and sulfate group, which results in a negative charge (Varki et al. 2022). *O*-glycosylation is found in the PTS domain (a domain rich in Pro, Thr, Ser) on most secreted and membrane-bound proteins (Varki et al. 2022). *O*-glycan is also typically negatively charged, promoting protein binding (Varki et al. 2022). The major glycoprotein families found in the vascular glycocalyx are various cell adhesion molecules (CAMs), including the selectin family, integrin family, and immunoglobulin superfamily (Varki et al. 2022; Reitsma et al. 2007) (Krieglstein and Granger 2001). Selectins are lectin-like molecules, which mediate leukocyte rolling on endothelial cells and platelet–leukocyte aggregation (Krieglstein and Granger 2001). E- and P-selectins are present on the endothelial surface, and their expression levels are increased by stimuli such as cytotoxins and oxidants (Krieglstein and Granger 2001). The integrin family of CAMs are transmembrane heterodimeric proteins that are composted of noncovalently bound α- and β-subunits (Takada et al. 2007). In humans, at least 18 α-units and 8 β-units are identified (Shimaoka and Springer 2003) and endothelial cells express at least 11 different integrins (Ruegg and Mariotti 2003). While a majority of those integrins mediate cell–matrix interactions, $\alpha_v\beta_3$ integrin is shown to be critical for activated platelet binding on the endothelial luminal surface (Ruegg and Mariotti 2003) (Bombeli et al. 1998). The immunoglobulin superfamily of glycoproteins are transmembrane proteins with luminal immunoglobulin-like domains (Krieglstein and Granger 2001) (Reitsma et al. 2007). This superfamily includes a broad range of molecules, and the most notable in endothelial cells are intercellular cell adhesion molecules-1 and -2 (ICAM-1 and -2), vascular cell adhesion molecule-1 (VCAM-1), and platelet-endothelial cell adhesion molecule-1 (PECAM)-1 (Krieglstein and Granger 2001). Both ICAM-1 and ICAM-2 are constitutively expressed and participate in leukocyte rolling during inflammation (Krieglstein and Granger 2001), but the protein expression of ICAM-2, a truncated form of ICAM-1, is not increased with inflammatory mediators such as tumor necrosis factor-α (TNFα) (de Fougerolles et al. 1991). VCAM-1 has very low expression at the basal level but is markedly upregulated with a cytokine challenge and plays a role in lymphocyte and monocyte adhesion (Krieglstein and Granger 2001). PECAM-1, similarly, medicates leukocyte cell adhesion, but, similar to ICAM-2, its expression is not altered upon cytokine stimulation (Henninger et al. 1997). Rather, PECAM-1 is shown to dampen cytokine levels during inflammation (Privratsky et al. 2012).

Nearly all vertebrate glycolipids are a subclass of glycolipids called glycosphingolipids, whose name has been derived from the enigmatic Egyptian Sphinx, as it was difficult to decipher the sphingosine structure for years (Varki et al. 2022). Glycosphingolipids are built on a ceramide lipid moiety that consists of long-chain amino-alcohol (sphingosine) in an amide linkage to a fatty acid and attached glycans (Varki et al. 2022). Structural and functional classifications of glycosphingolipids are based on glycans, and it is further subclassified based on the sugar charge, whether the glycans are sialylated or sulfated (Varki et al. 2022). On endothelial cells, various glycosphingolipids were identified (Kanda et al. 2004) and believed to play a role in cell–cell adhesion and signal transductions (Hakomori 2003). Many of the glycosphingolipids impart a negative charge through their acidic glycans, terminal sialic acid, or sulfate group (Hakomori 2003), contributing to the overall negative charge of the vascular glycocalyx.

With a negative charge and a hydrophilic nature, the glycocalyx also interacts with various plasma proteins, such as albumin and orosomucoid (Huxley and Curry 1991) and integrates these soluble factors into a substantial endothelial surface layer. Reduction of proteins on the endothelial surface alone is shown to result in a 31% reduction in glycocalyx thickness, demonstrating that glycosaminoglycans and proteins mutually support the glycocalyx structure (Cheng et al. 2016; Chappell et al. 2009a).

While the major constituents of the endothelial glycocalyx are known, it is difficult to understand the exact composition of its structure because (Xu and Shi 2014) glycocalyx compositions are heterogeneous depending on vessel function and (Lin and Davis 2023) the glycocalyx is not a static structure but is rather a dynamic matrix that reflects a balance between biosynthesis and removal (Lipowsky 2005).

4.1.1 Endothelial Glycocalyx Structure and Heterogeneity

The endothelial glycocalyx is an important structural component of the vasculature. The impressive size of this layer, however, has only been appreciated in the last 20 years owing to the advancements in imaging and fixation techniques. Recent imaging advances such as confocal microscopy, laser scanning microscopy, and two-photon microscopy have allowed for quantitative visualization of the glycocalyx (Kucharz et al. 2022; Megens et al. 2007; Twamley et al. 2021). Atomic force microscopy is used to measure the glycocalyx's thickness as well as its physical properties (Marsh and Waugh 2013) (Targosz-Korecka et al. 2017). To this date, electron microscopy is still widely used as an imaging method to visualize and measure glycocalyx thickness because it allows for direct visualization of the glycocalyx ultrastructure and does not depend on visualization of a particular component of the glycocalyx, such as the case for most ex vivo imaging techniques (Drost et al. 2023a). Historically, the drawback of electron microscopy has been the susceptibility to procedural artifacts, particularly alcohol dehydration that collapses the glycocalyx layer (Drost et al. 2023a; Ebong et al. 2011). However, recent electron microscopy approaches, coupled with improved fixation techniques, have detailed impressive sizes of the glycocalyx (Drost et al. 2023a; Ebong et al. 2011). Glycocalyx measurements in vivo without fixation can be achieved by the application of intravital microscopy in conjunction with zone exclusion with luminal fluorescent dextran, and of microparticle image velocimetry (Potter et al. 2009) (Schmidt et al. 2012), which led to the appreciation of in vivo glycocalyx manifestation, generally called the endothelial surface layer. By avoiding artifactual degradation during tissue fixation, measurements derived from in vivo imaging typically reveal a much thicker glycocalyx.

In addition to these technical challenges, the delicate and biologically dynamic nature of the glycocalyx makes it difficult to capture and interpret its status. Despite the challenges, heterogeneity in the endothelial glycocalyx has been recognized in recent years. While an endothelial glycocalyx exists on all vessels, its structure and compositions are markedly heterogeneous among vascular segments, reflecting the microenvironment specificity of glycocalyx function. Furthermore, significant endothelial glycocalyx heterogeneity exists at the organ level, vascular segment, and cellular levels.

At the organ level, heterogeneity can be observed when different organ microvascular beds are compared using the same imaging method. The endothelial glycocalyx is generally believed to occupy 20% of intravascular volume (Ince et al. 2016), yet one electron microscopy study found that this thickness varied substantially: pulmonary capillaries were observed to have only a thin glycocalyx layer occupying about 3.7% of luminal cross-sectional area, compared to cardiac (15.1%) and cerebral (40.1%) capillaries (Ando et al. 2018). However, this finding was seemingly contrasted by studies using intravital microscopy, which showed that the pulmonary microvascular glycocalyx was nearly three times thicker than cremasteric vessels (Schmidt et al. 2012; Han et al. 2016). Organ-specific heterogeneity of glycocalyx structures may also parallel that of capillary architecture. There are three major types of capillaries: continuous capillaries observed in organs such as the brain, lung, and heart; fenestrated capillaries in the kidney and intestinal mucosa; and sinusoid capillaries seen in the liver and spleen (Ono et al. 2017). Sinusoid hepatic capillaries are shown to have a much thinner glycocalyx layer (3.7% cross-sectional area) compared to continuous cardiac capillaries (13.6%) or fenestrated renal glomerular capillaries (16.7%) (Okada et al. 2017). The glycocalyx in fenestrated capillaries covers 60–80-nm pores and prevents large molecules from passing through (Suzuki et al. 2022), while 30–40-μm pores of sinusoid capillaries are not covered by a glycocalyx (Okada et al. 2017). This heterogeneity in the capillary glycocalyx is believed to reflect the functions of these vessels, such as gas exchange for pulmonary capillaries (Suzuki et al. 2022).

Additionally, there is substantial heterogeneity in the glycocalyx structure along different segments of the same vascular bed. Generally, the glycocalyx becomes thicker with increasing vessel size, as seen in the cerebral microcirculation in mice and in the sublingual microcirculation in humans (Kucharz et al. 2022; Rovas et al. 2019; Yoon et al. 2017).In the brain, it is observed that arterioles have an enriched glycocalyx layer compared to those of venules and capillaries (Kucharz et al. 2022). Observation of the lymphatic glycocalyx in humans shows that it is thinner than veins, and its morphology is described as discontinuous (Gianesini et al. 2023). Interestingly, removal of heparan sulfate from the lymphatic glycocalyx led to increased lymphatic drainage during acute inflammation, which suggests the beneficial effect of glycocalyx degradation during inflammation as opposed to its largely detrimental effect in other organs (Arokiasamy et al. 2019).

However, there may be substantial glycocalyx heterogeneity even within vessels of similar size (i.e., large arteries). Rodent aorta glycocalyx thickness is reported to be 0.27–0.67 μm (~35% luminal coverage) (Devaraj et al. 2009; Wiesinger et al. 2013; Targosz-Korecka et al. 2017). This is comparable to numerous reports of rodent capillary glycocalyx thickness, including myocardial capillaries (0.2–0.5 μm) (van den Berg et al. 2003; Okada et al. 2017), renal capillaries (0.8 μm) (Drost et al. 2023a; Drost et al. 2023b), skeletal muscle capillaries (0.4–0.5 μm) (Vink and Duling 1996), cerebral capillaries (0.42 μm) (Yoon et al. 2017), and cremaster muscle capillaries (0.5 μm) (Potter et al. 2009; Vink and Duling 1996). However, in mouse common carotid arteries, glycocalyx thickness is reported to be approximately 4.5 μm, occupying 66% of the luminal area (Megens et al. 2007; van den Berg et al. 2009). The reason for these substantial differences is unclear but may reflect the regional differences in flow patterns and shear stress. As such, within the internal carotid sinus region of the carotid bifurcation, the glycocalyx is about half the size of the common carotid artery, reflecting the disturbed flow pattern these vascular walls receive (van den Berg et al. 2009; van den Berg et al. 2006). This lack of a robust glycocalyx layer is speculated to be partly responsible for the development of atherosclerosis at this site (van den Berg et al. 2006, 2009).

Glycocalyx thickness is also heterogeneous at a cellular level. Heterogeneous glycocalyx distribution is reported in bovine lung microvascular cells, with a thicker layer at cell junctions and almost none at the nuclei (O'Callaghan et al. 2011).

Age and environmental forces also influence the endothelial glycocalyx. The glycocalyx is known to be thinner in older mice and humans. In aged mice, the glycocalyx in the mesentery and gastrocnemius microcirculation was 50–54% thinner compared to that in young mice, and the glycocalyx in the sublingual microcirculation in older humans was 33% thinner than that in younger humans (Machin et al. 2018). Similarly, the mesenteric lymphatic glycocalyx was around 100 nm in aged rats compared to their younger counterparts, in whom it was approximately 300 nm (Zolla et al. 2015). Diet also influences glycocalyx thickness, as shown in ApoE*3-Leiden mice, in which an atherogenic diet decreased glycocalyx thickness in the same region of the carotid artery (van den Berg et al. 2006). Sublingual glycocalyx thickness in humans is influenced by cigarette smoking, high-calorie meal intake, and coffee consumption (Eickhoff et al. 2020).

4.2 Part Two: Homeostatic Functions of the Endothelial Glycocalyx

The endothelial glycocalyx serves multiple functions critical to vascular homeostasis. Perhaps the most intuitive and important role of the endothelial glycocalyx lies in its support of endothelial barrier function. The negative charge of the glycocalyx establishes an oncotic gradient that limits protein infiltration and permeability (Curry and Adamson 2012); accordingly, the experimental loss of a major constituent, such as heparan sulfate, induces tissue edema (Jacob et al. 2007; Dull et al. 2012; Adamson 1990). Historically, vascular permeability had been defined by the Starling principle, which estimates the volume filtration rate per area based on the balance between hydrostatic and oncotic pressures and the filtration coefficient (porosity of the membrane) (Weinbaum et al. 2021). The discrepancies between this modeling and numerous biological observations, as well as the increased appreciation for the endothelial glycocalyx, led to the revised Starling principle (Weinbaum et al. 2021; Curry 2018), which accounts for the properties of the endothelial glycocalyx.

Another important function of the glycocalyx is to serve as a mechanosensor and transducer, essential for regulating vascular tone. A major constituent of the endothelial glycocalyx, heparan sulfate, is necessary for mechanosensing, as its removal leads to the loss of endothelial alignment to the direction of flow (Yao et al. 2007). In the presence of vascular flow, fluid shear stress is transduced by an intact glycocalyx into the endothelial cell, leading to the induction of endothelial nitric oxide synthase and production of nitric oxide (Florian et al. 2003; Pahakis et al. 2007) (Yen et al. 2015). This mechanotransduction of shear stress, which allows for autoregulation and accommodation of increasing rates of blood flow, is abolished with experimental degradation of heparan sulfate (Florian et al. 2003; Pahakis et al. 2007) (Yen et al. 2015).

Additionally, the endothelial glycocalyx shields endothelial surface adhesion molecules, limiting vascular adhesion of leukocytes and platelets. Loss of the endothelial glycocalyx exposes cell surface ICAM-1 and VCAM-1, promoting leukocyte adhesion (Schmidt et al. 2012; Constantinescu et al. 2003; Mulivor and Lipowsky 2002). Loss of the endothelial surface proteoglycan syndecan-1 induced pro-inflammatory endothelial phenotypic changes (Voyvodic et al. 2014). Platelet adhesion to endothelial cells, which initiates local coagulation and amplifies expression of adhesion molecules and pro-inflammatory chemokines, is attenuated by an intact glycocalyx (Ince et al. 2016; Chappell et al. 2014; Chappell et al. 2011).

The endothelial glycocalyx also maintains an anticoagulant vascular surface that opposes formation of intravascular clots. Several mechanisms contribute to this function, including heparan sulfate binding to antithrombin (Xu and Esko 2014), heparin cofactor II binding to dermatan sulfate (Tollefsen et al. 1983), and thrombomodulin on the cell surface. Heparan sulfate is assumed to play a major role, as it is believed to have "heparin-like" functions. Heparin, a widely used anticoagulant, is a form of heparan sulfate that is heavily sulfated compared to heparan sulfate, which facilitates binding and activation of antithrombin (Xu and Esko 2014). However, given the much lower degree of sulfation on heparan sulfate, the affinity of heparan sulfate to antithrombin is unclear. Indeed, antithrombin rarely binds to cell surface heparan sulfate (Xu and Esko 2014). Surprisingly, the little that is known about the impact of the endothelial surface heparan sulfate on coagulation is derived almost exclusively from endothelial cells in vitro (Dimitrievska et al. 2016), which have a strikingly aberrant glycocalyx structure (Chappell et al. 2009b), and further in vivo studies are warranted.

4.3 Part Three: Glycocalyx Degradation in Human Disease

4.3.1 Potential Pathogenic Mechanisms of Glycocalyx Degradation

Given the critical importance of an intact endothelial glycocalyx in vascular homeostasis, processes that cause glycocalyx degradation are likely to induce macro- and microvascular dysfunction, contributing to acute and chronic organ failure. The mechanisms underlying pathogenic glycocalyx degradation have been most extensively studied in acute illnesses such as sepsis or severe trauma/hemorrhage, potentially due to the relative ease of modeling acute insults (as opposed to chronic disease) in animals. In both acute and chronic illnesses, the circulating indices of glycocalyx damage are often distinct from those of endothelial cell death (Johansen et al. 2015; Piotti et al. 2021), suggesting that glycocalyx shedding may be a separate and thus modifiable event in vascular injury, occurring prior to loss of endothelial cell membrane integrity. As extensively reviewed elsewhere (Sullivan et al. 2021; Becker et al. 2015; Uchimido et al. 2019), multiple (and potentially complementary) "sheddases" have been implicated in the shedding of endothelial surface proteoglycans and their associated glycosaminoglycans during critical illnesses such as sepsis.

1. Heparan sulfate sheddases: Heparan sulfate, the most abundant glycosaminoglycan constituent of the endothelial glycocalyx (Reitsma et al. 2007), is degraded by only one known mammalian enzyme, i.e., heparanase-1 (referred to as "heparanase" in this review), a glucuronidase induced by endothelial-activating mediators such as TNFα (Schmidt et al. 2012) or angiopoietin-2 (Han et al. 2016; Lukasz et al. 2017; Drost et al. 2019). Heparanase activity and expression are increased during sepsis (Schmidt et al. 2012; Martin et al. 2015; Eustes et al. 2021) (in one diverging study (Matan et al. 2018)), leading to shedding of heparan sulfate fragments into the plasma of septic animals (Hofmann-Kiefer et al. 2009) and humans (Schmidt et al. 2014; Hippensteel et al. 2019a; Hippensteel et al. 2019b). Heparanase-mediated glycocalyx degradation is sufficient to induce pulmonary, glomerular, and mesenteric endothelial dysfunction in animal models of sepsis (Schmidt et al. 2012; Lygizos et al. 2013), leading to acute lung (Schmidt et al. 2012), kidney (Schmidt et al. 2016; Lygizos et al. 2013), and intestinal (Chen et al. 2015, 2017) injury, respectively. Accordingly, heparanase inhibitors have demonstrated benefit in preclinical models of sepsis (Schmidt et al. 2012; Lygizos et al. 2013; Chen et al. 2017). Intriguingly, heparanase-mediated heparan sulfate shedding is most apparent in animal models of (and in humans with) non-pulmonary sepsis (Schmidt et al. 2014), with relatively less shedding of heparan sulfate (Murphy et al. 2017; Schmidt et al. 2014) (and heparanase activity (Morris et al. 2015)) in sepsis arising from pulmonary infection (so-called "direct lung injury"). While non-pulmonary insults such as polytrauma and hemorrhage are similarly associated with heparan sulfate shedding into the plasma (Torres Filho et al. 2016), the role of heparanase as a trauma-induced sheddase is less well-studied. Furthermore, not all forms of trauma are associated with heparan sulfate shedding: both animal models and humans with traumatic brain injury demonstrated no increase in circulating heparan sulfate (Oshima et al. 2023), suggesting insult-specific heterogeneity of traumatic glycocalyx degradation.
2. Hyaluronic acid sheddases: Similar to heparan sulfate, animal (Byrne et al. 2018) and human (Nieuwdorp et al. 2009) endotoxemia models of sepsis induce shedding of hyaluronic acid into the circulation. Circulating hyaluronic acid levels in septic patients were associated with indices of organ dysfunction and disease severity (Yagmur et al. 2012). Urinary hyaluronic acid—a potential, albeit unproven, proxy for circulating hyaluronic acid—was found to predict sepsis mortality in humans (Schmidt et al. 2016). In contrast to heparan sulfate, the sheddases (either the host or pathogen) responsible for hyaluronic acid degradation are uncertain (Sullivan et al. 2021). Furthermore, elevated circulating hyaluronic acid may be influenced by sepsis-associated liver dysfunction, as hepatic endothelial cells mediate clearance of these glycans from the circulation (Eriksson et al. 1983). Finally, little is known about the size of circulating hyaluronic acid during sepsis, despite the critical relevance of fragment length to its pro- or anti-inflammatory effects (McKee et al. 1996; Jiang et al. 2005).
3. Proteoglycan sheddases: Proteoglycans can be cleaved from the endothelial surface by a variety of proteases, including host-derived matrix

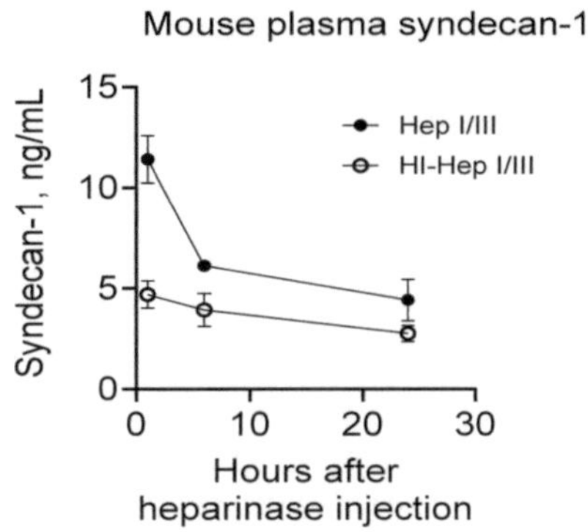

Fig. 4.2 Heparan sulfate degradation rapidly induces shedding of syndecan-1. Intravenous injection of 1 unit of heparinase I/III, a glucuronidase with no endogenous protease activity (Pahakis et al. 2007), into 8–12-week-old C57BL/6 mice leads to rapid shedding of syndecan-1 into the plasma. Heat-inactivated (HI) heparinase I/III serves as control. $n = 3–4$ mice/group; data provided as mean and standard error of the mean. $*p < 0.05$ by Mann–Whitney

metalloproteinases (MMPs) (Lipowsky 2011, 2012) and a disintegrin and metalloproteinase domain-containing proteins (ADAMs) (Yang et al. 2018). Furthermore, pathogens such as *Staphylococcus aureus* (Hayashida et al. 2009) and *Pseudomonas aeruginosa* (Damron et al. 2016; Spencer et al. 2010) can express proteases capable of proteoglycan cleavage. Intriguingly, the prior action of heparanase may augment (and even be sufficient to initiate, Fig. 4.2) protease-mediated proteoglycan shedding. Removal of heparan sulfate side chains exposes the proteoglycan core protein to proteases, potentially triggering secondary proteoglycan shedding after denudation of heparan sulfate side chains (Rangarajan et al. 2020). Furthermore, heparanase expression also induces host MMP expression via the activation of extracellular signal-regulated kinase (ERK) signaling (Rangarajan et al. 2020). These coordinated, multifaceted host processes of glycocalyx shedding suggest a unified response of a distressed endothelial cell. While the teleological benefit of such coordination is uncertain, it is possible that glycocalyx shedding facilitates host defense by enhancing leukocyte diapedesis to confront an invading pathogen, promoting local microthrombosis to prevent pathogen dissemination, and facilitating extravasation of antimicrobial and potentially host-reparative proteins.

In contrast to acute vascular injury, the glycocalyx aberrance characteristic of chronic vascular diseases (extensively reviewed elsewhere (Ahn et al. 2023)) such as atherosclerosis and diabetes is likely a consequence of the multiple deleterious effects of oxidative stress. Oxidized low-density lipoprotein (LDL) is sufficient to induce glycocalyx degradation in healthy animals in a superoxide-dependent mechanism (Constantinescu et al. 2001; Vink et al. 2000). Similarly, acute hyperglycemia is sufficient to disrupt the glycocalyx: Nieuwdorp et al. 2006) used 10 healthy male volunteers to demonstrate that acute hyperglycemia reduced whole-body glycocalyx volume (as quantified using the two-tracer technique) from 1.7 to 0.8 L, with concurrent rise in circulating hyaluronic acid. Similar glycocalyx loss was not induced by mannitol, an osmotic control for hyperglycemia. This glycocalyx loss could be inhibited by administration of *N*-acetylcysteine, again supporting a potential role for oxidative stress in hyperglycemia-induced glycocalyx degradation (Nieuwdorp et al. 2006).

Chronic glycocalyx impairment likely also reflects aberrant processes of glycocalyx synthesis and/or maintenance, which may be overcome by dietary supplementation of glycan precursors (Patil et al. 2022). Furthermore, as unidirectional fluid shear forces are necessary for proper glycocalyx synthesis (Siren et al. 2021), the aberrant shear forces observed in vascular regions (i.e., arterial bifurcations (van den Berg et al. 2006)) commonly affected by atherosclerosis may impair the homeostatic processes of glycocalyx synthesis and repair (Harding et al. 2018).

4.3.2 *Determining the Relevance of Glycocalyx Degradation to Human Disease*

4.3.2.1 Measuring Endothelial Glycocalyx Degradation in Humans

To investigate the impact of endothelial glycocalyx degradation on acute and chronic human diseases, it is necessary to quantify glycocalyx integrity within vascular beds relevant to the disease states studied. Several approaches have been used for such quantification, each with their own strengths and weaknesses:

1. Tracer dilution technique: This approach leverages the differences between two vascular tracers' ability to penetrate the endothelial glycocalyx. For example, by comparing the volume of distribution of a labeled small-weight dextran (capable of penetrating the glycocalyx) with that of labeled erythrocytes (incapable of penetrating the glycocalyx), the total endothelial glycocalyx volume can be calculated (Nieuwdorp et al. 2006, 2008). This approach, while attractive in its theoretical ability to quantify whole-body glycocalyx volume, relies upon several technical assumptions that have been criticized (Michel and Curry 2009). These limitations may be particularly problematic in the critically ill, as vascular leak may expand the volume of distribution of a low-weight dextran into the extravascular space. Furthermore, this approach does not allow specific insights into organ-specific glycocalyx degradation.
2. Intravital imaging of the sublingual microcirculation: In preclinical studies, the glycocalyx is best visualized using intravital microscopy, as tissue fixation and in vitro preparations often artifactually degrade the endothelial surface layer (Yang et al. 2013). Applying intravital microscopy approaches to humans, however, is limited by a paucity of vascular beds amenable to noninvasive imaging. The sublingual microcirculation is one such mucosal surface that can be noninvasively visualized using a handheld sidestream darkfield microscope (Colbert and Schmidt 2016). Using this technique (and automated image

processing software), glycocalyx thickness can be measured as the inverse of the so-called "perfused boundary region" (PBR)—the cross-sectional diameter of the microvessel that erythrocytes can access. Thus, as glycocalyx thickness is lost, the perfused boundary region increases in size. Accordingly, sepsis is associated with an increase in the perfused boundary region, which may be distinct from other indices of microcirculatory function (Rovas et al. 2019).

Consistent and accurate use of intravital microscopy as a bedside test of glycocalyx integrity can be technically challenging; for example, application of inappropriately high or low pressure by the intravital microscope to the sublingual surface can cause vascular dropout and limit the ability to capture high-quality images (Colbert and Schmidt 2016). Despite these potential limitations, perfused boundary region (PBR) measurements (as well as other intravital indices of microvascular function) have demonstrated robust interobserver consistency, even across multicenter studies (Eickhoff et al. 2020; Vellinga et al. 2015). However, additional concerns remain, including the relevance of the sublingual endothelial glycocalyx to organ failure occurring in other vascular beds, which are of more relevance to critical illness outcomes. Indeed, there exist substantial differences in glycocalyx composition both across and within the microcirculations of different organs, as noted previously in this review.

3. Circulating glycocalyx fragments: As the half-life of circulating heparan sulfate is short even in septic animals (Zhang et al. 2019), the detection of these fragments in plasma suggests ongoing glycocalyx degradation. While mass spectrometry is able to quantify both glycosaminoglycan fragment abundance and structure with high sensitivity and accuracy (Schmidt et al. 2014; Hippensteel et al. 2019a; Hippensteel et al. 2019b; Sun et al. 2015), this approach is currently expensive, technically laborious, and thus not yet practical for clinical use. Given the known association (Rangarajan et al. 2020) between heparanase and MMP induction, circulating heparan sulfate closely approximates circulating syndecan-1 (Hippensteel et al. 2019b). Thus, enzyme-linked immunosorbent assay (ELISA)-based quantification of syndecan-1 can serve as a less expensive alternative to mass spectrometry quantification of glycocalyx shedding. Finally, less expensive colorimetric assays, such as the dimethylmethylene blue (DMMB) assay of sulfated glycosaminoglycans, have shown promise in animal models of critical illness (Bastarache et al. 2022; Gómez et al. 2020), although concerns exist about their cross-reactivity to cell-free DNA released as part of neutrophil extracellular traps (Schmidt et al. 2016). Notably, whole-body quantification of circulating glycocalyx fragments does not allow identification of the vascular beds that are the source of these fragments. However, mass spectrometry's ability to define glycosaminoglycan sulfation patterns offers promise to discern organ-specific signatures of glycocalyx degradation (Schmidt et al. 2014; Oshima et al. 2019).

4.3.2.2 Is Endothelial Glycocalyx Degradation a Causal Mediator of Human Disease?

Human studies have observed an association of glycocalyx degradation with multiple different diseases. Whether such degradation is an independent (and potentially attributable) mediator of patient outcomes in these diseases, however, is less certain.

(a) Sepsis

Consistent with robust preclinical data implicating endothelial glycocalyx degradation as a direct mediator of septic organ injury, human studies (as summarized in a recent meta-analysis (Fernández-Sarmiento et al. 2023) as well as in other studies (Saoraya et al. 2021a; Steppan et al. 2011; Ostrowski et al. 2015a)) have demonstrated associations between syndecan-1 shedding and patient outcomes such as death, multi-organ dysfunction, and respiratory failure. Similarly, detection of circulating or urinary glycosaminoglycans has been associated with kidney failure (Schmidt et al. 2016), respiratory failure (Schmidt et al. 2014), coagulopathy (Ostrowski et al. 2015b; Ikeda et al. 2018), and/or death (Hippensteel et al. 2019b; Nelson et al. 2008, 2014) in human sepsis.

While endothelial glycocalyx degradation is commonly observed in human studies of sepsis, it is less clear whether it is simply a marker of vascular injury or an independent predictor of sepsis outcomes. Such determination requires large sample sizes—typically found within multicenter randomized controlled trials of sepsis therapeutics. For example, the ALBIOS (Albumin Italian Outcome Sepsis) study randomized 1818 patients with sepsis across 100 ICUs to treatment with 20% albumin or crystalloid, with albumin dosing titrated to maintain serum albumin >3 mg/dl (Caironi et al. 2014). A post hoc analysis of 956 ALBIOS subjects revealed that syndecan-1 shedding was an independent predictor of renal failure, coagulation failure, and mortality (Piotti et al. 2021). Accordingly, albumin supplementation—which had no impact on survival in the parent study—had no impact on syndecan-1 shedding in the associated sub-study, although it did suppress indices of endothelial cell damage (soluble vascular endothelial (VE)-cadherin) (Piotti et al. 2021). In contrast to ALBIOS, a sub-study of 1103 critically ill infected patients enrolled in the Procalcitonin and Survival Study (PASS), a randomized trial of procalcitonin-guided antibiotic escalation (Jensen et al. 2011), demonstrated that syndecan-1 was not an independent predictor of 90-day mortality in sepsis on multivariate analysis, although it did independently predict liver and renal failure by day 28 (Johansen et al. 2015). In contrast to syndecan-1, soluble thrombomodulin was an independent predictor of mortality and multisystem organ failure. The authors interpreted this discrepancy as evidence that syndecan-1 was distinct from markers of endothelial cell death (e.g., thrombomodulin release during membrane fragmentation). Finally, a sub-study of 167 patients with sepsis-induced acute respiratory distress syndrome enrolled in the CITRIS-ALI (Vitamin C Infusion for TReatment In Sepsis associated Acute Lung Injury) study of 4 days of treatment with ascorbic acid vs. placebo identified that baseline syndecan-1 and an increase in syndecan-1 over the first 48 h of the study both independently predicted all-cause mortality (Qiao et al. 2022).

Importantly, much remains to be learned about the pathogen determinants of glycocalyx degradation during sepsis. The COVID-19 pandemic demonstrated that viral sepsis is sufficient to induce endothelial glycocalyx degradation, mediated in part via the actions of heparanase (Druzak et al. 2023; Drost et al. 2022; Queisser et al. 2021; Buijsers et al. 2020). Intriguingly, Gram-negative bacteremia is associated with higher indices of glycocalyx degradation than Gram-positive bacteremia; the mechanisms underlying this difference are uncertain (Hippensteel et al. 2019b).

(b) Trauma

Multiple human studies have demonstrated that endothelial glycocalyx degradation is a robust predictor of patient outcomes in trauma. The Pragmatic, Randomized Optimal Platelet and Plasma Ratios (PROPPR) trial randomized 680 severely injured trauma patients to blood product resuscitation with a 1:1:1 or 1:1:2 ratio of platelets to fresh frozen plasma (FFP) to packed red blood cells (Holcomb et al. 2015). While the study was overall negative for its primary outcomes of 24-h and 30-day all-cause mortality, the 1:1:1 group had a lower risk of death by exsanguination at 24 h without any evidence of harm. Sub-studies of PROPPR demonstrated that syndecan-1 was independently associated with acute kidney injury (Dixon et al. 2023) and sepsis (Wei et al. 2018) on multivariate analysis. Multiple additional cohorts of trauma patients (Johansson et al. 2011a, b, 2017; Naumann et al. 2018) demonstrated that syndecan-1 is a quantitative index of the endotheliopathy of trauma (Gonzalez Rodriguez et al. 2017) and an independent predictor of trauma mortality.

(c) Ischemia–Reperfusion Injury

Ischemia–reperfusion injury has been implicated as the primary cause of multiple acute cardiac and vascular diseases. In a seminal paper, Rehm et al. demonstrated that cardiopulmonary bypass during major vascular surgery was associated with substantial (and rapid) elevations in the circulating indices of glycocalyx degradation (syndecan-1, heparan sulfate) in the absence of indices of cell death (Rehm et al. 2007). Urinary hyaluronic acid was an independent predictor of acute kidney injury in children undergoing cardiac surgery with cardiopulmonary bypass (Zarbock et al. 2014). Accordingly, patients undergoing off-pump cardiac surgery demonstrated less evidence of glycocalyx shedding than those receiving cardiopulmonary bypass (Bruegger et al. 2011).

Solid organ transplantation is similarly associated with ischemia–reperfusion injury to the allograft occurring during organ procurement, cold storage, and reperfusion. Intriguingly, circulating hyaluronic acid prior to lung transplantation (i.e., in potential organ donors) was associated with poor graft quality, and elevated syndecan-1 in lung transplant recipients was associated with primary graft dysfunction (Sladden et al. 2019). These findings are supported by large animal studies suggesting that glycocalyx degradation contributes to primary graft dysfunction (Rancan et al. 2018). Similar findings were observed with orthotopic liver transplantation (Schiefer et al. 2020).

Tissue injury during myocardial infarction may additionally be driven by ischemia–reperfusion injury to the myocardium; an accompanying loss of cardiac output could additionally drive systemic ischemia. Indeed, acute coronary syndrome has been associated with endothelial glycocalyx degradation, with parallel increases in systemic catecholamine tone (Ostrowski et al. 2013). Myocardial infarction-associated cardiogenic shock was associated with an early rise in circulating syndecan-1 that was associated with mortality on multivariate analysis (Jung et al. 2015).

Perhaps the most extreme form of ischemia–reperfusion injury is cardiac arrest, which subjects every vascular bed to potential glycocalyx degradation. Bro-Jeppesen et al. studied 163 patients with out-of-hospital cardiac arrest undergoing targeted temperature management (i.e., therapeutic hypothermia) (Bro-Jeppesen et al. 2016). Circulating syndecan-1 levels peaked at study entry; at 24 h, this index of glycocalyx degradation was associated with baseline interleukin (IL)-6 and an unshockable rhythm on univariate and multivariate analyses. Baseline syndecan-1 was not associated with the severity of organ injury at 24 or 72 h; however, the 72-h change in syndecan-1 was independently associated with mortality on multivariate analysis. Similar findings were observed in univariate analyses of other cohorts of out-of-hospital cardiac arrest (Grundmann et al. 2012).

(d) Chronic Vascular Diseases

In addition to acute illnesses, multiple chronic vascular diseases are associated with glycocalyx dysfunction and shedding (Ahn et al. 2023). Groen et al. used intravital microscopy of the sublingual microcirculation to assess glycocalyx thickness in 15 healthy young (median age: 24 years), 15 healthy aged (median age: 70 years), and 15 type II diabetic aged (median age: 70 years) men (Groen et al. 2014). Diabetic aged men demonstrated thinner glycocalyces than young healthy men, with older healthy men only demonstrating a nonsignificant trend toward thinner glycocalyces than young healthy men. Glycocalyx thickness was negatively associated with hemoglobin A1c (Hba1c) (Groen et al. 2014), and, accordingly, intensification of diabetic control (with a corresponding decrease in HbA1c) was associated with improvement in glycocalyx thickness (i.e., a drop in the perfused boundary region) and multiple indices of cardiovascular function (Lambadiari et al. 2019). Similarly, cigarette smokers demonstrated thinner glycocalyces than healthy controls; smoking cessation, accordingly, improved glycocalyx thickness, but these gains were lost in those who relapsed (Ikonomidis et al. 2017). Finally, HIV infection, a disease state associated with vascular injury, is characterized by increased syndecan-1 shedding; this shedding is predictive of renal dysfunction 5 years later (Cavalcante et al. 2023; Meneses et al. 2017). Glycocalyx thickness can be improved with the initiation of antiviral medications, independent of any change in circulating inflammatory markers (Fragkou et al. 2023).

4.3.3 *Randomized Human Trials of Putative Glycocalyx-Protective Therapies*

Given the independent associations of indices of endothelial glycocalyx degradation with patient outcomes, blocking glycocalyx shedding is an attractive therapeutic target in several acute and chronic vascular diseases. Accordingly, small randomized controlled trials have tested agents identified as being potentially glycocalyx-protective in preclinical models. These trials, largely underpowered to detect patient benefit, have had mixed results (Table 4.1). To provide sufficient statistical power, investigators have alternatively leveraged biobank patient samples collected from large randomized controlled trials to perform ancillary studies of any unforeseen benefit of the tested therapeutic on glycocalyx integrity. However, most therapies tested have multiple potential biological effects, and, as such, it is difficult to discern whether glycocalyx protection is responsible for an observed effect of a therapy on the measured outcomes.

Several classes of agents have been extensively tested as glycocalyx-protective therapeutics, generally with mixed results:

1. *Antioxidants*: Given their central role for oxidative stress in vascular disease as well as ischemia–reperfusion injury, antioxidants are attractive endothelial-protective treatments. Ascorbic acid (vitamin C) is an antioxidant that attracted brief, intense interest as a potential sepsis therapeutic. In patients enrolled in the CITRIS-ALI study of sepsis-associated acute respiratory distress syndrome (ARDS), randomization to ascorbic acid infusion decreased syndecan-1 shedding (Qiao et al. 2022), coincident with a potential (but not conclusive) improvement in survival as an exploratory secondary outcome in the parent trial (Fowler 3rd et al. 2019). Other studies of vitamin C in sepsis only revealed transient improvement in glycocalyx thickness (Belousoviene et al. 2023). In patients with chronic coronary artery disease, randomization to 1 month of treatment with the antioxidant hydroxytyrosol (a component of olive oil) was associated with improved sublingual endothelial glycocalyx thickness coincident with improved microvascular and (later) left ventricular diastolic function (Ikonomidis et al. 2023).
2. *Sheddase inhibitors*: Surprisingly, few trials have directly tested the impact of heparanase and/or MMP inhibitors on endothelial glycocalyx shedding in human disease. Three days of preoperative doxycycline, an MMP inhibitor with glycocalyx-protective effects in preclinical models (Mulivor and Lipowsky 2009), decreased circulating syndecan-1 and heparan sulfate after cardiopulmonary bypass for cardiac surgery, when compared to placebo (Gao et al. 2022). Coincident with this protection, doxycycline decreased postoperative albuminuria, myocardial injury, and lactic acid when compared to placebo. However, patients randomized to doxycycline demonstrated lower levels of glycocalyx degradation products even prior to surgery, raising some question about what processes were affected by doxycycline pretreatment. Heparanase inhibitors (such as low-molecular-weight heparin (Chen et al. 2023) and sulodexide

Table 4.1 Human studies of potential glycocalyx-protective therapeutics

Treatment	Disease state	No. of subjects	Impact on the endothelial glycocalyx
Antioxidants			
Ascorbic acid (Qiao et al. 2022)	Sepsis-induced ARDS	167	A sub-study of the CITRIS-ALI multicenter randomized controlled trial (Fowler 3rd et al. 2019). Randomization to ascorbic acid 50 mg/kg intravenously every 6 h decreased syndecan-1 at 48 h compared to placebo. Syndecan-1 was associated with the severity of hypoxemia and independently associated with 28-day mortality
Ascorbic acid (Belousoviene et al. 2023)	Sepsis	23	Randomization to ascorbic acid 50 mg/kg intravenously every 6 h transiently improved the perfused boundary region of sublingual microvessels after the first dose of ascorbic acid; no differences seen throughout the remainder of the study
Olive oil with hydroxytyrosol (Ikonomidis et al. 2023)	Chronic coronary artery disease	30	Subjects received four capsules containing olive oil (412.5 mg/cap) with hydroxytyrosol (2.5 mg/cap) or placebo daily for 1 month and then crossed over to alternative treatment. Olive oil/hydroxytyrosol improved sublingual glycocalyx thickness and other indices of microvascular function
Sheddase inhibitors			
Doxycycline (Gao et al. 2022)	Cardiopulmonary bypass for coronary artery bypass grafting/valve replacement	36	Doxycycline (100 mg/day for 3 days prior to surgery) decreased plasma syndecan-1, heparan sulfate, and MMP concentrations compared to placebo at the completion of cardiopulmonary bypass and arrival to the ICU. Attenuated syndecan-1 shedding was already apparent prior to surgery in the doxycycline group

(continued)

Table 4.1 (continued)

Treatment	Disease state	No. of subjects	Impact on the endothelial glycocalyx
Sulodexide (Broekhuizen et al. 2010)	Male patients with diabetes type II without macrovascular disease	20	A nonrandomized trial in which ten male patients with type II diabetes and 10 male healthy controls were treated with 8 weeks of sulodexide (200 mg/day, a heparanase/hyaluronidase inhibitor and a potential GAG substrate). Patients with diabetes had thinner sublingual and retinal microvascular glycocalyces at baseline than controls; this thinning improved with sulodexide. In healthy, male, age-matched controls, sulodexide had no effect on the sublingual glycocalyx but surprisingly decreased retinal glycocalyx thickness
Low-molecular-weight heparin (Chen et al. 2023)	Sepsis with acute gastrointestinal injury	95	Patients randomized to receive low-molecular-weight heparin (a known heparanase inhibitor) at doses sufficient to increase anti-Xa levels or usual care. Low-molecular-weight heparin decreased heparanase and syndecan-1 levels as well as improved indices of gastrointestinal injury. No survival or length of stay benefit
Prostacyclins			
Prostacyclin (Vigstedt et al. 2022)	Mechanically ventilated COVID-19 with endothelial injury (elevated thrombomodulin)	75	Patients randomized to continuous intravenous infusion of prostacyclin (1 ng/kg/min) or placebo for 72 h. A 24-h change in syndecan-1 blunted in the prostacyclin group; no effect on other vascular injury markers. A parent trial (Johansson et al. 2022) showed no statistically significant impact of prostacyclin on patient outcomes

(continued)

Table 4.1 (continued)

Treatment	Disease state	No. of subjects	Impact on the endothelial glycocalyx
Prostacyclin (Gybel-Brask et al. 2017)	Subarachnoid hemorrhage	89	Patients randomized to continuous infusion of prostacyclin (29 patients receiving 1 ng/kg/min; 30 patients receiving 2 ng/kg/min) or placebo (30 patients), starting at day 5 after subarachnoid hemorrhage and continuing to day 10. While syndecan-1 shedding increased over time, this was similar in all groups
Iloprost (Holmvang et al. 2012)	ST-elevation myocardial infarction	17	Patients randomized to 24 h of iloprost or saline infusion after percutaneous coronary intervention/eptifibatide. No impact of iloprost on the indices of glycocalyx shedding
Vasopressor support			
Norepinephrine (Damén et al. 2021)	Coronary artery bypass graft surgery	24	In all, 12 patients randomized to a higher mean arterial pressure goal (90 mmHg) by norepinephrine infusion; 12 patients randomized to lower goals (60 mmHg). No increase in syndecan-1 or HA shedding was noted in either group
Phenylephrine (Kaneko et al. 2020)	Abdominal surgery	52	In addition to an initial bolus of 8 ml/kg 6% hydroxyethyl starch and maintenance ringer's bicarbonate solution, 26 patients received continuous infusion of phenylephrine and 26 patients received a continuous infusion of saline. Hypertension managed by a decrease in phenylephrine infusion rate; hypotension managed by intravenous infusion of ephedrine or phenylephrine. No differences between groups in syndecan-1 or hyaluronic acid over a period of 30 min

(continued)

Table 4.1 (continued)

Treatment	Disease state	No. of subjects	Impact on the endothelial glycocalyx
Anti-inflammatories			
Hydrocortisone (Brettner et al. 2019)	Cardiac surgery with cardiopulmonary bypass	30	Patients randomized to hydrocortisone (100 mg intravenously over 10 min) prior to anesthesia induction demonstrated attenuated heparan sulfate shedding, but not syndecan-1 shedding, intraoperatively through 1 h post-cardiopulmonary bypass. Benefit was no longer apparent 4 h after the end of cardiopulmonary bypass. CRP levels were lower in the hydrocortisone-treated group. Patients in hydrocortisone group required more intraoperative FFP
Methylprednisolone (Xiang et al. 2022)	Laparoscopic gastrointestinal surgery	168	Patients randomized to 2 mg/kg methylprednisolone or placebo, administered 10 min before surgery. Methylprednisolone decreased syndecan-1 and heparan sulfate shedding at 24 and 72 h, respectively, and decreased the rate of postoperative delirium
Methylprednisolone (Lindberg-Larsen et al. 2017)	Total knee arthroplasty	70	Patients randomized to a single preoperative dose of methylprednisolone (125 mg intravenously) or saline. Methylprednisolone modestly decreased syndecan-1, soluble thrombomodulin, soluble E-selectin, VEGF, and CRP 24 h after surgery. Notably, substantial differences in baseline syndecan-1 between groups (control group >50% higher, due to four outliers)
Dexamethasone/ albumin (Yanase et al. 2021)	Colorectal, pancreatic, or liver surgery	72	Patients randomized to receive dexamethasone (16 mg) and albumin (100 ml 20% albumin)

(continued)

Table 4.1 (continued)

Treatment	Disease state	No. of subjects	Impact on the endothelial glycocalyx
			intravenously plus additional albumin in conjunction with as-needed crystalloid versus crystalloid only. The intervention group had no change in syndecan-1 but substantial decrease in shed heparan sulfate at postoperative day 0 (night of surgery) and postoperative day 1. Dexamethasone/albumin decreased CRP and the risk of postoperative complications
Lidocaine infusion (Pustetto et al. 2020)	Elective major abdominal surgery	40	Patients randomized to 1.5 mg/kg 1% lidocaine (as an anti-inflammatory agent) or saline bolus prior to anesthesia induction, followed by 2 mg/kg/hr. intravenous 1% lidocaine (or saline) infusion until surgery completion. Lidocaine had no impact on syndecan-1 shedding nor any index of vascular function (flow-mediated vasodilation, near-infrared spectroscopy). Lidocaine decreased stroke volume, cardiac index, and mean arterial pressure
Methotrexate/TNFα inhibition (Deyab et al. 2021)	Rheumatoid arthritis patients	39	A nonrandomized trial in which 19 methotrexate-naïve patients received methotrexate and 20 methotrexate nonresponders received TNFα inhibition. All patients were enrolled in the Norwegian PSARA (PSoriatic arthritis, Ankylosing spondylitis, Rheumatoid Arthritis) study (Deyab et al. 2017). Baseline syndecan-1 was independently associated with baseline CRP. Methotrexate treatment for 6 weeks was associated with a significant

(continued)

Table 4.1 (continued)

Treatment	Disease state	No. of subjects	Impact on the endothelial glycocalyx
			decrease in syndecan-1; this decrease was independent of changes in inflammatory indices. No benefit in methotrexate nonresponders treated with TNFα inhibitors
Intravenous fluid resuscitation practices			
Albumin (Piotti et al. 2021)	Septic shock	375	A sub-study of the ALBIOS trial (Caironi et al. 2014), in which patients were randomized to 20% albumin (to target serum albumin of 30 mg/dl) plus crystalloid vs. crystalloid alone for 28 days. Baseline syndecan-1 associated with 90-day mortality on multivariate analysis. Albumin had no impact on syndecan-1 or sphingosine-1-phosphate but decreased VE-cadherin
Bolus albumin (Bihari et al. 2019)	Healthy volunteers	6	A crossover study of different forms of bolus fluids: 30 ml/kg 0.9% saline, 30 ml/kg Hartmann's balanced crystalloid solution, 30 ml/kg 4% albumin, or 6 ml/kg of 20% albumin. Saline bolus caused syndecan-1 shedding and increased angiopoietin-2. Both saline and Hartmann's solution caused ultrasonographic evidence of lung edema as well as increased airway resistance
Hydroxyethyl starch (Kammerer et al. 2019)	Cystectomy	55	Patients randomized to intraoperative 6% 130/0.4 hydroxyethyl starch vs. intraoperative albumin. No differences between groups in the indices of glycocalyx degradation (syndecan-1, heparan sulfate, hyaluronic acid) in the first 3 days after operation. Glycocalyx

(continued)

Table 4.1 (continued)

Treatment	Disease state	No. of subjects	Impact on the endothelial glycocalyx
			degradation was not associated with coagulopathy as measured by rotational thromboelastometry
Hydroxyethyl starch (Liu et al. 2021)	Colorectal surgery	47	Patients randomized to three different stroke volume variation goals for fluid resuscitation immediately prior to surgery onset. If observed stroke volume variation was higher than the goal, patients would receive a bolus of 6% 130/0.4 hydroxyethyl starch. If stroke volume variation was lower than the target, no fluids other than maintenance lactated ringer's. Patients with lower stroke volume variation goals received more hydroxyethyl starch, had greater urine output, and had greater CO. however, this group had worsened indices of glycocalyx degradation (heparan sulfate, hyaluronic acid, syndecan-1) at the end of surgery. Patients who had lower stroke volume variation goals (i.e., those who received more fluids) had longer length of stay
Hydroxyethyl starch (Kim et al. 2017)	Off-pump coronary artery bypass surgery	120	Patients randomized to receive either 20 ml/kg hydroxyethyl starch 670/0.75 or crystalloid as an intraoperative fluid bolus, followed by as-needed crystalloid. The crystalloid group had higher fluid balance but no difference between groups in terms of glycocalyx degradation. No differences in microvascular or patient outcomes
Hydroxyethyl starch (Müller et al. 2016)	Severe sepsis	208	A sub-study of the Scandinavian Starch for Severe Sepsis/Septic Shock (6S) trial of fluid resuscitation with hydroxyethyl

(continued)

Table 4.1 (continued)

Treatment	Disease state	No. of subjects	Impact on the endothelial glycocalyx
			starch 130/0.42 or Ringer's acetate (Perner et al. 2012). No difference in syndecan-1 shedding between groups; no association of syndecan-1 with survival
Rate of intravenous fluids (Saoraya et al. 2021b)	Sepsis-induced hypovolemia	96	Randomization to 30 ml/kg/h vs. 10 ml/kg/h intravenous lactated ringer's for initial sepsis resuscitation had no impact on syndecan-1 at 6 h. Slower rates of infusion associated with less vasopressor support and respiratory failure at 6 h
Rate of intravenous fluids (Pouska et al. 2018)	Sepsis patients as well as patients undergoing elective major spinal surgery	50 spinal surgery, 16 sepsis	Patients were randomized to receive either a rapid fluid challenge (5 ml/kg, administered over 10 minutes using a pressure bag) or a slow fluid challenge (5 ml/kg, administered over 30 min using an infusion pump). In surgical patients, the randomized rate of fluid challenge had no impact on glycocalyx thickness. In septic patients, slow infusion improved glycocalyx thickness immediately after infusion but otherwise had no impact
Volume of intravenous fluids (Belavić et al. 2018)	Laparoscopic cholecystectomy	90	Randomized to three lactated ringer's infusion rates: 1 ml/kg/h intraoperatively continued for 6 h postoperatively; 5 ml/kg/h intraoperatively continued for 6 hours postoperatively; 15 ml/kg/h intraoperatively that continued at 10 ml/kg/h for 6 hpostoperatively. The highest-dose group (net fluid balance +2.7 L over 6 h) had increased syndecan-1 and hyaluronic acid both immediately and 6 hafter surgery, compared to the restrictive group. No differences in heparan sulfate

(continued)

Table 4.1 (continued)

Treatment	Disease state	No. of subjects	Impact on the endothelial glycocalyx
Bolus hypertonic saline (Smart et al. 2019)	ER patients with suspected sepsis	65	Patients randomized to receive 5 ml/kg 3% (hypertonic) saline or 10 ml/kg 0.9% (Normal) saline. Normal saline led to a transient decrease in syndecan-1 1 h after infusion; no difference between groups thereafter. Hypertonic saline increased osmolarity but had no effect on endothelial injury
Volatile anesthetics			
Desflurane (Oh et al. 2021)	Total knee arthroplasty with limb tourniquet	80	Randomization to intraoperative anesthesia with desflurane plus remifentanil vs. propofol plus remifentanil had no impact on syndecan-1
Sevoflurane (Kim et al. 2021)	Minimally invasive gastrectomy	136	Patients were randomized to sevoflurane/ remifentanil vs. propofol/ remifentanil. The propofol group had lower syndecan-1 at the end of the operation, but this difference was transient
Sevoflurane (Fang et al. 2021)	Cardiopulmonary bypass for elective heart valve surgery	51	Patients randomized to anesthesia with sevoflurane/ remifentanil vs. propofol/ remifentanil. Sevoflurane treatment led to less syndecan-1, heparan sulfate, and hyaluronic acid shedding in the period spanning aortic unclamping and 1 h after surgery completion; this benefit abated by 24 h
Sevoflurane (Kim et al. 2018)	Lung resection surgery with single-lung ventilation	78	Patients randomized to sevoflurane vs. propofol-based anesthesia. While single-lung ventilation was associated with a modest increase in the circulating indices of glycocalyx degradation, this was unaffected by anesthetic used

(continued)

Table 4.1 (continued)

Treatment	Disease state	No. of subjects	Impact on the endothelial glycocalyx
Sevoflurane (Nemme et al. 2020)	Open abdominal hysterectomy with rapid infusion of crystalloid	26	Patients randomized to sevoflurane vs. propofol-based anesthesia. No difference in the indices of endothelial glycocalyx shedding between randomized anesthesia groups
Sevoflurane (Maldonado et al. 2020)	Knee ligament reconstruction surgery (necessitating prolonged limb tourniquet)	16	Patients randomized to sevoflurane plus spinal anesthesia vs. propofol plus spinal anesthesia. The sevoflurane group had increased syndecan-1 at 90 min after tourniquet release; no difference at any earlier timepoint. No difference in heparan sulfate or thrombomodulin
Blood products			
Platelet transfusion (Benediktsson et al. 2019)	Thrombocytopenic patients with hematological malignancy	43	A nonrandomized study of 43 patients entailing 54 transfusion events. Syndecan-1 modestly increases immediately after platelet transfusion but returns to normal 1 h later and remains stable for the next 24 h. Changes in syndecan-1 did not correlate with transfusion-induced increases in blood platelet count
Fresh PRBC packed red blood cells (Baron-Stefaniak et al. 2019)	Critically ill patients receiving PRBC transfusion after cardiac surgery	11	Patients randomized to fresh (median 10 storage days) or standard issue (median 33 storage days) packed red blood cells. Standard issue packed red blood cells transiently increased pulmonary vascular resistance and pulmonary artery pressures. Transfusion-associated change in pulmonary artery pressures correlated with change in circulating syndecan-1

(continued)

Table 4.1 (continued)

Treatment	Disease state	No. of subjects	Impact on the endothelial glycocalyx
Ratio of platelets, plasma, and packed red blood cell transfusion (Dixon et al. 2023)	Trauma	599	Secondary analysis of the PROPPR (Pragmatic, Randomized Optimal Platelet and Plasma Ratios) trial (Holcomb et al. 2015), which randomized hemorrhaging trauma patients to 1:1:1 vs. 1:1:2 ratios of platelets:plasma: packed red blood cell transfusion. No impact of resuscitation strategy on the indices of glycocalyx degradation
Fresh frozen plasma (Straat et al. 2015)	Coagulopathic, non-bleeding critically ill patients prior to an invasive procedure	33	A nonrandomized study; patients received 12 ml/kg intravenous fresh frozen plasma, with blood drawn immediately prior and after transfusion. Syndecan-1, Von Willebrand factor, and TNFα levels decreased immediately after plasma transfusion. ADAMTS13 levels increased
Prehospital fresh frozen plasma (Gruen et al. 2020)	Severely injured trauma patients at risk of hemorrhagic shock	405	A sub-study of a randomized controlled trial of prehospital fresh frozen plasma in severely injured trauma patients (Sperry et al. 2018). In patients with the highest injury severity score, randomization to prehospital plasma decreased syndecan-1 shedding
Solvent/detergent-treated pooled plasma (Stensballe et al. 2018)	Emergency surgery for thoracic aortic dissection	44	Patients randomized to receive OctaplasLG (solvent/detergent-treated pooled plasma) versus normal fresh frozen plasma for all coagulation factor replacements. Patients randomized to OctaplasLG had more rapid normalization of syndecan-1 at 24 and 48 h compared to FFP. IL-6 levels were higher in the OctaplasLG group, paradoxically

(continued)

Table 4.1 (continued)

Treatment	Disease state	No. of subjects	Impact on the endothelial glycocalyx
Miscellaneous			
Tranexamic acid (Gruen et al. 2023)	Trauma with likely hemorrhage	766	A sub-study of the STAAMP (Study of Tranexamic Acid During Air Medical and Ground Prehospital Transport) trial of prehospital tranexamic acid (various dosing regimens) vs. placebo (Guyette et al. 2020). Syndecan-1 levels were significantly higher at all measured time points in non-surviving (30 d) patients. Prehospital transexamic acid decreased circulating syndecan-1 at 12 h in a dose-dependent manner; benefit abated by 24 h
Ischemic conditioning with brachial cuff inflation (Ikonomidis et al. 2021)	ST-elevation myocardial infarction (after percutaneous intervention)	278	Brachial cuff inflation (either one or two successive cycles of 5 min of 200-mmHg cuff pressure, then 5-min recovery) was associated with near-immediate (5 min later) improvement in the sublingual PBR 40 compared to no cuff inflation. One, but not two, cuff inflations associated with improved long-term cardiac outcomes
Spironolactone (Crompton et al. 2023)	Diabetic nephropathy	35	A sub-study of a randomized controlled trial of spironolactone vs. placebo added to usual care (including maximal ACE inhibition) (Mehdi et al. 2009). Spironolactone inhibited urinary MMP2, a glomerular glycocalyx sheddase implicated in animal models of diabetic nephropathy
Diabetic treatments (Ikonomidis et al. 2020)	Type II diabetes mellitus	160	Patients with type II diabetes already treated with metformin were randomized to insulin (40 patients), an

(continued)

Table 4.1 (continued)

Treatment	Disease state	No. of subjects	Impact on the endothelial glycocalyx
			SGLT2 inhibitor (emaglifozin, 40 patients), a GLP1 agonist (liraglutide, 40 patients), or their combination (liraglutide plus emaglifozin, 40 patients). After 12 months, patients treated with SGLT2 inhibitors, GLP1 agonists, or their combination had increased sublingual endothelial glycocalyx thickness on intravital microscopy compared to those treated with insulin, despite similar glycemic control. The improvement in glycocalyx thickness was the greatest in patients treated with combination therapy. Patients treated with GLP1 agonists or a combination of GLP1 agonists/SGLT2 inhibitors had improved indices of myocardial work. Patients treated with SGLT2 inhibitors or a combination of GLP1 agonists/SGLT2 inhibitors had improved indices of peripheral arterial stiffness
Respiratory muscle training (Campos et al. 2018)	Hemodialysis patients without pulmonary disease	41	2:1 randomization to thrice-weekly respiratory muscle training during hemodialysis sessions (29 subjects) or usual care (12 subjects) for 8 weeks. Training decreased circulating syndecan-1, angiopoietin-2, mean arterial pressure, and endothelin-1, while improving indices of pulmonary function
Targeted temperature management (Bro-Jeppesen et al. 2016)	Out-of-hospital cardiac arrest	163	A sub-study of the Targeted Temperature Management (TTM) trial (Nielsen et al. 2013). Patients were randomized to a temperature goal of 33 °C vs. 36 °C for

(continued)

Table 4.1 (continued)

Treatment	Disease state	No. of subjects	Impact on the endothelial glycocalyx
			24 h. No differences in glycocalyx degradation between groups. A 72-h change in syndecan-1 was associated with mortality on multivariate analysis
Diazoxide (Mennander et al. 2012)	Coronary artery bypass grafting surgery with cardiopulmonary bypass	13	Patients randomized to 50 μg/l diazoxide (a ATP-sensitive potassium channel) channel activator that mimics ischemic preconditioning) injected into the aortic root at the time of cross-clamping vs. no injection. Neither group differed in syndecan-1 or hyaluronic acid up to 1 hour after surgery

ARDS acute respiratory distress syndrome, *MMP* matrix metalloproteinase, *GAG* glycosaminoglycan, *TNFα* tumor necrosis factor-α, *CRP* C-reactive protein, *FFP* fresh frozen plasma, *VEGF* vascular endothelial growth factor, *VE*-cadherin, vascular endothelial cadherin, *HA* hyaluronan, *IL-6* interleukin-6, *ACE* angiotensin-converting enzyme, *SGLT2* sodium-glucose cotransporter 2, *GLP1* glucagon-like peptide-1, *ATP* adenosine

(Broekhuizen et al. 2010)) have also demonstrated benefit in acute (sepsis) and chronic (diabetes) vascular injury, respectively. However, both of these agents have multiple potential effects (e.g., anticoagulant effects of heparins and the potential role of sulodexide as a glycosaminoglycan substrate), making it difficult to ascertain whether the observed benefit was a direct consequence of heparanase inhibition.

3. *Blood products*: Fresh frozen plasma (FFP) is comprised of numerous potential glycocalyx-protective mediators such as erythrocyte-conditioned albumin (Adamson et al. 2014), apolipoprotein M (Christoffersen et al. 2011), sphingosine-1-phosphate (Zeng et al. 2014), and serine proteases such as antithrombin (Chappell et al. 2009a). These potential endothelial-protective benefits of FFP are most apparent in human trials of trauma resuscitation; for example, randomization of trauma patients at risk of hemorrhagic shock to prehospital FFP transfusion decreased indices of endothelial injury (Gruen et al. 2020). However, randomization of hemorrhaging trauma patients to a 1:1:1 ratio of platelets, FFP, and packed red blood cells had no impact on circulating glycocalyx fragments when compared to patients receiving 1:1:2 ratios of blood products (Dixon et al. 2023). In contrast to trauma, little is known about any durable endothelial effects of FFP transfusion in sepsis, with one nonrandomized observational study suggesting a modest, transient effect of FFP transfusion on syndecan-1 in non-bleeding critically ill patients (Straat et al. 2015).

4. *Intravenous fluids*: While intravenous fluids are essential for the resuscitation of critically ill patients, preclinical studies have suggested that the hemodynamic benefits of crystalloids may be offset by unanticipated injury to the endothelium (Byrne et al. 2018). Indeed, retrospective studies have largely (but not exclusively) suggested that volume overload may be associated with endothelial glycocalyx degradation (Hippensteel et al. 2019b; Saoraya et al. 2021a), although it is difficult to determine causality in the absence of randomized controlled trials specifically comparing the impact of different fluid resuscitation strategies on glycocalyx degradation. Small randomized studies of different rates of crystalloid infusion demonstrated little persistent impact on glycocalyx integrity (Saoraya et al. 2021b; Pouska et al. 2018), although high-volume intraoperative/postoperative fluid resuscitation strategies did induce increased circulating syndecan-1 in patients undergoing laparoscopic cholecystectomy (Belavić et al. 2018) or colorectal surgery (Liu et al. 2021). Colloids such as albumin (Piotti et al. 2021) and hydroxyethyl starch (Kammerer et al. 2019; Kim et al. 2017; Müller et al. 2016) have not demonstrated benefit or harm on glycocalyx integrity in critically ill and/or surgical patients. Similarly, hypertonic saline had little impact on glycocalyx integrity in patients with suspected sepsis (Smart et al. 2019). While there is interest in the potentially protective benefits of balanced crystalloids (as opposed to normal saline) in sepsis (Semler et al. 2018), a glycocalyx-protective effect of these balanced solutions as compared to normal saline is supported only by one small study performed in six healthy volunteers (Bihari et al. 2019).
5. *Volatile anesthetics*:. Inspired by preclinical studies suggesting an endothelial-protective effect of sevoflurane (Chappell et al. 2011; Annecke et al. 2010), multiple studies have compared this and other volatile anesthetics to intravenous anesthesia (i.e., propofol). Multiple small surgical studies in human patients, however, demonstrated either the absence (Oh et al. 2021; Kim et al. 2018; Nemme et al. 2020) or only transient (Kim et al. 2021; Fang et al. 2021) benefit of volatile anesthetics on glycocalyx integrity, with one (likely underpowered) study suggesting worsened glycocalyx shedding with sevoflurane (Maldonado et al. 2020).
6. *Anti-inflammatory agents*: Preclinical models have suggested a benefit of corticosteroids in endothelial glycocalyx protection (Chappell et al. 2007, 2010). Human studies, however, have been mixed. While 2 mg/kg intravenous methylprednisolone administered 10 minutes prior to laparoscopic gastrointestinal surgery decreased syndecan-1 and heparan sulfate shedding for 72 h postoperatively (Xiang et al. 2022), benefits were more modest in studies of cardiac surgery with cardiopulmonary bypass (Brettner et al. 2019), total knee arthroplasty (with limb tourniquet) (Lindberg-Larsen et al. 2017), or major abdominal surgery (Yanase et al. 2021).

In contrast to the acute vascular injury of major surgery, patients with chronic rheumatoid arthritis treated for 6 weeks with the disease-modifying (immunosuppressive) agent methotrexate decreased circulating syndecan-1 (Deyab et al. 2021). However, this potential glycocalyx protection occurred in the absence of any change in the circulating indices of inflammation, suggesting a mechanism of action independent of immunosuppression.

4.3.4 Looking to the Future: Glycocalyx Degradation as Predictive Enrichment in Clinical Trials

While multiple studies have suggested that the indices of endothelial glycocalyx degradation may independently predict outcomes in critical illness, relatively few have considered whether an individual patient's baseline severity of glycocalyx injury predicts that patient's response to an experimental therapy. This so-called "predictive enrichment"—in which an experimental agent is matched to patients at the greatest likelihood of benefitting from that agent—is a key concept in precision medicine (Beitler et al. 2021). By identifying patients with endothelial and/or endothelial glycocalyx injury at baseline, trialists (and, eventually, clinicians) may selectively target endothelial-protective treatments to those patients at the greatest likelihood of benefitting from these therapies. Relatively few studies, however, have considered baseline endothelial glycocalyx degradation as a factor that should guide treatment allocation. For example, Vigstedt et al. focused their study of prostacyclin in COVID-19 on patients with demonstrable endothelial injury (as quantified by increased circulating thrombomodulin); even with this predictive enrichment, prostacyclin did not improve endothelial glycocalyx shedding (Vigstedt et al. 2022). Gruen et al.'s sub-study of the PAMPer (Prehospital Air Medical Plasma) trial suggested that patients with baseline elevations in the circulating indices of glycocalyx degradation (as well as other markers of injury) were more likely to respond beneficially to prehospital plasma (Gruen et al. 2020). The development of rapid, point-of-care assays of glycocalyx integrity (such as intravital microscopy and/or the DMMB assay of sulfated glycosaminoglycans) may facilitate this rapid recognition of "endothelial-injured" patients that may benefit from targeted therapeutics. In addition, patient-level factors associated with baseline glycocalyx resilience (and thickness), such as sex/gender (Yuan et al. 2023), may help guide assignment of endothelial-protective therapies for those who require these therapies the most. Indeed, the greatest future impact of the endothelial glycocalyx in human disease may not be as a therapeutic target but as a quantifiable trait that could guide clinical decision-making in acute and chronic illnesses.

References

Adamson RH (1990) Permeability of frog mesenteric capillaries after partial pronase digestion of the endothelial glycocalyx. J Physiol 428:1–13

Adamson RH, Clark JF, Radeva M, Kheirolomoom A, Ferrara KW, Curry FE (2014) Albumin modulates S1P delivery from red blood cells in perfused microvessels: mechanism of the protein effect. Am J Physiol Heart Circ Physiol 306(7):H1011–H1017. https://doi.org/10.1152/ajpheart.00829.2013

Ahn SJ, Le Master E, Granados ST, Levitan I (2023) Impairment of endothelial glycocalyx in atherosclerosis and obesity. Curr Top Membr 91:1–19. https://doi.org/10.1016/bs.ctm.2023.02.001

Ando Y, Okada H, Takemura G et al (2018) Brain-specific ultrastructure of capillary endothelial Glycocalyx and its possible contribution for blood brain barrier. Sci Rep 8(1):17523. https://doi.org/10.1038/s41598-018-35976-2

Annecke T, Chappell D, Chen C et al (2010) Sevoflurane preserves the endothelial glycocalyx against ischaemia-reperfusion injury. Br J Anaesth 104(4):414–421. https://doi.org/10.1093/bja/aeq019

Arokiasamy S, King R, Boulaghrasse H et al (2019) Heparanase-dependent Remodeling of initial lymphatic Glycocalyx regulates tissue-fluid drainage during acute inflammation in vivo. Front Immunol 10:2316. https://doi.org/10.3389/fimmu.2019.02316

Baron-Stefaniak J, Leitner GC, Küntzel NKI et al (2019) Transfusion of standard-issue packed red blood cells induces pulmonary vasoconstriction in critically ill patients after cardiac surgery-a randomized, double-blinded, clinical trial. PLoS One 14(3):e0213000. https://doi.org/10.1371/journal.pone.0213000

Bastarache JA, Smith K, Jesse JJ et al (2022) A two-hit model of sepsis plus hyperoxia causes lung permeability and inflammation. Am J Phys Lung Cell Mol Phys 322(2):L273–L282. https://doi.org/10.1152/ajplung.00227.2021

Becker BF, Jacob M, Leipert S, Salmon AH, Chappell D (2015) Degradation of the endothelial glycocalyx in clinical settings: searching for the sheddases. Br J Clin Pharmacol. https://doi.org/10.1111/bcp.12629

Beitler JR, Thompson BT, Baron RM et al (2021) Advancing precision medicine for acute respiratory distress syndrome. Lancet Respir Med. https://doi.org/10.1016/s2213-2600(21)00157-0

Belavić M, Sotošek Tokmadžić V, Fišić E et al (2018) The effect of various doses of infusion solutions on the endothelial glycocalyx layer in laparoscopic cholecystectomy patients. Minerva Anestesiol 84(9):1032–1043. https://doi.org/10.23736/s0375-9393.18.12150-x

Belousoviene E, Pranskuniene Z, Vaitkaitiene E, Pilvinis V, Pranskunas A (2023) Effect of high-dose intravenous ascorbic acid on microcirculation and endothelial glycocalyx during sepsis and septic shock: a double-blind, randomized, placebo-controlled study. BMC Anesthesiol 23(1):309. https://doi.org/10.1186/s12871-023-02265-z

Benediktsson S, Kander T, Ostrowski SR, Johansson PI, Thomas OD, Schött U (2019) Platelet increment is not associated with endothelial damage in haematological patients: a prospective observational study. Scand J Clin Lab Invest 79(6):395–403. https://doi.org/10.1080/00365513.2019.1636403

Bihari S, Wiersema UF, Perry R et al (2019) Efficacy and safety of 20% albumin fluid loading in healthy subjects: a comparison of four resuscitation fluids. J Appl Physiol 126(6):1646–1660. https://doi.org/10.1152/japplphysiol.01058.2018

Bombeli T, Schwartz BR, Harlan JM (1998) Adhesion of activated platelets to endothelial cells: evidence for a GPIIbIIIa-dependent bridging mechanism and novel roles for endothelial intercellular adhesion molecule 1 (ICAM-1), alphavbeta3 integrin, and GPIbalpha. J Exp Med 187(3):329–339. https://doi.org/10.1084/jem.187.3.329

Brettner F, Chappell D, Nebelsiek T et al (2019) Preinterventional hydrocortisone sustains the endothelial glycocalyx in cardiac surgery. Clin Hemorheol Microcirc 71(1):59–70. https://doi.org/10.3233/ch-180384

Broekhuizen LN, Lemkes BA, Mooij HL et al (2010) Effect of sulodexide on endothelial glycocalyx and vascular permeability in patients with type 2 diabetes mellitus. Diabetologia 53(12):2646–2655. https://doi.org/10.1007/s00125-010-1910-x

Bro-Jeppesen J, Johansson PI, Hassager C et al (2016) Endothelial activation/injury and associations with severity of post-cardiac arrest syndrome and mortality after out-of-hospital cardiac arrest. Resuscitation 107:71–79. https://doi.org/10.1016/j.resuscitation.2016.08.006

Bruegger D, Schwartz L, Chappell D et al (2011) Release of atrial natriuretic peptide precedes shedding of the endothelial glycocalyx equally in patients undergoing on- and off-pump coronary artery bypass surgery. Basic Res Cardiol 106(6):1111–1121. https://doi.org/10.1007/s00395-011-0203-y

Buijsers B, Yanginlar C, de Nooijer A et al (2020) Increased plasma Heparanase activity in COVID-19 patients. Front Immunol 11:575047. https://doi.org/10.3389/fimmu.2020.575047

Byrne L, Obonyo NG, Diab SD et al (2018) Unintended consequences: fluid resuscitation worsens shock in an ovine model of Endotoxemia. Am J Respir Crit Care Med 198(8):1043–1054. https://doi.org/10.1164/rccm.201801-0064OC

Caironi P, Tognoni G, Masson S et al (2014) Albumin replacement in patients with severe sepsis or septic shock. N Engl J Med 370(15):1412–1421. https://doi.org/10.1056/NEJMoa1305727

Campos NG, Marizeiro DF, Florêncio ACL et al (2018) Effects of respiratory muscle training on endothelium and oxidative stress biomarkers in hemodialysis patients: a randomized clinical trial. Respir Med 134:103–109. https://doi.org/10.1016/j.rmed.2017.12.005

Cavalcante MG, Gomes P, Parente MSR et al (2023) Monitoring renal function in HIV patients without kidney disease using endothelial biomarkers: a prospective pilot study. AIDS Res Hum Retrovir 39(9):468–474. https://doi.org/10.1089/aid.2022.0113

Chappell D, Jacob M, Hofmann-Kiefer K et al (2007) Hydrocortisone preserves the vascular barrier by protecting the endothelial glycocalyx. Anesthesiology 107(5):776–784. https://doi.org/10.1097/01.anes.0000286984.39328.96

Chappell D, Jacob M, Hofmann-Kiefer K et al (2009a) Antithrombin reduces shedding of the endothelial glycocalyx following ischaemia/reperfusion. Cardiovasc Res 83(2):388–396. https://doi.org/10.1093/cvr/cvp097

Chappell D, Jacob M, Paul O et al (2009b) The glycocalyx of the human umbilical vein endothelial cell: an impressive structure ex vivo but not in culture. Circ Res 104(11): 1313–1317. https://doi.org/10.1161/CIRCRESAHA.108.187831

Chappell D, Dorfler N, Jacob M et al (2010) Glycocalyx protection reduces leukocyte adhesion after ischemia/reperfusion. Shock 34(2):133–139. https://doi.org/10.1097/SHK.0b013e3181cdc363

Chappell D, Heindl B, Jacob M et al (2011) Sevoflurane reduces leukocyte and platelet adhesion after ischemia-reperfusion by protecting the endothelial glycocalyx. Anesthesiology 115(3): 483–491. https://doi.org/10.1097/ALN.0b013e3182289988

Chappell D, Brettner F, Doerfler N et al (2014) Protection of glycocalyx decreases platelet adhesion after ischaemia/reperfusion: an animal study. Eur J Anaesthesiol 31(9):474–481. https://doi.org/10.1097/eja.0000000000000085

Chen S, Zhang X, Sun Y, Hu Z, Lu S, Ma X (2015) Unfractionated heparin attenuates intestinal injury in mouse model of sepsis by inhibiting heparanase. Int J Clin Exp Pathol 8(5):4903–4912

Chen S, He Y, Hu Z et al (2017) Heparanase mediates intestinal inflammation and injury in a mouse model of sepsis. J Histochem Cytochem 65(4):241–249. https://doi.org/10.1369/0022155417692536

Chen TT, Lv JJ, Chen L, Li M, Liu LP (2023) Heparanase inhibition leads to improvement in patients with acute gastrointestinal injuries induced by sepsis. World J Gastroenterol 29(35): 5154–5165. https://doi.org/10.3748/wjg.v29.i35.5154

Cheng MJ, Kumar R, Sridhar S, Webster TJ, Ebong EE (2016) Endothelial glycocalyx conditions influence nanoparticle uptake for passive targeting. Int J Nanomedicine 11:3305–3315. https://doi.org/10.2147/IJN.S106299

Christoffersen C, Obinata H, Kumaraswamy SB et al (2011) Endothelium-protective sphingosine-1-phosphate provided by HDL-associated apolipoprotein M. Proc Natl Acad Sci USA 108(23): 9613–9618. https://doi.org/10.1073/pnas.1103187108

Colbert JF, Schmidt EP (2016) Endothelial and microcirculatory function and dysfunction in sepsis. Clin Chest Med 37(2):263–275. https://doi.org/10.1016/j.ccm.2016.01.009

Constantinescu AA, Vink H, Spaan JA (2001) Elevated capillary tube hematocrit reflects degradation of endothelial cell glycocalyx by oxidized LDL. Am J Physiol Heart Circ Physiol 280(3): H1051–H1057. https://doi.org/10.1152/ajpheart.2001.280.3.H1051

Constantinescu AA, Vink H, Spaan JA (2003) Endothelial cell glycocalyx modulates immobilization of leukocytes at the endothelial surface. Arterioscler Thromb Vasc Biol 23(9):1541–1547. https://doi.org/10.1161/01.Atv.0000085630.24353.3d

Crompton M, Ferguson JK, Ramnath RD et al (2023) Mineralocorticoid receptor antagonism in diabetes reduces albuminuria by preserving the glomerular endothelial glycocalyx. JCI Insight 8(5). https://doi.org/10.1172/jci.insight.154164

Curry FE (2018) The molecular structure of the endothelial Glycocalyx layer (EGL) and surface layers (ESL) modulation of Transvascular exchange. Adv Exp Med Biol 1097:29–49. https://doi.org/10.1007/978-3-319-96445-4_2

Curry FE, Adamson RH (2012) Endothelial glycocalyx: permeability barrier and mechanosensor. Ann Biomed Eng 40(4):828–839. https://doi.org/10.1007/s10439-011-0429-8

Damén T, Saadati S, Forssell-Aronsson E et al (2021) Effects of different mean arterial pressure targets on plasma volume, ANP and glycocalyx-a randomized trial. Acta Anaesthesiol Scand 65(2):220–227. https://doi.org/10.1111/aas.13710

Damron FH, Oglesby-Sherrouse AG, Wilks A, Barbier M (2016) Dual-seq transcriptomics reveals the battle for iron during Pseudomonas aeruginosa acute murine pneumonia. Sci Rep 6:39172. https://doi.org/10.1038/srep39172

de Fougerolles AR, Stacker SA, Schwarting R, Springer TA (1991) Characterization of ICAM-2 and evidence for a third counter-receptor for LFA-1. J Exp Med 174(1):253–267. https://doi.org/10.1084/jem.174.1.253

Devaraj S, Yun JM, Adamson G, Galvez J, Jialal I (2009) C-reactive protein impairs the endothelial glycocalyx resulting in endothelial dysfunction. Cardiovasc Res 84(3):479–484. https://doi.org/10.1093/cvr/cvp249

Deyab G, Hokstad I, Whist JE et al (2017) Anti-rheumatic treatment is not associated with reduction of pentraxin 3 in rheumatoid arthritis, psoriatic arthritis and ankylosing spondylitis. PLoS One 12(2):e0169830. https://doi.org/10.1371/journal.pone.0169830

Deyab G, Reine TM, Vuong TT et al (2021) Antirheumatic treatment is associated with reduced serum Syndecan-1 in rheumatoid arthritis. PLoS One 16(7):e0253247. https://doi.org/10.1371/journal.pone.0253247

Dimitrievska S, Gui L, Weyers A et al (2016) New functional tools for Antithrombogenic activity assessment of live surface Glycocalyx. Arterioscler Thromb Vasc Biol 36(9):1847–1853. https://doi.org/10.1161/ATVBAHA.116.308023

Dixon A, Kenny JE, Buzzard L et al (2023) Acute respiratory distress syndrome, acute kidney injury, and mortality after trauma are associated with increased circulation of syndecan-1, soluble thrombomodulin, and receptor for advanced glycation end products. J Trauma Acute Care Surg. https://doi.org/10.1097/ta.0000000000004096

Drost CC, Rovas A, Kusche-Vihrog K et al (2019) Tie2 activation promotes protection and reconstitution of the endothelial Glycocalyx in human sepsis. Thromb Haemost 119(11): 1827–1838. https://doi.org/10.1055/s-0039-1695768

Drost CC, Rovas A, Osiaevi I et al (2022) Heparanase is a putative mediator of endothelial Glycocalyx damage in COVID-19 – a proof-of-concept study. Front Immunol 13:916512. https://doi.org/10.3389/fimmu.2022.916512

Drost CC, Unger A, Linke WA, Vink H, Kümpers P (2023a) The dynamics of the endothelial glycocalyx: a rare snapshot by electron microscopy. Angiogenesis 26(4):487–491. https://doi.org/10.1007/s10456-023-09880-x

Drost CC, Unger A, Linke WA, Vink H, Kumpers P (2023b) The dynamics of the endothelial glycocalyx: a rare snapshot by electron microscopy. Angiogenesis 26(4):487–491. https://doi.org/10.1007/s10456-023-09880-x

Druzak S, Iffrig E, Roberts BR et al (2023) Multiplatform analyses reveal distinct drivers of systemic pathogenesis in adult versus pediatric severe acute COVID-19. Nat Commun 14(1): 1638. https://doi.org/10.1038/s41467-023-37269-3

Dull RO, Cluff M, Kingston J et al (2012) Lung heparan sulfates modulate K(fc) during increased vascular pressure: evidence for glycocalyx-mediated mechanotransduction. Am J Physiol Lung Cell Mol Physiol 302(9):L816–L828. https://doi.org/10.1152/ajplung.00080.2011

Ebong EE, Macaluso FP, Spray DC, Tarbell JM (2011) Imaging the endothelial glycocalyx in vitro by rapid freezing/freeze substitution transmission electron microscopy. Arterioscler Thromb Vasc Biol 31(8):1908–1915. https://doi.org/10.1161/ATVBAHA.111.225268

Eickhoff MK, Winther SA, Hansen TW et al (2020) Assessment of the sublingual microcirculation with the GlycoCheck system: reproducibility and examination conditions. PLoS One 15(12): e0243737. https://doi.org/10.1371/journal.pone.0243737

Eriksson S, Fraser JRE, Laurent TC, Pertoft H, Smedsrød B (1983) Endothelial cells are a site of uptake and degradation of hyaluronic acid in the liver. Exp Cell Res 144(1):223–228. https://doi.org/10.1016/0014-4827(83)90458-5

Eustes AS, Campbell RA, Middleton EA et al (2021) Heparanase expression and activity are increased in platelets during clinical sepsis. J Thromb Haemost 19(5):1319–1330. https://doi.org/10.1111/jth.15266

Fang FQ, Sun JH, Wu QL et al (2021) Protective effect of sevoflurane on vascular endothelial glycocalyx in patients undergoing heart valve surgery: a randomised controlled trial. Eur J Anaesthesiol 38(5):477–486. https://doi.org/10.1097/eja.0000000000001429

Fernández-Sarmiento J, Molina CF, Salazar-Pelaez LM et al (2023) Biomarkers of Glycocalyx injury and endothelial activation are associated with clinical outcomes in patients with sepsis: a systematic review and meta-analysis. J Intensive Care Med 38(1):95–105. https://doi.org/10.1177/08850666221109186

Florian JA, Kosky JR, Ainslie K, Pang Z, Dull RO, Tarbell JM (2003) Heparan sulfate proteoglycan is a mechanosensor on endothelial cells. Circ Res 93(10):e136–e142. https://doi.org/10.1161/01.RES.0000101744.47866.D5

Fowler AA 3rd, Truwit JD, Hite RD et al (2019) Effect of vitamin C infusion on organ failure and biomarkers of inflammation and vascular injury in patients with sepsis and severe acute respiratory failure: the CITRIS-ALI randomized clinical trial. JAMA 322(13):1261–1270. https://doi.org/10.1001/jama.2019.11825

Fragkou PC, Ikonomidis I, Benas D et al (2023) Endothelial Glycocalyx integrity in treatment-naïve people living with HIV before and one year after antiretroviral treatment initiation. Viruses 15(7). https://doi.org/10.3390/v15071505

Funderburgh JL (2000) Keratan sulfate: structure, biosynthesis, and function. Glycobiology 10(10): 951–958. https://doi.org/10.1093/glycob/10.10.951

Gao W, Fang F, Xia TJ et al (2022) Doxycycline can reduce glycocalyx shedding by inhibiting matrix metalloproteinases in patients undergoing cardiopulmonary bypass: a randomized controlled trial. Microvasc Res 142:104381. https://doi.org/10.1016/j.mvr.2022.104381

Gianesini S, Rimondi E, Raffetto JD et al (2023) Human collecting lymphatic glycocalyx identification by electron microscopy and immunohistochemistry. Sci Rep 13(1):3022. https://doi.org/10.1038/s41598-023-30043-x

Gómez BI, Dubick MA, Schmidt EP, Shupp JW, Burmeister DM (2020) Plasma and urinary Glycosaminoglycans as evidence for Endotheliopathy in a swine burn model. J Surg Res 248: 28–37. https://doi.org/10.1016/j.jss.2019.11.006

Gonzalez Rodriguez E, Ostrowski SR, Cardenas JC et al (2017) Syndecan-1: a quantitative marker for the Endotheliopathy of trauma. J Am Coll Surg 225(3):419–427. https://doi.org/10.1016/j.jamcollsurg.2017.05.012

Groen BB, Hamer HM, Snijders T et al (2014) Skeletal muscle capillary density and microvascular function are compromised with aging and type 2 diabetes. J Appl Physiol 116(8):998–1005. https://doi.org/10.1152/japplphysiol.00919.2013

Gruen DS, Brown JB, Guyette FX et al (2020) Prehospital plasma is associated with distinct biomarker expression following injury. JCI Insight 5(8). https://doi.org/10.1172/jci.insight.135350

Gruen DS, Brown JB, Guyette FX et al (2023) Prehospital tranexamic acid is associated with a dose-dependent decrease in syndecan-1 after trauma: a secondary analysis of a prospective randomized trial. J Trauma Acute Care Surg 95(5):642–648. https://doi.org/10.1097/ta.0000000000003955

Grundmann S, Fink K, Rabadzhieva L et al (2012) Perturbation of the endothelial glycocalyx in post cardiac arrest syndrome. Resuscitation 83(6):715–720. https://doi.org/10.1016/j.resuscitation.2012.01.028

Guyette FX, Brown JB, Zenati MS et al (2020) Tranexamic acid during prehospital transport in patients at risk for Hemorrhage after injury: a double-blind, placebo-controlled, randomized clinical trial. JAMA Surg 156(1):11–20. https://doi.org/10.1001/jamasurg.2020.4350

Gybel-Brask M, Rasmussen R, Stensballe J, Johansson PI, Ostrowski SR (2017) Effect of delayed onset prostacyclin on markers of endothelial function and damage after subarachnoid hemorrhage. Acta Neurochir 159(6):1073–1078. https://doi.org/10.1007/s00701-017-3168-2

Haeger SM, Yang Y, Schmidt EP (2016) Heparan Sulfate in the developing, healthy, and injured lung. Am J Respir Cell Mol Biol 55(1):5–11. https://doi.org/10.1165/rcmb.2016-0043TR

Hakomori S (2003) Structure, organization, and function of glycosphingolipids in membrane. Curr Opin Hematol 10(1):16–24. https://doi.org/10.1097/00062752-200301000-00004

Han S, Lee SJ, Kim KE et al (2016) Amelioration of sepsis by TIE2 activation-induced vascular protection. Sci Transl Med 8(335):335ra55. https://doi.org/10.1126/scitranslmed.aad9260

Harding IC, Mitra R, Mensah SA, Herman IM, Ebong EE (2018) Pro-atherosclerotic disturbed flow disrupts caveolin-1 expression, localization, and function via glycocalyx degradation. J Transl Med 16(1):364. https://doi.org/10.1186/s12967-018-1721-2

Hayashida A, Bartlett AH, Foster TJ, Park PW (2009) Staphylococcus aureus beta-toxin induces lung injury through syndecan-1. Am J Pathol 174(2):509–518. https://doi.org/10.2353/ajpath.2009.080394

Henninger DD, Panés J, Eppihimer M et al (1997) Cytokine-induced VCAM-1 and ICAM-1 expression in different organs of the mouse. J Immunol 158(4):1825–1832

Hippensteel JA, Anderson BJ, Orfila JE et al (2019a) Circulating heparan sulfate fragments mediate septic cognitive dysfunction. J Clin Invest 129(4):1779–1784. https://doi.org/10.1172/jci124485

Hippensteel JA, Uchimido R, Tyler PD et al (2019b) Intravenous fluid resuscitation is associated with septic endothelial glycocalyx degradation. Crit Care 23(1):259. https://doi.org/10.1186/s13054-019-2534-2

Hofmann-Kiefer KF, Kemming GI, Chappell D et al (2009) Serum heparan sulfate levels are elevated in endotoxemia. Eur J Med Res 14(12):526–531. https://doi.org/10.1186/2047-783x-14-12-526

Holcomb JB, Tilley BC, Baraniuk S et al (2015) Transfusion of plasma, platelets, and red blood cells in a 1:1:1 vs a 1:1:2 ratio and mortality in patients with severe trauma: the PROPPR randomized clinical trial. JAMA 313(5):471–482. https://doi.org/10.1001/jama.2015.12

Holmvang L, Ostrowski SR, Dridi NP, Johansson P (2012) A single center, open, randomized study investigating the clinical safety and the endothelial modulating effects of a prostacyclin analog in combination with eptifibatide in patients having undergone primary percutaneous coronary intervention (PCI) for ST-segment elevation myocardial infarction. Prostaglandins Other Lipid Mediat 99(3–4):87–95. https://doi.org/10.1016/j.prostaglandins.2012.08.002

Huxley VH, Curry FE (1991) Differential actions of albumin and plasma on capillary solute permeability. Am J Phys 260(5 Pt 2):H1645–H1654. https://doi.org/10.1152/ajpheart.1991.260.5.H1645

Ihrcke NS, Wrenshall LE, Lindman BJ, Platt JL (1993) Role of heparan sulfate in immune system-blood vessel interactions. Immunol Today 14(10):500–505. https://doi.org/10.1016/0167-5699(93)90265-m

Ikeda M, Matsumoto H, Ogura H et al (2018) Circulating syndecan-1 predicts the development of disseminated intravascular coagulation in patients with sepsis. J Crit Care 43:48–53. https://doi.org/10.1016/j.jcrc.2017.07.049

Ikonomidis I, Marinou M, Vlastos D et al (2017) Effects of varenicline and nicotine replacement therapy on arterial elasticity, endothelial glycocalyx and oxidative stress during a 3-month smoking cessation program. Atherosclerosis 262:123–130. https://doi.org/10.1016/j.atherosclerosis.2017.05.012

Ikonomidis I, Pavlidis G, Thymis J et al (2020) Effects of glucagon-like Peptide-1 receptor agonists, sodium-glucose Cotransporter-2 inhibitors, and their combination on endothelial Glycocalyx, arterial function, and myocardial work index in patients with type 2 diabetes mellitus after 12-month treatment. J Am Heart Assoc 9(9):e015716. https://doi.org/10.1161/jaha.119.015716

Ikonomidis I, Vlastos D, Andreadou I et al (2021) Vascular conditioning prevents adverse left ventricular remodelling after acute myocardial infarction: a randomised remote conditioning study. Basic Res Cardiol 116(1):9. https://doi.org/10.1007/s00395-021-00851-1

Ikonomidis I, Katogiannis K, Chania C et al (2023) Association of hydroxytyrosol enriched olive oil with vascular function in chronic coronary disease. Eur J Clin Investig 53(7):e13983. https://doi.org/10.1111/eci.13983

Ince C, Mayeux PR, Nguyen T et al (2016) The endothelium in sepsis. Shock 45(3):259–270. https://doi.org/10.1097/SHK.0000000000000473

Jacob M, Bruegger D, Rehm M et al (2007) The endothelial glycocalyx affords compatibility of Starling's principle and high cardiac interstitial albumin levels. Cardiovasc Res 73(3):575–586. https://doi.org/10.1016/j.cardiores.2006.11.021

Jensen JU, Hein L, Lundgren B et al (2011) Procalcitonin-guided interventions against infections to increase early appropriate antibiotics and improve survival in the intensive care unit: a randomized trial. Crit Care Med 39(9):2048–2058. https://doi.org/10.1097/CCM.0b013e31821e8791

Jiang D, Liang J, Fan J et al (2005) Regulation of lung injury and repair by toll-like receptors and hyaluronan. Nat Med 11(11):1173–1179. https://doi.org/10.1038/nm1315

Johansen ME, Johansson PI, Ostrowski SR et al (2015) Profound endothelial damage predicts impending organ failure and death in sepsis. Semin Thromb Hemost 41(1):16–25. https://doi.org/10.1055/s-0034-1398377

Johansson PI, Stensballe J, Rasmussen LS, Ostrowski SR (2011a) A high admission syndecan-1 level, a marker of endothelial glycocalyx degradation, is associated with inflammation, protein C depletion, fibrinolysis, and increased mortality in trauma patients. Ann Surg 254(2):194–200. https://doi.org/10.1097/SLA.0b013e318226113d

Johansson PI, Sørensen AM, Perner A et al (2011b) Disseminated intravascular coagulation or acute coagulopathy of trauma shock early after trauma? An observational study. Crit Care 15(6): R272. https://doi.org/10.1186/cc10553

Johansson PI, Henriksen HH, Stensballe J et al (2017) Traumatic Endotheliopathy: a prospective observational study of 424 severely injured patients. Ann Surg 265(3):597–603. https://doi.org/10.1097/sla.0000000000001751

Johansson PI, Søe-Jensen P, Bestle MH et al (2022) Prostacyclin in intubated patients with COVID-19 and severe Endotheliopathy: a Multicenter, randomized clinical trial. Am J Respir Crit Care Med 205(3):324–329. https://doi.org/10.1164/rccm.202108-1855OC

Jung C, Fuernau G, Muench P et al (2015) Impairment of the endothelial glycocalyx in cardiogenic shock and its prognostic relevance. Shock 43(5):450–455. https://doi.org/10.1097/shk.0000000000000329

Kammerer T, Hulde N, Speck E et al (2019) Effects of balanced hydroxyethyl starch 6% (130/0.4) and albumin 5% on clot formation and glycocalyx shedding: subgroup analysis of a prospective randomized trial. Thromb Res 183:111–118. https://doi.org/10.1016/j.thromres.2019.10.020

Kanda T, Ariga T, Kubodera H et al (2004) Glycosphingolipid composition of primary cultured human brain microvascular endothelial cells. J Neurosci Res 78(1):141–150. https://doi.org/10.1002/jnr.20228

Kaneko T, Tatara T, Hirose M (2020) Effects of anaesthesia-induced hypotension and phenylephrine on plasma volume expansion by hydroxyethyl starch: a randomised controlled study. Acta Anaesthesiol Scand 64(5):620–627. https://doi.org/10.1111/aas.13548

Kim TK, Nam K, Cho YJ et al (2017) Microvascular reactivity and endothelial glycocalyx degradation when administering hydroxyethyl starch or crystalloid during off-pump coronary artery bypass graft surgery: a randomised trial. Anaesthesia 72(2):204–213. https://doi.org/10.1111/anae.13642

Kim HJ, Kim E, Baek SH et al (2018) Sevoflurane did not show better protective effect on endothelial glycocalyx layer compared to propofol during lung resection surgery with one lung ventilation. J Thorac Dis 10(3):1468–1475. https://doi.org/10.21037/jtd.2018.03.44

Kim NY, Kim KJ, Lee KY et al (2021) Effect of volatile and total intravenous anesthesia on syndecan-1 shedding after minimally invasive gastrectomy: a randomized trial. Sci Rep 11(1): 1511. https://doi.org/10.1038/s41598-021-81012-1

Krieglstein CF, Granger DN (2001) Adhesion molecules and their role in vascular disease. Am J Hypertens 14(6 Pt 2):44s–54s. https://doi.org/10.1016/s0895-7061(01)02069-6
Kucharz K, Janiurek MM, Christoffersen C, Lauritzen M (2022) Two-photon microscopy in vivo reveals brain vessel type-specific loss of glycocalyx caused by apoM/S1P signaling impairment. bioRxiv 2022:2022.04.11.487803. https://doi.org/10.1101/2022.04.11.487803
Lambadiari V, Pavlidis G, Kousathana F et al (2019) Effects of different antidiabetic medications on endothelial Glycocalyx, myocardial function, and vascular function in type 2 diabetic patients: one year follow-up study. J Clin Med 8(7). https://doi.org/10.3390/jcm8070983
Lin PK, Davis GE (2023) Extracellular matrix Remodeling in vascular disease: defining its regulators and pathological influence. Arterioscler Thromb Vasc Biol 43(9):1599–1616. https://doi.org/10.1161/atvbaha.123.318237
Lindberg-Larsen V, Ostrowski SR, Lindberg-Larsen M, Rovsing ML, Johansson PI, Kehlet H (2017) The effect of pre-operative methylprednisolone on early endothelial damage after total knee arthroplasty: a randomised, double-blind, placebo-controlled trial. Anaesthesia 72(10): 1217–1224. https://doi.org/10.1111/anae.13983
Lipowsky HH (2005) Microvascular rheology and hemodynamics. Microcirculation 12(1):5–15. https://doi.org/10.1080/10739680590894966
Lipowsky HH (2011) Protease activity and the role of the endothelial Glycocalyx in inflammation. Drug Discov Today Dis Models 8(1):57–62. https://doi.org/10.1016/j.ddmod.2011.05.004
Lipowsky HH (2012) The endothelial glycocalyx as a barrier to leukocyte adhesion and its mediation by extracellular proteases. Ann Biomed Eng 40(4):840–848. https://doi.org/10.1007/s10439-011-0427-x
Liu Y, Chen G, Gao J et al (2021) Effect of different levels of stroke volume variation on the endothelial glycocalyx of patients undergoing colorectal surgery: a randomized clinical trial. Exp Physiol 106(10):2124–2132. https://doi.org/10.1113/ep089348
Lukasz A, Hillgruber C, Oberleithner H et al (2017) Endothelial glycocalyx breakdown is mediated by angiopoietin-2. Cardiovasc Res 113(6):671–680. https://doi.org/10.1093/cvr/cvx023
Lygizos MI, Yang Y, Altmann CJ et al (2013) Heparanase mediates renal dysfunction during early sepsis in mice. Physiol Rep 1(6):e00153. https://doi.org/10.1002/phy2.153
Machin DR, Bloom SI, Campbell RA et al (2018) Advanced age results in a diminished endothelial glycocalyx. Am J Physiol Heart Circ Physiol 315(3):H531–H539. https://doi.org/10.1152/ajpheart.00104.2018
Maldonado F, Morales D, Gutiérrez R, Barahona M, Cerda O, Cáceres M (2020) Effect of sevoflurane and propofol on tourniquet-induced endothelial damage: a pilot randomized controlled trial for knee-ligament surgery. BMC Anesthesiol 20(1):121. https://doi.org/10.1186/s12871-020-01030-w
Marsh G, Waugh RE (2013) Quantifying the mechanical properties of the endothelial glycocalyx with atomic force microscopy. J Vis Exp 72:e50163. https://doi.org/10.3791/50163
Martin FA, Murphy RP, Cummins PM (2013) Thrombomodulin and the vascular endothelium: insights into functional, regulatory, and therapeutic aspects. Am J Phys Heart Circ Phys 304(12):H1585–H1597. https://doi.org/10.1152/ajpheart.00096.2013
Martin L, De Santis R, Koczera P et al (2015) The synthetic antimicrobial peptide 19-2.5 interacts with Heparanase and Heparan Sulfate in murine and human sepsis. PLoS One 10(11):e0143583. https://doi.org/10.1371/journal.pone.0143583
Matan M, King D, Peled E et al (2018) Heparanase level and procoagulant activity are reduced in severe sepsis. Eur J Haematol 100(2):182–188. https://doi.org/10.1111/ejh.12997
McKee CM, Penno MB, Cowman M et al (1996) Hyaluronan (HA) fragments induce chemokine gene expression in alveolar macrophages. The role of HA size and CD44. J Clin Invest 98(10): 2403–2413. https://doi.org/10.1172/jci119054
Megens RT, Reitsma S, Schiffers PH et al (2007) Two-photon microscopy of vital murine elastic and muscular arteries. Combined structural and functional imaging with subcellular resolution. J Vasc Res 44(2):87–98. https://doi.org/10.1159/000098259

Mehdi UF, Adams-Huet B, Raskin P, Vega GL, Toto RD (2009) Addition of angiotensin receptor blockade or mineralocorticoid antagonism to maximal angiotensin-converting enzyme inhibition in diabetic nephropathy. J Am Soc Nephrol 20(12):2641–2650. https://doi.org/10.1681/asn.2009070737

Meneses GC, Cavalcante MG, da Silva Junior GB et al (2017) Endothelial Glycocalyx damage and renal dysfunction in HIV patients receiving combined antiretroviral therapy. AIDS Res Hum Retrovir 33(7):703–710. https://doi.org/10.1089/aid.2016.0284

Mennander AA, Shalaby A, Oksala N et al (2012) Diazoxide may protect endothelial glycocalyx integrity during coronary artery bypass grafting. Scand Cardiovasc J 46(6):339–344. https://doi.org/10.3109/14017431.2012.717303

Michel CC, Curry FR (2009) Glycocalyx volume: a critical review of tracer dilution methods for its measurement. Microcirculation 16(3):213–219. https://doi.org/10.1080/10739680802527404

Mizumoto S, Yamada S (2022) The specific role of dermatan Sulfate as an instructive glycosaminoglycan in tissue development. Int J Mol Sci 23(13). https://doi.org/10.3390/ijms23137485

Morris A, Wang B, Waern I et al (2015) The role of heparanase in pulmonary cell recruitment in response to an allergic but not non-allergic stimulus. PLoS One 10(6):e0127032. https://doi.org/10.1371/journal.pone.0127032

Mulivor AW, Lipowsky HH (2002) Role of glycocalyx in leukocyte-endothelial cell adhesion. Am J Physiol Heart Circ Physiol 283(4):H1282–H1291. https://doi.org/10.1152/ajpheart.00117.2002

Mulivor AW, Lipowsky HH (2009) Inhibition of glycan shedding and leukocyte-endothelial adhesion in postcapillary venules by suppression of matrixmetalloprotease activity with doxycycline. Microcirculation 16(8):657–666. https://doi.org/10.3109/10739680903133714

Müller RB, Ostrowski SR, Haase N, Wetterslev J, Perner A, Johansson PI (2016) Markers of endothelial damage and coagulation impairment in patients with severe sepsis resuscitated with hydroxyethyl starch 130/0.42 vs ringer acetate. J Crit Care 32:16–20. https://doi.org/10.1016/j.jcrc.2015.11.025

Murphy LS, Wickersham N, McNeil JB et al (2017) Endothelial glycocalyx degradation is more severe in patients with non-pulmonary sepsis compared to pulmonary sepsis and associates with risk of ARDS and other organ dysfunction. Ann Intensive Care 7(1):102. https://doi.org/10.1186/s13613-017-0325-y

Naumann DN, Hazeldine J, Davies DJ et al (2018) Endotheliopathy of trauma is an on-scene phenomenon, and is associated with multiple organ dysfunction syndrome: a prospective observational study. Shock 49(4):420–428. https://doi.org/10.1097/shk.0000000000000999

Nelson A, Berkestedt I, Schmidtchen A, Ljunggren L, Bodelsson M (2008) Increased levels of glycosaminoglycans during septic shock: relation to mortality and the antibacterial actions of plasma. Shock 30(6):623–627. https://doi.org/10.1097/SHK.0b013e3181777da3

Nelson A, Berkestedt I, Bodelsson M (2014) Circulating glycosaminoglycan species in septic shock. Acta Anaesthesiol Scand 58(1):36–43. https://doi.org/10.1111/aas.12223

Nemme J, Krizhanovskii C, Ntika S, Sabelnikovs O, Vanags I, Hahn RG (2020) Hypervolemia does not cause degradation of the endothelial glycocalyx layer during open hysterectomy performed under sevoflurane or propofol anesthesia. Acta Anaesthesiol Scand 64(4):538–545. https://doi.org/10.1111/aas.13511

Nielsen N, Wetterslev J, Cronberg T et al (2013) Targeted temperature management at 33°C versus 36°C after cardiac arrest. N Engl J Med 369(23):2197–2206. https://doi.org/10.1056/NEJMoa1310519

Nieuwdorp M, Meuwese MC, Mooij HL et al (2008) Measuring endothelial glycocalyx dimensions in humans: a potential novel tool to monitor vascular vulnerability. J Appl Physiol 104(3):845–852. https://doi.org/10.1152/japplphysiol.00440.2007

Nieuwdorp M, van Haeften TW, Gouverneur MC et al (2006) Loss of endothelial glycocalyx during acute hyperglycemia coincides with endothelial dysfunction and coagulation activation in vivo. Diabetes 55(2):480–486. https://doi.org/10.2337/diabetes.55.02.06.db05-1103

Nieuwdorp M, Meuwese MC, Mooij HL et al (2009) Tumor necrosis factor-alpha inhibition protects against endotoxin-induced endothelial glycocalyx perturbation. Atherosclerosis 202(1):296–303. https://doi.org/10.1016/j.atherosclerosis.2008.03.024

O'Callaghan R, Job KM, Dull RO, Hlady V (2011) Stiffness and heterogeneity of the pulmonary endothelial glycocalyx measured by atomic force microscopy. Am J Physiol Lung Cell Mol Physiol 301(3):353–360. https://doi.org/10.1152/ajplung.00342.2010

Oh CS, Choi JM, Park EH et al (2021) Impact of Anesthetic agents on endothelial Glycocalyx injury during Total knee arthroplasty: Desflurane- vs. Propofol-based Anesthesia-a prospective randomized controlled trial. Biomed Res Int 2021:8880267. https://doi.org/10.1155/2021/8880267

Okada H, Takemura G, Suzuki K et al (2017) Three-dimensional ultrastructure of capillary endothelial glycocalyx under normal and experimental endotoxemic conditions. Crit Care 21(1):261. https://doi.org/10.1186/s13054-017-1841-8

Ono S, Egawa G, Kabashima K (2017) Regulation of blood vascular permeability in the skin. Inflamm Regen 37:11. https://doi.org/10.1186/s41232-017-0042-9

Oshima K, Han X, Ouyang Y et al (2019) Loss of endothelial sulfatase-1 after experimental sepsis attenuates subsequent pulmonary inflammatory responses. Am J Physiol Lung Cell Mol Physiol 317(5):L667–l677. https://doi.org/10.1152/ajplung.00175.2019

Oshima K, Siddiqui N, Orfila JE et al (2023) A role for decorin in improving motor deficits after traumatic brain injury. Matrix Biol. https://doi.org/10.1016/j.matbio.2023.12.005

Ostrowski SR, Sørensen AM, Windeløv NA et al (2012) High levels of soluble VEGF receptor 1 early after trauma are associated with shock, sympathoadrenal activation, glycocalyx degradation and inflammation in severely injured patients: a prospective study. Scand J Trauma Resusc Emerg Med 20:27. https://doi.org/10.1186/1757-7241-20-27

Ostrowski SR, Pedersen SH, Jensen JS, Mogelvang R, Johansson PI (2013) Acute myocardial infarction is associated with endothelial glycocalyx and cell damage and a parallel increase in circulating catecholamines. Crit Care 17(1):R32. https://doi.org/10.1186/cc12532

Ostrowski SR, Gaïni S, Pedersen C, Johansson PI (2015a) Sympathoadrenal activation and endothelial damage in patients with varying degrees of acute infectious disease: an observational study. J Crit Care 30(1):90–96. https://doi.org/10.1016/j.jcrc.2014.10.006

Ostrowski SR, Haase N, Müller RB et al (2015b) Association between biomarkers of endothelial injury and hypocoagulability in patients with severe sepsis: a prospective study. Crit Care 19(1): 191. https://doi.org/10.1186/s13054-015-0918-5

Pahakis MY, Kosky JR, Dull RO, Tarbell JM (2007) The role of endothelial glycocalyx components in mechanotransduction of fluid shear stress. Biochem Biophys Res Commun 355(1): 228–233. https://doi.org/10.1016/j.bbrc.2007.01.137

Patil NP, Gómez-Hernández A, Zhang F et al (2022) Rhamnan sulfate reduces atherosclerotic plaque formation and vascular inflammation. Biomaterials 291:121865. https://doi.org/10.1016/j.biomaterials.2022.121865

Perner A, Haase N, Guttormsen AB et al (2012) Hydroxyethyl starch 130/0.42 versus ringer's acetate in severe sepsis. N Engl J Med 367(2):124–134. https://doi.org/10.1056/NEJMoa1204242

Piotti A, Novelli D, Meessen J et al (2021) Endothelial damage in septic shock patients as evidenced by circulating syndecan-1, sphingosine-1-phosphate and soluble VE-cadherin: a substudy of ALBIOS. Crit Care 25(1):113. https://doi.org/10.1186/s13054-021-03545-1

Potter DR, Jiang J, Damiano ER (2009) The recovery time course of the endothelial cell glycocalyx in vivo and its implications in vitro. Circ Res 104(11):1318–1325. https://doi.org/10.1161/CIRCRESAHA.108.191585

Pouska J, Tegl V, Astapenko D, Cerny V, Lehmann C, Benes J (2018) Impact of intravenous fluid challenge infusion time on macrocirculation and endothelial Glycocalyx in surgical and critically ill patients. Biomed Res Int 2018:8925345. https://doi.org/10.1155/2018/8925345

Privratsky JR, Tilkens SB, Newman DK, Newman PJ (2012) PECAM-1 dampens cytokine levels during LPS-induced endotoxemia by regulating leukocyte trafficking. Life Sci 90(5–6): 177–184. https://doi.org/10.1016/j.lfs.2011.11.002

Pustetto M, Goldsztejn N, Touihri K, Engelman E, Ickx B, Van Obbergh L (2020) Intravenous lidocaine to prevent endothelial dysfunction after major abdominal surgery: a randomized controlled pilot trial. BMC Anesthesiol 20(1):155. https://doi.org/10.1186/s12871-020-01075-x

Qiao X, Kashiouris MG, L'Heureux M et al (2022) Biological effects of intravenous vitamin C on neutrophil extracellular traps and the endothelial Glycocalyx in patients with sepsis-induced ARDS. Nutrients 14(20). https://doi.org/10.3390/nu14204415

Queisser KA, Mellema RA, Middleton EA et al (2021) COVID-19 generates hyaluronan fragments that directly induce endothelial barrier dysfunction. JCI Insight 6(17). https://doi.org/10.1172/jci.insight.147472

Rancan L, Simón C, Sánchez Pedrosa G et al (2018) Glycocalyx degradation after pulmonary transplantation surgery. Eur Surg Res 59(3–4):115–125. https://doi.org/10.1159/000489492

Rangarajan S, Richter JR, Richter RP et al (2020) Heparanase-enhanced shedding of Syndecan-1 and its role in driving disease pathogenesis and progression. J Histochem Cytochem 68(12): 823–840. https://doi.org/10.1369/0022155420937087

Rehm M, Bruegger D, Christ F et al (2007) Shedding of the endothelial glycocalyx in patients undergoing major vascular surgery with global and regional ischemia. Circulation 116(17): 1896–1906. https://doi.org/10.1161/circulationaha.106.684852

Reitsma S, Slaaf DW, Vink H, van Zandvoort MA et al (2007) The endothelial glycocalyx: composition, functions, and visualization. Pflugers Arch 454(3):345–359. https://doi.org/10.1007/s00424-007-0212-8

Rizzo AN, Haeger SM, Oshima K et al (2022) Alveolar epithelial glycocalyx degradation mediates surfactant dysfunction and contributes to acute respiratory distress syndrome. JCI. Insight 7(2). https://doi.org/10.1172/jci.insight.154573

Rosenberg RD, Shworak NW, Liu J, Schwartz JJ, Zhang L (1997) Heparan sulfate proteoglycans of the cardiovascular system. Specific structures emerge but how is synthesis regulated? J Clin Invest 100(11 Suppl):S67–S75

Rovas A, Seidel LM, Vink H et al (2019) Association of sublingual microcirculation parameters and endothelial glycocalyx dimensions in resuscitated sepsis. Crit Care 23(1):260. https://doi.org/10.1186/s13054-019-2542-2

Ruegg C, Mariotti A (2003) Vascular integrins: pleiotropic adhesion and signaling molecules in vascular homeostasis and angiogenesis. Cell Mol Life Sci 60(6):1135–1157. https://doi.org/10.1007/s00018-003-2297-3

Saoraya J, Wongsamita L, Srisawat N, Musikatavorn K (2021a) Plasma syndecan-1 is associated with fluid requirements and clinical outcomes in emergency department patients with sepsis. Am J Emerg Med 42:83–89. https://doi.org/10.1016/j.ajem.2021.01.019

Saoraya J, Wongsamita L, Srisawat N, Musikatavorn K (2021b) The effects of a limited infusion rate of fluid in the early resuscitation of sepsis on glycocalyx shedding measured by plasma syndecan-1: a randomized controlled trial. J Intensive Care 9(1):1. https://doi.org/10.1186/s40560-020-00515-7

Schiefer J, Faybik P, Koch S et al (2020) Glycocalyx damage within human liver grafts correlates with graft injury and postoperative graft function after Orthotopic liver transplantation. Transplantation 104(1):72–78. https://doi.org/10.1097/tp.0000000000002838

Schmidt EP, Yang Y, Janssen WJ et al (2012) The pulmonary endothelial glycocalyx regulates neutrophil adhesion and lung injury during experimental sepsis. Nat Med 18(8):1217–1223. https://doi.org/10.1038/nm.2843

Schmidt EP, Li G, Li L et al (2014) The circulating glycosaminoglycan signature of respiratory failure in critically ill adults. J Biol Chem 289(12):8194–8202. https://doi.org/10.1074/jbc.M113.539452

Schmidt EP, Overdier KH, Sun X et al (2016) Urinary Glycosaminoglycans predict outcomes in septic shock and acute respiratory distress syndrome. Am J Respir Crit Care Med 194(4): 439–449. https://doi.org/10.1164/rccm.201511-2281OC

Semler MW, Self WH, Wanderer JP et al (2018) Balanced crystalloids versus saline in critically ill adults. N Engl J Med 378(9):829–839. https://doi.org/10.1056/NEJMoa1711584

Shimaoka M, Springer TA (2003) Therapeutic antagonists and conformational regulation of integrin function. Nat Rev Drug Discov 2(9):703–716. https://doi.org/10.1038/nrd1174

Siren EMJ, Luo HD, Bajaj S et al (2021) An improved in vitro model for studying the structural and functional properties of the endothelial glycocalyx in arteries, capillaries and veins. FASEB J 35(6):e21643. https://doi.org/10.1096/fj.201802376RRRR

Sladden TM, Yerkovich S, Grant M et al (2019) Endothelial Glycocalyx shedding predicts donor organ acceptability and is associated with primary graft dysfunction in lung transplant recipients. Transplantation 103(6):1277–1285. https://doi.org/10.1097/tp.0000000000002539

Smart L, Macdonald SPJ, Bosio E, Fatovich D, Neil C, Arendts G (2019) Bolus therapy with 3% hypertonic saline or 0.9% saline in emergency department patients with suspected sepsis: a pilot randomised controlled trial. J Crit Care 52:33–39. https://doi.org/10.1016/j.jcrc.2019.03.009

Spencer J, Murphy LM, Conners R, Sessions RB, Gamblin SJ (2010) Crystal structure of the LasA virulence factor from Pseudomonas aeruginosa: substrate specificity and mechanism of M23 metallopeptidases. J Mol Biol 396(4):908–923. https://doi.org/10.1016/j.jmb.2009.12.021

Sperry JL, Guyette FX, Brown JB et al (2018) Prehospital plasma during air medical transport in trauma patients at risk for Hemorrhagic shock. N Engl J Med 379(4):315–326. https://doi.org/10.1056/NEJMoa1802345

Stensballe J, Ulrich AG, Nilsson JC et al (2018) Resuscitation of Endotheliopathy and bleeding in thoracic aortic dissections: the VIPER-OCTA randomized clinical pilot trial. Anesth Analg 127(4):920–927. https://doi.org/10.1213/ane.0000000000003545

Steppan J, Hofer S, Funke B et al (2011) Sepsis and major abdominal surgery lead to flaking of the endothelial glycocalix. J Surg Res 165(1):136–141. https://doi.org/10.1016/j.jss.2009.04.034

Straat M, Müller MC, Meijers JC et al (2015) Effect of transfusion of fresh frozen plasma on parameters of endothelial condition and inflammatory status in non-bleeding critically ill patients: a prospective substudy of a randomized trial. Crit Care 19(1):163. https://doi.org/10.1186/s13054-015-0828-6

Sullivan RC, Rockstrom MD, Schmidt EP, Hippensteel JA (2021) Endothelial glycocalyx degradation during sepsis: causes and consequences. Matrix Biol Plus 12:100094. https://doi.org/10.1016/j.mbplus.2021.100094

Sun X, Li L, Overdier KH et al (2015) Analysis of total human urinary glycosaminoglycan disaccharides by liquid chromatography–tandem mass spectrometry. Anal Chem 87(12):6220–6227. https://doi.org/10.1021/acs.analchem.5b00913

Suzuki A, Tomita H, Okada H (2022) Form follows function: the endothelial glycocalyx. Transl Res 247:158–167. https://doi.org/10.1016/j.trsl.2022.03.014

Takada Y, Ye X, Simon S (2007) The integrins. Genome Biol 8(5):215. https://doi.org/10.1186/gb-2007-8-5-215

Targosz-Korecka M, Jaglarz M, Malek-Zietek KE et al (2017) AFM-based detection of glycocalyx degradation and endothelial stiffening in the db/db mouse model of diabetes. Sci Rep 7(1):15951. https://doi.org/10.1038/s41598-017-16179-7

Thorne RF, Legg JW, Isacke CM (2004) The role of the CD44 transmembrane and cytoplasmic domains in co-ordinating adhesive and signalling events. J Cell Sci 117(Pt 3):373–380. https://doi.org/10.1242/jcs.00954

Tollefsen DM, Pestka CA, Monafo WJ (1983) Activation of heparin cofactor II by dermatan sulfate. J Biol Chem 258(11):6713–6716

Torres Filho IP, Torres LN, Salgado C, Dubick MA (2016) Plasma syndecan-1 and heparan sulfate correlate with microvascular glycocalyx degradation in hemorrhaged rats after different resuscitation fluids. Am J Physiol Heart Circ Physiol 310(11):H1468–H1478. https://doi.org/10.1152/ajpheart.00006.2016

Twamley SG, Stach A, Heilmann H et al (2021) Immuno-electron and confocal laser scanning microscopy of the Glycocalyx. Biology (Basel) 10(5). https://doi.org/10.3390/biology10050402

Uchimido R, Schmidt EP, Shapiro NI (2019) The glycocalyx: a novel diagnostic and therapeutic target in sepsis. Crit Care 23(1):16. https://doi.org/10.1186/s13054-018-2292-6

van den Berg BM, Vink H, Spaan JA (2003) The endothelial glycocalyx protects against myocardial edema. Circ Res 92(6):592–594. https://doi.org/10.1161/01.RES.0000065917.53950.75

van den Berg BM, Spaan JA, Rolf TM, Vink H (2006) Atherogenic region and diet diminish glycocalyx dimension and increase intima-to-media ratios at murine carotid artery bifurcation. Am J Physiol Heart Circ Physiol 290(2):H915–H920. https://doi.org/10.1152/ajpheart.00051.2005

van den Berg BM, Spaan JA, Vink H (2009) Impaired glycocalyx barrier properties contribute to enhanced intimal low-density lipoprotein accumulation at the carotid artery bifurcation in mice. Pflugers Arch 457(6):1199–1206. https://doi.org/10.1007/s00424-008-0590-6

Varki A, Cummings RD, Esko JD et al (2022) Essentials of glycobiology, 4th edn. Cold Spring Harbor Laboratory Press, Cold Spring Harbor

Vellinga NA, Boerma EC, Koopmans M et al (2015) International study on microcirculatory shock occurrence in acutely ill patients. Crit Care Med 43(1):48–56. https://doi.org/10.1097/ccm.0000000000000553

Vigstedt M, Søe-Jensen P, Bestle MH et al (2022) The effect of prostacyclin infusion on markers of endothelial activation and damage in mechanically ventilated patients with SARS-CoV-2 infection. J Crit Care 69:154010. https://doi.org/10.1016/j.jcrc.2022.154010

Vink H, Duling BR (1996) Identification of distinct luminal domains for macromolecules, erythrocytes, and leukocytes within mammalian capillaries. Circ Res 79(3):581–589

Vink H, Constantinescu AA, Spaan JA (2000) Oxidized lipoproteins degrade the endothelial surface layer: implications for platelet-endothelial cell adhesion. Circulation 101(13):1500–1502. https://doi.org/10.1161/01.cir.101.13.1500

Voyvodic PL, Min D, Liu R et al (2014) Loss of syndecan-1 induces a pro-inflammatory phenotype in endothelial cells with a dysregulated response to atheroprotective flow. J Biol Chem 289(14):9547–9559. https://doi.org/10.1074/jbc.M113.541573

Wei S, Gonzalez Rodriguez E, Chang R, Holcomb JB, Kao LS, Wade CE (2018) Elevated Syndecan-1 after trauma and risk of sepsis: a secondary analysis of patients from the pragmatic, randomized optimal platelet and plasma ratios (PROPPR) trial. J Am Coll Surg 227(6):587–595. https://doi.org/10.1016/j.jamcollsurg.2018.09.003

Weinbaum S, Tarbell JM, Damiano ER (2007) The structure and function of the endothelial glycocalyx layer. Annu Rev Biomed Eng 9:121–167. https://doi.org/10.1146/annurev.bioeng.9.060906.151959

Weinbaum S, Cancel LM, Fu BM, Tarbell JM (2021) The Glycocalyx and its role in vascular physiology and vascular related diseases. Cardiovasc Eng Technol 12(1):37–71. https://doi.org/10.1007/s13239-020-00485-9

Wiesinger A, Peters W, Chappell D et al (2013) Nanomechanics of the endothelial glycocalyx in experimental sepsis. PLoS One 8(11):e80905. https://doi.org/10.1371/journal.pone.0080905

Xiang XB, Chen H, Wu YL, Wang K, Yue X, Cheng XQ (2022) The effect of preoperative methylprednisolone on postoperative delirium in older patients undergoing gastrointestinal surgery: a randomized, double-blind, placebo-controlled trial. J Gerontol A Biol Sci Med Sci 77(3):517–523. https://doi.org/10.1093/gerona/glab248

Xu D, Esko JD (2014) Demystifying heparan sulfate-protein interactions. Annu Rev Biochem 83:129–157. https://doi.org/10.1146/annurev-biochem-060713-035314

Xu J, Shi GP (2014) Vascular wall extracellular matrix proteins and vascular diseases. Biochim Biophys Acta 1842(11):2106–2119. https://doi.org/10.1016/j.bbadis.2014.07.008

Yagmur E, Koch A, Haumann M, Kramann R, Trautwein C, Tacke F (2012) Hyaluronan serum concentrations are elevated in critically ill patients and associated with disease severity. Clin Biochem 45(1–2):82–87. https://doi.org/10.1016/j.clinbiochem.2011.10.016

Yanase F, Tosif SH, Churilov L et al (2021) A randomized, Multicenter, open-label, blinded end point, phase 2, feasibility, efficacy, and safety trial of preoperative microvascular protection in patients undergoing major abdominal surgery. Anesth Analg 133(4):1036–1047. https://doi.org/10.1213/ane.0000000000005667

Yang Y, Yang G, Schmidt EP (2013) In vivo measurement of the mouse pulmonary endothelial surface layer. J Vis Exp 72:e50322. https://doi.org/10.3791/50322

Yang X, Meegan JE, Jannaway M, Coleman DC, Yuan SY (2018) A disintegrin and metalloproteinase 15-mediated glycocalyx shedding contributes to vascular leakage during inflammation. Cardiovasc Res 114(13):1752–1763. https://doi.org/10.1093/cvr/cvy167

Yao Y, Rabodzey A, Dewey CF Jr (2007) Glycocalyx modulates the motility and proliferative response of vascular endothelium to fluid shear stress. Am J Physiol Heart Circ Physiol 293(2): H1023–H1030. https://doi.org/10.1152/ajpheart.00162.2007

Yen W, Cai B, Yang J et al (2015) Endothelial surface glycocalyx can regulate flow-induced nitric oxide production in microvessels in vivo. PLoS One 10(1):e0117133. https://doi.org/10.1371/journal.pone.0117133

Yoon JH, Lee ES, Jeong Y (2017) In vivo imaging of the cerebral endothelial Glycocalyx in mice. J Vasc Res 54(2):59–67. https://doi.org/10.1159/000457799

Yuan L, Han J, van der Velden AIM et al (2023) Sex-specific association between microvascular health and coagulation parameters: The Netherlands epidemiology of obesity study. J Thromb Haemost 21(9):2585–2595. https://doi.org/10.1016/j.jtha.2023.06.001

Zarbock A, Meersch M, Van Aken H, Görlich D, Singbartl K (2014) Urinary hyaluronic acid as an early predictor of acute kidney injury after cardiac surgery. J Am Coll Cardiol 64(7):737–738. https://doi.org/10.1016/j.jacc.2014.05.034

Zeng Y, Adamson RH, Curry FR, Tarbell JM (2014) Sphingosine-1-phosphate protects endothelial glycocalyx by inhibiting syndecan-1 shedding. Am J Physiol Heart Circ Physiol 306(3):H363–H372. https://doi.org/10.1152/ajpheart.00687.2013

Zhang X, Han X, Xia K et al (2019) Circulating heparin oligosaccharides rapidly target the hippocampus in sepsis, potentially impacting cognitive functions. Proc Natl Acad Sci USA 116(19):9208–9213. https://doi.org/10.1073/pnas.1902227116

Zolla V, Nizamutdinova IT, Scharf B et al (2015) Aging-related anatomical and biochemical changes in lymphatic collectors impair lymph transport, fluid homeostasis, and pathogen clearance. Aging Cell 14(4):582–594. https://doi.org/10.1111/acel.12330

Chapter 5
Function and Plasticity of Microvascular Pericytes

Ariana D. Suarez-Martinez, Anthony Bruce, Ramon Castellanos-Sanchez, Tien Comlekoglu, David J. Csordas, Lilly DeCostanza, Kareem El-Ghazawi, Julie Leonard-Duke, Corrina Peachey, Daisy Zhu, Brant Isakson, Walter L. Murfee, and Shayn M. Peirce

Abstract Pericytes are mural cells that enwrap capillaries and are critically important in maintaining the health and homeostasis of the microcirculation. Although pericytes are most well-known for their role in providing stabilizing and supporting signals to the capillary endothelium, they also play a part in regulating immune cell recruitment, blood flow, and capillary permeability. Additionally, pericytes help modulate angiogenesis, and they have been shown to possess stem cell characteristics and pluripotency capabilities. Hence, the study of pericyte dynamics—how pericytes contribute to the growth and adaptation of the microvasculature, the signals that drive their phenotypic differentiation and the consequences thereof, and the alterations in microvascular function that ensue—has become a recent focus of research. An emerging body of literature suggests that pericyte dynamics in various tissues and organs throughout the body may be altered by diseases such as diabetes, Alzheimer's disease (AD), and fibrotic disease. Furthermore, alterations in pericyte dynamics likely contribute to the pathogenesis and progression of these and other diseases. This chapter summarizes the known functions of pericytes and how they adapt to disease, their origins and plasticity, and their potential to serve as therapeutic targets in different diseases. Throughout this chapter, we emphasize the experimental model systems that have been used to study pericytes as well as the challenges associated with identifying and isolating them. We conclude this chapter by highlighting some of the unanswered questions about pericyte identity, function, and dynamics.

A. D. Suarez-Martinez · R. Castellanos-Sanchez · T. Comlekoglu · D. J. Csordas
L. DeCostanza · K. El-Ghazawi · J. Leonard-Duke · C. Peachey · D. Zhu · W. L. Murfee
Department of Biomedical Engineering, University of Florida, Gainesville, FL, USA

A. Bruce · S. M. Peirce (✉)
Department of Biomedical Engineering, University of Virginia, Charlottesville, VA, USA
e-mail: smp6p@virginia.edu

B. Isakson
Department of Molecular Physiology and Biological Physics, University of Virginia, Charlottesville, VA, USA

R. F. Tuma et al. (eds.), *Handbook of Microcirculation*, Physiology in Health and Disease, https://doi.org/10.1007/978-3-031-99609-2_5

Keywords Pericyte dynamics · Cell heterogeneity · Cell fate · Phenotypic plasticity · Mural cells · Microvascular remodeling · Angiogenesis

5.1 Introduction to Pericytes

The entirety of the blood microcirculation is comprised of endothelial cells (ECs), which form the continuous inner lining of the vessel lumen through which blood flows, and mural cells, which include both vascular smooth muscle cells (SMCs) and pericytes. SMCs and pericytes enwrap the outer, or abluminal, surface of ECs and extend cellular protrusions through the basement membrane to establish cell–cell contacts with ECs. However, they reside in different locations along the microvascular tree. Pericytes enwrap capillaries, the smallest blood vessels in the body that have diameters ranging from 5 to 10 μm, while SMCs enwrap arterioles and venules, which have diameters ranging from approximately 11 to 250 μm. Moreover, the cell morphologies, or shapes, of pericytes also differ from those of SMCs in ways that are visually obvious by examining them under 100x (or higher) magnification. The classical pericyte morphology can be described as a soma, or a cell body, which appears like an octopus mantle that is tightly juxtaposed to the capillary basement membrane, extending filipodia along the length of the capillary and enwrapping it, like octopus tentacles (Fig. 5.1). The morphologies of SMCs depend on whether they are located on arterioles or venules and their respective roles in regulating vessel diameter. SMCs on arterioles perform the important function of regulating peripheral arterial resistance and blood flow to downstream capillaries by contracting to reduce arteriole diameter (i.e., "vasoconstriction") and relaxing to increase arteriole

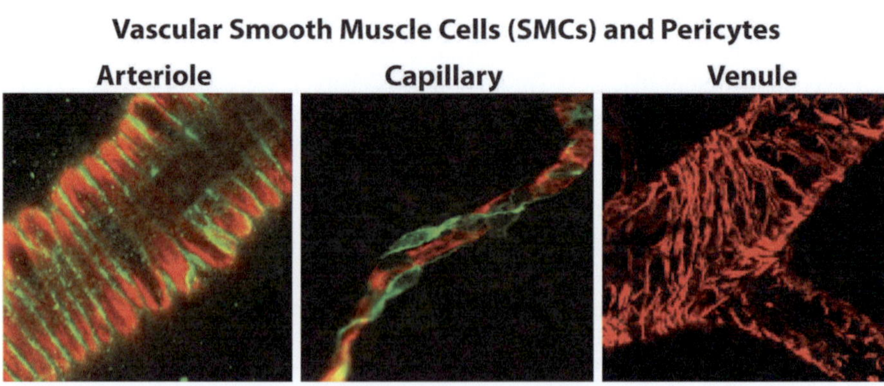

Fig. 5.1 Representative microscopic images of immunostained, vascularized windows in the rat mesentery showing the unique morphologies of SMCs in arterioles (left) vs. venules (right) and pericytes along capillaries (center). Red indicates αSMA in the arteriole and venule, which is expressed by SMCs. The red in the center indicates CD-31 immunostaining, an endothelial cell marker. The green in the arteriole and capillary indicates NG2 expressed by arteriolar SMCs and pericytes, respectively. *SMCs* smooth muscle cells, *αSMA* alpha-smooth muscle actin, *NG2* neuron–glia antigen 2

diameter (i.e., “vasodilation”), thereby decreasing or increasing blood flow to the downstream capillaries, respectively. Therefore, SMCs on arterioles are tightly aligned with one another, enwrapping the endothelium perpendicular to the lumen. This gives them an appearance of stacked donuts along the length of the arteriole. In contrast, SMCs on venules are not as contractile and venule diameter regulation is passive, which is reflected in the heterogeneous morphologies of SMCs on venules, appearing as a more webbed or “lace-like” structure (Fig. 5.1).

Pericytes were first discovered in 1873 by Charles Rouget as wrapping and contractile cells on the capillary wall of the frog hyaloid that he termed “Rouget cells” (Kelly-Goss et al. 2014; Ribatti et al. 2011). It was not until 1923 that the term “pericyte” was coined by Zimmermann to denote these mural cells on capillaries (Stapor et al. 2014). Traditionally, pericytes have been considered as immature SMCs (Nehls and Drenckhahn 1993). Their similar locations on the abluminal surface of ECs, the fact that they both extend protrusions through the basement membrane to make physical connections with the microvascular endothelium, and their expression of common markers, like alpha-smooth muscle actin (αSMA) and desmin, suggest a shared cell lineage (Nehls and Drenckhahn 1993). While pericytes can still be conservatively viewed as a mural cell subpopulation, their exact origin, plasticity (i.e., potential to differentiate into other cell types), and multifaceted roles in the regulation of microvascular homeostasis and during angiogenesis and microvascular remodeling remain elusive (Sims 2000; Gerhardt and Betsholtz 2003). Moreover, a growing body of evidence suggests that pericytes have context-dependent phenotypic plasticity, meaning that, in response to certain environmental stimuli, they can differentiate into other cell types and play important roles in the development and progression of disease, such as fibrosis, in various tissues and organs throughout the body (Ray et al. 2020; Hannan et al. 2021).

This chapter will provide an overview of the major areas of pericyte biology research and the associated driving questions. Pericyte adaptations in disease and their contributions to disease progression will be summarized in the context of targeting pericytes as a potential therapeutic strategy. Pericyte complexity will be highlighted by reflecting on the existing evidence supporting their potential to differentiate into other cell types and describing the experimental tools for conducting pericyte lineage-tracing studies in mouse models. Throughout this chapter, links to SMCs will be made to emphasize how the application of what we know about SMCs can provide a conceptual framework for advancing our current understanding of pericytes and motivating future research. This chapter will conclude with the identification of knowledge gaps and unanswered questions and with ideas for future studies that could be undertaken to fill the current voids in the literature.

5.2 Pericyte Functions and Adaptations in Disease

Pericytes, which reside on every capillary in all organs of the body (Fig. 5.2), have been shown to perform a multitude of different functions, some of which are organ-specific. For example, pericytes in the brain modulate capillary diameter and blood

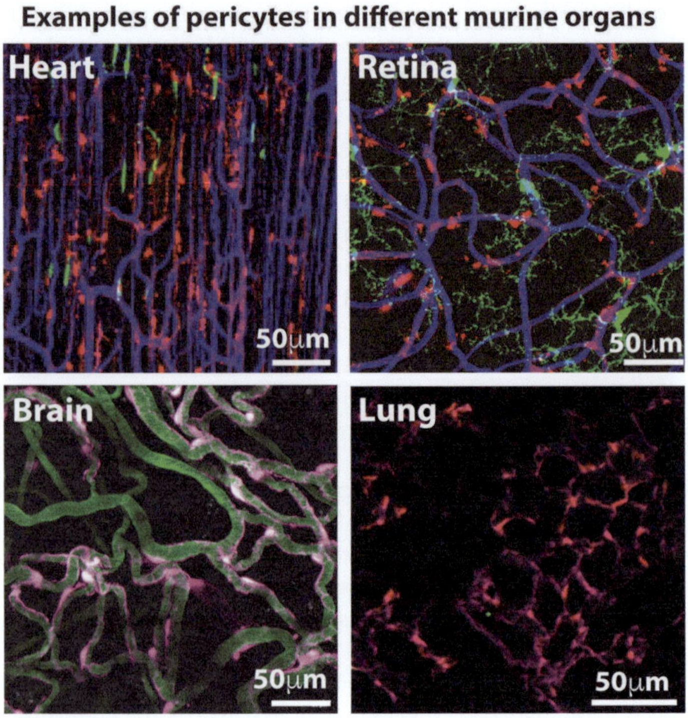

Fig. 5.2 Representative microscopic images of immunostained whole mount or histologically sectioned healthy murine organs: heart (sectioned), retina (whole-mount), brain (cranial window), and lung (sectioned). In each image, pericytes are labeled in red (but appears pink in the brain) and the capillaries they enwrap are shown in blue (heart, retina), green (brain), or purple (lung). Green-labeled cells are macrophages in the heart and microglia in the retina

flow (Hamilton et al. 2010; Gonzales et al. 2020; Eltanahy et al. 2021), whereas those in many different organs, including the brain, help stabilize capillaries (Saunders et al. 2006; Attwell et al. 2016; Alarcon-Martinez et al. 2021), regulate capillary permeability (Winkler et al. 2012; Hess et al. 2019), promote EC survival and proliferation (Bergers and Song 2005), and influence leukocyte recruitment (Pellowe et al. 2019). Different organs in the body have different ratios of pericytes to ECs. In the healthy brain and retina, for example, the pericyte to EC ratio is reported to be approximately 1:1, whereas in the skeletal or cardiac muscle, this ratio is closer to 1:50. These differing ratios reflect that pericytes have organ-specific functions, and they have also been shown to change their functionality in disease conditions (Murrant and Fletcher 2022). For example, in healthy conditions, liver pericytes store vitamin A, whereas in the brain, they have no functional storage abilities (Kostallari and Shah 2019). Moreover, in both organs, pericyte function changes during disease. In fibrotic liver disease, for example, pericytes no longer just store vitamin A but instead contribute to fibrosis by secreting extracellular matrix

(Kostallari and Shah 2019). In Alzheimer's disease (AD), pericytes play a more phagocytic role in which they clear amyloid-ß plaques through the low-density lipoprotein receptor-related protein 1 (LRP1) mechanism, thus impairing their ability to regulate capillary permeability in the blood–brain barrier (BBB) (Ma et al. 2018). SMCs also behave in functionally unique ways depending on their organ of residence and whether the organ is in a healthy or diseased state. In the healthy heart, for example, SMCs are normally highly contractile, but they dedifferentiate to take on a more proliferative or migratory phenotype under disease conditions (Zhuge et al. 2020). This section describes some of the key functions of pericytes, including their role in regulating capillary permeability, microvascular homeostasis and angiogenesis, and capillary blood flow, as they have been documented in the healthy microcirculation, and summarizes the major adaptations that pericytes undergo in disease settings.

5.2.1 Pericyte Regulation of Capillary Permeability

Throughout the body, pericytes play a role in regulating capillary permeability. In the brain, for example, pericytes are abundant, covering approximately 100% of the endothelium (Armulik et al. 2011) to help form the blood–brain barrier (BBB), a selective semipermeable membrane between the blood circulation and the central nervous system (CNS). The BBB ensures that neurons and other cells in the brain parenchyma are protected from pathogens (e.g., bacteria) and toxic substances by forming a physical boundary that confines harmful substances to the blood circulation. Brain pericytes mediate both endothelial transcytosis and endothelial tight junction opening and therefore play a critical role in determining which substances are allowed to cross the BBB (Quaegebeur et al. 2011).

When pericyte numbers are reduced, as was shown in pericyte-deficient mouse mutants, permeability of the BBB to water and to both low- and high-molecular-weight tracers is increased (Armulik et al. 2010). This was shown to be reversable by treating with the drug imatinib, which reduces endothelial transcytosis. Additionally, pericytes regulate BBB-specific gene expression patterns in ECs and induce polarization of the astrocyte end-feet that surround the brain blood vessels. Although these studies were conducted in pericyte-deficient mice, evidence from human cadaveric brains shows that pericyte numbers are reduced by 35–45% in the frontal lobe white matter of patients with post-stroke dementia, vascular dementia, and Alzheimer's disease compared to both post-stroke non-demented patients and healthy control subjects (Ding et al. 2020).

Consistent with their role in regulating capillary permeability in the brain, pericytes also regulate capillary permeability in other organs, such as the heart. Indeed, the loss of cardiac pericytes has been shown to increase heart capillary permeability. In mice treated with sunitinib, a platelet-derived growth factor receptor-β (PDGFR-β) inhibitor, pericyte reductions were accompanied by a 2.3-fold increase in capillary permeability relative to vehicle control-treated mice with normal pericyte numbers (Chintalgattu et al. 2013).

The molecular mechanisms by which pericytes regulate capillary permeability are an active area of investigation, and recent studies have suggested that both juxtacrine and paracrine signals are involved. For example, pericytes physically connect to ECs via *N*-cadherin, a cell surface protein expressed by both ECs and pericytes. Deletion of *N*-cadherin in either ECs or pericytes has been shown to increase capillary permeability in both the lung and brain, and this is accompanied by a reduction of vascular endothelial cadherin (VE-cadherin) accumulation in the adherens junctions that connect neighboring ECs. Furthermore, the dual Rac1/RhoA Rho guanine nucleotide exchange factor (GEF) Trio was shown to be a critical component of the *N*-cadherin adhesion complex and an activator of both Rac1 and RhoA signaling pathways in endothelial adherens junctions (Kruse et al. 2019). In acute lung injury, disruption of pericyte–EC interactions due to inflammation increases capillary permeability. Comparative genome-wide EC–pericyte cross talk analysis of healthy and inflamed lungs attributed these permeability increases to the impaired cross talk between endothelial nitric oxide and pericyte soluble guanylate cyclase (sGC) (He et al. 2023). sGC activation was shown to suppress cytoskeleton rearrangements in pericytes, leading to the stabilization of EC–pericyte interactions.

5.2.2 Pericyte Regulation of Microvascular Homeostasis and Angiogenesis

Pericytes provide both paracrine and juxtacrine support signals to ECs to ensure capillary maintenance and homeostasis. For example, the paracrine signals, angiopoietin-1 (Ang1) and − 2 (Ang-2), which are secreted by pericytes, ligate the Tie family of receptor tyrosine kinases expressed by ECs to mediate EC–pericyte interactions in both healthy and disease contexts (Partanen et al. 1992). When angiopoietin-1 (Ang-1) binds Tie2 on ECs, Tie2 phosphorylation and downstream signaling promotes vascular stability by promoting accumulation of VE-cadherin at adherens junctions and by reducing leukocyte adhesion and inflammation through downregulation of vascular cell adhesion molecule (VCAM), E-selectin, and intercellular adhesion molecule-1 (ICAM-1) expression (Sack et al. 2020). Sphingosine-1-phosphate (S1P) in the membranes of ECs promotes juxtacrine signaling between ECs and pericytes by upregulating *N*-cadherin expression in ECs (Paik et al. 2004). S1P signals through Rac to promote microtubule polymerization, resulting in trafficking of *N*-cadherin to polarized plasma membrane domains in ECs. *N*-cadherin in ECs binds to *N*-cadherin in neighboring pericytes, thereby enhancing the stability of EC–pericyte interactions (Paik et al. 2004). Moreover, studies that have injected anti-*N*-cadherin antibody (Gerhardt et al. 2000) or included short-interfering RNA in Matrigel plugs (Paik et al. 2004) underscore the functional importance for *N*-cadherin in stabilizing EC–pericyte connections.

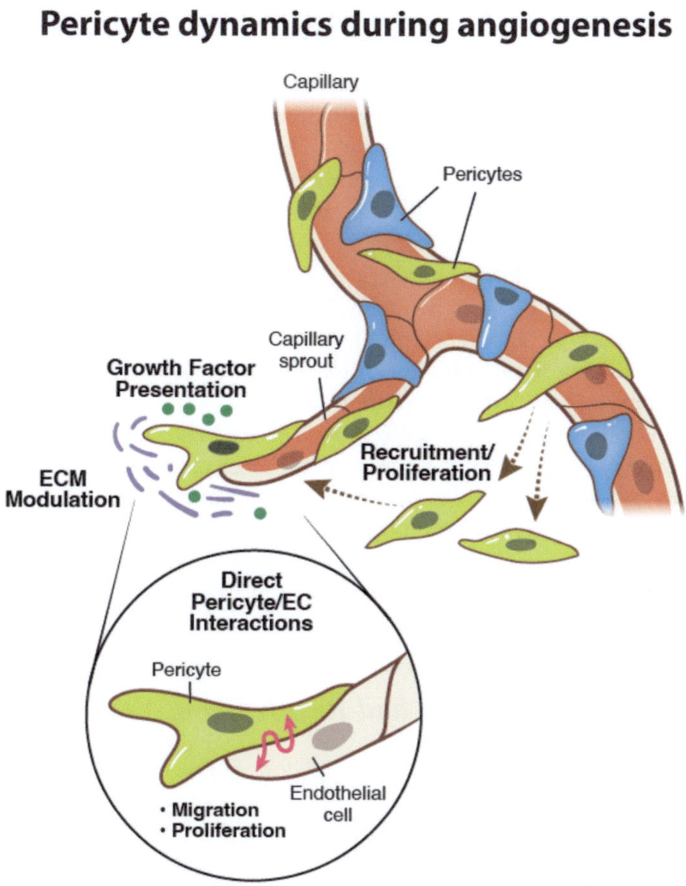

Fig. 5.3 Schematic overview of the major dynamics of pericytes during angiogenesis. Dynamics include pericyte recruitment, extracellular matrix modulation, direct cell-to-cell interactions, and paracrine signaling via secretion and binding of growth factors

Not surprisingly, given their important role in regulating EC stability, pericytes also participate in angiogenesis. So much so, that the field has often considered the coordination between pericytes and ECs as the key regulators of angiogenesis in both healthy and diseased contexts. During angiogenesis, pericyte dynamics are characterized by specific cell behaviors, such as: (1) recruitment to the abluminal surface of a newly forming vessel or capillary sprout, (2) extracellular matrix modulation through matrix metalloproteinase (MMP) secretion, (3) secretion of paracrine signals (e.g., growth factors and chemokines) that modulate EC and immune cell behaviors, and (4) direct interactions with ECs (Stapor et al. 2014; Gerhardt and Betsholtz 2003) (Fig. 5.3). The commonly accepted paradigm for pericyte involvement in angiogenesis, which is well-supported by the literature, is that pericytes are recruited by ECs that emerge as “tip cells” as they sprout from existing capillaries or postcapillary venules. ECs recruit pericytes by secreting platelet-derived growth factor-B (PDGF-B), which binds to the tyrosine kinase receptor PDGFR-β expressed by pericytes. Activation of PDGFR-β by PDGF-B induces pericyte proliferation and migration up the concentration gradient of PDGF-B in the direction of tip ECs (Hellström et al. 1999; Betsholtz 2004; Carmeliet and Jain 2011). After a pericyte has been recruited in this manner to a sprouting EC, it forms a physical connection with the EC, which initiates the assembly of the new capillary's basement membrane, beginning with the deposition of laminin, fibronectin, and collagen IV as the sprout elongates. The basement membrane is assembled and maintained by both ECs and pericytes. Moreover, once the sprout has connected to another capillary or capillary sprout, pericytes prevent proteolysis of matrix proteins by producing tissue inhibitor of metalloproteinase-3 (TIMP-3) to suppress further EC sprouting from the newly formed vessel (Stratman and Davis 2012). Pericyte secretion of certain diffusible growth factors can influence ECs and other interstitial cell types in the local environment. For example, pericytes can promote angiogenesis by secreting VEGF to signal EC sprouting. However, as discussed above, pericytes also secrete angiopoietin-1 (Ang-1), which binds to the Tie2 receptor expressed by ECs and induces capillary stabilization and maturation (Ribatti et al. 2011; Suri et al. 1996; Armulik et al. 2005).

5.2.3 Pericyte Contractility and Regulation of Vessel Diameter

While pericytes have been shown to be capable of contracting and regulating capillary diameters and blood flow in the brain and heart (Longden et al. 2023), far more is understood about the mechanisms underpinning SMC contraction in organs throughout the body. Therefore, given the close relationship between pericytes and SMCs, we will begin by summarizing how contraction is regulated in SMCs.

SMC contraction is facilitated by myosin light chain kinase (MLCK), which is activated by increases in cytosolic calcium. MLCK phosphorylates the regulatory light chain of myosin, enabling it to bind to actin and cause cellular contraction (Sweeney and Hammers 2018). Myosin light chain phosphatase (MLCP) dephosphorylates the regulatory light chain, thus inducing smooth muscle relaxation and therefore vasodilation. While this has been primarily studied in SMCs, evidence suggests this mechanism also facilitates pericyte contractility (Erdener et al. 2022).

SMCs are organized circumferentially around the vessel wall, enabling them to mediate efficient constriction of the vessel, but pericytes are organized along two axes. The primary processes of pericytes extend down the length of vessel, while the secondary processes are oriented circumferentially. Constriction of the secondary processes may modulate the luminal size and/or endothelial permeability (Dessalles et al. 2021). SMCs also express high levels of αSMA, while the expression of αSMA in pericytes is much lower and heterogeneous (Alarcon-Martinez et al. 2018, 2021). Both SMCs and pericytes express Myh11 as the predominant myosin isoform (Erdener et al. 2022).

Both myocardin and serum response factor (SRF) play important roles in regulating SMC contraction, and this is likely the case in pericytes. Many of the genes encoding the SMC contractile apparatus components (e.g., *Acta2*, *Myh11*, *Calponin, Sm22a*) contain CArG motifs, the binding site of SRF (Xia et al. 2017). SRF expression, however, is not unique to SMCs. The cell type-specific expression of these genes is regulated by myocardin, a coactivator of SRF unique to cardiac and smooth muscles (Wang et al. 2001). Increased expression of myocardin promotes a differentiated contractile SMC phenotype, while repression of myocardin promotes a dedifferentiated and proliferative SMC phenotype (Tang et al. 2008). In pericytes, although myocardin expression is much lower than it is in SMCs (Erdener et al. 2022), it may still play a role in pericyte contraction, and, SRF, which is expressed in pericytes, has already been shown to regulate pericyte migration and barrier function. As is the case in SMCs, pericyte expression of SRF has been reported to be dependent upon myocardin-related transcription factors (MRTFs) (Orlich et al. 2022).

In SMCs, purinergic signaling also regulates contraction and proliferation, and evidence suggests that this is true for pericytes as well. Purinergic receptors are made up of two groups. P1 receptors (A_1, A_{2A}, A_{2B}, and A_3), which respond to adenosine, and P2 receptors, which respond to adenosine triphosphate (ATP), adenosine diphosphate (ADP), uridine triphosphate (UTP), and/or uridine diphosphate (UDP). P2 receptors are further categorized into P2X (P2X1–7) and P2Y ($P2Y_1$, $P2Y_2$, $P2Y_4$, $P2Y_6$, $P2Y_{11}$, $P2Y_{12}$, $P2Y_{13}$, $P2Y_{14}$). The expression of these receptors in SMCs is heterogeneous between species and vascular beds, with P2X1 being the most consistently detected in SMCs (Burnstock and Ralevic 2014). Adenosine acts primarily as a vasodilator. In the coronary artery, this occurs through both endothelium-dependent and endothelium-independent pathways (Sousa and Diniz 2017; Cushing et al. 1991). SMC purinergic signaling is also an integral part of sympathetic vasoconstriction. ATP has been reported to be both a co-neurotransmitter with norepinephrine and released from SMCs in response to α-adrenergic stimulation. Upon α-adrenergic stimulation, ATP is released from SMC pannexin 1 (Panx1) channels acting in an autocrine manner to modulate the degree of vasoconstriction (Billaud et al. 2015; Dunaway et al. 2022). This occurs through P2Y receptors, while it appears that neuronal-derived ATP acts via P2X1 channels (Burnstock 2017; Vial and Evans 2002; Lamont et al. 2006; Angus and Wright 2015; Billaud et al. 2011; Prada et al. 2020). Purinergic signaling also controls the SMC phenotype. While low concentrations of ATP favor a contractile phenotype, higher concentrations favor a proliferative phenotype (Hogarth et al. 2004). This is primarily facilitated through P2Y receptors, which are more predominant on proliferative SMCs compared to contractile SMCs (Erlinge et al. 1998). Specifically, $P2Y_1$, $P2Y_2$, $P2Y_4$, and $P2Y_6$ receptor stimulation by ATP, UTP, and/or UDP has been shown to drive SMC proliferation (Hou et al. 2002; Harper et al. 1998; Erlinge 1998). In brain capillary pericytes, which play a role in regulating cerebral blood flow, extracellular ATP has been shown to cause pericyte contraction via stimulation of purinergic receptors and subsequent release of intracellular Ca^{2+}. In this context, pericyte contraction is mediated through $P2Y_1$ and $P2Y_2$ (Hørlyck et al. 2021). In summary, there is much to be learned about the molecular mechanisms of pericyte contraction, but inspecting the mechanisms of SMC contraction can provide potential pathways for future exploration of pericyte contraction in both health and disease. As an example of this potential impact, emerging work has substantiated a role for ion channels in pericyte subpopulation-specific contraction and regulation of local blood flow to tissue regions (Gonzales et al. 2020; Eltanahy et al. 2021; Klug et al. 2023).

5.3 Pericyte Origins, Identity, and Plasticity

In addition to altering their function in response to disease, pericytes also exhibit phenotypic plasticity and have the capacity to differentiate into other cell types, including fibroblasts and immune cells, as discussed below. These phenotypic changes are challenging to deduce given: (1) the spatial heterogeneity in their developmental origins, (2) their expression of markers that are also commonly expressed by other cell types, and (3) the lack of a unique pericyte marker. Hence, pericyte identity and plasticity can be most confidently documented by conducting experiments that utilize pericyte lineage-tracing approaches, which enable pericyte identification prior to the onset of disease and the tracking of pericytes throughout disease progression (Hannan et al. 2021; Hess et al. 2019; Corliss et al. 2020; Leonard-Duke et al. 2023). For example, the use of loxP-Cre systems to genetically remove genes specifically from mural cells in vivo has been present for nearly 30 years. Although several Cres have bene proposed, few have been demonstrated to be specific. For this reason, the Myh11 CreER^{T2+} from the Offerman group has been extensively used due to its strong specificity for SMCs and pericytes in a significant number of research studies (Wirth et al. 2008). However, the issue with this mouse was the inability to test the effects of smooth muscle cell genetic deletion in females because of the presence of Cre on the Y chromosome, and some groups reported a "leaky" Cre (non-inducible activation) in the gut smooth muscle. For this reason, two new smooth muscle cell-specific Cres have recently been published: from the Owens lab using *Myh11* as the Cre promoter (Deaton et al. 2022) and from the Miano lab using *Itga8* as the Cre promoter (Warthi et al. 2022). Both Cres demonstrate strong specificity for smooth muscle cells in male and female mice with a lack of "leakiness." This section will summarize what is known about the developmental origins of pericytes, how we identify them, and their phenotypic plasticity.

5.3.1 Pericyte Origins

Mural cells have been traced from the neural crest (ectoderm), the lateral plate mesoderm, the somites of the posterior mesoderm, and the mesothelium of their surrounding organs (Majesky 2007). Mural cells residing in vessels above the heart originate from the neural crest ectoderm, except for those cells residing in vessels in the posterior and dorsal brain, which originate from the posterior mesoderm tissue during development. Mural cells in the rest of the body originate from the mesoderm or mesothelium of their respective organs (Etchevers et al. 2001). Pericytes appear to share the same embryonic origins as vascular smooth muscle cells (Korn et al. 2002). Pericytes of the head, neck, and central nervous system (CNS) have been found to be neural crest in origin, and both mural cell types in the aorta can be traced back to the posterior somitic mesoderm (Martinez-Lemus 2012; Steucke et al. 2015; Latif et al. 2015).

5.3.2 Pericyte Identity

While various cell surface proteins have been used to identify pericytes, including αSMA, desmin, vimentin, PDGFR-β, and neuron–glia antigen 2 (NG2), none of these is unique to pericytes, making it difficult to distinguish pericytes from other vascular and nonvascular cell populations (Gerhardt and Betsholtz 2003; Nehls and Drenckhahn 1993; Kelly-Goss et al. 2014). One of the most used pericyte markers in recent years has been neuron–glia antigen-2 (NG2), also known as chondroitin sulfate proteoglycan-4 (CSPG-4) or melanoma chondroitin sulfate proteoglycan (MCSP). NG2 is a surface type I transmembrane proteoglycan protein of approximately 300 kDa that has functional intracellular and extracellular domains (Ampofo et al. 2017; Stallcup 2018; Pluschke et al. 1996). It has been used to identify pericytes in the angiogenic microvasculature during pre- and postnatal development (Grako and Stallcup 1995; Ozerdem et al. 2001), tumor angiogenesis (Chekenya et al. 2002), and in retinal and corneal angiogenesis models (Ozerdem et al. 2002). It is also expressed by most precursor cell types across many tissues during development. NG2 has been shown to promote motility and proliferation in ECs and pericytes by being an auxiliary receptor that enhances signaling via integrins (Grako et al. 1999; Makagiansar et al. 2004; Fukushi et al. 2004). NG2 has also been demonstrated to enhance the formation and maturation of EC zonula occludens-1 (ZO-1) intercellular junctions by activating integrin signaling (You et al. 2014).

The traditional view is that pericytes exist as a unique cell type; however, this view is now being challenged by the identification of transient pericyte subpopulations, particularly during angiogenesis (Fig. 5.4). For example, both NG2 and class III-β tubulin, another pericyte marker, transiently identify pericyte subpopulations in microvascular regions associated with capillary sprouting (Stapor et al. 2014).

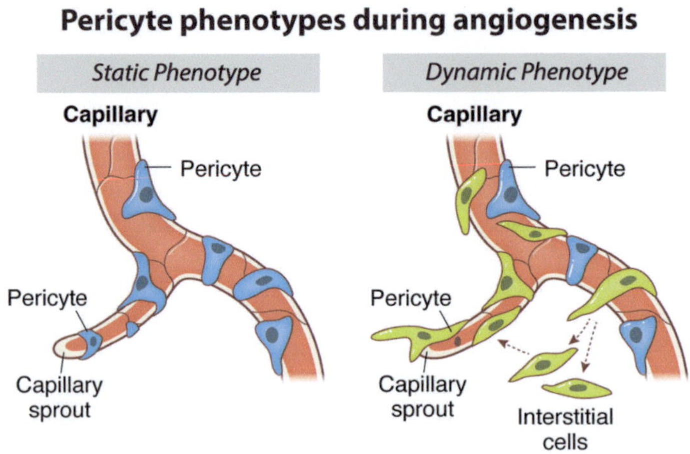

Fig. 5.4 Pericyte phenotypes change during angiogenesis. The discovery of pericyte subpopulations associated with capillary sprouting emphasize the temporal and spatially regulation of cell phenotype signatures. Realization of the transient cell types challenges a traditional view that all pericytes are the same

However, NG2 and class III-β tubulin are also expressed by neural cells and stem cell populations (Kelly-Goss et al. 2014), which contributes to the ambiguity about pericyte dynamics and subpopulations because pericytes can easily be confused with other cell types that express the same markers. The use of single-cell transcriptomics to identify and characterize pericyte subpopulations in multiple tissues and microvascular remodeling scenarios is producing new insights into pericyte heterogeneity (Muhl et al. 2020; He et al. 2018; Vanlandewijck et al. 2018), and the use of pericyte lineage-tracing models in mice is helping resolve the dynamic behaviors that pericytes exhibit during angiogenesis and capillary stabilization and characterize their pluripotency potential (Kelly-Goss et al. 2014; Payne et al. 2021).

5.3.3 *Pericyte Plasticity: PC/SMC-to-Fibroblast Phenotype Transitions*

Pericytes that interact with ECs can detach from the microvasculature during inflammation and disease (Hosaka et al. 2016; Chen et al. 2011; Wang et al. 2019; Lin et al. 2008). Detached pericytes migrate away from capillaries due to changes in the extracellular matrix (Chen et al. 2011; Wang et al. 2019) and levels of various growth factors (Hosaka et al. 2016). This has been observed in tumors (Hosaka et al. 2016), the choroid (Zhao et al. 2022), fibrotic kidneys (Lin et al. 2008; Yang et al. 2019; Humphreys et al. 2010), and the lung (Wang et al. 2019). As pericytes migrate away from capillaries, they downregulate the proteins involved in vascular stabilization (Lin et al. 2008) while upregulating fibroblastic proteins, including αSMA and NG2, as well as Col1a1 and other extracellular matrix proteins (Chen et al. 2011; Humphreys et al. 2010). These phenotypic changes are termed "pericyte-to-fibroblast transitions (PFTs)," though are often referred to as "pericyte-to-myofibroblast transitions (PMTs)" due to the resulting increase in the number of activated fibroblasts. These changes can be observed and quantified using lineage-tracing (Hosaka et al. 2016; Humphreys et al. 2010) and cell fate mapping techniques (Humphreys et al. 2010).

PMT involves the activation of molecular signaling pathways that drive EC detachment, migration, proliferation, and differentiation of pericytes into fibroblasts. The attachment of pericytes to the microvasculature is regulated by EC-derived PDGF. PDGF signaling involves four ligands, PDGF-A, PDGF-B, PDGF-C, and PDGF-DD, and two receptors, PDGFR-α and PDGFR-β (Kazlauskas 2017). As a class III receptor tyrosine kinase, PDGFR is autophosphorylated by ligand binding (Ribatti et al. 2011). The membrane protein, Notch1, regulates expression of PDGFR (Wang et al. 2019). Col1a1 and αSMA are upregulated when PDGFR is activated by ligation of PDGF-B. Transforming growth factor-beta 1 receptor (TGF-β1) and transforming growth factor-beta 2 receptor (TGF-β2) are also involved in modulating pericyte behaviors (Wu et al. 2013). For example, activation of TGF-β2 by its ligand TGF-β initiates the migration of pericytes away from capillaries after they detach from ECs. This behavior is modulated by the downstream Smad2/3 and Akt/mammalian target of rapamycin (Akt/mTOR) pathways in pericytes (Zhao et al. 2022). Activation of TGF-β1, in contrast, causes pericyte proliferation and contributes to the upregulation of fibroblast markers during PFT (Wu et al. 2013).

Various pharmacologically and/or genetically perturbing PDGF ligands and receptors have demonstrated their importance in PFT. For example, inhibiting PDGFR-α or PDGFR-β with anti-PDGFR-β antibodies or via genetic deletion attenuates PFT by decreasing the coverage of pericytes on capillaries (Hosaka et al. 2016; Zhao et al. 2022). This has been demonstrated in the fibrotic kidney (Chen et al. 2011). In contrast, PDGF-B knockdown increases coverage of pericytes on the tumor microvasculature and reduces fibroblast markers in the tumor (Hosaka et al. 2016; Wang et al. 2019). Conversely, PDGF-B overexpression in tumor environments reduces pericyte coverage, increases fibroblast markers, decreases pericyte markers, and enhances PFT (Hosaka et al. 2016).

Perturbations of TGF-β signaling also reveal the importance of this pathway in regulating PFT. For example, inhibition of TGF-β downregulates PDGF-B, causing reduced pericyte detachment, migration, and proliferation (Wu et al. 2013), while its overexpression enhances PFT (Zhao et al. 2022; Yang et al. 2019). TGF-β1 activation increases the transcription of the fibroblast markers αSMA and Col1a1 (Chen et al. 2011; Yang et al. 2019), while its inhibition attenuates PFT via decreased activation of Smad2 signaling (Humphreys et al. 2010; Wu et al. 2013).

5.3.4 *Pericyte Plasticity: PC/SMC-to-Immune Cell Phenotype Transitions*

Pericytes have long been described as having macrophage-like characteristics and exhibiting phagocytic behaviors. Due to the lack of a distinguishing marker to definitively identify pericytes, early studies identifying these cells as co-expressing canonical immune cell markers are convoluted (Navarro et al. 2016). The bulk of recent research has occurred in the brain, where pericytes are abundant and are an important component of the blood–brain barrier (BBB). Here, pericytes have been shown to play important roles in controlling neuroinflammation, including regulation of immune cell recruitment and infiltration, as well as confirming that pericytes have phagocytic abilities of their own (Jansson et al. 2014; Pieper et al. 2014; Rustenhoven et al. 2017). Experiments conducted with porcine brain capillary pericytes have shown that these pericytes are able to express major histocompatibility complex class 2 (MHC II) and CD68 (ED1) following stimulation with interferon gamma (IFNγ) (Smith et al. 2013). MHC II is important for foreign antigen presentation and is usually only found on antigen-presenting cells (APCs), while CD68 is a canonical macrophage marker, suggesting a transition of pericytes into macrophage-like cells.

In cases of ischemic stroke, pericytes seem to also be capable of transitioning into microglia-like cells. In a mouse model of stroke, PDGFR-β-expressing pericytes isolated from impacted tissues had increased levels of the stem cell marker nestin. PDGFR-β + cells have also been found to co-express ionized calcium-binding adaptor molecule 1 (Iba1), a marker that is expressed at high levels by microglia and macrophages. Importantly, PDGFR-β+/Iba1+ cells were also CD68-, implying that these cells originated from pericytes and not perivascular macrophages (Sakuma et al. 2016). Other studies have gone further and used lineage-tracing mice to show that Iba1+ cells originate from pericytes, as opposed to SMCs or fibroblasts, while others have shown that human brain-derived pericytes in hypoxic conditions express markers of microglia (Özen et al. 2014). Given the growing awareness that pericytes exhibit phenotypic plasticity in health and disease, there is an increasing need to identify the contexts, triggers, and molecular signals that dynamically modulate pericyte phenotypic changes in vivo.

5.3.5 *Extracellular Vesicles (EVs) in Modulating Pericyte Plasticity*

All cells secrete a heterogeneous array of membrane-enclosed vesicles to the extracellular space. Early research conducted in the 1980s on such vesicles concluded that they were destined for lysosomal degradation, often called cellular dust. In 1996, pioneering work by Raposo et al. (1996) documented, for the first time, that extracellular vesicles (EVs) played a role as modulators of the immune system.

Since then, thousands of papers have been published, pointing out the new role of EVs as a new form of cell-to-cell communication. We now know that the molecules contained within these EVs, contain proteins, growth factors, lipids, metabolites, and genetic material (mRNA, miRNA, and siRNA). These signaling molecules are transferred from cell to cell to modulate cellular function and plasticity. New evidence suggests that pericyte-derived EVs regulate endothelial cell function and plasticity, motivating the unexplored possibility that EC-derived EVs may signal reciprocally to modulate pericyte function and pericyte phenotypic differentiation. To date, pericyte-derived EVs have been shown to have a vasoprotective role in ECs against the harmful effects of hypoxia during spinal cord injury (Yin et al. 2022). However, some studies have suggested that EC-derived EVs can modulate the secretion of inflammatory growth factors by pericytes in vitro (Yamamoto et al. 2015). Moreover, the transcription factor KLF-2 (Kruppel-like factor 2) has been identified in SMC-derived EVs, which downregulate tight junctions in ECs and increase vascular permeability (Nehls and Drenckhahn 1993). This is interesting given that ECs overexpressing KLF-2 generate EVs capable of regulating SMC gene expression both in vitro and in vivo (Hergenreider et al. 2012). Indeed, EVs are an emerging area of research, and more studies are needed to determine how EC-derived EVs impact phenotypic differentiation of pericytes, and vice versa.

5.4 Consideration of Pericytes as Therapeutic Targets

Due to their important roles in modulating angiogenesis and capillary stabilization, which are often dysregulated during disease, pericytes have emerged as novel therapeutic targets and are receiving increasing attention for this reason (Lin et al. 2021). This section will describe how pericytes have been, and currently are being, considered as therapeutic targets in some of the major diseases afflicting today's society, including diabetic retinopathy, cardiovascular disease, neurodegenerative disease, and cancer.

As described above, pericytes are an integral component of the neurovascular unit in the retina (Abcouwer and Gardner 2014). The premature loss of pericytes is a pathological hallmark of diabetic retinopathy, which is a leading cause of vision loss in adults (Cogan et al. 1961). Several key molecular signaling pathways in pericytes have been linked to pericyte dropout in diabetic retinopathy, such as TGF-β, PDGF-B/PDGFR-β, and Ang-1/Tie 1. Consequently, all of these signaling pathways have been identified as potential therapeutic targets to prevent or delay vision loss (Arboleda-Velasquez et al. 2015). For example, Fovista (Opthotech Corp.), which inhibits PDGF signaling, made it into phase 3 clinical trials for treating age-related macular degeneration (AMD), which has many similarities to diabetic retinopathy, including pathological neovessel outgrowths. However, in the third and final phase 3 trial, Fovista failed to meet its primary goal and did not improve visual acuity at 12 months compared to Avastin monotherapy (Elvidge 2017). Moreover, activating Tie2 signaling with AKB-9778 (Aerpio Pharmaceuticals 2019), a selective inhibitor of vascular endothelial protein tyrosine phosphatase (VE-PTP), a membrane phosphatase in ECs that dephosphorylates Tie2, showed promise in treating nonproliferative diabetic retinopathy in a phase 1 clinical trial. However, twice daily subcutaneous administration of 15 mg AKB-9778 did not meet the primary end points of a phase 2 clinical trial, and the rates of progression to sight-threatening complications, including diabetic macular edema during the 48-week treatment period, were similar to that in placebo-treated patients (Genetic Engineering and Biotechnology News 2019). More recently, the drug Faricimab (Genentech), a bispecific antibody that targets Ang-2 and VEGF-A, has been approved by the Food and Drug Administration (FDA) in 2022 for the treatment of diabetic macular edema and neovascular AMD (Panos et al. 2023). The use of autologous adipose-derived pericyte progenitor cells, or mesenchymal stem cells, has also been explored as a means to repopulate the pool of retinal pericytes that are lost over time in diabetic retinopathy (Mendel et al. 2013; Cronk et al. 2015; Amos et al. 2008, 2011). While preclinical results in multiple murine models of diabetic retinopathy showed promise, the risks of injecting autologous, adipose-derived mesenchymal stem cells have proven to outweigh any potential benefits (Kuriyan et al. 2017).

Pericytes have also been suggested to play a role in cardiovascular disease, including ischemic heart disease, by reducing microvascular blood flow after ischemia, even after flow is restored to an upstream artery (Dabravolski et al. 2022). This has been attributed to contractile pericytes that constrict up to 40% of capillaries, thus reducing the volume of perfused blood within the affected region. Interestingly, capillary diameter reductions have been strongly co-localized with the location of pericytes on capillaries, and delivery of adenosine, a potent vasodilator, increased capillary diameters and perfusion by 57% (O'Farrell et al. 2017). Moreover, mice genetically engineered with mutations in PDGFR-β, are largely protected from capillary diameter constrictions in capillary segments containing pericytes, following the restoration of blood flow after ischemia (Methner et al. 2019). These studies support the use of adenosine in preventing pericyte constriction following coronary ischemic events (Le et al. 2022). Pericytes have also been shown to contract and inhibit cerebral reperfusion following an ischemic stroke (Hall et al. 2014). However, exogenous administration of tissue plasminogen activator (tPA) following ischemic stroke has proven to be therapeutic in restoring cerebral perfusion by decreasing pericyte coverage in areas of brain ischemia. Furthermore, Edaravone (brand name: Radicava, Mitsubishi Tanabe Pharma Corporation), a free radical scavenger, was demonstrated to relieve pericyte contraction and restore pericyte coverage (Deguchi et al. 2014).

Pericyte dropout and dysfunction have been implicated in neurodegenerative diseases, such as Alzheimer's disease (AD) and Parkinson's disease (PD) (Stevenson and Dieriks 2023; El-Ghazawi et al. 2024). It was shown that in AD, pericyte coverage of brain capillaries dropped by around 30% and that these microvascular impairments contributed to cognitive decline (Sengillo et al. 2013; Uemura et al. 2020). Furthermore, pericyte loss leads to accumulation of amyloid-β plaques, which is a characteristic pathological finding in AD (Sagare et al. 2013). Pericyte implantation into mouse models of AD demonstrated a reduction in amyloid-β plaques, opening the door to cell-based therapy as a promising modality for AD treatment (Tachibana et al. 2018). In Parkinson's disease, α-synuclein (α-syn), a small acidic protein that plays a role in vesicle endocytosis, membrane remodeling, and synaptic transmission, forms oligomer aggregates (Malmberg et al. 2020) and gives rise to Lewy bodies and Lewy neurites. Pericytes have been shown to efficiently endocytose and degrade fibrillar α-syn aggregates (Stevenson et al. 2022). Hence, enhancing lysosomal degradation processes in pericytes might be a therapeutic target for promoting α-syn degradation and slowing the progression of PD (Stevenson and Dieriks 2023).

Traditional therapeutic approaches targeting pericytes for the treatment of cancer have primarily focused on reducing the blood supply to tumors by targeting VEGF, most commonly (Moro et al. 2024). Bevacizumab, a human monoclonal antibody that binds to soluble VEGF isoforms to prevent the binding of VEGF with its receptors, was the first FDA-approved anti-VEGF therapy. However, despite showing efficacy in preclinical models of colorectal cancer, antiangiogenic drugs showed limited efficacy in treating solid tumors. One explanation for this is the fact that the tumor vasculature has decreased pericyte coverage, resulting in excessive capillary leakage and elevated interstitial fluid pressure (Goel et al. 2011). This counteracts the diffusion of drugs from the circulation into the tumor. Dr. Rakesh Jain introduced the concept of "normalizing" the tumor vasculature by using doses of antiangiogenic drugs, such as anti-VEGFR2, which retain pericytes on capillaries and improve tumor perfusion (Jain 2005). More recently, combining anti-VEGF therapy with conventional chemotherapy, such as intravenous 5-fluorouracil, has shown improved survival of patients with metastatic colon cancer patients compared to chemotherapy alone (Goel et al. 2011).

Pharmacologically targeting PDGFR-β with the drug imatinib, a receptor tyrosine kinase inhibitor, slows lymphoma growth in both human xenograft and murine allograft models by inducing apoptosis of tumor-associated PDGFR-β + pericytes and disrupting the tumor microvasculature (Ruan et al. 2013). However, clinical trials show that PDGFR-β inhibition with imatinib alone is ineffective, possibly because a reduction in pericyte coverage on tumor capillaries has been shown to enhance tumor metastasis (Cooke et al. 2012).

While these studies and others indicate that pharmacologically targeting pericytes, and/or leveraging them as a cell-based therapy, may yield therapeutic outcomes in different diseases, there remain numerous unresolved questions about the molecular targeting strategies, dosage, timing, and compensatory mechanisms that need to be determined in order to achieve clinical safety and efficacy. A comprehensive consideration of the broad array of organ-specific and nonspecific pericyte functions and the identity, plasticity, and adaptations of pericytes in different disease states can provide numerous strategies for designing pericyte-targeting therapies.

5.5 Future Directions and Conclusions

Future research will benefit from parallel but coordinated investigations of the function, identity, plasticity, and potential of pericytes as therapeutic targets across different diseases while remembering the historical viewpoint that pericytes are the less mature relatives of SMCs. Indeed, even in recent years, new insights from studies of SMCs have established pericytes as regulators of capillary blood flow via SMC-like mechanisms, including the involvement of ion channels in mediating pericyte contraction (Gonzales et al. 2020; Eltanahy et al. 2021; Klug et al. 2023; Hariharan et al. 2020). Just as the application of what we know about SMC biology can advance our understanding of pericytes, the growing bed of pericyte literature can inform the field's understanding of SMC biology. As already alluded to, pericytes can wrap and unwrap along capillaries during physiological angiogenesis and in response to fibrotic disease. SMCs can also unwrap from arterioles (Fig. 5.5) (Motherwell et al. 2017), motivating questions about how morphological remodeling of SMCs is regulated and how this remodeling alters SMC function.

The integration of different research areas, which are typically siloed according to the organ system (e.g., heart, brain, kidney, lung, skeletal muscle, eye, etc.), will also help provide novel insights that may not be realized by studying pericytes in unique organ contexts. Considering the diverse pericyte origins, multifaceted subpopulation identities, dynamic interactions with ECs and other cell types, robust pluripotency potential, and shared phenotypic lineages with other cell types leaves us with more questions than answers: Do interstitial cells differentiate into pericytes? Are unique pericyte subpopulations involved in regulating capillary sprouting, and are they

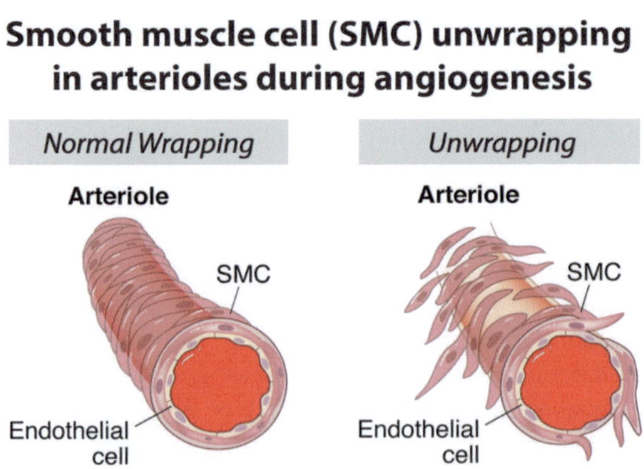

Fig. 5.5 Smooth muscle cell unwrapping along vasoactive arterioles can be associated with microvascular network growth scenarios characterized by angiogenesis. When considering the perivascular continuum, such potential morphological pattern changes can be viewed as an extension of downstream phenotypic and structural changes characteristic of pericyte dynamics. The observation of unwrapping provokes research questions related to how smooth muscle and pericyte function changes during remodeling scenarios

different from the pericyte subpopulations that regulate capillary permeability or constriction? How does pericyte function change over time, for example, with aging? Under what conditions can macrophages differentiate into pericytes, and vice versa? Do pericytes have the same developmental origin as neural support cells? All of these questions, and many more, remain to be answered in the context of both health and disease. It is obvious that we are just beginning to understand the multifaceted roles of pericytes in the microvasculature and beyond.

References

Abcouwer SF, Gardner TW (2014) Diabetic retinopathy: loss of neuroretinal adaptation to the diabetic metabolic environment. Ann N Y Acad Sci 1311:174–190. https://doi.org/10.1111/nyas.12412. PubMed PMID: 24673341; PMCID: PMC4154702

Aerpio Pharmaceuticals (2019) Aerpio shares plummet after Lead candidate fails phase IIb diabetic retinopathy trial. Genetic Engineering and Biotechnology News

Alarcon-Martinez L, Yilmaz-Ozcan S, Yemisci M, Schallek J, Kilic K, Can A, Di Polo A, Dalkara T (2018) Capillary pericytes express alpha-smooth muscle actin, which requires prevention of filamentous-actin depolymerization for detection. elife 7. https://doi.org/10.7554/eLife.34861. PubMed PMID: 29561727; PMCID: PMC5862523

Alarcon-Martinez L, Yemisci M, Dalkara T (2021) Pericyte morphology and function. Histol Histopathol 36(6):633–643. https://doi.org/10.14670/hh-18-314

Amos PJ, Shang H, Bailey AM, Taylor A, Katz AJ, Peirce SM (2008) IFATS collection: the role of human adipose-derived stromal cells in inflammatory microvascular remodeling and evidence of a perivascular phenotype. Stem Cells 26(10):2682–2690. https://doi.org/10.1634/stemcells.2008-0030. PubMed PMID: 18436860; PMCID: PMC2672107

Amos PJ, Mulvey CL, Seaman SA, Walpole J, Degen KE, Shang H, Katz AJ, Peirce SM (2011) Hypoxic culture and in vivo inflammatory environments affect the assumption of pericyte characteristics by human adipose and bone marrow progenitor cells. Am J Physiol Cell Physiol 301(6):C1378–C1388. https://doi.org/10.1152/ajpcell.00460.2010. PubMed PMID: 21865587; PMCID: PMC3233793

Ampofo E, Schmitt BM, Menger MD, Laschke MW (2017) The regulatory mechanisms of NG2/CSPG4 expression. Cell Mol Biol Lett 22:4. https://doi.org/10.1186/s11658-017-0035-3. PubMed PMID: 28536635; PMCID: PMC5415841

Angus JA, Wright CE (2015) ATP is not involved in alpha1-adrenoceptor-mediated vasoconstriction in resistance arteries. Eur J Pharmacol 769:162–166. https://doi.org/10.1016/j.ejphar.2015.11.013

Arboleda-Velasquez JF, Valdez CN, Marko CK, D'Amore PA (2015) From pathobiology to the targeting of pericytes for the treatment of diabetic retinopathy. Curr Diab Rep 15(2):573. https://doi.org/10.1007/s11892-014-0573-2. PubMed PMID: 25620405; PMCID: PMC5599150

Armulik A, Abramsson A, Betsholtz C (2005) Endothelial/pericyte interactions. Circ Res 97(6):512–523. https://doi.org/10.1161/01.RES.0000182903.16652.d7

Armulik A, Genové G, Mäe M, Nisancioglu MH, Wallgard E, Niaudet C, He L, Norlin J, Lindblom P, Strittmatter K, Johansson BR, Betsholtz C (2010) Pericytes regulate the blood-brain barrier. Nature 468(7323):557–561. https://doi.org/10.1038/nature09522

Armulik A, Mäe M, Betsholtz C (2011) Pericytes and the blood-brain barrier: recent advances and implications for the delivery of CNS therapy. Ther Deliv 2(4):419–422. https://doi.org/10.4155/tde.11.23

Attwell D, Mishra A, Hall CN, O'Farrell FM, Dalkara T (2016) What is a pericyte? J Cereb Blood Flow Metab 36(2):451–455. https://doi.org/10.1177/0271678x15610340. PubMed PMID: 26661200; PMCID: PMC4759679

Bergers G, Song S (2005) The role of pericytes in blood-vessel formation and maintenance. Neuro-Oncology 7(4):452–464. https://doi.org/10.1215/s1152851705000232. PubMed PMID: 16212810; PMCID: PMC1871727

Betsholtz C (2004) Insight into the physiological functions of PDGF through genetic studies in mice. Cytokine Growth Factor Rev 15(4):215–228. https://doi.org/10.1016/j.cytogfr.2004.03.005

Billaud M, Lohman AW, Straub AC, Looft-Wilson R, Johnstone SR, Araj CA, Best AK, Chekeni FB, Ravichandran KS, Penuela S, Laird DW, Isakson BE (2011) Pannexin1 regulates alpha1-adrenergic receptor- mediated vasoconstriction. Circ Res 109(1):80–85. https://doi.org/10.1161/CIRCRESAHA.110.237594. PubMed PMID: 21546608; PMCID: PMC3135971

Billaud M, Chiu YH, Lohman AW, Parpaite T, Butcher JT, Mutchler SM, DeLalio LJ, Artamonov MV, Sandilos JK, Best AK, Somlyo AV, Thompson RJ, Le TH, Ravichandran KS, Bayliss DA, Isakson BE (2015) A molecular signature in the pannexin1 intracellular loop confers channel activation by the alpha1 adrenoreceptor in smooth muscle cells. Sci Signal 8(364):ra17. https://doi.org/10.1126/scisignal.2005824. PubMed PMID: 25690012; PMCID: PMC4358815

Burnstock G (2017) Purinergic signaling in the cardiovascular system. Circ Res 120(1):207–228. https://doi.org/10.1161/CIRCRESAHA.116.309726

Burnstock G, Ralevic V (2014) Purinergic signaling and blood vessels in health and disease. Pharmacol Rev 66(1):102–192. https://doi.org/10.1124/pr.113.008029

Carmeliet P, Jain RK (2011) Molecular mechanisms and clinical applications of angiogenesis. Nature 473(7347):298–307. https://doi.org/10.1038/nature10144. PubMed PMID: 21593862; PMCID: PMC4049445

Chekenya M, Enger P, Thorsen F, Tysnes BB, Al-Sarraj S, Read TA, Furmanek T, Mahesparan R, Levine JM, Butt AM, Pilkington GJ, Bjerkvig R (2002) The glial precursor proteoglycan, NG2, is expressed on tumour neovasculature by vascular pericytes in human malignant brain tumours. Neuropathol Appl Neurobiol 28(5):367–380. https://doi.org/10.1046/j.1365-2990.2002.00412.x

Chen YT, Chang FC, Wu CF, Chou YH, Hsu HL, Chiang WC, Shen J, Chen YM, Wu KD, Tsai TJ, Duffield JS, Lin SL (2011) Platelet-derived growth factor receptor signaling activates pericyte-myofibroblast transition in obstructive and post-ischemic kidney fibrosis. Kidney Int 80(11): 1170–1181. https://doi.org/10.1038/ki.2011.208

Chintalgattu V, Rees ML, Culver JC, Goel A, Jiffar T, Zhang J, Dunner K Jr, Pati S, Bankson JA, Pasqualini R, Arap W, Bryan NS, Taegtmeyer H, Langley RR, Yao H, Kupferman ME, Entman ML, Dickinson ME, Khakoo AY (2013) Coronary microvascular pericytes are the cellular target of sunitinib malate-induced cardiotoxicity. Sci Transl Med 5(187):187ra69. https://doi.org/10.1126/scitranslmed.3005066. PubMed PMID: 23720580; PMCID: PMC3833098

Cogan DG, Toussaint D, Kuwabara T (1961) Retinal vascular patterns. IV Diabetic retinopathy. Arch Ophthalmol 66:366–378. https://doi.org/10.1001/archopht.1961.00960010368014

Cooke VG, LeBleu VS, Keskin D, Khan Z, O'Connell JT, Teng Y, Duncan MB, Xie L, Maeda G, Vong S, Sugimoto H, Rocha RM, Damascena A, Brentani RR, Kalluri R (2012) Pericyte depletion results in hypoxia-associated epithelial-to-mesenchymal transition and metastasis mediated by met signaling pathway. Cancer Cell 21(1):66–81. https://doi.org/10.1016/j.ccr.2011.11.024. PubMed PMID: 22264789; PMCID: PMC3999522

Corliss BA, Ray HC, Doty RW, Mathews C, Sheybani N, Fitzgerald K, Prince R, Kelly-Goss MR, Murfee WL, Chappell J, Owens GK, Yates PA, Peirce SM (2020) Pericyte bridges in homeostasis and hyperglycemia. Diabetes 69(7):1503–1517. https://doi.org/10.2337/db19-0471. PubMed PMID: 32321760; PMCID: PMC7306121

Cronk SM, Kelly-Goss MR, Ray HC, Mendel TA, Hoehn KL, Bruce AC, Dey BK, Guendel AM, Tavakol DN, Herman IM, Peirce SM, Yates PA (2015) Adipose-derived stem cells from diabetic mice show impaired vascular stabilization in a murine model of diabetic retinopathy.

Stem Cells Transl Med 4(5):459–467. https://doi.org/10.5966/sctm.2014-0108. PubMed PMID: 25769654; PMCID: PMC4414213

Cushing DJ, Brown GL, Sabouni MH, Mustafa SJ (1991) Adenosine receptor-mediated coronary artery relaxation and cyclic nucleotide production. Am J Phys 261(2 Pt 2):H343–H348. https://doi.org/10.1152/ajpheart.1991.261.2.H343

Dabravolski SA, Markin AM, Andreeva ER, Eremin II, Orekhov AN, Melnichenko AA (2022) Emerging role of pericytes in therapy of cardiovascular diseases. Biomed Pharmacother 156: 113928. https://doi.org/10.1016/j.biopha.2022.113928

Deaton RA, Bulut G, Serbulea V, Salamon A, Shankman LS, Nguyen AT, Owens GK (2022) A new autosomal Myh11-CreERT2 smooth muscle cell lineage tracing and gene knockout mouse model. Athero Thrombois Vasc Biol

Deguchi K, Liu N, Liu W, Omote Y, Kono S, Yunoki T, Deguchi S, Yamashita T, Ikeda Y, Abe K (2014) Pericyte protection by edaravone after tissue plasminogen activator treatment in rat cerebral ischemia. J Neurosci Res 92(11):1509–1519. https://doi.org/10.1002/jnr.23420. PubMed PMID: 24938625; PMCID: PMC4263311

Dessalles CA, Babataheri A, Barakat AI (2021) Pericyte mechanics and mechanobiology. J Cell Sci 134(6). https://doi.org/10.1242/jcs.240226

Ding R, Hase Y, Ameen-Ali KE, Ndung'u M, Stevenson W, Barsby J, Gourlay R, Akinyemi T, Akinyemi R, Uemura MT, Polvikoski T, Mukaetova-Ladinska E, Ihara M, Kalaria RN (2020) Loss of capillary pericytes and the blood-brain barrier in white matter in poststroke and vascular dementias and Alzheimer's disease. Brain Pathol 30(6):1087–1101. https://doi.org/10.1111/bpa.12888. PubMed PMID: 32705757; PMCID: PMC8018063

Dunaway LS, Billaud M, Macal E, Good ME, Medina CB, Lorenz U, Ravichandran KS, Koval M, Isakson BE (2022) Amount of pannexin 1 in smooth muscle cells regulates sympathetic nerve-induced vasoconstriction. Hypertension. In Press

El-Ghazawi K, Eyo UB, Peirce SM (2024) Brain microvascular Pericyte pathology linking Alzheimer's disease to diabetes. Microcirculation 31(7):e12877. https://doi.org/10.1111/micc.12877. PubMed PMID: 39222475; PMCID: PMC11471384

Eltanahy AM, Koluib YA, Gonzales A (2021) Pericytes: intrinsic transportation engineers of the CNS microcirculation. Front Physiol 12:719701. https://doi.org/10.3389/fphys.2021.719701. PubMed PMID: 34497540; PMCID: PMC8421025

Elvidge S (2017) Opthotech's Fovista crashes out in wet AMD. PharmaDive

Erdener SE, Kureli G, Dalkara T (2022) Contractile apparatus in CNS capillary pericytes. Neurophotonics 9(2):021904. https://doi.org/10.1117/1.NPh.9.2.021904. PubMed PMID: 35106320; PMCID: PMC8785978

Erlinge D (1998) Extracellular ATP: a growth factor for vascular smooth muscle cells. Gen Pharmacol 31(1):1–8. https://doi.org/10.1016/s0306-3623(97)00420-5

Erlinge D, Hou M, Webb TE, Barnard EA, Moller S (1998) Phenotype changes of the vascular smooth muscle cell regulate P2 receptor expression as measured by quantitative RT-PCR. Biochem Biophys Res Commun 248(3):864–870. https://doi.org/10.1006/bbrc.1998.9083

Etchevers HC, Vincent C, Le Douarin NM, Couly GF (2001) The cephalic neural crest provides pericytes and smooth muscle cells to all blood vessels of the face and forebrain. Development 128(7):1059–1068. https://doi.org/10.1242/dev.128.7.1059

Fukushi J, Makagiansar IT, Stallcup WB (2004) NG2 proteoglycan promotes endothelial cell motility and angiogenesis via engagement of galectin-3 and alpha3beta1 integrin. Mol Biol Cell 15(8):3580–3590. https://doi.org/10.1091/mbc.e04-03-0236. PubMed PMID: 15181153; PMCID: PMC491820

Gerhardt H, Betsholtz C (2003) Endothelial-pericyte interactions in angiogenesis. Cell Tissue Res 314(1):15–23. https://doi.org/10.1007/s00441-003-0745-x

Gerhardt H, Wolburg H, Redies C (2000) N-cadherin mediates pericytic-endothelial interaction during brain angiogenesis in the chicken. Dev Dyn 218(3):472–479

Goel S, Duda DG, Xu L, Munn LL, Boucher Y, Fukumura D, Jain RK (2011) Normalization of the vasculature for treatment of cancer and other diseases. Physiol Rev 91(3):1071–1121. https://doi.org/10.1152/physrev.00038.2010. PubMed PMID: 21742796; PMCID: PMC3258432

Gonzales AL, Klug NR, Moshkforoush A, Lee JC, Lee FK, Shui B, Tsoukias NM, Kotlikoff MI, Hill-Eubanks D, Nelson MT (2020) Contractile pericytes determine the direction of blood flow at capillary junctions. Proc Natl Acad Sci USA 117(43):27022–27033. https://doi.org/10.1073/pnas.1922755117. PubMed PMID: 33051294; PMCID: PMC7604512

Grako KA, Stallcup WB (1995) Participation of the NG2 proteoglycan in rat aortic smooth muscle cell responses to platelet-derived growth factor. Exp Cell Res 221(1):231–240. https://doi.org/10.1006/excr.1995.1371

Grako KA, Ochiya T, Barritt D, Nishiyama A, Stallcup WB (1999) PDGF (alpha)-receptor is unresponsive to PDGF-AA in aortic smooth muscle cells from the NG2 knockout mouse. J Cell Sci 112(Pt 6):905–915. https://doi.org/10.1242/jcs.112.6.905

Hall CN, Reynell C, Gesslein B, Hamilton NB, Mishra A, Sutherland BA, O'Farrell FM, Buchan AM, Lauritzen M, Attwell D (2014) Capillary pericytes regulate cerebral blood flow in health and disease. Nature 508(7494):55–60. https://doi.org/10.1038/nature13165. PubMed PMID: 24670647; PMCID: PMC3976267

Hamilton NB, Attwell D, Hall CN (2010) Pericyte-mediated regulation of capillary diameter: a component of neurovascular coupling in health and disease. Front Neuroenerg 2:2. https://doi.org/10.3389/fnene.2010.00005. PubMed PMID: 20725515; PMCID: PMC2912025

Hannan RT, Miller AE, Hung RC, Sano C, Peirce SM, Barker TH (2021) Extracellular matrix remodeling associated with bleomycin-induced lung injury supports pericyte-to-myofibroblast transition. Matrix Biol Plus 10:100056. https://doi.org/10.1016/j.mbplus.2020.100056. PubMed PMID: 34195593; PMCID: PMC8233458

Hariharan A, Weir N, Robertson C, He L, Betsholtz C, Longden TA (2020) The Ion Channel and GPCR toolkit of brain capillary Pericytes. Front Cell Neurosci 14:601324. https://doi.org/10.3389/fncel.2020.601324. PubMed PMID: 33390906; PMCID: PMC7775489

Harper S, Webb TE, Charlton SJ, Ng LL, Boarder MR (1998) Evidence that P2Y4 nucleotide receptors are involved in the regulation of rat aortic smooth muscle cells by UTP and ATP. Br J Pharmacol 124(4):703–710. https://doi.org/10.1038/sj.bjp.0701895. PubMed PMID: 9690862; PMCID: PMC1565449

He L, Vanlandewijck M, Mäe MA, Andrae J, Ando K, Del Gaudio F, Nahar K, Lebouvier T, Laviña B, Gouveia L, Sun Y, Raschperger E, Segerstolpe Å, Liu J, Gustafsson S, Räsänen M, Zarb Y, Mochizuki N, Keller A, Lendahl U, Betsholtz C (2018) Single-cell RNA sequencing of mouse brain and lung vascular and vessel-associated cell types. Sci Data 5:180160. https://doi.org/10.1038/sdata.2018.160. PubMed PMID: 30129931; PMCID: PMC6103262

He H, Yang W, Su N, Zhang C, Dai J, Han F, Singhal M, Bai W, Zhu X, Zhu J, Liu Z, Xia W, Liu X, Zhang C, Jiang K, Huang W, Chen D, Wang Z, He X, Kirchhoff F, Li Z, Liu C, Huan J, Wang X, Wei W, Wang J, Augustin HG, Hu J (2023) Activating NO-sGC crosstalk in the mouse vascular niche promotes vascular integrity and mitigates acute lung injury. J Exp Med 220(2). https://doi.org/10.1084/jem.20211422. PubMed PMID: 36350314; PMCID: PMC9984546

Hellström M, Kalén M, Lindahl P, Abramsson A, Betsholtz C (1999) Role of PDGF-B and PDGFR-beta in recruitment of vascular smooth muscle cells and pericytes during embryonic blood vessel formation in the mouse. Development 126(14):3047–3055. https://doi.org/10.1242/dev.126.14.3047

Hergenreider E, Heydt S, Tréguer K, Boettger T, Horrevoets AJ, Zeiher AM, Scheffer MP, Frangakis AS, Yin X, Mayr M, Braun T, Urbich C, Boon RA, Dimmeler S (2012) Atheroprotective communication between endothelial cells and smooth muscle cells through miRNAs. Nat Cell Biol 14(3):249–256. https://doi.org/10.1038/ncb2441

Hess DL, Kelly-Goss MR, Cherepanova OA, Nguyen AT, Baylis RA, Tkachenko S, Annex BH, Peirce SM, Owens GK (2019) Perivascular cell-specific knockout of the stem cell pluripotency gene Oct4 inhibits angiogenesis. Nat Commun 10(1):967. https://doi.org/10.1038/s41467-019-08811-z. PubMed PMID: 30814500; PMCID: PMC6393549

Hogarth DK, Sandbo N, Taurin S, Kolenko V, Miano JM, Dulin NO (2004) Dual role of PKA in phenotypic modulation of vascular smooth muscle cells by extracellular ATP. Am J Physiol Cell Physiol 287(2):C449–C456. https://doi.org/10.1152/ajpcell.00547.2003

Hørlyck S, Cai C, Helms HCC, Lauritzen M, Brodin B (2021) ATP induces contraction of cultured brain capillary pericytes via activation of P2Y-type purinergic receptors. Am J Physiol Heart Circ Physiol 320(2):H699–h712. https://doi.org/10.1152/ajpheart.00560.2020

Hosaka K, Yang Y, Seki T, Fischer C, Dubey O, Fredlund E, Hartman J, Religa P, Morikawa H, Ishii Y, Sasahara M, Larsson O, Cossu G, Cao R, Lim S, Cao Y (2016) Pericyte-fibroblast transition promotes tumor growth and metastasis. Proc Natl Acad Sci USA 113(38):E5618–E5627. https://doi.org/10.1073/pnas.1608384113. PubMed PMID: 27608497; PMCID: PMC5035870

Hou M, Harden TK, Kuhn CM, Baldetorp B, Lazarowski E, Pendergast W, Moller S, Edvinsson L, Erlinge D (2002) UDP acts as a growth factor for vascular smooth muscle cells by activation of P2Y(6) receptors. Am J Physiol Heart Circ Physiol 282(2):H784–H792. https://doi.org/10.1152/ajpheart.00997.2000

Humphreys BD, Lin SL, Kobayashi A, Hudson TE, Nowlin BT, Bonventre JV, Valerius MT, McMahon AP, Duffield JS (2010) Fate tracing reveals the pericyte and not epithelial origin of myofibroblasts in kidney fibrosis. Am J Pathol 176(1):85–97. https://doi.org/10.2353/ajpath.2010.090517. PubMed PMID: 20008127; PMCID: PMC2797872

Jain RK (2005) Normalization of tumor vasculature: an emerging concept in antiangiogenic therapy. Science 307(5706):58–62. https://doi.org/10.1126/science.1104819

Jansson D, Rustenhoven J, Feng S, Hurley D, Oldfield RL, Bergin PS, Mee EW, Faull RL, Dragunow M (2014) A role for human brain pericytes in neuroinflammation. J Neuroinflammation 11:104. https://doi.org/10.1186/1742-2094-11-104. PubMed PMID: 24920309; PMCID: PMC4105169

Kazlauskas A (2017) PDGFs and their receptors. Gene 614:1–7. https://doi.org/10.1016/j.gene.2017.03.003. PubMed PMID: 28267575; PMCID: PMC6728141

Kelly-Goss MR, Sweat RS, Stapor PC, Peirce SM, Murfee WL (2014) Targeting pericytes for angiogenic therapies. Microcirculation 21(4):345–357. https://doi.org/10.1111/micc.12107. PubMed PMID: 24267154; PMCID: PMC4079092

Klug NR, Sancho M, Gonzales AL, Heppner TJ, O'Brien RIC, Hill-Eubanks D, Nelson MT (2023) Intraluminal pressure elevates intracellular calcium and contracts CNS pericytes: role of voltage-dependent calcium channels. Proc Natl Acad Sci USA 120(9):e2216421120. https://doi.org/10.1073/pnas.2216421120. PubMed PMID: 36802432; PMCID: PMC9992766

Korn J, Christ B, Kurz H (2002) Neuroectodermal origin of brain pericytes and vascular smooth muscle cells. J Comp Neurol 442(1):78–88. https://doi.org/10.1002/cne.1423

Kostallari E, Shah VH (2019) Pericytes in the liver. Adv Exp Med Biol 1122:153–167. https://doi.org/10.1007/978-3-030-11093-2_9. PubMed PMID: 30937868; PMCID: PMC7137998

Kruse K, Lee QS, Sun Y, Klomp J, Yang X, Huang F, Sun MY, Zhao S, Hong Z, Vogel SM, Shin JW, Leckband DE, Tai LM, Malik AB, Komarova YA (2019) N-cadherin signaling via trio assembles adherens junctions to restrict endothelial permeability. J Cell Biol 218(1):299–316. https://doi.org/10.1083/jcb.201802076. PubMed PMID: 30463880; PMCID: PMC6314553

Kuriyan AE, Albini TA, Townsend JH, Rodriguez M, Pandya HK, Leonard RE 2nd, Parrott MB, Rosenfeld PJ, Flynn HW Jr, Goldberg JL (2017) Vision loss after Intravitreal injection of autologous "stem cells" for AMD. N Engl J Med 376(11):1047–1053. https://doi.org/10.1056/NEJMoa1609583. PubMed PMID: 28296617; PMCID: PMC5551890

Lamont C, Vial C, Evans RJ, Wier WG (2006) P2X1 receptors mediate sympathetic postjunctional Ca2+ transients in mesenteric small arteries. Am J Physiol Heart Circ Physiol 291(6):H3106–H3113. https://doi.org/10.1152/ajpheart.00466.2006

Latif N, Sarathchandra P, Chester AH, Yacoub MH (2015) Expression of smooth muscle cell markers and co-activators in calcified aortic valves. Eur Heart J 36(21):1335–1345. https://doi.org/10.1093/eurheartj/eht547

Le DE, Zhao Y, Kaul S (2022) Persistent coronary vasomotor tone during myocardial ischemia occurs at the capillary level and may involve Pericytes. Front Cardiovasc Med 9:930492. https://doi.org/10.3389/fcvm.2022.930492. PubMed PMID: 35811707; PMCID: PMC9263193

Leonard-Duke J, Bruce AC, Peirce SM, Taite LJ (2023) Variations in mechanical stiffness alter microvascular sprouting and stability in a PEG hydrogel model of idiopathic pulmonary fibrosis. Microcirculation 30(5–6):e12817. https://doi.org/10.1111/micc.12817. PubMed PMID: 37248193; PMCID: PMC10524245

Lin SL, Kisseleva T, Brenner DA, Duffield JS (2008) Pericytes and perivascular fibroblasts are the primary source of collagen-producing cells in obstructive fibrosis of the kidney. Am J Pathol 173(6):1617–1627. https://doi.org/10.2353/ajpath.2008.080433. PubMed PMID: 19008372; PMCID: PMC2626374

Lin A, Peiris NJ, Dhaliwal H, Hakim M, Li W, Ganesh S, Ramaswamy Y, Patel S, Misra A (2021) Mural cells: potential therapeutic targets to bridge cardiovascular disease and neurodegeneration. Cells 10(3). https://doi.org/10.3390/cells10030593. PubMed PMID: 33800271; PMCID: PMC7999039

Longden TA, Zhao G, Hariharan A, Lederer WJ (2023) Pericytes and the control of blood flow in brain and heart. Annu Rev Physiol 85:137–164. https://doi.org/10.1146/annurev-physiol-031522-034807. PubMed PMID: 36763972; PMCID: PMC10280497

Ma Q, Zhao Z, Sagare AP, Wu Y, Wang M, Owens NC, Verghese PB, Herz J, Holtzman DM, Zlokovic BV (2018) Blood-brain barrier-associated pericytes internalize and clear aggregated amyloid-β42 by LRP1-dependent apolipoprotein E isoform-specific mechanism. Mol Neurodegener 13(1):57. https://doi.org/10.1186/s13024-018-0286-0. PubMed PMID: 30340601; PMCID: PMC6194676

Majesky MW (2007) Developmental basis of vascular smooth muscle diversity. Arterioscler Thromb Vasc Biol 27(6):1248–1258. https://doi.org/10.1161/atvbaha.107.141069

Makagiansar IT, Williams S, Dahlin-Huppe K, Fukushi J, Mustelin T, Stallcup WB (2004) Phosphorylation of NG2 proteoglycan by protein kinase C-alpha regulates polarized membrane distribution and cell motility. J Biol Chem 279(53):55262–55270. https://doi.org/10.1074/jbc.M411045200

Malmberg M, Malm T, Gustafsson O, Sturchio A, Graff C, Espay AJ, Wright AP, El Andaloussi S, Lindén A, Ezzat K (2020) Disentangling the amyloid pathways: a mechanistic approach to etiology. Front Neurosci 14:256. https://doi.org/10.3389/fnins.2020.00256. PubMed PMID: 32372895; PMCID: PMC7186396

Martinez-Lemus LA (2012) The dynamic structure of arterioles. Basic Clin Pharmacol Toxicol 110(1):5–11. https://doi.org/10.1111/j.1742-7843.2011.00813.x. PubMed PMID: 21989114; PMCID: PMC4435689

Mendel TA, Clabough EB, Kao DS, Demidova-Rice TN, Durham JT, Zotter BC, Seaman SA, Cronk SM, Rakoczy EP, Katz AJ, Herman IM, Peirce SM, Yates PA (2013) Pericytes derived from adipose-derived stem cells protect against retinal vasculopathy. PLoS One 8(5):e65691. https://doi.org/10.1371/journal.pone.0065691. PubMed PMID: 23741506; PMCID: PMC3669216

Methner C, Mishra A, Golgotiu K, Li Y, Wei W, Yanez ND, Zlokovic B, Wang RK, Alkayed NJ, Kaul S, Iliff JJ (2019) Pericyte constriction underlies capillary derecruitment during hyperemia in the setting of arterial stenosis. Am J Physiol Heart Circ Physiol 317(2):H255–Hh63. https://doi.org/10.1152/ajpheart.00097.2019. PubMed PMID: 31125259; PMCID: PMC6732474

Moro M, Balestrero FC, Grolla AA (2024) Pericytes: jack-of-all-trades in cancer-related inflammation. Front Pharmacol 15:1426033. https://doi.org/10.3389/fphar.2024.1426033. PubMed PMID: 39086395; PMCID: PMC11288921

Motherwell JM, Azimi MS, Spicer K, Alves NG, Hodges NA, Breslin JW, Katakam PVG, Murfee WL (2017) Evaluation of arteriolar smooth muscle cell function in an ex vivo microvascular network model. Sci Rep 7(1):2195. https://doi.org/10.1038/s41598-017-02272-4. PubMed PMID: 28526859; PMCID: PMC5438412

Muhl L, Genové G, Leptidis S, Liu J, He L, Mocci G, Sun Y, Gustafsson S, Buyandelger B, Chivukula IV, Segerstolpe Å, Raschperger E, Hansson EM, Björkegren JLM, Peng XR, Vanlandewijck M, Lendahl U, Betsholtz C (2020) Single-cell analysis uncovers fibroblast heterogeneity and criteria for fibroblast and mural cell identification and discrimination. Nat Commun 11(1):3953. https://doi.org/10.1038/s41467-020-17740-1. PubMed PMID: 32769974; PMCID: PMC7414220

Murrant CL, Fletcher NM (2022) Capillary communication: the role of capillaries in sensing the tissue environment, coordinating the microvascular, and controlling blood flow. Am J Physiol Heart Circ Physiol 323(5):H1019–H1h36. https://doi.org/10.1152/ajpheart.00088.2022

Navarro R, Compte M, Álvarez-Vallina L, Sanz L (2016) Immune regulation by Pericytes: modulating innate and adaptive immunity. Front Immunol 7:480. https://doi.org/10.3389/fimmu.2016.00480. PubMed PMID: 27867386; PMCID: PMC5095456

Nehls V, Drenckhahn D (1993) The versatility of microvascular pericytes: from mesenchyme to smooth muscle? Histochemistry 99(1):1–12. https://doi.org/10.1007/bf00268014

O'Farrell FM, Mastitskaya S, Hammond-Haley M, Freitas F, Wah WR, Attwell D (2017) Capillary pericytes mediate coronary no-reflow after myocardial ischaemia. elife 6:6. https://doi.org/10.7554/eLife.29280. PubMed PMID: 29120327; PMCID: PMC5705208

Orlich MM, Dieguez-Hurtado R, Muehlfriedel R, Sothilingam V, Wolburg H, Oender CE, Woelffing P, Betsholtz C, Gaengel K, Seeliger M, Adams RH, Nordheim A (2022) Mural cell SRF controls Pericyte migration, vessel patterning and blood flow. Circ Res 131(4):308–327. https://doi.org/10.1161/CIRCRESAHA.122.321109. PubMed PMID: 35862101; PMCID: PMC9348820

Özen I, Deierborg T, Miharada K, Padel T, Englund E, Genové G, Paul G (2014) Brain pericytes acquire a microglial phenotype after stroke. Acta Neuropathol 128(3):381–396. https://doi.org/10.1007/s00401-014-1295-x. PubMed PMID: 24848101; PMCID: PMC4131168

Ozerdem U, Grako KA, Dahlin-Huppe K, Monosov E, Stallcup WB (2001) NG2 proteoglycan is expressed exclusively by mural cells during vascular morphogenesis. Dev Dyn 222(2):218–227. https://doi.org/10.1002/dvdy.1200

Ozerdem U, Monosov E, Stallcup WB (2002) NG2 proteoglycan expression by pericytes in pathological microvasculature. Microvasc Res 63(1):129–134. https://doi.org/10.1006/mvre.2001.2376

Paik JH, Skoura A, Chae SS, Cowan AE, Han DK, Proia RL, Hla T (2004) Sphingosine 1-phosphate receptor regulation of N-cadherin mediates vascular stabilization. Genes Dev 18(19):2392–2403. https://doi.org/10.1101/gad.1227804. PubMed PMID: 15371328; PMCID: PMC522989

Panos GD, Lakshmanan A, Dadoukis P, Ripa M, Motta L, Amoaku WM (2023) Faricimab: transforming the future of macular diseases treatment – a comprehensive review of clinical studies. Drug Des Devel Ther 17:2861–2873. https://doi.org/10.2147/dddt.S427416. PubMed PMID: 37746113; PMCID: PMC10516184

Partanen J, Armstrong E, Mäkelä TP, Korhonen J, Sandberg M, Renkonen R, Knuutila S, Huebner K, Alitalo K (1992) A novel endothelial cell surface receptor tyrosine kinase with extracellular epidermal growth factor homology domains. Mol Cell Biol 12(4):1698–1707. https://doi.org/10.1128/mcb.12.4.1698-1707.1992. PubMed PMID: 1312667; PMCID: PMC369613

Payne LB, Darden J, Suarez-Martinez AD, Zhao H, Hendricks A, Hartland C, Chong D, Kushner EJ, Murfee WL, Chappell JC (2021) Pericyte migration and proliferation are tightly synchronized to endothelial cell sprouting dynamics. Integr Biol (Camb) 13(2):31–43. https://doi.org/10.1093/intbio/zyaa027. PubMed PMID: 33515222; PMCID: PMC7919101

Pellowe AS, Sauler M, Hou Y, Merola J, Liu R, Calderon B, Lauridsen HM, Harris MR, Leng L, Zhang Y, Tilstam PV, Pober JS, Bucala R, Lee PJ, Gonzalez AL (2019) Endothelial cell-secreted MIF reduces pericyte contractility and enhances neutrophil extravasation. FASEB J

33(2):2171–2186. https://doi.org/10.1096/fj.201800480R. PubMed PMID: 30252532; PMCID: PMC6338650

Pieper C, Marek JJ, Unterberg M, Schwerdtle T, Galla HJ (2014) Brain capillary pericytes contribute to the immune defense in response to cytokines or LPS in vitro. Brain Res 1550:1–8. https://doi.org/10.1016/j.brainres.2014.01.004

Pluschke G, Vanek M, Evans A, Dittmar T, Schmid P, Itin P, Filardo EJ, Reisfeld RA (1996) Molecular cloning of a human melanoma-associated chondroitin sulfate proteoglycan. Proc Natl Acad Sci USA 93(18):9710–9715. https://doi.org/10.1073/pnas.93.18.9710. PubMed PMID: 8790396; PMCID: PMC38494

Prada MP, Syed AU, Reddy GR, Martin-Aragon Baudel M, Flores-Tamez VA, Sasse KC, Ward SM, Sirish P, Chiamvimonvat N, Bartels P, Dickson EJ, Hell JW, Scott JD, Santana LF, Xiang YK, Navedo MF, Nieves-Cintron M (2020) AKAP5 complex facilitates purinergic modulation of vascular L-type Ca(2+) channel CaV1.2. Nat Commun 11(1):5303. https://doi.org/10.1038/s41467-020-18947-y. PubMed PMID: 33082339; PMCID: PMC7575592

Quaegebeur A, Lange C, Carmeliet P (2011) The neurovascular link in health and disease: molecular mechanisms and therapeutic implications. Neuron 71(3):406–424. https://doi.org/10.1016/j.neuron.2011.07.013

Raposo G, Nijman HW, Stoorvogel W, Liejendekker R, Harding CV, Melief CJ, Geuze HJ (1996) B lymphocytes secrete antigen-presenting vesicles. J Exp Med 183(3):1161–1172. https://doi.org/10.1084/jem.183.3.1161. PubMed PMID: 8642258; PMCID: PMC2192324

Ray HC, Corliss BA, Bruce AC, Kesting S, Dey P, Mansour J, Seaman SA, Smolko CM, Mathews C, Dey BK, Owens GK, Peirce SM, Yates PA (2020) Myh11+ microvascular mural cells and derived mesenchymal stem cells promote retinal fibrosis. Sci Rep 10(1):15808. https://doi.org/10.1038/s41598-020-72875-x. PubMed PMID: 32978500; PMCID: PMC7519078

Ribatti D, Nico B, Crivellato E (2011) The role of pericytes in angiogenesis. Int J Dev Biol 55(3):261–268. https://doi.org/10.1387/ijdb.103167dr

Ruan J, Luo M, Wang C, Fan L, Yang SN, Cardenas M, Geng H, Leonard JP, Melnick A, Cerchietti L, Hajjar KA (2013) Imatinib disrupts lymphoma angiogenesis by targeting vascular pericytes. Blood 121(26):5192–5202. https://doi.org/10.1182/blood-2013-03-490763. PubMed PMID: 23632889; PMCID: PMC3695363

Rustenhoven J, Jansson D, Smyth LC, Dragunow M (2017) Brain Pericytes as mediators of Neuroinflammation. Trends Pharmacol Sci 38(3):291–304. https://doi.org/10.1016/j.tips.2016.12.001

Sack KD, Kellum JA, Parikh SM (2020) The angiopoietin-Tie2 pathway in critical illness. Crit Care Clin 36(2):201–216. https://doi.org/10.1016/j.ccc.2019.12.003. PubMed PMID: 32172809; PMCID: PMC8843037

Sagare AP, Bell RD, Zhao Z, Ma Q, Winkler EA, Ramanathan A, Zlokovic BV (2013) Pericyte loss influences Alzheimer-like neurodegeneration in mice. Nat Commun 4:2932. https://doi.org/10.1038/ncomms3932. PubMed PMID: 24336108; PMCID: PMC3945879

Sakuma R, Kawahara M, Nakano-Doi A, Takahashi A, Tanaka Y, Narita A, Kuwahara-Otani S, Hayakawa T, Yagi H, Matsuyama T, Nakagomi T (2016) Brain pericytes serve as microglia-generating multipotent vascular stem cells following ischemic stroke. J Neuroinflammation 13(1):57. https://doi.org/10.1186/s12974-016-0523-9. PubMed PMID: 26952098; PMCID: PMC4782566

Saunders WB, Bohnsack BL, Faske JB, Anthis NJ, Bayless KJ, Hirschi KK, Davis GE (2006) Coregulation of vascular tube stabilization by endothelial cell TIMP-2 and pericyte TIMP-3. J Cell Biol 175(1):179–191. https://doi.org/10.1083/jcb.200603176. PubMed PMID: 17030988; PMCID: PMC2064509

Sengillo JD, Winkler EA, Walker CT, Sullivan JS, Johnson M, Zlokovic BV (2013) Deficiency in mural vascular cells coincides with blood-brain barrier disruption in Alzheimer's disease. Brain Pathol 23(3):303–310. https://doi.org/10.1111/bpa.12004. PubMed PMID: 23126372; PMCID: PMC3628957

Sims DE (2000) Diversity within pericytes. Clin Exp Pharmacol Physiol 27(10):842–846. https://doi.org/10.1046/j.1440-1681.2000.03343.x

Smith AM, Graham ES, Feng SX, Oldfield RL, Bergin PM, Mee EW, Faull RL, Curtis MA, Dragunow M (2013) Adult human glia, pericytes and meningeal fibroblasts respond similarly to IFNy but not to TGFβ1 or M-CSF. PLoS One 8(12):e80463. https://doi.org/10.1371/journal.pone.0080463. PubMed PMID: 24339874; PMCID: PMC3855168

Sousa JB, Diniz C (2017) The Adenosinergic system as a therapeutic target in the vasculature: new ligands and challenges. Molecules 22(5). https://doi.org/10.3390/molecules22050752. PubMed PMID: 28481238; PMCID: PMC6154114

Stallcup WB (2018) The NG2 proteoglycan in Pericyte biology. Adv Exp Med Biol 1109:5–19. https://doi.org/10.1007/978-3-030-02601-1_2

Stapor PC, Sweat RS, Dashti DC, Betancourt AM, Murfee WL (2014) Pericyte dynamics during angiogenesis: new insights from new identities. J Vasc Res 51(3):163–174. https://doi.org/10.1159/000362276. PubMed PMID: 24853910; PMCID: PMC4149862

Steucke KE, Tracy PV, Hald ES, Hall JL, Alford PW (2015) Vascular smooth muscle cell functional contractility depends on extracellular mechanical properties. J Biomech 48(12):3044–3051. https://doi.org/10.1016/j.jbiomech.2015.07.029. PubMed PMID: 26283412; PMCID: PMC4592799

Stevenson TJ, Dieriks BV (2023) Wrapping up the role of pericytes in Parkinson's disease. Neural Regen Res 18(11):2395–2396. https://doi.org/10.4103/1673-5374.371362. PubMed PMID: 37282465; PMCID: PMC10360080

Stevenson TJ, Johnson RH, Savistchenko J, Rustenhoven J, Woolf Z, Smyth LCD, Murray HC, Faull RLM, Correia J, Schweder P, Heppner P, Turner C, Melki R, Dieriks BV, Curtis MA, Dragunow M (2022) Pericytes take up and degrade α-synuclein but succumb to apoptosis under cellular stress. Sci Rep 12(1):17314. https://doi.org/10.1038/s41598-022-20261-0. PubMed PMID: 36243723; PMCID: PMC9569325

Stratman AN, Davis GE (2012) Endothelial cell-pericyte interactions stimulate basement membrane matrix assembly: influence on vascular tube remodeling, maturation, and stabilization. Microsc Microanal 18(1):68–80. https://doi.org/10.1017/s1431927611012402. PubMed PMID: 22166617; PMCID: PMC3919655

Suri C, Jones PF, Patan S, Bartunkova S, Maisonpierre PC, Davis S, Sato TN, Yancopoulos GD (1996) Requisite role of angiopoietin-1, a ligand for the TIE2 receptor, during embryonic angiogenesis. Cell 87(7):1171–1180. https://doi.org/10.1016/s0092-8674(00)81813-9

Sweeney HL, Hammers DW (2018) Muscle contraction. Cold Spring Harb Perspect Biol 10(2). https://doi.org/10.1101/cshperspect.a023200. PubMed PMID: 29419405; PMCID: PMC5793755

Tachibana M, Yamazaki Y, Liu CC, Bu G, Kanekiyo T (2018) Pericyte implantation in the brain enhances cerebral blood flow and reduces amyloid-β pathology in amyloid model mice. Exp Neurol 300:13–21. https://doi.org/10.1016/j.expneurol.2017.10.023. PubMed PMID: 29106980; PMCID: PMC5745278

Tang RH, Zheng XL, Callis TE, Stansfield WE, He J, Baldwin AS, Wang DZ, Selzman CH (2008) Myocardin inhibits cellular proliferation by inhibiting NF-kappaB(p65)-dependent cell cycle progression. Proc Natl Acad Sci USA 105(9):3362–3367. https://doi.org/10.1073/pnas.0705842105. PubMed PMID: 18296632; PMCID: PMC2265140

Uemura MT, Maki T, Ihara M, Lee VMY, Trojanowski JQ (2020) Brain microvascular Pericytes in vascular cognitive impairment and dementia. Front Aging Neurosci 12:80. https://doi.org/10.3389/fnagi.2020.00080. PubMed PMID: 32317958; PMCID: PMC7171590

Vanlandewijck M, He L, Mäe MA, Andrae J, Ando K, Del Gaudio F, Nahar K, Lebouvier T, Laviña B, Gouveia L, Sun Y, Raschperger E, Räsänen M, Zarb Y, Mochizuki N, Keller A, Lendahl U, Betsholtz C (2018) A molecular atlas of cell types and zonation in the brain vasculature. Nature 554(7693):475–480. https://doi.org/10.1038/nature25739

Vial C, Evans RJ (2002) P2X(1) receptor-deficient mice establish the native P2X receptor and a P2Y6-like receptor in arteries. Mol Pharmacol 62(6):1438–1445. https://doi.org/10.1124/mol.62.6.1438

Wang D, Chang PS, Wang Z, Sutherland L, Richardson JA, Small E, Krieg PA, Olson EN (2001) Activation of cardiac gene expression by myocardin, a transcriptional cofactor for serum response factor. Cell 105(7):851–862. https://doi.org/10.1016/s0092-8674(01)00404-4

Wang YC, Chen Q, Luo JM, Nie J, Meng QH, Shuai W, Xie H, Xia JM, Wang H (2019) Notch1 promotes the pericyte-myofibroblast transition in idiopathic pulmonary fibrosis through the PDGFR/ROCK1 signal pathway. Exp Mol Med 51(3):1–11. https://doi.org/10.1038/s12276-019-0228-0. PubMed PMID: 30902967; PMCID: PMC6430797

Warthi G, Faulkner JL, Doja J, Ghanam AR, Gao P, Yang AC, Slivano OJ, Barris CT, Kress TC, Zawieja SD, Griffin SH, Xie X, Ashworth A, Christie CK, Bryant WB, Kumar A, Davis MJ, Long X, Gan L, de Chantemele EJB, Lyu Q, Miano JM (2022) Generation and comparative analysis of an Itga8-CreER (T2) mouse with preferential activity in vascular smooth muscle cells. Nat Cardiovasc Res 1(11):1084–1100. https://doi.org/10.1038/s44161-022-00162-1. PubMed PMID: 36424917; PMCID: PMC9681021

Winkler EA, Sengillo JD, Bell RD, Wang J, Zlokovic BV (2012) Blood-spinal cord barrier pericyte reductions contribute to increased capillary permeability. J Cereb Blood Flow Metab 32(10): 1841–1852. https://doi.org/10.1038/jcbfm.2012.113. PubMed PMID: 22850407; PMCID: PMC3463878

Wirth A, Benyo Z, Lukasova M, Leutgeb B, Wettschureck N, Gorbey S, Orsy P, Horvath B, Maser-Gluth C, Greiner E, Lemmer B, Schutz G, Gutkind JS, Offermanns S (2008) G12-G13-LARG-mediated signaling in vascular smooth muscle is required for salt-induced hypertension. Nat Med 14(1):64–68. https://doi.org/10.1038/nm1666

Wu CF, Chiang WC, Lai CF, Chang FC, Chen YT, Chou YH, Wu TH, Linn GR, Ling H, Wu KD, Tsai TJ, Chen YM, Duffield JS, Lin SL (2013) Transforming growth factor β-1 stimulates profibrotic epithelial signaling to activate pericyte-myofibroblast transition in obstructive kidney fibrosis. Am J Pathol 182(1):118–131. https://doi.org/10.1016/j.ajpath.2012.09.009. PubMed PMID: 23142380; PMCID: PMC3538028

Xia XD, Zhou Z, Yu XH, Zheng XL, Tang CK (2017) Myocardin: a novel player in atherosclerosis. Atherosclerosis 257:266–278. https://doi.org/10.1016/j.atherosclerosis.2016.12.002

Yamamoto S, Niida S, Azuma E, Yanagibashi T, Muramatsu M, Huang TT, Sagara H, Higaki S, Ikutani M, Nagai Y, Takatsu K, Miyazaki K, Hamashima T, Mori H, Matsuda N, Ishii Y, Sasahara M (2015) Inflammation-induced endothelial cell-derived extracellular vesicles modulate the cellular status of pericytes. Sci Rep 5:8505. https://doi.org/10.1038/srep08505. PubMed PMID: 25687367; PMCID: PMC4330530

Yang J, Wang M, Zhu F, Sun J, Xu H, Chong Lee Shin OL, Zhao Z, Pei G, Zhu H, Cao C, He X, Huang Y, Ma Z, Liu L, Wang L, Ning Y, Liu W, Xu G, Wang X, Zeng R, Yao Y (2019) Putative endothelial progenitor cells do not promote vascular repair but attenuate pericyte-myofibroblast transition in UUO-induced renal fibrosis. Stem Cell Res Ther 10(1):104. https://doi.org/10.1186/s13287-019-1201-5. PubMed PMID: 30898157; PMCID: PMC6429829

Yin GN, Shin TY, Ock J, Choi MJ, Limanjaya A, Kwon MH, Liu FY, Hong SS, Kang JH, Gho YS, Suh JK, Ryu JK (2022) Pericyte-derived extracellular vesicles-mimetic nanovesicles improves peripheral nerve regeneration in mouse models of sciatic nerve transection. Int J Mol Med 49(2). https://doi.org/10.3892/ijmm.2021.5073. PubMed PMID: 34935051; PMCID: PMC8711595

You WK, Yotsumoto F, Sakimura K, Adams RH, Stallcup WB (2014) NG2 proteoglycan promotes tumor vascularization via integrin-dependent effects on pericyte function. Angiogenesis 17(1): 61–76. https://doi.org/10.1007/s10456-013-9378-1. PubMed PMID: 23925489; PMCID: PMC3898355

Zhao Z, Zhang Y, Zhang C, Zhang J, Luo X, Qiu Q, Luo D, Zhang J (2022) TGF-β promotes pericyte-myofibroblast transition in subretinal fibrosis through the Smad2/3 and Akt/mTOR pathways. Exp Mol Med 54(5):673–684. https://doi.org/10.1038/s12276-022-00778-0. PubMed PMID: 35624154; PMCID: PMC9166792

Zhuge Y, Zhang J, Qian F, Wen Z, Niu C, Xu K, Ji H, Rong X, Chu M, Jia C (2020) Role of smooth muscle cells in cardiovascular disease. Int J Biol Sci 16(14):2741–2751. https://doi.org/10.7150/ijbs.49871. PubMed PMID: 33110393; PMCID: PMC7586427

Chapter 6
Physiology and Pathobiology of the Microvascular Endothelium

Richard W. Pierce and Jordan S. Pober

Abstract The microvasculature comprises blood vessels (<200 μm in diameter) that nourish and drain parenchymal and stromal cells throughout the body. These vessels are lined by a single-cell layer of endothelial cells (ECs) that share common features while exhibiting differing characteristics across tissues and vascular segments within the same tissue. In homeostasis, microvascular ECs actively regulate blood flow, maintain blood fluidity, control water, solute, and macromolecular transfer between blood and tissues, and modulate circulating immune cell recruitment and activation. Advances in single-cell transcriptional profiling have provided profound insights into the differences between ECs from specific organs and even within microvascular segments (i.e., arterioles, capillaries, and venules). This specialization of ECs in health reflects the broad diversity of capillary function to match the demands of specialized tissues. In disease, microvascular ECs may be activated by both paracrine and systemic factors to regulate blood flow, coagulation, permselectivity, and immune cell responses. This activation can occur in rapid (type I), acute (type II), and chronic (type III) phases, all of which aim to restore homeostasis. Endothelial dysfunction occurs when these processes become exaggerated or break down, thereby propagating pathology and potentially resulting in endothelial cell death. Recent profiling of microvascular ECs in health and disease has both deepened prior insights into pathological processes and identified new targets for clinical investigation.

Keywords Endothelial activation · Endothelial dysfunction · Coagulation · Permselectivity · Perfusion · Inflammation · Angiogenesis · Single-cell RNA sequencing · Autacoids · Cytokines

R. W. Pierce · J. S. Pober (✉)
Yale School of Medicine, New Haven, CT, USA
e-mail: jordan.pober@yale.edu

R. F. Tuma et al. (eds.), *Handbook of Microcirculation*, Physiology in Health and Disease, https://doi.org/10.1007/978-3-031-99609-2_6

6.1 Microvascular Endothelial Cells: Morphological Features and Markers

The microvasculature consists of arterioles, capillaries, and venules that are located within the tissues and organs they nourish and drain. These microvessels, which are <200 μm in diameter, are lined by a monolayer of epithelial cell-like, typically flat endothelial cells (ECs), which, in most organs, are the only noncirculating cells that directly interact with blood. Although it is now appreciated that microvascular ECs can vary significantly between different organs and even tissues within these organs, as well as between different vessel segments within the same organ, ECs have many shared features and express common "marker" proteins that separate them from other cell types, allowing them to be considered a unified organ system. Consistent with the concept of there being a common EC phenotype, unsupervised clustering of single-cell transcriptomics from many different organs has consistently grouped ECs together as a distinct cell population (Tabula Sapiens et al. 2022).

The commonalities between ECs in part arises from their largely shared developmental origin throughout the blood vascular system (Eichmann et al. 2005; Domigan and Iruela-Arispe 2012; Majesky 2018). During embryological development, the great vessels are formed from endothelial cell progenitors (angioblasts), which are generated within the embryonic blood islands, initially in the yolk sac and later in the aorta–mesonephros–gonad (AMG) region. Angioblasts migrate to specific sites within the embryo to form a series of interconnected capillary-like cords that spontaneously lumenize and recruit supporting mural cells, a process called vasculogenesis. The emerging great vessels, lined by ECs, give rise to EC sprouts that invade the developing organs where they mature to form the circulatory tree, a process known as developmental angiogenesis. In this way, most of the ECs lining the emerging microvasculature are derived from those lining the primordial great vessels. There are exceptions to this rule, in that within some organs, e.g., in the heart and kidneys, ECs lining segments of the microcirculation may arise from stromal progenitors that form microvessels, which anastomose with the invading vessel segments arising from the great vessels (Perez-Pomares et al. 2002; Sequeira Lopez and Gomez 2011). In post-embryological development, microvascular ECs may be generated in response to injury by angiogenesis from preexisting microvessels or by proliferation and differentiation of residual committed endothelial progenitor cells, now best referred to as endothelial colony-forming cells (ECFCs) to avoid confusion with pro-angiogenic macrophages, once inaccurately identified and labeled as "endothelial progenitor cells"; pro-angiogenic macrophages can express some EC marker proteins and promote but not persist in stable vessels (Basile et al. 2018). True ECFCs may be present in the circulation and/or reside in the adventitia of the vessel wall (Jolly et al. 2022).

A common morphological feature of microvascular ECs is that they are extremely thin, typically measuring <2 μm from the abluminal to luminal surface, except overlying a "nuclear bulge." However, they may cover an extensive surface area of 2500 μm^2 or more. (The high endothelial venules (HEVs) of the secondary or tertiary lymphoid organs are lined by plump ECs and are thus an exception to the flat shape of ECs; these altered ECs develop from flat venular ECs in response to signals from lymphoid tissue inducer cells.) (Ager 2017). Within individual ECs, much of the actin cytoskeleton is organized into dense peripheral bands of actin filaments that outline the lateral junctions of the cell, although ECs exposed to high shear stress, such as those lining arteries or arterioles, may also have longitudinal stress fibers of actin filaments (Gotlieb 1990). The microtubular system of ECs outwardly radiates from a single microtubular organizing center (MTOC) toward the lateral borders. The MTOC is typically found next to the centrally placed nucleus on the side nearest to the heart (i.e., the MTOC is located "upstream" vis-à-vis blood flow in the arteries and "downstream" in the veins) (Rogers et al. 1985). The actin cytoskeleton and the microtubular cytoskeleton are not co-localized. ECs also express a network of intermediate filaments that are principally composed of vimentin (Imcke et al. 1991).

The EC plasma membrane facing the blood is often highly enriched in flask-shaped membranous invaginations named caveolae (little caves) that are 50–100-nm in diameter (Palade and Bruns 1968). The lipid content of the caveolar membrane is enriched in cholesterol and glycosphingolipids (Liu and Schnitzer 1999). These lipid aggregates are stabilized on their cytosolic surface by cholesterol-binding scaffolding proteins called caveolin-1 and -2. Caveolin-1 knockout mice lack caveolae in their ECs (Frank et al. 2003). Caveolae may participate in vesicular transport of macromolecules across the EC monolayer and may also form clusters of various receptors and signaling molecules, facilitating receptor cross talk and internalization (Filippini and D'alessio 2020).

A characteristic organelle unique to the endothelium is the Weibel–Palade body (WPB) (Weibel and Palade 1964). These structures are actually specialized secretory vesicles that assume an elongated oblate shape due to storage of extremely large polymers of Von Willebrand factor (VWF) (Ewenstein et al. 1987), an adhesive protein for platelets that is synthesized only by ECs and megakaryocytes. The condensed polymers of VWF give the Weibel–Palade bodies a striated appearance by electron microscopy. EC-released polymers of VWF are fragmented in the plasma due to the action of ADAMTS13 and other circulating or cell-bound proteases and are therefore smaller than the polymers stored in WPBs (De Pablo-Moreno et al. 2022). VWF forms a complex with the hepatocyte-derived coagulation factor VIII and is sometimes referred to as a factor VIII-associated antigen. WPBs also contain a propeptide that is cleaved from mature VWF (called VWF-2) as well as several other proteins, including P-selectin, interleukin (IL)-8, endothelin-1, osteoprotegerin, and angiopoietin-2 (Nass et al. 2021).

Adjacent microvascular ECs in most organs form homotypic junctions at their borders to create a continuous EC lining. The junctional complexes of ECs are unique to this cell type (Bazzoni and Dejana 2004). Microvascular EC tight junctions are primarily formed by three transmembrane proteins: claudin-5, junctional adhesion molecule-1, and occludin, but, of these proteins, only claudin-5 appears to be specific for vascular tight junctions. The extracellular portions of claudin-5 form homophilic attachments between adjacent ECs and its intracellular portions interact with zona occludins-1 (ZO-1), a protein that connects tight junctions to the actin cytoskeleton. EC adherens junctions are formed by an EC-specific transmembrane protein vascular endothelial (VE)-cadherin (also known as cadherin 5). VE-cadherin, found at the edge of ECs, forms calcium-dependent homophilic connections with the VE-cadherin expressed by adjacent cells. The cytoplasmic portions of VE-cadherin associate with α- and β-catenins that connect these structures to the actin cytoskeleton. Since both claudin-5 and VE-cadherin are largely restricted to ECs, as noted above, these molecules can be used as markers of ECs in most vascular beds. In ECs, unlike in epithelial cell barriers, the tight and adherens junctions are spatially intermixed. ECs also develop gap junctions with adjacent cells that are principally formed by connexin 43. Microvascular ECs may also make contacts across the basement membrane with surrounding mural cells, i.e., smooth muscle cells (SMCs) or pericytes (PCs), forming hetero gap junctions (Segal 2005).

The luminal surface of ECs is coated with a hydrated gel known as the glycocalyx (Curry 2018). This layer is a complex mixture of glycosaminoglycans (heparan sulfate, chondroitin/dermatan sulfate, and hyaluronic acid, hyaluronan, and others) linked to proteoglycans expressed by ECs (transmembrane syndecans, membrane-bound glypicans and CD44, basement matrix-associated perlecans, and others). The composition of the glycocalyx may vary with anatomic location (Suzuki et al. 2022). Such differences may be functionally relevant as the glycocalyx has been associated with diverse processes such as glomerular filtration and blood–brain barrier maintenance. The thickness of the glycocalyx is estimated to be 500 nm. The glycocalyx is present in all vascular segments but is most pronounced in the microvasculature where it was first noted to alter the postcapillary tube hematocrit. Although a barrier, it does not seem to hinder the movement of blood cells and macromolecules. The glycocalyx serves to bind circulating plasma proteins to the EC luminal surface, including enzymes such as antithrombin-III, superoxide dismutase, lipoprotein lipase, and various chemokines. Breakdown and release of glycocalyx components is associated with microvascular dysfunction, and the composition of circulating products may identify the organ that is injured, but most studies have failed to establish a causal relationship between glycocalyx degradation and disruption of EC functions such as maintenance of permselectivity (Hahn et al. 2021).

In their anatomic location adjacent to flowing blood, ECs must resist desquamation mediated by shear forces. To do so, ECs synthesize and then attach to a basement membrane rich in laminins, type IV collagens, and proteoglycans (Davis and Senger 2005). Attachment is primarily mediated by integrins containing the β1 subunit, such as α1β1, α2β1, α5β1 and a6β1. CD44, a non-integrin receptor for hyaluronic acid (a glycosaminoglycan found on basement membrane proteoglycans) may also contribute to basement membrane attachment. Following injury or during vessel remodeling, ECs may also express αvβ3 integrin, which can interact with plasma- or EC-derived proteins deposited at sites of injury such as fibronectin of fibrin. When the cells are stressed, arteriolar endothelial cells may degranulate, depositing some of the VWF stored in their Weibel–Palade bodies on to their luminal surface or on to the exposed basement membrane. This is of pathophysiological relevance because platelets require VWF to adhere to ECs or to the basement membrane under arterial levels of high shear stress (Sixma and De Groot 1991).

VWF may also be bound by $\alpha_v\beta_3$ integrin, expressed on the surface of injured or angiogenic ECs, to strengthen their own attachment to the vessel wall.

In addition to resisting desquamation, ECs must also resist outward forces caused by hydrostatic or pulse pressure. To do so, they receive support from mural cells and the connective tissue these cells produce. In larger arterioles and venules, the vessel wall consists of an EC monolayer surrounded by mural cells, which are contractile SMCs that are circumferentially organized into ordered layers. The matrix formed by these cells in rich in type I collagen and elastin. Precapillary arterioles, capillaries, and postcapillary venules are instead supported by a discontinuous layer of longitudinally oriented PCs that reside within and contribute to synthesis of the EC basement membrane (Dessalles et al. 2021). PCs have points of direct contact with the abluminal surface of ECs and may be electrically coupled to each other by gap junctions. PC to EC ratios vary with vascular beds and can range from 1:1 (in the brain) to 1:100 (in the skeletal muscle).

Microvascular ECs exhibit heterogeneity along the vascular tree within individual organs as well as among the capillary beds formed within different organs. Specific examples of the latter will be discussed below. Here, we will describe some of the common characteristics of ECs that distinguish arteriolar, capillary, and venular segments within the same vascular bed. Arteriolar segments of the microvasculature range in diameter from 20 to 100 μm. Arteriolar ECs appear largely similar to each other regardless of the tissue in which they are located. Except at branch points, ECs on the arterial side of the circulation are strikingly elongated and align their long axis with the direction of flow (Gotlieb 1990). Individual ECs may measure about 50–100 μm in length and 10–20 μm in width. In vitro, EC elongation can be achieved by external application of shear stress in a flow chamber and is thus believed to arise from flowing blood in situ (Girard and Nerem 1995) where the shear force applied to arteriolar ECs may be ≥50 dynes/cm^2. Arteriolar ECs form tight, adherens and gap junctions with adjacent ECs. Since both claudin-5 and VE-cadherin are largely restricted to ECs, as noted above, these molecules can be used as markers of ECs in most vascular beds. Arteriolar ECs frequently contain numerous WPBs. Following exocytosis of WPBs, VWF can deposit on the luminal surface of ECs.

The ECs lining venules differ in some ways from those lining arterioles but are generally similar across different tissues. The average venular diameter tends to be larger than that of arterioles, ranging from 50 to 100 μm, and the numbers of supportive SMCs tend to be fewer. Elongation is much less evident in the venular side of the microvasculature where shear stress is reduced to as little as 1 dynes/cm^2. As a result, venular ECs are more polygonal but still retain their attenuated thickness. Venular ECs typically contain few WPBs, although they still constitutively synthesize and secrete VWF. The most proximal portion of the venule into which capillaries drain, known as the postcapillary venule, is the primary site of leukocyte extravasation during inflammatory processes. This segment of the microvasculature is still supported by PCs rather than SMCs, and the ECs lining the postcapillary venules form few if any tight junctions. Their intercellular junctions contain extensive invaginations of the plasma membrane, referred to as the lateral border recycling compartment, which contain multiple transmembrane proteins that interact with proteins on extravasating leukocytes. The best-characterized of these EC proteins, platelet endothelial cell adhesion molecule (PECAM-1 or CD31) is, as the name implies, shared with platelets. Other such molecules include CD99, polio virus receptor, and nectin-2 (Sullivan and Muller 2014).

In contrast to the relative conservation of features observed in the ECs lining arterioles and venules, capillary ECs vary more widely among different tissues, sometimes even between different regions of the same organ (Tse and Stan 2010). Continuous capillaries have small luminal diameters, sometimes <10 μM, and can be formed by a series of individual ECs that curve around to form a ring in which the edges are sealed by the same types of junctions that can connect one EC to another. Both intra- and inter-EC junctions can contain numerous tight junctions. Capillary ECs typically lack WPBs, although they still synthesize VWF. Many continuous capillaries form fenestrae, i.e., large pores covered by a "diaphragm-like" structure formed from a protein called plasmalemmal vesicle-associated protein (PLVAP, sometimes called PV-1 or, in humans only, PAL-E). Diaphragmed fenestrae may serve to facilitate gas exchange between blood and tissues while restricting passage of macromolecules and cells; antibodies recognizing PLVAP have been used to identify capillaries.

Various cellular biomarkers have traditionally been used to identify ECs, and these include many of the molecules described in the preceding paragraphs. Human ECs express blood group H (commonly recognized by Ulex europaeus agglutinin I), PECAM-1, the cell junctional molecules VE-cadherin and claudin-5, VWF, and P-selectin. Another widely used marker is CD146, once labeled a melanoma adhesion molecule; its function on the EC surface is proposed to be that of an adhesive receptor, but its putative ligands are not known. ECs also express E-selectin (CD62E), more specifically than they do P-selectin (CD62P), which is also expressed on platelets. However, unlike P-selectin, E-selectin is absent in resting ECs, but its synthesis is turned on when ECs are exposed to inflammatory cytokines in type IIa activation (see below). Of note, there are several relevant species differences in the expression of these commonly used markers. Mouse ECs do not react with Ulex europaeus agglutinin I but instead bind isolectin B4. Moreover, in mice, P-selectin is cytokine-inducible by the same inflammatory cytokines as E-selectin (Liu et al. 2010), whereas, in humans, P-selectin is basally expressed but sequestered within WPBs and is primarily regulated by mobilization to the cell surface concomitantly with secretion of VWF. Almost all of these blood vascular endothelial features are shared with one or more cell types, most often lymphatic endothelial cells and platelets, but a combination of these markers and the absence of other lineage determinants (such as CD45 for leukocytes) create a common signature. Lymphatic ECs are derived from blood ECs and, upon differentiation, lose some features of blood vascular ECs, such as inducible E-selectin expression, and acquire a number of new markers, but many characteristic lymphatic EC markers are expressed by some blood vascular ECs as well. Not surprisingly, lymphatic ECs are also heterogeneous, varying in part owing to the blood microvascular ECs from which they originate (Den Braanker et al. 2021). Two widely used markers that are expressed by blood vascular ECs and generally missing on lymphatic ECs in situ are the hyaluronic acid receptor CD44 and inducible expression of E-selectin. Lymphatic ECs often revert to a more blood vascular EC phenotype in culture.

6.2 Physiological Functions of ECs at Rest

The principal basal functions performed by vascular ECs are to maintain blood in a fluid state, to control local blood flow and vascular permeability, and to minimize interactions with leukocytes and platelets (Pierce et al. 2017b). As noted in the introduction, blood is normally contained entirely within the blood vascular system, a compartment lined entirely by ECs. Ex vivo, blood is highly unstable, spontaneously initiating coagulation when placed in contact with virtually any non-endothelial surface. ECs, throughout the vasculature, prevent coagulation by a combination of active and passive processes (Arnout et al. 2006). Coagulation is normally initiated by a protease, factor VIIa, that circulates at extremely low levels but whose activity is catalytically enhanced up to a million-fold by binding to a cell surface protein called tissue factor (TF). (The suffix "a" designates an active coagulation enzyme.) Consequently, factor VIIa is only **functionally** active when it binds to TF. The VIIa–TF complex cleaves and activates two additional proenzymes present in the circulation, converting factors IX–IXa and X–Xa. Factor IXa is a protease that, in concert with factor VIII, also activates X–Xa, augmenting the actions of the VIIa–TF complex. Factor Xa is a protease that, acting in concert with factor V, converts prothrombin (factor II) to thrombin. Thrombin, also known as factor IIa, is a protease that cleaves fibrinogen (factor I) into fibrin, and, fibrin, once formed, spontaneously polymerizes into an insoluble fibrin clot that may be stabilized by cross-links between lysine and glutamine side chains in a reaction catalyzed by factor XIII, which is an extracellular transglutaminase also known as fibrin-stabilizing factor. The enzymatic complexes involving factors IXa and VIII as well as Xa and V must assemble on a phospholipid surface containing phosphatidylserine to be fully active. The plasma membrane of a resting endothelium displays a luminal surface that does not favor assembly of these coagulation factors because healthy ECs, like most other cell types, sequester membrane phosphatidylserine on the inner leaflet of their plasma membrane lipid bilayer where it cannot interact with extracellular clotting factors. ECs, by forming a barrier to the passage of large plasma proteins, keep plasma clotting factors away from the extravascular tissues where procoagulant activities are constitutively expressed. Resting ECs express little, if any, tissue factor, and that which is expressed is sequestered (or "encrypted") intracellularly so that it cannot interact with factor VIIa. The net result is that coagulation is difficult to initiate and sustain in an environment encased by resting ECs.

These passive anticoagulant functions of ECs are supplemented by active ones. Resting ECs display tissue factor pathway inhibitor (TFPI) that antagonizes the factor VIIa–TF complex, preventing it from acting on factors IX and X. ECs also specifically inhibit thrombin cleavage of fibrinogen to fibrin in two important ways. First, as part of their glycocalyx, ECs synthesize and express heparan sulfate proteoglycans that share with anticoagulant heparin the ability to bind and activate antithrombin III. Active antithrombin III is a natural antiproteinase that can covalently bind to and inhibit thrombin activity. Second, ECs synthesize and express thrombomodulin (TM), a plasma membrane thrombin-binding protein. The active thrombin–TM complex is redirected away from cleavage of fibrinogen and instead acts as a proteolytic activator of protein C. Activated protein C, working in combination with protein S (which is synthesized by ECs), cleaves and inactivates factors V and VIII, proteins that serve as cofactors in the coagulation cascade. (This is why a mutant form of factor V, such as factor V Leyden, which still functions as a cofactor for factor Xa but cannot be cleaved by activated protein C, predisposes to thrombosis.) Furthermore, if thrombin does succeed in converting fibrinogen to fibrin despite these obstacles, ECs are a natural source of plasminogen activators, especially tissue-type plasminogen activators (tPAs), which activate circulating plasminogen to plasmin, an enzyme that cleaves fibrin and lyses clots or thrombi into soluble fragments. Cumulatively, the resting EC is a multifunctional anticoagulant machine.

As noted above, many of the reactions of the coagulation cascade must take place on a cell surface containing phosphatidylserine, and resting ECs normally present a cell surface to the bloodstream that lacks phosphatidylserine and thus is not supportive of coagulation. Activated platelets serve as a major alternative source of procoagulant lipid surfaces. Thus, in order to effectively block coagulation, it is also necessary to prevent platelet activation. The principal physiological activators of platelets include thrombin, adenosine triphosphate (ATP), and collagen. As we have noted, resting ECs prevent the activation of thrombin and thus platelet activation by this agent. They also express ectoenzymes that rapidly degrade ATP to inert adenosine monophosphate (AMP). Finally, the EC monolayer forms a barrier that keeps platelets away from collagen present in the tissues or basement membrane, preventing activation by this pathway as well.

ECs not only antagonize activators of platelets but also produce inhibitors of platelet activation. Nitric oxide (NO) gas may be released by ECs as arginine is converted to citrulline by an EC-specific isoform of nitric oxide synthase (NOS), called either NOS-III or endothelial NOS (e-NOS). NO may then diffuse into the platelet and activate soluble guanylate synthase, producing cyclic guanosine monophosphate (cGMP) that acts as an allosteric inhibitor of enzymes involved in platelet activation. The production of NO may be stimulated by shear forces to which the ECs, especially in arterioles, are routinely exposed.

A second major function of resting ECs, especially in the terminal arterioles of the microcirculation, is control of local blood flow to the tissues (Busse and Fleming 2006). ECs in precapillary arterioles regulate blood flow in response to changes in perfusion pressure (myogenic), tissue metabolism, and autocrine, paracrine, and endocrine signaling to alter the tone of mural cells like SMCs or PCs (Gutterman et al. 2016; Pierce et al. 2017b). Although microvascular ECs can be directly coupled to surrounding mural cells via gap junctions (Segal 2005), the principal control of tone is achieved by production of EC-derived diffusible mediators. Vasodilation and increased flow is typically achieved by release of NO. This is the same pathway that minimizes platelet activation, and it is not clear whether release of NO is biased toward either the luminal or abluminal surface. However, pharmacological blockade of e-NOS will raise blood pressure by reducing vasodilation (Shesely et al. 1996) and feeding of arginine to increase NO production will lower it (Shiraseb et al. 2022). At the same time, vasoconstriction and reduced flow may be mediated by secretion of several peptides. First is endothelin-1, which is derived from an endothelial cell-synthesized protein known as big endothelin and converted to its active form by an endothelial ectoenzyme known as endothelin-converting enzyme on the surface of ECs. Second is angiotensin-II, which is converted from its liver-derived plasma protein precursor angiotensinogen, by the sequential proteolytic actions of circulating renin (secreted by renal juxtaglomerular cells) and a second endothelial ectoenzyme, angiotensin-converting enzyme (ACE). Vasoconstrictors may also be tonically produced as suggested by the observation that blockade of angiotensin-II synthesis or inhibition of binding of angiotensin-II to its receptor can reduce blood pressure and increase local blood flow in normal individuals. The overall tone of the resistance vessels is regulated by the balance of vasodilatory and vasoconstrictive signals and their unique effects in the specific microvasculature.

Microvascular ECs in different organs have different mechanisms for regulating blood flow. Many mediators that vasoconstrict (angiotensin-II, endothelin-1, norepinephrine [via β_1 receptors], vasopressin, thromboxane A2, and serotonin) and vasodilate (NO, prostaglandins, bradykinin, histamine, epinephrine [via β_2 receptors], atrial naturetic peptide (ANP), and others) are produced by and/or act directly on ECs. However, the intricacies of organ-specific autoregulation are less well-understood. The pulmonary and cardiac microvasculature have lower β_1 receptor density and do not vasoconstrict in response to systemic epinephrine like the skin and gut microvasculature (Johnson 1998). Cardiac ECs are also exquisitely sensitive to changes in perfusion pressure, metabolism, hypoxemia, and even anemia (Goodwill et al. 2017). In the brain, autoregulation occurs in response to changes in blood pH, PaO_2, $PaCO_2$, and NO (Aaslid et al. 1989; Claassen et al. 2021). In the kidneys, autoregulation is primarily achieved through changes in renal arterial resistance modulated by the renin–angiotensin–aldosterone system (RAAS) and by tubuloglomerular feedback mechanisms (Burke et al. 2014). The RAAS system regulates blood pressure and blood volume by controlling the constriction and dilation of renal microvascular arterioles, while tubuloglomerular feedback mechanisms control renal blood flow by modulating the filtration rate of the glomerulus (Burke et al. 2014).

Regulation of vascular permeability, primarily across capillary ECs, is the third major function of the resting endothelium (Minshall and Malik 2006). Permeability is controlled by regulating both the paracellular and vesicular transport pathways. As we noted earlier, the permeability of the barrier that provides passage for macromolecules through the paracellular pathway varies across tissues and correlates with the extent of tight and adherens junctions formed between adjacent capillary ECs. This barrier is not static and is actively regulated to the tissue microenvironment. As will be discussed in subsequent sections of this chapter, loss of endothelial permselectivity and extravasation of plasma proteins is a characteristic component of inflammation. Tightening of the barrier may also occur. One specific agent known to increase barrier function is sphingosine-1-phosphate (Mcverry and Garcia 2004). The source of this mediator may be endothelial cells themselves, but it is also found in plasma (or activated platelet releasates). Within endothelial cells, sphingosine-1-phosphate is generated by enzymatic phosphorylation of sphingosine, which may be generated either by de novo synthesis or by enzymatic breakdown of membrane sphingomyelin by the sequential actions of neutral sphingomyelinase (which generates ceramide) and ceramidase. Sphingosine-1-phosphate acts on endothelial cells by binding to specific heterotrimeric G protein-coupled receptors (GPCRs), several of which are expressed by endothelial cells. In particular, engagement of the receptor SIP-1 (sometimes called SIPR-1) stabilizes inter-EC junctions, a response that may be mediated by activation of adenylate cyclase, leading to a rise in cAMP levels. This increase can activate the small GTPase Rap-1 through the GTP exchange factor (GEF) EPAC-1, which, in turn, triggers an increase in the activity of the small GTPase Rac-1. This leads to relaxation of the actin cytoskeleton and stabilization of tight and adherens junctions.

When the junctions are tightly closed to protein extravasation, especially in the central nervous system, protein transport across the capillary wall is conducted almost exclusively through vesicular transport pathways (Minshall and Malik 2006). As noted earlier, the principal vesicle system involved in transendothelial passage of proteins appears to be the caveolar system, although this has been disputed. The budding and trafficking of caveolae has been proposed to depend upon c-src-mediated phosphorylation of caveolin-1 and the associated protein dynamin-2. Transcellular vesicular transport is subject to regulation and may be increased by certain cytokines, notably vascular endothelial growth factor A (VEGF-A).

Resting ECs act to quiesce circulating leukocytes (Ley and Reutershen 2006). The absence of immune cell activation can be attributed to the fact that EC adhesion molecules found at sites of inflammation are either not expressed (like E-selectin or vascular cell adhesion molecule 1 [VCAM-1]), are expressed at low levels (like intracellular adhesion molecule 1 [ICAM-1]), or are sequestered internally (like P-selectin) in WPBs in the ECs of the resting venule. Resting ECs synthesize few activators of circulating leukocytes (like platelet-activating factor [PAF] or chemokines). Furthermore, constitutive NO production may dampen endothelial synthesis of these pro-inflammatory molecules (De Caterina et al. 1995), although basal expression is low even in e-NOS$^{-/-}$ mice (Kuhlencordt et al. 2004). Of greater relevance, constitutively produced NO in the venules and capillaries may also function as an inhibitor of leukocyte as well as platelet activation (Moncada 2006). Thus, as with coagulation, resting endothelial cells prevent inflammation by a combination of passive and active processes.

6.3 Identification of Unique Microvascular Endothelial Cell Populations

Whole transcriptome analysis, also known as RNA sequencing (RNA-seq), is a method to capture information about global changes in mRNA levels (Wang et al. 2009). This technique has been revolutionary in providing unbiased insights into cellular identity and adaptions. RNA-seq has been applied to patient samples to answer previously un-addressable questions (Byron et al. 2016). Refined in 2008, single-cell (sc) RNA-seq leverages the power of whole transcriptome analysis on individual cells to define cell populations and identify changes based on the transcriptional activity within these cell populations (Gawad et al. 2016). scRNA-seq has revolutionized our understanding of EC heterogeneity across different tissues and developmental phases. Using this new technology, collections of cells can be delineated into specific subpopulations expressing unique sets of genes, which can then be correlated with specialized functions. Indeed, when applied to organs, scRNA-seq revealed extensive heterogeneity among ECs, even within immediate proximity, and has identified new and key intermediate EC subtypes that arise during blood and lymphatic vessel development. As discussed below, the revolution of endothelial scRNA-seq allows the establishment of organ, and even entire human, atlases to redefine vascular cells in terms of distinctive molecular profiles (such as gene expression profiles) and to connect this information with classical cellular descriptions (such as location and morphology).

By the early 2010s, bulk RNA-seq methods had identified specific transcriptional profiles associated with the functional changes in labeled and sorted mouse ECs, including in the brain (*SPIB,* CD133), lung (*TBX3*, CD36), heart (*SPI*, CD36), liver (*ELF*, CD44), and bone marrow (*SFPI*, CD37) (Nolan et al. 2013). These studies have identified numerous novel transcriptional changes that prompted novel surface markers for EC development and organ and segment specialization. By the late 2010s, scRNA-seq techniques had sufficiently advanced to allow specific tissue-based EC profiling. These efforts resulted in the rise of publicly available bulk and single-cell RNA-seq transcriptional atlases, starting with Tabula Muris (https://tabula-muris.ds.czbiohub.org/) with the ambitious goal to transcriptionally profile every single cell in mice. Secondary EC-focused analysis of this evolving dataset suggested some ECs were more transcriptionally unique (e.g., liver, brain), whereas others (i.e., adipose, heart) shared more similar transcriptional profiles with ECs from other tissues and identified novel markers of tissue-specific ECs (Paik et al. 2020). These efforts culminated in the landmark effort by (Kalucka and colleagues 2020), profiling ECs from 11 different tissues in mice (heart, muscle ×2, intestine, colon, liver, spleen, brain, testis, kidney, and lung). This study discovered novel shared and unique marker genes for EC specialization and identified new EC subtypes within organs. Newly identified ECs included previously undescribed intestine ECs expressing *MADCAM1* in venular and *AQP7* as well as unexpectedly detectable numbers of proliferation cells across all organs. By clustering numerous ECs from different organs of the same species, this study also demonstrated that the ECs from each vascular bed transcriptionally resembled each other more than those from other vascular beds; however, numerous highly preserved (80–100% of ECs across tissues) marker genes for arterioles, venules, and lymphatic vessels have been discovered. Capillary ECs were more transcriptionally dissimilar based on tissue type, reinforcing the established paradigm that they exhibit a higher degree of plasticity, allowing them to adapt to each organ's microenvironment. Subsequent studies have identified markers of the arterial (clusterin, Crip1, Fibuilin-2, Sat1, Sema3G, Bmx, and Gastrokine-3), capillary (Rgcc, Sgk1, Sparc, Car4, and Mfsd2a), and venous (slc38a5, ApoE, Biglycan, Ctla2a, IL-6ST, ptgs1, and thymosin beta-10) vascular identities (Chavkin and Hirschi 2020).

Replication of these efforts in humans, both from isolated organs and within individuals, is redefining vascular biology in the 2020s and beyond. Ambitious projects through the Human Cell Atlas (https://www.humancellatlas.org/) and Chan Zuckerberg Initiative (https://tabula-sapiens-portal.ds.czbiohub.org/) are competing to collect transcriptional profiles of all human cells. These projects do not focus on ECs and sponsored investigations typically do not involve vascular biologists, so the precise EC segment and organ of origin is often mis-annotated. However, the data are publicly available for reanalysis. Studies focusing on ECs are challenging due to the inability to genetically label ECs beforehand and therefore rely on bootstrapped cellular annotation. However, through advances in bioinformatics analysis coupled with iterative manual curation by vascular biologists, there have been numerous successful efforts to directly link unique microvascular EC transcriptional profiles to specialized functions within vascular segments of organs. As listed in Table 6.1, the number of investigations incorporating single-cell RNA sequencing of ECs is exponentially increasing.

The revolutionary power of scRNA-seq on microvascular EC identity is perhaps most notably demonstrated in the lung. First in animals and later in humans, Gillich et al. (2020) identified five transcriptionally unique cellular populations, namely, artery, vein, lymphatic, and two types of alveolar capillaries, they dubbed either aerocytes (aCap, defined by *Apln*, *Ednrb*, and *Car4*) and general (gCap, defined by *Aplnr*). aCap were optimized for gas exchange and comprised the majority of alveolar capillaries, while gCap were anatomically intermingled and retained more stem cell-like features. More detailed analysis in the human lungs by Travaglini and later by Schupp detailed aCap as $vWF^{neg}/EMCN^{high}/EDNRB^{pos}$ and gCap as $vWF^{pos}/EMCN^{low}/EDN1^{pos}$ populations as well as identified separate bronchial and systemic arterial and venous populations.

Numerous other studies have eloquently detailed the transcriptional profiles of ECs in other organs, albeit with less correlation with specific function as the lung cell discussed above. Investigations incorporating scRNA-seq of noncancerous human tissue are listed in Table 6.1. Taken together, there are several organ-specific insights. ECs that comprise the blood–brain barrier are known to form tighter junctions with much more selective transcytosis. scRNA-seq studies on mice and humans support this notion, with increased junctional molecules (*OCLN, CLDN5, MARVELD2, LSR*), suppressed transcripts associated with transcytosis (*PLVAP, CAV1*), and increased Wnt/β-catenin signaling (*LEF1, AXIN2, FOXQ1*). In the

Table 6.1 Transcriptional markers of microvascular cells by organ

Microvascular bed	Investigations	Major findings
Heart	Litviňuková et al. (2020)	Single-cell and nuclei analyses of hearts from 15 donors identifying 10 EC subtypes with unique immune and angiogenic transcriptional profiles
		Basis for the human heart cell atlas
	Hu et al. (2021)	An atlas of non-diseased human cardiac arteries, identified pro-inflammatory (*CD31* + *ACKR1*+) and anti-inflammatory (*CD31* + *SULF1*+) EC subpopulations
Lung	Travaglini et al. (2020)	Identified nine subtypes of ECs, including two bronchial (*VWA1*+ *HSPG2*+), venular (*PLVAP*+), and cell cycle-associated (*MYC* + *HBEGF*+) and four distinct types of capillaries (two general capillaries and two aerocytes)
	Human Lung Cell Atlas (Schiller et al. 2019)	A single-cell database of more than 75,000 lung cells
	LungMap	A consortium-based open-access lung cell database. Identifies two types of lung capillary cells: Aplnr+ CAP1 (gCAPs) and Apln+ CAP2 (aCAPs)
	Adams et al. (2020)	An atlas of lung cells from patients with interstitial lung disease identifying ectopic peribronchial ECs
	Schupp et al. (2021)	An atlas of the lung vasculature from 73 patients identifying pan-endothelial, pan-vascular, and subpopulation-specific transcriptional profiles
	He et al. (2022a, b)	A cell atlas of human lung development from 5 to 22 postconception weeks identifying segment-specific transcriptional profiles of development, including capillary (*THY1*+ *CD24*+), arterial (*GRIA2* + *GJA5*+), and lymphatic (*PROX1* + *STAB1*+)
		The lung cell atlas is a part of the human lung cell atlas
Brain	Garcia et al. (2022)	A single nuclei atlas of the brain vasculature from biopsy samples identifying transcriptional profiles of arteries (*VEGFC*+, *BMX*+, *EFNB2*+), capillaries (*MFSD2A*+, *TFRC*+), and venules (*SLC38A5*+, *TSHZ2*+, *LRRC1*+)
	Winkler et al. (2022)	A single-cell atlas of the normal and malformed human brain vasculature identifying aberrant transcriptional profiles in arteriovenous malformation ECs
	Allen brain map	A single-cell atlas from numerous donated human brains
Kidney	Wu et al. (2018)	Single-cell analysis of kidney biopsies identifying four EC subtypes including a pro-angiogenic subset (*SEMA3G*+, *IGFBP3*+, *SERPINE2*+, and *AQP1*+)
		Kidney interactive transcriptomics

(continued)

Table 6.1 (continued)

Microvascular bed	Investigations	Major findings
	Young et al. (2018)	A single-cell atlas of fetal, pediatric, and adult kidneys identifying specific transcriptional profiles of the anatomic location of ECs (i.e., glomerular, descending/ascending limb, vasa recta, etc.)
	Combes et al. (2019)	A single-cell atlas of fetal kidney tissue identifying a single EC cluster with conserved markers
	Stewart et al. (2019)	Single-cell and spatial transcriptomic analyses of the human kidneys identifying specific profiles for glomerular, vasa recta, and peritubular capillary ECs
		The kidney cell atlas
	Hochane et al. (2019)	A single-cell atlas from fetal kidneys identifying a single EC cluster
		The human fetal kidney atlas
	Liao et al. (2020)	Data deposition of single-cell RNA sequencing from three human kidneys
	He et al. (2021)	Single-cell RNA sequencing of human kidney biopsies revealing (*Ehd3* + *Kdr*+) glomerular ECs
	Hansen et al. (2022)	A single-cell atlas from 56 human kidney biopsy samples identifying glomerular and two other EC populations
		The kidney precision medicine project
	The comprehensive single cell atlas of the human kidney	
Liver	Macparland et al. (2018)	A single-cell atlas of human livers revealing four EC subtypes: Periportal sinusoidal, central venous, portal arteriole, and venular ECs
		The human liver atlas
	Aizarani et al. (2019), Macparland et al. (2018)	A single-cell atlas from nine human livers identifying micro- and macrovascular EC subtypes
	Hou et al. (2021)	A single-cell and spatial RNA sequencing database of human livers from 8 to 17 postconception weeks identifying transcriptional profiles of developing ECs
	Wang et al. (2021)	A single-cell atlas of human liver tissue before and after 2 h of cold ischemia identifying seven EC transcriptionally unique populations
	Brancale and Vilarinho (2021)	A single-cell atlas of 28 human livers, resulting in the creation of the liver single cell atlas
	Wesley et al. (2022)	A single-cell atlas of deceased organ transplant donors and fetal tissues identifying progenitor ECs that share transcriptional profiles with stellate progenitor cells
	Guilliams et al. (2022)	A single-cell atlas of 19 healthy and obese liver biopsies identifying liver sinusoidal ECs
		The liver cell atlas
	Krämer et al. (2023)	A single-cell atlas from 19 healthy, 40 cirrhotic, and 8 non-cirrhotic donors identified sinusoidal Ecs

(continued)

Table 6.1 (continued)

Microvascular bed	Investigations	Major findings
Skin	Solé-Boldo et al. (2020)	Single-cell analysis of the human skin identifying vascular (*SELE, VWF, CDH5*) and lymphatic (*LYVE1, PROX1)* Ecs
	Li et al. (2021)	Single-cell RNA sequencing of the skin from 10 donors identified 7 subtypes: Arterioles, 2 types of capillaries, postcapillaries, venules, and 2 types of lymphatic ECs
	Zou et al. (2021)	Single-cell analysis of the human eyelid skin identifying endothelial cells
	He et al. (2022b)	Single-cell analysis of the skin from two donors revealed specific profiles for arterioles (*GJA5*), post-arteriole capillaries (*ASS1, S100A4*); pre-venular capillaries (*SOX17, PLAUR*) and venular (*EGR2, LRG1*) Ecs
Lymph nodes	Takeda et al. (2019)	Single-cell analysis of human lymph nodes reveals sic subtypes of ECs discriminated by *ACKR4, NT5E, CAV1, TNFRSF9, MARCO, LYVE1, MFAP4*, and *CLDN11* expression
	Abe et al. (2022)	Single-cell analysis of human lymph nodes identifying blood and lymphatic Ecs
Bone marrow	Jardine et al. (2021)	Single-cell analysis of fetal bone marrow reveals six EC subtypes discriminated by *KDR, CD34, CTSL, STAB2, SELENOP, VIM, PDGFB, UNC5B,* and *DLL4*
Spleen	No specific studies exist	

Kidney, liver, brain, bone marrow, spleen and lymph node. This could also go into a table

Allen Brain Map [Online]. Available: https://portal.brain-map.org/ [Accessed]

Comprehensive Single Cell Atlas of the Human Kidney [Online]. Available: https://chanzuckerberg.com/science/programs-resources/single-cell-biology/seednetworks/a-comprehensive-single-cell-atlas-of-the-human-kidney/ [Accessed]

Heart Cell Atlas [Online]. Available: https://www.heartcellatlas.org/ [Accessed]

Human fetal Kidney Atlas [Online]. Available: https://home.physics.leidenuniv.nl/~semrau/humanfetalkidneyatlas/ [Accessed]

Human Liver Atlas [Online]. Available: http://shiny.baderlab.org/HumanLiverAtlas/ [Accessed]

Human Lung Cell Atlas [Online]. Available: https://hlca.ds.czbiohub.org/ [Accessed]

Kidney Cell Atlas [Online]. Available: https://www.kidneycellatlas.org/ [Accessed]

Kidney Interactive Transcriptomics [Online]. Available: http://humphreyslab.com/SingleCell/ [Accessed]

Kidney Precision Medicine Project [Online]. Available: Kidney Precision Medicine Project [Accessed]

Liver Cell Atlas [Online]. Available: https://www.livercellatlas.org/index.php [Accessed]

Liver Single Cell Atlas [Online]. Available: http://liveratlas-vilarinholab.med.yale.edu/ [Accessed]

Lung Cell Atlas [Online]. Available: https://www.lungcellatlas.org/ [Accessed]

LungMAP [Online]. Available: https://lungmap.net/ [Accessed]

kidney, anatomic regional specialization is reflected in transcriptional profiling such as afferent arteriole ECs (*AQP1, Sox17, Cxcl12*), glomerular capillaries (*Ehd3, Mapt, Igfbp5*), efferent arterioles (*Plvap, Igfbp7, Mest*), peritubular capillaries (*Plvap, Igfbp3, Ehd3*), descending vasa recta (*Aqp1, Slc14a1, Fbln5*), and ascending vasa recta (*Nr2f2*). In the heart, at least ten distinct EC subtypes have been identified, including an arteriole, a venule, four capillaries, and four interstitial ECs with uniquely described subtypes, although the functional significance of these EC subtypes are not well-established. Similar scRNA-seq atlases exist for the liver, spleen, skin, and many other tissues, with more specific profiling being published each day. scRNA-seq is an ideal tool to describe the transcriptional differences between ECs.

6.4 Type I (Immediate) Activation of the Endothelium

Nearly any local or systemic corporeal perturbation elicits a response that involves the vasculature. Perhaps best-studied in ECs is local inflammation. Inflammation alters the basal state of ECs to favor three responses: increased local blood flow by dilation of the arterioles; increased leukocyte capture and extravasation at the venules; and leakage of plasma proteins at the venules. The changes in ECs that underlie these events are referred to as endothelial activation, commonly divided into type I (immediate, detectable within minutes, and independent of new protein synthesis) and type II (rapid, requires hours, and dependent upon new protein synthesis) (Pober and Cotran 1990). Type I activation, originally called stimulation, consists of a few stereotyped responses, to promote inflammation involving basally expressed GPCRs and signaling pathways. Many of these responses are triggered by autacoids, such as histamine, lipid mediators, and chemokines, or by enzymes, such as thrombin. Receptor signaling results from liberation of second messengers such as Gα•GTP and Gβγ subunits from a previously receptor-bound inactive heterotrimeric G protein complex. GPCRs that trigger type I activation, such as the histamine H1 receptor or the thrombin-activated protease-activated receptor (PAR) 1, are coupled to trimeric G proteins of the G_{q11} or $G_{12/13}$ type (Smolinska et al. 2022; Chandrabalan and Ramachandran 2021). These G protein subunits act as allosteric effectors that lead to phospholipase A_2 enzymatic liberation of arachidonic acid (AA) from plasma membrane lipids, phospholipase C release of inositol 3,4,5-trisphosphate-mediated elevations in cytosolic free calcium ions from activation of channels in the endoplasmic reticulum and plasma membrane, and activation of GEFs that activate the functions of small GTPases of the Rho family. Calcium ions, often through association and activation of the allosteric effector molecule, calmodulin, and activated Rho•GTP acting on Rho coiled kinases (ROCK), produce type I activation responses. Specific GPCRs, GEFs, and GTPase-activating proteins (GAPs, which cause small proteins to convert bound GTP to GTPase and thus lose activity) may vary among ECs of different microvascular beds and among segments of the same vascular bed. For example, histamine receptors are largely confined to postcapillary venules (Heltianu et al. 1982). Transcriptional profiling on

ECs from different segments and organs has recently revealed unique GAP and GEF patterns of expression among different ECs, which, like variation in receptor expression, can also result in differences in regulation in response to autocoid stimulation among microvascular segments and microvascular beds.

Type I activation of arteriolar ECs leads to production of mediators that relax the tone of mural cells, permitting increased blood flow to the site of type I activation. The principal mediators of inflammatory vasodilation are EC-derived NO and PGI_2 (Moncada 2006; Egan and Fitzgerald 2006). A rise in NO release above the basal level produced in resting ECs induced by laminar shear stress is initiated either when a calcium/calmodulin complex binds to and activates e-NOS or when e-NOS is phosphorylated at specific sites by kinases such as Akt, which is activated through a phosphatidylinositol 3-kinase pathway. PGI_2 is synthesized following a rise in cytosolic calcium-initiated activation of cellular phospholipase A_2 ($cPLA_2$), an enzyme that cleaves membrane phosphatidylcholine into arachidonic acid (AA) and lysophosphatidylcholine. AA is converted by prostaglandin H synthase 1, more commonly known as cyclooxygenase-1 (Cox-1), into PGH_2, and PGH_2 is then rapidly converted to PGI_2 by the actions of prostacyclin synthase. (In some instances, AA may also be converted into endothelium-derived hyperpolarizing factors (EDHF), which also mediates vasodilation.) The local vascular responses caused by vasodilation are responsible for the redness (rubor) and warmth (calor) associated with inflammation. Functionally, increased blood flow increases delivery of leukocytes to the tissue where they may be captured on the endothelial cell surface and induced to extravasate.

Some consequences of type I activation involve mobilization of WPBs, the contents of which are also influenced by the microvascular segment. Arteriolar ECs contain the highest density of WPBs within the microcirculation. At this site, calcium/calmodulin complexes formed in the ECs by GPCR signals initiate exocytosis of WPBs (Birch et al. 1994). Release of VWF by arteriolar ECs can create a surface that will allow platelets to adhere to the vessel wall through their glycoprotein 1 despite the high shear forces present within the arterial tree. The process of WPB exocytosis additionally involves *N*-ethylmaleimide sensitive factor (NSF), docking proteins associated with the cytosolic surface of the WPB membrane and of the plasma membrane (SNAPS), receptors for these proteins (SNARES), and small GTP-binding proteins of the Rab family. This process may be inhibited by NO, which nitrosylates and inactivates some of the WPB-associated docking proteins (Matsushita et al. 2003), thus providing a negative feedback loop.

The postcapillary venule is the principal site of leukocyte extravasation in most organs. In this location, ECs may also contain WPBs rich in adhesion molecules. Here, the most important consequence of WPB exocytosis is translocation of P-selectin to the endothelial plasma membrane where it may bind to ligands expressed on the surface of circulating leukocytes, initiating capture (tethering) and, in the face of continued shear force, rolling of leukocytes (principally neutrophils) on the EC luminal surface (Ley and Reutershen 2006). IL-8 contained within WPBs may also contribute to neutrophil activation (Middleton et al. 1997). At the same time, some of the lysophosphatidylcholine liberated by cPLA2 can be rapidly acetylated by an enzymatic reaction, generating an acyl form of platelet-activating factor (PAF) that is also displayed on the luminal membrane. PAF (despite the name) primarily acts on neutrophils, causing them to "activate" their surface integrins and rearrange their cytoskeleton from a round to a spread and motile form, similar to the actions of IL-8 (Louis et al. 2005).

The co-display of P-selectin and PAF (or IL-8) is referred to as juxtracrine signaling, converting rolling neutrophils to firmly attached and crawling forms, capable of diapedesis (Lorant et al. 1991). Attachment and crawling by neutrophils is principally mediated by two integrins, lymphocyte function-associated antigen-1 (LFA-1) (also known as $\alpha_L\beta_2$ or CD11aCD18) and Mac-1 (also known as $\alpha_M\beta_2$ CD11bCD18) (Ley and Reutershen 2006). PAF "activates" these integrins, a change that involves both increases in affinity due to a conformational change ("inside-out signaling") and increased plasma membrane expression. The endothelial cell surface ligands recognized by these neutrophil integrins are ICAM-1 (for LFA-1 only) and ICAM-2. ICAM-1 is present at low levels on the resting endothelium and may be slightly increased by type I activation, mobilizing preformed protein from an unknown storage compartment (Sugama et al. 1992). ICAM-2 is constitutively expressed at high levels.

The process of leukocyte diapedesis generally begins with crawling to the inter-endothelial junctions. Diapedesis involves interaction of the motile leukocytes with specific proteins concentrated within the lateral border recycling compartment that can be brough to the plasma membrane to form the channel through which leukocytes traverse. The process of mobilization of these proteins depends upon elevations in cAMP and activation of protein kinase A (Muller 2016).

Once through the endothelial layer, the leukocyte must then cross the basement membrane and migrate into the underlying interstitial connective tissue of the local organ. Among the principal matrix proteins at these sites are collagens, with which circulating neutrophils are poorly equipped to interact, lacking relevant receptors. A third crucial change that occurs in type I activation of the endothelium of the venules is the opening of the intercellular junctions, which, at this site, principally consist of VE-cadherin-dependent adherens junctions. The contraction of the actin cytoskeleton, mediated by dual signals of Rho/Rock inhibition of myosin light chain phosphatase and Ca^{+2}/calmodulin activation of myosin light chain kinase (Stevens et al. 2000), pulls apart the adherens junctions, allowing large plasma proteins, especially fibrinogen and fibronectin, to escape from the blood space and to modify the chemical composition of the basement membrane and interstitium. These molecules assemble in the tissue as a provisional matrix capable of supporting neutrophil attachment and migration mediated by certain β_1 and β_3 integrins expressed by these cells (Frangogiannis 2006).

A critical aspect of type I activation of ECs is that it is not sustained. This is because heterotrimeric GPCRs typically inactivate after a few minutes and then become desensitized to further stimulation. Consequently, type I activation only causes a transient leak of plasma proteins, producing edema and swelling (tumor) that quickly resolves. More sustained reactions and significant leukocyte accumulation and significant leukocyte accumulation require new gene transcription, a process described as type II activation.

6.5 Type IIa (Rapid) Activation of Endothelial Cells

Type II activation is slower in onset than type I activation but involves more prolonged responses, lasting hours to days (Pober et al. 2009). In much of the literature, "type II activation" has been used as a collective term for alterations in ECs that are involved in facilitating both innate and adaptive immune responses. Here, we will propose that these be subdivided into changes initiated in innate immunity (type IIa) and those associated with adaptive immunity (type IIb) as adaptive immune responses involve additional signals beyond those required for most innate immune responses. The prototypic mediators of type IIa activation associated with innate immune responses of ECs are pro-inflammatory cytokines, specifically tumor necrosis factor (TNF, sometimes designated as TNF-α) and interleukin-1 (which may be in the form of IL-1α and/or IL-1β, two homologous proteins with identical biological actions) (Le and Vilcek 1987). (Lymphotoxin [LT]-α acts on the same receptors as TNF and was once described as TNF-β, a now discarded nomenclature.) Although many cell types may synthesize these molecules, the most abundant source is mononuclear phagocytes responding to microbial products such as endotoxin (also known as lipopolysaccharide or LPS). TNF and IL-1 bind to distinct receptors expressed on ECs, but the major signaling pathways activated by these two cytokines converge. LPS and other microbial products, collectively designated as pathogen-associated molecular patterns (PAMPs) that mediate their effects through activation of toll-like receptors (TLRs), also signal through the same pathways as IL-1. PAMPs, which are major elicitors of TNF and IL-1 synthesis by mononuclear phagocytes, can directly act upon ECs through TLRs, but their effects on ECs are generally weaker than those elicited by these cytokines. As a result of the convergence signaling pathways within ECs, many different mediators produce similar changes in ECs. Almost all of these responses depend upon new gene transcription involving activation of a few specific transcription factors, most prominently nuclear factor-kappa B (NF-κB) (primarily involving NF-κB1/RelA homodimers) and various forms of activator protein 1 (AP-1). The capacity of these factors to initiate transcription of specific genes in a given cell type depends on whether the relevant binding sites in the enhancer region controlling that gene is accessible (i.e., "poised"), and this may vary among different microvascular ECs. Many of the genes whose transcription is activated by TNF or IL-1, as will be discussed shortly, contribute to inflammation. Although the actions of TNF and IL-1 extensively overlap, they can differ in two important ways. First, the signaling complex formed by the TNF receptor (specifically TNF receptor 1) can be modified after a lag of several hours by recruitment of new adapter proteins that activate programmed cell death by apoptosis or necroptosis. However, some of the NF-κB-induced genes can inhibit these death pathways, preventing EC death.

Second, TNF has a second receptor, TNFR2, whose signaling is less well-defined but appears to activate pro-survival and pro-growth pathways rather than inflammation. IL-17 may enhance some EC responses to IL-1 or TNF, but IL-17 effects on ECs in the absence of TNF or IL-1 are generally weak by comparison. Instead, IL-17 exerts its major vascular effects on PCs that may act in concert with ECs to promote acute inflammation (Liu et al. 2016).

Many of the target genes activated in ECs by TNF and/or IL-1 signaling are relevant for the induction of inflammation. One such gene is *Cox-2*; this enzyme catalyzes the same reaction as Cox-1, converting arachidonic acid to PGH_2, but is a higher output isoform, leading to much greater PGI_2 synthesis in response to type I agonists like histamine. TNF or IL-1 also initiates the de novo synthesis of tetrahydrobiopterin in ECs through induction of GTP cyclohydrolase I, the first enzyme in this biosynthetic pathway (Rosenkranz-Weiss et al. 1994). Tetrahydrobiopterin is an essential cofactor for e-NOS activity, so its synthesis by type II activated ECs serves to increase NO production. This enhanced capacity to make NO may be limited in duration because TNF destabilizes e-NOS mRNA through induction of a microRNA, so, 3 days after stimulation, the capacity of ECs to synthesize NO is actually decreased (Yoshizumi et al. 1993). E-selectin, which is induced by TNF and IL-1, plays a similar role as P-selectin, mediating the initial tethering and rolling of leukocytes. As noted earlier, P-selectin regulation in mice differs from that which occurs in humans and actually parallels E-selectin expression (Liu et al. 2016, p 26). Type IIa activated ECs also synthesize and display various chemokines, such as IL-8 (Kunkel et al. 1995), which bind to heparan sulfate proteoglycans on the EC luminal glycocalyx (Webb et al. 1993). Like PAF, these molecules, bind to heterotrimeric G protein-coupled receptors on the surface of leukocytes, and the combination of E-selectin and IL-8 on type IIa activated ECs can thus present a juxtracrine signal to neutrophils that is functionally similar to that provided by P-selectin and PAF on type I activated ECs. A notable difference is that syntheses of E-selectin and chemokines typically persist for hours rather than a few minutes. Consequently, neutrophil recruitment is much greater in settings of type IIa than type I activation.

An important characteristic of type IIa activation is that the molecules synthesized and expressed by type IIa activated ECs change with time (Briscoe et al. 1992). This can be clearly seen in the patterns of expression of leukocyte adhesion molecules by postcapillary venular ECs that initiate leukocyte capture and transendothelial migration. E-selectin is expressed most rapidly, reaching peak expression by 6–8 h that then declines; levels of ICAM-1 and VCAM-1 rise more slowly, showing a lag in the onset of several hours, peaking at about 24 h, and then are more sustained. Along with adhesion molecular is a rapid increase in chemokine transcription (i.e., CXCL-8, CXCL-10, CXCL-3, CXCL-5, and CXCL-6) (Lu et al. 2012). The changeover from high E-selectin/low VCAM-1 to low E-selectin/high VCAM-1 correlates with that in acute inflammation from neutrophil to monocyte cell recruitment. Changes in the expression patterns of leukocyte adhesion molecules follow changes in mRNA levels that are largely controlled by transcription, but posttranscriptional regulation by microRNAs has also been observed. As with many other properties of microvascular ECs, expression also varies with vascular beds (E-selectin may be more sustained in skin microvessels) and vascular segments. Venular cells are more likely to express E-selectin and VCAM-1 than are capillary or arterial ECs, whereas the opposite is true for ICAM-1. However, these patterns of segmental responses can be overwhelmed in vivo by the cytokine storm accompanying sepsis, causing ECs in the arterial circulation to express E-selectin (Redl et al. 1991). In EC cultures, high levels of shear stress, variations in matrix composition, or differences between two-dimensional (2D) and three-dimensional (3D) culture of ECs suggest that these differences in responses are at least partly controlled by environmental factors rather than by epigenetically fixed patterns. Consistent with the notion that segmental differences are not fixed fates are the observations that ECs isolated from the arterial or venous circulation rapidly, over hours to days, acquire a common in vitro phenotype differing from the arterial or venous patterns observed in situ and that ECs cultured from umbilical veins can acquire the microvascular features of arterioles, capillaries, and venules when allowed to reform vessels after transplantation into immunodeficient mouse hosts (Enis et al. 2005).

Recruitment of leukocytes follows a generally similar pattern in many different organs and is initiated by adhesion to activated ECs, but details may differ among organs (Margraf et al. 2019). Attached leukocytes transition from loose adhesion (with rolling on the EC luminal surface impelled by flowing blood) to firm adhesion (with cell flattening) induced by leukocyte activators such as EC-displayed PAF or chemokines. Flattened leukocytes then crawl to the sites of inter-EC junctions. The ECs of the postcapillary venule then play a more active role in the actual process of diapedesis. As leukocytes push through the EC monolayer as or near inter-EC junctions, additional EC adhesive molecules are brought to the regions of contact by mobilization of the lateral border recycling compartment. By forming tight connections between the ECs and leukocytes, this process minimizes the extent to which leukocyte diapedesis disrupts the barrier formed by the EC monolayer. Furthermore, as the leukocytes push through the EC monolayer, the ECs actively cover the channel through which leukocytes pass. These processes imply that ECs can sense the presence and movement of leukocytes. This may occur through clustering or tugging on EC adhesion molecules such as ICAM-1 or CD99, which are tethered to the EC cytoskeleton via linker proteins of the ezrin/radixin/moesin family (Watson et al. 2015).

The breaks that occur in the EC barrier during leukocyte diapedesis are rapidly healed and do not cause leakage (Martinelli et al. 2013). However, microvessels do become leaky as part of inflammation. Initially, this process is mediated by a set of responses through which the inter-EC junctions of the postcapillary venules are opened. When induced by IL-1 or TNF, opening of the junctions requires new protein synthesis arising from NF-κB-regulated genes. Many of these genes are Rho GAPs and GEFs, and, although these cytokines, do not induce a rise in cytosolic calcium ions, they still appear to increase myosin light chain phosphorylation mediated by ROCK inhibition of myosin light chain phosphatase MYP, similar to the transient responses we described in type I activation. A key difference may be related to the activation of RhoB, which can prevent Rac-1 from returning to the intercellular junction to reverse the leak (Marcos-Ramiro et al. 2016). Type IIa signaling causes a large increase in total RhoB and in its active, GTP-bound form. VE-cadherin is internalized and may be degraded as the leak persists. Inactivation of RhoB by a GAP appears to be critical for resolution of inflammatory leaks, which allows release of internalized Rac-1 to return to the plasma membrane. A genetic mutation in p190RhoGAP190B, encoded by *ARHGAP5*, is associated with a hereditary capillary leak syndrome episodically triggered (likely) by infections (Pierce et al. 2017c).

As in the case of type I activation, type IIa activation-induced leakiness allows for plasma proteins to enter the tissue and create a provisional matrix for leukocyte attachment, survival, and migration. The leakage of plasma proteins associated with type II activation is more sustained than that associated with type I activation, and sustained leakage of fibrinogen leads to the formation of a firm, fibrin-rich provisional matrix, accounting for the induration (hard swelling) associated with mononuclear cell-rich infiltrates as occurs in positive tuberculin skin tests. Over time, however, some degree of leakiness may involve capillaries rather than postcapillary venules, and we will return to this as a pathological process later in this chapter.

Type IIa activation also affects the normally anticoagulant and antiplatelet properties of ECs. TNF stimulation causes ECs to downregulate expression of both thrombomodulin (Lentz et al. 1991) and e-NOS (Yoshizumi et al. 1993). The latter effect, as noted above, may be counterbalanced for some time by the TNF-mediated increases in tetrahydrobiopterin levels (Rosenkranz-Weiss et al. 1994). Reduced NO may not only permit increased platelet activation but also enhanced exocytosis of WPBs, depositing both P-selectin and Von Willebrand factor on the EC surface (Lowenstein et al. 2005). At the same time that TNF or IL-1 reduces anticoagulant and antiplatelet activation, these cytokines also cause ECs to actively express a procoagulant protein in the form of tissue factor (Bevilacqua et al. 1986). CD40 signaling, which in many ways mimics those of IL-1 and TNF on endothelial cells, is actually more potent in this regard because in addition to activating NF-κB and AP-1, it also activates the transcription factor egr-1 (Bavendiek et al. 2002). Other procoagulant molecules such as fibrinogen-like protein 2 may also be upregulated in type II responses (Ghanekar et al. 2004).

The beneficial effect of pro-inflammatory responses involving leukocyte recruitment is to favor the eradication of infectious microbes. Procoagulant activities may similarly contribute to host defense by "walling off" infectious tissues from healthy ones, but the "fibrin cocoon" that keeps microbes in may also keep microbicidal leukocytes out. It has been observed that there is sequential activation of these responses in which procoagulant activities are delayed relative to pro-inflammatory ones. Such a time course may result because the loss of anticoagulant activities is slower than the induction of pro-inflammatory and procoagulant activities. This time course can be observed in the local Shwartzman reaction (Rothstein and Schreiber 1988). In this model of injury, a local injection of endotoxin or TNF must be administered 24 h prior to a systemic (i.e., intravenous) injection of the mediator. This sequence allows the decrease of anticoagulant activities like tissue factor pathway inhibitor, thrombomodulin, and e-NOS to occur prior to a reinduction of the procoagulant activities. Once the anticoagulant activates are removed, the procoagulant activities are essentially unopposed and thrombosis may ensue in response to the second dose of the inducing cytokine. The result of delaying coagulation is that it is only activated if the inflammatory response fails to eradicate the eliciting microbes.

6.6 Type IIb (Adaptive Immune-Mediated) Activation of Endothelial Cells

Type IIa activation with sequential recruitment of neutrophils and monocytes is often sufficient to eliminate the stimulus that triggered the inflammation in the first place. However, if the stimulus is not eradicated, then host defense may become amplified by involvement of the adaptive immune system. At this point, T and B cells, which specifically recognize precise molecular features of the stimulus, described as "antigens," adopt more potent effector mechanisms. Here, we propose that these changes, which are typically superimposed upon type IIa activated ECs to play a critical role in the recruitment of these cell populations (Abrahimi et al. 2015; Pober et al. 2017), either through pathways like those used by other leukocytes involving chemokines or through novel pathways involving presentation of antigens, be designated as type IIb activation.

T cells use a clonally expressed T-cell receptor (TCR) for antigens, formed by gene rearrangements during development in the thymus, and exit the thymus after having been selected to recognize peptides derived from non-self-proteins bound to cell surface proteins encoded by the major histocompatibility complex (MHC), called human leukocyte antigens (HLAs). HLA proteins fall into two structural categories: class I molecules, encoded by *HLA-A*, *-B*, and *-C* genes that present antigens to $CD8^{+}$ T cells, and class II molecules, encoded by *HLA-DR, -DP,* and -*DQ* genes that present antigens to $CD4^{+}$ T cells. Most microvascular ECs express both class I and class II HLA antigens in uniflamed tissues. T cell responses to antigens require that the antigen presenting cell provide additional antigen-independent signals, called costimulators, that also additional receptors on T cells. Naïve T cells, which have not yet encountered their specific antigen, and memory T cells, which have expanded and been altered by a prior encounter with their antigen, differ in the costimulatory signals they must see to become reactivated. ECs express costimulatory molecules that preferentially engage with a specific subset of memory T cells, called effector memory T cells, which are adapted for recruitment into peripheral tissues instead of the secondary lymphoid organs such as lymph nodes, the spleen, or intestinal Peyer's patches. Only effector memory T cells among circulating T-cell populations can extravasate in response to inflammatory chemokines. Alternatively, effector memory T cells that encounter their antigen displayed on the luminal surface of a microvascular EC can use that signal to initiate transendothelial migration and enter the tissue in an activated state. The interactions between T cell and ECs in chemokine- vs. antigen-based recruitment differ but overlap, and T cells that recognize their antigen become refractory to chemokine recruitment and utilize their antigen responses to guide their migration and activity (Manes et al. 2007). In the case of transplanted mouse hearts, T-cell activation by antigens within the microvessel lumen is required to initiate chemokine-based recruitment of leukocytes and bystander effector memory T cells that are not specific for the graft (Zhang and Lakkis 2015). For this reason, the earliest T cells that enter the tissue may be enriched in antigen-specific effector memory T cells and are sometimes referred to as "pioneer" T cells.

Like the induction of adhesion molecules, many of the costimulators that ECs use to activate effector memory T cells must be induced by cytokines and these are also involved in T-cell recruitment (Manes et al. 2022). In humans, the principal costimulatory of effector memory T cells that is constitutively expressed is LFA-3 (CD58), and a critical costimulatory molecule, which participates in transendothelial cell migration in response to antigens, is the (inducible T-cell costimulator ligand (ICOS-L), which is induced by IL-1 or TNF. In this setting, the T-cell receptor that recognizes ICOS-L is CD28; dendritic cells signal to T cell CD28 using other ligands (known as B7–1 or B7–2, also known as CD80 and CD86, respectively), but human microvascular ECs do not express these molecules. Both TNF and IL-1 can induce ECs to display complexes of IL-15 and IL-15Rα on their surface, which serves to enhance both T-cell and natural killer activation (Xie et al. 2020).

Whereas type IIa activation of microvascular ECs may be generally attributed to TNF or IL-1, signaling of cytokines like interferon (IFN)-γ or IL-4 activates NF-κB and AP-1 transcription factors, thus influencing the capacity of antigen-presenting cells to effectively participate in type IIb activation. These cytokines principally act through receptors that mediate Janus kinase/signal transducer and activator of transcription (JAK/STAT) signaling. IFN-γ generates activated STAT1 homodimers that induce secondary transcription factors that upregulate both HLA class I and class II expression as well as molecules required to process proteins into peptides and to load them on to HLA molecules so that they can be recognized by T cells. IFN-γ is also responsible for inducing EC synthesis of chemokines, such as monokine induced by gamma interferon (MiG, also designated as CXLC-9), IP-10 (CXCL-10), and (in humans) inducible T-cell alpha attractant (I-TAC also designated as CXCL-11), which are most active on bringing CXCR3-bearing effector memory T cells into the tissue in an antigen-independent manner. At the same time, IFN-γ also produces signals that can dampen a T-cell response, such as expression of the checkpoint inhibitors PD-L1 and PD-L2, which reduces T-cell activation by engaging with T-cell PD-1. This reflects a general principle of immunity, which is that the immune system may simultaneously be activated and regulated to prevent or limit damage to healthy tissues.

The types of EC responses promoted by IFN-γ are those that are most protective against viruses and intracellular bacteria, mediated by cytotoxic T lymphocytes (CTLs) and classically activated macrophages, respectively. This is sometimes referred to as type 1 immunity, and the $CD4^+$ T-helper cells that mediate this response, called Th1 cells, like $CD8^+$ CTL, are potent producers of IFN-γ. Th-2 immunity, which protects against multicellular parasites, is enhanced by IL-4 and IL-13 effects on ECs, is mediated by STAT6, and leads to recruitment of eosinophils and $CD4^+$ T helper cells, called Th2 cells, which make these cytokines. IL-17, produced by $CD4^+$ T cells that make IL-17 (called Th17 cells), is associated with neutrophilic responses to extracellular bacteria and preferentially alters ECs to recruit this cell type, but, as we have noted, IL-17 mostly acts on PCs than on ECs (Liu et al. 2016) (Pober et al. 2017) (Walch and Lakkis 2014).

In addition to responding to cytokines, ECs may respond to antibodies binding to EC surface proteins or to effector molecules generated by antibody binding to cell surface molecules (Belizna et al. 2006). Antibody-mediated clustering of certain cell surface proteins, such as class I HLA molecules, may induce some features of a type I inflammatory response such as exocytosis of WPBs (Yamakuchi et al. 2007) or ICAM-1 (Lawson et al. 2005). Although antibodies to certain EC proteins may also activate signaling pathways that lead to new gene transcription (Hickey et al. 2016), activation of complement is required to produce type II activation responses (Jane-Wit et al. 2013). Some responses may involve binding of C5a anaphylatoxin to relevant GPCRs on ECs (Albrecht et al. 2004), and the combination of antibody binding and C5a signals may lead to enzyme-mediated loss of anticoagulant heparan sulfates (Platt et al. 1991). Antibody binding to ECs can also lead to formation of membrane attack complexes (MACs) on the EC surface. Surprisingly, little cell death ensues, perhaps due to EC complement regulatory proteins. Moreover, ECs internalize into early endosomes where they initiate assembly of a signaling complex that both generates transcriptional changes through noncanonical forms of NF-κB and activates assembly of an NLRP3 inflammasome that leads to secretion of IL-1 and IL-18. Whereas IL-18 acts on T cells, IL-1 primarily feeds back on the ECs and induces type IIa activation responses (Xie et al. 2019).

6.7 Type III (Chronic) Activation of Endothelial Cells

If the adaptive immune response is unsuccessful in eliminating an inflammatory stimulus, the consequence may be chronic inflammation. Two important changes to microvascular ECs may occur in this setting. The first is angiogenesis or new blood vessel formation. Chronic inflammatory tissues often require an increase in blood flow to sustain the presence of inflammatory cells. This process probably shares many features with the angiogenesis of development, of tissue repair, and of tumor growth (Fischer et al. 2006) but may differ with regard to the sources and identities of the mediators involved. The principal angiogenic factors in the angiogenesis of chronic inflammation are not known with certainty, but many are likely derived from infiltrating leukocytes. In addition to well-described angiogenic factors, such as VEGF-A, which are associated with developmental angiogenesis, many inflammatory mediators also play a role in the angiogenesis of chronic inflammation (Benelli et al. 2006), either as stimulators or inhibitors. Certain chemokines, especially IL-8 (CXCL-8), have a pro-angiogenic effect, whereas others, such as IP-10 (CXCL-10), have an antiangiogenic effect (Mehrad et al. 2007). The latter may play a major part in the antiangiogenic actions of IFN-γ.

The second change associated with chronic inflammation is a remodeling of certain venules into a phenotype that resembles the high endothelial venules of secondary lymphoid organs like lymph nodes and gut-associated lymphoid tissue (Drayton et al. 2006). These changes, which involve the assumption of a "plump" or "tall" morphology, also involve the synthesis of sulfated glycoprotein ligands that can be recognized by L-selectin on leukocytes and the expression and/or display of certain chemokines involved in naïve T- and B-cell homing. Changes in the microvascular ECs lining postcapillary venules occur in the context of forming tertiary lymphoid structures, complete with segregated B- and T-cell areas, each with specialized accessory cells (like follicular dendritic cells in the B-cell areas and myeloid dendritic cells in the T-cell areas), and development of organized new lymphatics (Bery et al. 2022). Lymphotoxin-β, a heterotrimeric cell surface molecule containing a novel transmembrane protein complexed with two chains of lymphotoxin-α, appears to be a major inducer of the high endothelial venule (HEV) phenotype. The full consequences of HEV formation and tertiary lymphoid organogenesis are unknown, but it is likely that these stimuli play a role in epitope spreading, allowing the immune response to attack microbes in a broader manner that can avoid evasion.

6.8 Endothelial Dysfunction, Injury, Cytoprotection, and Death

Inflammatory and immune processes may damage microvascular ECs (Pober and Min 2006). ECs may be injured by the immune responses mediated by cytokines. TNF, in addition to activating ECs, may induce cell death, especially in combination with other cytokines, such as IFN-γ. The death pathway signals induced by TNF can be mimicked by other death-inducing ligands of the TNF superfamily, such as Fas or TNF-related apoptosis-inducing ligand (TRAIL). The best-described TNF-induced pathway of programmed cell death is apoptosis. This involves TNFR1-mediated assembly of a signaling complex that allows autocatalytic activation of procaspase 8 to active caspase 8. Caspase 8, in turn, proteolytically actives so-called executioner caspases that cause apoptosis. The caspase-activating complex requires several hours to assemble, and, in that interval, caspase 8 activation can be prevented by TNF-induced synthesis-inhibitory proteins. Among the best-characterized of these are the cFLICE inhibitory protein (c-FLIP), a protein that blocks TNF activation of caspase 8 (initially named FLICE), and the Bcl family member A2, which blocks mitochondrial release of cytochrome *c*, the initiator of caspase 9 activation, caused by Bax insertion into and oligomerization within the mitochondrial outer membrane. Even when TNF-induced apoptotic cell death is blocked, cells may still die. In this case, the death-inducing signaling pathway recruits and activates a kinase called receptor-interacting protein kinase (RIPK)-3 instead of procaspase 8. RIPK-3 phosphorylates a protein called mixed lineage kinase domain-like (MLKL), and phospho-MLKL forms oligomers that insert into the plasma membrane to form pores. These pores can overwhelm the pumps that keep water driven by osmotic gradients from entering a cell, resulting in cell swelling and ultimately cell bursting, a form of death called necroptosis. Activation of the necroptotic pathway may require proteolytic cleavage of TNFR1 mediated by the metalloproteinase ADAMS-17 (originally called TNFα-converting enzyme or TACE) (Bolik et al. 2022). Other pathways may also lead to EC injury and death. For example, ECs may be induced to assemble NLRP3 inflammasomes in response to the complement membrane attack complex (Xie et al. 2019). These structures, containing active caspase 1, can cleave the protein gasdermin D, the amino terminal fragments of which oligomerize and form pores in the plasma membrane. Normally, these structures allow processed IL-1β and IL-18 to escape from the cell, but excessive formation of gasdermin D pores leads to cell swelling and bursting, causing a form of programmed death called pyroptosis (Bai et al. 2020). ECs also appear to escape this fate, likely through expression of as yet incompletely characterized inhibitors that regulate this pathway. ECs are susceptible to injury by reactive oxygen intermediates (ROIs), such as those generated by neutrophils during ischemia/reperfusion injury. ROIs may also lead to NLRP3 inflammasome assembly and pyroptosis. Alternatively, ROIs may oxidize DNA and cause genomic stress, which, if excessive, triggers p53-induced apoptosis, or they may cause extensive iron-catalyzed lipid oxidation, a programmed death pathway called ferroptosis mediated by glutathione peroxidase 4 (*GPX4*) (Yuan et al. 2022).

Sublethal injuries may impair or render ECs incapable of performing their normal homeostatic functions, such as production of NO or maintenance of a permselective capillary barrier. Such changes are referred to as endothelial dysfunction. Dysfunctional changes may overlap with those of type II activation, and the terminology used to describe them may depend on whether the process appears to contribute to homeostasis, e.g., by supporting inflammation to eradicate infection, or lead to pathological changes, e.g., capillary leak and organ edema. The consequences of cell death may be more severe than those of dysfunction. One immediate consequence of EC death is often thrombosis. Apoptotic EC death can contribute to this by releasing plasma membrane-derived microparticles, providing a phosphatidylserine surface for assembly of the proteins of the coagulation cascade. Since these are potentially released from ECs exposed to type IIa activation, these vesicles may contain tissue factor, thus accelerating coagulation. Dying ECs also detach from their basement membrane. If such desquamation is sufficiently extensive that neighboring ECs cannot spread to cover the exposed basement membrane, this can serve as a signal for platelet adhesion, further exacerbating coagulation. In some cases, EC death may be part of homeostasis, for example, when eliminating infected cells. ECs are the primary target of some viruses, such as cytomegalovirus (CMV), and of other types of intracellular microbial infections, such as *Rickettsia* infection. In this case, ECs may be killed by CTLs, either by engaging with death receptors or by exocytosis of CTL granules containing perforin and granzymes that activate apoptosis in the target cell. Cell death by necroptosis or pyroptosis may lead to release of EC cytoplasmic or nuclear proteins as a consequence of cell bursting. Several specific molecules, e.g., histones or high mobility group B 1, which are often described as damage-associated molecular patterns (DAMPs), activate systemic inflammation, including acting on still viable ECs to produce type II activation.

Just as certain signals may render ECs more sensitive to injurious signals, others may make them less sensitive. One example of this are cytokines of the IL-6 family, especially IL-6 itself and IL-11 (Waxman et al. 2003). These cytokines bind to unique receptor heterodimers containing the gp130 coreceptor (encoded by *IL6ST*) signaling chain and activate JAK/STAT3 signaling. One important protective gene induced by this pathway is a protein called survivin (Kirkiles-Smith et al. 2004), which inhibits caspase activation and other cell death pathways. Other cytoprotective proteins are induced by TNF or IL-1, including various antioxidant enzymes and heme oxygenase-1 (HO-1) (Shaffer et al. 1990). HO-1 works, in part, by generating CO and the antioxidant bilirubin (Ryter and Choi 2007; Otterbein et al. 2003), both of which may reduce the damaging effects of ROIs. In the transplantation literature, changes that render ECs resistant to injury are referred to as accommodation, which can also be viewed as yet a different type of endothelial cell activation (Bach et al. 1997).

6.9 Clinical Assessment of Microvascular Activation and Dysfunction

As discussed above, the microvascular endothelium supports a wide array of functions essential for homeostasis and extends to within microns of all organ parenchymal cells. During acute illness, EC function may adapt to restore homeostasis, become dysfunctional to propagate the underlying pathology, or initiate death pathways with pathological consequences. EC dysfunction or death may clinically manifest as systemic inflammatory syndrome (leukocyte recruitment and inflammation), capillary leak (permeability), disseminated intravascular coagulation (coagulation), and microvascular capillary shunting (blood flow). All of these processes are clinically associated with organ failure in acutely ill humans (Pierce et al. 2017b). There are no accurate biomarkers for the diagnosis of EC activation, dysfunction, or death (Pierce et al. 2022). Further complicating the identification of a clinically useful biomarker is the exquisite organ and segmental specialization of microvascular ECs. Soluble adhesion molecules, such as ICAM-1, VCAM-1, and E-, L-, and P-selectins, are relevant to leukocyte adhesion and systemic inflammation. However, the physiological relevance of soluble adhesion molecules is not clear and most likely represents fragments of dead or sloughed ECs from any organ or vascular segment. Capillary leak is especially difficult to assess as capillary vascular beds are typically inaccessible, especially in acute ill people (Pierce et al. 2017a; Vermette et al. 2018). Sublingual intravital microscopy has been used to determine postcapillary hematocrit, as an indirect measure of fluid loss in the capillaries; however, this measure is not specific and reflects capillary leak in the tongue, not heart, lung, or brain capillaries that may be more relevant. Glycocalyx integrity is clinically correlated volume overload from a presumed capillary leak; however, factors inducing the release of glycocalyx products from ECs are poorly understood (Ando et al. 2018; Chappell et al. 2009). Further investigation is required to understand the mechanistic significance of sloughed glycocalyx components. Additionally, specific glycoprotein products may be preferentially expressed in specific vascular segments or organs. If confirmed, specific glycocalyx products may be able to directly correlate with specific organ injury. The markers of EC inflammation, modulation, and regulation of blood flow that have been studied are even more indirectly related to these functions. IL-6, VEGF, and a soluble form of one of its receptors (s-Flt1), which are well-established to be elevated in critical illness, are nonspecific pro-inflammatory cytokines that are secreted by ECs and many other cells. The effects of these cytokines on EC function are also poorly defined, and these molecules are unlikely to represent autocrine signaling. Soluble Fas (sFas) is believed to be antiapoptotic, while its ligand (sFasL) is proapoptotic, but these functions are not unique, or even best-studied, on ECs. Equally murky is the assessment of how microvascular EC regulation affects blood flow. The predominate class studied in this group are reactive nitrogen species believed to reflect the production, and therefor action, of nitric oxide (NO). Known as a potent vasodilator, NO is extremely short-lived and locally acting. Systemic levels of by-products of its degradation do not accurately reflect local blood flow regulation. Inhibitors of NO production also fall under this category and have not been demonstrated to be related to EC function. Finally, markers of coagulation are more specific to ECs, predominately in the lower flow capillaries and postcapillary venules. Functional markers, such as international normalized ratio (INR), prothrombin time (PT), partial thromboplastin time (PTT), and thromboelastography (TEG), could be coupled to levels of specific, EC-derived molecules and accurately reflect vascular function in this domain. However, these markers lack organ specificity. More investigation would be required to link specific functional coagulation profiles to circulating factor levels. These parameters would not provide information on other domains of EC function. Taken together, there are no single or combinations of endothelial-derived biomarkers to delineate EC activation, dysfunction, or death. Nearly all previously proposed biomarkers lack microvascular segment and organ specificity. Most biomarkers indirectly survey EC functions or rely on circulating markers of unknown biological significance.

The precision of scRNA-seq profiles of organ- and segment-specific microvascular ECs provide new possibilities for biomarker development. As these studies progress to detail EC activation and dysfunction or specific ECs in specific organs, new protein products may be identified with direct relevance to specific vascular responses. Alternatively, the specific genes could be targeted as relevant biomarkers. Cell-free unmethylated DNA, expected to be present when derived from genes expressed by ECs in some vascular beds and not others, may be able to provide a precise marker of organ-specific endothelial damage or death. New biomarkers would require clinical validation and would be able to link EC activation, damage, or death to specific organ failures.

6.10 Summary

Vascular ECs that line the entire circulatory system show common features and also tissue-specific differences, especially in the capillaries of the microcirculation. Microvascular ECs perform a number of constitutive functions, namely, regulation of blood flow (especially by arteriolar ECs), permselectivity (especially by capillary ECs), and inflammation (especially by venular ECs) and maintenance of blood fluidity (by all ECs). Compromise or loss of any of these functions constitutes endothelial dysfunction. Microvascular ECs may contribute to inflammation by acquisition of new properties, called activation. Early activation events may be protein synthesis-independent (type I activation) or protein synthesis-dependent (type II activation). Type II activation may evolve as the inflammatory process changes from innate (with type IIa activation) to adaptive immunity (with type IIb activation). Inflammation may injure ECs, contributing to dysfunction. ECs may also be activated to resist injury in the environment of inflammation processes. Chronic inflammation involves additional EC changes including angiogenesis and evolution into modified vessels similar to the high endothelial venules of lymph nodes. In conclusion, microvascular ECs are a remarkably plastic cell type, showing profound adaptations to promote host defense and resist injury.

References

Aaslid R, Lindegaard KF, Sorteberg W, Nornes H (1989) Cerebral autoregulation dynamics in humans. Stroke 20:45–52

Abe Y, Sakata-Yanagimoto M, Fujisawa M, Miyoshi H, Suehara Y, Hattori K, Kusakabe M, Sakamoto T, Nishikii H, Nguyen TB, Owada Y, Enomoto T, Sawa A, Bando H, Yoshida C, Tabata R, Terao T, Nakayama M, Ohshima K, Usuki K, Oda T, Matsue K, Chiba S (2022) A single-cell atlas of non-haematopoietic cells in human lymph nodes and lymphoma reveals a landscape of stromal remodelling. Nat Cell Biol 24:565–578

Abrahimi P, Liu R, Pober JS (2015) Blood vessels in Allotransplantation. Am J Transplant 15: 1748–1754

Adams TS, Schupp JC, Poli S, Ayaub EA, Neumark N, Ahangari F, Chu SG, Raby BA, Deiuliis G, Januszyk M, Duan Q, Arnett HA, Siddiqui A, Washko GR, Homer R, Yan X, Rosas IO, Kaminski N (2020) Single-cell RNA-seq reveals ectopic and aberrant lung-resident cell populations in idiopathic pulmonary fibrosis. Sci Adv 6:eaba1983

Ager A (2017) High endothelial Venules and other blood vessels: critical regulators of lymphoid organ development and function. Front Immunol 8:45

Aizarani N, Saviano A, Sagar N, Mailly L, Durand S, Herman JS, Pessaux P, Baumert TF, Grün DA (2019) Human liver cell atlas reveals heterogeneity and epithelial progenitors. Nature. 8;572 (7768):199–204. https://doi.org/10.1038/s41586-019-1373-2. https://pubmed.ncbi.nlm.nih.gov/31292543/

Albrecht EA, Chinnaiyan AM, Varambally S, Kumar-Sinha C, Barrette TR, Sarma JV, Ward PA (2004) C5a-induced gene expression in human umbilical vein endothelial cells. Am J Pathol 164:849–859

Ando Y, Okada H, Takemura G, Suzuki K, Takada C, Tomita H, Zaikokuji R, Hotta Y, Miyazaki N, Yano H, Muraki I, Kuroda A, Fukuda H, Kawasaki Y, Okamoto H, Kawaguchi T, Watanabe T, Doi T, Yoshida T, Ushikoshi H, Yoshida S, Ogura S (2018) Brain-specific ultrastructure of capillary endothelial Glycocalyx and its possible contribution for blood brain barrier. Sci Rep 8:17523

Arnout J, Hoylaerts MF, Lijnen HR (2006) Haemostasis. Handb Exp Pharmacol 176(II):1–42

Bach FH, Hancock WW, Ferran C (1997) Protective genes expressed in endothelial cells: a regulatory response to injury. Immunol Today 18:483–486

Bai B, Yang Y, Wang Q, Li M, Tian C, Liu Y, Aung LHH, Li PF, Yu T, Chu XM (2020) NLRP3 inflammasome in endothelial dysfunction. Cell Death Dis 11:776

Basile DP, Collett JA, Yoder MC (2018) Endothelial colony-forming cells and pro-angiogenic cells: clarifying definitions and their potential role in mitigating acute kidney injury. Acta Physiol (Oxf) 222

Bavendiek U, Libby P, Kilbride M, Reynolds R, Mackman N, Schonbeck U (2002) Induction of tissue factor expression in human endothelial cells by CD40 ligand is mediated via activator protein 1, nuclear factor kappa B, and Egr-1. J Biol Chem 277:25032–25039

Bazzoni G, Dejana E (2004) Endothelial cell-to-cell junctions: molecular organization and role in vascular homeostasis. Physiol Rev 84:869–901

Belizna C, Duijvestijn A, Hamidou M, Cohen Tervaert JW (2006) Antiendothelial cell antibodies in vasculitis and connective tissue disease. Ann Rheum Dis 65:1545–1550

Benelli R, Lorusso G, Albini A, Noonan DM (2006) Cytokines and chemokines as regulators of angiogenesis in health and disease. Curr Pharm Des 12:3101–3115

Bery AI, Shepherd HM, Li W, Krupnick AS, Gelman AE, Kreisel D (2022) Role of tertiary lymphoid organs in the regulation of immune responses in the periphery. Cell Mol Life Sci 79:359

Bevilacqua MP, Pober JS, Majeau GR, Fiers W, Cotran RS, Gimbrone MA Jr (1986) Recombinant tumor necrosis factor induces procoagulant activity in cultured human vascular endothelium: characterization and comparison with the actions of interleukin 1. Proc Natl Acad Sci USA 83: 4533–4537

Birch KA, Ewenstein BM, Golan DE, Pober JS (1994) Prolonged peak elevations in cytoplasmic free calcium ions, derived from intracellular stores, correlate with the extent of thrombin-stimulated exocytosis in single human umbilical vein endothelial cells. J Cell Physiol 160: 545–554

Bolik J, Krause F, Stevanovic M, Gandrass M, Thomsen I, Schacht SS, Rieser E, Muller M, Schumacher N, Fritsch J, Wichert R, Galun E, Bergmann J, Roder C, Schafmayer C, Egberts JH, Becker-Pauly C, Saftig P, Lucius R, Schneider-Brachert W, Barikbin R, Adam D, Voss M, Hitzl W, Kruger A, Strilic B, Sagi I, Walczak H, Rose-John S, Schmidt-Arras D (2022) Inhibition of ADAM17 impairs endothelial cell necroptosis and blocks metastasis. J Exp Med 219

Brancale J, Vilarinho S (2021) A single cell gene expression atlas of 28 human livers. J Hepatol 75: 219–220

Briscoe DM, Cotran RS, Pober JS (1992) Effects of tumor necrosis factor, lipopolysaccharide, and IL-4 on the expression of vascular cell adhesion molecule-1 in vivo. Correlation with CD3+ T cell infiltration. J Immunol 149:2954–2960

Burke M, Pabbidi MR, Farley J, Roman RJ (2014) Molecular mechanisms of renal blood flow autoregulation. Curr Vasc Pharmacol 12:845–858

Busse R, Fleming I (2006) Vascular endothelium and blood flow. Handb Exp Pharmacol 176 (II):43–78

Byron SA, Van Keuren-Jensen KR, Engelthaler DM, Carpten JD, Craig DW (2016) Translating RNA sequencing into clinical diagnostics: opportunities and challenges. Nat Rev Genet 17:257–271

Chandrabalan A, Ramachandran R (2021) Molecular mechanisms regulating proteinase-activated receptors (PARs). FEBS J 288:2697–2726

Chappell D, Jacob M, Paul O, Rehm M, Welsch U, Stoeckelhuber M, Conzen P, Becker BF (2009) The glycocalyx of the human umbilical vein endothelial cell: an impressive structure ex vivo but not in culture. Circ Res 104:1313–1317

Chavkin NW, Hirschi KK (2020) Single cell analysis in vascular biology. Front Cardiovasc Med 7:42

Claassen J, Thijssen DHJ, Panerai RB, Faraci FM (2021) Regulation of cerebral blood flow in humans: physiology and clinical implications of autoregulation. Physiol Rev 101:1487–1559

Combes AN, Zappia L, Er PX, Oshlack A, Little MH (2019) Single-cell analysis reveals congruence between kidney organoids and human fetal kidney. Genome Med 11

Curry FE (2018) The molecular structure of the endothelial Glycocalyx layer (EGL) and surface layers (ESL) modulation of Transvascular exchange. Adv Exp Med Biol 1097:29–49

Davis GE, Senger DR (2005) Endothelial extracellular matrix: biosynthesis, remodeling, and functions during vascular morphogenesis and neovessel stabilization. Circ Res 97:1093–1107

De Caterina R, Libby P, Peng HB, Thannickal VJ, Rajavashisth TB, Gimbrone MA Jr, Shin WS, Liao JK (1995) Nitric oxide decreases cytokine-induced endothelial activation. Nitric oxide selectively reduces endothelial expression of adhesion molecules and proinflammatory cytokines. J Clin Invest 96:60–68

De Pablo-Moreno JA, Serrano LJ, Revuelta L, Sanchez MJ, Liras A (2022) The vascular endothelium and coagulation: homeostasis, disease, and treatment, with a focus on the Von Willebrand factor and factors VIII and V. Int J Mol Sci 23

Den Braanker H, Van Stigt AC, Kok MR, Lubberts E, Bisoendial RJ (2021) Single-cell RNA sequencing reveals heterogeneity and functional diversity of lymphatic endothelial cells. Int J Mol Sci 22

Dessalles CA, Babataheri A, Barakat AI (2021) Pericyte mechanics and mechanobiology. J Cell Sci 134

Domigan CK, Iruela-Arispe ML (2012) Recent advances in vascular development. Curr Opin Hematol 19:176–183

Drayton DL, Liao S, Mounzer RH, Ruddle NH (2006) Lymphoid organ development: from ontogeny to neogenesis. Nat Immunol 7:344–353

Egan K, Fitzgerald GA (2006) Eicosanoids and the vascular endothelium. Handb Exp Pharmacol:189–211

Eichmann A, Li Y, Moyon D, Lenoble F, Purdanaud L, Breant C (2005) Vscular development: from precursor cells to branched arterial ad venous networks. Int J Dev Biol 49:259–267

Enis DR, Shepherd BR, Wang Y, Qasim A, Shanahan CM, Weissberg PL, Kashgarian M, Pober JS, Schechner JS (2005) Induction, differentiation, and remodeling of blood vessels after transplantation of Bcl-2-transduced endothelial cells. Proc Natl Acad Sci USA 102:425–430

Ewenstein BM, Warhol MJ, Handin RI, Pober JS (1987) Composition of the von Willebrand factor storage organelle (Weibel-Palade body) isolated from cultured human umbilical vein endothelial cells. J Cell Biol 104:1423–1433

Filippini A, D'alessio A (2020) Caveolae and lipid rafts in endothelium: valuable organelles for multiple functions. Biomolecules 10

Fischer C, Schneider M, Carmeliet P (2006) Principles and therapeutic implications of angiogenesis, vasculogenesis and arteriogenesis. Handb Exp Pharmacol 176:157–212

Frangogiannis NG (2006) Targeting the inflammatory response in healing myocardial infarcts. Curr Med Chem 13:1877–1893

Frank PG, Woodman SE, Park DS, Lisanti MP (2003) Caveolin, caveolae, and endothelial cell function. Arterioscler Thromb Vasc Biol 23:1161–1168

Garcia FJ, Sun N, Lee H, Godlewski B, Mathys H, Galani K, Zhou B, Jiang X, Ng AP, Mantero J, Tsai LH, Bennett DA, Sahin M, Kellis M, Heiman M (2022) Single-cell dissection of the human brain vasculature. Nature 603:893–899

Gawad C, Koh W, Quake SR (2016) Single-cell genome sequencing: current state of the science. Nat Rev Genet 17:175–188

Ghanekar A, Mendicino M, Liu H, He W, Liu M, Zhong R, Phillips MJ, Levy GA, Grant DR (2004) Endothelial induction of fgl2 contributes to thrombosis during acute vascular xenograft rejection. J Immunol 172:5693–5701

Gillich A, Zhang F, Farmer CG, Travaglini KJ, Tan SY, Gu M, Zhou B, Feinstein JA, Krasnow MA, Metzger RJ (2020) Capillary cell-type specialization in the alveolus. Nature 586:785–789

Girard PR, Nerem RM (1995) Shear stress modulates endothelial cell morphology and F-actin organization through the regulation of focal adhesion-associated proteins. J Cell Physiol 163: 179–193

Goodwill AG, Dick GM, Kiel AM, Tune JD (2017) Regulation of coronary blood flow. Compr Physiol 7:321–382

Gotlieb AI (1990) The endothelial cytoskeleton: organization in normal and regenerating endothelium. Toxicol Pathol 18:603–617

Guilliams M, Bonnardel J, Haest B, Vanderborght B, Wagner C, Remmerie A, Bujko A, Martens L, Thoné T, Browaeys R, De Ponti FF, Vanneste B, Zwicker C, Svedberg FR, Vanhalewyn T, Gonçalves A, Lippens S, Devriendt B, Cox E, Ferrero G, Wittamer V, Willaert A, Kaptein SJF, Neyts J, Dallmeier K, Geldhof P, Casaert S, Deplancke B, Ten Dijke P, Hoorens A, Vanlander A, Berrevoet F, Van Nieuwenhove Y, Saeys Y, Saelens W, Van Vlierberghe H, Devisscher L, Scott CL (2022) Spatial proteogenomics reveals distinct and evolutionarily conserved hepatic macrophage niches. Cell 185:379–396

Gutterman DD, Chabowski DS, Kadlec AO, Durand MJ, Freed JK, Ait-Aissa K, Beyer AM (2016) The human microcirculation: regulation of flow and beyond. Circ Res 118:157–172

Hahn RG, Patel V, Dull RO (2021) Human glycocalyx shedding: systematic review and critical appraisal. Acta Anaesthesiol Scand 65:590–606

Hansen J, Sealfon R, Menon R, Eadon MT, Lake BB, Steck B, Anjani K, Parikh S, Sigdel TK, Zhang G, Velickovic D, Barwinska D, Alexandrov T, Dobi D, Rashmi P, Otto EA, Rivera M, Rose MP, Anderton CR, Shapiro JP, Pamreddy A, Winfree S, Xiong Y, He Y, De Boer IH, Hodgin JB, Barisoni L, Naik AS, Sharma K, Sarwal MM, Zhang K, Himmelfarb J, Rovin B, El-Achkar TM, Laszik Z, He JC, Dagher PC, Valerius MT, Jain S, Satlin LM, Troyanskaya OG, Kretzler M, Iyengar R, Azeloglu EU (2022) A reference tissue atlas for the human kidney. Sci Adv 8:eabn4965

He B, Chen P, Zambrano S, Dabaghie D, Hu Y, Möller-Hackbarth K, Unnersjö-Jess D, Korkut GG, Charrin E, Jeansson M, Bintanel-Morcillo M, Witasp A, Wennberg L, Wernerson A, Schermer B, Benzing T, Ernfors P, Betsholtz C, Lal M, Sandberg R, Patrakka J (2021) Single-cell RNA sequencing reveals the mesangial identity and species diversity of glomerular cell transcriptomes. Nat Commun 12. https://doi.org/10.1038/s41467-021-22331-9

He P, Lim K, Sun D, Pett JP, Jeng Q, Polanski K, Dong Z, Bolt L, Richardson L, Mamanova L, Dabrowska M, Wilbrey-Clark A, Madissoon E, Tuong ZK, Dann E, Suo C, Goh I, Yoshida M, Nikolić MZ, Janes SM, He X, Barker RA, Teichmann SA, Marioni JC, Meyer KB, Rawlins EL (2022a) A human fetal lung cell atlas uncovers proximal-distal gradients of differentiation and key regulators of epithelial fates. Cell 185:4841–4860

He Y, Tacconi C, Dieterich LC, Kim J, Restivo G, Gousopoulos E, Lindenblatt N, Levesque MP, Claassen M, Detmar M (2022b) Novel blood vascular endothelial subtype-specific markers in human skin unearthed by single-cell transcriptomic profiling. Cells 11

Heltianu C, Simionescu M, Simionescu N (1982) Histamine receptors of the microvascular endothelium revealed in situ with a histamine-ferritin conjugate: characteristic high-affinity binding sites in venules. J Cell Biol 93:357–364

Hickey MJ, Valenzuela NM, Reed EF (2016) Alloantibody generation and effector function following sensitization to human leukocyte antigen. Front Immunol 7:30

Hochane M, Van Den Berg PR, Fan X, Bérenger-Currias N, Adegeest E, Bialecka M, Nieveen M, Menschaart M, S. M. Chuva De Sousa Lopes and S. Semrau. (2019) Single-cell transcriptomics reveals gene expression dynamics of human fetal kidney development. PLoS Biol 17:e3000152

Hou X, Yang Y, Li P, Zeng Z, Hu W, Zhe R, Liu X, Tang D, Ou M, Dai Y (2021) Integrating spatial Transcriptomics and single-cell RNA-seq reveals the gene expression Profling of the human embryonic liver. Front Cell Dev Biol 9

Hu Z, Liu W, Hua X, Chen X, Chang Y, Hu Y, Xu Z, Song J (2021) Single-cell transcriptomic atlas of different human cardiac arteries identifies cell types associated with vascular physiology. Arterioscler Thromb Vasc Biol 41:1408–1427

Imcke E, Ruszczak Z, Mayer-Da Silva A, Detmar M, Orfanos CE (1991) Cultivation of human dermal microvascular endothelial cells in vitro: immunocytochemical and ultrastructural characterization and effect of treatment with three synthetic retinoids. Arch Dermatol Res 283:149–157

Jane-Wit D, Manes TD, Yi T, Qin L, Clark P, Kirkiles-Smith NC, Abrahimi P, Devalliere J, Moeckel G, Kulkarni S, Tellides G, Pober JS (2013) Alloantibody and complement promote T cell-mediated cardiac allograft vasculopathy through noncanonical nuclear factor-kappaB signaling in endothelial cells. Circulation 128:2504–2516

Jardine L, Webb S, Goh I, Quiroga Londoño M, Reynolds G, Mather M, Olabi B, Stephenson E, Botting RA, Horsfall D, Engelbert J, Maunder D, Mende N, Murnane C, Dann E, Mcgrath J, King H, Kucinski I, Queen R, Carey CD, Shrubsole C, Poyner E, Acres M, Jones C, Ness T, Coulthard R, Elliott N et al (2021) Blood and immune development in human fetal bone marrow and down syndrome. Nature 598:327–331

Johnson M (1998) The beta-adrenoceptor. Am J Respir Crit Care Med 158:S146–S153

Jolly AJ, Lu S, Strand KA, Dubner AM, Mutryn MF, Nemenoff RA, Majesky MW, Moulton KS, Weiser-Evans MCM (2022) Heterogeneous subpopulations of adventitial progenitor cells regulate vascular homeostasis and pathological vascular remodelling. Cardiovasc Res 118: 1452–1465

Kirkiles-Smith NC, Mahboubi K, Plescia J, Mcniff JM, Karras J, Schechner JS, Altieri DC, Pober JS (2004) IL-11 protects human microvascular endothelium from alloinjury in vivo by induction of survivin expression. J Immunol 172:1391–1396

Krämer B, Nalin AP, Ma F, Eickhoff S, Lutz P, Leonardelli S, Goeser F, Finnemann C, Hack G, Raabe J, Tovinh M, Ahmad S, Hoffmeister C, Kaiser KM, Manekeller S, Branchi V, Bald T, Hölzel M, Hüneburg R, Nischalke HD, Semaan A, Langhans B, Kaczmarek DJ, Benner B, Lordo MR, Kowalski J, Gerhardt A, Timm J, Toma M, Mohr R, Türler A, Charpentier A, Van Bremen T, Feldmann G, Sattler A, Kotsch K, Abdallah AT, Strassburg CP, Spengler U, Carson WE, Mundy-Bosse BL, Pellegrini M, O'sullivan TE, Freud AG, Nattermann J (2023) Single-cell RNA sequencing identifies a population of human liver-type ILC1s. Cell Rep 42:111937

Kuhlencordt PJ, Rosel E, Gerszten RE, Morales-Ruiz M, Dombkowski D, Atkinson WJ, Han F, Preffer F, Rosenzweig A, Sessa WC, Gimbrone MA Jr, Ertl G, Huang PL (2004) Role of endothelial nitric oxide synthase in endothelial activation: insights from eNOS knockout endothelial cells. Am J Physiol Cell Physiol 286:C1195–C1202

Kunkel SL, Lukacs N, Strieter RM (1995) Expression and biology of neutropohil and endothelial cell-derived chemokines. Semin Cell Biol 6:327–336

Lawson C, Holder AL, Stanford RE, Smith J, Rose ML (2005) Anti-intercellular adhesion molecule-1 antibodies in sera of heart transplant recipients: a role in endothelial cell activation. Transplantation 80:264–271

Le J, Vilcek J (1987) Tumor necrosis factor and interleukin 1: cytokines with multiple overlapping biological activities. Lab Investig 56:234–248

Lentz SR, Tsiang M, Sadler JE (1991) Regulation of thrombomodulin by tumor necrosis factor-alpha: comparison of transcriptional and posttranscriptional mechanisms. Blood 77:542–550

Ley K, Reutershen J (2006) Leukocyte-endothelial interactions in health and disease. Handb Exp Pharmacol 176:97–133

Li Q, Zhu Z, Wang L, Lin Y, Fang H, Lei J, Cao T, Wang G, Dang E (2021) Single-cell transcriptome profiling reveals vascular endothelial cell heterogeneity in human skin. Theranostics 11:6461–6476

Liao J, Yu Z, Chen Y, Bao M, Zou C, Zhang H, Liu D, Li T, Zhang Q, Li J, Cheng J, Mo Z (2020) Single-cell RNA sequencing of human kidney. Scientific Data 7:4

Litviňuková M, Talavera-López C, Maatz H, Reichart D, Worth CL, Lindberg EL, Kanda M, Polanski K, Heinig M, Lee M, Nadelmann ER, Roberts K, Tuck L, Fasouli ES, Delaughter DM, Mcdonough B, Wakimoto H, Gorham JM, Samari S, Mahbubani KT, Saeb-Parsy K, Patone G, Boyle JJ, Zhang H, Zhang H, Viveiros A, Oudit GY, Bayraktar OA, Seidman JG, Seidman CE, Noseda M, Hubner N, Teichmann SA (2020) Cells of the adult human heart. Nature 588:466–472

Liu J, Schnitzer JE (1999) Analysis of lipids in caveolae. Methods Mol Biol 116:61–72

Liu Z, Miner JJ, Yago T, Yao L, Lupu F, Xia L, Mcever RP (2010) Differential regulation of human and murine P-selectin expression and function in vivo. J Exp Med 207:2975–2987

Liu R, Lauridsen HM, Amezquita RA, Pierce RW, Jane-Wit D, Fang C, Pellowe AS, Kirkiles-Smith NC, Gonzalez AL, Pober JS (2016) IL-17 promotes neutrophil-mediated immunity by activating microvascular Pericytes and not endothelium. J Immunol 197:2400–2408

Lorant DE, Patel KD, Mcintyre TM, Mcever RP, Prescott SM, Zimmerman GA (1991) Coexpression of GMP-140 and PAF by endothelium stimulated by histamine or thrombin: a juxtacrine system for adhesion and activation of neutrophils. J Cell Biol 115:223–234

Louis NA, Hamilton KE, Colgan SP (2005) Lipid mediator networks and leukocyte transmigration. Prostaglandins Leukot Essent Fatty Acids 73:197–202

Lowenstein CJ, Morrell CN, Yamakuchi M (2005) Regulation of Weibel-Palade body exocytosis. Trends Cardiovasc Med 15:302–328

Lu Z, Li Y, Jin J, Zhang X, Lopes-Virella MF, Huang Y (2012) Toll-like receptor 4 activation in microvascular endothelial cells triggers a robust inflammatory response and cross talk with mononuclear cells via interleukin-6. Arterioscler Thromb Vasc Biol 32:1696–1706

Macparland SA, Liu JC, Ma XZ, Innes BT, Bartczak AM, Gage BK, Manuel J, Khuu N, Echeverri J, Linares I, Gupta R, Cheng ML, Liu LY, Camat D, Chung SW, Seliga RK, Shao Z, Lee E, Ogawa S, Ogawa M, Wilson MD, Fish JE, Selzner M, Ghanekar A, Grant D, Greig P, Sapisochin G, Selzner N, Winegarden N, Adeyi O, Keller G, Bader GD, Mcgilvray ID (2018) Single cell RNA sequencing of human liver reveals distinct intrahepatic macrophage populations. Nat Commun 9

Majesky MW (2018) Vascular development. Arterioscler Thromb Vasc Biol 38:e17–e24

Manes TD, Shiao SL, Dengler TJ, Pober JS (2007) TCR signaling antagonizes rapid IP-10-mediated transendothelial migration of effector memory CD4+ T cells. J Immunol 178:3237–3243

Manes TD, Wang V, Pober JS (2022) Costimulators expressed on human endothelial cells modulate antigen-dependent recruitment of circulating T lymphocytes. Front Immunol 13:1016361

Marcos-Ramiro B, Garcia-Weber D, Barroso S, Feito J, Ortega MC, Cernuda-Morollon E, Reglero-Real N, Fernandez-Martin L, Duran MC, Alonso MA, Correas I, Cox S, Ridley AJ, Millan J (2016) RhoB controls endothelial barrier recovery by inhibiting Rac1 trafficking to the cell border. J Cell Biol 213:385–402

Margraf A, Ley K, Zarbock A (2019) Neutrophil recruitment: from model systems to tissue-specific patterns. Trends Immunol 40:613–634

Martinelli R, Kamei M, Sage PT, Massol R, Varghese L, Sciuto T, Toporsian M, Dvorak AM, Kirchhausen T, Springer TA, Carman CV (2013) Release of cellular tension signals self-restorative ventral lamellipodia to heal barrier micro-wounds. J Cell Biol 201:449–465

Matsushita K, Morrell CN, Cambien B, Yang SX, Yamakuchi M, Bao C, Hara MR, Quick RA, Cao W, O'rourke B, Lowenstein JM, Pevsner J, Wagner DD, Lowenstein CJ (2003) Nitric oxide regulates exocytosis by S-nitrosylation of N-ethylmaleimide-sensitive factor. Cell 115:139–150

Mcverry BJ, Garcia JG (2004) Endothelial cell barrier regulation by sphingosine 1-phosphate. J Cell Biochem 92:1075–1085

Mehrad B, Keane MP, Strieter RM (2007) Chemokines as mediators of angiogenesis. Thromb Haemost 97:755–762

Middleton J, Neil S, Wintle J, Clark-Lewis I, Moore H, Lam C, Auer M, Hub E, Rot A (1997) Transcytosis and surface presentation of IL-8 by venular endothelial cells. Cell 91:385–395

Minshall RD, Malik AB (2006) Transport across the endothelium: regulation of endothelial permeability. Handb Exp Pharmacol:107–144

Moncada S (2006) Nitric oxide and the vascular endothelium. Handb Exp Pharmacol 176:213–254

Muller WA (2016) Localized signals that regulate transendothelial migration. Curr Opin Immunol 38:24–29

Nass J, Terglane J, Gerke V (2021) Weibel Palade bodies: unique secretory organelles of endothelial cells that control blood vessel homeostasis. Front Cell Dev Biol 9:813995

Nolan DJ, Ginsberg M, Israely E, Palikuqi B, Poulos MG, James D, Ding BS, Schachterle W, Liu Y, Rosenwaks Z, Butler JM, Xiang J, Rafii A, Shido K, Rabbany SY, Elemento O, Rafii S (2013) Molecular signatures of tissue-specific microvascular endothelial cell heterogeneity in organ maintenance and regeneration. Dev Cell 26:204–219

Otterbein LE, Soares MP, Yamashita K, Bach FH (2003) Heme oxygenase-1: unleashing the protective properties of heme. Trends Immunol 24:449–455

Paik DT, Tian L, Williams IM, Rhee S, Zhang H, Liu C, Mishra R, Wu SM, Red-Horse K, Wu JC (2020) Single-cell RNA sequencing unveils unique transcriptomic signatures of organ-specific endothelial cells. Circulation 142:1848–1862

Palade GE, Bruns RR (1968) Structural modulations of plasmalemmal vesicles. J Cell Biol 37:633–649

Perez-Pomares JM, Carmona R, Gonzalez-Iriarte M, Atencia G, Wessels A, Munoz-Chapuli R (2002) Origin of coronary endothelial cells from epicardial mesothelium in avian embryos. Int J Dev Biol 46:1005–1013

Pierce R, Luckett PM, Faustino EVS (2017a) A survey of Pediatric critical care providers on the presence, severity, and assessment of capillary leak in critically ill children. J Pediatr Intensive Care 6:145–151

Pierce RW, Giuliano JS Jr, Pober JS (2017b) Endothelial cell function and dysfunction in critically ill children. Pediatrics 140

Pierce RW, Merola J, Lavik JP, Kluger MS, Huttner A, Khokha MK, Pober JS (2017c) A p190BRhoGAP mutation and prolonged RhoB activation in fatal systemic capillary leak syndrome. J Exp Med 214:3497–3505

Pierce RW, Giuliano JS, Whitney JE, Ouellette Y (2022) Endothelial dysfunction criteria in critically ill children: the PODIUM consensus conference. Pediatrics 149:S97–S102

Platt JL, Dalmasso AP, Lindman BJ, Ihrcke NS, Bach FH (1991) The role of C5a and antibody in the release of heparan sulfate from endothelial cells. Eur J Immunol 21:2887–2890

Pober JS, Cotran RS (1990) Cytokines and endothelial cell biology. Physiol Rev 70:427–451

Pober JS, Min W (2006) Endothelial cell dysfunction, injury and death. In: Moncada SHA (ed) Handbook of experimental pharmacology. Springer-Verlag, Berlin

Pober JS, Min W, Bradley JR (2009) Mechanisms of endothelial dysfunction, injury, and death. Annu Rev Pathol 4:71–95

Pober JS, Merola J, Liu R, Manes TD (2017) Antigen presentation by vascular cells. Front Immunol 8:1907

Redl H, Dinges HP, Buurman WA, Van Der Linden CJ, Pober JS, Cotran RS, Schlag G (1991) Expression of endothelial leukocyte adhesion molecule-1 in septic but not traumatic/hypovolemic shock in the baboon. Am J Pathol 139:461–466

Rogers KA, Mckee NH, Kalnins VI (1985) Preferential orientation of centrioles toward the heart in endothelial cells of major blood vessels is reestablished after reversal of a segment. Proc Natl Acad Sci USA 82:3272–3276

Rosenkranz-Weiss P, Sessa WC, Milstien S, Kaufman S, Watson CA, Pober JS (1994) Regulation of nitric oxide synthesis by proinflammatory cytokines in human umbilical vein endothelial cells. Elevations in tetrahydrobiopterin levels enhance endothelial nitric oxide synthase specific activity. J Clin Invest 93:2236–2243

Rothstein JL, Schreiber H (1988) Synergy between tumor necrosis factor and bacterial products causes hemorrhagic necrosis and lethal shock in normal mice. Proc Natl Acad Sci USA 85:607–611

Ryter SW, Choi AM (2007) Cytoprotective and anti-inflammatory actions of carbon monoxide in organ injury and sepsis models. Novartis Found Symp 280:165–175

Schiller HB, Montoro DT, Simon LM, Rawlins EL, Meyer KB, Strunz M, Vieira Braga FA, Timens W, Koppelman GH, Budinger GRS, Burgess JK, Waghray A, Van Den Berge M, Theis FJ, Regev A, Kaminski N, Rajagopal J, Teichmann SA, Misharin AV, Nawijn MC (2019) The human lung cell atlas: a high-resolution reference map of the human lung in health and disease. Am J Respir Cell Mol Biol 61:31–41

Schupp JC, Adams TS, Cosme C Jr, Raredon MSB, Yuan Y, Omote N, Poli S, Chioccioli M, Rose KA, Manning EP, Sauler M, Deiuliis G, Ahangari F, Neumark N, Habermann AC, Gutierrez AJ, Bui LT, Lafyatis R, Pierce RW, Meyer KB, Nawijn MC, Teichmann SA, Banovich NE, Kropski JA, Niklason LE, Pe'er D, Yan X, Homer RJ, Rosas IO, Kaminski N (2021) Integrated single-cell atlas of endothelial cells of the human lung. Circulation 144:286–302

Segal SS (2005) Regulation of blood flow in the microcirculation. Microcirculation 12:33–45

Sequeira Lopez ML, Gomez RA (2011) Development of the renal arterioles. J Am Soc Nephrol 22: 2156–2165

Shaffer JB, Treanor CP, Del Vecchio PJ (1990) Expression of bovine and mouse endothelial cell antioxidant enzymes following TNF-alpha exposure. Free Radic Biol Med 8:497–502

Shesely EG, Maeda N, Kim HS, Desai KM, Krege JH, Laubach VE, Sherman PA, Sessa WC, Smithies O (1996) Elevated blood pressures in mice lacking endothelial nitric oxide synthase. Proc Natl Acad Sci USA 93:13176–13181

Shiraseb F, Asbaghi O, Bagheri R, Wong A, Figueroa A, Mirzaei K (2022) Effect of l-arginine supplementation on blood pressure in adults: a systematic review and dose-response meta-analysis of randomized clinical trials. Adv Nutr 13:1226–1242

Single-Cell Transcriptome Atlas of Murine Endothelial Cells Cell (2020) Feb 20;180(4):764–779. e20. https://doi.org/10.1016/j.cell.2020.01.015. Epub 2020 Feb 13. https://pubmed.ncbi.nlm.nih.gov/32059779/

Sixma JJ, De Groot PG (1991) von Willebrand factor and the blood vessel wall. Mayo Clin Proc 66: 628–633

Smolinska S, Winiarska E, Globinska A, Jutel M (2022) Histamine: a mediator of intestinal disorders-a review. Meta 12

Solé-Boldo L, Raddatz G, Schütz S, Mallm J-P, Rippe K, Lonsdorf AS, Rodríguez-Paredes M, Lyko F (2020) Single-cell transcriptomes of the human skin reveal age-related loss of fibroblast priming. Commun Biol 3:188

Stevens T, Garcia JG, Shasby DM, Bhattacharya J, Malik AB (2000) Mechanisms regulating endothelial cell barrier function. Am J Physiol Lung Cell Mol Physiol 279:L419–L422

Stewart BJ, Ferdinand JR, Young MD, Mitchell TJ, Loudon KW, Riding AM, Richoz N, Frazer GL, Staniforth JUL, Vieira Braga FA, Botting RA, Popescu D-M, Vento-Tormo R, Stephenson E, Cagan A, Farndon SJ, Polanski K, Efremova M, Green K, Del Castillo Velasco-Herrera M, Guzzo C, Collord G, Mamanova L, Aho T, Armitage JN, Riddick ACP, Mushtaq I, Farrell S, Rampling D, Nicholson J, Filby A, Burge J, Lisgo S, Lindsay S, Bajenoff M, Warren AY, Stewart GD, Sebire N, Coleman N, Haniffa M, Teichmann SA, Behjati S, Clatworthy MR (2019) Spatiotemporal immune zonation of the human kidney. Science 365:1461–1466

Sugama Y, Tiruppathi C, Offakidevi K, Andersen TT, Fenton JW 2nd, Malik AB (1992) Thrombin-induced expression of endothelial P-selectin and intercellular adhesion molecule-1: a mechanism for stabilizing neutrophil adhesion. J Cell Biol 119:935–944

Sullivan DP, Muller WA (2014) Neutrophil and monocyte recruitment by PECAM, CD99, and other molecules via the LBRC. Semin Immunopathol 36:193–209

Suzuki A, Tomita H, Okada H (2022) Form follows function: the endothelial glycocalyx. Transl Res 247:158–167

Tabula Sapiens C, Jones RC, Karkanias J, Krasnow MA, Pisco AO, Quake SR, Salzman J, Yosef N, Bulthaup B, Brown P, Harper W, Hemenez M, Ponnusamy R, Salehi A, Sanagavarapu BA, Spallino E, Aaron KA, Concepcion W, Gardner JM, Kelly B, Neidlinger N, Wang Z, Crasta S, Kolluru S, Morri M, Pisco AO, Tan SY, Travaglini KJ, Xu C, Alcantara-Hernandez M, Almanzar N, Antony J, Beyersdorf B, Burhan D, Calcuttawala K, Carter MM, Chan CKF, Chang CA, Chang S, Colville A, Crasta S, Culver RN, Cvijovic I, D'amato G, Ezran C, Galdos FX, Gillich A, Goodyer WR, Hang Y, Hayashi A, Houshdaran S, Huang X, Irwin JC, Jang S, Juanico JV, Kershner AM, Kim S, Kiss B, Kolluru S, Kong W, Kumar ME, Kuo AH, Leylek R, Li B, Loeb GB, Lu WJ, Mantri S, Markovic M, Mcalpine PL, De Morree A, Morri M, Mrouj K, Mukherjee S, Muser T, Neuhofer P, Nguyen TD, Perez K, Phansalkar R, Pisco AO, Puluca N, Qi Z, Rao P, Raquer-Mckay H, Schaum N, Scott B, Seddighzadeh B, Segal J, Sen S, Sikandar S, Spencer SP, Steffes LC, Subramaniam VR, Swarup A, Swift M, Travaglini KJ, Van Treuren W, Trimm E, Veizades S, Vijayakumar S, Vo KC, Vorperian SK, Wang W, Weinstein HNW, Winkler J, Wu TTH, Xie J, Yung AR, Zhang Y, Detweiler AM, Mekonen H, Neff NF, Sit RV, Tan M, Yan J, Bean GR, Charu V, Forgo E, Martin BA, Ozawa MG, Silva O, Tan SY, Toland A, Vemuri VNP, Afik S, Awayan K, Botvinnik OB, Byrne A, Chen M, Dehghannasiri R, Detweiler AM, Gayoso A, Granados AA, Li Q, Mahmoudabadi G, Mcgeever A, De Morree A, Olivieri JE, Park M, Pisco AO, Ravikumar N, Salzman J, Stanley G, Swift M, Tan M, Tan W, Tarashansky AJ, Vanheusden R, Vorperian SK, Wang P, Wang S, Xing G, Xu C, Yosef N, Alcantara-Hernandez M, Antony J, Chan CKF, Chang CA, Colville A, Crasta S, Culver R, Dethlefsen L, Ezran C, Gillich A, Hang Y, Ho PY, Irwin JC, Jang S, Kershner AM, Kong W, Kumar ME, Kuo AH, Leylek R, Liu S, Loeb GB, Lu WJ, Maltzman JS, Metzger RJ, De Morree A, Neuhofer P, Perez K, Phansalkar R, Qi Z, Rao P, Raquer-Mckay H, Sasagawa K, Scott B, Sinha R, Song H, Spencer SP, Swarup A, Swift M, Travaglini KJ, Trimm E, Veizades S, Vijayakumar S, Wang B, Wang W, Winkler J, Xie J, Yung AR, Artandi SE, Beachy PA, Clarke MF, Giudice LC, Huang FW, Huang KC, Idoyaga J, Kim SK, Krasnow M, Kuo CS, Nguyen P, Quake SR, Rando TA, Red-Horse K, Reiter J, Relman DA, Sonnenburg JL, Wang B, Wu A, Wu SM, Wyss-Coray T (2022) The Tabula Sapiens: a multiple-organ, single-cell transcriptomic atlas of humans. Science 376:eabl4896

Takeda A, Hollmén M, Dermadi D, Pan J, Brulois KF, Kaukonen R, Lönnberg T, Boström P, Koskivuo I, Irjala H, Miyasaka M, Salmi M, Butcher EC, Jalkanen S (2019) Single-cell survey of human lymphatics unveils marked endothelial cell heterogeneity and mechanisms of homing for neutrophils. Immunity 51:561–572

Travaglini KJ, Nabhan AN, Penland L, Sinha R, Gillich A, Sit RV, Chang S, Conley SD, Mori Y, Seita J, Berry GJ, Shrager JB, Metzger RJ, Kuo CS, Neff N, Weissman IL, Quake SR, Krasnow MA (2020) A molecular cell atlas of the human lung from single-cell RNA sequencing. Nature 587:619–625

Tse D, Stan RV (2010) Morphological heterogeneity of endothelium. Semin Thromb Hemost 36: 236–245

Vermette D, Hu P, Canarie MF, Funaro M, Glover J, Pierce RW (2018) Tight junction structure, function, and assessment in the critically ill: a systematic review. Intensive Care Med Exp 6:37

Walch JM, Lakkis FG (2014) T-cell migration to vascularized organ allografts. Curr Opin Organ Transplant 19:28–32

Wang Z, Gerstein M, Snyder M (2009) RNA-Seq: a revolutionary tool for transcriptomics. Nat Rev Genet 10

Wang L, Li J, He S, Liu Y, Chen H, He S, Yin M, Zou D, Chen S, Luo T, Yu X, Wan X, Huang S, Guo Z, He X (2021) Resolving the graft ischemia-reperfusion injury during liver transplantation at the single cell resolution. Cell Death and Disease 12

Watson RL, Buck J, Levin LR, Winger RC, Wang J, Arase H, Muller WA (2015) Endothelial CD99 signals through soluble adenylyl cyclase and PKA to regulate leukocyte transendothelial migration. J Exp Med 212:1021–1041

Waxman AB, Mahboubi K, Knickelbein RG, Mantell LL, Manzo N, Pober JS, Elias JA (2003) Interleukin-11 and interleukin-6 protect cultured human endothelial cells from H2O2-induced cell death. Am J Respir Cell Mol Biol 29:513–522

Webb LM, Ehrengruber MU, Clark-Lewis I, Baggiolini M, Rot A (1993) Binding to heparan sulfate or heparin enhances neutrophil responsnes to interlukin 8. Proc Natl Acad Sci USA 90:7158–7162

Weibel ER, Palade GE (1964) New cytoplasmic components in arterial endothelia. J Cell Biol 23: 101–112

Wesley BT, Ross ADB, Muraro D, Miao Z, Saxton S, Tomaz RA, Morell CM, Ridley K, Zacharis ED, Petrus-Reurer S, Kraiczy J, Mahbubani KT, Brown S, Garcia-Bernardo J, Alsinet C, Gaffney D, Horsfall D, Tysoe OC, Botting RA, Stephenson E, Popescu D-M, Macparland S, Bader G, Mcgilvray ID, Ortmann D, Sampaziotis F, Saeb-Parsy K, Haniffa M, Stevens KR, Zilbauer M, Teichmann SA, Vallier L (2022) Single-cell atlas of human liver development reveals pathways directing hepatic cell fates. Nat Cell Biol 24:1487–1498

Winkler EA, Kim CN, Ross JM, Garcia JH, Gil E, Oh I, Chen LQ, Wu D, Catapano JS, Raygor K, Narsinh K, Kim H, Weinsheimer S, Cooke DL, Walcott BP, Lawton MT, Gupta N, Zlokovic BV, Chang EF, Abla AA, Lim DA, Nowakowski TJ (2022) A single-cell atlas of the normal and malformed human brain vasculature. Science 375

Wu H, Malone AF, Donnelly EL, Kirita Y, Uchimura K, Ramakrishnan SM, Gaut JP, Humphreys BD (2018) Single-cell transcriptomics of a human kidney allograft biopsy specimen defines a diverse inflammatory response. J Am Soc Nephrol 29:2069–2080

Xie CB, Qin L, Li G, Fang C, Kirkiles-Smith NC, Tellides G, Pober JS, Jane-Wit D (2019) Complement membrane attack complexes assemble NLRP3 Inflammasomes triggering IL-1 activation of IFN-gamma-primed human endothelium. Circ Res 124:1747–1759

Xie CB, Jiang B, Qin L, Tellides G, Kirkiles-Smith NC, Jane-Wit D, Pober JS (2020) Complement-activated interferon-gamma-primed human endothelium transpresents interleukin-15 to CD8+ T cells. J Clin Invest 130:3437–3452

Yamakuchi M, Kirkiles-Smith NC, Ferlito M, Cameron SJ, Bao C, Fox-Talbot K, Wasowska BA, Baldwin WM 3rd, Pober JS, Lowenstein CJ (2007) Antibody to human leukocyte antigen triggers endothelial exocytosis. Proc Natl Acad Sci USA 104:1301–1306

Yoshizumi M, Perrella M, Burnett JC Jr, Lee ME (1993) Tumor necrosis factor downregulates an endothelial nitic oxide synthase mRNA by shortening its half-life. Circ Res 73:205–209

Young MD, Mitchell TJ, Vieira Braga FA, Tran MGB, Stewart BJ, Ferdinand JR, Collord G, Botting RA, Popescu D-M, Loudon KW, Vento-Tormo R, Stephenson E, Cagan A, Farndon SJ, Del Castillo Velasco-Herrera M, Guzzo C, Richoz N, Mamanova L, Aho T, Armitage JN, Riddick ACP, Mushtaq I, Farrell S, Rampling D, Nicholson J, Filby A, Burge J, Lisgo S, Maxwell PH, Lindsay S, Warren AY, Stewart GD, Sebire N, Coleman N, Haniffa M, Teichmann SA, Clatworthy M, Behjati S (2018) Single-cell transcriptomes from human kidneys reveal the cellular identity of renal tumors. Science 361:594–599

Yuan W, Xia H, Xu Y, Xu C, Chen N, Shao C, Dai Z, Chen R, Tao A (2022) The role of ferroptosis in endothelial cell dysfunction. Cell Cycle 21:1897–1914

Zhang Q, Lakkis FG (2015) Memory T cell migration. Front Immunol 6:504

Zou Z, Long X, Zhao Q, Zheng Y, Song M, Ma S, Jing Y, Wang S, He Y, Esteban CR, Yu N, Huang J, Chan P, Chen T, Izpisua Belmonte JC, Zhang W, Qu J, Liu GH (2021) A single-cell transcriptomic atlas of human skin aging. Dev Cell 56:383–397

Chapter 7
The Cerebral Microcirculation

Ronald F. Tuma, Sean Salehi, and Ming Zhang

Abstract Under normal conditions, blood flow and pressure in the cerebral microcirculation are precisely controlled to match the needs of the brain. Blood vessels within the cerebral microcirculation are subject to a variety of control mechanisms. This chapter reviews both control mechanisms shared by other organ systems and those unique to the brain. Included among the control mechanisms reviewed are metabolic control, myogenic control, neural control, endothelial control and humoral control. The importance of communication among cells comprising the neurovascular unit in the regulation of cerebral blood flow is reviewed. The participation of astrocytes in cerebral vascular control is emphasized. The participation of cerebral microvessels in inflammatory responses is also reviewed. The unique properties of the glymphatic system are reviewed. The unique aspects of inflammatory responses in the brain are also reviewed.

Keywords Neurovascular unit · Blood–brain barrier · Astrocytes · Microglia · Pericytes · Glymphatic system · Glial cells

7.1 Historical Background

Understanding the brain and its function is an ancient goal. The observation that there are "pulsations of the brain during life" is published in the Smith Papyrus, one of the earliest written medical texts (seventh century BC). One of the complicating factors in the development of our understanding of the cerebral circulation was that

R. F. Tuma (✉)
Emeritus Professor of Physiology and Neurosurgery, Lewis Katz School of Medicine, Temple University, Philadelphia, PA, USA
e-mail: ronald.tuma@temple.edu

S. Salehi
Chicago Neurospine Surgery, Mokena, IL, USA

M. Zhang
Department of Pathology, Christiana Care Hospital, Newark, DE, USA

R. F. Tuma et al. (eds.), *Handbook of Microcirculation*, Physiology in Health and Disease, https://doi.org/10.1007/978-3-031-99609-2_7

the ancient Egyptians and early Greeks did not differentiate among tendons, blood vessels and nerves. The words neuron and neurology are derived from a Greek word whose original meaning was tendon. Some interesting insights were provided despite this confusion. In the sixth century BC, Pythagoras characterized the brain as the organ of reasoning. One of his pupils, Alcmaeon of Croton, recognized the movement of blood in the veins and the importance of blood for mental function. He hypothesized that sleep was caused by the "retreat" of blood from the brain into the great veins and that death occurred when this retreat was permanent. Unfortunately, these concepts were not perpetuated by Hippocrates and Aristotle, who believed that the brain was a large gland that functioned to cool and purify the blood. The brain was thought to release excess fluids and other materials by nasal discharge.

Galen (129–199), one of the most influential medical theoreticians, was unfortunately also responsible for the introduction of numerous erroneous ideas. He did, however, provide some interesting additional insights into the operation and function of the brain. These concepts dominated physiologic thought for the next millennium and a half. Galen presented the idea that the brain was nourished by both arterial and venous blood. He thought that venous blood originated in the liver and intestines and provided a "natural spirit" to the brain. According to his theory, the arteries delivered blood containing a "vital spirit" that came from air brought in through the trachea and transported to the lungs, pulmonary veins and the right ventricle, where it mixed with the blood. The brain then formed an "animal spirit" which was distributed to all of the muscles via the hollow nerves. The delivery or restriction of delivery of this animal spirit was responsible for controlling contraction of the muscles.

Progress was made during the Renaissance in clarifying the anatomic structure of the cerebral circulation. It is interesting that although he had even less insight into cardiovascular physiology than Galen, Leonardo da Vinci recognized that cervical compression could produce unconsciousness or irreversible damage if the compression lasted more than "the hundredth part of an hour" (Kety 1964).

Today, we recognize that precise homeostatic control of the interstitial fluid is absolutely critical for normal function of the central nervous system. The importance of this goal is reflected in the unique priority given to regulation of blood flow and permeability of blood vessels within the brain and spinal cord. Both the morphology and function of the CNS microcirculation are uniquely adapted to this objective.

An important change in our view of cerebral microvascular function occurred following the published introduction of the neurovascular unit as a conceptual framework in 2001 (Schaeffer and Iadecola 2021). Rather than considering the function of each cell type comprising the brain in isolation, the conceptual framework of the neurovascular unit recognizes that all of the various cell types in the brain are in intimate contact and have mutual influence on function. Included among the cell types forming the neurovascular unit are neurons, astrocytes, microglia, endothelial cells, pericytes and smooth muscle cells. The activity of the neurovascular unit coordinates numerous functions including blood–brain barrier function, metabolic regulation of blood flow, neurotransmission, synaptic plasticity and responses to ischemia and trauma. Single-cell transcriptomics have provided evidence of the segmental adapted diversity shown by components of the neurovascular unit (Schaeffer and Iadecola 2021).

7.2 Architecture

Specialized morphologic adaptations are evident when examining the unique way the arterial macrovessels interact in the regulation of vascular resistance in the central nervous system. In most organs of the body, the primary site of control of vascular resistance resides at the level of the arterioles within the microcirculation. In the central nervous system, large arterial vessels contribute much more to the control of vascular resistance. By the time blood enters intracranial arteries on the surface of the brain, vessels approximately 200 μm in diameter, pressure has decreased by over 50% (Heistad 2001). Because the cerebral arterioles are exposed to much smaller transmural pressure than the arterioles in other organs, intracranial arterioles have much thinner walls than those equivalent size vessels in other organs (Baumbach et al. 1988).

Pial arterial vessels course along the surface of the brain, where they contain numerous anastomoses. There are also AV shunts in both gray and white matter. The majority are small, with diameters <12 μm, however, a fewer number of larger shunts with diameters ranging from 12 to 160 μm. These vessels account for approximately 8–15% of cerebral blood flow when measured with 10 μm microspheres and < 2% when measured with 15 μm microspheres (Handbook). Penetrating arterioles branch perpendicularly from the pial arterioles, where they penetrate the parenchyma of the brain. These branches are relatively small in diameter. Penetrating arterioles are surrounded by a perivascular space (Virchow–Robin space) formed by the wall of the arterioles and a leptomeningeal sheath and containing cerebral spinal fluid. The Virchow–Robin space is, in reality, a continuation of the subarachnoid space. This space disappears where the glial limitans is formed by the astrocyte end-feet. The arterioles divide to form capillaries within the brain parenchyma. Capillary density is matched to the metabolic demand of the tissue supplied. Cerebral gray matter has approximately three times the capillary density of cerebral white matter (Woolsey et al. 1996). Within gray matter, capillary density is highest in the areas with the highest metabolic demand. Capillary density is greatest in the areas with highest synaptic density (Farkas and Luiten 2001). This relationship is reflected in the relative density of the capillary network of the cortical lamina, with lamina IV having the highest and lamina I having the next highest density (Hudetz 1997).

Only those capillaries on or near the surface of the brain are amenable to direct visualization by epifluorescence microscopy. Although confocal microscopy allows for visualization a greater depth, this technique is much more restricted in the number of vessels that can be studied simultaneously. Capillaries on the surface of the brain are of highly variable length and orientation. As a result, there are wide variations in transit times and path lengths in the cerebral capillary network (Hudetz

1997). Because capillaries within the parenchyma of the brain most often are branches of penetrating arterioles it is extremely difficulty to observe their origin. The degree of anastomoses among these capillaries varies with the species of animal. There is disaccord in reports of the degree to which the capillaries on the surface are perfused with flowing blood (Keyeux et al. 1995). Recently, most investigators have reported that, with the exception of random cessations of flow lasting only a few seconds, capillaries on the surface of the brain are continuously perfused with flowing blood (Seylaz et al. 1999; Pinard et al. 2000). Red blood cell velocity has been reported to range between 0.3 and 3.2 mm/s, with the majority of capillaries having red cell velocities between 0.4 and 1.8 mm/s (Hudetz 1997). Consideration must be given, however, to the effects of trauma and inflammation due to surgical preparation on perfused capillary density and red cell velocity. These effects have led to overestimation of capillary perfusion under resting conditions in other organs.

The presence of a glycocalyx, estimated to be approximately 90 nm thick, contributes resistance to blood flow through the capillary network. Alterations in arterial resistance normally compensate for this resistance, keeping flow closely matched to tissue need. Under conditions of maximal dilation, removal of the glycocalyx can increase cerebral blood flow by 30% (Vogel et al. 2000).

As is the case with other tissues, there is significant flux of oxygen and carbon dioxide across the walls of precapillary cerebral blood vessels. These losses are somewhat less than those seen in other tissues. Duling et al. (1979) measured a longitudinal decrease in intravascular oxygen tension from a mean of 98.5 mm Hg in first-order vessels (200–250 μm in diameter) down to less than 72.6 mm Hg in fourth-order arterioles (25 μm in diameter). The authors also point out that there should be a comparable flux of CO_2 across these vessels, which would contribute to elevation of CO_2 in the arterial microvessels and to the regulation of cerebral blood flow. The magnitude of this shunting is influenced by the velocity of the blood which provides a partial explanation for the enhancement of oxygen delivery to the brain following isovolemic hemodilution (Harrison 1989; Tuma et al. 1989; Lin et al. 1995; Hudetz et al. 1999).

The classical lymphatic architecture, with lymphatic vessels arranged in parallel with arterial and venous microvessels and terminating in lymphatic bulb found in other organs is not present in the brain. As a result, the central nervous system was described for many years as being immune privileged. It is now recognized, however, that other structures subserve this function. The glymphatic system includes a brain-wide network of perivascular pathways surrounding cerebral microvessels (Laman and Weller 2013; Sun et al. 2018). Brain interstitial fluid drains through the basement membrane surrounding arterioles and capillaries, finally emptying into cervical lymph nodes after passing along the walls of the carotid arteries. The restricted diameter of this pathway precluded the transport of dendritic cells. The rate of fluid movement and accumulation appears to be dynamically controlled by Aquaporin-4 through up and down regulation of these pores (Szczygielski et al. 2021). Aquaporin-4 function is responsive to both physiologic and pathophysiological conditions in the brain.

7.3 Blood–Brain Barrier

The first recorded observations of special permeability characteristics of brain microvessels were made by Biedl and Kraus in 1898 when they reported that bile acids were not neurotoxic when injected intravenously, but were neurotoxic when injected directly into the brain. A similar observation was made by Lewandowsky 2 years later, leading these investigators to postulate that the brain vessels had special permeability properties (Goldstein and Betz 1986). The concept of the existence of a "blood–brain barrier" is commonly attributed to studies conducted by Paul Erlich more than a century ago. Erlich found that some vital dyes, when injected into the vascular system were dispersed throughout almost all the tissues of the body. The one noteworthy exception was the brain. Erlich incorrectly interpreted this finding to be the result of a lack of affinity of brain tissue for the dye. Goldman, one of Erlich's students corrected this misconception by demonstrating, in later experiments, that when the dye was directly injected into the cerebral spinal fluid the brain was stained by the dye and did not stain other tissues of the body (Goldman 1909, 1913). These findings led to the use of the term blood–brain barrier to describe the extremely restrictive permeability of the cerebral blood vessels. It was not until the advent of the electron microscope, allowing for more detailed examination of the morphology of the cerebral microvessels, that the endothelial cells were verified to be the location of the blood–brain barrier. In a classic study, Reese and Krnovsky (1967) demonstrated that horseradish peroxides that was administered by intravenous injection was localized in the lumina of the cerebral microvessels and in vesicle within the endothelial cells but did not pass beyond the endothelium. The endothelial cell membrane has definite polarity, segregating proteins that form receptors and transporters between the luminal and abluminal sides of the membranes (Bundgaard 1986). Although the endothelial cells form the primary barrier to exchange, other components of the vessels wall play a critical role in the structure and function of the blood–brain barrier. Included in these important structures are the basal lamina, pericytes and astrocyte foot processes. Some investigators include neurons, in addition to the above mention structures, as components of a "functional neurovascular unit" that regulates blood–brain barrier permeability (Persidsky 1999).

There are a number of characteristics that distinguish cerebral endothelial cells from endothelial cells in other regions of the body. Specialized junctions between adjacent endothelial cells are responsible the selectivity of the blood–brain barrier to diffusion of charged molecules. These specialized junctions are formed by a junctional complex including both tight junctions and adherens junctions. This specialized complex creates the ultrastructural appearance of fusion of the outer leaflets of adjacent endothelial cells (Ballabh et al. 2004). The tight junctions are formed from three types of protein, claudins, occludins and junctional adhesion molecules. The primary seal of the tight junctions is formed by the binding of claudins in an endothelial cell membrane to claudins in the adjacent endothelial cell membrane (Nag 2003). Within the endothelial cell, claudins are connected to cytoskeletal actin.

Occludins, located in the same region as the claudins, are only found in the cerebral endothelial cells. The occludins are regulatory proteins that contribute to the increased electrical resistance across the blood–brain barrier, which can be as much as 40 times greater that the electrical resistance across endothelial cells in other regions (Furuse et al. 1999). Phosphorylation sites on occludin serine and threonine residues have a regulatory effect on barrier permeability (Sakakibara et al. 1997; Hirase et al. 2001). A group of accessory molecules (ZO-1 to 3, AF-6, 7H6 and cingulin) provide structural support and connections for claudins and occludins to structural molecules in the endothelial cells. Also located in the region of the tight junctions are several cytoplasmic signaling molecules that regulate the integrity of the tight junctions (Madara et al. 1992). It appears that claudins are the primary structural component of the tight junction and occludins enhance the tightness of the barrier (Persidsky 1999). The third component of the tight junction, the junctional adhesion molecules are members of the immunoglobulin superfamily with a single transmembrane domain. The junctional adhesion molecules interact with PECAM to regulate white blood cell migration (Martin-Padura et al. 1998).

The adherens junction component of the blood–brain barrier, located on the abluminal side of the endothelial cells, is formed from E, P and N cadherins. The cadherins are glycoproteins whose extracellular domain binds to the extracellular domain of cadherins in adjacent endothelial cells. This binding is calcium-dependent. The cadherins have an intercellular connection to α and γ catenins. The α and γ catenins bind to α, catenin which attaches to actin and vinculin (Gumbiner 1996). The mutual binding to actin by both the tight junctions and adherens junctions allows for interaction between these components of the blood–brain barrier (Matterand Balda 3003).

An additional factor contributing to the restrained movement of materials across the cerebral capillaries is the relative paucity of vesicles in cerebral endothelial cells. The density of vesicles in the cerebral endothelium is an order of magnitude less than the density in the endothelium in other organs (Stewart et al. 1987). Cerebral endothelium lack both fenestrations and transendothelial channels. Acute increases in blood pressure produce an increase in the number of pinocytotic vesicles in cerebral endothelium and a disruption of the blood–brain barrier (Mayhan 2001). The precise contribution of increased vesicular transport during hypertension to increased permeability of the cerebral vessels remains to be determined. It has also been demonstrated that focused ultrasound can induce increased vesicular transport in arteriolar microvessels and that this is associated with disruption of the blood–brain barrier (Sheikov et al. 2006).

The permeability of tight junctions is subject to regulation by cell signaling pathways. Alterations of intracellular free calcium concentration influences both signaling pathways that alter tight junction expression and distribution (Brown and Davis 2002). As mentioned previously, phosphorylation of both the transmembrane and associated intracellular proteins can alter tight junction function. The state of phosphorylation of these proteins is influenced by intercellular signaling molecules including inflammatory cytokines and growth factors (Persidsky 1999). G protein-coupled receptors provide an important pathway for regulatory modification of tight junctional permeability and influence the diapedesis of leukocytes into the brain (Adamson et al. 1999). Regulation of blood–brain barrier permeability occurs through the Rho signaling pathway by modification of cytoskeletal proteins. Initiation of this signaling pathway can be stimulated by cytokines and prostaglandins (Schmeck et al. 2003; Stamatovic et al. 2006). MCP-1 has been shown to be one of the important chemokines that activate this process (Stamatovic et al. 2005). Rab and Ras activation has also been shown to have influence on blood–brain barrier permeability (Stevenson 1999; Yamamoto et al. 1999).

Mitochondrial function may also be an important component of blood–brain barrier integrity, by providing the energy necessary to maintain ionic gradients across the blood–brain barrier. The density of mitochondria in the cerebral endothelium is significantly greater than that found in the endothelium of other tissues (Nag 2003).

Although the morphology of endothelial cells is uniform along vascular segments (Schaeffer and Iadecola 2021), there is considerable heterogeneity of function along these segments. This heterogeneity of is reflected by variations at the level of transcriptomes (Vanlandewijck et al. 2018).

Astrocytic end feet surround almost all of the brain capillary basement membranes. Cell culture studies have been reported to provide evidence that the development of the blood–brain barrier is not the result of intrinsic properties on the endothelial cells but were instead dependent upon communication between endothelial cells and astrocytes (Janzer and Raff 1987; Neuhaus et al. 1991; Rubin 1991). These results have been questioned by other investigators (Holash et al. 1993; Krum et al. 1997). While the precise role of and mechanism through which astrocytes contribute to the development and maintenance of the blood–brain barrier remain to be determined it is clear that there is a bidirectional communication between astrocytes and endothelial cells. Among the factors produced by astrocytes that can influence endothelial cells are TGFβ, glial cell-derived neurotropic growth factor, IL-6 and basic fibroblast growth factor (Sobue et al. 1999; Mi et al. 2001; Yoder 2002). Endothelial cell production of leukemia-inhibiting factor appears to influence astrocyte differentiation (Mi et al. 2001). The importance of astrocytes in the maintenance of blood–brain barrier function is highlighted by the fact that pial vessels, which have less interaction with astrocytes, have higher permeability than vessels located within the parenchyma (Abbott 2005). As will be discussed in a later section, astrocytes also have important influence on the control of vascular diameter and cerebral blood flow. There is accumulating evidence that in addition to astrocytes, other cells that form the neurovascular unit may contribute to maintenance of barrier function by brain endothelial cells. Pericytes are reported to be involved in the regulation of endothelial-astrocyte interactions (Ramsauer et al. 2002). Microglia have also been reported to modulate endothelial cell permeability (Abbott 2005).

Endothelial cells, pericytes and astrocytes are separated by a basement membrane synthesized by both endothelial cells and astrocytes. While the endothelial cells serve as the primary barrier to charged molecules, the basement membrane is also negatively charged and therefore forms a charge barrier between the blood and brain interstitial fluid, provides physical support, and regulates cell adhesion and migration

(Perlmutter and Chui 1990; Vorbrodt 1993). As in other vascular beds, the basement membrane surrounding the cerebral capillaries can be divided into three regions: the lamina rara externa, lamina densa and lamina rara interna (Farkas and Luiten 2001). The lamina densa, composed primarily of Collagen type IV, provides the major structural support provided by the basal lamina. Contained within the lamina rara are laminin and heparin sulfate proteoglycan, which promote cell adhesion (Perlmutter and Chui 1990). The basal lamina also serves as a reservoir for growth factors proteases, and protease inhibitors (Timpl 1994). In addition to serving as a physical anchor through the interaction of matrix proteins and integrins on endothelial cells, the cell interactions with the basement membrane can serve a signaling function and regulate expression of tight junction proteins (Savettieri et al. 2000; Tilling et al. 2002).

Although the contributions of the endothelial cells to restricting movement of materials from the vascular to the extravascular space is crucial for homeostasis, the cerebral endothelial cells can also play a critical role in signaling immune cells and facilitating their movement across the vessel wall. Because the brain is traditionally described as being "immune privileged" there is commonly a misunderstanding about the intricate immunological responses that occur under a variety of conditions. The more descriptive term of "immune modified" might provide a more accurate representation of the activities of the immune system in this area. Unique aspects of the inflammatory response in the CNS are the result of the properties of the blood–brain barrier and the absence of fully developed lymphatic antigen presenting sites in the CNS (Ransohoff 1999). As part of this intricate control of immune responses, cerebral endothelial cells can be stimulated to present a variety of adhesion molecules that promote rolling, adhesion and diapedesis of neutrophils, monocytes and lymphocytes. Included among the adhesion molecules expressed by cerebral endothelial cells are ICAM-1, VCAM-1, PECAM-1, E-selectin and P-selectin (Yang et al. 2005). Expression of these adhesion molecules is regulated by production of cytokines and chemokines by the brain (Wong and Dorovini-Zis 1992; Wong and Dorovini-Zis 1995; Easton and Dorovini-Zis 2001). Movement of leukocytes across the cerebral microvessels occurs through paracellular channels (Schenkel et al. 2004; Cullere et al. 2005). Passage through paracellular channels requires modification of tight junctions and adherens junctions. Changes in free intracellular Ca concentrations in response to white cell adhesion activate Rho GTPases, including Rac1, to attenuate adhesion between microvascular endothelial cells (Muller 2003; Burridge and Wennerberg 2004; Wittchen et al. 2005a; b). Rac1-mediated generation of ROS appears to mediate a decrease in adhesion forces under a number of pathologic conditions (Cook-Mills 2002; van Buul and Hordijk 2004; Wittchen et al. 2005a, b). Rho activation is also important in adhesion molecule expression on the surface of endothelial cells (Wojciak-Stothard et al. 1999).

The permeability restrictions imposed by the blood–brain barrier creates the need for transporters of essential molecules to which the endothelial cells are not permeable. Included among these molecules are glucose and amino acids that cannot be synthesized by the brain. Experiments conducted by Crone et al. demonstrated the extreme specificity of these transporters by showing that D-glucose is readily transported but that L-glucose is excluded from transport into the brain. He was the first to postulate that these transporters are located on the endothelial cells. There is an asymmetric distribution of transporters between the luminal and abluminal endothelial cell membranes. While transporters for large neutral amino acids are found on both membranes, sodium coupled active transporters for glycine, which can act as an inhibitory neurotransmitter, are found on the abluminal, but not the luminal endothelial cell membranes (Goldstein and Betz 1986).This group also demonstrated that potassium is actively transported into the cerebral endothelial cells where it diffuses down a concentration gradient into the blood.

In addition to transport of materials, metabolic activity of the endothelial cells also contributes to the control of distribution of materials across the blood–brain barrier. Conversion of substances, within the endothelial cells into compounds that cannot cross the cell membrane prevents them from entering the brain. Bertler et al. demonstrated this process in the handling of L-dopa. L-dopa is brought into the endothelial cell by large neutral amino acid transporters. However, upon entry into the endothelial cells, the L-dopa is enzymatically converted to, dopamine and DOPAC which are not permeable across the antiluminal membrane, effectively limiting transport of L-dopa into the brain.

The special properties of the blood–brain barrier are not only highly significant for homeostasis, but also have significant impact on drug delivery. This highly restrictive barrier can be of significant importance in determining the therapeutic effect of drugs whose delivery to the brain might or might not be desirable. Modulation of the blood–brain barrier by injection of hyperosmotic solutions into the carotid artery has been examined as a clinical strategy to allow entry of drugs that are normally excluded from the brain (Rapoport 2000, 2001). Other strategies for opening the barrier have also been explored (Doolittle et al. 2000; Kraemer et al. 2002). While it is expected that hydrophilic drugs would be excluded from delivery to the CNS, there are also numerous examples of lipophilic drugs that are also excluded (Golden and Pollack 2003). These compounds are excluded from the CNS by drug efflux transporters including P glycoprotein (Pgp), multidrug resistance proteins (MRPs) and the breast cancer resistance protein (BCRP) (Loscher and Potschka 2005a, b). These transporters have a wide variety of substrates, leading to the term "multidrug transporters" (Loscher and Potschka 2005a, b). The ability of these transporters to extrude drugs has obvious clinical implications for the treatment of CNS diseases. The majority of these transporters are members of the superfamily of ATP-binding cassette (ABC) proteins (Persidsky et al. 2006). The first member of this group of proteins to be extensively studied is the efflux transporter P-gp (Fricker and Miller 2004). Type 1 Pgp, the isoform found in human cerebral endothelial cells, is encoded by the MDR1 gene. It appears that Pgp may create a significant impediment to the delivery of drugs used to treat HIV and epilepsy (Persidsky et al. 2006). This transporter is predominantly located on the luminal side of brain endothelial cells in numerous mammalian species including primates, rats, mice cattle and pigs, allowing substrates for this transporter to be pumped back into the blood upon entry into the endothelial cells (Demeule et al. 2002; Schinkel and Jonker 2003; Sun et al. 2003). The activity of this transporter and changes associated with a number of disease processes, including ischemia, Alzheimer's disease, HIV, epilepsy, Parkinson's disease, and schizophrenia, are currently the subject of intense investigation (Marroni et al. 2003; Cirrito et al. 2005; Kortekaas et al. 2005; Loscher and Potschka 2005a, b; Volk et al. 2005; Spudich et al. 2006). P-gp is present not only on endothelial cells, but also on astrocytes and microglia (Lee and Bendayan 2004). In rodents, Type 1 Pgp is encoded by two genes (mdr1a and mdr1b) (Loscher and Potschka 2005a, b). Mdr1a is located in brain capillaries, whereas mdr1b is located primarily in brain parenchyma (Demeule et al. 2002).

The contribution of pericytes to the integrity and function of the blood–brain barrier is currently an area of intense interest. Pericytes are found to be present in almost all organs of the body, where they have been studied more extensively than they have in the brain. Unfortunately there is wide variability in the usage of this term. Some authors include all perivascular cells (Thomas 1999; Sims 2000), while others restrict this term to include cells surrounding capillaries and postcapillary venules. Traditionally, they are described as being enclosed in the vascular basal lamina and closely opposed to the abluminal side of endothelial cells (Bar and Budi Santoso 1984; Thomas 1999). Other authors have described a subpopulation of pericytes that are not completely surrounded by the basement membrane. The terms perivascular cell (50,72) (Graeber et al. 1992; Kida et al. 1993), perivascular microglia (Hickey et al. 1992) and fluorescent granular perithelial cells (Ookawara et al. 1996) have been used to describe these cells. Part of the difficulty in distinguishing among these cell types may be their common origin. There is general agreement that pericytes are of mesodermal origin (Sims 1986, 1991; Diaz-Flores et al. 1991). Macrophages circulating in the blood can transmigrate into the perivascular spaces to form perivascular macrophages. The rapidity and magnitude of this process is also variable and regulated.

Pericytes are heterogeneous in their structure and function. The three major functions that have been attributed to CNS pericytes are contractility, regulation of endothelial cell function and immune activity. All pericytes do not share the same functions. There is strong evidence that pericyte contraction can reduce capillary diameter through activation of purinergic receptors (Horlyck et al. 2021). There is debate as to whether they are involved in activity-induced dilation (Iadecola 2004; Fernandez-Klett et al. 2010; Hall et al. 2014; Hill et al. 2015) (Iadecola 2004). Based upon dynamic studies, Hartmann et al. (2021) concluded that capillary pericytes contract more slowly than mural cells on upstream arterioles and that they probably contribute more to regulation of basal perfusion and slow modulation of blood flow than upstream resistance (Heistad and Kontos 1983) and vessels (Hartmann et al. 2021) . The intimate contact between pericytes and endothelial cells allows for close communication between these two cell types. The pericytes have been reported, under some circumstances, to inhibit endothelial proliferation and to stabilize developing microvessels (Crocker et al. 1970; Shepro and Morel 1993; Hirschi and D'Amore 1996; Balabanov and Dore-Duffy 1998; Rucker et al. 2000). With proper stimulation, pericytes may, however, produce growth factors that stimulate endothelial proliferation (Folkman and D'Amore 1996; Takagi et al. 1996; Maisonpierre et al. 1997; Watanabe et al. 1997). The communication between perictyes and endothelial cells is bidirectional. Endothelial cells also have significant influence on pericyte function. It has been suggested that pericytes are involved in controlling endothelial permeability (Shepro and Morel 1993). A fraction of pericytes serve the function of macrophages in the central nervous system. One characteristic of macrophages shared by some pericytes is the presence of lysosomes and inclusion bodies (Murabe and Sano 1981; Mato et al. 1986; Farrell et al. 1987; Ookawara et al. 1996). Many pericytes have also been shown, by immunohistochemical staining, to contain numerous macrophage markers. The expression of these components varies with the conditions under which the cells are studied. Like macrophages, pericytes have been demonstrated to participate in phagocytic activity. This activity may be a component of blood–brain barrier function. Pericytes may also play an important role in the regulation of leukocyte transmigration across cerebral microvessels by releasing a number of immunoregulatory cytokines (Fabry et al. 1993). They play a very important role, as antigen presenting cells, in regulating lymphocyte movement into the brain.

When considering the BBB, it must be recognized that the transport of material into and out of the cerebral interstitial fluid depends upon the activity of two distinct barriers, the blood–brain barrier and the blood–cerebral spinal fluid barrier. The concentration of molecules in the interstitial fluid can be significantly different from those of the brain interstitial fluid.

The permeability of the blood–brain barrier is subject to modification by a number of vasoactive factors (Table 3 Abbott 2005) (Abbott 2000). Although there is some evidence of an increase in transcytosis in response to inflammatory vasoactive mediators (Joo 1993), the majority of studies support changes in permeability through tight junctions (Abbott and Revest 1991). There may be heterogeneity of responsiveness of cerebral vessels to vasoactive mediators. Pial vessels have both a higher baseline permeability and greater reactivity to modulators of permeability (Allt and Lawrenson 1997; Lawrenson et al. 1997). Although less responsive to inflammatory mediators than other vascular beds, the cerebral endothelial cells are, under appropriate circumstances, responsive to these molecules and permit inflammatory reactions to contribute to repair and healing. The ability of numerous inflammatory mediators to increase blood–brain barrier permeability is again a reflection of the fact that the CNS is immune modified rather than immune privileged, and that controlled inflammation may be of benefit under a number of circumstances (Perry et al. 1997; Hickey 2001). Even under normal conditions, a small percentage of brain microvessels ($<5\%$) may undergo tight junction opening (Plumb et al. 2002). This finding has led to the speculation that the cerebral vessels may normally undergo punctuate opening of the blood–brain barrier (Abbott 2005). From another perspective, the limited responsiveness of the cerebrovascular bed to inflammatory mediators may help to provide protection against white cell invasion and the resultant damage to neuronal structures.

A number of permeability-modifying agents have their effect through change in free calcium concentration in endothelial cells. Bradykinin, through stimulation of B_2 receptors, is able to induce mild increases in the permeability of the blood–brain barrier. Stimulation of the B_2 receptor has been reported to cause elevation on intracellular free calcium concentration, activation of phospholipase A_2 and phospholipase C, cause the release of arachidonic acid and initiate the production of free radicals (Abbott 2000). Bradykinin levels in the brain have been shown to increase following ischemia and cryoinjury (Maier-Hauff et al. 1984; Wahl et al. 1988). The increased permeability resulting from bradykinin production following ischemia has been demonstrated to be enhanced by IL-1β (Revest et al. 1991). Serotonin (5HT) is also elevated following stress and damage to the central nervous system (Sharma and Dey 1986a, b). As is the case with bradykinin, 5HT operates through elevation of intracellular free calcium in endothelial cells. The sources of 5HT include platelets, mast cells, endothelial cells and neurons (Abbott 2000). Histamine, released from histaminegic neurons, along with other sources, can increase blood–brain barrier permeability (Schilling and Wahl 1994). This effect is also associated with elevation of intracellular free calcium following stimulation of H_2 receptors when histamine is present in nanomolar concentrations. When the concentration of histamine increases into the micromolar range, H_1 receptor stimulation can reduce blood–brain barrier permeability (Sarker et al. 1998). While H_2 stimulation leads to an increase in intracellular calcium concentration, H_1 stimulation of receptors appears to cause an increase in cAMP. Elevations in intracellular calcium are also responsible for increased blood–brain barrier permeability following exposure adenosine nucleotides via activation of $P2Y_2$ purinoceptors (Revest et al. 1991; Purkiss et al. 1994; Nobles et al. 1995; Nobles and Abbott 1998).

There are also modifiers of blood–brain barrier that operate independent of changes in calcium concentration. Activation of alpha adrenergic receptors causes elevation of cAMP and activation of protein kinase A, while growth factors are reported to work through the MAPK signaling pathway to modify BBB permeability (Huber et al. 2001; Wolburg and Lippoldt 2002).

7.4 Blood Gases

The sensitivity of cerebral vessels to alterations in blood gasses also interacts with local control mechanisms. Cerebral resistance vessels dilate in response to hypercapnia and severe hypoxemia. These vessels are very responsive to changes in pCO_2. In addition to the direct effect of CO_2 on smooth muscle cells, effects on neuronal nitric oxide is thought to be contributing mechanisms (Iadecola 1992; Faraci et al. 2019). The cerebral vessels are less sensitive to changes in blood pO_2. Hypoxia does not increase cerebral blood flow until pO_2 is less than 50 mmHg (Lewis et al. 2014). This imparts less influence by changes in pO_2 under normal physiologic conditions.

7.5 Metabolic Control

Roy and Sherrington, in 1890 postulated that the chemical products of metabolism can cause variations in the caliber of cerebral blood vessels (Roy and Sherrington 1890). Although the precise mechanisms contributing to this "metabolic control" mechanism has been the subject of continued debate since the publication of this classic paper, it is clear that there is a precise and highly localized coupling between the metabolic requirements of the brain parenchyma and the magnitude of blood flow. This control mechanism predominates under normal conditions. Precise coupling is essential, since neurons rely almost exclusively on aerobic glycolysis and reductions in flow by as little as 40% can produce significant alterations in cerebral function. With localized increased metabolic activity, increased blood flow is limited to an intraparenchymal radii of approximately 250 μm around the active area.

The cerebral resistance vessels are responsive to a number of metabolic byproducts. Increasing interstitial fluid concentration of hydrogen ions, carbon dioxide, potassium and adenosine and decreasing the partial pressure of oxygen all result in a relaxation of vascular smooth muscle and dilation of the cerebral resistance vessels. There is continuing debate as to the relative importance of each of these metabolites in the normal regulation of cerebral flow.

A number of investigators have proposed that adenosine is the primary metabolite involved in the long term control of cerebral blood flow (Heistad and Kontos 1983). Adenosine is produced as a byproduct of ATP metabolism when oxygen delivery to the brain is insufficient to maintain energy stores. ATP is metabolized to adenosine which diffuses into the interstitial fluid. Adenosine is thought to stimulate A_2 receptors on smooth muscle cells (Ibayashi et al. 1991). This stimulation causes a reduction in cytosolic calcium vascular smooth muscle, resulting in vasodilation. Use of adenosine antagonists can block hypotension-induced vasodilation, demonstrating its contribution to coupling tissue oxygen demand with blood flow (Laudignon et al. 1991). It has also been shown, using cultured cortical astrocytes, that adenosine can activate purinoreceptors causing the release of NO, which could be an additional mechanism for regulation of cerebral blood flow (Janigro et al. 1996).

While not strictly a byproduct of metabolism, the concentration of potassium in the interstitial fluid of the brain increases when the frequency of neuronal depolarization increases. Potassium loss from the cytosol of neurons is small but causes significant increases in interstitial fluid concentration. Extracellular potassium concentration is regulated by both neuronal and glial mechanisms (Paulson and Newman 1987; Janigro et al. 1997). The extracellular potassium concentration normally ranges between 3 and 5 mM but increases to 10–12 mM with enhanced neuronal activation (Sykova 1983). Increases in interstitial potassium concentration in this range causes hyperpolarization of vascular smooth muscle through an inducible Na^+-K^+ electrogenic pump on the cell membrane and/or stimulation of inward rectifier potassium channels in the vascular smooth muscle of cerebral arteries and arterioles (McCarron and Halpern 1990a, b; Knot et al. 1996; Nguyen et al. 2000; Zaritsky et al. 2000). Inward rectifier potassium channels increase their open state probability when exposed to increases in extracellular potassium concentration in the range produced by increased neuronal activity. Either of these mechanisms can cause hyperpolarization of the vascular smooth muscle cells with a resultant closing of voltage gated calcium channels and relaxation of the muscle. Increases in potassium concentration greater than 15 mM would depolarize smooth muscle cells causing their constriction. However, concentration does not occur in the normal brain (Girouard et al. 2010). Endothelial cells to not appear to be involved with vasodilation in response to an increase in extracellular potassium concentration (Johnson et al. 1998). These investigators also report that inward rectifier potassium channels contribute to the resting tone of cerebral arteries. Inward rectifying potassium channels and sulfonylurea receptors are reported in feline arterioles to form ATP-sensitive potassium channels that require either L-arginine or L-lysine for agonists to open them (Knot et al. 1996).

Astrocytes express big potassium calcium activated channels that can contribute to the efflux of potassium from glial endfeet (Girouard et al. 2010).

Potassium initially received considerable attention because it was felt that its interstitial accumulation might precede that of adenosine and therefore explain the initial hyperemic response to elevations in metabolic rate. Others have however questioned whether the increase in potassium is rapid enough to explain the initial hyperemic response to an increase in metabolism (Lou et al. 1987). Potassium does not remain elevated during hypoxia-induced hyperemia and therefore may be important in initiating an increases in blood flow with other factors contributing to the maintenance of prolonged vasodilation under hypoxic conditions (Zauner et al. 2002).

Increased metabolism of the brain, if not matched by an increase in blood flow, also results in an increase in both CO_2 and hydrogen ion concentration. The prominent vasodilation of the cerebral microvessels in response to hypercapnia has been well recognized for many years. In humans, a 1 mmHg increase in blood pCO_2 produces a 2–4% increase in cerebral blood flow (Farkas and Luiten 2001). It has been demonstrated that CO_2 does not directly influence vascular smooth muscle, but rather operates indirectly through changes in hydrogen ion concentration (Busija and Heistad 1984). Hydrogen ions are unable to cross the blood–brain barrier, but CO_2 which is very lipid soluble, crosses the blood–brain barrier freely. The resultant decrease in pH in the extracellular fluid then causes dilation of the pial arterial vessels. Small pial arterial vessels dilate more than larger vessels, probably because the larger vessels are also influenced by reflex sympathetic stimulation that occurs during hypercapnia (Wei et al. 1980). With chronic hypercapnia or hypocapnia, the bicarbonate concentration of the CSF changes and blunts the effect of altered blood CO_2 on cerebral vessels. It has been hypothesized that the vasodilation caused by an increase in hydrogen ion may involve ATP sensitive potassium channels (Rosenblum 2003). It has also been suggested that increases in cerebral blood flow in response to alterations in pH may be modulated by nNOS production (Lindauer et al. 2001).

Hypoxia also causes dilation of the cerebral microvessels. The direct effects of alterations in blood oxygen tension are frequently misunderstood because hypoxia is often accompanied by hyperventilation. The resultant reduction in CO_2 tends to cause constriction of the cerebral microvessels masking the dilation resulting from hypoxia. When CO_2 is held constant, cerebral blood flow is linearly related to alterations in blood oxygen content. It is unlikely that changes in tissue PO_2 directly dilate vascular smooth muscle. One way in which changes in tissue oxygen tension may influence vascular diameter is through changes in adenosine production, as has been proposed in many tissues. Another mechanism that has been proposed to explain oxygen's effect on vascular diameter is through secondary changes in nitric oxide changes (Demchenko et al. 2002). According to this hypothesis, increases in tissue oxygen tension are associated with an increase in the production of superoxide, which inactivates nitric oxide. This theory is supported by studies of the response of mice that overexpress (+/+) or lack (−/−) extracellular superoxide dismutase. Mice that overexpress extracellular superoxide dismutase show significantly less increase in cerebrovascular resistance in response to increasing PO_2 than mice that do not express this enzyme. The highest activity level of extracellular superoxide dismutase is found in the wall of arterial vessel, where it serves to scavenge superoxide and preserve NO produced by endothelial cells. It appears from these studies that extracellular superoxide dismutase regulates the basal tone of cerebral resistance vessels and their response to increase PO_2. When PaO_2 falls below 60 mmHg, a significant vasodilation is elicited (Gupta et al. 1997). In cats and dogs, normal tissue oxygen tension has been reported to be approximately 28 mmHg. Despite the coupling mechanisms between changes in tissue oxygen tension and vascular resistance, it appears that changes in oxygen tension do not occur rapidly enough to explain the rapidity with which cerebral vascular resistance responds to a change in neuronal activity (Attwell and Iadecola 2002). As discussed below, changes in oxygen tension may influence coupling between astrocytes and smooth muscle cells in the control of vascular resistance.

Arachidonic acid derivatives are another class of compounds that appear to be involved in the regulation of blood flow during hypotension (Leffler and Busija 1987). Further evidence of the participation of prostaglandins is provided by the finding that indomethacin can interfere with cerebral autoregulation during hypotension (Leffler et al. 1987). Glutamate stimulation of astrocytes increases intracellular Ca^{2+}, which activates phospholipase A_2 leading to the production of arachidonic acid. The cyclooxygenase pathway, via COX2, forms vasodilator prostaglandins that contribute to functional hyperemia (Girouard and Iadecola 2006). There is still controversy as to whether calcium signaling in astrocytes precedes the rapid function hyperemia found in the brain (MacVicar and Newman 2015). Epoxyeicosatrienoic acids, formed via the p450 pathway, have also been proposed to contribute to functional hyperemia (Peng et al. 2002). Prostaglandins may also have an influence on NO production, providing an alternative mechanism for the regulation of cerebral blood flow (Najarian et al. 2000).

An explanation for the rapidity with which cerebral blood flow changes in response to an increase in neuronal activity remains a major goal. Fox and Raichle

(1986) studied the effect of somatosensory stimulation on cerebral blood flow, cerebral oxygen consumption and oxygen extraction in humans using PET. These investigators found that cerebral blood flow increased more than cerebral metabolism within 1–2 s after stimulation. The level of pO_2 actually increases rapidly above baseline during repetitive synaptic activity (Offenhauser et al. 2005; Devor et al. 2011). There are also a number of pathophysiologic situations such as hyperthermia and status epilepticus during which cerebral blood flow increases more than the metabolic requirements. Lou et al. (1987) postulated that a neuronal reflex is responsible for the initial vasodilation that accompanies an increase in metabolism. Separate responses are likely responsible for the initial increase in blood flow in response to an increase in metabolism and for the maintenance of flow with a sustained elevation in metabolic rate. As stated previously, adenosine is a likely candidate for the primary factor responsible for the sustained effect.

Conducted vasomotor responses, originating in cerebral penetrating arterioles, may also play a role in the coupling of blood flow to local changes in the metabolism of the brain (Cox et al. 1993; Dietrich et al. 1996; Nguyen et al. 2000; Horiuchi et al. 2002). It has been demonstrated that both ATP and adenosine stimulate conducted vasomotor responses in rats (Dietrich et al. 1996). Horuchi et al. 2002) report that activation of Na^+-K^+-ATPase acts as the generator for conducted vasodilation. It appears that both the endothelium and smooth muscle cells may serve as pathways for the conducted vasodilator response and that the pathway may depend upon the stimulus (Dietrich et al. 1996; Haas and Duling 1997; Bartlett and Segal 2000; Emerson and Segal 2000). Pathologic conditions such as ischemia/reperfusion and hypertension, that lead to endothelial dysfunction can impair local and conducted dilation in response to elevated potassium, interfering with the coupling of blood flow and metabolism in the brain (McCarron and Halpern 1990a, b; Marrelli et al. 1998).

While the mechanisms responsible for the exquisite coupling between metabolic requirements and blood flow are qualitatively similar to those occurring in other organs and tissues, one unique aspect of blood flow regulation in the CNS is derived from the unique contribution of astrocytes. Based upon the anatomic connections between astrocytes and blood vessels, Ramon y Cajal suggested that astrocytes might be involved in the regulation of blood flow (MacVicar and Newman 2015). It is now clear that these cells are intimately involved in intercellular communication among a variety of cell types, and that activation of astrocytes is an important component of numerous regulatory functions. The first speculations about the involvement of astrocytes in the regulation of cerebral blood flow were made primarily on the basis of morphology. Astrocytes processes surround both synapses microvessels and provide for a rapid way to transmit signals from neurons and synapses to the microvasculature. Recently is has been recognized that astrocyte processes have a minimal overlap. This allows each astrocyte to cover a specific area and positions them to couple this area with the appropriate microvessels regulating flow to this region (Volterra and Meldolesi 2005). The territory covered by a single astrocyte can include thousands of synapses (Bushong et al. 2002, 2004). This territory can, however, be subdivided into microdomains for more discrete regulation. The microdomains are formed by astrocytic lamellipodia and filopodia. These structures are mobile and dynamic (5). Microdomain signaling can occur without involving the entire cell (Simard et al. 2003; Hirrlinger et al. 2004; Mulligan and MacVicar 2004; Benediktsson et al. 2005). There are numerous ways through which astrocytes may influence microvascular resistance in the brain. It is though that the activity of astrocytes may provide at least a partial explanation for the rapidity with which changes in vascular resistance can respond to or even precede alteration in metabolism in the brain. Newman et al. (1984), Newman (1986) provided evidence that astrocytes may participate in a process known as potassium siphoning, in which the astrocytes control extracellular potassium concentrations and can rapidly transport potassium through their endfoot processes. This process provides a potential explanation for more rapid movement of potassium to the microvessels for the elicitation of dilation than could be afforded by the process of diffusion. This function is facilitated by the presence of voltage-dependent high-conductance potassium channels in perivascular astrocyte endfeet (Price et al. 2002).

Not only do astrocytes serve an important function as scavengers for neurotransmitters, they also have receptors for these transmitters that can activate the cells, causing an increase in intracellular calcium. Included among the neurotransmitters that are putative astrocyte activators are glutamate, GABA and possibly others (Anderson 2006). Astrocytes can also serve as a source of neurotransmitters (Fellin and Carmignoto 2004).

Changes in the intracellular calcium concentration of astrocytes can occur not only in response to the release of transmitters from neurons and also as the result of spontaneous excitation. This spontaneous excitation can remain confined to distal astrocytic processes or propagate both intracellularly and intercellularly (Parri et al. 2001; Nett et al. 2002; Nimmerjahn et al. 2004), creating a diverse potential for vascular control. Astrocytes can contribute to both dilation and constriction of arterial microvessels. Whether constriction or dilation occurs is dependent upon oxygen tension (Gordon et al. 2008; Gordon et al. 2011).

The first demonstration that electrical stimulation of perivascular astrocytes causes vasodilation was provided by Zonta et al. (2003a, b) utilizing cortical slices in vitro. These investigators also provided evidence that metabotropic glutamate receptors mediate this process, and that an elevation in astrocyte intracellular calcium concentration coincided with the time initiation of vasodilation (30 s). Studies utilizing brain slices as a model have demonstrated that astrocytes may either cause vasoconstriction or vasodilation, dependent upon conditions during stimulation. The changes in cytosolic calcium in astrocytes results in activation of PLA_2 leading to the formation of vasoactive eicosanoids (Filosa et al. 2004; Mulligan and MacVicar 2004; Attwell et al. 2010) (MacVicar and Newman 2015). Eicosanoids may play an important role as the paracrine mediating this process· Elevation of intracellular calcium in astrocytes as a result of neuronal stimulation appeared to induce vasodilation in brain slices through the release of PGE2 (Zonta et al. 2003a, b). When oxygen tension is low, lactate is also released, reducing the uptake of PGE2. There is a strong correlation between functional hyperemia and brain lactate concentration. There is also evidence that glutamate can induce the activation of astroctye cytochrome P450 epoxygenase that contribute to the production of epoxyeicosatrienoic acids (EETs) which have vasodilatory properties (Harder et al. 1998). This hypothesis is supported by the attenuation of functional hyperemia by inhibition of P450 epoxygenase activity (Alkayed et al. 1998). Increases in intracellular calcium concentration can also be mediate the release of 20-hydroxyeicosatetraenoic acid (20-HETE) causing vasoconstriction by inhibiting Ca^+-activated K^+ channels and by activating voltage-gated Ca + channels on vascular smooth muscle (Zou et al. 1996; Attwell et al. 2010). When oxygen tension increases the production of 20-HETE predominates, under conditions of reduced oxygen tension, EET production predominates. The tonic release of 20-HETE and ATP can not only contribute to acute changes but also to the maintenance of vascular tone. An in-vitro study demonstrated that elevation of astrocytic cytosolic calcium concentration could suppress vasomotion. Normal astrocytic regulation of blood flow is compromised by Alzheimer's disease (MacVicar and Newman 2015).

It is speculated that astrocytes may also contribute to basal vascular tone by the release of ATP which activates P2X2 receptors (Kur and Newman 2014).

Although capable of localized changes, astrocyte activation can also create propagating signals to neighboring astrocytes that can span hundreds of micrometers. It has been postulated that ATP and/or glutamate can act as an extracellular messenger for the propagation of these waves (Guthrie et al. 1999; Cohen et al. 2000; Schipke et al. 2002) (Anderson 2006).

It is also possible that astrocytes may have a mechanical influence cerebral capillary resistance. During potassium-induced cortical spreading depression, it has been postulated that the resultant swelling in astrocytic endfeet can compress the capillaries they surround, increasing resistance to the degree that capillary flow may cease in the affected area (Tomita et al. 2005). It has been proposed that changes in capillary diameter can have an influence on cerebral blood flow, with active capillary dilation contributing (Hall et al. 2014). It has been speculated that pericyte tone might also be influenced by mediators released from astrocytes (Hall et al. 2014; MacVicar and Newman 2015)

In addition to participating in the regulation of vascular resistance, astrocytes have a number of other important signaling functions that involve the cerebral microcirculation. They release chemokines that attract inflammatory cells (Babcock et al. 2003; Marella and Chabry 2004) and may be important in attracting neural stem cells curing repair processes (38). The term neurovascular unit has been used to describe the intimate developmental, structural and functional interrelationship among vascular cells, glial cells and neurons (Girouard and Iadecola 2006).

7.6 Neural Control

During the seventeenth century, Thomas Willis was the first to describe neural innervation of cerebral vessels. We now know that there are four components of the nervous system that have the potential to influence vascular diameter:

sympathetic nerves, parasympathetic nerves, the central pathways (including local interneurons) and sensory nerves (Gulbenkian et al. 2001). Sympathetic nerves innervating the cerebral vessels have their origin in the superior cervical ganglion and the locus coeruleus, while the parasympathetic fibers originate in the sphenopalatine, otic and internal carotid ganglia (Edvinsson et al. 1990a; b; Branston 1995). The possibilities for regulation of cerebrovascular resistance through neural control are extensive. In addition to the neurotransmitters noradrenaline and acetylcholine, cerebrovascular nerves also secrete other neurotransmitters including monoamines, polypeptides and gases (Table 7.1) (Gulbenkian et al. 2001). Over 15 vasoactive neurotransmitters have been demonstrated in axons innervating blood vessels in the Circle of Willis, including peptides and biogenic amines (Moskowitz et al. 1990).

Increased calcium influx resulting from glutamate activation of NMDA receptors on neurons by glutamate activates nNOS. The resultant release on nitric oxide by neurons contributes to dilation of cerebral resistance vessels. It is unclear whether this dilation is the direct effect of this NO on vascular smooth muscle, or the result of other pathways originating in astrocytes (Attwell et al. 2010).

Sympathetic nervous system effects on cerebral resistance vessels include both noradrenaline and neuropeptide Y and depending upon the species, somatostatin, calcitonin gene-related peptide and vasoactive intestinal peptide (Gulbenkian et al. 2001). The relative concentration of neurotransmitters in the sympathetic nervous system changes with age, resulting in a loss of neuropeptide Y and an increase in vasoactive intestinal peptide and calcitonin gene-related peptide. Neuropeptide Y is thought to active two different receptor subtypes: Y1 and Y2. The Y1 subtype is a postjunctional receptor in both vascular smooth muscle and endothelial cells, while the Y2 receptor is a presynaptic receptor that may limit noradrenaline release (Wahlestedt and Reis 1993). The sympathetic innervation of perforating arteries and small cortical arteries is relatively dense. The density of innervation decreases in smaller arterial vessels but is still present in arterioles as small as 15–20 mm. Most of the adrenergic nerves innervating cerebral vessels have their origin in cervical ganglia. However, the adrenergic nerves innervating pial vessels are hypothesized

Table 7.1 Vasoactive neurotransmitters

Nuclei and ganglia	Neurotransmitter
Subcortical	
Locus coeruleus	Noradrenaline
Raphe nucleus	5-HT
Basal forebrain	Ach
Thalamus	Glutamate
Extrinsic ganglia	
Superior cervical ganglia	NA NPY
Sphenopalatine/Otic ganglia	Ach VIP NOS PHI
Trigeminal ganglia	SP CGRP 5-HT NKA PACAP
Local interneurons	GABA VIP ach NOS NPY SOM

to primarily originate in the locus coeruleus and/or the hypothalamus (Duverger et al. 1987; Bekar et al. 2012).

While there is considerable evidence of the presence of 5-hydroxytryptamine containing nerve terminals innervating both large cerebral arteries and pial vessels, there is debate as to whether the 5-hydroxytryptamine is costored in noradrenergic sympathetic nerves or if these nerves are a separate population innervating these vessels (Lincoln 1995).

The response of the cerebral vessels to sympathetic stimulation differs significantly from the response of vessels in other organs of the body. Adrenergic receptors on the smooth muscle of the cerebral resistance vessels are less dense and less sensitive to stimulation compared to the adrenergic receptors on the resistance vessels in other organs of the body. The ED_{50} for norepinephrine is one hundred times greater for the basilar artery than for the saphenous artery. Adrenergic receptor activation in cerebral vessels also appears to lead to greater calcium influx from the extracellular fluid, which leads to interesting speculation about selective effects of calcium channel blockers on the cerebral resistance vessels (McCalden and Bevan 1981; Towart 1981). The results of sympathetic stimulation on cerebral vascular resistance are influenced by the fact that the larger cerebral vessels are much more responsive to stimulation than the small cerebral arterial vessels. The relative contribution of the large vs. small cerebral arterial vessels differs when systemic blood pressure changes. Under conditions of hypertension, larger cerebral vessels contribute more to vascular resistance in the brain, whereas under hypotensive conditions the smaller vessels have more impact on the control of vascular resistance. Since the sympathetic nerves have a greater influence over the large vessels, sympathetic stimulation increases resistance when blood pressure increases, but has little effect on cerebral vascular resistance when blood pressure is low. This is very beneficial with regard to homeostasis of the brain because sympathetic stimulation will help protect the capillaries from exposure to high pressures when blood pressure increases and therefore help maintain normal filtration rates. At the same time, because of the lack of effect on smaller cerebral arterioles, sympathetic stimulation contributes very little to cerebral vascular resistance when there is a reduction in blood pressure. Sympathetic nerves also seem to have a trophic effect on cerebral vessels, promoting vascular hypertrophy (Hart et al. 1980). This may have significant effect on long term autoregulatory adaptations of the cerebral vasculature.

The role of parasympathetic nerves in the regulation of cerebral blood flow is still unclear. Probable sites of origin of parasympathetic nerves innervating the cerebral blood vessels include sphenopalatine ganglion, the otic ganglia and internal carotid microganglia (Zhang et al. 1991). Parasympathetic terminals in the CNS costore Ach and vasoactive intestinal peptide (VIP). Parasympathetic nerves also release peptide histidine isoleucine (PHI) which is synthesized from the same precursor as VIP (Edvinsson et al. 1990a, b; Edvinsson 1991; Zhang et al. 1991; Arimura and Shioda 1995). VIP acts directly on smooth muscle cell VIP receptors, inducing an endothelial cell independent relaxation that has a greater effect on small arterial vessels than large arterial vessels (Edvinsson 1991). Studies conducted in cats indicate that VIP contributes to atropine-resistant neurogenic dilation of cerebral resistance vessels

(Bevan et al. 1986). In addition to VIP, nitric oxide may also function as a vasodilating neurotransmitter (Barroso et al. 1996). Cerebral vessels have a relatively dense innervation by NOS-immunoreactive nerve fibers (Nozaki et al. 1993; Yoshida et al. 1994; Goadsby et al. 1996). There is evidence that these fibers contribute to neurogenic vasodilation (Saito and Goto 1994). A subpopulation of nerve fibers containing VIP and NOS also contain PACAP, which has been shown to contribute to pial vasodilation (Uddman et al. 1993; Seki et al. 1995).

It has been speculated that the parasympathetic nervous system may participate in the development of collateral flow. This speculation is based on the observation that parasympathetic denervation increases infarct size in focal but not global ischemic models, indicating that flow may be better preserved in the penumbra when the parasympathetic nerves are intact (Moskowitz et al. 1990).

In addition to the direct effect of neurotransmitters released from sympathetic and parasympathetic nerves on blood vessels, the close approximation of nerve fibers (as little as 25 nm separation) from these two systems also creates the possibility that there may be direct axonal–axonal interactions involved in the control of cerebral vascular resistance (Lee 2002).

There is also evidence for the important contributions of sensory nerve fibers in the regulation of cerebral blood flow (Moskowitz et al. 1990; Macfarlane et al. 1991a, b). Included among the ganglia receiving projections from these sensory nerves are the trigeminal ganglia, internal carotid mini ganglia and cervical dorsal root ganglia. Peripheral ramifications of primary sensory (type C unmyelinated) neurons are thought to release a number of vasoactive neuropeptides. Two of the major vasoactive neuropeptides released by sensory neurons are substance P (SP) (Edvinsson et al. 1983) and calcitonin gene-related peptide (CGRP) (Edvinsson et al. 1987, 1998). These neurons also release a second tachykinin, neurokinin A (NKA) (Edvinsson et al. 1988). The relative density of innervation by nerves releasing these various neurotransmitters is species dependent. The relaxation caused by release of SP and NKA is dependent upon the presence of functioning endothelial cells, whereas vasorelaxation induced by CGRP release is both more potent and endothelial independent. PACAP also appears to function as a vasoactive neurotransmitter that is released from sensory nerves (Markowitz et al. 1987). A number of functions have been proposed for these sensory nerves. In animal studies, trigeminal ganglion lesion has been demonstrated to prolong the vasoconstrictor responses to elevated pH and $PGF2_{\alpha}$, and superfused norepinephrine. It has therefore been postulated that CGRP is involved in the restoration of normal vascular diameter under conditions of intense vasoconstriction such as subarachnoid hemorrhage and migraine (Edvinsson 1991, 2006). It has been demonstrated in a cat model that trigeminal ganglionectomy prolongs vasoconstriction caused by perivascular blood injection. The same investigator reports that there is depletion of CGRP but not of other perivascular neuropeptides in patients who have died following subarachnoid hemorrhage (Edvinsson et al. 1990a, b). The enhanced concentration of CGRP in the external jugular vein during migraine headache has also led to speculation that these sensory nerves may also be linked to this problem. These neurons may also play an important role in reactive hyperemia following transient cerebral ischemia.

Trigeminal ganglionectomy has been reported to reduce reactive hyperemia following 10 minutes of global ischemia in cats by almost 50% (Moskowitz et al. 1990; Macfarlane et al. 1991a, b). These investigators speculate that blocking this mechanism may be a useful strategy in reducing severe cortical hyperemia.

Although the vessels leading to, and on the surface of the brain, are innervated by nerves from multiple sources, the penetrating arterioles in the brain parenchyma, after the Virchow–Robin space has vanished, receive innervation from neurons located within the brain itself (Hamel 2006). The term "intrinsic innervation" is used to describe this component of vasoactive nerves.

The functional role of the central neural pathway, originating in subcortical basal areas is still poorly understood. Electrical stimulation of nuclei in this area can produce either increases or decreases in cortical blood flow, depending upon the area stimulated (Cauli et al. 2004). Receptors for the neurotransmitters released from these fibers exist on endothelial cells, vascular smooth muscle and astrocytes, allowing them to influence the entire neurovascular unit (Hamel 2006). There is evidence that the resistance alterations resulting from stimulation of the locus coeruleus is the result of adrenergic stimulation of astrocytes rather than a direct effect on the vascular wall (Mulligan and MacVicar 2004). Cholinergic nerve fibers that arise from the nucleus basalis of Meynert have been hypothesized to have a role in the control of vascular diameter in the cortex (Sato et al. 2004). In the rat, focal stimulation of these neurons increases ipsilateral cortical blood flow (Biesold et al. 1989; Adachi et al. 1990a, b). The dilation that results from these neurons involves intraparenchymal vessels but not the pial microvessels (Sato et al. 2004). It appears that the cortical vasodilation is mediated by both muscarinic and nicotinic cholinergic receptors. The vasodilation, mediated secondarily through an increased production of NO, could be produced by a variety of cells including endothelial cells, neurons and glial cells. Both cholinergic and serotonergic receptors have been identified on the perivascular end feet of astrocytes (Cohen et al. 1996; Luiten et al. 1996; Elhusseiny et al. 1999). Blood vessels in the hippocampus are also influenced by cholinergic fibers originating in the basal forebrain (Cao et al. 1989; Adachi et al. 1990a, b). In addition to the neurotransmitters discussed above, dopamine has been proposed to play an important role in the regulation of microvascular resistance and cerebral blood flow. Krimer et al. (1998) have demonstrated close apposition of fibers containing the dopamine transporter to penetrating arterioles and capillaries in the cerebral parenchyma. These nerves appear to innervate pericytes surrounding the capillaries. The potential to alter capillary resistance through neuronal activation of pericytes creates interesting potential for the local regulation of cerebral microvascular perfusion. Local application of dopamine on isolated cerebral arterial vessels causes constriction which may be mediated adrenergic or serotinergic receptors (Iadecola 1998).

The final potential role of nerves in the regulation of cerebral blood flow involves local interneurons. Initially it was shown that pial arterioles are innervated by nerve fibers from bipolar neurons within the cerebral cortex. These fibers were found to contain VIP and PHI. It is possible that the release of these peptides could elicit vasodilation when tissue metabolism increases (Zhang et al. 1991). Later studies have shown that depolarization of cortical GABA interneurons can induce either vasodilation or vasoconstriction. GABAnergic neurons that costored VIP or NO caused dilation, whole those that costore somatostatin cause contraction (Hamel 2006). These cortical interneurons receive afferent information from subcortical areas and can therefore integrate and relay information about changes in activity in these areas to elicit the appropriate vasomotor responses. A similar function is performed by interneurons (stellate cells) within the cerebellum. Depletion of cerebellar interneurons reduces cerebellar blood flow increases following sensory stimulation (Yang et al. 2000). Neurons from the ventral tegmental area also thought to be involved in the regulation of cerebral blood flow (Schaeffer and Iadecola 2021).

7.7 Myogenic Control

Cerebral resistance vessels are reported to have vigorous myogenic responsiveness (Ursino 1991). The myogenic response was first described by Bayliss (1902). Wall tension is the controlled parameter in this response. Following an increase or decrease in transmural pressure the myogenic response alters diameter to restore basal wall tension (Harder 1985; Harder et al. 1997). The mechanisms for myogenic control in the cerebral circulation appear to be the same as those controlling myogenic responses in other vascular beds. A number of mechanisms have been proposed to contribute to the initiation of the myogenic response. Included among these mechanisms are stretch activated ion channels, GPCRs, and membrane bound tissue necrosis factor, all of which contribute to increases in intracellular calcium levels in perivascular smooth muscle (Lidington et al. 2018). The myogenic response is dependent upon alterations in L-type calcium channels and calcium-activated potassium channels mediated by an arachidonic acid signal transduction pathway. 20-Hydroxyeicosatetraenoic acid (20-HETE) has been demonstrated to control the permeability of these channels (Harder et al. 1997).

7.8 Endothelial Control

As described previously, endothelial cells in the cerebral microcirculation are a key component of the blood–brain barrier, helping to protect the brain interstitial fluid from shifts in ionic concentration, and the concentration of other mediators that could adversely influence neuronal function. A second important function performed by endothelial cells is their contribution to the regulation of cerebral blood flow. In performing this function, the endothelial cells serve to signal vascular smooth muscle cells in response to changes in the blood circulating within the vessels.

The vasoactive mediators produced by cerebral endothelial cells are the same as those produced in other vascular beds throughout the body. Included among the vasodilators produced by endothelial cells are nitric oxide, endothelial hyperpolarizing factor, PGI_2 and PGE_2. Vasoconstrictors produced by endothelial cells include endothelin, thromboxane A_2 and $PGF_{2\alpha}$ (Andresen et al. 2006).

Numerous studies have demonstrated that the tonic release of nitric oxide under basal conditions, via the activity of eNOS (NOS3), is an important component of cerebral blood flow regulation (Tanaka et al. 1991; Prado et al. 1992; Wang et al. 1995; Atochin et al. 2003). The activity of eNOS in cerebral endothelial cells is regulated by intracellular calcium concentration, heat shock protein (HSP90), serine/threonine phosphorylation and tyrosine phosphorylation (Andresen et al. 2006). Cholinergic stimulation of eNOS in cerebral blood vessels is mediated by M(5) muscarinic receptors (Yamada et al. 2001). Estrogen is another factor that can enhance the tonic level of nitric oxide production (McNeill et al. 1999; Pelligrino et al. 2000; Santizo et al. 2000; Stirone et al. 2005). Interestingly, statins are another substance that can influence eNOS under both basal and pathologic conditions, in part explaining the protection from stroke provided by this drug. In addition to enhancing eNOS expression, they also reduce NADPH oxidase activity which increases NO availability (Wagner et al. 2000; Endres 2005).

A second vasodilator produced by endothelial cells is prostacyclin (PGI_2). PGI_2 is a metabolite of arachidonic acid produced via the cyclooxygenase (COX) pathway. There are three COX isoforms, all of which are found in cerebral endothelial cells (Chandrasekharan et al. 2002; Kis et al. 2003, 2004; Kis et al. 2005a, b). COX-1 is constitutively expressed, but can be upregulated. COX-2 is induced by inflammation. The release of PGI_2 by cerebral endothelial cells activates a G-protein-couple receptor on vascular smooth muscle cells, leading to an increase in intracellular CAMP and activates protein kinase A. This results in relaxation of the smooth muscle and vasodilation.

The endothelial hyperpolarizing factor (EDHF) pathway is a third mechanism through which endothelial cells can contribute to relaxation of cerebral microvessels (Golding et al. 2001; Golding et al. 2002; Bryan Jr. et al. 2005). Although the term EDHF implies a the production of a single vasoactive substance, EDHF has not as yet been clearly identified and may indeed represent a number of vasoactive agents released by endothelial cells. It is clear that this pathway is associated with endothelial cell and vascular smooth muscle cell hyperpolarization. The EDHF pathway can be initiated by agonist/receptor interactions and shear stress (McGuire et al. 2001; Golding et al. 2002). Increased endothelial intracellular calcium opens calcium activated potassium channels, hyperpolarizes endothelial cells and initiation of the EDHF pathway. The steps involved in the remainder of the signaling pathway remain to be identified. Among the substances that have been proposed to be involved in this mechanism of endothelial signaling of the vascular smooth muscle are: epoxyeicosatrienoic acids (EETS), potassium ions, gap junctions and hydrogen peroxide. The actions of the EDHF pathway are independent of NO and prostacyclin, and indeed may be upregulated as a protective mechanism when the NO production by endothelial cells becomes less effective under pathologic conditions (Bryan Jr. et al. 2005). It also has greater effect on arterial resistance vessels in the cerebral microcirculation that on larger arterial vessels (You et al. 1999). It is now speculated by a number of investigators that the EDHF mechanism may also contribute to tonic vasodilation of the cerebral microvessels.

The major constrictor produced by endothelial cells is endothelin (ET-1). Production of endothelin following head injury and ischemia/reperfusion injury contributes to an exacerbation of secondary injury associated with these insults. Endothelin, a peptide with a much longer half-life that the volatile free radical NO, is synthesized from inactive precursors in endothelial cells (Miyauchi and Masaki 1999). Endothelin can activate two receptors in the cerebral circulation ET_A and ET_B. These receptors are expressed on both endothelium and smooth muscle cells (Andresen et al. 2006). The predominant effect is vasoconstriction of both large and small arterial vessels in the brain through the ET_A receptor. Although activation of the ET_A receptor elicits vasoconstriction low concentrations ET-1 can cause dilation of pial arterioles (Faraci 1989).

Another potential role for endothelial cells in the regulation of cerebral blood flow is their participation in conducted upstream vasodilation. Although there is clear evidence of the participation of endothelial cells in conducted upstream vasodilation in a number of vascular beds, the degree to which this occurs in the cerebral vasculature remains to be determined.

7.9 Autoregulation

The autoregulatory ability of the cerebral resistance vessels, defined as the maintenance of constant blood flow in the face of changes in perfusion pressure, reflects the effectiveness of local control mechanisms. Cerebral blood flow remains constant when mean arterial pressure varies between 60 and 140 mmHg in the normal individual. Although intravascular pressure varies in small arteries and arterioles during alterations in systemic pressure, capillary pressure remains relatively constant (Faraci and Heistad 1990). When perfusion pressure exceeds the ability of the cerebral resistance vessels to autoregulate, there is a "breakthrough" of autoregulation that was interpreted to result from an inability of arterial vessels to resist the distending pressure. However, it has been proposed that this breakthrough of autoregulation may involve an active process, resulting from the activation of calcium-dependent potassium channels causing hyperpolarization and relaxation of vascular smooth muscle (Heistad 2001).

The autoregulatory range shifts to higher pressures under conditions of chronic hypertension, providing greater protection for cerebral microvessels from increased systemic pressure but less ability to compensate for reductions in systemic pressure. This shift in the range of autoregulation involves alterations in the morphology of the vessel wall, with an increase in wall/lumen ratio (Baumbach and Heistad 1983; Werber and Heistad 1984). While acute autoregulation in response to local regulatory mechanism is referred to as short-term autoregulation, chronic autoregulatory shifts resulting from morphologic changes in the vessel wall are often referred to a long-term autoregulation. Reversal of morphologic changes in resistance vessels is a slow process and must be taken into consideration during the treatment of chronic hypertension. The contribution of various arterial segments to the autoregulatory response appears to vary in different locations within the brain. During acute hypertensive episodes, the medulla appears to rely more on smaller arterial vessels for autoregulation of flow compared to the cerebrum (Faraci et al. 1987).

7.10 Humoral Regulation

The presence of the blood–brain barrier makes most regions of the brain significantly less responsive to humoral regulation than many other organs of the body. Despite these restrictions it is inaccurate to assume that humoral regulation is of no importance in the cerebral circulation. Specialized regions of the brain lacking a blood–brain barrier, such as the circumventricular regions and the choroids plexus, are influenced by circulating hormones such as vasopressin and angiotensin II (Faraci et al. 1990). Increased circulating levels of vasopressin can significantly reduce blood flow to the choroids plexus, resulting in an attenuation of cerebrospinal fluid production (Faraci et al. 1994).

Blood vessels of the choroids plexus are also responsive to angiotensin II. In addition to its effects on choroids plexus blood flow and cerebrospinal fluid formation, angiotensin II also influences functional hyperemia in the brain and the remodeling of cerebral blood vessels (Baumbach et al. 2003; Kazama et al. 2004). The constrictor effect on the mouse basilar artery (mean diameter 130 μm receptors through the Rho-kinase cell signaling pathway (Faraci et al. 2006) (Stenman and Edvinsson 2004). It has also been suggested that Angiotensin II may play a number of important roles in the deleterious cerebrovascular effects of hypertension (Lassegue and Griendling 2004; Saavedra 2005; Girouard et al. 2006). Angiotensin II-dependent hypertension can cause superoxide-mediated endothelial dysfunction and impaired neurovascular coupling (Kazama et al. 2004; Faraci et al. 2006).

While the majority of investigations report deleterious effects of Angiotensin II on the cerebral microcirculation there are also reports of a dilating effect of cerebral microvessels and potential protective effects against stroke (Takao et al. 2002). Using a photoelectric method, these investigators evaluated changes in cat cerebral parenchymal blood flow these investigators concluded that AT_1 receptors on parenchymal arterioles cause dilation of these vessels.

Vasopressin infusion causes a decrease in the resistance of large cerebral vessels. The constriction of small cerebral vessels in response to elevation in intravascular pressure in this arterial segment prevents any significant change in cerebral blood flow (Faraci et al. 1988). The dilation elicited by vasopressin infusion may be dependent upon endothelial-induced vasodilation (Faraci and Heistad 1990).

Estrogen is another circulating hormone that has been demonstrated to influence cerebral microvessels. There are conflicting reports concerning the effect of circulating estrogen on the normal cerebral microvascular bed (Littleton-Kearney et al. 2000). These investigators found that while the pial microvasculature responded superfusion with micromolar concentrations of 17B-estradiol in there was no response to alterations in circulating estrogen within the physiologic range under normal conditions. There are numerous reports of estrogen's protective effect during global and focal ischemia (Hurn et al. 1995; Alkayed et al. 1998; Pelligrino et al. 1998). Both eNOS synthase induction and an antioxidant effect are reported to contribute to this protection. The protective effect provided by circulation estrogen may be due to both direct vascular and perfusion independent effects (Wang et al. 1999). Estrogen has also been reported, through similar mechanisms, to provide a protective effect against head injury (Roof and Hall 2000a, b).

7.11 Inflammation

Inflammatory responses and the recruitment of inflammatory cells into the CNS are part of essential defense reactions and of pathological responses in a number of disease states. The involvement of the cerebral microcirculation in inflammatory reactions that exacerbate damage to the central nervous system has become an area of increasing interest over the past two decades. In addition to the damage to the CNS caused by trauma and/or ischemia, there is also a "secondary injury" occurring following trauma or reperfusion that can significantly enhance the magnitude of injury (Barone et al. 2002). Inflammatory responses within the CNS microcirculation form an important component of this secondary injury (Tuma et al. 1993, 1997). Modulation of these inflammatory responses has become an important therapeutic target in the treatment of stroke, trauma viral encephalitis and multiple sclerosis (del Zoppo et al. 2000; Legos et al. 2002).

Ischemic stroke usually involves an ischemic core of tissue that is often beyond salvage and a zone of tissue around the call called the "penumbra." The regulation and magnitude of blood flow through the microvasculature of the ischemic penumbra is a critical determinant of the eventual magnitude of the stroke (Barone et al. 2002).

Vasogenic edema, the result of changes in microvascular permeability following ischemia and trauma, is a well-recognized contributor to the secondary damage to the CNS and the target of considerable interest for many years. Ischemia causes a disruption of the blood–brain barrier through a number of mechanisms. Bradykinin, platelet-activating factor, thrombin, VEGF, matrix metalloproteinases and other proteases are all reported to contribute (Kamiya et al. 1993; Aschner et al. 1997; Heo et al. 1999; Zhang et al. 2000; Opdenakker et al. 2001; del Zoppo and Mabuchi 2003). Because of encapsulation in the bony structure of the cranium and spine, edema formation in the CNS can lead to elevation in intracranial pressure, compression of the cerebral vessels and further compromise of cerebral blood flow. Clinically this is commonly from changes in cerebral perfusion pressure (CCP) defined as mean arterial blood pressure minus ICP. CCP is the physiologic variable that defines the pressure gradient for blood flow across the brain and reductions in CCP are viewed to be one of the major management problems following injury to the CNS.

In addition to the paracrine substances mentioned above, ischemia and trauma cause the activation and production of a number of other signaling molecules that orchestrate cellular responses during secondary injury. Complement activation and the production of cytokines are important contributors. The complement cascade, when initiated following ischemia contributes to adhesion molecule upregulation, neutrophil chemotaxis, platelet activation and free radical production (Mocco et al. 2006). Early studies demonstrated that complement inhibition with cobra venom factor attenuated ischemia/reperfusion injury (Vasthare et al. 1998; Popovich and Jones 2003). More recently mice genetically deficient in selected complement proteins (Ciq, C3 and C5) demonstrated that C3 activation is the key component in complement-related inflammatory tissue injury (Mocco et al. 2006).

Many studies have been made to test the hypothesis that modulation of leukocyte participation in ischemia/reperfusion injury may influence the magnitude of the resultant damage to the brain. Initially this was done by rendering animals leukopenic prior to ischemia. Leukopenia was found to offer protection against ischemia/reperfusion injury in both focal (Dutka et al. 1989; Bednar et al. 1991) (Kochanek and Hallenbeck 1992; Hallenbeck 1997) and global ischemic models (Vasthare et al. 1990; Heinel et al. 1994). Since those initial studies, a number of investigators have tested a variety of substances designed to interfere with leukocyte adhesion and activation during reperfusion after ischemia. Antineutrophil antibodies were used in a rat focal ischemia model to provide evidence that circulating neutrophils are responsible in the initial surge of free radical production (Matsuo et al. 1995).

The expression of adhesion molecules on the endothelial surface after ischemia has been well documented. P selectin is the first to appear within the first hour following ischemia, followed by ICAM-1, ELAM-1 and E selectin. Antibodies which block the ability of neutrophils to bind to the endothelium, such as anti-rat ICAM-1, anti-CD11a and anti-CD18, have been found, in some studies, to have a protective effect for the ischemic brain (Matsuo et al. 1994). In a similar study, an anti-CD11b antibody reduced ischemic cell damage after transient but not after permanent focal occlusion in the rat (Jiang et al. 1994). Similar positive results were found in a rabbit model of cerebral emboli stroke (Bowes et al. 1995). Anti-ICAM-1 treatment produced a significant reduction in both the number of neutrophils and in the size of the infarct in both rat (Zhang et al. 1994) and rabbit, as did anti-Mac-1, which binds to CD11b/CD18 (Chopp et al. 1994). Bednar et al. were able to demonstrate that treatment with IB4, a monoclonal antibody directed against CD18, after a thromboembolic stroke in rabbits decreased intracranial pressure (Bednar et al. 1991). They were, however, unable to demonstrate any significant effect on infarct size. The recombinant neutrophil inhibitory factor (rNIF), a 41 kD hookworm glycoprotein which binds CD11b/CD18, has been shown to have neuroprotective abilities in rats with focal transient cerebral ischemia (Jiang et al. 1995). Soriano et al. showed that ICAM-1-deficient mice were less susceptible to cerebral ischemic injury (Soriano et al. 1996). A variety of techniques have also been used to demonstrate that blocking selectins is effective in reducing postischemic damage to the brain (Zhang et al. 1996; Goussev et al. 1998). These studies provide additional support for the hypothesis that modulation of leukocyte activity may provide a means to prevent exacerbation of ischemic damage during reperfusion. It is therefore clear that leukocyte rolling, adhesion and transmigration across the cerebral vasculature are important steps in the microvascular contribution to secondary injury following both ischemia (Weaver et al. 2002) and trauma (Popovich et al. 2002). Although adhesion molecule blockers have been shown to be an effective experimental method of reducing infarct volume following stroke (Mori et al. 1992; Jiang et al. 1998; Weaver et al. 2002), this has not proven to be an effective therapeutic tool in the clinic. This is because it appears the time postischemia that inhibitors of white cell migration remain effective is too short for effective treatment of most victims of stroke. Currently modulation of endogenous cannabinoid production and administration of exogenous cannabinoids and cannabinoid antagonists specifically targeted to either CB1 or CB2 receptors are being investigated for potential therapeutic use in the treatment of stroke (Zhang et al. 2007).

There is also growing interest in the hypothesis that while early invasion of the CNS by white cells contributes to injury, later activation and invasion by leukocytes may contribute to repair of damage following trauma (Lazarov-Spiegler et al. 1998; Ghirnikar et al. 2001; Popovich et al. 2002; Hauben and Schwartz 2003; Popovich and Jones 2003; Schwartz and Yoles 2005).

Cytokines/chemokines are another obvious target for the treatment of secondary injury following trauma and ischemia (Popovich 2000; Barone et al. 2002). The expression of these molecules plays a critical role in numerous factors contributing to this injury including: adhesion molecule expression, white cell and glial cell activation. TNF-α and IL-1β have been the focus of considerable interest following both stroke and traumatic brain injury. Chemokine receptor expression for CCR2, CCR3, CCR5, CCR10 and CCR4 has been demonstrated to increase following contusion injury of the spinal cord of the rat. There is an upregulation of IL-1β, TNF-α, IL-6 and M-CSF in the rat spinal cord (McTigue et al. 1998; Lee et al. 2000). The presence of both CXC-type chemokines (FRO alpha & IP-10) and CC-type chemokines (MCP-1 & MCP-5) has also been demonstrated in the injured spinal cord. It has been shown that infusion of a broad-spectrum chemokine inhibitor (vMIPII) directly into the spinal cord of rats decreases the invasion of inflammatory cell into the spinal cord and attenuates neuronal loss (Ghirnikar et al. 2000, 2001) Cytokine/chemokines can have both a proinflammatory and an anti-inflammatory role. Therefore, these molecules are involved in both exacerbation of injury and also repair, and modulation of the profile of cytokines/chemokines produced has become an area of significant interest.

Cannabinoids are another class of molecules that are increasingly recognized as important contributors to homeostatic mechanisms and inflammatory reactions in the central nervous system.

Experimental autoimmune encephalitis (EAE), an animal model of multiple sclerosis (MS), is a model of autoimmune disease resulting from abnormal immune responses in the central nervous system. MS lesions are characterized by chronic inflammation, microglial activation and myelin destruction mediated by the invasion of the CNS by immune cells including CD4-positive, CD8-positive T cells and monocytes. While it was previously though that the inflammatory response was primarily Th1 mediated, recent evidence indicates that Th17 cells may also play a prominent role. Despite the overwhelming evidence of the importance of white cells in the development of multiple sclerosis, there are still many unanswered questions concerning the mechanisms responsible for their invasion of the CNS parenchyma by leukocytes during the progress of this disease. It is clear that immune cells perform immunosurveillance in the CNS, but their mode of entry into the CNS may vary significantly with the kind of stimulus and CNS entry site (Engelhardt and Ransohoff 2005). These investigators have proposed that the molecular mechanisms involved in leukocyte–endothelial interactions may differ when comparing white matter to gray matter, the spinal cord and the brain and meningeal vessels versus parenchymal vessels. Under physiological conditions immunosurveillance in the CNS is restricted to the perivascular and subarachnoid spaces, with the lymphocytes excluded from the CNS parenchyma (Engelhardt and Ransohoff 2005).

Under quiescent conditions immunosurveillance may occur through lymphocytes invading the subarachnoid space via the choroids plexus and interacting with dendritic cells that normally reside there. P-selectins appear to play an important role in this process. During EAE, there is recruitment of dendritic cells to the choroids plexus and upregulation of VCAM-1 and ICAM-1 in addition to expression of MAdCAM-1 (Sasseville et al. 1992; Engelhardt et al. 2001; Engelhardt and Ransohoff 2005). Entry through the choroids plexus and reentry into the circulating blood allows for daily replenishment of lymphocytes in the CSF (Engelhardt and Ransohoff 2005).

Lymphocytes that had not been activated were thought to be unable penetrate the blood–brain barrier (Wekerle et al. 1986), although this has been suggested to occur during EAE (Zeine and Owens 1992; Engelhardt et al. 1998; McMahon et al. 2005). Studies utilizing intravital microscopic visualization of the mouse brain have also led to the conclusion that activated lymphocytes do not adhere to noninflamed microvessels (Piccio et al. 2002). In the presence of inflammatory stimuli, lymphoblasts are able to penetrate the blood–brain barrier where they encounter dendritic cells and macrophages in the perivascular spaces (Carrithers et al. 2000; Greter et al. 2005). Lymphoblasts that do not encounter the appropriate antigen do not remain in the central nervous system. When human lymphocytes taken from MS patients were injected into mice whose cerebral vasculature was inflamed by LPS injection, CD8+ lymphocytes were found to primarily interact with P-selectin glycoprotein ligand-1 (PSGL-1) to initiate rolling, while CD4+ T cells rolled via interactions with α4-integrins (Battistini et al. 2003). The α4 - integrins α4-integrins play a central role in migration across the blood–brain barrier under inflammatory conditions. E-selectin is also found to be involved in leukocyte rolling along inflamed cerebral microvessels (Carvalho-Tavares et al. 2000; Piccio et al. 2002). The integrin ligands ICAM-1 and VCAM-1 are both upregulated in EAE (Baron et al. 1993; Steffen et al. 1994). The involvement of ICAM-1 and its ligand LFA-1 in human multiple sclerosis is still a matter of debate (Baron et al. 1993; Cannella and Raine 1995; Kivisakk et al. 2003).

α4-Integrins not only play an important role in the adhesion of lymphocytes to pial vessels on the surface of the brain, but also in the direct capture of leukocytes in spinal cord microvessels. VCAM-1 also participates in this process (Vajkoczy et al. 2001). Once firmly adherent LFA-1 participates in the diapedesis of T lymphocyte microvessels supplying the spinal cord (Laschinger et al. 2002). Currently, there is growing interest in the potential use of cannabinoids to attenuate white cell invasion of the CNS and their participation in the development of EAE and MS. Chemokines, including CCL2 and CXCL13, are thought to be important signaling molecules involved in white cell invasion during EAE and multiple sclerosis (Krumbholz et al. 2006; Mahad et al. 2006). Selective agonists for the CB2 receptor have been shown to attenuate white cell rolling and adhesion and attenuate motor deficits in a mouse EAE model (Ni et al. 2004).

7.12 Age-Related Changes

Aging is associated with a loss of neurons, a decrease in glucose and oxygen consumption and compromised microvascular function within the brain. While there is still considerable debate as to which parameter causes a reduction in the other, there is increasing evidence for a bidirectional interaction associated with changes neuronal and microvascular structure, with neuronal tissue providing trophic influence on microvascular development, and microvessels having trophic influence on neuronal development (Leventhal et al. 1999; Riddle et al. 2003). Age-associated decrements in microvascular function may therefore not only result from, but also contribute to, a loss of neuronal function with age.

The evidence concerning age-related alterations in microvascular morphology is discordant, with reports of increased, decreased or no change in capillary density. Similarly, although rarefaction of the arterial system is generally associated with aging, there have been reports of no change, increased and decreased arteriolar density associated with the aging brain (Knox and Oliveira 1980; Bell and Ball 1981; Hutchins et al. 1996; Sonntag et al. 1997). There is evidence that aging is associated with a reduction in neuronal and microvascular plasticity in response to chronic stimuli in aged animals (Yamaguchi et al. 1988; Black et al. 1991; Szpak et al. 1999; Riddle et al. 2003). Sonntag and coworkers have provided evidence that decreased production of insulin-like growth factor (IGF-1) in aged animals may be a significant contributor to this reduced microvascular plasticity (Sonntag et al. 2000).

Descriptions of age-related changes in the morphology of the wall of blood vessels are more consistent. As would be expected, there is an age associated reduction in the elastic components of the vascular wall and an increase in collagen and the basement membrane (Hajdu et al. 1990). There are also reports of abnormal intracellular inclusions and flocculant deposits in the wall of microvessels (Knox et al. 1980; Keuker et al. 2000; Ritter et al. 2000; Uchida et al. 2000; Farkas and Luiten 2001). Basement membrane changes associated with aging include perivascular collagen deposits and thickening (Farkas and Luiten 2001) (Figure 4). These changes have significant implication for the regulation of blood flow and the blood–brain barrier.

There are numerous reports of age associated alterations in cerebral blood flow. While not always seen under baseline conditions, the responsiveness of cerebral vessels to a number of vasodilatory stimuli is diminished (Tomita et al. 1980; Vasthare et al. 1988; Mayhan et al. 1990; Jiang et al. 1992; Uchida et al. 2000; El-Assouad and Tayebati 2002). There is no difference between resting blood flow per gram of tissue in adult and senescent rats. However, following the administration of an anesthetic, cerebral blood flow in the senescent animals was significantly reduced compared to adult animals (Vasthare et al. 1988). Similarly, there is a reduction in the magnitude of reactive hyperemia following release of middle cerebral artery occlusion in senescent cats (Tomita et al. 1980). Diminished release of nitric oxide in response to a number of stimuli has been reported in aged animals (Mayhan et al. 1990). An enhanced response to vasoconstrictors has also been reported (Hajdu et al. 1993).

Alzheimer disease also results in a number of pathological alterations in the cerebral microvasculature. These abnormalities are visible with both light and electron microscopy. Under the light microscope, the microvessels appear irregular rather than smooth in contour (Scheibel 1987; Hashimura et al. 1991). With the use of the electron microscope, this was found to be the result of smooth muscle degeneration and swelling of astrocyte foot processes surrounding the vessels (Kimura et al. 1991; Yamashita et al. 1991). Endothelial cells have been shown to have reduced mitochondrial density (Stewart et al. 1992). These investigators also reported alterations in the number of tight junctions. Both factors implicate compromised function of the blood–brain barrier. The basement membrane also shows significant alterations in Alzheimer's disease (Figure 5). The contributing factors include accumulation of collagen, heparin sulfate proteoglycans, laminin and beta amyloid peptide. In addition to alterations in blood–brain barrier function, the presence of beta amyloid peptide in the wall of cerebral resistance vessels is reported to enhance vasoconstriction (Paris et al. 2000; Niwa et al. 2001). Amyloid-Beta has also been reported to negatively influence pericyte function. Leukocyte occlusion of capillary perfusion was found in a mouse model of Amyloid-beta accumulation (Schaeffer and Iadecola 2021). Higher resting Ca^{++} level has also been observed in Alzheimer's models, possibly contributing to enhanced constriction of cerebral arterioles (Kuchibhotla et al. 2009). Enhanced neuronal and microglial activity has also been reported to result from Alzheimer's disease (Terry et al. 1991; Wyss-Coray 2006; Venneti et al. 2008).

7.13 Conclusion

The structure and function of the cerebral microvessels are uniquely adapted to meet the specialized needs of the CNS. The intolerance of the CNS to even short periods of ischemia necessitates microvascular adaptations to maintain continuous adequate blood flow. The mechanisms contributing to the regulation of blood flow in the CNS are extremely complex. In addition to specialized adaptations of metabolic, myogenic and endothelial control mechanisms, there are unique aspects of neural control. Astrocytes have been proposed to make pivotal contributions to the rapid and exquisite coupling between changes in neuronal activity and regulation of cerebral blood flow. It also has become apparent that there is bidirectional communication among neurons, glia and microvessels which are now viewed to function as a neurovascular unit.

The sensitivity of neurons to alterations in the ionic composition of the extracellular fluid also places unique requirements for the control of vascular permeability. The structural and functional components of the BBB provide this protection with extremely intricate regulation of permeability. The permeability of hydrophilic substances is so limited that transporters are required to move most essential hydrophilic substances into and out of the CNS. There are also unique limitations to the movement of cells across the BBB. Although the CNS was for many years thought to be immune privileged, it is now recognized that the term "immune modified" is a more accurate description. While the movement of leukocytes is carefully restricted under normal conditions, it is clear that inflammatory responses and the invasion of the CNS by leukocytes play an important role in autoimmune disease and the exacerbation of damage following ischemia and trauma. The contribution of inflammatory regulating molecules such as cytokines and cannabinoids has become intense area of research interest. Alteration in both structure and the function of the cerebral microcirculation associated with aging and neurodegenerative disease leading to compromised function is also an area of interest.

References

Abbott NJ (2000) Inflammatory mediators and modulation of blood-brain barrier permeability. Cell Mol Neurobiol 20(2):131–147

Abbott NJ (2005) Dynamics of CNS barriers: evolution, differentiation, and modulation. Cell Mol Neurobiol 25(1):5–23

Abbott NJ, Revest PA (1991) Control of brain endothelial permeability. Cerebrovasc Brain Metab Rev 3(1):39–72

Adachi T, Biesold D et al (1990a) Stimulation of the nucleus basalis of Meynert and substantia innominata produces widespread increases in cerebral blood flow in the frontal, parietal and occipital cortices. Brain Res 514(1):163–166

Adachi T, Inanami O et al (1990b) Responses of regional cerebral blood flow following focal electrical stimulation of the nucleus basalis of Meynert and the medial septum using the [14C] iodoantipyrine method in rats. Neurosci Lett 112(2–3):263–268

Adamson P, Etienne S et al (1999) Lymphocyte migration through brain endothelial cell monolayers involves signaling through endothelial ICAM-1 via a rho-dependent pathway. J Immunol 162(5):2964–2973

Alkayed NJ, Harukuni I et al (1998) Gender-linked brain injury in experimental stroke. Stroke 29(1):159–165

Allt G, Lawrenson JG (1997) Is the pial microvessel a good model for blood-brain barrier studies? Brain Res Brain Res Rev 24(1):67–76

Andresen J, Shafi NI et al (2006) Endothelial influences on cerebrovascular tone. J Appl Physiol 100(1):318–327

Arimura A, Shioda S (1995) Pituitary adenylate cyclase activating polypeptide (PACAP) and its receptors: neuroendocrine and endocrine interaction. Front Neuroendocrinol 16(1):53–88
Aschner JL, Lum H et al (1997) Bradykinin- and thrombin-induced increases in endothelial permeability occur independently of phospholipase C but require protein kinase C activation. J Cell Physiol 173(3):387–396
Atochin DN, Demchenko IT et al (2003) Contributions of endothelial and neuronal nitric oxide synthases to cerebrovascular responses to hyperoxia. J Cereb Blood Flow Metab 23(10): 1219–1226
Attwell D, Iadecola C (2002) The neural basis of functional brain imaging signals. Trends Neurosci 25(12):621–625
Attwell D, Buchan AM et al (2010) Glial and neuronal control of brain blood flow. Nature 468(7321):232–243
Babcock AA, Kuziel WA et al (2003) Chemokine expression by glial cells directs leukocytes to sites of axonal injury in the CNS. J Neurosci 23(21):7922–7930
Balabanov R, Dore-Duffy P (1998) Role of the CNS microvascular pericyte in the blood-brain barrier. J Neurosci Res 53(6):637–644
Ballabh P, Braun A et al (2004) The blood-brain barrier: an overview: structure, regulation, and clinical implications. Neurobiol Dis 16(1):1–13
Bar T, Budi Santoso AW (1984) Identification of pericytes in the central nervous system by silver staining of the basal lamina. Cell Tissue Res 236(2):491–493
Baron J, Madri J et al (1993) Surface expression of alpha 4 integrin by CD4 T cells is required for their entry into brain parenchyma. J Exp Med 177(1):57–68
Barone FC, Tuma RF et al (2002) Brain inflammation, cytokines, and p38 MAP kinase signaling in stroke. In: Lin RCS (ed) Methods and new Frontiers in neuroscience. CRC Press, Boca Raton, pp 201–244
Barroso CP, Edvinsson L et al (1996) Nitroxidergic innervation of Guinea pig cerebral arteries. J Auton Nerv Syst 58(1–2):108–114
Bartlett IS, Segal SS (2000) Resolution of smooth muscle and endothelial pathways for conduction along hamster cheek pouch arterioles. Am J Physiol Heart Circ Physiol 278(2):H604–H612
Battistini L, Piccio L et al (2003) CD8+ T cells from patients with acute multiple sclerosis display selective increase of adhesiveness in brain venules: a critical role for P-selectin glycoprotein ligand-1. Blood 101(12):4775–4782
Baumbach GL, Heistad DD (1983) Effects of sympathetic stimulation and changes in arterial pressure on segmental resistance of cerebral vessels in rabbits and cats. Circ Res 52(5):527–533
Baumbach GL, Walmsley JG et al (1988) Composition and mechanics of cerebral arterioles in hypertensive rats. Am J Pathol 133(3):464–471
Baumbach GL, Sigmund CD et al (2003) Cerebral arteriolar structure in mice overexpressing human renin and angiotensinogen. Hypertension 41(1):50–55
Bayliss WM (1902) On the local reactions of the arterial wall to changes of internal pressure. J Physiol 28(3):220–231
Bednar MM, Raymond S et al (1991) The role of neutrophils and platelets in a rabbit model of thromboembolic. Stroke 22(1):44–50
Bekar LK, Wei HS et al (2012) The locus coeruleus-norepinephrine network optimizes coupling of cerebral blood volume with oxygen demand. J Cereb Blood Flow Metab 32(12):2135–2145
Bell MA, Ball MJ (1981) Morphometric comparison of hippocampal microvasculature in ageing and demented people: diameters and densities. Acta Neuropathol (Berl) 53(4):299–318
Benediktsson AM, Schachtele SJ et al (2005) Ballistic labeling and dynamic imaging of astrocytes in organotypic hippocampal slice cultures. J Neurosci Methods 141(1):41–53
Bevan JA, Buga GM et al (1986) In vitro evidence that vasoactive intestinal peptide is a transmitter of neuro-vasodilation in the head of the cat. Neuroscience 19(2):597–604
Biesold D, Inanami O et al (1989) Stimulation of the nucleus basalis of Meynert increases cerebral cortical blood flow in rats. Neurosci Lett 98(1):39–44

Black JE, Isaacs KR et al (1991) Usual vs. successful aging: some notes on experiential factors. Neurobiol Aging 12(4):325–328

Bowes MP, Rothlein R et al (1995) Monoclonal antibodies preventing leukocyte activation reduce experimental neurologic injury and enhance efficacy of thrombolytic therapy. Neurology 45(4): 815–819

Branston NM (1995) The physiology of the cerebrovascular parasympathetic innervation. Br J Neurosurg 9(3):319–329

Brown RC, Davis TP (2002) Calcium modulation of adherens and tight junction function: a potential mechanism for blood-brain barrier disruption after stroke. Stroke 33(6):1706–1711

Bryan RM Jr, You J et al (2005) Endothelium-derived hyperpolarizing factor: a cousin to nitric oxide and prostacyclin. Anesthesiology 102(6):1261–1277

Bundgaard M (1986) Pathways across the vertebrate blood-brain barrier: morphological view-points. Ann N Y Acad Sci 481:7–19

Burridge K, Wennerberg K (2004) Rho and Rac take center stage. Cell 116(2):167–179

Bushong EA, Martone ME et al (2002) Protoplasmic astrocytes in CA1 stratum radiatum occupy separate anatomical domains. J Neurosci 22(1):183–192

Bushong EA, Martone ME et al (2004) Maturation of astrocyte morphology and the establishment of astrocyte domains during postnatal hippocampal development. Int J Dev Neurosci 22(2): 73–86

Busija DW, Heistad DD (1984) Factors involved in the physiological regulation of the cerebral circulation. Rev Physiol Biochem Pharmacol 101:161–211

Cannella B, Raine CS (1995) The adhesion molecule and cytokine profile of multiple sclerosis lesions. Ann Neurol 37(4):424–435

Cao WH, Inanami O et al (1989) Stimulation of the septal complex increases local cerebral blood flow in the hippocampus in anesthetized rats. Neurosci Lett 107(1–3):135–140

Carrithers MD, Visintin I et al (2000) Differential adhesion molecule requirements for immune surveillance and inflammatory recruitment. Brain 123(Pt 6):1092–1101

Carvalho-Tavares J, Hickey MJ et al (2000) A role for platelets and endothelial selectins in tumor necrosis factor-alpha-induced leukocyte recruitment in the brain microvasculature. Circ Res 87(12):1141–1148

Cauli B, Tong XK et al (2004) Cortical GABA interneurons in neurovascular coupling: relays for subcortical vasoactive pathways. J Neurosci 24(41):8940–8949

Chandrasekharan NV, Dai H et al (2002) COX-3, a cyclooxygenase-1 variant inhibited by acetaminophen and other analgesic/antipyretic drugs: cloning, structure, and expression. Proc Natl Acad Sci USA 99(21):13926–13931

Chopp M, Zhang RL et al (1994) Postischemic administration of an anti-mac-1 antibody reduces ischemic cell damage after transient middle cerebral artery occlusion in rats. Stroke 25(4): 869–875

Cirrito JR, Deane R et al (2005) P-glycoprotein deficiency at the blood-brain barrier increases amyloid-beta deposition in an Alzheimer disease mouse model. J Clin Invest 115(11): 3285–3290

Cohen Z, Bonvento G et al (1996) Serotonin in the regulation of brain microcirculation. Prog Neurobiol 50(4):335–362

Cohen SA, Trikha M et al (2000) Potential future clinical applications for the GPIIb/IIIa antagonist, abciximab in thrombosis, vascular and oncological indications. Pathol Oncol Res 6(3):163–174

Cook-Mills JM (2002) VCAM-1 signals during lymphocyte migration: role of reactive oxygen species. Mol Immunol 39(9):499–508

Cox SB, Woolsey TA et al (1993) Localized dynamic changes in cortical blood flow with whisker stimulation corresponds to matched vascular and neuronal architecture of rat barrels. J Cereb Blood Flow Metab 13(6):899–913

Crocker DJ, Murad TM et al (1970) Role of the pericyte in wound healing. An ultrastructural study. Exp Mol Pathol 13(1):51–65

Cullere X, Shaw SK et al (2005) Regulation of vascular endothelial barrier function by Epac, a cAMP-activated exchange factor for rap GTPase. Blood 105(5):1950–1955
del Zoppo GJ, Mabuchi T (2003) Cerebral microvessel responses to focal ischemia. J Cereb Blood Flow Metab 23(8):879–894
del Zoppo G, Ginis I et al (2000) Inflammation and stroke: putative role for cytokines, adhesion molecules and iNOS in brain response to ischemia. Brain Pathol 10(1):95–112
Demchenko IT, Oury TD et al (2002) Regulation of the brain's vascular responses to oxygen. Circ Res 91(11):1031–1037
Demeule M, Regina A et al (2002) Drug transport to the brain: key roles for the efflux pump P-glycoprotein in the blood-brain barrier. Vasc Pharmacol 38(6):339–348
Devor A, Sakadzic S et al (2011) "overshoot" of O(2) is required to maintain baseline tissue oxygenation at locations distal to blood vessels. J Neurosci 31(38):13676–13681
Diaz-Flores L, Gutierrez R et al (1991) Microvascular pericytes: a review of their morphological and functional characteristics. Histol Histopathol 6(2):269–286
Dietrich HH, Kajita Y et al (1996) Local and conducted vasomotor responses in isolated rat cerebral arterioles. Am J Phys 271(3 Pt 2):H1109–H1116
Doolittle ND, Miner ME et al (2000) Safety and efficacy of a multicenter study using intraarterial chemotherapy in conjunction with osmotic opening of the blood-brain barrier for the treatment of patients with malignant brain tumors. Cancer 88(3):637–647
Duling BR, Kuschinsky W et al (1979) Measurements of the perivascular PO2 in the vicinity of the pial vessels of the cat. Pflugers Arch 383(1):29–34
Dutka AJ, Kochanek PM et al (1989) Influence of granulocytopenia on canine cerebral ischemia induced by air. Stroke 20(3):390–395
Duverger D, Edvinsson L et al (1987) Concentrations of putative neurovascular transmitters in major cerebral arteries and small pial vessels of various species. J Cereb Blood Flow Metab 7(4): 497–501
Easton AS, Dorovini-Zis K (2001) The kinetics, function, and regulation of P-selectin expressed by human brain microvessel endothelial cells in primary culture. Microvasc Res 62(3):335–345
Edvinsson L (1991) Innervation and effects of dilatory neuropeptides on cerebral vessels. New aspects. Blood Vessels 28(1–3):35–45
Edvinsson L (2006) Neuropeptide Y and the cerebral circulation. EXS 95:105–112
Edvinsson L, Rosendal-Helgesen S et al (1983) Substance P: localization, concentration and release in cerebral arteries, choroid plexus and dura mater. Cell Tissue Res 234(1):1–7
Edvinsson L, Ekman R et al (1987) Calcitonin gene-related peptide and cerebral blood vessels: distribution and vasomotor effects. J Cereb Blood Flow Metab 7(6):720–728
Edvinsson L, Brodin E et al (1988) Neurokinin a in cerebral vessels: characterization, localization and effects in vitro. Regul Pept 20(3):181–197
Edvinsson L, Delgado-Zygmunt T et al (1990a) Involvement of perivascular sensory fibers in the pathophysiology of cerebral vasospasm following subarachnoid hemorrhage. J Cereb Blood Flow Metab 10(5):602–607
Edvinsson L, Uddman R et al (1990b) Peptidergic innervation of the cerebral circulation. Role in subarachnoid hemorrhage in man. Neurosurg Rev 13(4):265–272
Edvinsson L, Mulder H et al (1998) Calcitonin gene-related peptide and nitric oxide in the trigeminal ganglion: cerebral vasodilatation from trigeminal nerve stimulation involves mainly calcitonin gene-related peptide. J Auton Nerv Syst 70(1–2):15–22
El-Assouad D, Tayebati SK (2002) Cholinergic innervation of pial arteries in senescent rats: an immunohistochemical study. Mech Ageing Dev 123(5):529–536
Elhusseiny A, Cohen Z et al (1999) Functional acetylcholine muscarinic receptor subtypes in human brain microcirculation: identification and cellular localization. J Cereb Blood Flow Metab 19(7):794–802
Emerson GG, Segal SS (2000) Endothelial cell pathway for conduction of hyperpolarization and vasodilation along hamster feed artery. Circ Res 86(1):94–100
Endres M (2005) Statins and stroke. J Cereb Blood Flow Metab 25(9):1093–1110

Engelhardt B, Ransohoff RM (2005) The ins and outs of T-lymphocyte trafficking to the CNS: anatomical sites and molecular mechanisms. Trends Immunol 26(9):485–495
Engelhardt B, Martin-Simonet MT et al (1998) Adhesion molecule phenotype of T lymphocytes in inflamed CNS. J Neuroimmunol 84(1):92–104
Engelhardt B, Wolburg-Buchholz K et al (2001) Involvement of the choroid plexus in central nervous system inflammation. Microsc Res Tech 52(1):112–129
Fabry Z, Fitzsimmons KM et al (1993) Production of the cytokines interleukin 1 and 6 by murine brain microvessel endothelium and smooth muscle pericytes. J Neuroimmunol 47(1):23–34
Faraci FM (1989) Effects of endothelin and vasopressin on cerebral blood vessels. Am J Phys 257(3 Pt 2):H799–H803
Faraci FM, Heistad DD (1990) Regulation of large cerebral arteries and cerebral microvascular pressure. Circ Res 66(1):8–17
Faraci FM, Mayhan WG et al (1987) Segmental vascular responses to acute hypertension in cerebrum and brain stem. Am J Phys 252(4 Pt 2):H738–H742
Faraci FM, Mayhan WG et al (1988) Effects of arginine vasopressin on cerebral microvascular pressure. Am J Phys 255(1 Pt 2):H70–H76
Faraci FM, Baumbach GL et al (1990) Cerebral circulation: humoral regulation and effects of chronic hypertension. J Am Soc Nephrol 1(1):53–57
Faraci FM, Kinzenbaw D et al (1994) Effect of endogenous vasopressin on blood flow to choroid plexus during hypoxia and intracranial hypertension. Am J Phys 266(2 Pt 2):H393–H398
Faraci FM, Lamping KG et al (2006) Cerebral vascular effects of angiotensin II: new insights from genetic models. J Cereb Blood Flow Metab 26(4):449–455
Faraci FM, Taugher RJ et al (2019) Acid-sensing ion channels: novel mediators of cerebral vascular responses. Circ Res 125(10):907–920
Farkas E, Luiten PG (2001) Cerebral microvascular pathology in aging and Alzheimer's disease. Prog Neurobiol 64(6):575–611
Farrell CR, Stewart PA et al (1987) Pericytes in human cerebral microvasculature. Anat Rec 218(4): 466–469
Fellin T, Carmignoto G (2004) Neurone-to-astrocyte signalling in the brain represents a distinct multifunctional unit. J Physiol 559(Pt 1):3–15
Fernandez-Klett F, Offenhauser N et al (2010) Pericytes in capillaries are contractile in vivo, but arterioles mediate functional hyperemia in the mouse brain. Proc Natl Acad Sci USA 107(51): 22290–22295
Filosa JA, Bonev AD et al (2004) Calcium dynamics in cortical astrocytes and arterioles during neurovascular coupling. Circ Res 95(10):e73–e81
Folkman J, D'Amore PA (1996) Blood vessel formation: what is its molecular basis? Cell 87(7): 1153–1155
Fox PT, Raichle ME (1986) Focal physiological uncoupling of cerebral blood flow and oxidative metabolism during somatosensory stimulation in human subjects. Proc Natl Acad Sci USA 83(4):1140–1144
Fricker G, Miller DS (2004) Modulation of drug transporters at the blood-brain barrier. Pharmacology 70(4):169–176
Furuse M, Sasaki H et al (1999) Manner of interaction of heterogeneous claudin species within and between tight junction strands. J Cell Biol 147(4):891–903
Ghirnikar RS, Lee YL et al (2000) Chemokine antagonist infusion attenuates cellular infiltration following spinal cord contusion injury in rat. J Neurosci Res 59(1):63–73
Ghirnikar RS, Lee YL et al (2001) Chemokine antagonist infusion promotes axonal sparing after spinal cord contusion injury in rat. J Neurosci Res 64(6):582–589
Girouard H, Iadecola C (2006) Neurovascular coupling in the normal brain and in hypertension, stroke, and Alzheimer disease. J Appl Physiol 100(1):328–335
Girouard H, Park L et al (2006) Angiotensin II attenuates endothelium-dependent responses in the cerebral microcirculation through nox-2-derived radicals. Arterioscler Thromb Vasc Biol 26(4): 826–832

Girouard H, Bonev AD et al (2010) Astrocytic endfoot Ca2+ and BK channels determine both arteriolar dilation and constriction. Proc Natl Acad Sci USA 107(8):3811–3816
Goadsby PJ, Uddman R et al (1996) Cerebral vasodilatation in the cat involves nitric oxide from parasympathetic nerves. Brain Res 707(1):110–118
Golden PL, Pollack GM (2003) Blood-brain barrier efflux transport. J Pharm Sci 92(9):1739–1753
Golding EM, You J et al (2001) Potentiated endothelium-derived hyperpolarizing factor-mediated dilations in cerebral arteries following mild head injury. J Neurotrauma 18(7):691–697
Golding EM, Marrelli SP et al (2002) Endothelium-derived hyperpolarizing factor in the brain: a new regulator of cerebral blood flow? Stroke 33(3):661–663
Goldman EE (1909) Die aussere und innere Sekretion des gesunden und kranken organismus imlichte der "vitalen farbung". Beitraege Klinishen Churgurie 64:192–265
Goldman EE (1913) Vitalfarbung am zentralnervensystme. Abhandlungen Preussischen Academie der Wissenschaften Physikalisch Mathematisch klasse 1:1–60
Goldstein GW, Betz AL (1986) The blood-brain barrier. Sci Am 255(3):74–83
Gordon GR, Choi HB et al (2008) Brain metabolism dictates the polarity of astrocyte control over arterioles. Nature 456(7223):745–749
Gordon GR, Howarth C et al (2011) Bidirectional control of arteriole diameter by astrocytes. Exp Physiol 96(4):393–399
Goussev AV, Zhang Z et al (1998) P-selectin antibody reduces hemorrhage and infarct volume resulting from MCA occlusion in the rat. J Neurol Sci 161(1):16–22
Graeber MB, Streit WJ et al (1992) Ultrastructural location of major histocompatibility complex (MHC) class II positive perivascular cells in histologically normal human brain. J Neuropathol Exp Neurol 51(3):303–311
Greter M, Heppner FL et al (2005) Dendritic cells permit immune invasion of the CNS in an animal model of multiple sclerosis. Nat Med 11(3):328–334
Gulbenkian S, Uddman R et al (2001) Neuronal messengers in the human cerebral circulation. Peptides 22(6):995–1007
Gumbiner BM (1996) Cell adhesion: the molecular basis of tissue architecture and morphogenesis. Cell 84(3):345–357
Gupta AK, Menon DK et al (1997) Non-invasive measurement of cerebral blood volume in volunteers. Br J Anaesth 78(1):39–43
Guthrie PB, Knappenberger J et al (1999) ATP released from astrocytes mediates glial calcium waves. J Neurosci 19(2):520–528
Haas TL, Duling BR (1997) Morphology favors an endothelial cell pathway for longitudinal conduction within arterioles. Microvasc Res 53(2):113–120
Hajdu MA, Heistad DD et al (1990) Effects of aging on mechanics and composition of cerebral arterioles in rats. Circ Res 66(6):1747–1754
Hajdu MA, McElmurry RT et al (1993) Effects of aging on cerebral vascular responses to serotonin in rats. Am J Phys 264(6 Pt 2):H2136–H2140
Hall CN, Reynell C et al (2014) Capillary pericytes regulate cerebral blood flow in health and disease. Nature 508(7494):55–60
Hallenbeck JM (1997) Cytokines, macrophages, and leukocytes in brain ischemia. Neurology 49(5 Suppl 4):S5–S9
Hamel E (2006) Perivascular nerves and the regulation of cerebrovascular tone. J Appl Physiol 100(3):1059–1064
Harder DR (1985) A cellular mechanism for myogenic regulation of cat cerebral arteries. Ann Biomed Eng 13(3–4):335–339
Harder DR, Lange AR et al (1997) Cytochrome P450 metabolites of arachidonic acid as intracellular signaling molecules in vascular tissue. J Vasc Res 34(3):237–243
Harder DR, Alkayed NJ et al (1998) Functional hyperemia in the brain: hypothesis for astrocyte-derived vasodilator metabolites. Stroke 29(1):229–234
Harrison MJ (1989) Influence of haematocrit in the cerebral circulation. Cerebrovasc Brain Metab Rev 1(1):55–67

Hart MN, Heistad DD et al (1980) Effect of chronic hypertension and sympathetic denervation on wall/lumen ratio of cerebral vessels. Hypertension 2(4):419–423

Hartmann DA, Berthiaume AA et al (2021) Brain capillary pericytes exert a substantial but slow influence on blood flow. Nat Neurosci 24(5):633–645

Hashimura T, Kimura T et al (1991) Morphological changes of blood vessels in the brain with Alzheimer's disease. Jpn J Psychiatry Neurol 45(3):661–665

Hauben E, Schwartz M (2003) Therapeutic vaccination for spinal cord injury: helping the body to cure itself. Trends Pharmacol Sci 24(1):7–12

Heinel LA, Rubin S et al (1994) Leukocyte involvement in cerebral infarct generation after ischemia and. Brain Res Bull 34(2):137–141

Heistad DD (2001) What's new in the cerebral microcirculation? Landis award lecture. Microcirculation 8(6):365–375

Heistad DD, Kontos HA (1983) Cerebral circulation. In: Shepherd JT, Abboud FM (eds) Handbood of physiology: the cardiovascular system, vol 3. American Physiological Society, Bethesda, pp 137–182

Heo JH, Lucero J et al (1999) Matrix metalloproteinases increase very early during experimental focal cerebral ischemia. J Cereb Blood Flow Metab 19(6):624–633

Hickey WF (2001) Basic principles of immunological surveillance of the normal central nervous system. Glia 36(2):118–124

Hickey WF, Vass K et al (1992) Bone marrow-derived elements in the central nervous system: an immunohistochemical and ultrastructural survey of rat chimeras. J Neuropathol Exp Neurol 51(3):246–256

Hill RA, Tong L et al (2015) Regional blood flow in the Normal and ischemic brain is controlled by arteriolar smooth muscle cell contractility and not by capillary Pericytes. Neuron 87(1):95–110

Hirase T, Kawashima S et al (2001) Regulation of tight junction permeability and occludin phosphorylation by Rhoa-p160ROCK-dependent and -independent mechanisms. J Biol Chem 276(13):10423–10431

Hirrlinger J, Hulsmann S et al (2004) Astroglial processes show spontaneous motility at active synaptic terminals in situ. Eur J Neurosci 20(8):2235–2239

Hirschi KK, D'Amore PA (1996) Pericytes in the microvasculature. Cardiovasc Res 32(4):687–698

Holash JA, Noden DM et al (1993) Re-evaluating the role of astrocytes in blood-brain barrier induction. Dev Dyn 197(1):14–25

Horiuchi T, Dietrich HH et al (2002) Mechanism of extracellular K+-induced local and conducted responses in cerebral penetrating arterioles. Stroke 33(11):2692–2699

Horlyck S, Cai C et al (2021) ATP induces contraction of cultured brain capillary pericytes via activation of P2Y-type purinergic receptors. Am J Physiol Heart Circ Physiol 320(2):H699–H712

Huber JD, Egleton RD et al (2001) Molecular physiology and pathophysiology of tight junctions in the blood-brain barrier. Trends Neurosci 24(12):719–725

Hudetz AG (1997) Blood flow in the cerebral capillary network: a review emphasizing observations with intravital microscopy. Microcirculation 4(2):233–252

Hudetz AG, Wood JD et al (1999) Effect of hemodilution on RBC velocity, supply rate, and hematocrit in the cerebral capillary network. J Appl Physiol 87(2):505–509

Hurn PD, Littleton-Kearney MT et al (1995) Postischemic cerebral blood flow recovery in the female: effect of 17 beta-estradiol. J Cereb Blood Flow Metab 15(4):666–672

Hutchins PM, Lynch CD et al (1996) The microcirculation in experimental hypertension and aging. Cardiovasc Res 32(4):772–780

Iadecola C (1992) Does nitric oxide mediate the increases in cerebral blood flow elicited by hypercapnia? Proc Natl Acad Sci USA 89(9):3913–3916

Iadecola C (1998) Neurogenic control of the cerebral microcirculation: is dopamine minding the store? Nat Neurosci 1(4):263–265

Iadecola C (2004) Neurovascular regulation in the normal brain and in Alzheimer's disease. Nat Rev Neurosci 5(5):347–360

Ibayashi S, Ngai AC et al (1991) Effects of topical adenosine analogs and forskolin on rat pial arterioles in vivo. J Cereb Blood Flow Metab 11(1):72–76
Janigro D, Wender R et al (1996) Adenosine-induced release of nitric oxide from cortical astrocytes. Neuroreport 7(10):1640–1644
Janigro D, Gasparini S et al (1997) Reduction of K+ uptake in glia prevents long-term depression maintenance and causes epileptiform activity. J Neurosci 17(8):2813–2824
Janzer RC, Raff MC (1987) Astrocytes induce blood-brain barrier properties in endothelial cells. Nature 325(6101):253–257
Jiang HX, Chen PC et al (1992) Age related alterations in the response of the pial arterioles to adenosine in the rat. Mech Ageing Dev 65(2–3):257–276
Jiang N, Zhang RL et al (1994) Anti-CD11b monoclonal antibody reduces ischemic cell damage after transient but not after permanenet MCA occlusion in the rat. Neurosci Res Commun 15: 85–93
Jiang N, Moyle M et al (1995) Neutrophil inhibitory factor is neuroprotective after focal ischemia in. Ann Neurol 38(6):935–942
Jiang N, Chopp M et al (1998) Neutrophil inhibitory factor treatment of focal cerebral ischemia in the rat. Brain Res 788(1–2):25–34
Johnson TD, Marrelli SP et al (1998) Inward rectifier potassium channels in the rat middle cerebral artery. Am J Phys 274(2 Pt 2):R541–R547
Joo F (1993) The role of second messenger molecules in the regulation of permeability in the cerebral endothelial cells. Adv Exp Med Biol 331:155–164
Kamiya T, Katayama Y et al (1993) The role of bradykinin in mediating ischemic brain edema in rats. Stroke 24(4):571–575
Kazama K, Anrather J et al (2004) Angiotensin II impairs neurovascular coupling in neocortex through NADPH oxidase-derived radicals. Circ Res 95(10):1019–1026
Kety SS (1964) The cerebral circulation. In: Fishman AP, Richards DW (eds) Circulation of the blood: men and ideas, vol 30. Oxford University Press, New York, pp 703–742
Keuker JI, Luiten PG et al (2000) Capillary changes in hippocampal CA1 and CA3 areas of the aging rhesus monkey. Acta Neuropathol (Berl) 100(6):665–672
Keyeux A, Ochrymowicz-Bemelmans D et al (1995) Induced response to hypercapnia in the two-compartment total cerebral blood volume: influence on brain vascular reserve and flow efficiency. J Cereb Blood Flow Metab 15(6):1121–1131
Kida S, Steart PV et al (1993) Perivascular cells act as scavengers in the cerebral perivascular spaces and remain distinct from pericytes, microglia and macrophages. Acta Neuropathol (Berl) 85(6): 646–652
Kimura T, Hashimura T et al (1991) Observations of microvessels in the brain with Alzheimer's disease by the scanning electron microscopy. Jpn J Psychiatry Neurol 45(3):671–676
Kis B, Snipes JA et al (2003) Putative cyclooxygenase-3 expression in rat brain cells. J Cereb Blood Flow Metab 23(11):1287–1292
Kis B, Snipes A et al (2004) Regional distribution of cyclooxygenase-3 mRNA in the rat central nervous system. Brain Res Mol Brain Res 126(1):78–80
Kis B, Snipes JA et al (2005a) Acetaminophen and the cyclooxygenase-3 puzzle: sorting out facts, fictions, and uncertainties. J Pharmacol Exp Ther 315(1):1–7
Kis B, Snipes JA et al (2005b) Acetaminophen-sensitive prostaglandin production in rat cerebral endothelial cells. Am J Physiol Regul Integr Comp Physiol 288(4):R897–R902
Kivisakk P, Mahad DJ et al (2003) Human cerebrospinal fluid central memory CD4+ T cells: evidence for trafficking through choroid plexus and meninges via P-selectin. Proc Natl Acad Sci USA 100(14):8389–8394
Knot HJ, Zimmermann PA et al (1996) Extracellular K(+)-induced hyperpolarizations and dilatations of rat coronary and cerebral arteries involve inward rectifier K(+) channels. J Physiol 492 (Pt 2):419–430
Knox CA, Oliveira A (1980) Brain aging in normotensive and hypertensive strains of rats. III. A quantitative study of cerebrovasculature. Acta Neuropathol (Berl) 52(1):17–25

Knox CA, Yates RD et al (1980) Effects of aging on the structural and permeability characteristics of cerebrovasculature in normotensive and hypertensive strains of rats. Acta Neuropathol (Berl) 51(1):1–13

Kochanek PM, Hallenbeck JM (1992) Polymorphonuclear leukocytes and monocytes/macrophages in the. Stroke 23(9):1367–1379

Kortekaas R, Leenders KL et al (2005) Blood-brain barrier dysfunction in parkinsonian midbrain in vivo. Ann Neurol 57(2):176–179

Kraemer DF, Fortin D et al (2002) Chemotherapeutic dose intensification for treatment of malignant brain tumors: recent developments and future directions. Curr Neurol Neurosci Rep 2(3): 216–224

Krimer LS, Muly EC 3rd et al (1998) Dopaminergic regulation of cerebral cortical microcirculation. Nat Neurosci 1(4):286–289

Krum JM, Kenyon KL et al (1997) Expression of blood-brain barrier characteristics following neuronal loss and astroglial damage after administration of anti-Thy-1 immunotoxin. Exp Neurol 146(1):33–45

Krumbholz M, Theil D et al (2006) Chemokines in multiple sclerosis: CXCL12 and CXCL13 up-regulation is differentially linked to CNS immune cell recruitment. Brain 129(Pt 1):200–211

Kuchibhotla KV, Lattarulo CR et al (2009) Synchronous hyperactivity and intercellular calcium waves in astrocytes in Alzheimer mice. Science 323(5918):1211–1215

Kur J, Newman EA (2014) Purinergic control of vascular tone in the retina. J Physiol 592(3): 491–504

Laman JD, Weller RO (2013) Drainage of cells and soluble antigen from the CNS to regional lymph nodes. J Neuroimmune Pharmacol 8(4):840–856

Laschinger M, Vajkoczy P et al (2002) Encephalitogenic T cells use LFA-1 for transendothelial migration but not during capture and initial adhesion strengthening in healthy spinal cord microvessels in vivo. Eur J Immunol 32(12):3598–3606

Lassegue B, Griendling KK (2004) Reactive oxygen species in hypertension; an update. Am J Hypertens 17(9):852–860

Laudignon N, Beharry K et al (1991) The role of adenosine in the vascular adaptation of neonatal cerebral blood flow during hypotension. J Cereb Blood Flow Metab 11(3):424–431

Lawrenson JG, Reid AR et al (1997) Molecular characteristics of pial microvessels of the rat optic nerve. Can pial microvessels be used as a model for the blood-brain barrier? Cell Tissue Res 288(2):259–265

Lazarov-Spiegler O, Rapalino O et al (1998) Restricted inflammatory reaction in the CNS: a key impediment to axonal regeneration? Mol Med Today 4(8):337–342

Lee TJ (2002) Sympathetic modulation of nitrergic neurogenic vasodilation in cerebral arteries. Jpn J Pharmacol 88(1):26–31

Lee G, Bendayan R (2004) Functional expression and localization of P-glycoprotein in the central nervous system: relevance to the pathogenesis and treatment of neurological disorders. Pharm Res 21(8):1313–1330

Lee YL, Shih K et al (2000) Cytokine chemokine expression in contused rat spinal cord. Neurochem Int 36(4–5):417–425

Leffler CW, Busija DW (1987) Prostanoids and pial arteriolar diameter in hypotensive newborn pigs. Am J Phys 252(4 Pt 2):H687–H691

Leffler CW, Busija DW et al (1987) Effect of therapeutic dose of indomethacin on the cerebral circulation of newborn pigs. Pediatr Res 21(2):188–192

Legos JJ, Tuma RF et al (2002) Pharmacological interventions for stroke: failures and future. Expert Opin Investig Drugs 11(5):603–614

Leventhal C, Rafii S et al (1999) Endothelial trophic support of neuronal production and recruitment from the adult mammalian subependyma. Mol Cell Neurosci 13(6):450–464

Lewis NC, Messinger L et al (2014) Effect of acute hypoxia on regional cerebral blood flow: effect of sympathetic nerve activity. J Appl Physiol (1985) 116(9):1189–1196

Lidington D, Kroetsch JT et al (2018) Cerebral artery myogenic reactivity: the next frontier in developing effective interventions for subarachnoid hemorrhage. J Cereb Blood Flow Metab 38(1):17–37

Lin SZ, Chiou TL et al (1995) Hemodilution accelerates the passage of plasma (not red cells) through cerebral microvessels in rats. Stroke 26(11):2166–2171

Lincoln J (1995) Innervation of cerebral arteries by nerves containing 5-hydroxytryptamine and noradrenaline. Pharmacol Ther 68(3):473–501

Lindauer U, Kunz A et al (2001) Nitric oxide from perivascular nerves modulates cerebral arterial pH reactivity. Am J Physiol Heart Circ Physiol 281(3):H1353–H1363

Littleton-Kearney MT, Agnew DM et al (2000) Effects of estrogen on cerebral blood flow and pial microvasculature in rabbits. Am J Physiol Heart Circ Physiol 279(3):H1208–H1214

Loscher W, Potschka H (2005a) Drug resistance in brain diseases and the role of drug efflux transporters. Nat Rev Neurosci 6(8):591–602

Loscher W, Potschka H (2005b) Role of drug efflux transporters in the brain for drug disposition and treatment of brain diseases. Prog Neurobiol 76(1):22–76

Lou HC, Edvinsson L et al (1987) The concept of coupling blood flow to brain function: revision required? Ann Neurol 22(3):289–297

Luiten PG, de Jong GI et al (1996) Ultrastructural localization of cholinergic muscarinic receptors in rat brain cortical capillaries. Brain Res 720(1–2):225–229

Macfarlane R, Moskowitz MA et al (1991a) Postischemic cerebral blood flow and neuroeffector mechanisms. Blood Vessels 28(1–3):46–51

Macfarlane R, Tasdemiroglu E et al (1991b) Chronic trigeminal ganglionectomy or topical capsaicin application to pial vessels attenuates postocclusive cortical hyperemia but does not influence postischemic hypoperfusion. J Cereb Blood Flow Metab 11(2):261–271

MacVicar BA, Newman EA (2015) Astrocyte regulation of blood flow in the brain. Cold Spring Harb Perspect Biol 7(5)

Madara JL, Parkos C et al (1992) The movement of solutes and cells across tight junctions. Ann N Y Acad Sci 664:47–60

Mahad D, Callahan MK et al (2006) Modulating CCR2 and CCL2 at the blood-brain barrier: relevance for multiple sclerosis pathogenesis. Brain 129(Pt 1):212–223

Maier-Hauff K, Baethmann AJ et al (1984) The kallikrein-kinin system as mediator in vasogenic brain edema. Part 2: studies on kinin formation in focal and perifocal brain tissue. J Neurosurg 61(1):97–106

Maisonpierre PC, Suri C et al (1997) Angiopoietin-2, a natural antagonist for Tie2 that disrupts in vivo angiogenesis. Science 277(5322):55–60

Marella M, Chabry J (2004) Neurons and astrocytes respond to prion infection by inducing microglia recruitment. J Neurosci 24(3):620–627

Markowitz S, Saito K et al (1987) Neurogenically mediated leakage of plasma protein occurs from blood vessels in dura mater but not brain. J Neurosci 7(12):4129–4136

Marrelli SP, Johnson TD et al (1998) Altered function of inward rectifier potassium channels in cerebrovascular smooth muscle after ischemia/reperfusion. Stroke 29(7):1469–1474

Marroni M, Marchi N et al (2003) Vascular and parenchymal mechanisms in multiple drug resistance: a lesson from human epilepsy. Curr Drug Targets 4(4):297–304

Martin-Padura I, Lostaglio S et al (1998) Junctional adhesion molecule, a novel member of the immunoglobulin superfamily that distributes at intercellular junctions and modulates monocyte transmigration. J Cell Biol 142(1):117–127

Mato M, Aikawa E et al (1986) Tridimensional observation of fluorescent granular perithelial (FGP) cells in rat cerebral blood vessels. Anat Rec 215(4):413–419

Matsuo Y, Onodera H et al (1994) Correlation between myeloperoxidase-quantified neutrophil accumulation. Stroke 25(7):1469–1475

Matsuo Y, Kihara T et al (1995) Role of neutrophils in radical production during ischemia and reperfusion. J Cereb Blood Flow Metab 15(6):941–947

Mayhan WG (2001) Regulation of blood-brain barrier permeability. Microcirculation 8(2):89–104

Mayhan WG, Faraci FM et al (1990) Effects of aging on responses of cerebral arterioles. Am J Phys 258(4 Pt 2):H1138–H1143

McCalden TA, Bevan JA (1981) Sources of activator calcium in rabbit basilar artery. Am J Phys 241(2):H129–H133

McCarron JG, Halpern W (1990a) Impaired potassium-induced dilation in hypertensive rat cerebral arteries does not reflect altered Na+,K(+)-ATPase dilation. Circ Res 67(4):1035–1039

McCarron JG, Halpern W (1990b) Potassium dilates rat cerebral arteries by two independent mechanisms. Am J Phys 259(3 Pt 2):H902–H908

McGuire JJ, Ding H et al (2001) Endothelium-derived relaxing factors: a focus on endothelium-derived hyperpolarizing factor(s). Can J Physiol Pharmacol 79(6):443–470

McMahon EJ, Bailey SL et al (2005) Epitope spreading initiates in the CNS in two mouse models of multiple sclerosis. Nat Med 11(3):335–339

McNeill AM, Kim N et al (1999) Chronic estrogen treatment increases levels of endothelial nitric oxide synthase protein in rat cerebral microvessels. Stroke 30(10):2186–2190

McTigue DM, Tani M et al (1998) Selective chemokine mRNA accumulation in the rat spinal cord after contusion injury. J Neurosci Res 53(3):368–376

Mi H, Haeberle H et al (2001) Induction of astrocyte differentiation by endothelial cells. J Neurosci 21(5):1538–1547

Miyauchi T, Masaki T (1999) Pathophysiology of endothelin in the cardiovascular system. Annu Rev Physiol 61:391–415

Mocco J, Mack WJ et al (2006) Complement component C3 mediates inflammatory injury following focal cerebral ischemia. Circ Res 99(2):209–217

Mori E, del Zoppo GJ et al (1992) Inhibition of polymorphonuclear leukocyte adherence suppresses no-reflow. Stroke 23(5):712–718

Moskowitz MA, Macfarlane R et al (1990) Neurogenic control of the cerebral circulation during global ischemia. Stroke 21(11 Suppl):168–171

Muller WA (2003) Leukocyte-endothelial-cell interactions in leukocyte transmigration and the inflammatory response. Trends Immunol 24(6):327–334

Mulligan SJ, MacVicar BA (2004) Calcium transients in astrocyte endfeet cause cerebrovascular constrictions. Nature 431(7005):195–199

Murabe Y, Sano Y (1981) Morphological studies on neuroglia. I. Electron microscopic identification of silver-impregnated glial cells. Cell Tissue Res 216(3):557–568

Nag S (2003) Morphology and molecular properties of cellular components of normal cerebral vessels. Methods Mol Med 89:3–36

Najarian T, Marrache AM et al (2000) Prolonged hypercapnia-evoked cerebral hyperemia via K(+) channel- and prostaglandin E(2)-dependent endothelial nitric oxide synthase induction. Circ Res 87(12):1149–1156

Nett WJ, Oloff SH et al (2002) Hippocampal astrocytes in situ exhibit calcium oscillations that occur independent of neuronal activity. J Neurophysiol 87(1):528–537

Neuhaus J, Risau W et al (1991) Induction of blood-brain barrier characteristics in bovine brain endothelial cells by rat astroglial cells in transfilter coculture. Ann N Y Acad Sci 633:578–580

Newman EA (1986) High potassium conductance in astrocyte endfeet. Science 233(4762):453–454

Newman EA, Frambach DA et al (1984) Control of extracellular potassium levels by retinal glial cell K+ siphoning. Science 225(4667):1174–1175

Nguyen TS, Winn HR et al (2000) ATP-sensitive potassium channels may participate in the coupling of neuronal activity and cerebrovascular tone. Am J Physiol Heart Circ Physiol 278(3):H878–H885

Ni X, Geller EB et al (2004) Win 55212-2, a cannabinoid receptor agonist, attenuates leukocyte/endothelial interactions in an experimental autoimmune encephalomyelitis model. Mult Scler 10(2):158–164

Nimmerjahn A, Kirchhoff F et al (2004) Sulforhodamine 101 as a specific marker of astroglia in the neocortex in vivo. Nat Methods 1(1):31–37

Niwa K, Porter VA et al (2001) A beta-peptides enhance vasoconstriction in cerebral circulation. Am J Physiol Heart Circ Physiol 281(6):H2417–H2424
Nobles M, Abbott NJ (1998) Modulation of the effects of extracellular ATP on [Ca2+]i in rat brain microvacular endothelial cells. Eur J Pharmacol 361(1):119–127
Nobles M, Revest PA et al (1995) Characteristics of nucleotide receptors that cause elevation of cytoplasmic calcium in immortalized rat brain endothelial cells (RBE4) and in primary cultures. Br J Pharmacol 115(7):1245–1252
Nozaki K, Moskowitz MA et al (1993) Possible origins and distribution of immunoreactive nitric oxide synthase-containing nerve fibers in cerebral arteries. J Cereb Blood Flow Metab 13(1): 70–79
Offenhauser N, Thomsen K et al (2005) Activity-induced tissue oxygenation changes in rat cerebellar cortex: interplay of postsynaptic activation and blood flow. J Physiol 565(Pt 1): 279–294
Ookawara S, Mitsuhashi U et al (1996) Study on distribution of pericyte and fluorescent granular perithelial (FGP) cell in the transitional region between arteriole and capillary in rat cerebral cortex. Anat Rec 244(2):257–264
Opdenakker G, Van den Steen PE et al (2001) Gelatinase B functions as regulator and effector in leukocyte biology. J Leukoc Biol 69(6):851–859
Paris D, Town T et al (2000) Beta-amyloid vasoactivity and proinflammation in microglia can be blocked by cGMP-elevating agents. Ann N Y Acad Sci 903:446–450
Parri HR, Gould TM et al (2001) Spontaneous astrocytic Ca2+ oscillations in situ drive NMDAR-mediated neuronal excitation. Nat Neurosci 4(8):803–812
Paulson OB, Newman EA (1987) Does the release of potassium from astrocyte endfeet regulate cerebral blood flow? Science 237(4817):896–898
Pelligrino DA, Santizo R et al (1998) Cerebral vasodilating capacity during forebrain ischemia: effects of chronic estrogen depletion and repletion and the role of neuronal nitric oxide synthase. Neuroreport 9(14):3285–3291
Pelligrino DA, Ye S et al (2000) Nitric-oxide-dependent pial arteriolar dilation in the female rat: effects of chronic estrogen depletion and repletion. Biochem Biophys Res Commun 269(1): 165–171
Peng X, Carhuapoma JR et al (2002) Suppression of cortical functional hyperemia to vibrissal stimulation in the rat by epoxygenase inhibitors. Am J Physiol Heart Circ Physiol 283(5): H2029–H2037
Perlmutter LS, Chui HC (1990) Microangiopathy, the vascular basement membrane and Alzheimer's disease: a review. Brain Res Bull 24(5):677–686
Perry VH, Anthony DC et al (1997) The blood-brain barrier and the inflammatory response. Mol Med Today 3(8):335–341
Persidsky Y (1999) Model systems for studies of leukocyte migration across the blood - brain barrier. J Neurovirol 5(6):579–590
Persidsky Y, Ramirez SH et al (2006) Blood-brain barrier: structural components and function under physiologic and pathologic conditions. J Neruoimmune Pharmacology 1:223–236
Piccio L, Rossi B et al (2002) Molecular mechanisms involved in lymphocyte recruitment in inflamed brain microvessels: critical roles for P-selectin glycoprotein ligand-1 and heterotrimeric G(i)-linked receptors. J Immunol 168(4):1940–1949
Pinard E, Engrand N et al (2000) Dynamic cerebral microcirculatory changes in transient forebrain ischemia in rats: involvement of type I nitric oxide synthase. J Cereb Blood Flow Metab 20(12): 1648–1658
Plumb J, McQuaid S et al (2002) Abnormal endothelial tight junctions in active lesions and normal-appearing white matter in multiple sclerosis. Brain Pathol 12(2):154–169
Popovich PG (2000) Immunological regulation of neuronal degeneration and regeneration in the injured spinal cord. Prog Brain Res 128:43–58
Popovich PG, Jones TB (2003) Manipulating neuroinflammatory reactions in the injured spinal cord: back to basics. Trends Pharmacol Sci 24(1):13–17

Popovich PG, Guan Z et al (2002) The neuropathological and behavioral consequences of intraspinal microglial/macrophage activation. J Neuropathol Exp Neurol 61(7):623–633
Prado R, Watson BD et al (1992) Endothelium-derived nitric oxide synthase inhibition. Effects on cerebral blood flow, pial artery diameter, and vascular morphology in rats. Stroke 23(8): 1118–1123
Price DL, Ludwig JW et al (2002) Distribution of rSlo Ca2+-activated K+ channels in rat astrocyte perivascular endfeet. Brain Res 956(2):183–193
Purkiss JR, West D et al (1994) Stimulation of phospholipase C in cultured microvascular endothelial cells from human frontal lobe by histamine, endothelin and purinoceptor agonists. Br J Pharmacol 111(4):1041–1046
Ramsauer M, Krause D et al (2002) Angiogenesis of the blood-brain barrier in vitro and the function of cerebral pericytes. FASEB J 16(10):1274–1276
Ransohoff R (1999) Mechanisms of inflammation in MS tissue: adhesion molecules and chemokines. J Neuroimmunol 98(1):57–68
Rapoport SI (2000) Osmotic opening of the blood-brain barrier: principles, mechanism, and therapeutic applications. Cell Mol Neurobiol 20(2):217–230
Rapoport SI (2001) Advances in osmotic opening of the blood-brain barrier to enhance CNS chemotherapy. Expert Opin Investig Drugs 10(10):1809–1818
Reese TS, Karnovsky MJ (1967) Fine structural localization of a blood-brain barrier to exogenous peroxidase. J Cell Biol 34(1):207–217
Revest PA, Abbott NJ et al (1991) Receptor-mediated changes in intracellular [Ca2+] in cultured rat brain capillary endothelial cells. Brain Res 549(1):159–161
Riddle DR, Sonntag WE et al (2003) Microvascular plasticity in aging. Ageing Res Rev 2(2): 149–168
Ritter LS, Orozco JA et al (2000) Leukocyte accumulation and hemodynamic changes in the cerebral microcirculation during early reperfusion after stroke. Stroke 31(5):1153–1161
Roof RL, Hall ED (2000a) Estrogen-related gender difference in survival rate and cortical blood flow after impact-acceleration head injury in rats. J Neurotrauma 17(12):1155–1169
Roof RL, Hall ED (2000b) Gender differences in acute CNS trauma and stroke: neuroprotective effects of estrogen and progesterone. J Neurotrauma 17(5):367–388
Rosenblum WI (2003) ATP-sensitive potassium channels in the cerebral circulation. Stroke 34(6): 1547–1552
Roy CS, Sherrington CS (1890) On the regulation of the blood supply of the brain. J Physiol 11:85–108
Rubin LL (1991) The blood-brain barrier in and out of cell culture. Curr Opin Neurobiol 1(3): 360–363
Rucker HK, Wynder HJ et al (2000) Cellular mechanisms of CNS pericytes. Brain Res Bull 51(5): 363–369
Saavedra JM (2005) Brain angiotensin II: new developments, unanswered questions and therapeutic opportunities. Cell Mol Neurobiol 25(3–4):485–512
Saito A, Goto K (1994) Vasodilator innervation of small cerebral arteries of Guinea pigs. J Auton Nerv Syst 49(Suppl):S59–S62
Sakakibara A, Furuse M et al (1997) Possible involvement of phosphorylation of occludin in tight junction formation. J Cell Biol 137(6):1393–1401
Santizo RA, Anderson S et al (2000) Effects of estrogen on leukocyte adhesion after transient forebrain ischemia. Stroke 31(9):2231–2235
Sarker MH, Easton AS et al (1998) Regulation of cerebral microvascular permeability by histamine in the anaesthetized rat. J Physiol 507(Pt 3):909–918
Sasseville VG, Newman WA et al (1992) Elevated vascular cell adhesion molecule-1 in AIDS encephalitis induced by simian immunodeficiency virus. Am J Pathol 141(5):1021–1030
Sato A, Sato Y et al (2004) Activation of the intracerebral cholinergic nerve fibers originating in the basal forebrain increases regional cerebral blood flow in the rat's cortex and hippocampus. Neurosci Lett 361(1–3):90–93

Savettieri G, Di Liegro I et al (2000) Neurons and ECM regulate occludin localization in brain endothelial cells. Neuroreport 11(5):1081–1084
Schaeffer S, Iadecola C (2021) Revisiting the neurovascular unit. Nat Neurosci 24(9):1198–1209
Scheibel AB (1987) Alterations of the cerebral capillary bed in the senile dementia of Alzheimer. Ital J Neurol Sci 8(5):457–463
Schenkel AR, Mamdouh Z et al (2004) Locomotion of monocytes on endothelium is a critical step during extravasation. Nat Immunol 5(4):393–400
Schilling L, Wahl M (1994) Opening of the blood-brain barrier during cortical superfusion with histamine. Brain Res 653(1–2):289–296
Schinkel AH, Jonker JW (2003) Mammalian drug efflux transporters of the ATP binding cassette (ABC) family: an overview. Adv Drug Deliv Rev 55(1):3–29
Schipke CG, Boucsein C et al (2002) Astrocyte Ca2+ waves trigger responses in microglial cells in brain slices. FASEB J 16(2):255–257
Schmeck B, Brunsch M et al (2003) Rho protein inhibition blocks cyclooxygenase-2 expression by proinflammatory mediators in endothelial cells. Inflammation 27(2):89–95
Schwartz M, Yoles E (2005) Macrophages and dendritic cells treatment of spinal cord injury:from the bench to the clinic. Acta Neurochir Suppl 93:147–150
Seki Y, Suzuki Y et al (1995) The effects of pituitary adenylate cyclase-activating polypeptide on cerebral arteries and vertebral artery blood flow in anesthetized dogs. Eur J Pharmacol 275(3): 259–266
Seylaz J, Charbonne R et al (1999) Dynamic in vivo measurement of erythrocyte velocity and flow in capillaries and of microvessel diameter in the rat brain by confocal laser microscopy. J Cereb Blood Flow Metab 19(8):863–870
Sharma HS, Dey PK (1986a) Influence of long-term immobilization stress on regional blood-brain barrier permeability, cerebral blood flow and 5-HT level in conscious normotensive young rats. J Neurol Sci 72(1):61–76
Sharma HS, Dey PK (1986b) Probable involvement of 5-hydroxytryptamine in increased permeability of blood-brain barrier under heat stress in young rats. Neuropharmacology 25(2): 161–167
Sheikov N, McDannold N et al (2006) Brain arterioles show more active vesicular transport of blood-borne tracer molecules than capillaries and venules after focused ultrasound-evoked opening of the blood-brain barrier. Ultrasound Med Biol 32(9):1399–1409
Shepro D, Morel NM (1993) Pericyte physiology. FASEB J 7(11):1031–1038
Simard M, Arcuino G et al (2003) Signaling at the gliovascular interface. J Neurosci 23(27): 9254–9262
Sims DE (1986) The pericyte—a review. Tissue Cell 18(2):153–174
Sims DE (1991) Recent advances in pericyte biology--implications for health and disease. Can J Cardiol 7(10):431–443
Sims DE (2000) Diversity within pericytes. Clin Exp Pharmacol Physiol 27(10):842–846
Sobue K, Yamamoto N et al (1999) Induction of blood-brain barrier properties in immortalized bovine brain endothelial cells by astrocytic factors. Neurosci Res 35(2):155–164
Sonntag WE, Lynch CD et al (1997) Decreases in cerebral microvasculature with age are associated with the decline in growth hormone and insulin-like growth factor 1. Endocrinology 138(8): 3515–3520
Sonntag WE, Lynch C et al (2000) The effects of growth hormone and IGF-1 deficiency on cerebrovascular and brain ageing. J Anat 197(Pt 4):575–585
Soriano SG, Lipton YF et al (1996) Intercellular adchesion molecule-1-deficient mice are less susceptible to cerebral ischemia-reperfusion injury. Ann Neurol 39:618–624
Spudich A, Kilic E et al (2006) Inhibition of multidrug resistance transporter-1 facilitates neuroprotective therapies after focal cerebral ischemia. Nat Neurosci 9(4):487–488
Stamatovic SM, Shakui P et al (2005) Monocyte chemoattractant protein-1 regulation of blood-brain barrier permeability. J Cereb Blood Flow Metab 25(5):593–606

Stamatovic SM, Dimitrijevic OB et al (2006) Protein kinase Calpha-RhoA cross-talk in CCL2-induced alterations in brain endothelial permeability. J Biol Chem 281(13):8379–8388
Steffen BJ, Butcher EC et al (1994) Evidence for involvement of ICAM-1 and VCAM-1 in lymphocyte interaction with endothelium in experimental autoimmune encephalomyelitis in the central nervous system in the SJL/J mouse. Am J Pathol 145(1):189–201
Stenman E, Edvinsson L (2004) Cerebral ischemia enhances vascular angiotensin AT1 receptor-mediated contraction in rats. Stroke 35(4):970–974
Stevenson BR (1999) Understanding tight junction clinical physiology at the molecular level. J Clin Invest 104(1):3–4
Stewart PA, Magliocco M et al (1987) A quantitative analysis of blood-brain barrier ultrastructure in the aging human. Microvasc Res 33(2):270–282
Stewart PA, Hayakawa K et al (1992) A morphometric study of the blood-brain barrier in Alzheimer's disease. Lab Investig 67(6):734–742
Stirone C, Boroujerdi A et al (2005) Estrogen receptor activation of phosphoinositide-3 kinase, akt, and nitric oxide signaling in cerebral blood vessels: rapid and long-term effects. Mol Pharmacol 67(1):105–113
Sun H, Dai H et al (2003) Drug efflux transporters in the CNS. Adv Drug Deliv Rev 55(1):83–105
Sun BL, Wang LH et al (2018) Lymphatic drainage system of the brain: a novel target for intervention of neurological diseases. Prog Neurobiol 163-164:118–143
Sykova E (1983) Extracellular K+ accumulation in the central nervous system. Prog Biophys Mol Biol 42(2–3):135–189
Szczygielski J, Kopanska M et al (2021) Cerebral microcirculation, perivascular unit, and Glymphatic system: role of Aquaporin-4 as the gatekeeper for water homeostasis. Front Neurol 12:767470
Szpak GM, Lechowicz W et al (1999) Border zone neovascularization in cerebral ischemic infarct. Folia Neuropathol 37(4):264–268
Takagi H, King GL et al (1996) Identification and characterization of vascular endothelial growth factor receptor (Flt) in bovine retinal pericytes. Diabetes 45(8):1016–1023
Takao M, Kobari M et al (2002) Dilatation of cerebral parenchymal vessels mediated by angiotensin type 1 receptor in cats. Neurosci Lett 318(2):108–112
Tanaka K, Gotoh F et al (1991) Inhibition of nitric oxide synthesis induces a significant reduction in local cerebral blood flow in the rat. Neurosci Lett 127(1):129–132
Terry RD, Masliah E et al (1991) Physical basis of cognitive alterations in Alzheimer's disease: synapse loss is the major correlate of cognitive impairment. Ann Neurol 30(4):572–580
Thomas WE (1999) Brain macrophages: on the role of pericytes and perivascular cells. Brain Res Brain Res Rev 31(1):42–57
Tilling T, Engelbertz C et al (2002) Expression and adhesive properties of basement membrane proteins in cerebral capillary endothelial cell cultures. Cell Tissue Res 310(1):19–29
Timpl R (1994) Proteoglycans of basement membranes. EXS 70:123–144
Tomita M, Gotoh F et al (1980) "low perfusion hyperemia" following middle cerebral arterial occlusion in cats of different age groups. Stroke 11(6):629–636
Tomita M, Schiszler I et al (2005) Initial oligemia with capillary flow stop followed by hyperemia during K+-induced cortical spreading depression in rats. J Cereb Blood Flow Metab 25(6): 742–747
Towart R (1981) The selective inhibition of serotonin-induced contractions of rabbit cerebral vascular smooth muscle by calcium-antagonistic dihydropyridines. An investigation of the mechanism of action of nimodipine. Circ Res 48(5):650–657
Tuma RF, White JV et al (eds) (1989) The role of Hemodilution in optimal patient care. W. Zuckschwerdt Verlag, Munchen
Tuma RF, Vashtare US et al (1993) White cell involvement in cerebral ischemia and reperfusion injury. In: Tomita M, Mchedlishvili G, Rosenblum WI, Fukuuchi Y (eds) Microcirculatory stasis in the brain. Elsevier Science Publishers, Amsterdam, pp 123–131

Tuma RF, Vasthare US et al (1997) Hypertonic saline administration attenuates spinal cord injury. J Trauma 42(5 Suppl):S54–S60
Uchida S, Suzuki A et al (2000) Effects of age on cholinergic vasodilation of cortical cerebral blood vessels in rats. Neurosci Lett 294(2):109–112
Uddman R, Goadsby PJ et al (1993) PACAP, a VIP-like peptide: immunohistochemical localization and effect upon cat pial arteries and cerebral blood flow. J Cereb Blood Flow Metab 13(2): 291–297
Ursino M (1991) Mechanisms of cerebral blood flow regulation. Crit Rev Biomed Eng 18(4): 255–288
Vajkoczy P, Laschinger M et al (2001) Alpha4-integrin-VCAM-1 binding mediates G protein-independent capture of encephalitogenic T cell blasts to CNS white matter microvessels. J Clin Invest 108(4):557–565
van Buul JD, Hordijk PL (2004) Signaling in leukocyte transendothelial migration. Arterioscler Thromb Vasc Biol 24(5):824–833
Vanlandewijck M, He L et al (2018) A molecular atlas of cell types and zonation in the brain vasculature. Nature 554(7693):475–480
Vasthare US, Irion GL et al (1988) Differential effects of anesthetic agents on regional blood flow and central hemodynamic parameters in rats. Drug Dev Res 14(1):59–67
Vasthare US, Heinel LA et al (1990) Leukocyte involvement in cerebral ischemia and reperfusion injury. Surg Neurol 33(4):261–265
Vasthare US, Barone FC et al (1998) Complement depletion improves neurological function in cerebral ischemia. Brain Res Bull 45(4):413–419
Venneti S, Wang G et al (2008) The positron emission tomography ligand DAA1106 binds with high affinity to activated microglia in human neurological disorders. J Neuropathol Exp Neurol 67(10):1001–1010
Vogel J, Sperandio M et al (2000) Influence of the endothelial glycocalyx on cerebral blood flow in mice. J Cereb Blood Flow Metab 20(11):1571–1578
Volk H, Potschka H et al (2005) Immunohistochemical localization of P-glycoprotein in rat brain and detection of its increased expression by seizures are sensitive to fixation and staining variables. J Histochem Cytochem 53(4):517–531
Volterra A, Meldolesi J (2005) Astrocytes, from brain glue to communication elements: the revolution continues. Nat Rev Neurosci 6(8):626–640
Vorbrodt AW (1993) Morphological evidence of the functonal polarization of brain microvascular endothelium. In: Pardridge WM (ed) The blood-brain barrier. Cellular and molecular biology. Raven, New York, pp 137–164
Wagner AH, Kohler T et al (2000) Improvement of nitric oxide-dependent vasodilatation by HMG-CoA reductase inhibitors through attenuation of endothelial superoxide anion formation. Arterioscler Thromb Vasc Biol 20(1):61–69
Wahl M, Unterberg A et al (1988) Mediators of blood-brain barrier dysfunction and formation of vasogenic brain edema. J Cereb Blood Flow Metab 8(5):621–634
Wahlestedt C, Reis DJ (1993) Neuropeptide Y-related peptides and their receptors – are the receptors potential therapeutic drug targets? Annu Rev Pharmacol Toxicol 33:309–352
Wang Q, Pelligrino DA et al (1995) The role of neuronal nitric oxide synthase in regulation of cerebral blood flow in normocapnia and hypercapnia in rats. J Cereb Blood Flow Metab 15(5): 774–778
Wang Q, Santizo R et al (1999) Estrogen provides neuroprotection in transient forebrain ischemia through perfusion-independent mechanisms in rats. Stroke 30(3):630–637
Watanabe S, Morisaki N et al (1997) Cultured retinal pericytes stimulate in vitro angiogenesis of endothelial cells through secretion of a fibroblast growth factor-like molecule. Atherosclerosis 130(1–2):101–107
Weaver M, Leshley K et al (2002) LEX032, a novel recombinant serpin, protects the brain after transient focal ischemia. Microvasc Res 63(3):327–334

Wei EP, Kontos HA et al (1980) Dependence of pial arteriolar response to hypercapnia on vessel size. Am J Phys 238(5):697–703
Wekerle H, Linington C et al (1986) Cellular immune reactivity within the CNS. Trends Neurosci 9: 271–277
Werber AH, Heistad DD (1984) Effects of chronic hypertension and sympathetic nerves on the cerebral microvasculature of stroke-prone spontaneously hypertensive rats. Circ Res 55(3): 286–294
Wittchen ES, van Buul JD et al (2005a) Trading spaces: rap, Rac, and rho as architects of transendothelial migration. Curr Opin Hematol 12(1):14–21
Wittchen ES, Worthylake RA et al (2005b) Rap1 GTPase inhibits leukocyte transmigration by promoting endothelial barrier function. J Biol Chem 280(12):11675–11682
Wojciak-Stothard B, Williams L et al (1999) Monocyte adhesion and spreading on human endothelial cells is dependent on rho-regulated receptor clustering. J Cell Biol 145(6):1293–1307
Wolburg H, Lippoldt A (2002) Tight junctions of the blood-brain barrier: development, composition and regulation. Vasc Pharmacol 38(6):323–337
Wong D, Dorovini-Zis K (1992) Upregulation of intercellular adhesion molecule-1 (ICAM-1) expression in primary cultures of human brain microvessel endothelial cells by cytokines and lipopolysaccharide. J Neuroimmunol 39(1–2):11–21
Wong D, Dorovini-Zis K (1995) Expression of vascular cell adhesion molecule-1 (VCAM-1) by human brain microvessel endothelial cells in primary culture. Microvasc Res 49(3):325–339
Woolsey TA, Rovainen CM et al (1996) Neuronal units linked to microvascular modules in cerebral cortex: response elements for imaging the brain. Cereb Cortex 6(5):647–660
Wyss-Coray T (2006) Inflammation in Alzheimer disease: driving force, bystander or beneficial response? Nat Med 12(9):1005–1015
Yamada M, Lamping KG et al (2001) Cholinergic dilation of cerebral blood vessels is abolished in M(5) muscarinic acetylcholine receptor knockout mice. Proc Natl Acad Sci USA 98(24): 14096–14101
Yamaguchi S, Kobayashi S et al (1988) Effect of aging on collateral circulation via pial anastomoses in cats. Gerontology 34(4):157–164
Yamamoto T, Harada N et al (1999) In vivo interaction of AF-6 with activated Ras and ZO-1. Biochem Biophys Res Commun 259(1):103–107
Yamashita K, Miyakawa T et al (1991) Vascular changes in the brains with Alzheimer's disease. Jpn J Psychiatry Neurol 45(1):79–84
Yang G, Huard JM et al (2000) Stellate neurons mediate functional hyperemia in the cerebellar molecular layer. J Neurosci 20(18):6968–6973
Yang L, Froio RM et al (2005) ICAM-1 regulates neutrophil adhesion and transcellular migration of TNF-alpha-activated vascular endothelium under flow. Blood 106(2):584–592
Yoder EJ (2002) Modifications in astrocyte morphology and calcium signaling induced by a brain capillary endothelial cell line. Glia 38(2):137–145
Yoshida K, Okamura T et al (1994) Histological and functional studies on the nitroxidergic nerve innervating monkey cerebral, mesenteric and temporal arteries. Jpn J Pharmacol 65(4):351–359
You J, Johnson TD et al (1999) Functional heterogeneity of endothelial P2 purinoceptors in the cerebrovascular tree of the rat. Am J Phys 277(3 Pt 2):H893–H900
Zaritsky JJ, Eckman DM et al (2000) Targeted disruption of Kir2.1 and Kir2.2 genes reveals the essential role of the inwardly rectifying K(+) current in K(+)-mediated vasodilation. Circ Res 87(2):160–166
Zauner A, Daugherty WP et al (2002) Brain oxygenation and energy metabolism: part I-biological function and pathophysiology. Neurosurgery 51(2):289–301
Zeine R, Owens T (1992) Direct demonstration of the infiltration of murine central nervous system by Pgp-1/CD44high CD45RB(low) CD4+ T cells that induce experimental allergic encephalomyelitis. J Neuroimmunol 40(1):57–69

Zhang ET, Mikkelsen JD et al (1991) Prepro-vasoactive intestinal polypeptide-derived peptide sequences in cerebral blood vessels of rats: on the functional anatomy of metabolic autoregulation. J Cereb Blood Flow Metab 11(6):932–938

Zhang RL, Chopp M et al (1994) Anti-ICAM-1 antibody reduces ischemic cell damage after transient middle cerebral artery occlusion in the rat. Neurology 44(9):1747–1751

Zhang RL, Chopp M et al (1996) E-selectin in focal cerebral ischemia and reperfusion in the rat. J Cereb Blood Flow Metab 16(6):1126–1136

Zhang ZG, Zhang L et al (2000) VEGF enhances angiogenesis and promotes blood-brain barrier leakage in the ischemic brain. J Clin Invest 106(7):829–838

Zhang M, Martin BR et al (2007) Cannabinoid CB(2) receptor activation decreases cerebral infarction in a mouse focal ischemia/reperfusion model. J Cereb Blood Flow Metab 27:1387

Zonta M, Angulo MC et al (2003a) Neuron-to-astrocyte signaling is central to the dynamic control of brain microcirculation.[see comment]. Nat Neurosci 6(1):43–50

Zonta M, Sebelin A et al (2003b) Glutamate-mediated cytosolic calcium oscillations regulate a pulsatile prostaglandin release from cultured rat astrocytes. J Physiol 553(Pt 2):407–414

Zou AP, Fleming JT et al (1996) 20-HETE is an endogenous inhibitor of the large-conductance ca (2+)-activated K+ channel in renal arterioles. Am J Phys 270(1 Pt 2):R228–R237

Chapter 8
Regulation of Coronary Microvascular Resistance in Health and Disease

Song Yi Shin, Adam G. Goodwill, Vahagn Ohanyan, Liya Yin, Feng Dong, Ping Zhang, Leah Choban, Catherine Nicole Domingo, Sarah Godel, Emily Kodash, Anna Martin, Taylor Yamamoto, Christian Bittner, and William Chilian

Abstract This chapter discusses the regulation of microvascular resistance and responses in the coronary circulation. This chapter is divided into seven sections: Introduction and Background, Metabolic Regulation of Coronary Microvascular Resistance, Endothelial and Myogenic Mechanisms, Neurohumoral Control, Cardiac Lymphatic Microcirculation, Pathophysiological Disturbances, and Summary and Future Directions. Each section contains a recent view about the key factors and mechanisms that underscore the control of the coronary microvascular resistance.

Keywords Coronary Circulation · Coronary Microcirculation · Coronary Blood Flow

8.1 Introduction and Background

8.1.1 General Perspectives

The previous version of this chapter was published in 2008, and the current authors' intention is to build upon the previous version with updated information in metabolic control mechanisms, a new section on the history of research in the coronary microcirculation, and the introduction of more pathophysiology. While there continues to be progress in our understanding of the regulation of blood flow to the heart by the coronary microcirculation, gaps in our knowledge remain. As one example, science has known the composition of the human genome for over two decades, but one cannot state unequivocally what factors connect coronary blood flow to cardiac

S. Y. Shin · A. G. Goodwill (✉) · V. Ohanyan · L. Yin · F. Dong · P. Zhang · L. Choban · C. N. Domingo · S. Godel · E. Kodash · A. Martin · T. Yamamoto · C. Bittner · W. Chilian (✉)
Department of Biomedical Sciences, Northeast Ohio Medical University, Rootstown, OH, USA
e-mail: agoodwill@neomed.edu; ekondash@neomed.edu; wchilian@neomed.edu

R. F. Tuma et al. (eds.), *Handbook of Microcirculation*, Physiology in Health and Disease, https://doi.org/10.1007/978-3-031-99609-2_8

aerobic metabolism—perhaps the most important function of the coronary microcirculation, vis-à-vis metabolic dilation or metabolic hyperemia. An imbalance in this relationship can result in demand-induced ischemia leading to cardiac dysfunction and even necrosis. The microcirculation of the heart is tasked with the important function of maintaining the proper balance between oxygen supply (product of flow and oxygen content) and oxygen demands (myocardial oxygen consumption). To accomplish this task, the resistance vessels of the heart must integrate inputs from many intrinsic and extrinsic vasodilator and vasoconstrictor signals. It is our goal to discuss these myriad inputs in the context of their actions on coronary blood flow and coronary resistance vessels. We have also extended our discussion to pathophysiological disturbances which alter some regulatory processes within the coronary microcirculation. Although the regulation of permeability is a critical component of substrate delivery to the working heart, we have not discussed the regulation of permeability or exchange processes in this chapter. However, we refer to some of the advances made in this area in the section on Chronology of Research in the Coronary Microcirculation. The principal focus of this chapter is directed at the regulation of coronary blood flow by the coronary microcirculation.

There are a variety of mechanisms that contribute to the regulation of coronary blood flow and thus tone of coronary resistance vessels, including the metabolic activity of the myocardium, myogenic activity of coronary blood vessels, neurohumoral factors, and endothelium-dependent stimuli. Of these influences, metabolic control mechanisms are undoubtedly the most important, because they link flow to metabolism and are involved in ischemic vasodilation. Myogenic and endothelial regulatory mechanisms exert influences on coronary microvascular diameters through the transduction of intravascular pressure (stretch) and flow (shear stress) and may participate in the integration of the response of the microvascular network in the metabolic control of coronary blood flow. In myogenic responses, coronary arterioles constrict and dilate in response to increased and decreased intraluminal pressure, respectively (Kuo et al. 1988). In flow-induced responses, increased shear stress on the vascular wall elicits endothelium-dependent dilation of coronary microvessels (Kuo et al. 1990a). Shear stress-induced dilation shares some similarities with endothelium-dependent dilation induced by agonists, but also some differences. Neurohumoral mechanisms seem to preferentially modulate tone in small arteries, but in the absence of autoregulatory adjustments, these factors can play a major role on tone in small arterioles.

Before we elaborate on the various regulatory factors that control tone in the coronary microcirculation, it is important to provide the perspective of the locations of vascular resistance within the heart (Fig. 8.1). Chilian et al. (1986) estimated the distribution of resistance from microvascular pressure measurements in the beating heart under basal experimental conditions (with tone intact in an anesthetized open chest preparation and during dipyridamole-induced dilation). In this situation, the proportion of the dissipated pressure (between various points in the microcirculation) as a percentage of the total pressure drop (difference between the aorta and the right atrium) approximates the amount of resistance in the vascular segment. Under baseline conditions small arteries and arterioles constitute about 30 and 40% of total

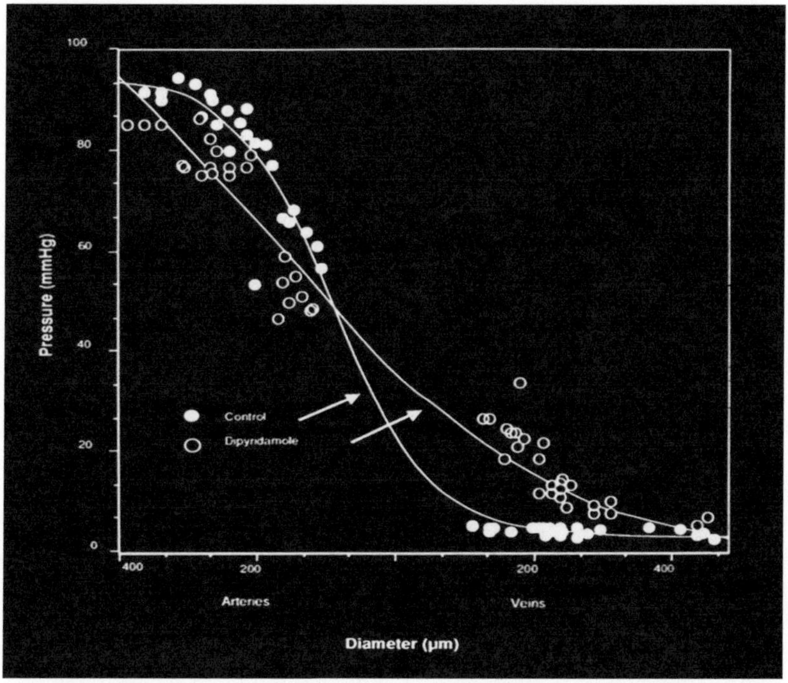

Fig. 8.1 The distribution of pressures throughout the left coronary circulation under basal, control conditions (with tone), and during dipyridamole-induced coronary vasodilation. Dipyridamole produced a 5- to 6-fold increase in flow, and despite this huge increase in flow, the pressure drop across the small arteries and arterioles decreased by half. This implies that resistance decreased by 10- to 12-fold during dipyridamole-induced vasodilation. (Adapted from Chilian et al. 1989a)

resistance, respectively (Fig. 8.1). During intense vasodilation, resistance in the arteriolar fraction drops considerably (pressure drop is halved when flow increases 6-fold, indicating a 12-fold drop in resistance). This drop in resistance is most striking in the arterioles, which then results in a redistribution of resistance from small arterioles to upstream small arteries and downstream venules. Thus, depending on the particular condition, arterioles, small arteries, and even venules can play a role in the control of total coronary resistance. It is important to note that under normal conditions, arterioles comprise the bulk of resistance, and the combination of small arteries and arterioles constitute the majority of resistance to flow in the coronary circulation.

Although the focus of this review is on regulatory mechanisms in the coronary microcirculation, it is important to highlight some regional differences in the distribution and regulation of resistance across the wall of the left ventricle, i.e., a comparison of the endocardial vs. epicardial microcirculations. The most resistive segment to blood flow (>60%) resides in microvessels less than 150 μm in diameter (Chilian et al. 1986). However, it is important to highlight that this observation was made in the epicardial microcirculation, and the distribution of resistance and pressures in the endocardial microcirculation is different. In vitro studies have shown that arteriolar pressures are significantly lower in the subendocardium than in the subepicardium (Chilian 1991a), and adenosine produces greater dilation of subendocardial microvessels and redistributes flow toward the subendocardium under normal perfusion pressure (Zhang et al. 2000). However, reduction of coronary pressure causes a relative increase in subepicardial blood flow and a preferential increase in subepicardial perfusion in response to adenosine (Zhang et al. 2000). These studies not only suggest a differential adenosine response across the ventricular wall but also implicate a possible influence of luminal pressure on adenosine responsiveness. In vivo studies of coronary circulation also showed that a reduction in coronary microvascular pressure, as a result of a proximal coronary stenosis, causes a redistribution of perfusion with a greater blood flow in the subepicardium than in the subendocardium (Bache and Schwartz 1982).

Although some of the regional differences in subepicardial and subendocardial flow regulation may be related to the distribution of resistances (both vascular and extravascular), inherent differences in vascular regulation may also play a role. Quillen and Harrison (1992) demonstrated using isolated buffer-perfused vessels that subendocardial coronary arterioles have an increased sensitivity to adenosine and forskolin, when compared to subepicardial arterioles of comparable size (Quillen and Harrison 1992). This difference did not occur for sodium nitroprusside or the endothelium-dependent vasodilators bradykinin, ADP, or A23187. By using this in vitro approach, they were able to avoid confounding factors such as differences in blood flow and washout of metabolites between epicardial and endocardial layers. This intrinsic difference may be an adaptation to the increased prevalence to ischemia, and hence increased release of adenosine and other metabolites in the subendocardium versus the subepicardium (Zhang et al. 2000).

The intrinsic variation of vascular responsiveness, in a pressure-dependent manner, across the ventricular wall may contribute to the redistribution of transmural flow during a reduction in coronary perfusion pressure (Zhang et al. 2000). Coronary subepicardial and subendocardial arterioles exhibit identical vascular responsiveness to K_{ATP} channel activators, adenosine and pinacidil. However, by activating smooth muscle K_{ATP} channels, vascular dilation to these agonists is enhanced at lower intraluminal pressures. Since coronary microvascular pressure is not homogeneously distributed across the ventricular wall, lower intraluminal pressure in the subendocardium would potentially recruit blood flow to this region by adenosine (Zhang et al. 2000). However, during coronary arterial stenosis or occlusion, the release of adenosine associated with the reduction of arteriolar pressure in the subendocardium could redistribute blood flow in this area. Therefore, the inherent pressure gradient in the coronary microcirculation across the ventricular wall may be an important determinant of transmural flow in vivo during resting conditions or under metabolic stress.

In the normal heart, blood flow to the subendocardium is either equal to or slightly greater than that to the subepicardium. During exercise, sympathetic activation of vascular β-adrenoceptors can contribute to coronary vasodilation (Gorman et al. 2000a; Miyashiro and Feigl 1993), and the subendocardial flow is slightly redistributed toward the subepicardium (Hein et al. 2004). For healthy individuals, this usually does not pose a grave situation, but for those with cardiovascular complications such as coronary stenosis, the redistribution of flow to the subepicardium is accentuated during exercise.

8.1.2 Chronology of Research in the Coronary Microcirculation

Figure 8.2 illustrates the chronology of research in the coronary microcirculation. This figure includes reports focusing on morphometric parameters as well as studies directed at understanding the control of blood flow and vascular tone. The first report of measurements of microvascular parameters in the coronary circulation was published in 1928 (Wearn et al. 1928). In this article, the author used microscopic evaluation of tissue sections from the hearts of cats, rabbits, and humans. The author observed higher capillary densities in cardiac tissue than previously reported by Krogh for skeletal muscle and concluded that the heart has a "very rich blood supply." The capillary densities in the ventricles were higher than in the atria. In

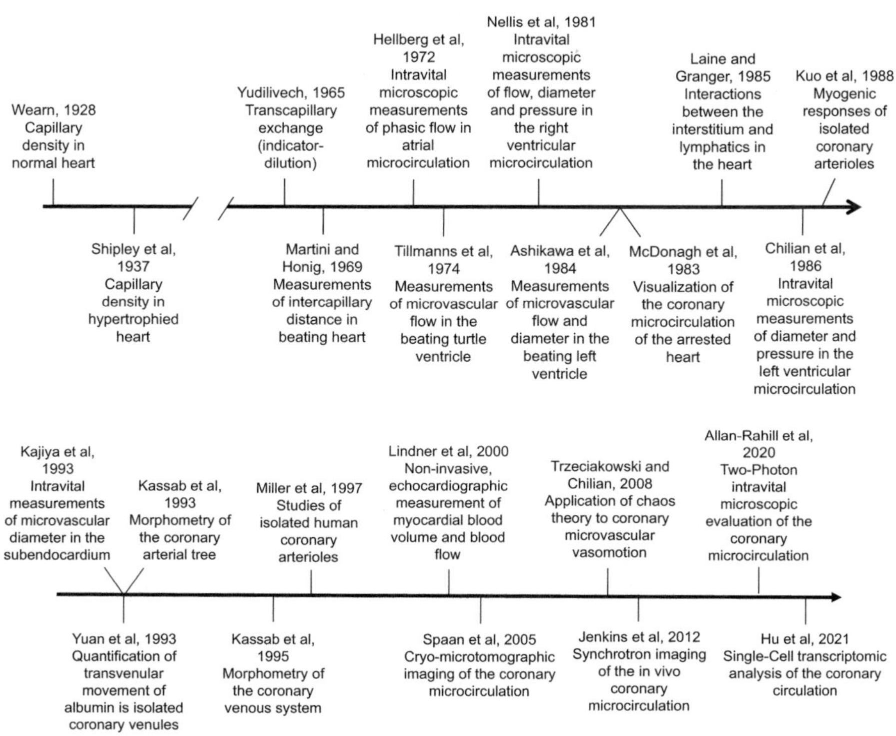

Fig. 8.2 Chronology of research in the coronary microcirculation

1937, Shipley et al. noted a decrease in capillary density in hypertrophied ventricles (Shipley et al. 1937). These authors noted that "capillaries do not multiply" during cardiac enlargement, which could impede exchange of metabolic substrates. The authors speculated that the gravity of this issue is not known but requires more physiological measurements. It is humbling to think that over eight decades the impact of capillary rarefaction in cardiac hypertrophy is still an unanswered question.

Although research in morphometric attributes of the coronary microcirculation, e.g., capillary density, continued, the next advance centered around the implementation of indicator-dilution techniques to understand capillary and solute exchange in the heart. Specifically, Yudilevich and Martindejulian documented transvascular exchange for potassium, sodium, and iodide in the heart (Yudilevich and Martindejulian 1965). This use of indicator-dilution techniques for the study of solute exchange led to a vista of investigations by multiple laboratories studying exchange processes in the heart, as well as transit times of solutes and gases, and capillary permeability-surface area products (Bassingthwaighte et al. 1986; Rose et al. 1977; Rose and Goresky 1976; Duran and Yudilevich 1974, 1978; Duran et al. 1973; Katz and Feigl 1987; Roth and Feigl 1981).

The next advance involved direct visualization of the microcirculation of the beating heart. Martini and Honig used "stop-motion" cinematography of the epicardial surface of the beating rat heart to measure intercapillary distances at varying arterial PO_2 (Martini and Honig 1969). Their results supported the concept of capillary reserve as the intercapillary distance was reduced during hypoxia, suggesting that not every capillary is being perfused at every moment. This finding was also supported by the observations made with indicator dilution techniques that reported an increase in coronary permeability-surface area with vasodilation (Duran et al. 1973). Advances in direct visualization of the beating heart were made by some of the pioneering work of Richard Bing and his colleagues (Hellberg et al. 1972; Bing et al. 1971, 1972a, b; Tillmanns et al. 1974). In these papers (Hellberg et al. 1972; Tillmanns et al. 1974), high-speed cinematography was used in conjunction with intravital microscopy to measure phasic flow velocities in the atria and left ventricle. Nellis et al. built upon these approaches using an innovative system in which stroboscopic illumination was synchronized with the heart such that a flash would occur at the same point in each cardiac cycle (Nellis et al. 1981). When viewed through a microscope, the heart would appear motionless, and a camera synchronized with the strobe allowed high-resolution images to be captured. Figure 8.3 (top) shows the details of this preparation in which the right ventricle was penetrated with a fiber optic light source and the tissue was transilluminated for visualization. Nellis and colleagues also implemented a program in which the strobe flash would advance 1/100 of a cardiac cycle per beat, which then engendered the heart to be viewed in slow motion. One-hundred heartbeats then constituted a cardiac cycle. They used this approach to synchronize a 3-dimensional electromagnetic micromanipulator to cardiac motion and via micropuncture techniques, measured pressures in coronary arterioles and venules in the beating right ventricle (Fig. 8.3, bottom). A few years later, Ashikawa and colleagues devised a different approach to

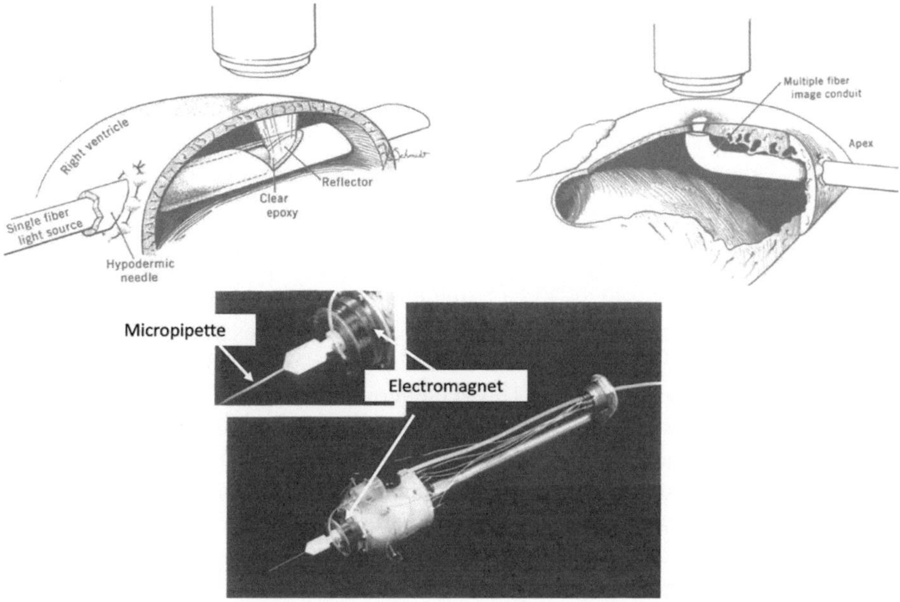

Fig. 8.3 Components of the electromechanical micromanipulator and fiber optic transillumination enable measurements of pressure and diameter of the beating heart. Top: Diagrams of the preparations used by Nellis and colleagues for transillumination of the right ventricle for stroboscopic visualization of the right ventricular coronary microcirculation of the beating heart. The preparation on the top right was also used for the microvascular pressure measurements in the beating heart using the 3-dimensional electromagnetic micromanipulator. (Bottom; adapted from Nellis et al. 1981)

compensate for cardiac motion—a floating microscope objective that moved in conjunction with the beating heart (Kanatsuka et al. 1990a; Ashikawa et al. 1984, 1986). They used this approach to measure phasic flow velocities and diameters in epicardial vessels of the left ventricle.

In 1983, McDonagh developed a different approach to studying the coronary microcirculation—the potassium chloride arrested heart (McDonagh et al. 1984). Despite the criticisms of the physiological relevance of such a preparation, it was used to understand some of the dynamics of leukocyte diapedesis in the coronary microcirculation as well as the influences of ischemia-reperfusion injury (Ritter et al. 2011; Gale et al. 2001; Reynolds and McDonagh 1994; McDonagh and Roberts 1986). Laine and Granger made measurements of the interactions between the interstitium and lymphatic circulation in the heart (Laine and Granger 1985). Their interesting report suggested that a mechanism employed by the heart to limit edema during increased venous pressures is a striking increase in interstitial pressures. A year later, Chilian et al. (1986) reported the distribution of microvascular pressures of the beating left ventricle under baseline conditions (pentobarbital anesthesia) and during coronary vasodilation with papaverine and dipyridamole (Chilian et al. 1986, 1989a). These authors based their approach on the Nellis technique, but adapted their stroboscopic system to video (rather than cinematography) and made measurements in the left ventricular coronary microcirculation.

In 1988, a completely different manner of studying the coronary microcirculation was reported, vis-à-vis, the study of isolated coronary arterioles (Kuo et al. 1988). These investigators (Kuo et al. 1988) adapted techniques for the study of arterioles from other organ systems to the heart. Their initial report detailed differences in myogenic responses between arterioles isolated from the subepicardium and subendocardium of pig hearts. These investigators utilized this preparation to study a variety of functional attributes of the coronary microcirculation including shear stress-induced vasodilation and longitudinal gradients in arterial and arteriolar responses to metabolites, myogenic stimuli, and shear stress (Kuo et al. 1990b, 1995). Although this in vitro approach was used by many laboratories, a significant advance was made by the group of David Gutterman. Specifically, his group started to study microvessels isolated from the human heart (discarded atrial appendages following cardiopulmonary bypass) in 1997 (Miller et al. 1997). Although there is a continual problem regarding this approach about a suitable control group (patients undergoing valve replacements without coronary disease were used as controls for those with coronary disease), this group has made important contributions to our understanding of the pathophysiological consequences of vascular disease extending into the microcirculation of the heart (Freed et al. 2014).

In 1993, three important advances were reported. The laboratory of Fumihiko Kajiya reported visualization of the endocardial coronary microcirculation of the beating heart (Yada et al. 1993). Their preparation involved the use of an image-transferring conduit surrounded by a ring of fiberoptic illuminators and a silicone ring at the end of the conduit (Fig. 8.4). When the ring was pressed against the

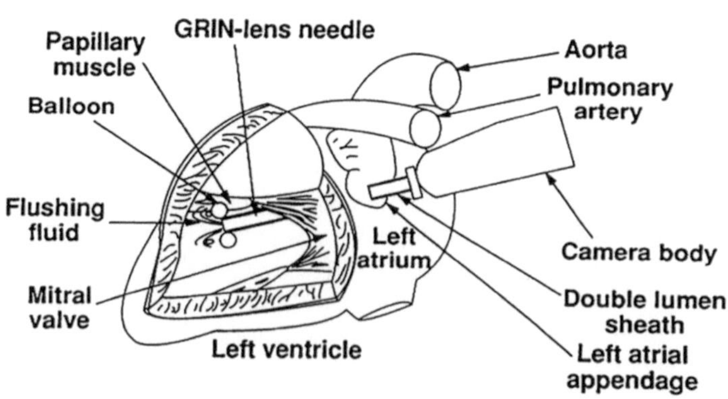

Fig. 8.4 Experimental preparation to obtain access of the subendocardial microcirculation using a gradient index lens needle (GRIN). The needle probe is inserted into the left ventricular lumen via access through the left atrium and mitral valve. The probe has a balloon on the end which, when inflated, prevents the tip of the probe from contacting the subendocardium. The balloon can create a small pocket when positioned “against” the subendocardium, and blood is flushed from the tip of the probe to enable visualization of the subendocardial microcirculation. (Adapted from Yada et al. 1993)

endocardium and a small cannula allowed the flushing of blood from the area within the ring, the endocardial microcirculation could be visualized. A criticism of the previous studies of the in vivo coronary microcirculation was that the measurements were confined to the epicardial surface, but this system circumvented that problem. Of note, the dynamics of flow velocity in microvessels from the subendocardium were very different than those reported in the subepicardium with a significant component of systolic retrograde flow in the subendocardium (Toyota et al. 2005). We would be remiss to not point out a caveat of all of the preparations involving in vivo analysis of the coronary microcirculation. Every approach employs some stabilization of the heart to prevent the "Z" motion, i.e., the "up and down" motion of the ventricle occurring during the normal cycles of contraction/relaxation. Undoubtedly, these procedures affect some attribute of mechanical forces and interpretations should always be made in the context of this issue.

The second major advance of 1993 was the adaptation of isolated vessel techniques to quantify the permeability of postcapillary venules in the heart (Yuan et al. 1993). This preparation enabled the discovery that increases in venular flow velocity, without changes in pressure, increase venular permeability (Yuan et al. 1992) and also engendered the study of inflammatory cell-endothelial dynamics in coronary exchange vessels (Yuan et al. 1995). The third advance of 1993 pertained to the detailed morphometric analysis of the entire coronary arterial and capillary circulation (Kassab et al. 1993). This detailed, painstaking analysis revealed some striking findings, such as capillaries do not necessarily stem from the last order of arterioles; rather they can branch from vessels one and even two branch points from the precapillary vessel.

In the twenty-first century, many of the advances have focused on imaging modalities using contrast echocardiography to noninvasively measure flow and vascular volume in tissue (Lindner et al. 2000), synchrotron imaging to noninvasively measure the diameter of coronary resistance vessels (Jenkins et al. 2012), and two-photon microscopy to visualize diameters and flows in the beating heart (Allan-Rahill et al. 2020). Two other advances in the twenty-first century are worth noting. In 2005, Spaan et al. utilized the technique of cryomicrotomography to reconstruct the coronary vasculature (Spaan et al. 2005). This technique was even used to visualize and reconstruct the coronary microcirculation of the human heart, which was found to contain a significant number of subendocardial collateral vessels in heart failure (van den Wijngaard et al. 2010). Finally, in 2008, Trzeciakowski and Chilian measured an attribute of the coronary microcirculation that they termed "flowmotion" (Trzeciakowski and Chilian 2008). Flowmotion was the "back and forth" flow that occurred in an arterial-arterial anastomosis, which was measured by following a "wavefront" of fluorescence. Modeling of this behavior revealed a chaotic pattern—we remind the reader that chaos is not random, it pertains to complexity. Importantly, if the coronary microcirculation follows such a nonlinear control scheme, then models based on linearity will never accurately model the behavior of the microcirculation of the heart.

Perhaps, in the next decade, the use of artificial intelligence and more sophisticated imaging modalities will enable an even better understanding of the regulation of the microcirculation of the heart. We opine that a more complete understanding of the complex control mechanisms embedded within the coronary microcirculation will stem from observations similar to those by Hu et al. (2021), who reported results from single-cell analysis of large coronary arteries. Although microvessels were not studied, the use of such an analysis, coupled with spatial genomics, will likely yield a far better understanding of the different phenotypes of endothelial cells and vascular smooth muscle in the coronary microcirculation. We believe that if we understand the myriad phenotypes of vascular cells, a better understanding of control mechanisms will advance.

8.2 Metabolic Regulation of Coronary Microvascular Resistance

8.2.1 General Concepts

Since the previous version of this chapter, there has been progress toward understanding the connection of coronary blood flow to myocardial metabolism. Since the bulk of coronary vascular resistance resides in arterioles and small arteries—measurements of flow reflect tone in these resistance vessels. Despite this progress, it is humbling to note that the identity of the human genome has been known for over two decades, yet one cannot state the precise mechanisms by which blood flow is coupled to metabolism in the heart. To study metabolic dilation of the coronary microvasculature, investigators have used both in vivo experiments, in which changes in myocardial oxygen balance have been used, and in vitro experiments in which isolated cardiac myocytes have been used in combination with isolated coronary arterioles.

The heart requires a continuous supply of oxygen to sustain normal aerobic metabolism required for the production of ATP to enable cardiac contraction, maintenance of ion gradients, etc. Oxygen is required, even under resting conditions, as the myocardium extracts 75–80% of the available oxygen from the blood, leaving very little oxygen extraction reserve. Since oxygen extraction is near maximal, to augment oxygen delivery, blood flow to the working heart must be increased. This process of coupling blood flow to metabolism is designated by several terms: metabolic dilation, metabolic hyperemia, and/or functional hyperemia. For the sake of brevity in the remainder of this discussion, we will use the term metabolic dilation. For the heart to sustain normal function, coronary blood flow must be tightly matched to the metabolic demands of the myocardium on a moment-to-moment basis (Alders et al. 2004), and any increase in myocardial oxygen demand (~10 mL O_2/min/100 g myocardium at rest), which can increase up to fivefold during exercise, needs to be met by a similar increase in myocardial oxygen supply. The ability of the coronary resistance vessels to dilate in response to increments in myocardial oxygen demand is extremely important to maintain an adequate oxygen supply. If the oxygen supply is inadequate, the affected region stops beating and there could be irreversible tissue necrosis. It is therefore essential that cardiac myocytes, the beneficiaries of coronary blood flow, communicate their needs to the coronary resistance vasculature, to fine-tune the regulation of coronary flow and to direct this flow to the regions of the myocardium that require it most.

8.2.2 *Metabolic Hyperemia: Putative Metabolites and Effectors*

The high level of basal oxygen extraction requires an increase in coronary blood flow to accommodate any major increase in myocardial oxygen demand. Since mean aortic pressure increases only by 10–20% and average extravascular compressive forces increase during exercise, it is clear that an increase in coronary blood flow must be accommodated by a decrease in coronary vascular resistance. However, the traditional picture of a homogeneous array of resistance vessels that act principally in response to local myocardial needs, but can also respond weakly to systemic vasomotor influences, is incorrect. Metabolic regulation of coronary vascular resistance occurs principally in coronary arterioles (<150 μm in diameter) to tightly couple oxygen demand and supply, whereas the small arteries do not appear to respond to local metabolic influences because they are distant from the parenchymal source of metabolites and these vessels are less sensitive to putative metabolites (Kuo et al. 1995). Figure 8.5 represents the changes in diameter of epicardial vessels

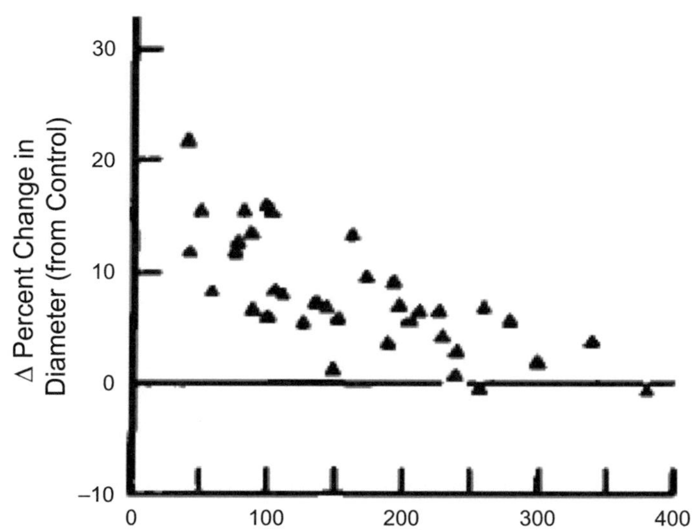

Fig. 8.5 Responses of coronary arteries and arterioles to increased cardiac metabolism produced by pacing. Responses to increased metabolism are plotted as a percent change in their diameter from the baseline, control diameter. (Adapted from Kanatsuka et al. 1989)

in response to increased myocardial oxygen consumption produced by pacing (Kanatsuka et al. 1989). Note the smallest vessels responded most robustly, but even small arteries as large as 300–350 μm in diameter showed some degree of vasodilation. We believe this represents a coordinated response by the entire network in which metabolic stimuli dilate the smallest vessels while myogenic and flow (shear stress)-dependent stimuli lead to upstream vasodilation as suggested by our group many years ago (Kuo et al. 1995; Jones et al. 1993a, 1995a). In this model, during increased myocardial oxygen consumption, the production of vasoactive metabolites produces dilation of the smallest coronary arterioles. This dilation would decrease the pressure upstream in larger arterioles, which would recruit these vessels through myogenic vasodilation. Dilation of these two segments would lead to a decrease in overall resistance, which would increase flow through small arteries and recruit these vessels to dilate through flow-mediated vasodilation. Although this scheme is hypothetical based on the reactivity of metabolic, myogenic and flow-dependent stimuli in the coronary microcirculation, it is a plausible explanation.

Originally, metabolic control was thought of as an increased production of a vasodilator by the cardiac myocytes in response to an increased myocardial metabolism, that diffused through the interstitial space to the coronary vascular smooth muscle, reacted with a receptor, and caused vasodilation. Proposed mediators of metabolic vasodilation include adenosine (resulting from ATP hydrolysis), bradykinin, CO_2 and $[H^+]$, H_2O_2, and potassium. However, none of these vasodilators fulfilled the requirement that blocking its production or its receptor abolishes metabolic coronary vasodilation.

Interestingly, recent studies have focused on the role of vasoconstrictors in metabolic control of coronary blood flow (even though this appears enigmatic!). Besides the increased production of a vasodilator, decreased production of a vasoconstrictor could also result in coronary vasodilation, thereby augmenting coronary blood flow. Vasoconstrictors that have been proposed to play a role in metabolic regulation are endothelin and angiotensin-II. Thus, several substances have been proposed to play a role in the process of metabolic regulation. However, unraveling of the exact mechanism of metabolic vasodilation has been difficult, likely because blockade of a single pathway results in compensatory increments in the contribution of other pathways (redundancy in design). Thus, it is now widely accepted (but this may be incorrect) that there is no single substance responsible for metabolic regulation. Rather, many substances act in concert with the coronary vasculature to match coronary blood flow to the local demands of the myocardium. Several of these substances will be discussed below; however, two methods of measuring metabolic control of the coronary vasculature will be described first.

A sensitive way to study alterations in coronary vascular tone in relation to myocardial metabolism in vivo is the myocardial oxygen balance, i.e., the balance between myocardial oxygen demand and myocardial oxygen supply. This balance is reflected in relationship between coronary venous PO_2 and myocardial oxygen demand as is explained below. An increase in coronary resistance vessel tone will limit coronary blood flow and hence myocardial oxygen supply at a given level of myocardial oxygen consumption, forcing the myocardium to increase its myocardial oxygen extraction (in order to meet myocardial oxygen demand), which results in a lower coronary venous PO_2. Conversely, a decrease in resistance vessel tone increases myocardial oxygen supply at a given level of myocardial oxygen consumption resulting in an increased coronary venous PO_2. The coronary venous PO_2 or coronary venous oxygen saturation thus represents an index of myocardial tissue oxygenation (i.e., the balance between myocardial oxygen supply and oxygen demand), which is determined by the coronary resistance vessel tone (Fig. 8.5). The advantage of measuring metabolic control this way is that one can measure directly in the system of interest, in a physiologically highly relevant in vivo system.

It is difficult to identify the main contributors to metabolic regulation in vivo because metabolic regulation encompasses redundant systems that act in concert to modulate vascular tone; inhibition of one pathway may be compensated by altered output of the others. Unless more pathways are blocked simultaneously, metabolic regulation appears to be maintained (Ishibashi et al. 1998). Also, changes in production of vasoactive substances are difficult to measure. The endothelium forms a two-way barrier between the bloodstream and the myocardial interstitium and can actively metabolize and produce vasoactive and cardioactive substances. Therefore, concentrations of vasoactive substances as measured in the coronary sinus do not necessarily reflect concentrations in the interstitium. To circumvent some of these problems, an in vitro system has been developed. In this system, isolated cardiac myocytes are used that can be metabolically stimulated by electrical field stimulation at different rates. The supernatant of these myocytes can then be transferred to isolated pressurized coronary arterioles and their vasodilator or vasoconstrictor response can be measured. Quiescent myocytes consume approximately 20 nL O_2/min per 100,000 cells (Merkus et al. 2002). Oxygen consumption increases dramatically when the myocytes are electrically stimulated and depends on contraction rate. Stimulation of myocytes at 100 and 200 beats per minute results in a 20- and 30-fold increases in oxygen consumption respectively (Merkus et al. 2002). The main disadvantage of this in vitro system is that by removing the myocytes and coronary arterioles from their natural environment, the close proximity between vessels and myocytes is lost. Thereby, the possibility exists that some short-lived vasoactive factors such as nitric oxide and carbon monoxide will not be preserved. Furthermore, it is difficult to mimic in vivo concentrations of endogenous substances. The changes in vascular diameter in response to administration of the supernatant are, therefore, qualitative rather than quantitative. However, an advantage of the in vitro system is that oxygen supply to the myocytes is independent of the coronary circulation (diffusion of oxygen in the bath), which circumvents a concern related to the in vivo experiments. For example, if one blocked the actions or production of a metabolic dilator in vivo, then it is likely ischemia would ensure, and there may be compensatory ischemic dilation, that may be mediated by a different set of dilators. Thus, the study itself may be impacting the observations, which can lead to erroneous conclusions.

Adenosine has been proposed as one of the messengers that couples myocardial oxygen demand to the vasomotor tone of the coronary resistance vessels (Berne and

Rubio 1974; Olsson et al. 1992). Adenine nucleotides do not cross the cell membrane of cardiac myocytes, but adenosine formed from the action of nucleotide phosphorylase on AMP can be transported out of myocytes into the interstitial space (Sparks Jr. and Bardenheuer 1986). Upon entering the interstitial space, adenosine can interact with A_2 receptors on coronary vascular smooth muscle to produce vasodilation and an increase in coronary blood flow (Olsson et al. 1992). Studies in dogs have demonstrated that endogenous adenosine production is not mandatory for maintaining resting coronary blood flow. Thus, studies in dogs have generally failed to demonstrate an effect of intracoronary adenosine deaminase (Saito et al. 1981; Kroll and Feigl 1985; Hanley et al. 1986), or adenosine receptor blockade (Schutz et al. 1977; Giles and Wilcken 1977), on basal coronary flow, although some studies have reported an increase in coronary vascular resistance (Martin et al. 1991), or a decrease in resting coronary venous $P\text{O}_2$ (Tune et al. 2000a), following adenosine receptor blockade. In doses that caused marked inhibition of exogenous adenosine-induced coronary vasodilation, intracoronary adenosine deaminase or intravenous 8-phenyl-theophylline (8PT) had no effect on resting coronary blood flow, oxygen extraction or coronary venous $P\text{O}_2$ in awake dogs (Bache et al. 1988). In the canine heart in vivo, adenosine release has been reported to be enhanced during exercise (Watkinson et al. 1979; Bacchus et al. 1982), although recent work suggests that interstitial adenosine concentrations may not increase significantly during exercise (Tune et al. 2000a). Furthermore, demonstration of an essential role of adenosine in mediating exercise-induced coronary vasodilation requires that interruption of the adenosine effect interfere with exercise-induced coronary vasodilation. Yet, the shift in the relationship between myocardial oxygen consumption and coronary venous $P\text{O}_2$ is parallel, indicating that the vasodilator effect of adenosine does not increase during exercise in either dogs (Ishibashi et al. 1998; Tune et al. 2000b), swine (Merkus et al. 2003a), or humans (Edlund and Sollevi 1995; Edlund et al. 1989, 1995). Similarly, when the supernatant of isolated myocytes is transferred to isolated coronary arterioles, blockade of A_2 receptors with 8PT decreases the vasodilator properties of the supernatant but to an equal amount in stimulated and unstimulated cardiac myocytes (Merkus et al. 2002). Taken together, these findings suggest that adenosine may contribute to a basal offset point of coronary vasomotor tone, but does not contribute to the increase in blood flow during metabolic hyperemia.

Endothelin (ET)-1 is one of the most potent coronary vasoconstrictor agent known to date. It is produced in endothelial cells by cleavage of its nonvasoactive precursors, preproendothelin and big ET (Rubanyi and Polokoff 1994). The ET receptors are located both on the endothelium and on vascular smooth muscle. Binding of ET to ET_B receptors on the endothelium leads to production of NO and prostacyclin, which induce vasodilation, whereas binding of ET to the ET_A and ET_B receptors on vascular smooth muscle leads to vasoconstriction (Rubanyi and Polokoff 1994). Administration of exogenous ET causes ET_B-mediated vasodilation at low doses but constriction at high doses, indicating that the ET_B receptor on the endothelium is more sensitive to ET than the receptors on vascular smooth muscle (Rubanyi and Polokoff 1994). Measurements of ET levels in blood yield concentrations in the picomolar range, while receptor sensitivities are in the nanomolar range (Frelin and Guedin 1994). However, reports on the role of endogenous ET have shown that, despite its low plasma concentrations and most likely due to its abluminal secretion, ET contributes to vascular tone in the coronary circulation of humans, swine, and dogs (Merkus et al. 2002; Takamura et al. 2000; Merkus et al. 2003b; Halcox et al. 2001).

A role for ET in regulating resistance vessel tone during acute exercise has been proposed. Short-term exercise does not result in changes in plasma ET-levels (Takamura et al. 2000; Lenz et al. 1998; Haitsma et al. 2001), although small increases have been reported as well (Maeda et al. 1994; Ahlborg et al. 1995). Since vasodilators can modify the response of coronary microvessels to ET (Merkus et al. 2000), withdrawal of the vasoconstrictor influence of ET could contribute to exercise-induced vasodilation. The first study that investigated the role of ET in the regulation of coronary vasomotor tone during exercise found that the effect of ET_A/ET_B blockade on coronary vascular tone tended to decrease in dogs during incremental levels of exercise (Takamura et al. 2000). In a study in swine that were exercised on a motor-driven treadmill under control conditions and after the ET_A-receptor blockade, it was observed that ET_A-receptor blockade caused coronary vasodilation under resting conditions, evidenced by an increase in coronary venous $P\text{O}_2$. Interestingly, the effects of the ET_A receptor antagonist waned at increasing levels of exercise (Fig. 8.6), indicating that the vasoconstrictor influence of ET is blunted during increased myocardial metabolic demand, as occurs during exercise (Merkus et al. 2003b). Moreover, the ET_B receptor was found to exert a tonic vasodilator influence on the porcine coronary vasculature. These findings are in accordance with a study in dogs in vivo (Gorman et al. 2005) as well as with in vitro findings that the ET-vasoconstrictor influence on coronary arterioles is modified by the cardiomyocytes according to their metabolic status, so that at higher pacing rates myocytes inhibit the vasoconstrictor influence of ET (Merkus et al. 2002).

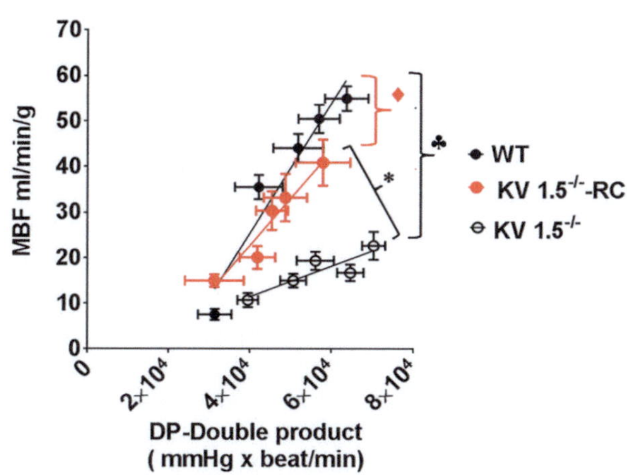

Fig. 8.6 Relationships between cardiac work (DP) and myocardial blood flow (MBF) in WT, Kv1.5$^{-/-}$ (male and female), and Kv1.5$^{-/-}$, SM22-Tet-Kv1.5 mice (Kv1.5$^{-/-}$RC). (Adapted from Ohanyan et al. 2015)

Although a decrease in local ET release, for example due to an increased NO production during exercise, could theoretically have contributed to metabolic coronary vasodilation, this is less likely. Thus, neither coronary arterial nor coronary venous ET-levels changed during exercise in either swine (Merkus et al. 2003b) or dogs (Takamura et al. 2000). There are several mechanisms that may account for the exercise-induced modulation of the effects of ET. First, interstitial adenosine levels may increase during exercise, which can decrease the sensitivity of the vasculature to ET (Merkus et al. 2000). Second, NO production increases during exercise, which can directly modulate the binding of ET to the ET_A receptor as well as the receptor-signal transduction (Goligorsky et al. 1994; Wiley and Davenport 2001). Thus, it is possible that ET_A receptor sensitivity may be decreased during exercise through an increase in NO, adenosine or both, thereby facilitating metabolic vasodilation.

Recent work on coronary metabolic dilation has focused on the effectors of this process, i.e., those components of the system that lead to the vasodilation. Within this context, much of the progress has centered around potassium channels that mediate the connection of flow to metabolism. There is some debate about whether the connection is mediated by a metabolite that opens the potassium channel or whether the connection is mediated by an electrical connection between cardiac myocytes and the vasculature. In particular, Zhao et al. proposed during increased myocardial metabolism, a decrease in cardiac myocyte ATP levels would lead to the opening of cardiac myocyte K_{ATP} channels (Zhao et al. 2020). In an elegant series of experiments, the authors found that hyperpolarization of cardiac myocytes can be conducted to electrically connected cells, such as vascular smooth muscle and endothelial cells. The cardiac hyperpolarization decreased intracellular calcium in vascular smooth muscle, leading to relaxation, vasodilation of resistance vessels and increased flow (in their case reduced perfusion pressure in an artery feeding the papillary muscle during constant flow perfusion). This article attracted a commentary by Grainger and Santana, entitled metabolic-electrical control of coronary blood flow (Grainger and Santana 2020). This hypothesis is unique but suffers from some limitations that we are compelled to mention. First, it is based on the assumption that a decrease in ATP occurs in cardiac myocytes during increased metabolism. This decrease in ATP is the event leading to the opening of the cardiac K_{ATP} channels. However, evidence that detracts from the novelty of the hypothesis shows that cardiac ATP levels are maintained during heightened aerobic metabolism produced by pacing (Dunn et al. 1978; Allison and Holsinger Jr. 1977). Also in the aerobically working heart, most reports show very small changes in coronary sinus PO_2 indicating that oxygen balance (delivery equals consumption) is intact (Merkus et al. 2002; Tune et al. 2004), such that there is not a hypoxic error signal, e.g., a reduction in tissue ATP, that needs to be corrected by increased flow. Also, Tune et al. reported that K_{ATP} channel blockade decreased resting cardiac venous PO_2, but did not affect myocardial oxygen balance during exercise (Tune et al. 2001). Another reason for our skepticism about the viability of the electro-metabolic coupling hypothesis relates to the fact that the electrical connections between cardiac myocytes and adjacent cells would conduct current from all sources, not just K_{ATP} channels. Although the current conducted by the K_{ATP} channels is large in cardiac myocytes, there are three major potassium currents and 18 different types of potassium channels in cardiac myocytes (Jeevaratnam et al. 2018). It seems improbable that only the K_{ATP} channel would have conducted current; thus, the authors are skeptical of this hypothesis. Another component that fuels our skepticism about electro-metabolic coupling are numerous observations showing that isolated cardiac myocytes, when stimulated to beat at high rates produce sufficient quantities of metabolites to induce vasodilation of isolated arterioles (Merkus et al. 2002; Guarini et al. 2016; Saitoh et al. 2006).

Another type of conducted response that could contribute to the connection of flow to metabolism would be responses conducted along the vascular tree. In the brain, this mechanism appears to be well documented and appears to enable fine-tuning of metabolic control (Longden et al. 2017). In this paradigm, neuronal activity induces an increase in extracellular potassium, via depolarization and repolarization, activating endothelial inward rectifying potassium channels. This hyperpolarization response is conducted retrograde (upstream) to arterioles and via myoendothelial junctions results in arteriolar hyperpolarization and vasodilation. This mechanism makes sense for the brain where discrete sites of neuronal activity can cause local dilation in a confined area. In the coronary circulation, this type of local fine-tuning makes less sense because *all* cardiac myocytes are active with every beat (under normal conditions). However, there is evidence for Kir channels in coronary arterioles and arteries (Quayle et al. 1996; Dora et al. 2022). Moreover, Rivers et al. (2001) observed conducted responses in isolated coronary arterioles that were blocked by barium-chloride, which reportedly antagonizes Kir channels (Robertson et al. 1996). Conducted responses may facilitate upstream recruitment of arterioles and small arteries to facilitate a coordinated network response to increased metabolic demand, but whether this response is part of the coronary microvascular response to heightened metabolism remains unresolved.

Advances in our understanding of the role of specific ion channels as effectors that connect flow to metabolism stemmed from our work in the early 2000s (Saitoh et al. 2006; Rogers et al. 2006, 2007). We advanced the hypothesis that coronary blood flow is connected to metabolism via mitochondrial production of H_2O_2. Such coupling is a feed-forward mechanism as it does not rely on an error signal, such as tissue hypoxia, for production of the metabolite. Rather, the production of H_2O_2 is a by-product of aerobic metabolism from electrons that leak from mitochondrial complexes to form superoxide ($O_2^{\cdot -}$), which is rapidly converted into H_2O_2 by the actions of superoxide dismutase. H_2O_2 opens Kv channels via a redox reaction since the channel opening was inhibited by the reductant, dithiothreitol. These findings led to several studies in the literature pertaining to the role of Kv channels in coronary regulation (Goodwill et al. 2016; Berwick et al. 2013; Dick et al. 2008). However, a limitation of these studies pertained to the use of pharmacological antagonists, which frequently have off-target effects, and are typically not as specific as originally thought. As one example, the drug 4-aminopyridine is used to block voltage-gated potassium channels with the thought that the drug is preferential for Kv channel families 1–4, but is now known to antagonize members of each of the

12 families of Kv channels (Rodriguez-Rangel et al. 2020). Thus, to understand the role of specific ion channels, a more refined approach was needed.

To circumvent the issue of ion channel nonspecificity, investigators have advanced their approaches to include the study of mice, enabling the manipulation of mouse genetics to rigorously study the roles of specific ion channels in coronary metabolic dilation. To this end, Ohanyan et al. interrogated the role of Kv1.5 channels in coronary metabolic dilation (Ohanyan et al. 2015). The foundation for this work was based on reports that the Kv1.5 channel had redox and oxygen-sensing properties (Archer et al. 2000, 2004), which would render it a suitable link between myocardial metabolism and coronary blood flow. Ohanyan et al. found that the coupling between metabolism (as indicated by the product of heart rate and mean arterial pressure; or the triple product of heart rate, arterial pressure, and stroke volume) was substantially altered in mice null for Kv1.5 channels (Fig. 8.6). It is important to mention that cardiac work was increased by varying amounts of i.v. norepinephrine. Because the genetically modified mice had a global knockout of Kv1.5 channels (all tissues would lack the channel), the authors also incorporated a double transgenic on the knockout background that enabled re-expression of the ion channel in smooth muscle under the control of a tet-on promoter (Kv1.5$^{-/-}$RC [Kv1.5-null mouse with reconstituted Kv1.5 channel]). Treatment with a tetracycline antibiotic (in drinking water or food) expresses the channel in smooth muscle. As shown in Fig. 8.6 in red, the re-expression of Kv1.5 channels in smooth muscle rescued the impaired coronary dilation, suggesting that the presence of the Kv1.5 channel in coronary smooth muscle is critically important for the coupling of coronary blood flow and cardiac metabolism. The coupling of flow to metabolism in the heart was important for tissue oxygenation during a cardiac stress test by i.v. norepinephrine to increase cardiac work (Fig. 8.7). This figure shows measurements of left ventricular PO_2 by L-band electron paramagnetic resonance under basal conditions and during norepinephrine infusion. Note in wild-type mice, cardiac

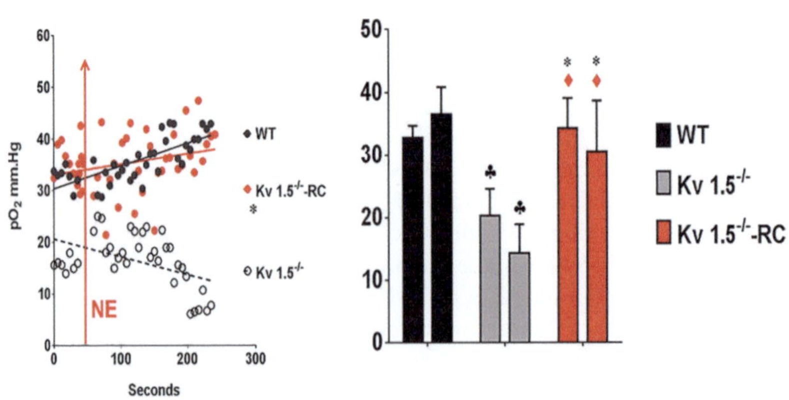

Fig. 8.7 Myocardial tissue oxygenation (PO_2) in WT, Kv1.5$^{-/-}$(male and female), and Kv1.5$^{-/-}$, SM22-Tet-Kv1.5 mice (Kv1.5$^{-/-}$RC) during increased metabolic demands. (Adapted from Ohanyan et al. 2015)

tissue oxygenation is maintained during the stress test, but oxygenation drops in the null mice to a low of 6 mmHg. The Kv1.5-RC mice (in red) had a similar response as wild-type mice with oxygenation maintained during the stress test; again, indicating the importance of the Kv1.5 channel in the connection of coronary blood flow to myocardial metabolism. One question pertaining to this result is whether a tissue PO_2 would be too low to sustain aerobic metabolism and function. Although this question was not directly answered, insight was provided by the observation that during the cardiac stress test, arterial pressure was maintained in the wild-type and Kv1.5-RC groups but fell considerably in the Kv1.5-null mice. This suggests the fall of tissue oxygenation during the stress test is due to compromised cardiac pump function because of insufficient blood supply and oxygen delivery that could not sustain normal aerobic metabolism.

The role of the Kv1.5 channel in myocardial metabolic hyperemia was further established in experiments aimed at studying mechanisms of channel regulation and the role of regulatory subunits in the connection between metabolism and flow. Ohanyan et al. reported that the Kvβ2 subunit was critical in the connection of coronary blood flow to cardiac work (Ohanyan et al. 2021). They observed in mice null for the Kvβ2 subunit, an impairment of metabolic dilation compared to wild-type animals, i.e., for a given increase in cardiac work produced by i.v. norepinephrine, the increase in blood flow was far less in the null animals than in wild types (Fig. 8.8). Interestingly, mice null for the Kvβ1.1 subunit showed enhanced metabolic hyperemia, i.e., for a certain workload, blood flow was higher in Kvβ1.1$^{-/-}$ mice compared to wild-type mice (Fig. 8.8). Another important aspect of this work pertained to an inability of the mice to sustain cardiac pump function during the norepinephrine stress test (Fig. 8.9), similar to the finding reported by Ohanyan et al. (2015). It is important to mention that in this latter paper, Ohanyan

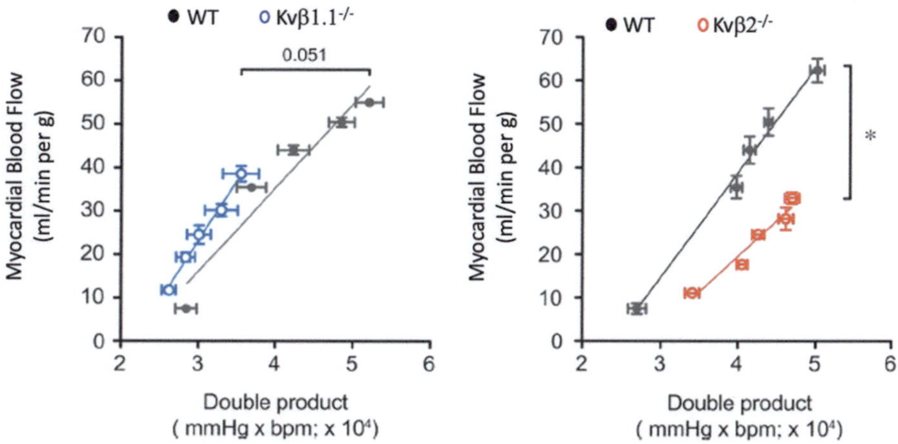

Fig. 8.8 Plots of myocardial blood flow during increases in cardiac metabolism (double product of heart rate x mean arterial pressure) induced by i.v. norepinephrine in wild-type (WT) mice and those null for the Kvβ1.1 or Kvβ2 subunits of the Kv1.5 channel. (Adapted from Ohanyan et al. 2021)

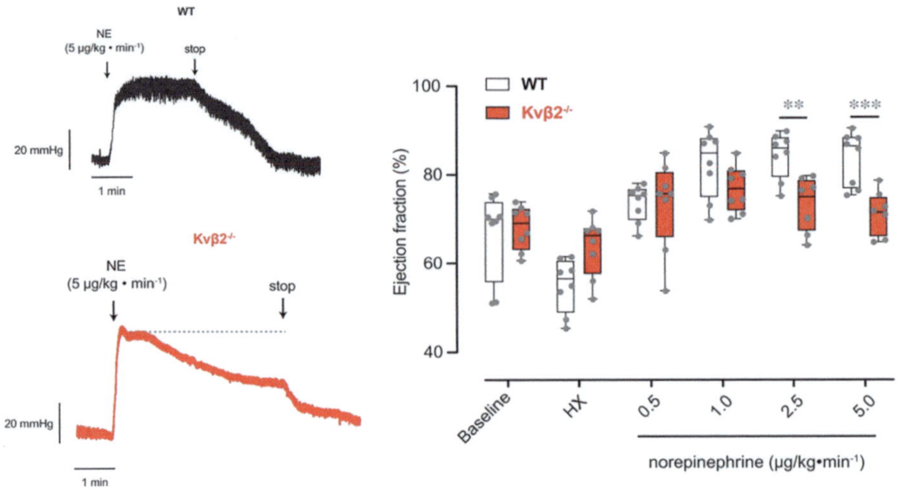

Fig. 8.9 Aortic pressure (left panels) and left ventricular ejection fraction (right panel) during i.v. norepinephrine in wild-type (WT) mice and those null for the Kvβ2 subunit of the Kv1.5 channel. Note, during the norepinephrine stress test, wild-type (WT) animals maintained high arterial pressure and cardiac function were increased and stable from baseline. However, in the Kvβ2-null mice, neither arterial pressure nor ejection fraction was maintained during the stress test. (Adapted from Ohanyan et al. 2021)

and colleagues studied mice null for the alpha subunit (pore-forming subunit) of the Kv1.5 channel. The figure illustrates the deterioration of cardiac pump function in the Kvα2$^{-/-}$ mice during the norepinephrine stress test, which contrasts with the response of the wild-type mice to the i.v. norepinephrine where cardiac function was enhanced and maintained. The result of such deterioration of cardiac function was a decrease in arterial pressure (after an initial increase) in the null mice that contrasted to the wild-type mice where arterial pressure was maintained at a high level during the stress test.

A missing "piece" into the role of Kv1.5 channels in facilitating the connection of coronary blood flow to cardiac metabolism/work, i.e., coronary metabolic dilation or metabolic hyperemia pertains to how this ion channel is regulated. Dwenger et al. helped unravel this conundrum by showing the redox regulation of the channel (Dwenger et al. 2022). These investigators found that the NAD(P)(H) reductase activity of the channel affected the redox state of vascular myocytes during changes in cardiac metabolism as viewed by mass spectrometric imaging of cardiac tissue specimens obtained under the various conditions. This alteration in redox state (increase in the NADH:NAD$^+$ ratio) potentiated Kv1 channel activity in a Kvα2-dependent manner. Elimination of the Kvα2 subunit reductase activity through a point mutation in the subunit eliminated the redox sensitivity of the channel. These investigators also observed the changes in this smooth muscle response were regulated by products produced by electrically stimulated cardiac myocytes as stimulation of the smooth muscle directly did not cause any change in the redox state of the cell.

The synthesis of this work on the Kv1.5 channel suggests strongly it plays a key role in coronary metabolic dilation. The authors would like to offer some perspective on key missing elements in the complete understanding of metabolic hyperemia in the heart. Specifically, the precise mediators that connect flow to metabolism are still not fully known or understood. Obviously other mechanisms and processes play a role as deletion of the channel, the Kvα2 subunit, or elimination of the Kvα2 reductase activity does not eliminate all metabolic hyperemia—the connection of flow to metabolism is attenuated, but not eliminated. There must be other processes and effectors that facilitate the connection of myocardial blood flow to cardiac work. As an example, Ohanyan et al. reported that Kv1.3 channels play a role, although a lesser role than Kv1.5 channels in connecting coronary blood flow to cardiac metabolism (Ohanyan et al. 2017). This is one example of the redundancy that must occur for such a fundamental process as metabolic hyperemia in the heart.

We would be remiss to not mention that many other dilators and constrictors have been proposed (Table 8.1), but the distillation of all results leads to the inescapable conclusion that the identity (or identities) of causal factors for metabolic regulation remain unknown. Table 8.1 summarizes findings that have supported (and some refuted) the role of many suspected metabolic dilators or the signaling scheme by which dilation is produced. We wish we could say there exists some consensus, but to be completely frank, the unequivocal evidence supposedly supporting or refuting is open to debate. For example, some investigators have dismissed the adenosine hypothesis because calculated (based on a model) interstitial levels of adenosine are too low to be involved in metabolic dilation, yet we offer the caveat that the measurement has not been made only the calculation. Other results supposedly refuting the involvement or supporting a particular ion channel are made with pharmacological antagonists that affect other channels. For example, glibenclamide and tetraethylammonium are used to block K_{ATP} and K_{Ca} channels, respectively, but both antagonists block other K channels such as Kv. There is growing, but yet preliminary, evidence that reactive oxygen species may play a role in metabolic dilation (Zhang et al. 2004) and coronary autoregulation (Yada et al. 2003), but these studies are still preliminary. A recent study of trying to identify a "best fit" predictive model for metabolic hyperemia in the heart identified a feed-forward system as the best fit (Gharahi et al. 2022) and control schemes relying on a negative feedback system were not nearly as predictive. The authors opine this makes sense as imagine a negative feedback scheme involving the production of adenosine to produce metabolic dilation. The heart would literally have to be driven into an oxygen imbalance to produce vasodilatory metabolites. During exercise and metabolic hyperemia (under normal conditions), the heart is not ischemic, and as mentioned previously, ATP levels do not fall (Dunn et al. 1978). However, we emphasize that there still remains no consensus as to the metabolites, processes, and effectors that link flow and metabolism.

Table 8.1 Putative metabolic vasodilators and signaling mediators/effectors

Vasodilator	References
Hydrogen peroxide	Zhang et al. (2004), Yada et al. (2003)
Adenosine	Watkinson et al. (1979), Miller et al. (1979), Rubio et al. (1974), Berne et al. (1971)
Vascular myocyte redox status	Dwenger et al. (2022), Saitoh et al. (2007)
Norepinephrine (as a feed-forward dilator)	Miyashiro and Feigl (1993), Gorman et al. (2000a, b)
ATP, ADP, AMP	Farias 3rd et al. (2005), Gorman et al. (2010)
Endothelin-1 (inversely proportional to metabolic dilation)	Merkus et al. (2002)
Protons	Ishizaka and Kuo (1996), Ishizaka et al. (1999)
Bradykinin	Merkus et al. (2002), Prasad et al. (1999)
Potassium ion	Rivers et al. (2001)
Osmolality	Vlahakes et al. (1989), Ishizaka and Kuo (1997)
G Protein mediation	Sato et al. (2002), Tanikawa et al. (2000)
Effector	
K_{ATP} channels as effectors	Zhao et al. (2020), Takeda et al. (2006)
K_{Ca} channel as effectors	Merkus et al. (2006)
Kv channels as effectors	Ohanyan et al. (2015, 2021), Dwenger et al. (2022), Ohanyan et al. (2017), Rogers et al. (2006), Dwenger et al. (2018)
Kir channels as effectors	Rivers et al. (2001), Knot et al. (1996)
Contrarian evidence	
Evidence against K_{ATP} channels	Tune et al. (2001), Richmond et al. (1999), Duncker et al. (1993)
Evidence against adenosine	Tune et al. (2000a, b; 2001)
Evidence against nitric oxide	Tune et al. (2001)

8.2.3 Autoregulation

The discussion pertaining to ischemic vasodilation will be less than the preceding section on metabolic hyperemia. In part, this relates to a far greater consensus on factors contributing to ischemic vasodilation in the heart. Our focus will center on coronary autoregulation during reductions in coronary perfusion pressure. Autoregulation is defined as the proclivity of a circulation to maintain constant blood flow during changes in perfusion pressure, without changes in metabolic demand. For organ systems other than the heart, this would occur during a reduction in aortic pressure. However, for the heart, the situation is more complicated because a reduction in aortic pressure also would reduce cardiac work and flow would decrease because of lower metabolic hyperemia. This is not to indicate that autoregulation is not important for the heart, but it is more applicable to our understanding of flow regulation in the setting of coronary disease. As a coronary stenosis grows pressure in the coronary artery downstream from the stenosis falls, but flow is maintained until the stenosis reaches a critical level. This critical level

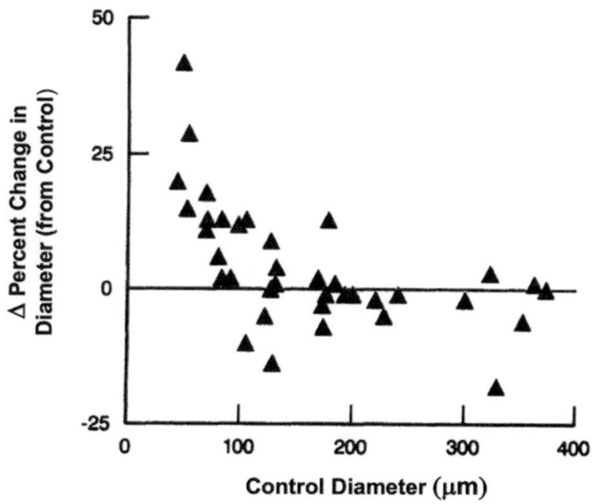

Fig. 8.10 Responses of coronary arteries and arterioles to decreases coronary perfusion pressure. Responses to the decreased perfusion pressure were plotted as a percent change in their diameter from the baseline, control diameter. (Adapted from Kanatsuka et al. 1990b)

likely coincides with the lower limit of autoregulation, below which perfusion becomes pressure-dependent. Importantly it is the autoregulatory process that maintains coronary blood flow despite the presence of a stenosis. In a sense, this important aspect of flow regulation is one of the conundrums in clinical cardiology making most patients with coronary disease asymptomatic, because autoregulatory adjustments maintain flow during the advancement of occlusive disease.

The coronary autoregulatory response to reductions in perfusion pressure principally involve adjustments in diameter in the smallest resistance vessels in the coronary circulation (Kanatsuka et al. 1990b; Chilian and Layne 1990) with small arteries serving as a passive player in this response as they showed decreases in diameter during reductions in pressure. Figure 8.10 illustrates the responses of epicardial coronary arterioles and small arteries to reductions of coronary perfusion pressure (reduced to ~60 mmHg from a control of ~90 mmHg. As mentioned, the smallest arterioles (those less than 100 μm in diameter) show the greatest dilation (with the extent of dilation greater in the smaller vessels). Small arteries between 100 and 200 μm show a variable response with some dilating and others constricting, and those small arteries above 200 μm either show no response or a passive constriction. What is unknown is the contribution of metabolic and myogenic processes to the dilation at reduced perfusion pressures, although it is noteworthy to add that coronary arterioles do possess the capability of reacting to intraluminal pressure and likely stretch (Kuo et al. 1988, 1990a, b, 1995).

In contrast to the feed-forward process that appears to best predict metabolic hyperemia in the heart, ischemic vasodilation is best fit by a negative feedback scheme involving tissue hypoxia during ischemia and the consequential production of ischemic vasodilatory metabolites by metabolic pathways reliant on oxygen metabolism (Warne et al. 2023). Such a negative feedback was elegantly illustrated in the previous study in which authors studied coronary autoregulation during reductions in perfusion pressure under normoxia and hypoxemia. Induction of hypoxemia increased autoregulatory gain (the extent to which flow is restored to baseline following reductions in perfusion pressure). This result expands on the concept presented many years ago by Dole and Nuno (1986) who reported that coronary autoregulation was intimately related to coronary sinus/venous PO_2 with reductions in PO_2 increasing autoregulatory regulatory gain (a term used to describe the ability of the circulation to adjust to the control value before the pressure change) and elevations producing the opposite response. Despite the observations and acceptance that oxygen tension is involved in coronary autoregulation (Warne et al. 2023; Dole and Nuno 1986; Drake-Holland et al. 1984), there are some disagreements about the role of various effectors and putative metabolites. Kiel et al. (2018) argued that the level of oxygenation is not critical, rather it is the level of vasomotor tone. However, the majority of the literature support the idea that autoregulation of coronary blood flow is related to the correction of a flow deficit during reductions in perfusion pressure and that a single substance cannot explain both processes (Laird et al. 1981).

One of the classical mediators proposed for modulating coronary autoregulation is adenosine (Berne 1980). Adenosine, the production of ATP hydrolysis and the conversion of ADP to AMP and finally to adenosine during tissue hypoxia occurring when oxygen delivery is limited. Adenosine is a potent coronary vasodilator and also demonstrates a distinct preference for small coronary arterioles as shown by Kuo et al. (1995). Yet, there are many challenges to the adenosine hypothesis for coronary autoregulation. Stepp et al. found that coronary sinus levels of adenosine only increased when coronary perfusion pressure dropped to 60 mmHg or less (Stepp et al. 1997). Because autoregulation had occurred at the lesser pressure reductions without any change in adenosine, the authors concluded that adenosine is not required for most autoregulatory adjustments. Komaru et al. also found that dilation of epicardial coronary arterioles during pressure reductions was not compromised by adenosine receptor antagonism or superfusion of the epicardium with adenosine deaminase (Komaru et al. 1994). Another report that challenged the idea that purine metabolism may contribute to coronary autoregulation reported that antagonism of P2Y (Kuo et al. 1988) receptors did not affect coronary pressure-flow relations during reductions in perfusion pressure. Another report further complicates these observations! Yada et al. reported that during reductions in coronary perfusion pressure, inhibition of nitric oxide synthase blocked dilation of epicardial small arteries but not of small arterioles. However, combined administration of a nitric oxide synthase inhibitor and catalase (to degrade hydrogen peroxide) attenuated dilation of coronary arterioles to reductions in perfusion pressure. Combined administration of catalase and an adenosine receptor antagonist blocked coronary arteriolar vasodilation to the reductions in perfusion pressure (Yada et al. 2003). These investigators supported a role for adenosine in the autoregulatory adjustments of the coronary vasculature to reductions in perfusion pressure. Taken together, there is yet no consensus for or against the role of adenosine in coronary autoregulation. However, given the importance of this vasomotor adjustment the authors opine that further investigations are needed to finally understand the role of adenosine and purinergic receptor signaling in this important process.

Similar to the role of adenosine in coronary autoregulation, the role of K_{ATP} channels in coronary autoregulation is also controversial. Stepp et al. reported that antagonism of these channels induced coronary constriction (reduced flow at a certain pressure) but did not affect autoregulation (Stepp et al. 1997). In contrast, Dankelman observed that K_{ATP} antagonism did not affect the steady state adjustment in flow during reductions in perfusion pressure, but they did observe a decreased rate of flow adjustment (fourfold) leading them to conclude these channels play a role in autoregulation (Dankelman et al. 1994). Although the conclusions of these two studies are different, both groups made the same observation with the measurements of the steady-state flow after the pressure reduction. The Dankelman study however also evaluated another variable—the time it took for the flow adjustment and found that K_{ATP} channel blockage lengthened the time necessary for flow to return to a steady state. Another report observed that dilation of epicardial coronary arterioles to reductions in perfusion pressure was abolished by glibenclamide, which suggests that K_{ATP} channels play a role in this response (Komaru et al. 1991a).

8.3 Endothelial and Myogenic Mechanisms of Coronary Regulation

8.3.1 Endothelium-Dependent Vasodilation

The vascular endothelium is a single cell layer forming the inner lining of blood vessels and serves many functions in vascular homeostasis and tone. The endothelium is a very heterogeneous structure throughout the vasculature in terms of function. The vascular endothelium plays a key role in the local regulation of vascular tone by the release of vasodilator substances (i.e., endothelium-derived relaxing factor (EDRF = nitric oxide, NO) and prostacyclin) and vasoconstrictor substances (i.e., thromboxane A2, free radicals, or endothelin) (Mullane and Pinto 1987). Using either agents like acetylcholine or changes in flow to stimulate the release of EDRF (NO), clinical studies have revealed the importance of EDRF in both basal and stimulated control of vascular tone in large epicardial coronary arteries and in the coronary microcirculation (Mullane and Pinto 1987). Morphological studies have shown the presence of tight junctions and gap junctions between adjacent endothelial cells. These cell-cell junctions are important in the control of endothelial functions as a preselective barrier between the blood and underlying tissues, and also central for protecting the vessel wall from platelet deposition and thrombus formation (Mullane and Pinto 1987). Coronary microvascular endothelium plays a major role in the modulation of coronary blood flow. The use of endothelium-dependent agonists to elicit microvascular vasodilation is often used as an index of the functional status of the endothelium (Quillen et al. 1990; Chilian et al. 1993). In large coronary arteries, or vessels from other vascular beds, agonists such as acetylcholine, serotonin or substance P have been shown to elicit endothelium-dependent vasodilation (Tschudi et al. 1991; Lamping et al. 1989). Furthermore, since the synthesis of NO by endothelium requires the presence of L-arginine, analogues of L-arginine are often used to inhibit the endothelium-dependent dilation of coronary vessels (Tschudi et al. 1991; Jones et al. 1995b). Substances other than NO, however, also play a role in endothelium-mediated dilation of coronary resistance vessels. For example, inhibition of the L-arginine pathway in porcine coronary resistance vessels (300 μm) inhibits basal release of NO and α_2-adrenergic-mediated release, but only partially inhibits stimulation by serotonin, and has no effect on bradykinin-stimulated release (Tschudi et al. 1991).

Endothelial control mechanisms are not exclusively confined to NO. Other endothelium-dependent dilators have been described to mediate both agonist-dependent and flow-mediated responses. However, under most circumstances, these other dilator systems are apparent when the production of NO by eNOS is impaired. Nonetheless, depending on the model, the species, and the experimental condition, these dilators, including epoxyeicosatrienoic acids or related products of cytochrome P450, prostacyclin, and hydrogen peroxide, can serve as endothelium-derived relaxing factors (Nishikawa et al. 2000; Huang et al. 2001, 2005; Miura et al. 2003). Interestingly, they have different mechanisms of dilation from those mediated by NO, and may even work on different segments of the vasculature.

In the last two decades, endothelium-dependent dilation in response to an increase in flow has been demonstrated in large coronary arteries from a variety of species in vitro and in vivo and has been shown to be endothelium-dependent (Jones et al. 1993b). Control of coronary blood flow requires a balance between flow-dependent properties of the large coronary microvessels upstream and constriction of the small vessels downstream, e.g., basal metabolism-dependent tone (Cornelissen et al. 2002). Flow-dependent dilation is mediated by shear stress-induced production of NO (Cornelissen et al. 2002; Jones et al. 1993b). Coronary flow control is the result of the modulation of tone in resistance arteries.

Intact heart studies have provided indirect evidence that NO mediates flow-dependent dilation in the coronary microcirculation. In isolated hearts with controlled flow perfusion, coronary microvascular flow-dependent dilation is apparent as a secondary fall in perfusion pressure after a step increase in flow (and pressure). The L-arginine analogues increase coronary microvascular resistance in most isolated heart preparations, indicating a tonic dilating role for basal, probably flow-dependent, EDRF release (Jones et al. 1993b). However, the L-arginine analogues have not always been found to increase coronary vascular resistance in vivo. Although these studies provide circumstantial evidence that coronary microvessels exhibit flow-dependent vasoactive responses, this cannot be unequivocally demonstrated in the intact heart, since microvascular tone is always influenced by alterations in myocardial oxygen consumption when coronary pressure or flow are altered (Jones et al. 1993b).

Flow-dependent dilation of coronary arteries has been shown to be blocked by the stereo-specific analogues of L-arginine, which inhibit the activity of the Ca^{2+}-dependent nitric oxidase synthase enzyme, indicating a causal role for NO (Jones et al. 1993b). The decrease in coronary vascular resistance during reactive hyperemia is in part due to the increase in flow velocity and shear stress associated with increased blood flow, and dependent on the presence of an intact endothelium. To eliminate the confounding influences of metabolic changes associated with reactive hyperemia, Kuo et al. (1990a) have demonstrated that following removal of the endothelium from isolated coronary arterioles, flow-induced dilation was lost, while myogenic activity was preserved. Also, similar to other regulatory mechanisms of the coronary microcirculation, flow, or shear-dependent vasodilation is heterogeneously distributed throughout the coronary microvasculature, with the predominant site being large arterioles greater than 100 μm in diameter (Kuo et al. 1995).

Flow-dependent dilation has been demonstrated in isolated porcine coronary venules 40–80 μm in diameter (Kuo et al. 1992a). The venules developed spontaneous tone and dilated progressively in response to step increases in flow, while pressure was held constant. As was the case for arteriolar flow-dependent dilation, the venular flow-dependent dilation was also endothelium-dependent, mediated by NO or a similar substance. Flow-dependent dilation of coronary venules would tend to limit the rise in capillary pressure that follows the greater decrease in arteriolar compared to venous resistance and may thus limit the increase in capillary permeability to water and the possibility of myocardial edema. Flow also modulates coronary venular permeability, a major factor involved in the delivery of nutrients to the myocardium. Since permeability changes passively with pressure, changes in permeability may serve to link both pressure- and flow-dependent vascular control mechanisms to myocardial metabolism, and therefore it ultimately explains the metabolic control of coronary blood flow (Jones et al. 1993b).

The role of integrins in mediating flow-induced changes in vascular tone is suggested by several studies in which flow-induced vascular responses are inhibited by integrin blockade. Integrin-binding peptides containing Arg-Gly-Asp (RGD) inhibit this flow-induced vasodilation, suggesting that integrin signaling is pivotal for the mechanosensory process involved in shear stress mechanotransduction mechanisms and the production of NO (Muller et al. 1997).

In the coronary arterioles, increases in flow-induced shear stress elicit a vasodilation that is dependent on the pathway involving integrins, focal adhesion kinase, src, PI3 kinase, and Akt, leading to phosphorylation of endothelial nitric oxide synthase and the release of NO (Koshida et al. 2005). This renders dilation to flow or shear stress largely independent of changes in cytosolic calcium because the phosphorylation of eNOS by Akt induces activation independent of calcium, which has been confirmed experimentally (Muller et al. 1999).

An important question pertains to the importance of flow- or shear stress-dependent dilatory mechanisms in vivo. Stepp and Chilian have approached this problem by estimating shear stress throughout the intact coronary microcirculation and found that shear stress was poorly regulated in the microcirculation compared to small arteries in which shear stress was well regulated (Stepp et al. 1999). This was slightly different than the predictions of in vitro results in which sensitivity to shear stress was highest in large arterioles around 100 μm in diameter (Kuo et al. 1995), but in the absence of the parenchymal influences through the production of metabolites it is not surprising that the responses differ among locations. This may also help explain observations that myriad interactions occur between myogenic and agonist-induced changes in tone on the responsiveness of flow-dependent dilation (Kuo et al. 1991; Stepp et al. 2001).

8.3.2 Myogenic Control Mechanisms

The myogenic response is defined either as constriction of a vessel in response to an increase in transmural pressure, or stretch, or as dilation in response to a decrease in pressure (Jones et al. 1993b). The myogenic responsiveness of coronary arterioles is truly myogenic, meaning it arises from an inherent response in the smooth muscle myocytes, and myogenic reactions are generally unaffected by careful endothelial removal (Kuo et al. 1988). Figure 8.11 (Kuo et al. 1990a) shows that endothelial removal in 12 vessels did not affect the active pressure-diameter relations. All

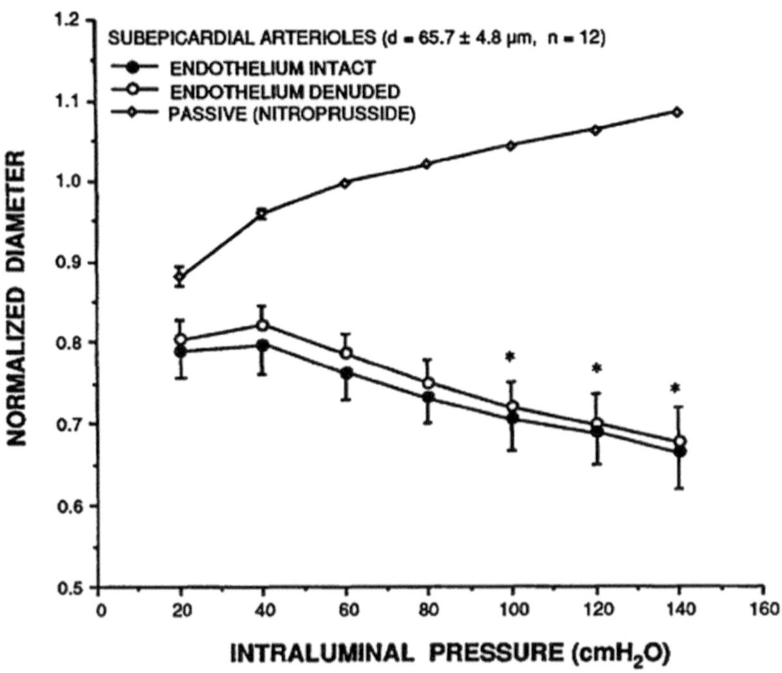

Fig. 8.11 Pressure diameter relations of subepicardial arterioles before (closed circle) and after (open circle) mechanical denudation of endothelium. There is no statistical difference in myogenic response before and after mechanical denudation. Lumen diameters were normalized to diameter at a pressure of 60 cm H_2O in presence of nitroprusside (10^{-4} M). Average luminal diameter (d) in physiological saline-albumin solution at 60 cm H_2O is shown. Vertical bars denote mean ± SEM. *Significantly different from diameter at 60 cm H_2O pressure with or without endothelium. (Reprinted with permission from the American Physiological Society from Kuo et al. 1990a)

12 vessels constricted and dilated at high (>60 cm H_2O) and low pressures, respectively, and none of these vessels dilated in response to bradykinin. In 4 of 16 vessels, however, the myogenic response was abolished after mechanical denudation, spontaneous tone was eliminated, and their contractile response to acetylcholine was significantly reduced. This suggests that excessive abrasion could produce vascular smooth muscle damage. After administration of nitroprusside, all vessels dilated and behaved passively during changes in intraluminal pressure (Fig. 8.11). This supports the hypothesis that the myogenic response of small coronary arterioles does not depend on the presence of an intact, functional endothelium.

Fluctuations in intraluminal pressure mediating myogenic reactions may contribute to vasomotor adaptations in coronary tone (DeFily 1998). The myogenic theory of autoregulation states that an intrinsic property of the blood vessel, or more specifically, vascular smooth muscle regulates vascular tone in response to changes in intraluminal pressure. The myogenic response of small coronary arterioles does not depend on the presence of an intact, functional endothelium (Kuo et al. 1990a). This is an important observation, albeit a negative one, concerning the accumulating evidence describing the complex interactions between the endothelium and vascular smooth muscle. Specifically, the endothelium has been shown to: (1) modulate the vasoactivity of several autacoids; (2) was reported to mediate flow-induced vasodilation of large arteries; and (3) was implicated in stretch-induced and pressure-dependent vascular constrictor responses. Indeed, denudation of the endothelium does not alter the myogenic response of isolated arterioles (Kuo et al. 1990a).

The vascular myogenic mechanism is thought to be responsible for the development of spontaneous vascular tone and for the constriction of a blood vessel in response to intravascular pressure elevation and dilation in response to pressure reduction. Theoretically, the maintenance of a tonic level of vascular tone via the myogenic response is physiologically important for the production of a background vasomotor tone against which vasodilators can work and for the regulation of blood flow and capillary pressure (Martinez-Lemus et al. 2003). The myogenic response varies with the initial diameter of the vessel, such that coronary vessels larger than 150–300 μm in diameter do not exhibit significant myogenic activity (Kuo et al. 1988, 1990a). Similarly, in vivo autoregulatory dilation in response to a reduction in coronary intraluminal pressure occurs primarily in coronary arterioles smaller than 100–150 μm in diameter (Chilian and Layne 1990)' (Kanatsuka et al. 1990a). This size dependence has also been demonstrated in isolated human coronary arterioles (Miller et al. 1997).

There is also a transmural variation in the myogenic activity of coronary arterioles. The myogenic activity of isolated subepicardial coronary arterioles has been demonstrated to be more active than that of subendocardial arterioles, at both low and high pressures (Kuo et al. 1988). This suggests that the reduced autoregulatory capacity of the subendocardium, previously attributed to differences in ventricular wall tension, myocardial metabolism, and other factors, may be due to intrinsic differences in the vascular smooth muscle between the inner subendocardial and outer subepicardial layers of the ventricular wall.

An important question pertains to the contribution of the myogenic response to in vivo adjustments in coronary tone during autoregulation, and we must admit that the answer is still unknown. We believe the myogenic response in vivo plays a minor role because of the observation of Dole and Nuno (1986). These investigators studied the role of myocardial oxygen tension (via coronary sinus PO_2) on autoregulation. A modest increase in oxygenation completely abolished autoregulation. This casts doubt on the myogenic contributions because the extent of the pressure drop (and the myogenic response is elicited by changes in pressure and stretch) increased oxygenation was the same as that during control conditions, yet the autoregulatory responses were very different. If the myogenic response was pivotal then the response should be unaffected by the oxygenation, which was not the observation. However, the myogenic response may still be involved in certain adjustments where changes in intravascular pressures are involved in the vasomotor adjustments, which will be described in the following section.

8.3.3 Interaction of Pressure- and Flow-Dependent Control

Important roles for myogenic and flow-dependent responses in the coronary circulation are quite compatible with dominant metabolic control of coronary blood flow (Jones et al. 1993b). These responses will amplify metabolic dialation if they enable recruitment of vessels far upstream of the smallest arterioles in closest contact with the myocardial parenchyma (Jones et al. 1993b). Coronary arterioles between 80 and 150 μm are more sensitive to flow than smaller arterioles or than larger arteries (Kuo et al. 1995). However, small arterioles (<100 μm) exhibit the greatest myogenic responsiveness, and the smallest arterioles appear to be the most sensitive to metabolic stimuli (Zhang et al. 2000). Thus, smooth muscle tone in different sized coronary microvessels is controlled by predominantly different mechanisms. Spatial variations in regulatory responses are likely to be important in the intact microvascular network, in which metabolic, myogenic, and endothelial control mechanisms may be integrated as series-coupled resistances (Zhang et al. 2000).

Pressure-and flow-dependent control mechanisms might interact locally if they simultaneously influence the diameter of an individual microvessel, and they might interact by changing the pressure and flow in distant microvessels in the same network. The local interactions between myogenic constriction and dialation and endothelium-dependent dialation have been evaluated in isolated subepicardial coronary arterioles 40–80 μm in diameter (Zhang et al. 2000). Within certain physiological ranges of pressure and flow, myogenic constriction and flow-dependent dialation were additive. Functional competition between myogenic constriction and flow-dependent dialation was apparent also as diminished flow-dependent dialation with increasing myogenic tone, and diminished myogenic constriction in the presence of flow (Zhang et al. 2000).

Figure 8.12 (Kuo et al. 1991) shows examples of the interaction of pressure- and flow-induced responses in isolated coronary arterioles. The effect of flow on

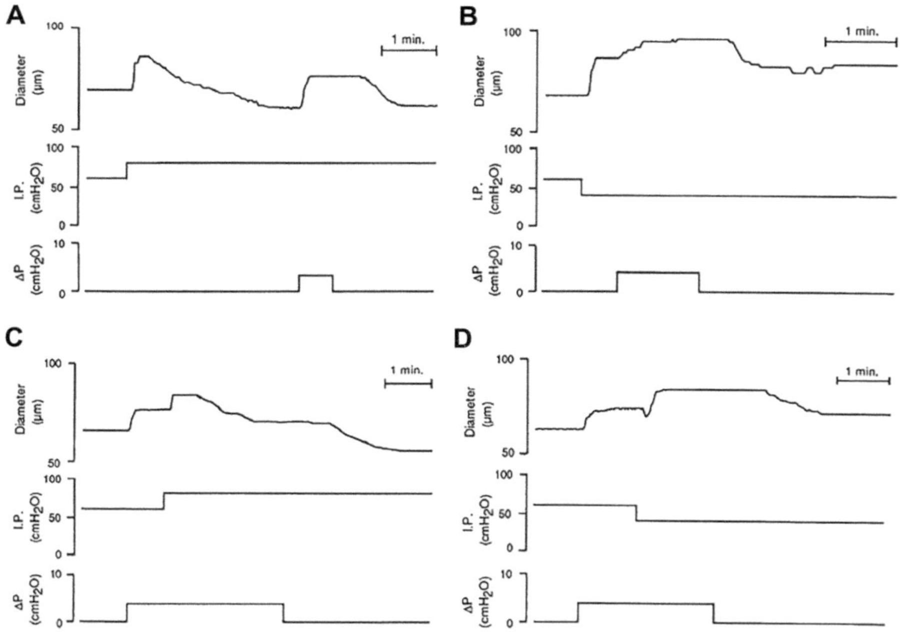

Fig. 8.12 Interactions between myogenic responses (changes in internal pressure [I.P.]) and flow-dependent dilation (changes in the pressure drop across the arteriole [ΔP]) of coronary arterioles. (**a**). Myogenic constriction was inhibited by an increase in flow. (**b**). Myogenic dilation was potentiated by flow. (**c**). Flow-induced dilation was attenuated by elevating pressure. (**d**). Flow-induced dilation was potentiated by lowering pressure. (Reprinted with permission from the American Physiological Society from reference Kuo et al. 1991)

pressure-induced constriction and dilation is shown in Fig. 8.12a, b, respectively. The pressure changes shown here caused submaximal constriction or dilation. When intraluminal pressure (IP) was increased from 60 to 80 cmH_2O, the arteriole initially distended but regained tone and constricted further from its original diameter of 69 to 61 μm (Fig. 8.12a). This pressure-induced myogenic constriction was reversed by flow (when ΔP was increased from 0 to 4 cmH_2O), with diameter increasing from 61 to 76 μm. When flow was stopped ($\Delta P = 0$ cmH_2O), vasoconstriction occurred, and the diameter gradually returned to its preflow value (Fig. 8.12a). Figure 8.12b shows that a myogenic dilation (IP decreased from 60 to 40 cmH_2O) was potentiated by an increase in flow, producing vasoconstriction from its initial value of 69 to 96 μm. Upon the cessation of flow, the vessel returned to the diameter that occurred during the myogenic dilation (from 96 to 85 μm). Figure 8.12c, d shows the effects of pressure-induced responses on vessels with flow. Under control conditions ($IP = 60$ cmH_2O, $\Delta P = 0$ cmH_2O), vasodilation occurred when flow was initiated ($\Delta P = 4$ cmH_2O), but the increase in diameter was attenuated by elevating luminal pressure ($IP = 80$ cmH_2O) (Fig. 8.12c). In contrast, myogenic relaxation (IP decreased from 60 to 40 cmH_2O) potentiated the vasodilatory effect of flow (Fig. 8.12)

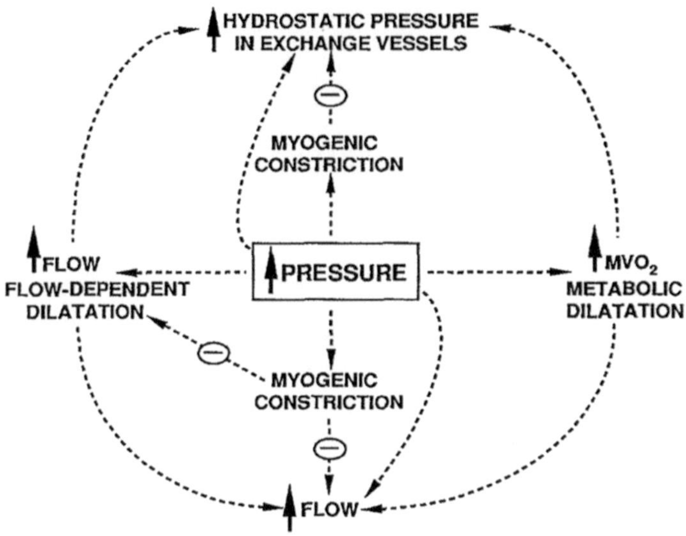

Fig. 8.13 Conceptual interactions between myogenic constriction and flow-dependent dilation of during an increase in aortic pressure. (Reprinted with permission from Basic Research in Cardiology from reference Jones et al. 1993b)

Figure 8.13 (Jones et al. 1993b) illustrates the relevance of the coronary microvascular response to increases in aortic pressure, for instance during the sympathetic activation of exercise. Flow-dependent dilation will occur when aortic pressure is increased, because the arteriolar flow is increased by the increased coronary pressure drop across the coronary circulation but may be opposed by simultaneous myogenic constriction in response to the increase in coronary pressure. Opposition by myogenic constriction may be advantageous as a balancing mechanism minimizing the increase in tissue perfusion and increase in pressure in exchange vessels, thus protecting the myocardium from edema (Jones et al. 1993b)

The major importance of pressure- and flow-dependent mechanisms of coronary blood flow control under normal conditions is likely to be their contribution to the matching of coronary blood flow to the metabolic demands of the myocardium (Jones et al. 1993b). According to this scheme, the smallest arterioles dilate during increased metabolic demand, lowering microvascular resistance so that myocardial perfusion increases. This metabolic dialation will decrease the upstream arteriolar pressure, stimulating the myogenic dilation of slightly larger arterioles and, consequently, a further decrease in resistance. Increased flow in upstream arterioles (around 80–150 μm diameter) stimulates flow-dependent dilation and recruits even larger arteries to dilate (Jones et al. 1993b). Flow-dependent dilation of upstream arterioles may have two advantageous functions in this situation: (1) to further reduce the overall resistance to flow, and (2) to increase the pressure and flow transmitted to downstream arterioles so that they retain tone and, thus, their sensitivity to further changes in metabolic demand. This second possibility is attributed to flow-dependent dialation in the role of maintaining a normal distribution of coronary microvascular resistance during functional hyperemia (Jones et al. 1993b).

In summary, the responses of the coronary microvasculature to pressure and flow may facilitate coronary blood flow to change with metabolic need in an integrated manner. Blunted responses to either pressure or flow will have adverse consequences in disease. Abnormal coronary microvascular responses to flow have been demonstrated in a number of disease states, including atherosclerosis. Kuo et al. have demonstrated loss of endothelial responses to flow in isolated coronary arterioles from pigs with dietary hypercholesterolemia and coronary atherosclerosis (Kuo et al. 1992b). Normal flow-dependent dilation was restored in these microvessels by L-arginine, the substrate for the nitric oxide synthase enzyme. These findings provided evidence that atherosclerosis induces a major abnormality of the synthesis or release of NO, or a similar endothelium-derived relaxing factor (EDRF)-type substance in coronary microvascular endothelial cells. Loss of EDRF release in response to agonists also occurs after ischemia and reperfusion and may be prevented by preconditioning (DeFily and Chilian 1993). In addition to the clinical relevance of all these observations to patients with atherosclerosis, they may also be relevant to the coronary endothelial dysfunctions documented in patients presenting with many different cardiovascular pathologies.

8.4 Neurohumoral Control General Concepts

Innervation of the coronary vasculature by both the parasympathetic and sympathetic nervous systems has been documented in both animals and humans (Hirsch and Borghard-Erdle 1961; Woollard 1926). Both sympathetic and parasympathetic nerve fibers are located within the vascular wall of large arteries as well as resistance arteries and arterioles (Hirsch and Borghard-Erdle 1961) with more dense innervation being present in resistance arteries and arterioles. However, assessment of the direct influence of parasympathetic and sympathetic stimulation on coronary vascular resistance is complicated by secondary changes in resistance that occur in response to changes in myocardial demand initiated by activation of myocardial sympathetic and parasympathetic nerve endings. A variety of techniques have been employed in coronary studies to evaluate parasympathetic or sympathetic control of the coronary vasculature in the absence of neural influences on the myocardium and subsequent changes in heart rate, myocardial contractility, and perfusion pressure.

8.4.1 Sympathetic Control

Sympathetic control of the coronary circulation is exerted through both α-adrenergic and β-adrenergic effects (Gorman et al. 2000a; Miyashiro and Feigl 1993; Jackson et al. 1987; Trivella et al. 1990; Jones and Gwirtz 1990; Huang and Feigl 1988;

Chilian 1990; Holtz 1990); however, the direct effects of sympathetic stimulation of coronary vascular receptors are often masked by the metabolic vasodilation which occurs in response to sympathetic stimulation of the heart. Data obtained during conditions in which metabolic effects of sympathetic stimulation are controlled indicate that both α-adrenergic and β-adrenergic regulation of coronary vascular resistance may be important during varying physiological conditions. Investigations employing techniques to evaluate both mRNA and protein expression of α-adrenergic and β-adrenergic receptors indicate that these receptors are present in the coronary microcirculation although the distribution of receptor subtypes is heterogeneous (Muntz et al. 1985, 1986; Vatner et al. 1986; Hein et al. 2004).

The contribution of α-adrenergic activation to regulation of coronary microvascular tone appears minimal at rest (Griggs Jr. et al. 1984; Chilian et al. 1981; Gwirtz and Stone 1984); however, during conditions of increased myocardial oxygen demand created by sympathetic stimulation (Mohrman and Feigl 1978) and exercise (Jones and Gwirtz 1990; Duncker et al. 1994; Gwirtz et al. 1986; Gwirtz and Stone 1981; Dai et al. 1989), experimental evidence indicates that α-adrenergic constriction competes with autoregulatory mechanisms and plays a significant role in determining microvascular resistance. Similarly, the influence of α-adrenergic constriction is unmasked when autoregulatory mechanisms are mitigated by coronary stenosis (Jones and Gwirtz 1990; Ishibashi et al. 1997; Laxson et al. 1989; Seitelberger et al. 1988). Although the primary effect of α-adrenergic activation of coronary vessels is vasoconstriction of vascular smooth muscle, coronary α-adrenergic-mediated vasoconstriction has been difficult to demonstrate due to competing metabolic vasodilatory influences that occur secondary to augmented myocardial oxygen consumption. Coronary vasoconstriction in α-receptor-blocked hearts can be elicited by injection of phenylephrine and other α_1-receptor agonists. Mohrman and Feigl 1978) measured coronary blood flow during conditions of increased myocardial metabolism created by sympathetic stimulation before and after α-adrenergic blockade. They found that at the same level of myocardial oxygen consumption, blood flow was significantly higher in the presence of α-adrenergic blockade. These results indicated that the adrenergic constriction competed with metabolic vasodilation, thus reducing oxygen delivery to the myocardium. Studies in exercising animals have shown that α-adrenergic constriction restricts coronary blood flow during exercise (Duncker et al. 1994; Gwirtz et al. 1986; Gwirtz and Stone 1981; Dai et al. 1989; Bache et al. 1991; Heyndrickx et al. 1982). During conditions such as hypotension (Jones and Gwirtz 1990; Jones et al. 1983, 1986; Liang and Jones 1985) and myocardial ischemia (Jones and Gwirtz 1990; Seitelberger et al. 1988; Heusch and Deussen 1983), in which myocardial oxygen delivery is impaired, adrenergic constriction competes with vasodilation produced by local metabolic mechanisms. Administration of α-adrenergic blocking agents increases coronary blood flow during ischemia indicating persistent α-adrenergic constriction which may, in fact, aggravate the ischemic condition (Jones and Gwirtz 1990; Seitelberger et al. 1988). Similarly, coronary blood flow increases during hypotension following α-adrenergic blockade (Jones et al. 1983, 1986; Liang and Jones 1985). The increase in blood flow that occurs after removal of α-adrenergic constriction is accompanied by an increase in myocardial oxygen consumption suggesting that α-adrenergic tone limits myocardial oxygenation as well as coronary flow.

Huang and Feigl (1988) proposed the possibility that adrenergic coronary constriction present during exercise may be more pronounced in the subepicardial layers, promoting the maintenance of blood flow to the vulnerable subendocardium. These investigators blocked α-receptors in a distinct area of the myocardium and then measured regional myocardial perfusion during exercise. They observed that the endocardial-to-epicardial ratio of myocardial blood flow was better maintained in the unblocked region of the myocardium. The distribution of α-adrenoceptors across the ventricular wall is relatively uniform (Saffitz 1989); thus, Huang and Feigl (1988) proposed that increased production of adenosine might attenuate release of norepinephrine in the subendocardium. More recently, it has been postulated that α-adrenoceptor-induced vasoconstriction stiffens intramyocardial arterial vessels, and decreases intramyocardial capacitance, thus reducing the wasteful flow oscillations that occur during the cardiac cycle (Morita et al. 1997). Thus, although the precise physiological role of the adrenergic constriction present during conditions of heightened physiological stress remains to be determined, it is apparent that α-adrenoceptor blockade results in greater coronary blood flow and increased myocardial performance during exercise or ischemia.

A number of studies have focused on determining whether the adrenergic constriction in the coronary circulation which persists during conditions of increased myocardial demand and limited oxygen delivery is mediated by α_1- or α_2-adrenergic receptors. Gwirtz et al. 1986) determined that the adrenergic constriction that remains during exercise is mediated by α_1-adrenergic receptors. They injected prazosin, a specific α_1-antagonist, into the coronary circulation of dogs during submaximal exercise and found an increase in circumflex blood flow that was accompanied by an increase in left ventricular function (an increase in the rate of segmental shortening and in dP/dt max). These results have been confirmed by Dai et al. (1989) who reported that coronary blood flow in running dogs was increased by blockade with prazosin. Combined blockade with both prazosin and idazoxin, an α_2-antagonist, did not produce any further increase in coronary blood flow at the same level of exercise. In contrast to these results, Seitelberger and colleagues (1988) reported that α_2-adrenergic blockade attenuated myocardial ischemia in running dogs. The reason for the differences in these studies is not completely clear, but may be related to the presence of ischemia produced by stenosis in the study of Seitelberger et al. (1988). No ischemia was present in the studies by Gwirtz et al. (1986) and Dai et al. (1989); the measurements were made during conditions of submaximal exercise.

Varying results have also been reported concerning the contribution of α_1- and α_2-mediated constriction during conditions of myocardial ischemia. Seitelberger et al. (1988) determined that myocardial dysfunction produced by ischemia during exercise was attenuated by the blockade of α_2-adrenergic receptors with idazoxin. Heusch and Deussen (1983) found that stimulation of cardiac sympathetic nerves produced an increase in resistance in coronary resistance distal to a severe stenosis.

Blockade with the α_2-antagonist, rauwolscine, but not with prazosin prevented the increase in resistance suggesting that the effect was mediated by α_2-adrenergic constriction.

Some of the differences found in the studies described above may be related to the heterogeneous distribution of adrenergic receptors within the coronary vascular tree. These studies have considered the coronary resistance vasculature as a homogenous unit; however, even within the microcirculation, a heterogeneous distribution of the effects of α-adrenergic activation is present (Chilian 1990; Chilian et al. 1989b). Large epicardial coronary arteries constrict in response to α_1-receptor agonists (Heusch et al. 1984); however, the primary sites of α-adrenergic regulation in the coronary circulation appear to be in small arteries and large arterioles. In contrast to large epicardial arteries, there appears to be a functional distribution of both α_1- and α_2-adrenergic receptors in coronary resistance vessels (Heusch et al. 1984; Chilian 1991b). Activation of α-receptors by exogenous norepinephrine infusion produces constriction of coronary arterioles greater than 100 μm in diameter and a simultaneous dilation of arterioles less than 100 μm in diameter (Chilian 1990, 1991b; Chilian et al. 1989b). These diverse effects are likely related to pressure changes created by the constriction of larger arterioles. If the upstream vessels (>100 μm) constrict this will reduce pressure in the smaller downstream arterioles. The decrease in pressure may then trigger a myogenic dilation in these smaller vessels. In contrast, if perfusion pressure is maintained below physiological levels (hypoperfusion), autoregulatory responses to pressure changes can be eliminated and both α_1- and α_2-adrenergic vasoconstriction can be unmasked in both large (>100 μm) and small coronary arterioles (<100 μm) (Chilian 1991b).

α-Adrenergic receptors that mediate vasodilation are also present on the endothelium of coronary arterial vessels. The selective α_2-agonist, clonidine, relaxes isolated epicardial coronary arteries precontracted with prostaglandin $F_{2\alpha}$ and resistance arteries precontracted with acetylcholine (Richard et al. 1990; Nakayama et al. 1988; Tschudi et al. 1991). Although intense activation of α-adrenergic receptors has been consistently shown to produce vasoconstriction in the coronary circulation, some work has suggested that α_2-adrenergic receptors may have dual effects in the coronary circulation. Threshold activation of α_2-adrenergic receptors causes vasodilation of the coronary vasculature (Hori et al. 1988). Maximal dilation of the coronary circulation in response to either exogenous or endogenous adenosine can be augmented by α_2-adrenergic receptor stimulation with low doses of clonidine (Hori et al. 1989). In the beating heart, the vasoconstrictor responses of small arteries and arterioles to α_1- and α_2-agonists were potentiated by the L-arginine analogs, N^G-nitro-L-arginine and N^G-nitro-L-arginine-methyl ester suggesting that endothelium-dependent release of NO occurs simultaneously with α-adrenergic activation and may compete with the adrenergic constriction (Jones et al. 1993c). Under most conditions, this dilation to α-adrenergic agonists appears to be masked by predominant constrictor effects; however, in exercising dogs, α_2-adrenergic coronary vasoconstriction is unmasked only when adrenergic stimulation of NO production is blocked with an arginine analogue (Ishibashi et al. 1997). Furthermore, it is possible that under pathological conditions, such as in atherosclerosis in which endothelial

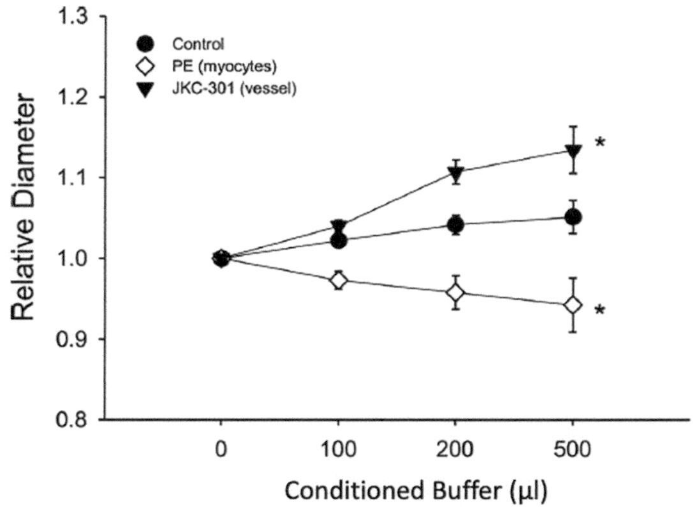

Fig. 8.14 The requisite role that cardiac myocytes have in coronary α-adrenergic constriction. Administration of phenylephrine (PE) to isolated cardiac myocytes results in the release of a product (or products) that produce endothelial-mediated constriction of isolated coronary vessels (after administration of the buffer bathing the cardiac myocytes (conditioned buffer)). The endothelin-A antagonist, JKC-301, blocked the constriction. (Adapted from Merkus et al. 2005)

function is impaired (Baumgart et al. 1999), the loss of α_2-adrenoceptor-mediated, endothelium-dependent vasodilation may contribute to exaggerated adrenergic constriction.

Paradoxically, it has been shown by Jones et al. (1995c) that isolated coronary arterioles do not respond to stimulation with α-adrenergic agonists, despite experimental conditions similar to those in which venules demonstrate marked constriction. The disparity in experimental results found using in vivo and in vitro preparations of the coronary microcirculation may be due to the indirect effects of α-adrenergic stimulation in the heart. Endothelin antagonists block α_1-adrenergic constriction of coronary arterioles in the intact beating heart (DeFily et al. 1999), and cardiac myocytes stimulated with phenylephrine produce vasoactive substances (including endothelin) capable of constricting isolated coronary arterioles (Tiefenbacher et al. 1998). Figure 8.14 shows that coronary arterioles constrict to supernatant derived from myocytes treated with phenylephrine. This constriction was blocked by the addition of the endothelin receptor antagonist, JKC-301 to the isolated arterioles (Merkus et al. 2005). Although the effects of phenylephrine on the isolated arterioles are not shown, direct administration of an α_1-adrenergic agonist to an isolated coronary arteriole does not cause direct constriction, which is similar to the observations of Jones et al. (1993b). Autoradiographic studies indicate that the density of α_1-adrenergic receptors on cardiac myocytes far exceeds that of coronary arterioles (Muntz et al. 1985); thus, α_1-adrenoceptor-stimulated production of vasoactive substances by cardiac myocytes constitutes a likely mechanism for sympathetic constriction in the coronary microcirculation. It is important to emphasize that

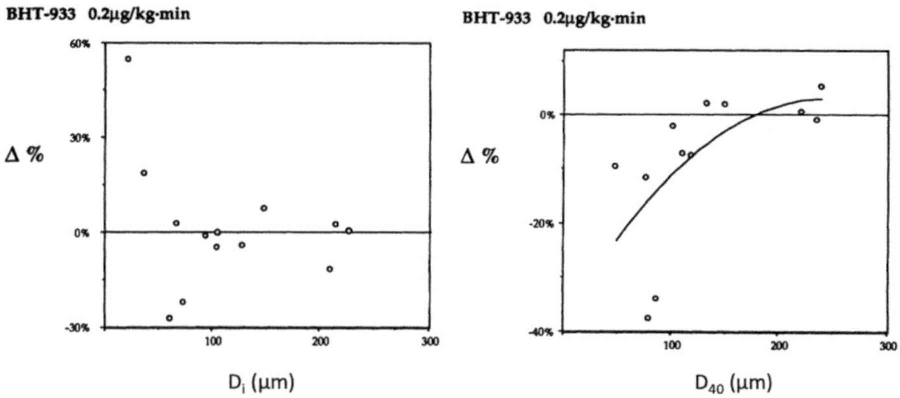

Fig. 8.15 The responses of epicardial coronary arterial and arteriolar microvessels to the α_2-adrenergic agonist BHT-933 under basal control conditions (D_i) and after a reduction in coronary perfusion pressure to 40 mmHg (D_{40}). The response is presented as a percent change of the diameter of the vessel from that measured at D_i or at $D_{40.}$ (Adapted from Chilian 1991b)

isolated arterioles appear refractory to stimulation with adrenergic agonists, and experimental evidence indicates that adrenergic modulation of coronary microvascular tone occurs through indirect mechanisms that are principally governed by myocardial demand, which was shown by Merkus et al. (2002).

Considered together, existing data indicate that α-adrenergic control of the coronary circulation is important during both physiological and pathophysiological conditions. The predominant effect of adrenergic stimulation of the coronary circulation is to increase vascular tone. Under various conditions that are associated with an increase in sympathetic stimulation, blockade of adrenergic receptors results in an increase in coronary blood flow. Within the coronary microcirculation, α-adrenergic constriction can be demonstrated in both small arteries and in arterioles; however, under normal conditions in which the coronary vascular bed retains its ability to autoregulate, small arterioles escape from adrenergic constriction and show substantial vasodilation. Figure 8.15 illustrates this concept (Chilian 1991a, b). Under control conditions in which autoregulatory mechanisms remain intact the α_2-agonist, BHT-933, constricts large coronary microvessels but in many small arterioles autoregulatory dilation predominates over the vasoconstrictor effects. Significant constriction to α_2-adrenergic stimulation is present in both large and small arterioles when autoregulation is prevented during hypoperfusion at 40 mm Hg. Although adrenergic vasoconstriction can occur throughout the coronary circulation, functional responses to adrenergic stimulation are heterogeneous when tone is intact. The heterogeneous responses to adrenergic stimulation result mainly from the heterogeneous distribution of adrenergic receptors and modulation of adrenergic constriction by other controlling factors such as autoregulatory and endothelium-dependent responses.

Radioligand binding studies indicate that as with α-adrenergic receptors, the distribution of α-adrenergic receptors is heterogeneous within the coronary circulation. The ratio of α_1-/α_2- adrenergic receptors in large vessels is almost 2:1 (Vatner et al. 1986), whereas a predominance of α_2-adrenergic receptors is found in resistance vessels (Murphree and Saffitz 1988). Reverse transcription-polymerase chain reaction and immunohistochemistry indicate the presence of α_2-adrenergic receptors in coronary arterioles, with a markedly greater level of both mRNA and protein expression of α_2-adrenergic receptors in subepicardial as compared to subendocardial arterioles (Hein et al. 2004). Intracoronary infusion of isoproterenol increases coronary blood flow (Sekiguchi et al. 1992). Isolated porcine coronary arterioles relax in response to both norepinephrine and epinephrine (Nakayama et al. 1988; Quillen et al. 1992). Recent data indicate that norepinephrine produces dilation of human coronary arterioles through activation of α_2-adrenoceptors on vascular smooth muscle (Sun et al. 2002). Direct application of isoproterenol causes dilation of coronary microvessels in both in vivo and in vitro experimental preparations (Hein et al. 2004; Sekiguchi et al. 1992; Wang et al. 1994). In isolated porcine coronary arterioles, isoproterenol-induced vasodilation was inhibited by the α_2-adrenoceptor blocker, ICI-118,551, but was insensitive to treatment with the α_1-adrenergic antagonist, atenolol (Hein et al. 2004).

Blockade of coronary β_2-adrenoceptors decreases coronary blood flow during exercise in both dogs and pigs (Gorman et al. 2000a; Duncker et al. 1998; DiCarlo et al. 1988). Feed-forward β-adrenergic vasodilation has been demonstrated in the coronary circulation in response to intracoronary norepinephrine and during exercise (Gorman et al. 2000a; Miyashiro and Feigl 1993; Duncker et al. 1998). Gorman et al. (2005) showed that in exercising dogs, specific α-adrenoceptor blockade increased coronary blood flow, whereas combined blockade of β- and α-adrenoceptors resulted in a decrease in coronary blood flow. If feed-forward β-adrenergic vasodilation contributes to the exercise-induced increase in coronary blood flow, then coronary venous oxygen tension should be lower after specific blockade of β-adrenoceptors during exercise at an equivalent myocardial oxygen consumption (Gorman et al. 2000a). Figure 8.16 (Gorman et al. 2000b) shows the relationship between coronary venous oxygen tension and myocardial oxygen consumption at rest and during three levels of exercise. The difference in slopes between β-adrenoceptor blockade and combined α- and β-adrenoceptor blockade demonstrates feed-forward β-adrenergic vasodilation. The steep decline in coronary venous oxygen tension as myocardial oxygen consumption increases indicates that myocardial oxygen supply-to-consumption ratio is reduced by blockade of β-adrenergic vasodilation.

8.4.2 Parasympathetic Control

Cholinergic innervation of small coronary arteries has been demonstrated through acetylcholinesterase staining (Denn and Stone 1976). Early studies of the effects of parasympathetic stimulation were inconclusive, and attempts to demonstrate parasympathetic vasodilation produced negative results mainly because the negative chronotropic and inotropic effects of vagal stimulation were not controlled (Feigl

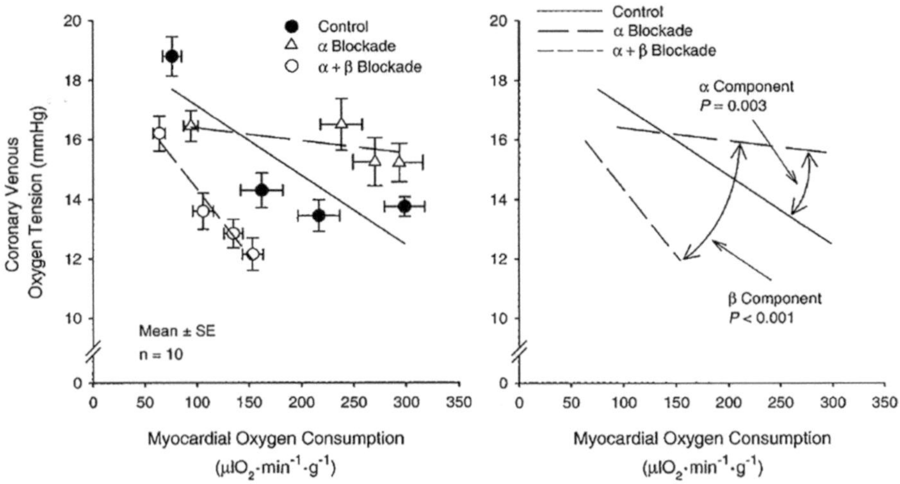

Fig. 8.16 Blockade of feed-forward β-adrenergic dilation of the coronary circulation reduces the myocardial oxygen supply-to-consumption ratio in exercising dogs. Note, α-adrenergic blockade shifts the relationship upward, meaning this constrictor mechanism decreased venous PO_2 at a given level of oxygen consumption. Note the opposite effect when β- and α-adrenergic blockades were combined—coronary venous PO_2 decreased for a given level of oxygen consumption. This indicates that there was beta-adrenergic dilation during exercise. (Adapted from Gorman et al. 2000a)

1969, 1983). However, Feigl (1969) reported that vagal stimulation resulted in an increase in coronary blood flow in the dog heart. The effects of vagal stimulation on heart rate, contractility, and myocardial oxygen consumption were controlled by constant pacing and did not account for the increase in blood flow. Additionally, atropine blocked the increase in coronary blood flow produced by vagal stimulation, demonstrating the involvement of muscarinic receptors. Both in vitro and in vivo studies have shown that acetylcholine produces coronary vasodilation in dogs (Cox et al. 1983), rabbits, baboons, and goats (Feigl et al. 1990). However, the vasoactive effects of acetylcholine are mediated through both the endothelium and the vascular smooth muscle, and a biphasic response has been observed in some experimental models (Knight et al. 1991). Acetylcholine-induced responses of the coronary microcirculation are species- and dose-dependent (Feigl et al. 1990; Knight et al. 1991; Van Winkle and Feigl 1989). Species differences are likely related to differences in distribution of receptors on the endothelium and smooth muscle, and the strength of direct constriction of vascular smooth muscle relative to endothelium-dependent release of NO or other vasodilator substances. In species that display coronary dilation to acetylcholine, removal of the endothelium converts the vasodilatory effect of acetylcholine to vasoconstriction (Kalsner 1989). In pigs and cattle, acetylcholine acts as a vasoconstrictor in coronary vessels presumably due to a lack of endothelial muscarinic receptors (Kalsner 1989; Young et al. 1987, 1988; Cowan and McKenzie 1990). In several species, low doses of acetylcholine produce dilation of coronary arterioles, whereas high doses induce a constrictor response (Feigl et al. 1990; Knight et al. 1991; Van Winkle and Feigl 1989).

Isolated porcine coronary arterioles (Kuo et al. 1988), resistance arteries (Nakayama et al. 1988), and epicardial arteries (Myers et al. 1991) all display similar sensitivity in their vasoconstrictor response to acetylcholine. Acetylcholine produces similar vasodilation in isolated canine coronary arteries and arterioles (Myers et al. 1989). Lamping et al. (1992) found similar results in a beating heart preparation. In their study, both endogenous acetylcholine released in response to vagal stimulation and exogenous acetylcholine produced dilation of arterial vessels ranging from 50 to 400 μm in diameter. The dilation produced by acetylcholine was similar in all vessel sizes. Using the same beating heart preparation, Komaru et al. (1991b) also reported that acetylcholine produced dilation in all sizes of coronary microvessels (50–250 μm in diameter). However, they found that acetylcholine produced greater dilation in vessels less than 120 μm in diameter. These investigators also found differences in the mechanisms by which acetylcholine produced vasodilation in large and small coronary arterioles; inhibition of NO formation eliminated the vasodilatory response to acetylcholine in large arterioles but only partially inhibited acetylcholine-induced vasodilation in small arterioles. Isolated porcine coronary resistance arteries constrict in response to acetylcholine, but inhibition of NO synthase antagonizes the constriction produced by acetylcholine (Myers et al. 1991). In human coronary arterioles, responsiveness to acetylcholine has been reported to differ dramatically in atrial and ventricular vessels (Miller Jr. et al. 1998). Atrial vessels constrict to acetylcholine whereas isolated ventricular arterioles display vasodilation. The results of this human study must be interpreted with caution because the data were collected from a large sample of patients of diverse ages and various underlying cardiovascular diseases. Collectively, data from animal and human studies suggest that, like other control mechanisms, the vasoactive effects of cholinergic stimulation vary between segments of the coronary circulation.

The role of the parasympathetic nervous system in controlling coronary blood flow under physiological conditions has not been clearly established. The contribution of parasympathetic activity to the control of resting coronary blood flow has not been considered in species where acetylcholine exerts a predominant vasodilatory effect. Presumably the contribution of parasympathetic vasodilation is small compared to the vasodilation exerted by metabolic mechanisms, especially since increased metabolic vasodilation would presumably be accompanied by withdrawal of vagal tone. However, it is possible that under resting conditions there is a tonic vasodilatory effect exerted by parasympathetic release of acetylcholine. In dogs and baboons, vagal stimulation during ventricular pacing causes coronary vasodilation that is countered by a metabolic reduction of coronary flow (Van Winkle and Feigl 1989). In the pig, where acetylcholine produces constriction of large coronary arteries (Myers et al. 1991), resistance arteries (Nakayama et al. 1988) and arterioles (Myers et al. 1991; Kuo et al. 1990a), Cowan and McKenzie (1990) reported that neither cholinergic blockade with atropine nor vagal ligation altered resting coronary blood flow, indicating a lack of parasympathetic influence on basal coronary tone.

8.4.3 Nonadrenergic, Noncholinergic Control

Perivascular sensory nerves release both calcitonin gene-related peptide (CGRP) and substance P. Varicose nerve fibers that display immunoreactivity for both CGRP and substance are present in epicardial and intramyocardial arteries and coronary arterioles (Saetrum Opgaard et al. 1997; Gulbenkian et al. 1990; Coupe et al. 1990). Stimulation of sensory fibers in the myocardium by metabolites produced during ischemia can trigger release of these neuropeptides (Franco-Cereceda et al. 1987a, 1989); however, it is less clear how efferent release of either substance P or CGRP contributes to the determination of coronary vascular resistance at rest or during nonpathological conditions. However, these neuropeptides can also affect cholinergic release of acetylcholine, resulting in both chronotropic and inotropic effects in the myocardium. Thus, the direct effects of these neuropeptides on the coronary resistance vasculature have proved difficult to establish in vivo. Although the endogenous release of these peptides may be triggered by ischemic conditions (Franco-Cereceda et al. 1987a, 1989), data indicating that they play significant role in determination of coronary vascular resistance are lacking.

CGRP elicits endothelium-independent vasodilation of both coronary arteries and arterioles (Gulbenkian et al. 1990; Sekiguchi et al. 1994; Franco-Cereceda et al. 1987b), but evidence that its vasodilatory influence plays a significant role in determining coronary microvascular resistance at rest or ischemia is lacking. Although ischemia triggers the release of CGRP (Franco-Cereceda et al. 1987a, 1989), superfusion with a CGRP receptor antagonist had no effect on arteriolar tone during ischemia in the beating dog heart (Sekiguchi et al. 1994). Under normal, resting conditions, blockade with α-CGRP (Gorman et al. 2000a; Miyashiro and Feigl 1993; Hein et al. 2004; Wearn et al. 1928; Shipley et al. 1937; Yudilevich and Martindejulian 1965; Bassingthwaighte et al. 1986; Rose et al. 1977; Rose and Goresky 1976; Duran and Yudilevich 1974, 1978; Duran et al. 1973; Katz and Feigl 1987; Roth and Feigl 1981; Martini and Honig 1969; Hellberg et al. 1972; Bing et al. 1971, 1972a, b; Tillmanns et al. 1974; Nellis et al. 1981; Kanatsuka et al. 1990a; Ashikawa et al. 1984, 1986; McDonagh et al. 1984; Ritter et al. 2011; Gale et al. 2001; Reynolds and McDonagh 1994; McDonagh and Roberts 1986; Laine and Granger 1985) did not alter systemic hemodynamics or coronary blood flow in either anesthetized rats or conscious dogs (Shen et al. 2001). Thus, although exogenous CGRP produces strong vasodilation of coronary arterial vessels, endogenous CGRP does not contribute significantly to maintenance of coronary microvascular tone.

Substance P is a potent dilator of coronary arterioles (Kuo et al. 1995; Muller et al. 1999; Miller Jr. et al. 1998). Substance P-induced dilation is substantially greater in coronary arterioles (<100 μm ID) than in small arteries (>150 μm ID) isolated from porcine hearts (Kuo et al. 1995), indicating the heterogeneity of its role in determining coronary microvascular resistance. Early studies report that the predominant effect of substance P in the intact coronary circulation is to produce vasodilation (Maxwell 1968; Brum et al. 1986; Ezra et al. 1986); however, Nakamura et al. reported that substance P elicited constriction of resistance vessels in the hearts of conscious dogs (Nakamura et al. 1990) both before and after ganglionic blockade. These investigators found that bolus injections of substance P produced a biphasic response of dilation followed by sustained constriction. The difference in the results of this study and those of previous studies may be related to the dose and intravenous administration of substance P, and the balance between endothelium-dependent vasodilation and constriction of vascular smooth muscle. In isolated pig coronary arteries, dilation of substance P is transient, and removal of the endothelium completely abolishes the relaxation (Franco-Cereceda et al. 1987b). High doses of substance P cause constriction of smooth muscle in rat portal vein (Hellstrand and Jarhult 1980). In a later study, intracoronary administration of substance P also resulted in a significant reduction of coronary blood flow in conscious dogs (Nakamura et al. 1991). In rabbit hearts studied with a Langendorff preparation, infusion of substance P significantly decreased coronary vascular resistance (Maley et al. 1994). Although the vasoactive effects of substance P in the coronary microcirculation have been documented in several species, the role of endogenous substance P in the determination of coronary microvascular resistance at rest or varying physiological conditions remains to be determined.

8.4.4 Humoral Control

Various humoral substances that are present in circulating blood or produced in peripheral tissues exert vasoactive effects on the coronary microvasculature. Many of these agonists produce heterogeneous effects within the coronary circulation due to variations in the density of receptor subtypes and due to stimulation of both endothelium-dependent dilation and direct constriction of vascular smooth muscle. Serotonin, a humoral agent produced by activated platelets, produces heterogeneous effects in the coronary circulation. Serotonin constricts large coronary arteries but has also been reported to increase total coronary blood flow (Lamping et al. 1985; Cappelli-Bigazzi et al. 1991). In large arteries, serotonin produces constriction that is modulated by endothelium-dependent vasodilation, evidenced by the fact that removal of the endothelium results in augmentation of vasoconstriction in response to serotonin (Lamping et al. 1985). Serotonin also constricts small coronary arteries (150–300 μm) (Lamping et al. 1989). In contrast, serotonin produces marked dilation of coronary arterioles less than 100 μm in diameter, presumably due to a predominant effect on the endothelium in these smaller vessels (Lamping et al. 1989). In agreement with these previous findings, Lamping et al. recently showed that dilation of small microvessels is predominantly mediated by NO release from the endothelium via 5-HT_1 receptor stimulation, whereas serotonin-induced constriction of coronary microvessels is mediated by 5-HT_2 receptors (Lamping 1997).

The vasoactive effects of vasopressin also vary within the coronary circulation. In isolated epicardial arteries, vasopressin produces vasodilation (Myers et al. 1989). In the beating heart, diameter measurements of epicardial vessels ranging in size from approximately 50 to 300 μm showed that vasopressin caused slight dilation of vessels greater than 100 μm in diameter while it constricted vessels less than 90 μm in diameter. These diameter changes were accompanied by an overall increase in vascular resistance and a decrease in coronary blood flow (Lamping et al. 1989). The diversity of these responses could be related to autoregulatory adjustments. The decrease in resistance in large arteries due to the direct vasodilatory effects of vasopressin may produce an increase in downstream pressure promoting myogenic constriction of resistance vessels. Alternatively, the dilation of large vessels may be due to endothelium-dependent vasodilation which is absent in smaller vessels. Vasopressin produces endothelium-dependent relaxation of isolated coronary arteries, whereas isolated resistance arteries constrict in response to vasopressin (Myers et al. 1989). Vasopressin-induced constriction of coronary resistance arteries is modulated by endothelial release of NO, and augmented in the presence of endothelial dysfunction (Sellke and Quillen 1992). These data support the notion that the disparate effects of vasopressin in large and small coronary vessels are due to a differential contribution of endothelium-mediated mechanisms.

In the intact heart, angiotensin II has been reported to produce an increase (Needleman et al. 1975), decrease (Heyndrickx et al. 1976), or biphasic changes (transient decrease followed by an increase) (Porsti et al. 1993) in coronary blood flow. Similarly, the direct effects of Angiotensin II on vascular tone of coronary microvessels also appear to be diverse. In canine and porcine epicardial arteries, angiotensin II produces modest vasoconstriction (Zhang et al. 2003). Myers et al. (1994) reported that canine coronary arterioles preconstricted with endothelin dilated in response to Angiotensin II, whereas porcine coronary arterioles were unresponsive. Recently, isolated porcine coronary arterioles have been reported to constrict to low concentrations and dilate to high concentrations of angiotensin II (Zhang et al. 2003). mRNA transcripts for both angiotensin II type 1 (AT_1) receptors and angiotensin II type 2 (AT_2) receptors were detected in porcine coronary arterioles, with the level of expression for AT_2 receptor mRNA being approximately twofold higher than that for AT_1 receptor mRNA. Blockade of the AT_1 receptor with losartan eliminated the angiotensin II-induced constriction in porcine coronary arterioles; treatment with the AT_2 receptor antagonist, PD123319, abolished coronary arteriolar vasodilation (Zhang et al. 2003). Thus, angiotensin II appears to exert both vasodilatory and vasoconstrictor effects in the coronary microcirculation through activation of specific angiotensin II receptor subtypes. Diverse responses may be related to the distribution of receptor subtypes within the coronary vascular tree, and species differences in receptor subtype expression.

8.5 Cardiac Lymphatic Microcirculation

8.5.1 Introduction to the Lymphatic System

The lymphatic system is a highly complex and often underappreciated component of the human body. However, if it fails, many functions in the body would be compromised. From its role in immune function to its critical involvement in maintaining fluid balance, the lymphatic system plays a vital role in overall homeostasis in circulatory systems. While the circulatory system, with its beating heart and coursing blood, garners most attention, the lymphatic system operates quietly in the background, serving as a key partner in maintaining homeostasis. The lymphatic system is a complicated web of vessels, nodes, and pumps designed to perform several essential functions, including immune defense, fluid balance, and nutrient transport.

Interstitial fluid is the body fluid between blood vessels and cells and represents the exchange of soluble nutrients and wastes as blood circulates through the body. In the process of exchange, fluids are filtered through capillary pores and clefts from the bloodstream (due to the balance of capillary and tissue, oncotic and hydrostatic pressures, there is a net positive pressure driving fluid filtration from capillaries into tissue), and it is crucial that the leaked fluid must return to the blood to maintain adequate volume in the cardiovascular system. This is resolved by the lymphatic system, which comprises five primary conduits and an organ: lymphatic capillaries, collecting lymphatic vessels, lymph nodes, lymph trunk and ducts. This section provides a comprehensive examination of the cardiac lymphatic system, delving into its structure, functions, and clinical implications related to microcirculation and serves as a thorough resource to those seeking a deeper understanding of this intricate network.

8.5.2 Cardiac Lymphatics and Pathology

The heart possesses a robust lymphatic network responsible for draining extracellular fluid and waste from the myocardium with each cardiac contraction. This lymphatic system propels lymph through collecting vessels, spanning from the apex to the base of the heart, ultimately directing it toward draining lymph nodes located around the aortic arch. Cardiac lymphatics, specialized to adapt to the heart's unique microenvironment, play a vital role in maintaining an ideal milieu for facilitating electrical conduction and promoting cardiomyocyte force generation (Harris et al. 2023). Anatomically, cardiac lymphatics are distributed through coronary arteries (Symbas et al. 1963) and extended throughout all layers including the endocardium, valves, myocardium, and epicardium (Loukas et al. 2011). Aligned with the architecture of lymphatic vessels observed in other organs, the cardiac lymphatic network comprises initial lymphatic vessels with button-like junctions of endothelial cells and their high permeabilities, which subsequently drain into larger collecting lymphatics equipped with valves (Klaourakis et al. 2021). While cardiac lymphatics adhere to the general anatomical and functional principles of the lymphatic system outlined earlier, cardiac collecting vessels lack muscle layers compared to collecting vessels in other organs; thus, cardiac contraction is the driving force of lymph propulsion (Itkin et al. 2021). Another variation found in heart lymphatics is that they uniquely form a mesh-like network known as a plexus, which is composed of both initial and precollecting vessels. While plexus formations are frequently observed in the epicardium near the heart's apex, their precise biological function remains undisclosed (Ratajska et al. 2014; Sacchi et al. 1999). However, it is plausible that they facilitate efficient removal of fluid and immune cells at a high throughput rate (Miller et al. 1961).

The cardiac lymphatic network supports normal cardiac function and aids in the healing process of the myocardium following injury (Miller 2011; Cui 2010a). Cardiac lymphatics transport lymphatic fluid from the myocardium and surrounding structures to larger lymphatic vessels, eventually returning it to the bloodstream. Lymphatic drainage from the heart typically follows the coronary vasculature, with lymphatic vessels converging into larger trunks and ducts (Xu et al. 2007). Dysfunction of coronary lymphatics and reduced flow rate can contribute to acute and chronic pathological conditions such as myocardial edema, inflammation, and fibrosis, impairing cardiac output (Cui 2010a, b). Specifically adapted to the cardiac microenvironment, the roles of cardiac lymphatics in cardiovascular disease and the resolution of myocardial edema are currently emerging. Several studies have revealed that altered lymphatic functions are closely related to different stages of cardiovascular diseases (Balasubbramanian et al. 2019; Henri et al. 2016; Lim et al. 2013). This is characterized by a loss of ability to efficiently absorb and remove fluids, contributing to fluid accumulation in the heart or myocardial edema. Myocardial edema is often a manifestation of underlying cardiovascular diseases or conditions that affect the structure and function of the heart. Wang et al. (2009) found that impaired cardiac lymphatic function affected LV function, reducing ejection fraction following 90 days post lymphatic obstruction in rabbits (Wang et al. 2009). A study with lymphatic vessel endothelial hyaluronan receptor 1 (LYVE1)-null mice, which lack lymphatic endothelium phenotype, showed a decreased capacity for wound healing, and reduced cardiac output, indicating that initial or precollecting lymphatics play a crucial role in wound recovery and cardiac function (Vieira et al. 2018). Following the cardiac injury, lymphangiogenesis is induced in response to compensate for the loss of cardiomyocytes and heal the fibrotic tissues (Kim et al. 2014; Kinashi et al. 2013; Suzuki et al. 2012). Klotz et al. reported that MI significantly promoted a lymphangiogenic response in the mouse heart and demonstrated the effect of lymphangiogenesis on cardiac injury. In the study, they observed increased levels of vascular endothelial growth factor receptor-3 (VEGFR-3) 24 h to 21 days postmyocardial infarction, and that lymphangiogenesis was enhanced with the treatment of vascular endothelial growth factor-C, showing improved ejection fraction (Klotz et al. 2015). The improvement of cardiac function is associated with decreased inflammation (Smart et al. 2011) and attenuated edema due in part to the VEGF-C-induced lymphangiogenesis (Karkkainen et al. 2001). By contrast, it has been shown that blocking VEGFR signaling deteriorates inflammation and suppresses lymphangiogenesis (Mäkinen et al. 2001), thus indicating a pivotal role of VEGF in cardiac recovery.

8.6 Pathophysiological Disturbances

8.6.1 Concepts of Coronary Disease

Canonically, coronary artery disease is equated to remodeling of the epicardial coronary vasculature via progressive obstruction of the lumen due to the inward growth of a stenosis. Although these large arteries play nearly no physiological role in the regulation of coronary blood flow (under normal conditions), they function as vascular conduits providing a permissive action wherein all meaningful blood supply to the myocardium must first pass through the epicardial coronaries to gain access to the coronary capillary vasculature for gas and nutrient exchange. Ultimately, the volume flow through any coronary artery is regulated by the downstream microvascular circulation (i.e., resistance vessels). In conditions where the coronary artery narrows due to a stenosis, flow is maintained through the coronary tree via autoregulatory actions of the coronary microcirculation. Biophysically, epicardial narrowing results in diminished downstream perfusion pressure resulting in coronary microvascular dilation which, in turn, restores coronary flows to control values. This phenomenon operates on a physiological continuum until epicardial narrowing becomes so great as to become flow-limiting.

A previous review from our group, published in 2011, detailed the history of syndrome X (Chilian et al. 1993). This syndrome, first described by Likoff et al. (1967) and expanded on in subsequent studies by Cannon 3rd et al. (1983, 1985, 1987), was characterized by anginal pain in the absence of epicardial stenosis. These early investigations described phenomena of exertional ECG changes reflective of myocardial ischemia (ST elevations) that were exacerbated during vasoconstrictive maneuvers. Moreover, more focused studies demonstrated that these patients had diminished dilator responses. Several subsequent decades of research concluded that individuals with syndrome X were afflicted with a condition wherein anginal pain was induced resulting from an inability of the microcirculation to adequately dilate to myocardial demand, reflected as an impaired dilator reserve. Whether "microvascular angina" was a genuine phenomenon was contested from the initial description of syndrome X in the late 1960s until 2017 when Bairey Merz et al. published the results of a working group of The Cardiovascular Disease in Women Committee of the American College of Cardiology, the National Heart, Lung, and Blood Institute, American Heart Association, and European Society of Cardiology (Bairey Merz et al. 2017). This working group was the first to develop consensus criteria for Ischemia and No Obstructive Coronary Artery Disease (INOCA) for which they provided a detailed roadmap toward characterizing the patient condition of a "syndrome of symptoms and signs suggesting ischemic heart disease but found to have no obstructed coronary arteries." Since that 2017 publication, the scientific community has largely accepted the existence and seriousness of INOCA as reflected by the roughly 250 studies that have been published in the last 7 years.

It is important to note how coronary microvascular dysfunction (CMD), also known as INOCA, and/or Myocardial Ischemia and No Obstructive Coronary Artery

Disease (MINOCA) are defined. These conditions are defined on the basis of reduced coronary vasodilator reserve *in the absence of obstructive coronary disease*. As an aside, most patients with CMD have some degree of large vessel disease, but the lesions are minor and not viewed as obstructive (Bairey Merz et al. 2017). Coronary vasodilator reserve is the ability of the circulation to increase coronary blood flow (during maximal dilation) above basal flow. As an example, coronary reserve of 3 indicates that during maximal dilation, coronary blood flow increased threefold above baseline. However, there is not a consensus value that is used to define microvascular dysfunction with some groups using values such as 2.32 (Pepine et al. 2010), while others use values ranging from 2.0 to 2.5 (Sara et al. 2016; Shah et al. 2018). Many groups propose a coronary reserve values of >2.5 as normal, and less than 2.5 (in the absence of obstructive disease) as indicating CMD (Marroquin et al. 2005). The cause(s) of CMD is (are) not completely understood. Endothelial cell inflammation and systemic inflammation does not appear to play a role in CMD (Marroquin et al. 2005).

There are likely many contributing factors to CMD and some of our preclinical work may provide insight into the complexity of this issue. Specifically, in a model of obesity, insulin resistance, and type 2 diabetes, the coupling of coronary blood flow to myocardial metabolism is blunted compared to normal, lean animals (Guarini et al. 2016). To determine if this adverse effect was due to altered production of vasoactive metabolites by cardiac myocytes or due to alterations in microvascular reactivity a bioassay system was developed. This bioassay was based on isolating cardiac myocytes and coronary arterioles from obese and lean rats. The myocytes were electrically stimulated to increase metabolism for 20 minutes and the bathing buffer was removed and administered in aliquots to isolated arterioles (Fig. 8.17). The rationale for this was to collect the buffer after it had been "conditioned" by vasoactive metabolites released by the working cardiac myocytes. Vasodilation of arterioles from lean rats was significantly less using conditioned buffer from obese rat myocytes compared to lean myocytes. This implies that at least part of deficient metabolic dilation relates to inadequate amounts and perhaps moieties of metabolites from myocytes isolated from the obese rats. Moreover, vasodilation to conditioned buffer from the lean control myocytes was significantly less in arterioles from the obese rats compared to those from the leans. This result suggests that alterations in vessel function may also play a role in the compromised dilation. Thus, microvascular dysfunction may stem from a cardiac and/or vascular issue.

The consequences of CMD are serious. Although most attention focuses on the adverse outcomes of obstructive coronary artery disease (also known as coronary heart disease [CHD]), e.g., a myocardial infarction occurs about every 40 seconds (data from American Heart Association), the overall negative impact of CMD is roughly equivalent to CHD (Bairey Merz et al. 2006; Gulati et al. 2009). Although CMD is less common in males than females, if it occurs in males the adverse outcomes are similar to those in females (Murthy et al. 2012).

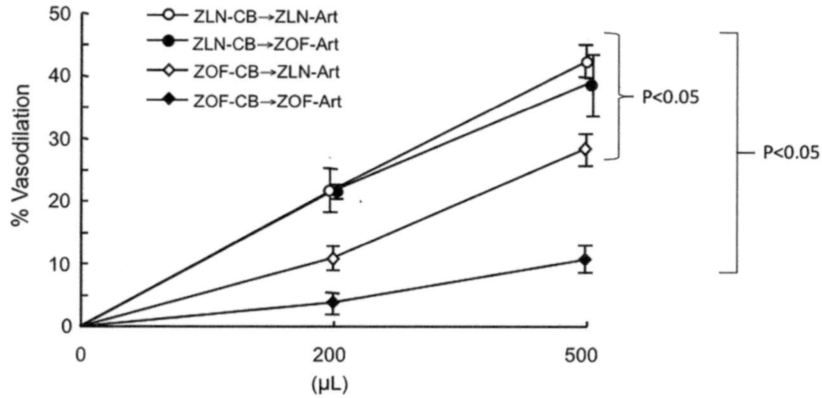

Fig. 8.17 Vasodilation of arterioles (Art) to conditioned buffer (CB) from Zucker obese fatty (ZOF) and Zucker lean (ZLN) rats. Note vasodilation was compromised (ZOF < ZLN) in arterioles from ZOF rats, and to conditioned buffer from ZOF cardiac myocytes. (Adapted from Guarini et al. 2016)

8.6.2 Pathophysiology of Endothelial Dysfunction

The vascular endothelium plays an essential role in barrier function for the vascular wall, as an interface between vascular and humoral niches, and as a regulator of vasomotor tone in response to varied stimuli. It borders on impossible to provide an exhaustive list of the factors known to be secreted by the vascular endothelium and to which the vascular smooth muscle is responsive. For the purpose of this review, we focus on those factors that are most well characterized as dysregulated in disease states and identified in the etiology of coronary microvascular dysfunction. From a vasomotor perspective, endothelial dysfunction can be defined as reduced dilator response to endothelium-dependent agonists and/or altered endothelial responses resulting from a nonphysiologic balance of vasodilator and vasoconstrictor mediators. Owing to the dual responsibility of barrier function and vasomotor tone regulator, any disruption of vascular endothelial function can have significant impact on the function of the coronary tree. It is, therefore, unsurprising that endothelial dysfunction is causally involved in the microvascular pathology of heart disease.

Arguably the most well-known and well-studied endothelium-dependent vasodilator is nitric oxide (NO). Coronary NO release has been shown to be stimulated by numerous receptor pathways. Notable endothelium-dependent agonists include bradykinin, histamine, serotonin, and substance P (Bohlen 2015; Broten et al. 1992; Chiu and Chien 2011; Durand and Gutterman 2013; Liu and Gutterman 2009; Moncada et al. 1991; Van Bibber et al. 1995). NO is also proportionally released in response to increases in shear (Stepp et al. 1999; Liu and Gutterman 2009; Canty Jr. and Schwartz 1994; Stepp et al. 2001). Once released, NO freely diffuses into the vascular smooth muscle layer and induces a vasodilator response through cyclic guanosine monophosphate-dependent relaxation of vascular smooth muscle.

Interestingly, while the capacity of NO to induce agonist-induced vasodilator responses in the coronary microcirculation is uncontested, the physiological role of this mediator remains equivocal. Regardless of whether NO significantly mediates coronary vascular responses under physiological conditions, reductions in NO bioavailability serve as a sentinel of overall endothelial function. At the level of the epicardial coronaries, reductions in NO have been shown to promote inflammation, coronary plaque formation/growth/rupture and clot formation (Loftus 2011; Khambata et al. 2017). In these same conduit vessels, NO has been shown to contribute to resting vessel diameter (Canty Jr. and Schwartz 1994; Berdeaux et al. 1994; Chu et al. 1990, 1991; Egashira et al. 1996). Throughout the coronary vascular tree NO has been shown to mediate shear stress-dependent dilator responses under normal conditions (Kuo et al. 1993, 1995; Stepp et al. 1999, 2001) but the regulation of diameter by shear stress is less important in coronary arterioles than small arteries (Stepp et al. 1999). The result of this differential response is that, when tested in vivo, inhibition of NO produces a uniform, downward shift in coronary flow across a physiological continuum of myocardial demand (Tune et al. 2000a; Altman et al. 1994; Bernstein et al. 1996; Duncker and Bache 1994; Ishibashi et al. 1998; Merkus et al. 2004). These data are consistent with what would be predicted by loss of the tonic vasodilation due to shear stress-induced production of NO in the coronary vasculature. What then, is the role of NO in coronary microvascular pathophysiology? Jones et al. proposed that the coronary tree is differentially responsive to NO as a means to prevent excess shear stress, thereby preserving coronary flow reserve in resistance vasculature (Jones et al. 1993a). Loss of that protection could contribute to the loss of flow reserve seen in heart disease. However, we emphasize that the role of NO in the modulation of coronary vasomotor tone is far less in magnitude than metabolic dilation and autoregulation.

While the literature lacks compelling evidence in support of the prominent role of NO for the physiological regulation of coronary vascular tone, there is overwhelming evidence that the loss of NO bioavailability is reflective of the presence and magnitude of disease (Tune et al. 2017; Bonetti et al. 2003; Cai and Harrison 2000; De Vriese et al. 2000; Koltai et al. 1997; Picchi et al. 2006; Zhang et al. 2006). Perhaps a better question than what does loss of NO cause, is what causes NO loss. If the loss of NO itself is not reflexively deleterious, perhaps the loss of NO bioavailability provides insight into disrupted cellular processes whose effects go beyond NO. Early investigations into disease associated dysregulation of NO focused extensively on the availability of substrates, cofactors, and key enzymes. Experimental and preclinical supplementation with the essential eNOS cofactor, tetrahydrobiopterin, has been shown to reduce the magnitude of endothelial dysfunction in several different studies. However, despite this promising early observation, no clinical value has even been found from this approach. Perhaps less encouraging, attempting to drive NO production through supplementation with the precursor, L-arginine, showed promise in early animal studies (Drexler et al. 1991; Boger et al. 1998) that were not reproducible in human trials. Contrary to these negative findings, recent studies have repeatedly demonstrated the link between reductions in NO bioavailability and increases in NO scavenging by oxidant species

(Heitzer et al. 2001; Zalba et al. 2001). It is known that increased oxidative stress in the form of superoxide can scavenge NO by converting NO to peroxynitrite ($ONOO^-$). Production of superoxide is known to be increased in disease conditions that predispose individuals to cardiovascular disease (e.g., obesity, diabetes mellitus, hypertension, metabolic syndrome), and although superoxide production can be the result of decreased tetrahydrobiopterin availability, the negative findings with supplementation would seem to lack support of eNOS cofactors as causative in most disease conditions. Recent work by Kiyooka et al. has provided compelling evidence that changes to mitochondrial structure and function may cause enhanced superoxide production leading to NO scavenging and endothelial dysfunction (Kiyooka et al. 2022). Using a Zucker Obese Fatty rat model, the researchers observed that diminished endothelium-dependent vasodilator responses in obese animals coincided with mitochondrial strand lesions and reduced endothelial mitochondrial complex activities. In this same manuscript, they were able to demonstrate causality, as experimental enhancement of strand damage using mitochondrial directed exonucleases further blunted endothelium-dependent dilator responses. Administration of a strand repairing exonuclease fully rescued dilator responses to the same magnitude as found in age-matched lean counterparts. This work provides a direct mechanistic link between endothelial mitochondrial health and NO bioavailability. So, while NO may only play a small and compensable role in the regulation of vascular tone, its absence is likely indicative of impaired mitochondrial function in the coronary vasculature. The ultimate implications of this mitochondrial functional impairment remain an area for future investigation.

Metabolites of arachidonic acid have also been implicated in the health and disease of the coronary circulation. Unlike NO where studies of substrate availability were common and supplementation studies have been performed, the literature has never concerned itself with arachidonic acid bioavailability as it is found in the plasma membrane of nearly every cell in the body (Stillwell 2016). In circumstances where arachidonic acid is metabolized, three major metabolic pathways are available for substrate creation: cyclooxygenase, lipoxygenase, or the cytochrome P-450 pathway (Wang et al. 2021). Of the three major pathways for arachidonic acid metabolism, the cyclooxygenase pathway has received the greatest attention largely due to the observation that daily aspirin therapy may mitigate myocardial infarctions. First described in the 1950s by Laurence Craven (1950, 1953, 1956), daily aspirin therapy was studied over the course of the next 40 years without significant mechanistic insight until the 1970s when Vane identified that aspirin was capable of inhibiting synthesis of one of the major cyclooxygenase metabolites, prostaglandins (Vane 1971), ultimately sharing a Nobel Prize in 1982 with Bergstrom and Samuelsson for this discovery (Bergstrom et al. 1965; Bergstrom and Samuelsson 1962; Frolich 1982). Patrono built on this work in the 1980s demonstrating that aspirin was able to permanently inhibit thromboxane by cyclooxygenase (Patrono et al. 1985). These observations fueled interest in the role of arachidonic acid metabolism in the maintenance of coronary health. Ultimately, it was established that the inhibition of cyclooxygenase activity provided an antithrombotic action that helped to protect individuals on low-dose aspirin therapies from coronary thrombosis. Whether these same metabolites played a significant role in vasomotor tone required further investigation.

These same cyclooxygenase metabolites, whose actions can determine the probability of coronary thrombosis, turned out to also have the capacity to regulate coronary vascular tone. Arachidonic acid breakdown by cyclooxygenase yields a variety of vasoactive metabolites, nonexhaustively including the vasoconstrictors prostaglandin H_2, prostaglandin $F_{2\alpha}$ and thromboxane A_2 (described in Section *Role of Vasoconstrictors*) and prostaglandin I_2 (described immediately hereafter). Prostaglandin I_2, sometimes referred to as prostacyclin or PGI_2, diffuses from the vascular endothelium to interact with vascular smooth muscle cells where it binds its cognate receptor (Li et al. 2021). Once bound, a cAMP-dependent pathway acts on K^+ channels to elicit vasodilator responses that have been documented in both animal models and humans (Durand and Gutterman 2013; Beyer and Gutterman 2012; Feletou and Vanhoutte 2009). With the knowledge that these clinically relevant biomolecules could elicit vasomotor response as a foundation, a number of investigations were undertaken to examine whether there was a role for arachidonic acid metabolites in coronary health or disease. In 1984, Dai et al. established that while cyclooxygenase inhibition was able to attenuate arachidonate-stimulated increases in coronary flow, that same blockade had no effect in their canine model at rest or across a range of treadmill-induced cardiac stress (Dai and Bache 1984). The observation that cyclooxygenase inhibition has little if any effect on coronary flow has been repeated in both animal models and humans by multiple other laboratories (Edlund et al. 1989; Harlan et al. 1978; Hintze and Kaley 1977). These observations, however, are made in conditions of cardiovascular health. When the balance between coronary oxygen delivery and cardiac metabolic demand is perturbed as is seen in hypoxia (Afonso et al. 1974), anoxia (Block et al. 1974), and myocardial ischemia (Alexander et al. 1975), increased prostaglandin release occurs. Indeed, in conditions of coronary artery disease, cyclooxygenase inhibition elicits a vasoconstrictor response (Friedman et al. 1981) supporting a role for prostanoids in the maintenance of appropriate vascular tone in disease states. Adding a level of complexity, in skeletal muscle vascular beds, there has been an observed shift from dilator prostanoids to constrictor prostanoids that occurs concomitant with the dyslipidemia (Goodwill et al. 2008a) or metabolic syndrome (Goodwill et al. 2008b). This shift may very well occur within and contribute to coronary vascular dysfunction secondary to obesity and type II diabetes. Whether this shift has consequence specific to the coronary vasculature will require further study. What is clear is that, while little evidence exists for a role of prostanoids in the regulation of vascular tone in conditions of health, their role becomes apparent with the progression of coronary artery disease.

It was observed that, even when NO action and cyclooxygenase activity were maximally inhibited, a sustained endothelial-dependent hyperpolarization remained (Ellinsworth et al. 2016; Fleming et al. 1996; Gutterman et al. 2005, 2016; Nishikawa et al. 1999). It is from this observation that we get the final group of endothelium-dependent vasodilators, endothelium-dependent hyperpolarizing factors (EDHFs). For the last 30 years, numerous groups have worked to identify the specific secreted factors responsible for this persistent dilator response. The list of EDHFs is lengthy but a few have risen to the top as potentially more significant players than others. These include epoxyeicosatrienoic acids (EETs; cytochrome P-450 metabolites of arachidonic acid), H_2O_2, and K^+ (Ellinsworth et al. 2016; Batenburg et al. 2004a, b; Edwards et al. 1998; Gebremedhin et al. 1998; Liu et al. 2011; Matoba and Shimokawa 2003; Matoba et al. 2000, 2002; Randall et al. 1996). Currently, the role of any of these mediators in a specific coronary vasomotor response remains contested. The greatest body of evidence for an EDHF-derived physiological mediator of flow exists for H_2O_2. The Gutterman laboratory has consistently advanced our knowledge of the capacity of this reactive oxygen species to potentially tie flow to metabolism (Durand and Gutterman 2013; Ellinsworth et al. 2016; Gutterman et al. 2016). In recent work, they have further demonstrated a role in disease. Examining coronary arterioles from human subjects, these investigators observed that NO-dependent flow-mediated dilator responses are compromised in patients with coronary artery disease. In its place, flow-mediated responses become reliant on mitochondrially generated H_2O_2 (Chabowski et al. 2023), and this conversion appears to be mediated by ceramides (Freed et al. 2014). It is noteworthy to add that the source of H_2O_2 is somewhat controversial as Jugilon found in a preclinical model of diabetes that the source of hydrogen peroxide was cytosolic (Juguilon et al. 2022). Despite these differences, both studies show that during endothelial dysfunction, there is a switch from NO to H_2O_2 as the principal mediator of endothelium-dependent dilation. These data illustrate that endothelial-dependent dilator responses are likely important in normal processes as the system has selected to allow for compensatory pathways in disease states. If flow-mediated responses could occur in the absence of endothelial signaling and/or if these responses were not important for organ level homeostasis, it is improbable that a compensation system would be in place.

To summarize the role of endothelium-dependent dilators in the etiology of coronary microvascular dysfunction, it is safe to say that endothelium-dependent dilators play a role in health and disease. Which specific dilators are relevant to each condition, remains undetermined. Based on the variety of endothelium-dependent dilators and their often-overlapping functions, it seems apparent that the system has been designed to preserve endothelial regulation of vasomotor tone whenever possible. Adding a layer of complexity, our system appears to change emphasis over time with certain dilators more relevant in youth than in age. This remains an area of significant need for investigation with ample opportunities for discovery.

8.6.3 Role of Vasoconstrictors

Apart from the attenuation of vasodilator activity, an increase in the production of endothelial vasoconstrictors could well be involved in the sequelae of various disorders concerning the microcirculation. Much as NO is the gold standard for endothelium-dependent dilators, endothelin-1 is the textbook example of an endothelium-dependent constrictor. Interestingly, endothelin-1 has oppositional effects dependent on which of its cognate receptors it binds. Binding to endothelial ETB receptors stimulates NO and prostacyclin production (Folta et al. 1989; Traverse et al. 1996) eliciting a vasodilator response. By contrast, activation of smooth muscle ETA or ETB receptors produces powerful vasoconstriction which can last up to an hour (Gorman et al. 2005; Rubanyi and Polokoff 1994; Yanagisawa et al. 1988). When comparing the relative strengths of each receptor, it becomes readily apparent that the system favors ETA signaling, resulting in vasoconstriction. This is supported by studies where endothelin-1 was administered directly into an epicardial coronary resulting in complete cessation of coronary blood flow suggesting arteriolar spasm (Gorman et al. 2005).

potent constrictor actions, endothelin-1 can induce total cessation of coronary flow and would appear to have role in microvascular pathologies characterized by coronary insufficiency. Despite this, numerous reports show that blockade of endothelin receptors has little if any impact on either coronary flow or cardiac metabolism (Gorman et al. 2005; Merkus et al. 2002, 2003b, 2006; Takamura et al. 2000). However, we would be remiss not to mention that Merkus et al. observed a role for endothelin at low levels of myocardial oxygen consumption, which waned as oxygen consumption increased (Merkus et al. 2002). While additional studies have been performed to establish whether endothelin might serve as a metabolic dilator, those are inconclusive to date. The evidence for the role of endothelin-1 in microvascular disease is more compelling than the evidence for its role in health. Endothelial endothelin production has been shown to be increased in atherosclerosis, hyperlipidemia and hypertension (Schiffrin 2005). In animal models of chronic coronary artery stenosis, microvessels distal to the stenotic epicardial coronaries were shown to have a greater sensitivity to endothelin-1 than what was demonstrated in control counterparts (Sorop et al. 2008). This supports the idea that endothelin-1 may limit coronary blood flow, particularly under conditions of physiological stress. In line with this, coronary flow was shown to improve with ETA receptor blockade in both collateralized and noncollateralized vessels from a model of slowly progressing myocardial ischemia (Symons et al. 1997). These data, while not conclusive, are the basis for an argument that endothelin-1 may be more relevant in pathology than in physiology. However, studies investigating the effects of endothelin antagonists have only shown moderate effects at best (Rich and McLaughlin 2003). Nonetheless, the results of preclinical data support the premise that a more nuanced investigation of endothelin-1's role in cardiovascular disease may be merited.

Similar to endothelins, constrictor prostanoids have been shown to be elevated with disease. Specifically, prostaglandin H_2 is known to be elevated in conditions of hypertension. End products of cyclooxygenase metabolism, prostaglandin $F_{2\alpha}$, and thromboxane A_2 have been shown to produce marked vasoconstriction in the coronary circulation (Ammar Jr. et al. 2000; Feng et al. 2011; Kadokami et al. 1996; Rigel and Shetty 1997). While there is little if any evidence of a role for these constrictors in health, each of these compounds has been implicated in coronary vasospasm (Houston and Vanhoutte 1986; Konidala and Gutterman 2004; Shepherd and Vanhoutte 1986). Additional vasomotor contributions to disease remain poorly defined. However, these compounds do contribute to cardiovascular disease in non-microvascular ways, briefly detailed below.

8.7 Summary and Future Directions

The preceding discussion hopefully elaborated the idea that much is known about the regulation of the coronary microcirculation under physiological and pathophysiological conditions. Unfortunately some of our knowledge is negative, vis-à-vis, we know what is not known! Despite being in the twenty-first century with the completed sequence of the human genome, specific knowledge about causal factors for autoregulation, ischemic dilation, and metabolic hyperemia remain unidentified. We have discussed many possibilities and contributing factors for these responses, e.g., myogenic mechanisms, shear stress-dependent dilation, hydrogen peroxide, etc., but to date there is no consensus about a specific substance or substances and effectors for autoregulation and metabolic hyperemia. If medical science will ever understand coronary microvascular disease it is imperative that basic science has a more complete understanding of regulation of the coronary microcirculation. The uniqueness of this vascular bed render generalized conclusions about vascular control mechanism suspect. As one example, the coronary circulation constricts to α-adrenergic agonists, but as pointed out, this process involves activation of cardiac myocyte a-adrenoceptors and the release of substances to stimulate endothelin-1 production by the vasculature. The authors know of no other circulation in the body with such a mechanism. Future experiments should take advantage of recent advances in technology, e.g., mass spectroscopy for identification of metabolites and signaling molecules that are causally involved in fundamental control process in the regulation of coronary blood flow, two-photon imaging of the beating heart, single cell genomics to understand the myriad phenotype of vascular cells. Although cardiac lymphatics generally conform to the general anatomic and functional role of the lymphatic system inherit in most organ systems, they also display unique attributes that renders them distinct from lymphatic beds in other organs. Cardiac lymphatics appear to have adapted to the microenvironment of the beating heart, and their role in cardiovascular disease and myocardial edema remains an open question.

Finally, we predict that a better understanding of the mechanisms and factors that are responsible for regulation of the coronary microcirculation will undoubtedly lead to advances in the treatment of many vascular and cardiac diseases characterized by abnormal microvascular behavior.

References

Afonso S, Bandow GT, Rowe GG (1974) Indomethacin and the prostaglandin hypothesis of coronary blood flow regulation. J Physiol 241:299–308

Ahlborg G, Weitzberg E, Lundberg J (1995) Metabolic and vascular effects of circulating endothelin-1 during moderately heavy prolonged exercise. J Appl Physiol 78:2294–2300

Alders DJ, Groeneveld AB, de Kanter FJ, van Beek JH (2004) Myocardial O2 consumption in porcine left ventricle is heterogeneously distributed in parallel to heterogeneous O2 delivery. Am J Physiol Heart Circ Physiol 287:H1353–H1361

Alexander RW, Kent KM, Pisano JJ, Keiser HR, Cooper T (1975) Regulation of postocclusive hyperemia by endogenously synthesized prostaglandins in the dog heart. J Clin Invest 55:1174–1181

Allan-Rahill NH, Lamont MRE, Chilian WM, Nishimura N, Small DM (2020) Intravital microscopy of the beating murine heart to understand cardiac leukocyte dynamics. Front Immunol 11:92

Allison TB, Holsinger JW Jr (1977) Transmural metabolic gradients in the normal dog left ventricle: effect of right atrial pacing. Am J Phys 233:H217–H221

Altman JD, Kinn J, Duncker DJ, Bache RJ (1994) Effect of inhibition of nitric oxide formation on coronary blood flow during exercise in the dog. Cardiovasc Res 28:119–124

Ammar RF Jr, Gutterman DD, Brooks LA, Dellsperger KC (2000) Impaired dilation of coronary arterioles during increases in myocardial O(2) consumption with hyperglycemia. Am J Physiol Endocrinol Metab 279:E868–E874

Archer SL, Weir EK, Reeve HL, Michelakis E (2000) Molecular identification of O2 sensors and O2-sensitive potassium channels in the pulmonary circulation. Adv Exp Med Biol 475:219–240

Archer SL, Wu XC, Thebaud B, Nsair A, Bonnet S, Tyrrell B, McMurtry MS, Hashimoto K, Harry G, Michelakis ED (2004) Preferential expression and function of voltage-gated, O2-sensitive K+ channels in resistance pulmonary arteries explains regional heterogeneity in hypoxic pulmonary vasoconstriction: ionic diversity in smooth muscle cells. Circ Res 95:308–318

Ashikawa K, Kanatsuka H, Suzuki T, Takishima T (1984) A new microscopic system for the continuous observation of the coronary microcirculation in the beating canine left ventricle. Microvasc Res 28:387–394

Ashikawa K, Kanatsuka H, Suzuki T, Takishima T (1986) Phasic blood flow velocity pattern in epimyocardial microvessels in the beating canine left ventricle. Circ Res 59:704–711

Bacchus AN, Ely SW, Knabb RM, Rubio R, Berne RM (1982) Adenosine and coronary blood flow in conscious dogs during normal physiological stimuli. Am J Phys 243:H628–H633

Bache RJ, Schwartz JS (1982) Effect of perfusion pressure distal to a coronary stenosis on transmural myocardial blood flow. Circulation 65:928–935

Bache RJ, Dai XZ, Schwartz JS, Homans DC (1988) Role of adenosine in coronary vasodilation during exercise. Circ Res 62:846–853

Bache RJ, Homans DC, Dai XZ (1991) Adrenergic vasoconstriction limits coronary blood flow during exercise in hypertrophied left ventricle. Am J Physiol 260:H1489–H1494

Bairey Merz CN, Shaw LJ, Reis SE, Bittner V, Kelsey SF, Olson M, Johnson BD, Pepine CJ, Mankad S, Sharaf BL, Rogers WJ, Pohost GM, Lerman A, Quyyumi AA, Sopko G, Investigators W (2006) Insights from the NHLBI-Sponsored Women's Ischemia Syndrome Evaluation (WISE) Study: Part II: gender differences in presentation, diagnosis, and outcome with regard to gender-based pathophysiology of atherosclerosis and macrovascular and microvascular coronary disease. J Am Coll Cardiol 47:S21–S29

Bairey Merz CN, Pepine CJ, Walsh MN, Fleg JL (2017) Ischemia and no obstructive coronary artery disease (INOCA): developing evidence-based therapies and research agenda for the next decade. Circulation 135:1075–1092

Balasubbramanian D, Lopez Gelston CA, Rutkowski JM, Mitchell BM (2019) Immune cell trafficking, lymphatics and hypertension. Br J Pharmacol 176:1978–1988

Bassingthwaighte JB, Chinard FP, Crone C, Goresky CA, Lassen NA, Reneman RS, Zierler KL (1986) Terminology for mass transport and exchange. Am J Physiol 250:H539–H545
Batenburg WW, de Vries R, Saxena PR, Danser AH (2004a) L-S-nitrosothiols: endothelium-derived hyperpolarizing factors in porcine coronary arteries? J Hypertens 22:1927–1936
Batenburg WW, Popp R, Fleming I, de Vries R, Garrelds IM, Saxena PR, Danser AH (2004b) Bradykinin-induced relaxation of coronary microarteries: S-nitrosothiols as EDHF? Br J Pharmacol 142:125–135
Baumgart D, Haude M, Gorge G, Liu F, Ge J, Grosse-Eggebrecht C, Erbel R, Heusch G (1999) Augmented alpha-adrenergic constriction of atherosclerotic human coronary arteries. Circulation 99:2090–2097
Berdeaux A, Ghaleh B, Dubois-Rande JL, Vigue B, Drieu La Rochelle C, Hittinger L, Giudicelli JF (1994) Role of vascular endothelium in exercise-induced dilation of large epicardial coronary arteries in conscious dogs. Circulation 89:2799–2808
Bergstrom S, Samuelsson B (1962) Isolation of prostaglandin E1 from human seminal plasma. Prostaglandins and related factors. 11. J Biol Chem 237:3005–3006
Bergstrom S, Carlson LA, Ekelund LG, Oro L (1965) Cardiovascular and metabolic response to infusions of prostaglandin E1 and to simultaneous infusions of noradrenaline and prostaglandin E1 in man. Prostaglandin and related factors 35. Acta Physiol Scand 64:332–339
Berne RM (1980) The role of adenosine in the regulation of coronary blood flow. Circ Res 47:807–813
Berne RM, Rubio R (1974) Regulation of coronary blood flow. Adv Cardiol 12:303–317
Berne RM, Rubio R, Dobson JG Jr, Curnish RR (1971) Adenosine and adenine nucleotides as possible mediators of cardiac and skeletal muscle blood flow regulation. Circ Res 28(Suppl 1): 115
Bernstein RD, Ochoa FY, Xu X, Forfia P, Shen W, Thompson CI, Hintze TH (1996) Function and production of nitric oxide in the coronary circulation of the conscious dog during exercise. Circ Res 79:840–848
Berwick ZC, Dick GM, O'Leary HA, Bender SB, Goodwill AG, Moberly SP, Owen MK, Miller SJ, Obukhov AG, Tune JD (2013) Contribution of electromechanical coupling between Kv and Ca v1.2 channels to coronary dysfunction in obesity. Basic Res Cardiol 108:370
Beyer AM, Gutterman DD (2012) Regulation of the human coronary microcirculation. J Mol Cell Cardiol 52:814–821
Bing RJ, Wayland H, Rickart A, Hellberg K (1971) Studies on the coronary microcirculation by direct visualization. G Ital Cardiol 1:401–408
Bing RJ, Rickart A, Hellberg K, Wayland H (1972a) Direct visualization of coronary microcirculation. Kardiologiia 12:18–24
Bing RJ, Hellberg K, Wayland H (1972b) Microcirculation of the heart. Adv Exp Med Biol 22:253–266
Block AJ, Poole S, Vane JR (1974) Modification of basal release of prostaglandins from rabbit isolated hearts. Prostaglandins 7:473–486
Boger RH, Bode-Boger SM, Thiele W, Creutzig A, Alexander K, Frolich JC (1998) Restoring vascular nitric oxide formation by L-arginine improves the symptoms of intermittent claudication in patients with peripheral arterial occlusive disease. J Am Coll Cardiol 32:1336–1344
Bohlen HG (2015) Nitric oxide and the cardiovascular system. Compr Physiol 5:808–823
Bonetti PO, Lerman LO, Lerman A (2003) Endothelial dysfunction: a marker of atherosclerotic risk. Arterioscler Thromb Vasc Biol 23:168–175
Broten TP, Miyashiro JK, Moncada S, Feigl EO (1992) Role of endothelium-derived relaxing factor in parasympathetic coronary vasodilation. Am J Phys 262:H1579–H1584
Brum JM, Go VL, Sufan Q, Lane G, Reilly W, Bove AA (1986) Substance P distribution and effects in the canine epicardial coronary arteries. Regul Pept 14:41–55
Cai H, Harrison DG (2000) Endothelial dysfunction in cardiovascular diseases: the role of oxidant stress. Circ Res 87:840–844

Cannon RO 3rd, Watson RM, Rosing DR, Epstein SE (1983) Angina caused by reduced vasodilator reserve of the small coronary arteries. J Am Coll Cardiol 1:1359–1373
Cannon RO 3rd, Leon MB, Watson RM, Rosing DR, Epstein SE (1985) Chest pain and "normal" coronary arteries – role of small coronary arteries. Am J Cardiol 55:50B–60B
Cannon RO 3rd, Schenke WH, Leon MB, Rosing DR, Urqhart J, Epstein SE (1987) Limited coronary flow reserve after dipyridamole in patients with ergonovine-induced coronary vasoconstriction. Circulation 75:163–174
Canty JM Jr, Schwartz JS (1994) Nitric oxide mediates flow-dependent epicardial coronary vasodilation to changes in pulse frequency but not mean flow in conscious dogs. Circulation 89:375–384
Cappelli-Bigazzi M, Nuno DW, Lamping KG (1991) Evidence of a role for compounds derived from arginine in coronary response to serotonin in vivo. Am J Physiol 261:H404–H409
Chabowski DS, Hughes WE, Hockenberry JC, LoGiudice J, Beyer AM, Gutterman DD (2023) Lipid phosphate phosphatase 3 maintains NO-mediated flow-mediated dilatation in human adipose resistance arterioles. J Physiol 601:469–481
Chilian WM (1990) Adrenergic vasomotion in the coronary microcirculation. Basic Res Cardiol 85 (Suppl 1):111–120
Chilian WM (1991a) Microvascular pressures and resistances in the left ventricular subepicardium and subendocardium. Circ Res 69:561–570
Chilian WM (1991b) Functional distribution of alpha 1- and alpha 2-adrenergic receptors in the coronary microcirculation. Circulation 84:2108–2122
Chilian WM, Layne SM (1990) Coronary microvascular responses to reductions in perfusion pressure. Evidence for persistent arteriolar vasomotor tone during coronary hypoperfusion. Circ Res 66:1227–1238
Chilian WM, Boatwright RB, Shoji T, Griggs DM Jr (1981) Evidence against significant resting sympathetic coronary vasoconstrictor tone in the conscious dog. Circ Res 49:866–876
Chilian WM, Eastham CL, Marcus ML (1986) Microvascular distribution of coronary vascular resistance in beating left ventricle. Am J Physiol 251:H779–H788
Chilian WM, Layne SM, Klausner EC, Eastham CL, Marcus ML (1989a) Redistribution of coronary microvascular resistance produced by dipyridamole. Am J Phys 256:H383–H390
Chilian WM, Layne SM, Eastham CL, Marcus ML (1989b) Heterogeneous microvascular coronary alpha-adrenergic vasoconstriction. Circ Res 64:376–388
Chilian WM, Kuo L, DeFily DV, Jones CJ, Davis MJ (1993) Endothelial regulation of coronary microvascular tone under physiological and pathophysiological conditions. Eur Heart J 14 (Suppl I):55–59
Chiu JJ, Chien S (2011) Effects of disturbed flow on vascular endothelium: pathophysiological basis and clinical perspectives. Physiol Rev 91:327–387
Chu A, Chambers DE, Lin CC, Kuehl WD, Cobb FR (1990) Nitric oxide modulates epicardial coronary basal vasomotor tone in awake dogs. Am J Phys 258:H1250–H1254
Chu A, Chambers DE, Lin CC, Kuehl WD, Palmer RM, Moncada S, Cobb FR (1991) Effects of inhibition of nitric oxide formation on basal vasomotion and endothelium-dependent responses of the coronary arteries in awake dogs. J Clin Invest 87:1964–1968
Cornelissen AJM, Dankelman J, VanBavel E, Spaan JAE (2002) Balance between myogenic, flow-dependent, and metabolic flow control in coronary arterial tree: a model study. Am J Physiol Heart Circ Physiol 282:H2224–H2237
Coupe MO, Mak JC, Yacoub M, Oldershaw PJ, Barnes PJ (1990) Autoradiographic mapping of calcitonin gene-related peptide receptors in human and Guinea pig hearts. Circulation 81:741–747
Cowan CL, McKenzie JE (1990) Cholinergic regulation of resting coronary blood flow in domestic swine. Am J Physiol 259:H109–H115
Cox DA, Hintze TH, Vatner SF (1983) Effects of acetylcholine on large and small coronary arteries in conscious dogs. J Pharma Exp Ther 225:764–769

Craven LL (1950) Acetylsalicylic acid, possible preventive of coronary thrombosis. Ann West Med Surg 4:95

Craven LL (1953) Experiences with aspirin (Acetylsalicylic acid) in the nonspecific prophylaxis of coronary thrombosis. Miss Valley Med J 75:38–44

Craven LL (1956) Prevention of coronary and cerebral thrombosis. Miss Valley Med J 78:213–215

Cui Y (2010a) Impact of lymphatic vessels on the heart. Thorac Cardiovasc Surg 58:1–7

Cui Y (2010b) The role of lymphatic vessels in the heart. Pathophysiology 17:307–314

Dai XZ, Bache RJ (1984) Effect of indomethacin on coronary blood flow during graded treadmill exercise in the dog. Am J Phys 247:H452–H458

Dai XZ, Sublett E, Lindstrom P, Schwartz JS, Homans DC, Bache RJ (1989) Coronary flow during exercise after selective alpha 1- and alpha 2-adrenergic blockade. Am J Physiol 256:H1148–H1155

Dankelman J, Van der Ploeg CP, Spaan JA (1994) Glibenclamide decelerates the responses of coronary regulation in the goat. Am J Phys 266:H1715–H1721

De Vriese AS, Verbeuren TJ, Van de Voorde J, Lameire NH, Vanhoutte PM (2000) Endothelial dysfunction in diabetes. Br J Pharmacol 130:963–974

DeFily DV (1998) Control of microvascular resistance in physiological conditions and reperfusion. J Mol Cell Cardiol 30:2547–2554

DeFily DV, Chilian WM (1993) Preconditioning protects coronary arteriolar endothelium from ischemia-reperfusion injury. Am J Phys 265:H700–H706

DeFily DV, Nishikawa Y, Chilian WM (1999) Endothelin antagonists block alpha1-adrenergic constriction of coronary arterioles. Am J Physiol 276:H1028–H1034

Denn MJ, Stone HL (1976) Automic innervation of dog coronary arteries. J Appl Physiol 41:30–35

DiCarlo SE, Blair RW, Bishop VS, Stone HL (1988) Role of beta 2-adrenergic receptors on coronary resistance during exercise. J Appl Physiol 64:2287–2293

Dick GM, Bratz IN, Borbouse L, Payne GA, Dincer UD, Knudson JD, Rogers PA, Tune JD (2008) Voltage-dependent K+ channels regulate the duration of reactive hyperemia in the canine coronary circulation. Am J Physiol Heart Circ Physiol 294:H2371–H2381

Dole WP, Nuno DW (1986) Myocardial oxygen tension determines the degree and pressure range of coronary autoregulation. Circ Res 59:202–215

Dora KA, Lin J, Borysova L, Beleznai T, Taggart M, Ascione R, Garland C (2022) Signaling and structures underpinning conducted vasodilation in human and porcine intramyocardial coronary arteries. Front Cardiovasc Med 9:980628

Drake-Holland AJ, Laird JD, Noble MI, Spaan JA, Vergroesen I (1984) Oxygen and coronary vascular resistance during autoregulation and metabolic vasodilation in the dog. J Physiol 348: 285–299

Drexler H, Zeiher AM, Meinzer K, Just H (1991) Correction of endothelial dysfunction in coronary microcirculation of hypercholesterolaemic patients by L-arginine. Lancet 338:1546–1550

Duncker DJ, Bache RJ (1994) Inhibition of nitric oxide production aggravates myocardial hypoperfusion during exercise in the presence of a coronary artery stenosis. Circ Res 74:629–640

Duncker DJ, Van Zon NS, Altman JD, Pavek TJ, Bache RJ (1993) Role of K+ATP channels in coronary vasodilation during exercise. Circulation 88:1245–1253

Duncker DJ, Van Zon NS, Crampton M, Herrlinger S, Homans DC, Bache RJ (1994) Coronary pressure-flow relationship and exercise: contributions of heart rate, contractility, and alpha 1-adrenergic tone. Am J Physiol 266:H795–H810

Duncker DJ, Stubenitsky R, Verdouw PD (1998) Autonomic control of vasomotion in the porcine coronary circulation during treadmill exercise: evidence for feed-forward beta-adrenergic control. Circ Res 82:1312–1322

Dunn RB, McDonough KM, Griggs DM (1978) Cardioacceleration by atrial pacing and transmural metabolite levels in the canine left ventricle. Proc Soc Exp Biol Med 158:230–234

Duran WN, Yudilevich DL (1974) Capillary and cellular barriers to ouabain transport in the heart. Microvasc Res 7:84–88

Duran WN, Yudilevich DL (1978) Estimate of capillary permeability coefficients of canine heart to sodium and glucose. Microvasc Res 15:195–205
Duran WN, Alvarez OA, Yudilevich DL (1973) Influence of maximal vasodilatation on glucose and sodium blood-tissue transport in canine heart. Microvasc Res 6:347–359
Durand MJ, Gutterman DD (2013) Diversity in mechanisms of endothelium-dependent vasodilation in health and disease. Microcirculation 20:239–247
Dwenger MM, Ohanyan V, Navedo MF, Nystoriak MA (2018) Coronary microvascular Kv1 channels as regulatory sensors of intracellular pyridine nucleotide redox potential. Microcirculation 25
Dwenger MM, Raph SM, Reyzer ML, Lisa Manier M, Riggs DW, Wohl ZB, Ohanyan V, Mack G, Pucci T, Moore JB, Hill BG, Chilian WM, Caprioli RM, Bhatnagar A, Nystoriak MA (2022) Pyridine nucleotide redox potential in coronary smooth muscle couples myocardial blood flow to cardiac metabolism. Nat Commun 13:2051
Edlund A, Sollevi A (1995) Theophylline increases coronary vascular tone in humans: evidence for a role of endogenous adenosine in flow regulation. Acta Physiol Scand 155:303–311
Edlund A, Sollevi A, Wennmalm A (1989) The role of adenosine and prostacyclin in coronary flow regulation in healthy man. Acta Physiol Scand 135:39–46
Edlund A, Conradsson T, Sollevi A (1995) A role for adenosine in coronary vasoregulation in man. Effects of theophylline and enprofylline. Clin Physiol 15:623–636
Edwards G, Dora KA, Gardener MJ, Garland CJ, Weston AH (1998) K+ is an endothelium-derived hyperpolarizing factor in rat arteries. Nature 396:269–272
Egashira K, Katsuda Y, Mohri M, Kuga T, Tagawa T, Kubota T, Hirakawa Y, Takeshita A (1996) Role of endothelium-derived nitric oxide in coronary vasodilatation induced by pacing tachycardia in humans. Circ Res 79:331–335
Ellinsworth DC, Sandow SL, Shukla N, Liu Y, Jeremy JY, Gutterman DD (2016) Endothelium-derived hyperpolarization and coronary vasodilation: diverse and integrated roles of epoxyeicosatrienoic acids, hydrogen peroxide, and gap junctions. Microcirculation 23:15–32
Ezra D, Laurindo FR, Eimerl J, Goldstein RE, Peck CC, Feuerstein G (1986) Tachykinin modulation of coronary blood flow. Eur J Pharmacol 122:135–138
Farias M 3rd, Gorman MW, Savage MV, Feigl EO (2005) Plasma ATP during exercise: possible role in regulation of coronary blood flow. Am J Physiol Heart Circ Physiol 288:H1586–H1590
Feigl EO (1969) Parasympathetic control of coronary blood flow in dogs. Circ Res 25:509–519
Feigl EO (1983) Coronary physiology. Physiol Rev 63:1–205
Feigl EO, Van Winkle DM, Miyashiro JK (1990) Cholinergic vasodilatation of coronary resistance vessels in dogs, baboons and goats. Blood Vessels 27:94–105
Feletou M, Vanhoutte PM (2009) EDHF: an update. Clin Sci (Lond) 117:139–155
Feng J, Liu Y, Chu LM, Clements RT, Khabbaz KR, Robich MP, Bianchi C, Sellke FW (2011) Thromboxane-induced contractile response of human coronary arterioles is diminished after cardioplegic arrest. Ann Thorac Surg 92:829–836
Fleming I, Bauersachs J, Busse R (1996) Paracrine functions of the coronary vascular endothelium. Mol Cell Biochem 157:137–145
Folta A, Joshua IG, Webb RC (1989) Dilator actions of endothelin in coronary resistance vessels and the abdominal aorta of the Guinea pig. Life Sci 45:2627–2635
Franco-Cereceda A, Saria A, Lundberg JM (1987a) Ischaemia and changes in contractility induce release of calcitonin gene-related peptide but not neuropeptide Y from the isolated perfused Guinea-pig heart. Acta Physiol Scand 131:319–320
Franco-Cereceda A, Rudehill A, Lundberg JM (1987b) Calcitonin gene-related peptide but not substance P mimics capsaicin-induced coronary vasodilation in the pig. Eur J Pharmacol 142: 235–243
Franco-Cereceda A, Saria A, Lundberg JM (1989) Differential release of calcitonin gene-related peptide and neuropeptide Y from the isolated heart by capsaicin, ischaemia, nicotine, bradykinin and ouabain. Acta Physiol Scand 135:173–187

Freed JK, Beyer AM, LoGiudice JA, Hockenberry JC, Gutterman DD (2014) Ceramide changes the mediator of flow-induced vasodilation from nitric oxide to hydrogen peroxide in the human microcirculation. Circ Res 115:525–532

Frelin C, Guedin D (1994) Why are circulating concentrations of endothelin-1 so low? Cardiovasc Res 28:1613–1622

Friedman PL, Brown EJ Jr, Gunther S, Alexander RW, Barry WH, Mudge GH Jr, Grossman W (1981) Coronary vasoconstrictor effect of indomethacin in patients with coronary-artery disease. N Engl J Med 305:1171–1175

Frolich JC (1982) The 1982 Nobel Prize in medicine: prostaglandins. Dtsch Med Wochenschr 107: 1932–1934

Gale SC, Hokama JY, Ritter LS, Gorman GD, Copeland JG, McDonagh PF (2001) Pentoxifylline reduces coronary leukocyte accumulation early in reperfusion after cold ischemia. Ann Thorac Surg 71:1305–1311

Gebremedhin D, Harder DR, Pratt PF, Campbell WB (1998) Bioassay of an endothelium-derived hyperpolarizing factor from bovine coronary arteries: role of a cytochrome P450 metabolite. J Vasc Res 35:274–284

Gharahi H, Figueroa CA, Tune JD, Beard DA (2022) Multiscale model of the physiological control of myocardial perfusion to delineate putative metabolic feedback mechanisms. J Physiol 600: 1913–1932

Giles RW, Wilcken DE (1977) Reactive hyperaemia in the dog heart: inter-relations between adenosine, ATP, and aminophylline and the effect of indomethacin. Cardiovasc Res 11:113–121

Goligorsky MS, Tsukahara H, Magazine H, Andersen TT, Malik AB, Bahou WF (1994) Termination of endothelin signaling: role of nitric oxide. J Cell Physiol 158:485–494

Goodwill AG, Stapleton PA, James ME, D'Audiffret AC, Frisbee JC (2008a) Increased arachidonic acid-induced thromboxane generation impairs skeletal muscle arteriolar dilation with genetic dyslipidemia. Microcirculation 15:621–631

Goodwill AG, James ME, Frisbee JC (2008b) Increased vascular thromboxane generation impairs dilation of skeletal muscle arterioles of obese Zucker rats with reduced oxygen tension. Am J Physiol Heart Circ Physiol 295:H1522–H1528

Goodwill AG, Noblet JN, Sassoon D, Fu L, Kassab GS, Schepers L, Herring BP, Rottgen TS, Tune JD, Dick GM (2016) Critical contribution of KV1 channels to the regulation of coronary blood flow. Basic Res Cardiol 111:56

Gorman MW, Tune JD, Richmond KN, Feigl EO (2000a) Feedforward sympathetic coronary vasodilation in exercising dogs. J Appl Physiol 89:1892–1902

Gorman MW, Tune JD, Richmond KN, Feigl EO (2000b) Quantitative analysis of feedforward sympathetic coronary vasodilation in exercising dogs. J Appl Physiol 89:1903–1911

Gorman MW, Farias M 3rd, Richmond KN, Tune JD, Feigl EO (2005) Role of endothelin in alpha-adrenoceptor coronary vasoconstriction. Am J Physiol Heart Circ Physiol 288:H1937–H1942

Gorman MW, Rooke GA, Savage MV, Jayasekara MP, Jacobson KA, Feigl EO (2010) Adenine nucleotide control of coronary blood flow during exercise. Am J Physiol Heart Circ Physiol 299: H1981–H1989

Grainger N, Santana LF (2020) Metabolic-electrical control of coronary blood flow. Proc Natl Acad Sci USA 117:8231–8233

Griggs DM Jr, Chilian WM, Boatwright RB, Shoji T, Williams DO (1984) Evidence against significant resting alpha-adrenergic coronary vasoconstrictor tone. Fed Proc 43:2873–2877

Guarini G, Kiyooka T, Ohanyan V, Pung YF, Marzilli M, Chen YR, Chen CL, Kang PT, Hardwick JP, Kolz CL, Yin L, Wilson GL, Shokolenko I, Dobson JG Jr, Fenton R, Chilian WM (2016) Impaired coronary metabolic dilation in the metabolic syndrome is linked to mitochondrial dysfunction and mitochondrial DNA damage. Basic Res Cardiol 111:29

Gulati M, Cooper-DeHoff RM, McClure C, Johnson BD, Shaw LJ, Handberg EM, Zineh I, Kelsey SF, Arnsdorf MF, Black HR, Pepine CJ, Merz CN (2009) Adverse cardiovascular outcomes in women with nonobstructive coronary artery disease: a report from the Women's Ischemia

Syndrome Evaluation Study and the St James Women Take Heart Project. Arch Intern Med 169: 843–850
Gulbenkian S, Edvinsson L, Saetrum Opgaard O, Wharton J, Polak JM, David-Ferreira JF (1990) Peptide-containing nerve fibres in Guinea-pig coronary arteries: immunohistochemistry, ultrastructure and vasomotility. J Auton Nerv Syst 31:153–167
Gutterman DD, Miura H, Liu Y (2005) Redox modulation of vascular tone: focus of potassium channel mechanisms of dilation. Arterioscler Thromb Vasc Biol 25:671–678
Gutterman DD, Chabowski DS, Kadlec AO, Durand MJ, Freed JK, Ait-Aissa K, Beyer AM (2016) The human microcirculation: regulation of flow and beyond. Circ Res 118:157–172
Gwirtz PA, Stone HL (1981) Coronary blood flow and myocardial oxygen consumption after alpha adrenergic blockade during submaximal exercise. J Pharm Exp Ther 217:92–98
Gwirtz PA, Stone HL (1984) Coronary vascular response to adrenergic stimulation in exercise-conditioned dogs. J Appl Physiol Respir Environ Exerc Physiol 57:315–320
Gwirtz PA, Overn SP, Mass HJ, Jones CE (1986) Alpha 1-adrenergic constriction limits coronary flow and cardiac function in running dogs. Am J Physiol 250:H1117–H1126
Haitsma DB, Bac D, Raja N, Boomsma F, Verdouw PD, Duncker DJ (2001) Minimal impairment of myocardial blood flow responses to exercise in the remodeled left ventricle early after myocardial infarction, despite significant hemodynamic and neurohumoral alterations. Cardiovasc Res 52:417–428
Halcox JP, Nour KR, Zalos G, Quyyumi AA (2001) Coronary vasodilation and improvement in endothelial dysfunction with endothelin ET(A) receptor blockade. Circ Res 89:969–976
Hanley FL, Grattan MT, Stevens MB, Hoffman JI (1986) Role of adenosine in coronary autoregulation. Am J Phys 250:H558–H566
Harlan DM, Rooke TW, Belloni FL, Sparks HV (1978) Effect of indomethacin on coronary vascular response to increased myocardial oxygen consumption. Am J Phys 235:H372–H378
Harris NR, Balint L, Dy DM, Nielsen NR, Mendez HG, Aghajanian A, Caron KM (2023) The ebb and flow of cardiac lymphatics: a tidal wave of new discoveries. Physiol Rev 103:391–432
Hein TW, Zhang C, Wang W, Kuo L (2004) Heterogeneous beta2-adrenoceptor expression and dilation in coronary arterioles across the left ventricular wall. Circulation 110:2708–2712
Heitzer T, Schlinzig T, Krohn K, Meinertz T, Munzel T (2001) Endothelial dysfunction, oxidative stress, and risk of cardiovascular events in patients with coronary artery disease. Circulation 104:2673–2678
Hellberg K, Wayland H, Rickart AL, Bing RJ (1972) Studies on the coronary microcirculation by direct visualization. Am J Cardiol 29:593–597
Hellstrand P, Jarhult J (1980) Effects of nine different gastrointestinal polypeptides on vascular smooth muscle in vitro. Acta Physiol Scand 110:89–94
Henri O, Pouehe C, Houssari M, Galas L, Nicol L, Edwards-Lévy F, Henry JP, Dumesnil A, Boukhalfa I, Banquet S, Schapman D, Thuillez C, Richard V, Mulder P, Brakenhielm E (2016) Selective stimulation of cardiac lymphangiogenesis reduces myocardial edema and fibrosis leading to improved cardiac function following myocardial infarction. Circulation 133:1484–1497
Heusch G, Deussen A (1983) The effects of cardiac sympathetic nerve stimulation on perfusion of stenotic coronary arteries in the dog. Circ Res 53:8–15
Heusch G, Deussen A, Schipke J, Thamer V (1984) Alpha 1- and alpha 2-adrenoceptor-mediated vasoconstriction of large and small canine coronary arteries in vivo. J Cardiovasc Pharmacol 6: 961–968
Heyndrickx GR, Boettcher DH, Vatner SF (1976) Effects of angiotensin, vasopressin, and methoxamine on cardiac function and blood flow distribution in conscious dogs. Am J Physiol 231:1579–1587
Heyndrickx GR, Muylaert P, Pannier JL (1982) alpha-Adrenergic control of oxygen delivery to myocardium during exercise in conscious dogs. Am J Physiol 242:H805–H809
Hintze TH, Kaley G (1977) Prostaglandins and the control of blood flow in the canine myocardium. Circ Res 40:313–320

Hirsch E, Borghard-Erdle A (1961) The innervation of the human heart. Arch Pathol 71:384–407
Holtz J (1990) Alpha-adrenoceptor subtypes in the coronary circulation. Basic Res Cardiol 85 (Suppl 1):81–95
Hori M, Kitakaze M, Tamai J, Koretsune Y, Iwai K, Iwakura K, Kagiya T, Kitabatake A, Inoue M, Kamada T (1988) Alpha 2-adrenoceptor activity exerts dual control of coronary blood flow in canine coronary artery. Am J Physiol 255:H250–H260
Hori M, Kitakaze M, Tamai J, Iwakura K, Kitabatake A, Inoue M, Kamada T (1989) Alpha 2-adrenoceptor stimulation can augment coronary vasodilation maximally induced by adenosine in dogs. Am J Physiol 257:H132–H140
Houston DS, Vanhoutte PM (1986) Serotonin and the vascular system. Role in health and disease, and implications for therapy. Drugs 31:149–163
Hu Z, Liu W, Hua X, Chen X, Chang Y, Hu Y, Xu Z, Song J (2021) Single-cell transcriptomic atlas of different human cardiac arteries identifies cell types associated with vascular physiology. Arterioscler Thromb Vasc Biol 41:1408–1427
Huang AH, Feigl EO (1988) Adrenergic coronary vasoconstriction helps maintain uniform transmural blood flow distribution during exercise. Circ Res 62:286–298
Huang A, Wu Y, Sun D, Koller A, Kaley G (2001) Effect of estrogen on flow-induced dilation in NO deficiency: role of prostaglandins and EDHF. J Appl Physiol 91:2561–2566
Huang A, Sun D, Jacobson A, Carroll MA, Falck JR, Kaley G (2005) Epoxyeicosatrienoic acids are released to mediate shear stress-dependent hyperpolarization of arteriolar smooth muscle. Circ Res 96:376–383
Ishibashi Y, Duncker DJ, Bache RJ (1997) Endogenous nitric oxide masks alpha 2-adrenergic coronary vasoconstriction during exercise in the ischemic heart. Circ Res 80:196–207
Ishibashi Y, Duncker DJ, Zhang J, Bache RJ (1998) ATP-sensitive K^+ channels, adenosine, and nitric oxide-mediated mechanisms account for coronary vasodilation during exercise. Circ Res 82:346–359
Ishizaka H, Kuo L (1996) Acidosis-induced coronary arteriolar dilation is mediated by ATP-sensitive potassium channels in vascular smooth muscle. Circ Res 78:50–57
Ishizaka H, Kuo L (1997) Endothelial ATP-sensitive potassium channels mediate coronary microvascular dilation to hyperosmolarity. Am J Phys 273:H104–H112
Ishizaka H, Gudi SR, Frangos JA, Kuo L (1999) Coronary arteriolar dilation to acidosis: role of ATP-sensitive potassium channels and pertussis toxin-sensitive G proteins. Circulaiton 99:558–563
Itkin M, Rockson SG, Burkhoff D (2021) Pathophysiology of the lymphatic system in patients with heart failure: JACC state-of-the-art review. J Am Coll Cardiol 78:278–290
Jackson CV, Pope TK, Lucchesi BR (1987) Coronary artery vasodilation in the canine: physiological and pharmacological roles of beta-adrenergic receptors. J Cardiovasc Pharmacol 10:196–204
Jeevaratnam K, Chadda KR, Huang CL, Camm AJ (2018) Cardiac potassium channels: physiological insights for targeted therapy. J Cardiovasc Pharmacol Ther 23:119–129
Jenkins MJ, Edgley AJ, Sonobe T, Umetani K, Schwenke DO, Fujii Y, Brown RD, Kelly DJ, Shirai M, Pearson JT (2012) Dynamic synchrotron imaging of diabetic rat coronary microcirculation in vivo. Arterioscler Thromb Vasc Biol 32:370–377
Jones CE, Gwirtz PA (1990) Alpha 1-adrenergic coronary constriction during exercise and ischemia. Basic Res Cardiol 85(Suppl 1):177–192
Jones CE, Farrell TA, Ator R (1983) Evidence that a coronary alpha-adrenergic tone limits myocardial blood flow and oxygenation in acute hemorrhagic hypotension. Circ Shock 11: 329–340
Jones CE, Liang IY, Maulsby MR (1986) Cardiac and coronary effects of prazosin and phenoxybenzamine during coronary hypotension. J Pharm Exp Ther 236:204–211
Jones CJ, Kuo L, Davis MJ, Chilian WM (1993a) Distribution and control of coronary microvascular resistance. Adv Exp Med Biol 346:181–188

Jones CJH, Kuo L, Davis MJ, Chilian WM (1993b) Myogenic and flow-dependent control mechanisms in the coronary microcirculation. Basic Res Cardiol 88:2–10

Jones CJ, DeFily DV, Patterson JL, Chilian WM (1993c) Endothelium-dependent relaxation competes with alpha 1- and alpha 2-adrenergic constriction in the canine epicardial coronary microcirculation. Circulation 87:1264–1274

Jones CJ, Kuo L, Davis MJ, Chilian WM (1995a) Regulation of coronary blood flow: coordination of heterogeneous control mechanisms in vascular microdomains. Cardiovasc Res 29:585–596

Jones CJH, Kuo L, Davis MJ, Chilian WM, DeFily DV (1995b) Role of nitric oxide in the coronary microvascular responses to adenosine and increased metabolic demand. Circulation 91:1807–1813

Jones CJ, Kuo L, Davis MJ, Chilian WM (1995c) α-adrenergic responses of isolated canine coronary microvessels. Basic Res Cardiol 90:61–69

Juguilon C, Wang Z, Wang Y, Enrick M, Jamaiyar A, Xu Y, Gadd J, Chen CW, Pu A, Kolz C, Ohanyan V, Chen YR, Hardwick J, Zhang Y, Chilian WM, Yin L (2022) Mechanism of the switch from NO to H(2)O(2) in endothelium-dependent vasodilation in diabetes. Basic Res Cardiol 117:2

Kadokami T, Egashira K, Kuwata K, Fukumoto Y, Kozai T, Yasutake H, Kuga T, Shimokawa H, Sueishi K, Takeshita A (1996) Altered serotonin receptor subtypes mediate coronary microvascular hyperreactivity in pigs with chronic inhibition of nitric oxide synthesis. Circulation 94: 182–189

Kalsner S (1989) Cholinergic constriction in the general circulation and its role in coronary artery spasm. Circ Res 65:237–257

Kanatsuka H, Lamping KG, Eastham CL, Dellsperger KC, Marcus ML (1989) Comparison of the effects of increased myocardial oxygen consumption and adenosine on the coronary microvascular resistance. Circ Res 65:1296–1305

Kanatsuka H, Lamping KG, Eastham CL, Marcus ML (1990a) Heterogeneous changes in epimyocardial microvascular size during graded coronary stenosis. Evidence of the microvascular site for autoregulation. Circ Res 66:389–396

Kanatsuka H, Ashikawa K, Komaru T, Suzuki T, Takishima T (1990b) Diameter change and pressure-red blood cell velocity relations in coronary microvessels during long diastoles in the canine left ventricle. Circ Res 66:503–510

Karkkainen MJ, Jussila L, Ferrell RE, Finegold DN, Alitalo K (2001) Molecular regulation of lymphangiogenesis and targets for tissue oedema. Trends Mol Med 7:18–22

Kassab GS, Rider CA, Tang NJ, Fung YC (1993) Morphometry of pig coronary arterial trees. Am J Phys 265:H350–H365

Katz SA, Feigl EO (1987) Little carbon dioxide diffusional shunting in coronary circulation. Am J Phys 253:H614–H625

Khambata RS, Ghosh SM, Rathod KS, Thevathasan T, Filomena F, Xiao Q, Ahluwalia A (2017) Antiinflammatory actions of inorganic nitrate stabilize the atherosclerotic plaque. Proc Natl Acad Sci USA 114:E550–E559

Kiel AM, Goodwill AG, Baker HE, Dick GM, Tune JD (2018) Local metabolic hypothesis is not sufficient to explain coronary autoregulatory behavior. Basic Res Cardiol 113:33

Kim H, Kataru RP, Koh GY (2014) Inflammation-associated lymphangiogenesis: a double-edged sword? J Clin Invest 124:936–942

Kinashi H, Ito Y, Mizuno M, Suzuki Y, Terabayashi T, Nagura F, Hattori R, Matsukawa Y, Mizuno T, Noda Y, Nishimura H, Nishio R, Maruyama S, Imai E, Matsuo S, Takei Y (2013) TGF-β1 promotes lymphangiogenesis during peritoneal fibrosis. J Am Soc Nephrol 24:1627–1642

Kiyooka T, Ohanyan V, Yin L, Pung YF, Chen YR, Chen CL, Kang PT, Hardwick JP, Yun J, Janota D, Peng J, Kolz C, Guarini G, Wilson G, Shokolenko I, Stevens DA, Chilian WM (2022) Mitochondrial DNA integrity and function are critical for endothelium-dependent vasodilation in rats with metabolic syndrome. Basic Res Cardiol 117:3

Klaourakis K, Vieira JM, Riley PR (2021) The evolving cardiac lymphatic vasculature in development, repair and regeneration. Nat Rev Cardiol 18:368–379

Klotz L, Norman S, Vieira JM, Masters M, Rohling M, Dubé KN, Bollini S, Matsuzaki F, Carr CA, Riley PR (2015) Cardiac lymphatics are heterogeneous in origin and respond to injury. Nature 522:62–67

Knight DR, Shen YT, Young MA, Vatner SF (1991) Acetylcholine-induced coronary vasoconstriction and vasodilation in tranquilized baboons. Circ Res 69:706–713

Knot HJ, Zimmermann PA, Nelson MT (1996) Extracellular K(+)-induced hyperpolarizations and dilatations of rat coronary and cerebral arteries involve inward rectifier K(+) channels. J Physiol 492(Pt 2):419–430

Koltai MZ, Hadhazy P, Posa I, Kocsis E, Winkler G, Rosen P, Pogatsa G (1997) Characteristics of coronary endothelial dysfunction in experimental diabetes. Cardiovasc Res 34:157–163

Komaru T, Lamping KG, Eastham CL, Dellsperger KC (1991a) Role of ATP-sensitive potassium channels in coronary microvascular autoregulatory responses. Circ Res 69:1146–1151

Komaru T, Lamping KG, Eastham CL, Harrison DG, Marcus ML, Dellsperger KC (1991b) Effect of an arginine analogue on acetylcholine-induced coronary microvascular dilatation in dogs. Am J Physiol 261:H2001–H2007

Komaru T, Lamping KG, Dellsperger KC (1994) Role of adenosine in vasodilation of epimyocardial coronary microvessels during reduction in perfusion pressure. J Cardiovasc Pharmacol 24:434–442

Konidala S, Gutterman DD (2004) Coronary vasospasm and the regulation of coronary blood flow. Prog Cardiovasc Dis 46:349–373

Koshida R, Rocic P, Saito S, Kiyooka T, Zhang C, Chilian WM (2005) Role of focal adhesion kinase in flow-induced dilation of coronary arterioles. Arterioscler Thromb Vasc Biol 25:2548–2553

Kroll K, Feigl EO (1985) Adenosine is unimportant in controlling coronary blood flow in unstressed dog hearts. Am J Phys 249:H1176–H1187

Kuo L, Davis MJ, Chilian WM (1988) Myogenic activity in isolated subepicardial and subendocardial coronary arterioles. Am J Physiol 255:H1558–H1562

Kuo L, Chilian WM, Davis MJ (1990a) Coronary arteriolar myogenic response is independent of endothelium. Circ Res 66:860–866

Kuo L, Davis MJ, Chilian WM (1990b) Endothelium-dependent, flow-induced dilation of isolated coronary arterioles. Am J Physiol 259:H1063–H1070

Kuo L, Chilian WM, Davis MJ (1991) Interaction of pressure- and flow-induced responses in porcine coronary resistance vessels. Am J Physiol Heart Circ Physiol 261:H1706–H1715

Kuo L, Arko F, Chilian WM, Davis MJ (1992a) Nitrovasodilator-mediated flow-induced dilation in isolated porcine coronary venules. HFASEB J 6:A1752

Kuo L, Davis MJ, Cannon MS, Chilian WM (1992b) Pathophysiological consequences of atherosclerosis extend into the coronary microcirculation: restoration of endothelium-dependent responses by L-arginine. Circ Res 70:465–476

Kuo L, Arko F, Chilian WM, Davis MJ (1993) Coronary venular responses to flow and pressure. Circ Res 72:607–615

Kuo L, Davis MJ, Chilian WM (1995) Longitudinal gradients for endothelium-dependent and -independent vascular responses in the coronary microcirculation. Circulation 92:518–525

Laine GA, Granger HJ (1985) Microvascular, interstitial, and lymphatic interactions in normal heart. Am J Phys 249:H834–H842

Laird JD, Breuls PN, van der Meer P, Spaan JA (1981) Can a single vasodilator be responsible for both coronary autoregulation and metabolic vasodilation? Basic Res Cardiol 76:354–358

Lamping KG (1997) Response of native and stimulated collateral vessels to serotonin. Am J Physiol 272:H2409–H2415

Lamping KG, Marcus ML, Dole WP (1985) Removal of the endothelium potentiates canine large coronary artery constrictor responses to 5-hydroxytryptamine in vivo. Circ Res 57:46–54

Lamping KG, Kanatsuka H, Eastham CL, Chilian WM, Marcus ML (1989) Nonuniform vasomotor responses of the coronary microcirculation to serotonin and vasopressin. Circ Res 65:343–351
Lamping KG, Chilian WM, Eastham CL, Marcus ML (1992) Coronary microvascular response to exogenously administered and endogenously released acetylcholine. Microvasc Res 43:294–307
Laxson DD, Dai XZ, Homans DC, Bache RJ (1989) The role of alpha 1- and alpha 2-adrenergic receptors in mediation of coronary vasoconstriction in hypoperfused ischemic myocardium during exercise. Circ Res 65:1688–1697
Lenz T, Nadansky M, Gossmann J, Oremek G, Geiger H (1998) Exhaustive exercise-induced tissue hypoxia does not change endothelin and big endothelin plasma levels in normal volunteers. Am J Hypertens 11:1028–1031
Li Z, Luo W, Fang S, Chen X, Lin T, Zhou S, Zhang L, Yang W, Li Z, Ye J, Wang J, Liu P, Li Z (2021) Prostacyclin facilitates vascular smooth muscle cell phenotypic transformation via activating TP receptors when IP receptors are deficient. Acta Physiol (Oxf) 231:e13555
Liang IY, Jones CE (1985) Alpha 1-adrenergic blockade increases coronary blood flow during coronary hypoperfusion. Am J Physiol 249:H1070–H1077
Likoff W, Segal BL, Kasparian H (1967) Paradox of normal selective coronary arteriograms in patients considered to have unmistakable coronary heart disease. N Engl J Med 276:1063–1066
Lim HY, Thiam CH, Yeo KP, Bisoendial R, Hii CS, McGrath KC, Tan KW, Heather A, Alexander JS, Angeli V (2013) Lymphatic vessels are essential for the removal of cholesterol from peripheral tissues by SR-BI-mediated transport of HDL. Cell Metab 17:671–684
Lindner JR, Villanueva FS, Dent JM, Wei K, Sklenar J, Kaul S (2000) Assessment of resting perfusion with myocardial contrast echocardiography: theoretical and practical considerations. Am Heart J 139:231–240
Liu Y, Gutterman DD (2009) Vascular control in humans: focus on the coronary microcirculation. Basic Res Cardiol 104:211–227
Liu Y, Bubolz AH, Mendoza S, Zhang DX, Gutterman DD (2011) H2O2 is the transferrable factor mediating flow-induced dilation in human coronary arterioles. Circ Res 108:566–573
Loftus I (2011) Mechanisms of plaque rupture. In: Fitridge R, Thompson M (eds) Mechanisms of vascular disease: a reference book for vascular specialists. University of Adelaide Press, Adelaide
Longden TA, Dabertrand F, Koide M, Gonzales AL, Tykocki NR, Brayden JE, Hill-Eubanks D, Nelson MT (2017) Capillary K(+)-sensing initiates retrograde hyperpolarization to increase local cerebral blood flow. Nat Neurosci 20:717–726
Loukas M, Abel N, Shane Tubbs R, Grabska J, Birungi J, Anderson RH (2011) The cardiac lymphatic system. Clin Anat 24:684–691
Maeda S, Miyauchi T, Goto K, Matsuda M (1994) Alteration of plasma endothelin-1 by exercise at intensities lower and higher than ventilatory threshold. J Appl Physiol 77:1399–1402
Mäkinen T, Jussila L, Veikkola T, Karpanen T, Kettunen MI, Pulkkanen KJ, Kauppinen R, Jackson DG, Kubo H, Nishikawa S, Ylä-Herttuala S, Alitalo K (2001) Inhibition of lymphangiogenesis with resulting lymphedema in transgenic mice expressing soluble VEGF receptor-3. Nat Med 7: 199–205
Maley RH, Vrettos AM, Arden WA, Chein S, Salley RK, Gross DR (1994) Coronary vascular and myocardial effects of substance P in hypercholesterolemic rabbits. Neuropeptides 27:95–103
Marroquin OC, Kip KE, Mulukutla SR, Ridker PM, Pepine CJ, Tjandrawan T, Kelsey SF, Mankad S, Rogers WJ, Merz CN, Sopko G, Sharaf BL, Reis SE (2005) Inflammation, endothelial cell activation, and coronary microvascular dysfunction in women with chest pain and no obstructive coronary artery disease. Am Heart J 150:109–115
Martin SE, Tidmore WC, Patterson RE (1991) Adenosine receptor blockade with 8-p-sulfophenyltheophylline aggravates coronary constriction. Am J Phys 260:H1753–H1759
Martinez-Lemus LA, Wu X, Wilson E, Hill MA, Davis GE, Davis MJ, Meininger GA (2003) Integrins as unique receptors for vascular control. J Vasc Res 40:211–233

Martini J, Honig CR (1969) Direct measurement of intercapillary distance in beating rat heart in situ under various conditions of O 2 supply. Microvasc Res 1:244–256

Matoba T, Shimokawa H (2003) Hydrogen peroxide is an endothelium-derived hyperpolarizing factor in animals and humans. J Pharmacol Sci 92:1–6

Matoba T, Shimokawa H, Nakashima M, Hirakawa Y, Mukai Y, Hirano K, Kanaide H, Takeshita A (2000) Hydrogen peroxide is an endothelium-derived hyperpolarizing factor in mice. J Clin Invest 106:1521–1530

Matoba T, Shimokawa H, Kubota H, Morikawa K, Fujiki T, Kunihiro I, Mukai Y, Hirakawa Y, Takeshita A (2002) Hydrogen peroxide is an endothelium-derived hyperpolarizing factor in human mesenteric arteries. Biochem Biophys Res Commun 290:909–913

Maxwell GM (1968) Actions of substance P on the general, pulmonary, and coronary haemodynamics and metabolism of intact dogs. Br J Pharmacol 32:514–522

McDonagh PF, Roberts DJ (1986) Prevention of transcoronary macromolecular leakage after ischemia-reperfusion by the calcium entry blocker nisoldipine. Direct observations in isolated rat hearts. Circ Res 58:127–136

McDonagh PF, Niven AT, Roberts D (1984) Direct visualization of the coronary microcirculation for pharmacologic and physiologic studies. Microvasc Res 28:180–196

Merkus D, Stepp DW, Jones DW, Nishikawa Y, Chilian WM (2000) Adenosine preconditions against endothelin-induced constriction of coronary arterioles. Am J Phys 279:H2593–H2597

Merkus D, Duncker DJGM, Chilian WM (2002) Metabolic regulation of coronary vascular tone-role of endothelin-1. Am J Physiol Heart Circ Physiol 283:H1915–H1921

Merkus D, Haitsma DB, Fung TY, Assen YJ, Verdouw PD, Duncker DJ (2003a) Coronary blood flow regulation in exercising swine involves parallel rather than redundant vasodilator pathways. Am J Physiol Heart Circ Physiol 285:H424–H433

Merkus D, Houweling B, Mirza A, Boomsma F, van den Meiracker AH, Duncker DJ (2003b) Contribution of endothelin and its receptors to the regulation of vascular tone during exercise is different in the systemic, coronary and pulmonary circulation. Cardiovasc Res 59:745–754

Merkus D, Houweling B, Zarbanoui A, Duncker DJ (2004) Interaction between prostanoids and nitric oxide in regulation of systemic, pulmonary, and coronary vascular tone in exercising swine. Am J Physiol Heart Circ Physiol 286:H1114–H1123

Merkus D, Brzezinska AK, Zhang C, Saito S, Chilian WM (2005) Cardiac myocytes control release of endothelin-1 in coronary vasculature. Am J Physiol Heart Circ Physiol 288:H2088–H2092

Merkus D, Sorop O, Houweling B, Hoogteijling BA, Duncker DJ (2006) KCa+ channels contribute to exercise-induced coronary vasodilation in swine. Am J Physiol Heart Circ Physiol 291: H2090–H2097

Miller AJ (2011) The grossly invisible and generally ignored lymphatics of the mammalian heart. Med Hypotheses 76:604–606

Miller FJ Jr, Dellsperger KC, Gutterman DD (1998) Pharmacologic activation of the human coronary microcirculation in vitro: endothelium-dependent dilation and differential responses to acetylcholine. Cardiovasc Res 38:744–750

Miller AJ, Pick R, Katz LN (1961) Lymphatics of the mitral valve of the dog. Demonstration and discussion of the possible significance. Circ Res 9:1005–1009

Miller WL, Belardinelli L, Bacchus A, Foley DH, Rubio R, Berne RM (1979) Canine myocardial adenosine and lactate production, oxygen consumption, and coronary blood flow during stellate ganglia stimulation. Circ Res 45:708–718

Miller FJJ, Dellsperger KC, Gutterman DD (1997) Myogenic constriction of human coronary arterioles. Am J Phys 273:H257–H264

Miura H, Bosnjak JJ, Ning G, Saito T, Miura M, Gutterman DD (2003) Role for hydrogen peroxide in flow-induced dilation of human coronary arterioles. Circ Res 92:e31–e40

Miyashiro JK, Feigl EO (1993) Feedforward control of coronary blood flow via coronary beta-receptor stimulation. Circ Res 73:252–263

Mohrman DE, Feigl EO (1978) Competition between sympathetic vasoconstriction and metabolic vasodilation in the canine coronary circulation. Circ Res 42:79–86

Moncada S, Palmer RM, Higgs EA (1991) Nitric oxide: physiology, pathophysiology, and pharmacology. Pharmacol Rev 43:109–142

Morita K, Mori H, Tsujioka K, Kimura A, Ogasawara Y, Goto M, Hiramatsu O, Kajiya F, Feigl EO (1997) Alpha-adrenergic vasoconstriction reduces systolic retrograde coronary blood flow. Am J Physiol 273:H2746–H2755

Mullane KM, Pinto A (1987) Endothelium, arachidonic acid, and coronary vascular tone. Fed Proc 46:54–62

Muller JM, Chilian WM, Davis MJ (1997) Integrin signaling transduces shear stress-dependent vasodilation of coronary arterioles. Circ Res 80:320–326

Muller JM, Davis MJ, Kuo L, Chilian WM (1999) Changes in coronary endothelial cell Ca2+ concentration during shear stress- and agonist-induced vasodilation. Am J Physiol 276:H1706–H1714

Muntz KH, Garcia C, Hagler HK (1985) alpha 1-Receptor localization in rat heart and kidney using autoradiography. Am J Physiol 249:H512–H519

Muntz KH, Meyer L, Gadol S, Calianos TA (1986) Alpha-2 adrenergic receptor localization in the rat heart and kidney using autoradiography and tritiated rauwolscine. J Pharm Exp Ther 236: 542–547

Murphree SS, Saffitz JE (1988) Delineation of the distribution of beta-adrenergic receptor subtypes in canine myocardium. Circ Res 63:117–125

Murthy VL, Naya M, Foster CR, Gaber M, Hainer J, Klein J, Dorbala S, Blankstein R, Di Carli MF (2012) Association between coronary vascular dysfunction and cardiac mortality in patients with and without diabetes mellitus. Circulation 126:1858–1868

Myers PR, Banitt PF, Guerra R Jr, Harrison DG (1989) Characteristics of canine coronary resistance arteries: importance of endothelium. Am J Physiol 257:H603–H610

Myers PR, Banitt PF, Guerra R Jr, Harrison DG (1991) Role of the endothelium in modulation of the acetylcholine vasoconstrictor response in porcine coronary microvessels. Cardiovasc Res 25:129–137

Myers PR, Katwa LC, Tanner M, Morrow C, Guarda E, Parker JL (1994) Effects of angiotensin II on canine and porcine coronary epicardial and resistance arteries. J Vasc Res 31:338–346

Nakamura Y, Parent R, Lavallee M (1990) Opposite effects of substance P on conductance and resistance coronary vessels in conscious dogs. Am J Physiol 258:H565–H573

Nakamura Y, Parent R, Lavallee M (1991) Disparate effects of substance P on systemic and coronary beds in conscious dogs. Circulation 84:300–312

Nakayama K, Osol G, Halpern W (1988) Reactivity of isolated porcine coronary resistance arteries to cholinergic and adrenergic drugs and transmural pressure changes. Circ Res 62:741–748

Needleman P, Marshall GR, Sobel BE (1975) Hormone interactions in the isolated rabbit heart. Synthesis and coronary vasomotor effects of prostaglandins, angiotensin, and bradykinin. Circ Res 37:802–808

Nellis SH, Liedtke AJ, Whitesell L (1981) Small coronary vessel pressure and diameter in an intact beating rabbit heart using fixed-position and free-motion techniques. Circ Res 49:342–353

Nishikawa Y, Stepp DW, Chilian WM (1999) In vivo location and mechanism of EDHF-mediated vasodilation in canine coronary microcirculation. Am J Phys 277:H1252–H1259

Nishikawa Y, Stepp DW, Chilian WM (2000) Nitric oxide exerts feedback inhibition on EDHF-induced coronary arteriolar dilation in vivo. Am J Physiol Heart Circ Physiol 279:H459–H465

Ohanyan V, Yin L, Bardakjian R, Kolz C, Enrick M, Hakobyan T, Kmetz J, Bratz I, Luli J, Nagane M, Khan N, Hou H, Kuppusamy P, Graham J, Fu FK, Janota D, Oyewumi MO, Logan S, Lindner JR, Chilian WM (2015) Requisite role of Kv1.5 channels in coronary metabolic dilation. Circ Res 117:612–621

Ohanyan V, Yin L, Bardakjian R, Kolz C, Enrick M, Hakobyan T, Luli J, Graham K, Khayata M, Logan S, Kmetz J, Chilian WM (2017) Kv1.3 channels facilitate the connection between metabolism and blood flow in the heart. Microcirculation 24

Ohanyan V, Raph SM, Dwenger MM, Hu X, Pucci T, Mack G et al (2021) Myocardial blood flow control by oxygen sensing vascular Kvbeta proteins. Circ Res 128:738–751

Olsson RA, Bunger R, Spaan JA (1992) Coronary circulation. In: Fozzard HA, Haber E, Jennings RB, Katz AM, Morgan HE (eds) The heart and cardiovascular system. Raven Press, New York, pp 1393–1426

Patrono C, Ciabattoni G, Patrignani P, Pugliese F, Filabozzi P, Catella F, Davi G, Forni L (1985) Clinical pharmacology of platelet cyclooxygenase inhibition. Circulation 72:1177–1184

Pepine CJ, Anderson RD, Sharaf BL, Reis SE, Smith KM, Handberg EM, Johnson BD, Sopko G, Bairey Merz CN (2010) Coronary microvascular reactivity to adenosine predicts adverse outcome in women evaluated for suspected ischemia results from the National Heart, Lung and Blood Institute WISE (Women's Ischemia Syndrome Evaluation) study. J Am Coll Cardiol 55:2825–2832

Picchi A, Gao X, Belmadani S, Potter BJ, Focardi M, Chilian WM, Zhang C (2006) Tumor necrosis factor-alpha induces endothelial dysfunction in the prediabetic metabolic syndrome. Circ Res 99:69–77

Porsti I, Hecker M, Bassenge E, Busse R (1993) Dual action of angiotensin II on coronary resistance in the isolated perfused rabbit heart. Naunyn Schmiedeberg's Arch Pharmacol 348:650–658

Prasad A, Husain S, Quyyumi AA (1999) Abnormal flow-mediated epicardial vasomotion in human coronary arteries is improved by angiotensin-converting enzyme inhibition: a potential role of bradykinin. J Am Coll Cardiol 33:796–804

Quayle JM, Dart C, Standen NB (1996) The properties and distribution of inward rectifier potassium currents in pig coronary arterial smooth muscle. J Physiol 494(Pt 3):715–726

Quillen JE, Harrison DG (1992) Vasomotor properties of porcine endocardial and epicardial microvessels. Am J Physiol Heart Circ Physiols 262:H1143–H1148

Quillen JE, Sellke FW, Brooks LA, Harrison DG (1990) ischemia-reperfusion impairs endothelium-dependent relaxation of coronary microvessels but does not affect large arteries. Circulaiton 82:586–594

Quillen J, Sellke F, Banitt P, Harrison D (1992) The effect of norepinephrine on the coronary microcirculation. J Vasc Res 29:2–7

Randall MD, Alexander SP, Bennett T, Boyd EA, Fry JR, Gardiner SM, Kemp PA, McCulloch AI, Kendall DA (1996) An endogenous cannabinoid as an endothelium-derived vasorelaxant. Biochem Biophys Res Commun 229:114–120

Ratajska A, Gula G, Flaht-Zabost A, Czarnowska E, Ciszek B, Jankowska-Steifer E, Niderla-Bielinska J, Radomska-Lesniewska D (2014) Comparative and developmental anatomy of cardiac lymphatics. Scient World J 2014:183170

Reynolds JM, McDonagh PF (1994) Platelets do not modulate leukocyte-mediated coronary microvascular damage during early reperfusion. Am J Phys 266:H171–H181

Rich S, McLaughlin VV (2003) Endothelin receptor blockers in cardiovascular disease. Circulation 108:2184–2190

Richard V, Tanner FC, Tschudi M, Luscher TF (1990) Different activation of L-arginine pathway by bradykinin, serotonin, and clonidine in coronary arteries. Am J Physiol 259:H1433–H1439

Richmond KN, Tune JD, Gorman MW, Feigl EO (1999) Role of K+ATP channels in local metabolic coronary vasodilation. Am J Phys 277:H2115–H2123

Rigel DF, Shetty SS (1997) A novel model of conduit coronary constriction reveals local actions of endothelin-1 and prostaglandin F2alpha. Am J Phys 272:H2054–H2064

Ritter L, Davidson L, Henry M, Davis-Gorman G, Morrison H, Frye JB, Cohen Z, Chandler S, McDonagh P, Funk JL (2011) Exaggerated neutrophil-mediated reperfusion injury after ischemic stroke in a rodent model of type 2 diabetes. Microcirculation 18:552–561

Rivers RJ, Hein TW, Zhang C, Kuo L (2001) Activation of barium-sensitive inward rectifier potassium channels mediates remote dilation of coronary arterioles. Circulation 104:1749–1753

Robertson BE, Bonev AD, Nelson MT (1996) Inward rectifier K+ currents in smooth muscle cells from rat coronary arteries: block by Mg2+, Ca2+, and Ba2+. Am J Phys 271:H696–H705

Rodriguez-Rangel S, Bravin AD, Ramos-Torres KM, Brugarolas P, Sanchez-Rodriguez JE (2020) Structure-activity relationship studies of four novel 4-aminopyridine K(+) channel blockers. Sci Rep 10:52

Rogers PA, Dick GM, Knudson JD, Focardi M, Bratz IN, Swafford AN Jr, Saitoh S, Tune JD, Chilian WM (2006) H2O2-induced redox-sensitive coronary vasodilation is mediated by 4-aminopyridine-sensitive K+ channels. Am J Physiol Heart Circ Physiol 291:H2473–H2482

Rogers PA, Chilian WM, Bratz IN, Bryan RM Jr, Dick GM (2007) H2O2 activates redox- and 4-aminopyridine-sensitive Kv channels in coronary vascular smooth muscle. Am J Physiol Heart Circ Physiol 292:H1404–H1411

Rose CP, Goresky CA (1976) Vasomotor control of capillary transit time heterogeneity in the canine coronary circulation. Circ Res 39:541–554

Rose CP, Goresky CA, Bach GG (1977) The capillary and sarcolemmal barriers in the heart an exploration of labeled water permeability. Circ Res 41:515–533

Roth AC, Feigl EO (1981) Diffusional shunting in the canine myocardium. Circ Res 48:470–480

Rubanyi GM, Polokoff MA (1994) Endothelins: molecular biology, biochemistry, pharmacology, physiology, and pathophysiology. Pharmacol Rev 46:325–415

Rubio R, Wiedmeier VT, Berne RM (1974) Relationship between coronary flow and adenosine production and release. J Mol Cell Cardiol 6:561–566

Sacchi G, Weber E, Aglianò M, Cavina N, Comparini L (1999) Lymphatic vessels of the human heart: precollectors and collecting vessels. A morpho-structural study. J Submicrosc Cytol Pathol 31:515–525

Saetrum Opgaard O, Gulbenkian S, Edvinsson L (1997) Innervation and effects of vasoactive substances in the coronary circulation. Eur Heart J 18:1556–1568

Saffitz JE (1989) Distribution of alpha 1-adrenergic receptors in myocytic regions and vasculature of feline myocardium. Am J Physiol 257:H162–H169

Saito D, Steinhart CR, Nixon DG, Olsson RA (1981) Intracoronary adenosine deaminase reduces canine myocardial reactive hyperemia. Circ Res 49:1262–1267

Saitoh S, Zhang C, Tune JD, Potter B, Kiyooka T, Rogers PA, Knudson JD, Dick GM, Swafford A, Chilian WM (2006) Hydrogen peroxide: a feed-forward dilator that couples myocardial metabolism to coronary blood flow. Arterioscler Thromb Vasc Biol 26:2614–2621

Saitoh S, Kiyooka T, Rocic P, Rogers PA, Zhang C, Swafford A, Dick GM, Viswanathan C, Park Y, Chilian WM (2007) Redox-dependent coronary metabolic dilation. Am J Physiol Heart Circ Physiol 293:H3720–H3725

Sara JD, Lennon RJ, Ackerman MJ, Friedman PA, Noseworthy PA, Lerman A (2016) Coronary microvascular dysfunction is associated with baseline QTc prolongation amongst patients with chest pain and non-obstructive coronary artery disease. J Electrocardiol 49:87–93

Sato K, Komaru T, Shioiri H, Takeda S, Takahashi K, Kanatsuka H, Shirato K (2002) Vasodilator signals from the ischemic myocardium are transduced to the coronary vascular wall by pertussis toxin-sensitive G proteins: a new experimental method for the analysis of the interaction between the myocardium and coronary vessels. J Am Coll Cardiol 39:1859–1865

Schiffrin EL (2005) Vascular endothelin in hypertension. Vasc Pharmacol 43(1):19–29

Schutz W, Zimpfer M, Raberger G (1977) Effect of aminophylline on coronary reactive hyperaemia following brief and long occlusion periods. Cardiovasc Res 11:507–511

Seitelberger R, Guth BD, Heusch G, Lee JD, Katayama K, Ross J Jr (1988) Intracoronary alpha 2-adrenergic receptor blockade attenuates ischemia in conscious dogs during exercise. Circ Res 62:436–442

Sekiguchi N, Kanatsuka H, Komaru T, Akai K, Sato K, Wang Y, Sugi M, Ashikawa K, Takishima T (1992) Effects of alpha and beta adrenergic blockade on coronary arterial microvessels in the beating canine heart. Cardiovasc Res 26:415–421

Sekiguchi N, Kanatsuka H, Sato K, Wang Y, Akai K, Komaru T, Takishima T (1994) Effect of calcitonin gene-related peptide on coronary microvessels and its role in acute myocardial ischemia. Circulation 89:366–374

Sellke FW, Quillen JE (1992) Altered effects of vasopressin on the coronary circulation after ischemia. J Thor Cardiovasc Surg 104:357–363

Shah SJ, Lam CSP, Svedlund S, Saraste A, Hage C, Tan RS, Beussink-Nelson L, Ljung Faxen U, Fermer ML, Broberg MA, Gan LM, Lund LH (2018) Prevalence and correlates of coronary microvascular dysfunction in heart failure with preserved ejection fraction: PROMIS-HFpEF. Eur Heart J 39:3439–3450

Shen YT, Pittman TJ, Buie PS, Bolduc DL, Kane SA, Koblan KS, Gould RJ, Lynch JJ Jr (2001) Functional role of alpha-calcitonin gene-related peptide in the regulation of the cardiovascular system. J Pharma Exp Ther 298:551–558

Shepherd JT, Vanhoutte PM (1986) Mechanisms responsible for coronary vasospasm. J Am Coll Cardiol 8:50A–54A

Shipley RA, Shipley LJ, Wearn JT (1937) The capillary supply in normal and hypertrophied hearts of rabbits. J Exp Med 65:29–42

Smart N, Bollini S, Dubé KN, Vieira JM, Zhou B, Davidson S, Yellon D, Riegler J, Price AN, Lythgoe MF, Pu WT, Riley PR (2011) De novo cardiomyocytes from within the activated adult heart after injury. Nature 474:640–644

Sorop O, Merkus D, de Beer VJ, Houweling B, Pistea A, McFalls EO, Boomsma F, van Beusekom HM, van der Giessen WJ, VanBavel E, Duncker DJ (2008) Functional and structural adaptations of coronary microvessels distal to a chronic coronary artery stenosis. Circ Res 102:795–803

Spaan JA, ter Wee R, van Teeffelen JW, Streekstra G, Siebes M, Kolyva C, Vink H, Fokkema DS, VanBavel E (2005) Visualisation of intramural coronary vasculature by an imaging cryomicrotome suggests compartmentalisation of myocardial perfusion areas. Med Biol Eng Comput 43:431–435

Sparks HV Jr, Bardenheuer H (1986) Regulation of adenosine formation by the heart. Circ Res 58: 193–201

Stepp DW, Kroll K, Feigl EO (1997) K+ATP channels and adenosine are not necessary for coronary autoregulation. Am J Phys 273:H1299–H1308

Stepp DW, Nishikawa Y, Chilian WM (1999) Regulation of shear stress in the canine coronary microcirculation. Circulation 100:1555–1561

Stepp DW, Merkus D, Nishikawa Y, Chilian WM (2001) Nitric oxide limits coronary vasoconstriction by a shear stress-dependent mechanism. Am J Physiol Heart Circ Physiol 281:H796–H803

Stillwell W (2016) An introduction to biological membranes: composition, structure and function. Academic Press is an Imprint of Elsevier, Amsterdam; Boston, p 579

Sun D, Huang A, Mital S, Kichuk MR, Marboe CC, Addonizio LJ, Michler RE, Koller A, Hintze TH, Kaley G (2002) Norepinephrine elicits beta2-receptor-mediated dilation of isolated human coronary arterioles. Circulation 106:550–555

Suzuki Y, Ito Y, Mizuno M, Kinashi H, Sawai A, Noda Y, Mizuno T, Shimizu H, Fujita Y, Matsui K, Maruyama S, Imai E, Matsuo S, Takei Y (2012) Transforming growth factor-β induces vascular endothelial growth factor-C expression leading to lymphangiogenesis in rat unilateral ureteral obstruction. Kidney Int 81:865–879

Symbas PN, Cooper T, Gantner GA Jr, Willman VL (1963) Lymphatic drainage of the heart: effects of experimental interruption of Lymphatics. Surg Forum 14:254–256

Symons JD, Rendig SV, Fu LW, Longhurst JC (1997) Endothelin-1 limits increases in blood flow to native and collateral-dependent myocardium. Am J Phys 273:R41–R48

Takamura M, Parent R, Cernacek P, Lavallee M (2000) Influence of dual ET(A)/ET(B)-receptor blockade on coronary responses to treadmill exercise in dogs. J Appl Physiol 89:2041–2048

Takeda S, Komaru T, Takahashi K, Sato K, Kanatsuka H, Kokusho Y, Shirato K, Shimokawa H (2006) Beating myocardium counteracts myogenic tone of coronary microvessels: involvement of ATP-sensitive potassium channels. Am J Physiol Heart Circ Physiol 291:H3050

Tanikawa T, Kanatsuka H, Koshida R, Tanaka M, Sugimura A, Kumagai T, Miura M, Komaru T, Shirato K (2000) Role of pertussis toxin-sensitive G protein in metabolic vasodilation of coronary microcirculation. Am J Physiol Heart Circ Physiol 279:H1819–H1829

Tiefenbacher CP, DeFily DV, Chilian WM (1998) Requisite role of cardiac myocytes in coronary alpha1-adrenergic constriction. Circulation 98:9–12

Tillmanns H, Ikeda S, Hansen H, Sarma JS, Fauvel JM, Bing RJ (1974) Microcirculation in the ventricle of the dog and turtle. Circ Res 34:561–569

Toyota E, Ogasawara Y, Hiramatsu O, Tachibana H, Kajiya F, Yamamori S, Chilian WM (2005) Dynamics of flow velocities in endocardial and epicardial coronary arterioles. Am J Physiol Heart Circ Physiol 288:H1598–H1603

Traverse JH, Judd D, Bache RJ (1996) Dose-dependent effect of endothelin-1 on blood flow to normal and collateral-dependent myocardium. Circulation 93:558–566

Trivella MG, Broten TP, Feigl EO (1990) Beta-receptor subtypes in the canine coronary circulation. Am J Physiol 259:H1575–H1585

Trzeciakowski J, Chilian WM (2008) Chaotic behavior of the coronary circulation. Med Biol Eng Comput 46:433–442

Tschudi M, Richard V, Bühler FR, Lüscher TF (1991) Importance of endothelium-derived nitric oxide in porcine coronary resistance arteries. Am J Physiol 260:H13–H20

Tune JD, Richmond KN, Gorman MW, Feigl EO (2000a) Role of nitric oxide and adenosine in control of coronary blood flow in exercising dogs. Circulation 101:2942–2948

Tune JD, Richmond KN, Gorman MW, Olsson RA, Feigl EO (2000b) Adenosine is not responsible for local metabolic control of coronary blood flow in dogs during exercise. Am J Physiol Heart Circ Physiol H278:H74–H84

Tune JD, Richmond KN, Gorman MW, Feigl EO (2001) K(ATP)(+) channels, nitric oxide, and adenosine are not required for local metabolic coronary vasodilation. Am J Physiol Heart Circ Physiol 280:H868–H875

Tune JD, Gorman MW, Feigl EO (2004) Matching coronary blood flow to myocardial oxygen consumption. J Appl Physiol 97:404–415

Tune JD, Goodwill AG, Sassoon DJ, Mather KJ (2017) Cardiovascular consequences of metabolic syndrome. Transl Res 183:57–70

Van Bibber R, Traub O, Kroll K, Feigl EO (1995) EDRF and norepinephrine-induced vasodilation in the canine coronary circulation. Am J Phys 268:H1973–H1981

van den Wijngaard JP, van Horssen P, ter Wee R, Coronel R, de Bakker JM, de Jonge N, Siebes M, Spaan JA (2010) Organization and collateralization of a subendocardial plexus in end-stage human heart failure. Am J Physiol Heart Circ Physiol 298:H158–H162

Van Winkle DM, Feigl EO (1989) Acetylcholine causes coronary vasodilation in dogs and baboons. Circ Res 65:1580–1593

Vane JR (1971) Inhibition of prostaglandin synthesis as a mechanism of action for aspirin-like drugs. Nat New Biol 231:232–235

Vatner DE, Knight DR, Homcy CJ, Vatner SF, Young MA (1986) Subtypes of beta-adrenergic receptors in bovine coronary arteries. Circ Res 59:463–473

Vieira JM, Norman S, Villa Del Campo C, Cahill TJ, Barnette DN, Gunadasa-Rohling M, Johnson LA, Greaves DR, Carr CA, Jackson DG, Riley PR (2018) The cardiac lymphatic system stimulates resolution of inflammation following myocardial infarction. J Clin Invest 128: 3402–3412

Vlahakes GJ, Giamber SR, Rothaus KO, Powell WJ Jr (1989) Hyperosmotic mannitol and collateral blood flow to ischemic myocardium. J Surg Res 47:438–446

Wang SY, Friedman M, Johnson RG, Weintraub RM, Sellke FW (1994) Adrenergic regulation of coronary microcirculation after extracorporeal circulation and crystalloid cardioplegia. Am J Physiol 267:H2462–H2470

Wang YL, Wang XH, Liu YL, Kong XQ, Wang LX (2009) Cardiac lymphatic obstruction impairs left ventricular function and increases plasma endothelin-1 and angiotensin II in rabbits. Lymphology 42:182–187

Wang B, Wu L, Chen J, Dong L, Chen C, Wen Z, Hu J, Fleming I, Wang DW (2021) Metabolism pathways of arachidonic acids: mechanisms and potential therapeutic targets. Signal Transduct Target Ther 6:94

Warne CM, Essajee SI, Tucker SM, Figueroa CA, Beard DA, Dick GM, Tune JD (2023) Oxygen-sensing pathways below autoregulatory threshold act to sustain myocardial oxygen delivery during reductions in perfusion pressure. Basic Res Cardiol 118:12
Watkinson WP, Foley DH, Rubio R, Berne RM (1979) Myocardial adenosine formation with increased cardiac performance in the dog. Am J Phys 236:H13–H21
Wearn JT et al (1928) The extent of the capillary bed of the heart. J Exp Med 47:273–290
Wiley KE, Davenport AP (2001) Nitric oxide-mediated modulation of the endothelin-1 signalling pathway in the human cardiovascular system. Br J Pharmacol 132:213–220
Woollard HH (1926) The innervation of the heart. J Anat Physiol 60:345–373
Xu X, Lin H, Lv H, Zhang M, Zhang Y (2007) Adventitial lymphatic vessels – an important role in atherosclerosis. Med Hypotheses 69:1238–1241
Yada T, Hiramatsu O, Kimura A, Goto M, Ogasawara Y, Tsujioka K, Yamamori S, Ohno K, Hosaka H, Kajiya F (1993) In vivo observation of subendocardial microvessels of the beating porcine heart using a needle-probe videomicroscope with a CCD camera. Circ Res 72:939–946
Yada T, Shimokawa H, Hiramatsu O, Kajita T, Shigeto F, Goto M, Ogasawara Y, Kajiya F (2003) Hydrogen peroxide, an endogenous endothelium-derived hyperpolarizing factor, plays an important role in coronary autoregulation in vivo. Circulation 107:1040–1045
Yanagisawa M, Kurihara H, Kimura S, Goto K, Masaki T (1988) A novel peptide vasoconstrictor, endothelin, is produced by vascular endothelium and modulates smooth muscle Ca2+ channels. J Hypertens Suppl 6:S188–S191
Young MA, Knight DR, Vatner SF (1987) Autonomic control of large coronary arteries and resistance vessels. Prog Cardiovasc Dis 30:211–234
Young MA, Knight DR, Vatner SF (1988) Parasympathetic coronary vasoconstriction induced by nicotine in conscious calves. Circ Res 62:891–895
Yuan Y, Granger HJ, Zawieja DC, Chilian WM (1992) Flow modulates coronary venular permeability by a nitric oxide-related mechanism. Am J Phys 263:H641–H646
Yuan Y, Chilian WM, Granger HJ, Zawieja DC (1993) Permeability to albumin in isolated coronary venules. Am J Phys 265:H543–H552
Yuan Y, Mier RA, Chilian WM, Zawieja DC, Granger HJ (1995) Interaction of neutrophils and endothelium in isolated coronary venules and arterioles. Am J Phys 268:H490–H498
Yudilevich D, Martindejulian P (1965) Potassium, sodium, and iodide transcapillary exchange in the dog heart. Am J Phys 208:959–967
Zalba G, San Jose G, Moreno MU, Fortuno MA, Fortuno A, Beaumont FJ, Diez J (2001) Oxidative stress in arterial hypertension: role of NAD(P)H oxidase. Hypertension 38:1395–1399
Zhang C, Hein TW, Kuo L (2000) Transmural difference in coronary arteriolar dilation to adenosine: effect of luminal pressure and K(ATP) channels. Am J Physiol Heart Circ Physiol 279:H2612–H2619
Zhang C, Hein TW, Wang W, Kuo L (2003) Divergent roles of angiotensin II AT1 and AT2 receptors in modulating coronary microvascular function. Circ Res 92:322–329
Zhang C, Brzezinska AK, Chilian WM (2004) Hydrogen peroxide is a coronary vasodilator linking myocardial metabolism to coronary blood flow. Circulation 110:72–74
Zhang C, Xu X, Potter BJ, Wang W, Kuo L, Michael L, Bagby GJ, Chilian WM (2006) TNF-alpha contributes to endothelial dysfunction in ischemia/reperfusion injury. Arterioscler Thromb Vasc Biol 26:475–480
Zhao G, Joca HC, Nelson MT, Lederer WJ (2020) ATP- and voltage-dependent electro-metabolic signaling regulates blood flow in heart. Proc Natl Acad Sci USA 117:7461–7470

Chapter 9
Ocular Circulation

Leopold Schmetterer and Munirah Ismail

Abstract The eye has a complex vascular supply with the retinal circulation oxygenating the inner retina including the ganglion cells and the choroidal circulation oxygenating the outer retina including the photoreceptors. Over the last 50 years, significant progress has been made for better understanding the physiology of ocular circulation including autoregulation and neurovascular coupling. Studying the microvasculature in vivo is easier than in other organs because one has a direct view on the vessels through the transparent ocular media. As such, techniques have been developed that are based on optical imaging allowing to study several aspects of retinal and choroidal blood flow. With the commercialization of optical coherence tomography angiography, there is a renewed interest in the role of retinal and choroidal vasculature in ocular disease, because the technique is widely available to clinicians. This leads to a better understanding of the role of perfusion abnormalities in diseases such as glaucoma, age-related macular degeneration, and diabetic retinopathy.

Keywords Retinal blood flow · Choroidal blood flow · Microvasculature · Ocular imaging · Retina · Glaucoma

---

L. Schmetterer (✉)
Singapore Eye Research Institute, Singapore, Singapore

SERI-NTU Advanced Ocular Engineering (STANCE), Singapore, Singapore

Academic Clinical Program, Duke-NUS Medical School, Singapore, Singapore

School of Chemical and Biomedical Engineering, Nanyang Technological University, Singapore, Singapore

Department of Clinical Pharmacology, Medical University Vienna, Vienna, Austria

Center for Medical Physics and Biomedical Engineering, Medical University Vienna, Vienna, Austria

Institute of Molecular and Clinical Ophthalmology, Basel, Switzerland
e-mail: leopold.schmetterer@medunwien.ac.at

M. Ismail
Singapore Eye Research Institute, Singapore, Singapore

R. F. Tuma et al. (eds.), *Handbook of Microcirculation*, Physiology in Health and Disease, https://doi.org/10.1007/978-3-031-99609-2_9

9.1 The Vascular System of the Eye

The vasculature of the eye, which delivers oxygen and metabolites, is responsible for waste material removal and ensures thermoregulation of the retina (Parver et al. 1980). The pumping action of the heart establishes a pressure gradient from the arteries to the arterioles, the capillaries, the venules, and the veins. The arterioles are the main site of vascular resistance, and accordingly, these vessels are the major site of pressure drop within the vascular system. This comes with a conversion from pulsatile flow into steady flow even though blood flow in capillaries and veins shows some degree of pulsatility.

In humans, the retina receives vascular supply from two vascular beds that branch off from the ophthalmic artery. The ophthalmic artery does, however, also supply other tissues including the meninges, face, and upper nose. The inner retina, including the retinal ganglion cells and the retinal nerve fiber layer, obtains its oxygen supply via the retinal circulation. The outer retina is avascular in healthy subjects. It is supplied by the choroid via diffusion of oxygen. The choriocapillaris layer is located within approximately 30 μm below the retinal pigment epithelium (RPE) to allow for sufficient amount of oxygen to reach the photoreceptors.

The retinal vessels are branches of the central retinal artery (CRA), which runs inferior to the optic nerve within its dural sheath. The diameter of the CRA is in the order of 150 to 500 μm (Dorner et al. 2002a; Pemp et al. 2013; Baldoncini et al. 2019). The CRA and its branches are functionally arterioles and therefore constitute the major site of vascular resistance. These vessels are, however, usually referred to as arteries, and we follow this terminology in this review paper. The retinal vasculature is visible on fundus photographs with arteries appearing less bright than veins because of their higher proportion of oxygenated hemoglobin (Fig. 9.1). In approximately 25% of humans, a cilioretinal artery is present, which provides additional vascular supply to the inner retina. This artery belongs anatomically to the posterior ciliary artery system, and it branches either from the choroid or directly from a posterior ciliary artery. In most cases, this vessel is located at the temporal edge of the optic nerve head.

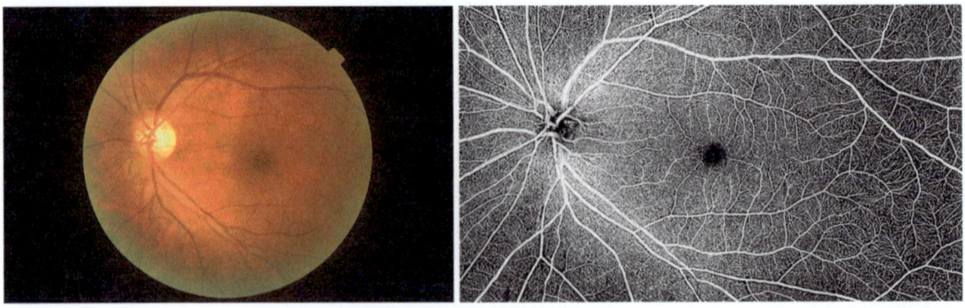

Fig. 9.1 Color fundus photography and OCT angiography in a healthy subject showing the retinal vasculature

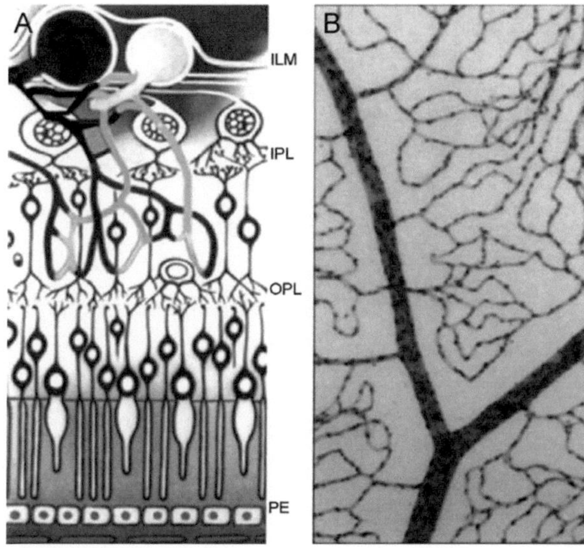

Fig. 9.2 Capillary plexus in the inner retina. (**a**) Schematic representation. (**b**) Tryptic digest of a flatmounted monkey retina, stained with hematoxylin and eosin. Note the broad capillary-free zone around the artery. (Reproduced with permission from Pournaras et al. 2008)

In Fig. 9.1, retinal arteries can be seen surrounded by a capillary-free zone that can be visualized in vivo by OCTA and adaptive optics scanning laser ophthalmoscopy (AOSLO) (Arthur et al. 2019; Balaratnasingam et al. 2018). In nonhuman primates, histological studies indicate that this zone is in the order of 30 and 70 μm (Fig. 9.2) (Okada and Ohta 1994). The tissue adjacent to the retinal branch arteries is oxygenated directly from these larger vessels. The central retinal vein (CRV) runs parallel to the CRA within the optic sheath. It drains the retinal blood into the cavernous sinus. The retinal vasculature is an end organ and as such all the blood delivered via the arterial system is drained via the venous system. The branching pattern of retinal veins shows similarities with the arterial angioarchitecture as seen from Fig. 9.1. The retinal capillaries have a diameter between 3.5 and 6 μm. Their walls contain endothelial cells, pericytes, and basement membranes. In contrast to other vascular beds, the retina does not contain precapillary sphincters. Hence, the retinal capillaries show continuous perfusion. The retinal capillary network is arranged in four layers (Fig. 9.3). The superficial plexus is located in the nerve fiber and ganglion cell layer. It contains the larger retinal branch arteries and veins that run in parallel to the nerve fibers as well as the microvessels, which form a dense capillary network. The intermediate plexus contains tortuous capillaries that are arranged in a lobular way (Chan et al. 2012). This plexus is located in the inner nuclear layer (INL) that is in proximity to the amacrine cells. The deep vascular bed consists of dense capillary vessels that are arranged in a complex network and located at the outermost side of the INL (Chan et al. 2012). The radial peripapillary capillaries are located close to the nerve fiber layer and are limited to the region around the optic nerve head nerve head (Henkind 1967; Yu et al. 2014). This is functionally required, because the nerve fiber bundles are thick in this region and have high metabolic demand to maintain ion pumping (Snodderly et al. 1992). The

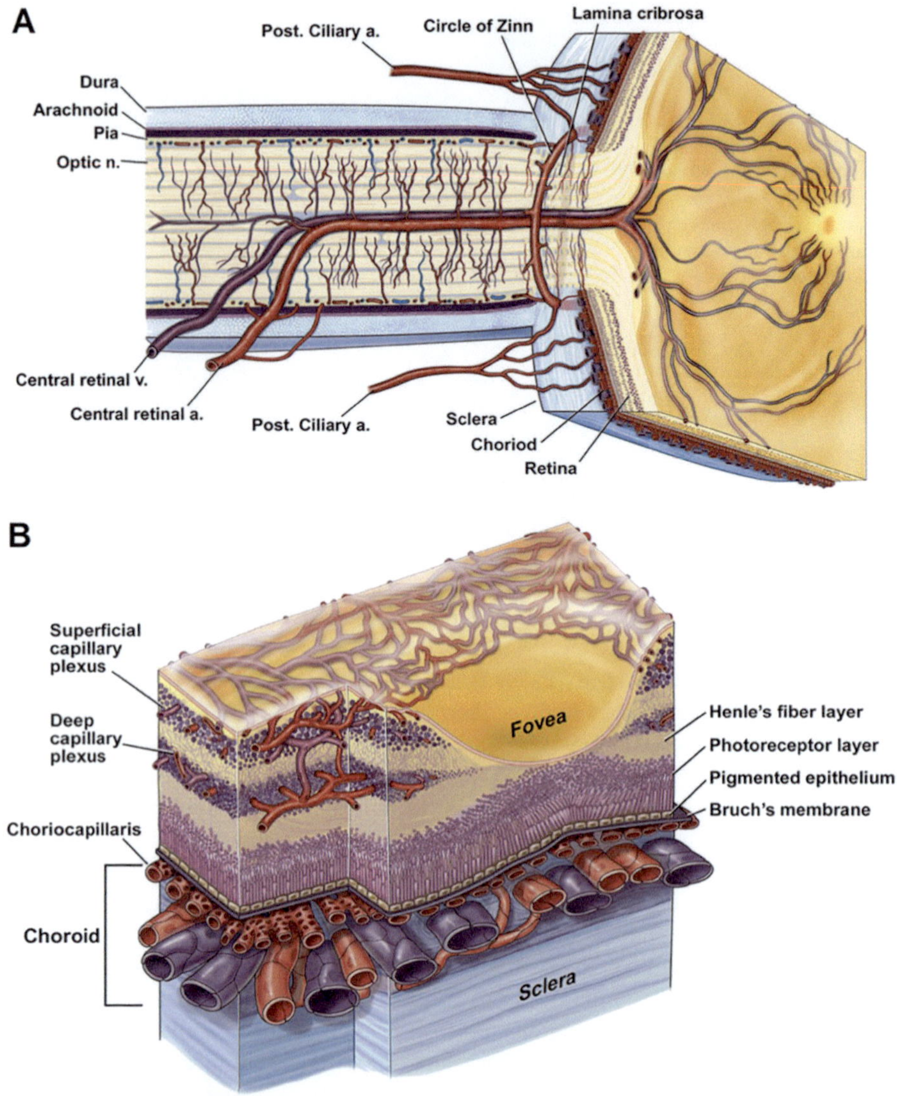

Fig. 9.3 Anatomy of ocular circulation (a—artery, b—vein, n—nerve). (**a**), Cut away drawing along the superior-inferior axis of the human eye through the optic nerve, showing the vascular supply to the retina and choroid. (**b**), Drawing showing vasculature of the retina and choroid. (Reproduced with permission from Anand-Apte and Hollyfield 2009)

superficial and deep networks are connected by anastomotic capillaries that branch at angles close to 90 degrees.

Retinal vessels have 1 to 7 layers of smooth muscle cells that control vascular tone. The capillaries of the retina show a remarkably high density of pericytes. Traditionally, a ratio of 1:1 between pericytes and endothelial cells has been reported

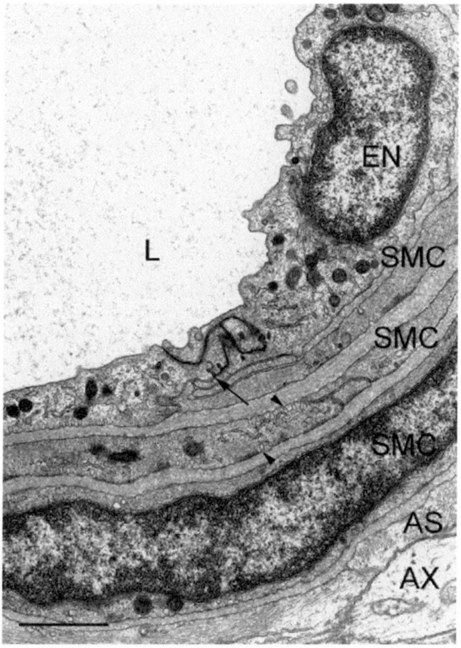

Fig. 9.4 Inner blood–retinal barrier. Electron micrograph of an artery in the inner pig retina. Tissue was fixed in the presence of tannic acid to enhance fluid filling of the intercellular endothelial cleft. The vessel wall is composed of the endothelium (EN) and three layers of smooth muscle cells (SMC), with numerous caveolae (arrowheads). Astrocytic processes (AS) enwrap the vessel. Note high contrast in intercellular cleft and vesicular pits up to the site, where a tight junctional element is located (arrow). AX, axon. Bar, 1 mm. (Reproduced with permission from Pournaras et al. 2008)

(Lutty et al. 2012). A recent study of the human retina, however, showed a 94.5% frequency of pericyte coverage on human retinal capillaries (Chan-Ling et al. 2011). In other organs, the endothelial to pericyte ratio in normal tissues vary between 1:1 and 10:1, while pericyte coverage of the endothelial abluminal surface is typically between 70% and 10% (Geevarghese and Herman 2014). Endothelial cells are lined in parallel to the blood vessels. The blood vessels share their basement membrane with pericytes and endothelial cells (Fig. 9.4). The endothelial cells have tight-cell junctions, which form the blood–retinal barrier.

The choroidal circulation gets its main vascular supply from the long and short ciliary arteries (CSAs). This is part of the uveal system that also supplies the iris and ciliary body. The number of short posterior CSAs shows high interindividual variability typically ranging between 6 and 12. They penetrate the sclera in the proximity of the optic nerve and then branch to form the arterioles and capillaries. The choroid consists of five layers including Bruch's membrane, three vascular layers as well as the suprachoroidea. The outermost vascular layer is Haller's layer that includes the largest choroidal arteries and veins. Sattler's layer contains smaller arteries and veins. Anatomically there is no clear border between Haller's and Sattler's layer. The choriocapillaris consists of highly anastomosed capillaries and is between 7 and 10 μm thick. The choriocapillaris is arranged in a lobular arrangement. It is thickest in the macular area with capillary diameters in the order of 16 to 20 μm. In the equatorial region, the vessel diameters become wider (20–50 μm), but vessels are sparser. The choriocapillaris gradually thins toward the equatorial region. These capillaries are fenestrated, which allows for molecular

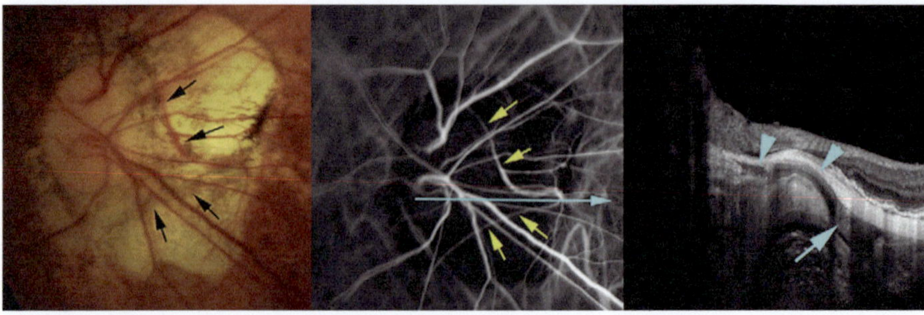

Fig. 9.5 Detection of the path from the retrobulbar short posterior ciliary artery to the Zinn–Haller ring by optical coherence tomography (OCT). (Left) Fundus photograph of the left eye of a 55-year-old man with high myopia. A large temporal conus is observed. Blood vessels suggesting the Zinn–Haller ring are seen within the conus (arrows). (Middle) Indocyanine green angiographic finding at 1 minute after dye injection showing the temporal part of the Zinn–Haller ring with a triangular shape (arrows). A lateral short posterior ciliary artery enters the Zinn–Haller ring at its most horizontally distant point. A blue line shows the scanned line by OCT. (Right) Horizontal OCT scan along the blue line in the middle image shows that the short posterior ciliary artery enters the sclera (arrow) and courses toward the Zinn–Haller ring intrasclerally (along the arrowheads). (Reproduced with permission from Ohno-Matsui et al. 2013)

transport between the retina and choroid. The choriocapillaris lobules drain into venules that join the larger venules of the outer conduit layer. The 4–5 vortex veins penetrate the sclera at the equator.

The optic nerve head has its own complex vascular supply that shows significant interindividual variability in humans (Hayreh 2001). In most cases, the surface layer of the ONH is supplied by the CRA and, as such, is part of the retinal vascular system. In cases with a cilio-retinal artery, the corresponding sector of the surface layer usually obtains its blood supply from this vessel. The prelaminar region is located below the surface nerve fiber layer, which extends toward the lamina cribrosa. Deeper layers of ONH vasculature supply the lamina cribrosa as well as the retrolaminar regions. The pre- and postlaminar vessels receive their blood supply from the PCAs and the peripapillary choroid. In majority of cases, the medial and lateral short posterior ciliary arteries form the circle Zinn and Haller around the optic nerve, which can be seen ophthalmoscopically in some subjects, particularly in those with high myopia (Fig. 9.5).

The vasculature of the anterior segment gets its main supply from the anterior ciliary arteries, but the long posterior ciliary arteries and anastomotic connections from the anterior choroid contribute to this supply as well. The anterior ciliary arteries (usually two) travel in parallel to the extraocular muscles pierce the sclera near the posterior pole and then travel anteriorly between the sclera and choroid to join the major arterial circle of the iris. The major arterial circle of the iris branches off into smaller arteries to supply the iris and ciliary body. The venous drainage from the anterior segment is also directed into the vortex veins. The cornea is not vascularized in healthy humans. The limbus, however, contains a ring of vasculature that contributes to corneal oxygenation. Whereas the conjunctiva is richly vascularized, the sclera contains only a few blood vessels, because of its low metabolic demand. The anterior uveal system includes the iris and the ciliary body. The principal blood supply of the iris is the major arterial circle, which enters the iris through its root. The iris blood vessels are radially oriented. The ciliary body is also supplied by the major arterial circle, which branches into the anterior parts of the ciliary processes and forms a capillary bed.

9.2 Oxygenation and Blood Flow of the Retina

There are fundamental differences in the oxygenation of the inner and outer retina. The outer retina is avascular and the photoreceptors depend on oxygen supply from the choroid, which is delivered via diffusion. The inner retina including the two nuclear layers, the two plexiform layers, and the nerve fiber layer receive oxygen mainly from the retinal vessels (Linsenmeier and Zhang 2017). Oxygen profiles in the retina differ between light and dark-adapted conditions. In the dark-adapted retina, a trough in oxygen tension (PO_2) is seen at approximately 80% retinal depth in a variety of species (Fig. 9.6). This is related to the high oxygen consumption of photoreceptor inner segments. Since gradients in the PO_2 curve from both sides can be observed, both the choroidal and retinal circulation contribute to photoreceptor oxygenation with most of the oxygen diffusing from the choroid. In the light-adapted retina, the retinal oxygen profile looks different (Fig. 9.6). While there is little change in choroidal PO_2 (Linsenmeier and Braun 1992; Braun and Linsenmeier 1995), the minimum PO_2 in the outer retina is higher than in the dark-adapted retina. The contribution of the retinal circulation to photoreceptor oxygenation is minimal or completely absent in the light-adapted retina.

In the inner retina, the average PO_2 in the deeper region is slightly lower than in the very superficial region in dark-adapted animals (Braun and Linsenmeier 1995). This difference between the deeper region and superficial region disappears when the retina is light adapted (Braun and Linsenmeier 1995), because some choroidal O_2

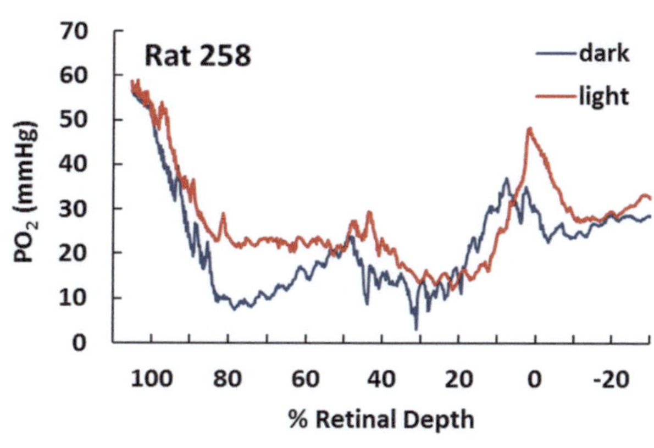

Fig. 9.6 Dark-adapted (blue) and light-adapted (red) retinal PO2 profiles from a Long–Evans rat. (Reprinted with permission from Linsenmeier and Zhang 2017)

diffuses from the choroid toward the inner retina. In contrast to the outer retina, inner retinal PO_2 is almost independent of the illumination level of the retina (Linsenmeier and Zhang 2017). Since the oxygen consumption of the vitreous is low, vitreal PO_2 close to the retina is a good indicator of inner retinal PO_2 and is easier to measure. However, this may not be true in pathological conditions such as diabetic retinopathy.

The photoreceptors also consume glucose via anaerobic glycolysis leading to a considerable accumulation of lactate (Wrinkler 1995). In contrast, anaerobic glycolysis is absent in the inner retina. Under light-adapted conditions, the photoreceptors are solely oxygenated by the choroid. In darkness, however, the PO_2 reaches 0 mmHg at the proximal side of the inner segments, which means a change in the gradient of PO_2 toward inner segment. This implies that, in darkness, some of the oxygen to the photoreceptors is delivered from the retinal circulation (Linsenmeier and Zhang 2017). Photoreceptors consume significant amount of oxygen that is mainly used for maintenance of the dark current, and the generation of guanosine triphosphate (Haugh-Scheidt et al. 1995).

Akin to any other vascular bed, blood flow in the eye is determined by ocular perfusion pressure (OPP) and vascular resistance (R). As such, volumetric blood flow can be written according to the Hagen–Poiseuille equation as $Q = OPP/R$. R increases linearly with length of the vessel segment and the viscosity of blood and is inversely related to the fourth power of the radius of the vessel. The equation is based on a model of incompressible and Newtonian fluids resulting in laminar flow. Whereas the Hagen–Poiseuille equation may be well applicable to flow in the larger retinal vessels, it does, however, not apply to microvascular flow because it is nonlaminar and the viscosity of blood changes with the diameter of the vessel. There is a pronounced decrease in viscosity as the vessel's diameter decreases to ensure energy-efficient perfusion (Fahraeus–Lindqvist effect). The effective viscosity in retinal microvessels is dependent on the number of erythrocytes and leukocytes and may be increased in conditions with increased leukocyte counts (Lasta et al. 2012a; Told et al. 2012, 2014).

Retinal blood flow is characterized by a high vascular resistance and accordingly, blood flow rate is low. Using noninvasive technology, the retinal blood flow rate has been measured to be approximately 40 μL/min in healthy subjects (Riva et al. 1985; Doblhoff-Dier et al. 2014; Garhofer et al. 2012; Tan et al. 2015; Baumann et al. 2011; Haindl et al. 2016). The diameter of larger retinal branch veins and arteries can be up to 200 μm, which makes the vessel functionally venules and arterioles. There is an almost linear relationship between retinal vessel diameter and retinal blood velocities in both arteries and veins (Riva et al. 1985; Garhofer et al. 2012) as predicted from Murray's law (Fig. 9.7). Blood velocities in retinal arteries are, however, higher than in veins, because of the equation of continuity. Blood velocities in larger retinal vessels are in the order of 6 to 25 mm/s. In the capillaries, velocities are in the range between 1 and 4 mm/s and exhibit quite significant spatial and temporal variations (Warner et al. 2021). Recently, resistance and blood flow in the retina have been modeled using retinal vessel calibers, fractal dimension, perfusion pressure, and population-based hematocrit values (Pappelis et al. 2020). Retinal

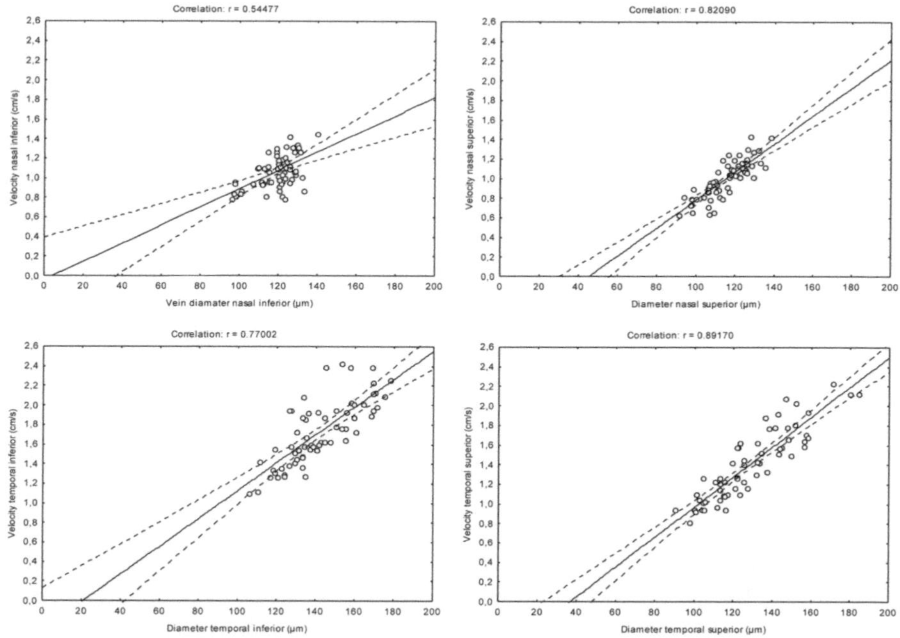

Fig. 9.7 Linear correlation analysis between mean blood velocities and vessel diameters for the largest venules in each quadrant in healthy subjects ($P < 0.001$ each). Data are shown separately for the nasal inferior, nasal superior, temporal inferior, and temporal superior quadrant ($n = 64$). (Reprinted from with permission from Garhofer et al. 2012)

blood flow decreases with age, which is most likely due to an age-related loss of retinal ganglion cells and reduced metabolic demand (Bata et al. 2019a). Oxygen extraction in retinal vessels is accordingly low. The arteriovenous oxygen difference is high, in the order of 35–45% (Lasta et al. 2012b; Geirsdottir et al. 2012; Jani et al. 2014; Stefánsson et al. 2019; Wang et al. 2022a). Total retinal oxygen extraction is accordingly in the range of 2.0 to 2.5 µL (O_2)/min (Werkmeister et al. 2015; Fondi et al. 2017). Choroidal blood flow is much higher than retinal blood flow because the vascular resistance is low. In humans, there is no technique to measure absolute choroidal blood flow. Data in nonhuman primates indicate that the absolute flow rate is approximately 600 µL/min (Alm and Bill 1973). In humans, pulsatile ocular blood flow has been estimated as 600–1000 µL/min (Yang et al. 1997; Silver et al. 1989; Berisha et al. 2010), but the ratio between pulsatile and nonpulsatile flow is unknown. The arteriovenous O_2 difference cannot be measured in humans, but it is considered to be very small with a value of approximately 1% as measured in cats (Alm and Bill 1972). There is some evidence for gender differences in ocular blood flow, but the evidence for this is inconclusive (Schmidl et al. 2015a).

9.3 Techniques for Measuring Blood Flow

Measurement of blood flow in the human eye is not an easy task. Multiple techniques have been developed, but none of the techniques has gained wide-spread clinical application for measuring blood flow (Schmetterer and Garhofer 2007; Schmetterer et al. 2012). In recent years, OCT angiography and laser speckle flowgraphy have become commercially available, but only the former has gained widespread clinical use. Some of the older historical methods will only be mentioned briefly in this article.

9.3.1 Invasive Technology in Experimental Animals

9.3.1.1 Microsphere Technology

The microsphere method for measuring blood flow can be used in multiple organs and absolute values of regional blood flow can be obtained. When microspheres are injected into the systemic circulation, the number of entrapped microspheres is proportional to the blood flow through the tissues. In the eye, plastic radioactive microspheres (Granstam and Granstam 1999; Nilsson and Alm 2012), fluorescent microspheres (Nork et al. 2006; Shih et al. 2013), and colored microspheres have been used (Wang et al. 2007a; Told et al. 2016). Whereas in principle many organs can be studied simultaneously, reliable measurements of local blood flow require optimization of the size and number of microspheres. This also makes it difficult to study retina and choroid in the same experiment, because of the differences in capillary lumen. As with all other methods in experimental animal studies, care has to be taken such that the anesthesia and the experimental procedures do not influence the results. Anesthetics can either influence the results directly because of the effects of the anesthetic drugs on vascular tone and/or vascular regulation including neuronal input, but also via the fluctuations of arterial blood pressure during the course of the experiments. An obvious advantage of the microsphere method is the lack of surgical intervention. However, the animals cannot be followed longitudinally because the technique is terminal.

9.3.1.2 Hydrogen Clearance

Hydrogen clearance is based on the detection of the rate of clearance of an inert gas from the tissue of interest. This technique has been used in the eye by several investigators to study perfusion in different tissues (Yu et al. 1991; Stefánsson et al. 1988; Sugiyama et al. 1996). This technique is, however, not easily applied to the eye because the condition that the tissue has to be perfused uniformly in a three-dimensional volume is neither fulfilled for the choroid nor for the retina.

Hydrogen can either be administered via lung inhalation, intra-arterial injection or local routes. The hydrogen is detected via an intraocularly placed microelectrode that allows for the measurement of arrival and clearance of the gas. Placement of the microelectrodes can be done in proximity of either larger retinal vessels or vascularized tissue resulting in different clearance curves. The blood flow values are usually reported in units of mL/min/100 g tissue.

9.3.1.3 Oxygen-Sensitive Electrodes

Techniques that quantify oxygen can be divided into those that quantify the partial pressure of free oxygen in tissue (PO_2) and those that measure the saturation of oxygen bound to hemoglobin (SaO_2). SaO_2 and PO_2 are interrelated in a nonlinear way and can be interconverted except in conditions of very high oxygenation when SaO_2 is almost 100%. Oxygen-sensitive electrodes work based on polarography. When a noble metal is polarized at about −0.7 V relative to a reference electrode, it becomes a cathode where oxygen is chemically reduced (Linsenmeier and Zhang 2017). This is associated with a current that is proportional to the PO_2 in the tissue. Different oxygen electrodes have been used in the vitreous humor (Alder et al. 1991; Pournaras et al. 1990; Yu et al. 1990), the retina (Alder et al. 1991; Birol et al. 2007; Pournaras et al. 1989), and the optic nerve head (Ahmed et al. 1994; Cour et al. 2000; Pedersen et al. 2006). The technique provides high spatial resolution for measuring tissue PO_2.

9.3.2 *Noninvasive Technology in Humans*

9.3.2.1 Dye-Based Angiography

Fluorescein and indocyanine green angiography are important tools for visualizing blood flow in clinics. There have been multiple approaches to quantify retinal and choroidal blood flow based on this technique (Ben-Sira et al. 1973; Jung et al. 1983; Wolf et al. 1994; Duijm et al. 1996). The most advanced approach is to combine video fluorescein angiography based on scanning laser ophthalmoscopy with digital image processing (Gabel et al. 1988; Arend et al. 1991). Early studies quantified retinal perfusion by dye dilution technique, but time resolution was low with images captured at a frame rate of 1 to 2 Hz (Hitchings and Spaeth 1977; Riva et al. 1978). Parameters obtained were arm-retina time and the arteriovenous passage time (Hitchings and Spaeth 1977; Riva et al. 1978), but only semiquantitative measurements of perfusion were obtained. More advanced approaches use arteriovenous passage time and mean dye velocity using the early phase of the angiograms (Jung et al. 1983; Preußner et al. 1983). Another approach is to directly visualize retinal capillary microcirculation in perifoveal capillaries (Funatsu et al. 2006; Noma et al. 2009) providing information on the macular microcirculation. Whereas dye-based angiography still has an important role in clinical decision-making, quantification of blood flow based on the technology has lost its importance. The invasive nature of dye administration and the potential risks associated with it makes it difficult to apply the technology for research purposes.

9.3.2.2 Measurement of Retinal Vessel Diameters

Measuring retinal vessel diameters or retinal vessel calibers has been widely used as biomarkers for both ocular and systemic disease (Sun et al. 2009; Ikram et al. 2013a; Guo et al. 2020). There is no clear definition of these parameters since different techniques may measure different anatomical features. This can either include the vessel wall, can refer to the diameter as measured from the two inner sides of the vessel walls or can be defined as the blood column covered by erythrocytes excluding the erythrocyte-free plasma zone close to the vessel wall. Retinal vessel diameters can be extracted from any high-quality fundus images. Parameters that are reported include central retinal arterial equivalent (CRAE), central retinal venular equivalent (CRVE), and arteriovenous ratio (Fig. 9.8). Several formulas for summarizing retinal vessel diameters have been proposed (Knudtson et al. 2003; Heitmar et al. 2015). A commercial fundus-camera based system is available, specifically targeting retinal vessel diameter analysis as well as the response of retinal vessels to flicker light stimulation (Garhofer et al. 2010; Heitmar et al. 2010). In terms of blood flow, the information is limited as no information on retinal blood velocity is available. In patients with systemic hypertension, for instance, CRAE is reduced indicating arteriolar vasoconstriction but CRVE is increased which is most likely due to inflammation (Ding et al. 2014). Whether hypertension is associated with altered volumetric blood flow based on these measurements is, however, unclear. Due to the equation of continuity, it is, however, obvious that the ratio of retinal blood velocity between arteries and veins is shifted toward higher values, because the retina is an end organ and total arterial blood flow equals total retinal venous blood flow.

9.3.2.3 Laser Doppler Velocimetry

The technique is based on the optical Doppler effect. Laser light that hits moving erythrocytes is shifted in frequency and this frequency shift is dependent on the velocity of the scatterers. First measurements in retinal vessels were performed in the 1970s to measure the blood velocity (Riva et al. 1972; Tanaka et al. 1974). Since the technique does not provide any depth resolution, a continuous spectrum of Doppler shifts is detected because of the velocity profile within the vessel. The maximum Doppler shift corresponds to the maximum flow velocity in the center of the vessel. Quantification of blood velocity is based on the assumption of a parabolic velocity profile, which is fulfilled in case of laminar flow, but not at vessel bifurcations. Absolute measurement of retinal blood velocity is, however, difficult because

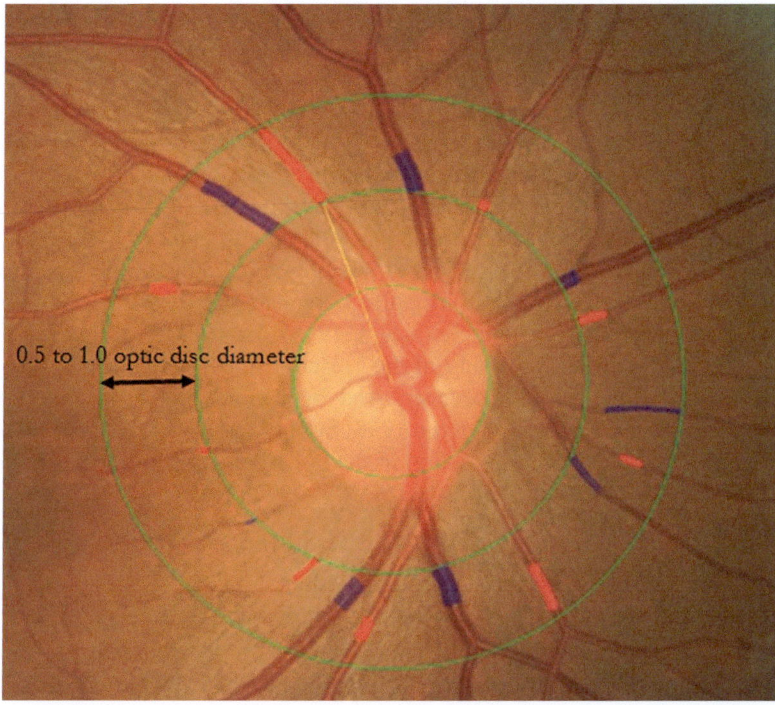

Fig. 9.8 Retinal fundus photograph assessed quantitatively by the IVAN software (University of Wisconsin). The measured area of retinal vascular parameters is standardized as the region from 0.5 to 1.0 disc diameters away from the disc margin. Retinal arteriolar (red) and venular (blue) calibers are summarized as CRAE and CRVE, respectively, from retinal fundus photograph. CRAE and CRVE were defined based on the revised Knudtson–Parr–Hubbard formula. (Reproduced with permission from Ikram et al. 2013b)

the angle between the laser beam and the retinal vessel needs to be known. One way to overcome this problem is to detect the scattered laser light in two distinct directions, a technique called bidirectional laser Doppler velocimetry (Riva et al. 1979, 1981). Combing this technology with measurement of vessel diameter provided the first results for total retinal blood flow in humans (Riva et al. 1985).

9.3.2.4 Laser Doppler Flowmetry

The technique is related to laser Doppler velocimetry, but in contrast to single blood vessel, vascularized tissue is illuminated. Based on a scattering theory of light in tissue, average blood velocity, blood volume and blood flow can be derived (Bonner and Nossal 1990). The technology has been used to study optic nerve head (Riva et al. 1992) and subfoveal choroidal (Riva et al. 1994a) blood flow in humans. No absolute values of blood flow can be obtained and as such, the method is mainly suitable for measuring blood flow changes in response to stimuli. The sampling depth of laser Doppler flowmetry is difficult to determine in vivo. Studies in nonhuman primates indicate that laser Doppler flowmetry technique mainly samples blood flow changes in the superficial layers of the optic nerve head and is less influenced by perfusion in deeper regions (Riva et al. 1994a). When the laser beam is directed toward the fovea, the signal arises from the choroidal microvasculature, because the retina lacks vessels in this region. When the laser beam is directed toward other retinal locations, the signal is also largely influenced by the choroidal perfusion (Polska et al. 2002). Mapping of the microcirculation of the human retina can be achieved by scanning laser Doppler flowmetry based on a scanning laser ophthalmoscope (Michelson and Schmauss 1995), but this technique is highly sensitive to motion and the scanning of the beam itself can induce a Doppler shift. Laser Doppler flowmetry has also been used by various groups in experimental animals (Kiel and Shepherd 1992; Koss 1998; Strohmaier et al. 2011), and much of our knowledge on blood flow regulation in the eye stems from these experiments.

9.3.2.5 Color Doppler Imaging

Color Doppler imaging is based on the acoustical Doppler effect. It combines B-scan ultrasonography for structural imaging with velocity extraction based on the Doppler shift of the wave (Williamson and Harris 1996; Dimitrova and Kato 2010; Stalmans et al. 2011). In the eye, this technique is mainly used for the visualization of blood velocities in the following retrobulbar vessels: ophthalmic artery (OA), posterior ciliary arteries (PCAs), and the central retinal artery (CRA) as well as other vessels including retrobulbar veins (Fig. 9.9). The typical frequencies of the ultrasound probes are in the order of 5 to 10 MHz to achieve sufficient penetration. Doppler shift of the ultrasound wave does not only depend on the blood velocity but also on the Doppler angle. The direction of the sampling gate therefore needs to be adjusted based on the anatomy of the artery. While this is usually achieved without problems in the CRA and the OA, it is more difficult in the PCAs. Peak systolic and end diastolic velocities (PSV, EDV) are extracted from the time slope of the Doppler shift. Mean flow velocity (MFV) can be calculated as the time mean of the spectral outline over a heart cycle. In addition, a resistance index (RI = (PSV − EDV)/PSV) can be calculated as a measure of vascular resistance distal to these vascular beds. A study in healthy subjects indicates, however, that this is a poor measure of vascular resistance in the retina (Polska et al. 2001). This technique is not capable of measuring volumetric blood flow because its resolution is not sufficient to resolve the vessel diameter with the required precision. Since the ultrasound probe is put on the closed eye lid, care has to be taken such that the intraocular pressure is not increased due to the force induced on the eye.

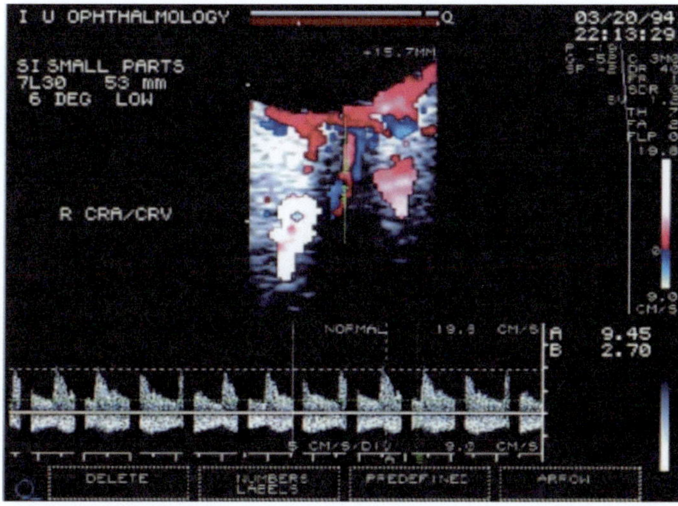

Fig. 9.9 Color Doppler image of the central retinal artery and vein taken with a 7.5 MHz linear probe (Siemens Quantum 2000 system). The Doppler shifted spectrum (time velocity curve) is displayed at the bottom of the image. Red and blue pixels represent blood movement toward and away from the transducer, respectively. (Reproduced with permission from Chung et al. 1999)

9.3.2.6 Laser Speckle Flowgraphy

Laser speckle flowgraphy measures blood flow using laser speckle statistics. Laser speckles arise from interference of coherent light at rough scatterers. The speckle pattern which appears under the illumination of laser irradiation can only be described statistically. When objects are moving, for instance red blood cells in the blood stream, the structure of the pattern varies. The higher the variation, the higher the velocity of the scatterers. Analysis of blood flow is based on autocorrelation techniques. Early approaches using laser speckle flowgraphy only allowed for semiquantitative analysis of perfusion (Briers and Fercher 1982). Later, instruments were developed that allow for quantification of blood flow at the posterior pole of the eye (Tamaki et al. 1994, 1995; Sugiyama et al. 2010). Nowadays, several commercial systems are available.

In most systems, a near infrared laser is used as light source coupled to a fundus camera and a digital charge-coupled device camera. The most widely used output parameter of laser speckle technology is mean blur rate (MBR), which is a measure of relative blood flow velocity and is expressed in arbitrary units (au). When a scan is started, 118 images are captured at a rate of 30 frames per second, resulting in a measurement time of 4 s. From the 118 acquired images, a color-coded map is calculated, which depicts the distribution of perfusion.

The system can either be used to study perfusion in vascularized tissue or to study blood flow in retinal vessels. For perfusion measurements, the ONH area is selected by positioning an ellipsoid region of interest at the ONH margin. Thereafter, larger

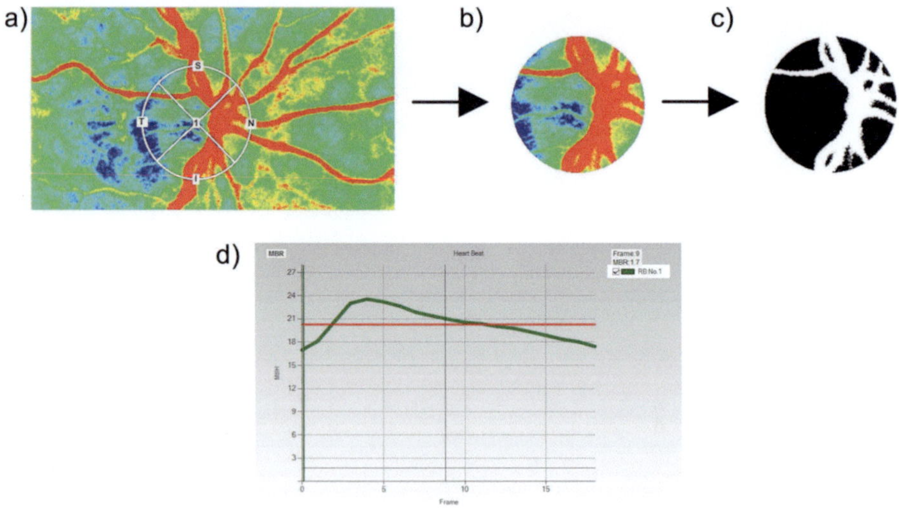

Fig. 9.10 Laser speckle flowgraphy (LSFG) scan of the optic nerve head (ONH) area. (**a**) The ONH margin is to be delineated using an ellipsoid region of interest with variable size and radii. (**b**) The selected area representing the entire optic disc is taken into account for further analysis using the "vessel extraction" function. (**c**) The "vessel extraction" function distinguishes between areas of visible surface vessels (white) and ONH tissue areas (black). (**d**) Typical pulse-waveform curve (green) with a steep incline during the systolic phase and a flatter decline during the diastolic phase. The red line indicates the mean level of mean blur rate (MBR). (Reproduced with permission from Luft et al. 2016a)

retinal vessels are automatically detected by thresholding according to their higher blood velocities. Thus, MBR can be determined for the total ONH area, for vessel area or for tissue area (Fig. 9.10).

It is obvious that the speckle contrast in the tissue is not only affected by the velocity of the red blood cells but also by the number of moving scatterers. Hence, there is a longstanding discussion whether the technique measures flow, velocity, or some quantity related to volumetric blood flow (Briers et al. 2013). A model relating three-dimensional microvascular blood flow in tissue to measured autocorrelation values is currently lacking. The technique has been validated in a variety of animal studies and it has been shown that measurements of laser speckle flowgraphy correlate well with values obtained using either the microsphere method or the hydrogen gas clearance method (Sugiyama et al. 1996; Aizawa et al. 2014). As with laser Doppler flowmetry, the penetration depth of the technique is not completely understood. When the laser is directed toward vascularized tissue of the retina, a sum signal between retinal and choroidal blood flow is measured. Animal data indicate that the retinal contribution is in the order of 10% (Isono et al. 1997), which is in agreement with data from laser Doppler flowmetry studies. The results may be influenced by pigmentation and as such, results between different ethnicities may not be comparable. The majority of data with this technology has been obtained in Japan (Kobayashi et al. 2022; Enomoto et al. 2021), but the technique was also successfully applied in Caucasians (Luft et al. 2016a). When measurements are done in retinal vessels, the correlation with data as obtained with Doppler OCT was not good, particularly at higher arterial flow rates because of saturation effects (Luft et al. 2016b).

9.3.2.7 Doppler OCT

Doppler OCT is a functional extension of OCT (Leitgeb et al. 2014). The first description of this technique was realized based on a time domain OCT system (Chen et al. 1997). In Fourier domain OCT, the Doppler shift of light can be quantified by analyzing the phase of the complex OCT signal (White et al. 2003; Leitgeb et al. 2003, 2004). Similar to laser Doppler velocimetry, measurement of blood velocity requires knowledge of the incident angle of the laser beam. Multibeam systems for the illumination of the retina were found to overcome this angle ambiguity (Haindl et al. 2016; Pedersen et al. 2007; Werkmeister et al. 2008; Dai et al. 2013; Szegedi et al. 2020a; Trasischker et al. 2013; Blatter et al. 2013a). Combining these measurements with diameter measurements of retinal vessels by either using fundus photography or phase tomograms of OCT, allows for measurement of total retinal blood flow (Doblhoff-Dier et al. 2014; Haindl et al. 2016; Hosseinaee et al. 2018; Fondi et al. 2016). Other approaches to overcome the angle ambiguity were also realized. The use of bi-directional Doppler cross sections perpendicular to the illumination plane allows for calculation of retinal blood flow independent of the Doppler angle (Blatter et al. 2013b). Dual-plane scanning patterns have also been used to determine the angle between blood flow and the scanning beam (Wang et al. 2007b). When using 3D Doppler data sets, en face images can be used to extract absolute blood velocity as well as total retinal blood flow without knowledge of the Doppler angles (Fig. 9.11) (Baumann et al. 2011; Tayyari et al. 2014). Another approach is to study retinal vessels at relatively high Doppler angle which is automatically detected by auto-alignment (Tani et al. 2017). Recently, it has been proposed to use the forward-scattering signal instead of the backward scattering signal because it is insensitive to vessel orientation (Nam et al. 2022).

Doppler OCT has also been used to reconstruct the velocity vector field from measured phase data allowing for the visualization of velocity profiles at retinal bifurcations (Fig. 9.12) (Aschinger et al. 2015). Line scanning protocols of multichannel OCT were employed to investigate venous pulsatile caliber oscillations and flow pulsatility (Wartak et al. 2019).

9.3.2.8 OCT Angiography

OCT angiography is a technique that is closely related to Doppler OCT (Spaide et al. 2018; Kashani et al. 2017; Ang et al. 2018a; Chen and Wang 2017). The first approach to noninvasively visualize the vasculature in the human eye based on

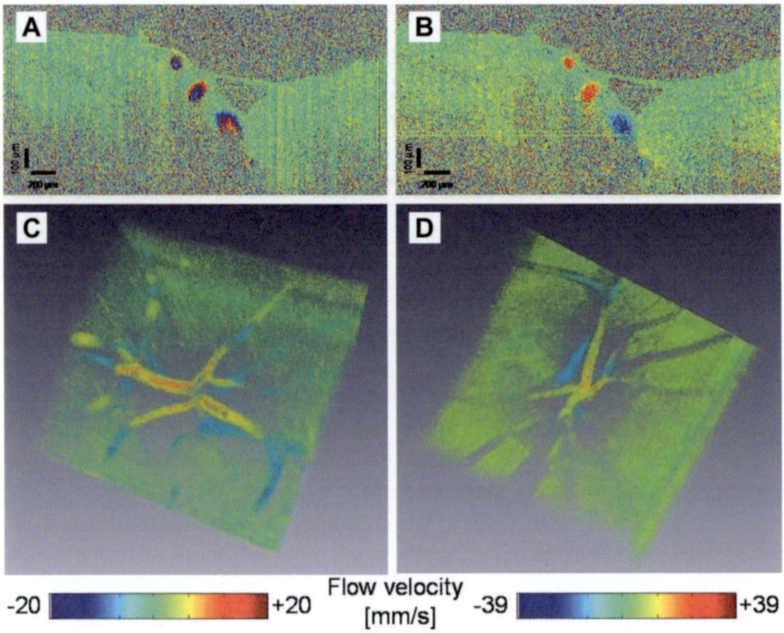

Fig. 9.11 Volumetric Doppler OCT imaging of retinal vasculature. (**a**) Doppler OCT B-scan image at 100 kHz close to the optic disk. (**b**) Corresponding Doppler OCT image at 200 kHz. Arterial and venous vessels in the papilla can be distinguished in red and blue color. Note that the color scale encodes the ranges of ±19.7 mm/s and ± 39.5 mm/s for 100 kHz and 200 kHz, respectively. 3D renderings of volumetric data sets at 100 kHz (**c**) and 200 kHz (**d**) show the three-dimensional structure of the retinal arteries and veins branching in the optic disk. (Reproduced with permission from Baumann et al. 2011)

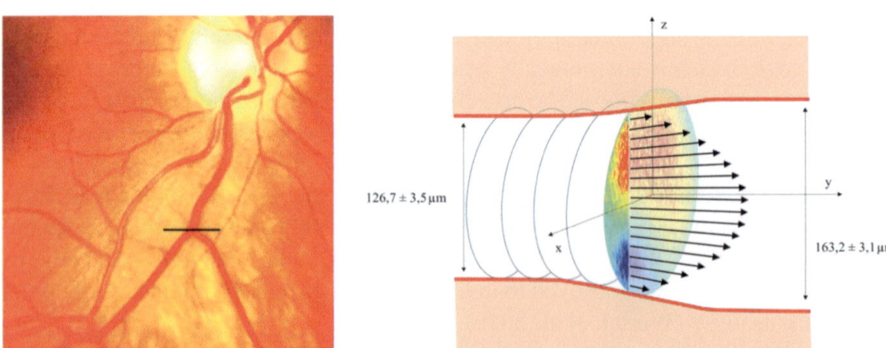

Fig. 9.12 Fundus image of a venous convergence site with the black bar marking the scanning position. Reconstructed velocity vector field at the scanned position of a convergence: the red line depicts the vessel wall geometry. (Adapted with permission from Aschinger et al. 2015)

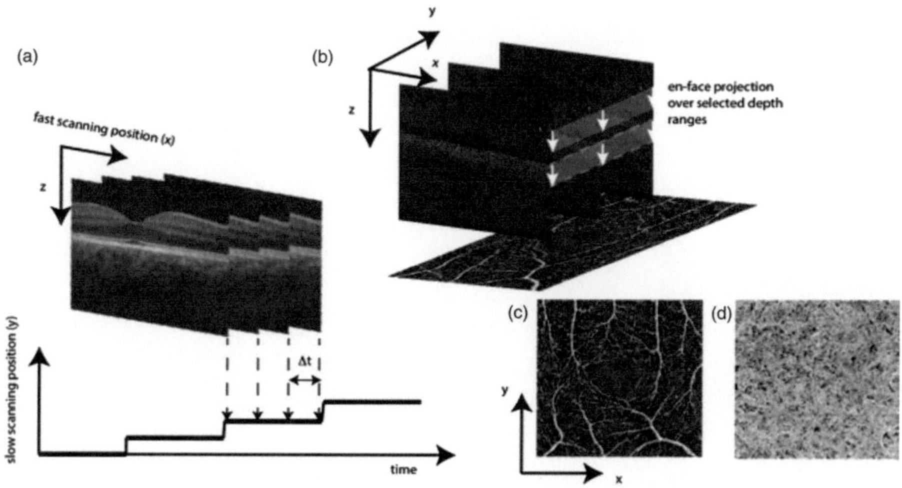

Fig. 9.13 (**a**) Recording scheme for OCT (OCT) angiography. (**b**) Calculated 3-D OCT angiography. (**c**) Taking the en face projection by, e.g., plotting the maximum intensity results in comprehensive en face angiography maps. (**d**) Dense choroidal vasculature obtained by integration over dashed depth region in (**b**). (Reproduced with permission from Wolfgang et al. 2014)

OCT was based on Doppler phase as well as variance or power of the phase analysis (Makita et al. 2006). Later, multiple algorithms and scan protocols were proposed based on logarithmic intensity and speckle contrast imaging. Split spectrum amplitude decorrelation (SSADA) improves signal to noise and bulk eye motion artifacts (Jia et al. 2012). Optical microangiography (OMAG) is another approach that includes both amplitude and phase of the OCT signal (An and Wang 2008).

Nowadays, repeated B-scans are used to detect motion contrast thereby visualizing the ocular vasculature (Fig. 9.13). Nonmoving tissues will not produce an autocorrelation signal in repeated B-scans but moving objects produce a signal that is dependent on the blood speed. Based on this mode of function, it is obvious that the technique is sensitive to eye motion and motion artifacts are common (Spaide et al. 2015). Segmentation of OCT angiograms is usually done to extract either two or three retinal vascular plexuses as well as the choriocapillaris (Fig. 9.14).

OCT angiography has also been applied to the anterior segment of the eye (Ang et al. 2018b; Lee et al. 2019; Siddiqui and Yin 2019). This technique has several advantages over other vascular anterior segment imaging modalities including rapid image acquisition, noninvasiveness, absence of leakage, 3D visualization, and imaging in cases of corneal opacification. This technology has been used to study the vasculature of the cornea in pathological conditions, the iris, sclera, episclera, and conjunctiva.

OCT angiography has recently renewed the interest of many clinicians in ocular blood flow because it is readily available as a functional extension of OCT platforms. The technology allows for the noninvasive visualization of the vasculature with

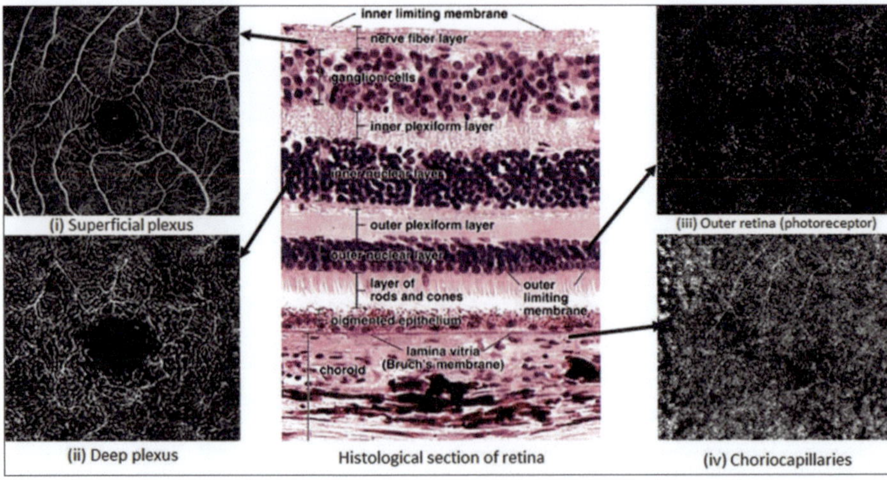

Fig. 9.14 The location of different en face zones in relation to histology of the human retina. The four en face zones include (i) the superficial plexus, the capillary network in ganglion cell layer and nerve fiber layer; (ii) the deep plexus, a network of capillaries in the inner plexiform layer with offshoot of 55 μm; (iii) the outer retina (photoreceptors); and (iv) the choriocapillaris (choroid) with offshoot of 30 μm. (Reproduced with permission from Chalam 2016)

unprecedented 3D resolution. This has led to the discovery of novel insights into the pathophysiology of retinal and choroidal vascular disease and also to the standard use of the technology in clinical care. A variety of quantitative metrics have been proposed for the analysis of OCT angiograms (Tan et al. 2020a; Yao et al. 2020). The foveal avascular zone (FAZ) can be separately analyzed for each capillary plexus. The vessel density is based on the skeletonized vessel map and is calculated as the ratio of vessel area to image area. A one-pixel line represents the vessel thereby excluding the effect of vessel diameter. Perfusion density is calculated as the ratio of the binarized perfusion area to the entire imaged area. Alternative metrics such as fractal dimension (Corvi et al. 2018) or complexity index (Chu et al. 2016) were also calculated from OCT angiograms.

Since the choriocapillaris is highly vascularized, the nonperfused area is usually used as an outcome measure after binarization of the image (Spaide et al. 2018). These nonperfused regions are called flow voids, flow deficits or signal voids. The size and number of such flow voids can provide information on choroidal integrity (Spaide 2016). The B-scan rate, the A-scan rate, the scan direction, the oversampling ratio as well as the lateral and axial resolution need to be considered when calculating vascular metrics based on OCT angiography (Tan et al. 2019, 2020a). Magnification is another important factor to consider. The field of view is a function of axial eye length and several approaches were published on how to correct for this (Bennett et al. 1994). This obviously has a major effect on quantification of FAZ size, which in turn also affects measures such as perfusion density and vessel density.

Whenever quantitative metrics are calculated it is important to understand the area that is being analyzed. Generally, data for vessel density or perfusion density are not comparable between different OCT systems, because each vendor uses its own algorithm (Corvi et al. 2018). There is currently no consensus on an optimal approach for this problem. Moreover, it needs to be considered that none of the parameters currently extracted from commercial OCT angiography is a direct measure of blood flow. Hence, any conclusions related to quantitative perfusion of the retina and/or choroid are unjustified.

Nonetheless, there are several approaches to quantify capillary blood velocities based on OCT angiography approaches. Such techniques require high OCT A-scan rates to allow for estimation of retinal blood flow speeds. The basis of this idea is that OCT angiography signals saturate with long interscan times and measuring the threshold at which this saturation occurs allows for quantification of blood velocities (Tokayer et al. 2013). Using OCTA signals measured at two different interscan time points, variable interscan time-analysis (VISTA) has allowed for the characterization of choroidal blood flow alterations in patients with age-related macular degeneration (AMD) with geographic atrophy (GA) (Choi et al. 2015). Using the ratio of these en face OCTA signals, mapping of relative blood flow velocities is attained (Ploner et al. 2016). Recently a next generation VISTA algorithm evaluating the temporal autocorrelation decay was introduced (Hwang et al. 2023). This gives rise to the estimation of retinal blood velocities at the individual capillary segment level by fitting the OCT angiography to a temporal autocorrelation model (Fig. 9.15).

9.3.2.9 Spectroscopic Oximetry

Noninvasive retinal oxygen saturation measurements were first done in 1965 based on absorption profiles of hemoglobin that depend on SO_2 (Hickam et al. 1963). This technique has been improved by using digital imaging (Beach et al. 1999) and automated software solutions (Hardarson et al. 2006). Commercial systems for noninvasive spectroscopic oximetry use a two-wavelength approach (Stefánsson et al. 2019; Shughoury et al. 2020). These two images are taken concomitantly so that the influence of eye movements is eliminated (Fig. 9.16). Validation of this technique is not easy, because of the lack of a gold standard method. One approach to study the accuracy is to compare SO_2 values obtained using oximetry in retinal arteries to systemic SO_2 values in patients with systemic hypoxemia (Traustason et al. 2011; Palkovits et al. 2013). The two-wavelength approach may, however, be influenced by retinal pigmentation and vessel diameter and therefore requires careful calibration (Hammer et al. 2008). Application of such calibration values to eyes from different ethnicities is, however, not possible because of differences in pigmentation levels (Beach et al. 1999). Another factor influencing the results is blood velocity (Jeppesen and Bek 2019). To overcome these limitations, several authors have used multiwavelength or hyperspectral approaches for retinal oximetry (Johnson et al. 2007; Desjardins et al. 2016; Kaluzny et al. 2017; Shahidi et al. 2017; DePaoli et al. 2019). Alternatively, white light optical coherence tomography can also be used to

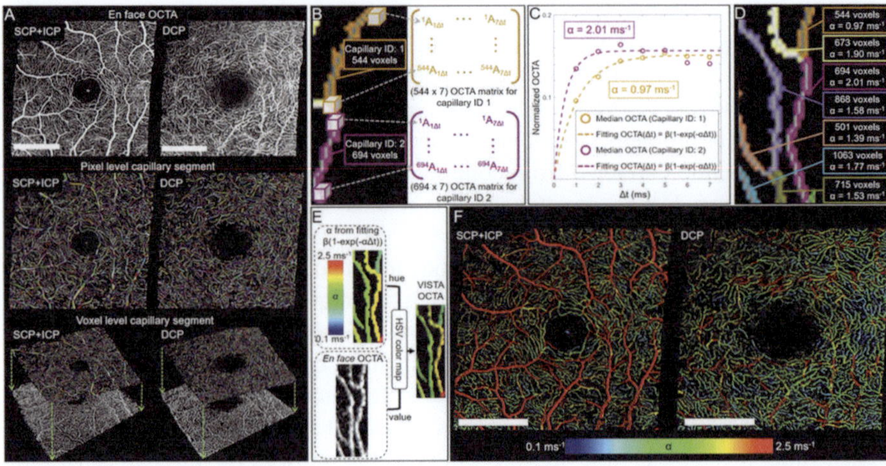

Fig. 9.15 VISTA pipeline for temporal autocorrelation decay (α) evaluation at the capillary level. (**a**) Voxel level capillary identification. En face OCT angiography (top) is transformed into a 2D graph (middle). The capillary ID from the 2D graph is projected onto a 3D vessel mask (bottom) and implemented separately for SCP + ICP and DCP. (**b**) Spatial compilation of normalized OCT angiography. The spatially compiled OCT angiography of each capillary can be represented with an OCT angiography matrix. (**c**) Temporal autocorrelation decay (α) evaluation of two capillaries shown in (**b**). (**d**) Number of voxels and corresponding α in the capillaries. (E) Visualization scheme using an HSV (Hue, Saturation, Value) colormap to generate VISTA OCT angiography. (**f**) VISTA OCT angiography of SCP + ICP and DCP. Scale bars: 1 mm. (Reproduced with permission from Hwang et al. 2023)

study retinal SO_2 (Wang et al. 2022a; Chen et al. 2017).This technique has also been extended toward the retinal microcirculation (Pi et al. 2020). Monte Carlo models have recently been developed that dealt with the impact of scattering, blood volume fraction and lens yellowing (Akitegetse et al. 2022) and will hopefully improve the reliability of this technique in the future.

9.4 Regulation of Ocular Blood Flow

9.4.1 Local Metabolic Control

Metabolic control of blood flow is based on the interaction between parenchymal cells and smooth muscle cells. In the retina and the optic nerve head, the key regulator is oxygen. During conditions of hypoxia vasodilator signals are activated to increase retinal perfusion (Fallon et al. 1987; Strenn et al. 1997; Cheng et al. 2016; Rose et al. 2016). In both animals and humans, this increase in perfusion is, however, not associated with an increase in inner retinal oxygen consumption as long as the systemic PO_2 is above 35 mmHg (Alm and Bill 1972; Enroth-Cugell et al. 1980; Palkovits et al. 2014a). Little is known about the mechanisms of oxygen-induced

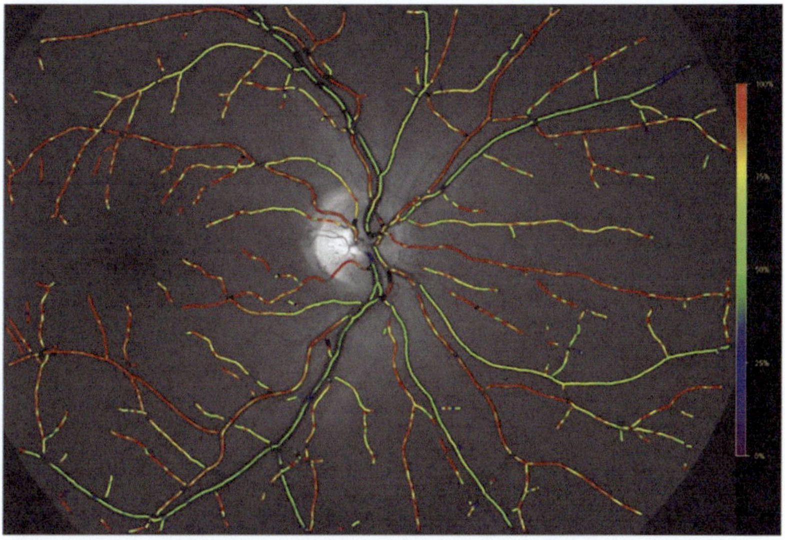

Fig. 9.16 Retinal oximetry image in a healthy subject. The color scale on the right-hand side shows oxygen saturation. The arterioles are generally red indicating oxygen saturation in the 90–100% range and the venules are generally green/yellow, indicating oxygen saturation in the 50–65% range. The oximeter software excludes vessels below 6 pixels (56 μm) from analysis. The variability in the measurement increases with smaller vessel diameters and this cut off is used to ensure high quality data. The excluded vessel segments are labeled gray on the image. (Reproduced with permission from Stefánsson et al. 2019)

vasodilatation in the retina, but nitric oxide (NO) and prostaglandins appear to be involved in this process (Nagaoka et al. 2002; Kringelholt et al. 2013). During hyperoxia, strong retinal vasoconstriction is observed, which is associated with a pronounced reduction in blood flow (Fig. 9.17) (Cheng et al. 2016; Kiss et al. 2002; Werkmeister et al. 2012; Gilmore et al. 2005; Harris et al. 1996; Riva et al. 1986a). OCT angiography metrics also decrease during hyperoxia, but it needs to be kept in mind that the technique does not measure blood flow (Pechauer et al. 2015; Xu et al. 2016; Hommer et al. 2021). The vasoconstrictor response to hyperoxia involves endothelin-1 as well as the arachidonic acid metabolites thromboxane and 20-HETE (Zhu et al. 1998; Song et al. 2016; Dallinger et al. 2000; Takagi et al. 1996). In contrast to blood flow in the retina, perfusion in the choroid is almost independent on systemic PO_2 (Schmetterer et al. 1996, 2000; Geiser et al. 2000; Kergoat and Faucher 1999; Gallice et al. 2017). During systemic hyperoxia, large amounts of oxygen diffuse from the choroid into the retina, thereby increasing inner retinal PO_2 (Pournaras et al. 1989; Linsenmeier and Yancey 1989; Palkovits et al. 2014b).

Perfusion in both the retina and the choroid is strongly dependent on partial pressure of carbon dioxide (Fig. 9.17) (PCO_2) (Harris et al. 1996; Geiser et al. 2000; Gallice et al. 2017; Venkataraman et al. 2005; Shahidi et al. 2014; Dorner et al. 2002b; Schmetterer et al. 1995; Luksch et al. 2002). In the brain, it has been demonstrated that reduced extracellular pH, is mainly responsible for vasodilation

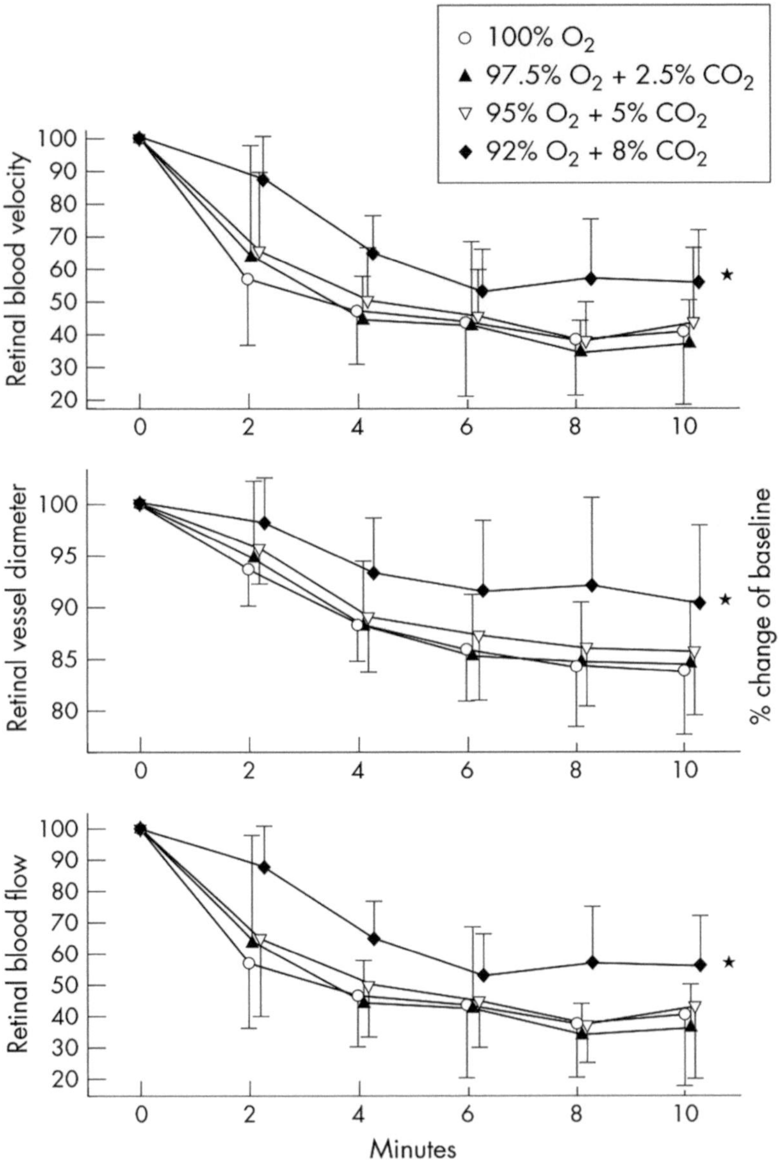

Fig. 9.17 Time course of retinal venous diameter, retinal blood velocity, and retinal blood flow during breathing of different mixtures of O2/CO2 in healthy subjects. Data are presented as means (SD) ($n = 12$). Asterisks indicate significant differences versus pure oxygen breathing. (Reproduced with permission from Schmetterer et al. 1996)

in response to local changes in PCO_2 (Kontos et al. 1977). In the retina, this has not been confirmed, but the most likely effect is the diffusion of CO_2 across the blood–retinal barrier, which results in reduced pH in the extracellular space. Activation of acid-sensing ion channel-1A by extracellular acidosis seems to be a key mechanism in CO_2-induced vasodilation in brain (Faraci et al. 2019). The cerebral blood flow response in the brain is dependent on nitric oxide (NO) (Okamoto et al. 1997; Iadecola and Zhang 1996), a mechanism that also been confirmed for the ocular vasculature (Schmetterer et al. 1997a).

The metabolic theory of blood flow regulation assumes that blood flow is adapted via a feedback loop in response to changes in PO_2, PCO_2, pH, and other vasoactive mediators. This may for instance play a role in hypoxic states such as early diabetes when retinal blood flow increases (Fondi et al. 2017). It is, however, unlikely that such metabolic mechanisms play an important role in blood flow regulation during changes in arterial blood pressure or neuronal stimulation, because of their relatively slow mechanism of action.

9.4.2 *Autoregulation in Response to Changes in Perfusion Pressure*

Autoregulation refers to the ability of a vascular bed to maintain its blood flow during changes in perfusion pressure (Pournaras et al. 2008). In its strict sense, it refers to isolated vascular beds. Changes in vascular tone during in vivo experiments are not only due to the change in perfusion pressure but also due to potential changes in the neural or humoral input. This is unlikely to occur in the retina because the retinal vasculature lacks autonomic innervation (McDougal and Gamlin 2015; Bek 2018; Bergua et al. 2013). In the choroid, however, the vessels are richly innervated (McDougal and Gamlin 2015), and in vivo responses to pressure stimuli are at least partially mediated via neural mechanisms (Klooster et al. 1996; Lütjen-Drecoll 2006; Reiner et al. 2018).

Another issue relates to the calculation of perfusion pressure. This is defined as the difference in arterial minus venous pressure. In the retina, neither arterial nor venous pressure can be measured noninvasively in vivo. Noninvasive plethysmographic measurements of retinal venous pressure have been performed (Meyer-Schwickerath et al. 1995; Stodtmeister et al. 2022), but the results are dependent on cerebrospinal fluid pressure (Lo et al. 2021; Morgan et al. 2016). Hence, the vast majority of autoregulation studies have estimated ocular perfusion pressure (OPP) as mean arterial pressure (MAP)—intraocular pressure (IOP) in the supine position and as 2/3 MAP—IOP in the sitting position to account for the drop in pressure between the level of the eye and the level of the heart (Costa et al. 2014; Leske 2009; Grover and Budenz 2011; Schmidl et al. 2011a, 2015b). While this may be a good approximation, there are several limitations to this approach. On the one hand, the arterial tree from the heart to the eye represents a vascular resistance, which is associated with a drop in pressure. Hence, the arterial pressure at the level of the central retinal artery will be lower than at the level of the heart. On the other hand, retinal venous pressure is slightly higher than IOP, in order to allow blood to exit the eye ball toward the venous circulation (Glucksberg and Dunn 1993). The same also holds true for the choroidal veins (Mäepea 1992). As such, the calculated OPP will always overestimate the actual OPP by a couple of mmHg. There are two principal methods to change OPP, either via modification of the arterial side or the venous side or a combination of both. Multiple techniques have been used to study this behavior. The majority of studies investigated autoregulation either during an increase in blood pressure or an increase in IOP.

Myogenic mechanisms play a key role in autoregulation. Myogenic tone is defined as vasoconstriction that occurs in an isolated blood vessel at a constant pressure (Claassen et al. 2021). Similar to the brain, retinal arteries respond to changes in transmural pressure, which refers to the pressure within the vessel minus outside the vessel, with development of myogenic tone (Barabas et al. 2020). When arterial blood pressure is decreased, myogenic tone is reduced and vessel diameter increases. An increase in blood pressure increases myogenic tone thereby reducing vascular diameter. These changes in vessel diameter are associated with changes in vascular resistance that keep blood flow constant within certain limits of ocular perfusion pressure.

With increases in BP, a reduction in vessel diameter helps to normalize wall tension. This concept is supported by the fact that similar phenotypic changes are seen in isolated vessels studied under pressurized conditions in vitro, where blood flow is absent. The vascular myogenic response is intrinsic to vascular muscle cells and experiments in cerebral vessels have shown that the mechanism is still functioning when endothelial cells are removed (Faraci et al. 1989a; Knot and Nelson 1995). The mechanisms regulating myogenic response in brain and retina seem to be closely related (Barabas et al. 2020). When transmural pressure increases, cation channels on the surface of the vascular smooth muscle cells are activated in response to the mechanical stress (Welsh et al. 2002). This leads to cell membrane potential depolarization which induces vasoconstriction due to the increase in voltage-dependent Ca influx (Wu and Davis 2001; Kur et al. 2013, 2014). Transient receptor potential channels contribute to myogenic signaling in the retinal vasculature with TRPV2 being the most important regulator (McGahon et al. 2016; Thébault 2021). While myogenic mechanisms in vascular smooth muscle cells are intrinsic, additional factors related to endothelial-, metabolic-, neural-, and immune-related signaling, have also been identified (Claassen et al. 2021).

A schematic drawing of autoregulation in response to changes in ocular perfusion pressure is shown in Fig. 9.18. Two examples are shown, one in which autoregulation is present over an OPP range of 40 mmHg and another in which the autoregulatory range is lower, only spanning over an OPP range of 20 mmHg. The autoregulatory range is also termed autoregulatory plateau. The lower limit of autoregulation is defined at the OPP value at which blood flow starts to decline. The OPP value at which blood flow starts to increase when OPP is increased is called the upper limit of autoregulation. This autoregulatory capacity is achieved by the adaptation of vascular tone during changes in transmural pressure as discussed above. It is often stated that the upper and lower limit of autoregulation represent maximum vasoconstriction and maximum vasodilatation, respectively. Experiments

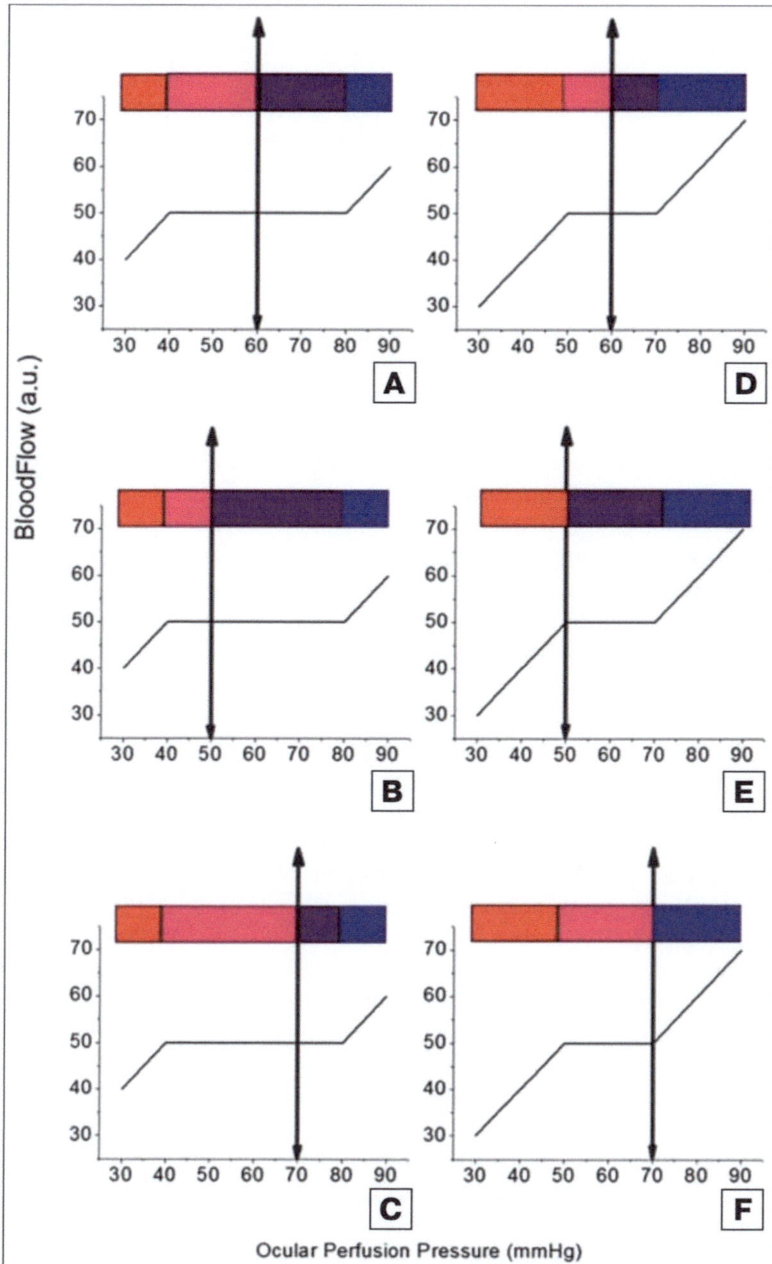

Fig. 9.18 Theoretical model of autoregulation. Red bars indicate the ischemic region, pink bars the region of autoregulatory reserve toward lower perfusion pressure, purple bars the region of autoregulatory reserve toward lower ocular perfusion pressure (OPP), and blue bars the region of hyperperfusion. (**a–c**) A person with an autoregulatory plateau of 40 mmHg. The autoregulatory reserve depends on the baseline OPP. At lower baseline OPP (**b**), the autoregulatory reserve toward lower OPPs is reduced. At higher OPP (**c**), the autoregulatory reserve toward lower OPPs is increased, but the autoregulatory reserve towards higher OPPs is reduced. (**d–f**) A person with an

in healthy subjects, however, have shown that even below the limit of lower autoregulation, retinal vessels still exhibit flicker-induced vasodilatation (Garhöfer et al. 2003).

Failure of autoregulatory capacity may have severe clinical consequences. When the OPP falls below the lower level of autoregulation, tissue will be at risk for ischemia and hypoxia. When the OPP exceeds the upper level of autoregulation, there is a risk for bleeding. Several common eye diseases are associated with abnormal autoregulatory ranges, which are considered to play a role in their pathophysiology.

In the retina, autoregulation has been proven in a wide variety of studies and most likely reflects, to a large degree, myogenic blood flow regulation. During an isometric exercise-induced increase in OPP, retinal blood flow in humans stays constant until an increase in the order of 30–40% (Robinson et al. 1986; Dumskyj et al. 1996; Witkowska et al. 2017). Retinal blood flow autoregulation has also been shown to occur when OPP is decreased for instance during an experimental increase in IOP (Fig. 9.19) (Riva et al. 1986b; Puchner et al. 2020).

In older textbooks, it is frequently mentioned that the choroid is a strictly passive vascular bed with a linear pressure–flow relationship. This is, to large degree, based on animal experiments using the microsphere technology (Alm and Bill 1972). Newer studies have, however, proven that the pressure–flow relationship is nonlinear in vivo (Fig. 9.20) (Kiel and Shepherd 1992; Kiel 1994; Kiel and van Heuven 1995). In humans, the choroid has been shown to react to both; a decrease (Riva et al. 1997a; Schmidl et al. 2012) and an increase in OPP (Schmidl et al. 2012; Calzetti et al. 2018; Riva et al. 1997b; Lovasik et al. 2003; Kiss et al. 2001). Animal studies support the concept that myogenic mechanisms play a role in this response (Kiel 1994), but there is currently no study available investigating myogenic tone in choroidal arterioles. A strong modulating function of neural components to this response has been proven, which may be responsible for most of the vascular tone changes during perfusion pressure challenges (Reiner et al. 2018; Li et al. 2021; Zagvazdin et al. 2000).

As mentioned above, the vascular anatomy of the ONH is complex. Little is known about potential differences between the superficial and deep vascular components of the ONH in term of pressure autoregulation. Most of the data on ONH autoregulation used either laser Doppler flowmetry or laser speckle flowgraphy, and therefore, data are limited to the superficial vascular bed. With this technology, autoregulatory behavior was seen during an increase (Witkowska et al. 2017; Schmidl et al. 2012; Bata et al. 2019b; Movaffaghy et al. 1998; Boltz et al. 2013a) as well as a decrease (Hashimoto et al. 2017a; Iwase et al. 2018; Pillunat et al. 1997; Riva et al. 1997c) in OPP. Two studies have compared autoregulatory changes in the

Fig. 9.18 (continued) abnormally reduced autoregulatory plateau of 20 mmHg. When baseline OPP is 60 mmHg (**d**), both the reserves toward higher and lower OPPs are reduced. When baseline OPP is only 50 mmHg, there is no reserve toward ischemia left (**e**), whereas no reserve toward hyperperfusion is left when OPP is 70 mmHg (**f**)

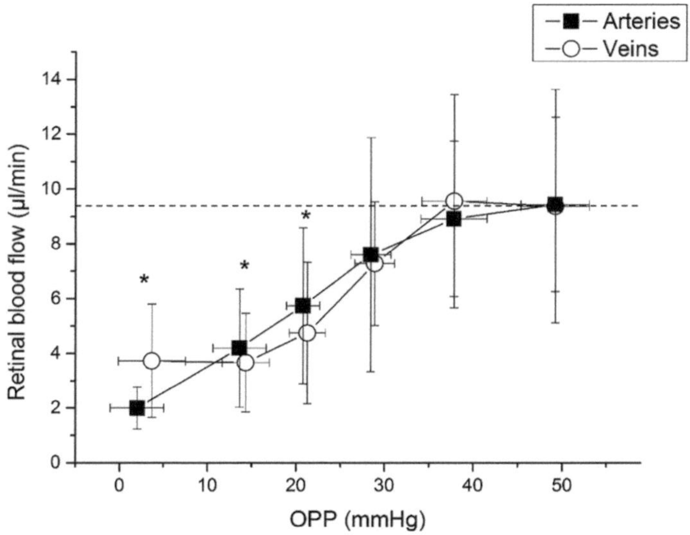

Fig. 9.19 Pressure–flow relationship for retinal arteries and veins determined by categorized OPP and retinal blood flow during application of the suction cup to increase IOP in healthy subjects. Data were sorted into groups of 15 values, each according to ascending OPP. Data are presented as mean ± SD ($n = 15$). *Significant changes versus baseline for retinal arteries and veins. (Reproduced with permission from Puchner et al. 2020)

ONH and choroid during standardized stimuli by either using laser speckle flowgraphy (Kiyota et al. 2018) or laser Doppler flowmetry (Schmidl et al. 2012) and found distinct differences between the two vascular beds (Fig. 9.21).

An important difference between the autoregulation in the brain and the eye is that in the eye, venous pressure can become very high when IOP is high. There is now evidence from a wide array of animal and human studies that blood flow reacts differently if either the arterial or the venous pressure is modulated (Fig. 9.22) (Kiel and van Heuven 1995; Tani et al. 2014; Wang et al. 2015; Boltz et al. 2013b; Polska et al. 2007; Popa-Cherecheanu et al. 2019). In humans, the regulation of choroidal and optic nerve head blood flow during combined changes in systemic blood pressure and IOP is dependent on the absolute values of BP and IOP (Boltz et al. 2013b; Polska et al. 2007; Popa-Cherecheanu et al. 2019). Blood flow in both vascular beds is less sensitive to changes in arterial blood pressure than in IOP. Interestingly, this finding is compatible with the myogenic mechanism being involved. During an artificial increase in IOP, venous pressure will increase with an associated increase in transmural pressure gradient. As mentioned above, the myogenic theory is based on the concept that changes in transmural pressure cause smooth muscle relaxation when OPP is lowered. Hence, ChBF may exhibit much less autoregulatory response during an IOP-induced drop in OPP, because it is paralleled by a decrease in perfusion pressure gradient. When OPP is changed via the arterial side, however, the full myogenic response is initiated due to the increase in transmural pressure. This concept is also supported by a study in healthy subjects

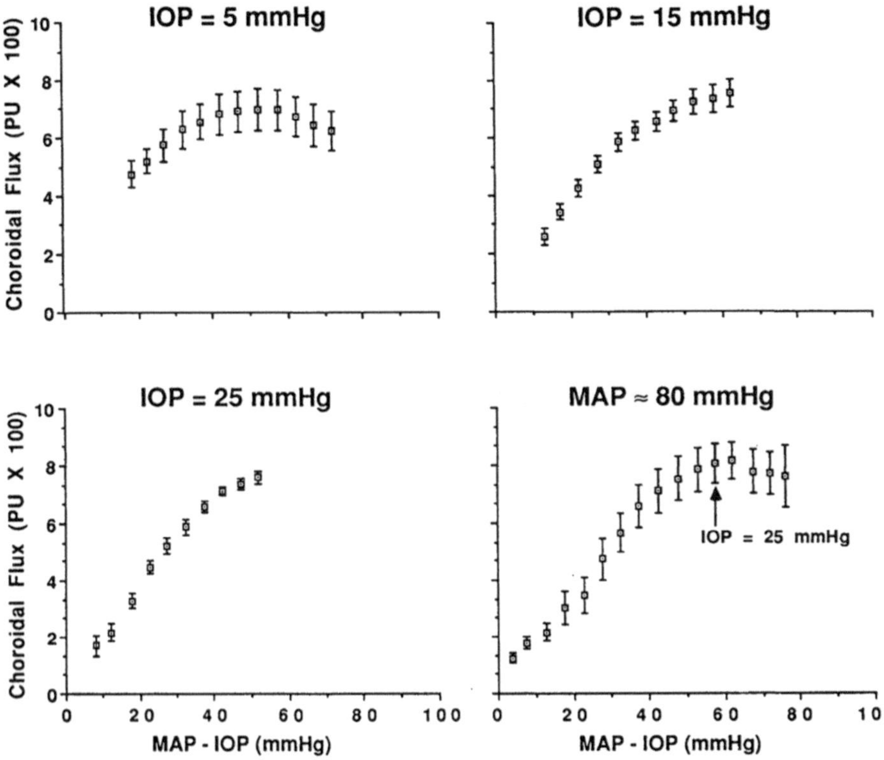

Fig. 9.20 Choroidal pressure–flow relationships in the rabbit when arterial pressure was decreased at intraocular pressures (IOP) of 5, 15, and 25 mmHg, and when intraocular pressure was increased at the prevailing mean arterial pressure (80 mmHg). At the lowest IOP, choroidal blood flow was well maintained until the perfusion pressure fell below approximately 40 mmHg. Raising the IOP to 15 and 25 mmHg shifted the upper portion of the pressure–flow curves upward, so the curves became progressively more linear. When IOP was the manipulated variable, blood flow was pressure-independent until intraocular pressure was raised above approximately 25 mmHg. (Reproduced with permission from Kiel and Shepherd 1992)

in which the IOP was lowered pharmacologically with an associated increase in the autoregulatory plateau (Boltz et al. 2011).

Several studies have found that autoregulation of ocular vascular beds is altered by pharmacological intervention. A role of NO in autoregulation of all ocular vascular beds has been shown in a wide variety of animal and human studies (Fig. 9.23) (Tani et al. 2014; Koss 1999; Schmetterer and Polak 2001; Okuno et al. 2002; Schmidl et al. 2013; Orgül et al. 1999; Kiel 1999; Simader et al. 2009; Luksch et al. 2003). Interestingly, NO synthase inhibition alters the choroidal but not the ONH pressure–flow relationship when IOP is increased, again, highlighting the differences in the mechanisms between choroid and ONH (Boltz et al. 2011; Simader et al. 2009). Modulating the endothelin system has also been proven to change the ocular pressure/flow relationship mediated via the endothelin-A receptor

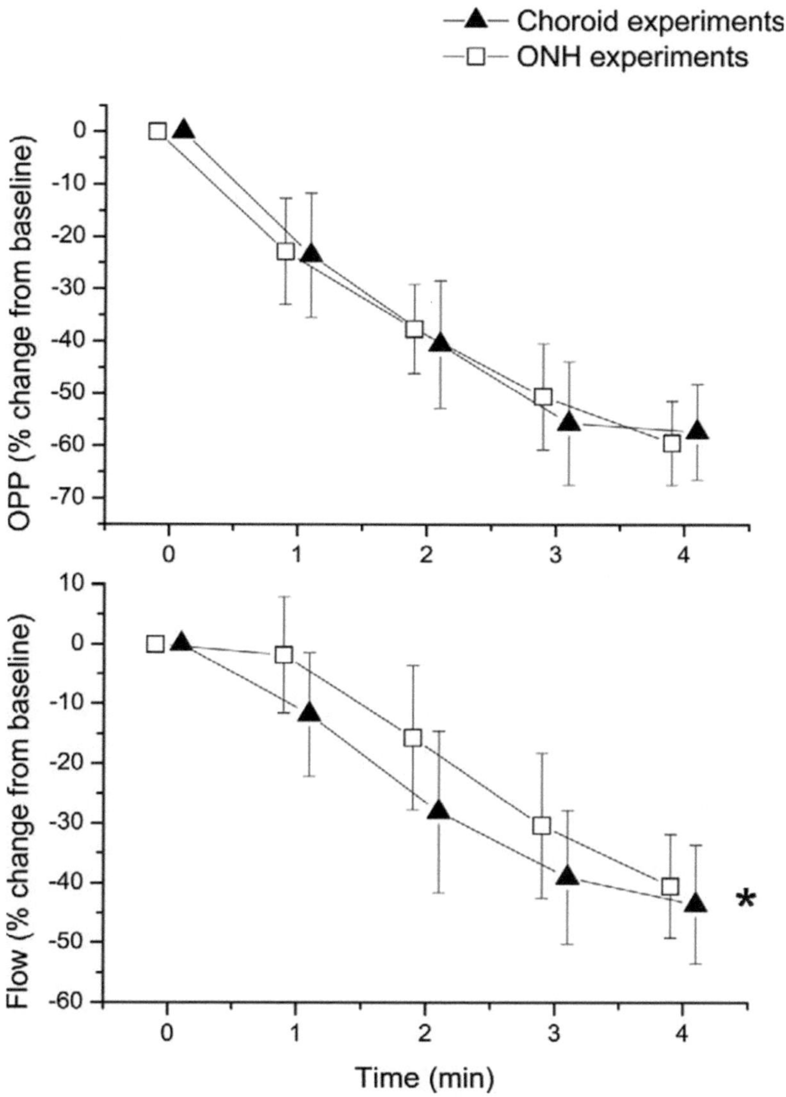

Fig. 9.21 Effect of an IOP increase on OPP and flow in healthy subjects. Data are shown separately for the artificial IOP increase for choroidal ($n = 24$, solid up triangles) and ONH experiments ($n = 24$, open squares). Data are presented as percentage change from baseline (means ± SD). *Significant differences between the choroidal and ONH experiments. (Reproduced with permission from Schmidl et al. 2012)

(Fig. 9.24) (Kiel 2000; Boltz et al. 2013c; Luksch et al. 2006; Fuchsjäger-Mayrl et al. 2003). Other factors that have been shown to modulate the autoregulatory pressure response in the eye include adenosine (Tani et al. 2014), NMDA receptors (Tani

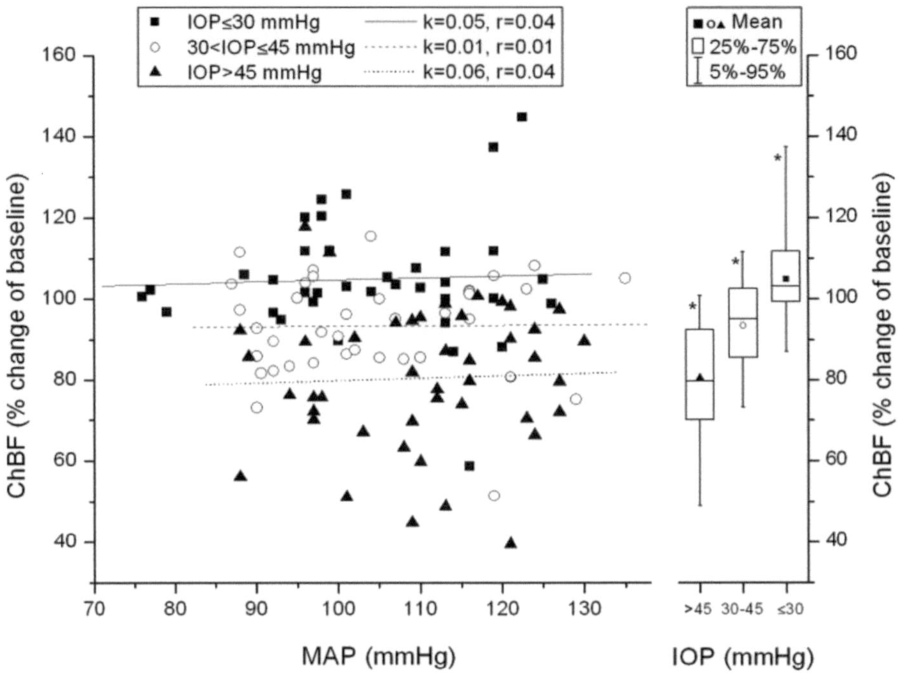

Fig. 9.22 Mean arterial pressure–choroidal blood flow (MAP–ChBF) relationship during combined experimental IOP increase and isometric exercise using categorized data according to IOP in healthy subjects. All MAP and ChBF pairs, except baseline, were sorted into three groups according to IOP (group 1: IOP $\leq$ 30 mmHg; group 2: 30 $>$ IOP $\leq$ 45 mmHg; group 3: IOP $>$ 45 mmHg). Left: regression analysis for each group, performed separately. Right: box-and-whisker plots, indicating mean ChBF at different IOPs independent of MAP. *Significant differences between these groups. *k*, slope of the regression line; *r*, correlation coefficient. (Reproduced with permission from Polska et al. 2007)

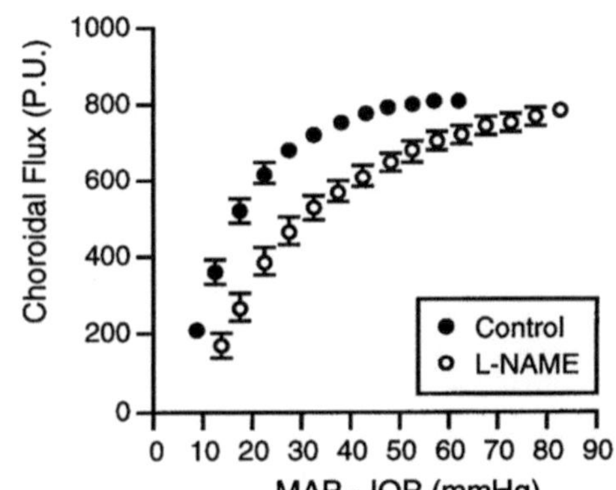

Fig. 9.23 Effect of inhibiting nitric oxide synthase (L-NAME, 10 mbg kg^{-4}, i.v.). pressure–flow curves in the rabbit choroid. L-NAME-reduced group mean choroidal flow over the entire perfusion pressure range was examined. (Reproduced with permission from Kiel 1999)

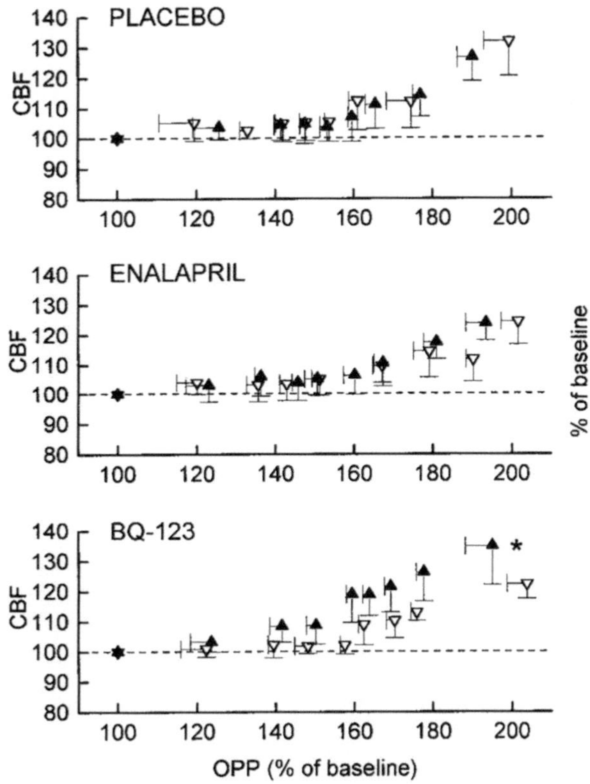

Fig. 9.24 Pressure–flow relationship between ocular perfusion pressure (OPP) and choroidal blood flow (CBF) during isometric exercise in healthy subjects. Data are shown for the angiotensin converting enzyme inhibitor enalapril and for the endothelin receptor antagonist BQ-123. The first period of squatting was without the drug (baseline; open triangles). The second squatting period was during administration of placebo, BQ-123, or enalapril (filled triangles). The mean and the lower limits of the 95% confidence interval are shown ($n = 12$). (Reproduced with permission from Fuchsjäger-Mayrl et al. 2003)

et al. 2014), alpha- and beta-adrenergic blockade (Kiel and Lovell 1996), and calcium channel blockade (Kiel 1999).

In perfused and pressurized brain slices, it was postulated that astrocytes are involved in modulating myogenic tone during changes in pressure (Kim et al. 2015). This concept is supported by rabbit experiments in which animals treated with the gliotoxic L-2-aminoadipic acid (LAA) showed reduced autoregulatory reserve in the ONH (Shibata et al. 2012).

9.4.3 Neurovascular Coupling

When neurons are stimulated with light, the retinal vessels vasodilate. This response is due to a fundamental mechanism termed functional hyperemia that describes the increase in blood flow when neurons are activated. This phenomenon has been described in detail in the brain but also occurs in the retina (Attwell et al. 2010). The increase in blood flow and vessel diameter during flicker stimulation has been described by many authors in both animals (Riva et al. 1991; Toi and Riva 1995; Buerk et al. 1995; Wang and Bill 1997; Srienc et al. 2010; Kornfield and Newman

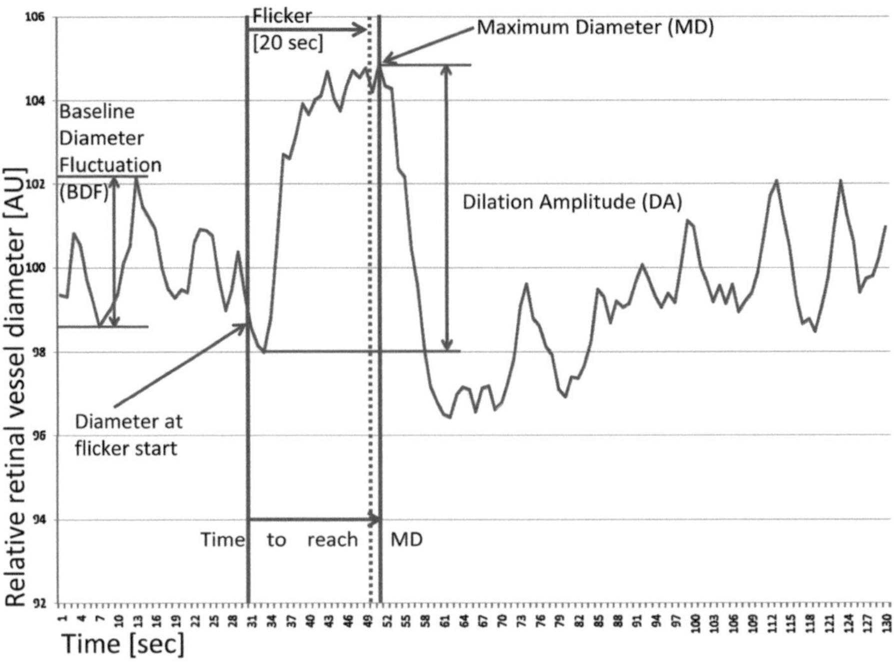

Fig. 9.25 The retinal diameter response to flicker provocation. Illustrated are changes in a retinal vessel as it is stimulated by flickering light. On the far left, BDF defines the fluctuation in the baseline diameter. The central section shows an increase in relative vessel diameter due to flickering light. This response provides the MD, the DA, and the time to reach MD. The right side shows the responses of the vessel as it recovers. (Reproduced with permission from Heitmar et al. 2010)

2014; Werkmeister et al. 2014; Albanna et al. 2018; Tamplin et al. 2022) and humans (Fig. 9.25) (Formaz et al. 1997; Riva et al. 2001; Falsini et al. 2002; Garhüfer et al. 2002, 2004a; Polak et al. 2002; Riva et al. 2004a; Wang et al. 2011; Zhong et al. 2012; Son et al. 2016; Duan et al. 2016; Aschinger et al. 2017; Fondi et al. 2018; Warner et al. 2020; Nesper et al. 2019; Kallab et al. 2021a). This is paralleled by an increase in retinal oxygen extraction (Felder et al. 2015; Palkovits et al. 2015). This relaxation can either be mediated via smooth muscle cells or via the retinal pericytes that also show contractile tone in the retina (Anderson 1996; Erdener et al. 2022).

In the brain, it has been shown that NO and prostaglandins are involved in the process of neurovascular coupling (Lindauer et al. 1999; Lacroix et al. 2015). Results for NO involvement have also been confirmed in the retina (Kondo et al. 1997; Buerk et al. 1996; Dorner et al. 2003a). Whereas these substances are produced in neurons and directly diffuse through the vessels, another mechanism plays a key role in brain neurovascular coupling. Astroglial cells act as intermediaries in neurovascular signaling (Metea and Newman 2006; Iadecola 2017). According to this model, astrocytes are stimulated via neurotransmitters such as glutamate and ATP which are produced in neurons (Fig. 9.26). This induces an

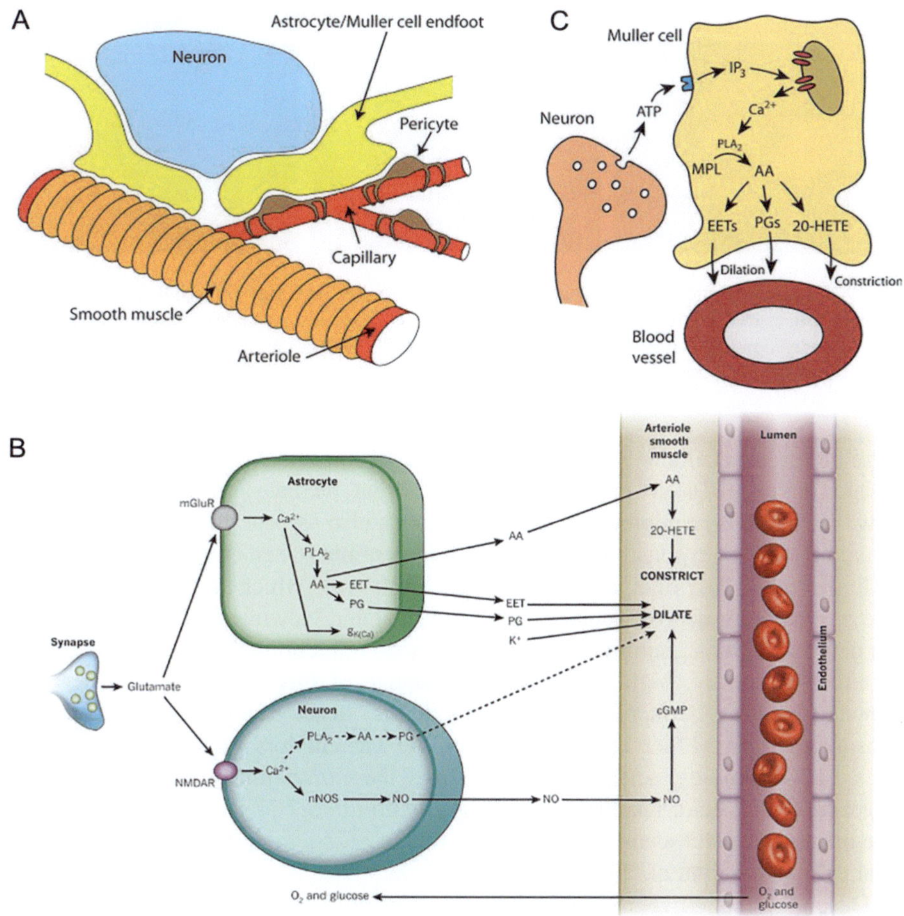

Fig. 9.26 Cellular and molecular mechanisms of neurovascular coupling. (**a**) Cells of the neurovascular unit regulate blood flow in the brain and retina. The diameters of arteries and arterioles are controlled by smooth muscle cells, while capillary diameter is controlled by pericytes. Astroglial cells in the brain and Müller glial cells in the retina surround arterioles and capillaries and their contractile cells. Signaling molecules released from neurons, and glial cell endfeet control the contractile state of smooth muscle cells and pericytes. (**b**) Signaling pathways that mediate neurovascular coupling in the brain. Glutamate released from neurons acts on NMDA receptors in neurons (NMDAR) to raise intracellular Ca2+ levels, causing neuronal nitric oxide synthase (nNOS) to release NO, dilating smooth muscle cells. Raised Ca^{2+} may also (dashed line) generate arachidonic acid (AA) from phospholipase A2 (PLA2), which is converted to prostaglandins (PG) that dilate vessels. Glutamate also raises Ca^{2+} levels in astrocytes by activating metabotropic glutamate receptors (mGluR), generating AA and three types of AA metabolites: prostaglandins and epoxyeicosatrienoic acids (EETs), which dilate vessels, and 20-hydroxy-eicosatetraenoic acid (20-HETE), which constricts vessels. (**c**) Neurovascular coupling in the retina. Adenosine triphosphate (ATP) released from neurons stimulates purinergic receptors on Müller cells, leading to the production of IP3 and the release of Ca2+ from internal stores. Ca2+ activates PLA2, which converts membrane phospholipids (MPL) to AA which is subsequently metabolized to the vasodilators PGs and EETs, and to the vasoconstrictor 20-HETE. (Reproduced with permission from Nippert and Newman 2020)

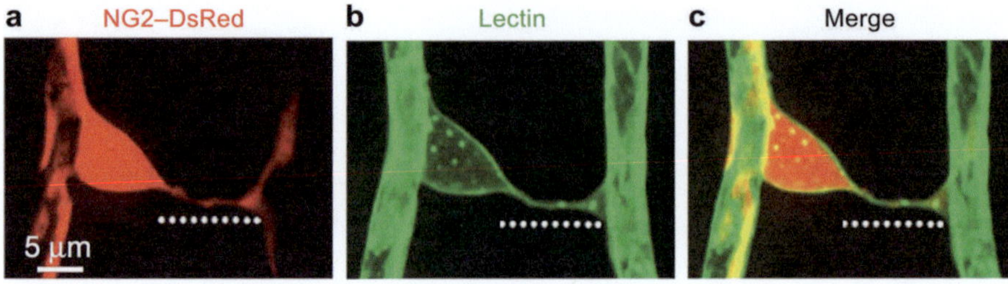

Fig. 9.27 Interpericyte tunneling nanotubes connect retinal pericytes on distal capillaries in a soma-to-process configuration. Examination of retinas from mice that express red fluorescent protein under control of the NG2 (Cspg4) promoter (NG2–DsRed) and Lectin. (Reproduced with permission from Alarcon-Martinez et al. 2020)

increase in intracellular Ca^{2+} activating phospholipase A2 and other enzymes (Mishra et al. 2016) to produce vasodilators that mediate the increase in blood flow. In addition, endothelium-derived factors also appear to contribute to the signaling that its responsible for functional hyperemia (Chen et al. 2014).

In the retina, the principal glial cell is the Müller cell, which also has an important role in mediating neurovascular coupling (Metea and Newman 2007). Blockade of the neuron to glia signaling causes a pronounced reduction of the response to light stimulation thereby, highlighting the importance of this mechanism (Metea and Newman 2006). The increase in intracellular Ca^{2+} leads to a release of vasoactive substances from Müller cells that include the vasodilators prostaglandin E2 and epoxyeicosatrienoic acids as well as the vasoconstrictor 20-hydroxy-eicosatetraenoic acid, resulting in net vasodilation under physiological conditions. The main role of NO seems to be the regulation of production of these vasodilators in glial cells (Attwell et al. 2010).

During systemic hyperoxia, the hyperemic response in the retina is preserved in the rat (Mishra et al. 2011), but increased in humans (Palkovits et al. 2014c). Another potential mechanism involved in neurovascular coupling is the release of K+ from Müller cells (Newman et al. 1984). Potassium ion concentration increases in the vitreous humor close to the optic nerve head as measured with microelectrodes during flicker stimulation in cats (Buerk et al. 1995). However, K^+ siphoning does not play a role in functional hyperemic response in the retina (Metea et al. 2007).

Recently, nanotube-like processes that connect two pericytes with retinal capillaries systems were found to play a role in neurovascular coupling (Alarcon-Martinez et al. 2020) in the retina (Fig. 9.27). These processes have an open-ended proximal side and a closed-ended terminal that connects the pericyte processes via gap junctions. This allows intracellular Ca^{2+} signaling mediating communication between adjacent pericytes. Pericytes rely on the processes to control neurovascular coupling. The role of caveolin-1 in flicker-induced retinal vasodilatation was also established. Caveolin-1 knockout mice show reduced response of optic nerve head blood flow to flicker stimulation (Fig. 9.28) (Loo et al. 2021).

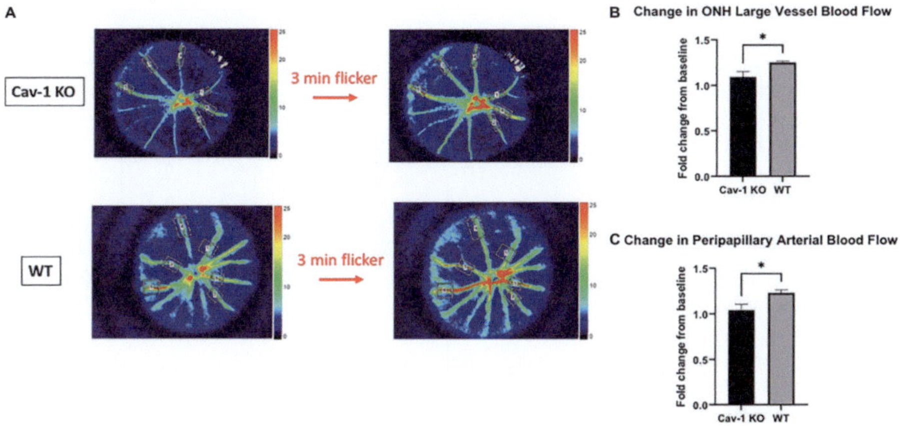

Fig. 9.28 Blood flow changes after 3 min of light flicker stimulus indicate differences in the neurovascular coupling of response of Caveolin-1 knockout (KO) compared to wild-type mice (WT). (**a**) Representative LSFG images taken at the optic nerve head in Caveolin-1 KO and WT mice. (**b**) Caveolin-1 KO show a lesser increase in large vessel blood flow at the optic nerve head after flicker stimulus, compared to WT. (**c**) Caveolin-1 KO show a lesser increase in peripapillary arterial blood flow after flicker stimulus, compared to WT, $*P < 0.05$. (Reproduced with permission from Loo et al. 2021)

9.5 Endothelial Factors

The endothelium plays a key role in regulating vascular tone (Ashby and Mack 2021; Busse and Fleming 2006; Haefliger et al. 1994a). Arteries, arterioles, capillaries, venules, and veins are lined with endothelial cells. Endothelial cells produce both vasodilators and vasoconstrictors that regulate blood flow. Endothelial dysfunction as observed in diseases such as diabetes alters retinal blood flow and contributes to macro- and microvascular damage (De Vriese et al. 2000; Shi and Vanhoutte 2017). Many mechanisms contribute to regulation of vascular tone by endothelial cells. These can roughly be divided into two main components: release of signaling molecules that modify vascular tone (Furchgott and Vanhoutte 1989; Godo and Shimokawa 2017) and spread of electrical signals from endothelial cells to smooth muscles (Aydin et al. 1991).

The most important endothelium-derived vasodilator is NO (Moncada 1994). NO is produced from the amino acid l-arginine via three isoforms of NO synthase (NOS), NOS1, NOS2, and NOS3. NOS1 and NOS3 are constitutive, and NOS2 is inducible, playing an important role in pathological states that are associated with alteration in the NO system. NO is a small gaseous mediator that can easily diffuse through tissue, similar to oxygen. Since NO cannot be stored in vivo, it only affects the cells that are in close proximity to the production site. A wide variety of studies have proven that inhibition of NO synthase reduces basal vascular tone in the ocular vasculature (Fig. 9.29) (Donati et al. 1995, 2003a; Haefliger et al. 1992, 1994b; Laspas et al. 2014; Mann et al. 1995; Schmetterer et al. 1997b; Luksch et al. 2000).

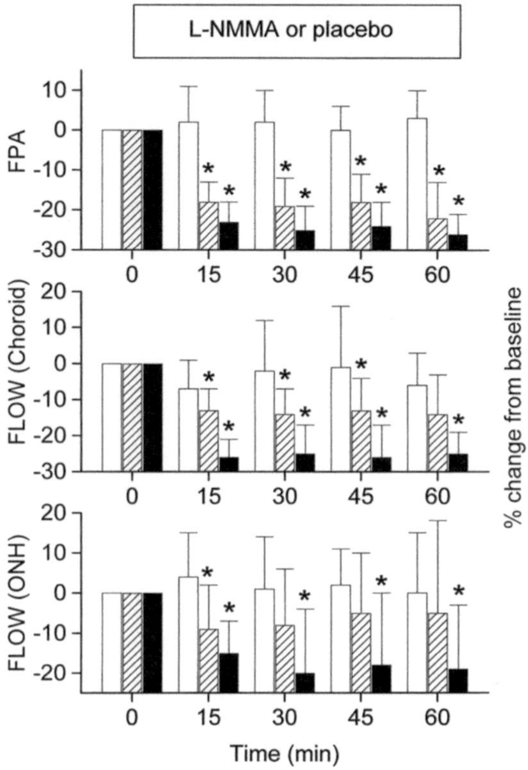

Fig. 9.29 Percent change in fundus pulsation amplitude (FPA); blood flow in the choroid, FLOW (Choroid); and blood flow in the optic nerve head, FLOW (ONH) after administration of L-NMMA (hatched bars: 3 mg/kg over 5 min followed by 30 μg/kg per minute over 55 min; solid bars: 6 mg/kg over 5 min followed by 60 μg/kg per minute over 55 min) or placebo (hollow bars). Data are presented as mean ± SD ($n = 12$). Asterisks indicate significant effects of L-NMMA versus baseline as calculated from the absolute values. (Reproduced with permission from Luksch et al. 2000)

NO has a wide array of functions in the vasculature of choroid, retina as well as ONH and appears to be involved in vascular dysregulation as observed in some diseases (Schmetterer and Polak 2001; Toda and Nakanishi-Toda 2007; Erdinest et al. 2021; Toda et al. 2010). Other endothelium-derived vasodilators that affect vascular tone includes potassium ion (Buerk et al. 1995; Metea and Newman 2007; Needham et al. 2014; Kur et al. 2012) and an endothelium-derived hyperpolarizing factors that has not been identified so far (Delaey et al. 2007; Haefliger et al. 2001; Delaey and van de Voorde 2000).

Another important system regulating vascular tone is the endothelin family. Three structurally different endothelin isoforms, endothelin-1, endothelin-2, and endothelin-3 have been identified, with endothelin-1 as being one of the most potent vasoconstrictors known (Mateo and De ArtiÑAno 1997; Barton and Yanagisawa 2019). Circulating levels of endothelin-1 are low, and its main role is to act as an autocrine/paracrine mediator. Endothelin-1 is a potent endothelium-derived vasoconstrictor that controls vascular tone in many organs including the eye (Granstam et al. 1993; Nishimura et al. 1996; Polak et al. 2001, 2003; Schmetterer et al. 1997c; Kida et al. 2018; Kawamura et al. 2002; Chakravarthy et al. 1992; Bursell et al. 1995; Hétu et al. 2013). The vascular effects of endothelin-1 are mediated by two receptor subtypes called endothelin A and endothelin B receptors. In the eye,

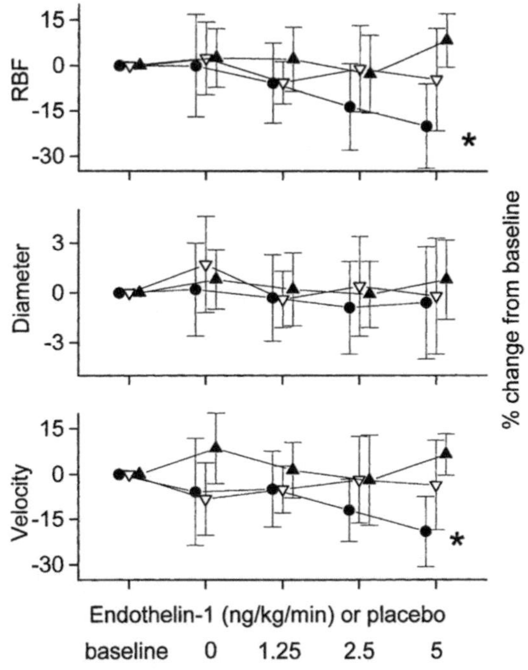

Fig. 9.30 The effect of endothelin-1 and/or the endothelin receptor antagonist BQ-123 infusion on retinal blood flow (RBF), retinal VD (diameter), and retinal blood velocity (velocity). Three study days were scheduled for each subject ($n = 12$): endothelin-1 /placebo (solid circles), placebo/BQ-123 (solid up triangles) and endothelin-1 /BQ-123 (open down triangles). Data are presented as means ± SD. (Reproduced with permission from Polak et al. 2003)

endothelin A receptors are localized at blood vessels, whereas endothelin B receptor are mainly found on neural and glial tissues (MacCumber and D'Anna 1994). The endothelin A mediates vasoconstriction and stimulation of the endothelin B receptor induces vasodilatation (Fig. 9.30).

Arachidonic acid metabolites control vascular tone via multiple metabolites (Chawengsub et al. 2009; Zhou et al. 2021) including prostaglandins, thromboxane A2, and prostacyclin. The majority of receptors that mediate the vasoactive actions of the arachidonic acid metabolites is present at the posterior pole of the eye (Abran et al. 1995, 1997). Ocular levels of prostanoids are higher in the perinatal than in the adult eye (Hardy et al. 2000) indicating an important role of prostaglandins in the regulation of blood flow in newborn animals. As such, prostanoids appear to be involved in the hypoxic processes that lead to retinopathy of prematurity (Hardy et al. 2000). In the newborn animal, prostanoids play a role in autoregulatory processes (Hardy et al. 1994). In adults, this effect may be less pronounced, but prostaglandin E2 is produced during an increase in perfusion pressure, partially mediating autoregulatory vasoconstriction (Hardy et al. 1994). This is, however, controversial because some studies have reported that prostaglandin E2 acts as a potent vasodilator in both the retina and the choroid (Mori et al. 2007). Controversial data have also been reported for prostaglandin E2 (Kitanishi et al. 2001; Pournaras et al. 1978) and prostaglandin F2α (Astin et al. 1994; Stjernschantz et al. 2000; Ohkubo and Chiba 1987). Similar to other vascular beds, thromboxane is a potent vasoconstrictor (Faraci et al. 1989b; Torring et al. 2014).

Shear stress regulates endothelial function, thereby influencing vascular tone (Bilotta and Abramov 1989). Flow-mediated vasodilatation in large system arteries is well established (Thijssen et al. 2011). In smaller vessels, this phenomenon is more difficult to measure because of the lower shear stress levels due to lower blood velocities and smaller vessel diameters. Techniques to measure flow velocity such as laser Doppler velocimetry allow for calculation of shear stress in retinal arteries based on theoretical velocity profiles (Nagaoka et al. 2002; Nagaoka and Yoshida 2006). Although Doppler OCT technologies allow for measurement of velocity profiles (Aschinger et al. 2015; Yazdanfar et al. 2003; Logean et al. 2000), this technology has not yet been used to calculate shear stress values. High shear stress caused by increased retinal blood velocities compromise the inner blood retinal barrier function (Molins et al. 2019), thereby establishing a link between perfusion abnormalities and leakage as seen in diabetes (Cunha-Vaz 2016).

9.5.1 Neural Control of Blood Flow

Due to the rich innervation of the choroid by parasympathetic, sympathetic and trigeminal sensory nerves, choroidal blood flow is under neural control (Reiner et al. 2018). Facial parasympathetic nerves regulate choroidal blood flow as shown by several animal studies. Mediators such as NO, acetylcholine and vasoactive intestinal polypeptide (VIP) cause vasodilation in the ocular vasculature. It has, for instance, been shown that administration of VIP causes an increase in choroidal blood flow (Nilsson and Bill 1984). Activation of preganglionic input to the pterygopalatine ganglion by facial nerve stimulation increases choroidal blood flow (Stjernschantz and Bill 1980; Nilsson et al. 1985; Nilsson 2000). Direct activation of the superior salivatory nucleus induces choroidal vasodilation and an increase in blood flow (Steinle et al. 2000). Centrally mediated autonomic reflexes also play a role in the change of choroidal vascular resistance during changes in blood pressure (Fig. 9.31) (Reiner et al. 2011).

The choroid is also innervated by sympathetic noradrenergic nerve fibers that originate from the superior cervical ganglion (Lütjen-Drecoll 2006). The sympathetic neurotransmitter noradrenaline induces pronounced vasoconstriction in the choroid (Gherezghiher et al. 1991). Cervical sympathetic stimulation causes an increase in choroidal vascular resistance (Steinle et al. 2000; Riva et al. 1994b; Alm 1977). The sympathetic input appears to exert a small vasoconstrictor tone in the choroid under basal conditions (Reiner et al. 2018). The sympathetic input into the choroid may also regulate choroidal vascular tone during elevated blood pressure, maintaining constant blood flow (Bill 1985). This is supported by experiments showing that superior cervical ganglion removal impairs choroidal autoregulation during changes in perfusion pressure (Reiner et al. 2018). Sensory nerve fibers from the trigeminal ganglion also innervate the choroid (McDougal and Gamlin 2015; Corvetti et al. 1988), and have been proposed to be involved in the thermoregulation of choroidal blood flow (Reiner et al. 2018).

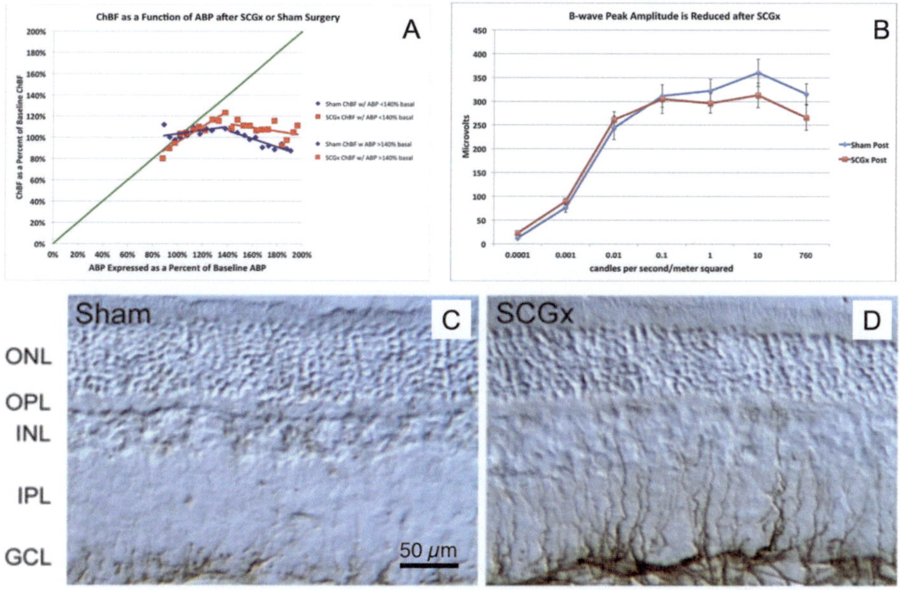

Fig. 9.31 At 2–3 months postsuperior cervical ganglion removal (SCGx), choroidal baroregulation during high arterial blood pressure (ABP) was impaired. Image A shows a plot of choroidal blood flow (ChBF) as function of ABP for sham rat eyes ($n = 14$) and superior cervical ganglion removal in rat eyes ($n = 14$). As ABP rapidly rose above baseline after L-NAME administration, ChBF in sham eyes remained relatively stable but followed ABP in superior cervical ganglion removal eyes. After ABP had stabilized at an elevated level, ChBF in superior cervical ganglion removal eyes declined toward baseline but remained elevated compared to sham eyes. The impairment in ChBF baroregulation was associated with a deficit in the flash-evoked scotopic b-wave ERG peak, which was significantly reduced for superior cervical ganglion removal eyes compared to control eyes (**b**). Additionally, GFAP immunolabeling of Müller cells was increased in retina by 3 weeks after superior cervical ganglion removal (**c, d**). The immunolabeled Müller cell processes in superior cervical ganglion removal eyes traversed the IPL and some extended into the INL (**c**). By contrast, in control retinas, GFAP labeling of Müller cell processes did not extend much beyond the GCL (**d**). (Reproduced with permission from Reiner et al. 2018)

9.5.2 *Humoral Control of Blood Flow*

Adenosine is a nucleoside which is present in the entire body and exerts strong vasodilator effects (Zhang et al. 2021). Animal studies have proven that adenosine increases blood flow in new born piglets (Gidday and Park 1993a) as well as in adult cats (Portellos et al. 1995). In humans, adenosine increases retinal and choroidal blood flow (Polska et al. 2003). The nucleoside is involved in hypoxia-induced vasodilatation and retinal autoregulation in new born animals (Gidday and Park 1993b) but does not modulate the choroidal pressure/flow curve in the human choroid (Schmidl et al. 2011b). Recent evidence suggests that the vasoactive effect of adenosine triphosphate differs in vessels at different branching levels (Dons-Jensen et al. 2019; Ernst et al. 2022).

Glucose has important vascular effects impairing endothelium-dependent relaxation (Guo et al. 2000; Kito et al. 2018). In the eye, hyperglycemia induces vasodilator effects in the retina and the choroid (Sullivan et al. 1990; Bursell et al. 1996; Luksch et al. 2001; Pemp et al. 2010). In other vascular beds, this vasodilator effect has been attributed to changes in osmotic load (Hoffman et al. 1999). Alternatively, the vasodilator responses to hyperglycemia may be linked to an increased ratio of NADH/NAD+, because of an increased rate of reduction of NAD+ to NADH (Williamson et al. 1993; Ido et al. 2001). In the eye, insulin also causes vasodilatation (Schmetterer et al. 1997d; Polak et al. 2000) which is additive to glucose vasodilator action (Luksch et al. 2001). The increase in choroidal blood flow is reduced after NOS inhibition (Schmetterer et al. 1997d) and is enhanced by coadministration of L-arginine (Dallinger et al. 2003).

9.5.3 Altered Blood Flow in Ocular Disease

9.5.3.1 Glaucoma

Multiple cross-sectional studies have shown that glaucoma is associated with reduced retinal and optic nerve head blood flow (Yoshioka et al. 2021; Inada et al. 1986; Hwang et al. 2012; Deokule et al. 2010; Yamada et al. 2019; Resch et al. 2011; Jeon et al. 2022). There is a long-standing debate on whether this is a cause or consequence of the disease. It has been argued that a loss of retinal ganglion cells and their axons will result in decreased oxygen demand and capillaries may die as a result of this reduced need. On the other hand, theories have been formulated suggesting that ischemia may promote apoptosis of retinal ganglion cells, thereby increasing the risk of visual field loss (Osborne et al. 1999; Cherecheanu et al. 2013). Most likely, both phenomena are true because evidence has been accumulated for both mechanisms.

The first hypothesis has been supported by several recent studies indicating a strong spatial association between retinal perfusion and visual field defects (Figs. 9.32 and 9.33) (Wong et al. 2020, 2021; Yaoeda et al. 2003; Kallab et al. 2022; Calzetti et al. 2022). Although longitudinal studies on the sequence of events are lacking, the general believe is that the glaucoma-related loss of RGCs leads to localized capillary dropout. The second hypothesis is currently supported by a wide variety of studies. Color Doppler imaging studies have shown that low blood velocities in the arteries supplying the ocular vasculature are associated with future visual field progression (Satilmis 2003; Galassi et al. 2003; Zeitz et al. 2006; Siesky et al. 2016). These results were confirmed by other studies using either laser Doppler flowmetry (Zink et al. 2003), OCT angiography (Jeon et al. 2019; Wang et al. 2022b) or laser speckle flowgraphy (Fig. 9.34) (Kiyota et al. 2021, 2022). Moreover, there is evidence that both high and low blood pressures are risk factors for glaucoma incidence and progression (Leeman and Kestelyn 2019; Costa et al. 2009; Skrzypecki et al. 2019). Nocturnal hypotension is also a risk factor for visual field

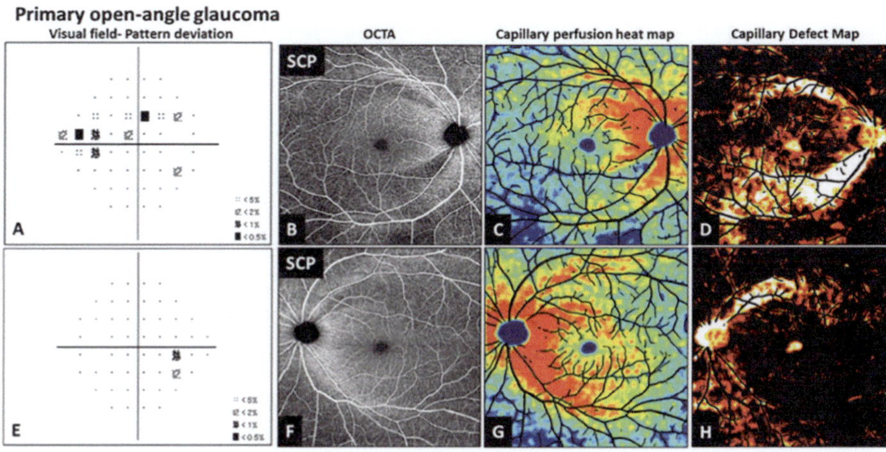

Fig. 9.32 Use of normative capillary perfusion maps in quantifying capillary defects in glaucoma (POAG) eyes. A, E) Visual fields in POAG. B, F) 12 × 12 mm OCTA images in POAG eyes (Male, Age: 38, SSI: 10). The original grayscale OCTA images (second column) were processed to create capillary perfusion maps (third column). Their comparisons with age-matched normative capillary maps (fourth column) showed both the loci and severities of the capillary defects. (Reproduced with permission from Tan et al. 2021a)

Fig. 9.33 Focal structure–function relationship in glaucoma. Factors from the multivariate mixed-effects modeling at test points from the 24–2 visual field (VF) for (**a**) focal retinal nerve fiber layer (FNL) thickness and (**b**) focal capillary perfusion density (FCL). Green indicates a test location in which the corresponding parameter was significantly associated with VF loss. Locations located nearer fixation (0°, 0°) were more significantly influenced by FCD than FNL thickness. Correlations for nasal VF test locations were poor for both FNL thickness and FCD. One VF test location at eccentricity (9°, −3°) was excluded due to the limited arc of the FNL thickness defined by the trajectories for that location. (Reproduced with permission from Wong et al. 2020)

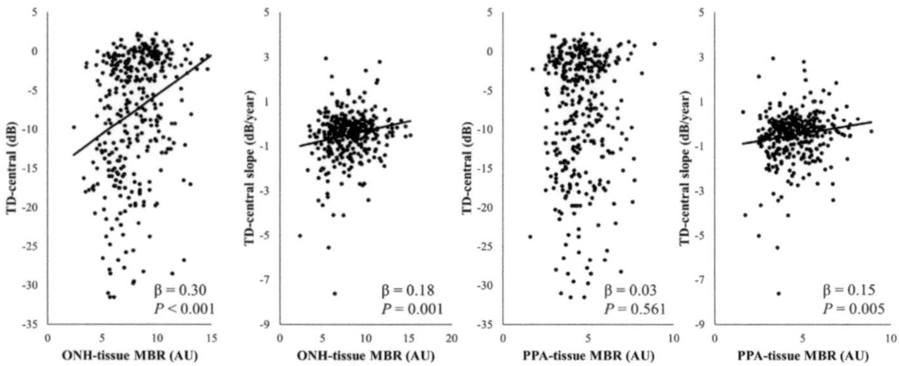

Fig. 9.34 Scatterplots showing comparisons between temporal optic nerve head (ONH)-tissue mean blur rate (MBR) and temporal peripapillary chorioretinal atrophy (PPA)-tissue MBR, total deviation (TD)-central, and TD-central slope in patients with glaucoma. Temporal ONH-tissue MBR was positively correlated with both TD-central and TD-central slope ($ß = 0.30$, $P < 0.001$; $ß = 0.18$, $P < 0.001$, respectively). Temporal PPA-tissue MBR was correlated only with TD-central slope, not with TD-central ($ß = 0.15$, $P < 0.005$; $ß = 0.03$, $P = 0.561$, respectively). AU = arbitrary units. (Reproduced with permission from Kiyota et al. 2020)

deterioration, most likely causing ischemic episodes during sleep (Charlson et al. 2014). Additionally, cardiovascular disease has been identified as an important risk factor for glaucoma progression (Marshall et al. 2021).

Evidence for altered autoregulation in glaucoma has been obtained from multiple studies. Group correlation between blood pressure or OPP and ocular hemodynamic parameters has been reported in multiple studies (Marjanović et al. 2016; Fuchsjäger-Mayrl et al. 2004; Garhöfer et al. 2010; Gherghel et al. 2000; Plange et al. 2008). A higher degree of correlation between perfusion pressure and blood flow is indicative of vascular dysregulation in such studies. This abnormal correlation in glaucoma is normalized after pharmacological reduction of IOP, in agreement with the idea that IOP has the main influence on the pressure/flow relationship in the eye (Fuchsjäger-Mayrl et al. 2010).

Multiple studies have reported abnormal blood flow responses to changes in OPP in glaucoma (Schmidl et al. 2011a). These studies can be grouped into investigations where OPP was reduced (Quaranta et al. 1994; Grunwald et al. 1984) and those where IOP was increased (Galambos et al. 2006; Evans et al. 1999). Evidence has also shown that the response to changes in OPP shows a wide variability among glaucoma patients (Bata et al. 2019b; Feke and Pasquale 2008). A few studies did not find any difference between the autoregulatory behavior between glaucoma patients and healthy controls, but the change in OPP induced in these studies was most likely too small to reach the lower or upper limit of autoregulation (Weigert et al. 2005; Pournaras et al. 2004; Mursch-Edlmayr et al. 2019). Abnormal autoregulation was also found in a study investigating the variations of ONH blood flow during diurnal variations in OPP (Sehi et al. 2005).

Altered autoregulation in glaucoma is linked to pericyte dysfunction and loss of interpericyte tunneling nanotubes (Alarcon-Martinez et al. 2022). In addition, it has been shown in healthy subjects that reduced ganglion cell thickness in healthy subjects is associated with both high blood pressure and abnormal blood flow autoregulation, which may make this group of subjects more susceptible to glaucomatous damage (Pappelis and Jansonius 2021). This is also compatible with findings in a nonhuman primate model indicating that abnormal autoregulation is at least partially a consequence of ganglion cell loss (Cull et al. 2015).

A reduced hyperemic response to stimulation with diffuse luminance flicker in glaucoma has been consistently reported using several techniques to assess blood flow (Riva et al. 2004b; Garhöfer et al. 2004b; Gugleta et al. 2012, 2013; Mroczkowska et al. 2013). This is compatible with results showing an abnormal oxygenation response to flicker stimulation in glaucoma patients (Al Zoubi et al. 2022). One study did not report a difference between flicker-evoked blood flow changes as measured with laser speckle flowgraphy in glaucoma patients as compared to healthy controls, but the study was most likely underpowered (Mursch-Edlmayr et al. 2020). A longitudinal study did not find an association between reduced functional hyperemia and structural progression in glaucoma patients, but again the number of included patients and the duration of follow up was too low to draw any meaningful conclusions (Waldmann et al. 2016). Glaucoma has also been associated with endothelial dysfunction (Resch et al. 2009a) and changes in the endothelin system (Chauhan 2008; Prasanna et al. 2011; Shoshani et al. 2012; Yorio et al. 2002). Evidence for endothelial dysfunction has been obtained from systemic (Henry et al. 1999; Buckley et al. 2002; Bukhari et al. 2016; Cellini et al. 2012; Su et al. 2008; Bojic et al. 2022) as well as from ocular vascular stimulation tests (Polak et al. 2007). Dual inhibition of endothelin receptors increases ocular blood flow in patients with glaucoma (Resch et al. 2009b). In patients with pseudoexfoliation glaucoma, an accumulation of pseudoexfoliative microfibrillar material leads to vascular endothelial dysfunction (Bourouki et al. 2019).

9.5.3.2 Diabetes and Diabetic Retinopathy

Retinal vascular changes are well documented in diabetes and diabetic retinopathy (DR) including dysfunction of the blood retinal barrier associated with the development of diabetic macular edema (Bhagat et al. 2009). Other key features include vascular basement membrane thickening (Roy et al. 2010) and a progressive loss of pericytes and vascular endothelial cells (Simó et al. 2018). During the course of this disease, vasodegeneration occurs, which is clinically seen as vascular dropout (Tonade et al. 2016). This capillary dropout and retinal ischemia trigger the release of growth factors including vascular endothelial growth factors (VEGF), which then leads to proliferative DR.

There are conflicting results of whether retinal blood flow is increased or decreased in early stages of DR (Fondi et al. 2017; Bursell et al. 1996; Pemp et al. 2010; Khuu et al. 2017; Patel et al. 1992; Pemp and Schmetterer 2008; Ludovico et al. 2003; Kohner et al. 1975; Kohner 1976; Burgansky-Eliash et al. 2012; Omae et al. 2015; Findl et al. 2000) or animal models of DR (Bursell et al. 1996; Muir et al. 2012). This may be related to differences in the technology used to assess blood flow but also due to different phenotypes. It also needs to be considered that there may be a redistribution of blood flow in different vascular plexuses of the retina. Hence, results may depend largely on the technology that is used form measuring retinal perfusion. Finally, it is important to mention that both insulin and glucose may influence the results due to their vasodilator actions as mentioned above (Luksch et al. 2001).

OCT angiography has enabled the study of important aspects of vascular alterations in diabetes and DR (Borrelli et al. 2021; Boned-Murillo et al. 2022; Chua et al. 2020a; Lee and Rosen 2016; Johannesen et al. 2019; Tey et al. 2019; Kannenkeril et al. 2022). Features that can be studied with OCT angiography include microaneurysms, neovascularization, peripheral retinal nonperfusion, foveal avascular zone alterations, as well as quantitative changes in perfusion and vessel density (Borrelli et al. 2021; Chua et al. 2020a; Tey et al. 2019; Monteiro-Henriques et al. 2022). Widefield OCT angiography is nowadays the preferred method to identify peripheral capillary nonperfusion (Fig. 9.35) (Hirano et al. 2018; Zhang et al. 2018; Tan et al. 2020b; Guo et al. 2018). Interestingly, widefield OCT angiography has also been shown to provide improved detection of retinal neovascularization that is not detected clinically (You et al. 2020; Khalid et al. 2021; Vaz-Pereira et al. 2021; Lu et al. 2022). Quantitative OCT angiography metrics predict the risk of DR progression and may therefore add to traditional risk factors in the management of DR (Sun et al. 2019).

OCT angiography can also be used to study choroidal changes in diabetes. Significant loss of choriocapillaris has been described histologically (Johnson et al. 2005; Lutty 2017), but in vivo studies on choroidal blood flow in diabetes have been sparse (Nagaoka et al. 2004). Analysis of OCT angiography slabs has again proven that pronounced vascular changes are present in the diabetic choroid (Dodo et al. 2017; Li et al. 2019; Tan et al. 2022a; Wang et al. 2022c, d; Liu et al. 2022). It has been shown that retina inflammatory processes appear to trigger endothelial cell damage in the choroid (Brinks et al. 2022). Nevertheless, the role of the choroid and its blood flow in the development and progression of DR remains poorly understood.

Diabetes is characterized by abnormal retinal and optic nerve head vascular autoregulation (Sinclair et al. 1982; Rassam et al. 1995; Mehlsen et al. 2011; Hashimoto et al. 2017b). The mechanism underlying this effect is not well characterized. Obviously, loss of pericytes (Trost et al. 2019) as well as endothelial dysfunction are candidates (Schmetterer and Polak 2001; Wong et al. 2013) for being involved in the loss of autoregulatory capacity. Uncoupling of gap junctions is also a candidate mechanism for disrupting autoregulation in response to changes in ocular perfusion pressure (Shibata et al. 2011).

Functional hyperemia is reduced in the diabetic retina in a stage-dependent manner (Fig. 9.36) (Garhöfer et al. 2004c; Nguyen et al. 2009; Lecleire-Collet et al. 2011; Sörensen et al. 2016; Mandecka et al. 2007; Pemp et al. 2009a; Lasta et al. 2013; Garhöfer et al. 2020). Correlation between reduced blood flow responses

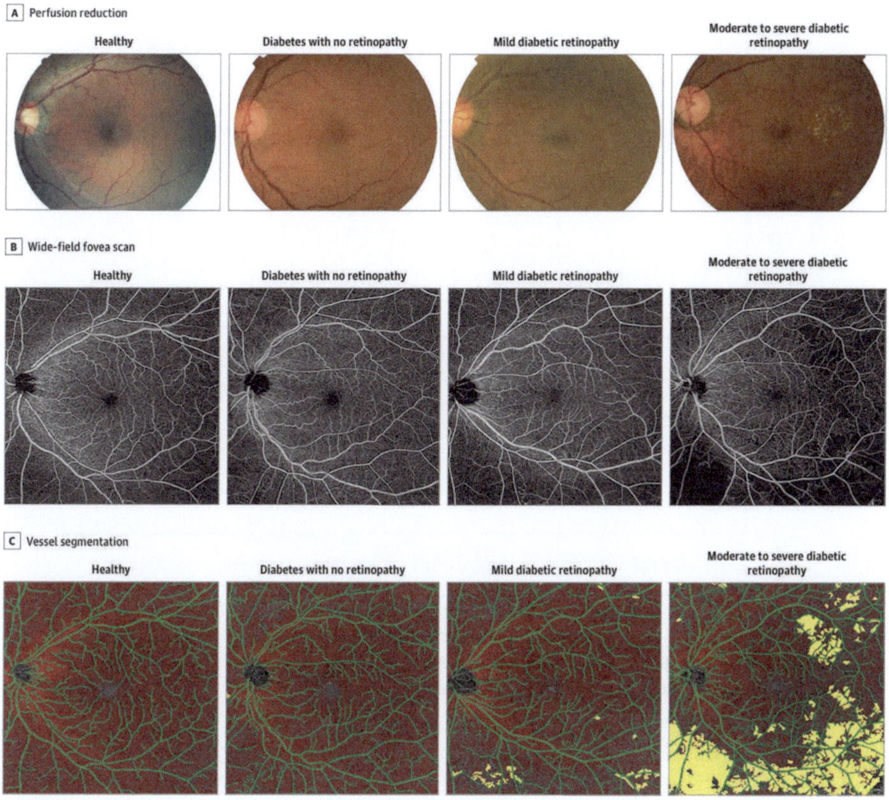

Fig. 9.35 Representative wide-field OCT angiography images of eyes with different diabetic retinopathy severities. Larger retinal vessel are segmented in green, areas of nonperfusion in yellow. (Reproduced with permission from Tan et al. 2020b)

to flicker stimulation and endothelial dysfunction has been observed in diabetes (Pemp et al. 2009b; Schirutschke et al. 2021). Interestingly, experimentally-induced hyperglycemia also impacts neurovascular coupling in the retina (Dorner et al. 2003b) which is compatible with results showing reduced flicker-induced retinal vascular reactivity in subjects with impaired glucose tolerance (Patel et al. 2012). Abnormal retinal neurovascular coupling has also been observed in animal models of DR elucidating some of the mechanisms underlying this effect (Hanaguri et al. 2021; Mishra and Newman 2010, 2012). The upregulation of NOS2 in retinal neurons and glial cells appears to play an important role in affecting neurovascular coupling in diabetes. The mechanism for the NO-induced reduction in neurovascular coupling is believed to include the blockade of production of vasoactive factors released from Müller cells (Nippert and Newman 2020). This mechanism is supported by results showing that the NOS2 inhibitor aminoguanidine restores neurovascular coupling in the diabetic rat retina (Mishra and Newman 2012).

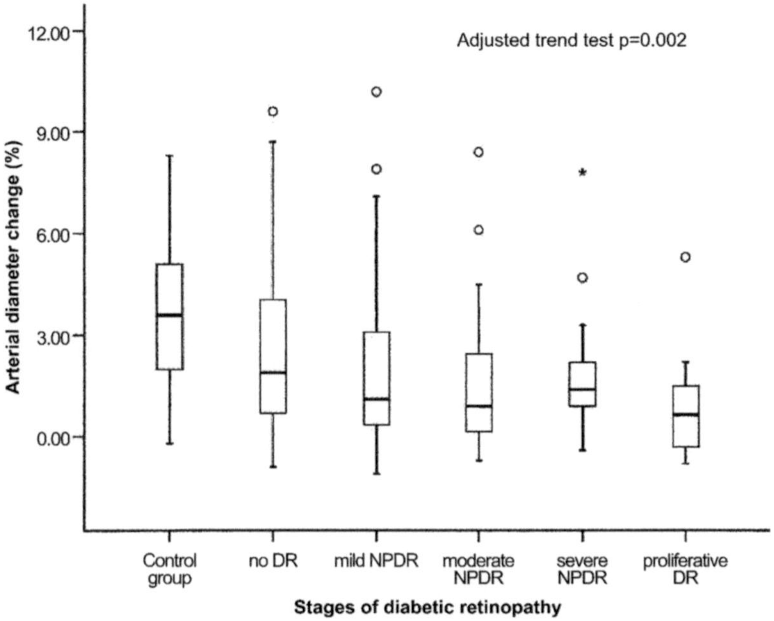

Fig. 9.36 Arterial diameter changes in response to stimulation with diffuse luminance flicker at different stages of diabetic retinopathy (DR). (Reproduced with permission from Mandecka et al. 2007)

Endothelial dysfunction is a key feature of diabetes (Shi and Vanhoutte 2017; Schiekofer et al. 2000; Triggle and Ding 2010) and can also be detected at the level of ocular vasculature (Toda and Nakanishi-Toda 2007; Schmetterer et al. 1997e; Wright et al. 2020). Alterations in the endothelial surface layer, specifically in the glycocalyx, seem to play a role in the retinal distribution of red blood cells, explaining reduced blood flow in the deep capillary plexus (Harris et al. 2019).

9.5.3.3 Age-Related Macular Degeneration

Hypoxia and ischemia have been implicated in the pathophysiology of age-related macular degeneration to trigger angiogenesis and interplay with inflammation (Bressler 2009). This is supported by studies that have associated hypoxia-inducible transcription factors (HIFs) with choroidal neovascularization (CNV) (Mammadzada et al. 2020). As expected, choroidal blood flow is reduced in age-related macular degeneration (Pemp and Schmetterer 2008; Grunwald et al. 1998; Metelitsina et al. 2006; Chen et al. 2001; Üretmen et al. 2003; Ciulla et al. 2002). Patients with AMD also show abnormal choroidal autoregulation (Metelitsina et al. 2010). With increasing severity of the disease, choroidal blood flow gradually decreases (Berenberg et al. 2012; Grunwald et al. 2005). Low choroidal blood flow has been shown to be a risk factor for the development of

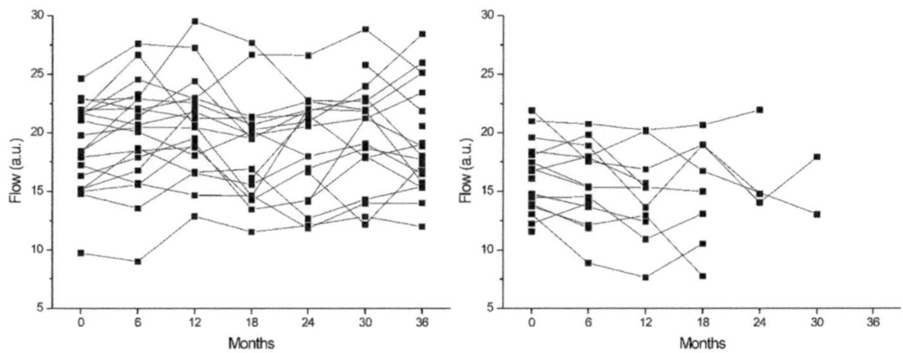

Fig. 9.37 Time course of choroidal blood flow (FLOW) during the 36-month observation period. Left: the patients with no choroidal neovascularization (CNV) in the study eye; right: the patients with CNV. The development of CNV was the end point, after which no further measurements were obtained. (Reproduced with permission from Boltz et al. 2010)

CNV (Fig. 9.37) (Metelitsina et al. 2008; Boltz et al. 2010; Xu et al. 2010). The relationship between delayed patchy choroidal filling and pseudodrusen was also established (Zhou et al. 2017).

Choroidal vascularity index is defined as the ratio of vascular area to the total choroidal area in OCT images (Betzler et al. 2022). Choroidal vascularity index was consistently found to be lower in AMD eyes compared to healthy eyes (Koh et al. 2017; Wei et al. 2017; Keenan et al. 2020). The parameter is associated with the rate of geographic atrophy enlargement in patients with AMD (Giannaccare et al. 2020; Sacconi et al. 2022).

With the introduction of OCT angiography, our understanding of disease progression has increased (Spaide et al. 2018). Evidence has indicated that ischemic features of the choriocapillaris are related to the progression of early AMD, CNV and geographic atrophy. Eyes with unilateral Type 3 CNV show more pronounced choriocapillaris nonperfusion areas as compared to nonexudative fellow eyes (Borrelli et al. 2018). This is in agreement with other findings showing that CNV vascular complexity is correlated with choriocapillaris flow deficits adjacent to the CNV area (Alagorie et al. 2020; Nesper et al. 2021).

Locally there is a large overlap between areas of macular atrophy and areas of choriocapillaris dropout in AMD (Takasago et al. 2019). More importantly, abnormalities in choriocapillaris perfusion in the macular area contribute to disease progression in eyes with geographic atrophy (Fig. 9.38) (Shi et al. 2021; You et al. 2021; Corvi et al. 2021a, 2021b; Corradetti et al. 2021). This is compatible with histological studies showing that choriocapillaris breakdown precedes retinal degeneration (Biesemeier et al. 2014). In addition, choriocapillaris flow deficits predict the development and enlargement of drusen (Nassisi et al. 2019).

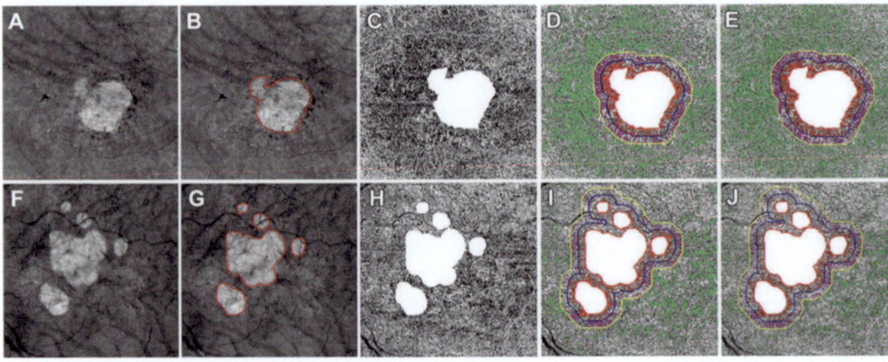

Fig. 9.38 Choriocapillaris (CC) analysis in two eyes with geographic atrophy (GA) secondary to nonexudative age-related macular degeneration imaged using swept source OCTA at the baseline visit using 6 × 6 mm scan patterns. (**a** and **f**): GA lesions identified as a bright area on en face structure images using a custom slab from 64 μm to 400 μm under Bruch's membrane (BM) at the baseline visit from two eyes. (**b** and **g**): Manually outlined GA lesions at the baseline visits for the two eyes (red outlines). (**c** and **h**): CC en face flow images using a 15 μm thick slab with inner boundary located 4 μm under BM. The area of GA was masked as white and excluded from CC analysis. (**d** and **i**): CC flow deficits (FDs) analyzed in different regions when using a global threshold method. White: The area of GA lesions which was excluded from CC FDs analysis. R1: CC FDs were colored coded as red in R1 region extending from 0 μm to 300 μm outside of GA border. R2: CC FDs were colored coded as purple in R2 region extending from >300 μm to 600 μm outside of GA border. R3: CC FDs were colored coded as green in the R3 region correspond to the total scan region minus the GA region, R1 and R2. E and J: CC FDs analyzed in different regions when using a local threshold method (Phansalkar method with a window radius of 3 pixels = 18 μm). White: The area of GA lesions which was excluded from CC FDs analysis. R1: CC FDs were colored coded as red CC FDs in R1 region extending from 0 μm to 300 μm outside of GA border. R2: CC FDs were colored coded as purple in R2 region extending from >300 μm to 600 μm outside of GA border. R3: CC FDs were colored coded as green in the R3 region corresponds to the total scan region minus the GA region, R1 and R2. (Reproduced with permission from Shi et al. 2021)

9.5.3.4 Systemic Disease

Systemic hypertension is a risk factor for major age-related eye diseases including AMD (Katsi et al. 2015), DR (Yamazaki et al. 2018), and glaucoma (Nislawati et al. 2021). As such, it is thought that high blood pressure amplifies perfusion problems that are associated with these diseases. As expected, optic nerve head blood flow autoregulation is altered in patients with systemic hypertension (Fig. 9.39) (Hashimoto et al. 2017c). Given the mechanisms of autoregulation as described in Fig. 9.18, one would assume that the upper limit of autoregulation will be reached earlier if the starting point is shifted toward higher levels. OCT angiography studies have revealed that hypertension is associated with microvascular changes in both the choroid and the retina (Fig. 9.40) (Monteiro-Henriques et al. 2022; Zeng et al. 2022; Tan et al. 2021b, 2022b; Anjos et al. 2021; Nolde et al. 2021; Sun et al. 2020; Chua et al. 2019a; b, 2021). Patients with hypertension who exhibit signs of left ventricular remodeling show altered retinal capillary density at the superficial retinal plexus on

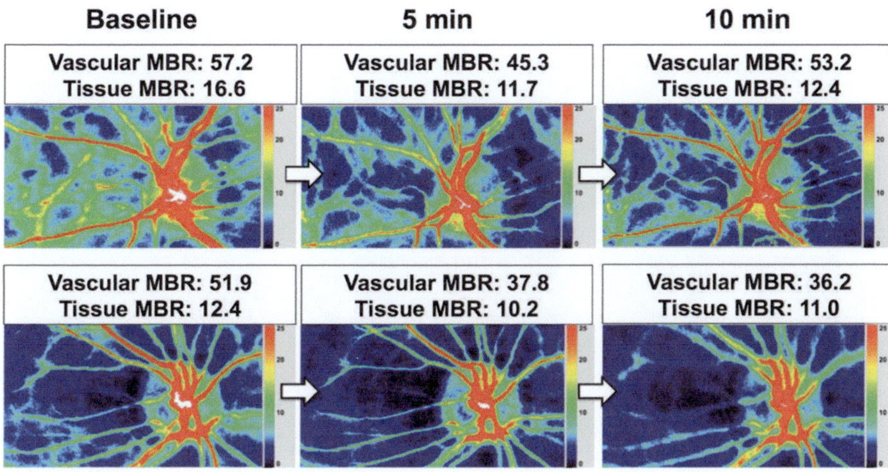

Fig. 9.39 Illustrative laser speckle flowgraphy images from subjects in the control and hypertension and hyperlipidemia groups. (Top) Laser speckle flowgraphy color maps of optic nerve head (ONH) blood flow obtained during vitreous surgery to treat an epiretinal membrane in a 67-year-old woman without systemic disorders. The images shown were obtained at baseline and 5 and 10 mins after the elevation of intraocular pressure by approximately 15 mmHg. The vascular and tissue mean blur rates are presented in the upper panels of each figure part. (Bottom) Laser speckle flowgraphy color maps of ONH blood flow obtained in a 69-year-old woman with hypertension and hyperlipidemia during vitreous surgery to treat a macular hole. (Reproduced with permission from Hashimoto et al. 2017c)

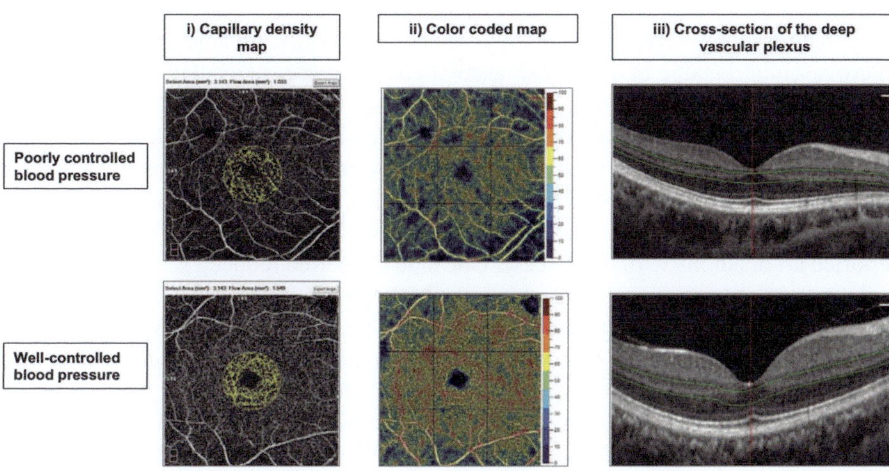

Fig. 9.40 Capillary density map of the macular region showing retinal microvasculature of participants with (**a**) poorly controlled blood pressure (less dense) and (**b**) well controlled blood pressure (more dense). (i) Capillary density extracted map with a circular (diameter of 1.0 mm centered on the fovea) measurement region defined. (ii) Capillary density color-coded map. (iii) Deep vascular plexus (slab boundary of 15–70 mm below the inner plexiform layer). (Reproduced with permission from Chua et al. 2019b)

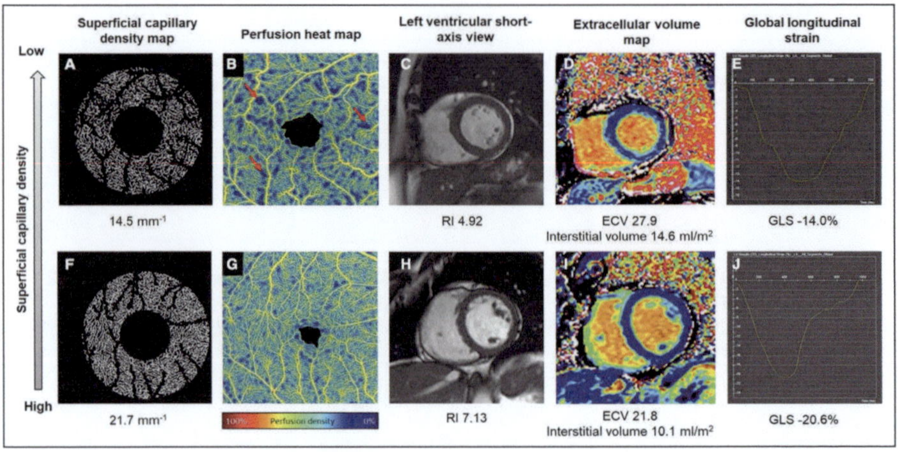

Fig. 9.41 Lower superficial capillary density in a patient with hypertension (**a** and **b**; 14.5 mm^{-1}) with (**c**) lower remodeling index, (**d**) higher interstitial volume and extracellular volume fraction, and (**e**) worse global longitudinal strain as compared with a denser superficial capillary density in a patient with hypertension (**f** and **g**; 21.7 mm^{-1}) with (**h**) higher remodeling index, (**i**) lower interstitial volume, and (**j**) better global longitudinal strain. Presence of larger nonperfused area (blue regions demarcated by red arrows) (**b**) can be seen in patient with adverse cardiac remodeling markers as compared with (**c**) a patient with favorable cardiac remodeling markers. Foveal avascular zone area (in black; **b** and **g**) and large vessels (seen as a black outline of vessels in **a** and **f**) were excluded from the calculation of capillary density. ECV indicates extracellular volume fraction; GLS, global longitudinal strain; and RI, remodeling index. (Reproduced from Chua et al. 2022)

OCT angiography. This provides a unique opportunity of studying the role of microvascular damage in the cardiovascular consequences of systemic hypertension ((Chua et al. 2022), Fig. 9.41).

Ocular perfusion defects have also been detected in neurodegenerative diseases of the brain. Two studies have reported that retinal venular narrowing is associated with Alzheimer's disease (Frost et al. 2013; Cheung et al. 2014). In addition, these studies investigated parameters such as fractal dimensions, tortuosity, and branching and reported associations with the disease. One study reported reduced retinal blood flow and oxygen extractions in patients with Alzheimer's disease in line with OCT angiography findings (Szegedi et al. 2020b). In the longitudinal Rotterdam Study, however, retinal vascular features were not associated with incident dementia (Schrijvers et al. 2012). Multiple studies using OCT angiography have detected retinal microvascular changes in Alzheimer's disease and mild cognitive impairment (Fig. 9.42) (Yeh et al. 2022; Chalkias et al. 2022; Jin et al. 2021; Zabel et al. 2021; Rifai et al. 2021; Wang et al. 2021; Chua et al. 2020b; Wu et al. 2020). In multiple sclerosis, the pattern appears to be more complex, because increased and decreased blood flows in the retina have been indicated (Wang et al. 2018; Kallab et al. 2021b; Chen et al. 2021; Svrčinová et al. 2020; Liu et al. 2019; Wang et al. 2014; Akarsu et al. 2004). This may be related to the level of inflammation in these patients, which may cause vasodilation and counteract the state of vasoconstriction that is usually associated with neurodegenerative diseases.

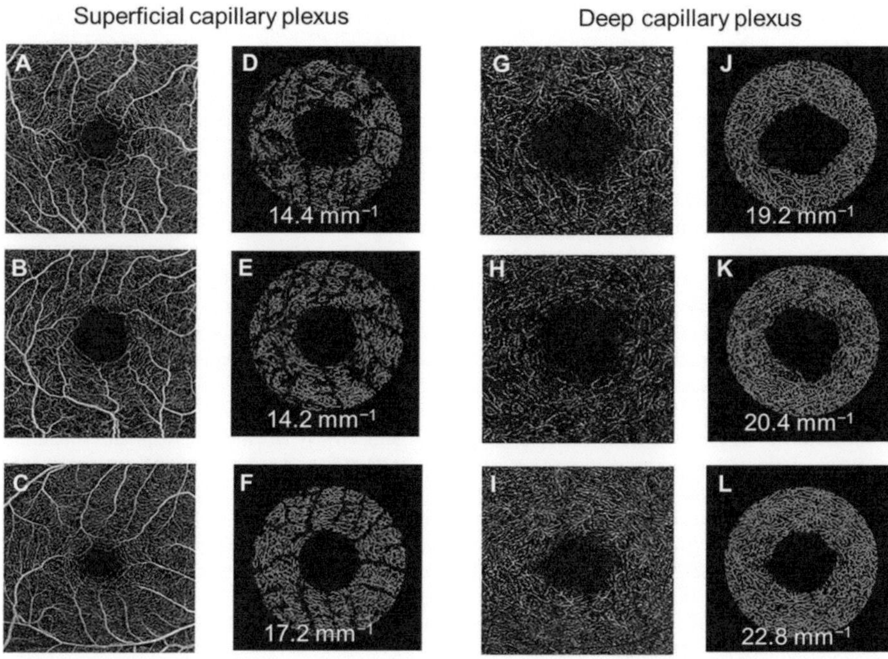

Fig. 9.42 Optical coherence tomography angiography (OCTA) images of the superficial (**a**–**c**) and deep (**g**–**i**) capillary plexuses were extracted from the OCTA machines. (**d**–**f**, **j**–**l**) Vessel density maps of the macular annulus region showing retinal microvasculature of participants with Alzheimer's disease (AD; **d**, **j**), mild cognitive impairment (MCI; **e**, **k**), and controls (**f**, **l**). AD participants showed a decrease in vessel densities in both plexuses compared to controls. MCI participants showed a decrease vessel density only in superficial capillary plexus and not the deep capillary plexus. (Reproduced with permission from Chua et al. 2020b)

References

Abran D et al (1995) Characterization and ontogeny of PGE2 and PGF2α receptors on the retinal vasculature of the pig. Prostaglandins 50(5):253–267

Abran D et al (1997) Characterization and regulation of prostaglandin E2 receptor and receptor-coupled functions in the choroidal vasculature of the pig during development. Circ Res 80(4): 463–472

Ahmed J, Linsenmeier RA, Dunn R (1994) The oxygen distribution in the prelaminar optic nerve head of the cat. Exp Eye Res 59(4):457–466

Aizawa N et al (2014) Laser speckle and hydrogen gas clearance measurements of optic nerve circulation in albino and pigmented rabbits with or without optic disc atrophy. Invest Ophthalmol Vis Sci 55(12):7991–7996

Akarsu C, Tan FU, Kendi T (2004) Color Doppler imaging in optic neuritis with multiple sclerosis. Graefes Arch Clin Exp Ophthalmol 242(12):990–994

Akitegetse C et al (2022) Monte-Carlo simulation and tissue-phantom model for validation of ocular oximetry. Biomed Opt Express 13(5):2929–2946

Al Zoubi H et al (2022) Optic disc blood perfusion and oxygenation in glaucoma. Graefes Arch Clin Exp Ophthalmol 260:3587

Alagorie AR et al (2020) Quantitative assessment of choriocapillaris flow deficits surrounding choroidal neovascular membranes. Retina 40(11):2106–2112

Alarcon-Martinez L et al (2020) Interpericyte tunnelling nanotubes regulate neurovascular coupling. Nature 585(7823):91–95

Alarcon-Martinez L et al (2022) Pericyte dysfunction and loss of interpericyte tunneling nanotubes promote neurovascular deficits in glaucoma. Proc Natl Acad Sci USA 119(7):e2110329119

Albanna W et al (2018) Non-invasive evaluation of neurovascular coupling in the murine retina by dynamic retinal vessel analysis. PLoS One 13(10):e0204689

Alder VA, Yu D-Y, Cringle SJ (1991) Vitreal oxygen tension measurements in the rat eye. Exp Eye Res 52(3):293–299

Alm A (1977) The effect of sympathetic stimulation on blood flow through the uvea, retina and optic nerve in monkeys (Macaca irus). Exp Eye Res 25(1):19–24

Alm A, Bill A (1972) The oxygen supply to the retina, I. Effects of changes in intraocular and arterial blood pressures, and in arterial PO2 and PCO2 on the oxygen tension in the vitreous body of the cat. Acta Physiol Scand 84(2):261–274

Alm A, Bill A (1973) Ocular and optic nerve blood flow at normal and increased intraocular pressures in monkeys (Macaca irus): a study with radioactively labelled microspheres including flow determinations in brain and some other tissues. Exp Eye Res 15(1):15–29

An L, Wang RK (2008) In vivo volumetric imaging of vascular perfusion within human retina and choroids with optical micro-angiography. Opt Express 16(15):11438–11452

Anand-Apte B, Hollyfield JG (2009) Developmental anatomy of the retinal and choroidal vasculature. In: Besharse J, Bok D (eds) Encyclopedia of the eye. Academic Press; Elsevier Books, London, p 9e15

Anderson DR (1996) Glaucoma, capillaries and pericytes. 1. Blood flow regulation. Ophthalmologica 210(5):257–262

Ang M et al (2018a) Optical coherence tomography angiography: a review of current and future clinical applications. Graefes Arch Clin Exp Ophthalmol 256(2):237–245

Ang M et al (2018b) Anterior segment optical coherence tomography. Prog Retin Eye Res 66:132–156

Anjos R et al (2021) Application of optical coherence tomography angiography macular analysis for systemic hypertension. A systematic review and meta-analysis. Am J Hypertens 35(4): 356–364

Arend O et al (1991) Retinal microcirculation in patients with diabetes mellitus: dynamic and morphological analysis of perifoveal capillary network. Br J Ophthalmol 75(9):514

Arthur E et al (2019) Distances from capillaries to arterioles or venules measured using OCTA and AOSLO. Invest Ophthalmol Vis Sci 60(6):1833–1844

Aschinger GC et al (2015) Blood flow velocity vector field reconstruction from dual-beam bidirectional Doppler OCT measurements in retinal veins. Biomed Opt Express 6(5):1599–1615

Aschinger GC et al (2017) Effect of diffuse luminance flicker light stimulation on total retinal blood flow assessed with dual-beam bidirectional Doppler OCT. Invest Ophthalmol Vis Sci 58(2): 1167–1178

Ashby JW, Mack JJ (2021) Endothelial control of cerebral blood flow. Am J Pathol 191(11): 1906–1916

Astin M, Stjernschantz J, Selén G (1994) Role of nitric oxide in PGF2α-induced ocular hyperemia. Exp Eye Res 59(4):401–408

Attwell D et al (2010) Glial and neuronal control of brain blood flow. Nature 468(7321):232–243

Aydin F, Rosenblum WI, Povlishock JT (1991) Myoendothelial junctions in human brain arterioles. Stroke 22(12):1592–1597

Balaratnasingam C et al (2018) Comparisons between histology and optical coherence tomography angiography of the periarterial capillary-free zone. Am J Ophthalmol 189:55–64

Baldoncini M et al (2019) Microsurgical anatomy of the central retinal artery. World Neurosurg 130:e172–e187

Barabas P et al (2020) Chapter seven –h ion channels and myogenic activity in retinal arterioles. In: Jackson WF (ed) Current topics in membranes. Academic Press, pp 187–226
Barton M, Yanagisawa M (2019) Endothelin: 30 years from discovery to therapy. Hypertension 74(6):1232–1265
Bata AM et al (2019a) Age-related decline of retinal oxygen extraction in healthy subjects. Invest Ophthalmol Vis Sci 60(8):3162–3169
Bata AM et al (2019b) Optic nerve head blood flow regulation during changes in arterial blood pressure in patients with primary open-angle glaucoma. Acta Ophthalmol 97(1):e36–e41
Baumann B et al (2011) Total retinal blood flow measurement with ultrahigh speed swept source/ Fourier domain OCT. Biomed Opt Express 2(6):1539–1552
Beach JM et al (1999) Oximetry of retinal vessels by dual-wavelength imaging: calibration and influence of pigmentation. J Appl Physiol 86(2):748–758
Bek T (2018) Is lack of autonomic nerves in retinal vessels a protection from electrical stimulation generated by activity in visual neurons? Acta Ophthalmol 96(2):e263–e263
Bennett AG, Rudnicka AR, Edgar DF (1994) Improvements on Littmann's method of determining the size of retinal features by fundus photography. Graefes Arch Clin Exp Ophthalmol 232(6): 361–367
Ben-Sira I, Riva CE, Roberts W (1973) Fluorophotometric recording of fluorescein dilution curves in human retinal vessels. Invest Ophthalmol Vis Sci 12(4):310–312
Berenberg TL et al (2012) The association between drusen extent and foveolar choroidal blood flow in age-related maculare degeneration. Retina 32(1):25–31
Bergua A et al (2013) Innervation pattern of the preocular human central retinal artery. Exp Eye Res 110:142–147
Berisha F et al (2010) A study comparing ocular pressure pulse and ocular fundus pulse in dependence of axial eye length and ocular volume. Acta Ophthalmol 88(7):766–772
Betzler BK et al (2022) Choroidal vascularity index: a step towards software as a medical device. Br J Ophthalmol 106(2):149
Bhagat N et al (2009) Diabetic macular edema: pathogenesis and treatment. Surv Ophthalmol 54(1): 1–32
Biesemeier A et al (2014) Choriocapillaris breakdown precedes retinal degeneration in age-related macular degeneration. Neurobiol Aging 35(11):2562–2573
Bill A (1985) Some aspects of the ocular circulation. Friedenwald lecture. Invest Ophthalmol Vis Sci 26(4):410–424
Bilotta J, Abramov I (1989) Orientation and direction tuning of goldfish ganglion cells. Vis Neurosci 2(1):3–13
Birol G et al (2007) Oxygen distribution and consumption in the macaque retina. Am J Physiol Heart Circ Physiol 293(3):H1696–H1704
Blatter C et al (2013a) Dove prism based rotating dual beam bidirectional Doppler OCT. Biomed Opt Express 4(7):1188–1203
Blatter C et al (2013b) Angle independent flow assessment with bidirectional Doppler optical coherence tomography. Opt Lett 38(21):4433–4436
Bojic L et al (2022) Brachial flow – mediated dilation and carotid intima – media thickness in glaucoma patients. BMC Ophthalmol 22(1):275
Boltz A et al (2010) Choroidal blood flow and progression of age-related macular degeneration in the fellow eye in patients with unilateral choroidal neovascularization. Invest Ophthalmol Vis Sci 51(8):4220–4225
Boltz A et al (2011) Effect of latanoprost on choroidal blood flow regulation in healthy subjects. Invest Ophthalmol Vis Sci 52(7):4410–4415
Boltz A et al (2013a) Optic nerve head blood flow autoregulation during changes in arterial blood pressure in healthy young subjects. PLoS One 8(12):e82351
Boltz A et al (2013b) Regulation of optic nerve head blood flow during combined changes in intraocular pressure and arterial blood pressure. J Cereb Blood Flow Metab 33(12):1850–1856

Boltz A et al (2013c) Role of endothelin-a receptors in optic nerve head red cell flux regulation during isometric exercise in healthy humans. Am J Physiol Heart Circ Physiol 304(1):H170–H174

Boned-Murillo A et al (2022) Optical coherence tomography angiography in diabetic patients: a systematic review. Biomedicine 10(1):88

Bonner RF, Nossal R (1990) Principles of laser Doppler flowmetry. In: Shepherd AP, Öberg PA (eds) Laser-Doppler blood flowmetry. Kluwer Academic Publishers, Boston, pp 57–72

Borrelli E et al (2018) Reduced choriocapillaris flow in eyes with type 3 neovascularization and age-related macular degeneration. Retina 38(10):1968–1976

Borrelli E et al (2021) Optical coherence tomography angiography in diabetes. Asia-Pacific J Ophthalmol 10(1):20–25

Bourouki E et al (2019) Pseudoexfoliative glaucoma, endothelial dysfunction, and arterial stiffness: the role of circulating apoptotic endothelial microparticles. J Glaucoma 28(8):749–755

Braun RD, Linsenmeier RA (1995) Retinal oxygen tension and the electroretinogram during arterial occlusion in the cat. Invest Ophthalmol Vis Sci 36(3):523–541

Bressler SB (2009) Introduction: understanding the role of angiogenesis and antiangiogenic agents in age-related macular degeneration. Ophthalmology 116(10):S1–S7

Briers JD, Fercher AF (1982) Retinal blood-flow visualization by means of laser speckle photography. Invest Ophthalmol Vis Sci 22(2):255–259

Briers D et al (2013) Laser speckle contrast imaging: theoretical and practical limitations. J Biomed Opt 18(6):066018

Brinks J et al (2022) Exploring the choroidal vascular labyrinth and its molecular and structural roles in health and disease. Prog Retin Eye Res 87:100994

Buckley C et al (2002) Systemic vascular endothelial cell dysfunction in normal pressure glaucoma. Br J Ophthalmol 86(2):227

Buerk DG, Riva CE, Cranstoun SD (1995) Frequency and luminance-dependent blood flow and K+ ion changes during flicker stimuli in cat optic nerve head. Invest Ophthalmol Vis Sci 36(11): 2216–2227

Buerk DG, Riva CE, Cranstoun SD (1996) Nitric oxide has a vasodilatory role in cat optic nerve head during flicker stimuli. Microvasc Res 52(1):13–26

Bukhari SMI et al (2016) Microvascular endothelial function and severity of primary open angle glaucoma. Eye 30(12):1579–1587

Burgansky-Eliash Z et al (2012) Increased retinal blood flow velocity in patients with early diabetes mellitus. Retina 32(1):112–119

Bursell SE et al (1995) The in vivo effect of endothelins on retinal circulation in nondiabetic and diabetic rats. Invest Ophthalmol Vis Sci 36(3):596–607

Bursell SE et al (1996) Retinal blood flow changes in patients with insulin-dependent diabetes mellitus and no diabetic retinopathy. Invest Ophthalmol Vis Sci 37(5):886–897

Busse R, Fleming I (2006) Vascular endothelium and blood flow. In: Moncada S, Higgs A (eds) The vascular endothelium II. Springer, Berlin/Heidelberg, pp 43–78

Calzetti G et al (2018) Assessment of choroidal blood flow using laser speckle flowgraphy. Br J Ophthalmol 102(12):1679

Calzetti G et al (2022) Measuring optic nerve head perfusion to monitor glaucoma: a study on structure–function relationships using laser speckle flowgraphy. Acta Ophthalmol 100(1):e181–e191

Cellini M et al (2012) Endothelin-1 plasma levels and vascular endothelial dysfunction in primary open angle glaucoma. Life Sci 91(13):699–702

Chakravarthy U et al (1992) The effect of endothelin 1 on the retinal microvascular pericyte. Microvasc Res 43(3):241–254

Chalam KV, Sambhav K (2016) Optical coherence tomography angiography in retinal diseases. J Ophthal Vis Res 11(1):84

Chalkias E et al (2022) Differentiating degenerative from vascular dementia with the help of optical coherence tomography angiography biomarkers. Healthcare (Basel) 10(3):539

Chan G et al (2012) Quantitative morphometry of perifoveal capillary networks in the human retina. Invest Ophthalmol Vis Sci 53(9):5502–5514

Chan-Ling T et al (2011) Role of CD44+ stem cells in mural cell formation in the human choroid: evidence of vascular instability due to limited pericyte ensheathment. Invest Ophthalmol Vis Sci 52(1):399–410

Charlson ME et al (2014) Nocturnal systemic hypotension increases the risk of glaucoma progression. Ophthalmology 121(10):2004–2012

Chauhan BC (2008) Endothelin and its potential role in glaucoma. Can J Ophthalmol 43(3): 356–360

Chawengsub Y, Gauthier KM, Campbell WB (2009) Role of arachidonic acid lipoxygenase metabolites in the regulation of vascular tone. Am J Physiol Heart Circ Physiol 297(2):H495–H507

Chen C-L, Wang RK (2017) Optical coherence tomography based angiography [invited]. Biomed Opt Express 8(2):1056–1082

Chen Z et al (1997) Optical Doppler tomographic imaging of fluid flow velocity in highly scattering media. Opt Lett 22(1):64–66

Chen S-J et al (2001) Pulsatile ocular blood flow in asymmetric exudative age related macular degeneration. Br J Ophthalmol 85(12):1411

Chen BR et al (2014) A critical role for the vascular endothelium in functional neurovascular coupling in the brain. J Am Heart Assoc 3(3):e000787

Chen S et al (2017) Retinal oximetry in humans using visible-light optical coherence tomography [invited]. Biomed Opt Express 8(3):1415–1429

Chen Q et al (2021) Retinal microvascular and neuronal function in patients with multiple sclerosis: 2-year follow-up. Mult Scler Relat Disord 56:103314

Cheng RW et al (2016) Relationship between retinal blood flow and arterial oxygen. J Physiol 594(3):625–640

Cherecheanu AP et al (2013) Ocular perfusion pressure and ocular blood flow in glaucoma. Curr Opin Pharmacol 13(1):36–42

Cheung CY-L et al (2014) Microvascular network alterations in the retina of patients with Alzheimer's disease. Alzheimers Dement 10(2):135–142

Choi W et al (2015) Ultrahigh-speed, swept-source optical coherence tomography angiography in nonexudative age-related macular degeneration with geographic atrophy. Ophthalmology 122(12):2532–2544

Chu Z et al (2016) Quantitative assessment of the retinal microvasculature using optical coherence tomography angiography. J Biomed Opt 6:066008

Chua J et al (2019a) Impact of systemic vascular risk factors on the choriocapillaris using optical coherence tomography angiography in patients with systemic hypertension. Sci Rep 9(1):5819

Chua J et al (2019b) Impact of hypertension on retinal capillary microvasculature using optical coherence tomographic angiography. J Hypertens 37(3):572–580

Chua J et al (2020a) Optical coherence tomography angiography in diabetes and diabetic retinopathy. J Clin Med 9(6):1723

Chua J et al (2020b) Retinal microvasculature dysfunction is associated with Alzheimer's disease and mild cognitive impairment. Alzheimer's Res Ther 12(1):161

Chua J et al (2021) Choriocapillaris microvasculature dysfunction in systemic hypertension. Sci Rep 11(1):4603

Chua J et al (2022) Relationship of quantitative retinal capillary network and myocardial remodeling in systemic hypertension. J Am Heart Assoc 11(6):e024226

Chung HS et al (1999) Progress in measurement of ocular blood flow and relevance to our understanding of glaucoma and age-related macular degeneration. Prog Retin Eye Res 18(5): 669–687

Ciulla TA et al (2002) Choroidal perfusion perturbations in non-neovascular age related macular degeneration. Br J Ophthalmol 86(2):209

Claassen JAHR et al (2021) Regulation of cerebral blood flow in humans: physiology and clinical implications of autoregulation. Physiol Rev 101(4):1487–1559
Corradetti G et al (2021) Scotopic microperimetric sensitivity and inner choroid flow deficits as predictors of progression to nascent geographic atrophy. Br J Ophthalmol 105(11):1584
Corvetti G, Pignocchino P, Sisto Daneo L (1988) Distribution and development of substance P immunoreactive axons in the chick cornea and uvea. Basic Appl Histochem 32(1):187–192
Corvi F et al (2018) Reproducibility of vessel density, fractal dimension, and foveal avascular zone using 7 different optical coherence tomography angiography devices. Am J Ophthalmol 186: 25–31
Corvi F et al (2021a) Topography of choriocapillaris flow deficit predicts development of neovascularization or atrophy in age-related macular degeneration. Graefes Arch Clin Exp Ophthalmol 259(10):2887–2895
Corvi F et al (2021b) Choriocapillaris flow deficits as a risk factor for progression of age-related macular degeneration. Retina 41(4):686–693
Costa VP, Arcieri ES, Harris A (2009) Blood pressure and glaucoma. Br J Ophthalmol 93(10):1276
Costa VP et al (2014) Ocular perfusion pressure in glaucoma. Acta Ophthalmol 92(4):e252–e266
Cour MI et al (2000) Optic nerve oxygen tension: effects of intraocular pressure and dorzolamide. Br J Ophthalmol 84(9):1045
Cull G et al (2015) Compromised optic nerve blood flow and autoregulation secondary to neural degeneration. Invest Ophthalmol Vis Sci 56(12):7286–7292
Cunha-Vaz J (2016) Mechanisms of retinal fluid accumulation and blood-retinal barrier breakdown. Dev Ophthalmol 58:11–20
Dai C et al (2013) Absolute retinal blood flow measurement with a dual-beam doppler optical coherence tomography. Invest Ophthalmol Vis Sci 54(13):7998–8003
Dallinger S et al (2000) Endothelin-1 contributes to hyperoxia-induced vasoconstriction in the human retina. Invest Ophthalmol Vis Sci 41(3):864–869
Dallinger S et al (2003) Vasodilator effects of l-arginine are stereospecific and augmented by insulin in humans. Am J Physiol Endocrinol Metab 284(6):E1106–E1111
De Vriese AS et al (2000) Endothelial dysfunction in diabetes. Br J Pharmacol 130(5):963–974
Delaey C, van de Voorde J (2000) Regulatory mechanisms in the retinal and choroidal circulation. Ophthalmic Res 32(6):249–256
Delaey C et al (2007) The endothelium-derived hyperpolarising factor (EDHF) in isolated bovine choroidal arteries. Exp Eye Res 84(6):1067–1073
Deokule S et al (2010) Association of visual field severity and parapapillary retinal blood flow open-angle glaucoma. J Glaucoma 19(5):293–298
DePaoli DT et al (2019) Convolutional neural networks for spectroscopic analysis in retinal oximetry. Sci Rep 9(1):11387
Desjardins M et al (2016) Preliminary investigation of multispectral retinal tissue oximetry mapping using a hyperspectral retinal camera. Exp Eye Res 146:330–340
Dimitrova G, Kato S (2010) Color Doppler imaging of retinal diseases. Surv Ophthalmol 55(3): 193–214
Ding J et al (2014) Retinal vascular caliber and the development of hypertension: a meta-analysis of individual participant data. J Hypertens 32(2):225
Doblhoff-Dier V et al (2014) Measurement of the total retinal blood flow using dual beam Fourier-domain Doppler optical coherence tomography with orthogonal detection planes. Biomed Opt Express 5(2):630–642
Dodo Y et al (2017) Clinical relevance of reduced decorrelation signals in the diabetic inner choroid on optical coherence tomography angiography. Sci Rep 7(1):5227
Donati G et al (1995) Nitric oxide controls arteriolar tone in the retina of the miniature pig. Invest Ophthalmol Vis Sci 36(11):2228–2237
Dons-Jensen A et al (2019) The diameter of retinal arterioles is unaffected by intravascular administration of the adenosine a2A receptor agonist regadenoson in normal persons. Biomedicine Hub 4(2):1–10

Dorner GT et al (2002a) Calculation of the diameter of the central retinal artery from noninvasive measurements in humans. Curr Eye Res 25:341–345
Dorner GT et al (2002b) Response of retinal blood flow to CO2-breathing in humans. Eur J Ophthalmol 12(6):459–466
Dorner GT et al (2003a) Nitric oxide regulates retinal vascular tone in humans. Am J Physiol Heart Circ Physiol 285(2):H631–H636
Dorner GT et al (2003b) Hyperglycemia affects flicker-induced vasodilation in the retina of healthy subjects. Vis Res 43(13):1495–1500
Duan A et al (2016) Evidence of flicker-induced functional hyperaemia in the smallest vessels of the human retinal blood supply. PLoS One 11(9):e0162621
Duijm HFA et al (1996) Study of choroidal blood flow by comparison of SLO fluorescein angiography and microspheres. Exp Eye Res 63(6):693–704
Dumskyj MJ et al (1996) Autoregulation in the human retinal circulation: assessment using isometric exercise, laser Doppler velocimetry, and computer-assisted image analysis. Microvasc Res 51(3):378–392
Enomoto N et al (2021) Characterization of laser speckle flowgraphy pulse waveform parameters for the evaluation of the optic nerve head and retinal circulation. Sci Rep 11(1):6847
Enroth-Cugell C, Goldstick TK, Linsenmeier RA (1980) The contrast sensitivity of cat retinal ganglion cells at reduced oxygen tensions. J Physiol 304(1):59–81
Ento Key (2025) Ocular Circulation, https://entokey.com/ocular-circulation-4/
Erdener S, Küreli G, Dalkara T (2022) Contractile apparatus in CNS capillary pericytes. Neurophotonics 9(2):021904
Erdinest N et al (2021) Nitric oxide interaction with the eye. Vision (Basel) 5(2):29
Ernst C, Aalkjær C, Bek T (2022) ATP induced calcium signaling activity in perivascular cells differ at different vascular branch levels in the porcine retina. Microvasc Res 139:104256
Evans DW et al (1999) Glaucoma patients demonstrate faulty autoregulation of ocular blood flow during posture change. Br J Ophthalmol 83(7):809
Fallon TJ, Maxwell DL, Kohner EM (1987) Autoregulation of retinal blood flow in diabetic retinopathy measured by the blue-light entoptic technique. Ophthalmology 94(11):1410–1415
Falsini B, Riva CE, Logean E (2002) Flicker-evoked changes in human optic nerve blood flow: relationship with retinal neural activity. Invest Ophthalmol Vis Sci 43(7):2309–2316
Faraci FM, Baumbach GL, Heistad DD (1989a) Myogenic mechanisms in the cerebral circulation. J Hypertens 7(4):S61–S64
Faraci FM et al (1989b) Atherosclerosis potentiates constrictor responses of cerebral and ocular blood vessels to thromboxane in monkeys. Stroke 20(2):242–247
Faraci FM et al (2019) Acid-sensing ion channels. Circ Res 125(10):907–920
Feke GT, Pasquale LR (2008) Retinal blood flow response to posture change in glaucoma patients compared with healthy subjects. Ophthalmology 115(2):246–252
Felder AE et al (2015) Inner retinal oxygen extraction fraction in response to light flicker stimulation in humans. Invest Ophthalmol Vis Sci 56(11):6633–6637
Findl O et al (2000) Ocular haemodynamics and colour contrast sensitivity in patients with type 1 diabetes. Br J Ophthalmol 84(5):493
Fondi K et al (2016) Measurement of retinal vascular caliber from optical coherence tomography phase images. Invest Ophthalmol Vis Sci 57(9):OCT121–OCT129
Fondi K et al (2017) Retinal oxygen extraction in individuals with type 1 diabetes with no or mild diabetic retinopathy. Diabetologia 60(8):1534–1540
Fondi K et al (2018) Evaluation of flicker induced hyperemia in the retina and optic nerve head measured by laser speckle flowgraphy. PLoS One 13(11):e0207525
Formaz F, Riva CE, Geiser M (1997) Diffuse luminance flicker increases retinal vessel diameter in humans. Curr Eye Res 16(12):1252–1257
Frost S et al (2013) Retinal vascular biomarkers for early detection and monitoring of Alzheimer's disease. Transl Psychiatry 3(2):e233–e233

Fuchsjäger-Mayrl G et al (2003) Role of endothelin-1 in choroidal blood flow regulation during isometric exercise in healthy humans. Invest Ophthalmol Vis Sci 44(2):728–733
Fuchsjäger-Mayrl G et al (2004) Ocular blood flow and systemic blood pressure in patients with primary open-angle glaucoma and ocular hypertension. Invest Ophthalmol Vis Sci 45(3): 834–839
Fuchsjäger-Mayrl G et al (2010) Effect of dorzolamide and timolol on ocular pressure: blood flow relationship in patients with primary open-angle glaucoma and ocular hypertension. Invest Ophthalmol Vis Sci 51(3):1289–1296
Funatsu H et al (2006) Tracing method in the assessment of retinal capillary blood flow velocity by fluorescein angiography with scanning laser ophthalmoscope. Jpn J Ophthalmol 50(1):25–32
Furchgott RF, Vanhoutte PM (1989) Endothelium-derived relaxing and contracting factors. FASEB J 3(9):2007–2018
Gabel VP, Birngruber R, Nasemann J (1988) The scanning laser ophthalmoscope and its use as a fluorescein angiography instrument. Fortschritte Ophthalmologie: Zeitschrift der Deutschen Ophthalmologischen Gesellschaft, 1988(0723–8045 (Print))
Galambos P et al (2006) Compromised autoregulatory control of ocular hemodynamics in glaucoma patients after postural change. Ophthalmology 113(10):1832–1836
Galassi F et al (2003) Ocular hemodynamics and glaucoma prognosis: a color Doppler imaging study. Arch Ophthalmol 121(12):1711–1715
Gallice M et al (2017) Hypoxic, hypercapnic, and hyperoxic responses of the optic nerve head and subfoveal choroid blood flow in healthy humans. Invest Ophthalmol Vis Sci 58(12):5460–5467
Garhöfer G et al (2003) Flicker light–induced vasodilatation in the human retina: effect of lactate and changes in mean arterial pressure. Invest Ophthalmol Vis Sci 44(12):5309–5314
Garhöfer G et al (2004a) Diffuse luminance flicker increases blood flow in major retinal arteries and veins. Vis Res 44(8):833–838
Garhöfer G et al (2004b) Response of retinal vessel diameters to flicker stimulation in patients with early open angle glaucoma. J Glaucoma 13(4):340–344
Garhöfer G et al (2004c) Reduced response of retinal vessel diameters to flicker stimulation in patients with diabetes. Br J Ophthalmol 88(7):887
Garhofer G et al (2010) Use of the retinal vessel analyzer in ocular blood flow research. Acta Ophthalmol 88(7):717–722
Garhöfer G et al (2010) Retrobulbar blood flow velocities in open angle glaucoma and their association with mean arterial blood pressure. Invest Ophthalmol Vis Sci 51(12):6652–6657
Garhofer G et al (2012) Retinal blood flow in healthy young subjects. Invest Ophthalmol Vis Sci 53(2):698–703
Garhöfer G et al (2020) Retinal neurovascular coupling in diabetes. J Clin Med 9(9):2829
Garhüfer G et al (2002) Influence of diffuse luminance flicker on choroidal and optic nerve head blood flow. Curr Eye Res 24(2):109–113
Geevarghese A, Herman IM (2014) Pericyte-endothelial crosstalk: implications and opportunities for advanced cellular therapies. Transl Res 163(4):296–306
Geirsdottir A et al (2012) Retinal vessel oxygen saturation in healthy individuals. Invest Ophthalmol Vis Sci 53(9):5433–5442
Geiser MH et al (2000) Response of choroidal blood flow in the foveal region to hyperoxia and hyperoxia-hypercapnia. Curr Eye Res 21(2):669–676
Gherezghiher T, Okubo H, Koss MC (1991) Choroidal and ciliary body blood flow analysis: application of laser Doppler flowmetry in experimental animals. Exp Eye Res 53(2):151–156
Gherghel D et al (2000) Relationship between ocular perfusion pressure and retrobulbar blood flow in patients with glaucoma with progressive damage. Am J Ophthalmol 130(5):597–605
Giannaccare G et al (2020) Choroidal vascularity index quantification in geographic atrophy using binarization of enhanced-depth imaging optical coherence tomographic scans. Retina 40(5): 960–965
Gidday JM, Park TS (1993a) Microcirculatory responses to adenosine in the newborn pig retina. Pediatr Res 33(6):620–627

Gidday JM, Park TS (1993b) Adenosine-mediated autoregulation of retinal arteriolar tone in the piglet. Invest Ophthalmol Vis Sci 34(9):2713–2719

Gilmore ED et al (2005) Retinal arteriolar diameter, blood velocity, and blood flow response to an isocapnic hyperoxic provocation. Am J Physiol Heart Circu Physiol 288(6):H2912–H2917

Glucksberg MR, Dunn R (1993) Direct measurement of retinal microvascular pressures in the live, anesthetized cat. Microvasc Res 45(2):158–165

Godo S, Shimokawa H (2017) Endothelial functions. Arterioscler Thromb Vasc Biol 37(9):e108–e114

Granstam E, Granstam SO (1999) Regulation of uveal and retinal blood flow in STZ-diabetic and non-diabetic rats; involvement of nitric oxide. Curr Eye Res 19(4):330–337

Granstam E, Wang L, Bill A (1993) Vascular effects of endothelin-1 in the cat; modification by indomethacin and l-NAME. Acta Physiol Scand 148(2):165–176

Grover DS, Budenz DL (2011) Ocular perfusion pressure and glaucoma. Int Ophthalmol Clin 51(3): 19–25

Grunwald JE et al (1984) Retinal autoregulation in open-angle glaucoma. Ophthalmology 91(12): 1690–1694

Grunwald JE et al (1998) Foveolar choroidal blood flow in age-related macular degeneration. Invest Ophthalmol Vis Sci 39(2):385–390

Grunwald JE et al (2005) Reduced foveolar choroidal blood flow in eyes with increasing AMD severity. Invest Ophthalmol Vis Sci 46(3):1033–1038

Gugleta K et al (2012) Dynamics of retinal vessel response to flicker light in glaucoma patients and ocular hypertensives. Graefes Arch Clin Exp Ophthalmol 250(4):589–594

Gugleta K et al (2013) Retinal neurovascular coupling in patients with glaucoma and ocular hypertension and its association with the level of glaucomatous damage. Graefes Arch Clin Exp Ophthalmol 251(6):1577–1585

Guo X et al (2000) High glucose impairs endothelium-dependent relaxation in rabbit aorta. Acta Pharmacol Sin 21(2):169–173

Guo Y et al (2018) MEDnet, a neural network for automated detection of avascular area in OCT angiography. Biomed Opt Express 9(11):5147–5158

Guo S et al (2020) Association between caliber of retinal vessels and cardiovascular disease: a systematic review and meta-analysis. Curr Atheroscler Rep 22(4):16

Haefliger IO, Flammer J, Lüscher TF (1992) Nitric oxide and endothelin-1 are important regulators of human ophthalmic artery. Invest Ophthalmol Vis Sci 33(7):2340–2343

Haefliger IO et al (1994a) The vascular endothelium as a regulator of the ocular circulation: a new concept in ophthalmology? Surv Ophthalmol 39(2):123–132

Haefliger IO, Zschauer A, Anderson DR (1994b) Relaxation of retinal pericyte contractile tone through the nitric oxide-cyclic guanosine monophosphate pathway. Invest Ophthalmol Vis Sci 35(3):991–997

Haefliger IO et al (2001) Endothelium-dependent vasoactive modulation in the ophthalmic circulation. Prog Retin Eye Res 20(2):209–225

Haindl R et al (2016) Total retinal blood flow measurement by three beam Doppler optical coherence tomography. Biomed Opt Express 7(2):287–301

Hammer M et al (2008) Retinal vessel oximetry-calibration, compensation for vessel diameter and fundus pigmentation, and reproducibility. J Biomed Opt 5:054015

Hanaguri J et al (2021) Retinal blood flow dysregulation precedes neural retinal dysfunction in type 2 diabetic mice. Sci Rep 11(1):18401

Hardarson SH et al (2006) Automatic retinal oximetry. Invest Ophthalmol Vis Sci 47(11): 5011–5016

Hardy P et al (1994) Free radicals in retinal and choroidal blood flow autoregulation in the piglet: interaction with prostaglandins. Invest Ophthalmol Vis Sci 35(2):580–591

Hardy P et al (2000) Oxidants, nitric oxide and prostanoids in the developing ocular vasculature: a basis for ischemic retinopathy. Cardiovasc Res 47(3):489–509

Harris A et al (1996) Laser Doppler flowmetry measurement of changes in human optic nerve head blood flow in response to blood gas perturbations. J Glaucoma 5(4):258–265
Harris NR et al (2019) Blood flow distribution and the endothelial surface layer in the diabetic retina. Biorheology 56:181–189
Hashimoto R et al (2017a) Autoregulation of optic nerve head blood flow induced by elevated intraocular pressure during vitreous surgery. Curr Eye Res 42(4):625–628
Hashimoto R et al (2017b) Impaired autoregulation of blood flow at the optic nerve head during vitrectomy in patients with type 2 diabetes. Am J Ophthalmol 181:125–133
Hashimoto R et al (2017c) Impairment of autoregulation of optic nerve head blood flow during vitreous surgery in patients with hypertension and hyperlipidemia. Graefes Arch Clin Exp Ophthalmol 255(11):2227–2235
Haugh-Scheidt LM, Griff ER, Linsenmeier RA (1995) Light-evoked oxygen responses in the isolated toad retina. Exp Eye Res 61(1):73–81
Hayreh SS (2001) The blood supply of the optic nerve head and the evaluation of it — myth and reality. Prog Retin Eye Res 20(5):563–593
Heitmar R et al (2010) Continuous retinal vessel diameter measurements: the future in Rretinal vessel assessment? Invest Ophthalmol Vis Sci 51(11):5833–5839
Heitmar R, Kalitzeos AA, Panesar V (2015) Comparison of two formulas used to calculate summarized retinal vessel calibers. Optom Vis Sci 92(11):1085
Henkind P (1967) Radial peripapillary capillaries of the retina. I. Anatomy: human and comparative. Br J Ophthalmol 51(2):115
Henry E et al (1999) Peripheral endothelial dysfunction in normal pressure glaucoma. Invest Ophthalmol Vis Sci 40(8):1710–1714
Hétu S et al (2013) Assessment of retinal and choroidal blood flow changes using laser Doppler flowmetry in rats. Curr Eye Res 38(1):158–167
Hickam JB, Frayser R, Ross JC (1963) A study of retinal venous blood oxygen saturation in human subjects by photographic means. Circulation 27(3):375–385
Hirano T et al (2018) Wide-field en face swept-source optical coherence tomography angiography using extended field imaging in diabetic retinopathy. Br J Ophthalmol 102(9):1199
Hitchings RA, Spaeth GL (1977) Fluorescein angiography in chronic simple and low-tension glaucoma. Br J Ophthalmol 61(2):126
Hoffman RP et al (1999) Hyperglycemia without hyperinsulinemia produces both sympathetic neural activation and vasodilation in normal humans. J Diabetes Complicat 13(1):17–22
Hommer N et al (2021) Effect of hyperoxia and hypoxia on retinal vascular parameters assessed with optical coherence tomography angiography. Acta Ophthalmol 100:1–8
Hosseinaee Z et al (2018) Comparative study of optical coherence tomography angiography and phase-resolved doppler optical coherence tomography for measurement of retinal blood vessels caliber. Transl Vis Sci Technol 7(4):18–18
Hwang JC et al (2012) Relationship among visual field, blood flow, and neural structure measurements in glaucoma. Invest Ophthalmol Vis Sci 53(6):3020–3026
Hwang Y et al (2023) Retinal blood flow speed quantification at the capillary level using temporal autocorrelation fitting OCTA [invited]. Biomed Opt Express 14(6):2658–2677
Iadecola C (2017) The neurovascular unit coming of age: a journey through neurovascular coupling in health and disease. Neuron 96(1):17–42
Iadecola C, Zhang F (1996) Permissive and obligatory roles of NO in cerebrovascular responses to hypercapnia and acetylcholine. Am J Phys 271(4):R990–R1001
Ido Y et al (2001) NADH: sensor of blood flow need in brain, muscle, and other tissues. FASEB J 15(8):1419–1421
Ikram MK et al (2013a) Retinal vascular caliber measurements: clinical significance, current knowledge and future perspectives. Ophthalmologica 229(3):125–136
Ikram MK et al (2013b) Retinal vascular caliber as a biomarker for diabetes microvascular complications. Diabetes Care 36(3):750–759

Inada E et al (1986) Histamine antagonists and d-tubocurarine-induced hypotension in cardiac surgical patients. Clin Pharmacol Ther 40(5):575–580
Isono H et al (1997) Analysis of choroidal blood flow by laser speckle flowgraphy. Nippon Ganka Gakkai Zasshi 101(8):684–691
Iwase T et al (2018) Evaluation of optic nerve head blood flow in response to increase of intraocular pressure. Sci Rep 8(1):17235
Jani PD et al (2014) Normative values and predictors of retinal oxygen saturation. Retina 34(2): 394–401
Jeon SJ et al (2019) Association of retinal blood flow with progression of visual field in glaucoma. Sci Rep 9(1):16813
Jeon SJ et al (2022) Macular blood flow and pattern electroretinogram in normal tension glaucoma. J Clin Med 11(7):1790
Jeppesen SK, Bek T (2019) The retinal oxygen saturation measured by dual wavelength oximetry in larger retinal vessels is influenced by the linear velocity of the blood. Curr Eye Res 44(1):46–52
Jia Y et al (2012) Split-spectrum amplitude-decorrelation angiography with optical coherence tomography. Opt Express 20(4):4710–4725
Jin Q et al (2021) A systematic review and meta-analysis of retinal microvascular features in Alzheimer's disease. Front Aging Neurosci 13:1–11
Johannesen SK et al (2019) Optical coherence tomography angiography and microvascular changes in diabetic retinopathy: a systematic review. Acta Ophthalmol 97(1):7–14
Johnson MA et al (2005) Ocular structure and function in an aged monkey with spontaneous diabetes mellitus. Exp Eye Res 80(1):37–42
Johnson W et al (2007) Snapshot hyperspectral imaging in ophthalmology. J Biomed Opt 1:014036
Jung F et al (1983) Quantification of characteristic blood-flow parameters in the vessels of the retina with a picture analysis system for video-fluorescence angiograms: initial findings. Graefes Arch Clin Exp Ophthalmol 221(3):133–136
Kallab M et al (2021a) Plexus-specific effect of flicker-light stimulation on the retinal microvasculature assessed with optical coherence tomography angiography. Am J Physiol Heart Circ Physiol 320(1):H23–H28
Kallab M et al (2021b) Retinal oxygen metabolism and haemodynamics in patients with multiple sclerosis and history of optic neuritis. Front Neurosci 15
Kallab M et al (2022) Combining vascular and nerve fiber layer thickness measurements to model glaucomatous focal visual field loss. Ann N Y Acad Sci 1511(1):133–141
Kaluzny J et al (2017) Bayer filter snapshot hyperspectral fundus camera for human retinal imaging. Curr Eye Res 42(4):629–635
Kannenkeril D et al (2022) Retinal capillary damage is already evident in patients with hypertension and prediabetes and associated with HbA1c levels in the nondiabetic range. Diabetes Care 45(6):1472–1475
Kashani AH et al (2017) Optical coherence tomography angiography: a comprehensive review of current methods and clinical applications. Prog Retin Eye Res 60:66–100
Katsi VK et al (2015) Essential hypertension in the pathogenesis of age-related macular degeneration: a review of the current evidence. J Hypertens 33(12):2382–2388
Kawamura H et al (2002) Endothelin-induced changes in the physiology of retinal pericytes. Invest Ophthalmol Vis Sci 43(3):882–888
Keenan TD et al (2020) Choroidal thickness and vascularity vary with disease severity and subretinal drusenoid deposit presence in nonadvanced age-related macular degeneration. Retina 40(4):632–642
Kergoat H, Faucher C (1999) Effects of oxygen and carbogen breathing on choroidal hemodynamics in humans. Invest Ophthalmol Vis Sci 40(12):2906–2911
Khalid H et al (2021) Widefield optical coherence tomography angiography for early detection and objective evaluation of proliferative diabetic retinopathy. Br J Ophthalmol 105(1):118
Khuu LA et al (2017) Aqueous humor endothelin-1 and total retinal blood flow in patients with non-proliferative diabetic retinopathy. Eye 31(10):1443–1450

Kida T et al (2018) Data on the involvement of endothelin-1 (ET-1) in the dysregulation of retinal veins. Data Brief 21:59–62
Kiel JW (1994) Choroidal myogenic autoregulation and intraocular pressure. Exp Eye Res 58(5): 529–543
Kiel JW (1999) Modulation of choroidal autoregulation in the rabbit. Exp Eye Res 69(4):413–429
Kiel JW (2000) Endothelin modulation of choroidal blood flow in the rabbit. Exp Eye Res 71(6): 543–550
Kiel JW, Lovell MO (1996) Adrenergic modulation of choroidal blood flow in the rabbit. Invest Ophthalmol Vis Sci 37(4):673–679
Kiel JW, Shepherd AP (1992) Autoregulation of choroidal blood flow in the rabbit. Invest Ophthalmol Vis Sci 33(8):2399–2410
Kiel JW, van Heuven WA (1995) Ocular perfusion pressure and choroidal blood flow in the rabbit. Invest Ophthalmol Vis Sci 36(3):579–585
Kim KJ et al (2015) Astrocyte contributions to flow/pressure-evoked parenchymal arteriole vasoconstriction. J Neurosci 35(21):8245
Kiss B et al (2001) Ocular hemodynamics during isometric exercise. Microvasc Res 61(1):1–13
Kiss B et al (2002) Retinal blood flow during hyperoxia in humans revisited: concerted results using different measurement techniques. Microvasc Res 64(1):75–85
Kitanishi K et al (2001) Liposomal prostaglandin E1 enhances optic nerve head blood flow in cats. J Ocular Pharma Ther 17(2):115–122
Kito K et al (2018) Endothelium-dependent vasodilation in the cerebral arterioles of rats deteriorates during acute hyperglycemia and then is restored by reducing the glucose level. J Anesth 32(4): 531–538
Kiyota N et al (2018) The impact of intraocular pressure elevation on optic nerve head and choroidal blood flow. Invest Ophthalmol Vis Sci 59(8):3488–3496
Kiyota N et al (2020) Progression in open-angle glaucoma with myopic disc and blood flow in the optic nerve head and peripapillary chorioretinal atrophy zone. Ophthalmology Glaucoma 3(3): 202–209
Kiyota N et al (2021) Time-course changes in optic nerve head blood flow and retinal nerve fiber layer thickness in eyes with open-angle glaucoma. Ophthalmology 128(5):663–671
Kiyota N et al (2022) The relationship between choroidal blood flow and glaucoma progression in a Japanese study population. Jpn J Ophthalmol 66:425
Klooster J et al (1996) Sympathetic innervation of the rat choroid: an autoradiographic tracing and immunohistochemical study. Ophthalmic Res 28(1):36–43
Knot HJ, Nelson MT (1995) Regulation of membrane potential and diameter by voltage-dependent K+ channels in rabbit myogenic cerebral arteries. Am J Phys 269(1):H348–H355
Knudtson MD et al (2003) Revised formulas for summarizing retinal vessel diameters. Curr Eye Res 27(3):143–149
Kobayashi T et al (2022) Characteristics of laterality in the optic nerve head microcirculation obtained by laser speckle flowgraphy in healthy subjects. Graefes Arch Clin Exp Ophthalmol 260:1–7
Koh LHL et al (2017) Choroidal vascular changes in age-related macular degeneration. Acta Ophthalmol 95(7):e597–e601
Kohner EM (1976) The problems of retinal blood flow in diabetes. Diabetes 25(2 SUPPL):839–844
Kohner EM et al (1975) The retinal blood flow in diabetes. Diabetologia 11(1):27–33
Kondo M, Wang L, Bill A (1997) The role of nitric oxide in hyperaemic response to flicker in the retina and optic nerve in cats. Acta Ophthalmol Scand 75(3):232–235
Kontos HA, Raper AJ, Patterson JL (1977) Analysis of vasoactivity of local pH, PCO2 and bicarbonate on pial vessels. Stroke 8(3):358–360
Kornfield TE, Newman EA (2014) Regulation of blood flow in the retinal trilaminar vascular network. J Neurosci 34(34):11504
Koss MC (1998) Role of nitric oxide in maintenance of basal anterior choroidal blood flow in rats. Invest Ophthalmol Vis Sci 39(3):559–564

Koss MC (1999) Functional role of nitric oxide in regulation of ocular blood flow. Eur J Pharmacol 374(2):161–174

Kringelholt S, Holmgaard K, Bek T (2013) Relaxation of porcine retinal arterioles during acute hypoxia in vitro depends on prostaglandin and NO synthesis in the perivascular retina. Curr Eye Res 38(9):965–971

Kur J, Newman EA, Chan-Ling T (2012) Cellular and physiological mechanisms underlying blood flow regulation in the retina and choroid in health and disease. Prog Retin Eye Res 31(5): 377–406

Kur J et al (2013) Ca2+ sparks promote myogenic tone in retinal arterioles. Br J Pharmacol 168(7): 1675–1686

Kur J et al (2014) Role of ion channels and subcellular Ca2+ signaling in arachidonic acid–induced dilation of pressurized retinal arterioles. Invest Ophthalmol Vis Sci 55(5):2893–2902

Lacroix A et al (2015) COX-2-derived prostaglandin E2 produced by pyramidal neurons contributes to neurovascular coupling in the rodent cerebral cortex. J Neurosci 35(34):11791

Laspas P et al (2014) Role of nitric oxide synthase isoforms for ophthalmic artery reactivity in mice. Exp Eye Res 127:1–8

Lasta M et al (2012a) Effects of increased white blood cell count on retinal perfusion during hyperoxia-induced vasoconstriction. Microvasc Res 83(2):126–130

Lasta M et al (2012b) Reproducibility of retinal vessel oxygen saturation measurements in healthy young subjects. Acta Ophthalmol 90(8):e616–e620

Lasta M et al (2013) Neurovascular dysfunction precedes neural dysfunction in the retina of patients with type 1 diabetes. Invest Ophthalmol Vis Sci 54(1):842–847

Lecleire-Collet A et al (2011) Evaluation of retinal function and flicker light-induced retinal vascular response in normotensive patients with diabetes without retinopathy. Invest Ophthalmol Vis Sci 52(6):2861–2867

Lee J, Rosen R (2016) Optical coherence tomography angiography in diabetes. Curr Diab Rep 16(12):123

Lee WD et al (2019) Optical coherence tomography angiography for the anterior segment. Eye and Vision 6(1):4

Leeman M, Kestelyn P (2019) Glaucoma and blood pressure. Hypertension 73(5):944–950

Leitgeb RA et al (2003) Real-time assessment of retinal blood flow with ultrafast acquisition by color Doppler Fourier domain optical coherence tomography. Opt Express 11(23):3116–3121

Leitgeb RA et al (2004) Real-time measurement of in vitro flow by Fourier-domain color Doppler optical coherence tomography. Opt Lett 29(2):171–173

Leitgeb RA et al (2014) Doppler optical coherence tomography. Prog Retin Eye Res 41:26–43

Leske MC (2009) Ocular perfusion pressure and glaucoma: clinical trial and epidemiologic findings. Curr Opin Ophthalmol 20(2):73

Li L et al (2019) Quantitative analysis of retinal and choroid capillary ischaemia using optical coherence tomography angiography in type 2 diabetes. Acta Ophthalmol 97(3):240–246

Li C et al (2021) Role of the superior salivatory nucleus in parasympathetic control of choroidal blood flow and in maintenance of retinal health. Exp Eye Res 206:108541

Lindauer U et al (1999) Nitric oxide: a modulator, but not a mediator, of neurovascular coupling in rat somatosensory cortex. Am J Phys 277(2):H799–H811

Linsenmeier RA, Braun RD (1992) Oxygen distribution and consumption in the cat retina during normoxia and hypoxemia. J Gen Physiol 99(2):177–197

Linsenmeier RA, Yancey CM (1989) Effects of hyperoxia on the oxygen distribution in the intact cat retina. Invest Ophthalmol Vis Sci 30(4):612–618

Linsenmeier RA, Zhang HF (2017) Retinal oxygen: from animals to humans. Prog Retin Eye Res 58:115–151

Liu Y et al (2019) Retinal tissue perfusion in patients with multiple sclerosis. Curr Eye Res 44(10): 1091–1097

Liu T et al (2022) Retinal and choroidal vascular perfusion and thickness measurement in diabetic retinopathy Ppatients by the swept-source optical coherence tomography angiography. Front Med 9
Lo L et al (2021) Non-invasive measurement of intracranial pressure through application of venous ophthalmodynamometry. In: 2021 43rd annual international conference of the IEEE Engineering in Medicine & Biology Society (EMBC)
Logean E, Schmetterer LF, Riva CE (2000) Optical Doppler velocimetry at various retinal vessel depths by variation of thesource coherence length. Appl Opt 39(16):2858–2862
Loo JH et al (2021) Loss of caveolin-1 impairs light flicker-induced neurovascular coupling at the optic nerve head. Front Neurosci 15
Lovasik JV et al (2003) Choroidal blood flow during exercise-induced changes in the ocular perfusion pressure. Invest Ophthalmol Vis Sci 44(5):2126–2132
Lu ES et al (2022) Detection of neovascularisation in the vitreoretinal interface slab using widefield swept-source optical coherence tomography angiography in diabetic retinopathy. Br J Ophthalmol 106(4):534
Ludovico J et al (2003) Alterations of retinal capillary blood flow in preclinical retinopathy in subjects with type 2 diabetes. Graefes Arch Clin Exp Ophthalmol 241(3):181–186
Luft N et al (2016a) Ocular blood flow measurements in healthy white subjects using laser speckle flowgraphy. PLoS One 11(12):e0168190
Luft N et al (2016b) Measurements of retinal perfusion using laser speckle flowgraphy and Doppler optical coherence tomography. Invest Ophthalmol Vis Sci 57(13):5417–5425
Luksch A et al (2000) Effects of systemic NO synthase inhibition on choroidal and optic nerve head blood flow in healthy subjects. Invest Ophthalmol Vis Sci 41(10):3080–3084
Luksch A et al (2001) Glucose and insulin exert additive ocular and renal vasodilator effects on healthy humans. Diabetologia 44(1):95–103
Luksch A et al (2002) Effect of inhalation of different mixtures of O2 and CO2 on retinal blood flow. Br J Ophthalmol 86(10):1143
Luksch A et al (2003) Role of NO in choroidal blood flow regulation during isometric exercise in healthy humans. Invest Ophthalmol Vis Sci 44(2):734–739
Luksch A et al (2006) ETa-receptor blockade, but not ACE inhibition, blunts retinal vessel response during isometric exercise. Am J Physiol Heart Circ Physiol 290(4):H1693–H1698
Lütjen-Drecoll E (2006) Choroidal innervation in primate eyes. Exp Eye Res 82(3):357–361
Lutty GA (2017) Diabetic choroidopathy. Vis Res 139:161–167
Lutty GA, Bhutto I, McLeod DS (2012) Anatomy of the ocular vasculatures. In: Schmetterer L, Kiel JW (eds) Ocular blood flow, pp 3–21
MacCumber MW, D'Anna SA (1994) Endothelin receptor-binding subtypes in the human retina and choroid. Arch Ophthalmol 112(9):1231–1235
Mäepea O (1992) Pressures in the anterior ciliary arteries, choroidal veins and choriocapillaris. Exp Eye Res 54(5):731–736
Makita S et al (2006) Optical coherence angiography. Opt Express 14(17):7821–7840
Mammadzada P, Corredoira PM, André H (2020) The role of hypoxia-inducible factors in neovascular age-related macular degeneration: a gene therapy perspective. Cell Mol Life Sci 77(5):819–833
Mandecka A et al (2007) Influence of flickering light on the retinal vessels in diabetic patients. Diabetes Care 30(12):3048–3052
Mann RM et al (1995) Nitric oxide and choroidal blood flow regulation. Invest Ophthalmol Vis Sci 36(5):925–930
Marjanović I et al (2016) Relationship between blood pressure and retrobulbar blood flow in dipper and nondipper primary open-angle glaucoma patients. Eur J Ophthalmol 26(6):588–593
Marshall H et al (2021) Cardiovascular disease predicts structural and functional progression in early glaucoma. Ophthalmology 128(1):58–69
Mateo AO, De ArtiÑAno AA (1997) Highlights on endothelins: a review. Pharmacol Res 36(5): 339–351

McDougal DH, Gamlin PD (2015) Autonomic control of the eye. In: Comprehensive physiology, pp 439–473
McGahon MK et al (2016) TRPV2 channels contribute to stretch-activated cation currents and myogenic constriction in retinal arterioles. Invest Ophthalmol Vis Sci 57(13):5637–5647
Mehlsen J et al (2011) Lack of effect of short-term treatment with amlodipine and Lisinopril on retinal autoregulation in normotensive patients with type 1 diabetes and mild diabetic retinopathy. Acta Ophthalmol 89(8):764–768
Metea MR, Newman EA (2006) Glial cells dilate and constrict blood vessels: a mechanism of neurovascular coupling. J Neurosci 26(11):2862
Metea MR, Newman EA (2007) Signalling within the neurovascular unit in the mammalian retina. Exp Physiol 92(4):635–640
Metea MR, Kofuji P, Newman EA (2007) Neurovascular coupling is not mediated by potassium siphoning from glial cells. J Neurosci 27(10):2468
Metelitsina TI et al (2006) Effect of systemic hypertension on foveolar choroidal blood flow in age related macular degeneration. Br J Ophthalmol 90(3):342
Metelitsina TI et al (2008) Foveolar choroidal circulation and choroidal neovascularization in age-related macular degeneration. Invest Ophthalmol Vis Sci 49(1):358–363
Metelitsina TI et al (2010) Effect of isometric exercise on choroidal blood flow in patients with age-related macular degeneration. Br J Ophthalmol 94(12):1629
Meyer-Schwickerath R et al (1995) Central retinal venous outflow pressure. Graefes Arch Clin Exp Ophthalmol 233(12):783–788
Michelson G, Schmauss B (1995) Two dimensional mapping of the perfusion of the retina and optic nerve head. Br J Ophthalmol 79(12):1126
Mishra A, Newman EA (2010) Inhibition of inducible nitric oxide synthase reverses the loss of functional hyperemia in diabetic retinopathy. Glia 58(16):1996–2004
Mishra A, Newman E (2012) Aminoguanidine reverses the loss of functional hyperemia in a rat model of diabetic retinopathy. Front Neuroenerg:3
Mishra A, Hamid A, Newman EA (2011) Oxygen modulation of neurovascular coupling in the retina. Proc Natl Acad Sci USA 108(43):17827–17831
Mishra A et al (2016) Astrocytes mediate neurovascular signaling to capillary pericytes but not to arterioles. Nat Neurosci 19(12):1619–1627
Molins B et al (2019) Shear stress modulates inner blood retinal barrier phenotype. Exp Eye Res 187:107751
Moncada S (1994) Nitric oxide. J Hypertens 12(10):S35–S39
Monteiro-Henriques I, Rocha-Sousa A, Barbosa-Breda J (2022) Optical coherence tomography angiography changes in cardiovascular systemic diseases and risk factors: a review. Acta Ophthalmol 100(1):e1–e15
Morgan WH, Hazelton ML, Yu D-Y (2016) Retinal venous pulsation: expanding our understanding and use of this enigmatic phenomenon. Prog Retin Eye Res 55:82–107
Mori A et al (2007) Stimulation of prostanoid IP and EP2 receptors dilates retinal arterioles and increases retinal and choroidal blood flow in rats. Eur J Pharmacol 570(1):135–141
Movaffaghy A et al (1998) Blood flow in the human optic nerve head during isometric exercise. Exp Eye Res 67(5):561–568
Mroczkowska S et al (2013) Primary open-angle glaucoma vs normal-tension glaucoma: the vascular perspective. JAMA Ophthalmology 131(1):36–43
Muir ER, Rentería RC, Duong TQ (2012) Reduced ocular blood flow as an early indicator of diabetic retinopathy in a mouse model of diabetes. Invest Ophthalmol Vis Sci 53(10): 6488–6494
Mursch-Edlmayr AS et al (2019) Differences in optic nerve head blood flow regulation in normal tension glaucoma patients and healthy controls as assessed with laser speckle flowgraphy during the water drinking test. J Glaucoma 28(7):649–654

Mursch-Edlmayr AS et al (2020) Comparison of neurovascular coupling between normal tension glaucoma patients and healthy individuals with laser speckle flowgraphy. Curr Eye Res 45(11): 1438–1442

Nagaoka T, Yoshida A (2006) Noninvasive evaluation of wall shear stress on retinal microcirculation in humans. Invest Ophthalmol Vis Sci 47(3):1113–1119

Nagaoka T et al (2002) The effect of nitric oxide on retinal blood flow during hypoxia in cats. Invest Ophthalmol Vis Sci 43(9):3037–3044

Nagaoka T et al (2004) Alteration of choroidal circulation in the foveal region in patients with type 2 diabetes. Br J Ophthalmol 88(8):1060

Nam AS, Braaf B, Vakoc BJ (2022) Using the dynamic forward scattering signal for optical coherence tomography based blood flow quantification. Opt Lett 47(12):3083–3086

Nassisi M et al (2019) Choriocapillaris flow impairment predicts the development and enlargement of drusen. Graefes Arch Clin Exp Ophthalmol 257(10):2079–2085

Needham M et al (2014) The role of K+ and cl− channels in the Rregulation of retinal arteriolar Ttone and blood flow. Invest Ophthalmol Vis Sci 55(4):2157–2165

Nesper PL et al (2019) Hemodynamic response of the three macular capillary plexuses in dark adaptation and flicker stimulation using optical coherence tomography angiography. Invest Ophthalmol Vis Sci 60(2):694–703

Nesper PL, Ong JX, Fawzi AA (2021) Exploring the relationship between multilayered choroidal neovascularization and choriocapillaris flow deficits in AMD. Invest Ophthalmol Vis Sci 62(3): 12–12

Newman EA, Frambach DA, Odette LL (1984) Control of extracellular potassium levels by retinal glial cell K+ siphoning. Science 225(4667):1174–1175

Nguyen TT et al (2009) Flicker light–induced retinal vasodilation in diabetes and diabetic retinopathy. Diabetes Care 32(11):2075–2080

Nilsson SFE (2000) The significance of nitric oxide for parasympathetic vasodilation in the eye and other orbital tissues in the cat. Exp Eye Res 70(1):61–72

Nilsson SFE, Alm A (2012) Determination of ocular blood flows with the microsphere method. In: Schmetterer L, Kiel J (eds) Ocular blood flow. Springer, Berlin/Heidelberg, pp 25–47

Nilsson SFE, Bill A (1984) Vasoactive intestinal polypeptide (VIP): effects in the eye and on regional blood flows. Acta Physiol Scand 121(4):385–392

Nilsson SFE, Linder J, Bill A (1985) Characteristics of uveal vasodilation produced by facial nerve stimulation in monkeys, cats and rabbits. Exp Eye Res 40(6):841–852

Nippert AR, Newman EA (2020) Regulation of blood flow in diabetic retinopathy. Vis Neurosci 37: E004

Nishimura K et al (1996) Effects of endothelin-1 on optic nerve head blood flow in cats. J Ocular Pharmacol Ther 12(1):75–83

Nislawati R et al (2021) Role of hypertension as a risk factor for open-angle glaucoma: a systematic review and meta-analysis. BMJ Open Ophthalmol 6(1):e000798

Nolde JM et al (2021) Capillary vascular density in the retina of hypertensive patients is associated with a non-dipping pattern independent of mean ambulatory blood pressure. J Hypertens 39(9): 1826–1834

Noma H et al (2009) Macular microcirculation in hypertensive patients with and without branch retinal vein occlusion. Acta Ophthalmol 87(6):638–642

Nork TM et al (2006) Measurement of regional choroidal blood flow in rabbits and monkeys using fluorescent microspheres. Arch Ophthalmol 124(6):860–868

Ohkubo H, Chiba S (1987) Responses of isolated canine ophthalmic and ciliary arteries to vasoactive substances. Jpn J Ophthalmol 31(4):627–634

Ohno-Matsui K, Kasahara K, Moriyama M (2013) Detection of zinn-haller arterial ring in highly myopic eyes by simultaneous indocyanine green gngiography and optical coherence tomography. Am J Ophthalmol 155(5):920–926

Okada S, Ohta Y (1994) Microvascular architecture of the retina in the Japanese monkey (Macaca fuscata fuscata). Anthropol Sci 102(Supplement):139–153

Okamoto H et al (1997) Neuronal NOS-derived NO plays permissive role in cerebral blood flow response to hypercapnia. Am J Phys 272(1):H559–H566

Okuno T et al (2002) Evidence that nitric oxide is involved in autoregulation in optic nerve head of rabbits. Invest Ophthalmol Vis Sci 43(3):784–789

Omae T, Nagaoka T, Yoshida A (2015) Relationship between retinal blood flow and serum adiponectin concentrations in patients with type 2 diabetes mellitus. Invest Ophthalmol Vis Sci 56(6):4143–4149

Orgül S, Gugleta K, Flammer J (1999) Physiology of perfusion as it relates to the optic nerve head. Surv Ophthalmol 43:S17–S26

Osborne NN et al (1999) Neuroprotection in relation to retinal ischemia and relevance to glaucoma. Surv Ophthalmol 43:S102–S128

Palkovits S et al (2013) Measurement of retinal oxygen saturation in patients with chronic obstructive pulmonary disease. Invest Ophthalmol Vis Sci 54(2):1008–1013

Palkovits S et al (2014a) Regulation of retinal oxygen metabolism in humans during graded hypoxia. Am J Physiol Heart Circ Physiol 307(10):H1412–H1418

Palkovits S et al (2014b) Retinal oxygen metabolism during normoxia and hyperoxia in healthy subjects. Invest Ophthalmol Vis Sci 55(8):4707–4713

Palkovits S et al (2014c) Effect of increased oxygen tension on flicker-induced vasodilatation in the human retina. J Cereb Blood Flow Metab 34(12):1914–1918

Palkovits S et al (2015) Relation of retinal blood flow and retinal oxygen extraction during stimulation with diffuse luminance flicker. Sci Rep 5(1):18291

Pappelis K, Jansonius NM (2021) U-shaped effect of blood pressure on structural OCT metrics and retinal perfusion in ophthalmologically healthy subjects. Invest Ophthalmol Vis Sci 62(12):5–5

Pappelis K, Choritz L, Jansonius NM (2020) Microcirculatory model predicts blood flow and autoregulation range in the human retina: in vivo investigation with laser speckle flowgraphy. Am J Physiol Heart Circ Physiol 319(6):H1253–H1273

Parver LM, Auker C, Carpenter DO (1980) Choroidal blood flow as a heat dissipating mechanism in the macula. Am J Ophthalmol 89(5):641–646

Patel V et al (1992) Retinal blood flow in diabetic retinopathy. BMJ 305(6855):678–683

Patel SR et al (2012) Abnormal retinal vascular reactivity in individuals with impaired glucose tolerance: a preliminary study. Invest Ophthalmol Vis Sci 53(9):5102–5108

Pechauer AD et al (2015) Optical coherence tomography angiography of peripapillary retinal blood flow response to hyperoxia. Invest Ophthalmol Vis Sci 56(5):3287–3291

Pedersen DB et al (2006) Optic nerve pH and PO2: the effects of carbonic anhydrase inhibition, and metabolic and respiratory acidosis. Acta Ophthalmol Scand 84(4):475–480

Pedersen CJ et al (2007) Measurement of absolute flow velocity vector using dual-angle, delay-encoded Doppler optical coherence tomography. Opt Lett 32(5):506–508

Pemp B, Schmetterer L (2008) Ocular blood flow in diabetes and age-related macular degeneration. Can J Ophthalmol 43(3):295–301

Pemp B et al (2009a) Reduced retinal vessel response to flicker stimulation but not to exogenous nitric oxide in type 1 diabetes. Invest Ophthalmol Vis Sci 50(9):4029–4032

Pemp B et al (2009b) Correlation of flicker-induced and flow-mediated vasodilatation in patients with endothelial dysfunction and healthy volunteers. Diabetes Care 32(8):1536–1541

Pemp B et al (2010) Retinal blood flow in type 1 diabetic patients with no or mild diabetic retinopathy during euglycemic clamp. Diabetes Care 33(9):2038–2042

Pemp B et al (2013) Calculation of central retinal artery diameters from non-invasive ocular haemodynamic measurements in type 1 diabetes patients. Acta Ophthalmol 91:1755–3768

Pi S et al (2020) Retinal capillary oximetry with visible light optical coherence tomography. Proc Natl Acad Sci USA 117(21):11658–11666

Pillunat LE et al (1997) Autoregulation of human optic nerve head circulation in response to increased intraocular pressure. Exp Eye Res 64(5):737–744

Plange N et al (2008) Prolonged retinal arteriovenous passage time is correlated to ocular perfusion pressure in normal tension glaucoma. Graefes Arch Clin Exp Ophthalmol 246(8):1147–1152

Ploner SB et al (2016) Toward quantitative optical coherence tomography angiography: visualizing blood flow speeds in ocular pathology using variable interscan time analysis. Retina 36:S118
Polak K et al (2000) Effects of insulin on retinal and pulsatile choroidal blood flow in humans. Arch Ophthalmol 118(1):55–59
Polak K et al (2001) Effect of endothelin and BQ123 on ocular blood flow parameters in healthy subjects. Invest Ophthalmol Vis Sci 42(12):2949–2956
Polak K, Schmetterer L, Riva CE (2002) Influence of flicker frequency on flicker-induced changes of retinal vessel diameter. Invest Ophthalmol Vis Sci 43(8):2721–2726
Polak K et al (2003) Regulation of human retinal blood flow by endothelin-1. Exp Eye Res 76(5): 633–640
Polak K et al (2007) Altered nitric oxide system in patients with open-angle glaucoma. Arch Ophthalmol 125(4):494–498
Polska E et al (2001) RI in central retinal artery as assessed by CDI does not correspond to retinal vascular resistance. Am J Physiol Heart Circ Physiol 280(4):H1442–H1447
Polska E et al (2002) Measurements in the peripheral retina using LDF and laser interferometry are mainly influenced by the choroidal circulation. Curr Eye Res 24(4):318–323
Polska E et al (2003) Effects of adenosine on intraocular pressure, optic nerve head blood flow, and choroidal blood flow in healthy humans. Invest Ophthalmol Vis Sci 44(7):3110–3114
Polska E et al (2007) Regulation of choroidal blood flow during combined changes in intraocular pressure and arterial blood pressure. Invest Ophthalmol Vis Sci 48(8):3768–3774
Popa-Cherecheanu A et al (2019) Regulation of choroidal blood flow during isometric exercise at different levels of intraocular pressure. Invest Ophthalmol Vis Sci 60(1):176–182
Portellos M et al (1995) Effects of adenosine on ocular blood flow. Invest Ophthalmol Vis Sci 36(9):1904–1909
Pournaras C, Tsacopoulos M, Chapuis P (1978) Studies on the role of prostaglandins in the regulation of retinal blood flow. Exp Eye Res 26(6):687–697
Pournaras CJ et al (1989) Diffusion of O2 in the retina of anesthetized miniature pigs in normoxia and hyperoxia. Exp Eye Res 49(3):347–360
Pournaras CJ et al (1990) Experimental retinal branch vein occlusion in miniature pigs induces local tissue hypoxia and vasoproliferative microangiopathy. Ophthalmology 97(10):1321–1328
Pournaras CJ et al (2004) Regulation of optic nerve head blood flow in normal tension glaucoma patients. Eur J Ophthalmol 14(3):226–235
Pournaras CJ et al (2008) Regulation of retinal blood flow in health and disease. Prog Retin Eye Res 27(3):284–330
Prasanna G, Krishnamoorthy R, Yorio T (2011) Endothelin, astrocytes and glaucoma. Exp Eye Res 93(2):170–177
Preußner PR et al (1983) Quantitative measurement of retinal blood flow in human beings by application of digital image-processing methods to television fluorescein angiograms. Graefes Arch Clin Exp Ophthalmol 221(3):110–112
Puchner S et al (2020) Changes in retinal blood flow in response to an experimental increase in IOP in healthy participants as assessed with Doppler optical coherence tomography. Invest Ophthalmol Vis Sci 61(2):33–33
Quaranta L et al (1994) The effect of increased intraocular pressure on pulsatile ocular blood flow in low tension glaucoma. Surv Ophthalmol 38:S177–S182
Rassam S, Patel V, Kohner E (1995) The effect of experimental hypertension on retinal vascular autoregulation in humans: a mechanism for the progression of diabetic retinopathy. Exp Physiol 80(1):53–68
Reiner A et al (2011) Age-related impairment in choroidal blood flow compensation for arterial blood pressure fluctuation in pigeons. Invest Ophthalmol Vis Sci 52(10):7238–7247
Reiner A et al (2018) Neural control of choroidal blood flow. Prog Retin Eye Res 64:96–130
Resch H et al (2009a) Endothelial dysfunction in glaucoma. Acta Ophthalmol 87(1):4–12
Resch H et al (2009b) Effect of dual endothelin receptor blockade on ocular blood flow in patients with glaucoma and healthy subjects. Invest Ophthalmol Vis Sci 50(1):358–363

Resch H et al (2011) Correlation of optic disc morphology and ocular perfusion parameters in patients with primary open angle glaucoma. Acta Ophthalmol 89(7):e544–e549
Rifai OM et al (2021) The application of optical coherence tomography angiography in Alzheimer's disease: a systematic review. Alzheimers Dement 13(1):e12149
Riva C et al (1972) Laser Doppler measurements of blood flow in capillary tubes and retinal arteries. Investig Ophthalmol 11(11):936–944
Riva CE, Feke GT, Ben-Sira I (1978) Fluorescein dye-dilution technique and retinal circulation. Am J Phys 234(3):H315–H322
Riva CE et al (1979) Bidirectional LDV system for absolute measurement of blood speed in retinal vessels. Appl Opt 18(13):2301–2306
Riva CE et al (1981) Fundus camera based retinal LDV. Appl Opt 20(1):117–120
Riva CE et al (1985) Blood velocity and volumetric flow rate in human retinal vessels. Invest Ophthalmol Vis Sci 26(8):1124–1132
Riva CE, Pournaras CJ, Tsacopoulos M (1986a) Regulation of local oxygen tension and blood flow in the inner retina during hyperoxia. J Appl Physiol 61(2):592–598
Riva CE, Grunwald JE, Petrig BL (1986b) Autoregulation of human retinal blood flow. An investigation with laser Doppler velocimetry. Invest Ophthalmol Vis Sci 27(12):1706–1712
Riva CE et al (1991) Flicker evoked increase in optic nerve head blood flow in anesthetized cats. Neurosci Lett 128(2):291–296
Riva CE et al (1992) Laser Doppler flowmetry in the optic nerve. Exp Eye Res 55(3):499–506
Riva CE et al (1994a) Choroidal blood flow in the foveal region of the human ocular fundus. Invest Ophthalmol Vis Sci 35(13):4273–4281
Riva CE et al (1994b) Local choroidal blood flow in the cat by laser Doppler flowmetry. Invest Ophthalmol Vis Sci 35(2):608–618
Riva CE et al (1997a) Effect of acute decreases of perfusion pressure on choroidal blood flow in humans. Invest Ophthalmol Vis Sci 38(9):1752–1760
Riva CE et al (1997b) Choroidal blood flow during isometric exercises. Invest Ophthalmol Vis Sci 38(11):2338–2343
Riva CE et al (1997c) Autoregulation of human optic nerve head blood flow in response to acute changes in ocular perfusion pressure. Graefes Arch Clin Exp Ophthalmol 235(10):618–626
Riva CE, Falsini B, Logean E (2001) Flicker-evoked responses of human optic nerve head blood flow: luminance versus chromatic modulation. Invest Ophthalmol Vis Sci 42(3):756–762
Riva CE, Logean E, Falsini B (2004a) Temporal dynamics and magnitude of the blood flow response at the optic disk in normal subjects during functional retinal flicker-stimulation. Neurosci Lett 356(2):75–78
Riva CE et al (2004b) Flicker-evoked response measured at the optic disc rim is reduced in Ocular hypertension and early glaucoma. Invest Ophthalmol Vis Sci 45(10):3662–3668
Robinson F et al (1986) Retinal blood flow autoregulation in response to an acute increase in blood pressure. Invest Ophthalmol Vis Sci 27(5):722–726
Rose K, Kulasekara SI, Hudson C (2016) Intervisit repeatability of retinal blood oximetry and total retinal blood flow under varying systemic blood gas oxygen saturations. Invest Ophthalmol Vis Sci 57(1):188–197
Roy S et al (2010) Vascular basement membrane thickening in diabetic retinopathy. Curr Eye Res 35(12):1045–1056
Sacconi R et al (2022) Choroidal vascularity index is associated with geographic atrophy progression. Retina 42(2):381–387
Satilmis MUC et al (2003) Rate of progression of glaucoma correlates with retrobulbar circulation and intraocular pressure. Am J Ophthalmol 135(5):664–669
Schiekofer S et al (2000) Endothelial dysfunction in diabetes mellitus. In seminars in thrombosis and hemostasis, vol 26. Thieme Medical Publishers, Inc, p 503
Schirutschke H et al (2021) Comparative study of microvascular function: forearm blood flow versus dynamic retinal vessel analysis. Clin Physiol Funct Imaging 41(1):42–50

Schmetterer L, Garhofer G (2007) How can blood flow be measured? Surv Ophthalmol 52(6): S134–S138
Schmetterer L, Polak K (2001) Role of nitric oxide in the control of ocular blood flow. Prog Retin Eye Res 20(6):823–847
Schmetterer L et al (1995) The effect of hyperoxia and hypercapnia on fundus pulsations in the macular and optic disc region in healthy young men. Exp Eye Res 61(6):685–690
Schmetterer L et al (1996) The effect of inhalation of different mixtures of O2 and CO2 on ocular fundus pulsations. Exp Eye Res 63(4):351–355
Schmetterer L et al (1997a) Role of NO in the O2 and CO2 responsiveness of cerebral and ocular circulation in humans. Am J Phys 273(6):R2005–R2012
Schmetterer L et al (1997b) The effect of systemic nitric oxide-synthase inhibition on ocular fundus pulsations in man. Exp Eye Res 64(3):305–312
Schmetterer L et al (1997c) Effects of endothelin-1 (ET-1) on ocular hemodynamics. Curr Eye Res 16(7):687–692
Schmetterer L et al (1997d) Renal and ocular hemodynamic effects of insulin. Diabetes 46(11): 1868–1874
Schmetterer L et al (1997e) Nitric oxide and ocular blood flow in patients with IDDM. Diabetes 46(4):653–658
Schmetterer L et al (2000) A comparison between laser interferometric measurement of fundus pulsation and pneumotonometric measurement of pulsatile ocular blood flow 2. Effects of changes in pCO2 and pO2 and of isoproterenol. Eye 14(1):46–52
Schmetterer L, Kiel JW (2012) Ocular blood flow. Springer, Berlin, Heidelberg
Schmidl D, Garhofer G, Schmetterer L (2011a) The complex interaction between ocular perfusion pressure and ocular blood flow – relevance for glaucoma. Exp Eye Res 93(2):141–155
Schmidl D et al (2011b) Role of adenosine in the control of choroidal blood flow during changes in ocular perfusion pressure. Invest Ophthalmol Vis Sci 52(8):6035–6039
Schmidl D et al (2012) Comparison of choroidal and optic nerve head blood flow regulation during changes in ocular perfusion pressure. Invest Ophthalmol Vis Sci 53(8):4337–4346
Schmidl D et al (2013) Role of nitric oxide in optic nerve head blood flow regulation during isometric exercise in healthy humans. Invest Ophthalmol Vis Sci 54:1964–1970. https://doi.org/10.1167/iovs.12-11406
Schmidl D et al (2015a) Gender differences in ocular blood flow. Curr Eye Res 40(2):201–212
Schmidl D et al (2015b) Der okuläre perfusionsdruck und seine bedeutung für das glaukom. Klin Monatsbl Augenheilkd 232(02):141–146
Schrijvers EMC et al (2012) Retinopathy and risk of dementia. Rotterdam Study 79(4):365–370
Sehi M et al (2005) Anterior optic nerve capillary blood flow response to diurnal variation of mean ocular perfusion pressure in early untreated primary open-angle glaucoma. Invest Ophthalmol Vis Sci 46(12):4581–4587
Shahidi AM et al (2014) Assessment of total retinal blood flow using Doppler Fourier domain optical coherence tomography during systemic hypercapnia and hypocapnia. Physiol Rep 2(7): e12046
Shahidi AM et al (2017) Retinal oxygen saturation in patients with primary open-angle glaucoma using a non-flash hypespectral camera. Curr Eye Res 42(4):557–561
Shi Y, Vanhoutte PM (2017) Macro- and microvascular endothelial dysfunction in diabetes. J Diabetes 9(5):434–449
Shi Y et al (2021) Correlations between choriocapillaris and choroidal measurements and the growth of geographic atrophy using swept source OCT imaging. Am J Ophthalmol 224:321–331
Shibata M et al (2011) Disruption of gap junctions may be involved in impairment of autoregulation in optic nerve head blood flow of diabetic rabbits. Invest Ophthalmol Vis Sci 52(5):2153–2159
Shibata M et al (2012) Involvement of glial cells in the autoregulation of optic nerve head blood flow in rabbits. Invest Ophthalmol Vis Sci 53(7):3726–3732

Shih Y-YI et al (2013) Quantitative retinal and choroidal blood flow during light, dark adaptation and flicker light stimulation in rats using fluorescent microspheres. Curr Eye Res 38(2):292–298
Shoshani YZ et al (2012) Endothelin and its suspected role in the pathogenesis and possible treatment of glaucoma. Curr Eye Res 37(1):1–11
Shughoury A et al (2020) Retinal oximetry in glaucoma: investigations and findings reviewed. Acta Ophthalmol 98(6):559–571
Siddiqui Y, Yin J (2019) Anterior segment applications of optical coherence tomography angiography. Semin Ophthalmol 34(4):264–269
Siesky B et al (2016) Reductions in retrobulbar and retinal capillary blood flow strongly correlate with changes in optic nerve head and retinal morphology over 4 years in open-angle glaucoma patients of african descent compared with patients of european descent. J Glaucoma 25(9): 750–757
Silver DM et al (1989) Estimation of pulsatile ocular blood flow from intraocular pressure. Acta Ophthalmol 67(S191):25–29
Simader C et al (2009) Role of NO in the control of choroidal blood flow during a decrease in ocular perfusion pressure. Invest Ophthalmol Vis Sci 50(1):372–377
Simó R, Stitt AW, Gardner TW (2018) Neurodegeneration in diabetic retinopathy: does it really matter? Diabetologia 61(9):1902–1912
Sinclair SH et al (1982) Retinal vascular autoregulation in diabetes mellitus. Ophthalmology 89(7): 748–750
Skrzypecki J et al (2019) Blood pressure and glaucoma: at the crossroads between cardiology and ophthalmology. Cardiol J 26(1):8–12
Snodderly DM, Weinhaus RS, Choi JC (1992) Neural-vascular relationships in central retina of macaque monkeys (Macaca fascicularis). J Neurosci 12(4):1169
Son T et al (2016) Optical coherence tomography angiography of stimulus evoked hemodynamic responses in individual retinal layers. Biomed Opt Express 7(8):3151–3162
Song Y et al (2016) Glial endothelin-1 regulates retinal blood flow during hyperoxia in cats. Invest Ophthalmol Vis Sci 57(11):4962–4969
Sörensen BM et al (2016) Prediabetes and type 2 diabetes are associated with generalized microvascular dysfunction. Circulation 134(18):1339–1352
Spaide RF (2016) Choriocapillaris flow features follow a power law distribution: implications for characterization and mechanisms of disease progression. Am J Ophthalmol 170:58–67
Spaide RF, Fujimoto JG, Waheed NK (2015) Image artifacts in optical coherence tomography angiography. Retina 35(11):2163–2180
Spaide RF et al (2018) Optical coherence tomography angiography. Prog Retin Eye Res 64:1–55
Srienc A, Kurth-Nelson Z, Newman E (2010) Imaging retinal blood flow with laser speckle flowmetry. Front Neuroenerg 2:128
Stalmans I et al (2011) Use of colour Doppler imaging in ocular blood flow research. Acta Ophthalmol 89(8):e609–e630
Stefánsson E, Wagner HG, Seida M (1988) Retinal blood flow and its autoregulation measured by intraocular hydrogen clearance. Exp Eye Res 47(5):669–678
Stefánsson E et al (2019) Retinal oximetry: metabolic imaging for diseases of the retina and brain. Prog Retin Eye Res 70:1–22
Steinle JJ, Krizsan-Agbas D, Smith PG (2000) Regional regulation of choroidal blood flow by autonomic innervation in the rat. Am J Physiol Regul Integr Comp Physiol 279(1):R202–R209
Stjernschantz J, Bill A (1980) Vasomotor effects of facial nerve stimulation: noncholinergic vasodilation in the eye. Acta Physiol Scand 109(1):45–50
Stjernschantz J et al (2000) Microvascular effects of selective prostaglandin analogues in the eye with special reference to latanoprost and glaucoma treatment. Prog Retin Eye Res 19(4): 459–496
Stodtmeister R et al (2022) Measurement of the retinal venous pressure with a new instrument in healthy subjects. Graefes Arch Clin Exp Ophthalmol 260(4):1237–1244

Strenn K et al (1997) Reproducibility and sensitivity of scanning laser Doppler flowmetry during graded changes in PO2. Br J Ophthalmol 81(5):360
Strohmaier C et al (2011) A novel, microscope based, non-invasive laser Doppler flowmeter for choroidal blood flow assessment. Exp Eye Res 92(6):545–551
Su W-W et al (2008) Glaucoma is associated with peripheral vascular endothelial dysfunction. Ophthalmology 115(7):1173–1178
Sugiyama T et al (1996) Measurement of optic nerve head circulation: comparison of laser speckle and hydrogen clearance methods. Jpn J Ophthalmol 40(3):339–343
Sugiyama T et al (2010) Use of laser speckle flowgraphy in ocular blood flow research. Acta Ophthalmol 88(7):723–729
Sullivan PM et al (1990) Retinal blood flow during hyperglycemia. A laser Doppler velocimetry study. Invest Ophthalmol Vis Sci 31(10):2041–2045
Sun C et al (2009) Retinal vascular caliber: systemic, environmental, and genetic associations. Surv Ophthalmol 54(1):74–95
Sun Z et al (2019) OCT angiography metrics predict progression of diabetic retinopathy and development of diabetic macular edema: a prospective study. Ophthalmology 126(12): 1675–1684
Sun C et al (2020) Systemic hypertension associated retinal microvascular changes can be detected with optical coherence tomography angiography. Sci Rep 10(1):9580
Svrčinová T et al (2020) Changes in oxygen saturation and the retinal nerve fibre layer in patients with optic neuritis associated with multiple sclerosis in a 6-month follow-up. Acta Ophthalmol 98(8):841–847
Szegedi S et al (2020a) Repeatability and reproducibility of total retinal blood flow measurements using bi-directional doppler OCT. Transl Vis Sci Technol 9(7):34–34
Szegedi S et al (2020b) Anatomical and functional changes in the retina in patients with Alzheimer's disease and mild cognitive impairment. Acta Ophthalmol 98(7):e914–e921
Takagi C et al (1996) Endothelin-1 action via endothelin receptors is a primary mechanism modulating retinal circulatory response to hyperoxia. Invest Ophthalmol Vis Sci 37(10): 2099–2109
Takasago Y et al (2019) Macular atrophy findings by optical coherence tomography angiography compared with fundus autofluorescence in treated exudation age-related macular degeneration. Retina 39(2):296–302
Tamaki Y et al (1994) Noncontact, two-dimensional measurement of retinal microcirculation using laser speckle phenomenon. Invest Ophthalmol Vis Sci 35(11):3825–3834
Tamaki Y et al (1995) Non-contact, two-dimensional measurement of tissue circulation in choroid and optic nerve head using laser speckle phenomenon. Exp Eye Res 60(4):373–383
Tamplin MR et al (2022) Measuring hyperemic response to light flicker stimulus using continuous laser speckle flowgraphy in mice. Exp Eye Res 216:108952
Tan O et al (2015) En face Doppler total retinal blood flow measurement with 70 kHz spectral optical coherence tomography. J Biomed Opt 6:066004
Tan B et al (2019) Quantitative analysis of choriocapillaris in non-human primates using swept-source optical coherence tomography angiography (SS-OCTA). Biomed Opt Express 10(1): 356–371
Tan B et al (2020a) Approaches to quantify optical coherence tomography angiography metrics. Ann Transl Med 8(18):1205
Tan B et al (2020b) Quantitative microvascular analysis with wide-field optical coherence tomography angiography in eyes with diabetic retinopathy. JAMA Netw Open 3(1):e1919469–e1919469
Tan B et al (2021a) Developing a normative database for retinal perfusion using optical coherence tomography angiography. Biomed Opt Express 12(7):4032–4045
Tan W et al (2021b) The application of optical coherence tomography angiography in systemic hypertension: a meta-analysis. Front Med 8:1–15

Tan B et al (2022a) Combining retinal and choroidal microvascular metrics improves discriminative power for diabetic retinopathy. Br J Ophthalmol 107:1–7

Tan W et al (2022b) The new era of retinal imaging in hypertensive patients. Asia-Pacific J Ophthalmol 11(2):149–159

Tanaka T, Riva C, Ben-Sira I (1974) Blood velocity measurements in human retinal vessels. Science 186(4166):830–831

Tani T et al (2014) Autoregulation of retinal blood flow in response to decreased ocular perfusion pressure in cats: comparison of the effects of increased intraocular pressure and systemic hypotension. Invest Ophthalmol Vis Sci 55(1):360–367

Tani T et al (2017) Repeatability and reproducibility of retinal blood flow measurement using a doppler optical coherence tomography flowmeter in healthy subjects. Invest Ophthalmol Vis Sci 58(7):2891–2898

Tayyari F et al (2014) Variability and repeatability of quantitative, fourier-domain optical coherence tomography Doppler blood flow in young and elderly healthy subjects. Invest Ophthalmol Vis Sci 55(12):7716–7725

Tey KY et al (2019) Optical coherence tomography angiography in diabetic retinopathy: a review of current applications. Eye and Vision 6(1):37

Thébault S (2021) Minireview: insights into the role of TRP channels in the retinal circulation and function. Neurosci Lett 765:136285

Thijssen DHJ et al (2011) Assessment of flow-mediated dilation in humans: a methodological and physiological guideline. Am J Physiol Heart Circ Physiol 300(1):H2–H12

Toda N, Nakanishi-Toda M (2007) Nitric oxide: ocular blood flow, glaucoma, and diabetic retinopathy. Prog Retin Eye Res 26(3):205–238

Toda N, Imamura T, Okamura T (2010) Alteration of nitric oxide-mediated blood flow regulation in diabetes mellitus. Pharmacol Ther 127(3):189–209

Toi VV, Riva CE (1995) Variations of blood flow at optic nerve head induced by sinusoidal flicker stimulation in cats. J Physiol 482(1):189–202

Tokayer J et al (2013) Blood flow velocity quantification using split-spectrum amplitude-decorrelation angiography with optical coherence tomography. Biomed Opt Express 4(10): 1909–1924

Told R et al (2012) Effects of increased white blood cell count on endothelin-induced vasoconstriction in healthy subjects. Exp Eye Res 97(1):49–54

Told R et al (2014) Interaction between leukocytes and erythrocytes in the human retina: effects of pentoxifylline on hyperoxia-induced vasoconstriction during increased neutrophil counts. Microvasc Res 92:85–90

Told R et al (2016) Total retinal blood flow in a nonhuman primate optic nerve transection model using dual-beam bidirectional Doppler FD-OCT and microsphere method. Invest Ophthalmol Vis Sci 57(3):1432–1440

Tonade D, Liu H, Kern TS (2016) Photoreceptor cells produce inflammatory mediators that contribute to endothelial cell death in diabetes. Invest Ophthalmol Vis Sci 57(10):4264–4271

Torring MS, Aalkjaer C, Bek T (2014) Constriction of porcine retinal arterioles induced by endothelin-1 and the thromboxane analogue U46619 in vitro decreases with increasing vascular branching level. Acta Ophthalmol 92(3):232–237

Trasischker W et al (2013) In vitro and in vivo three-dimensional velocity vector measurement by three-beam spectral-domain Doppler optical coherence tomography. J Biomed Opt 18(11): 116010

Traustason S et al (2011) Retinal oxygen saturation in patients with systemic hypoxemia. Invest Ophthalmol Vis Sci 52(8):5064–5067

Triggle CR, Ding H (2010) A review of endothelial dysfunction in diabetes: a focus on the contribution of a dysfunctional eNOS. J Am Soc Hypertens 4(3):102–115

Trost A et al (2019) Pericytes in the retina. In: Birbrair A (ed) Pericyte biology in different organs. Springer International Publishing, Cham, pp 1–26

Üretmen Ö et al (2003) Color Doppler imaging of choroidal circulation in patients with asymmetric age-related macular degeneration. Ophthalmologica 217(2):137–142

Vaz-Pereira S, Morais-Sarmento T, Engelbert M (2021) Update on optical coherence tomography and optical coherence tomography angiography imaging in proliferative diabetic retinopathy. Diagnostics (Basel) 11(10):1869

Venkataraman ST et al (2005) The impact of hypercapnia on retinal capillary blood flow assessed by scanning laser Doppler flowmetry. Microvasc Res 69(3):149–155

Waldmann NP et al (2016) The prognostic value of retinal vessel analysis in primary open-angle glaucoma. Acta Ophthalmol 94(6):e474–e480

Wang L, Bill A (1997) Effects of constant and flickering light on retinal metabolism in rabbits. Acta Ophthalmol Scand 75(3):227–231

Wang L et al (2007a) Microspheres method for ocular blood flow measurement in rats: size and dose optimization. Exp Eye Res 84(1):108–117

Wang Y et al (2007b) In vivo total retinal blood flow measurement by Fourier domain Doppler optical coherence tomography. J Biomed Opt 12(4):041215

Wang Y et al (2011) Flicker-induced changes in retinal blood flow assessed by Doppler optical coherence tomography. Biomed Opt Express 2(7):1852–1860

Wang X et al (2014) Optical coherence tomography angiography of optic nerve head and parafovea in multiple sclerosis. Br J Ophthalmol 98(10):1368

Wang L, Cull GA, Fortune B (2015) Optic nerve head blood flow response to reduced ocular perfusion pressure by alteration of either the blood pressure or intraocular pressure. Curr Eye Res 40(4):359–367

Wang L et al (2018) Microvascular blood flow velocities measured with a retinal function imager: inter-eye correlations in healthy controls and an exploration in multiple sclerosis. Eye and Vision 5(1):29

Wang X et al (2021) Decreased retinal vascular density in Alzheimer's disease (AD) and mild cognitive impairment (MCI): an optical coherence tomography angiography (OCTA) study. Front Aging Neurosci 12:12

Wang J et al (2022a) A baseline study of oxygen saturation in parafoveal vessels using visible light optical coherence tomography. Front Med 9:1–7

Wang YM et al (2022b) Optical coherence tomography angiography metrics predict normal tension glaucoma progression. Acta Ophthalmol 100

Wang W et al (2022c) Choriocapillaris perfusion assessed using swept source optical coherence tomographic angiography and the severity of diabetic retinopathy. Br J Ophthalmol 107:320163

Wang W et al (2022d) Choriocapillaris flow deficit and the risk of referable diabetic retinopathy: a longitudinal SS-OCTA study. Br J Ophthalmol 107:320704

Warner RL et al (2020) Full-field flicker evoked changes in parafoveal retinal blood flow. Sci Rep 10:16051. https://doi.org/10.1038/s41598-020-73032-0

Warner RL et al (2021) Measuring temporal and spatial variability of red blood cell velocity in human retinal vessels. Invest Ophthalmol Vis Sci 62(14):29–29

Wartak A et al (2019) Investigating spontaneous retinal venous pulsation using Doppler optical coherence tomography. Sci Rep 9(1):4237

Wei X et al (2017) Choroidal vascularity index: a novel optical coherence tomography based parameter in patients with exudative age-related macular degeneration. Retina 37(6):1120–1125

Weigert G et al (2005) Effects of moderate changes in intraocular pressure on ocular hemodynamics in patients with primary open-angle glaucoma and healthy controls. Ophthalmology 112(8): 1337–1342

Welsh DG et al (2002) Transient receptor potential channels regulate myogenic tone of resistance arteries. Circ Res 90(3):248–250

Werkmeister RM et al (2008) Bidirectional Doppler Fourier-domain optical coherence tomography for measurement of absolute flow velocities in human retinal vessels. Opt Lett 33(24): 2967–2969

Werkmeister RM et al (2012) Response of retinal blood flow to systemic hyperoxia as measured with dual-beam bidirectional Doppler fourier-domain optical coherence tomography. PLoS One 7(9):e45876

Werkmeister R et al (2014) Measurement of retinal blood flow in the rat by combining Doppler Fourier-domain optical coherence tomography with fundus imaging. J Biomed Opt 19(10): 106008

Werkmeister RM et al (2015) Retinal oxygen extraction in humans. Sci Rep 5(1):15763

White BR et al (2003) In vivo dynamic human retinal blood flow imaging using ultra-high-speed spectral domain optical Doppler tomography. Opt Express 11(25):3490–3497

Williamson TH, Harris A (1996) Color Doppler ultrasound imaging of the eye and orbit. Surv Ophthalmol 40(4):255–267

Williamson JR et al (1993) Hyperglycemic pseudohypoxia and diabetic complications. Diabetes 42(6):801–813

Witkowska KJ et al (2017) Optic nerve head and retinal blood flow regulation during isometric exercise as assessed with laser speckle flowgraphy. PLoS One 12(9):e0184772

Wolf S, Arend O, Reim M (1994) Measurement of retinal hemodynamics with scanning laser ophthalmoscopy: reference values and variation. Surv Ophthalmol 38:S95–S100

Wolfgang D et al (2014) Optical coherence tomography today: speed, contrast, and multimodality. J Biomed Opt 19(7):071412

Wong VHY et al (2013) Susceptibility of streptozotocin-induced diabetic rat retinal function and ocular blood flow to acute intraocular pressure challenge. Invest Ophthalmol Vis Sci 54(3): 2133–2141

Wong D et al (2020) Focal structure–function relationships in primary open-angle glaucoma using OCT and OCT-A measurements. Invest Ophthalmol Vis Sci 61(14):33–33

Wong D et al (2021) Combining OCT and OCTA for focal structure–function modeling in early primary open-angle glaucoma. Invest Ophthalmol Vis Sci 62(15):8–8

Wright WS et al (2020) Retinal physiology and circulation: effect of diabetes. In: Comprehensive physiology, pp 933–974

Wrinkler BS (1995) A quantitative assessment of glucose metabolism in the isolated rat retina. In: Christen Y, Doly M, Droy-Lefaix MT (eds) Les Seminaires Ophtalmologiques D'ipsen. Elsevier Science, New York, pp 78–96

Wu X, Davis MJ (2001) Characterization of stretch-activated cation current in coronary smooth muscle cells. Am J Physiol Heart Circ Physiol 280(4):H1751–H1761

Wu J et al (2020) Retinal microvascular attenuation in mental cognitive impairment and Alzheimer's disease by optical coherence tomography angiography. Acta Ophthalmol 98(6): e781–e787

Xu W et al (2010) Association of risk factors for choroidal neovascularization in age-related macular degeneration with decreased foveolar choroidal circulation. Am J Ophthalmol 150(1):40–47

Xu H et al (2016) Microcirculatory responses to hyperoxia in macular and peripapillary regions. Invest Ophthalmol Vis Sci 57(10):4464–4468

Yamada Y et al (2019) The relationship between interocular asymmetry of visual field defects and optic nerve head blood flow in patients with glaucoma. J Glaucoma 28(3):231–237

Yamazaki D, Hitomi H, Nishiyama A (2018) Hypertension with diabetes mellitus complications. Hypertens Res 41(3):147–156

Yang YC et al (1997) Pulsatile ocular blood flow measurements in healthy eyes: reproducibility and reference values. J Glaucoma 6(3):175–179

Yao X et al (2020) Quantitative optical coherence tomography angiography: a review. Exp Biol Med 245(4):301–312

Yaoeda K et al (2003) Relationship between optic nerve head microcirculation and visual field loss in glaucoma. Acta Ophthalmol Scand 81(3):253–259

Yazdanfar S, Rollins AM, Izatt JA (2003) In vivo imaging of human retinal flow dynamics by color Doppler optical coherence tomography. Arch Ophthalmol 121(2):235–239

Yeh T-C, Kuo C-T, Chou Y-B (2022) Retinal microvascular changes in mild cognitive impairment and Alzheimer's disease: a systematic review, meta-analysis, and meta-regression. Front Aging Neurosci 14:1–17

Yorio T, Krishnamoorthy R, Prasanna G (2002) Endothelin: is it a contributor to glaucoma pathophysiology? J Glaucoma 11(3):259–270

Yoshioka T et al (2021) Retinal blood flow reduction in normal-tension glaucoma with single-hemifield damage by Doppler optical coherence tomography. Br J Ophthalmol 105(1):124

You QS et al (2020) Detection of clinically unsuspected retinal neovascularization with wide-field optical coherence tomography angiography. Retina 40(5):891–897

You QS et al (2021) Geographic atrophy progression is associated with choriocapillaris flow deficits measured with optical coherence tomographic angiography. Invest Ophthalmol Vis Sci 62(15):28–28

Yu DY, Cringle SJ, Alder VA (1990) The response of rat vitreal oxygen tension to stepwise increases in inspired percentage oxygen. Invest Ophthalmol Vis Sci 31(12):2493–2499

Yu DY, Alder VA, Cringle SJ (1991) Measurement of blood flow in rat eyes by hydrogen clearance. Am J Phys 261(3):H960–H968

Yu PK, Cringle SJ, Yu D-Y (2014) Correlation between the radial peripapillary capillaries and the retinal nerve fibre layer in the normal human retina. Exp Eye Res 129:83–92

Zabel P et al (2021) Quantitative assessment of retinal thickness and vessel density using optical coherence tomography angiography in patients with Alzheimer's disease and glaucoma. PLoS One 16(3):e0248284

Zagvazdin Y, Fitzgerald MEC, Reiner A (2000) Role of muscarinic cholinergic transmission in edinger-westphal nucleus-induced choroidal vasodilation in pigeon. Exp Eye Res 70(3): 315–327

Zeitz O et al (2006) Glaucoma progression is associated with decreased blood flow velocities in the short posterior ciliary artery. Br J Ophthalmol 90(10):1245

Zeng R et al (2022) Retinal microvasculature and vasoreactivity changes in hypertension using optical coherence tomography-angiography. Graefes Arch Clin Exp Ophthalmol 260:1–11

Zhang Q et al (2018) Ultra-wide optical coherence tomography angiography in diabetic retinopathy. Quant Imaging Med Surg 8(8):743–753

Zhang Y et al (2021) Adenosine and adenosine receptor-mediated action in coronary microcirculation. Basic Res Cardiol 116(1):22

Zhong Z et al (2012) Local flicker stimulation evokes local retinal blood velocity changes. J Vis 12(6):3–3

Zhou Q et al (2017) Association between pseudodrusen and delayed patchy choroidal filling in the comparison of age-related macular degeneration treatments trials. Acta Ophthalmol 95(6):e518–e520

Zhou Y et al (2021) Effects of arachidonic acid metabolites on cardiovascular health and disease. Int J Mol Sci 22(21):12029

Zhu YUN, Park TS, Gidday JM (1998) Mechanisms of hyperoxia-induced reductions in retinal blood flow in newborn pig. Exp Eye Res 67(3):357–369

Zink JM et al (2003) Association between lower optic nerve laser Doppler blood volume measurements and glaucomatous visual field progression. Br J Ophthalmol 87(12):1487

Chapter 10
Microcirculation in Wound Healing

Harvey N. Mayrovitz and Kawaiola Aoki

Abstract The microcirculation has a substantial role in wound healing, as it provides oxygen and proliferative/immunological mediators necessary to the healing process. In this chapter, we discuss the concept of the wound as a temporary added "organ" due to its increased metabolic needs and how the microcirculation functions as a reserve to support healing. We provide a brief overview of the normal healing process, including the factors driving angiogenesis, and address the differences in the microarchitecture of the skin's circulation. While microcirculation is a highly plastic system, we consider aspects that may contribute to dysfunctional perfusion to the skin, such as aging and other pathologies, and their consequences in delayed healing and the development of ulcers. This chapter will provide an in-depth explanation of the structure and physiology of microcirculation in the paradigm of the healing process and address the role of microcirculation in some of the most common chronic and often difficult to heal wounds including venous, arterial, neuropathic, diabetic, and pressure ulcers.

Keywords Venous ulcers · Arterial ulcers · Neuropathic ulcers · Diabetic ulcers · Pressure ulcers · Skin aging · Wound healing and blood flow · Microvasculature and wounds · Leg ulcers · Venous insufficiency

10.1 Introduction

To heal a skin wound or ulcer initially to a certain extent requires adequate blood flow to meet the increased demands of the metabolizing wound tissue. In this sense, as and if the wound goes on to heal, it may be compared to a temporarily added

H. N. Mayrovitz (✉)
Dr. Kiran C Patel College of Allopathic Medicine, Nova Southeastern University, Davie, FL, USA
e-mail: mayrovit@nova.edu

K. Aoki
Dr. Kiran C Patel College of Osteopathic Medicine, Nova Southeastern University, Davie, FL, USA

R. F. Tuma et al. (eds.), *Handbook of Microcirculation*, Physiology in Health and Disease, https://doi.org/10.1007/978-3-031-99609-2_10

"organ." However, if healing is significantly delayed or prevented, a more-or-less permanently added "organ" taxes the localized microcirculatory system with the continued added demand for blood flow. Most of the information that we have regarding the relationship between wounds, their healing and blood flow is derived from measurements and outcomes of wounds that are present on the skin, and of those, most are wounds on the skin of the lower extremities. Blood flow and the microcirculation, in one form or another, play a role in wounds of all etiologies that include arterial ulcers, venous ulcers, and neuropathic ulcers with the latter most notably, but not exclusively, observed in persons with diabetes mellitus (DM).

For arterial ulcers, the causative factor in the development of the ulcer and its difficulty of healing is usually, if not always, inadequate nutritional skin microcirculation, resulting in tissue ischemia as causation and inadequate microvascular reserve to support the healing of chronic ischemic ulcers. The deficit in microvascular reserve may be attributable to structural or functional issues such as may be present in some patients with DM, but in general the main factor is an overall deficient blood flow associated with the presence of peripheral arterial disease (PAD). The situation is quite different in the case of venous ulcers in which skin ulceration is generally secondary to chronic venous insufficiency (CVI) that causes pressure induced injury in superficial lower extremity veins followed by skin breakdown and the occurrence of the venous ulcer. Although full details on the microcirculatory involvement are still not completely clarified, plugging of capillaries via activated leukocytes is at least thought to be involved causing diminished skin nutritional microcirculation. An important nonleg-related ulcer is the so-called decubitus or pressure ulcer. Its linkage to blood flow and more directly to deficits in skin and subcutaneous microcirculation is attributable to nonrelieved skin tissue pressure and shear forces that cause microvascular perfusion deficits sufficient to cause skin break down and ulcer development.

In this chapter, these wounds and others are considered in detail especially with respect to linkages and roles of blood flow and the microcirculation in either or both wound development and healing. Prior to that, a review of the normal wound healing process is described, and the angiogenesis process is presented to provide context. Factors considered in the angiogenesis process are the roles of growth factors, tissue oxygen levels, and the extravascular matrix involvement. Thereafter, further details related to the skin microcirculation are presented followed by microcirculatory aspects of wound healing in specific conditions including aging and ulcers classified as venous, arterial, diabetic and pressure ulcers.

10.2 Normal Wound Healing Process: Overview of Phases

Following injury, most skin lesions heal rapidly, an ability that is vital to human survival and maintaining the protective barrier that lines the body's surfaces and cavities (Hay et al. 2014). Wound healing is an intricate progression of reactions and interactions between cells and mediators, affected by many intrinsic and extrinsic factors (Broughton 2nd et al. 2006). The wound healing process is divided into four overlapping stages, and an interruption during any of the stages can lead to a spectrum of pathologies: From prolonged inflammation and an inability to re-epithelialize the wound causing chronic ulcers; an overgrowth of granulation tissue leading to pyogenic granulomas; to an excessive fibrotic response generating hypertrophic scars and keloids (Sun et al. 2014). These four stages—hemostasis, inflammation, proliferation, and remodeling—are each characterized by distinct molecular, cellular, and physiological events described in brief below: (Sun et al. 2014; Gurtner et al. 2008).

Hemostasis begins after injury to stop blood loss. Local mediators trigger vasoconstriction in vascular smooth muscle cells to reduce blood flow. Platelets and the coagulation cascade form a fibrin clot, which acts as a reservoir for cytokines and growth factors and, in later stages, a fibronectin-rich scaffold for a provisional extracellular matrix (ECM) (Martin 1997; Li et al. 2007). These platelet-derived mediators and bacterial byproducts act as chemotactic cues to recruit immune cells, initiating inflammation hours after injury (Ross and Odland 1968).

During the inflammatory stage, local vasodilation allows plasma-like fluid to leak into the tissue space. The influx of this fluid transports nutrients, antibodies, and substances such as histamine, serotonin, proteolytic enzymes, kinins, prostaglandins, and cells into the wound bed (Phillips 2000). The wound is infiltrated by immune cells: First by neutrophils, which kill bacteria and degrade the damaged matrix; then monocytes within 24 h. These monocytes then become macrophages which phagocytose microbes and remove debris. Macrophages release growth factors and chemokines, such as platelet-derived growth factor (PDGF), fibroblast growth factor (FGF), vascular endothelial growth factor (VEGF), and Transforming Growth factor-a and -b (TGF-a and TGF-b), which assists in the transition to the proliferation stage (Koh and Dipietro 2011; Werner and Grose 2003).

In this stage, released chemotactic agents induce cells' to mobilize, proliferate, and differentiate (Phillips 2000). Stem cells migrate into the wound bed from the basal layer, as well as a significant contribution from remnants of dermal appendages—such as the residual stumps of hair follicles (Martin 1997). These stem cells and keratinocytes proliferate and migrate to achieve re-epithelialization and coverage of the wound (Phillips 2000; Li et al. 2007). Fibroplasia is the accumulation and proliferation of fibroblasts and their production of collagen and ground matrix proteins forming the granulation tissue. This early ECM includes collagens, proteoglycans, and elastin and fills the wound space acting as scaffolding and contact guidance for migrating cells and supporting angiogenesis (Woodley 1985).

Toward the end of the proliferative stage, the fibroblasts differentiate into actin-rich, contractile myofibroblasts, which pull together the wound edges (Tomasek et al. 2002). The final remodeling stage involves the conversion of the dermis from type III to type I collagen and removing cells from earlier stages. This tightly controlled synthesis of new collagen and lysis of old collagen is mainly achieved through the actions of matrix metalloproteinases (MMPs) (Visse and Nagase 2003). Even during normal physiological wound healing, the end-product is neither aesthetically nor functionally perfect. The connective tissue scar that reconstitutes the wound consists of poorly reorganized collagen in dense parallel bundles compared to the basketweave configuration of unwounded tissue. Additionally, hair bulbs or sweat glands do not regenerate if the injury is too deep into the dermis, and there are no remnants of these epidermal appendages (Martin 1997).

10.3 Angiogenesis and Microcirculation in Wound Healing

Angiogenesis occurs throughout all four overlapping stages of wound healing and is initiated immediately after tissue injury. New capillaries sprout from pre-existing vessels and penetrate the wound bed forming a microvasculature network and are visible in the wound bed 3–5 days after injury (Tonnesen et al. 2000). The granulation tissue that forms during the proliferative phase acts as a matrix for proliferating blood vessels, migrating fibroblasts, and new collagen (Folkman 1995; Rees et al. 1999). These new proliferating capillaries are vital to tissue regeneration by bringing in oxygen and micronutrients and removing catabolic waste products (O'connor et al. 2000; Tonnesen et al. 2000). Angiogenesis is regulated by complex growth factor–receptor, cell–cell, and cell–matrix reactions in an orderly cascade of molecular and cellular events (Honnegowda et al. 2015).

10.3.1 Growth Factors in Angiogenesis

Regulation of vascular growth is a balance between proangiogenic and antiangiogenic factors present throughout the body. When these angiogenic stimulators and inhibitors are at a physiological balance, vascular growth is suppressed (Hanahan and Weinberg 2011). Antiangiogenic factors include angiostatin, tissue inhibitors of matrix metalloproteinase-2 (TIMP-2), thrombospondin-1 (TSP-1), endostatin, platelet factor-4 (PF-4), and interferon $\alpha/\beta/\gamma$. These inhibitory factors circulate in the bloodstream at low physiological levels and are stored in the ECM surrounding the blood vessels (Honnegowda et al. 2015).

Immediately following injury, proangiogenic stimuli are released into the wound bed, including thrombin, fibrinogen fragments, thymosin-β_4, and growth factors. This initiates angiogenesis and shifts the balance toward vascular growth (Majima et al. 2000; Pugh and Ratcliffe 2003; Semenza 2002). Tissue damage releases basic fibroblast growth factor (FGF-2) from previously intact cells (Matsuoka et al. 2006). FGF-2 and acidic fibroblast growth factor (FGF-1) are synthesized by inflammatory cells and dermal fibroblasts involved in wound healing. FGF-1 and -2 are released from the ECM and bind to the high-affinity protein family of transmembrane tyrosine kinases FGF receptor-1 (FGFR-1), expressed on endothelial cells of different origins, and FGFR-2, expressed under some circumstances (Presta et al. 2005).

Whereas FGF-1 and -2 that are released from endothelial cells act in an autocrine manner, both mechanisms modulate cell proliferation, migration, protease production, receptor expression, and gap–junction communication (Presta et al. 2005; Barrientos et al. 2008). The highly expressed αvβ3 integrin on endothelial cells during angiogenesis and its interaction with FGF-2 augments the cell's mitogenic activity and cell adhesion to ECM, thereby allowing sustenance for neovascularization (Rusnati et al. 1997).

Platelets and inflammatory cells in the bloodstream also store growth factors. During bleeding and hemostasis, thrombin in the wound upregulates vascular endothelial growth factor (VEGF) (Tsopanoglou and Maragoudakis 1999). Platelets release platelet-derived growth factor (PDGF), transforming growth factor (TGF-α and TGF-β), VEGF, FGF-2, platelet-derived endothelial growth factor, and angiopoietin-1 (Ang-1), which then stimulates endothelial proliferation and migration (Hellberg et al. 2010; Li et al. 2001; Pintucci et al. 2002). Also, macrophages provide a source of cytokines to amplify the angiogenesis cascade and stimulate fibroplasia and neovascularization (Tonnesen et al. 2000). Macrophages and monocytes release PDGF, VEGF, Ang-1, TGF-α, FGF-2, IL-8, and TNF-α during the inflammatory phase. PDFG, VEGF, and FGF-2 act with a synergistic ability to vascularize tissues (Grimm et al. 2009).

10.3.2 Role of Hypoxia

Hypoxia is the driving force for vascular proliferation. The hypoxic gradient between injured and healthy tissue promotes the gene expression of HIF-1α and triggers VEGF production (Grimm et al. 2009; Acker and Plate 2003). Expression of COX-2 during the inflammatory stage also leads to the production of VEGF and other promoters of angiogenesis. VEGF that is present in wound tissue and exudate, increases vascular permeability and transwall hydraulic conductivity (Acker and Plate 2003; Howdieshell et al. 2003). The leakage of fibrinogen and fibronectin through these fenestrations is essential to constructing the provisional ECM (Leonardi et al. 2003). Hypoxia also triggers endothelial cell production of nitric oxide (NO), promoting vasodilation and improving local blood flow (Smith Jr. et al. 2009).

10.3.3 Extracellular Matrix Degradation and Vascular Stabilization

When angiogenic growth factors bind to cell surface receptors of pre-existing venules (parent vessels), this growth factor receptor binding activates signaling pathways within these endothelial cells. Activated endothelial cells release proteolytic enzymes, such as thrombin-exposed endothelial cells releasing release gelatinase A (MMP-2), which dissolves the basement membrane of the parent vessels (Nguyen et al. 2001). Endothelial cells sprout outward through the basement membrane and migrate into the wound bed facilitated by the actions of integrins (αvβ3, αvβ5, and αvβ1). The tissue matrix is dissolved ahead of the sprouting vessel by the MMPs (Honnegowda et al. 2015).

While these proteases break down damaged tissue matrix, proangiogenic stimulators are released, such as fragment E, yielded from fibrin's cleavage, which stimulates angiogenesis directly and enhances VEGF and FGF-2 (Bootle-Wilbraham et al. 2001). As capillary sprouts develop, they digest endothelial cells and infiltrate the ECM after penetrating the vascular basement membrane and continue to branch and form networks (Morgan et al. 2007). These vascular sprouts form tubular channels that connect to vascular loops, which then differentiate into afferent (arterial) and efferent (venous) limbs (Honnegowda et al. 2015).

Vascular stability is controlled by Ang-1 and tyrosine kinase with immunoglobulin-like and EGF-like domains 2 (Tie-2) in the wound bed. Ang-1 binding to its receptor, Tie-2, on activated endothelial cells produces PDGF and recruits mural cells, smooth muscle cells and pericytes (Inoki et al. 2002; Ma et al. 2007; Korff et al. 2001). Stabilizing the vascular architecture allows for blood flow to begin in the mature vessel (Honnegowda et al. 2015). At the terminal stages of healing, growth factor levels decline in the wound, and angiogenesis is suppressed (Kumar et al. 2009). Pericytes secrete an inhibitory form of activated TGF-β that impedes vascular proliferation (Korff et al. 2001; Darland and D'amore 2001). Endostatin, a cleavage product of collagen XVIII, is present surrounding the vascular basement membrane and inhibits wound vascularity (Lange-Asschenfeldt et al. 2001).

10.4 Microcirculatory Aspects of Wound Healing in Skin

Skin circulation has two primary functions: Regulating body temperature and delivering nutrients and oxygen for the skin cell metabolism (Awan et al. 2011). The microvasculature architecture is not homogeneous, as differences depend on anatomical and topographic characteristics specific to the skin area (Braverman 1983). These microcirculatory areas have been classified according to four features: (1) presence/absence of arteriovenous anastomoses (AVAs); (2) structure and location of end arteriole blocking devices; (3) structure of valve-containing venules; and (4) microvessel tissue relationships (Curri 1990). Most skin has nutritive perfusion provided through capillaries. However, specific sites also have arterioles and venules that directly communicate with each other. These AVA provide low resistance pathways from arteriole to vein through which blood flow is modulated to meet thermal demands (Zweifach 1987; Hales et al. 1985). Some anatomical regions, such as the sole and dorsum of the foot, fingertips, and nail folds do not contain these AVAs (Rendell et al. 1993).

The anatomical arrangement of the cutaneous microcirculation is organized in two horizontal plexuses: A deep (lower) plexus containing arterioles and venules starting at the dermal–subcutaneous junction, and a superficial (upper) plexus within the papillary dermis located 1–1.5 mm below the skin surface. These two plexuses are structurally different and connected through paired ascending arterioles and descending venules (Braverman 2000). Autonomic nerves regulate perfusion in the deep plexus, primarily allowing AVA to serve a thermoregulatory function. AVAs act as regulatory capillary flow devices, have a much greater vasodilatory response to heat, and are also affected by emotions (Braverman et al. 1992; Braverman 2000; Charkoudian 2003; Mork et al. 2002). At the dermal–subcutaneous junction, collecting veins, the presence of bi-cusped valves prevent retrograde blood flow (Braverman 2000). The superficial plexus is predominantly composed of smaller diameter (10–20 μm) postcapillary venules that contribute to the capillary loops of the dermal papillae. Arterial capillaries rise to form the dermal papillary loops representing the nutritive component of the skin microcirculation. These capillaries are sites of inflammatory cell emigration, histamine-induced vascular permeability, and deposition of immune complexes in vasculitis (Braverman 2000).

A primary function of skin papillary capillaries is to deliver O_2 and nutrients to maintain and promote epithelial cell proliferation (Sundheim et al. 2017). Pericytes are effector cells involved in regulating some nutritive capillary blood flow via activation of their contractile proteins (actin and myosin) (Attwell et al. 2016; Birbrair et al. 2015). Perfusion in papillary capillaries is regulated locally based on metabolic needs (Mork et al. 2002). Pericytes constrict or dilate in response to PCO_2, pH, and local humoral factors (endothelin and NO) (Chen and Anderson 1997). The O_2 delivery capacity depends on the product of microvascular blood flow and oxygen extraction. Papillary oxygen extraction is high compared to other tissues, limiting the need for a high blood flow (Awan et al. 2011).

Following skin injury or wounding, the skin healing process is associated with an increase in the local metabolic rate that facilitates cellular proliferation and production of the extra cellular matrix. During the repair process, a critical function of this regulatory system is to maintain O_2 delivery for the expansion of stem cells at the epidermal basement membrane. The O_2 is delivered via noninnervated nutritive papillary skin capillaries from the superficial plexus (Braverman 1997). Increased blood perfusion in the deep thermoregulatory plexus occurs between 30 and 60 min after trauma, but papillary capillary nutritive perfusion does not appear to change (Wikslund et al. 2021). As long as the O_2 supply is sufficient, pericytes regulating papillary nutritive perfusion remain contracted, narrowing the diameter of nutritive capillaries. However, beyond about an hour, pericytes experience signals of increased metabolic need, and in response to the acidic extracellular pH and elevated PCO_2, will relax, thereby promoting increased capillary network blood perfusion (Attwell et al. 2016; Birbrair et al. 2015; Chen and Anderson 1997). Blood flow is thus increased in the superficial plexus via ascending arterioles from the deep plexus facilitating increased nutritive perfusion.

Angiogenesis is triggered by multiple cellular signals that occur early in the wound healing process and continues throughout. The plasticity of the microcirculatory system permits a quick remodeling process that can occur in hours to a few days (Ricard and Simons 2015; Wikslund et al. 2022). After about 24 h, papillary nutritive capillaries are observed to be elongated and dilated, and the superficial vascular plexuses are more visible (Wikslund et al. 2022). As a consequence, the increase in capillary surface area allows for a commensurate increase in diffusion capacity that aids the repair process and corresponds to the rise in the tissue's metabolic rate for stem cell production of new epidermal cells (Wikslund et al. 2022).

During this process, the density of papillary capillaries and erythrocyte velocities in capillaries near the injury site (2 mm distant) are increased (Wikslund et al. 2021). The repair zone, based on the Krogh model and the maximal O_2 diffusion distance, extends up to 1 mm from the injury site (Forster et al. 2017). This zone with increased metabolic and O_2 needs is met by an increased number of O_2-carrying erythrocytes passing through the area and reduced diffusion distances. Beyond 2 mm from the injury site, oxygen may be supplied by erythrocytes in a mechanism analogous to upstream flow-mediated dilation that increases microcirculation to and within the sites with increased needs (Markos et al. 2013).

10.5 Microcirculatory Aspects of Wound Healing in Specific Conditions

10.5.1 Aging and Associated Skin Changes

The aging of the skin is influenced by intrinsic factors related to simple chronological aging effects that are independent of environmental insults and also by extrinsic factors that are related to the cumulative insults of environmental exposure, such as UV radiation (Fisher et al. 1997). The net effect of the these combined aging processes is a diminished ability to maintain homeostasis leading to a progressive loss of function of the skin barrier and increased vulnerability to environmental insults (Gilchrest et al. 1994). At the dermal–epidermal junction, the rete ridges are flattened with senescence. Thus, in this skin area, where the papillary dermis maintains contact with the epidermis, flattening the dermal–epidermal junction results in decreased surface contact between these two layers and gives the appearance of atrophy (Kurban and Bhawan 1990; Montagna and Carlisle 1990). In addition to the decline in epidermal and dermal thickness and composition, most resident cells are reduced in number and there is an associated diminished skin microcirculation (Bentov and Reed 2014; Montagna and Carlisle 1990). The skin microcirculation shows impaired regulation, aspects of inflammatory response changes, decreased numbers of progenitor cells, and declines in circulatory mediators (Bentov and Reed 2014). These age-associated changes impede the microvasculature's ability to thermoregulate and respond to injury, thereby increasing the chance of tissue hypoperfusion that negatively impacts wounds from reaching the angiogenic stage of repair (Bentov and Reed 2014; Gould et al. 2015).

The literature presents conflicting data, with most showing an age-related decrease in angiogenesis (Holm-Pedersen and Viidik 1972; Swift et al. 1999) and a few showing an increase (Passaniti et al. 1992). Possible mechanisms of impaired angiogenesis are attributed to impaired endothelial cell function and reduced VEGF expression (Rivard et al. 1999). In animal models, delayed wound capillary ingrowth has been partially attributed to reduced angiogenic factors FGF, VEGF, and TGF-β (Rivard et al. 1999; Swift et al. 1999). Additionally, aged endothelial cells secrete less nitric oxide, likely decreasing vasodilation potential (Rivard et al. 1999) and diminishing microvascular reserve.

During wound healing, the ECM of older adults demonstrates prolonged inflammation, increased MMP and elastase expression, reduced TGF-β, and a weakened cellular response (Gurtner et al. 2008). The atrophic skin of the older adult cannot adapt quickly to the mechanical demands of an injury causing the healing response to be prolonged and blunted with amplified inflammation and differences in signal transduction resulting in inferior ECM production (Edwards et al. 2013; Gould et al. 2015). Accordingly, re-epithelialization, collagen synthesis, and angiogenesis exhibit an age-related delay (Swift et al. 1999). Dermal lymphatic drainage has been shown to be decreased in aging skin causing wounds to be less able to clear pathogens and have diminished wound contraction potential (Gosain and Dipietro 2004).

During the proliferative phase of wound healing, older individuals exhibit a 50% increase in time for keratinocytes to migrate from the basal layer to the surface compared to younger individuals (Gilchrest et al. 1982). The number and size of dermal fibroblasts decrease in advanced age (Plisko and Gilchrest 1983), and these fibroblasts exhibit a diminished response to growth factors and capacity to replicate (Plisko and Gilchrest 1983; West 1994). To assemble a new collagen framework, the ECM must first be degraded by MMP during remodeling. Angiogenesis also requires MMP activity to allow for the invasion of endothelial cells (Reed et al. 2000). Older adults have been shown to have higher expression of proteases and higher activity levels, and these proteases, particularly MMP2, are elevated in older postmenopausal women. Thus, estrogen replacement therapy can stimulate the migration and proliferation of keratinocytes and the elaboration of the matrix (Ashcroft et al. 2002). Lastly, in photoaged skin, the MMPs activated by UV radiation produce disorganized collagen fibrils, causing an accumulation of abnormal elastin-containing materials (Fisher et al. 1997). Due to comorbidities associated with older individuals, including vascular disease, venous insufficiency, unrelieved pressure, or DM, this population is more susceptible to chronic wounds. The sequelae of these conditions whether venous leg ulcers, diabetic foot ulcers, arterial ulcers, or pressure ulcers, disproportionately afflict older adults (Gould et al. 2015).

Since blood flow adequacy is an important determinant of wound healing efficiency, it is useful to further consider the potential impacts of age-related changes in skin blood flow. Early work using the hairless mouse skin model demonstrated age-related changes based on laser Doppler perfusion measurements (Mayrovitz

1992). In humans, early measurements of skin blood flow in deltoid skin using ^{133}Xenon washout methods in 65 men aged 20–70 pointed to a 40% skin blood flow reduction over this age range potentially impacting the skin's ability to adapt to stressors (Tsuchida 1993). Laser Doppler methods have also been used to demonstrate reduced microvascular vasodilatory reserve in older vs. younger persons. For example, using direct skin heating as a stimulus, increases in forearm skin blood perfusion were less in men aged 60–79 years than in men between the ages of 20 and 39 years (Richardson 1989). A similar reduction in heat-induced hyperemia was observed in the foot dorsum of elder men vs. younger men that was attributed to a reduction in vascularity of the capillary network (Weiss et al. 1992). A similar conclusion was reported based on laser Doppler measurements of heat-induced hyperemia in a small group of young and older men (Tolino and Wilkin 1988). Further work by Inoue and coworkers compared responses of passive heating of young (20–25 years) vs. older (64–76 years) men and showed a reduced heat-induced skin perfusion increase in skin of the back, chest, forearm and thigh subsequent to passive heating (Inoue et al. 1998). Further evidence for an age-related reduction in the maximum achievable skin blood flow was provided by a sustained one-hour heating protocol that used venous occlusion plethysmography (Rooke et al. 1994).

Given the importance of adequate blood flow increases during the wound healing process, it is likely that the age-related decline in vasodilatory reserve is relevant to slowed or blunted wound healing in some patients. Contrastingly, in at least one study, the postischemic forearm skin hyperemic response was not found to be different based on a three-minute occlusion-release protocol in a small group of 10 young and 10 older subjects (Kelly et al. 1995). More recently, an age-related differential response in finger pad blood perfusion of the third finger was observed during simulated Braille reading with the index finger (Murata et al. 2021). During this procedure, finger pad perfusion decreased in both young (20–24 years) and older (64–70 years) subjects, but a significantly greater reduction was observed in the younger persons. Since it had been established that the finger blood perfusion is subject to neurovascular control (Mayrovitz and Groseclose 2002a, b, 2005; Mayrovitz et al. 2005), it is likely that the age-related features are at least in part attributable to such neurovascular differences. An example of the neurovascular assessment using the response to a deep and rapid inspiratory gasp is illustrated in Fig. 10.1. However, the decreased skin blood perfusion and decreased pulse pressure that accompanies the inspiratory gasp reflex as shown in Fig. 10.1 have not as yet been evaluated with respect to possible age-related differences.

However, the potential role of age-related changes in neurovascular functioning as a contributor to delayed or absent healing has been suggested with possible linkages to sensory nerve function (Ardron et al. 1991). These workers demonstrated a reduced axon reflex vasodilation to be present in 15 elderly patients who were afflicted with chronic venous ulcers, thereby potentially implicating age-related changes in skin C-fiber function impacting skin blood perfusion. Although a major function of these skin sensory nerves is pain signal transmission, they also release

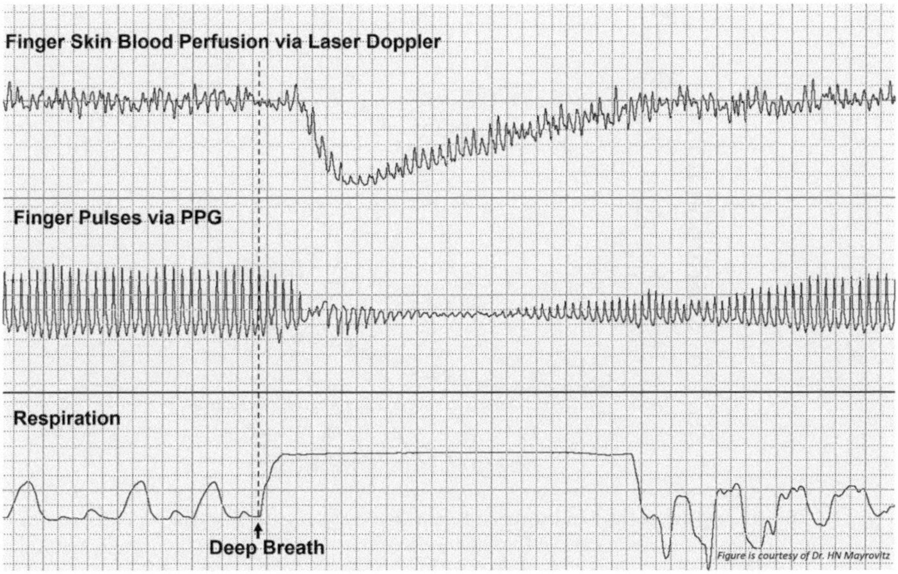

Fig. 10.1 Illustrating the response to an inspiratory gasp held for 25 s
The finger pad skin blood perfusion (upper panel) falls dramatically during a rapid inspiratory gasp, but during the breath holding, it is observed to recover. A similar change is noted in the pulse plethysmographs recording (middle panel) of the index finger pressure pulses of the same hand but with a longer time to recover. Upward movement on the respiration tracing indicates inspiration. Horizontal scale is 1 sec/div. (Figure is provided as a courtesy of Dr. HN Mayrovitz)

substances that impact microvessel vasodilatory states thereby affecting blood flow and, in that way, may impact wound healing.

A potential role of sensory afferent fiber function deficit with increasing age is consistent with the age-related trend in their laser Doppler perfusion response to capsaicin challenge (Munce and Kenney 2003). In this study of a group of eight men between the ages of 65 and 80, the vasodilatory response of forearm skin blood perfusion to capsaicin was significantly less than observed in men either 40–55 years or 18–30 years. When sensory nerve-related vasodilation was blocked using a local anesthetic, the age-related difference in heat-induced vasodilation between young vs. older men was reported to be lost (Tew et al. 2011). This further supports a role for an age-related sensory nerve changes that may impact wound healing via limitations on maximal vasodilation. However, direct linkages have yet to be confirmed experimentally. The use of such localized heating in combination with assessments of skin blood perfusion responses is useful to evaluate microvascular blood flow reserve and its potential limitations. Figure 10.2 illustrates a typical response to local skin heating as influenced by the presence of significant peripheral arterial disease in only the left leg of a patient. The much-reduced microvascular reserve would have a negative impact on wound healing.

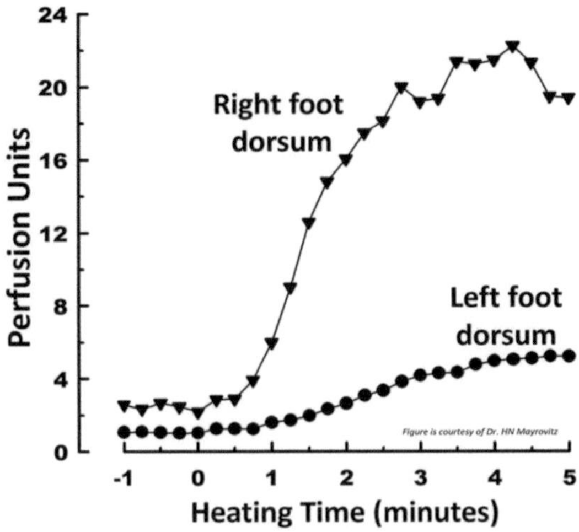

Fig. 10.2 Illustrating the normal and reduced microvascular perfusion response to local heating Skin blood perfusion measured via laser Doppler on foot dorsum of a patient with hemodynamically significant peripheral arterial disease in the left leg only. Heating was local to a skin temperature of 42 °C and maintained. The significantly reduced microvascular perfusion reserve is evident by the differences between the right and left dorsum responses. This would have a negative impact on the healing of a metabolically active wound if present. (Figure is provided as a courtesy of Dr. HN Mayrovitz)

10.5.2 Pressure Ulcers

10.5.2.1 Major Features and Blood Flow Involvement

The Centers for Medicare and Medicaid Services (CMS) has referred to pressure ulcers as describing a chronic skin lesion mainly caused by excess pressure over bony prominences that occlude blood flow (Horn et al. 2015). Although such ulcers tend to develop in elderly and also in persons with spinal cord injury (Harrow and Mayrovitz 2014), they can occur in anyone subject to sufficiently large and unrelieved pressure. Interfaces between the body surface at points of contact at bony prominences such as the trochanter, heel, malleolus, and sacral region are particularly vulnerable to the development of pressure ulcers. Use of thermal scanning of the sacral region to predict sacral ulcer development in hospital admitted patients failed to be sufficient selective (Mayrovitz et al. 2018).

However, measurements of sacral skin blood perfusion via laser Doppler imaging reveal a high level of normal skin blood perfusion in resting healthy persons compared to other nearby and distant skin sites (Mayrovitz et al. 2002). Deprivation of this blood flow during sustained supine lying is likely one factor in pressure ulcer development. Indeed, laser Doppler imaging of acquired sacral ulcers demonstrates increased blood perfusion at the center of the ulcer (Huiming et al. 2021). Measurements of skin blood perfusion and its response to local heating in periulcer regions have demonstrated elevated preheat values and augmented hyperemia in comparison

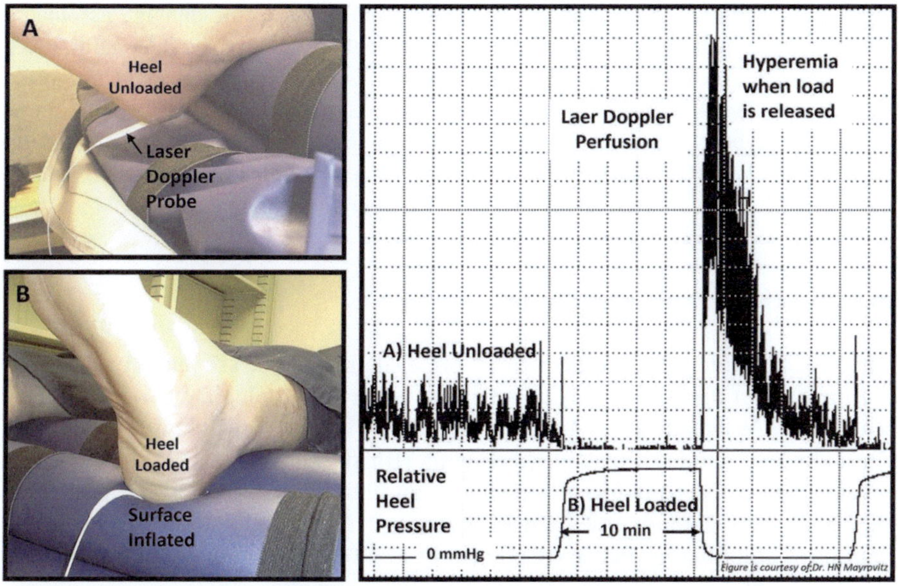

Fig. 10.3 Normal hyperemic response to heel loading and off-loading
A flat laser Doppler probe is placed on the heel in (**a**) with the heel unloaded since the support surface cell deflated. In (**b**), the surface is inflated to load the heel and the blood perfusion is seen to drop to unmeasurable levels as shown in the right panel. Upon deflation, after a loading time of just 10 min, a large hyperemic response is noted that is normal. In persons with compromised vascular and microvascular status, the hyperemic response is much diminished or absent. (Figure is provided as a courtesy of Dr. HN Mayrovitz)

to normal skin but these skin regions have a significantly reduced oxygen as determined by transcutaneous oxygen measurements (Schubert 2000). The important potential role of altered microvascular perfusion as a main source of the genesis of pressure ulcers has been studied by determining the blood flow deficits during tissue loading and subsequent unloading on the foot heel, a common site for pressure ulcer development (Mayrovitz et al. 1997; Mayrovitz and Smith 1998, 2002; Mayrovitz et al. 1999, 2003a, b, 2004). A normal hyperemic response to heel loading on a support surface and its release is shown in Fig. 10.3 demonstrating that even rather short durations of compression-related ischemia are followed by significant hyperemia in normal tissue.

10.5.2.2 Tissue Injury and Risk Factors

Studies of the details of pressure ulcers demonstrate injuries to skin and underlying tissue due to sustained external pressure and shear forces thought to be attributable to internal tissue deformation (Gebhardt 2002; Bouten et al. 2003). This distortion or deformation is caused by compression and/or shear between the skeleton and support surface (e.g., bed or chair) that leads to localized acute damage associated with the tissue ischemia, and potentially subsequent tissue necrosis (Gebhardt 2002; Gefen 2009a, b). Therefore, this type of injury is commonly found in immobile patients on the skin and soft tissue that as was previously noted occurs most frequently at bony areas of the body, such as the occiput, trochanters, sacrum, malleoli, and heel (Gebhardt 2002). Other common locations include ischial, patella, and pretibial ulcers (Vasconez et al. 1977; Cannon and Cannon 2004).

10.5.2.3 Generalized Risk Factors

It has been reported that pressure ulcers are a significant disease burden worldwide and the highest disability index compared to other dermatological conditions (Hay et al. 2014). As already noted, predisposing factors include loss of movement, loss of sensation, and failure of reactive hyperemia (Bhattacharya and Mishra 2015). The development of pressure ulcers depends on an individual's health status and tissue tolerance. Age, comorbidities (such as Type 2 DM), and nutritional status are significant contributing factors (Coleman et al. 2014) with two-thirds of pressure ulcers occurring in those aged 60–80 years (Leblebici et al. 2007).

Normally as tissue distortion occurs and causes ischemia, CNS signals of discomfort and pain stimulate protective movements to relieve the pressure. However, immobility or a loss of sensation may prevent signals from being communicated to restore circulation to the area, resulting in the development of a pressure ulcer (Bliss 1998). Even in healthy patients, ischemic responses occur with positions such as supine, lateral lying, and high sitting, corresponding to pressures between skin and support surface from 30 to 90 mm Hg (Bader and Worsley 2018). Thus, underlying factors in individual anatomy (tissue structure and geometry) and physiology (microcirculation and nutrition factors) are generalized factors involved (Coleman et al. 2014).

10.5.2.4 Tissue Damage Causation

Prolonged external pressure of sufficient magnitude on skin occludes capillary perfusion resulting in ischemic soft tissue injury, skin breakdown, ulcer development and possible necrosis depending on the amount of pressure and its duration of action. Clinical experience in nursing home facilities and in hospital settings shows that despite proper positional changing, pressure ulcers still occur (Tsuji et al. 2005). Two different damage mechanisms, both of which depend on tissue ischemia, have been put forward to account for the pathogenesis of pressure ulcers. The first is ischemic damage due to interruption of perfusion to tissue that causes direct cell damage (Bouten et al. 2003). The other is ischemic damage caused by decreased nutrients and oxygen and increased toxic metabolites in the tissue. It has been reported that this latter process occurs at relatively low internal tissue strains and can take several hours to develop (Soetens et al. 2019b). Contrastingly, direct cell damage would be caused by higher pressures and greater internal strains and can happen within tens of minutes. The concept is that the development of a pressure ulcer can be associated with both the magnitude and duration of tissue deformation—high deformation for short durations and low deformation for long periods (Bouten et al. 2003) with the type of injury dependent on the characteristic of the tissue loading pattern (Oomens et al. 2015). Continuous interface pressure monitoring of patients with spinal cord injuries are consistent with this concept (Fryer et al. 2022).

10.5.2.5 Tissue Pressure and Distortion as a Cause of Ischemia

It has been reported that tissues can sustain pressure of around 30–32 mm Hg for a short duration, but pressures exceeding this will cause occlusion of the microcirculation (Gefen 2009a, b). However, these values are in part based on what was assumed to be an average arterial-end capillary intravascular pressure, and thus, larger values would cause sufficient compression to cause capillary occlusion. However, this concept does not take into account the tissue structure and its properties that lay between the pressurized skin surface and the underlying vessels to be compressed. Further, such values do not intrinsically include the fact that skin surface shear and underlying tissue strain may be equally or more important in affecting the underlying tissue and vessels therein (Oomens et al. 2010).

The manner in which surface forces are transmitted to and through underlying tissue to impact the microcirculation is complex and in very much need of new research efforts. The impact of support surface dynamic pressure patterns on underlying microcirculatory parameters are also of interest (Chai et al. 2017). There is also the possibility that the focus on "compressing" capillaries has hidden the potential involvement of the lymphatic vessels that have a much lower internal pressure. It would appear that sustained occlusion of the underlying lymphatic vessels should be considered as a potential contributor to some forms of pressure ulcer development. Some data supports this concept (Gray et al. 2016) and points to the variability of thresholds among patients (Worsley et al. 2020). Despite the caveats of this paragraph, much work in related areas has been done and concepts put forward that help clarify the factors involved as they relate to the involvement of microvascular perfusion.

For example, the potential involvement of venous vessels, that have a low intravascular pressure, and possible effects on the subpapillary thermoregulatory A/V shunts has been described (Ambrozy et al. 2013). Additionally, the concept and potential role of tissue creep that allows soft tissues to accommodate an external load thereby reducing the surface external stress but in so doing, increasing internal tissue strain (Dodd and Gross 1991). Such strain and associated tissue distortion, if large enough, decreases blood flow via a variety of possible mechanisms (Callam et al. 1985). Distortion-related vascular kinking can compromise tissue blood perfusion causing ischemia and possibly necrosis within hours (Bhattacharya and Mishra 2015). Further, such internal shearing forces tend to distort and occlude blood flow more easily due to bending or kinking vessels rather than direct compression (Goossens et al. 1997).

10.5.2.6 Other Issues and Considerations

Differences in local tissue structure and microvasculature lead to differential abilities to withstand distortion forces and hence different involvements of the localized microcirculation. For example, foot soles have a thin, soft tissue covering blood vessels that are adapted for weight bearing and foot soles rarely develop pressure ulcers. The sacrum and ischial tuberosities do not have such well-adapted blood vessels. Even though these regions have a thicker layer of soft tissue, they can develop pressure ischemia even under light compression.

Depending on the pressure application duration, its relief is normally associated with a reactive hyperemia allowing rapid restoration of O_2 and waste product removal in previously ischemic tissue. A failure of the reactive hyperemic cycle or a major deficit in it, will prevent the tissues from recovering from the pressure-induced ischemia and, eventual permanent damage. In addition, following periods of ischemia, cellular injury resulting from reperfusion may occur when blood flow is restored. Nutrient and oxygen-deprived tissue may reduce metabolism to preserve function. When perfusion is restored, high levels of cytotoxic reactive oxygen species may be produced that exceed the capacity of the free radical scavenging mechanism (Tsuji et al. 2005). It is likely that ischemia-reperfusion injury, in addition to ischemic necrosis, contributes to the formation of some pressure ulcers (Salcido et al. 1994; Tsuji et al. 2005).

In addition to hyperemic responses to prior ischemia, some blood vessels exhibit pressure-induced-vasodilation (PIV) in which a pressure applied to skin causes cutaneous microvessels to vasodilate thereby potentially mitigating ischemic effects (Fromy et al. 2000). However, once a threshold level is exceeded, the PIV response doesn't compensate for compression pressures that would normally be experienced. At these higher pressures (as low as 60 mmHg), complete microvascular occlusion has been observed (Worsley et al. 2020). Similar pressures were shown to cause local ischemia and inflammation and accumulation of metabolites due to anaerobic metabolism (Allman 1997; Soetens et al. 2019a; Knight et al. 2001). Additionally, persons with an already compromised microcirculation are at greater risk of a pressure ulcer and once it occurs may require a longer recovery period (Bogie et al. 1995).

10.5.3 Venous Ulcers

10.5.3.1 Incidence, Location, and Factors

About 80% of lower extremity wounds are venous ulcers with nearly 95% of them located in the gaiter area, somewhere between ankle and knee, but often closer to the malleolus than knee. An example of a venous ulcer located in the gaiter area is shown in Fig. 10.4. This figure demonstrates some of its common features such as being a shallow ulcer with irregular margins often with surrounding hyperpigmentation. The predilection for the gaiter area may be related to reported differences in the veno-arterial vasoconstrictor response evaluated by assessing the skin perfusion reduction from supine to standing (Bull et al. 1995).

Factors involved in venous ulcer development include venous reflux and venous hypertension due to incompetence of deep and communicating vein valves and thrombosis of deep vein segments (Gschwandtner and Ehringer 2001). Also, CVI is often present and associated with venous drainage obstruction and increased venous pressure and reflux due to arteriovenous fistulas. Previous leg injuries, deep vein thrombosis, phlebitis, and older age are among the risk factors for the development of venous leg ulcers (Nelson et al. 2000; Collins and Seraj 2010). Approximately 1.5% of Americans have venous ulcers with a female to male ratio of near 1.6 to 1. About 20% of those developing venous ulcers do so prior to age 40 and about 40% of patients that develop venous ulcers have a history of deep vein thrombosis and have a diagnosis of CVI. Although the evolution of skin ulcers

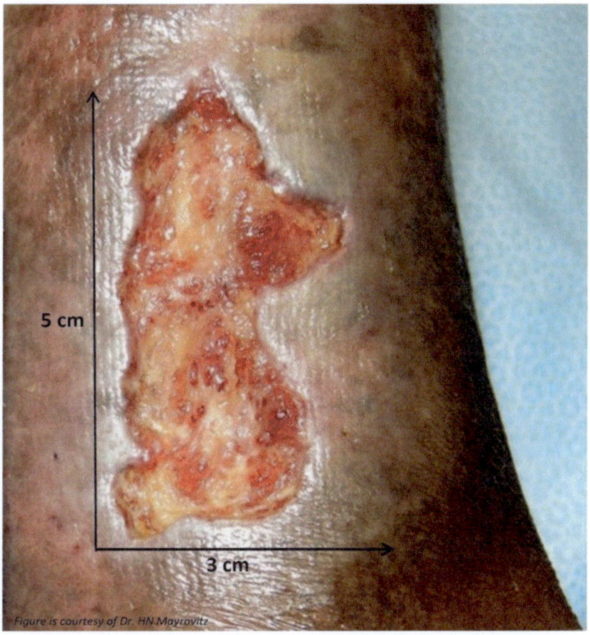

Fig. 10.4 Venous ulcer located on the lateral gaiter area
These ulcers typically have an irregular shape and characteristic wound bed granulation tissue and surrounding tissue hyperpigmentation. (Figure is provided as a courtesy of Dr. HN Mayrovitz)

from venous hypertension is not fully understood; contributory factors include inflammatory processes, intercellular and vascular adhesion molecule upregulation, protein rich edema, leukocyte trapping, oxygen deprivation, and microcirculatory deficits (Robles-Tenorio et al. 2022).

10.5.3.2 Microvascular Involvement

Although the precise pathophysiology of venous ulcer development is unclear, various theories have been described (Vasudevan 2014), and early reviews have considered multiple factors as contributory (Shami et al. 1992). For example, the "white cell trapping" theory describes a release of free radicals that result in tissue death as a consequence of venous hypertension (Brem et al. 2004). And as noted, microcirculatory deficits due to increased activation of platelets, monocytes and neutrophils leading to microvascular aggregation and microvascular entrapment of neutrophils has been reported (Robles-Tenorio et al. 2022).

10.5.3.3 Microvascular Linkages to CVI

It has been proposed that the link to CVI may be that the microvasculature can no longer fully regulate cutaneous blood flow in skin areas severely affected by the CVI (Junger et al. 1996b). Accordingly, the venous congestion leads to capillary and postcapillary hypertension during ambulation that is manifest in the skin as edema, hyperpigmentation, induration, white atrophy, and skin ulcers (Junger et al. 2000). Earlier work by this group described a lack of cutaneous microvascular reserve that prevents blood flow from matching sudden changes in tissue demand due to orthostasis (Junger et al. 1996a). Since cutaneous microangiopathy appears to precede the development of trophic skin alteration in the presence of chronic venous congestion, and there appears to be a close correlation between the extent of the clinical manifestation and severity of microangiopathy (Junger et al. 1996a, b), it is likely that cutaneous microcirculation forms the link between venous hemodynamics and congestive dermatoses.

In mild CVI, morphological changes in the capillaries, such as moderate dilatations and increased tortuosity, can be seen via capillary microscopy. The higher the ambulatory venous hypertension, the more enlarged and tortuous the capillaries (Fagrell 1995). The increased diameters of pericapillary spaces create a halo effect, that indicates increased transcapillary filtration, that varies in size along the outer border of the dermal papillae, thereby creating a cobblestone appearance (Speiser and Bollinger 1991; Junger et al. 2000). In severe CVI, many or most capillaries are occluded by white blood cells and hemorrhagic disturbances, leading to decreased capillary density (Coleridge Smith et al. 1988). In areas of white atrophy, there is an avascular field in which few or no capillaries are found (Franzeck et al. 1984).

Compression therapy is the mainstay of treatment for CVI, which has been shown to improve most subjective symptoms in patients. Clinically effective compression therapy reverses the pathogenic processes of CVI on the microcirculation and interstitial architecture by improving nutritive blood flow in congested areas (Abu-Own et al. 1994; Christopoulos et al. 1991; Leu et al. 1993). One mechanism by which compression therapy improves the situation is by recruiting previously under perfused capillaries thereby increasing capillary density and nutritive blood flow and by promoting angiogenesis. An improvement in nutritive perfusion generally accompanies improved healing of venous ulcers. In the first 2 weeks, changes in the capillary density can be seen to have prognostic significance. Increased capillary density suggests accelerated healing within approximately 6 weeks of therapy, while a lack of angiogenesis is related to delayed healing (Junger et al. 2000).

10.5.3.4 Normal and Abnormal Calf Pump Function

Normally the venous calf muscle pump helps propel blood in leg veins against gravity toward the heart (Gschwandtner and Ehringer 2001), with each calf muscle contraction compressing the deep veins forcing blood forward with function venous valves preventing backflow into superficial veins. The pressure gradient in the deep veins and the presence of functional valves in the communicating veins, allows blood from superficial veins to flow normally (Back et al. 1995). However, with dysfunction or failure of the valves each contraction of the leg muscles propagates an associated pressure peak into the superficial venous system. The ambulatory venous hypertension is propagated into the nutritive capillaries, and these high-pressure peaks eventually deteriorate the microcirculation of the skin—initially through capillary dilatation, followed by gradual rarefaction (Junger et al. 2000).

These events and the blood flow dynamics that occur are schematically illustrated in Fig. 10.5 in which the normally low-pressure superficial veins become exposed to the high pressures induced by the reverse flow pathways associated with incompetent valves. Concomitant with the tissue injury are inflammatory processes, increased vascular permeability and edema and or lymphedema. In Fig. 10.6, two examples of venous ulcers are shown in which skin blood perfusion assessed via laser Doppler methods is being measured in the periulcer region of two patients. The initial blood perfusion measurement is made at a skin temperature of 35 °C and then locally heated to 44 °C with responses as shown in Fig. 10.7. The responses shown in Fig. 10.7 demonstrate a common finding for ulcers of venous origin; an elevated periulcer basal resting flow with little if any microvascular reserve when stimulated with heat as shown in part B, but with normal responses in healthy control skin as shown in part A. It has been reported that in patients in which the microvascular reserve is relatively maintained healing of venous ulcers is expedited (Mlacak et al. 2005).

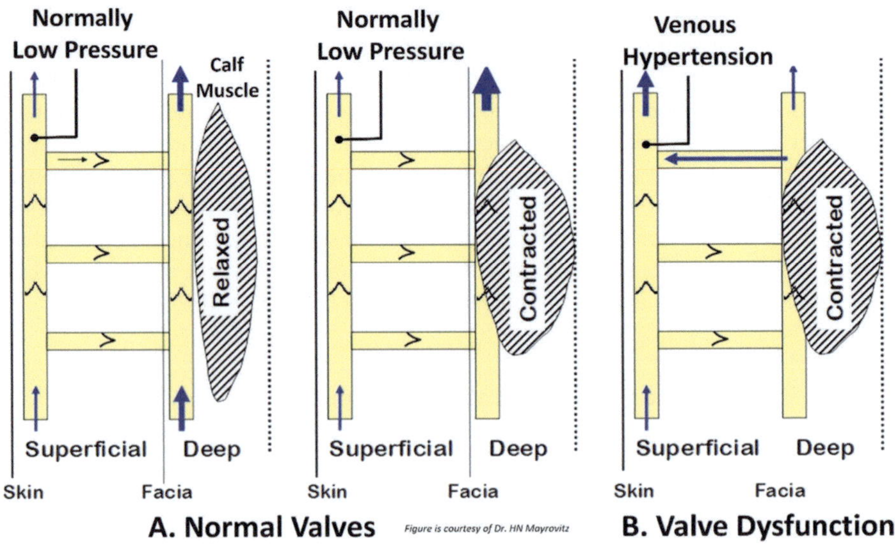

Fig. 10.5 Schematic of the impact and hemodynamics of incompetent venous valves
The normally low pressure experienced by the superficial veins is subject to high pressures in the presence of the valve incompetency as shown in part B. This elevated pressure is not well tolerated and causes venous injury that triggers a sequence of events that may lead to the development of a venous ulcer. (Figure is provided as a courtesy of Dr. HN Mayrovitz)

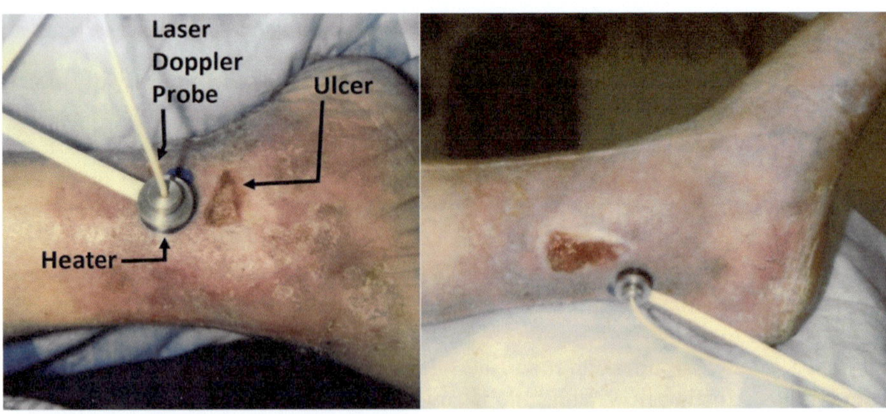

Fig. 10.6 Illustrating periwound skin blood perfusion measurement in venous ulcers
A laser Doppler probe is fitted through a concentric hole in the heater that is in contact with skin. Localized heating produces an increase in microvascular perfusion in healthy skin but with a differ pattern in periwound skin as shown in Fig. 10.7. (Figure is provided as a courtesy of Dr. HN Mayrovitz)

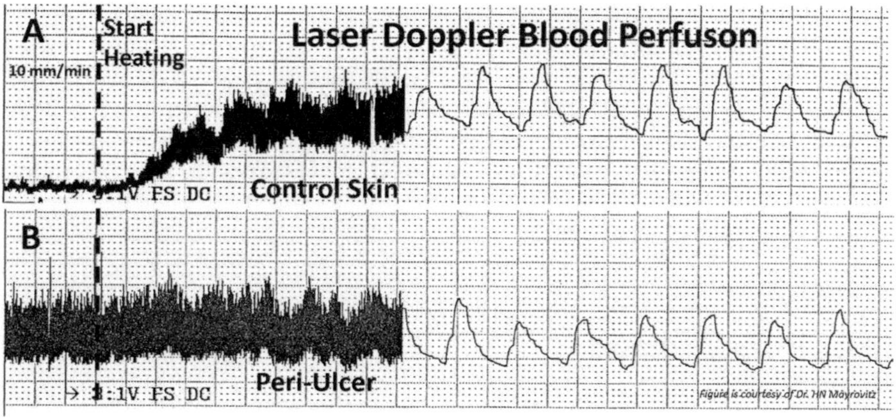

Fig. 10.7 Skin blood perfusion responses to heating in healthy vs. periulcer skin
The responses show a normal response to localized heating (**a**) and a common finding associated with venous ulcers (**b**). In B, an elevated periulcer basal resting perfusion is noted with little if any microvascular reserve when stimulated with heat. Contrastingly, in control skin as shown in part (**a**) a normal active hyperemia is noted in response to the heating. (Figure is provided as a courtesy of Dr. HN Mayrovitz)

10.5.3.5 The Wound and the Periwound Skin

Within the wound bed of venous ulcers, the capillary density is reported to be heterogeneous (Burnand et al. 1981; Leu 1991). Almost no capillaries are seen in areas without granulation tissue. These areas have poor nutritive microcirculation contributing to the formation of venous ulcers but moderate subpapillary blood flow. This moderate subpapillary perfusion may likely facilitate future wound healing and the construction of granulation tissue (Gschwandtner and Ehringer 2001).

As venous ulcers heal, they form granulation tissue seen grossly as erythematous areas within the ulcers (Ambrozy et al. 2013), as may be visualized in Fig. 10.4. In these areas, a few capillary sprouts are found embedded in small edematous pockets (Leu 1991; Burnand et al. 1981; Gschwandtner et al. 1999). This stage of healing is characterized by a distinct increase in subpapillary perfusion and increased appearance of visible capillaries. Nutritive perfusion in or adjacent to venous ulcers is lower than in normal skin areas although is reported to be higher than in ulcer areas without granulation tissue (Gschwandtner and Ehringer 2001). Thus, subpapillary perfusion appears to be crucial not only for thermoregulation but also for wound healing at this stage (Ambrozy et al. 2013). Capillary density is significantly improved but still low compared to noncompromised tissue. Additionally, these capillaries appear long and dilated, unlike the typical glomerular-like morphology in patients with chronic hypertension (Ambrozy et al. 2013; Gschwandtner et al. 1999, 2001).

At the scarring stage, the microvascular pattern of the skin differs from that observed in previously intact periulcer skin (Gschwandtner et al. 1999, 2001) and has glomerular-shaped capillaries characteristic of what is seen in CVI (Ambrozy et al. 2013). The nutritive capillary density was found to be relatively high and similar to that seen in the early stages of venous ulcers prior to the ulcer developing granulation tissue (Ambrozy et al. 2013, Gschwandtner et al. 1999, 2001). The capillary density was considered to be moderate (20–30 capillaries/mm^2) in comparison to healthy skin (Lamah et al. 1996), although it is unclear if this capillary density is due to neoangiogenesis or contraction of the tissues. In healed periulcer skin, there is high nutritive and moderate subpapillary blood perfusion that appears close in value to normal skin (Konecny et al. 1987).

10.5.4 Arterial Ulcers

Arterial ulcers represent about 5% of all leg ulcers and are commonly located on the leg or foot and represent about 10–20% of all nonhealing lower extremity ulcers (Mekkes et al. 2003). The most prevalent predisposing condition for the development of an arterial ulcer is advanced peripheral arterial disease (PAD) present within arteries that supply the lower leg and foot (Spentzouris and Labropoulos 2009). Skin breakdown and ulceration is thus attributable to inadequate microvascular perfusion causing ischemic and hypoxic effects on skin and subcutaneous tissue (Grey et al. 2006). These ulcers will often occur in areas subject to pressure or trauma such as at the malleolus or posterior heel or the toes leading to necrosis and the need for amputation as illustrated in Fig. 10.8. Arterial ulcers are difficult to heal in the absence of an adequate restoration of blood flow, which itself may be more difficult to accomplish due to the presence of comorbid conditions (Greer et al. 2012). In

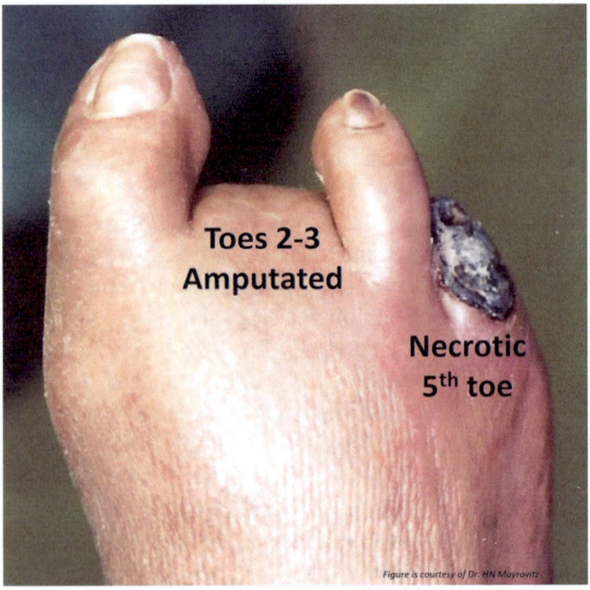

Fig. 10.8 Illustrating aspects of an arterial ulcer A patient with critical ischemia due to significant PAD in whom toes 2–3 were previously amputated and toe 5 is necrotic. (Figure is provided as a courtesy of Dr. HN Mayrovitz)

patients such as illustrated in Fig. 10.8, assessments of the response to local periulcer heating that is used to measure microvascular perfusion reserve are similar to that illustrated for the left foot dorsum in Fig. 10.2 or in more extreme cases, with even less or an absent perfusion increase. The presence of this microcirculatory deficit is accompanied by abnormally low levels of tissue oxygen levels as assessed by transcutaneous oxygen tension measurements (Ueno et al. 2010; Cina et al. 1984; Mayrovitz and Larsen 1994). Patients with PAD who also have DM and a foot ulcer are at greater risk of amputation (Azhar et al. 2021). However, that risk depends on the revascularization approach used (Butt et al. 2019) and on the extent of the PAD (Apelqvist et al. 2011). An epidemiological study reported that about 0.5% of the Korean population had both PAD and a diabetic foot ulcer (Chun et al. 2019).

10.5.5 Wounds in the Diabetic Patient

Persons with DM are generally at increased risk of developing skin ulcerations in part related to the presence of neuropathy, ischemia, and poor glycemic control (Clayton Jr. and Elasy 2009), although the development of DM foot ulcers has been reported to be less with improved glucose control that also improves microcirculation (Rathsman et al. 2014). The higher likelihood of PAD and the presence of microvascular deficits in DM (Thiruvoipati et al. 2015) increase the chances of ischemia, tissue to breakdown, and ulcer formation. In diabetic patients with PAD, the nature and distribution of leg arterial stenoses causing the ischemia is varied (Graziani et al. 2007) but even without the presence of significant PAD, foot ulcers in diabetic patients are often more difficult to heal for reasons that include reduced blood flow (Dinh et al. 2011; Catrina and Zheng 2016), wound oxygen deficits (Okonkwo and Dipietro 2017), and infection. Although wound healing in patients with DM tends to be inferior in the presence of reduced microcirculation (Lowry et al. 2017), the relationship between microvascular assessments and wound healing is not always easy to predict (Mennes et al. 2021).

In some patients with DM, it takes less local pressure to reduce skin blood flow in regions of bony prominence thereby laying the groundwork for ulcerations at these sites. When sensory neuropathy is present, normal pressure/pain signals are diminished or absent, thereby removing warning of developing tissue injury. Most of these types of ulcers develop on the foot, with plantar ulcers often associated with neuropathy. An example of such an ulcer is shown in Fig. 10.9. In these cases, elimination of elevated foot pressures combined with standard wound care are indicated. Statistics suggest that about 15–25% of persons with DM will get a foot ulcer (Yazdanpanah et al. 2018) with an annual incidence rate of 2–4% (Crawford et al. 2011). Diabetic-related nonhealing ulcers account for 140,000 extremity amputations per year in the USA (Mizelle Jr. 2021) and an annual amputation incidence rate between 0.5% and 0.8% (number of amputations per patient-year).

In terms of mechanisms and processes, DM causes structural and functional metabolic alterations to the arteriolar and microcirculatory systems, especially in

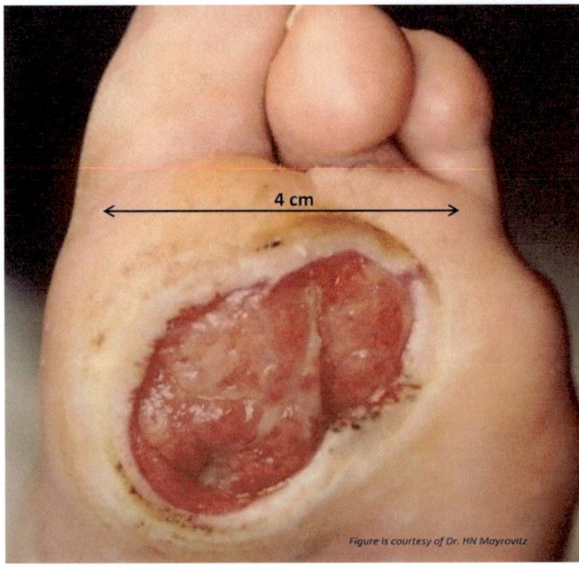

Fig. 10.9 An example of a plantar diabetic neuropathic ulcer. (Figure is provided as a courtesy of Dr. HN Mayrovitz)

the lower extremities of poorly controlled blood sugar patients (Cohen 1993; Nathan 1993; Raskin et al. 1983). In the capillary circulation, lumen size is reduced, and the vessels have increased stiffness (Jaap et al. 1996; Rayman et al. 1995). The increased stiffness is due to a thickened basement membrane and arteriolar hyalinosis, as well as glycosylation and formation of nonenzymatic advanced glycosylation end products (AGEs) (Hile and Veves 2003), thus limiting the vessel's ability to vasodilate and eventually autoregulate (Tooke 1995). Simultaneous measurements of transcutaneous oxygen, laser Doppler perfusion, cardiac output, and leg blood flow in 60 patients with and 60 patients without DM led to the conclusion that there was a direct linkage between a diabetic-related deficit in tissue oxygen and a submaximal microvascular vasodilatory reserve with little dependence on other circulatory factors(Mayrovitz and Larsen 1996). Other work indicated elevated levels of skin tissue water although apparently not significantly related to HbA1c values in patients with DM (Mayrovitz et al. 2013, 2016). Such alterations are likely to impact wounds when present in persons with DM.

Indeed, diabetic complications have been associated with impaired vasodilatory capacity. Nitric oxide (NO)-dependent smooth muscle vasodilation has been shown to be abnormal in patients with diabetic foot ulcers, and endothelium-dependent and -independent vasodilation is abnormal in diabetic neuropathy (Korzon-Burakowska and Edmonds 2006). However, in early work, it has been reported that within the arteriolar circulation, blood flow may be normal or even increased (Parving et al. 1983). It was posited that these characteristic changes are caused by endothelial injury attributable to increased microvascular pressure and shear force in leg microcirculation and leads to an injury response and proliferation of extravascular matrix proteins (Ajjam et al. 1985; Tilton et al. 1985). The thickened membranes impair leukocyte transwall emigration, thereby increasing the susceptibility of the diabetic foot to infection, and additionally reduce the needed hyperemic response to injury (Flynn and Tooke 1992; Rayman et al. 1986). As nutritive perfusion of the skin plays a role in ulcer development and wound healing, these changes are a major concern in diabetic patients who have microvascular deficits (Korzon-Burakowska and Edmonds 2006).

As discussed, PAD and peripheral neuropathy are both major causes of a patient developing a diabetic ulcer and contribute to poor healing outcomes (Hinchliffe et al. 2016; Prompers et al. 2008). However, in many patients with microvascular deficits, there is no corresponding clinical evidence of lower extremity macrovascular disease. However, dysfunctional vessels of the microvasculature and impairment of the nerve-axon reflex may be concomitant abnormalities contributing to the neuro-ischemic diabetic limb—leading to ulceration and impaired wound healing (Hile and Veves 2003; Quigley et al. 1991). It has been shown that revascularization significantly improves, but does not completely resolve, these impairments (Arora et al. 2002). However, microangiopathy as a direct cause of ulcer development has not been fully confirmed, and in fact, one study did not find a relationship between microvascular abnormalities and wound healing (Korzon-Burakowska and Edmonds 2006).

10.6 Conclusion

Proper wound healing is inseparable from a functional microcirculation, as adequate localized blood flow and metabolic waste removal are necessary to meet the increased demands of healing tissue. Normal wound healing consists of four overlapping phases: Hemostasis, inflammation, proliferation, and remodeling, and microcirculation participation is to varying degrees present at each phase. Throughout these phases, angiogenesis occurs to supply nutrients and O_2 to wounded tissues, with the microcirculation delivering growth factors and other needed substrates. Skin has nutritive blood perfusion (via the superficial plexus) regulated by metabolic needs with some areas supplied by arteriovenous anastomoses (the deep plexus) largely under autonomic control. Following wounding, the latter vasodilate to provide increased microvascular blood flow to support the wound healing process including increased nutritive perfusion to support stem cell expansion and production of new epidermal cells. Angiogenesis occurs rapidly, thereby increasing microcirculatory potential to meet the needs of healing wound.

Aging affects skin's composition and its microcirculation. Age-associated changes include decreased blood flow, impaired ability to adapt to stressors (including thermoregulation and vasodilatory reserve), prolonged inflammation, decreased progenitor cells and mediators, increased migration time for keratinocytes, and increased matrix metalloproteinase activity. All these factors can contribute to delayed and dysfunctional wound healing. Derangements of the microcirculation can also contribute to the development of wounds, including arterial, venous, pressure, and diabetic (neuropathic) ulcers. In these pathologies, microvascular perfusion is impeded in some way: Peripheral artery disease causes occlusion and ischemia in arterial ulcers; venous congestion, which may contribute to ischemic and inflammatory conditions in venous ulcers; external pressure causes direct damage and ischemia in pressure ulcers; metabolic and structural alterations impair vasodilatory capacity and contributing to possible infectious processes in diabetic ulcers. In summary, wounds can be viewed as an added "organ" that taxes the microcirculatory system with continued demand for blood flow. These wounds exemplify the effects of dysfunctional microcirculation when the increased needs cannot be met, not only impeding healing but taxing the system to such a degree that the skin structure cannot be maintained.

References

Abu-Own A, Shami SK, Chittenden SJ, Farrah J, Scurr JH, Smith PD (1994) Microangiopathy of the skin and the effect of leg compression in patients with chronic venous insufficiency. J Vasc Surg 19:1074–1083

Acker T, Plate KH (2003) Role of hypoxia in tumor angiogenesis-molecular and cellular Angiogenic crosstalk. Cell Tissue Res 314:145–155

Ajjam ZS, Barton S, Corbett M, Owens D, Marks R (1985) Quantitative evaluation of the dermal vasculature of diabetics. Q J Med 54:229–239

Allman RM (1997) Pressure ulcer prevalence, incidence, risk factors, and impact. Clin Geriatr Med 13:421–436

Ambrozy E, Waczulikova I, Willfort A, Bohler K, Cauza K, Ehringer H, Heinz G, Koppensteiner R, Maric S, Gschwandtner ME (2013) Healing process of venous ulcers: the role of microcirculation. Int Wound J 10:57–64

Apelqvist J, Elgzyri T, Larsson J, Londahl M, Nyberg P, Thorne J (2011) Factors related to outcome of Neuroischemic/ischemic foot ulcer in diabetic patients. J Vasc Surg 53(1582–8):E2

Ardron ME, Helme RD, Mckernan S (1991) Microvascular skin responses in elderly people with varicose leg ulcers. Age Ageing 20:124–128

Arora S, Pomposelli F, Logerfo FW, Veves A (2002) Cutaneous microcirculation in the neuropathic diabetic foot improves significantly but not completely after successful lower extremity revascularization. J Vasc Surg 35:501–505

Ashcroft GS, Mills SJ, Ashworth JJ (2002) Ageing and wound healing. Biogerontology 3:337–345

Attwell D, Mishra A, Hall CN et al (2016) What is a Pericyte? J Cereb Blood Flow Metab 36:451–455

Awan ZA, Haggblad E, Wester T, Kvernebo MS, Halvorsen PS, Kvernebo K (2011) Diffuse reflectance spectroscopy: systemic and microvascular oxygen saturation is linearly correlated and hypoxia leads to increased spatial heterogeneity of microvascular saturation. Microvasc Res 81:245–251

Azhar A, Basheer M, Abdelgawad MS, Roshdi H, Kamel MF (2021) Prevalence of peripheral arterial disease in diabetic foot ulcer patients and its impact in limb salvage. Int J Low Extrem Wounds *22:518*

Back TL, Padberg FT Jr, Araki CT, Thompson PN, Hobson RW 2nd (1995) Limited range of motion is a significant factor in venous ulceration. J Vasc Surg 22:519–523

Bader DL, Worsley PR (2018) Technologies to monitor the health of loaded skin tissues. Biomed Eng Online 17:40

Barrientos S, Stojadinovic O, Golinko MS, Brem H, Tomic-Canic M (2008) Growth factors and cytokines in wound healing. Wound Repair Regen 16:585–601
Bentov I, Reed MJ (2014) Anesthesia, microcirculation, and wound repair in aging. Anesthesiology 120:760–772
Bhattacharya S, Mishra RK (2015) Pressure ulcers: current understanding and newer modalities of treatment. Indian J Plast Surg 48:4–16
Birbrair A, Zhang T, Wang ZM, Messi ML, Mintz A, Delbono O (2015) Pericytes at the intersection between tissue regeneration and pathology. Clin Sci (Lond) 128:81–93
Bliss MR (1998) Hyperaemia. J Tissue Viability 8:4–13
Bogie KM, Nuseibeh I, Bader DL (1995) Early progressive changes in tissue viability in the seated spinal cord injured subject. Paraplegia 33:141–147
Bootle-Wilbraham CA, Tazzyman S, Thompson WD, Stirk CM, Lewis CE (2001) Fibrin fragment E stimulates the proliferation, migration and differentiation of human microvascular endothelial cells in vitro. Angiogenesis 4:269–275
Bouten CV, Oomens CW, Baaijens FP, Bader DL (2003) The etiology of pressure ulcers: skin deep or muscle bound? Arch Phys Med Rehabil 84:616–619
Braverman IM (1983) The role of blood vessels and lymphatics in cutaneous inflammatory processes: an overview. Br J Dermatol 109(Suppl 25):89–98
Braverman IM (1997) The cutaneous microcirculation: ultrastructure and microanatomical organization. Microcirculation 4:329–340
Braverman IM (2000) The cutaneous microcirculation. J Investig Dermatol Symp Proc 5:3–9
Braverman IM, Schechner JS, Silverman DG, Keh-Yen A (1992) Topographic mapping of the cutaneous microcirculation using two outputs of laser-Doppler Flowmetry: flux and the concentration of moving blood cells. Microvasc Res 44:33–48
Brem H, Kirsner RS, Falanga V (2004) Protocol for the successful treatment of venous ulcers. Am J Surg 188:1–8
Broughton G 2nd, Janis JE, Attinger CE (2006) Wound healing: an overview. Plast Reconstr Surg 117:1e-S–32e-S
Bull R, Ansell G, Stanton AW, Levick JR, Mortimer PS (1995) Normal cutaneous microcirculation in gaiter zone (ulcer-susceptible skin) versus nearby regions in healthy young adults. Int J Microcirc Clin Exp 15:65–74
Burnand KG, Whimster I, Clemenson G, Thomas ML, Browse NL (1981) The relationship between the number of capillaries in the skin of the venous ulcer-bearing area of the lower leg and the fall in foot vein pressure during exercise. Br J Surg 68:297–300
Butt T, Lilja E, Orneholm H, Apelqvist J, Gottsater A, Eneroth M, Acosta S (2019) Amputation-free survival in patients with diabetes mellitus and peripheral arterial disease with heel ulcer: open versus endovascular surgery. Vasc Endovasc Surg 53:118–125
Callam MJ, Ruckley CV, Harper DR, Dale JJ (1985) Chronic ulceration of the leg: extent of the problem and provision of care. Br Med J (Clin Res Ed) 290:1855–1856
Cannon BC, Cannon JP (2004) Management of pressure ulcers. Am J Health Syst Pharm 61:1895–1905. Quiz 1906-7
Catrina S-B, Zheng X (2016) Disturbed hypoxic responses as a pathogenic mechanism of diabetic foot ulcers. Diabetes Metab Res Rev 32:179–185
Chai CY, Sadou O, Worsley PR, Bader DL (2017) Pressure signatures can influence tissue response for individuals supported on an alternating pressure mattress. J Tissue Viability 26:180–188
Charkoudian N (2003) Skin blood flow in adult human thermoregulation: how it works, when it does not, and why. Mayo Clin Proc 78:603–612
Chen Q, Anderson DR (1997) Effect of CO2 on intracellular Ph and contraction of retinal capillary Pericytes. Invest Ophthalmol Vis Sci 38:643–651
Christopoulos DC, Nicolaides AN, Belcaro G, Kalodiki E (1991) Venous hypertensive microangiopathy in relation to clinical severity and effect of elastic compression. J Dermatol Surg Oncol 17:809–813

Chun DI, Kim S, Kim J, Yang HJ, Kim JH, Cho JH, Yi Y, Kim WJ, Won SH (2019) Epidemiology and burden of diabetic foot ulcer and peripheral arterial disease in Korea. J Clin Med 8
Cina C, Katsamouris A, Megerman J, Brewster DC, Strayhorn EC, Robison JG, Abbott WM (1984) Utility of transcutaneous oxygen tension measurements in peripheral arterial occlusive disease. J Vasc Surg 1:362–371
Clayton W Jr, Elasy TA (2009) A review of the pathophysiology, classification, and treatment of foot ulcers in diabetic patients. Clin Diabetes 27:52–58
Cohen RA (1993) Dysfunction of vascular endothelium in diabetes mellitus. Monogr Am Heart Assoc 87:V67–V76
Coleman S, Nixon J, Keen J, Wilson L, Mcginnis E, Dealey C, Stubbs N, Farrin A, Dowding D, Schols JM, Cuddigan J, Berlowitz D, Jude E, Vowden P, Schoonhoven L, Bader DL, Gefen A, Oomens CW, Nelson EA (2014) A new pressure ulcer conceptual framework. J Adv Nurs 70: 2222–2234
Coleridge Smith PD, Thomas P, Scurr JH, Dormandy JA (1988) Causes of venous ulceration: a new hypothesis. Br Med J (Clin Res Ed) 296:1726–1727
Collins L, Seraj S (2010) Diagnosis and treatment of venous ulcers. Am Fam Physician 81:989–996
Crawford F, Mccowan C, Dimitrov BD, Woodburn J, Wylie GH, Booth E, Leese GP, Bekker HL, Kleijnen J, Fahey T (2011) The risk of foot ulceration in people with diabetes screened in community settings: findings from a cohort study. QJM 104:403–410
Curri SB (1990) Microvascular anatomy of the skin and its appendages. Phlebologie 43:407–430
Darland DC, D'amore PA (2001) Tgf Beta is required for the formation of capillary-like structures in three-dimensional Cocultures of 10t1/2 and endothelial cells. Angiogenesis 4:11–20
Dinh T, Elder S, Veves A (2011) Delayed wound healing in diabetes: considering future treatments. Diabetes Manag 1:509–519
Dodd KT, Gross DR (1991) Three-dimensional tissue deformation in subcutaneous tissues overlying bony prominences may help to explain external load transfer to the Interstitium. J Biomech 24:11–19
Edwards H, Finlayson K, Courtney M, Graves N, Gibb M, Parker C (2013) Health service pathways for patients with chronic leg ulcers: identifying effective pathways for facilitation of evidence based wound care. BMC Health Serv Res 13:86
Fagrell B (1995) Vital microscopy and the pathophysiology of deep venous insufficiency. Int Angiol 14:18–22
Fisher GJ, Wang ZQ, Datta SC, Varani J, Kang S, Voorhees JJ (1997) Pathophysiology of premature skin aging induced by ultraviolet light. N Engl J Med 337:1419–1428
Flynn MD, Tooke JE (1992) Aetiology of diabetic foot ulceration: a role for the microcirculation? Diabet Med 9:320–329
Folkman J (1995) Seminars in medicine of the Beth Israel hospital, Boston. Clinical applications of research on angiogenesis. N Engl J Med 333:1757–1763
Forster JC, Harriss-Phillips WM, Douglass MJ, Bezak E (2017) A review of the development of tumor vasculature and its effects on the tumor microenvironment. Hypoxia (Auckl) 5:21–32
Franzeck UK, Bollinger A, Huch R, Huch A (1984) Transcutaneous oxygen tension and capillary morphologic characteristics and density in patients with chronic venous incompetence. Circulation 70:806–811
Fromy B, Abraham P, Saumet JL (2000) Progressive calibrated pressure device to measure cutaneous blood flow changes to external pressure strain. Brain Res Brain Res Protoc 5:198–203
Fryer S, Caggiari S, Major D, Bader DL, Worsley PR (2022) Continuous pressure monitoring of inpatient spinal cord injured patients: implications for pressure ulcer development. Spinal Cord
Gebhardt K (2002) Pressure ulcer prevention. Part 1. Causes of pressure ulcers. Nurs Times 98:41–44
Gefen A (2009a) Reswick and Rogers pressure-time curve for pressure ulcer risk. Part 1. Nurs Stand 23:64–68
Gefen A (2009b) Reswick and Rogers pressure-time curve for pressure ulcer risk. Part 2. Nurs Stand 23:40–44

Gilchrest BA, Murphy GF, Soter NA (1982) Effect of chronologic aging and ultraviolet irradiation on Langerhans cells in human epidermis. J Invest Dermatol 79:85–88
Gilchrest BA, Garmyn M, Yaar M (1994) Aging and photoaging affect gene expression in cultured human keratinocytes. Arch Dermatol 130:82–86
Goossens RH, Snijders CJ, Holscher TG, Heerens WC, Holman AE (1997) Shear stress measured on beds and wheelchairs. Scand J Rehabil Med 29:131–136
Gosain A, Dipietro LA (2004) Aging and wound healing. World J Surg 28:321–326
Gould L, Abadir P, Brem H, Carter M, Conner-Kerr T, Davidson J, Dipietro L, Falanga V, Fife C, Gardner S, Grice E, Harmon J, Hazzard WR, High KP, Houghton P, Jacobson N, Kirsner RS, Kovacs EJ, Margolis D, Mcfarland Horne F, Reed MJ, Sullivan DH, Thom S, Tomic-Canic M, Walston J, Whitney JA, Williams J, Zieman S, Schmader K (2015) Chronic wound repair and healing in older adults: current status and future research. J Am Geriatr Soc 63:427–438
Gray RJ, Voegeli D, Bader DL (2016) Features of lymphatic dysfunction in compressed skin tissues – implications in pressure ulcer aetiology. J Tissue Viability 25:26–31
Graziani L, Silvestro A, Bertone V, Manara E, Andreini R, Sigala A, Mingardi R, De Giglio R (2007) Vascular involvement in diabetic subjects with ischemic foot ulcer: a new morphologic categorization of disease severity. Eur J Vasc Endovasc Surg 33:453–460
Greer N, Foman N, Dorrian J, Fitzgerald P, Macdonald R, Rutks I, Wilt T (2012) Advanced wound care therapies for non-healing diabetic, venous, and arterial ulcers: a systematic review, Washington
Grey JE, Harding KG, Enoch S (2006) Venous and arterial leg ulcers. BMJ 332:347–350
Grimm D, Bauer J, Schoenberger J (2009) Blockade of neoangiogenesis, a new and promising technique to control the growth of malignant tumors and their metastases. Curr Vasc Pharmacol 7:347–357
Gschwandtner ME, Ehringer H (2001) Microcirculation in chronic venous insufficiency. Vasc Med 6:169–179
Gschwandtner ME, Ambrozy E, Fasching S, Willfort A, Schneider B, Bohler K, Gaggl U, Ehringer H (1999) Microcirculation in venous ulcers and the surrounding skin: findings with capillary microscopy and a laser Doppler imager. Eur J Clin Investig 29:708–716
Gschwandtner ME, Ambrozy E, Maric S, Willfort A, Schneider B, Bohler K, Gaggl U, Ehringer H (2001) Microcirculation is similar in ischemic and venous ulcers. Microvasc Res 62:226–235
Gurtner GC, Werner S, Barrandon Y, Longaker MT (2008) Wound repair and regeneration. Nature 453:314–321
Hales JR, Jessen C, Fawcett AA, King RB (1985) Skin ava and capillary dilatation and constriction induced by local skin heating. Pflugers Arch 404:203–207
Hanahan D, Weinberg RA (2011) Hallmarks of cancer: the next generation. Cell 144:646–674
Harrow JH, Mayrovitz HN (2014) Subepidermal moisture surrounding pressure ulcers in persons with a spinal cord injury: a pilot study. J Spinal Cord Med 37:719–728
Hay RJ, Johns NE, Williams HC, Bolliger IW, Dellavalle RP, Margolis DJ, Marks R, Naldi L, Weinstock MA, Wulf SK, Michaud C et al (2014) The global burden of skin disease in 2010: an analysis of the prevalence and impact of skin conditions. J Invest Dermatol 134:1527–1534
Hellberg C, Ostman A, Heldin CH (2010) Pdgf and vessel maturation. Recent Results Cancer Res 180:103–114
Hile C, Veves A (2003) Diabetic neuropathy and microcirculation. Curr Diab Rep 3:446–451
Hinchliffe RJ et al (2016) Iwgdf guidance on the diagnosis, prognosis and management of peripheral artery disease in patients with foot ulcers in diabetes. Diabetes Metab Res Rev 32 (Suppl 1):37–44
Holm-Pedersen P, Viidik A (1972) Tensile properties and morphology of healing wounds in young and old rats. Scand J Plast Reconstr Surg 6:24–35
Honnegowda TM, Kumar P, Udupa EGP, Kumar S, Kumar U, Rao P (2015) Role of angiogenesis and Angiogenic factors in acute and chronic wound healing. Plast Aesthet Res 2:243–249
Horn SD, Barrett RS, Fife CE, Thomson B (2015) A predictive model for pressure ulcer outcome: the wound healing index. Adv Skin Wound Care 28:560–572

Howdieshell TR, Webb WL et al (2003) Inhibition of inducible nitric oxide synthase results in reductions in wound vascular endothelial growth factor expression, granulation tissue formation, and local perfusion. Surgery 133:528–537

Huiming G, Yuming W, Mingliang Y, Changbin L, Qiuchen H, Jianjun L (2021) Study on the characteristics of microcirculation in the site of pressure ulcer in patients with spinal cord injury. Sci Prog 104:368504211028726

Inoki I, Shiomi T, Hashimoto G, Enomoto H, Nakamura H, Makino K, Ikeda E, Takata S, Kobayashi K, Okada Y (2002) Connective tissue growth factor binds vascular endothelial growth factor (Vegf) and inhibits Vegf-induced angiogenesis. FASEB J 16:219–221

Inoue Y, Shibasaki M, Hirata K, Araki T (1998) Relationship between skin blood flow and sweating rate, and age related regional differences. Eur J Appl Physiol Occup Physiol 79:17–23

Jaap AJ, Shore AC, Stockman AJ, Tooke JE (1996) Skin capillary density in subjects with impaired glucose tolerance and patients with type 2 diabetes. Diabet Med 13:160–164

Junger M, Hahn M, Klyscz T, Rassner G (1996a) Influence of healing on the disturbed blood flow regulation in venous ulcers. Vasa 25:341–348

Junger M, Klyscz T, Hahn M, Rassner G (1996b) Disturbed blood flow regulation in venous leg ulcers. Int J Microcirc Clin Exp 16:259–265

Junger M, Steins A, Hahn M, Hafner HM (2000) Microcirculatory dysfunction in chronic venous insufficiency (cvi). Microcirculation 7:S3–S12

Kelly RI, Pearse R, Bull RH, Leveque JL, De Rigal J, Mortimer PS (1995) The effects of aging on the cutaneous microvasculature. J Am Acad Dermatol 33:749–756

Knight SL, Taylor RP, Polliack AA, Bader DL (2001) Establishing predictive indicators for the status of loaded soft tissues. J Appl Physiol 90:2231–2237

Koh TJ, Dipietro LA (2011) Inflammation and wound healing: the role of the macrophage. Expert Rev Mol Med 13:E23

Konecny U, Ehringer H, Jung M, Koppensteiner R, Minar E, Stumpflen A, Kolbl M (1987) Mapping the capillary density of hands and feet in healthy Probands. Vasa Suppl 20:113–116

Korff T, Kimmina S, Martiny-Baron G, Augustin HG (2001) Blood vessel maturation in a 3-dimensional spheroidal Coculture model: direct contact with smooth muscle cells regulates endothelial cell quiescence and abrogates Vegf responsiveness. FASEB J 15:447–457

Korzon-Burakowska A, Edmonds M (2006) Role of the microcirculation in diabetic foot ulceration. Int J Low Extrem Wounds 5:144–148

Kumar I, Staton CA, Cross SS, Reed MW, Brown NJ (2009) Angiogenesis, vascular endothelial growth factor and its receptors in human surgical wounds. Br J Surg 96:1484–1491

Kurban RS, Bhawan J (1990) Histologic changes in skin associated with aging. J Dermatol Surg Oncol 16:908–914

Lamah M, Chaudhry H, Mortimer PS, Dormandy JA (1996) Repeatability of Intravital Capillaroscopic measurement of capillary density. Int J Microcirc Clin Exp 16:23–29

Lange-Asschenfeldt B, Velasco P, Streit M, Hawighorst T, Pike SE, Tosato G, Detmar M (2001) The angiogenesis inhibitor Vasostatin does not impair wound healing at tumor-inhibiting doses. J Invest Dermatol 117:1036–1041

Leblebici B, Turhan N, Adam M, Akman MN (2007) Clinical and epidemiologic evaluation of pressure ulcers in patients at a university hospital in Turkey. J Wound Ostomy Continence Nurs 34:407–411

Leonardi R, Caltabiano M, Pagano M, Pezzuto V, Loreto C, Palestro G (2003) Detection of vascular endothelial growth factor/vascular permeability factor in periapical lesions. J Endodont 29:180–183

Leu HJ (1991) Morphology of chronic venous insufficiency – light and electron microscopic examinations. Vasa 20:330–342

Leu AJ, Yanar A, Geiger M, Franzeck UK, Bollinger A (1993) Microangiopathy in chronic venous insufficiency before and after Sclerotherapy and compression treatment: results of a one-year follow-up study. Phlebology 8:99–106

Li JJ, Huang YQ, Basch R, Karpatkin S (2001) Thrombin induces the release of Angiopoietin-1 from platelets. Thromb Haemost 85:204–206
Li J, Chen J, Kirsner R (2007) Pathophysiology of acute wound healing. Clin Dermatol 25:9–18
Lowry D, Saeed M, Narendran P, Tiwari A (2017) The difference between the healing and the nonhealing diabetic foot ulcer: a review of the role of the microcirculation. J Diabetes Sci Technol 11:914–923
Ma J, Wang Q, Fei T, Han JD, Chen YG (2007) Mcp-1 mediates Tgf-Beta-induced angiogenesis by stimulating vascular smooth muscle cell migration. Blood 109:987–994
Majima M, Hayashi I, Muramatsu M, Katada J, Yamashina S, Katori M (2000) Cyclo-Oxygenase-2 enhances basic fibroblast growth factor-induced angiogenesis through induction of vascular endothelial growth factor in rat sponge implants. Br J Pharmacol 130:641–649
Markos F et al (2013) What is the mechanism of flow-mediated arterial dilatation. Clin Exp Pharmacol Physiol 40:489–494
Martin P (1997) Wound healing – aiming for perfect skin regeneration. Science 276:75–81
Matsuoka H, Sisson TH, Nishiuma T, Simon RH (2006) Plasminogen-mediated activation and release of hepatocyte growth factor from extracellular matrix. Am J Respir Cell Mol Biol 35: 705–713
Mayrovitz HN (1992) Age and site variability of skin blood perfusion in the hairless mouse ear determined by laser Doppler Flowmetry. Int J Microcirc Clin Exp 11:297–306
Mayrovitz HN, Groseclose EE (2002a) Inspiration-induced vascular responses in finger dorsum skin. Microvasc Res 63:227–232
Mayrovitz HN, Groseclose EE (2002b) Neurovascular responses to sequential deep inspirations assessed via laser-Doppler perfusion changes in dorsal finger skin. Clin Physiol Funct Imaging 22:49–54
Mayrovitz HN, Groseclose EE (2005) Inspiration-induced Vasoconstrictive responses in dominant versus non-dominant hands. Clin Physiol Funct Imaging 25:69–74
Mayrovitz HN, Larsen PB (1994) Standard and near-surface laser-Doppler perfusion in foot dorsum skin of diabetic and nondiabetic subjects with and without coexisting peripheral arterial disease. Microvasc Res 48:338–348
Mayrovitz HN, Larsen PB (1996) Functional microcirculatory impairment: a possible source of reduced skin oxygen tension in human diabetes mellitus. Microvasc Res 52:115–126
Mayrovitz HN, Sims N (2002) Effects of different cyclic pressurization and relief patterns on heel skin blood perfusion. Adv Skin Wound Care 15:158–164
Mayrovitz HN, Sims N (2004) Effects of support surface relief pressures on heel skin blood flow in persons with and without diabetes mellitus. Adv Skin Wound Care 17:197–201
Mayrovitz HN, Smith J (1998) Heel-skin microvascular blood perfusion responses to sustained pressure loading and unloading. Microcirculation 5:227–233
Mayrovitz HN, Smith J, Delgado M, Regan MB (1997) Heel blood perfusion responses to pressure loading and unloading in women. Ostomy Wound Manage 43:16–20
Mayrovitz HN, Macdonald J, Smith JR (1999) Blood perfusion Hyperaemia in response to graded loading of human heels assessed by laser-Doppler imaging. Clin Physiol 19:351–359
Mayrovitz HN, Sims N, Taylor MC (2002) Sacral skin blood perfusion: a factor in pressure ulcers? Ostomy Wound Manage 48(34–8):40–42
Mayrovitz HN, Sims N, Dribin L (2003a) Heel skin Hyperaemia: direct compression versus vascular occlusion. Clin Physiol Funct Imaging 23:354–359
Mayrovitz HN, Sims N, Taylor MC, Dribin L (2003b) Effects of support surface relief pressures on heel skin blood perfusion. Adv Skin Wound Care 16:141–145
Mayrovitz HN, Groseclose EE, King D (2005) No effect of 85 Mt permanent magnets on laser-Doppler measured blood flow response to inspiratory gasps. Bioelectromagnetics 26:331–335
Mayrovitz HN, Mcclymont A, Pandya N (2013) Skin tissue water assessed via tissue dielectric constant measurements in persons with and without diabetes mellitus. Diabetes Technol Ther 15:60–65

Mayrovitz HN, Volosko I, Sarkar B, Pandya N (2016) Arm, leg, and foot skin water in persons with diabetes mellitus (Dm) in relation to Hba1c assessed by tissue dielectric constant (Tdc) technology measured at 300 Mhz. J Diabetes Sci Technol
Mayrovitz HN, Spagna PE, Taylor MC (2018) Sacral skin temperature assessed by thermal imaging: role of patient vascular attributes. J Wound Ostomy Continence Nurs 45:17–21
Mekkes JR, Loots MA, Van Der Wal AC, Bos JD (2003) Causes, investigation and treatment of leg ulceration. Br J Dermatol 148:388–401
Mennes OA, Van Netten JJ, Van Baal JG, Slart R, Steenbergen W (2021) The association between foot and ulcer microcirculation measured with laser speckle contrast imaging and healing of diabetic foot ulcers. J Clin Med 10
Mizelle RM Jr (2021) Diabetes, race, and amputations. Lancet 397:1256–1257
Mlacak B, Blinc A, Gale N, Ivka B (2005) Microcirculation disturbances in patients with venous ulcer before and after healing as assessed by laser Doppler flux-Metry. Arch Med Res 36:480–484
Montagna W, Carlisle K (1990) Structural changes in ageing skin. Br J Dermatol 122(Suppl 35): 61–70
Morgan MR, Humphries MJ, Bass MD (2007) Synergistic control of cell adhesion by Integrins and Syndecans. Nat Rev Mol Cell Biol 8:957–969
Mork C, Kvernebo K, Asker CL, Salerud EG (2002) Reduced skin capillary density during attacks of Erythromelalgia implies arteriovenous shunting as Pathogenetic mechanism. J Invest Dermatol 119:949–953
Munce TA, Kenney WL (2003) Age-specific skin blood flow responses to acute capsaicin. J Gerontol A Biol Sci Med Sci 58:304–310
Murata J, Murata S, Kodama T, Nakano H, Soma M, Nakae H, Satoh Y, Kogo H, Umeki N (2021) Age-related changes in the response of finger skin blood flow during a braille character discrimination task. Healthcare (Basel) 9:9
Nathan DM (1993) Long-term complications of diabetes mellitus. N Engl J Med 328:1676–1685
Nelson EA, Bell-Syer SE, Cullum NA (2000) Compression for preventing recurrence of venous ulcers. Cochrane Database Syst Rev Cd002303
Nguyen M, Arkell J, Jackson CJ (2001) Human endothelial gelatinases and angiogenesis. Int J Biochem Cell Biol 33:960–970
O'connor DS, Schechner JS, Adida C, Mesri M, Rothermel AL, Li F, Nath AK, Pober JS, Altieri DC (2000) Control of apoptosis during angiogenesis by Survivin expression in endothelial cells. Am J Pathol 156:393–398
Okonkwo UA, Dipietro LA (2017) Diabetes and wound angiogenesis. Int J Mol Sci 18
Oomens CW, Loerakker S, Bader DL (2010) The importance of internal strain as opposed to Interface pressure in the prevention of pressure related deep tissue injury. J Tissue Viability 19: 35–42
Oomens CW, Bader DL, Loerakker S, Baaijens F (2015) Pressure induced deep tissue injury explained. Ann Biomed Eng 43:297–305
Parving HH, Viberti GC, Keen H, Christiansen JS, Lassen NA (1983) Hemodynamic factors in the genesis of diabetic Microangiopathy. Metabolism 32:943–949
Passaniti A, Taylor RM, Pili R, Guo Y, Long PV, Haney JA, Pauly RR, Grant DS, Martin GR (1992) A simple, quantitative method for assessing angiogenesis and antiangiogenic agents using reconstituted basement membrane, heparin, and fibroblast growth factor. Lab Investig 67: 519–528
Phillips SJ (2000) Physiology of wound healing and surgical wound care. ASAIO J 46:S2–S5
Pintucci G, Froum S, Pinnell J, Mignatti P, Rafii S, Green D (2002) Trophic effects of platelets on cultured endothelial cells are mediated by platelet-associated fibroblast growth Factor-2 (Fgf-2) and vascular endothelial growth factor (Vegf). Thromb Haemost 88:834–842
Plisko A, Gilchrest BA (1983) Growth factor responsiveness of cultured human fibroblasts declines with age. J Gerontol 38:513–518

Presta M, Dell'era P, Mitola S, Moroni E, Ronca R, Rusnati M (2005) Fibroblast growth factor/fibroblast growth factor receptor system in angiogenesis. Cytokine Growth Factor Rev 16:159–178

Prompers L, Schaper N, Apelqvist J, Edmonds M, Jude E, Mauricio D, Uccioli L, Urbancic V, Bakker K, Holstein P, Jirkovska A, Piaggesi A, Ragnarson-Tennvall G, Reike H, Spraul M, Van Acker K, Van Baal J, Van Merode F, Ferreira I, Huijberts M (2008) Prediction of outcome in individuals with diabetic foot ulcers: focus on the differences between individuals with and without peripheral arterial disease. The Eurodiale Study. Diabetologia 51:747–755

Pugh CW, Ratcliffe PJ (2003) Regulation of angiogenesis by hypoxia: role of the Hif system. Nat Med 9:677–684

Quigley FG, Faris IB, Duncan HJ (1991) A comparison of Doppler ankle pressures and skin perfusion pressure in subjects with and without diabetes. Clin Physiol 11:21–25

Raskin P, Pietri AO, Unger R, Shannon WA Jr (1983) The effect of diabetic control on the width of skeletal-muscle capillary basement membrane in patients with type I diabetes mellitus. N Engl J Med 309:1546–1550

Rathsman B, Jensen-Urstad K, Nystrom T (2014) Intensified insulin treatment is associated with improvement in skin microcirculation and Ischaemic foot ulcer in patients with type 1 diabetes mellitus: a Long-term follow-up study. Diabetologia 57:1703–1710

Rayman G, Williams SA, Spencer PD, Smaje LH, Wise PH, Tooke JE (1986) Impaired microvascular Hyperaemic response to minor skin trauma in type I diabetes. Br Med J (Clin Res Ed) 292: 1295–1298

Rayman G, Malik RA, Sharma AK, Day JL (1995) Microvascular response to tissue injury and capillary ultrastructure in the foot skin of type I diabetic patients. Clin Sci (Lond) 89:467–474

Reed MJ, Corsa AC, Kudravi SA, Mccormick RS, Arthur WT (2000) A deficit in collagenase activity contributes to impaired migration of aged microvascular endothelial cells. J Cell Biochem 77:116–126

Rees M, Hague S, Oehler MK, Bicknell R (1999) Regulation of endometrial angiogenesis. Climacteric 2:52–58

Rendell MS, Kelly ST, Bamisedun O, Luu T, Finney DA, Knox S (1993) The effect of increasing temperature on skin blood flow and red cell deformability. Clin Physiol 13:235–245

Ricard N, Simons M (2015) When it is better to regress: dynamics of vascular pruning. PLoS Biol 13:E1002148

Richardson D (1989) Effects of age on cutaneous circulatory response to direct heat on the forearm. J Gerontol 44:M189–M194

Rivard A, Fabre JE, Silver M, Chen D, Murohara T, Kearney M, Magner M, Asahara T, Isner JM (1999) Age-dependent impairment of angiogenesis. Circulation 99:111–120

Robles-Tenorio A, Lev-Tov H, Ocampo-Candiani J (2022) Venous leg ulcer. Statpearls, Treasure Island (Fl)

Rooke GA, Savage MV, Brengelmann GL (1994) Maximal skin blood flow is decreased in elderly men. J Appl Physiol 77:11–14

Ross R, Odland G (1968) Human wound repair. Ii. Inflammatory cells, epithelial-mesenchymal interrelations, and Fibrogenesis. J Cell Biol 39:152–168

Rusnati M, Tanghetti E, Dell'era P, Gualandris A, Presta M (1997) Alphavbeta3 integrin mediates the cell-adhesive capacity and biological activity of basic fibroblast growth factor (Fgf-2) in cultured endothelial cells. Mol Biol Cell 8:2449–2461

Salcido R, Donofrio JC, Fisher SB, Legrand EK, Dickey K, Carney JM, Schosser R, Liang R (1994) Histopathology of pressure ulcers as a result of sequential computer-controlled pressure sessions in a fuzzy rat model. Adv Wound Care 7:23–26

Schubert V (2000) The influence of local heating on skin microcirculation in pressure ulcers, monitored by a combined laser Doppler and transcutaneous oxygen tension probe. Clin Physiol 20:413–421

Semenza G (2002) Signal transduction to hypoxia-inducible factor 1. Biochem Pharmacol 64:993–998

Shami SK, Shields DA, Scurr JH, Smith PD (1992) Leg ulceration in venous disease. Postgrad Med J 68:779–785

Smith RS Jr, Gao L, Bledsoe G, Chao L, Chao J (2009) Intermedin is a new Angiogenic growth factor. Am J Physiol Heart Circ Physiol 297:H1040–H1047

Soetens JFJ, Worsley PR, Bader DL, Oomens CWJ (2019a) Investigating the influence of intermittent and continuous mechanical loading on skin through non-invasive sampling of Il-1alpha. J Tissue Viability 28:1–6

Soetens JFJ, Worsley PR, Herniman JM, Langley GJ, Bader DL, Oomens CWJ (2019b) The expression of anaerobic metabolites in sweat and sebum from human skin subjected to intermittent and continuous mechanical loading. J Tissue Viability 28:186–193

Speiser DE, Bollinger A (1991) Microangiopathy in mild chronic venous incompetence (Cvi): morphological alterations and increased Transcapillary diffusion detected by fluorescence Videomicroscopy. Int J Microcirc Clin Exp 10:55–66

Spentzouris G, Labropoulos N (2009) The evaluation of lower-extremity ulcers. Semin Intervent Radiol 26:286–295

Sun BK, Siprashvili Z, Khavari PA (2014) Advances in skin grafting and treatment of cutaneous wounds. Science 346:941–945

Sundheim LK, Sporastoyl AH, Wester T, Salerud G, Kvernebo K (2017) Acute skin trauma induces hyperemia, but superficial papillary nutritive perfusion remains unchanged. Microcirculation 24

Swift ME, Kleinman HK, Dipietro LA (1999) Impaired wound repair and delayed angiogenesis in aged mice. Lab Investig 79:1479–1487

Tew GA, Klonizakis M, Moss J, Ruddock AD, Saxton JM, Hodges GJ (2011) Role of sensory nerves in the rapid cutaneous vasodilator response to local heating in young and older endurance-trained and untrained men. Exp Physiol 96:163–170

Thiruvoipati T, Kielhorn CE, Armstrong EJ (2015) Peripheral artery disease in patients with diabetes: epidemiology, mechanisms, and outcomes. World J Diabetes 6:961–969

Tilton RG, Faller AM, Burkhardt JK, Hoffmann PL, Kilo C, Williamson JR (1985) Pericyte degeneration and acellular capillaries are increased in the feet of human diabetic patients. Diabetologia 28:895–900

Tolino A, Wilkin JK (1988) Aging and cutaneous vascular thermoregulation responses. J Invest Dermatol 90:613

Tomasek JJ, Gabbiani G, Hinz B, Chaponnier C, Brown RA (2002) Myofibroblasts and Mechanoregulation of connective tissue Remodelling. Nat Rev Mol Cell Biol 3:349–363

Tonnesen MG, Feng X, Clark RA (2000) Angiogenesis in wound healing. J Investig Dermatol Symp Proc 5:40–46

Tooke JE (1995) Microvascular function in human diabetes. A physiological perspective. Diabetes 44:721–726

Tsopanoglou NE, Maragoudakis ME (1999) On the mechanism of thrombin-induced angiogenesis. Potentiation of vascular endothelial growth factor activity on endothelial cells by up-regulation of its receptors. J Biol Chem 274:23969–23976

Tsuchida Y (1993) The effect of aging and arteriosclerosis on human skin blood flow. J Dermatol Sci 5:175–181

Tsuji S, Ichioka S, Sekiya N, Nakatsuka T (2005) Analysis of ischemia-reperfusion injury in a microcirculatory model of pressure ulcers. Wound Repair Regen 13:209–215

Ueno H, Fukumoto S, Koyama H, Tanaka S, Maeno T, Murayama M, Otsuka Y, Mima Y, Kawaguchi Y, Shoji T, Inaba M, Nishizawa Y (2010) Regions of arterial stenosis and clinical factors determining transcutaneous oxygen tension in patients with peripheral arterial disease. J Atheroscler Thromb 17:858–869

Vasconez LO, Schneider WJ, Jurkiewicz MJ (1977) Pressure sores. Curr Probl Surg 14:1–62

Vasudevan B (2014) Venous leg ulcers: pathophysiology and classification. Indian Dermatol Online J 5:366–370

Visse R, Nagase H (2003) Matrix Metalloproteinases and tissue inhibitors of Metalloproteinases: structure, function, and biochemistry. Circ Res 92:827–839

Weiss M, Milman B, Rosen B, Eisenstein Z, Zimlichman R (1992) Analysis of the diminished skin perfusion in elderly people by laser Doppler Flowmetry. Age Ageing 21:237–241
Werner S, Grose R (2003) Regulation of wound healing by growth factors and cytokines. Physiol Rev 83:835–870
West MD (1994) The cellular and molecular biology of skin aging. Arch Dermatol 130:87–95
Wikslund LK, Amundsen VS, Kvernebo AK, Standal OK, Kvernebo K (2021) Skin trauma rapidly induces thermoregulatory plexus hyperemia, while an increased nutritive papillary capillary function can be detected after 24 H. Microcirculation E12735
Wikslund LK, Kaljusto ML, Amundsen VS, Kvernebo K (2022) Microvascular remodeling following skin injury. Microcirculation E12755
Woodley DT (1985) Cutaneous wound healing: a model for cell-matrix interactions. J Am Acad Dermatol 12:420–433
Worsley PR, Crielaard H, Oomens CWJ, Bader DL (2020) An evaluation of dermal microcirculatory occlusion under repeated mechanical loads: implication of lymphatic impairment in pressure ulcers. Microcirculation 27:E12645
Yazdanpanah L, Shahbazian H, Nazari I, Arti HR, Ahmadi F, Mohammadianinejad SE, Cheraghian B, Hesam S (2018) Incidence and risk factors of diabetic foot ulcer: a population-based diabetic foot cohort (Adfc study)-two-year follow-up study. Int J Endocrinol 2018: 7631659
Zweifach BW (1987) Design of the microcirculation. Proc Chin Acad Med Sci Peking Union Med Coll 2:225–227

Chapter 11
Sickle Cell Anemia

Manu O. Platt, Hannah Song Lee, and Liana Hatoum

Abstract Sickle cell anemia (SCA) is a homozygous recessive genetic disorder affecting nearly 100,000 people in the United States and millions worldwide. A mutation in the β-globin subunit of hemoglobin causes polymerization into rigid fibers after deoxygenation within red blood cells (RBCs), deforming them into a "sickle" shape. Sickled RBCs cause blockage of vessels in postcapillary venules, ischemia, and pain. Accumulation of membrane damage causes hemolysis of the RBCs, reducing the number of circulating RBCs and anemia. Hemolysis also stimulates chronic vascular damage after releasing heme into plasma, increased systemic levels of inflammatory cytokines, and subsequent consequential effects that damage multiple tissues and organ systems. Sickling of RBCs happens under deoxygenated conditions, beginning in the capillaries where gas exchange occurs, and vaso-occlusions form in the postcapillary venules. This makes microcirculation a prime region for studies to understand pathological complications of sickle cell anemia. Though the initiating cause is the mutation in β-globin protein, the resultant pathological processes are numerous. This chapter will discuss underlying mechanisms, describe links between a mutation in red blood cells altering cell behavior in the microcirculation and systemic damage to shortening of life spans for individuals with SCA, and identify targets for treatment and current therapies.

Keywords Sickle cell anemia · Endothelial cells · Genetic disease · Blood flow · Vaso-occlusion

M. O. Platt (✉)
Wallace H. Coulter Department of Biomedical Engineering, Georgia Institute of Technology and Emory University, Atlanta, GA, USA

National Institute of Biomedical Imaging and Bioengineering, National Institutes of Health, Bethesda, USA
e-mail: Manu.platt@nih.gov

H. Song Lee · L. Hatoum
Wallace H. Coulter Department of Biomedical Engineering, Georgia Institute of Technology and Emory University, Atlanta, GA, USA
e-mail: Hannah.lee2@nih.gov; liana.hatoum@emory.edu

R. F. Tuma et al. (eds.), *Handbook of Microcirculation*, Physiology in Health and Disease, https://doi.org/10.1007/978-3-031-99609-2_11

11.1 Molecular Consequence of the Sickle Cell Mutation

Sickle cell anemia is the most common form of a large group of hemoglobinopathies caused by mutations of the beta-globin subunit of hemoglobin that in general are called sickle cell disease (SCD). Normal hemoglobin A (HbA) is a tetramer that is formed from two α-globin subunits and two β-globin subunits that self-assemble into a hemoglobin molecule. Hemoglobin has the key responsibility of binding oxygen and releasing carbon dioxide in the pulmonary circulation, returning to the heart to be pumped out into the systemic circulation, to provide oxygen to the tissues of the body and carry away the carbon dioxide back to the lungs to be exhaled. Sickle hemoglobin (HbS) differs from HbA because of the mutated β-globin S. The mutated allele contains an adenine to thymine mutation that, after translation, encodes for a valine in place of a glutamic acid as the sixth amino acid in β-globin protein. Substitution of valine, a hydrophobic amino acid, for glutamic acid, a negatively charged amino acid, changes the structural properties of the hemoglobin tetramer. After deoxygenation, hemoglobin undergoes a conformational change that places the hydrophobic valines on the surface of the protein creating a thermodynamically unfavorable situation. To sequester the valines from water, hemoglobin molecules self-assemble into stiff HbS polymers that can completely deform the RBC membrane.

HbS polymerization is the functional cause of all the pathologies of sickle cell anemia (Steinberg 1998; Pawliuk et al. 2001; Vekilov 2007). As the HbS polymers form fibers, these fibers then stretch across the RBC and begin to deform its membrane. It is important to remember that RBCs are highly deformable to enable their squeezing through capillaries to maximize the surface area to volume ratios for maximal gas exchange. To be highly deformable, mature RBCs have ejected their nucleus, which, in other cells is the most rigid part of the cell. In the absence of the nucleus, the stiff HbS polymers are able to highly deform the RBC membrane into the canonical “sickle” shape. Biophysical studies of the self-assembly of HbS molecules into fibrils and fibers have been characterized (Turner et al. 2003) as well as mechanics of the HbS polymer, buckling length, and force exerted on the RBC membrane (Wang et al. 2002). Beyond deforming the RBC, HbS polymers change the rheological properties of the cell, increasing viscosity of the membrane, and their oblong shapes cause cells to get stuck in the microcirculation resulting vaso-occlusion (Connes et al. 2016).

Vicious cycle for RBC sickling has been described as a process beginning in the capillaries during deoxygenation. Deoxygenation causes HbS polymerization which increases viscosity of the RBC membrane, slowing down the flow velocity and increasing RBC transit time in the capillary. The longer the time in the capillary, the greater amount of oxygen that can be delivered to tissue but also increasing deoxygenation of the HbS which increases polymerization and continues to slow down the transit time in a vicious cycle (Barabino et al. 2010a). Delay times in HbS polymerization have also been identified in that the rates of polymerization can, in general, be longer than the time for the RBCs to transit through the capillary; RBCs

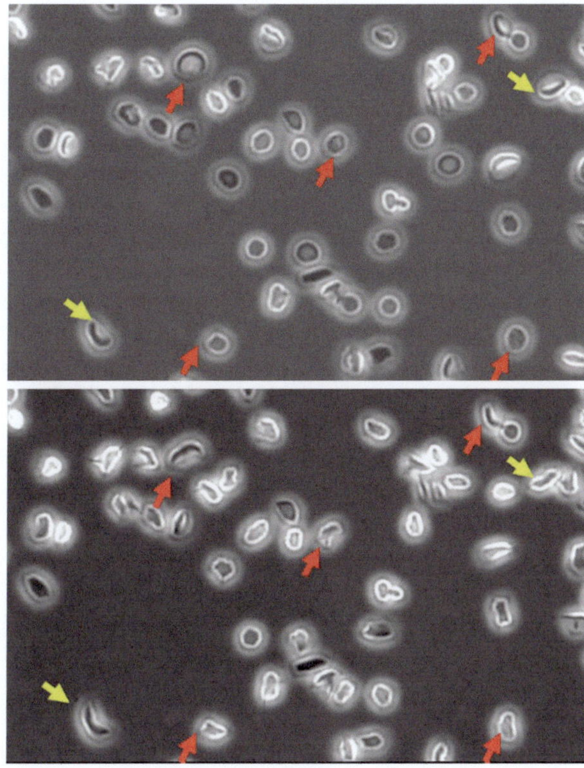

Fig. 11.1 Microscope images of red blood cells from a person with sickle cell anemia (HbSS) captured with video. Still images are shown from (**a**) RBCs under ambient oxygen levels and shown at (**b**) 2% O_2 levels for deoxygenated conditions. Red arrows indicate reversible sickling as indicated by cell deformability based on oxygen levels. Yellow arrows indicate irreversibly sickled cells that retain their shape under oxygenated and deoxygenated conditions

might pass without sickling (Kaul et al. 2009). In the video capture images of Fig. 11.1, some of the RBCs do not sickle over the same time as others. However, if other factors that increase the time to traverse the capillaries exist, such as white blood cells that are less deformable or greater blood viscosity, then a prolonged time would lead to greater sickling and even reduce the ability of the RBC to make it through the narrow capillaries (Kaul et al. 2009). This would initiate a vaso-occlusion. The rate of this polymerization is proportional to HbS concentrations inside of the RBC (to the 34th power) (Bunn 1997; Noguchi et al. 1988; Brittenham et al. 1985).

11.1.1 Other Mutations Cause Variations of Sickle Cell Disease

The HbSS homozygous mutation is the focus of this chapter, but there are other mutations in the β-globin allele that are of the family of sickle cell diseases that will be discussed here briefly. If the β-globin allele has a mutation from glutamic acid (acidic) to lysine (basic), it is referred to as βC allele or HbC. If an individual inherits a βS and a βC mutation, then they have the HbSC genotype which changes the structure of hemoglobin but causes a milder hemolytic anemia and less complications compared to sickle cell anemia with HbSS (Sundd et al. 2019).

Other mutations in hemoglobin could result in alpha or beta-thalassemias. A null β-globin allele that results in one of the chromosomes producing no β-globin subunits causes Hbβ0 (Figueiredo 2015), and sickle cell/β-thalassemia is divided either HbSβ0 or HbSβ+ depending when sickle β-globin is combined with either no β-globin or decreased amount of normal β-globin protein. This form produces some normal HbA so is usually milder than HbSS for its clinical presentation. In other genetic mutations that affect hemoglobin is the expression of fetal Hb (HbF). This is a hemoglobin tetramer of two α-globin subunits and two γ-globin subunits (Platt et al. 1994). Hereditary persistence of fetal hemoglobin (HPFH) is a condition that maintains expression of HbF on during childhood to adulthood (Forget 1998; Kutlar et al. 1984) when normally γ-globin gene is silenced within weeks after birth (Bard 1975). Fetal hemoglobin interferes with HbS polymerization and reduces the severity of SCA for those individuals that also have high expression of HbF (Steinberg and Sebastiani 2012).

11.1.2 Red Blood Cell Damage Caused by Sickle Hemoglobin Polymerization

Vaso-occlusive crises (VOC) will be discussed in further detail in sect. II, but it is important to discuss the consequences of HbS polymerization on the red blood cells themselves. These repeated events of HbS polymerization, membrane stretching, and sickling under deoxygenated conditions were followed by polymer "melting," or depolymerization, once the HbS binds oxygen again in the venous side of pulmonary circulation. These cycles lead to severe damage and permanent alterations to the RBC membrane and its function. The accumulated damage of RBCs during sickling and resolution was first characterized by Fabry and Nagel showing a bimodal distribution of hemoglobin concentrations in RBCs using density gradient separation on blood from people with HbSS (Fabry and Nagel 1982). Mean corpuscular hemoglobin concentration (MCHC) expressed as g/dL is the measure of red blood cell density, and it varies among the population of RBCs within a person with SCA. This is due to the membrane damage that occurs with each sickling cycle that causes membrane permeability and damage to transporter ions (Joiner and Franco 2001; Frenette and Atweh 2007). This causes loss of water to the external environment, but protein does not exit the cell which increases the HbS concentration per cell, measured as the MCHC. Membrane deformation and erythrocyte dehydration will continue until ultimately the membrane is viscous, damaged, and stiff characterized as an irreversibly sickled cell (Smith 2nd et al. 1982; Evans and Mohandas 1987; Nash et al. 1988; Kuypers 2014).

Kaul and others described the four major populations of red blood cells in people with sickle cell anemia based on their density in 1983 (Kaul et al. 1983) and their progression from the reticulocytes in fraction 1 to the irreversibly sickled cells, and the most dense, in fraction 4. Fraction 2 contained the discocytic cells with MCHC levels of AA RBCs, more dense than the lighter reticulocytes that are still immature RBCs that do not have fully developed RBC membranes with the biconcave shape. Fraction 3 contains very dense, irregularly shaped discocytes with higher MCHCs than fraction 2 between the range of 30 g/dL from fraction 1 but less than 45 g/dL of fraction 4 (Kaul et al. 1983). With the reminder that the more dense cells will polymerize HbS faster at the same deoxygenation level, fractions 3 and 4 will generate more sickle polymers, intracellularly, faster than fractions 1 and 2. However, because of the greater membrane deformability of fractions 1 and 2 due to reduced viscosity and damage accumulation, those cells are more likely to sickle and change shape under deoxygenation, but fractions 3 (high-density discocytes) and fraction 4 (irreversibly sickled cells) will not deform as readily, but are more likely to rupture or lyse due to HbS polymerization. This hemolysis of the RBCs causes the anemia, this genetic disease is named for, and leading to systemic hemoglobin levels from 6 to 11 g/dL (Kato et al. 2006; Connes et al. 2014; Nouraie et al. 2013). There is variability in the life span of the RBCs in sickle cell anemia from reticulocytes to irreversibly sickled cells being reduced from 120 days in AA RBCs to 9–10 days for SS RBCs (Kaul et al. 1983). Damage to the RBCs also causes phosphatidylserine exposure to the extracellular face of the membrane, a marker for cell removal to phagocytes. Hemolysis occurs by extravascular phagocytosis by macrophages with only one-third occurring through intravascular hemolysis (Crosby 1955).

11.2 Vaso-occlusions

Vaso-occlusions are blocked blood vessels that can cause great pain known as the sickle crisis. For a long while, the cause was believed to be just aggregated sickled red blood cells in the microvasculature, but these vaso-occlusions are now known to be the result of more complex phenomena involving interactions and adhesions among red blood cells, leukocytes, platelets, and the endothelial cells that comprise the inner lining of the vasculature (Fig. 11.2). Evidence exists that the painful vaso-occlusions are caused by more than just sickling (Embury 2004) Sickling is associated with deoxygenation, but people with sickle cell disease do not get hypoxia-induced bursts of sickling that cause a VOC. Since the full volume of blood is pumped through the heart once per minute, than any red blood cell would be deoxygenated once per minute meaning that sickling is a constant process even without forming a vaso-occlusion. There is no correlation between painful vaso-occlusions and the number of dense red blood cells, which, as discussed above, are those most likely to polymerize (Nagel 1993). In the capillaries and other small vessels, the red blood cells pass in a single-file (Eaton and Hofrichter 1990). It has been argued that sickling is not necessary and sufficient to cause the occlusions and

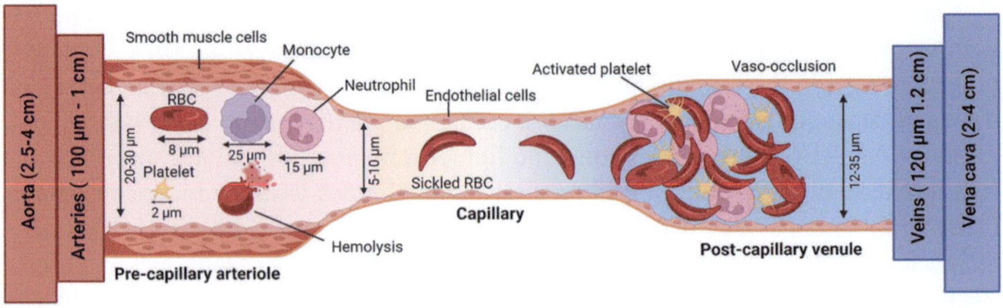

Fig. 11.2 Schematic to compare blood cell diameters with blood vessel diameters and localization of sickling. Sickling of RBCs happens in the capillaries under deoxygenated conditions. Circulating blood cells including RBCs, monocytes, neutrophils, and platelets interact and adhere to each other and to the endothelial cells contributing to vaso-occlusions which form in the postcapillary venules. Diameters not drawn to scale. (Created with BioRender.com)

that sickling may be secondary to another initiating factor (Embury 2004), while others argue that the initiation of vaso-occlusion is highly stochastic (Eaton and Hofrichter 1990).

Evidence indicates that VOC is frequently initiated by environmental stimuli such as infection, hypoxia, dehydration, and by inflammatory stimuli (Rees et al. 2010; Novelli and Gladwin 2016). Damaged, sickle red blood cells also have greater adhesivity to the endothelium which is another factor for VOC initiation (Kaul et al. 2009). Initiation may also take place over time and space with a number of nucleating sites susceptible to these environmental and inflammatory stimuli, and in the clinic, there are few data that identify the actual location of vaso-occlusion for the patient in a sickle crisis (Embury 2004). Several sites of vaso-occlusion have been identified in humans and animal models:

- small vessels of autopsy and biopsy specimens (Boros et al. 1976)
- capillaries of the nailbed (Lipowsky et al. 1987)
- forearm skin of subjects with sickle cell disease (Rodgers et al. 1984)
- large arteries of sickle cell patients with strokes (Stockman et al. 1972)
- capillaries and postcapillary venules in the rat mesocecum perfused with human sickle cells (Kaul et al. 1989a)
- mucosal intestinal tissues of transgenic knockout sickle cell mice (Embury et al. 1999; Embury et al. 2004)
- arteriolar-capillary bifurcations and venular junctions in microvascular flow models (Kaul et al. 1989b)

11.2.1 Endothelial Cells and Vaso-occlusion

Endothelial cells line the vascular walls and provide the surfaces for the adhesions to initiate accumulation of red and white blood cells to form the vaso-occlusion. Endothelial cells are not just passive on the surface but respond to inflammatory stimuli to express cell adhesion molecules to promote further cell adhesion. Chronic tissue damage, caused by cycles of ischemia-reperfusion, by stiffened/sickled RBCs to tissue beds, plus the anemia, and by free heme released after red blood cell hemolysis and resultant circulating inflammatory stimuli affects endothelial cells.

A number of cytokine stimuli are circulating in the plasma of people with sickle cell disease: tumor necrosis factor-α (TNF-α), interleukin-1, interferon-γ, IL-2, IL-2 receptor (Embury 2004). Activated endothelial cells also produce their own cytokines in response to being activated: IL-1β, IL-6, and tumor necrosis factor (TNF), which lead to a chronic inflammatory state (Sundd et al. 2019). In response to these stimuli, endothelial cells upregulate expression of cell adhesion molecules vascular cell adhesion protein 1 (VCAM1) (Kaul et al. 2000; Setty and Stuart 1996), intercellular adhesion molecule 1 (ICAM1) (Hines et al. 2003), and P- and E-selectins that increase binding and adhesion of red and white blood cells (Setty and Stuart 1996; Keegan et al. 2012a). Integrin αVβ3 is another prominent endothelial receptor (Kaul et al. 2000; Zennadi et al. 2004). Thrombospondin and von Willebrand factor are bridging factors that can bind to endothelial cells and to RBCs connecting the two cell types through an indirect binding (Manodori et al. 1998; Wick et al. 1987). Endothelial cells are also activated by the contact with the red blood cells, monocytes (Keegan et al. 2012b) and free heme and hemoglobin released during hemolysis (Hsu et al. 2007).

11.2.2 Red Blood Cells and Vaso-occlusion

HbSS red blood cells are more adhesive to endothelial cells than HbAA normal red blood cells (Hebbel 1997; Wagner et al. 2004), due to their accumulation of membrane damage from repetitive sickling cycles. By some estimates, each cell sickles an average of almost 100 times a day (Embury 2004). Adhesion molecules are overexpressed on the surface of sickling red blood cells and can promote RBC binding to the upregulated cell adhesion molecules on the endothelial cell surface in response to the chronic damage and inflammation of sickle cell disease as described in the above section. Very late antigen 4 (VLA-4) is a reticulocyte-specific adhesion molecule also known as α4β1 integrin, and CD36 and BCAM (basal cell adhesion molecule) are some of these molecules on the RBC (Sundd et al. 2019). The less dense RBCs in fractions 1 and 2, reticulocytes and the deformable discocytes, are even more adhesive than the irreversibly sickled red cells (Sugihara et al. 1992; Barabino et al. 1987). This was confirmed with parallel plate flow chambers but also in experiments where oxygenated HbSS red cells were infused into rat microvasculature experimental models where adhesion was highest in postcapillary venules (Kaul et al. 1989a). There is an inverse relationship between vascular diameter and adhesion, with smaller the diameter, the higher the adhesion (Zennadi et al. 2007). Once adhered, these HbSS red cells can remain bound under the wall shear rates of postcapillary venules (200–600 sec^{-1}) (Kaul et al. 1995). Other studies have confirmed that HbSS red cells can adhere to endothelial cells under arterial levels of shear stress of 10 dynes/cm^2 and higher (Grabowski 1987; La Celle 1986; Smith and La Celle 1986).

The hemolysis of red blood cells releases heme and hemoglobin to induce ROS generation by endothelial cells, scavenging of nitric oxide (NO) which further damages endothelial cells, and promotes vascular narrowing which increases probability of vaso-occlusions forming. Damaged RBCs prior to hemolysis also expose phosphatidylserine on their extracellular plasma membrane surface (Setty et al. 2002), which may also contribute to RBC-endothelial or RBC-white blood cell adhesions, and activation of coagulation pathways that would further close off the lumen and promote the VOC (Kuypers 2007; Kuypers and de Jong 2004).

11.2.3 Leukocytes, Platelets, and Vaso-occlusion

Chronic, systemic damage caused by sickling red blood cells activates leukocytes and inflammation. Monocytes and neutrophils are among the main subcategories of leukocytes that are elevated in people with SCA and their adhesion to endothelial cells and to red blood cells can exacerbate vaso-occlusion formation. Tumor necrosis factor-α TNFα stimulates endothelium, upregulates surface expression of vascular adhesion molecules, and induces red and white blood cell adhesion (Makis et al. 2000). Peripheral blood monocytes from people with SCD had 139% more (TNFα) per cell than normal monocytes (Belcher et al. 2000; Zennadi et al. 2008) and interferon gamma (IFNγ) is elevated and promotes monocyte activation (Collison et al. 2015). There are also increased numbers of circulating monocytes in people with SCD (Conran et al. 2007). Neutrophils are also activated in SCA and use selectins produced by stimulated endothelium to initiate binding (Zhang et al. 2016). With their upregulation, increased adhesion to the vascular endothelium, and release of other inflammatory cytokines, it is no surprise that there are correlations between leukocyte numbers and increased morbidity and mortality for people with SCA (Miller et al. 2000; Elmariah et al. 2014). Mouse models have confirmed the roles of circulating leukocytes (Belcher et al. 2003) and neutrophils in vaso-occlusions by adhering to endothelial cells and then being able to bind SS red blood cells. Together these cells more rapidly close off the lumen, occlude it, and reduce blood flow (Zhang et al. 2016).

Platelets also play a part in sickle cell vaso-occlusions and have been found in neutrophil-RBC aggregates in the blood (Dominical et al. 2014). P-selectin and activated αIIbβ3 integrin (Zhang et al. 2016) can mediate these adhesions that promote aggregate formation. Thrombospondin (Sugihara et al. 1992), CXCL4, and β-thromboglobulin are other platelet-derived biological markers elevated in SCA (Hebbel and Key 2016). Platelets also release other cytokines to activate other immune cells contributing in an indirect, nonbinding capacity (Davila et al. 2015).

11.2.4 Thrombosis, Embolisms, and Fat Emboli

Sickle cell anemia (SCA) facilitates a state of chronic coagulation activation, also known as hypercoagulation, where most parts of the coagulation cascade are altered (Manwani et al. 2019; Sparkenbaugh and Pawlinski 2017) causing formation of clots and cellular aggregations. This coagulation also contributes to the sickle pain crisis (Manwani et al. 2019; Ataga and Key 2007a; Nasimuzzaman and Malik 2019; Noubouossie et al. 2016; Piel et al. 2017) by promoting vaso-occlusions in microcirculation. Some mechanisms that have been identified are increased plasma concentrations of fibrinopeptide A (FpA) (Leichtman and Brewer 1978; Ekwere et al. 2014), a biomarker of clotting activation that is cleaved from fibrinogen during fibrin polymerization, and could be cleaved by other proteases beyond thrombin (Ferrall-Fairbanks et al. 2018). Increased platelet and tissue factor activation has been observed, along with decreased levels of anticoagulant proteins and abnormal fibrinolysis (Noubouossie et al. 2016). Anticoagulants and antiplatelet therapeutics have not helped prevent or treat hypercoagulation and vaso-occlusive conditions in SCD patients (Manwani et al. 2019; Ataga and Key 2007b; Nardo-Marino et al. 2020; Roberts et al. 2018; Telen et al. 2016; Biemond et al. 2019; Charneski and Congdon 2010). Anticoagulant approaches also bring the risk of hemorrhage and bleeding risks, which have limited their studies clinically (Nardo-Marino et al. 2020), as there are microtears and miniclots throughout the macro- and microvasculature of people with SCA due to the stiffened, sickled red cell damage to thin vessels.

When coagulation cascades lead to clots and thrombi in the circulation, the risk of them causing an embolism and blocking blood flow is elevated. Of particular concern for people with sickle cell anemia and those without, is the dangerous pulmonary embolism where pulmonary circulation is blocked. Thrombotic clots and fat and marrow emboli can form and block pulmonary microcirculation (Vichinsky et al. 1994). An additional risk for people with SCA is that thrombosis can cause acute chest syndrome; autopsies have shown platelet thrombi in the lung microvasculature of these patients that succumbed to ACS (Anea et al. 2016).

11.3 Acute Chest Syndrome

Acute chest syndrome is a leading cause of death for people with SCA, and is also a frequent cause for hospitalization (Novelli and Gladwin 2016). ACS affects both adults and children, but more common in children. Its severity increases with age putting adults a risk of $>10\%$ being fatal or causing multiorgan failure or neurological damage (Vichinsky et al. 1997). ACS is considered a complication of VOC because it typically occurs 2–3 days after vaso-occlusive pain crisis (Vichinsky et al. 2000). There are multifactorial mechanisms that may initiate the pulmonary injury including a pulmonary infarction, embolism, fat embolism, and infection (Vichinsky et al. 2000), with infection being the most common trigger in children (Vichinsky et al. 1997). The proinflammatory state of people with SCD due to the chronic damage and cytokine release discussed earlier can amplify immune responses to pathogens that present in the airway (DeBaun and Strunk 2016; Willen et al. 2018). Taken together with the hyperreactive bronchioles, asthma, and chronic inflammation in the airways exacerbate the risk of a heightened risk of airway closure (Leong et al. 1997). Both elevated platelet counts and massive drops in platelet counts are linked to poor outcomes in ACS (Anea et al. 2016; Chaturvedi et al. 2016; Mekontso Dessap et al. 2011); platelets play a part in vaso-occlusions as described earlier with cytokine release and binding in heterocellular aggregates of RBCs, neutrophils, and monocytes.

To counter these responses, antibiotics are given to patients to fight off the infections, bronchodilators and oxygen are provided, and usually a red blood cell transfusion is indicated (Howard et al. 2015). Exchange transfusion is when the patient's blood is drained or removed before adding in the donor AA RBCs to lower the level of HbS in the circulation to below 30% of total hemoglobin while keeping the total hemoglobin below 10 g/dL (Kassim et al. 2015). Delaying transfusion runs the risk of causing a hyperhemolytic episode with destruction of the patient's RBCs and any transfused cells (Gardner et al. 2015). Steroids can also be prescribed to reduce the acute inflammation (Howard et al. 2015; Bernini et al. 1998), but are limited to life-threatening events with ACS. While beneficial acutely, steroids are associated with an increased risk of recurrence of ACS when the steroids have stopped (Ogunlesi et al. 2014).

11.4 Strokes, Blood Flow, and Large Artery Complications

11.4.1 Strokes

One of the most devastating vascular complications for people with sickle cell anemia is the increased risk and mortality due to ischemic and hemorrhagic strokes. Strokes are arterial diseases and arteries carry oxygenated blood away from the heart to deliver oxygen to tissues, so not as subject to sickling caused by deoxygenation. When there's a blocked artery in the cerebral circulation, then oxygen is not being delivered to neurons in the brain, causing death and loss of both cognitive and motor functions. Children with SCA have a 300-fold higher risk of acute stroke than other children, and by 45 years of age, one in four adults with SCA will have had a stroke (Ohene-Frempong et al. 1998). Of the children, 11% will have a major stroke with paralysis, and 33% will suffer from silent strokes causing impaired mental function and cognitive abilities (Hillery and Panepinto 2004; Adams et al. 1998). Most strokes are caused by large artery remodeling confirmed by autopsies from SCD stroke cases indicating fragmented elastic lamina, intimal hyperplasia, and luminal occlusion (Platt et al. 2006, 2007; Ross 1999). Another unique feature of microvascular generation that happens in people with SCA is moyamoya where collateral vessels form around occluded cerebral arteries in the Circle of Willis (Merkel et al. 1978). These vessels are at risk of being leaky and causing intracranial bleeding (Dobson et al. 2002).

11.4.2 Hemodynamics and Stroke

Sickled red blood cells, increased cell density, and increased cell viscosities can change viscosity of the whole blood. Heterocellular aggregations of RBCs, WBCs, and platelets also increase the whole blood viscosity. Fluid mechanics have linked disturbed blood flow profiles with expression of cell adhesion molecules by endothelial cells, recruitment of monocytes binding to endothelium (Sorescu et al. 2004), reactive oxygen species generation (Hwang et al. 2003), and vascular wall remodeling through proteases (Platt et al. 2006, 2007), extracellular matrix deposition, and lipid deposition (Ross 1999). Altered hemodynamics in children and adults with SCA have been linked to clinical interventions and as indicators of underlying problems. Increased time averaged mean blood flow velocity in the internal carotid artery, the middle cerebral artery, or the anterior cerebral artery is an indicator of high risk for stroke and indicates the use of regular blood transfusions to mitigate this risk (Adams et al. 1998, 2004). This finding was from the STOP clinical trial that used transcranial Doppler to measure blood flow in the cerebrovasculature, and if a velocity of 200 cm/s or greater was measured, then the child was placed on a regimen of monthly blood transfusions. This was one of the successful clinical trials that was ended prematurely because those children in the placebo arm, measuring velocities >200 cm/s but only kept on standard care were suffering strokes and mortality (Adams et al. 1998, 2004).

Mechanisms underlying the increased velocities are still being determined and may be due to a number of reasons. The anemia of people with SCA will cause them to have higher velocities (Prohovnik et al. 2009; Adams et al. 1990). The hematocrit, number of RBCs per volume of blood, is 45% on average, but for people with SCA, it is around 25%. Yet, even with less RBCs per unit volume, because the HbSS RBCs are more damaged, dense, and viscous, the whole blood viscosity remains the same between AA and SS individuals, in general (Barabino et al. 2010a). It has been postulated that partial vaso-occlusions or aggregations of adhered cell aggregates to the endothelial wall can form stenoses in the lumen that cause disturbed flow poststenosis, but also increases the velocity of the fluid while in the stenosed region (Barabino et al. 2010a). Cerebral angiograms of children with SCA revealed bumpy microfeatures in the scans that were not detected in AA angiograms that were prepared by the same algorithms (Rivera et al. 2016), and indicate that changes to the anatomy of cerebral vessels were sufficient to increase blood flow velocities.

11.4.3 Microcirculation Involvement in Stroke

While transcranial Doppler may detect velocities in the larger arteries and provide clinical indication of stroke risks, cerebral small-vessel vasculopathy is another important area. Lesions in the small vessels are more common in the cerebrovasculature of people with SCA that rises from 13% as a toddler to 50% by adulthood (Wang et al. 2008; Bernaudin et al. 2011; Calvet et al. 2017; Kassim et al. 2016). Intracerebral arterioles with lesions were identified in autopsy studies (Koshy et al. 1990). This hypoperfusion of the microcirculation caused by these lesions and stenoses can cause hyperperfusion of the macrocirculation. Explained differently, occlusions in the capillaries and postcapillary venules from increasing viscosity as the HbS deoxygenates and polymerizes reduces the perfusion and slows the velocity. In the larger arteries, anemia and stenoses cause higher velocities and greater volumetric blood flow rates. This has been named the perfusion paradox of SCD (Nath et al. 2004), and is another of the conundrums that require a systems view and approach to solve problems with this complicated genetic disease.

11.5 Treatments for Sickle Cell Anemia

The goal to treat the disease is to prevent the formation of the sickle hemoglobin polymer. This can be accomplished (1) by reducing the amount of HbS in the blood through blood transfusion, (2) small-molecule drugs to (a) prevent self-assembly of the sickle hemoglobin polymer by increasing fetal hemoglobin expression using hydroxyurea, (b) prevent sickling by hemoglobin retention of oxygen with voxelotor, and (c) block binding molecules that promote cell adhesions, and (3) replacing HSCs that are genetically encoded to produce HbS via hematopoietic stem cell transplantation (Kassim and DeBaun 2014).

11.5.1 Small-Molecule Sickle Cell Medications

11.5.1.1 Hydroxyurea

Hydroxyurea is a drug that increases fetal hemoglobin (HbF) expression to mitigate sickle polymer formation (McGann and Ware 2015). A former chemotherapeutic agent, hydroxyurea is toxic to rapidly dividing cells such as proliferating progenitor cells in the bone marrow to produce RBC precursors with HbS. After their death, the slower proliferating RBC precursors with HbF begin to comprise a larger percentage of RBCs (Platt 2008). The inhibitory effect of hydroxyurea and HbF on HbS polymerization was first shown in 1995 with a reduction in painful crises in patients on the treatment group (Charache et al. 1995). Hydroxyurea also reduces cerebral blood flow velocities measured with transcranial Doppler (Zimmerman et al. 2007) with HbF expression reducing RBC deformability and subsequent damage (Barabino et al. 2010b). Hydroxyurea reduces incidences of vaso-occlusive crises and mortality with excellent safety (Charache et al. 1995). In higher-income countries, up to greater than 60% of SCA patients may be on hydroxyurea (Ataga et al. 2017), but this percentage is almost nil in most African countries which see a greater number of people born and living with sickle cell anemia (McGann et al. 2016). With favorable outcomes from human clinical trials, more infants, toddlers, and children with SCA are being prescribed hydroxyurea (Wang et al. 2011; Heeney et al. 2016).

11.5.1.2 Voxelotor

Voxelotor is an FDA-approved drug as an inhibitor of HbS polymerization by binding to the N-terminus of α chain of hemoglobin resulting in tighter binding of oxygen (Metcalf et al. 2017). By holding onto oxygen longer, this decreases sickling of RBCs, since being less likely to release the oxygen reduces the rate of undergoing the valine-exposing conformational change that causes HbS polymerization. RBC membrane viscosity is then reduced with less damage accumulating (Dufu et al. 2018). In phase 3 clinical trial (HOPE study), voxelotor did decrease markers of hemolysis, but there was no difference for vaso-occlusive events between treatment and placebo groups (Howard et al. 2021).

11.5.1.3 Crizanlizumab

Crizanlizumab is an FDA-approved, humanized monoclonal antibody that binds to P-selectin on platelets and endothelial cells to block binding of red blood cells, neutrophils, monocytes, and accumulation of the hetero-cellular aggregates that contribute to vaso-occlusive crisis events. Its efficacy has been shown to decrease sickle cell-related pain crises than placebo (Ataga et al. 2017).

11.5.2 Blood Transfusions

The hematopoietic stem cells in the bone marrow, as do all the cells in a person living with sickle cell disease, contain the HbS mutation in their genomic DNA. RBCs differentiated from these stem cells will all contain this mutant hemoglobin. Transfusions of HbAA or HbAS containing RBCs have been used therapeutically to reduce the amount of HbS circulating. An RBC containing HbAA survives for about 120 days, while an HbSS RBC survives on average 9–10 days with some range of variability. Therefore, if a person with sickle cell disease is transfused with HbAA or HbAS RBCs, then they should survive longer than the HbSS RBCs, but the bone marrow is producing a steady supply of new RBCs that eventually will overtake the number of HbAA or HbAS RBCs, requiring regular blood transfusions. Blood transfusions are beneficial in that they improve microvascular flow with less sickling RBCs to occlude microvasculature, and there is reduced endothelial injury and inflammation (Hyacinth et al. 2014, 2015). There are limits to chronic blood transfusions in that the cost of testing the blood products for infectious agents and blood-borne pathogens makes them available mostly in high-resource countries. Adverse effects are also consequences of chronic blood transfusions; iron overload and an immune response to the donor's blood are major issues. Iron chelating drugs have been used to assist with iron overload and patients are matched with repeat donors to mitigate the risk of alloimmunization. Altogether, those on chronic blood transfusions have a higher quality of life related to their health outcomes and reduced risk of some of the pathophysiological complications of SCD (Beverung et al. 2015).

11.5.3 Bone Marrow Transplants/Hematopoietic Stem Cell Transplantation

Currently, the only functional cure for SCA is hematopoietic stem cell transplantation (HSCT) which was previously known as bone marrow transplantation. The principle is that the HSCs resident in the bone marrow that contain the HbSS mutation are destroyed by chemotherapeutic or radiation strategies. Modified strategies now use reduced-intensity conditioning regimens (Hsieh et al. 2014). Then donor HSCs that are human leukocyte antigen (HLA)-matched but are HbAA or HbAS genotypes are infused, will engraft in the marrow space, and reconstitute the peripheral blood with RBCs that do not sickle under deoxygenated conditions. This has been prioritized for younger patients with severe symptoms (i.e., strokes, ACS, frequent sickle crises). With advancement in technologies, more adults have been able to successfully be functionally cured of SCA. Systemic tissue and organ damage is not corrected by the HSCT, but the hope is that progression of that damage will not continue since the primary cause of RBC sickling will have been corrected.

This solution has not been available to every patient. Many patients do not have a suitable HLA-matched donor, close relative or other person. This increases the risk of graft vs. host disease (GVHD) after HSCT which occurs even 1 out of 3 times for the best sibling HLA matches. With new advances in biomedical approaches, the average rate of graft failure has been lowered to just 2% (Gluckman et al. 2017), and acute and chronic GVHD limited to 14% or lower (Walters et al. 2016). Clinical trials have begun expanding donor pools to haploidentical matches that share 50% of the HLA antigens, and though there is an increased risk of graft rejection (Saraf et al. 2016), this has provided some options to a broader population that was forced to continue without a curative therapy. Still a relatively new therapeutic option for SCA, approximately 2000 people have undergone HSCT for SCA with a 90% survival for the United States and Europe (Gluckman et al. 2017; Walters et al.

2016). Notice that this therapy is still limited to high-income countries (Gluckman 2013) and is missing a large swath of people living with SCA.

11.6 Conclusion

The singularity of that point mutation from HbA to HbS occurs in one protein that is only expressed in one type of cell, but the red blood cells travel throughout the macro- and microcirculation to all organs and all vascular beds putting the entire system under stress and damage. As more is learned about sickle cell anemia and its complications, the symptoms are being addressed, while genome editing and advanced hematopoietic stem cell transplants seek to solve the problem at its root. The best solutions will also have applicability across cultures and across economic resources to address this global genetic disease and extend lives.

References

Adams RJ, Nichols FT, Aaslid R, Mckie VC, Mckie K, Carl E et al (1990) Cerebral vessel stenosis in sickle-cell disease – criteria for detection by transcranial Doppler. Am J Pediat Hematol 12(3):277–282

Adams RJ, McKie VC, Hsu L, Files B, Vichinsky E, Pegelow C et al (1998) Prevention of a first stroke by transfusions in children with sickle cell anemia and abnormal results on transcranial Doppler ultrasonography. N Engl J Med 339(1):5–11

Adams RJ, Brambilla DJ, Granger S, Gallagher D, Vichinsky E, Abboud MR et al (2004) Stroke and conversion to high risk in children screened with transcranial Doppler ultrasound during the STOP study. Blood 103(10):3689–3694. https://doi.org/10.1182/blood-2003-08-2733

Anea CB, Lyon M, Lee IA, Gonzales JN, Adeyemi A, Falls G et al (2016) Pulmonary platelet thrombi and vascular pathology in acute chest syndrome in patients with sickle cell disease. Am J Hematol 91(2):173–178. https://doi.org/10.1002/ajh.24224

Ataga KI, Key NS (2007a) Hypercoagulability in sickle cell disease: new approaches to an old problem. Hematology Am Soc Hematol Educ Program 2007:91–96

Ataga KI, Key NS (2007b) Hypercoagulability in sickle cell disease: new approaches to an old problem. Hematology 2007(1):91–96. https://doi.org/10.1182/asheducation-2007.1.91

Ataga KI, Kutlar A, Kanter J, Liles D, Cancado R, Friedrisch J et al (2017) Crizanlizumab for the prevention of pain crises in sickle cell disease. N Engl J Med 376(5):429–439. https://doi.org/10.1056/NEJMoa1611770

Barabino GA, McIntire LV, Eskin SG, Sears DA, Udden M (1987) Endothelial cell interactions with sickle cell, sickle trait, mechanically injured, and normal erythrocytes under controlled flow. Blood 70(1):152–157

Barabino GA, Platt MO, Kaul DK (2010a) Sickle cell biomechanics. Annu Rev Biomed Eng 12: 345–367. https://doi.org/10.1146/annurev-bioeng-070909-105339

Barabino GA, Platt MO, Kaul DK (2010b) Sickle cell biomechanics. Annu Rev Biomed Eng 12. https://doi.org/10.1146/annurev-bioeng-070909-105339

Bard H (1975) The postnatal decline of hemoglobin F synthesis in normal full-term infants. J Clin Invest 55(2):395–398. https://doi.org/10.1172/JCI107943

Belcher JD, Marker PH, Weber JP, Hebbel RP, Vercellotti GM (2000) Activated monocytes in sickle cell disease: potential role in the activation of vascular endothelium and vaso-occlusion. Blood 96(7):2451–2459

Belcher JD, Bryant CJ, Nguyen J, Bowlin PR, Kielbik MC, Bischof JC et al (2003) Transgenic sickle mice have vascular inflammation. Blood 101(10):3953–3959. https://doi.org/10.1182/blood-2002-10-3313

Bernaudin F, Verlhac S, Arnaud C, Kamdem A, Chevret S, Hau I et al (2011) Impact of early transcranial Doppler screening and intensive therapy on cerebral vasculopathy outcome in a newborn sickle cell anemia cohort. Blood 117(4):1130–1140. https://doi.org/10.1182/blood-2010-06-293514

Bernini JC, Rogers ZR, Sandler ES, Reisch JS, Quinn CT, Buchanan GR (1998) Beneficial effect of intravenous dexamethasone in children with mild to moderately severe acute chest syndrome complicating sickle cell disease. Blood 92(9):3082–3089

Beverung LM, Strouse JJ, Hulbert ML, Neville K, Liem RI, Inusa B et al (2015) Health-related quality of life in children with sickle cell anemia: impact of blood transfusion therapy. Am J Hematol 90(2):139–143. https://doi.org/10.1002/ajh.23877

Biemond BJ, Tombak A, Kilinc Y, Al-Khabori M, Abboud MR, Nafea M et al (2019) Efficacy and safety of sevuparin, a novel non-anti-coagulant heparinoid, in patients with acute painful vaso-occlusive crisis; a global, multicenter double-blind, randomized, placebo-controlled phase 2 trial (TVOC01). Blood 134:614

Boros L, Thomas C, Weiner WJ (1976) Large cerebral vessel disease in sickle cell anaemia. J Neurol Neurosurg Psychiatry 39(12):1236–1239

Brittenham GM, Schechter AN, Noguchi CT (1985) Hemoglobin S polymerization: primary determinant of the hemolytic and clinical severity of the sickling syndromes. Blood 65(1):183–189

Bunn HF (1997) Pathogenesis and treatment of sickle cell disease. N Engl J Med 337(11):762–769. https://doi.org/10.1056/NEJM199709113371107

Calvet D, Tuilier T, Mele N, Turc G, Habibi A, Abdallah NA et al (2017) Low fetal hemoglobin percentage is associated with silent brain lesions in adults with homozygous sickle cell disease. Blood Adv 1(26):2503–2509. https://doi.org/10.1182/bloodadvances.2017005504

Charache S, Terrin ML, Moore RD, Dover GJ, Barton FB, Eckert SV et al (1995) Effect of hydroxyurea on the frequency of painful crises in sickle cell anemia. Investigators of the multicenter study of Hydroxyurea in sickle cell anemia. N Engl J Med 332(20):1317–1322. https://doi.org/10.1056/NEJM199505183322001

Charneski L, Congdon HB (2010) Effects of antiplatelet and anticoagulant medications on the vasoocclusive and thrombotic complications of sickle cell disease: a review of the literature. Am J Health Syst Pharm 67(11):895–900. https://doi.org/10.2146/ajhp090229

Chaturvedi S, Ghafuri DL, Glassberg J, Kassim AA, Rodeghier M, DeBaun MR (2016) Rapidly progressive acute chest syndrome in individuals with sickle cell anemia: a distinct acute chest syndrome phenotype. Am J Hematol 91(12):1185–1190. https://doi.org/10.1002/ajh.24539

Collison JL, Carlin LM, Eichmann M, Geissmann F, Peakman M (2015) Heterogeneity in the Locomotory behavior of human monocyte subsets over human vascular endothelium in vitro. J Immunol 195(3):1162–1170. https://doi.org/10.4049/jimmunol.1401806

Connes P, Lamarre Y, Waltz X, Ballas SK, Lemonne N, Etienne-Julan M et al (2014) Haemolysis and abnormal haemorheology in sickle cell anaemia. Br J Haematol 165(4):564–572. https://doi.org/10.1111/bjh.12786

Connes P, Alexy T, Detterich J, Romana M, Hardy-Dessources MD, Ballas SK (2016) The role of blood rheology in sickle cell disease. Blood Rev 30(2):111–118. https://doi.org/10.1016/j.blre.2015.08.005

Conran N, Saad STO, Costa FF, Ikuta T (2007) Leukocyte numbers correlate with plasma levels of granulocyte-macrophage colony-stimulating factor in sickle cell disease. Ann Hematol 86(4):255–261. https://doi.org/10.1007/s00277-006-0246-6

Crosby WH (1955) The metabolism of hemoglobin and bile pigment in hemolytic disease. Am J Med 18(1):112–122. https://doi.org/10.1016/0002-9343(55)90208-4
Davila J, Manwani D, Vasovic L, Avanzi M, Uehlinger J, Ireland K et al (2015) A novel inflammatory role for platelets in sickle cell disease. Platelets 26(8):726–729. https://doi.org/10.3109/09537104.2014.983891
DeBaun MR, Strunk RC (2016) The intersection between asthma and acute chest syndrome in children with sickle-cell anaemia. Lancet 387(10037):2545–2553. https://doi.org/10.1016/S0140-6736(16)00145-8
Dobson SR, Holden KR, Nietert PJ, Cure JK, Laver JH, Disco D et al (2002) Moyamoya syndrome in childhood sickle cell disease: a predictive factor for recurrent cerebrovascular events. Blood 99(9):3144–3150
Dominical VM, Samsel L, Nichols JS, Costa FF, McCoy JP Jr, Conran N et al (2014) Prominent role of platelets in the formation of circulating neutrophil-red cell heterocellular aggregates in sickle cell anemia. Haematologica 99(11):e214–e217. https://doi.org/10.3324/haematol.2014.108555
Dufu K, Patel M, Oksenberg D, Cabrales P (2018) GBT440 improves red blood cell deformability and reduces viscosity of sickle cell blood under deoxygenated conditions. Clin Hemorheol Microcirc 70(1):95–105. https://doi.org/10.3233/CH-170340
Eaton WA, Hofrichter J (1990) Sickle cell hemoglobin polymerization. Adv Protein Chem 40:63–279. https://doi.org/10.1016/s0065-3233(08)60287-9
Ekwere T, Ebele U, Ekanem M, Ogunro P (2014) Relevance of fibrinolytic protein (D-dimer) and fibrinopeptide a as markers of sickle cell anaemia vaso-occlusive crisis. Niger Postgrad Med J 21(3):225–230
Elmariah H, Garrett ME, De Castro LM, Jonassaint JC, Ataga KI, Eckman JR et al (2014) Factors associated with survival in a contemporary adult sickle cell disease cohort. Am J Hematol 89(5):530–535. https://doi.org/10.1002/ajh.23683
Embury SH (2004) The not-so-simple process of sickle cell vasoocclusion. Microcirculation 11(2):101–113. https://doi.org/10.1080/10739680490278277
Embury SH, Mohandas N, Paszty C, Cooper P, Cheung ATW (1999) In vivo blood flow abnormalities in the transgenic knockout sickle cell mouse. J Clin Invest 103(6):915–920
Embury SH, Matsui NM, Ramanujam S, Mayadas TN, Noguchi CT, Diwan BA et al (2004) The contribution of endothelial cell P-selectin to the microvascular flow of mouse sickle erythrocytes in vivo. Blood 104(10):3378–3385. https://doi.org/10.1182/blood-2004-02-0713
Evans EA, Mohandas N (1987) Membrane-associated sickle hemoglobin: a major determinant of sickle erythrocyte rigidity. Blood 70(5):1443–1449
Fabry ME, Nagel RL (1982) Heterogeneity of red cells in the sickler: a characteristic with practical clinical and pathophysiological implications. Blood Cells 8(1):9–15
Ferrall-Fairbanks MC, West DM, Douglas SA, Averett RD, Platt MO (2018) Computational predictions of cysteine cathepsin-mediated fibrinogen proteolysis. Protein Sci 27(3):714–724. https://doi.org/10.1002/pro.3366
Figueiredo MS (2015) The compound state: Hb S/beta-thalassemia. Rev Bras Hematol Hemoter 37(3):150–152. https://doi.org/10.1016/j.bjhh.2015.02.008
Forget BG (1998) Molecular basis of hereditary persistence of fetal hemoglobin. Ann N Y Acad Sci 850:38–44. https://doi.org/10.1111/j.1749-6632.1998.tb10460.x
Frenette PS, Atweh GF (2007) Sickle cell disease: old discoveries, new concepts, and future promise. J Clin Invest 117(4):850–858. https://doi.org/10.1172/JCI30920
Gardner K, Hoppe C, Mijovic A, Thein SL (2015) How we treat delayed haemolytic transfusion reactions in patients with sickle cell disease. Br J Haematol 170(6):745–756. https://doi.org/10.1111/bjh.13494
Gluckman E (2013) Allogeneic transplantation strategies including haploidentical transplantation in sickle cell disease. Hematology Am Soc Hematol Educ Program 2013:370–376. https://doi.org/10.1182/asheducation-2013.1.370

Gluckman E, Cappelli B, Bernaudin F, Labopin M, Volt F, Carreras J et al (2017) Sickle cell disease: an international survey of results of HLA-identical sibling hematopoietic stem cell transplantation. Blood 129(11):1548–1556. https://doi.org/10.1182/blood-2016-10-745711

Grabowski EF (1987) Sickle erythrocytes adhere to endothelial cell monolayers (ECM'S) exposed to flowing blood. Prog Clin Biol Res 240:167–179

Hebbel RP (1997) Adhesive interactions of sickle erythrocytes with endothelium. J Clin Invest 100(11 Suppl):S83–S86

Hebbel RP, Key NS (2016) Microparticles in sickle cell anaemia: promise and pitfalls. Br J Haematol 174(1):16–29. https://doi.org/10.1111/bjh.14112

Heeney MM, Hoppe CC, Abboud MR, Inusa B, Kanter J, Ogutu B et al (2016) A multinational trial of Prasugrel for sickle cell Vaso-occlusive events. N Engl J Med 374(7):625–635. https://doi.org/10.1056/NEJMoa1512021

Hillery C, Panepinto J (2004) Pathophysiology of stroke in sickle cell disease. Microcirculation 11(2):195–208. https://doi.org/10.1080/10739680490278600

Hines PC, Zen Q, Burney SN, Shea DA, Ataga KI, Orringer EP et al (2003) Novel epinephrine and cyclic AMP-mediated activation of BCAM/Lu-dependent sickle (SS) RBC adhesion. Blood 101(8):3281–3287. https://doi.org/10.1182/blood-2001-12-0289

Howard J, Hart N, Roberts-Harewood M, Cummins M, Awogbade M, Davis B et al (2015) Guideline on the management of acute chest syndrome in sickle cell disease. Br J Haematol 169(4):492–505. https://doi.org/10.1111/bjh.13348

Howard J, Ataga KI, Brown RC, Achebe M, Nduba V, El-Beshlawy A et al (2021) Voxelotor in adolescents and adults with sickle cell disease (HOPE): long-term follow-up results of an international, randomised, double-blind, placebo-controlled, phase 3 trial. Lancet Haematol 8(5):e323–ee33. https://doi.org/10.1016/S2352-3026(21)00059-4

Hsieh MM, Fitzhugh CD, Weitzel RP, Link ME, Coles WA, Zhao X et al (2014) Nonmyeloablative HLA-matched sibling allogeneic hematopoietic stem cell transplantation for severe sickle cell phenotype. JAMA 312(1):48–56. https://doi.org/10.1001/jama.2014.7192

Hsu LL, Champion HC, Campbell-Lee SA, Bivalacqua TJ, Manci EA, Diwan BA et al (2007) Hemolysis in sickle cell mice causes pulmonary hypertension due to global impairment in nitric oxide bioavailability. Blood 109(7):3088–3098

Hwang J, Saha A, Boo YC, Sorescu GP, McNally JS, Holland SM et al (2003) Oscillatory shear stress stimulates endothelial production of O2- from p47phox-dependent NAD(P)H oxidases, leading to monocyte adhesion. J Biol Chem 278(47):47291–47298

Hyacinth HI, Adams RJ, Voeks JH, Hibbert JM, Gee BE (2014) Frequent red cell transfusions reduced vascular endothelial activation and thrombogenicity in children with sickle cell anemia and high stroke risk. Am J Hematol 89(1):47–51. https://doi.org/10.1002/ajh.23586

Hyacinth HI, Adams RJ, Greenberg CS, Voeks JH, Hill A, Hibbert JM et al (2015) Effect of chronic blood transfusion on biomarkers of coagulation activation and thrombin generation in sickle cell patients at risk for stroke. PLoS One 10(8):e0134193. https://doi.org/10.1371/journal.pone.0134193

Joiner CH, Franco RS (2001) The activation of KCL cotransport by deoxygenation and its role in sickle cell dehydration. Blood Cells Mol Dis 27(1):158–164. https://doi.org/10.1006/bcmd.2000.0365

Kassim AA, DeBaun MR (2014) The case for and against initiating either hydroxyurea therapy, blood transfusion therapy or hematopoietic stem cell transplant in asymptomatic children with sickle cell disease. Expert Opin Pharmacother 15(3):325–336. https://doi.org/10.1517/14656566.2014.868435

Kassim AA, Galadanci NA, Pruthi S, DeBaun MR (2015) How I treat and manage strokes in sickle cell disease. Blood 125(22):3401–3410. https://doi.org/10.1182/blood-2014-09-551564

Kassim AA, Pruthi S, Day M, Rodeghier M, Gindville MC, Brodsky MA et al (2016) Silent cerebral infarcts and cerebral aneurysms are prevalent in adults with sickle cell anemia. Blood 127(16):2038–2040. https://doi.org/10.1182/blood-2016-01-694562

Kato GJ, McGowan V, Machado RF, Little JA, Jt T, Morris CR et al (2006) Lactate dehydrogenase as a biomarker of hemolysis-associated nitric oxide resistance, priapism, leg ulceration, pulmonary hypertension, and death in patients with sickle cell disease. Blood 107(6):2279–2285. https://doi.org/10.1182/blood-2005-06-2373

Kaul DK, Fabry ME, Windisch P, Baez S, Nagel RL (1983) Erythrocytes in sickle-cell-anemia are heterogeneous in their rheological and hemodynamic characteristics. J Clin Invest 72(1):22–31

Kaul DK, Fabry ME, Nagel RL (1989a) Microvascular sites and characteristics of sickle cell adhesion to vascular endothelium in shear flow conditions: pathophysiological implications. Proc Natl Acad Sci USA 86(9):3356–3360. https://doi.org/10.1073/pnas.86.9.3356

Kaul DK, Fabry ME, Nagel RL (1989b) Erythrocytic and vascular factors influencing the microcirculatory behavior of blood in sickle cell anemia. Ann N Y Acad Sci 565:316–326. https://doi.org/10.1111/j.1749-6632.1989.tb24179.x

Kaul DK, Fabry ME, Costantini F, Rubin EM, Nagel RL (1995) In vivo demonstration of red cell-endothelial interaction, sickling and altered microvascular response to oxygen in the sickle transgenic mouse. J Clin Invest 96(6):2845–2853. https://doi.org/10.1172/JCI118355

Kaul DK, Tsai HM, Liu XD, Nakada MT, Nagel RL, Coller BS (2000) Monoclonal antibodies to alphaVbeta3 (7E3 and LM609) inhibit sickle red blood cell-endothelium interactions induced by platelet-activating factor. Blood 95(2):368–374

Kaul DK, Finnegan E, Barabino GA (2009) Sickle red cell-endothelium interactions. Microcirculation 16(1):97–111. https://doi.org/10.1080/10739680802279394

Keegan PM, Wilder CL, Platt MO (2012a) Tumor necrosis factor alpha stimulates cathepsin K and V activity via juxtacrine monocyte-endothelial cell signaling and JNK activation. Mol Cell Biochem 367(1–2):65–72. https://doi.org/10.1007/s11010-012-1320-0

Keegan PM, Surapaneni S, Platt MO (2012b) Sickle cell disease activates peripheral blood mononuclear cells to induce cathepsins k and v activity in endothelial cells. Anemia 2012: 201781. https://doi.org/10.1155/2012/201781

Koshy M, Thomas C, Goodwin J (1990) Vascular lesions in the central nervous system in sickle cell disease (neuropathology). J Assoc Acad Minor Phys 1(3):71–78

Kutlar A, Gardiner MB, Headlee MG, Reese AL, Cleek MP, Nagle S et al (1984) Heterogeneity in the molecular basis of three types of hereditary persistence of fetal hemoglobin and the relative synthesis of the G gamma and a gamma types of gamma chain. Biochem Genet 22(1–2):21–35. https://doi.org/10.1007/BF00499284

Kuypers FA (2007) Membrane lipid alterations in hemoglobinopathies. Hematology Am Soc Hematol Educ Program 2007:68–73. https://doi.org/10.1182/asheducation-2007.1.68

Kuypers FA (2014) Hemoglobin s polymerization and red cell membrane changes. Hematol Oncol Clin North Am 28(2):155–179. https://doi.org/10.1016/j.hoc.2013.12.002

Kuypers FA, de Jong K (2004) The role of phosphatidylserine in recognition and removal of erythrocytes. Cell Mol Biol 50(2):147–158

La Celle PL (1986) Alterations by leukocytes of erythrocyte flow in microchannels. Blood Cells 12(1):179–189

Leichtman DA, Brewer GJ (1978) Elevated plasma levels of fibrinopeptide a during sickle cell anemia pain crisis–evidence for intravascular coagulation. Am J Hematol 5(3). https://doi.org/10.1002/ajh.2830050303

Leong MA, Dampier C, Varlotta L, Allen JL (1997) Airway hyperreactivity in children with sickle cell disease. J Pediatr 131(2):278–283. https://doi.org/10.1016/s0022-3476(97)70166-5

Lipowsky HH, Sheikh NU, Katz DM (1987) Intravital microscopy of capillary hemodynamics in sickle cell disease. J Clin Invest 80(1):117–127. https://doi.org/10.1172/JCI113036

Makis AC, Hatzimichael EC, Bourantas KL (2000) The role of cytokines in sickle cell disease. Ann Hematol 79(8):407–413

Manodori AB, Matsui NM, Chen JY, Embury SH (1998) Enhanced adherence of sickle erythrocytes to thrombin-treated endothelial cells involves interendothelial cell gap formation. Blood 92(9):3445–3454

Manwani D, Davila J, Minniti CP (2019) Coagulopathy in sickle cell disease. Elsevier:757–760

McGann PT, Ware RE (2015) Hydroxyurea therapy for sickle cell anemia. Expert Opin Drug Saf 14(11):1749–1758. https://doi.org/10.1517/14740338.2015.1088827

McGann PT, Tshilolo L, Santos B, Tomlinson GA, Stuber S, Latham T et al (2016) Hydroxyurea therapy for children with sickle cell anemia in sub-Saharan Africa: rationale and design of the REACH trial. Pediatr Blood Cancer 63(1):98–104. https://doi.org/10.1002/pbc.25705

Mekontso Dessap A, Deux JF, Abidi N, Lavenu-Bombled C, Melica G, Renaud B et al (2011) Pulmonary artery thrombosis during acute chest syndrome in sickle cell disease. Am J Respir Crit Care Med 184(9):1022–1029. https://doi.org/10.1164/rccm.201105-0783OC

Merkel KH, Ginsberg PL, Parker JC, Post MJ (1978) Cerebrovascular disease in sickle cell anemia: a clinical, pathological and radiological correlation. Stroke 9(1):45–52

Metcalf B, Chuang C, Dufu K, Patel MP, Silva-Garcia A, Johnson C et al (2017) Discovery of GBT440, an orally bioavailable R-state stabilizer of sickle cell hemoglobin. ACS Med Chem Lett 8(3):321–326. https://doi.org/10.1021/acsmedchemlett.6b00491

Miller ST, Sleeper LA, Pegelow CH, Enos LE, Wang WC, Weiner SJ et al (2000) Prediction of adverse outcomes in children with sickle cell disease. N Engl J Med 342(2):83–89. https://doi.org/10.1056/NEJM200001133420203

Nagel RL (1993) Sickle cell anemia is a multigene disease: sickle painful crises, a case in point. Am J Hematol 42(1):96–101. https://doi.org/10.1002/ajh.2830420119

Nardo-Marino A, Brousse V, Rees D (2020) Emerging therapies in sickle cell disease. Br J Haematol 190. https://doi.org/10.1111/bjh.16504

Nash GB, Johnson CS, Meiselman HJ (1988) Rheologic impairment of sickle RBCs induced by repetitive cycles of deoxygenation-reoxygenation. Blood 72(2):539–545

Nasimuzzaman M, Malik P (2019) Role of the coagulation system in the pathogenesis of sickle cell disease. Blood Adv 3(20):3170–3180. https://doi.org/10.1182/bloodadvances.2019000193

Nath KA, Katusic ZS, Gladwin MT (2004) The perfusion paradox and vascular instability in sickle cell disease. Microcirculation 11(2):179–193. https://doi.org/10.1080/10739680490278592

Noguchi CT, Rodgers GP, Serjeant G, Schechter AN (1988) Levels of fetal hemoglobin necessary for treatment of sickle cell disease. N Engl J Med 318(2):96–99. https://doi.org/10.1056/NEJM198801143180207

Noubouossie D, Key NS, Ataga KI (2016) Coagulation abnormalities of sickle cell disease: relationship with clinical outcomes and the effect of disease modifying therapies. Blood Rev 30(4):245–256. https://doi.org/10.1016/j.blre.2015.12.003

Nouraie M, Lee JS, Zhang Y, Kanias T, Zhao X, Xiong Z et al (2013) The relationship between the severity of hemolysis, clinical manifestations and risk of death in 415 patients with sickle cell anemia in the US and Europe. Haematologica 98(3):464–472. https://doi.org/10.3324/haematol.2012.068965

Novelli EM, Gladwin MT (2016) Crises in sickle cell disease. Chest 149(4):1082–1093. https://doi.org/10.1016/j.chest.2015.12.016

Ogunlesi F, Heeney MM, Koumbourlis AC (2014) Systemic corticosteroids in acute chest syndrome: friend or foe? Paediatr Respir Rev 15(1):24–27. https://doi.org/10.1016/j.prrv.2013.10.004

Ohene-Frempong K, Weiner SJ, Sleeper LA, Miller ST, Embury S, Moohr JW et al (1998) Cerebrovascular accidents in sickle cell disease: rates and risk factors. Blood 91(1):288–294

Pawliuk R, Westerman KA, Fabry ME, Payen E, Tighe R, Bouhassira EE et al (2001) Correction of sickle cell disease in transgenic mouse models by gene therapy. Science 294(5550):2368–2371. https://doi.org/10.1126/science.1065806

Piel FB, Steinberg MH, Rees DC (2017) Sickle cell disease. N Engl J Med 376(16):1561–1573. https://doi.org/10.1056/nejmra1510865

Platt OS (2008) Hydroxyurea for the treatment of sickle cell anemia. N Engl J Med 358(13):1362–1369. https://doi.org/10.1056/NEJMct0708272

Platt OS, Brambilla DJ, Rosse WF, Milner PF, Castro O, Steinberg MH et al (1994) Mortality in sickle cell disease. Life expectancy and risk factors for early death. N Engl J Med 330(23):1639–1644. https://doi.org/10.1056/NEJM199406093302303

Platt MO, Ankeny RF, Jo H (2006) Laminar shear stress inhibits cathepsin L activity in endothelial cells. Arterioscler Thromb Vasc Biol 26(8):1784–1790

Platt MO, Ankeny RF, Shi GP, Weiss D, Vega JD, Taylor WR et al (2007) Expression of cathepsin K is regulated by shear stress in cultured endothelial cells and is increased in endothelium in human atherosclerosis. Am J Physiol Heart Circ Physiol 292(3):H1479–H1486

Prohovnik I, Hurlet-Jensen A, Adams R, De Vivo D, Pavlakis SG (2009) Hemodynamic etiology of elevated flow velocity and stroke in sickle-cell disease. J Cereb Blood Flow Metab 29(4): 803–810. https://doi.org/10.1038/jcbfm.2009.6

Rees DC, Williams TN, Gladwin MT (2010) Sickle-cell disease. Lancet 376(9757):2018–2031. https://doi.org/10.1016/S0140-6736(10)61029-X

Rivera CP, Veneziani A, Ware RE, Platt MO (2016) Original research: sickle cell anemia and pediatric strokes: computational fluid dynamics analysis in the middle cerebral artery. Exp Biol Med (Maywood) 241(7):755–765. https://doi.org/10.1177/1535370216636722

Roberts MZ, Gaskill GE, Kanter-Washko J, Kyle TR, Jones BC, Bohm NM (2018) Effectiveness and safety of oral anticoagulants in patients with sickle cell disease and venous thromboembolism: a retrospective cohort study. J Thromb Thrombolysis 45(4):512–515. https://doi.org/10.1007/s11239-018-1637-y

Rodgers GP, Schechter AN, Noguchi CT, Klein HG, Nienhuis AW, Bonner RF (1984) Periodic microcirculatory flow in patients with sickle-cell disease. N Engl J Med 311(24):1534–1538. https://doi.org/10.1056/NEJM198412133112403

Ross R (1999) Atherosclerosis – an inflammatory disease. N Engl J Med 340(2):115–126

Saraf SL, Oh AL, Patel PR, Jalundhwala Y, Sweiss K, Koshy M et al (2016) Nonmyeloablative stem cell transplantation with Alemtuzumab/low-dose irradiation to cure and improve the quality of life of adults with sickle cell disease. Biol Blood Marrow Transplant 22(3): 441–448. https://doi.org/10.1016/j.bbmt.2015.08.036

Setty BN, Stuart MJ (1996) Vascular cell adhesion molecule-1 is involved in mediating hypoxia-induced sickle red blood cell adherence to endothelium: potential role in sickle cell disease. Blood 88(6):2311–2320

Setty BN, Kulkarni S, Stuart MJ (2002) Role of erythrocyte phosphatidylserine in sickle red cell-endothelial adhesion. Blood 99(5):1564–1571. https://doi.org/10.1182/blood.v99.5.1564

Smith CM 2nd, Krivit W, White JG (1982) The irreversibly sickled cell. Am J Pediatr Hematol Oncol 4(3):307–315

Smith BD, La Celle PL (1986) Erythrocyte-endothelial cell adherence in sickle cell disorders. Blood 68(5):1050–1054

Sorescu GP, Song H, Tressel SL, Hwang J, Dikalov S, Smith DA et al (2004) Bone morphogenic protein 4 produced in endothelial cells by oscillatory shear stress induces monocyte adhesion by stimulating reactive oxygen species production from a nox1-based NADPH oxidase. Circ Res 95(8):773–779

Sparkenbaugh E, Pawlinski R (2017) Prothrombotic aspects of sickle cell disease. J Thromb Haemost 15(7):1307–1316. https://doi.org/10.1111/jth.13717

Steinberg MH (1998) Pathophysiology of sickle cell disease. Baillieres Clin Haematol 11(1): 163–184. https://doi.org/10.1016/s0950-3536(98)80074-7

Steinberg MH, Sebastiani P (2012) Genetic modifiers of sickle cell disease. Am J Hematol 87(8): 795–803. https://doi.org/10.1002/ajh.23232

Stockman JA, Nigro MA, Mishkin MM, Oski FA (1972) Occlusion of large cerebral vessels in sickle-cell anemia. N Engl J Med 287(17):846–849

Sugihara K, Sugihara T, Mohandas N, Hebbel RP (1992) Thrombospondin mediates adherence of CD36+ sickle reticulocytes to endothelial cells. Blood 80(10):2634–2642

Sundd P, Gladwin MT, Novelli EM (2019) Pathophysiology of sickle cell disease. Annu Rev Pathol 14:263–292. https://doi.org/10.1146/annurev-pathmechdis-012418-012838

Telen MJ, Batchvarova M, Shan S, Bovee-Geurts PH, Zennadi R, Leitgeb A et al (2016) Sevuparin binds to multiple adhesive ligands and reduces sickle red blood cell-induced vaso-occlusion. Br J Haematol 175(5):935–948

Turner MS, Briehl RW, Ferrone FA, Josephs R (2003) Twisted protein aggregates and disease: the stability of sickle hemoglobin fibers. Phys Rev Lett 90(12):128103

Vekilov PG (2007) Sickle-cell haemoglobin polymerization: is it the primary pathogenic event of sickle-cell anaemia? Br J Haematol 139(2):173–184. https://doi.org/10.1111/j.1365-2141.2007.06794.x

Vichinsky E, Williams R, Das M, Earles AN, Lewis N, Adler A et al (1994) Pulmonary fat embolism: a distinct cause of severe acute chest syndrome in sickle cell anemia. Blood 83(11):3107–3112

Vichinsky EP, Styles LA, Colangelo LH, Wright EC, Castro O, Nickerson B (1997) Acute chest syndrome in sickle cell disease: clinical presentation and course. Cooperative study of sickle cell disease. Blood 89(5):1787–1792

Vichinsky EP, Neumayr LD, Earles AN, Williams R, Lennette ET, Dean D et al (2000) Causes and outcomes of the acute chest syndrome in sickle cell disease. National Acute Chest Syndrome Study Group. N Engl J Med 342(25):1855–1865. https://doi.org/10.1056/NEJM200006223422502

Wagner MC, Eckman JR, Wick TM (2004) Sickle cell adhesion depends on hemodynamics and endothelial activation. J Lab Clin Med 144(5):260–267. https://doi.org/10.1016/j.lab.2004.08.004

Walters MC, De Castro LM, Sullivan KM, Krishnamurti L, Kamani N, Bredeson C et al (2016) Indications and results of HLA-identical sibling hematopoietic cell transplantation for sickle cell disease. Biol Blood Marrow Transplant 22(2):207–211. https://doi.org/10.1016/j.bbmt.2015.10.017

Wang JC, Turner MS, Agarwal G, Kwong S, Josephs R, Ferrone FA et al (2002) Micromechanics of isolated sickle cell hemoglobin fibers: bending moduli and persistence lengths. J Mol Biol 315(4):601–612

Wang WC, Pavlakis SG, Helton KJ, McKinstry RC, Casella JF, Adams RJ et al (2008) MRI abnormalities of the brain in one-year-old children with sickle cell anemia. Pediatr Blood Cancer 51(5):643–646. https://doi.org/10.1002/pbc.21612

Wang WC, Ware RE, Miller ST, Iyer RV, Casella JF, Minniti CP et al (2011) Hydroxycarbamide in very young children with sickle-cell anaemia: a multicentre, randomised, controlled trial (BABY HUG). Lancet 377(9778):1663–1672. https://doi.org/10.1016/S0140-6736(11)60355-3

Wick TM, Moake JL, Udden MM, Eskin SG, Sears DA, McIntire LV (1987) Unusually large von Willebrand factor multimers increase adhesion of sickle erythrocytes to human endothelial cells under controlled flow. J Clin Invest 80(3):905–910. https://doi.org/10.1172/JCI113151

Willen SM, Rodeghier M, Strunk RC, Bacharier LB, Rosen CL, Kirkham FJ et al (2018) Aeroallergen sensitization predicts acute chest syndrome in children with sickle cell anaemia. Br J Haematol 180(4):571–577. https://doi.org/10.1111/bjh.15076

Zennadi R, Hines PC, De Castro LM, Cartron JP, Parise LV, Telen MJ (2004) Epinephrine acts through erythroid signaling pathways to activate sickle cell adhesion to endothelium via LW-alphavbeta3 interactions. Blood 104(12):3774–3781. https://doi.org/10.1182/blood-2004-01-0042

Zennadi R, Moeller BJ, Whalen EJ, Batchvarova M, Xu K, Shan S et al (2007) Epinephrine-induced activation of LW-mediated sickle cell adhesion and vaso-occlusion in vivo. Blood 110(7):2708–2717. https://doi.org/10.1182/blood-2006-11-056101

Zennadi R, Chien A, Xu K, Batchvarova M, Telen MJ (2008) Sickle red cells induce adhesion of lymphocytes and monocytes to endothelium. Blood 112(8):3474–3483

Zhang D, Xu C, Manwani D, Frenette PS (2016) Neutrophils, platelets, and inflammatory pathways at the nexus of sickle cell disease pathophysiology. Blood 127(7):801–809. https://doi.org/10.1182/blood-2015-09-618538

Zimmerman SA, Schultz WH, Burgett S, Mortier NA, Ware RE (2007) Hydroxyurea therapy lowers transcranial Doppler flow velocities in children with sickle cell anemia. Blood 110(3):1043–1047. https://doi.org/10.1182/blood-2006-11-057893

Chapter 12
Microvascular Effects of Cannabinoids

Ronald F. Tuma, Hongbo Li, Zachary W. Reichenbach, Sara Jane Ward, and Melanie Elliott

Abstract The use of cannabinoids, commonly referred to as marijuana, for both recreational and therapeutic purposes has been one of the most hotly debated topics in medicine during the past two decades. This chapter reviews the political and economic factors influencing perceptions of cannabinoids and provides a detailed description of the actions of cannabinoids at the level of the microcirculation. The importance of the endocannabinoid system as a major homeostatic system is reviewed. The effects of CB2 receptor activation, providing anti-inflammatory protection without causing a euphoric effect, are described. Potential therapeutic targets, including treatment of stroke, head trauma, traumatic brain injury, spinal cord injury, and multiple sclerosis, are also explored.

Keywords Cannabinoid receptor 1 (CB1) · Cannabinoid receptor 2 (CB2) · Cannabidiol (CBD) · Phytocannabinoids · Tetrahydrocannabinol · Endocannabinoid system

R. F. Tuma (✉)
Emeritus Professor of Physiology and Neurosurgery, Lewis Katz School of Medicine, Temple University, Philadelphia, PA, USA
e-mail: ronald.tuma@temple.edu

H. Li · S. J. Ward
Center for Substance Abuse Research, Lewis Katz School of Medicine, Temple University, Philadelphia, PA, USA

Z. W. Reichenbach
Department of Medicine and Center for Substance Abuse Research, Lewis Katz School of Medicine, At Temple University, Philadelphia, PA, USA

M. Elliott
Department of Neuroscience, Sidney Kimmel Medical College, Thomas Jefferson University, Philadelphia, PA, USA

R. F. Tuma et al. (eds.), *Handbook of Microcirculation*, Physiology in Health and Disease, https://doi.org/10.1007/978-3-031-99609-2_12

12.1 Introduction

The term cannabinoid is derived from cannabis sativa, the name given to the plant by the famous Swedish botanist Karl Von Linnea (Linnaeus) around 1753. The term cannabis is derived from either a Greek or an Arabic word for hemp, and sativa refers to cultivated.

An understanding of modern perceptions concerning cannabinoids (or as popularly known marijuana) requires familiarity with the history of human use of this compound. Despite the intensity of current controversies over cannabis, it should be understood that human use of this plant has been extensive and varied over the centuries. Cannabis plants are one of the first plants to be cultivated by humans as long ago as 12,000 BC. The plant probably originated in the Mongolian steps and was brought to various parts of the world by migratory groups over the centuries. There is a controversy as to whether there is just one species of cannabis or if there are three different species (sativa, indica, ruderalis). Many botanists speculate that the three variants are a single species grown in different milieu. It has been used for food production of rope, paper, clothing, and sails, and for religious ceremony and recreational purposes. The fact that the fibers from cannabis are superior for the production of rope made the cultivation of the plant extensive at a time when sailing ships had great military and commercial importance. As a result, many royal proclamations forced farmers to grow cannabis, including early settlers at Jamestown. Both President George Washington and Thomas Jefferson grew cannabis, although there is no evidence it was ever smoked or eaten by them.

Chinese texts from 2727 BC report cannabis use for medicinal purposes, where it was mixed with wine to provide analgesia. A number of European medical texts from the sixteenth and seventeenth century list therapeutic uses for cannabis. The medicinal use of cannabis was introduced to British medicine by the physician William O′ Shaughnessy in the mid-1800s after observing its use in India. In 1890, Sir John Russell Reynolds, Queen Victoria's personal physician, stated that *"Indian hemp, when pure and administered carefully, is one of the most valuable medicines we possess."* He is also quoted as stating "In almost all painful maladies, I have found Indian hemp by far the most useful drug."

During the mid-nineteenth century, cannabis was listed in the pharmacopoeia and was one of the most widely dispensed medications by pharmacies in the United States (Russo 2016). At the time, cannabis preparations were produced with a variety of formulations, often including a number of psychoactive components such as opiates. This contributed to the effects of these other components being ascribed to cannabis. Despite the popularity of cannabis in the nineteenth century, perceptions began to change during the early twentieth century. The changes resulted from both racism and economic factors. Following the Mexican Revolution of 1910, there was a large influx of Mexican immigrants into the southwestern United States. These immigrants were unwelcome and the subject of racial prejudice. Many of the immigrants from Mexico brought with them the habit of using cannabis, and cannabis use became associated with this population. The term marijuana was

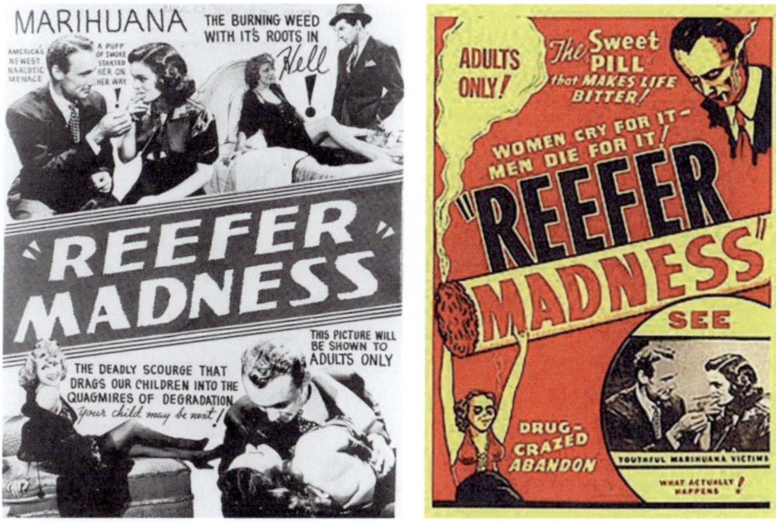

Fig. 12.1 Posters for the film Reefer Madness distributed with the support of the Federal Narcotics Bureau designed to demonize marijuana

used to replace the term cannabis in part to associate its use with Hispanic people. A number African-American musicians also used cannabis and this was an additional a source of prejudice. The most strident effort to vilify cannabis was initiated by Harry J. Anslinger**,** the newly appointed head of the Federal Narcotics Bureau (Booth 2003). Anslinger viewed his bureau as being in competition with other government agencies such as the FBI for funding, and sought a cause that would inspire public support, He focused his attention on marijuana and maintained a long ardent campaign to demonize its use. He used extremely hyperbolic descriptions of the ill effects of marijuana and supported these statements with racial overtones:

the primary reason to outlaw marijuana is its effect on the degenerate races.
Reefer makes darkies think they are as good as white men
You smoke a joint and you are likely to kill your brother.
Marijuana is the most violence –causing drug in the history of mankind.

A film titled Refer Madness was distributed across the United States to support these ideas further suggest that use of marijuana by white women would lead to their becoming sex slaves for dark skinned people (Fig. 12.1)**.** The majority of the medical community did not support these ideas but they persisted in public opinion. In 1944, Fiorella LaGuardia, the Mayor of New York, commissioned a panel of experts to produce a report evaluating these claims. The report disagreed with the claims made but was ignored. Political influence on the perceptions of marijuana use persisted through the Nixon administration. According to John Erlichman, Nixon's policy advisor, "Look, we understood we couldn't make it illegal to be young or poor or black in the United States, but we could criminalize their common pleasure. We understood that drugs were not the health problem we were making them out to be, but it was such a perfect issue for the Nixon White House that we couldn't resist it." "We knew we couldn't make it illegal to be either against the war or black, but by getting the public to associate the hippies with marijuana and blacks with heroin, and then criminalize both heavily, we could disrupt those communities. We could arrest their leaders, raid their homes break up their meetings, and vilify them night after night on the evening news. Did we know we were lying about the drugs? Of course we did." The criminalization of cannabinoid use was to continue through a number of future administrations. The result of these long standing pressures was to place a stigma on cannabinoid use, including medicinal cannabinoids. A second unfortunate result was the classification of marijuana as a Schedule 1 drug, even though this was opposed by the medical community. Schedule 1 drugs are drugs deemed to have high addictive potential and no medicinal use. Although there is a substantial body of basic science studies of cannabinoids, there is still a severe paucity of clinical trials as a result of the Schedule 1 classification. An opinion in opposition to this classification is summarized by Lester Grinspoon, MD, Emeritus Professor of Psychiatry, Harvard Medical School: *"If marijuana were a new discovery rather than a well-known substance carrying cultural and political baggage it would be hailed as a wonder drug."* This is not to say, as will be discussed below, there are not untoward effects of cannabinoid use and whether they will reach their potential as "wonder drugs" remains to be determined.

Currently, the THC content of many cannabis plants today greatly exceeds the content found in plants from previous decades. Currently the content of THC in cannabis plants can be up to 20 fold higher than the varieties grown in the 1970s (Pacher et al. 2018). Synthetic cannabinoids and designer drugs sold for recreational purposes may have as much as 200 times the THC content and may function as full CB1 agonists giving them much greater potency. This may explain multiple recent reports of severe adverse reaction, including cardiovascular effects.

12.1.1 Definition

The definition of cannabinoids is somewhat cumbersome. Cannabinoids are a class of diverse chemical compounds acting on receptors in cells and include phytocannabinoids (compounds from the cannabis plant), endocannabinoids (produced endogenously in animals), and compounds which mimic the actions of phytocannabinoids or have a similar structure. The cannabis plant produces well over 100 different cannabinoids along with other biologically active compounds including terpenes and flavonoids.

The first cannabinoids to be isolated and structurally identified were cannabidiol (CBD) and tetrahydrocannabinol (THC). CBD was first isolated by Roger Adams in 1940. The structures of CBD and tetrahydrocannabinol (THC) were determined by Professor Rafael Mechoulam, considered to be the father of modern cannabinoid research in 1963 (Sondheimer et al. 1964; Mechoulam and Hanus 2000). THC is the cannabinoid primarily responsible for the euphoric effects of marijuana, whereas CBD does not produce euphoria but has other important biologic effects.

12.1.2 *The Endocannabinoid System*

As mentioned previously, cannabinoids are endogenously produced in humans and other animals and are a component of the endocannabinoid system. The endocannabinoid system includes the enzymes responsible for the production of endocannabinoids, endocannabinoids, cannabinoid receptors and the enzymes responsible for their degradation (Fig. 12.2). The first cannabinoid receptor (CB1) was identified in the laboratory of Dr. Allyn Howlett (Devane et al. 1988). A second cannabinoid receptor (CB2) was identified in 1995. Cannabinoid receptors are located within the myocardium and in both vascular smooth muscle and endothelial cells along with circulating blood cells (Munro et al. 1993; Blazquez et al. 2003; Zoratti et al. 2003; Mestre et al. 2006; Rajesh et al. 2007b). All of these components are currently targets for therapeutic intervention (Pacher et al. 2005a). CB1 and CB2 receptors are the most thoroughly studied endocannabinoid receptors. These receptors share approximately 44% homology in structure but differ significantly in both distribution and effect. While the CB1 receptor is widely distributed in neurons, it is actually the most prolific coupled receptor in the brain, and the CB2 receptor is primarily located on immune cells (Cabral and Dove Pettit 1998; Cabral et al. 2001; Cabral and Staab 2005), endothelial cells and cardiac myocytes. CB1 receptors are particularly concentrated on both GABA-releasing neurons (Inhibitory neurons) and glutamate releasing neurons. Activation of CB1 receptors leads to retrograde suppression of neurotransmitter release, which may be excitatory or inhibitory depending on the location in the brain (Fig. 12.3). CB1 receptors are also located in other regions of the body and are expressed on immune cells and endothelial cells

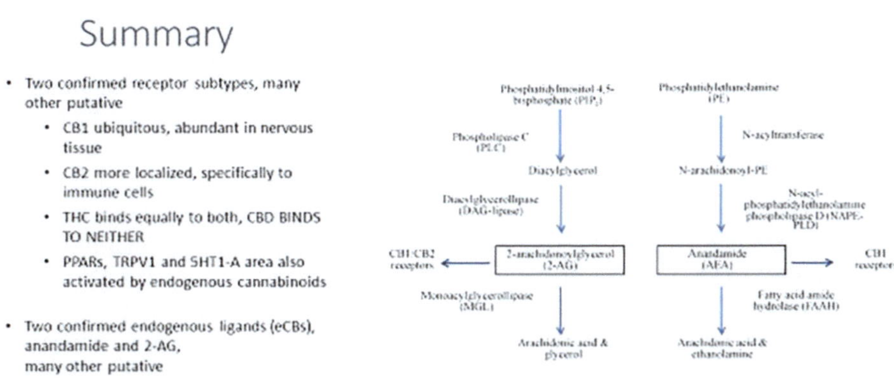

Fig. 12.2 An outline of the steps involved in the synthesis and degradation of 2-AG and AEA. It is important to notice that the byproduct of degradation of both compounds results in the production of arachidonic acid, the precursor of prostaglandins

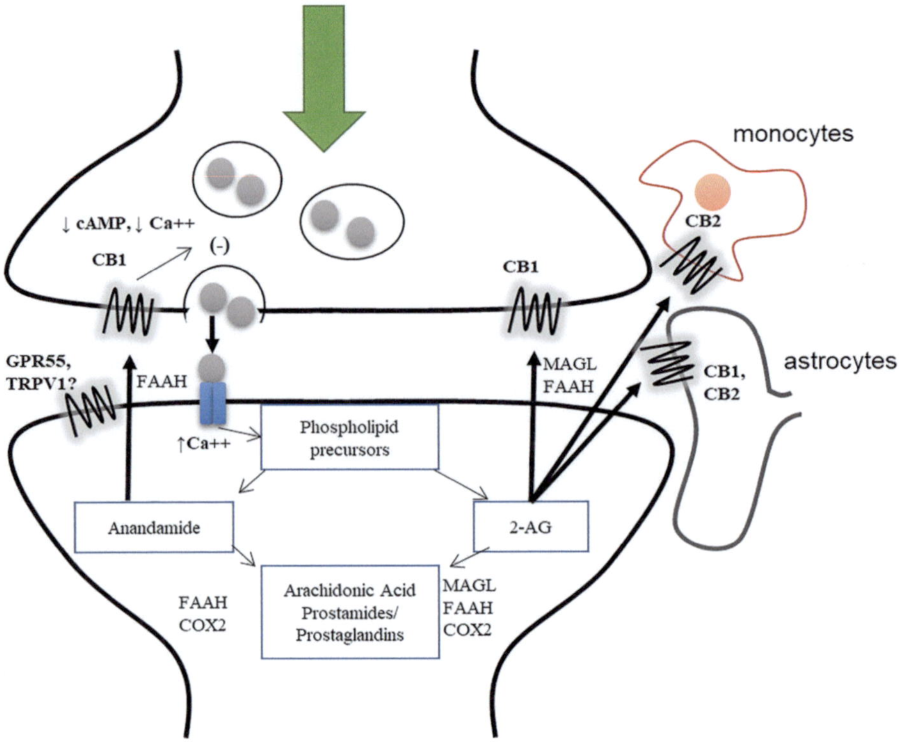

Fig. 12.3 Cannabinoids function of retrograde neurotransmitters in addition to influencing surrounding cells

(Gebremedhin et al. 1999; Liu et al. 2000). Cannabinoid receptors located on afferent and efferent neurons in the autonomic nervous system can have significant impact on cardiovascular function. While it is often mistakenly stated that there are no CB2 receptors in the brain, this refers to the fact that there are very few CB2 receptors on neurons. There are however CB2 receptors on microglia and other immune cells located within the brain. CB2 receptors are also located in endothelial cells, smooth muscle cells and immune cells in the brain and other parts of the body. Stimulation of the CB2 receptor has been shown to modulate immune cells activities and inflammatory responses. CB_2 is a G_i protein-coupled receptor and its activation triggers a series of signal transduction pathways, which eventually lead to either up- or downregulation of gene transcription (Klein et al. 2001). They primarily signal through the MAPK pathway including ERK, JNK, and p38 (Rom and Persidsky 2013). In most cases, the genes involved are coded for proinflammatory cytokines (Klein et al. 2001). Cannabinoid receptors inhibit adenylyl cyclase activation through activation of $G_{i/o}$ protein activation and can also modulate ion channel function. In addition, they can engage G-protein-independent pathways through β−arrestin pathways (Howlett et al. 2002; Pertwee et al. 2010; Delgado-Peraza et al. 2016; Soethoudt et al. 2017).

Endocannabinoids have been demonstrated to be synthesized in microvascular endothelium, smooth muscle cells, platelets, monocytes, macrophage foam cells, lymphocytes and dendritic cells (Sugiura et al. 1998; Galiegue et al. 1995; Deutsch et al. 1997; Rajesh et al. 2008, 2010). Endogenous cannabinoids are not stored in vesicles, but are produced upon demand, as occurs following increases in intracellular calcium. Currently the most well understood endocannabinoids are arachidonoyl ethanolamide (AEA) and 2-arachidonoylglycerol (2-AG). Differences in the enzymes responsible for their production, receptor activation and degradation are likely of significance for endocannabinoid regulation (Fig. 12.2). AEA functions as a partial agonist for the CB1 receptor, whereas 2-AG is a full agonist for both CB1 and CB2. Since AEA is only a partial agonist for CB1 it has the potential to limit CB1 activation by 2-AG through competitive affinity for the CB1 receptor (Di Marzo and De Petrocellis 2012). The endogenous cannabinoids are rapidly degraded by enzymes including fatty-acid 2-amide hydrolase (FAAH), monoglyceride lipase (MGL), and cyclooxygenase 2 (prostaglandin G/H synthase)(Yu et al. 1997; Kozak et al. 2000; Cravatt et al. 2001; Dinh et al. 2002). Inhibition of these enzymes is being explored as potential therapeutic target. As can be seen in Fig. 12.2, the degradation of the endogenous cannabinoids into arachidonic acid provides a direct connection with the eicosanoid system. The magnitude of contributions of the eicosanoid system to exogenous and endocannabinoid effects may be an important factor that remains to be determined.

Endocannabinoids have been demonstrated to induce both vasodilation and vasoconstriction (Pacher et al. 2005a, 2006; Montecucco and Di Marzo 2012; Stanley and O'Sullivan 2014; Benyo et al. 2016). In vivo studies have demonstrated that CB1 activation contributes to a decrease in myocardial contractility and blood pressure through a direct effect on myocytes, blood vessels, and the autonomic nervous system (Pacher et al. 2004, 2005a, 2018; Batkai and Pacher 2009). CB2 receptor activation, on the other hand, does not have direct hemodynamic effect and provides attenuation of cardiovascular dysfunction under a number of pathologic conditions. Anandamide has been shown to induce sustained vasodilation and a decrease in resistance in arterial vessels by activation of CB1R (Pacher et al. 2004). In addition, it also appears to mediate transient vasodilation in mesenteric and possibly other vascular beds through activation of TRP channels and perhaps what is termed the abnormal cannabidiol receptor, which has been postulated to be the GPCR GPR18 (Bondarenko 2014; Sierra et al. 2018). Indeed, it has been speculated that anandamide may be EDHF (Randall et al. 1996; Randall and Kendall 1997; Randall and Kendall 1998) or facilitate the actions of EDRF (Jarai et al. 1999). The vasodilation induced by anandamide may have both an endothelial and a nonendothelial component (Chaytor et al. 1999; Wagner et al. 1999). An indirect mechanism for anandamide-induced vasodilation has also been proposed to be through activation of TRPV1 receptors on sensory neurons which in turn release CGRP (Zygmunt et al. 1999). In addition, anandamide may, under certain conditions, cause a temporary vasoconstriction in certain microvascular beds such as the spleen (Wagner et al. 2001).

There is significant evidence that endocannabinoids play an important role in the hypotension associated with hemorrhagic, endotoxic, and cardiogenic shock (Wagner et al. 1997; Varga et al. 1998; Godlewski et al. 2004). These effects appear to contribute to survival since mortality is increased following blockade of CB1 receptors (Wagner et al. 1997, 2001) and survival is improved following treatment with THC or HU-210 (Varga et al. 1998; Wagner et al. 2003) . Although initial studies were focused on the activation of CB1 receptors to provide protection from excitotoxic injury, ischemia/reperfusion injury, and CNS autoimmune diseases

(Nagayama et al. 1999; Parmentier-Batteur et al. 2002; Muthian et al. 2004), later it has become recognized that the anti-inflammatory effect of cannabinoids may be very important in this in the management of these problems. An added factor is that unlike CB1 receptors, CB2 receptor activation is not associated with euphoric effects.

12.1.3 Multiple Sclerosis/EAE

Definitive observational evidence of the effect of cannabinoids on leukocyte-endothelial interactions in the microcirculation was provided by studies of the effect of cannabinoid treatment on experimental autoimmune encephalomyelitis (EAE), a mouse model used to replicate many of the sequential pathologic changes that occur in with multiple sclerosis (Ni et al. 2004). EAE models can serve to define and characterize the sequential events leading to CNS autoimmune inflammation (Becher et al. 2006).

The specific characteristics of disease progression during EAE are dependent upon the method used to induce an autoimmune reaction against myelin (Steinman 1996).

Immunization of C57BL/6 mice with myelin oligodendrocyte protein (MOG) results in the development of a chronic form of EAE. Immunization of SJL/J mice with proteolipid protein (PLP) results in the development of a remitting/relapsing form of EAE. EAE can also be induced by an adoptive transfer technique, in which myelin antigen-specific T cells are injected into naïve recipients. As illustrated in Fig. 12.3, there are numerous potential sites of action that could be affected by the selective CB2R agonist O-1966 when treating EAE.

It was hypothesized that, since the CB1 receptor inhibits synaptic transmission, its activation could be used to control clinical EAE including spasticity, tremor, and paralysis (Lyman et al. 1989; Wirguin et al. 1994; Baker et al. 2000). Later studies demonstrated that administration of competitive endocannabinoid uptake inhibitors and selective inhibitors of FAAH also attenuated spasticity in a chronic relapsing model of EAE (Baker et al. 2001).

In the first investigation of microvascular changes following cannabinoid administration, mouse myelin oligodendrocyte protein and pertussis toxin were used to induce EAE. Using in vivo intravital microscopy, leukocyte/endothelial interactions were evaluated via a cranial window implanted 2 days before induction of EAE. The

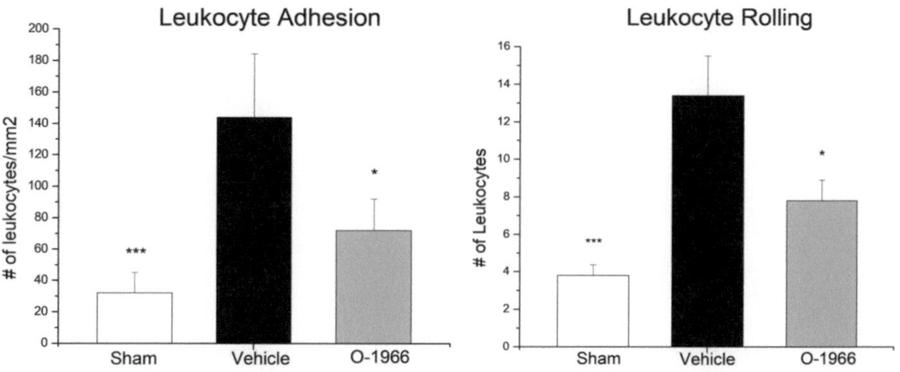

Fig. 12.4 In vivo administration of O-1966 reduces endogenous leukocyte rolling and adhesion. C57BL/ 6 mice were immunized with MOG 35–55. One group was treated with O-1966 (1 mg/kg) on days 7, 11, and 15. The control group received vehicle. Sham—nonimmunized mice. Leukocyte rolling and adhesion to pial microvessels was measured on day 15 by intravital microscopy (cranial window) as previously described. $^{*}p < 0.05$, $^{***}p < 0.001$ (Ni et al. 2004)

results demonstrated that EAE increases leukocyte rolling and firm adhesion to microvessels in the brain, and that this increased leukocyte/endothelial interaction can be attenuated by administration of WIN 55212–2 (Ni et al. 2004). WIN 55212–2, an agonist for both CB1 and CB2 receptors. Later selective CB1 and CB2 antagonist were given individually to differentiate between the effects of CB1 and CB2 activation. Surprisingly, use of the selective antagonists for the CB1 receptor (SR 141716A) and the CB2 receptor (SR144528) in this study demonstrated that the cannabinoid's inhibitory effects on leukocyte/endothelial interactions are mediated by activating CB2 receptors, not through activation of the CB1 receptor (Ni et al. 2004). A later study investigation of selective activation of the CB2 receptor was achieved by administration of the highly selective compound O-1966. The selectivity of this compound for the CB2 receptor as compared to the CB1 receptor was verified by ^{35}S-GTPγS binding (Wiley et al. 2002). Investigations of the effects of O-1966 on leukocyte rolling and adhesion and EAE progression demonstrated that administration of O-1966 significantly reduced leukocyte rolling and adhesion (Fig. 12.4). It was also demonstrated that administration of O-1966 also improved motor function in the chronic EAE model (C57BL/6/MOG), in the remitting–relapsing model (SJL/J/PLP), and in an adoptive transfer model (Fig. 12.5). Marez et al. confirmed the role of the CB2 receptor by showing that showing that encephalitogenic T cells from CB2-deficient mice were more aggressive in terms of CNS infiltration and caused more severe EAE (Maresz et al. 2007a, b).

Later studies were continued to elucidate the mechanisms of action through which a selective CB2 agonist interfered with the progression of EAE. Cellular autoimmune responses, with central contribution by T cells, play an essential role in the development and progression of MS immunopathology (Steinman 1996). EAE models have helped to elucidate the sequential events leading to CNS autoimmune

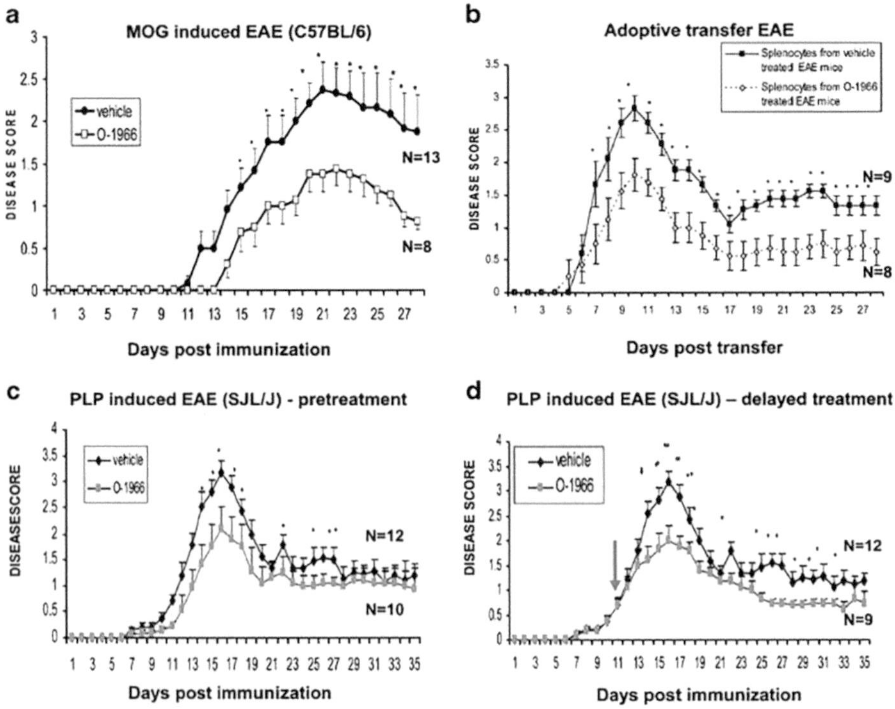

Fig. 12.5 Effects of O-1966 administration in models of EAE. a C57BL/6 mice were immunized with MOG 35–55 /CFA and pertussis toxin (days 0 and + 2). One group received O-1966 (1 mg/kg) on day 7 and every fourth day thereafter (up to day 28). The control group received vehicle. $^*p < 0.01$. b One group of C57BL/6 mice was immunized with MOG 35–55 /CFA, and a second group was immunized and treated with O-1966 on days 5 and 9. On day 11 postimmunization, the mice were sacrificed, and splenocytes were restimulated ex vivo with MOG 35–55 in the presence of IL-2 (2 U/mL), anti-IFNγ Ab (10 μg/mL), and rIL-12 (5 ng/mL). Four days later, splenic CD4+ T cells (1×10^7) were injected i.p. into naïve C57BL/6 recipients. $^*p < 0.05$. c SJL/ J mice were immunized with PLP 139–151 /CFA and pertussis toxin (day 0). One group was treated with O-1966 (1 mg/kg) on day 7 and every fourth day thereafter (up to day 28). The control group received vehicle. $^*p < 0.01$. d SJL/J mice were immunized with PLP 139–151 /CFA and pertussis toxin (day 0). One group was treated with O-1966 (1 mg/kg) on day 11 after development of clinical symptoms and every fourth day thereafter (up to day 27). The control group received vehicle. $^*p < 0.05$. Disease scores were assigned on the basis of motor function as follows: 0, no abnormality; 1, limp tail or hind limb weakness (waddling gait); 2, limp tail and hind limb weakness; 3, partial hind limb paralysis; 4, complete hind limb paralysis; and 5, moribund. Disease scores were compared using ANOVA with repeated measures (Kong et al. 2014)

inflammation (Becher et al. 2006). The first step occurs with priming of encephalitogenic CD4+ T cells in the periphery followed by recruitment of activated T cells across the microcirculation into the CNS, and finally reactivation in the perivascular space. The effector phase is characterized by inflammatory cell migration into parenchyma, microglial activation, and destruction of oligodendrocytes (Kong et al. 2014). Activation of naïve T cells requires cognate contact with dendritic

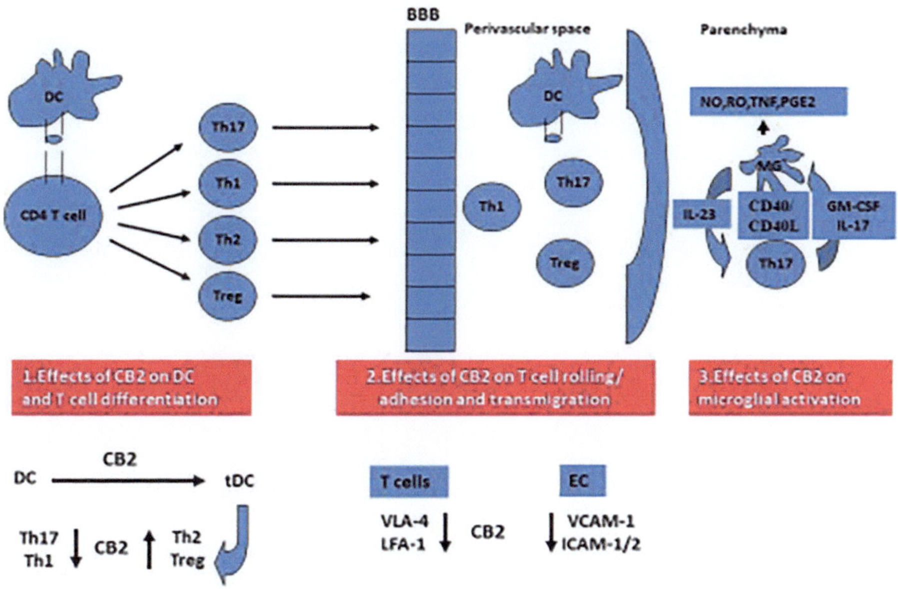

Fig. 12.6 The proposed sites of CB_2 agonist action in EAE include peripheral and perivascular dendritic cells, T cells, endothelial cells, and microglia

cells (DC), which present the processed antigen in the context of MHC II and costimulatory molecules (CD40, CD80, CD86). At this stage, depending on the nature of DC, i.e., stimulatory or tolerogenic, CD4+ T cells could follow two alternative pathways, differentiating into either proinflammatory Th17/Th1 effectors or anti-inflammatory Th2/Treg cells. Following activation and maturation into effectors, T cells migrate to the CNS, cross the blood–brain barrier (BBB), and are reactivated by perivascular antigen-presenting cells. It has been shown that perivascular DC are the major players in encephalitogenic T-cell reactivation (Hickey and Kimura 1988; Archambault et al. 2005; Greter et al. 2005; McMahon et al. 2005, 2006; Kong et al. 2014). Reactivated T cells penetrate deeper into the CNS parenchyma where they activate resident microglia to secrete oxygen radicals, nitric oxide, proinflammatory cytokines, and chemokines, and further promote local inflammation and attract additional inflammatory cells. Activated T cells upregulate IL-23 expression and release from microglia (Becher et al. 2006), and IL-23 maintains the function of the Th17 effector cells that play a critical role in the pathogenesis of EAE/MS, the Th17 effectors. Although microglia do not function as APCs to allow T-cell entry in the CNS, they play a critical role in the effector phase, and blocking microglial activation of CB2 receptors represses EAE (Heppner et al. 2005; Kong et al. 2014). All of these steps are subject to regulation by CB2 receptor agonists (Fig. 12.6). Previously published reports indicated that delta(9)-tetrahydrocannabinol (THC), a mixed CB1/CB2 receptor agonist, and the CB2 agonist JWH-133 shift the Th1/Th2 balance in favor of Th2, presumably through

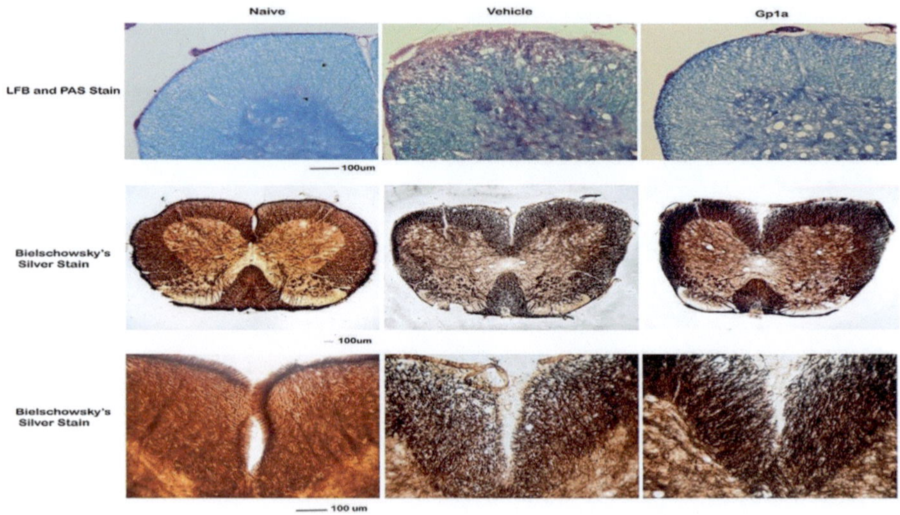

Fig. 12.7 Administration of the selective CB2 agonist Gp1a significantly attenuated axonal/nerve fiber degeneration as evidenced by luxol fast blue, periodic acid-Schiff, and Bielschowsky staining of mouse spinal cords following the development of EAE

the inhibition of IL-12 production by antigen-presenting cells (Yuan et al. 2002; Correa et al. 2005; Lu et al. 2006). Another selective CB2 agonist (GP1a) has been shown to modulate DC and T-cell differentiation (Kong et al. 2014), inducing a shift in T-cell differentiation from Th1/Th17 to Th2/Treg. This leads to a reduction in clinical scores in during EAE and reduced demyelination and axonal loss (Kong et al. 2014) Fig. 12.7. This compound attenuated accumulation of pathogenic immune cells in the CNS. This effect was associated with reductions in the expression of CNS and T-cell chemokine receptors, chemokines and adhesion molecules. These finding are supported by reports that CB1 and CB2 receptor deficient mice develop more sever EAE (Pryce et al. 2003; Maresz et al. 2007a, b; Palazuelos et al. 2008).

An additional effect of selective CB2 receptor activation is the reduction of dendritic cell migration, primarily through inhibition of matrix metalloproteinase (MMP9). CB2 receptor activation on dendritic cells resulted in a reduction of cAMP levels with subsequent decease in ERK activation and reduced binding of c-Fof and c-Jun to the MMP-9 promoter activator protein (Adhikary et al. 2012). The role of MMP9 in dendritic cell migration to inflammatory sites and their diapedesis out of capillaries has been demonstrated a number of investigators showing that MMP-9 is necessary for degradation of extracellular matrix through cleavage of collagen IV, allowing for the migration of DCs and macrophages (Ichiyasu et al. 2004; Chabot et al. 2006; Zozulya et al. 2007).

12.1.4 Stroke

Based on the hypothesis that modifying CB1 receptor activity could influence excitotoxicity following cerebral ischemia, the initial studies of cannabinoid influence on stroke focused on the CB1 receptor. Mixed results were reported, with some indicating that CB1 receptor activation was protective while others indicated that blocking the CB1 receptor was protective (Nagayama et al. 1999; Muthian et al. 2004). CB1-induced hypothermia may have contributed to observations of neuroprotection. In addition, an investigation utilizing CB_1 knockout mice showed an increase in infarct size compared with wild-type animals (Parmentier-Batteur et al. 2002). The possibility that CB1 embryonic deletion may lead to abnormal brain development must be taken into consideration when interpreting these results. However, in a separate investigation utilizing the CB_1 receptor antagonists SR141716 and LY320135, blockade of this receptor was found to reduce infarct volume in a mouse MCAO model (Muthian et al. 2004). The CB1 antagonist SR141716A was also found to be neuroprotective in a mouse model of permanent ischemia. The precise role of the CB1 receptor following ischemia and reperfusion remains an open question.

While the ultimate treatment for ischemia is the restoration of blood flow, unfortunately reperfusion also causes additional tissue damage, referred to a secondary injury, mediated by reactive oxygen and reactive nitrogen species upon reperfusion, as well as to the rapid activation of a variety of proinflammatory genes (Pacher et al. 2005b, 2007). Leukocytes play an important role in this process. Reperfusion injury rapidly expands damage caused by the initial ischemia and attenuation of inflammatory processes presents potential targets for attenuation of secondary injury.

Having previously demonstrated the contribution of leukocyte activation and penetration across the blood–brain barrier to the exacerbation of injury following stroke (Vasthare et al. 1990; Heinel et al. 1994), and based on previous studies demonstrating activation of the CB_2-attenuated inflammation in models of multiple sclerosis, traumatic brain injury, and Alzheimer's disease (Schwenkreis and Tegenthoff 2003; Walter et al. 2003; Ni et al. 2004; Walter and Stella 2004; Jackson et al. 2005), studies were conducted to evaluate the role of selective CB2 receptor agonists in modulating ischemic injury following stroke. In these studies, stroke was induced by transient occlusion of the middle cerebral artery of mice created by blocking the origin of the vessel with the tip of a ligature, and later removing the ligature to allow for reperfusion. Changes in microcirculatory function were observed through a closed cranial window (Zhang et al. 2007). Observations of the cerebral microvessels were made prior to ischemia, 1 h following ischemia and 24 h following ischemia. It was shown that within minutes of ischemia, cerebral vascular endothelium is activated and leukocytes begin to roll on inflamed endothelial cells in venules. Once activated, leukocytes begin to firmly adhere to endothelial cells. This process is mediated by series of interactions between cell adhesion molecules expressed on both leukocytes and endothelial cells. Cell adhesion molecules, such as intracellular adhesion molecule-1, can be highly induced during the early stage of ischemia by proinflammatory cytokines such as tumor necrosis factor-α (Kishimoto and Rothlein 1994; Rothlein et al. 1994; Kataoka et al. 2004). After cerebral ischemia, neutrophils are the first leukocytes to infiltrate at the site of inflammation and monocytes are subsequently recruited. Leukocytes activation and migration have been implicated as primary contributors to ischemia/reperfusion injury. In addition to their role in physical obstruction of capillaries, they participate in inflammatory responses and cause brain tissue damage by various mechanisms. Proinflammatory cytokines (tumor necrosis factor-α and interleukin-1β), secreted by leukocytes, not only activate vascular endothelial cells and amplify inflammatory response but also directly induce neuronal injury (Wood 2003). Matrix metalloproteases (MMPs), secreted by macrophages, can degrade the constituents of basal lamina and thus contribute to blood–brain barrier disruption and cerebral edema during stroke (Maier et al. 2004). These studies highlight the involvement of white blood cells in exacerbating ischemic injury, and the protection offered by interfering with the ability of white blood cells to adhere to endothelial cells and undergo diapedesis into the brain (Vasthare et al. 1990; Heinel et al. 1994; Connolly Jr. et al. 1996; Kanemoto et al. 2002; Weaver et al. 2002).

Leukocyte rolling and adhesion are prerequisites for transmigration across microvessels an invasion of the parenchyma. Administration of the selective CB2 agonist ischemia resulted in significant attenuation of leukocyte rolling and adhesion (Fig. 12.8). The reduction in leukocyte/endothelial interactions occurred with both pretreatment and post treatment, with the largest effect with pretreatment. There is also considerable evidence that there is a significant increase in endogenous cannabinoid production and cannabinoid receptors (Sugiura et al. 2001; Patel et al. 2005; Hillard 2008) following ischemia. While it is often argued that pretreatment with a cannabinoid would not have clinical relevance for the treatment of stroke, it should be recognized that there are numerous medical procedures associated with a significant risk for stroke, and pretreatment could be useful in this context. Consistent with the effects on leukocyte rolling and adhesion was the finding that administration of the selective CB2 receptor agonist resulted in a significant decrease in infarct size, as measured by triphenyltetrazolium chloride staining 24 hours after ischemia (Fig. 12.9). The improvement seen following administration of the CB2 receptor selective agonists was not the result of significant changes in blood flow to the ischemic region during occlusion. Treatment with selective CB2 agonists resulted in a significant improvement in neurologic function. Neurologic (motor) function was assessed using the standard five point deficit scale with higher numbers reflecting higher levels of dysfunction and the ability of the mice to run on a RotaRod. Later studies revealed that the inhibition of leukocyte rolling and adhesion was the result of a downregulation of adhesion molecules on both circulating leukocytes and endothelial cells. Treatment with the selective CB2 agonist significantly decreased ICAM and Mac1 expression and enhanced blood–brain barrier protection as evidenced by a reduction in Evans blue extravasation following ischemia. Activated Mac-1 on leukocytes binds to ICAM-1 on endothelial cells causing firm adhesion (Kishimoto and Rothlein 1994; Lindsberg et al. 1996). In addition, iNOS

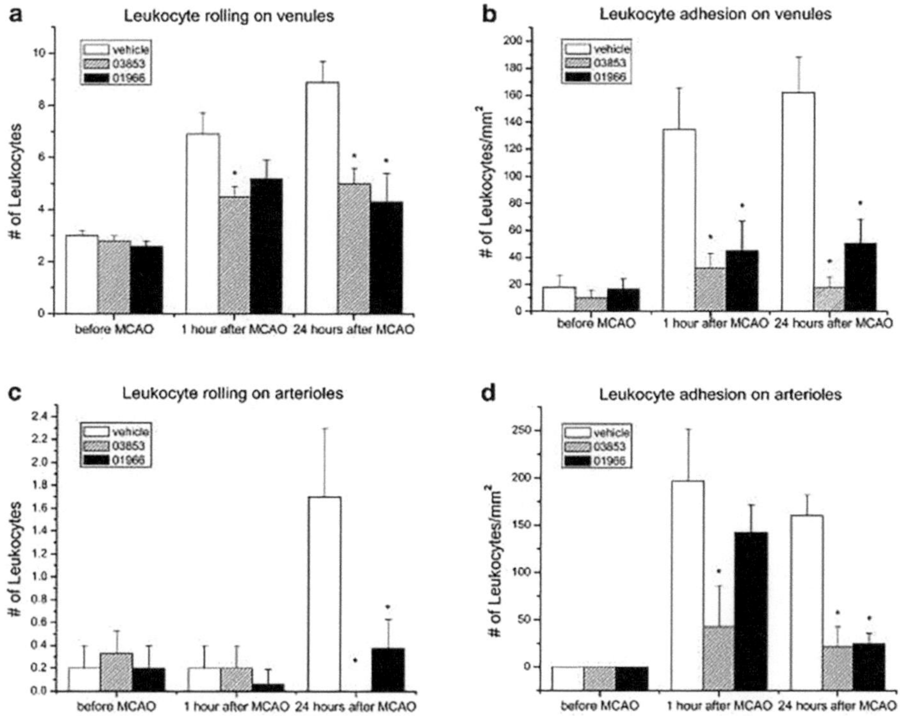

Fig. 12.8 In vivo administration of CB_2 receptor agonists reduces leukocyte rolling and adhesion during cerebral ischemic/reperfusion injury. C57BL/6 mice were subjected to middle cerebral artery occlusion and reperfusion as described (Zhang et al. 2007). One group received O-1966, a second group received the CB_2 receptor agonist O-3853, and the control group received vehicle. Leukocyte rolling and adhesion to cranial venules and arterioles before, 1 hour, and 24 h after MCAO was determined by intravital microscopy as described previously. $*p < 0.05$. (Zhang et al. 2007)

transcription and NO production in macrophages can be largely inhibited by CB2 activation (Berdyshev 2000; Klein and Cabral 2006). Both proinflammatory cytokines and NO are neurotoxic, leading to neuronal death during stroke. CB2 stimulation is also able to inhibit antigen-presenting cell activity, decrease antibody production from B lymphocytes, and down regulate inflammatory cytokine production (Klein and Cabral 2006). All these investigations indicate that activation of the CB2 receptor may provide a potent mechanism for interference with inflammatory response. The finding that CB2 knockout mice had much larger infarctions and decreased motor function following cerebral ischemia supports the concept that CB2R activation is protective for ischemia/reperfusion injury (Zhang et al. 2009).

Preliminary data indicated that the combination of a CB2 agonist with a CB1 antagonist provided an additive effect in protecting the brain from ischemia/reperfusion injury. These studies also demonstrated that the combination of a CB2 agonist with a CB1 antagonist may elicit dilation of small arterioles and therefore improve collateral flow (Fig. 12.10). The CB2 agonist/CB1 antagonist combination was also

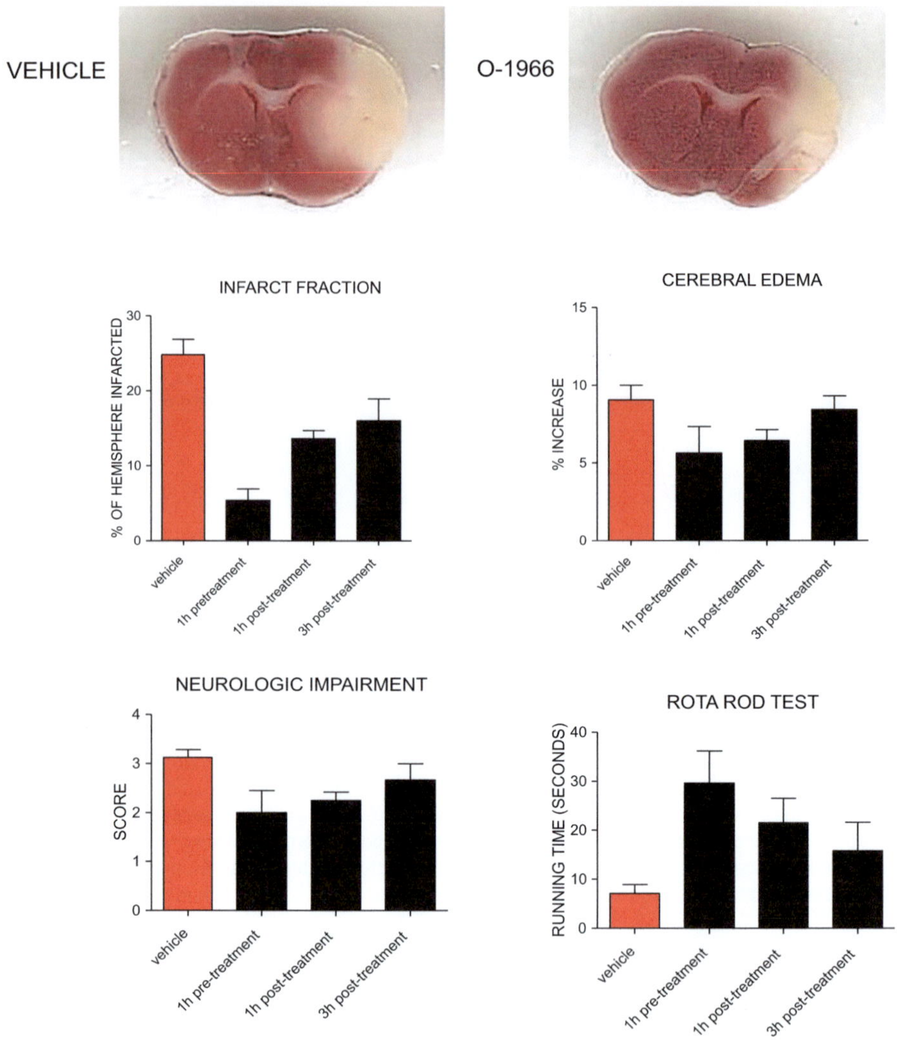

Fig. 12.9 Administration of the selective CB2 agonist O-1966 decreased infarct fraction as measured by TTC, cerebral edema, and neurologic impairment while improving the ability of the mice to run on the Rota Rod. (Zhang et al. 2007)

found to have additive protective effects following ischemia. Treatment with this combination resulted in a reduction in infarct size (Zhang et al. 2008).

Very soon after publication of results describing the microvascular effects of selective CB2 agonist on the brain microvasculature during and after ischemia, similar microvascular effects were demonstrated in other organs. In an elegant study involving ischemia/reperfusion injury to the liver, Pacher's group found results very similar to those described in the brain (Pacher and Hasko 2008).

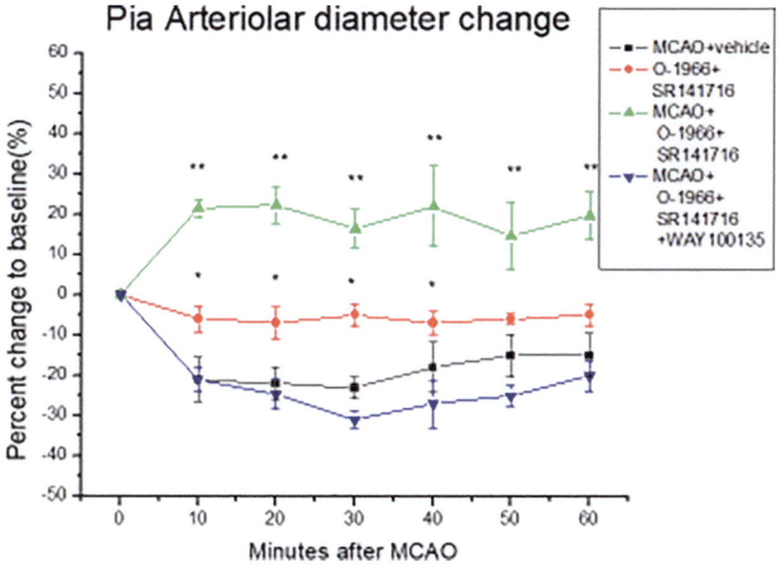

Fig. 12.10 Effects of treatment with the selective CB2 agonist (O-1966), the selective CB1 antagonist (SR141716) administered singly and in combination. The combination of the selective CB2 agonist with the selective CB1 antagonist resulted in an increase in pial arteriolar diameter the occlusion period

Through mechanisms similar to those described for cerebral ischemia/reperfusion injury, circulating, and infiltrating polymorphonuclear cells, Kupffer cells and inflammatory cytokines/chemokines play an important role in liver ischemia/reperfusion (Ohkohchi et al. 1999; Jaeschke 2006). As was the case for cerebral ischemia/reperfusion injury, stimulation of CB2R receptor activity in the liver was shown to have significant influence on outcome following ischemia/reperfusion injury through modification of inflammatory responses attenuating secondary inflammatory damage following reperfusion. Included among the effects were a reduction in inflammatory cell invasion, MIP-2, tissue lipid peroxidation, adhesion molecule expression, and adhesion of neutrophils to liver sinusoidal cells (Pacher and Hasko 2008). The role of CB2 receptors in protecting the liver from ischemia/reperfusion injury are further confirmed by the finding that $CB2^{-/-}$ knockout mice demonstrate increased damage and inflammatory responses following the injury.

There are also multiple similar effects at the level of the coronary microcirculation that contribute to the protective effect of endogenous and exogenous cannabinoids following myocardial ischemia/reperfusion injury (Tuma and Steffens 2012; Steffens and Pacher 2012). Both ex vivo and in vivo studies have demonstrated the protective effect of CB2 receptor activation in reducing microvascular damage in the heart following ischemia reperfusion injury. It is interesting to speculate about the potential therapeutic effect of combining 2CB2 agonists with thrombolytic agents to protect from reperfusion injury. While thrombolytic therapy is effective in restoring blood flow after acute stroke and myocardial infarction, the endothelial dysfunction is actually enhanced by thrombolytic agents (Suzuki et al. 2009; Docagne et al. 2015). Activation of CB2 receptors may contribute to protecting endothelial cells and maintaining barriers to vascular permeability.

A number of studies have provided evidence that endocannabinoids contribute to ischemic preconditioning, the phenomenon where exposure of a tissue to brief periods of ischemia imparts a degree of protection for the tissue during a more extended period of ischemia (Yellon and Downey 2003; Pacher et al. 2006). The phenomenon of ischemic preconditioning of the myocardium appears to be dependent upon activation of the CB2 receptor (Krylatov et al. 2002a, b). CB2 receptors on endothelial and circulating inflammatory cells are thought to contribute to this phenomenon (Jaeschke 2006; Pacher and Hasko 2008). The effects of CB1 activation on myocardial infarction are less clear.

The anti-inflammatory effect of CB2 receptor activation also appears to protect the central nervous system from mechanical trauma. For the purpose of these studies, controlled cortical impact to the CNS was produced by a computer-controlled piston, carefully monitored for both impact velocity and depth of penetration. Rodent studies have demonstrated a significant reduction in inflammatory responses in animals treated with a CB2 agonist following controlled impact studies of both spinal cord injury and traumatic brain injury. Traumatic brain injury causes an increase in TNF-alpha protein, and increase in ICAM-1 expression, microglial activation, inflammatory cell invasion, iNos production, and an increase in blood–brain barrier permeability. A central component of this response in inhibition of the conversion of microglia into an inflammatory phenotype (Elliott et al. 2011; Amenta et al. 2012a, b, 2014).Administration of CB2 agonists attenuated all of these responses. In addition, administration of a selective CB2 agonist also decreased perivascular substance P accumulation (Fig. 12.11). Locomotor and exploratory behavior were also significantly improved (Elliott et al. 2011).

Following a similar pattern, administration of selective CB2 agonists also attenuated mechanical injury to the spinal cord. Mice treated with the selective CB2 agonist O-1966 had improved motor function in a model of controlled compression of the spinal cord (Baty et al. 2008). The most notable finding was a remarkable improvement in bladder function in the CB2 treated animals. In a later study, mice treated with a combination of the selective CB2 agonist (O-1966) and the CB1 antagonist SR141716 had a better recovery than mice treated with either agent alone (Heller et al. 2009). Adhikary et al. evaluated changes in specific inflammatory responses, including the generation of free radicals, proinflammatory cytokines, and chemokines, white blood cell invasion, and activation of resident inflammatory cells, which are major components of secondary injury (Adhikary et al. 2011). CXCL-1 and CCL-2 are proinflammatory and can function as chemoattractants for neutrophils and monocytes, respectively, while CXCL9 and CXCL11 may function as T-cell attractants. Recently, IL-17-producing T cells have been recognized as contributing to the exacerbation of central nervous system (CNS) injury. Also modified were Toll-like receptor 2 (TLR2) and TLR4 expressions which are proposed to function in the regulation of proinflammatory cytokines following spinal cord injury (Kigerl et al. 2007) (Table 12.1).

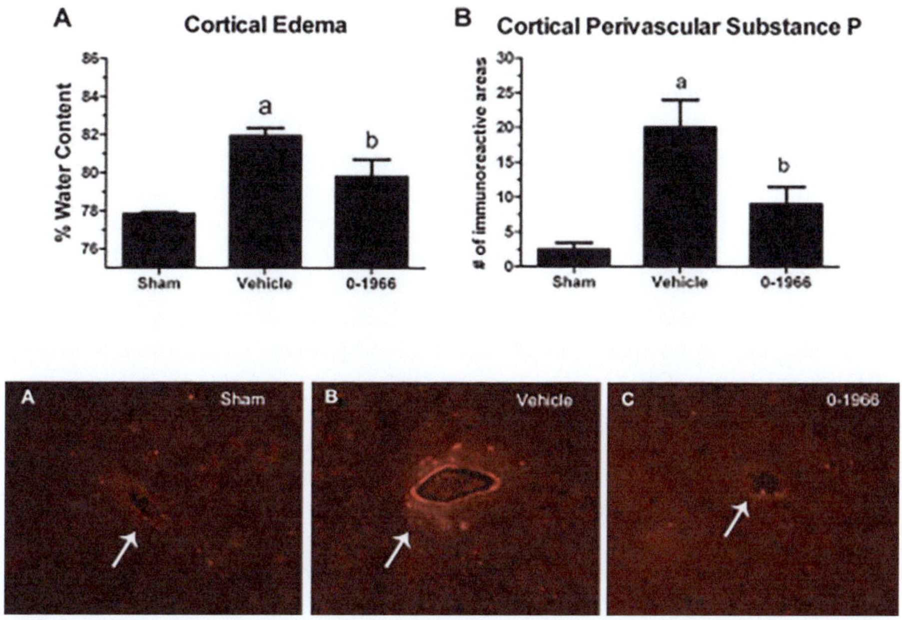

Fig. 12.11 Cortical edema and the accumulation of perivascular substance P following controlled cortical impact were significantly attenuated by administration of the selective CB2 agonist O-1966. (Elliott et al. 2011)

Table 12.1 Diseases influenced by the endocannabinoid system

Organ	Diseases
Cardiovascular	Myocardial infarction Stroke Atherosclerosis Restenosis
CNS	Multiple sclerosis Depression and anxiety Traumatic brain injury Schizophrenia
Liver	Hepatitis Fibrosis Steatosis
Metabolic disease	Obesity Insulin resistance Pancreatitis Insulin secretion Adipose function
Bone	Osteoporosis Abnormal bone formation
Gastrointestinal	IBD
Cancer	Inhibition of some type of tumors
Reproductive system	Fertility Endometriosis
Autoimmune diseases	Arthritis Organ transplant

12.1.5 Cannabidiol

Although not a component of the endocannabinoid system, a compound that has recently garnered significant attention is cannabidiol (CBD). CBD may however increase the concentration of endogenous cannabinoids (Maione et al. 2011; Leishman et al. 2018). Cannabidiol is one of the major chemicals found in the cannabis plant. Unlike THC, CBD does not have a euphoric effect and therefore some advantage for use as a therapeutic agent. Also, consumption of high levels of THC has been speculated to be associated with heart attack and stroke. However, CBD does not appear to cause any adverse cardiovascular effects. Under basal conditions, there is no significant change in blood pressure or heart rate following CBD administration. CBD has been reported to have antioxidant, anti-inflammatory, and immunomodulatory effects. While many of the actions of CBD are similar to CB2 agonists, the effects of CBD are not primarily through activation of CB1 or CB2 receptors. The mechanisms of action of CBD have not been fully elucidated and there are multiple targets for CBD activation. Among the receptors influencing microvascular function and thought to be modulated by CBD are TRPV1–4 receptors and 5-HT_{1A}, 5-$HT_{2A,}$ and PPAR-γ receptors. CBD has been shown interact with numerous other receptors, but in many cases only at very high concentrations. In addition, CBD may influence adenosine, noradrenalin, dopamine and serotonin uptake (Kicman and Toczek 2020). It has been reported that CBD can modulate the actions of THC. Since it can modulate arachidonic metabolism, CBD can also influence to production of prostaglandins (McPartland et al. 2015; Ligresti et al. 2016).

Although CBD administration does not evoke hypotension or increased heart rate when administered in vivo, CBD has been shown to have vasodilator properties under physiological and pathological conditions (Stanley et al. 2015; Baranowska-Kuczko et al. 2020). PPAR-γ receptors may participate in CBD-induced vasodilation (Baranowska-Kuczko et al. 2020). Modification of intracellular proteins may be responsible for the vasodilatory and anti-inflammatory effects of CBD (Kicman and Toczek 2020).

The microvascular effects of CBD have been reported to present therapeutic potential for its use in a wide variety of disorders involving ischemia/reperfusion injury as occurs with myocardial infarction and stroke. As is the case with selective CB2 agonists, administration of CBD prior to ischemia/reperfusion injury of the myocardium reduced infarct size in the myocardium (Durst et al. 2007; Walsh et al. 2010; Feng et al. 2015; Gonca and Darici 2015). A decrease in arrhythmias following ischemia/reperfusion injury in animals treated with CBD was attributed to inhibition of release of arrhythmogenic substances from platelets and adenosine stimulation of A1 adenosine receptors. It has been suggested that inhibition of adenosine uptake plays a major role in providing cardioprotection from inflammatory damage (Xu et al. 2005; Carrier et al. 2006). Administration of cannabidiol also resulted in reduced leukocyte infiltration across the microcirculation.

Cannabidiol also protected the brain in mouse and rat models of middle cerebral artery occlusion/reperfusion (Mishima et al. 2005; Hayakawa et al. 2007, 2008; Khaksar and Bigdeli 2017). In these studies, CBD was found to be protective when given before and also following ischemia. Administration of CBD also increased cerebral blood flow during the occlusion period, and improved microvascular perfusion for 1 h after reperfusion. The 5-HT_{1A} receptor was found to contribute to this response (Mishima et al. 2005). Activation of the CB2 and adenosine receptors has also been reported to participate in its neuroprotective response (Pazos et al. 2013; Ceprian et al. 2019). Evidence was also provided that CBD administration given days after stroke improved outcome (Hayakawa et al. 2009). The anti-inflammatory effect of cannabidiol administration is evidenced by reduced neutrophil invasion of the brain. These beneficial effects extend to the treatment of perinatal-hypoxic-ischemic encephalopathy (Alvarez et al. 2008; Pazos et al. 2013; Barata et al. 2019).

There is also evidence that the microvascular effects of CBD may also have a role in treating sepsis-related encephalitis. In a model of murine encephalitis evoked by lipopolysaccharide administration, CBD attenuated the increased expression of TNFα, COX-2, inducible NOS expression, and oxidative stress markers, helping to preserve the blood–brain barrier (Ruiz-Valdepenas et al. 2011). CBD can influence a number of blood cell functions. CBD increases phospholipase A_2 expression and lipoxygenase production in platelets and inhibits adenosine or epinephrine stimulated platelet aggregation (White and Tansik 1980).

Protection from ischemia/reperfusion injury of the kidneys and liver are also a potential targets for Cannabinoid therapy. CBD reduced inflammation in both organs following ischemia/reperfusion injury (Fouad and Jresat 2011; Mukhopadhyay et al. 2011; Fouad et al. 2012). In the liver, this was associated with a reduction in ICAM-1, VCM-1, and reduced neutral infiltration.

As a result of its anti-inflammatory actions, cannabidiol can attenuate microvascular damage caused by diabetes. In coronary vessels, diabetic complications can cause endothelial cells to increase superoxide generation, NF-kB activation, inducible NOS expression, and adhesion and diapedesis of monocytes. All of these proinflammatory changes are attenuated by CBD (Rajesh et al. 2007a). CBD also enhances vasodilation in diabetic animals (Kicman and Toczek 2020).

Dosing issues with selective CB2 agonists and CBD are complex, as has been demonstrated by many investigators in that midrange doses are often more effective than either higher or lower doses. This results in an inverted U shape dose–response curve, or even in the case of CBD, an N-shaped dose–response curve. This can be attributed to off target activation of receptors at higher doses (Ferraro et al. 2001; Pegorini et al. 2006).

12.2 Summary

Although employed for medicinal purposes over thousands of years, the therapeutic use of cannabinoids for medicinal has received remarkable resurgence of interest over the last few decades. The social, political and racial biases that have contributed to the classification of cannabinoids as Schedule 1 drugs have severely limited the ability to perform clinical trials and therefore our understanding of both the therapeutic potential and the potential adverse effects of medicinal cannabinoids. Despite the limitations imposed on clinical trials, multiple studies have been conducted at the basic science level in experimental animals. Revelations of the importance of the endocannabinoid system as a major homeostatic system has enhanced our appreciation of the therapeutic potential of manipulation of this system in multiple organs. The widespread effect of cannabinoids in multiple organs is derived in large part from its anti-inflammatory action at the microvascular level. There are also multiple targets for cannabinoids at the microvascular level that can modulate inflammatory responses. Since both CB2 selective agents and CBD can be administered without inducing a euphoric effect, it enhances their attractiveness as therapeutic agents.

References

Adhikary S, Li H et al (2011) Modulation of inflammatory responses by a cannabinoid-2-selective agonist after spinal cord injury. J Neurotrauma 28(12):2417–2427

Adhikary S, Kocieda VP et al (2012) Signaling through cannabinoid receptor 2 suppresses murine dendritic cell migration by inhibiting matrix metalloproteinase 9 expression. Blood 120(18): 3741–3749

Alvarez FJ, Lafuente H et al (2008) Neuroprotective effects of the nonpsychoactive cannabinoid cannabidiol in hypoxic-ischemic newborn piglets. Pediatr Res 64(6):653–658

Amenta PS, Jallo JI et al (2012a) A cannabinoid type 2 receptor agonist attenuates blood-brain barrier damage and neurodegeneration in a murine model of traumatic brain injury. J Neurosci Res 90(12):2293–2305

Amenta PS, Starke RM et al (2012b) Successful treatment of a traumatic carotid pseudoaneurysm with the pipeline stent: case report and review of the literature. Surg Neurol Int 3:160

Amenta PS, Jallo JI et al (2014) Cannabinoid receptor type-2 stimulation, blockade, and deletion alter the vascular inflammatory responses to traumatic brain injury. J Neuroinflammation 11:191

Archambault AS, Sim J et al (2005) Defining antigen-dependent stages of T cell migration from the blood to the central nervous system parenchyma. Eur J Immunol 35(4):1076–1085

Baker DP, Croxford G, Brown JL, Pertwee P, Huffman RG, Layward JW (2000) Cannabinoids control spasticity and tremor in a multiple sclerosis model. Nature 404:84–87

Baker DP, Croxford G, Brown JL, Pertwee P, Makriyannis RG, Khanolkar A, Layward A, Fezza L, Bisogno F, Di Marzo T (2001) Endocannabinoids control spasticity in a multiple sclerosis model. FASEB J 15(2):300–302

Baranowska-Kuczko M, Kozlowska H et al (2020) Vasodilatory effects of cannabidiol in human pulmonary and rat small mesenteric arteries: modification by hypertension and the potential pharmacological opportunities. J Hypertens 38(5):896–911

Barata L, Arruza L et al (2019) Neuroprotection by cannabidiol and hypothermia in a piglet model of newborn hypoxic-ischemic brain damage. Neuropharmacology 146:1–11

Batkai S, Pacher P (2009) Endocannabinoids and cardiac contractile function: pathophysiological implications. Pharmacol Res 60(2):99–106

Baty DE, Zhang M et al (2008) Cannabinoid CB2 receptor activation attenuates motor and autonomic function deficits in a mouse model of spinal cord injury. Clin Neurosurg 55:172–177

Becher B, Bechmann I et al (2006) Antigen presentation in autoimmunity and CNS inflammation: how T lymphocytes recognize the brain. J Mol Med 84(7):532–543

Benyo Z, Ruisanchez E et al (2016) Endocannabinoids in cerebrovascular regulation. Am J Physiol Heart Circ Physiol 310(7):H785–H801

Berdyshev EV (2000) Cannabinoid receptors and the regulation of immune response. Chem Phys Lipids 108(1–2):169–190

Blazquez C, Casanova ML et al (2003) Inhibition of tumor angiogenesis by cannabinoids. FASEB J 17(3):529–531

Bondarenko AI (2014) Endothelial atypical cannabinoid receptor: do we have enough evidence? Br J Pharmacol 171(24):5573–5588

Cabral GA, Dove Pettit DA (1998) Drugs and immunity: cannabinoids and their role in decreased resistance to infectious disease. J Neuroimmunol 83(1–2):116–123

Cabral GA, Staab A (2005) Effects on the immune system. Handb Exp Pharmacol 168:385–423

Cabral GA, Harmon KN et al (2001) Cannabinoid-mediated inhibition of inducible nitric oxide production by rat microglial cells: evidence for CB1 receptor participation. Adv Exp Med Biol 493:207–214

Carrier EJ, Auchampach JA et al (2006) Inhibition of an equilibrative nucleoside transporter by cannabidiol: a mechanism of cannabinoid immunosuppression. Proc Natl Acad Sci USA 103(20):7895–7900

Ceprian M, Vargas C et al (2019) Cannabidiol administration prevents hypoxia-ischemia-induced Hypomyelination in newborn rats. Front Pharmacol 10:1131

Chabot V, Reverdiau P et al (2006) CCL5-enhanced human immature dendritic cell migration through the basement membrane in vitro depends on matrix metalloproteinase-9. J Leukoc Biol 79(4):767–778

Chaytor AT, Martin PE et al (1999) The endothelial component of cannabinoid-induced relaxation in rabbit mesenteric artery depends on gap junctional communication. J Physiol 520(Pt 2): 539–550

Connolly ES Jr, Winfree CJ et al (1996) Cerebral protection in homozygous null ICAM-1 mice after middle cerebral artery occlusion. Role of neutrophil adhesion in the pathogenesis of stroke. J Clin Invest 97(1):209–216

Correa F, Mestre L et al (2005) Activation of cannabinoid CB2 receptor negatively regulates IL-12p40 production in murine macrophages: role of IL-10 and ERK1/2 kinase signaling. Br J Pharmacol 145(4):441–448

Cravatt BF, Demarest K et al (2001) Supersensitivity to anandamide and enhanced endogenous cannabinoid signaling in mice lacking fatty acid amide hydrolase. Proc Natl Acad Sci USA 98(16):9371–9376

Delgado-Peraza F, Ahn KH et al (2016) Mechanisms of biased beta-Arrestin-mediated signaling downstream from the cannabinoid 1 receptor. Mol Pharmacol 89(6):618–629

Devane WA, Dysarz FA 3rd et al (1988) Determination and characterization of a cannabinoid receptor in rat brain. Mol Pharmacol 34(5):605–613

Di Marzo V, De Petrocellis L (2012) Why do cannabinoid receptors have more than one endogenous ligand? Philos Trans R Soc Lond Ser B Biol Sci 367(1607):3216–3228

Dinh TP, Carpenter D et al (2002) Brain monoglyceride lipase participating in endocannabinoid inactivation. Proc Natl Acad Sci USA 99(16):10819–10824

Docagne F, Parcq J et al (2015) Understanding the functions of endogenous and exogenous tissue-type plasminogen activator during stroke. Stroke 46(1):314–320

Durst R, Danenberg H et al (2007) Cannabidiol, a nonpsychoactive cannabis constituent, protects against myocardial ischemic reperfusion injury. Am J Physiol Heart Circ Physiol 293(6): H3602–H3607

Elliott MB, Tuma RF et al (2011) Acute effects of a selective cannabinoid-2 receptor agonist on neuroinflammation in a model of traumatic brain injury. J Neurotrauma 28(6):973–981

Feng Y, Chen F et al (2015) Pharmacologic effects of Cannabidiol on acute Reperfused myocardial infarction in rabbits: evaluated with 3.0T cardiac magnetic resonance imaging and histopathology. J Cardiovasc Pharmacol 66(4):354–363

Ferraro L, Tomasini MC et al (2001) The cannabinoid receptor agonist WIN 55,212-2 regulates glutamate transmission in rat cerebral cortex: an in vivo and in vitro study. Cereb Cortex 11(8): 728–733

Fouad AA, Jresat I (2011) Therapeutic potential of cannabidiol against ischemia/reperfusion liver injury in rats. Eur J Pharmacol 670(1):216–223

Fouad AA, Al-Mulhim AS et al (2012) Cannabidiol treatment ameliorates ischemia/reperfusion renal injury in rats. Life Sci 91(7–8):284–292

Gebremedhin D, Lange AR et al (1999) Cannabinoid CB1 receptor of cat cerebral arterial muscle functions to inhibit L-type Ca2+ channel current. Am J Phys 276(6 Pt 2):H2085–H2093

Godlewski G, Malinowska B et al (2004) Presynaptic cannabinoid CB(1) receptors are involved in the inhibition of the neurogenic vasopressor response during septic shock in pithed rats. Br J Pharmacol 142(4):701–708

Gonca E, Darici F (2015) The effect of cannabidiol on ischemia/reperfusion-induced ventricular arrhythmias: the role of adenosine A1 receptors. J Cardiovasc Pharmacol Ther 20(1):76–83

Greter M, Heppner FL et al (2005) Dendritic cells permit immune invasion of the CNS in an animal model of multiple sclerosis. Nat Med 11(3):328–334

Hayakawa K, Mishima K et al (2007) Delayed treatment with cannabidiol has a cerebroprotective action via a cannabinoid receptor-independent myeloperoxidase-inhibiting mechanism. J Neurochem 102(5):1488–1496

Hayakawa K, Mishima K et al (2008) Cannabidiol prevents a post-ischemic injury progressively induced by cerebral ischemia via a high-mobility group box1-inhibiting mechanism. Neuropharmacology 55(8):1280–1286

Hayakawa K, Irie K et al (2009) Therapeutic time window of cannabidiol treatment on delayed ischemic damage via high-mobility group box1-inhibiting mechanism. Biol Pharm Bull 32(9): 1538–1544

Heinel LA, Rubin S et al (1994) Leukocyte involvement in cerebral infarct generation after ischemia and. Brain Res Bull 34(2):137–141

Heller JE, Baty DE et al (2009) The combination of selective inhibition of the cannabinoid CB1 receptor and activation of the cannabinoid CB2 receptor yields improved attenuation of motor and autonomic deficits in a mouse model of spinal cord injury. Clin Neurosurg 56:84–92

Heppner FL, Greter M et al (2005) Experimental autoimmune encephalomyelitis repressed by microglial paralysis. Nat Med 11(2):146–152

Hickey WF, Kimura H (1988) Perivascular microglial cells of the CNS are bone marrow-derived and present antigen in vivo. Science 239(4837):290–292

Hillard CJ (2008) Role of cannabinoids and endocannabinoids in cerebral ischemia. Curr Pharm Des 14(23):2347–2361

Howlett AC, Barth F et al (2002) International Union of Pharmacology. XXVII. Classification of cannabinoid receptors. Pharmacol Rev 54(2):161–202

Ichiyasu H, McCormack JM et al (2004) Matrix metalloproteinase-9-deficient dendritic cells have impaired migration through tracheal epithelial tight junctions. Am J Respir Cell Mol Biol 30(6): 761–770

Jackson SJ, Diemel LT et al (2005) Cannabinoids and neuroprotection in CNS inflammatory disease. J Neurol Sci 233(1–2):21–25

Jaeschke H (2006) Mechanisms of liver injury. II. Mechanisms of neutrophil-induced liver cell injury during hepatic ischemia-reperfusion and other acute inflammatory conditions. Am J Physiol Gastrointest Liver Physiol 290(6):G1083–G1088

Jarai Z, Wagner JA et al (1999) Cannabinoid-induced mesenteric vasodilation through an endothelial site distinct from CB1 or CB2 receptors. Proc Natl Acad Sci USA 96(24):14136–14141

Kanemoto Y, Nakase H et al (2002) Effects of anti-intercellular adhesion molecule-1 antibody on reperfusion injury induced by late reperfusion in the rat middle cerebral artery occlusion model. Neurosurgery 51(4):1034

Kataoka H, Kim SW et al (2004) Leukocyte-endothelium interactions during permanent focal cerebral ischemia in mice. J Cereb Blood Flow Metab 24(6):668–676

Khaksar S, Bigdeli MR (2017) Correlation between Cannabidiol-induced reduction of infarct volume and inflammatory factors expression in ischemic stroke model. Basic Clin Neurosci 8(2):139–146

Kicman A, Toczek M (2020) The effects of Cannabidiol, a non-intoxicating compound of cannabis, on the cardiovascular system in health and disease. Int J Mol Sci 21(18)

Kigerl KA, Lai W et al (2007) Toll-like receptor (TLR)-2 and TLR-4 regulate inflammation, gliosis, and myelin sparing after spinal cord injury. J Neurochem 102(1):37–50

Kishimoto TK, Rothlein R (1994) Integrins, ICAMs, and selectins: role and regulation of adhesion molecules in neutrophil recruitment to inflammatory sites. Adv Pharmacol 25:117–169

Klein TW, Cabral GA (2006) Cannabinoid-induced immune suppression and modulation of antigen-presenting cells. J Neuroimmune Pharmacol 1:50–64

Klein TW, Newton CA et al (2001) Cannabinoids and the immune system. Pain Res Manag 6(2): 95–101

Kong W, Li H et al (2014) Selective CB2 receptor activation ameliorates EAE by reducing Th17 differentiation and immune cell accumulation in the CNS. Cell Immunol 287(1):1–17

Kozak KR, Rowlinson SW et al (2000) Oxygenation of the endocannabinoid, 2-arachidonylglycerol, to glyceryl prostaglandins by cyclooxygenase-2. J Biol Chem 275(43): 33744–33749

Krylatov AV, Uzhachenko RV et al (2002a) Endogenous cannabinoids improve myocardial resistance to arrhythmogenic effects of coronary occlusion and reperfusion: a possible mechanism. Bull Exp Biol Med 133(2):122–124

Krylatov AV, Uzhachenko RV et al (2002b) Anandamide and R-(+)-methanandamide prevent development of ischemic and reperfusion arrhythmia in rats by stimulation of CB2-receptors. Eksp Klin Farmakol 65(3):6–9

Leishman E, Manchanda M et al (2018) Cannabidiol's upregulation of N-acyl Ethanolamines in the central nervous system requires N-acyl Phosphatidyl ethanolamine-specific phospholipase D. Cannabis Cannabinoid Res 3(1):228–241

Ligresti A, De Petrocellis L et al (2016) From Phytocannabinoids to cannabinoid receptors and endocannabinoids: pleiotropic physiological and pathological roles through complex pharmacology. Physiol Rev 96(4):1593–1659

Lindsberg PJ, Carpen O et al (1996) Endothelial ICAM-1 expression associated with inflammatory cell response in human ischemic stroke. Circulation 94(5):939–945

Liu J, Gao B et al (2000) Functional CB1 cannabinoid receptors in human vascular endothelial cells. Biochem J 346(Pt 3):835–840

Lu T, Newton C et al (2006) Cannabinoid treatment suppresses the T-helper cell-polarizing function of mouse dendritic cells stimulated with legionella pneumophila infection. J Pharmacol Exp Ther 319(1):269–276

Lyman WD, Sonett JR et al (1989) Delta 9-tetrahydrocannabinol: a novel treatment for experimental autoimmune encephalomyelitis. J Neuroimmunol 23(1):73–81

Maier CM, Hsieh L et al (2004) Matrix metalloproteinase-9 and myeloperoxidase expression: quantitative analysis by antigen immunohistochemistry in a model of transient focal cerebral ischemia. Stroke 35(5):1169–1174

Maione S, Piscitelli F et al (2011) Non-psychoactive cannabinoids modulate the descending pathway of antinociception in anaesthetized rats through several mechanisms of action. Br J Pharmacol 162(3):584–596

Maresz K, Pryce G et al (2007a) Direct suppression of CNS autoimmune inflammation via the cannabinoid receptor CB1 on neurons and CB2 on autoreactive T cells. Nat Med 13(4):492–497

Maresz K, Pryce G et al (2007b) Direct suppression of CNS autoimmune inflammation via the cannabinoid receptor CB(1) on neurons and CB(2) on autoreactive T cells. Nat Med 13(4): 492–497

McMahon EJ, Bailey SL et al (2005) Epitope spreading initiates in the CNS in two mouse models of multiple sclerosis. Nat Med 11(3):335–339

McMahon EJ, Bailey SL et al (2006) CNS dendritic cells: critical participants in CNS inflammation? Neurochem Int 49(2):195–203

McPartland JM, Duncan M et al (2015) Are cannabidiol and Delta(9) -tetrahydrocannabivarin negative modulators of the endocannabinoid system? A systematic review. Br J Pharmacol 172(3):737–753

Mechoulam R, Hanus L (2000) A historical overview of chemical research on cannabinoids. Chem Phys Lipids 108(1–2):1–13

Mestre L, Correa F et al (2006) The synthetic cannabinoid WIN 55,212-2 increases COX-2 expression and PGE2 release in murine brain-derived endothelial cells following Theiler's virus infection. Biochem Pharmacol 72(7):869–880

Mishima K, Hayakawa K et al (2005) Cannabidiol prevents cerebral infarction via a serotonergic 5-hydroxytryptamine1A receptor-dependent mechanism. Stroke 36(5):1077–1082

Montecucco F, Di Marzo V (2012) At the heart of the matter: the endocannabinoid system in cardiovascular function and dysfunction. Trends Pharmacol Sci 33(6):331–340

Mukhopadhyay P, Rajesh M et al (2011) Cannabidiol protects against hepatic ischemia/reperfusion injury by attenuating inflammatory signaling and response, oxidative/nitrative stress, and cell death. Free Radic Biol Med 50(10):1368–1381

Munro S, Thomas KL et al (1993) Molecular characterization of a peripheral receptor for cannabinoids. Nature 365(6441):61–65

Muthian S, Rademacher DJ et al (2004) Anandamide content is increased and CB1 cannabinoid receptor blockade is protective during transient, focal cerebral ischemia. Neuroscience 129(3): 743–750

Nagayama T, Sinor AD et al (1999) Cannabinoids and neuroprotection in global and focal cerebral ischemia and in neuronal cultures. J Neurosci 19(8):2987–2995

Ni X, Geller EB et al (2004) Win 55212-2, a cannabinoid receptor agonist, attenuates leukocyte/endothelial interactions in an experimental autoimmune encephalomyelitis model. Mult Scler 10(2):158–164

Ohkohchi N, Shibuya H et al (1999) Kupffer's cells modulate neutrophile activity by superoxide anion and tumor necrosis factor-delta in reperfusion injury of liver transplantation-mechanisms of radical generation and reperfusion injury after cold ischemia. Transplant Proc 31(1–2): 1055–1058

Pacher P, Hasko G (2008) Endocannabinoids and cannabinoid receptors in ischaemia-reperfusion injury and preconditioning. Br J Pharmacol 153(2):252–262

Pacher P, Batkai S et al (2004) Haemodynamic profile and responsiveness to anandamide of TRPV1 receptor knock-out mice. J Physiol 558(Pt 2):647–657

Pacher P, Batkai S et al (2005a) Cardiovascular pharmacology of cannabinoids. Handb Exp Pharmacol(168):599–625

Pacher P, Schulz R et al (2005b) Nitrosative stress and pharmacological modulation of heart failure. Trends Pharmacol Sci 26(6):302–310

Pacher P, Batkai S et al (2006) The endocannabinoid system as an emerging target of pharmacotherapy. Pharmacol Rev 58(3):389–462

Pacher P, Beckman JS et al (2007) Nitric oxide and peroxynitrite in health and disease. Physiol Rev 87(1):315–424

Pacher P, Steffens S et al (2018) Cardiovascular effects of marijuana and synthetic cannabinoids: the good, the bad, and the ugly. Nat Rev Cardiol 15(3):151–166

Palazuelos J, Davoust N et al (2008) The CB(2) cannabinoid receptor controls myeloid progenitor trafficking: involvement in the pathogenesis of an animal model of multiple sclerosis. J Biol Chem 283(19):13320–13329

Parmentier-Batteur S, Jin K et al (2002) Increased severity of stroke in CB1 cannabinoid receptor knock-out mice. J Neurosci 22(22):9771–9775
Patel S, Carrier EJ et al (2005) The postmortal accumulation of brain N-arachidonylethanolamine (anandamide) is dependent upon fatty acid amide hydrolase activity. J Lipid Res 46(2):342–349
Pazos MR, Mohammed N et al (2013) Mechanisms of cannabidiol neuroprotection in hypoxic-ischemic newborn pigs: role of 5HT(1A) and CB2 receptors. Neuropharmacology 71:282–291
Pegorini S, Zani A et al (2006) Vanilloid VR1 receptor is involved in rimonabant-induced neuroprotection. Br J Pharmacol 147(5):552–559
Pertwee RG, Howlett AC et al (2010) International Union of Basic and Clinical Pharmacology. LXXIX. Cannabinoid receptors and their ligands: beyond CB(1) and CB(2). Pharmacol Rev 62(4):588–631
Pryce G, Ahmed Z et al (2003) Cannabinoids inhibit neurodegeneration in models of multiple sclerosis. Brain 126(Pt 10):2191–2202
Rajesh M, Mukhopadhyay P et al (2007a) Cannabidiol attenuates high glucose-induced endothelial cell inflammatory response and barrier disruption. Am J Physiol Heart Circ Physiol 293(1): H610–H619
Rajesh M, Pan H et al (2007b) Cannabinoid-2 receptor agonist HU-308 protects against hepatic ischemia/reperfusion injury by attenuating oxidative stress, inflammatory response, and apoptosis. J Leukoc Biol 82(6):1382–1389
Randall MD, Kendall DA (1997) Involvement of a cannabinoid in endothelium-derived hyperpolarizing factor-mediated coronary vasorelaxation. Eur J Pharmacol 335(2–3):205–209
Randall MD, Kendall DA (1998) Anandamide and endothelium-derived hyperpolarizing factor act via a common vasorelaxant mechanism in rat mesentery. Eur J Pharmacol 346(1):51–53
Randall MD, Alexander SP et al (1996) An endogenous cannabinoid as an endothelium-derived vasorelaxant. Biochem Biophys Res Commun 229(1):114–120
Rom S, Persidsky Y (2013) Cannabinoid receptor 2: potential role in immunomodulation and neuroinflammation. J Neuroimmune Pharmacol 8(3):608–620
Rothlein R, Kishimoto TK et al (1994) Cross-linking of ICAM-1 induces co-signaling of an oxidative burst from mononuclear leukocytes. J Immunol 152(5):2488–2495
Ruiz-Valdepenas L, Martinez-Orgado JA et al (2011) Cannabidiol reduces lipopolysaccharide-induced vascular changes and inflammation in the mouse brain: an intravital microscopy study. J Neuroinflammation 8(1):5
Russo EB (2016) Current therapeutic cannabis controversies and clinical trial design issues. Front Pharmacol 7:309
Schwenkreis P, Tegenthoff M (2003) Therapeutic use of cannabinoids in neurology. Schmerz 17(5):367–373
Sierra S, Luquin N et al (2018) The endocannabinoid system in cardiovascular function: novel insights and clinical implications. Clin Auton Res 28(1):35–52
Soethoudt M, Grether U et al (2017) Cannabinoid CB(2) receptor ligand profiling reveals biased signalling and off-target activity. Nat Commun 8:13958
Sondheimer F, Mechoulam R et al (1964) The synthesis of 19-hydroxy-10-alpha-testosterone. Tetrahedron 20(11):2473–2485
Stanley C, O'Sullivan SE (2014) Vascular targets for cannabinoids: animal and human studies. Br J Pharmacol 171(6):1361–1378
Stanley CP, Hind WH et al (2015) Cannabidiol causes endothelium-dependent vasorelaxation of human mesenteric arteries via CB1 activation. Cardiovasc Res 107(4):568–578
Steffens S, Pacher P (2012) Targeting cannabinoid receptor CB(2) in cardiovascular disorders: promises and controversies. Br J Pharmacol 167(2):313–323
Steinman L (1996) Multiple sclerosis: a coordinated immunological attack against myelin in the central nervous system. Cell 85(3):299–302
Sugiura T, Kodaka T et al (1998) Detection of an endogenous cannabimimetic molecule, 2-arachidonoylglycerol, and cannabinoid CB1 receptor mRNA in human vascular cells: is

2-arachidonoylglycerol a possible vasomodulator? Biochem Biophys Res Commun 243(3): 838–843

Sugiura T, Yoshinaga N et al (2001) Rapid generation of 2-arachidonoylglycerol, an endogenous cannabinoid receptor ligand, in rat brain after decapitation. Neurosci Lett 297(3):175–178

Suzuki Y, Nagai N et al (2009) Tissue-type plasminogen activator (t-PA) induces stromelysin-1 (MMP-3) in endothelial cells through activation of lipoprotein receptor-related protein. Blood 114(15):3352–3358

Tuma and Steffens (2012) Targeting the endocannabinoid system to limit myocardial and cherebral ischemic and reperfusion injury. Curr Pharm Biotechnol Curr Pharm Biotechnol 13(1):46–58

Varga K, Wagner JA et al (1998) Platelet- and macrophage-derived endogenous cannabinoids are involved in endotoxin-induced hypotension. FASEB J 12(11):1035–1044

Vasthare US, Heinel LA et al (1990) Leukocyte involvement in cerebral ischemia and reperfusion injury. Surg Neurol 33(4):261–265

Wagner JA, Varga K et al (1997) Activation of peripheral CB1 cannabinoid receptors in haemorrhagic shock. Nature 390(6659):518–521

Wagner JA, Varga K et al (1999) Mesenteric vasodilation mediated by endothelial anandamide receptors. Hypertension 33(1 Pt 2):429–434

Wagner JA, Hu K et al (2001) Endogenous cannabinoids mediate hypotension after experimental myocardial infarction. J Am Coll Cardiol 38(7):2048–2054

Wagner JA, Hu K et al (2003) CB(1) cannabinoid receptor antagonism promotes remodeling and cannabinoid treatment prevents endothelial dysfunction and hypotension in rats with myocardial infarction. Br J Pharmacol 138(7):1251–1258

Walsh SK, Hepburn CY et al (2010) Acute administration of cannabidiol in vivo suppresses ischaemia-induced cardiac arrhythmias and reduces infarct size when given at reperfusion. Br J Pharmacol 160(5):1234–1242

Walter L, Stella N (2004) Cannabinoids and neuroinflammation. Br J Pharmacol 141(5):775–785

Walter L, Franklin A et al (2003) Nonpsychotropic cannabinoid receptors regulate microglial cell migration. J Neurosci 23(4):1398–1405

Weaver M, Leshley K et al (2002) LEX032, a novel recombinant serpin, protects the brain after transient focal ischemia. Microvasc Res 63(3):327–334

White HL, Tansik RL (1980) Effects of delta 9-tetrahydrocannabinol and cannabidiol on phospholipase and other enzymes regulating arachidonate metabolism. Prostaglandins Med 4(6): 409–417

Wiley JL, Beletskaya ID et al (2002) Resorcinol derivatives: a novel template for the development of cannabinoid CB(1)/CB(2) and CB(2)-selective agonists. J Pharmacol Exp Ther 301(2): 679–689

Wirguin I, Mechoulam R et al (1994) Suppression of experimental autoimmune encephalomyelitis by cannabinoids. Immunopharmacology 28(3):209–214

Wood PL (2003) Neuroinflammation: mechanisms and management. Humana Press, Totowa

Xu Z, Mueller RA et al (2005) Cardioprotection with adenosine A2 receptor activation at reperfusion. J Cardiovasc Pharmacol 46(6):794–802

Yellon DM, Downey JM (2003) Preconditioning the myocardium: from cellular physiology to clinical cardiology. Physiol Rev 83(4):1113–1151

Yu M, Ives D et al (1997) Synthesis of prostaglandin E2 ethanolamide from anandamide by cyclooxygenase-2. J Biol Chem 272(34):21181–21186

Yuan M, Kiertscher SM et al (2002) Delta 9-tetrahydrocannabinol regulates Th1/Th2 cytokine balance in activated human T cells. J Neuroimmunol 133(1–2):124–131

Zhang M, Martin BR et al (2007) Cannabinoid CB(2) receptor activation decreases cerebral infarction in a mouse focal ischemia/reperfusion model. J Cereb Blood Flow Metab 27:1387

Zhang M, Martin BR et al (2008) Modulation of the balance between cannabinoid CB(1) and CB (2) receptor activation during cerebral ischemic/reperfusion injury. Neuroscience 152(3): 753–760

Zhang M, Adler MW et al (2009) CB(2) receptor activation attenuates microcirculatory dysfunction during cerebral ischemic/reperfusion injury. Microvasc Res 78:86

Zoratti C, Kipmen-Korgun D et al (2003) Anandamide initiates ca(2+) signaling via CB2 receptor linked to phospholipase C in calf pulmonary endothelial cells. Br J Pharmacol 140(8): 1351–1362

Zozulya AL, Reinke E et al (2007) Dendritic cell transmigration through brain microvessel endothelium is regulated by MIP-1alpha chemokine and matrix metalloproteinases. J Immunol 178(1):520–529

Zygmunt PM, Petersson J et al (1999) Vanilloid receptors on sensory nerves mediate the vasodilator action of anandamide. Nature 400(6743):452–457

Chapter 13
Tumor Microcirculation: Abnormalities and Normalization

Dai Fukumura

Abstract Blood and lymphatic vessels in tumors are structurally and functionally abnormal resulting in aberrant and immunosuppressive tumor microenvironment (TME). We postulated that judicious doses of antiangiogenic therapy, such as anti-vascular endothelial growth factor (VEGF) therapies, could transiently normalize tumor vasculature—improvement of vessel morphology and function, closer to normal vessels but not completely normal—and hence TME. In both preclinical models and clinical settings, normalization of vasculature and TME has shown improvement of the delivery and efficacy of various therapeutic agents including chemotherapy and immunotherapy. Immunotherapy, such as immune checkpoint blockade (ICB), has revolutionized cancer therapy recently. It now emerged as a major therapeutic modality in oncology and is providing hope for cure to a fraction of cancer patients. However, currently, the majority of cancer patients do not derive benefit from these treatments. Aberrantly elevated levels of proangiogenic factors, such as VEGF and angiopoietin-2 (Ang-2), significantly contribute to the tumor vascular abnormalities and resultant TME. Judicious use of agents targeting these pathways can transiently normalize vasculature and TME, improve therapeutic responsiveness, increase the infiltration of immune effector cells into tumors, and convert the intrinsically immunosuppressive TME to an immune-supportive TME. Thus, judiciously combining antiangiogenic therapies and immunotherapies might enhance the efficacy of immunotherapy. In this chapter, we outline the roles of VEGF and Ang-2 in tumor immune evasion and progression and discuss the evidence indicating that antiangiogenic agents can normalize the TME. We also introduce the application of normalization strategies for CAR-T-cell immunotherapy as well as systemic conditions compromising (obesity) or improving (exercise) vessels/TME. Overall, this chapter discusses that antiangiogenic agents could be combined with immune checkpoint blockade and other treatments/conditions to potentially improve patient outcomes.

D. Fukumura (✉)
Edwin L. Steele Laboratories, Department of Radiation Oncology, Massachusetts General Hospital and Harvard Medical School, Boston, MA, USA
e-mail: dfukumura@mgh.harvard.edu

R. F. Tuma et al. (eds.), *Handbook of Microcirculation*, Physiology in Health and Disease, https://doi.org/10.1007/978-3-031-99609-2_13

Keywords Tumor vasculature · Tumor microenvironment · Vascular normalization · Vascular endothelial growth factor · Angiopoietin · Obesity · Renin-angiotensin system · Exercise · Pancreatic cancer · Breast cancer

13.1 The Abnormal Tumor Vasculature

In normal tissues/organs, blood vessels are well differentiated and form functional vessel networks. Blood vessels are indispensable for the delivery of nutrients and oxygen as well as the removal of waste products. Furthermore, blood vessels play an important role in maintaining homeostasis and providing passages to recruit various immune cells from systemic circulation to the target tissues. On the other hand, blood and lymphatic vessels in tumors are structurally and functionally abnormal (Fig. 13.1) as compared to those in normal counterparts (Jain 2014). These abnormal vessels have impaired perfusion, resulting in hypoxia and a low pH in tumor tissues (Jain 2014). Tumor blood vessels are leaky, and lymphatic drainage is dysfunctional in tumors, together, leading to elevated interstitial fluid pressure in the tumor microenvironment (TME) (Jain 2013). Tumor cells can overcome such harsh conditions in TME and exhibit a survival advantage over normal cells via various mechanisms (Jain 2014). The abnormal vessels and impaired perfusion also hinder

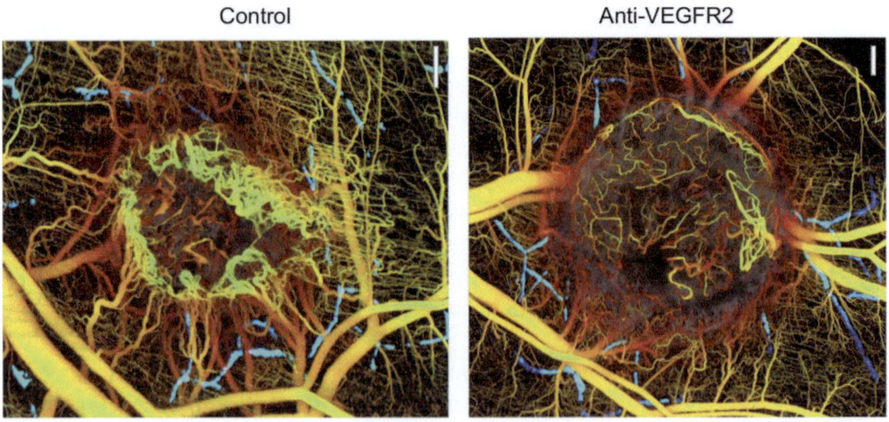

Fig. 13.1 Doppler-OCT image demonstrates normalization of murine tumor vasculature during anti-VEGF receptor 2 (VEGFR2) treatment. Depth projection of the vasculature in a murine mammary carcinoma (MCaIV) in the dorsal skinfold chamber of a SCID mouse is shown 5 days after treatment with anti-VEGFR2 antibody (DC101) as imaged by Doppler OCT (Vakoc et al. 2009). Superficial vessels are brighter (yellowish), whereas deeper (up to 4 mm) vessels are darker (reddish). Tumor vessels are heterogeneous, dilated and tortuous. Control (*left*) and treated (*right*) tumors show strikingly different vascular morphologies. Anti-VEGFR2-treated tumors show smaller vessel diameter, reduced density and less tortuousity, suggesting the vascular normalizing effect of the treatment. The tumor vessels are surrounded by a bed of relatively normal capillary networks. The lymphatic vascular networks are also identifiable (blue) for both tumors. Vertical scale bar, 500 μm. (Adapted from Vakoc et al. 2009)

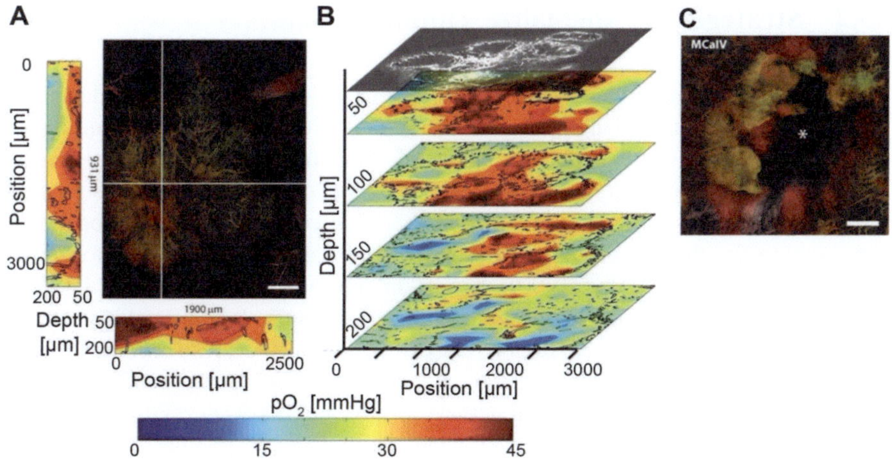

Fig. 13.2 Simultaneous angiography and oxygen tension measurements reveal the abnormal function of tumor vessels and the dependence of oxygen tension on the distance to the nearest vessel. (**A**) The large panel shows a maximum intensity depth projection of MCaIV tumor angiography with an oxygen tension heat map to the left and bottom of the angiography image. Images were taken by Multiphoton Phosphorescence Quenching Microscopy System. The depths of the heat maps go from 50 μm to 200 μm and are taken from the planes noted by the white lines in the angiography image. In the angiography, color denotes depth (from deep to shallow: green, red, and white). The color scale of the oxygen heat maps is shown in the bottom panel (blue indicates 0 mmHg oxygen tension while red indicates 45 mmHg). Black outlines on the oxygen heat maps indicate the morphometry of the tumor vessels. Scale bar 500 μm in all panels. (**B**) Oxygen tension maps from 50 μm, 100 μm, 150 μm, and 200 μm deep within the tumor shown in (A). The angiographic depth slice at 50 μm is shown above the 50 μm deep oxygen heat map for comparison. Black lines on the oxygen heat maps indicate the morphometry of the tumor vessels. (**C**) Angiography of a MCaIV tumor. Asterisk (*) indicates area lacking perfused vessels. (Adapted from Martin et al. 2023)

the delivery of cytotoxic drugs and immune cells from the circulation into tumors. Even if these drugs and immune cells could enter the tumor, abnormal TME, such as hypoxia and low pH, interferes with the antitumor activity of these therapeutic molecules and cells (Jain 2014). Various cytokines produced by diverse cell types in response to TME, such as hypoxia and low pH, cause immunosuppression (Fig. 13.2). Abnormal leaky vessels also facilitate the intravasation of metastatic tumor cells and serve as a conduit for systemic dissemination. Tumor blood vessels also sustain a favorable niche for cancer stem cells (Carmeliet and Jain 2011a; Jain et al. 1996). Our work and that of others have revealed that, in comparison with non-malignant tissues, each component of the TME is abnormal and heterogeneous, and this abnormality fuels tumor progression and treatment resistance (Jain 2013, 2014). Thus, normalization of the TME could improve the outcomes of patients treated with the standard of care and emerging anticancer therapies, including chemotherapy, radiation therapy, molecularly targeted therapies, and immunotherapy (Palucka and Coussens 2016).

13.2 Strategies to Normalize Tumor Vessels

Abnormal tumor vasculature and TME are largely due to the imbalance of proangiogenic versus antiangiogenic signaling (Jain 2005, 2014; Jain et al. 2014). In normal tissues, this balance is exquisitely maintained to ensure the normal morphology and function of blood vessels. However, this balance tips toward proangiogenesis, resulting in immature and abnormal vessels during carcinogenesis. Indeed, vessels already become abnormal in the early phase of carcinogenesis (Hagendoorn et al. 2006). In 2001, to restore this balance, we proposed to use antiangiogenic agents and demonstrated that a judicious dose of antiangiogenic therapy transiently improves tumor vessels structurally and functionally (Jain 2001). We referred to this phenomenon as "normalization" as the vessels become closer to normal vessels but, nonetheless, not perfectly normal (Fig. 13.3) (Fukumura et al. 2018). To test our vascular normalization hypothesis, we studied

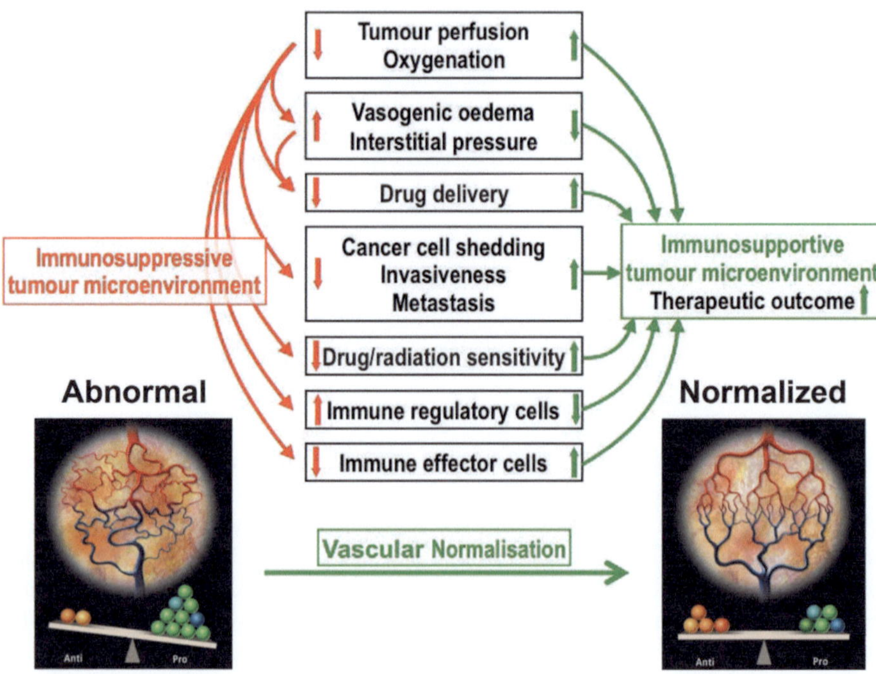

Fig. 13.3 Vascular normalization reprograms TME from immunosuppressive to immunosupportive. The structural and functional abnormalities in tumor vessels lead to insufficient blood flow/perfusion, resulting in impaired delivery of oxygen/nutrients, drugs, and immune cells. Hypoxic condition desensitizes tumors to cytotoxic therapies such as chemotherapy and radiation therapy. Furthermore, these conditions in the TME increase immunosuppressive M2 phenotype and Tregs and decrease immune effector T cells, leading to compromised antitumor immune response. Restoration of pro−/antiangiogenic factor balance can induce vascular normalization, decrease many of these adverse effects, and enhance the delivery and efficacy of chemo-, radio- and immunotherapies. (Adapted from Fukumura et al. 2018)

the effects of agents targeting VEGF signaling in both preclinical models and patients with antiangiogenic therapies. VEGF signaling has been the major target of antiangiogenic therapies in patients. A number of preclinical and clinical studies from our own institution and others have provided evidence in support of the normalization hypothesis for a number of malignancies (Jain 2014; Schmittnaegel et al. 2017; Batchelor et al. 2007; Chen and Mellman 2013). We first asked a cardinal question—the effect of vascular normalization on the concomitantly administered cytotoxic therapies such as chemotherapy and radiation therapy (Willett et al. 2004). We then found that vascular normalization also converts TME phenotype from immunosuppressive to immunosupportive, thereby improving the outcomes of vaccine-based anticancer immunotherapy (Huang et al. 2012). Subsequently, many other laboratories provided further evidence of vascular normalization by multiple anti-VEGF agents in various preclinical models, as well as in different immunotherapies (Jain 2014; Chen and Mellman 2013). A major challenge with anti-VEGF therapy-induced normalization is its transient nature. Duration (window) of normalization is relatively short (days to weeks) (Winkler et al. 2004). It is dependent on the tumor and dose of the anti-VEGF agent used (Huang et al. 2012). In our attempts to extend the normalization window in preclinical cancer models, we discovered that angiopoietin/TIE-2 signaling mediates the recruitment of pericytes to the tumor vessels (Winkler et al. 2004), and that ectopic expression of Ang-2 (negative ligand of TIE2) can abrogate the benefit of anti-VEGF therapy (Chae et al. 2010). Moreover, concomitant blockade of Ang-2 and VEGF extends both the window of normalization and the survival benefit, compared with the inhibition of either pathway alone, in part, by reprogramming the immunosuppressive TME (Kloepper et al. 2016; Peterson et al. 2016). Interestingly, in glioblastoma (GBM) patients, circulating Ang-2 levels rebound following significant initial decrease by anti-VEGF therapies (Batchelor et al. 2010). Furthermore, Ang-2 levels have been reported to be elevated or increased in patients with melanoma who have an unfavorable response to. Importantly, Ang-2 levels are high in melanoma patients with poor response to ICB (Wu et al. 2017) and are predictive marker of ICB response (Wu et al. 2017).

13.2.1 Optimal Dose of Anti-VEGF Therapy for Vessel Normalization

Anti-VEGF therapy induces normalization of tumor vasculature via pruning of immature blood vessels (Winkler et al. 2004; Tong et al. 2004). The normalization process is transient and reversible, in a tumor-dependent and dose-dependent manner, and lasts for around a week in mice and possibly for a few months in humans (Jain 2014; Batchelor et al. 2007; Winkler et al. 2004; Batchelor et al. 2010; Akbari et al. 2022). The window of vessel normalization ends when an excessive number of vessels are pruned or when tumors switch to the use of alternative angiogenic pathways or alternative mechanisms of blood vessel recruitment, such as vessel co-option (Winkler et al. 2004; Griveau et al. 2018). Notably, high doses of antiangiogenic agents can result in a short normalization window by causing excessive amounts of vessel pruning, thus exacerbating hypoxia and acidosis in the TME (Jain 2014). High doses of anti-VEGF agents can also lead to increased deposition of extracellular matrix that, together with hypoxia promotes the infiltration of immunosuppressive and/or protumor immune cells, such as monocytic and granulocytic myeloid-derived suppressor cells (MDSCs) (Jung et al. 2017a, b; Rahbari et al. 2016). These results suggest that high-dose anti-VEGF therapy can limit the benefits of concomitantly administered therapies such as chemotherapy and induces immunosuppressive TME (Rahbari et al. 2016). In mouse models of CRC, high-dose anti-VEGF treatment can also result in the recruitment of immunosuppressive, nonclassical, Ly6C^{low} monocytes that also confer resistance to anti-VEGF therapy, thus setting up a vicious cycle (Jung et al. 2017a, b). Furthermore, treatment-resistant cancer stem cells reside in the hypoxic niche, and excessive antiangiogenic therapy can cause hypoxia and thereby support these cells (Conley et al. 2012). In contrast, low doses of antiangiogenic agents have the potential to induce prolonged vessel normalization, thereby reducing tumor hypoxia and acidosis (Jain 2014; Huang et al. 2012), and also reduce the number of stem cells, hence avoiding or delaying therapy resistance. In fact, the results of two retrospective studies have shown that low-dose bevacizumab in combination with chemoradiotherapy confers a greater survival benefit than high-dose bevacizumab in patients with glioblastoma (Levin et al. 2015, 2017; Lorgis et al. 2012; Lu-Emerson et al. 2015). Hence, to realize the full potential of vascular normalization, the dose and schedule of anti-VEGF agents might need to be tailored.

13.2.2 ANG-2 Induces Resistance to Vessel Normalization (with Anti-VEGF Therapy)

Similar to other anticancer drugs (such as molecularly targeted agents), tumors inevitably develop resistance to anti-VEGF agents, and one key mediator of resistance is hypoxia (Jain 2014). Several molecular and cellular mechanisms of escape from antiangiogenic therapies have been studied that includes activation of alternative angiogenic pathways, recruitment of proangiogenic myeloid cells, increased vessel maturation (increased pericyte coverage of tumor blood vessels, tumor cell invasion, and alternative modes of vessel recruitment (Jain 2014; Carmeliet and Jain 2011a; Bergers and Hanahan 2008). Among these, we identified the activation of the Ang-2/TIE-2 axis as an escape mechanism from anti-VEGF therapy. Ang-1 mediates vessel maturation in preclinical models (Winkler et al. 2004), whereas overexpression of Ang-2 compromises the benefit of anti-VEGF therapy by hampering vessel normalization (Chae et al. 2010). Ang-2 is coexpressed with VEGF in many human solid tumor. Circulating Ang-2 transiently decrease during the window of normalization and return to baseline levels when tumors progress after anti-VEGF therapy in cancer patients (Batchelor et al. 2010). Tumor-derived VEGF can increase EC expression of Ang-2 that promotes endothelial activation, tumor progression and lung metastasis (Minami et al. 2013). Finally, Ang-2 is a candidate for predictive and/or prognostic marker of therapeutic responsiveness to anticancer treatments, including immunotherapy using immune-checkpoint inhibitors.

13.2.3 Cotargeting of Ang-1 and Ang-2 Impairs Beneficial Effects of Ang-2 Targeting

Therapeutic agents targeting the Ang-2/TIE-2 axis led a number of preclinical models to tumor regression. While Ang-2 is an antagonistic TIE-2 ligand, Ang-1 is an agonistic ligand of TIE-2. Dual targeting of Ang-1 and Ang-2 may counteract each other. Indeed, the clinical trials of Ang-1/2 dual targeting agents such as trebananib showed disappointing outcomes. In a phase II trial involving patients with CRC, Trebananib combined with chemotherapeutics or sorafenib did not improve the median PFS or overall survival of CRCs, metastatic gastroesophageal cancers and metastatic RCCs in multiple Phase II studies (Peeters et al. 2013; Eatock et al. 2013; Rini et al. 2012). Interestingly, in a phase III trial, trebananib plus paclitaxel significantly improved survival of patients with ascites at baseline only suggesting possible vessel normalizing effects of trebananib (Monk et al. 2016). On the other hand, the failure of trebananib plus PEGylated liposomal doxorubicin in improving ovarian cancer treatments suggests vascular normalization of dual anti-Ang-1/Ang-2 blockade is insuffficient (Reardon et al. 2017). This dual blockade approach can provide tumor control in some preclinical models, although the degree of benefit and the underlying mechanism of action seem to be context-dependent (Bergers and Hanahan 2008; Liu et al. 2016; Holopainen et al. 2012; De Palma et al. 2005).

13.2.4 Cotargeting VEGF and Ang-2 Improves Vascular Normalization and Immune TME

Because of the complementary roles of VEGF and Ang-2 in tumor angiogenesis, dual VEGF/Ang-2 blockade prolonged survival compared with anti-VEGF or anti-Ang-2 monotherapy in mouse models of glioblastoma (Kloepper et al. 2016; Peterson et al. 2016), melanoma (Schmittnaegel et al. 2017), pancreatic neuroendocrine tumors (PNETs) (Schmittnaegel et al. 2017), and metastatic (Schmittnaegel et al. 2017; Wu et al. 2016), or early stage (resected) breast cancer as indicated by reduced metastatic dissemination and micrometastatic growth (Wu et al. 2016; Kienast et al. 2013). These findings suggest that the benefit of dual antiangiogenic therapy is dependent on the tumor type (or model used), the stage of disease progression, and probably the specific characteristics of the TME. Treatment with an anti-VEGF/Ang-2 bispecific antibody (A2V) resulted in the increased extravasation and perivascular accumulation of both CD4+ and CD8+ T cells in metastatic breast cancer, PNETs, or melanoma (Schmittnaegel et al. 2017). T-cell accumulation was confined to tumor areas with normalized vessels (Schmittnaegel et al. 2017). Increased production of IFNγ and greater expression of the activation marker CD69 were observed on cytotoxic T lymphocytes (CTLs) in comparison to monotherapy with neutralizing antibodies to either Ang-2 or VEGF (Schmittnaegel et al. 2017). Reprogramming of protumor, M2-like tumor associated macrophages (TAMs) to an anticancer M1 phenotype is a major contributor to the therapeutic efficacy of dual antiangiogenic therapy targeting VEGF and Ang-2, independent of the tumor model used (Schmittnaegel et al. 2017; Kloepper et al. 2016; Peterson et al. 2016). In GBMs, macrophages/microglia constitute up to 40% of the tumor mass (Charles et al. 2011). Macrophage depletion with an anti-CSF-1 antibody abrogated the benefits of dual VEGF/Ang-2 in GBMs (Peterson et al. 2016). These approaches could induce an extended window of vessel normalization and reprogramming of immune TME toward an immunostimulatory phenotype. We also hypothesize that the judicious use of low doses of anti-VEGF agents (Levin et al. 2017; Kabbinavar et al. 2003; Willett et al. 2005) might further improve dual VEGF/Ang-2 targeting. Nevertheless, careful dose selection for dual antiangiogenic treatments in the clinic would be crucial to prevent excessive vessel pruning. However, based on previous studies (Schmittnaegel et al. 2017; Kloepper et al. 2016; Peterson et al. 2016), dual Ang-2/VEGF blockade alone may not induce durable clinical responses. Specifically, dual antiangiogenic therapy with the anti-VEGF/Ang-2 antibody A2V led to upregulation of PD-L1 on ECs and tumor cells in multiple mouse tumor models, resulting in immunosuppression (Schmittnaegel et al. 2017). Notwithstanding, this resistance mechanism could be counteracted by concomitant treatment with an anti-PD-L1 antibody, resulting in improved tumor control (Schmittnaegel et al. 2017). The benefits of dual VEGF/Ang-2 inhibition may also be tumor dependent. It seems that dual VEGF/Ang-2 blockade might be particularly effective against tumors of the CNS. Glioblastomas are particularly hypoxic, which poses a considerable challenge to therapy (Jain 2014; Lu-Emerson et al. 2015), and normalizing brain tumor blood vessels for an extended period of time might help overcome this obstacle. In addition, the high abundance of macrophages in glioblastomas (Charles et al. 2011) could potentially render these tumors particularly sensitive to dual VEGF/Ang-2 inhibition, owing to reprogramming of macrophages toward an anticancer phenotype, as observed in preclinical models (Kloepper et al. 2016; Peterson et al. 2016). Taken together, these data indicate that the successful translation of dual anti-VEGF/Ang-2 therapy will require careful mechanistic studies and biomarker-driven clinical trial designs.

Alternatively, targeting of nonredundant vessel-normalization pathways by inhibiting the renin–angiotensin system is another way to potentially further extend the window of normalization (Chauhan et al. 2013; Pinter and Jain 2017) as discussed in Sect. 13.5. In line with this hypothesis, the results of a retrospective study demonstrated that patients with recurrent glioblastoma treated with bevacizumab who used angiotensin receptor blockers (ARBs) had prolonged overall survival compared with those who did not use ARBs (Levin et al. 2017).

13.3 Vessel Normalization Can Improve Immunotherapy

Immunotherapy, especially immune checkpoint blockade (ICB), has revolutionized cancer treatment and brought hope for a cure to cancer patients with previously incurable diseases. However, the majority of these patients still do not benefit from this approach. The success of ICB immunotherapy requires the recruitment, expansion, and effective anticancer activity of CTLs in the TME. The accumulation of T cells is a multistep process. First, T cells must interact with tumor vascular endothelial cells (ECs) via cell adhesion molecules (CAMs), such as E-selectin, ALCAM-1, ICAM-1, and VCAM-1, expressed on the luminal surface of ECs lining blood vessels. These CAMs slow down circulating CTLs, making them roll and adhere to the inner vessel wall and then extravasate across the endothelial barriers and vessel wall. Dual anti-VEGF/Ang-2 therapy with A2V has shown to upregulate CAM expression during the window of vascular normalization and can thereby facilitate the accumulation of anticancer CTLs in preclinical models of various types of tumors (Schmittnaegel et al. 2017). Low doses of an anti-VEGFR2 antibody have also shown enhanced T-cell infiltration in tumors and anticancer efficacy—during the window of normalization (Huang et al. 2012; Shrimali et al. 2010). It should be noted that upregulation of PD-L1 has been demonstrated on both ECs and tumor cells following treatment with anti-VEGFR2 therapy (Allen et al. 2017). This provides a strong rationale supporting the combination of vascular-normalizing agents with interventions that alleviate functional blockade of T cells, such as the combination of ICBs and anti-Ang-2/VEGF antibody A2V in various cancer models (Bergers and Hanahan 2008). Interestingly, ICB of CTLA-4 and PD-1 can normalize blood vessels in multiple models of transplanted breast and other tumors (Tian et al. 2017). The vascular normalization effect observed in these studies was attributed to increased $CD4^+$ type-1 T helper (T_H1) cell accumulation and anticancer activity (Tian et al. 2017). Hence, the effects of T_H1 cells on vascular normalization in the context of ICB need to be further validated in other preclinical models. Whether ICB-mediated vessel normalization is induced across a range of tumor types remains to be determined. If found to be broadly applicable, this mechanism provides another rationale for the pursuit of combinatorial approaches using antiangiogenic agents and immune-checkpoint inhibitors, as these therapies target nonredundant pathways for vessel normalization and abrogation of immunosuppression in the TME.

13.4 Low-Dose Anti-VEGF Therapy Improves CAR-T-Cell Delivery and Efficacy

Chimeric antigen receptor (CAR)-T cell therapy has transformed the treatment of multiple types of hematological cancers. However, its efficacy against solid tumors has been limited. (Rafiq et al. 2020; Young et al. 2022) This may be because, unlike hematological cancers, delivery of CAR-T cells is significant hurdle for solid tumor treatment. Immunosuppressive TME also hinders the efficacy of antitumor activity of CAR-T cells which manage to infiltrate into solid tumors. As discussed above, antiangiogenic therapies such as anti-VEGF therapies can normalize tumor vessels. Vascular normalization can improve the delivery of CD8+ T cells and the efficacy of immunotherapy in solid tumors. In fact, the US FDA has approved multiple different combinations of anti-VEGF agents and ICB for solid tumors including liver, kidney, lung and endometrial cancers. We have recently shown that anti-VEGF therapy can improve the delivery and efficacy of CAR-T cells in orthotopic GBM models in immunocompetent mice (Dong et al. 2023). To this end, we developed EGFRvIII expressing murine GBM cell lines. EGFRvIII is one of the most common neoantigens in human GBM and EGFRvIII targeting CAR T cells have been developed and tested in the clinic. To determine if anti-VEGF antibody (B20) treatment-induced vascular normalization increased EGFRvIII-CAR-T-cell infiltration in the GBM TME, we performed intravital microscopy of murine orthotopic GBMs with or without B20 treatment. We found that B20 treatment improved CAR-T-cell infiltration and distribution throughout the GBM TME, delayed tumor growth, and prolonged survival of GBM-bearing mice compared to EGFRvIII-CAR-T-cell therapy alone (Fig. 13.4). Flow cytometry analysis confirmed that anti-VEGF treatment increases accumulation of EGFRvIII-CAR-T cells as well as their

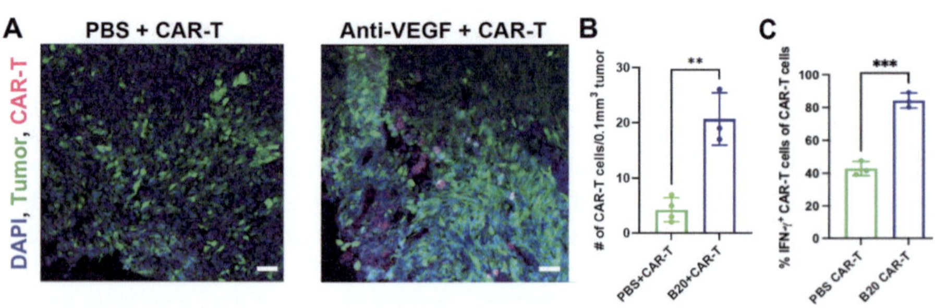

Fig. 13.4 Anti-VEGF treatment improves the infiltration of EGFRvIII-CAR-T cells and their function. (**A**) Intratumoral distribution of EGFRvIII-CAR-T cells, fluorescently labeled with CellTracker™ Red CMTPX Dye (shown in magenta), inside of GFP-GSC005 GBM tumors (green) imaged by multiphoton microscopy. Images were taken 24 hours after the injection of CAR-T cells (shown in pink. Scale bars = 50 μm. (**B**) Measurement of CAR-T cell number in GFP-GSC005 tumors treated with PBS or anti-VEGF antibody, B20 (2.5 mg/kg). (**C**) Percentage of IFNγ+ CAR T cells inside of GSC005 GBM tumors measured by flow cytometry. Error bars show ± SEM. Statistical analysis was performed using Student's *t* test. **$p < 0.01$ and ***p <0.001. (Adapted from Dong et al. 2023)

proliferation and cytotoxic activity in GBMs. Furthermore, the combined therapy facilitated recruitment of antitumor immune cells such as dendritic cells and CD8+ T cells. With regard to long-term therapeutic efficacy in murine GBM tumor models, the combination of B20 with EGFRvIII-CAR-T cells significantly improved the survival benefit of GBM-bearing mice while monotherapy with either B20 or CAR-T cells showed only minor improvement in the animal survival. These findings provide compelling data and a rationale for clinical evaluation of anti-VEGF agents with CAR T cells for cancer patients which is currently being tested in clinical trials (Akbari et al. 2022).

13.5 Role of Obesity in Cancers: Pancreatic Cancer

Obesity is a pandemic worldwide— about 70% of the adult population in the United States are overweight or obese (Ogden et al. 2014)—and the majority of cancer patients have excess weight at diagnosis (Li et al. 2009). Excess body weight worsens the outcome of cancer patients (Li et al. 2009; Yuan et al. 2013; Calle et al. 2003; McWilliams et al. 2010; Smits and van Geenen 2011). Thus, understanding why obesity is associated with a worse prognosis would lead to novel treatments and enhancement of current therapies.

Pancreatic cancer is the fourth-leading cause of cancer-associated death worldwide, with an overall five-year survival rate of 7% (American Cancer Society: Cancer Facts and Figures 2015). The risk of pancreatic cancer is about 50% greater for individuals with obesity, particularly those with increased abdominal adiposity (Genkinger et al. 2011). We studied clinically relevant orthotopic and genetically engineered mouse models (GEMMs) of PDAC.

13.5.1 Obesity Promotes Pancreatic Tumor Progression

Obesity promotes tumor initiation and progression across various tumor models (Incio et al. 2016a; Zyromski et al. 2009; Bracci 2012). We studied the effects of obesity on PDACs using a high-fat diet (HFD)-induced obesity (DIO) model as well as genetic models of obesity such as *ob/ob* and *db/db*. We found that obese animals tended to develop tumors earlier than lean mice. Obesity accelerated the growth of implanted tumors (Incio et al. 2016b). Obese mice presented with increased tumor weight, metastatic dissemination, and local invasion. These data confirmed the tumor-promoting effect of obesity in PDACs. PDAC is a highly desmoplastic cancer characterized by activated pancreatic stellate cells (PSCs) and an excessive accumulation of extracellular matrix (ECM) (Dvorak 1986; Whatcott et al. 2015). We found that obesity further promotes desmoplasia in PDACs (Incio et al. 2016b). In addition to molecular interactions between cancer cells, stromal cells (e.g., PSCs), and the ECM, these components produce physical forces (solid stress—pressure from solid tissue components). Solid stress compresses tumor blood vessels, causing poor and heterogeneous tumor perfusion (Stylianopoulos et al. 2012; Chauhan et al. 2014). These mechanically induced alterations in vascular perfusion create a formidable barrier to the delivery and efficacy of cytotoxic therapies such as chemotherapeutics and radiation, leading to poorer treatment outcomes (Chauhan et al. 2013; Olive et al. 2009; Alvarez et al. 2013; Hidalgo and Von Hoff 2012; Endrich et al. 1979). The hypoxia that results from dysfunctional tumor vessels and decreased perfusion also causes pro-tumor immune cell recruitment, cytokine production, inflammation, and, ultimately, fibrosis (Costa et al. 2007; Hosogai et al. 2007; Incio and Soares 2013).

13.5.2 Obesity Induces a Steatotic and Fibrotic Microenvironment in PDACs

We next determined whether the abundance of fibrotic adipocyte-rich areas in tumors from obese mice led to an overall increase in tumor fibrosis. In obese mice, hypertrophic adipocytes and associated fibrosis are observed in the visceral adipose tissue. We also found more and larger adipocytes in the tumor microenvironment. Furthermore, Masson's trichrome staining revealed abundant fibrosis in tumor areas enriched in adipocytes or alongside adjacent visceral adipose tissues. Using second harmonic generation (SHG) imaging, we found that obesity significantly increased the SHG signal, which represents fibrillar collagen, in orthotopically grown PDACs. Consistent with the increase in the fibrotic reaction in adipocyte-rich regions in tumors, the expression of α smooth muscle actin (αSMA) also increased in these regions in tumors in obese mice (Fig. 13.6a). Taken together, we found that tumors in the obese setting are enriched in enlarged adipocytes, activated PSCs, and collagen I (Incio et al. 2016b).

13.5.3 Obesity-Aggravated Desmoplasia Is Associated with Impaired Vascular Perfusion, Intratumoral Drug Delivery, and Treatment Efficacy

We have reported that desmoplasia reduces the delivery and efficacy of chemotherapy in PDAC by decreasing vessel perfusion (Chauhan et al. 2013). Here, we tested if obesity-aggravated desmoplasia reduced the efficacy of chemotherapy. Indeed, we found that in obese animals, the percentage of perfused vessels per total tumor area (Fig. 13.5a, b). To determine if reduced perfusion impairs the delivery of chemotherapeutic agents, we measured the uptake of 5-fluorouracil (5-FU), an approved cytotoxic agent used in the treatment of PDAC (Neoptolemos et al. 2012). Obesity significantly decreased both the uptake and efficacy of 5-FU in orthotopic PDAC

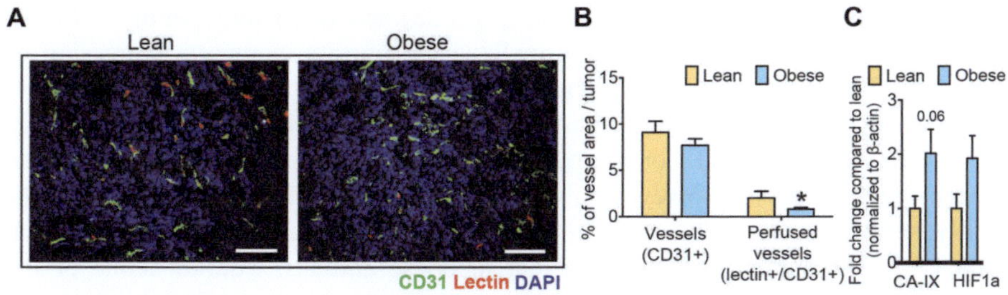

Fig. 13.5 Obesity-aggravated desmoplasia is associated with impaired perfusion, oxygenation, drug delivery and resistance to chemotherapy. Effect of diet-induced obesity (DIO) on tumor perfusion. (**A**) Representative immunohistochemistry images of CD31(+) vessels and lectin in PAN02 tumors. Scale bars: 200 μm. (**B**) Quantification of total (CD31+) and lectin-positive (CD31/lectin+) vessel area in PAN02 tumors. Obese mice presented with decreased perfusion ($n = 3$–12 tumors/group). (**C**) Effect of DIO on tumor hypoxia. Obesity decreased delivery of the chemotherapeutic agent. Consequentry, 5-FU was less effective in preventing tumor growth in DIO animals than in lean. Data are shown as mean ± SEM. *P* values were determined by the Student *t* test unless otherwise stated. *, $p < 0.05$. (Adapted from Incio et al. 2016)

models. Consistent with this finding, we observed an increase in hypoxia markers in tumors from obese mice (Fig. 13.5c). Collectively, these data show that, in addition to directly promoting PDAC growth and metastasis, obesity reduces vascular perfusion, drug delivery, and the efficacy of 5-FU chemotherapy (Incio et al. 2016b).

13.5.4 The Inhibition of Angiotensin II Type-1 Receptor (AT1) Signaling Reduces Obesity-Induced Desmoplasia, Tumor Growth, and Increases the Response to Chemotherapy

Adipocytes abundantly express the AT1 (Biogps 2015; Sun et al. 2013), a major profibrotic pathway that becomes activated in a proinflammatory environment (Biogps 2015; Sun et al. 2013). We have shown that tumor desmoplasia results in part from the activation of PSCs through AT1 signaling (Chauhan et al. 2013). We found that obesity activated AT1 downstream pathways such as p38, ERK, AKT, pS6, and 4EBP1 (Chauhan et al. 2013; Yang et al. 2009; Hama et al. 2004, 2006). In addition, several target genes of the AT1 pathway, including *Col1a2, Tgfβ1, and matrix metalloproteinases* genes, were upregulated in tumors in obesity. To test if the inhibition of AT1 signaling could reverse desmoplasia in obese mice, we used the angiotensin II receptor 1 blocker (ARB) losartan and mice deficient in AT1 (*Agtr1a*$^{-/-}$) to inhibit AT1 signaling. We found that losartan treatment significantly decreased the SHG signal for fibrillar collagen and reduced collagen I immunofluorescence. Losartan normalized the obesity-induced expression of desmoplasia-related markers (αSMA, decorin, Smad-2, p-P38) and ECM remodeling (MMP9). We confirmed that AT1 deletion normalized the obesity-increased expression of vimentin. We then determined whether AT1 blockade could improve the response to chemotherapy, particularly in obese mice with desmoplastic tumors. In lean mice, the genetic deletion of AT1 or the losartan treatment had no effect on tumor size in mouse PDAC models. However, in obese mice, AT1 blockade resulted in significantly smaller tumors. In conclusion, we found that obesity is associated with increased AT1 signaling in tumors; that in obese animals, losartan alone or the genetic blockade of AT1 has direct antitumor effects; and that AT1 inhibition can also enhance the efficacy of 5-FU chemotherapy.

13.5.5 Tumor-Associated Neutrophils Mediate Obesity-Induced PSC Activation and Tumor Progression

The fibrotic phenotype in obesity is largely the consequence of a persistent proinflammatory state, which is characterized by the production of cytokines by hypoxic and dysfunctional hypertrophic adipocytes and the recruitment of immune cells (Hosogai et al. 2007; Incio and Soares 2013; Kanasaki and Koya 2011). In addition, we have previously described increased inflammation and immune cell recruitment in PDACs implanted in obese mice (Incio et al. 2016a). Hence, we next determined whether adipocyte-associated inflammation was responsible for the increased tumor desmoplasia and accelerated tumor growth observed in obese animals. Flow cytometric analysis revealed that obesity promotes the infiltration of myeloid cells, especially Ly6G(+) tumor-associated neutrophils (TANs) in orthotopic PDAC models in obese mice. TAN depletion (TAN-D) with an anti-Ly6G antibody reverted the increased tumor growth in obese mice. This effect only occurred when TAN depletion was initiated on day 1 but not on day 7 of the experiment, indicating the relative importance of TANs in tumor progression at an early stage. We next determined whether TANs could affect desmoplasia. TANs preferentially accumulate near activated PSCs suggesting a potential paracrine crosstalk between TANs and PSCs. TAN depletion in PDACs in obese animals decreased the number and activation of PSCs. We also observed reduced expression of AT1, collagen I, and MMP9 by TAN depletion. In addition, TAN depletion significantly increased the vascular surface area and the number of perfused vessels in the PDACs. Our data suggests that obesity increases TAN recruitment in PDAC, which mediates the activation of PSCs and tumor growth.

13.5.6 IL-1ß Mediates Obesity-Induced Tumor-Associated Neutrophil Infiltration and PSC Activation in PDACs

We next determined which inflammatory cytokines mediate obesity-induced fibrotic processes and TAN infiltration in the tumor microenvironment. We found that in PDACs, obesity significantly increased the expression of IL1-ß by five fold and somewhat IL-6, TNF-a, IL-12, or CXCL1. Not surprisingly, IL-1ß was abundantly expressed by adipocytes in adipocyte-rich areas where activated PSCs predominate (Fig. 13.6a). Antibody neutralization of IL-1ß significantly decreased TAN infiltration in obese PDACs. In addition, the decreased TAN recruitment was associated with an increase in CD8(+) cytotoxic T-cell population (decreased in obese mice) and a decrease in regulatory T cells (Tregs) (Fig. 13.6b). Similar to TAN depletion effects, IL-1ß inhibition in obese animals significantly reduced the PDAC growth as well as the expression levels of αSMA and vimentin (the EMT marker). In addition to adipocytes, IL-1ß was also expressed in about 70% of TANs themselves, and TAN depletion reduced tumor IL-1ß levels suggesting an autocrine mechanism Furthermore, TAN depletion significantly reduced the levels of CXCL1, indicating

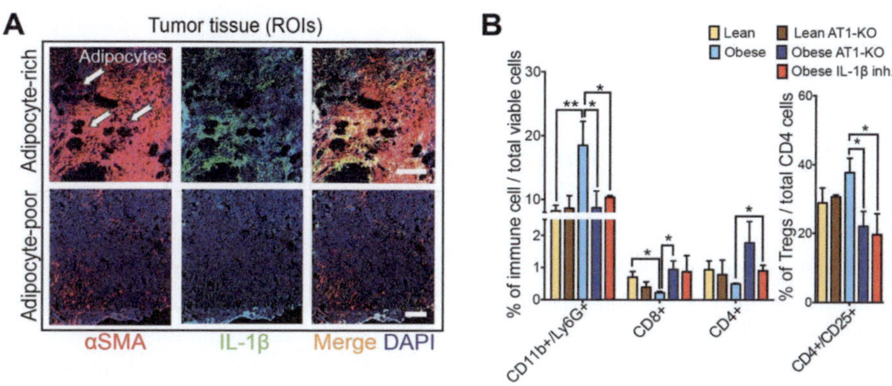

Fig. 13.6 The adipose microenvironment promotes TAN infiltration and fibrosis via IL-1ß. Effect of obesity on cytokine expression in PAN02 tumors. Multiplex protein revealed that PAN02 tumors from obese mice had increased expression of IL-1ß. (**A**) Representative images of αSMA and IL-1ß expression in adipocyte-rich and poor regions. Scale bars: 200 μm. IL-1ß was abundantly expressed in adipocytes and PSCs in the adipocyte-rich areas where PSCs predominate. (**B**) Effect of IL-1ß and AT1 blockade on immune cell profile. Quantification normalized by total viable cells or total CD4 cells ($n = 3$–6 tumors/group). Obesity promoted an increase in TANs, a decrease in CD8(+) lymphocytes, and a strong tendency for increased Tregs. An anti-IL1ß-neutralizing antibody or genetic AT1 blockade decreased CD45(+)CD11b(+)Ly6G(+) TAN infiltration while recovering CD4+ and CD8+ T cells and decreasing Tregs in obese mice (one-way ANOVA, $n = 3$–6 tumors/group). Data are shown as mean ± SEM. *P* values were determined by the Student *t* test unless otherwise stated. *, $p < 0.05$; **, $p < 0.01$. (Adapted from Incio et al. 2016)

that this cytokine may also play a role in obesity-induced inflammation. In conclusion, our findings suggest IL-1ß is involved in obesity-induced TAN recruitment, immunosuppressive microenvironment, and PDAC progression.

13.5.7 AT1 Blockade Can Reciprocally Inhibit Inflammation and Immune Cell Infiltration in Obesity

We observed that αSMA-positive PSCs also abundantly (~70% of PSCs) expressed IL-1ß. Since IL-1ß recruits TANs, which localize in close proximity to PSCs, we examined whether targeting PSCs could also interfere with IL-1ß production and TAN recruitment, particularly in obese mice. Indeed, in obese mice, prevention of PSC activation by losartan-inhibition of AT1 signaling decreased IL-1ß levels in PDACs. In addition, the genetic deletion of AT1 decreased TAN recruitment in PDACs in obese but not lean mice. Consequently, recapitulating IL-1ß inhibition, AT1 blockade increased CD8(+) T cells and reduced Tregs in tumors from obese but not lean mice. The effects of AT1 blockade on the expression of IL-1ß and immune cell recruitment are consistent with the prevention of obesity-induced tumor growth and indicate that AT1 signaling mediates obesity-augmented tumor progression via modulation of the immune microenvironment.

13.5.8 Obese PDAC Patients Have a Worse Prognosis: AT1 Blockade May Counteract It

Importantly, cancer lesions in obese mice and patients have an increased adipocyte content (Zyromski et al. 2009; Hori et al. 2014). Of clinical relevance, the interaction of cancer cells with adipocytes—both in the form of accumulation of fat in the pancreas and at the invasive tumor front into the local adipose tissue—is associated with worse outcomes in PDAC patients (Smits and van Geenen 2011; Hefetz-Sela and Scherer 2013). We analyzed data collected from patients ($n = 715$) admitted to MGH with a diagnosis of PDAC. The average BMI of patients was 26.419. Overweight and obesity (BMI 25–35) is associated with worse survival compared with normal weight (BMI <25) patients (33% decrease in median survival days). Obesity associated with more advanced disease at baseline and dramatically decreased survival. The survival advantage of chemotherapy or tumor resection was effectively reduced by obesity. Finally, we found in patients that were treated with ARBs or angiotensin-converting enzyme inhibitors (ACEis), obesity-associated worse survival did not occur. In conclusion, our preclinical models successfully predicted an aggravated disease in obese patients, a reduced response to therapy, and an improvement with AT1 inhibition.

13.6 Role of Obesity in Cancers: Breast Cancer

13.6.1 Obesity Impedes Anti-VEGF Therapy in Breast Cancers (BCs)

Antivascular endothelial growth factor (VEGF) therapy has failed to improve survival in patients with breast cancer (BC). In metastatic breast cancer (BC) in particular, the FDA revoked the approval of the vascular endothelial growth factor (VEGF) antibody bevacizumab due to a lack of overall survival benefit in phase III trials (Lohmann and Chia 2012).

Proinflammatory and alternative angiogenic factors can sustain angiogenesis and tumor progression despite VEGF blockade (Jain 2013; Carmeliet and Jain 2011a; Jain et al. 2009; Duda et al. 2013; Carmeliet and Jain 2011b). Many of these same factors are also increased in obesity, *where the rate of adipose tissue expansion exceeds that of angiogenesis, resulting in inadequate tissue vascularization and hypoxia* (Incio and Soares 2013; Ye et al. 2007; Khan et al. 2013; Fukumura et al. 2016). In response to hypoxia, cytokines and chemokines (adipokines) are released, causing inflammation and the recruitment of additional proinflammatory and proangiogenesis mediators (Costa et al. 2007; Hosogai et al. 2007; Gati et al. 2014; Hursting 2012; Lijnen 2008). Obesity is associated with hypoxic adipose tissues, including those in the breast, resulting in increased production of some of the aforementioned factors. Obesity promotes resistance to anti-VEGF therapy in BCs via systemic as well as local overproduction of inflammatory and alternative angiogenic factors. Hence, we hypothesized that obesity could contribute to anti-VEGF therapy's lack of efficacy. Adipose tissue makes up a large proportion of BC, with cancer-associated adipocytes (CAAs) being the predominant cell population in the stromal compartment of BCs (Wang et al. 2012).

13.6.2 Diet-Induced Obesity Attenuates the Effect of Anti-VEGF Therapy on BC Progression in Mouse Models

In obesity-accelerated tumor growth in the early stages of tumor development, tumors in obese mice become 50% larger in size compared to those in lean mice (~150 vs. 100 mm^3) (Incio et al. 2018). Treatment with antimouse VEGF antibody (B20–4.1.1) was more effective at inhibiting tumor growth in lean mice as compared to obese mice (Fig. 13.7c). Nine days of anti-VEGF treatment reduced the murine BC tumor volumes in lean mice by 50–60%. Whereas in obese mice, it was only 10–25% (Fig. 13.7c). To determine whether the reduced sensitivity of tumors in obesity to anti-VEGF treatment was due to diet or body weight, we switched some low-fat diet mice to a high-fat diet. There is no difference in tumor growth in low-fat to low-fat diet mice as compared to low-fat to high-fat diet mice, indicating that body

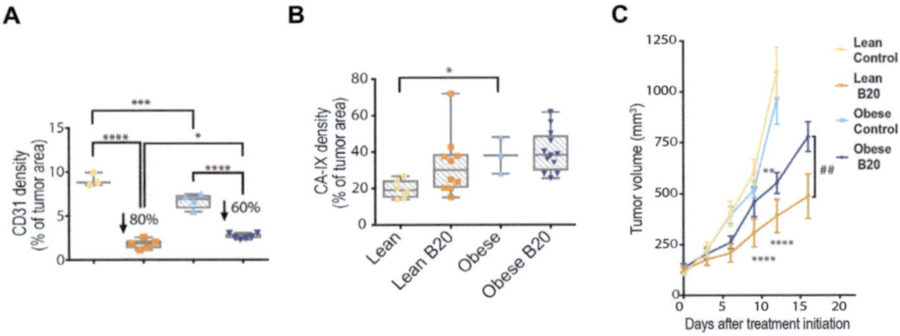

Fig. 13.7 Anti-VEGF therapy is less effective in reducing tumor vessel density and growth in obese mice. (**A**) Quantification of CD31+ expression (% of viable tumor area). (**B**) Quantification of hypoxia (% of CA-IX expression over DAPI viable area) in whole E0771 tumors from lean and obese animals treated with anti-VEGF antibody (B20) or control IgG. Significant differences in A-B using two-way ANOVA with post hoc test for multiple comparisons are indicated, $*p < 0.05$, $***p < 0.001$, $****p < 0.0001$. Data in A–B are shown as individual values plus box plots with min, max, and median values. (**C**) Growth curves of E0771 tumors. Tumors grown in obese versus lean mice were treated with B20 or control IgG (E0771: $n = 3$ animals in obese control group, 6 animals for other groups). Significant differences using two-way ANOVA with post hoc test for multiple comparisons are indicated. $**p < 0.01$, $****p < 0.0001$ control vs. B20 in lean or obese settings; $\#\# p < 0.01$, lean B20 vs. obese B20. Data are shown as mean ± SEM

weight, rather than diet, accounted for the reduced sensitivity of tumors in obesity to anti-VEGF treatment. Taken together, these data indicate that obesity promotes BC growth, and anti-VEGF therapy is less effective in inhibiting angiogenesis and BC progression in obese mice.

13.6.3 The Adipocyte-Rich and Hypoxic Tumor Microenvironment in Obese Mice Is Associated with Reduced Response to Anti-VEGF Therapy

To determine why anti-VEGF antibody (B20) failed to control tumors from obese animals to the same extent as in lean animals, we first assessed the effect of B20 on tumor vasculature. Similar to patients with obesity, orthotopic BCs from obese mice presented with lower vessel density (Fig. 13.7a) and higher expression of hypoxia markers, namely CA-IX and glucose transporter 1 (GLUT-1) were highly expressed in obese tumors as compared to tumors from lean mice and retained their expression even in the presence of anti-VEGF therapy (Fig. 13.7b). Conversely, in lean mice, the expression of the active forms of these mediators was relatively low and appeared to decrease after anti-VEGF therapy. These findings may account for the increased tumor burden in B20-treated obese mice compared to B20-treated lean mice (Fig. 13.7c). Furthermore, the hypovascularity and hypoxia in tumors from obese animals were associated with increased presence of CAAs. We further confirmed the presence of CAAs by detecting various adipocyte markers such as perilipin, Oil Red O staining, caveolin, and fatty acid synthase (FAS) (Mottagui-Tabar et al. 2003). Not surprisingly, in obese tumors, we observed increases in both adipocyte size and number. Moreover, adipocyte-rich tumor areas had lower vessel density and were hypoxic. The expression of proliferative markers was particularly pronounced in tumor areas abundant in hypertrophic CAAs (Dirat et al. 2011). In addition, when tumor-bearing animals were treated with B20, tumor tissue in adipocyte-rich regions remained viable (nonnecrotic), whereas extensive necrosis was observed in adipocyte-poor areas, suggesting a protective effect of adipose stroma in tumors.

In subsequent studies, we found that BC patients with obesity harbored increased systemic concentrations of interleukin 6 (IL-6) and/or fibroblast growth factor 2 (FGF-2), and their tumor vasculature was less sensitive to anti-VEGF treatment (Incio et al. 2018). Mouse models revealed that obesity impairs the effects of anti-VEGF therapy on angiogenesis, tumor growth, and metastasis. In one murine BC model, obesity was associated with increased IL-6 production from adipocytes and myeloid cells within tumors. IL-6 blockade abrogated the obesity-induced resistance to anti-VEGF therapy in primary and metastatic sites by directly impacting tumor cell proliferation, normalizing tumor vasculature, alleviating hypoxia, and reducing immunosuppression. Similarly, in another mouse model, obesity was associated with increased FGF-2. Normalization of FGF-2 signaling by metformin or specific FGFR inhibition decreased vessel density and restored tumor.

The lack of stratification by obesity status despite the high prevalence of obesity among cancer patients, including BCs, may explain the lack of overall survival benefits with anti-VEGF therapy in BCs and other treatments. Our work reveals strategies to improve the clinical outcome of BC. Stratification of patients based on obesity status as well as obesity-related markers and inflammatory or angiogenic factors will allow a more personalized approach to cancer treatment. It would be worthwhile to evaluate prospectively whether targeting specific inflammatory or angiogenic factors can overcome obesity-induced resistance to anti-VEGF and other treatments such as conventional therapies and novel molecularly targeted therapies.

13.7 Role of Exercise in Cancers

13.7.1 Exercise Normalizes Tumor Vasculature and TME

Exercise is the standard of care for obesiy and is generally recommended for cancer patients. Exercise does have beneficial effects on cancer, such as improvement of cancer prevention and overall survival of cancer patients (Holmes et al. 2005; Iyengar and Jones 2019; Williams 2014; Cannioto et al. 2020). Exercise is a

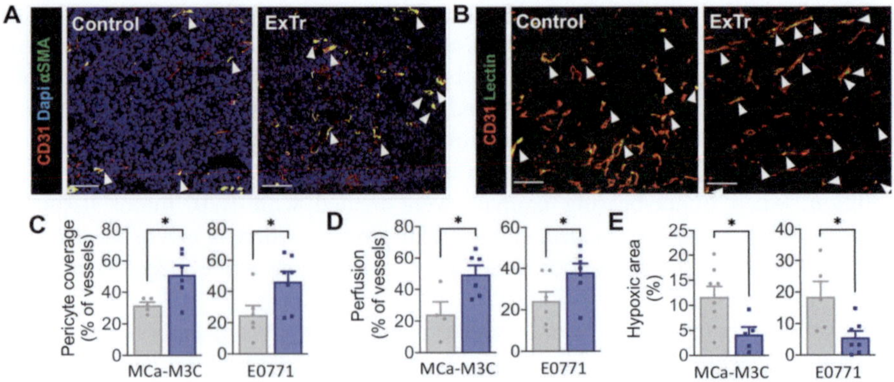

Fig. 13.8 Exercise training normalizes tumor blood vessels and reduces hypoxia. Immunofluorescence quantification was done in tumor sections to determine vessel normalization and hypoxia. (**A**, **B**) Representative images from E0771 model. Red indicates vessels (CD31); Green indicates pericytes (αSMA) or perfused vessels (lectin). White arrows highlight co-staining. Scale bar, 100 μm. (**C**, **D**) Corresponding quantitative data. **C**, Vessel normalization, characterized by the fraction of vessels with pericyte coverage ($CD31^{+}\alpha SMA^{+}$). **D**, Perfused vessels, expressed as a fraction of vessels ($CD31^{+}$) also positive for lectin. **E**, Hypoxic area (pimonidazole); $n = 4–8$ (MCa-M3C) and $n = 5–7$ (E0771). Statistical differences between groups done using unpaired Student's *t* test. Data presented as mean ± SEM. (Adapted from Gomes dos Santos et al. 2021)

physiological intervention and affects the tumor microenvironment (TME) (McCullough et al. 2014; Betof et al. 2015; Schadler et al. 2016; Pedersen et al. 2016; Miller et al. 2016; Rundqvist et al. 2020). To precisely study the role of exercise in tumors, we prescribed an individualized ExTr program so as to maintain the same relative exercise intensity. Indeed, we found that exercise can increase tumor perfusion, normalize tumor vasculature, decrease tumor growth, and potentiate concomitantly administered cancer treatments in preclinical breast cancer models (Gomes-Santos et al. 2021). It has been shown that tumor perfusion and oxygenation increase from the very first exercise session (McCullough et al. 2014; Miller et al. 2016; Garcia et al. 2016). Mean arterial pressure elevates during exercise, which increases tumor perfusion due to the lack of autonomic regulation (McCullough et al. 2014; Zlotecki et al. 1995). Importantly, exercise training (ExTr) promotes tumor vessel normalization (Fig. 13.8). ExTr improves the function of these vessels, as evidenced by an increased fraction of perfused vessels and reduced hypoxia, rendering tumors less aggressive (Incio et al. 2018) (Fig. 13.8). This may partly explain the association of ExTr with reduced tumor burden (Betof et al. 2015; Pedersen et al. 2016) and suggests ExTr as a form of vascular normalization therapy. More importantly, normalized tumor vasculature and microenvironment reprograms the immunosuppressive TME and enhances antitumor immunity (Huang et al. 2012). An improved vascular network favors adaptive antitumor immunity by promoting trafficking and re-localization of tumor-infiltrating lymphocytes (TILs), leading to improved outcomes of various immunotherapies, including immune checkpoint blockade (ICB) (Schmittnaegel et al. 2017; Huang et al. 2012; Tian et al. 2017; Ashcraft et al. 2019; Zheng et al. 2018; Chen et al. 2019).

13.7.2 Exercise Improves Antitumor Immunity

Limited T-cell recruitment and function in the TME correlates with poor prognosis in BC (Chung et al. 2017; Denkert et al. 2018). An abnormal TME, caused in part by hypoperfusion, results in the exclusion of $CD8^+$ T cells and reduces the impact of immunotherapies such as ICB. Vascular normalization induced by antiangiogenic therapies enhances trafficking of $CD8^+$ T cells in murine BC models (Schmittnaegel et al. 2017; Tian et al. 2017; Zheng et al. 2018; Chen et al. 2019). Among TILs, $CD8^+$ T cells predict overall treatment response and overall survival in patients with BC (Liu et al. 2017; Wang et al. 2017; Savas et al. 2018). Indeed, we found that ExTr increased the infiltration of $CD8^+$ T cells into the BC TME using flow cytometry (Fig. 13.9a). We profiled the immune TME in BCs during ExTr and found that ExTr consistently boosted $CD8^+$ T-cell infiltration (Fig. 13.9a). Consistent with flow cytometry analysis, we found that ExTr increased the intratumoral distribution of $CD8^+$ T cells within the TME (Fig. 13.9b). On the other hand, we did not see any changes in other cell populations potentially relevant to antitumor immunity (Fig. 13.9a), such as CD4+ T cells, Tregs, tumor-associated macrophages (TAMs), DCs, or natural killer cells (NK) cells. Furthermore, CD8+ T-cell depletion completely abrogated ExTr-induced tumor growth delay, indicating that $CD8^+$ T cells' tumor-infiltration is necessary for the antitumor effect of ExTr. Then, we determined the immune phenotype of $CD8^+$ T cells in BCs in response to ExTr. The proliferation marker Ki67 was elevated in ExTr groups (Fig. 13.9c). In addition, ExTr increased the fraction of effector $CD8^+$ T cells expressing granzyme B

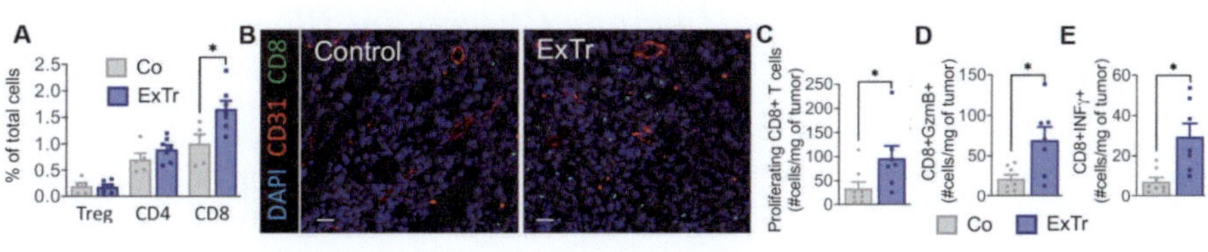

Fig. 13.9 Exercise training improves infiltration and effector function of $CD8^+$ T cells in breast tumors. (**A**) Flow cytometric analysis of tumor infiltrating $TCR^+CD4^+FOXP3^+$ (Tregs), TCR^+CD4^+, and TCR^+CD8^+ T cells, assessed as percentage of total cells in breast tumors. (**B**) Representative immunofluorescence image [red indicates vessels ($CD31^+$) and green indicates $CD8^+$ T cells] Scale bar, 25 μm. (C–E) Number and function of $CD8^+$ T-cells. (C) Proliferating ($Ki67^+$) $CD8^+$ T cells. (**D**) Granzyme B^+ and (E) $IFN\gamma^+$ $CD8^+$ T cells. For flow cytometry, 2 to 3 independent experiments were performed, and the results are representative of one single experiment ($n = 6$–10 mice per group). Statistical differences comparing groups by unpaired Student's *t* test. Data presented as mean ± SEM. (Adapted from Gomes dos Santos et al. 2021)

(Fig. 13.9d) and IFNγ (Fig. 13.9e). **Moderate intensity is optimal for exercise-induced antitumor effects:** Based on the recommendation for cardiovascular exercise. We initially used moderate-vigor exercise. To determine optimal exercise intensity for various anti-tumor effects, we compared tumor vasculature, immune profile and tumor growth in tumors with different intensities of exercise – Low-intensity (~30% of maximum), Moderate-intensity (60% of maximum), or High-intensity (~90% of maximum), or Control (Co, untrained). Indeed, moderate intensity exercise promotes most beneficial effects for tumor vascular normalization, anti-tumor immunity and inhibition of tumor progression (Gomes-Santos et al. 2024). Thereafter, we continue to study the effects of moderate intensity exercise.

13.7.3 CXCL9/11-CXCR3 Pathway Mediates the Antitumor Effect of Exercise

To dissect the mechanism of ExTr-induced $CD8^+$ T-cell infiltration, we conducted a proteome profile array of bulk tumors, and found ExTr significantly increased specific ligands for the CXCR3 receptor such as CXCL9 and CXCL11 chemokines. CXCR3 signaling mediates the recruitment of $CD8^+$ T cells into tumors (House et al. 2020; Tokunaga et al. 2018; Moon et al. 2018; Luo et al. 2019). We also observed that ExTr increased $CD8^+$ T cells expressing CXCR3 (more than 2-3-fold) in the breast cancers. We then tested functional role of CXCR3 in antitumor effect of ExTr using CXCR3 deficient mice. The number of $CD8^+$ T cells after ExTr in E0771 tumors grown in $Cxcr3^{-/-}$ mice was ten-fold lower than that observed in wild-type mice (Gomes-Santos et al. 2021). Finally, $Cxcr3^{-/-}$ mice lost benefit of ExTr on antitumor activity.

13.7.4 Exercise Improves Immune Checkpoint Blockade Therapy in BC

Limited T-cell recruitment and function in the TME correlates with poor prognosis in BCs (Chung et al. 2017; Denkert et al. 2018). Among TILs, $CD8^+$ T cells predict overall treatment response and overall survival in patients with BC (Liu et al. 2017; Wang et al. 2017; Savas et al. 2018). A varying degree of infiltrating $CD8^+$ T cells express multiple immune checkpoints (e.g., coexpression of programmed cell death protein 1 (PD-1) and cytotoxic T-lymphocyte-associated protein 4 (CTLA-4)), which cause these cells to become exhausted (Beavis et al. 2018). ICB mitigates $CD8^+$ T-cell exhaustion by targeting these receptors and is approved for a subset of triple-negative BC patients (Zhang and Plitas 2020).

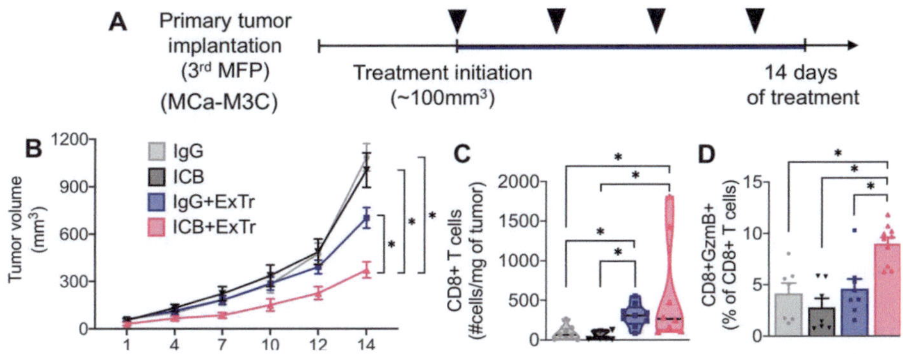

Fig. 13.10 Exercise training sensitizes breast cancer to immune checkpoint blockade. (**A**) Experimental design of the study. When MCa-M3C tumors reached ~100 mm^3, mice underwent ExTr (60% of Vmax ExTr, for 14 days) or kept sedentary and were treated with anti-PD-1 plus anti-CTLA-4 (ICB) or isotype control IgG (i.p., every 4 days; black arrowheads). (**B**) ICB effects on tumor volume. (**C–D**), TCR$^+$CD8$^+$ T-cell infiltration and function. (C) Absolute number of CD8$^+$ T cells. (**D**) Relative number of granzyme B$^+$CD8$^+$ T cells. Statistical significance was determined by Kruskal–Wallis (**B**) and two-way ANOVA (C–D). Data presented as mean ± SEM (**B**, **D**) or median ± interquartile range and distribution (violin plot, C). (Adapted from Gomes dos Santos et al. 2021)

An abnormal TME, caused in part by hypoperfusion, results in the exclusion of CD8$^+$ T cells and reduces the impact of immunotherapies such as ICB. We previously demonstrated that vascular normalization could increase immune cells' infiltration into tumors. Subsequently, multiple studies have confirmed these findings in several tumor models (Schmittnaegel et al. 2017; Huang et al. 2012; Tian et al. 2017; Zheng et al. 2018). To test this in the context of ExTr, we treated mice with either ICB (anti-PD-1 plus anti-CTLA-4) or ExTr alone or a combination of ICB and ExTr when tumors were established (Fig. 13.10a). We found no effect of ICB alone in the orthotopic BCs (Fig. 13.10b). On the other hand, the addition of ICB to ExTr exhibited superior BC tumor control, delaying tumor growth, reducing the final tumor volume, and further enhancing the antitumor effect of ExTr (Fig. 13.10b–d). ExTr and ICB seem to have a synergistic effect in the recruitment and function of CTLs as the number of intratumoral CD8$^+$ T cells and granzyme B$^+$ CD8$^+$ T cells were significantly increased with ExTr plus ICB (Fig. 13.10c, d).

Interestingly, ExTr prevents cancer-induced chronic fatigue, as evidenced by increased exercise capacity (Gomes-Santos et al. 2021). More importantly, ExTr was associated with reduced recurrence rate and metastatic burden in the metastatic MCa-M3C model (Gomes-Santos et al. 2021).

We also determined the effects of ExTr on tumor transcriptional signature using RNA-seq analysis (Gomes-Santos et al. 2021). ExTr induced metabolic reprogramming by reducing glucose uptake/insulin pathways and upregulated several gene sets associated with oxidative metabolism, such as oxidative phosphorylation (OXPHOS). It also enhanced gene expression linked to cytokine activation and downregulated several pathways involved with immunosuppression in BCs (Chen et al. 2019; Gunderson et al. 2020; Peranzoni et al. 2018). This molecular reprogramming suggests a less glycolytic and less immunosuppressive tumor, which is consistent with vessel normalization.

Altogether, exerecise intervention in established tumors can normalize tumor vasculature and TME, induce antitumor immunity (increased $CD8^+$ T-cell infiltration and function), and thus, improve control of refractory tumors when combined with ICB.

Acknowledgments The work of the author is supported in part by US NIH National Institute of Health grants (P01CA080124; R01-CA096915, R01-CA208205, R01-CA259253, R01-NS118929, R21-CA302986.

This article was prepared based on the following publications.

1. Fukumura D et al (2018) Enhancing cancer immunotherapy using antiangiogenics: opportunities and challenges. Nat Rev Clin Oncol 15:325–340. https://doi.org/10.1038/nrclinonc.2018.29
2. Dong X et al (2023) Anti-VEGF therapy improves EGFR-vIII-CAR-T cell delivery and efficacy in syngeneic glioblastoma models in mice. J Immunother Cancer 11. https://doi.org/10.1136/jitc-2022-005583
3. Incio J et al (2016) Obesity-induced inflammation and desmoplasia promote pancreatic cancer progression and resistance to chemotherapy. Cancer Discov 6:852–869. https://doi.org/10.1158/2159-8290.CD-15-1177
4. Incio J et al (2018) Obesity promotes resistance to anti-VEGF therapy in breast cancer by up-regulating IL-6 and potentially FGF-2. Sci Transl Med 10. https://doi.org/10.1126/scitranslmed.aag0945
5. Gomes-Santos IL et al (2021) Exercise training preserves myocardial strain and improves exercise tolerance in doxorubicin-induced cardiotoxicity. Front Cardiovasc Med 8:605993. https://doi.org/10.3389/fcvm.2021.605993

References

Akbari P et al (2022) Directing CAR T cells towards the tumor vasculature for the treatment of solid tumors. Biochim Biophys Acta Rev Cancer 1877:188701. https://doi.org/10.1016/j.bbcan.2022.188701

Allen E et al (2017) Combined antiangiogenic and anti-PD-L1 therapy stimulates tumor immunity through HEV formation. Sci Transl Med 9. https://doi.org/10.1126/scitranslmed.aak9679

Alvarez R et al (2013) Stromal disrupting effects of nab-paclitaxel in pancreatic cancer. Br J Cancer 109:926–933. https://doi.org/10.1038/bjc.2013.415

American Cancer Society: Cancer Facts and Figures 2015. http://www.cancer.org/acs/groups/content/@editorial/documents/document/acspc-044552.pdf

Ashcraft KA, Warner AB, Jones LW, Dewhirst MW (2019) Exercise as adjunct therapy in cancer. Semin Radiat Oncol 29:16–24. https://doi.org/10.1016/j.semradonc.2018.10.001

Batchelor TT et al (2007) AZD2171, a pan-VEGF receptor tyrosine kinase inhibitor, normalizes tumor vasculature and alleviates edema in glioblastoma patients. Cancer Cell 11:83–95. https://doi.org/10.1016/j.ccr.2006.11.021

Batchelor TT et al (2010) Phase II study of cediranib, an oral pan-vascular endothelial growth factor receptor tyrosine kinase inhibitor, in patients with recurrent glioblastoma. J Clin Oncol 28: 2817–2823. https://doi.org/10.1200/JCO.2009.26.3988

Beavis PA et al (2018) Dual PD-1 and CTLA-4 Checkpoint blockade promotes antitumor immune responses through CD4(+)Foxp3(−) cell-mediated modulation of CD103(+) dendritic cells. Cancer Immunol Res 6:1069–1081. https://doi.org/10.1158/2326-6066.CIR-18-0291

Bergers G, Hanahan D (2008) Modes of resistance to anti-angiogenic therapy. Nat Rev Cancer 8: 592–603

Betof AS et al (2015) Modulation of murine breast tumor vascularity, hypoxia and chemotherapeutic response by exercise. J Natl Cancer Inst 107. https://doi.org/10.1093/jnci/djv040

Biogps (2015). http://biogps.org/#goto=genereport&id=185

Bracci PM (2012) Obesity and pancreatic cancer: overview of epidemiologic evidence and biologic mechanisms. Mol Carcinog 51:53–63. https://doi.org/10.1002/mc.20778

Calle EE, Rodriguez C, Walker-Thurmond K, Thun MJ (2003) Overweight, obesity, and mortality from cancer in a prospectively studied cohort of U.S. adults. N Engl J Med 348:1625–1638. https://doi.org/10.1056/NEJMoa021423

Cannioto RA et al (2020) Physical activity before, during and after chemotherapy for high-risk breast cancer: relationships with survival. J Natl Cancer Inst 113. https://doi.org/10.1093/jnci/djaa046

Carmeliet P, Jain RK (2011a) Molecular mechanisms and clinical applications of angiogenesis. Nature 473:298–307. https://doi.org/10.1038/nature10144

Carmeliet P, Jain RK (2011b) Principles and mechanisms of vessel normalization for cancer and other angiogenic diseases. Nat Rev Drug Discov 10:417–427. https://doi.org/10.1038/nrd3455

Chae SS et al (2010) Angiopoietin-2 interferes with anti-VEGFR2-induced vessel normalization and survival benefit in mice bearing gliomas. Clin Cancer Res 16:3618–3627. https://doi.org/10.1158/1078-0432.CCR-09-3073

Charles NA, Holland EC, Gilbertson R, Glass R, Kettenmann H (2011) The brain tumor microenvironment. Glia 59:1169–1180. https://doi.org/10.1002/glia.21136

Chauhan VP et al (2013) Angiotensin inhibition enhances drug delivery and potentiates chemotherapy by decompressing tumour blood vessels. Nat Commun 4:2516. https://doi.org/10.1038/ncomms3516

Chauhan VP et al (2014) Compression of pancreatic tumor blood vessels by hyaluronan is caused by solid stress and not interstitial fluid pressure. Cancer Cell 26:14–15

Chen DS, Mellman I (2013) Oncology meets immunology: the cancer-immunity cycle. Immunity 39:1–10. https://doi.org/10.1016/j.immuni.2013.07.012

Chen IX et al (2019) Blocking CXCR4 alleviates desmoplasia, increases T-lymphocyte infiltration, and improves immunotherapy in metastatic breast cancer. Proc Natl Acad Sci USA 116:4558–4566. https://doi.org/10.1073/pnas.1815515116

Chung W et al (2017) Single-cell RNA-seq enables comprehensive tumour and immune cell profiling in primary breast cancer. Nat Commun 8:15081. https://doi.org/10.1038/ncomms15081

Conley SJ et al (2012) Antiangiogenic agents increase breast cancer stem cells via the generation of tumor hypoxia. Proc Natl Acad Sci USA 109:2784–2789. https://doi.org/10.1073/pnas.1018866109

Costa C, Incio J, Soares R (2007) Angiogenesis and chronic inflammation: cause or consequence? Angiogenesis 10:149–166. https://doi.org/10.1007/s10456-007-9074-0

De Palma M et al (2005) Tie2 identifies a hematopoietic lineage of proangiogenic monocytes required for tumor vessel formation and a mesenchymal population of pericyte progenitors. Cancer Cell 8:211–226. https://doi.org/10.1016/j.ccr.2005.08.002

Denkert C et al (2018) Tumour-infiltrating lymphocytes and prognosis in different subtypes of breast cancer: a pooled analysis of 3771 patients treated with neoadjuvant therapy. Lancet Oncol 19:40–50. https://doi.org/10.1016/S1470-2045(17)30904-X

Dirat B et al (2011) Cancer-associated adipocytes exhibit an activated phenotype and contribute to breast cancer invasion. Cancer Res 71:2455–2465. https://doi.org/10.1158/0008-5472.CAN-10-3323

Dong X et al (2023) Anti-VEGF therapy improves EGFR-vIII-CAR-T cell delivery and efficacy in syngeneic glioblastoma models in mice. J Immunother Cancer 11. https://doi.org/10.1136/jitc-2022-005583

Duda DG, Munn LL, Jain RK (2013) Can we identify predictive biomarkers for antiangiogenic therapy of cancer using mathematical modeling? J Natl Cancer Inst 105:762–765. https://doi.org/10.1093/jnci/djt114

Dvorak HF (1986) Tumors: wounds that do not heal. Similarities between tumor stroma generation and wound healing. N Engl J Med 315:1650–1659. https://doi.org/10.1056/NEJM198612253152606

Eatock MM et al (2013) Phase II randomized, double-blind, placebo-controlled study of AMG 386 (trebananib) in combination with cisplatin and capecitabine in patients with metastatic gastro-oesophageal cancer. Ann Oncol 24:710–718. https://doi.org/10.1093/annonc/mds502

Endrich B, Reinhold HS, Gross JF, Intaglietta M (1979) Tissue perfusion inhomogeneity during early tumor growth in rats. J Natl Cancer Inst 62:387–395

Fukumura D, Incio J, Shankaraiah RC, Jain RK (2016) Obesity and cancer: an angiogenic and inflammatory link. Microcirculation 23:191–206. https://doi.org/10.1111/micc.12270

Fukumura D, Kloepper J, Amoozgar Z, Duda DG, Jain RK (2018) Enhancing cancer immunotherapy using antiangiogenics: opportunities and challenges. Nat Rev Clin Oncol 15:325–340

Garcia E et al (2016) Blood flow responses to mild-intensity exercise in ectopic vs. orthotopic prostate tumors; dependence upon host tissue hemodynamics and vascular reactivity. J Appl Physiol 121:15–24. https://doi.org/10.1152/japplphysiol.00266.2016

Gati A et al (2014) Obesity and renal cancer: role of adipokines in the tumor-immune system conflict. Onco Targets Ther 3:e27810. https://doi.org/10.4161/onci.27810

Genkinger JM et al (2011) A pooled analysis of 14 cohort studies of anthropometric factors and pancreatic cancer risk. Int J Cancer 129:1708–1717. https://doi.org/10.1002/ijc.25794

Gomes-Santos IL et al (2021) Exercise training preserves myocardial strain and improves exercise tolerance in doxorubicin-induced cardiotoxicity. Front Cardiovasc Med 8:605993. https://doi.org/10.3389/fcvm.2021.605993

Griveau A, Seano G, Jain RK, Rowitch DH (2018) A glial signature and Wnt7 signaling regulate glioma-vascular interactions and tumor microenvironment. Cancer Cell:874

Gunderson AJ et al (2020) TGFbeta suppresses CD8(+) T cell expression of CXCR3 and tumor trafficking. Nat Commun 11:1749. https://doi.org/10.1038/s41467-020-15404-8

Hagendoorn J et al (2006) Onset of abnormal blood and lymphatic vessel function and interstitial hypertension in early stages of carcinogenesis. Cancer Res 66:3360–3364

Hama K et al (2004) Angiotensin II stimulates DNA synthesis of rat pancreatic stellate cells by activating ERK through EGF receptor transactivation. Biochem Biophys Res Commun 315: 905–911

Hama K et al (2006) Angiotensin II promotes the proliferation of activated pancreatic stellate cells by Smad7 induction through a protein kinase C pathway. Biochem Biophys Res Commun 340: 742–750

Hefetz-Sela S, Scherer PE (2013) Adipocytes: impact on tumor growth and potential sites for therapeutic intervention. Pharmacol Ther 138:197–210. https://doi.org/10.1016/j.pharmthera.2013.01.008

Hidalgo M, Von Hoff DD (2012) Translational therapeutic opportunities in ductal adenocarcinoma of the pancreas. Clin Cancer Res 18:4249–4256. https://doi.org/10.1158/1078-0432.ccr-12-1327

Holmes MD, Chen WY, Feskanich D, Kroenke CH, Colditz GA (2005) Physical activity and survival after breast cancer diagnosis. JAMA 293:2479–2486. https://doi.org/10.1001/jama.293.20.2479

Holopainen T et al (2012) Effects of angiopoietin-2-blocking antibody on endothelial cell-cell junctions and lung metastasis. J Natl Cancer Inst 104:461–475. https://doi.org/10.1093/jnci/djs009

Hori M et al (2014) Association of pancreatic Fatty infiltration with pancreatic ductal adenocarcinoma. Clin Transl Gastroenterol 5:e53. https://doi.org/10.1038/ctg.2014.5

Hosogai N et al (2007) Adipose tissue hypoxia in obesity and its impact on adipocytokine dysregulation. Diabetes 56:901–911

House IG et al (2020) Macrophage-derived CXCL9 and CXCL10 are required for antitumor immune responses following immune checkpoint blockade. Clin Cancer Res 26:487–504. https://doi.org/10.1158/1078-0432.CCR-19-1868

Huang Y et al (2012) Vascular normalizing doses of antiangiogenic treatment reprogram the immunosuppressive tumor microenvironment and enhance immunotherapy. Proc Natl Acad Sci USA 109:17561–17566. https://doi.org/10.1073/pnas.1215397109

Hursting SD (2012) Minireview: the year in obesity and cancer. Mol Endocrinol 26:1961–1966. https://doi.org/10.1210/me.2012-1283

Incio J, Soares R (2013) In: Gonzalez-Perez RR, Rueda BR (eds) Tumor angiogenesis regulators. CRC

Incio J et al (2016a) PlGF/VEGFR-1 signaling promotes macrophage polarization and accelerated tumor progression in obesity. Clin Cancer Res 22. https://doi.org/10.1158/1078-0432.CCR-15-1839

Incio J et al (2016b) Obesity-induced inflammation and desmoplasia promote pancreatic cancer progression and resistance to chemotherapy. Cancer Discov 6:852–869. https://doi.org/10.1158/2159-8290.CD-15-1177

Incio J et al (2018) Obesity promotes resistance to anti-VEGF therapy in breast cancer by up-regulating IL-6 and potentially FGF-2. Sci Transl Med 10. https://doi.org/10.1126/scitranslmed.aag0945

Iyengar NM, Jones LW (2019) Development of exercise as interception therapy for cancer: a review. JAMA Oncol 5. https://doi.org/10.1001/jamaoncol.2019.2585

Jain RK (2001) Normalizing tumor vasculature with anti-angiogenic therapy: a new paradigm for combination therapy. Nat Med 7:987–989. https://doi.org/10.1038/nm0901-987

Jain RK (2005) Normalization of tumor vasculature: an emerging concept in antiangiogenic therapy. Science 307:58–62. https://doi.org/10.1126/science.1104819

Jain RK (2013) Normalizing tumor microenvironment to treat cancer: bench to bedside to biomarkers. J Clin Oncol 31:2205–2218. https://doi.org/10.1200/JCO.2012.46.3653

Jain RK (2014) Antiangiogenesis strategies revisited: from starving tumors to alleviating hypoxia. Cancer Cell 26:605–622. https://doi.org/10.1016/j.ccell.2014.10.006

Jain RK et al (1996) Leukocyte-endothelial adhesion and angiogenesis in tumors. Cancer Metastasis Rev 15:195–204

Jain RK et al (2009) Biomarkers of response and resistance to antiangiogenic therapy. Nat Rev Clin Oncol 6:327–338

Jain RK, Martin JD, Stylianopoulos T (2014) The role of mechanical forces in tumor growth and therapy. Annu Rev Biomed Eng 16:321–346. https://doi.org/10.1146/annurev-bioeng-071813-105259

Jung K et al (2017a) Targeting CXCR4-dependent immunosuppressive Ly6Clow monocytes improves antiangiogenic therapy in colorectal cancer. Proc Natl Acad Sci USA 114:10455–10460. https://doi.org/10.1073/pnas.1710754114

Jung K et al (2017b) Ly6Clo monocytes drive immunosuppression and confer resistance to anti-VEGFR2 cancer therapy. J Clin Invest 127:3039–3051. https://doi.org/10.1172/JCI93182

Kabbinavar F et al (2003) Phase II, randomized trial comparing bevacizumab plus fluorouracil (FU)/leucovorin (LV) with FU/LV alone in patients with metastatic colorectal cancer. J Clin Oncol 21:60–65. https://doi.org/10.1200/JCO.2003.10.066

Kanasaki K, Koya D (2011) Biology of obesity: lessons from animal models of obesity. J Biomed Biotechnol 2011:197636. https://doi.org/10.1155/2011/197636

Khan S, Shukla S, Sinha S, Meeran SM (2013) Role of adipokines and cytokines in obesity-associated breast cancer: therapeutic targets. Cytokine Growth Factor Rev 24:503–513. https://doi.org/10.1016/j.cytogfr.2013.10.001

Kienast Y et al (2013) Ang-2-VEGF-A CrossMab, a novel bispecific human IgG1 antibody blocking VEGF-A and Ang-2 functions simultaneously, mediates potent antitumor, antiangiogenic, and antimetastatic efficacy. Clin Cancer Res 19:6730–6740. https://doi.org/10.1158/1078-0432.CCR-13-0081

Kloepper J et al (2016) Ang-2/VEGF bispecific antibody reprograms macrophages and resident microglia to anti-tumor phenotype and prolongs glioblastoma survival. Proc Natl Acad Sci USA 113:4476–4481. https://doi.org/10.1073/pnas.1525360113

Levin VA et al (2015) Impact of bevacizumab administered dose on overall survival of patients with progressive glioblastoma. J Neuro-Oncol 122:145–150. https://doi.org/10.1007/s11060-014-1693-x

Levin VA, Chan J, Datta M, Yee JL, Jain RK (2017) Effect of angiotensin system inhibitors on survival in newly diagnosed glioma patients and recurrent glioblastoma patients receiving chemotherapy and/or bevacizumab. J Neuro-Oncol 134. https://doi.org/10.1007/s11060-017-2528-3

Li D et al (2009) Body mass index and risk, age of onset, and survival in patients with pancreatic cancer. JAMA 301:2553–2562. https://doi.org/10.1001/jama.2009.886

Lijnen HR (2008) Angiogenesis and obesity. Cardiovasc Res 78:286–293. https://doi.org/10.1093/cvr/cvm007

Liu C, Workman CJ, Vignali DA (2016) Targeting regulatory T cells in tumors. FEBS J 283:2731–2748. https://doi.org/10.1111/febs.13656

Liu S et al (2017) Role of cytotoxic tumor-infiltrating lymphocytes in predicting outcomes in metastatic HER2-positive breast cancer: a secondary analysis of a randomized clinical trial. JAMA Oncol 3:e172085. https://doi.org/10.1001/jamaoncol.2017.2085

Lohmann AE, Chia S (2012) Patients with metastatic breast cancer using bevacizumab as a treatment: is there still a role for it? Curr Treat Options in Oncol 13:249–262. https://doi.org/10.1007/s11864-012-0181-9

Lorgis V et al (2012) Relation between bevacizumab dose intensity and high-grade glioma survival: a retrospective study in two large cohorts. J Neuro-Oncol 107:351–358. https://doi.org/10.1007/s11060-011-0748-5

Lu-Emerson C et al (2015) Lessons from anti-vascular endothelial growth factor and anti-vascular endothelial growth factor receptor trials in patients with glioblastoma. J Clin Oncol 33:1197–1213. https://doi.org/10.1200/JCO.2014.55.9575

Luo R et al (2019) Cisplatin facilitates radiation-induced abscopal effects in conjunction with PD-1 checkpoint blockade through CXCR3/CXCL10-mediated T-cell recruitment. Clin Cancer Res 25:7243–7255. https://doi.org/10.1158/1078-0432.CCR-19-1344

McCullough DJ, Stabley JN, Siemann DW, Behnke BJ (2014) Modulation of blood flow, hypoxia, and vascular function in orthotopic prostate tumors during exercise. J Natl Cancer Inst 106: dju036. https://doi.org/10.1093/jnci/dju036

McWilliams RR et al (2010) Obesity adversely affects survival in pancreatic cancer patients. Cancer 116:5054–5062. https://doi.org/10.1002/cncr.25465

Miller A et al (2016) Perfusion pressure is a critical determinant of the intratumoral extravasation of oncolytic viruses. Mol Ther 24:306–317. https://doi.org/10.1038/mt.2015.219

Minami T et al (2013) The calcineurin-NFAT-angiopoietin-2 signaling axis in lung endothelium is critical for the establishment of lung metastases. Cell Rep 4:709–723. https://doi.org/10.1016/j.celrep.2013.07.021

Monk BJ et al (2016) Final results of a phase 3 study of trebananib plus weekly paclitaxel in recurrent ovarian cancer (TRINOVA-1): long-term survival, impact of ascites, and progression-free survival-2. Gynecol Oncol 143:27–34. https://doi.org/10.1016/j.ygyno.2016.07.112

Moon EK et al (2018) Intra-tumoral delivery of CXCL11 via a vaccinia virus, but not by modified T cells, enhances the efficacy of adoptive T cell therapy and vaccines. Onco Targets Ther 7: e1395997. https://doi.org/10.1080/2162402X.2017.1395997

Mottagui-Tabar S et al (2003) Evidence for an important role of perilipin in the regulation of human adipocyte lipolysis. Diabetologia 46:789–797. https://doi.org/10.1007/s00125-003-1112-x

Neoptolemos JP et al (2012) Effect of adjuvant chemotherapy with fluorouracil plus folinic acid or gemcitabine vs observation on survival in patients with resected periampullary adenocarcinoma: the ESPAC-3 periampullary cancer randomized trial. JAMA 308:147–156. https://doi.org/10.1001/jama.2012.7352

Ogden CL, Carroll MD, Kit BK, Flegal KM (2014) Prevalence of childhood and adult obesity in the United States, 2011-2012. JAMA 311:806–814. https://doi.org/10.1001/jama.2014.732

Olive KP et al (2009) Inhibition of Hedgehog signaling enhances delivery of chemotherapy in a mouse model of pancreatic cancer. Science 324:1457–1461

Palucka AK, Coussens LM (2016) The basis of oncoimmunology. Cell 164:1233–1247. https://doi.org/10.1016/j.cell.2016.01.049

Pedersen L et al (2016) Voluntary running suppresses tumor growth through epinephrine- and IL-6-dependent NK cell mobilization and redistribution. Cell Metab 23:554–562. https://doi.org/10.1016/j.cmet.2016.01.011

Peeters M et al (2013) A randomised, double-blind, placebo-controlled phase 2 study of trebananib (AMG 386) in combination with FOLFIRI in patients with previously treated metastatic colorectal carcinoma. Br J Cancer 108:503–511. https://doi.org/10.1038/bjc.2012.594

Peranzoni E et al (2018) Macrophages impede CD8 T cells from reaching tumor cells and limit the efficacy of anti-PD-1 treatment. Proc Natl Acad Sci USA 115:E4041–E4050. https://doi.org/10.1073/pnas.1720948115

Peterson TE et al (2016) Dual inhibition of Ang-2 and VEGF receptors normalizes tumor vasculature and prolongs survival in glioblastoma by altering macrophages. Proc Natl Acad Sci USA 113:4470–4475. https://doi.org/10.1073/pnas.1525349113

Pinter M, Jain RK (2017) Targeting the renin-angiotensin system to improve cancer treatment: Implications for immunotherapy. Sci Transl Med 9. https://doi.org/10.1126/scitranslmed.aan5616

Rafiq S, Hackett CS, Brentjens RJ (2020) Engineering strategies to overcome the current roadblocks in CAR T cell therapy. Nat Rev Clin Oncol 17:147–167. https://doi.org/10.1038/s41571-019-0297-y

Rahbari NN et al (2016) Anti-VEGF therapy induces ECM remodeling and mechanical barriers to therapy in colorectal cancer liver metastases. Sci Transl Med 8:360ra135. https://doi.org/10.1126/scitranslmed.aaf5219

Reardon DA et al (2017) Phase 2 and biomarker study of trebananib, an angiopoietin-blocking peptibody, with and without bevacizumab for patients with recurrent glioblastoma. Cancer 124. https://doi.org/10.1002/cncr.31172

Rini B et al (2012) AMG 386 in combination with sorafenib in patients with metastatic clear cell carcinoma of the kidney: a randomized, double-blind, placebo-controlled, phase 2 study. Cancer 118:6152–6161. https://doi.org/10.1002/cncr.27632

Rundqvist H et al (2020) Cytotoxic T-cells mediate exercise-induced reductions in tumor growth. elife 9. https://doi.org/10.7554/eLife.59996

Savas P et al (2018) Single-cell profiling of breast cancer T cells reveals a tissue-resident memory subset associated with improved prognosis. Nat Med 24:986–993. https://doi.org/10.1038/s41591-018-0078-7

Schadler KL et al (2016) Tumor vessel normalization after aerobic exercise enhances chemotherapeutic efficacy. Oncotarget 7:65429–65440. https://doi.org/10.18632/oncotarget.11748

Schmittnaegel M et al (2017) Dual angiopoietin-2 and VEGFA inhibition elicits antitumor immunity that is enhanced by PD-1 checkpoint blockade. Sci Transl Med 9. https://doi.org/10.1126/scitranslmed.aak9670

Shrimali RK et al (2010) Antiangiogenic agents can increase lymphocyte infiltration into tumor and enhance the effectiveness of adoptive immunotherapy of cancer. Cancer Res 70:6171–6180. https://doi.org/10.1158/0008-5472.CAN-10-0153

Smits MM, van Geenen EJ (2011) The clinical significance of pancreatic steatosis. Nat Rev Gastroenterol Hepatol 8:169–177. https://doi.org/10.1038/nrgastro.2011.4

Stylianopoulos T et al (2012) Causes, consequences, and remedies for growth-induced solid stress in murine and human tumors. Proc Natl Acad Sci USA 109:15101–15108. https://doi.org/10.1073/pnas.1213353109

Sun K, Tordjman J, Clement K, Scherer PE (2013) Fibrosis and adipose tissue dysfunction. Cell Metab 18:470–477. https://doi.org/10.1016/j.cmet.2013.06.016

Tian L et al (2017) Mutual regulation of tumour vessel normalization and immunostimulatory reprogramming. Nature 544:250–254. https://doi.org/10.1038/nature21724

Tokunaga R et al (2018) CXCL9, CXCL10, CXCL11/CXCR3 axis for immune activation – a target for novel cancer therapy. Cancer Treat Rev 63:40–47. https://doi.org/10.1016/j.ctrv.2017.11.007

Tong RT et al (2004) Vascular normalization by vascular endothelial growth factor receptor 2 blockade induces a pressure gradient across the vasculature and improves drug penetration in tumors. Cancer Res 64:3731–3736

Wang YY et al (2012) Adipose tissue and breast epithelial cells: a dangerous dynamic duo in breast cancer. Cancer Lett 324:142–151. https://doi.org/10.1016/j.canlet.2012.05.019

Wang K, Shen T, Siegal GP, Wei S (2017) The CD4/CD8 ratio of tumor-infiltrating lymphocytes at the tumor-host interface has prognostic value in triple negative breast cancer. Hum Pathol 69: 110–117. https://doi.org/10.1016/j.humpath.2017.09.012

Whatcott CJ et al (2015) Desmoplasia in primary tumors and metastatic lesions of pancreatic cancer. Clin Cancer Res 21. https://doi.org/10.1158/1078-0432.CCR-14-1051

Willett CG et al (2004) Direct evidence that the VEGF-specific antibody bevacizumab has antivascular effects in human rectal cancer. Nat Med 10:145–147. https://doi.org/10.1038/nm988

Willett CG et al (2005) Surrogate markers for antiangiogenic therapy and dose-limiting toxicities for bevacizumab with radiation and chemotherapy: continued experience of a phase I trial in rectal cancer patients. J Clin Oncol 23:8136–8139. https://doi.org/10.1200/JCO.2005.02.5635

Williams PT (2014) Significantly greater reduction in breast cancer mortality from post-diagnosis running than walking. Int J Cancer 135:1195–1202. https://doi.org/10.1002/ijc.28740

Winkler F et al (2004) Kinetics of vascular normalization by VEGFR2 blockade governs brain tumor response to radiation: role of oxygenation, angiopoietin-1, and matrix metalloproteinases. Cancer Cell 6:553–563. https://doi.org/10.1016/j.ccr.2004.10.011

Wu FT et al (2016) Efficacy of Cotargeting Angiopoietin-2 and the VEGF pathway in the adjuvant postsurgical setting for early breast, colorectal, and renal cancers. Cancer Res 76:6988–7000. https://doi.org/10.1158/0008-5472.CAN-16-0888

Wu X et al (2017) Angiopoietin-2 as a biomarker and target for immune checkpoint therapy. Cancer Immunol Res 5:17–28. https://doi.org/10.1158/2326-6066.CIR-16-0206

Yang F, Chung AC, Huang XR, Lan HY (2009) Angiotensin II induces connective tissue growth factor and collagen I expression via transforming growth factor-beta-dependent and -independent Smad pathways: the role of Smad3. Hypertension 54:877–884

Ye J, Gao Z, Yin J, He Q (2007) Hypoxia is a potential risk factor for chronic inflammation and adiponectin reduction in adipose tissue of ob/ob and dietary obese mice. Am J Physiol Endocrinol Metab 293:E1118–E1128. https://doi.org/10.1152/ajpendo.00435.2007

Young RM, Engel NW, Uslu U, Wellhausen N, June CH (2022) Next-generation CAR T-cell therapies. Cancer Discov 12:1625–1633. https://doi.org/10.1158/2159-8290.CD-21-1683

Yuan C et al (2013) Prediagnostic body mass index and pancreatic cancer survival. J Clin Oncol 31: 4229–4234. https://doi.org/10.1200/JCO.2013.51.7532

Zhang JQ, Plitas G (2020) Immunotherapeutic strategies in breast cancer: a clinical update. J Surg Oncol 123. https://doi.org/10.1002/jso.26287

Zheng X et al (2018) Increased vessel perfusion predicts the efficacy of immune checkpoint blockade. J Clin Invest 128:2104–2115. https://doi.org/10.1172/JCI96582

Zlotecki RA, Baxter LT, Boucher Y, Jain RK (1995) Pharmacologic modification of tumor blood flow and interstitial fluid pressure in a human tumor xenograft: network analysis and mechanistic interpretation. Microvasc Res 50:429–443. https://doi.org/10.1006/mvre.1995.1069

Zyromski NJ et al (2009) Obesity potentiates the growth and dissemination of pancreatic cancer. Surgery 146:258–263. https://doi.org/10.1016/j.surg.2009.02.024

Chapter 14
Microvascular Dysfunction Related to COVID-19 and Systemic Inflammation

Dominga Lapi, Martina Chiurazzi, Espedita Muscariello, and Antonio Colantuoni

Abstract The coronavirus disease 2019 (COVID-19) pandemic began in Wuhan, China, on December 2019, caused by severe acute respiratory distress syndrome coronavirus (SARS-CoV-2), a new coronavirus related to bat-origin SARS-like viruses. It quickly spread globally due to its high transmissibility, leading to a wide range of disease severities, from asymptomatic to fatal, among certain populations, such as the elderly and those with pre-existing morbidities, being more vulnerable. Over time, SARS-CoV-2 developed multiple variants, including Alpha, Beta, Delta and Omicron, with Omicron showing the most mutations and highest spread. The disease course has been divided into three phases: the initial or first phase, characterized by virus invasion and replication, with clinical mild symptoms; the second phase, with immunity responses and predominant respiratory clinical symptoms; the third or last phase, usually in 10–20% of cases progressing to systemic inflammation or hyperinflammation phase, eventually resulting in intensive care unit treatment. The three phases have presented consistent overlap in many patients. However, the virus primarily affects the vascular system, causing "angiocentric inflammation" characterized by endothelial dysfunction, inflammation and intravascular clotting, which can lead to multiorgan failure. The virus enters cells through angiotensin-converting enzyme 2 (ACE2) receptors, triggering immune responses and inflammatory damage, and progresses through stages: viral replication, pulmonary inflammation and hyperinflammation, which can lead to acute respiratory distress syndrome (ARDS). COVID-19 also causes the cytokine storm, characterized by an overactive immune response that worsens tissue damage. Additionally, the virus affects platelets, neutrophils and the complement system, leading to intravascular clotting and further inflammation. COVID-19 is increasingly

D. Lapi (✉)
Department of Biology, University of Pisa, Pisa, Italy
e-mail: dominga.lapi@unipi.it

M. Chiurazzi · E. Muscariello · A. Colantuoni (✉)
Department of Clinical Medicine and Surgery, Federico II University Medical School, Naples, Italy
e-mail: anto.colanti@gmail.coms

R. F. Tuma et al. (eds.), *Handbook of Microcirculation*, Physiology in Health and Disease, https://doi.org/10.1007/978-3-031-99609-2_14

recognized as a systemic vascular disease, affecting microcirculation, and in some individuals, it leads to long-term effects known as long COVID.

The fight against the virus requires ongoing monitoring of new variants and a deeper understanding of the disease's pathophysiological processes across its various stages. The disease's focus on the vascular system calls for further research into the interactions between endothelial cells and pericytes, both under normal conditions and during viral infection, to clarify their role in disease progression. Additionally, improving viscometric techniques to analyze the blood of COVID-19 patients could provide crucial insights into the effects on microcirculation. Advancing in vivo assessments of microcirculation through sublingual methods could also enhance understanding of the mechanisms behind microvascular damage in COVID-19 patients.

Keywords COVID-19 · SARS-CoV-2 · Systemic vascular disease · Acute respiratory distress syndrome · Angiocentric inflammation · Systemic inflammation · Endothelial dysfunction · Microcirculation

14.1 Introduction

14.1.1 SARS-CoV-2 Pandemic

In Wuhan, China, there were the first cases of a new viral respiratory infection on December 2019. The virus was identified as a novel coronavirus, causing severe acute respiratory distress syndrome (SARS), and named SARS-CoV-2, family Coronaviridae, genus Betacoronavirus, subgenus Sarbecovirus, the causative agent of coronavirus disease 2019 (COVID-19). SARS-CoV-2 has been shown to be closely related to two bat-derived SARS-like coronaviruses: bat-SL-CoVZC45 and bat-SL-CoVZXC21. Another coronavirus, SARS-CoV, caused the SARS outbreak in China during 2002–2003. The Middle East Respiratory Syndrome (MERS) was detected in June 2012 in a man in Saudi Arabia, who died of acute pneumonia and renal failure, caused by a coronavirus (MERS-CoV) isolated from his sputum. Several cases were reported in Jordan and in UK in the same period (De Wit et al. 2016).

SARS-CoV2 is highly contagious as evidenced by its spread to nearly all countries worldwide within a very short time. Three years after the pandemic, it has now become clear that SARS-CoV-2 infection can result in highly heterogeneous disease. COVID-19 ranges from asymptomatic to deadly, with variable severity according to age, concomitant condition and ethnicity.

The morbidity and mortality of COVID-19 have changed the world despite the advanced research working toward its management and prevention. However, the SARS-CoV-2 virus continued to evolve into multiple variants. Therefore, the World Health Organization (WHO) has renamed the variants as variant of concern (VOC), such as Omicron (B.1.1.529) and variant of interest (VOI). VOC are variants with increased transmissibility or detrimental changes in COVID-19 epidemiology. VOI are variants with genetic changes that are predicted or are known to affect viral characteristics (Zhang et al. 2022a, b). A September 2021 WHO report stated that VOI include C.37 (Lambda) and B.1.621 (Mu).

Different variants (VOC) have been identified over time and geography: Alpha (B.1.1.7) in the United Kingdom, Beta (B.1.351) in South Africa, Gamma (P.1) in Brazil, Delta (B.1.617.2) in India. The Omicron variant (B.1.1.529) was identified in Botswana and South Africa on November 24 2021 (Da Silva et al. 2023).

Omicron has the highest number of mutations compared to other variants discovered to date, with over 30 mutations located within the spike protein (S-protein), a critical component in determining the infectivity and antigenicity of the virus. In addition, 15 of these mutations are located in the receptor-binding domain (RBD) of S-protein, which are believed to lead to breakthrough infections or reinfections. Surveillance shows that certain sublineages identified of the Omicron variant have shown increased prevalence. As a result, there is a significant concern of Omicron having higher transmission and hospitalization rates, significantly reduced vaccine effectiveness (VE), development of more severe disease and increased mortality compared with other variants.

Up today, the WHO reported 2,178,394 deaths in Europe and 6,713,093 deaths worldwide.

Global efforts in research, including the pathophysiology of COVID-19, have been undertaken, with a rapid and rarely seen surge of publications related to a single disease, within months (Chams et al. 2020).

What is known, indeed, about the pathogenesis, epidemiology and treatment/prevention strategies regarding SARS-CoV-2 infection has expanded immensely during the pandemic. In an effort to help guide targeted public health strategies and medical interventions, an important area of research is working to clarify the specific chronic medical conditions associated with worse outcomes from SARS-CoV-2 infection. At this moment (February 2023), 336,368 entries for COVID-19 are reported in PUB MED.

It is interesting that COVID-19 is associated with a myriad of clinical manifestations, compared with other diseases. It can affect the lungs, heart, kidneys, central nervous system, gastrointestinal system and interfere with the blood clotting processes. Therefore, many studies have been aimed at finding the "common multiple denominator" or what is the structure that unifies all the tissues and organs on which the virus acts. To date, growing evidence supports endothelial dysfunction as the unified key mechanism in the pathogenesis of COVID-19 (Xu et al. 2023; Ambrosino et al. 2022).

The vascular endothelium, the outermost layer of blood vessels, represents a dynamic interface between the blood and the various tissues/organs and maintains tissue homeostasis. Physiological functions of the endothelium include fine control of vascular tone, tissue hemostasis, barrier integrity, inflammation, oxidative stress, vascular permeability, and structural and functional integrity. Numerous viral species, such as dengue, Ebola and cytomegalovirus can infect endothelial cells (ECs) and cause endothelial dysfunction (Konradt and Hunter 2018). COVID-19 primarily injures the vascular endothelium, characterizing the early stage with a disrupted vessel regulation (Nicosia et al. 2021).

Histopathological observations demonstrate that COVID-19 is a (micro)vascular and endothelial disease in which endothelial dysfunction plays a fundamental role (Otifi and Adiga 2022; Nägele et al. 2020; Martínez-Salazar et al. 2022). Direct or indirect mechanism after SARS-CoV-2 infection and the consequent endothelialitis/endotheliopathy triggers multiple instances of endothelial dysfunction, including altered vascular tone, oxidative stress, inflammation/leukocyte adhesion, endothelial–mesenchymal transition (EndoMT), mitochondria dysfunction, virus-induced senescence, cytokine storm, and coagulopathy. Prevalent data indicate that the clinical manifestations of severe COVID-19 result from systemic inflammation and thrombosis.

14.2 From the Virus Entry to Systemic Inflammation

COVID-19 spreading and pandemic have been facilitated by human-to human transmission through respiratory tract droplets from an infected person to other people, by cough or sneeze droplets or direct contact with respiratory secretions (Jayaweera et al. 2020). The disease course has been divided into three phases: the initial or first phase, characterized by virus invasion and replication, with clinical mild symptoms; the second phase with immunity responses and predominant respiratory clinical symptoms; the third or last phase, usually in 10–20% of cases, progressing to a systemic inflammation or hyperinflammation phase, eventually resulting in intensive care unit treatment (Aguilar et al. 2020; Bartoli et al. 2021). The three phases have presented consistent overlap in many patients; however, the sequence of clinical and pathophysiological events has been useful to define the complex mechanisms of the viral infection.

The fundamental point is represented by the entry of the virus into the host cells. SARS-CoV-2 presents a single-stranded positive-sense RNA genome, which is one of the largest genome among RNA viruses (Alexandersen et al. 2020). The genome encodes four structural proteins; spike (S), envelope (E), membrane (M) and nucleocapsid (N) (Troyano-Hernáez et al. 2021). The S protein recognizes the host cell receptor on the target cells: the S protein constitutes homotrimer on the viral surface with a crown-like appearance. The virus entry is due to the cleavage by furin at S1/S2 boundary, resulting in the formation of S1 subunit for receptor binding and S2 for membrane fusion. A further cleavage at the S2' site on cell surface by type 2 transmembrane serine protease (TMPRSS2) or by cathepsin in endosome is essential for exposition of fusion peptide to start the fusion of membrane. TMPRSS2 is able to cleave S at residues located immediately upstream of Fusion Protein (FP), facilitating FP to insert into the host cell membrane and mediate the viral entry at cell surface (Zhang et al. 2022a, b; Mahoney et al. 2021; Bestle et al. 2020).

TMPRSS2 is expressed in epithelial cells of respiratory tract, heart, prostate and gastrointestinal tracts. In vitro studies indicate that inhibition of TMPRSS2 by camostat mesylate reduced the entry of SARS-CoV2 in the human airway cells, suggesting that TMPRSS2 play a role for SARS-CoV2 infection of respiratory system (Hoffmann et al. 2020).

Another way to entry is represented by endocytosis by Clathrin-mediated mechanism or other pathways as reported in several studies (Jackson et al. 2022). This mechanism of virus entry has been suggested to participate in the amplification of SARS-CoV2 infection in extrapulmonary organs after the first infection of respiratory epithelium involving the TMPRSS2 mediated membrane fusion (Hoffmann and Pöhlmann 2021).

Moreover, it has been reported that S protein expressed on infected cell surface cooperates with ACE2 receptor on neighboring cells and induces the syncytia formation, contributing to the pathology of SARS-CoV2 infection. In COVID-19 autopsies, multinucleated giant cells were found in lung tissues, indicating that syncytia could facilitate viral transmission from the primary infection sites enabling multiple viral genomes to be transferred into single cells (Buchrieser et al. 2020; Nejat et al. 2023).

The SARS-CoV-2 viral particles enter cells and release the virus genome (RNA) into the cytoplasm. The viruses, once inside, release their genome and transform the cell into a "virus factory." They are able to use all the replication apparatus of the host cells, managing to reproduce their RNA and all the viral proteins needed to build the new viruses, which numerous will be assembled inside the cell until they cause the cellular death with the release of the new SARS-CoV-2 viral particles. These will then be able to infect other cells and, above all, other individuals (Jackson et al. 2022; Zhang et al. 2021).

The first phase of infection is characterized by infiltration of lung (usually prevalent) or gastrointestinal or other organs parenchyma with SARS-CoV2 proliferation (Fraga-Silva et al. 2021; Suberviola and Fito 2022). The clinical symptoms are mild and correspond to the first response by innate immune system with monocyte/macrophage infiltration. The sequence of viral infection mechanisms is characterized by vasodilation, microvascular permeability and leukocyte recruitment. These microvascular changes progress in the second phase, causing usually lung involvement and hypoxia with respiratory dysfunction. COVID-19 patients present an acute respiratory distress syndrome (ARDS), characterized by slight reduction in pulmonary compliance. Therefore, these patients are relatively "ease to ventilate." However, the progression to ARDS is accompanied by multiple pathophysiological mechanisms, such as marked vascular endothelial damage with lung edema and microvascular thrombosis resulting in increased respiratory dead space, requiring supportive respiratory treatment in patients. COVID-19 ARDS manifest itself unpredictably, due to different disease phase and the variable contribution of host response, physiological reserve and comorbidities (Romagnoli et al. 2020; Aslan et al. 2021; Gibson et al. 2020).

The patients progress to the third phase when the inflammatory response increases (hyperinflammation phase), characterized by increased production of cytokines (cytokine storm), multiple intravascular coagulopathies and multiorgan failure (MOF) (Gao et al. 2021; Silva et al. 2023a, b). The first reports on COVID-19 indicated a dramatic increase in cytokines, such as (Interleukin) IL-6, IL-1β, IL-2,

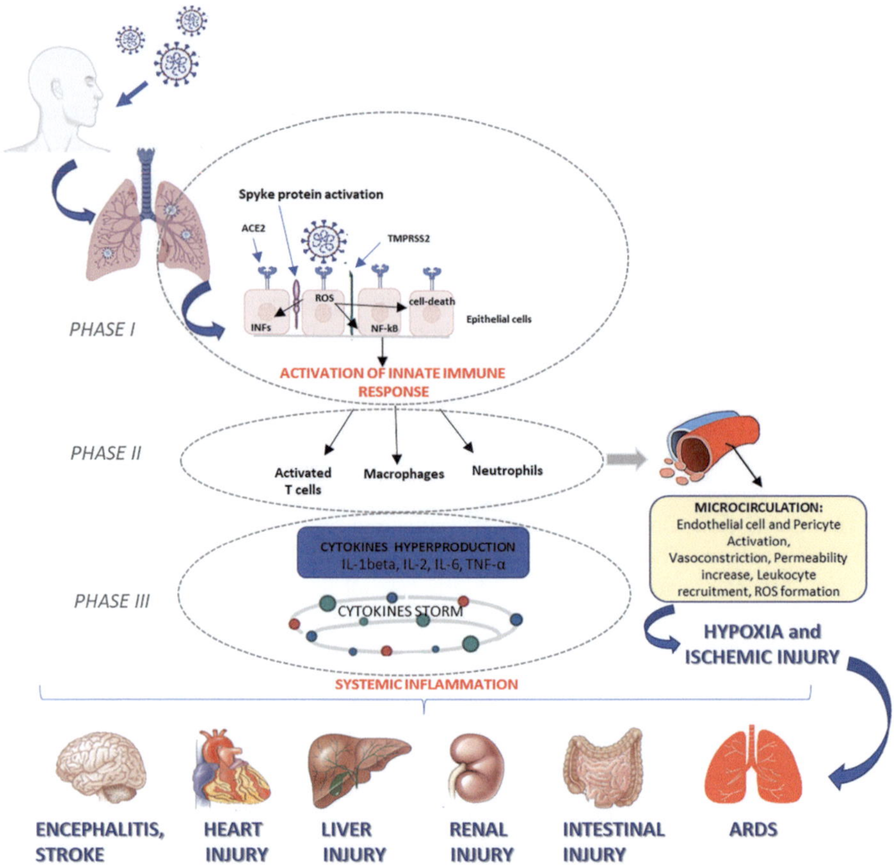

Fig. 14.1 Scheme of the main phases of the COVID-19 infection, from the entry of the virus at the level of the endothelial cells to the multi-organ damage
PHASE I: SARS-CoV-2 entry into the host and interaction of its S protein with ACE2 by TMPRSS2 promoting endocytosis to infect respiratory endothelial cells
PHASE II: immune cells activation that cause: impairment to the microcirculation and an uncontrolled immune cell activation (cytokine storm) which distinguishes *PHASE III*. The systemic inflammation determines: *MODS* multiorgan dysfunction syndrome and *ARDS* acute respiratory distress syndrome

IL-7, tumor necrosis factor (TNF)-α, macrophage inflammatory protein (MIP) 1α, C-reactive protein (CRP) and several others (Costela-Ruiz et al. 2020; Tang et al. 2020). Viral injuries, huge increase in cytokine release and several molecules induced by inflammatory vascular damage facilitate systemic microvascular inflammation, with further endothelial activation and prothrombotic conditions. All these conditions result in acute pulmonary thrombosis, ischemic stroke, myocardial infarction, kidney, liver and heart failure, systemic arterial thrombosis complicated by septic shock and secondary bacterial infections. These complications have resulted in a great number of fatalities worldwide. A schematic presentation of COVID-19 course is reported in Fig. 14.1.

14.2.1 *The Role of ACE2: Imbalance of the Renin–Angiotensin System*

Angiotensin-converting enzyme 2 (ACE2), the functional receptor of SARS-CoV-2, plays a crucial role in the pathogenesis of COVID-19, as it provides viral entry into human cells (Kuba et al. 2021; Beyerstedt et al. 2021). The viral spike (S) protein of SARS-CoV-2 binds to the ACE2 receptor. SARS-CoV-2 has a marked affinity for ACE2 and this binding leads to host cell entry of the virus in concert with S-protein priming by the host cell protease TMPRSS2.

The renin–angiotensin system (RAS) can finely tune the amount of Ang II, the main active RAS component, exerting its effects mainly via Angiotensin II type 1 receptors (AT1R). Major effects of Ang II include vasoconstriction, renal sodium reabsorption and potassium excretion, aldosterone synthesis stimulation, blood pressure elevation, induction of inflammatory and profibrotic pathways (Ekholm and Kahan 2021; Benigni et al. 2010). ACE2 cleaves Ang II to angiotensin (1–7), which exerts vasodilating, anti-inflammatory and antifibrotic effects through binding to the Mas receptor (Ni et al. 2020). In addition, ACE2 cleaves Ang I into angiotensin (1–9), which is in turn converted into angiotensin (1–7) by ACE.

It was observed that ACE2 was highly expressed on lung alveolar epithelial cells and small intestinal epithelial cells, consistent with potential routes of viral transmission of SARS-CoV-2, as both respiratory and gastrointestinal systems share interfaces with the external environment. Additionally, ACE2 was present on vascular endothelial cells and smooth muscle cells in all organs studied. Therefore, the interplay between ACE1 and ACE2 is fundamental to facilitate the balance of constriction and dilation induced, respectively, by Ang II and Ang 1–7 (Lapi et al. 2021; Gheblawi et al. 2020), a balance that is known to modulate the microvascular functions in all organs of living organisms. Consequently, this ACE/ACE2 balance plays a crucial role in determining the availability of different angiotensin peptides and hence the balance between proinflammatory and profibrotic and anti-inflammatory and antifibrotic pathways (Lapi et al. 2021).

Since the RAS is based on a dynamic equilibrium between two opposite axes, down-regulation of the enzyme responsible for converting Ang II into Ang (1–7) implicates an enhanced activity of RAS classical axis. The most accredited hypothesis is that the upregulation of this classical RAS arm might be partially responsible for the deleterious pathophysiological mechanisms of COVID-19, in association with viral activation of the innate and adaptive immune response (Ni et al. 2020).

Therefore, the RAS is formed by two opposite counter-regulatory arms: the classical one, composed by ACE/Angiotensin II (Ang II)/AT1 receptor (AT1R), and the alternative anti-inflammatory arm, comprising ACE2/Ang (1–7)/Mas Receptor (MasR) (Masaru and Horiuchi 2009). Consequently, it has been suggested that ACE2 targeting by SARS-CoV-2 is not just a casual route of infection, but a causal factor of the disease itself: COVID-19 is, at least in part, the consequence of RAS imbalance, due to an exaggerated activation of the classical arm (ACE/Ang II/AT1R), leading to pulmonary injury, hematological alterations and hyperinflammatory state. Hence, according to the RAS imbalance hypothesis, the RAS pathway disruption is at the center of COVID-19 pathophysiological mechanisms. Numerous results, reported up to date, support the hypothesis that Ang II could play a role in inducing alterations in the microvascular networks of different organs. Conversely, Angiotensin 1–7 appears to trigger cascades of events preserving microvascular networks from damage.

Recently, in support of this hypothesis Lapi et al. (2021) observed a determining role of ACE/ACE2 imbalance during cerebral hypoperfusion–reperfusion injury. In this experimental study, it has been shown that under baseline conditions of perfusion Ang II does not affect pial microvascular networks. Conversely, reducing the perfusion of pial vessels with the bilateral occlusion of carotid arteries for 30 min and recovering the perfusion for 60 min (in vivo hypoperfusion–reperfusion model), Ang II significantly increases the damage of the cerebral microcirculation, characterized by exacerbated microvascular permeability, increased leukocyte adhesion to venular walls and ROS production, marked reduction in capillary perfusion. On the other hand, in the same hypoperfusion–reperfusion conditions, Ang 1–7 has totally opposite effects, preventing the microvascular alterations triggered by Ang II (Lapi et al. 2021). In this way, these data support the suggested competition between the RAS end final molecules, which are effective in exacerbating or preventing vascular damage under the pathophysiological conditions induced by hypoperfusion and reperfusion. Similar data have been reported by Arroia and coworkers studying the different effects of Ang II and Ang 1–7 in ischemic stroke (Arroja et al. 2016). The effects of Ang 1–7 during hypoperfusion and reperfusion were exclusively due to the activation of the MAS receptor. Its inhibition, indeed, completely abolished the protective effects of angiotensin 1–7, while the inhibition of the AT2R did not interfere with the protection exerted by Ang 1–7. The effects of diminazene aceturate, a stimulator of ACE2, both systemically and locally administered, indicate the key role of the ACE2 in balancing the overall RAS mechanisms, facilitating the catabolism of Ang II and the formation of Ang 1–7. By stimulating the activity of ACE2, diminazene aceturate DIZE) was able to preserve pial microcirculation from damage induced by hypoperfusion and reperfusion in the rat model. The local effect triggered by DIZE was impressive, because it was able to protect pial microcirculation after local application of Ang II under the condition of hypoperfusion and reperfusion. Therefore, the local balance between ACE and ACE2 enzymes could play a key role in facilitating or inhibiting the effects of Ang II on microvascular networks under the conditions of hypoperfusion and reperfusion.

Finally, it is worthwhile to consider the role of ACE and ACE2 in SARS-CoV2 infection. Taking into account the widespread presence of ACE and ACE2 in several organs, it has been hypothesized that the imbalance between the two arms of RAS may contribute to SARS-CoV2 infection. Audit and coworkers, as well as Lanza and colleagues (Oudit et al. 2009; Lanza et al. 2020), indeed, suggested that SARS-CoV2 disease (COVID-19), through ACE2 involvement in the intracellular entry of the virus, may be effective in inducing a prevalence of Ang II compared to the amount of Ang 1–7. Consequently, the imbalance in ACE and ACE2 activity, with the accumulation of Ang II, may induce the activation of AT1R with consequent microvascular damage.

The RAS imbalance was suggested by Lanza et al. (2020) as effective in pulmonary damage in severe acute distress syndromes. In particular, in mice models of SARS-CoV induced lung injury, there was a reduced expression of ACE2 on lung cell membranes; moreover, there was an increase in pulmonary Ang II levels after infection with Spike (S318–510)-Fc. The results indicated that mice had lung edema as well as increase in lung neutrophil infiltration and microvascular permeability (Kuba et al. 2021). Interestingly, in another model of lung damage, induced by hyperoxia, the mice were administered with DIZE (Diminazene Aceturate), an ACE2 agonist, and compared with mice treated with MLN-4760, an ACE2 inhibitor.

The findings indicated that lung ACE2 expression decreased as well as the activity of the enzyme during hyperoxia treatment. Moreover, there was enhanced Ang II activity compared with that of Ang (1–7). The treatment with DIZE during hyperoxia induced an increase in ACE 2 expression and activity, resulting in decreased ratio between Ang II and Ang (1–7). On the other hand, MLN-4760, inhibitor of ACE2, administered at the beginning of hyperoxia treatment, caused a decrease in ACE2 expression and activity, as well as an increase of Ang II proportion compared to Ang (1–7). Therefore, DIZE was effective in reducing the lung damage, the response to inflammation, the formation of oxygen-derived free radicals caused by hyperoxia. On the contrary, MLN-4760 was effective in increasing the lung damage (Fang et al. 2019).

In conclusion, Ang II and Ang 1–7 are effective in inducing opposite effects in different physiopathological conditions, and the imbalance between Ang II and Ang 1–7 could induce a prevalence of AT1R activation, which facilitates microvascular changes and redox stress with consequent organ implication. Therefore, the imbalance between ACE/Ang II/AT1R and ACE2/Ang (1–7)/MasR axes results in multiple organ dysfunction and uncontrolled inflammatory response.

14.2.2 The Pulmonary Involvement: Hypoxia and Acute Respiratory Distress Syndrome

In several studies, most common reported symptoms of COVID-19 are fever, cough (60–80%), shortness of breath (53–80%) and changes in taste (ageusia) and smell (anosmia) (64–80%). However, the clinical symptoms characterizing COVID-19 patients are a viral pneumonia and hypoxemic respiratory failure resulting in acute respiratory distress syndrome (ARDS) in 10–20% of patients, the most common reason for Intensive Care Unit (ICU) admission (Long et al. 2022). Bacterial and fungal coinfection affect up to 8% of patients. Concurrent viral infections are common, with 20.7% of COVID-19 patients infected with one other virus. The spreading of pandemic has been accompanied by extrapulmonary presentations and complications. All organs have been affected by SARS-CoV2 infection: myocardial disease occur in 20% of COVID-19 patients, represented by acute coronary syndrome, dysrhythmias, heart failure and myocarditis. Some studies suggest that cardiac involvement is present even in patients with minimal or no symptoms. Up to 80% of patients experience neurologic symptoms during the course of COVID-19. Gastrointestinal (GI) symptoms are common, with one third of patients with COVID-19 presenting first with GI symptoms: loss of appetite (40%) and diarrhea (50%). Dermatologic manifestations are present in up to 20% of cases: erythema or urticaria. The hematologic alterations are common in critical COVID-19 patients, with venous thromboembolism event in 31% of cases, taking into account that the multiple intravascular coagulation is one of the results induced by systemic inflammation.

In most cases of COVID-19 infection, the first step is the contact of the virus with the first airway cells, due to the virus-full droplets from infected people inspired by healthy subjects. Several studies have investigated the in vitro mechanisms of nose cells infection by SARS-CoV2 virus. Nasal airways are pseudostratified epithelia, including multiciliated apical epithelial, basal and mucus-producing goblet (or secretory) cells. The mucus layer is atop the ciliary brush and an underlying periciliary layer (PCL) keeps mucus away from the epithelia, but permits ciliary beating. Mucus traps virus particles, swept to the laryngopharynx by ciliated epithelial cells (CECs), constituting 80% of nasal epithelium with about 300 cilia beating in coordinated manner. Infectious particles are eliminated by coughing or swallowing. The PCL plays the key role to block the viral entry. Nasal and upper airways CECs are the main targets for the COVID-19 infections. SARS-CoV2 receptors, ACE2 with TMPRSS2, are localized in cilia of CECs. The infection is a two-step process: viral particles bind the ACE2 receptor on the surface of airways cilia. Cilia facilitated their transport through the PCL mucus layer. Only a few ciliated human nasal epithelial cells (HNEs) are infected at the beginning. Within 24 hours SARS-CoV2 virus is able to induce elongation of highly branched microvilli by stimulation of protein kinases, especially p21-activated kinases 1 and 4 (PAK1/4). These kinases facilitate the virus exit across the PCL layer accompanied by lateral spread to other region: it is suggested by mucociliary transport (Wu et al. 2022). Infection spreading from upper to lower airways can be facilitated by the changes in epithelium structure: the height and complexity gradually decrease along the respiratory tract, from the tall pseudostratified ciliated columnar cells interspersed with goblet cells to the tin pneumocytes. Taking into the account that ACE 2 receptor expression is well represented along the whole respiratory tract, it is reasonable to hypothesize that pneumocytes can be more easily attacked than nasal cells by SARS-CoV2 (Gelardi et al. 2021).

The dynamics of nasopharynx epithelial differentiation upon SARS-CoV2 infection have been assessed in the very accurate study by Chua and coworkers (Chua et al. 2020). They studied 5 healthy controls and 19 COVID-19 patients (8: moderate; 11 critical). They observed that basal cells differentiate into secretory cells, further differentiating into ciliated cells. Under the viral infection, the basal cells are able to differentiate in ciliated cells. Moreover, they found that immature secretory cells differentiate into ciliated cells, mediated by Interferon G-responsive cells. It is really interesting that the ACE2 epithelial target cells increased three-fold upon infection when compared to the cells in healthy controls. Secretory and ciliated cells contained the highest fraction of SARS-CoV2 infected cells, corresponding to ACE2+ cells overrepresented within the epithelial cells compared to the immune cells (58–98-fold enrichment). Moreover, ACE2 expression is interferon-responsive, because ACE2 gene in human airway has been proven to be induced by interferon, secreted by immune cells, such as Cytotoxic T Lymphocytes (CTLs).

The uptake of the virus by the cell stimulates the activation of metalloproteinase (ADAM 17) (a disintegrin and metalloproteinase 17) also known as tumor necrosis factor alpha (TNF-α) convertase (TACE). This enzyme regulates the conversion of TNFα of membrane in the soluble form, which has greater inflammatory effects. The metalloproteinases (MMPs) are a family of zinc-dependent enzymes that can degrade most extracellular matrix proteins. To date, 23 MMPs have been identified in humans (Raeeszadeh-Sarmazdeh et al. 2020), all of which are released as inactive zymogens (preproenzymes) secreted from cells into the extracellular space. Subsequently, MMPs are processed proteolytically or by modifying the thiol group by oxidation for activation (Wang et al. 2022).

Therefore, COVID-19 causes an overexpression of MMP, such as MMP-1, and vascular endothelial growth factor (VEGF)-A. MMP-1 is an interstitial collagenase capable of degrading collagen types I, II and III and plays a crucial role in vascular remodeling and diseases (Huang et al. 2023). It is necessary for tissue development and maintenance, but its overexpression promotes hyperactivation of MMP-1/PAR1 signaling, which can increase the expression of VEGF receptor 2, decrease vascular endothelial cell function and lead to excessive recruitment and activation of inflammatory cells (Quintero-Fabián et al. 2019). The 6-month follow-up after hospital discharge of patients with COVID-19 revealed that approximately half of the patients with moderate or severe COVID-19 developed impaired pulmonary diffusion and early fibrotic changes in association with elevated MMP-1. MMP-1 may play a role in tissue damage associated with SARS-CoV-2 infection and MMP-1 levels may be a prognostic biomarker of COVID-19.

MMP2, a constitutively expressed MMP in the brain is involved in central nervous system pathology, and MMP9 is a major inducible MMP released in neuroinflammatory responses. MMP-2 and MMP-9 levels are significantly elevated in the cerebrospinal fluid of patients with COVID-19 who have neurological syndrome compared to those without neurological syndrome (Mohammadhosayni et al. 2021). In addition, patients with respiratory failure due to COVID-19 have elevated MMP-9 in the circulation; overexpression of MMP-2 and MMP-8 caused lipid peroxidation, which resulted in intensive destruction of lung tissue in severe COVID-19 cases (Huang et al. 2023). Regulation of MMP expression and activity is complex. Mmp2−/− mice display more pronounced interstitial and perivascular fibrosis in response to angiotensin II (Hardy et al. 2018). SARS-CoV-2 binds to ACE 2 for intracellular invasion, thereby increasing vasoconstriction and inflammation.

Furthermore, the infected epithelial cells secrete a large amount of chemokines, effective in recruiting and activating immune cells populations at the site of virus attack. In patients with moderate COVID-19, there is a higher expression of the chemokine-ligand encoding genes compared to healthy controls: CXCL1, CXCL3, CXCL6, CXCL16 and CXCL17, known to recruit neutrophils, T cells (CTLs and CD4T/Treg) and mast cells. Ciliated cells express CCL15, known to facilitate the inflow of monocytes/macrophages and neutrophils through the CCR1 receptors. It is worth noting that the expression of chemokines and their receptors increases markedly in critical COVID-19 patients, indicating a massive recruitment of immune cells in the sites of inflammation, with higher activation of non-resident macrophages (nrMa), monocyte-derived macrophages (moMa) and cytotoxic T lymphocytes (CTL). nrMa were characterized by increase in expression of chemokine encoding genes CCL2 (encoding monocyte chemoattractant protein-1, MCP1), CCL3 (macrophage inflammatory protein-1 alpha, MIP1α), CCL20, CXCL1 and CXCL3. Moreover, these cells expressed the proinflammatory cytokines, such as interleukin 1-β (IL-1β), IL-8, IL-18 and tumor necrosis factor α (TNFα) at higher levels in critical COVID-19 patients.

However, comparing the viral load, measured by the total count of viral reads and viral copy number between higher airway mucosa and bronchial lavage (BL) samples, BL presented higher values, with increased number of positive macrophages and CTLs, characterized by stronger inflammatory signature, especially represented by augmented expression of Caspase 3 (CASP3), the effector of both immune cell types.

The virus infection, therefore, is characterized by cell entry and virus replication, with corresponding immune native and adaptive cell response: extravasation of immune cells and recruitment of blood monocytes and neutrophils to the infected tissue. Microvascular endothelial cells play a key role in this mechanism, because virus infection activates endothelial cells that facilitate the extravasation of immune cells. In the patients affected by COVID-19, it has been possible to identify immune cells in the upper respiratory tract: prevalent neutrophils are accompanied by resident macrophages, dendritic cells, nonresident macrophages, monocyte-derived macrophages, regulatory T cells, cytotoxic T cells, natural killer T cells, proliferating NKT cells, natural B cells and mast cells. The impressive cellular extravasation is linked to the impact of chemokines and proinflammatory cytokines, such as IL-1β and TNFα, known to increase microvascular permeability, on endothelial cells, that are stimulated to contract and release Weibel–Palade bodies substances, such as von Willebrand multimer factor and thromboxane A and B. Moreover, the endothelial cells bring to the surface P-selectin able to facilitate leukocyte adhesion and extravasation. Therefore, the activation and contraction of endothelial cells contribute to the increase in permeability: taking into the account the very large surface of alveolar pneumocytes (T1 and 2) and the corresponding endothelial layer in capillaries, the lower airway anatomical structures are under very strong virus-induced turmoil. The alveolar edema and extravasation of immune cells affect the gas exchange at alveolar levels, with increasing reduction of hemoglobin oxygenation and sequential hypoxemia. This condition can be blocked by immune response and/or therapeutical treatment. On the other hand, in critical COVID-19 patients, hypoxemia is accompanied by intravascular coagulation and lung infiltrates with respiratory failure exiting in Acute Respiratory Distress Syndrome (ARDS), requiring Intensive Care Unity admission.

Hypoxia appears to play a key role in the worsening of lung infection due to SARS-CoV-2. Data reported in COVID-19 patients by Ackermann and colleagues (2020) indicate diffuse alveolar damage (DAD) with necrosis of alveolar lining cells, pneumocytes type 2 hyperplasia and linear intra-alveolar fibrin deposition. In four of seven cases, the changes were focal with mild interstitial edema. In three cases, there were homogeneous fibrin deposit and marked interstitial edema with early intra-alveolar organization. They observed a distorted vascularity with structurally deformed capillaries: the elongated capillaries showing sudden changes in diameter and the presence of intussusceptive pillars within their lumen. They found ultra-structural damage to the endothelium, as well as the presence of intracellular SARS-CoV-2; moreover, they observed nearly total occlusion of precapillary and postcapillary vessels. In another paper, Ackermann and colleagues (2023) describe the complex three-dimensional phenomenon of bronchopulmonary shunting in COVID-19 and the excessive neovascularization with a complete remodeling of lung circulation (Fig. 14.2). These data support the pivotal role of hypoxia during COVID-19, effective in inducing the activation of hypoxia-inducible factors (HIF) resulting in increased production of VEGF in COVID-19 patients and sequential increased microvascular permeability and activation of neoangiogenesis. Moreover, HIF-1 expression is induced by viral infection and obesity, exacerbating the production of proinflammatory cytokines. Therefore, increasing hypoxia amplifies the COVID-19 pathogenetic mechanisms, contributing to hyperinflammation and coagulopathy, such as diffuse intravascular coagulation (Somers et al. 2020). HIFs, indeed, promote the expression of tissue factor (TF) and plasminogen activator inhibitor-1 (PAI-1), while suppressing the expression of tissue factor pathway inhibitor (TPFI), thrombomodulin and protein S, facilitating the development of intravascular coagulation. Hypoxia activates macrophages and other cells to synthesize and release proinflammatory cytokines, such as TNF-α and IL-6. Furthermore, HIF-1 facilitates the expression of inflammasome NLRP3 in a variety of inflammasome cells linked to IL-1 production and pyroptosis. Finally, hypoxia or hypoperfusion facilitate the proinflammatory effects of Ang II, through activation of ATR1, triggering increase in microvascular permeability, ROS formation and profibrotic mechanisms, due to the imbalance between ACE/ACE2 activity in COVID-19 patients.

The worsening of clinical symptoms due to the lung viral infection and correlated angiocentric inflammation, with significant reduction in alveolar gas exchanges and arterial blood oxygen pressure, results in acute respiratory distress syndrome (ARDS).

ARDS is an acute life threatening inflammatory lung injury manifested by hypoxia and stiff lungs due to increased pulmonary vascular permeability and almost always requiring mechanical ventilation support. ARDS was first described by Ashbaugh and colleagues in 1967 (Bernard et al. 1994) and represents an acute response to diverse provoking trigger factors and etiologies, resulting in bilateral lung opacities on radiography and hypoxemia. Numerous studies have addressed the

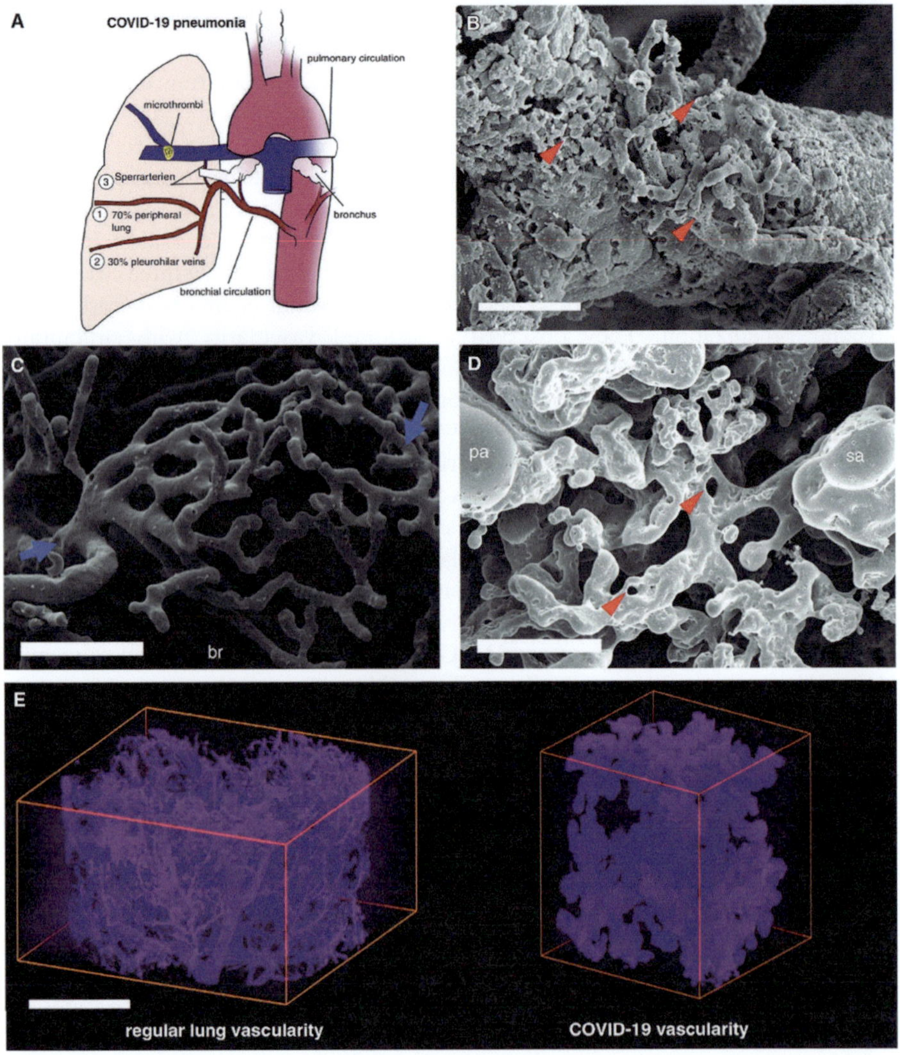

Fig. 14.2 Intrapulmonary shunting via bronchial circulation in coronavirus disease (COVID-19) pneumonia. In (**a**) scheme of the bronchial circulation arises primarily direct from the thoracic aorta and intercostal arteries and, under normal conditions, constitutes less than 2% volume of the left cardiac output. However, the bronchial flow can significantly exceed the regular hemodynamics upon hypoxic and inflammatory stimulus. In (**b**) Scanning electron micrograph of a microvascular corrosion cast demonstrates the expansion of vasa vasorum of an intralobular artery by intussusceptive angiogenesis (red arrowheads) in the lung tissue of a patient who succumbed to severe COVID-19 pneumonia. Scale bar, 100 μm. In (**c**) Scanning electron micrograph of a microvascular corrosion cast demonstrates the peribronchial plexus in healthy control lungs (blue arrows). Scale bar, 200 μm. In (**d**) Scanning electron micrograph of a microvascular corrosion cast reveals the cross-circulation in COVID-19 lung tissue between a Sperrarterie (or blocked artery) and a lobular artery in a secondary pulmonary lobule, which is characterized by plexiform-like vascular proliferations by intussusceptive angiogenesis (red arrowheads). Scale bar, 200 μm. In (**e**) Three-dimensional evaluation of microvascular corrosion casts by Synchrotron radiation X-ray tomographic microscopy illustrating the irregular, hierarchical vascularity with dilated, blunt-like vessels in COVID-19 pneumonia. Scale bar, 100 μm. br = bronchiole, pa = lobular artery; sa = Sperrarterie (blocked artery). (Reprinted with permission of the American Thoracic Society. Copyright © 2023 American Thoracic Society. All rights reserved Ackermann et al. 2023)

various clinical aspects of the syndrome, its pathogenesis, risk factors and treatment. However, despite the intense research, only few effective therapies for ARDS have been postulated, including the lung protection strategies.

The American-European Consensus Conference (AECC, 1994) defined ARDS as the acute onset of hypoxemia (arterial partial pressure of oxygen to fraction of inspired oxygen (PaO_2/FIO_2] $\leq$ 200 mm Hg) with bilateral infiltrates on frontal chest radiograph, with no evidence of left atrial hypertension and acute lung injury (ALI). ALI was defined using similar criteria, but having $PaO_2/FIO_2 \leq 300$ mm Hg. Over the years, with the ongoing research on this topic, issues regarding the validity and reliability of this definition emerged. A panel of experts assembled in 2011 (an initiative of the European Society of Intensive Care Medicine endorsed by the American Thoracic Society and the Society of Critical Care Medicine) and developed the Berlin Definition of ARDS (Bordes et al. 2014; Ranieri et al. 2012).

The severity of the hypoxemia defines the severity of ARDS: (1) Mild ARDS—The PaO_2/FiO_2 is >200 mmHg, but ≤300 mmHg, on a ventilator with a positive end-expiratory pressure (PEEP) or continuous positive airway pressure ≥ 5 cm H_2O. (2) Moderate ARDS—The PaO_2/FiO_2 is >100 mmHg, but ≤200 mmHg, on a ventilator with a PEEP ≥5 cm H_2O. (3) Severe ARDS—The PaO_2/FiO_2 is ≤100 mmHg on a ventilator with a PEEP ≥5 cm H_2O.

Several factors are involved in the progression of ARDS, such as proinflammatory cytokines, facilitating the extravasation of neutrophils into the alveoli, impairing microvascular endothelium as well as the alveolar epithelium, resulting in pulmonary edema (Yang et al. 2021). These events result in alterations of gas exchange, diminution of lung compliance. At the same time, there is a rise in lung blood pressure with changes in lung hemodynamics. COVID-19 associated ARDS has been hypothesized to be different compared to the common ARDS, taking into the account that COVID-19 patients present hypoxemia, correlated to compliant lung compared to classical ARDS. Therefore, typical protective ventilatory settings might not be indicated in patients with COVID-19-related ARDS. It has been observed that infection caused by SARS-CoV-2 induces alterations in lung microvascular endothelial cells. Moreover, lung epithelial layers present marked changes, correlated to the blood vessel extension in pulmonary circulation. Endothelial cell activation facilitates the extravasation of immuno-competent cells with consequent activation of the mechanisms inducing inflammatory cascade and coagulation. These processes can cause barrier damage, gas exchange dysfunction and coagulation activation. In COVID-19 patients, several pathological characteristics have been reported: from lung edema and vascular congestion up to focal intrapulmonary hemorrhage as well as infarction.

It has been reported that SARS-CoV-2 infection promotes endothelialitis in lung microvasculature; in particular, intracellular viral bodies have been detected in autoptic specimen of lung tissues. Postmortem studies have shown that the lung vascular system of people affected by COVID-19 exhibited distinctive features. The predominant vascular changes including endothelial cell injury, pulmonary capillary microthrombosis and hyperplasia are the main features of the patients with COVID-19, with consequent dissimilarities in the mechanisms, involved in development of the pathological conditions correlated to dissimilarities in the reaction to the therapy, when this acute syndrome has been compared with typical ARDS.

It has been reported that COVID-19 patients show a characteristic chest CT with ground-glass opacity present in both lungs with a peripheral dissemination. However, the chest CT did not show the ARDS features present in the typical ARDS. In particular, pulmonary compliance has been defined as normal in several persons affected by COVID-19, presenting the acute respiratory distress syndrome. Moreover, ARDS induced by several factors is different compared to the syndrome described in COVID-19 patients (Li and Ma 2020). It has been observed that in typical ARDS, the lung compliance, under static conditions, deteriorated according to the degree of alterations in pulmonary functions, in direct correlation: increasing hypoxemia caused a decrease in compliance. However, in COVID-19 patients, the compliance under static conditions did not change, even though increased hypoxemia was present during ARDS. It is reasonable to suggest that many other factors could affect hypoxemia, that is related to alveolar edema or obstruction: characteristic in typical ARDS.

In an accurate study, Gattinoni and colleagues (2020) described the variations of clinical and physiopathological conditions in the patients affected by COVID-19, presenting ARDS. They suggested two phenotypes, according to compliance or elastance, the degree of ventilation and perfusion of the lung, the pulmonary weight and the degree of pulmonary recruitment: type Low (with low values of the parameters) and type H (with high values of parameters). The type H appeared to be more close to the typical ARDS, with worsening clinical conditions. The suggestion was that patients get L type at the beginning of the main disease, then they get H type, indicating the increase in the severity of the disease and in pulmonary conditions. Several studies have reported this progression of the disease, promoting the classification by CRITICAL CARE COMMUNITY in subtypes of ARDS induced by COVID-19. It is worth noting that the COVID-19 patients affected by ARDS can present different responses to the therapy. In general, ARDS caused by COVID-19 is characterized by pulmonary alveoli flooding and edema, inducing changes in gas exchange and hypoxemia, with marked changes in endothelial functions and endothelialitis, effective in inducing intravascular clotting and occlusion of vasculature. Therefore, COVID-19-induced ARDS present peculiar characteristics, when compared with typical ARDS: endothelium inflammation with consequent intravascular coagulation. This sequence of events is correlated to hypoxemia, characterized by intrapulmonary shunt with consistent ventilation of portions of the lungs scarcely irritated (dead space).

Grasselli and colleagues reviewed 26 studies for a total of 14,075 critically ill COVID-19 patients, requiring mechanical ventilation. The median ratio of partial pressure of arterial oxygen to fraction of inspired oxygen (PaO_2/FiO_2) was in the range from 102 to 198 mm/Hg, while the positive end expiratory pressure (PEEP) values ranged from 9 to 16.5 cmH_2O and static respiratory system compliance values ranged between 24 and 49 mL/cmH_2O (Grasselli et al. 2021). It is interesting to note the prone position was widely used in COVID-19 patients, ranging from 17% to 81% to facilitate blood oxygenation. This rate was higher than previously reported in classical "ARDS." However, in another study, Grasselli and coworkers (2020) recruited 71 COVID-19 patients, compared with a data set of non-COVID-19-related classical ARDS obtained from the physiological database ($n = 269$) used in the creation of the Berlin definition and the database of the LUNG-SAFE study ($n = 3022$). Median static compliance of the respiratory system was 28% higher in COVID-19 patients [$n = 297$, 41 mL/cm H_2O (33–52)] than in those with classical ARDS [$n = 960$, 32 mL/cm H_2O (25–43)]. However, only 17 (6%) of 297 COVID-19 patients with ARDS had compliance greater than the 95th percentile of the patients with classical ARDS. Moreover, static compliance decreased as PaO_2/FiO_2 decreased in patients with classical ARDS, while it remained unchanged in COVID-19 patients. According to D-dimer (the soluble fibrin-degradation product, marker of activation of coagulation and fibrinolysis) blood concentration and static compliance, the patients were divided in four groups: (1) High D-dimer, Low Compliance, HDLC (71 patients, 27%): with D-dimer concentrations greater than median in COVID-19 ARDS (1880 ng/mL) and static compliance equal or less than the median (41 mL/cm H_2O); (2) Low D-dimers, High Compliance, LDHC (67 patients, 26%); (3) Low D-Dimers, Low Compliance, LDLC (60 patients, 23%); (4) High D-dimers, High Compliance, HDHC (63 patients, 24%). Patients with D-dimer concentration equal or less than median had normal perfusion scans regardless of compliance. Conversely, 15 of 16 patients with D-dimer concentration greater than the median had bilateral diffuse areas of hypoperfusion consistent with the presence of thrombi or emboli. 28 day mortality was 36% (93 of 261 patients): 40 of 71 (56%) patients in HDLC group. The Authors conclude that the patients with COVID-19-related ARDS have a form of injury similar to classical ARDS: when increased parenchymal damage (low static compliance) and increased D-dimer concentrations occur together, mortality is extremely high.

14.2.3 The Cytokine Storm

The cytokine storm has been defined as a hyperinflammatory state secondary to the production of cytokines by a dysregulated immune system (Zhang et al. 2022a, b). This condition has been first described in 1993 for the graft-versus-host disease following allogeneic hematopoietic stem cell transplantation. Moreover, in 2010 it has been reported to define the adverse syndrome developed after chimeric antigen receptor (CAR) T-cell therapy (Fajgenbaum and June 2020). The cytokine release syndrome described a similar condition induced by infusion of muromonab and CD3 (OKT3, antibody anti-CD3). Cytokine storm has been defined as a life-threatening systemic inflammatory syndrome involving elevated levels of circulating cytokines and immune-cell hyperactivation, triggered by pathogens or other conditions, such as sepsis, autoinflammatory disease, primary and secondary hemophagocytic lymphohistiocytosis and multicentric Castleman's disease. The last disease was described in 1954 and 1956 (Castleman and Towne 1954; Castleman et al. 1956) and is characterized by lymphadenopathy with high serum levels of IL-6.

The cytokine storm or the cytokine release syndrome has been widely discussed, because of disagreement about the mechanisms interfering with an appropriate inflammatory response. Previous studies indicate that severe acute respiratory syndrome coronavirus 1 (SARS-CoV-1) and Middle East Respiratory Syndrome Coronavirus (MERS) were effective in inducing a cytokine storm. A great number of studies from the beginning of SARS-CoV-2 pandemic have documented that COVID-19 patients have increased levels of inflammatory chemokines and cytokines. When the alveolar epithelial cells are infected there is an interesting cross-talk with the immune cells, resulting in intensive induction and activation of innate and adaptive immune responses. The chemokines and cytokines detected in COVID-19 patients have been reported in several studies. The list is very long: Interleukin 6 (IL-6), IL-1β, IL-2, IL-10, tumor necrosis factor-related apoptosis-inducing ligand (TRAIL), IL-15, IL-8/CXCL8, IL-12, IL-5, IL-17, CD163, tumor necrosis factor-β (TNF-β), IL-3, IL-4, IL-13, CXCL2/SDF-1 (stromal cell-derived factor-1), sTNFR1, leukocyte inhibitory factor (LIF), macrophage inflammatory proteins (MIPs), IL-7, IL-9, regulated upon activation normal T cell expressed and presumably secreted (RANTES)/CCL5, CCL7, CCL8, CCL13, CCL17, CXCL10/IP10 and CXCL11, which are associated with the poor prognosis of COVID-19. There is increased secretion of MIP-1α, MIP-1β, transforming growth factor-alpha (TGF-α), monocyte chemoattractant protein-1 (MCP-1/CCL2), TNF-α, interleukin 1-α (IL-1α), nerve growth factor-beta (NGF-β), basic fibroblast growth factor (FGFb), IFN-α2, granulocyte-colony stimulating factor (G-CSF), tumor necrosis factor receptor superfamily 9 (4-1BB/TNFRS9/CD137), granulocyte macrophage-colony stimulating factor (GM-CSF), neurite growth promoting factor 2 (Midkine), IL-21, Flt-3 ligand (FLT3, growth factor of hematopoietic lineages), chemokine ligand 28 (CCL28), Fas/TNFSF6 ligand (FasL), IL-25, IL-23, CD40 ligand (CD40L)/CD154/TNFSF5, CXCL14/BRAK (breast- and kidney-expressed chemokine), IL-31, programmed death ligand 1 (PD-L1/B7-H1), IFN-γ and IFN-γ-inducible protein 10 (IP-10) (Silva et al. 2023a, b; Zanza et al. 2022).

Moreover, inflammatory infiltrates within different tissues in COVID-19 patients have been reported in bioptic and autoptic samples. These data suggest that in the majority of cases, the disease represents a self-limiting flu-like syndrome. In a minority of patients (10–20%), suggested predisposed subjects, the infection of lung cells, in particular of type 2 pneumocytes, can be effective in inducing a rich inflammatory cell infiltrate, characterized by neutrophils, macrophages, CD8+ and CD4+ T lymphocytes, with consequent marked release of cytokines leading to bilateral pneumonia and ARDS resulting in multiorgan damage (Domingo et al. 2020). The mechanism suggested in the development of cytokine storm during SARS-CoV2 infection is the inadequate immune response to the virus infection. Blanco-Melo and colleagues have observed an impaired capacity to produce interferons by infected cells, reducing the host immune response against the viral attach. On the other hand, there is a huge increase in neutrophil- and macrophage-recruitment chemokines release by the infected cells (Blanco-Melo et al. 2020). Moreover, in COVID-19 patients, autoantibodies against several immunomodulatory proteins have been detected: in particular, anti-type 1 interferon antibodies, associated to critical disease and death (Bastard et al. 2021). In summary, in the first phase of infection, the innate immune response is unable to clear the infected cells: the virus replication continues. Ly and colleagues have observed that SARS-CoV-2 is able to replicate inside the alveolar macrophage (Lv et al. 2021). In the second phase, the immune system, recovering the ability to fight the virus, produces an excessive reaction. The activation of inflammasomes in macrophages, especially the NLRP3, contributes to the exacerbation of the inflammatory process by recruiting more signature of inflammatory mediators (Silva et al. 2023a, b).

A role in the cytokine storm resulting in systemic inflammation has the activation of the complement system, as highlighted by Afzali and colleagues (2022). The complement system is the host immune system's first response to pathogens. The complement is constituted by over 30 proteins organized in the 3 activation pathways: the classical, the lectin and the alternative, converging on the cleavage of central component C3 by enzyme complexes: the classical/lectin, and the alternative C3 convertases. The generated activation products (C3a, C4a, C3b and C4b) are able to eliminate pathogens through opsonization, to attract and activate neutrophils and macrophage, to enhance humoral immunity and T-cell response.

SARS-CoV-2 N protein has been shown to activate mannose-binding protein-associated serine protease 2 through MBL (a C-type lectin) and increases C4b deposition in vitro (Noris et al. 2020). Moreover, in lung tissue of COVID-19 patients there were deposits of mannose-binding protein-associate serine protease 2 and C4. These data indicate that both MBL opsonization and deposition of C3 and C4 onto virions participate to SARS-CoV-2 neutralization. Furthermore, studies on other viruses, such as influenza A virus H1N1, indicate that the lectin pathway could facilitate the excessive inflammatory response associated to viral infection. It is worth noting that both natural and induced antibodies form immune complexes with viral antigens, able to activate C1 and promote complement activation via the classical pathway. This process is accompanied by C3b deposition on the viral envelope, interfering with virus attachment to susceptible cells and facilitating the phagocytosis by macrophages.

In COVID-19-associated tissue inflammatory injury the terminal pathway, constituting the common endpoint of the 3 complement activation cascades, may play a role. The binding of C3b to either the classical/lectin or the alternative C3 convertase is able to form the C5 convertase, that cleaves C5, generating C5a and C5b. C5a is defined anaphylatoxin, a potent stimulating factor involved in exacerbating inflammatory reactions. C5b facilitates the formation of C5b-9, able to insert itself into cell membranes causing cell injury and dysfunction. C5a promotes monocyte and neutrophil attraction, aggregation and activation to generate an oxidative burst with the release of ROS, exerting a critical role in virus-induced lung damage and mortality. The interaction between C5a and C5aR on endothelial cells causes the upregulation of adhesion molecules, facilitating the adhesion of leukocytes and promoting their transmigration into lung parenchyma. C5a triggers mast cell degranulation, initiates cytokine storm and increases vascular permeability.

SARS-CoV-2 can, therefore, activate the complement system either directly through the lectin pathways and/or the alternative pathway. However, the endothelium injury and dysfunction promoting inflammation and thromboinflammation, i.e., inflammation associated with coagulation and thrombosis, are effective in activating the complement system. The viral Spike and Nucleocapsid proteins are recognized by lectin pathways components, activating the complement. Moreover, IgG and IgM antibodies directed against the biding domain of S protein activate the complement system. The S protein can interfere with the alternative pathway of complement activation by binding heparan sulfate and competing with factor H, a negative regulator of complement activation. In COVID-19 patients, high circulating levels of C5a and C5b-9, the terminal activation fragment, correlate with clinical severity. Noris and colleagues (2020) suggest that C5a directly interacts with its C5aR on endothelial cells, causing profound perturbation on the physiologically thromboresistant endothelium, including upregulation of tissue factor and loss of thrombomodulin, which activates the coagulation and exocytosis of P-selectin and ultra-large multimers of von Willebrand Factor (vWF), facilitating platelets adhesion and aggregation. C5b-9 induces endothelial activation and dysfunction, including the expression of tissue factor, adhesion molecules and the release of chemokines and platelets-activating factor, able to promote inflammation, to increase microvascular permeability and to trigger coagulation process. C5b-9 is a powerful platelet agonist inducing the secretion of storage granules and the release of procoagulant platelets microparticles.

It is interesting to note that different authors suggest, as above reported, that the renin–angiotensin–aldosterone system (RAAS) may play a role in the cytokine storm. As above discussed, the main receptor used by SARS-CoV2 to enter the human cells is angiotensin-converting enzyme 2 (ACE2). The imbalance between ACE and ACE2, due to the reduction of ACE2 availability, can induce an increase in angiotensin II activity, through ATR1, increasing the proinflammatory, prothrombotic and profibrotic effects, that are counterbalanced by Angiotensin 1–7. The prevalence of angiotensin II could facilitate the proinflammatory mechanism and facilitate the cytokine storm.

The activation of infected and immune cells is accompanied by a marked increase in reactive oxygen species (ROS) production, known to induce oxidative stress, exacerbating the cytokine storm and facilitating tissues and organ damage.

Finally, the impact of infection on pulmonary circulation and microcirculation causes the activation of endothelial cells and pericytes, with the release of several factors, contributing to the cytokine storm. In critical COVID-19 patients, the cytokine storm and lung or other organ damage result in systemic inflammation with a prothrombotic state and hypercoagulability, ending in multiorgan failure.

14.2.4 The Systemic Inflammation and Multiorgan Failure

14.2.4.1 Endothelium and Endothelialitis

The virus replication, insufficiently opposed by innate immune responses, promotes the development of an acute inflammation of endothelial cells in the pulmonary microcirculation accompanied by the cytokine storm, luminal plugging of inflammatory cells and intravascular thrombosis. Hadjadj and coworkers (2020) carried out an integrated immune analysis on a cohort of 50 COVID-19 patients with various disease severity. A distinct phenotype was observed in severe and critical patients, consisting of a highly impaired interferon (IFN) type I response, characterized by no IFN-β and low IFN production and activity, which was associated with a persistent blood viral load and an exacerbated inflammatory response. Inflammation was partially driven by the transcriptional factor nuclear factor-κB (NF-κB) and characterized by increased tumor necrosis factor-α and interleukin-6 production and signaling.

The endothelial cell inflammation or endothelialitis is facilitated by pre-existing morbidities, such as diabetes, arterial hypertension and cardiovascular diseases, and can extend into the systemic circulation, leading to systemic inflammation and resulting in multiorgan failure. Early reports from China and Europe indicate that the severity of patients' condition and mortality increased exponentially above the age of 60 years, suggesting that human aging facilitates SARS-CoV-2 infection (Vrints et al. 2020).

Ackermanm and colleagues (2020) compared seven lungs obtained during autopsy from patients who died from COVID-19 with seven lungs from patients died from ARDS secondary to influenza A (H1N1). Ten age-matched uninfected lungs were the controls. The histological pattern in lungs from COVID-19 and influenza patients was diffuse alveolar damage (DAD) with perivascular T-cell infiltration. However, in COVID-19 patients, the lungs presented distinctive vascular features, consisting in severe endothelial injury associated with the presence of intracellular virus and disrupted cell membranes. Moreover, pulmonary vessels displayed widespread thrombosis with microangiopathy. Alveolar capillary microthrombi were prevalent in COVID-19 patients, accompanied by new vessel growth, through a mechanism of intussusceptive angiogenesis. It is worth noting that COVID-19 lungs showed increased expression of ACE2 in alveolar epithelial cells and endothelial cells, compared with the scarce expression in control lungs, while ACE2 positive lymphocytes were present in perivascular tissue and in the alveoli, but not in controls. CD3-positive, CD4-positive and CD8-positive T cells were detected in precapillary and postcapillary vessel walls (diameter 200–400 μm). CD15-positive neutrophils were adjacent to the alveolar epithelial lining. Moreover, 79 inflammation-related genes, out of 249 examined, were differentially regulated only in specimens from COVID-19 patients. Impressive is the distribution of thrombi in pulmonary arteries (diameter: 1–2 mm) without complete luminal obstruction, in alveolar capillaries (fibrin thrombi) and in postcapillary venules (diameter: <1 mm). By three-dimensional micro-computerized-tomography (mCT) of the pulmonary specimens, the lungs of COVID-19 patients presented nearly total occlusion of precapillary and postcapillary vessels. It is interesting to note that transmission electron microscopy of the COVID-19 endothelium demonstrated ultrastructural damage to the endothelium, as well as the presence of intracellular SARS-CoV-2. The virus was identified also in extracellular space. All the data reported in this accurate study prompted the Authors to define "angiocentric" the inflammation induced by SARS-CoV2. Therefore, the endothelium is the main target of virus infection. Endothelial cells and SARS-CoV-2 have an intimate relationship according to Cunha Barbosa and colleagues (2021). A lot of studies and comments have highlighted the central role of the endothelial cells in the pathogenetic mechanisms of COVID-19. The involvement of endothelial cells by virus infection causes the release of many factors, such as vWF, effective in triggering intravascular coagulation, as observed by Ackermann and colleagues in the lung of COVID-19 patients. Growing evidence of endothelial cell involvement has been reported from the spreading of COVID-19 pandemic, such as the circulating endothelial cells (Guervilly et al. 2020) and circulating markers of endotheliopathy, such as angiopoietin-2, follistatin-like protein 3 (FSTL3) and plasminogen activator inhibitor-1 (PAI-1) and of angiogenesis (vascular endothelium growth factor: VEGF-A; platelet-derived growth factor: PDGF-AA and PGGF-AB/BB) (Pine et al. 2020). Moreover, increases in circulating vWF and soluble thrombomodulin were correlated to severity of SARS-CoV-2 infections (Goshua et al. 2020). The crucial contribution of endotheliopathy and thrombocytopathy to COVID-19 thromboinflammation has been reviewed by Gu and colleagues, stimulating collaborative efforts to target both thrombocytopathy and endotheliopathy (Gu et al. 2020). Interesting data are reported in two studies, where plasma proteomics and eicosanoid metabolomics have been employed by Roh and colleagues (2022) and Voiriot and coworkers (2022), respectively.

The study by Roh and colleagues was carried out in age-matched non-COVID-control subjects ($n = 26$) and in COVID-19 patients with moderate disease [hospitalized, but not requiring intensive care unit (ICU) admission, $n = 25$] and COVID-19 patients with severe disease (requiring ICU care, $n = 29$). The mean age of the participants was 64 ± 11 years with 40% woman. A parallel experimental study was carried out in hamsters uninfected and SARS-CoV-2 infected ($n = 3$/group). In this very interesting study, 4730 distinct human proteins were analyzed (4996 were the analytes). The first impressive result indicated that in age-matched subjects, the senescence-associated secretory phenotype (SASP-1), a marker of biological aging, was the most highly enriched process. Moreover, using a more comprehensive SASP-2 set including all published SAPS proteins, the results show that SAPS-2 was the most significant upregulated process, associated with cardiac involvement. In infected hamsters, the same enrichment of SAPS-1 and SAPS-2 was found in lung transcriptional profiles compared with age-matched noninfected hamsters, showing that SARS-CoV-2 infection can induce these age-related senescence genes. It is worth noting that in COVID-19 patients, extracellular histones, markers of neutrophil extracellular traps and von Willebrand factor (vWF) were the most elevated circulating proteins: the increase of vWF was by 341%. It is interesting that 171 plasma proteins were differently expressed between the two groups of COVID-19 patients. The most down-regulated protein (44% reduction) was a Disintegrin and Metalloproteinase with a Thrombospondin type 1 motif, member 13 (ADAMTS13), a secreted protease inhibiting thrombosis by cleaving vWF multimers. It was found that vWF/ADAMTS13 ratios increased with increasing COVID-19 severity. ADAMTS13 levels were inversely correlated to D-dimer levels, indicating that the overall thrombotic conditions are higher in COVID-19 patients with decreased levels of ADAMTS13. It is interesting that in SARS-CoV-2 infected hamsters, ADAMTS13 gene expression was reduced by 67% compared to controls. The lung of infected animals presented vascular thrombus formation and small vessels fibrin deposition. There was a decrease in mRNA levels resulting in reduced synthesis of the antithrombotic protein. This was accompanied by a 3.8-fold increase in vWF expression. Furthermore, investigating the involvement of transforming growth factor beta (TGFβ) superfamily in the mechanisms of COVID-19 cardiovascular injury, it was found that FSTL3, among 526 proteins differently expressed, was positively correlated with N-terminal of brain natriuretic peptide (NT-proBNP), the marker of heart failure, and cardiac troponin (TnT), the marker of myocardial necrosis. However, the data indicate that activins and TGF-β increased the expression of FSTL3 in cardiomyocytes and the percentage of senescent cardiomyocytes in culture. TGF-β1 was increased by 74% in COVID-19 patients as well as in lungs of SARS-CoV-2 infected hamsters.

All these data support the endothelial cell damage induced by SARS-CoV-2 infection first at pulmonary levels. Afterward, the release of prothrombotic factors and the down regulation of ADAMTS13 facilitate massive pulmonary intravascular thrombosis, with concomitant release of angiogenic factors. The lung vascular turmoil can be arrested by immune responses: innate and adaptative. However, the effect of SARS.CoV-2 on innate immune response and the effects on the senescence-associated secretory phenotype impair the host defense response, causing a replication of SARS-CoV-2 resulting in systemic inflammation and multiorgan failure.

Voiriot and colleagues studied 76 COVID-19 patients admitted to the ICU, with a mean age of 64 years (range 52–70) and 71% male, dividing them into two groups based on 28-day outcomes. Forty-five patients (59%) had a poor outcome. The patients underwent fiberoptic bronchoscopy with bronchoalveolar lavage (BAL). Blood and BAL fluid (BALF) were analyzed using cytological and biochemical protocols. Additionally, an eicosanoid metabolomic analysis was performed on serum and BALF cell-free supernatants. The results were noteworthy because TxB2 (a derivative of TXA2) and 20-HETE (20-Hydroxy- 5Z,8X,11Z,14Z-hydroxyeicosatetraenoic acid) levels in BALF, were increased accompanied by increased VEGF. These values were higher in COVID-19 patients with poor outcome, when compared with their counterparts. TxB2 has been shown to play a key role in neutrophil adhesion and endothelium activation in experimental models of acute lung injury (Zarbock et al. 2006). Moreover, TxA increases almost 15 times in the experimental model of rat pial microvascular hypoperfusion–reperfusion injury (Lapi et al. 2020). Furthermore, 20-HETE has been proven to promote endothelial dysfunction through the NF-kB signaling, effective in inducing the transcription of angiotensin-converting enzyme (Garcia et al. 2016). The eicosanoid signature of vascular injury was detected also in the systemic compartment, because of higher blood concentration of cytochrome P450-dependent lipids, such as 8,9 EET (8,9-Epoxy-5Z,11Z,14Z-eicosatrienoic acid) and 11,12 EET (11,12-Epoxy-5Z,8Z,14Z-eicosatrienoic acid), known to promote arteriolar vasoconstriction, in COVID-19 patients with a poor 28 day outcome compared to their counterparts. It was found a very high pulmonary-to-systemic gradient for IL-8, a key neutrophil chemoattractant, which was ten times more concentrated in BALF than in blood, suggesting that pulmonary vascular injury induced by SARS-CoV-2 infection causes leakage of alveolar fluid and a loss of lung compartmentalization. Hyperpermeability is effective in promoting the extravasation of proinflammatory mediators from alveoli to the systemic compartment, leading to marked systemic inflammation response resulting in multiorgan failure. The results reported by Hadjadj and colleagues (2020) on the higher blood level of IL-6 compared to low level of RNA suggest high pulmonary production followed by massive release into the systemic compartment, especially in the severe COVID-19 patients.

The vascular dysfunction or endotheliopathy, however, is still debated and new several data suggest a "growing role for pericytes" (McQuaid and Montagne 2022). Avolio and coworkers suggest that the spike protein of SARS-CoV-2 can disrupt human cardiac pericytes function through CD147 receptor-mediated signaling and promote a noninfective mechanism of COVID-19 microvascular disease. The effects are secretion of proinflammatory molecules typically involved in cytokine storm and production of proapoptotic factors causing endothelial cell death (Avolio et al. 2021). Furthermore, new organoid models have been implemented to study the pericyte function under SARS-CoV-2 infection. Lu Wang and colleagues studied the SARS-CoV-2 infection in a human three-dimensional neural-perivascular "assembloid" with pericyte-like cells (PLCs) integrated into a cortical organoid (Wang et al. 2021). Before infection, PCLs facilitate astrocytic maturation and production of basement membrane components. Under SARS-CoV-2 infection, the PLCs served as viral replication hubs, with virus spreading to astrocytes and mediating type I interferon transcriptional responses, resulting in astrocyte death.

Khan and coworkers (2022) have observed the presence of the Spike glycoprotein in NG2-positive (proteoglycan) pericytes of lung tissues obtained postmortem from five patients (out of eight) who died from COVID-19, while the S protein was undetectable in two age-matched non-COVID-19 control lung autopsies. Moreover, the expression of VE-cadherin (CD144), as a marker of microvascular permeability, results decreased in nonventilated COVID-19 patients, most markedly increased in patients with macro- and micro-thrombosis compared with controls. Furthermore, in lung capillaries of COVID-19 patients there were platelet recruitment and vWF deposition, while in lung large vessels there were thrombi rich in vWF and fibrin with low content in platelets. The Authors implemented a 3D human vascular organoid, with vascular endothelium (CD144+) and pericytes (CD140b+). Organoids were exposed to SARS-CoV-2 and assessed for Spike proteins after 48 hours. Colocalization analysis indicates a strong presence of the Spike in pericytes with lower levels detected in endothelial cells. There was increased apoptosis of endothelial cells and pericytes: higher in pericytes than in endothelial cells. It is worth noting that SARS-CoV-2 infection progressively changed the expression and distribution of NG2 and decreased CD144 expression, due to apoptosis of pericytes, characterized by their preferential uptake of the virus. This mechanism was associated with the disruption of the vessel architecture, detachment of pericytes and loss of vascular integrity. Finally, the Spike protein (S) and Nucleocapsid protein (N) decreased the number of endothelial cells, in the same model, but were unable to activate the endothelial cells.

All data indicate that the SARS-CoV-2 infection is facilitated by ACE2 receptor in epithelial cells. The replication of virus can be blocked by immune response, in most cases. However, the deficiency of the native immune response may facilitate the progression of the pathological mechanisms with replication in the pericytes and endothelial cells, with activation of chemokine and cytokine production. This process is accompanied by increase in microvascular permeability and recruitment of immune and adaptative cells, producing a lot of molecules able to trigger a cytokine storm. The activation of pericytes and endothelial cells triggers intravascular coagulation resulting in systemic inflammation up to the multiorgan failure, requiring Intensive Care Unit.

14.3 Platelets and Thrombogenesis

Platelets are sprouting from megakaryocytes as "disk-shaped structures" without nucleus, usually released in the bone marrow. Moreover, some studies have demonstrated that the lungs may contribute to platelet production by resident megakaryocytes actively releasing platelets into the circulation (Weyrich and Zimmerman 2013; Lefrançais et al. 2017). The platelets are present in the blood for up to 10 days and are eliminated from circulation usually in the spleen (Kaplan and Saba 1978; Kuter 1996). Platelets are involved in the protection of vascular wall integrity. In particular, the platelets are flowing near by the vessel wall in the circulation, according to their physical properties and contribute to maintain blood components within the vessel lumen (Holinstat 2017; Tomaiuolo et al. 2017) (Fig. 14.1). It is worth noting that the intraluminal distribution of the platelets facilitates their participation in hemostasis after injury to the vascular wall (Holinstat 2017). After endothelial disruption, platelets rapidly adhere to the exposed subendothelial extracellular matrix (ECM) through interactions mediated by several surface receptors. Key among these are the collagen-binding receptors integrin $\alpha2\beta1$ and glycoprotein VI (GPVI), as well as the glycoprotein Ib-V-IX complex, which binds von Willebrand factor (vWF) anchored to the subendothelial surface (Holinstat 2017).

In addition to their well-recognized function in maintaining hemostasis and contributing to thrombosis, platelets are increasingly being acknowledged for their broader biological roles. Emerging evidence suggests that platelets play unique and active roles in modulating the immune response and participating in inflammatory processes. von Hundelshausen and Weber (2007) proposed that platelets exert distinctive functions in modulating immune responses and facilitating inflammatory reactions. These findings suggest that platelets are not merely passive elements in coagulation but also serve as key immunomodulatory cells, interacting with leukocytes, endothelial cells and soluble mediators to influence both innate and adaptive immune pathways.

Upon activation, platelets exhibit a marked upregulation of adhesive surface molecules, most notably P-selectin and the lysosomal marker CD63 (Mizurini et al. 2021). P-selectin, in particular, plays a central role in mediating interactions between activated platelets and immune cells—such as lymphocytes, neutrophils, and monocytes—that express P-selectin glycoprotein ligand-1.

This interaction facilitates the formation of platelet-leukocyte microaggregates and promotes critical immune processes, including leukocyte tethering, rolling, and eventual firm adhesion at sites of vascular injury or inflammation (Diacovo et al. 1994).

In addition to P-selectin, several other surface receptors expressed on platelets have been implicated in mediating interactions with the immune system. These include the integrin αIIbβ3 (Oki et al. 2006), CD40 ligand (Prasad et al. 2003), intercellular adhesion molecule-2 (ICAM-2) (Diacovo et al. 1994), junctional adhesion molecules (JAMs) (Prota et al. 2003), and a range of chemokine receptors (Murphy et al. 2000). Collectively, these molecules facilitate platelet communication with various immune cells, highlighting the multifaceted role of platelets beyond their classical involvement in maintaining vascular integrity and mediating hemostasis and thrombosis. It is now well recognized that platelets also contribute actively to complex mechanisms of immune and inflammatory responses. These functions are particularly evident in pathological conditions such as SARS-CoV-2 infection, where activated platelets contribute to thrombotic obstruction in small vessels, exacerbating disease severity.

Autopsy studies have played a pivotal role in reshaping our understanding of COVID-19, revealing it not only as a respiratory illness but also as a disease with profound angiogenic and vascular involvement. Detailed pathological examinations of lung tissues from individuals who succumbed to the infection frequently demonstrated widespread microvascular injury, characterized by fibrin deposition and accumulation of activated platelets obstructing small pulmonary vessels. In addition to these microvascular changes, thrombotic events were also observed in larger pulmonary arteries in certain cases, underscoring the multifocal nature of COVID-19-associated coagulopathy and its contribution to respiratory failure (Jonigk et al. 2022). Furthermore, in hospitalized patients with severe COVID-19, a high incidence (about 69%) of venous thromboembolism (VTE) has been reported. This percentage remained high despite thromboprophylaxis treatment. Lodigiani and colleagues found that the incidence of VTE (despite thromboprophylaxis) was 27.6% in ICU COVID-19 patients and 6.6% in the general ward, while the rate of stroke was 2.5% and acute coronary syndrome 1.1% (Lodigiani et al. 2020). Patients with COVID-19 pneumonia exhibit coagulation abnormalities, most commonly elevated levels of fibrinogen and D-dimer, often with mild thrombocytopenia

(Wool and Miller 2020; Levi et al. 2020). Several studies have confirmed that an elevated D-dimer is associated with a higher mortality rate. A subset of COVID-19 patients can have abnormally short prothrombin time (PT) and activated partial thromboplastin time (aPTT), often related to elevated Factor VIII (FVIII) (Lin et al. 2021). In more severely affected COVID-19 patients, a disseminated intravascular coagulopathy (DIC)-like state can develop with relatively mild prolongation of the PT and aPTT, while fibrinogen remains normal/elevated (Asakura and Ogawa 2020).

Three stages of COVID-19-associated coagulopathy have been proposed:

1. elevated D-dimer,
2. elevated D-dimer together with mildly prolonged PT/INR, aPTT and mild thrombocytopenia,
3. critical illness and laboratory results progressing toward classic DIC.

In severe cases of COVID-19 (58–95%), a mild thrombocytopenia has been detected, and the platelet count was only 23×10^9/L to 31×10^9/L lower compared with nonsevere COVID-19 patients (Kossiva et al. 2021). It has been suggested that such severely ill patients with systemic immune and coagulation activation maintain reasonable platelet counts because of marked compensatory platelet production response. Severe thrombocytopenia is only rarely reported in COVID-19 patients, for instance, in association with an immune thrombocytopenic purpura-like state. Viral infection can be associated with thrombocytopenia due to a variety of causes. Hypoproliferative thrombocytopenia is observed at later stages of viral infection and is generally mediated via enhanced platelet clearance/destruction. Viruses can also interact with megakaryocytes and reduce platelet synthesis (Raadsen et al. 2021).

Platelets play an important role in inflammatory signaling as well as in infectious response. By combining thrombotic and immune recruitment functions, platelets may help focus hemostasis and immune responses against potential infectious agents to prevent microbial invasion. Platelets also appear to play a role in recruiting and activating circulating leukocytes to the endothelial surface, leading to white blood cell diapedesis. The interactions between endothelial cells, platelets, and leukocytes play a critical role in the procoagulant effect of viral infections (Filaly et al. 2023).

Several mechanisms of COVID-19-associated thrombocytopenia have been suggested with a possible anti-infective effect, to prevent viremic spread via the bloodstream. Several data confirm that there are significant intravascular platelet aggregates in COVID-19 autopsy lung specimens, located primarily in the interalveolar capillaries and smaller vessels. Surprisingly, none of the COVID-19 patients assessed was thrombocytopenic. An increased mean platelet volume implies an increase in circulating young platelets and is the body's response to thrombocytopenia (Delshad et al. 2021). The number of ribosomes is higher in large platelets and they incorporate more amino acids, suggesting a higher potential for protein synthesis. Larger platelets have increased hemostatic potential, bind more fibrinogen and have greater levels of phosphorylation after thrombin stimulation than smaller platelets (Handtke and Thiele 2020). COVID-19 patients were found to have significantly larger mean platelet volume (MPV) than critically ill non-COVID-19 patients matched for platelet count (11.6 vs. 10.5 fL). Moreover, in COVID-19 patients reticulated platelets were increased: these are immature platelets with a high granule content, residual mRNA, and increased mean volume compared with older circulating platelets. The reticulated platelet count is positively associated with cardiovascular risk and mortality. A further feature is that the COVID-19 is associated with the increased production of large immature platelets, as megakaryocytes respond to increased platelet consumption. Interestingly, COVID-19 is associated with increased numbers of immature platelets even at normal platelet counts. As immature platelets are known to be more functional, this could be another mechanism for increased clotting events in COVID-19 (Barale et al. 2021).

In keeping with these findings that COVID-19 patients have larger and less mature platelets (parameters associated with increased platelet reactivity), Ranucci and colleagues (2020) showed that, upon admission to the ICU, a group of intubated COVID-19 patients with normal platelet counts had an increased platelet contribution to clot strength according to a viscoelastic analysis.

Zhang and colleagues have reported their detection of ACE2 expression by platelets, the direct stimulatory effects of SARS-CoV-2 or its Spike protein on platelets and the ability of recombinant human ACE2 protein or anti-Spike monoclonal antibody to inhibit Spike protein-induced platelet activation (Zhang et al. 2023). Cappelletto and colleagues found that Spike protein was active as a sole agonist or, even more effectively, by enhancing the function of known platelet activators, such as increasing Ca^{2+} flux, phosphatidylserine externalization on the platelet outer membrane, activating Transmembrane protein scramblase TMEM16F and thrombin generation (Cappelletto et al. 2023). Hottz and coworkers (2020) have further demonstrated increased platelet–monocyte interaction and associated tissue factor expression by monocytes in severe COVID-19 patients. Traces of SARS-CoV-2 mRNA were detected in isolated platelets by RT-qPCR and virions have been depicted within platelet sections by electron microscopy techniques. SARS-CoV-2 mRNA-positive platelet samples were more often collected from older COVID-19 patients compared with younger COVID-19 patients. These findings are within the overall picture that SARS-CoV-2 infections are in general more severe in older patients, who are more likely to have a negative outcome.

However, there are numerous alternative receptors that could bind SARS-CoV-2 and promote entry into platelets: CD147 (basigin, EMMPRIN), CD26, integrins and endosomal toll-like receptor-mediated pathway. CD147 is a glycosylated membrane protein, expressed in several blood cells including platelets and has been described as a possible entry vector for SARS-CoV-1, HIV, and measles (Fenizia et al. 2021). The SARS-CoV-2 spike protein contains an RGD motif (arginine–glycine–aspartate) in the receptor-binding domain near the distal dip, which is frequently utilized by viruses (e.g., herpes simplex virus or Epstein–Barr virus) to bind integrin proteins. A homologous KGD sequence has been detected in the SARS-CoV-1 spike protein (Rohlfing et al. 2021). The RGD domain is identical to the RGD sequence in the α-chain of fibrin. Platelets express and present several integrin proteins on their cellular surface, such as integrin $\alpha IIb\beta 3$ that binds to the RGD sequence within fibrin. These integrins, therefore, could allow virion binding to the surface of platelets or even enable their entry into the cytoplasm. However, the presence of SARS-CoV-2 mRNA and virus particles in platelets could also be explained by SARS-CoV-2 infection of megakaryocytes (Zaid and Guessous 2022). Megakaryocytes have an overlapping receptor stock. Therefore, the earlier-described entry vectors can be applied to megakaryocytes as well as platelets. In addition, megakaryocytes are able to engulf hematopoietic cells and thus take up virus particle. mRNA and virions could consequently be diverted from megakaryocytes into platelets during the shedding of the cell fragments into the blood stream. Moreover, platelets have the ability to assimilate proteins, vesicles and other cell components by themselves. Hence, platelets could fuse with extracellular vesicles containing virus particles or mRNA shed from infected cell, such as epithelial cells. It has been recognized that patients with severe acute respiratory failure show enhanced pathophysiological signs of inflamed endothelium in the pulmonary microcirculation (endothelialitis). Activation of platelets is a major trigger for endothelial dysfunction. Enhanced platelet activation and subsequent endothelial reactions are likely to promote microcirculatory arrest in organ failure. In addition, the tissue damage caused by the virus especially in the lung epithelium can further activate platelets. Tissue damage factors, such as vWF, fibrinogen, PAI-1, soluble thrombomodulin and angiopoietin 2, are elevated in plasma from COVID-19 patients, facilitating the development of coagulopathies.

14.4 Monocytes and Macrophages: The Inflammasome NLRP3

Monocytes and macrophages exhibit remarkable plasticity, allowing them to dynamically adjust their gene expression profiles in response to signals from their surrounding microenvironment or systemic signals. This process enables these cells to adopt functionally distinct phenotypes: the proinflammatory M1 and the anti-inflammatory M2 states. M1 cells are primarily involved in host defense mechanisms, producing high levels of proinflammatory cytokines and reactive oxygen species to combat pathogens. In contrast, M2 cells contribute to the resolution of inflammation by promoting tissue remodeling, repair, and the restoration of homeostasis.

It is worth noting that this adaptation of monocytes and macrophages is not a fixed or terminal state; it can occur at various stages of the inflammatory response and is reversible depending on the evolving local contest. The innate immune response to inflammation is largely mediated by a family of evolutionarily conserved intracellular pattern recognition receptors known as NOD-like receptors (NLRs), or nucleotide-binding domain and leucine-rich repeat-containing proteins.

These receptors are predominantly expressed in monocytes and macrophages, where they play a critical role in detecting pathogen-associated molecular patterns (PAMPs) and danger-associated molecular patterns (DAMPs) within the cytosol initiating inflammatory signaling cascades (Awad et al. 2017). Upon activation, certain NLRs, such as NLRP3, assemble into multiprotein complexes known as inflammasomes. These complexes promote the activation of caspase-1 (Swanson et al. 2019; Man and Kanneganti 2016), which in turn leads to the maturation and secretion of proinflammatory cytokines such as IL-1β and IL-18, as well as the induction of pyroptotic cell death. Through these mechanisms, NLRs and their associated inflammasomes are essential for initiating and sustaining inflammatory responses, particularly during the early stages of infection and tissue injury.

NLRs are multidomain cytosolic proteins formed by a central nucleotide-binding domain (NACHT), an N-terminal effector domain (such as a pyrin domain or CARD), and a C-terminal region enriched in leucine-rich repeats (LRRs). The LRRs are thought to contribute to autoregulation by maintaining the receptor in an inactive conformation until a triggering signal is encountered (Mason et al. 2012). NLRs are classified into several subfamilies, including NLRPs, NLRCs, and NLRBs (also known as NAIPs) by differences in their domain composition.

Among the NLR family members, several have been identified as inflammasome-forming sensors, including NLRP1, NLRP3 up to NLRP12, NLRC4, and NAIP. These proteins act as molecular scaffolds for the assembly of inflammasomes, multiprotein complexes that activate caspase-1 and promote the maturation and secretion of proinflammatory cytokines such as interleukin-1β (IL-1β) and interleukin-18 (IL-18).

Particularly, the NLRP3 inflammasome has garnered significant attention due to its well-established role in human disease. Mutations in the NLRP3 gene are causally linked to a group of rare, inherited autoinflammatory disorders collectively known as cryopyrin-associated periodic syndromes (CAPS) (Booshehri and Hoffman 2019). These syndromes are characterized by uncontrolled inflammasome activation and excessive IL-1β release from peripheral blood mononuclear cells (PBMCs), leading to recurrent systemic inflammation and tissue damage.

Therefore, macrophages have a crucial role in innate immune response during viral infections. As mentioned above, two types of monocyte-derived macrophages are developed: classically activated macrophages (M1) and alternatively activated macrophages (M2). M1 macrophages with proinflammatory role are stimulated by tumor necrosis alpha (TNF-α), INF-γ, and pathogen-associated molecular patterns (PAMPs), while M2 macrophages with anti-inflammatory role are activated by IL-13 and IL-4 (Kerneur et al. 2022). Activation of M1 macrophages increases the serum level of the pteridine neopterin (NPT).

Moreover, T cells through TNF-α, natural killer (NK) cells through interferon gamma (INF-γ) and viral infections through pathogen-associated molecular patterns (PAMPs) activate classical macrophage (M1) for the release of neopterin (NPT), which inhibits release of adhesion molecules (Al-kuraishy et al. 2021). As a consequence, a high release of NPT leads to the negative feedback inhibition of macrophage activity with macrophage polarization induction from proinflammatory M1 to the anti-inflammatory M2 type (Ross et al. 2021).

NPT is regarded as a marker of cellular immune activation and T helper 1 (Th1) type 1 immune response. NPT is different from biopterins, which are pterin derivative and act as cofactor for endogenous enzymes, mainly amino acid hydroxylase, for synthesis of different molecules such as nitric oxide (Al-kuraishy et al. 2021). Briefly, NPT is an oxidized form of dihydroneopterin during antioxidant reactions, where high NPT levels in serum and other biological fluids are linked to an elevated production of reactive oxygen species (ROS) and induction of oxidative stress (OS) during intense activation of cellular immunity. Various types of viral infections are able to activate the formation and release of INF-γ from infected cells, where INF-γ stimulates macrophages to produce and release NPT. Therefore, NPT concentrations reflect an underlying cell-mediated immune activation.

It has been suggested that SARS-CoV-2 infection leads to endothelialitis which provokes the release of proinflammatory cytokines. As a consequence, endothelial inflammation activates the production of IL-6 and adhesion molecules, like ICAM-1, MCP-1 and VCAM-1 that facilitate T cells and monocyte adhesions and infiltrations into vascular neointima with secretion of TNF-α (Ahmad et al. 2022). Furthermore, SARS-CoV-2 infection is associated with macrophage activation with acquisition of proinflammatory phenotype (M1) and activation of NF-κB and Nod-like receptor pyrin 3 (NLRP3) inflammasomes leading to hyper-inflammatory status (Sefik et al. 2022). Different studies have shown that NPT exert anti-inflammatory effects through suppressing NF-κB signaling and NLRP3 inflammasomes. The inhibitory effect of NPT on the NLRP3 inflammasomes is mediated through activation of anti-inflammatory IL-10 and nuclear translocation of transcription factor-2 (Nrf-2). As well, a prolonged NPT effect accelerates the development of M2 over M1 leading to an anti-inflammatory status to counterbalance the M1-mediated proinflammatory response (Al-kuraishy et al. 2021).

The latest structural models of mature SARS-CoV-2 peptides indicated that the virus produces all NLRP3 inflammasome-activating proteins (Freeman and Swartz 2020). SARS-CoV2 activates, indeed, the inflammasome and creates a phenomenon called pyroptosis as a form of inflammatory programmed cell death pathway associated with the release of IL-1β, especially in the epithelial cells of the respiratory tract (Freeman and Swartz 2020).

Bellmann-Weiler and colleagues have studied 115 COVID-19 patients to assess if neopterin, either alone or in combination of other established markers of inflammation, such as C reactive protein (CRP), procalcitonin (PCT) or IL-6 might be useful to predict the clinical course of hospitalized patients with SARS-CoV2 infection. The working hypothesis was that neopterin is exclusively produced by monocytic cells in response to stimulation by the T-helper cell type 1 (Th1)-derived cytokine IFN-γ, thereby providing information on the degree of cell-mediated immune activation in COVID-19 patients. Conversely, IL-6 determination does not provide information on its source, as multiple cells and tissues can produce this protein during cytokine storm in SARS-CoV-2 infection. It was found that higher serum levels (>45 nmol/L) of neopterin were associated to increased risk of mechanical ventilation, ICU treatment and adverse outcomes, supporting the hypothesis that more advanced T-cell-triggered immune activation is a major driver of adverse outcomes. The increase in serum NPT and IL-6 indicate that an activated IFN-γ pathway and a stimulated acute phase response, as reflected by IL-6, predict the subsequent clinical course (Bellmann-Weiler et al. 2021).

Moreover, NPT level in cerebrospinal fluid (CSF) is elevated in patients with severe SARS-CoV-2 infection with neurological disorders suggesting a remarkable pattern of CSF inflammation and central nervous system (CNS) pathobiology in COVID-19-induced brain injury.

Therefore, inflammasomes are key regulators of proinflammatory cytokine secretion and play a key role in COVID-19 pathological mechanisms. The accurate study by Sefik and colleagues has clarified several mechanisms triggered by SARS-CoV-2 involving the activation of inflammasome and inflammasome-related genes (Sefik et al. 2022). They utilized the humanized mouse model of COVID-19 with a human immune system (MISTRG6-hACE2) and studied the involvement of macrophages in the acute and chronic stages of the disease. Sustained interferon (IFN) response and viral RNA persisting for months characterize the chronic phase of the infection. They demonstrate that SARS-CoV-2 infection and replication in lung-resident human macrophages is a critical driver of the disease. In response to infection mediated by CD16 and ACE2 receptors, human macrophages are effective in activating inflammasomes, releasing IL-1 and IL-8 and undergoing pyroptosis. These mechanisms contribute to the hyperinflammation of the lungs. The inhibition of the NLRP3 inflammasome pathway reverses the lung inflammatory changes and results in the release of virus by the infected macrophages. Therefore, the role of inflammasomes is to oppose host infection by SARS-CoV-2 by production of inflammatory cytokines and to suicide by pyroptosis to prevent a productive viral cycles. The results indicate that resident alveolar macrophages are activated at the peak of infection, followed by an inflammatory response, characterized by infiltrating monocytes and monocyte-derived macrophages. The differentiating macrophages maintain their inflammatory features and remain activated during the infection. It is worth noting that macrophages present enriched interferon-stimulated genes, such as IFITM3, IFITM and IRF7. These genes were suppressed by administration of antibodies against the receptor of Interferon A (anti-IFNAR2) and a substance blocking the viral replication (Redemsivir). Interestingly, this treatment did not affect the release of Interferon-γ produced by cytotoxic T cells, as the first response of immune system. All aspects of inflammasome activation have been detected, from assembling of specific NOD-like receptors (NLRs) and adaptor molecules to activate caspases, the central effectors proteins. When the inflammatory complexes are formed, active caspase-1 cleaves and activates the proinflammatory IL-1 family cytokines, IL-1β and IL-8, usually elevated in COVID-19 patients. Finally, pyroptosis was characterized by detecting LDH and gasdermin D (GSDMD), a substrate of active Caspase-1 and pore-forming executer of pyroptosis in the serum of infected mice.

In conclusion, the role of inflammasome in COVID-19 hyperinflammation is important. It is interesting that NLRP3 inflammasome inhibitors, such as flavonoid compounds, can be a potential treatment for severe inflammation, which is the main life-threatening condition in patients with COVID-19 (Ali Saeedi-Boroujeni and Mahmoudian-Sani 2021). However, the increased virus production observed upon inflammasome blockade could pose a significant risk to benefit of wholesale inhibition of the pathway (Sefik et al. 2022). Therefore, further studies are required to clarify all the stages of SARS-CoV-2 infection.

14.5 Neutrophils and Neutrophil Extracellular Trap (NET)

Neutrophils play a key role in the innate immune response. Differentiated in the bone marrow from the granulocyte–monocyte progenitor cells (Akashi et al. 2000; Görgens et al. 2013), mature neutrophils are characterized by the presence of granules in their cytosol and a multilobulated nucleus (Campbell et al. 1995). Neutrophils are released into the circulation, where they can be recruited to sites of both sterile and pathogen-induced injury by activated endothelial cells, resident immune cells and injured epithelial cells (de Oliveira et al. 2016). When they reach the site of injury, engagement of neutrophil surface receptors with proinflammatory signals, such as bacterial components, fungal β-glucan or cytokines, triggers a signaling cascade resulting in enhanced neutrophil effector functions (i.e., activation) (Mayadas et al. 2014).

Neutrophils have several immune functions, including phagocytosis, reactive oxygen species (ROS) production and degranulation, as well as the formation and release of sticky web-like structures composed of decondensed chromatin filaments called neutrophil extracellular traps (NETs). These structures are decorated with histones and neutrophil granule proteins, such as lactoferrin, cathepsins, neutrophil elastase (NE) and myeloperoxidase (MPO), as well as cytoplasmic and cytoskeletal proteins. Mitochondrial DNA is found in NETs that are effective in immobilizing pathogens, limit their dissemination and enable their killing by antimicrobial proteins. Although the components of NETs have bactericidal activity, excessive NET formation has proinflammatory characteristics and contributes to host cell injury. Thus, NETs have also been discussed as potentially key factors in different diseases, such as diabetes, thrombosis and lung injury.

The seminal work of Brinkmann et al. (2004) first reported the ability of activated neutrophils to release nuclear DNA into the extracellular environment, where it can trap and neutralize pathogens in a process termed NETosis: a special form of programmed cell death of neutrophils. NETosis is induced by a range of microbial stimuli and proinflammatory mediators [i.e., interleukins: IL-8 and IL-1β]. The increased generation of reactive oxygen species (ROS) is a crucial intracellular process that causes NETosis. Recent evidence shows that NETosis is classified into two categories: (i) "suicidal" NETosis, which occurs in the presence of NADPH oxidase activity, and (ii) "vital" NETosis, which occurs without NADPH oxidase activity.

(i) In the "suicidal" NETosis the fundamental prerequisite is the stimulation of the neutrophils which determines an increase of calcium ions released from the endoplasmic reticulum. Sequentially, protein-kinase C (PKC) is activated and the NADPH oxidase (NOX) complex is formed. The activated complex induces the production of ROS, such as O_2, H_2O_2 and HOCl. ROS are important in the rearrangement of the cytoskeleton and glycolytic ATP production, which are required for NET formation (Ackermann et al. 2021). Early during NET formation, granular NE and MPO translocate to the nucleus and drive nuclear and chromatin decondensation. Peptidylarginine deiminase 4 (PAD 4) contributes to chromatin decondensation by histone hypercitrullination. Citrullination unleashes energy of coiled DNA leading to catapult-like ejection of NETs, after licensing calpain to further decondense nuclei before extra-cellular trap release. PAD 4 retains enzymatic activity in the extracellular environment and modifies proteins: those of extracellular matrix and coagulation factors. In addition, cathepsin C (CatC) is required for the activation of NE and other serine proteases. During NET formation, NE cleaves gasdermin D (GSDMD) a molecule centrally involved in pyroptosis, facilitating granule and plasma membrane permeabilization. NET formation is promoted by caspase-11-dependent activation of GSDMD.

Sprenkeler and colleagues (2022) revealed that a complete and active actin polymerization process as well as active myosin II are critical for DNA release during NETosis. Cyclin-dependent kinases (CDK) also play a key role in NET production; NETs are not produced until CDK4 and CDK6 are activated. It has been suggested that mitochondrial DNA (mtDNA), rather than nuclear DNA, may constitute the raw material of NETs in some diseases, including SLE and trauma. Furthermore, CO_2 and HCO_3- levels may influence NETosis efficiency and pH may alter the effectiveness of glycolysis in neutrophils regarding the "suicide" NETosis (Huang et al. 2022). (ii) A different type of NETosis is "vital" or NOX-independent pathway. In contrast to the previously described "suicidal" NETosis, the "vital" NETosis takes approximately 30 min, while the "suicidal" NETosis lasts 3–4 h. Furthermore, the cause of these two forms of NETosis is also quite different. "Vital" NETosis can be stimulated by activated platelets, microorganisms and complement proteins. Calcium ions enter neutrophils after activation, which is regulated by small conductance potassium channel member three (SK3). The influx of Ca2+ activates PAD4 to aid in H3 citrullination. As a consequence, the electrostatic bond between histone and DNA weakens, causing chromatin decondensation. Vesicles transport these chromatins, together with histones and granular proteins, out of the neutrophil. The neutrophils are still alive after "vital" NETosis and NOX is not needed during the process. Recent data indicate that cleavage of the histone n-terminal may be used to determine the type of NETosis; this is because NE cleaves the n-terminal tails of core histones during "suicidal" NETosis but not "vital" NETosis.

The crucial point for NETosis is the presence of receptors on the cell membrane. Neutrophils, which are at the forefront of recognizing and killing pathogens, have a variety of receptors that are vital to the human immune system: both innate and adaptive immunity. A vast number of receptors are also activation involved in the NETosis process. However, it is worth noting that NETs exert both pro- and anti-inflammatory effects, which have been defined as context-dependent. The proinflammatory effects comprehend the induction of type I Interferons (IFNs), proinflammatory cytokines, NLRP3 inflammasome promotion of adaptative immune responses, damage to the endothelium and immunothrombosis. NET aggregation can occlude vessels and ducts in various organs and promote organ damage. On the other hand, aggregation of NETs can promote trapping and cleavage of proinflammatory mediators by NET-bound proteases, leading to downregulation of inflammatory responses and resolution of inflammation.

The neutrophil receptors are numerous and differently activated. Pattern recognition receptors (PRRs) are a kind of recognition molecules that are mostly expressed on the surface and endosome of cells and can recognize one or more pathogen-related molecular patterns (PAMPs). There are several types of PRRs found on neutrophils, including toll-like receptors (TLRs), C-type lectin receptors (CLRs), RIG-I-like receptors and nucleotide-binding oligomerization domain-like receptors (NLRs). All TLRs, NLRs, and CLRs receptors play a role in the NETosis process.

Complement receptors (CRs) on the cell surface can mediate several biological effects, such as phagocytosis, immune regulation, adhesion, clearance of immune complexes and inflammation, by engaging with the active fragments generated during complement activation. There are several CRs expressed on lymphoid and myeloid cells, and some of them, including CR4, CR3 (Mac-1 or CD11b/CD18), CR1 (CD35) and CR5, are linked to NETosis. Viruses may also activate neutrophils through CRs. SARS-CoV-2 activates compliment and complement regulators, causing accumulation of C5a and C5b-9 in the blood of COVID-19 patients, Complement deposition has been detected in microvessels, occasionally in proximity to SARS-CoV-2 glycoproteins. Complement activation triggers NETs also in COVID-19.

Fc receptors are receptors for the C-terminal of immunoglobulin (Ig) Fc. After Ig binds to an antigen, the Fc segment of the antibody becomes allosteric and binds to the Fc receptor on the cell membrane, resulting in various biological effects. As a result, Fc receptors are vital in regulating immune function and each type of Ig has a corresponding Fc receptor. Human neutrophils express two Fc to recognize IgG molecules, including FcgRIIa (CD32a) and FcgRIIIb (CD16b). Previous findings indicated that immune complexes (ICs) could induce NET formation; however, opposite results have been reported on this mechanism. Therefore, further studies are required to clarify all the mechanisms involved.

Chemokine receptors, expressed on the cell membranes of immune and endothelial cells, are a type of G protein-coupled receptor that mediates chemokine activity. Chemokine receptors can recruit neutrophils to sites of infection, trauma and abnormal proliferation through chemokine interactions. CRs are divided into four main subfamilies: CXCR, CCR, CX3CR and XCR. Among them, CXCR1, CXCR2, CXCR4 and CXCR7 are thought to be associated with NETosis.

The increase in number of neutrophils (neutrophilia) has been directly related to severe COVID-19 course. The NET-remnants, such as circulating cell-free DNA, citrullinated H3 or MPO-DNA complexes, are found in circulation of patients with severe COVID-19. Moreover, neutrophil–platelet aggregates and

(continued)

NETs loaded with tissue factor (TF) are present in the blood of severe COVID-19 patients. The excessive NET formation may also cause direct vascular injury, resulting in organ damage as observed with histopathology studies, showing NET-based immunothrombosis. Thrombotic complications contribute to morbidity and mortality in severe COVID-19 patients, Thrombosis, indeed, affects both the arterial and venous circulation, leading to acute coronary syndrome, stroke, deep vein thrombosis, pulmonary embolism and microvascular thrombosis (Ackermann et al. 2021).

Lung autopsies from COVID-19 patients revealed widespread occlusion of small pulmonary vessels by aggregated NETs (Middleton et al. 2020). NET-rich thrombi, platelets and fibrin were observed in the lung, heart and kidney (Leppkes et al. 2020). It is worth noting that immature and low-density neutrophils predominate in severe COVID-19. It has been suggested that hypogranular neutrophils produced during emergency myelopoiesis have a higher propensity to release NET. However, the cooperation of numerous components in NETs, including platelets, endothelial cells, coagulation factors, and inorganic polyphosphate, is required for NET-induced intravascular coagulation. In particular, the major components of NETs, such as genomic DNA and citrullinated histone H3, have been suggested as coagulation inducers. When cell free-DNA from human neutrophils spikes into plasma, it triggers thrombin generation by binding to factor XII (FXII), which is enhanced by diminished anticoagulant factors and poor fibrinolysis. Complements also stimulate tissue factor production and interact with the platelet/NETs/thrombin axis, making them important players in COVID-19 immunothrombosis. Furthermore, histones boost thrombin synthesis in a dose-dependent manner, in a process that relies on activatable platelets as well as platelet activation indirectly, but not by platelet tissue factor. Other components, such as neutrophil elastase and cathepsin G via protease-activated receptors, and neutrophil elastase can also degrade tissue factor pathway inhibitors to promote coagulation. NETs and histones may also stimulate platelets to be procoagulant by upregulating the expression of P-selectin.

In addition, the cytotoxic action of the enzymes and histones that are released from NETs can promote endothelial cell death and endothelial dysfunction, which is considered a critical pathogenic mechanism of organ injury in COVID-19 patients. NET production, indeed, promotes lung epithelial cell death and intravascular thrombus formation, crucial for inducing acute lung injury in COVID-19. Moreover, due to the presence of abundant pulmonary vascular neutrophils and dynamically released platelets, NET-induced immunothrombosis is more likely to occur, resulting in occlusion of the microvasculature and ischemic injury. Autopsies of patients who have died from COVID-19 have also revealed occlusive thrombi inside the pulmonary

(continued)

vasculature. These autopsies have also shown that incidences of alveolar capillary microthrombi in COVID-19 patients are significantly higher than in influenza patients. NETs may also contribute to renal failure and liver injury in COVID-19 patients due to their prothrombotic activity. Numerous microthrombi in the hepatic sinusoids as well as ischemic-type hepatic necrosis are found in the postmortem livers of COVID-19 patients. In particular, NE can further cleave the spike protein and thus activates the fusogenic peptide of SARS-CoV-2 spike protein S2. These findings suggest that the proteolytic activity of neutrophil-derived enzymes may modulate membrane fusion of the virus. In conclusion, NETs directly activate endothelial cells, induce endothelial to mesenchymal transition and apoptotic endothelial cell death. Thus, NETs compromise endothelial integrity and barrier function and promote endothelial dysfunction.

14.6 Microvascular Investigations by Sublingual Microcirculation Videomicroscopy, Nail Fold Videocapillaroscopy, Laser Doppler Perfusion Monitoring and Laser Speckle Contrast Imaging in COVID-19 Patients

Several studies indicate that coronavirus disease (COVID-19) is a systemic disease due to the vascular endothelium injury with different clinical manifestations from severe and acute to completely asymptomatic course (Otifi and Adiga 2022). Lasting effects and long-term sequelae could persist after the acute infection and may be due to persistent viral infection and microvascular damage. The microcirculation has been defined the core of COVID-19 pathogenetic mechanisms and direct observations of microvascular involvement have been reported, according to noninvasive and reliable methods for microcirculation status monitoring, such as sublingual microcirculation videomicroscopy or nailfold videocapillaroscopy or optical coherence tomography angiography (OCTA). Moreover, interesting data have been obtained by laser-Doppler perfusion monitoring and laser speckle contrast imaging. These methods of in vivo microvascular investigations are difficult to be utilized during a worldwide pandemic, triggered by a very aggressive virus, such as SARS-CoV-2. However, many interesting and suggesting results have been reported.

Damiani and colleagues (2020) studied 12 COVID-19 patients (10 males, age 56 $\pm$ 10 years), admitted to an Intensive Care Unit (ICU, Ancona, Italy). They evaluated the sublingual microcirculation by means of incident dark field videomicroscopy (Cytocam, Braedius Medical, Amsterdam, NL). They found that for vessels smaller than 20 μm in diameter, the total vessel density (TVD) was 15.3

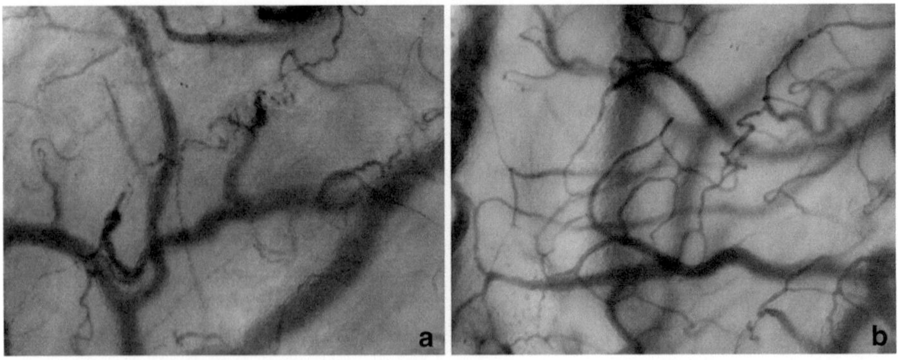

Fig. 14.3 Sublingual microcirculation of two patients with SARS-COV2 severe pneumonia. Patient (**a**) showed a significantly lower perfused vessel density as compared to patient (**b**). PaO2/FiO2 ratio was 74 mmHg in patient (**a**) and 247 mmHg in patient (**b**). (Reprinted with permission of Copyright © 2022 Elsevier Inc. All rights reserved. Damiani et al. 2020)

(14.5–17.1) mm/mm^2, lower than 20.4 ± 3.7 mm/mm^2 detected in control patients with pneumonia not affected by SARS-CoV2. Furthermore, the perfused vessel density (PVD) was 14.9 (14.1–16.9) mm/mm^2 compared with 19.3 ± 4.0 mm/mm^2; the percentage of perfused vessels (PPV) was 97.3 (95.1–98.8)% compared to 98.3 (95.4–100)%, microvascular flow index (MFI) was 2.9 (2.6–3) versus 3 (2.7–3.0) and flow heterogeneity index was 0.3 (0–0.4) versus 0 (0.0–0.2) evaluated in pneumonic control non COVID-19 patients. D-dimer levels were inversely correlated with PVD, TVD and PaO_2/FiO_2 ($p < 0.016$, 0.032, and 0.047, respectively). In particular, it was observed that in two COVID-19 patients a significantly lower perfused vessel density was correlated with higher D-dimer levels, supporting a link between coagulopathy and microvascular perfusion disturbances in patients with SARS-CoV-2 severe pneumonia (Fig. 14.3) (Damiani et al. 2020). Moreover, the same group of Scientists studied nine patients with severe COVID-19 pneumonia in a retrospective data collection. The patients received veno-venous extracorporeal membrane oxygenation (VV-ECMO), as a rescue treatment for severe cases of acute respiratory failure, refractory to conventional mechanical ventilation, neuromuscular blockade and prone positioning. They reported the sublingual microcirculation alterations in vessels with a diameter less than 20 μm: TVD 16.81 (14.46–18.6) mm/mm^2, PVD 15.3 (14.09–17.96) mm/mm^2, PPV 94.85 (93.82–97.79)%, MFI 2.5 (2.5–2.92)% and HI 0.4 (0.18–0.4). In these patients they found that TVD and PVD were inversely correlated to D-dimer levels: 1428 (628.25–5743.75) ng/mL, supporting previous results. Moreover, these microvascular parameters were inversely related to aspartate aminotransferase (AST): 54 (33.5–97.259 U/L and alanine aminotransferase (ALT): 85.5 (38.75–172.75) U/L. These data indicate that VV-ECMO per se was not able to prevent microvascular alterations, effective in damaging liver function (Carsetti et al. 2020).

Bottari and colleagues have reported the sublingual microcirculation analysis (SMA) in three children with severe SARS-CoV-2 infection, who required admission to the pediatric intensive care unit (PICU). The first patient, a 14-year-old boy, with a relapsed bone lymphoblastic B lymphoma was admitted to the PICU for treatment of acute SARS-CoV-2 infection with hypoxic respiratory failure. The clinical picture of acute respiratory distress syndrome (ARDS) required orotracheal intubation and invasive mechanical ventilation. The SMA data were: TVD 18.5 mm/mm^2, PVD 9.31 mm/mm^2, PPV 79.3% MFI 1.3 and HI 2.25. The D-dimer level was 2.65 μg/mL, demonstrating a severe COVID-19, resulting in bilateral pneumotorax and fatal outcome. The two patients, very young: a 20-month-old boy and a 15-day-old newborn, had better SMA values and lower D-dimer levels and were discharged from the PICU. In particular, the MFI < 2.5 correlated with an adverse outcome (Bottari et al. 2022).

Interesting data on sublingual microcirculation are reported by Ospina-Tascón and colleagues (2017), who observed that in 42 patients [(24 male, age 48 (33–63) years] with moderate and severe ARDS, due to SARS-CoV-2 infection, the percentage of small vessel perfused (PPV, diameter < 20 μm) and the microcirculation flow index (MFI) were inversely related only to the dead space ventilation (V_D/V_T). In this case, the mismatch between ventilation and blood oxygenation in the lung plays a key role in the circulation and microcirculation response to the SARS-CoV-2 infection. It is worth noting that the Authors reported the data of functional capillary density and total capillary density as number of vessels/mm^2, while other Authors (Damiani et al. 2020) indicate the length of vessels/mm^2, The comparison of data is really difficult.

Favaron and colleagues studied sublingual microcirculation in 34 COVID-19 patients (28 males; age; 59.18 ± 11.7 years) compared with 33 healthy volunteers (15 males, age 45.8 ± 12.1 years). All the COVID-19 patients were invasively mechanically ventilated. The main results in these patients indicate that TVD increased 22.8 ± 1.5 vs. 20.0 ± 3.3 in Controls; as well as functional capillary density (FCD), PPV: 97.6 ± 2.1% vs. 94.6 ± 6.5%, red blood cell velocity: 362 ± 48 vs. 306 ± 53 μm s^{-1}. Moreover, there was the increase in capillary hematocrit (5.3 ± 1.3 vs. 4.7 ± 0.8) and in capillary-hematocrit to systemic-hematocrit ratio (0.18 ± 0.0 vs. 0.11 ± 0.0) accompanied by increase in the number of leukocytes in microvessels (17.6 ± 6.7 vs. 5.2 ± 2.3). Furthermore, red blood cell microaggregates were observed and quantified in 22 COVID-19 patients compared with 2 healthy controls. In three patients, microaggregates obstructed capillary flow. In the patients D-dimer concentration levels were 2.5 (1.1–4.5) mg/L, accompanied by increase in C-reactive protein, procalcitonin, IL-6 and leukocytosis (Favaron et al. 2021).

Kanoore Edul and colleagues investigated sublingual macrocirculation in 27 COVID-19 patients [20 males, age 60 (51–63) years], intubated and mechanically ventilated. Total and perfused vascular density (21.9 ± 3.9 mm/mm^2 vs. 16.7 ± 1.6; 21.0 ± 3.5 vs. 16.6 ± 1.6 mm/mm^2, respectively) and heterogeneity flow index (0.91 ± 0 0.24 vs. 0.04 ± 0.03) were high compared with those of healthy volunteers (Kanoore et al. 2021). On the contrary, the proportion of perfused vessels (0.96 ± 0.03 versus 1.00 ± 0.00), microvascular flow index (2.79 ± 0.10 vs. 2.97 ± 0.03) and red blood cell velocity

(1124 ± 161 vs. 1331 ± 190 μm s^{-1}) were reduced. D-dimer levels [1.30 (0.58–2.93) μg/mL] were not correlated with either total or perfused vascular density. However, evaluating the microvessels (diameter < 20 μm), they found an increased number of small vessels with stopped, intermittent or slow flow.

Rovas and colleagues investigated the sublingual microvascular dysfunction in COVID-19 patients in a very accurate paper: the MISTIC study (Rovas et al. 2021). They observed and reported data on capillaries with a diameter of 4–7 μm and glycocalyx dimensions, as perfused boundary region (PBR). The results are fundamental to understand the angiocentric mechanism of COVID-19 turmoil. They found a tremendous decrease in vascular density in the diameter classes 4–10 μm in mechanically ventilated (MV) patients compared to controls. A less pronounced pattern was observed in patients nonmechanically ventilated (nMV). Moreover, the capillary density loss in the range D 4–6 μm vessels significantly correlated with D-dimer levels [4.83 (2–10.30) mg/L]. It is worth noting that the microvessel density in the diameter range > 10 μm was not different among the nMV patients [9, 2 females, age 64 (53–77) years], MW patients [14, 1 female, age 61 (55–67) and healthy controls [15, 8 females; age 35 (28–66) years]. Furthermore, in MV patients the capillary red blood cell velocity was significantly lower compared to healthy individuals [81.7 (74.3–91.9) μm/s vs. 106.4 (95.5–121.0) μm/s, while nMV patients had normal V_{RBC} [99.4 (86.8–118.0) μm/s]. The glycocalyx damage was evaluated by means of perfused boundary region (PBR), highest in MV COVID-19 patients vs. nMV patients and healthy controls [2.44 (2.28–2.55) μm vs. 2.16 (2.01–2.24) μm and 2.24 (2.12–2.23) μm, respectively. These data were supported by high levels of Hyaluronic acid and syndecan-1 in MV COVID-19 patients correlated positively with D-dimer levels and inversely with ADAMTS 13 antigen levels, the antithrombotic metalloproteinase which cleaves vWF multimers after their release from activated endothelium. The increased detected markers of endothelial dysfunction were: Angpt-2 and soluble TIE2 (angiopoietin-1 receptor), VGF-A and its endogenous inhibitor sFlt-1, Procalcitonin and TNF-α. It is interesting to note that higher PBR values and reduced ADAMTS13 antigen levels were associated with mortality. These data indicate that SARS-CoV-2 infection is characterized by endothelial activation, glycocalyx damage and severe capillary impairment, due to intravascular clotting, causing a vascular multisystemic disease. The involvement of glycocalyx in COVID-19 mechanisms of endothelium injury has been reported in several studies (Astapenko et al. 2022; Lambadiari et al. 2021). The key role of microvascular damage is played by capillary derangement, as shown by Irina Osiaevi and colleagues, who reported that in long COVID-19 syndrome, i.e., in patients suffering from long term sequelae and persistent symptoms even sixth months after the initial infection, there was persistent capillary rarefaction. They found that the dimensions of glycocalyx, in 27 patients, with persistent symptoms after >12 weeks from recovery of SARS-CoV-2 infection, were comparable to those of healthy controls. However, they report a significant decrease of vascular density, affecting exclusively very small capillaries; the decrease was by 45.16% in microvessels with diameter of 5 μm, by 35.60% in 6 μm diameter, and by 22.79% in 7 μm diameter microvessels. These data were comparable with those of critically ill COVID-19 patients. Moreover, the data indicate that the microvessels affected did not respond adequately to local variations of tissue metabolic demand (Osiaevi et al. 2023).

Furthermore, Druzak and colleagues report that in mechanistic in-vitro studies, using microvasculature-on-chip devices, the plasma from critically ill COVID-19 adult patients induces fibrinogen-dependent red blood cell aggregation, that mechanically damage the microvascular glycocalyx. The plasma from non-COVID sepsis patients causes less significant alterations in the overall blood rheology. The pediatric acute COVID-19 patients present alterations associated with cytokine upregulation (Druzak et al. 2023). These results are supported by the observation of Choi and colleagues, who studied in 5621 hospitalized COVID-19 patients the estimated blood viscosity (eBV) utilizing the Walburn–Schneck model. They report that one-centipoise increase in estimated high-shear rate blood viscosity (eHSBV) and estimated low-shear rate blood viscosity (eLSBV) were associated with a 36.0% and 7.0% increase in death, respectively. The method requires hematocrit, albumin and total proteins for the evaluation of the blood viscosity (Choi et al. 2022).

The microvascular rarefication has been found in retinal microcirculation, investigated by Optical Coherence Tomography Angiography (OCTA). The data indicate that the foveal, parafoveal and whole image vessel density in the deep capillary plexus was statistically lower in patients recovered from COVID-19 compared to healthy controls (Kazantzis et al. 2023). The meta-analysis carried out by the Authors included 12 studies on patients recovered from SARS-CoV-2 infection compared with healthy controls. On the other hand, the vessel density in the superficial capillary plexus did not show any significant difference between the two groups.

The nailfold microcirculation, investigated by nailfold capillaroscopy in COVID-19 pneumonia patients, was described by Natalello and colleagues, who recruited 82 patients, (age: 58.8 ± 13.2 years; male 68.3%); 28 during the hospitalization and 54 after recovery and hospital discharge. The microvessel alterations were common to both groups: pericapillary edema, enlarged capillaries, sludge flow and reduced capillary density. Acute COVID-19 patients, when compared to recovered patients presented a higher prevalence of hemosiderin deposits as results of micro-hemorrhage and micro-thrombosis, sludge flow and pericapillary edema. The recovered patients showed a higher prevalence of enlarged capillaries, loss of capillaries and empty dermal papillae (Natalello et al. 2021). Sulli and colleagues report the videocapillaroscopic microvascular changes in 61 COVID-19 survivors (age: 58 ± 13 m years; mean days from recovery: 104 ± 53) divided in two subgroups, according to the severity of active infection: 34 with past mild disease and 27 with past severe disease. These COVID-19 survivor patients were compared with 31 primary Raynaud's phenomenon (PRP) patients (age: 59 ± 15 years; disease duration: 11 ± 10 years) and 30 control subjects (CNT) (age: 58 ± 13 years). The nailfold capillary number per linear millimeter was significantly reduced in severe and mild COVID-19 survivors than in both PRP and CNT subjects (Fig. 14.4). The isolated microhemorrhages were more frequent in PRP patients than in COVID-19 survivors (Sulli et al. 2022).

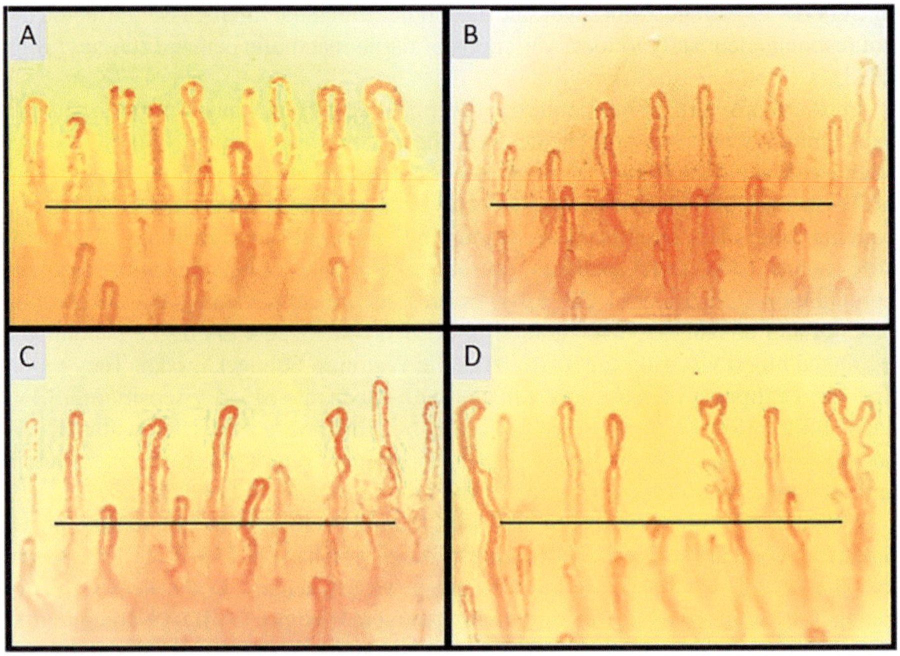

Fig. 14.4 Images of Nailfold videocapillaroscopy: in (**a**) it was showed a control (capillaries with regular array and density; in (**b**) patient with primary Raynaud's phenomenon (capillaries hairpin shaped and crossing); in (**c**) patient with moderate Covid-19 (reduced number of capillaries, tortuous and crossed) and in (**d**) patient with severe Covid-19 (reduced number of capillaries with abnormal shape). Horizontal bar is one millimeter long. (Reprinted with permission of Copyright © 2022 Elsevier Inc. All rights reserved. Sulli et al. 2022)

Cutaneous microcirculation has been studied in COVID-19 patients utilizing the Laser_Doppler Perfusion Monitoring (LDPM) or Laser Speckle Contrast Imaging. Sabioni and colleagues recruited 32 patients with mild or severe COVID-19 (25% S-Covid; age: 58.2 ± 14.1 and 65.6 ± 6.5 years, male 54% and 100%, respectively) and 14 controls (age: 56.3 ± 9.6 years; male 43%). The S-COVID patients required the ICU admission and mechanical ventilation. The basal flow was similar between the M-COVID patients and the healthy volunteers, but was higher in S-COVID patients than in M-COVID patients and healthy volunteers. The peak microvascular vasodilator response, induced by local thermal hyperemia (LTH) or local heating of the laser probe to 44 °C for 20 min, was markedly reduced in both patients groups compared to healthy volunteers. Moreover, the flow during LTH was lower in S-COVID patients than in M-COVID patients. Furthermore, the percent increases in microvascular flow, from baseline to the peak flow during LTH, were markedly decreased in both patients groups compared to the healthy volunteers. In S-COVID patients, there was higher concentration of cytokines: IL-1α, IL-2Rα, IL-6, IL-15, IL-17, IFNα2 and IL-2p40 (Sabioni et al. 2021). It is interesting that Sabioni and colleagues investigated the microvascular response to LTH in patients recovered from acute COVID-19, followed for 12–15 months after the recovery. The results indicate that after 12–15 months from acute SARS-CoV-2 infections, the microvascular response to local thermal hyperemia in COVID-19 patients was not different compared to the values obtained in healthy volunteers (Sabioni et al. 2023).

However, recent observations show that a fairly large proportion of patients who have recovered from a SARS-CoV-2 infection subsequently suffer long-term effects of the disease (Zharkikh et al. 2023). These include symptoms such as weakness, breathlessness, chest and joint pain, confusion, memory and concentration problems (so-called "brain fog"), mood changes, etc. These and other symptoms can persist for months after the disease itself and significantly reduce patients' quality of life (De Luca et al. 2022). These disorders are referred to as "long COVID" or post-COVID syndrome. Current research is largely focused on the acute stage of SARS-CoV-2, but ongoing monitoring of the long-term effects of the disease is also necessary.

It is worth noting that Zharkikh and colleagues evaluated the cutaneous microcirculation in one patient, an 18-years-old male, who was accidentally infected with SARS-CoV-2. They utilized the laser-Doppler flowmetry (LDF) to record signals of blood flow oscillations, analyzed by wavelet transform to calculate their amplitude-frequency spectra. The LDF measurements were carried out 10 days before the onset of the disease and during 26 days after the recovery. Moreover, 23 patients who had long COVID symptoms for a prolonged period of time after the recovery were investigated by LDF measurements, with probe on the wrists or fingers and shins or toes. The power spectral analysis permits to differentiate five frequency bands, corresponding to endothelial (0.005–0.021 Hz), neurogenic (0.021–0.052 Hz), myogenic (0.052–0.145 Hz), respiratory (0.145–0.6 Hz) and cardiac (0.6–2.0 Hz)-correlated mechanisms of microvascular regulations. Comparing the tracings before and after the infections, in the young COVID-19 patients, the myogenic component (0.052–0.0145 Hz) decreased after the infection in the wrist-derived oscillations. In the shins, the respiratory (0.145–0.6 Hz) and cardiac (0.6–2 Hz) components decreased during the recovery after the infection. In 23 COVID-19 patients (3 with severe and 20 with moderate infection, age: 58 ± 9 years) compared with 13 aged-matched healthy volunteers, the spectral density of microvascular oscillations was reduced after the recovery from infection, indicating lower values of microcirculation blood flow compared to controls, but there was an increase in the spectral density of neurogenic-correlated band, accompanied by increase in spectral density of respiratory and cardiac bands. The overall increase in oscillations of microvascular blood flow suggests a recovery of the mechanisms regulating the peripheral blood distribution. However, the authors did not differentiate the nitric oxide dependent or independent components, engulfed in the first band of oscillations, useful for understanding endothelial involvement in the pathogenetic mechanism of the disease (Zharkikh et al. 2023).

Gille-Johnson and colleagues investigated microvascular function in 23 patients with critically ill COVID-19 through bedside assessment of skin perfusion and microvascular reactivity in response to vasoactive drugs by laser speckle contrast imaging (LSCI). Speckle contrast analysis with colors ranging from blue (no perfusion) to red (high perfusion) showed basal microvascular perfusion. They found that patients with critical COVID-19 had significantly higher basal skin perfusion and reduced responses to both endothelial-dependent and endothelial-independent vasodilatory drugs compared with non-COVID controls, matched for age, sex BMI, and comorbidities. During the postinfectious phase, at least 3 months after disease onset, basal skin perfusion was still significantly higher in surviving patients compared with controls, indicating that critical COVID-19 is associated with long-term microvascular dysregulation (Gille-Johnson et al. 2022).

Finally, Romanowska-Kocejko and colleagues investigated the response to brachial artery occlusion for 3 min, utilizing the flow-mediated skin fluorescence (FMSF), a noninvasive optical technique, based on cutaneous nicotinamide adenine dinucleotide (NADH) fluorescence intensity registration. In particular, the level of NADH fluorescence corresponds to the balance of mitochondrial oxidation-reduction processes occurring in the tissue, reflected by the balance between the oxidized form of the coenzyme (NAD^+) and its reduced form (NADH). They compared the responses of a 30-year-old healthy subject and a 30-year-old post-COVID patient. They found that the ischemic response or increase in NADH fluorescence, during brachial artery occlusion, was higher in healthy subject than in post COVID patient, as well as the hyperemic response or the relative maximal decrease in NADH fluorescence, during reperfusion phase (Romanowska-Kocejko et al. 2022).

Hellmann and colleagues studied a 36-year-old man, with unremarkable medical history, referred to their cardiology out-patient clinic because of reduced exercise tolerance, shortness of breath and fatigue, which had been continuing for about 2 months after COVID-19. The molecular test performed in December 2020 using the real-time reverse transcriptase–polymerase chain reaction yielded a positive result. However, besides general muscle weakness, the patient did not manifest any serious symptoms of infection. On admission, the levels of D-dimer, B-type natriuretic peptide, and cardiac troponin were within reference ranges. On 2-dimensional echocardiography, normal left ventricular ejection fraction (65%) was noted, without regional dysfunction. Subsequently microcirculation measurements were carried out on the forearm during and following brachial artery occlusion, simultaneously by LSCI (PeriCam PSI System, Perimed, Järfälla, Sweden) and FMSF (Angionica Ltd., Łódź, Poland). Speckle contrast analysis with colors ranging from blue (no perfusion) to red (high perfusion) showed basal microvascular perfusion. A relatively weak postocclusive reactive hyperemic response was detected by LSCI, suggesting a reduced endothelium-dependent vasodilatation. Likewise, using FMSF, modest oscillations were detected as well as slightly reduced ischemic and hyperemic responses (Hellmann et al. 2021).

The reported data are preliminary, but suggestive for noninvasive monitoring of microvascular impairment in COVID-19 and post-COVID patients.

14.7 Future Perspectives

After the first SARS-CoV-2 spreading pandemic, several highly mutated forms of the virus emerged, indicating the complex interplay of virus biology in the context of changing human immunity due to both vaccination and prior infection. However, there are complicated relationships among virus antigenicity, transmission and virulence, which have unpredictable implications for the future development and disease burden of COVID-19 (Carabelli et al. 2023).

The fight against the virus requires a systematic survey of new variants and the improvement of the knowledge on the pathophysiological mechanisms that characterize the different phases of the disease. The angiocentric core of the disease implies new researches on the relationship between endothelial cell and pericytes under physiological conditions and under virus infection, to clarify the role of these cells in the progression of the disease from the initial virus entry to the systemic inflammation. Furthermore, viscometric methods could be improved to characterize blood of COVID-19 patients to obtain essential data on rheology and resulting impact on microcirculation. At the same time, improvement in in vivo evaluation of microcirculation by sublingual methods could increase the knowledge of the mechanisms of microvascular damage in COVID-19 patients.

References

Ackermann M, Verleden SE, Kuehnel M, Haverich A, Welte T, Laenger F, Jonigk D (2020) Pulmonary vascular Endothelialitis, thrombosis, and angiogenesis in Covid-19. N Engl J Med 383:120–128. https://doi.org/10.1056/NEJMoa2015432

Ackermann M, Anders HJ, Bilyy R, Bowlin GL, Daniel C, De Lorenzo R, Herrmann M (2021) Patients with COVID-19: in the dark-NETs of neutrophils. Cell Death Differ 28:3125–3139. https://doi.org/10.1038/s41418-021-00805-z

Ackermann M, Tafforeau P, Wagner WL, Walsh CL, Werlein C, Kühnel MP, Jonigk DD (2023) The bronchial circulation in COVID-19 pneumonia. Am J Respir Crit Care Med 205:1. https://doi.org/10.1164/rccm.202103-0594IM

Afzali B, Noris M, Lambrecht BN, Kemper C (2022) The state of complement in COVID-19. Nat Rev Immunol 22:77–84. https://doi.org/10.1038/s41577-021-00665-1

Aguilar RB, Hardigan P, Mayi B, Sider D, Piotrkowski J, Mehta JP, Dev J, Seijo Y, Camargo AL, Andux L, Hagen K, Hernandez MB (2020) Current understanding of COVID-19 clinical course and investigational treatments. Front Med (Lausanne) 21(7):555301. https://doi.org/10.3389/fmed.2020.555301

Ahmad F, Kannan M, Ansaric AW (2022) Role of SARS-CoV-2 -induced cytokines and growth factors in coagulopathy and thromboembolism. Cytokine Growth Factor Rev 63:58–68. https://doi.org/10.1016/j.cytogfr.2021.10.007

Akashi K, Traver D, Miyamoto T, Weissman IL (2000) A clonogenic common myeloid progenitor that gives rise to all myeloid lineages. Nature 404:193–197. https://doi.org/10.1038/35004599

Alexandersen S, Chamings A, Bhatta TR (2020) SARS-CoV-2 genomic and subgenomic RNAs in diagnostic samples are not an indicator of active replication. Nat Commun 11. https://doi.org/10.1038/s41467-020-19883-7

Ali Saeedi-Boroujeni A, Mahmoudian-Sani MR (2021) Anti-inflammatory potential of quercetin in COVID-19 treatment. J Inflamm (Lond) 28(18):3. https://doi.org/10.1186/s12950-021-00268-6

Al-kuraishy HM, Al-Gareeb AI, Alzahrani KJ, Cruz-Martins N, Batiha GES (2021) The potential role of neopterin in Covid-19: a new perspective. Mol Cell Biochem 476:4161–4166. https://doi.org/10.1007/s11010-021-04232-z

Ambrosino P, Calcaterra IL, Mosella M, Formisano R, D'Anna SE, Bachetti T, Marcuccio G, Galloway B, Mancini FP, Papa A, Motta A, Di Minno MND, Maniscalco M (2022) Endothelial dysfunction in COVID-19: a unifying mechanism and a potential therapeutic target. Biomedicine 10:812. https://doi.org/10.3390/biomedicines10040812

Arroja MMC, Emma Reid M, McCabe C (2016) Therapeutic potential of the renin angiotensin system in ischaemic stroke. Exp Trans Stroke Med 8:8. https://doi.org/10.1186/s13231-016-0022-1

Asakura H, Ogawa H (2020) COVID-19-associated coagulopathy and disseminated intravascular coagulation. Int J Hematol 113:45–57. https://doi.org/10.1007/s12185-020-03029-y

Aslan A, Aslan C, Zolbanin NM, Jafari R (2021) Acute respiratory distress syndrome in COVID-19: possible mechanisms and therapeutic management. Pneumonia (Nathan) 6(13):14. https://doi.org/10.1186/s41479-021-00092-9

Astapenko D, Tomasova A, Ticha A, Hyspler R, Chua HS, Manzoor M, Cerny V (2022) Endothelial glycocalyx damage in patients with severe COVID-19 on mechanical ventilation—a prospective observational pilot study. Clin Hemorheol Microcirc 81:205–219. https://doi.org/10.3233/CH-221401

Avolio E, Carrabba M, Milligan R, Williamson MK, Beltrami AP, Gupta K (2021) The SARS-CoV-2 spike protein disrupts human cardiac pericytes function through CD147 receptor-mediated signalling: a potential non-infective mechanism of COVID-19 microvascular disease. Clin Sci (Lond) 22(135):2667–2689. https://doi.org/10.1042/CS20210735

Awad F, Assrawi E, Jumeau C, Georgin-Lavialle S, Cobret L, Duquesnoy P, Karabina A (2017) Impact of human monocyte and macrophage polarization on NLR expression and NLRP3 inflammasome activation. PLoS One 12:1–18

Barale C, Melchionda E, Morotti A, Russo I (2021) Prothrombotic phenotype in COVID-19: focus on platelets. Int J Mol Sci 22:13638. https://doi.org/10.3390/ijms222413638

Barbosa LC, Gonçalves TL, de Araujo LP, de Oliveira Rosario LV, Ferrer VP (2021) Endothelial cells and SARS-CoV-2: an intimate relationship. Vasc Pharmacol 137:106829. https://doi.org/10.1016/j.vph.2021.106829

Bartoli A, Gabrielli F, Alicandro T, Nascimbeni F, Andreone P (2021) COVID-19 treatment options: a difficult journey between failed attempts and experimental drugs. Intern Emerg Med 16:281–308. https://doi.org/10.1007/s11739-020-02569-9

Bastard P, Gervais A, Le Voyer T, Rosain J, Philippot Q, Manry J, Casanova JL (2021) Autoantibodies neutralizing type I IFNs are present in ~ 4% of uninfected individuals over 70 years old and account for ~ 20% of COVID-19 deaths. Sci Immunol 19(6):eabl4340. https://doi.org/10.1126/sciimmunol.abl4340

Bellmann-Weiler R, Lanser L, Burkert F, Seiwald S, Fritsche G, Wildner S, Weiss G (2021) Neopterin predicts disease severity in hospitalized patients with COVID-19. *Open forum*. Infect Dis 8:ofaa521. https://doi.org/10.1093/ofid/ofaa521

Benigni A, Cassis P, Remuzzi G (2010) Angiotensin II revisited: new roles in inflammation, immunology and aging. EMBO Mol Med 2:247–257. https://doi.org/10.1002/emmm.201000080

Bernard GR, Artigas A, Brigham KL et al (1994) The American-European consensus conference on ARDS. Definitions, mechanisms, relevant outcomes, and clinica trial coordination. Am J Respir Crit Care Med 149:818–824. https://doi.org/10.1164/ajrccm.149.3.7509706

Bestle D, Heindl MR, Limburg H et al (2020) TMPRSS2 and furin are both essential for proteolytic activation of SARS-CoV-2 in human airway cells. Life Sci Alliance 23;3(9):e202000786. https://doi.org/10.26508/lsa.202000786

Beyerstedt S, Casaro EB, Bevilaqua Rangel E (2021) COVID-19: angiotensin-converting enzyme 2 (ACE2) expression and tissue susceptibility to SARS-CoV-2 infection. Eur J Clin Microbiol Infect Dis 40:905–919. https://doi.org/10.1007/s10096-020-04138-6

Blanco-Melo D, Nilsson-Payant BE, Liu WC, Uhl S, Hoagland D, Møller R, Jordan TX, tenOever BR (2020) Imbalanced host response to SARS-CoV-2 drives development of COVID-19. Imbalanced host response to SARS-CoV-2 drives development of COVID-19. Cell 28(181): 1036–1045.e9. https://doi.org/10.1016/j.cell.2020.04.026

Booshehri LM, Hoffman HM (2019) CAPS and NLRP3. J Clin Immunol 39. https://doi.org/10.1007/s10875-019-00638-z

Bordes J, Lacroix G, Esnault P, Goutorbe P, Cotte J, Dantzer E, Meaudre E (2014) Comparison of the Berlin definition with the American European consensus definition for acute respiratory distress syndrome in burn patients. Burns 40:562–567

Bottari G, Damiani E, Confalone V, Scorcella C, Casarotta E, Gandolfo C, Stoppa F, Cecchetti C, Donati A (2022) Microvascular dysfunction in pediatric patients with SARS-COV-2 pneumonia: report of three severe cases. Microvasc Res 141:104312. https://doi.org/10.1016/j.mvr.2022.104312

Brinkmann V, Reichard U, Goosmann C, Fauler B, Uhlemann Y, Zychlinsky A (2004) Neutrophil extracellular traps kill bacteria. Science 303:1532–1535. https://doi.org/10.1126/science.1092385

Buchrieser J, Dufloo J, Hubert M, Monel B, Planas D, Rajah MM, Schwartz O (2020) Syncytia formation by SARS-CoV-2-infected cells. EMBO J 39:e106267. https://doi.org/10.15252/embj.2020106267

Campbell MS, Lovell MA, Gorbsky GJ (1995) Stability of nuclear segments in human neutrophils and evidence against a role for microfilaments or microtubules in their genesis during differentiation of HL60 myelocytes. J Leukoc Biol 58:659–666. https://doi.org/10.1002/jlb.58.6.659

Cappelletto A, Allan HE, Crescente M, Schneider E, Bussani R, Ali H, Secco I, Vodret S, Simeone R, Mascaretti L, Zacchigna S, Warner TD, Giacca M (2023) SARS-CoV-2 spike protein activates TMEM16F-mediated platelet procoagulant activity. Front Cardiov Med 9. https://doi.org/10.3389/fcvm.2022.1013262

Carabelli AM, Peacock TP, Thorne LG, Harvey WT, Hughes J, Robertson DL (2023) SARS-CoV-2 variant biology: immune escape, transmission and fitness. Nat Rev Microbiol 21:162–177

Carsetti A, Damiani E, Casarotta E, Scorcella C, Domizi R, Montomoli J, Donati A (2020) Sublingual microcirculation in patients with SARS-CoV-2 undergoing veno-venous extracorporeal membrane oxygenation. Microvasc Res 132:1040640

Castleman B, Towne VW (1954) Case records of the Massachusetts General Hospital; weekly clinicopathological exercises; founded by Richard C Cabot. N Engl J Med 251:396–400. https://doi.org/10.1056/NEJM195409022511008

Castleman B, Iverson L, Menendez VP (1956) Localized mediastinal lymph-node hyperplasia resembling thymoma. Cancer 9:822–830

Chams N, Chams S, Badran R, Shams A, Araji A, Raad M, Mukhopadhyay S, Stroberg E, Duval EJ, Barton LM, Hajj Hussein I (2020) COVID-19: a multidisciplinary review. Front Public Health 29(8):383. https://doi.org/10.3389/fpubh.2020.00383

Choi D, Waksman O, Shaik A, Mar P, Chen O, Cho DJ, Rosenson RS (2022) Association of Blood Viscosity with Mortality among Patients Hospitalized with COVID-19. J Am Coll Cardiol 80: 316–328. https://doi.org/10.1016/j.jacc.2022.04.060

Chua RL, Lukassen S, Trump S, Hennig BP, Wendisch D, Pott F, Eils R (2020) COVID-19 severity correlates with airway epithelium–immune cell interactions identified by single-cell analysis. Nat Biotechnol 38:970–979

Costela-Ruiz VJ, Illescas-Montes R, Puerta-Puerta JM, Ruiz C, Melguizo-Rodrígueza L (2020) SARS-CoV-2 infection: the role of cytokines in COVID-19 disease. Cytokine Growth Factor Rev 54:62–75. https://doi.org/10.1016/j.cytogfr.2020.06.001

Da Silva SJR, Kohl A, Pena L, Pardee K (2023) Recent insights into SARS-CoV-2 omicron variant. Rev Med Virol 33:e2373. https://doi.org/10.1002/rmv.2373

Damiani E, Carsetti A, Casarotta E, Scorcella C, Domizi R, Adrario E, Donati A (2020) Microvascular alterations in patients with SARS-COV-2 severe pneumonia. Ann Intensive Care 10: 60. https://doi.org/10.1186/s13613-020-00680-w

De Luca R, Bonanno M, Calabrò RS (2022) Psychological and cognitive effects of Long COVID: a narrative review focusing on the assessment and rehabilitative approach. J Clin Med 11:6554. https://doi.org/10.3390/jcm11216554

de Oliveira S, Rosowski EE, Huttenlocher A (2016) Neutrophil migration in infection and wound repair: going forward in reverse. Nat Rev Immunol 27(16):378–391. https://doi.org/10.1038/nri.2016.49

De Wit E, van Doremalen N, Falzarano DVJ (2016) SARS and MERS: recent insights into emerging coronaviruses. Nat Rev Microbiol 14:523–534

Delshad M, Safaroghli-Azar A, Pourbagheri-Sigaroodi A, Poopak B, Shokouhi S, Bashashb D (2021) Platelets in the perspective of COVID-19; pathophysiology of thrombocytopenia and its implication as prognostic and therapeutic opportunity. Int Immunopharmacol 99(107995):21. https://doi.org/10.1016/j.intimp.2021.107995

Diacovo TG, deFougerolles AR, Bainton DF, Springer TA (1994) A functional integrin ligand on the surface of platelets: intercellular adhesion molecule-2. J Clin Invest 94:1243–1251. https://doi.org/10.1172/JCI117442

Domingo P, Mur I, Pomar V, Corominas H, Casademont J, de Benito N (2020) The four horsemen of a viral apocalypse: the pathogenesis of SARS-CoV-2 infection (COVID-19). EBioMedicine 58:102887. https://doi.org/10.1016/j.ebiom.2020.102887

Druzak S, Iffrig E, Roberts BR, Zhang T, Fibben KS, Sakurai Y, Maier CL (2023) Multiplatform analyses reveal distinct drivers of systemic pathogenesis in adult versus pediatric severe acute COVID-19. Nat Commun 14:1638

Ekholm M, Kahan T (2021) The impact of the renin-angiotensin-aldosterone system on inflammation, coagulation, and Atherothrombotic complications, and to aggravated COVID-19. Front Pharmacol 12:640185. https://doi.org/10.3389/fphar.2021.640185

Fajgenbaum DC, June CH (2020) Cytokine Storm. N Engl J Med 383:2255–2273. https://doi.org/10.1056/NEJMra2026131

Fang Y, Gao F, Liu Z (2019) Angiotensin-converting enzyme 2 attenuates inflammatory response and oxidative stress in hyperoxic lung injury by regulating NF-κB and Nrf2 pathways. QJM 112:914–924. https://doi.org/10.1093/qjmed/hcz206

Favaron E, Ince C, Hilty M, Ergin B, van der Zee P, Uz Z, Endeman H (2021) Capillary leukocytes, microaggregates, and the response to hypoxemia in the microcirculation of coronavirus disease 2019 patient. Crit Care Med 49:661–670. https://doi.org/10.1097/CCM.0000000000004862

Fenizia C, Galbiati S, Vanetti C, Vago R, Clerici M, Tacchetti C, Daniele T (2021) SARS-CoV-2 entry: at the crossroads of CD147 and ACE2. Cells 10:1434. https://doi.org/10.3390/cells10061434

Filaly HE, Mabrouk M, Atifi F, Guessous F, Akarid K, Merhi Y, Zaid Y (2023) Dissecting platelet's role in viral infection: a double-edged effector of the immune system. Int J Mol Sci 24:2009. https://doi.org/10.3390/ijms24032009

Fraga-Silva TF, Maruyama SR, Sorgi CA, de Sousa Russo EM, Morais Fernandes AP, de Barros Cardoso CR, Bonato LD (2021) COVID-19: integrating the complexity of systemic and pulmonary immunopathology to identify biomarkers for different outcomes. Front Immunol 11:599736. https://doi.org/10.3389/fimmu.2020.599736

Freeman TL, Swartz TH (2020) Targeting the NLRP3 Inflammasome in severe COVID-19. Front Immunol 11:1518. https://doi.org/10.3389/fimmu.2020.01518

Gao XM, Xu G, Wang B, Liu C (2021) Cytokine storm syndrome in coronavirus disease 2019: a narrative review. JIM Virtual Issue: COVID-19 289:147–161. https://doi.org/10.1111/joim.13144

Garcia V, Shkolnik B, Milhau L, Falck JR, Laniado M (2016) 20-HETE activates the transcription of angiotensin-converting enzyme via nuclear factor-κB translocation and promoter binding. J Pharmacol Exp Ther 356:525–533. https://doi.org/10.1124/jpet.115.229377

Gattinoni L, Chiumello D, Caironi P, Busana M, Romitti F, Brazzi L, Camporota L (2020) COVID-19 pneumonia: different respiratory treatments for different phenotypes? Intensive Care Med 46: 1099–1102. https://doi.org/10.1007/s00134-020-06033-2

Gelardi M, Notargiacomo M, Trecca EMC, Cassano M, Ciprandi G (2021) COVID-19: the difference between the nose and the lung. Monaldi Arch Chest Dis 91:1708. https://doi.org/10.23750/abm.v90i2-S.8103

Gheblawi M, Wang K, Viveiros A, Nguyen Q, Zhong JC, Turner AJ, Raizada MK, Grant MB, Oudit GY (2020) Angiotensin-converting enzyme 2: SARS-CoV-2 receptor and regulator of the renin-angiotensin system. Circ Res 126:1456–1474. https://doi.org/10.1161/CIRCRESAHA.120.317015

Gibson PG, Qin L et al (2020) COVID-19 acute respiratory distress syndrome (ARDS): clinical features and differences from typical pre-COVID-19 ARDS. Med J Aust 213:54–56. https://doi.org/10.5694/mja2.50674

Gille-Johnson P, Kessler I, Tehrani S (2022) Impaired microvascular function in patients with critical COVID-19. Crit Care Explor 4:e0643. https://doi.org/10.1097/CCE.0000000000000643

Görgens A, Radtke S, Möllmann M, Cross M, Dürig J, Horn PA, Giebel B (2013) Revision of the human hematopoietic tree: granulocyte subtypes derive from distinct hematopoietic lineages. Cell Rep 30(3):1539–1552. https://doi.org/10.1016/j.celrep.2013.04.025

Goshua G, Pine AB, Meizlish ML, Chang CH, Zhang H, Bahel P, Lee AI (2020) Endotheliopathy in COVID-19-associated coagulopathy: evidence from a single-Centre, cross-sectional study. Lancet Haematol. 7:e575–e582. https://doi.org/10.1016/S2352-3026(20)30216-7

Grasselli G, Tonetti T, Protti A, Langer T, Girardis M, Bellani G, Ranieri VM (2020) Pathophysiology of COVID-19-associated acute respiratory distress syndrome: a multicentre prospective observational study. Lancet Respir Med 8:1201–1208. https://doi.org/10.1016/S2213-2600(20)30370-2

Grasselli G, Cattaneo E, Florio G, Ippolito M, Zanella A, Cortegiani A, Einav S (2021) Mechanical ventilation parameters in critically ill COVID-19 patients: a scoping review. Crit Care 20(25): 115. https://doi.org/10.1186/s13054-021-03536-2

Gu SX, Tyagi T, Jain K, Gu VW, Lee SH, Hwa JM, Hwa J (2020) Thrombocytopathy and endotheliopathy: crucial contributors to COVID-19 thromboinflammation. Nat Rev Cardiol 18:124–209. https://doi.org/10.1038/s41569-020-00469-1

Guervilly C, Burtey S, Sabatier F, Cauchois R, Lano G, Abdili E, Dignat-George F (2020) Circulating endothelial cells as a marker of endothelial injury in severe COVID -19. J Infect Dis 222:1789–1793. https://doi.org/10.1093/infdis/jiaa528

Hadjadj J, Yatim N, Barnabei L, Corneau A, Boussier J, Smith N, Terrier B (2020) Impaired type I interferon activity and inflammatory responses in severe COVID-19 patients. Science 7(369): 718–724. https://doi.org/10.1126/science.abc6027

Handtke S, Thiele T (2020) Large and small platelets—(when) do they differ? J Thromb Haemost 18:1256–1267. https://doi.org/10.1111/jth.14788

Hardy E, Hardy-Sosa A, Fernandez-Patron C (2018) MMP-2: is too low as bad as too high in the cardiovascular system? Review: *integrative cardiovascular physiology and pathophysiology*. Am J Physiol Heart Circ Physiol 315:H1332–H1340. https://doi.org/10.1152/ajpheart.00198.2018.0363-6135/18

Hellmann M, Romanowska-Kocejko M, Żarczyńska-Buchowiecka M, Dudziak M (2021) Subtle microvascular endothelial impairment in a young patient recovered from mild COVID-19. Pol Arch Intern Med 29(131):561–562. https://doi.org/10.20452/pamw.15938

Hoffmann M, Pöhlmann S (2021) How SARS-CoV-2 makes the cut. Nat Microbiol 6:828–829

Hoffmann M, Kleine-Weber H, Schroeder S, Krüger N, Herrler T, Erichsen S, Schiergens TS, Pöhlmann S (2020) SARS-CoV-2 cell entry depends on ACE2 and TMPRSS2 and is blocked by a clinically proven protease inhibitor. Cell 181(2):271–280. https://doi.org/10.1016/j.cell.2020.02.052

Holinstat M (2017) Normal platelet function. Cancer Metastasis Rev 36:195–198. https://doi.org/10.1007/s10555-017-9677-x
Hottz ED, Azevedo-Quintanilha IC, Palhinha L, Teixeira L, Barreto EA, Pão CRR, Righy C, Franco S, Souza TML, Kurtz P, Bozza FA, Bozza PT (2020) Platelet activation and platelet-monocyte aggregate formation trigger tissue factor expression in patients with severe COVID-19. Blood 10(136):1330–1341. https://doi.org/10.1182/blood.2020007252
Huang J, Hong W, Wan M, Zheng L (2022) Molecular mechanisms and therapeutic target of NETosis in diseases. MedComm 19(3):e162. https://doi.org/10.1002/mco2.162
Huang JJ, Wang CW, Liu Y, Zhang YY, Yang NB, Yu YC, Jiang Q, Song QF, Qian GQ (2023) Role of the extracellular matrix in COVID-19. World J Clin Cases 6:73–83. https://doi.org/10.12998/wjcc.v11.i1.73
Jackson CB, Farzan M, Chen B, Choe H (2022) Mechanisms of SARS-CoV-2 entry into cells. Nat Rev Mol Cell Biol 23:3–20. https://doi.org/10.1038/s41580-021-00418-x
Jayaweera M, Perera H, Gunawardana B, Manatunge J (2020) Transmission of COVID-19 virus by droplets and aerosols: a critical review on the unresolved dichotomy. Environ Res 188:109819. https://doi.org/10.1016/j.envres.2020.109819
Jonigk D, Werlein C, Acker T, Boor P (2022) Organ manifestations of COVID-19: what have we learned so far (not only) from autopsies? Springer Virchows Archiv 481:139–159. https://doi.org/10.1007/s00428-022-03319-2
Kanoore E, Vanina S, Caminos E, Juan F, Ferrara G, Estenssoro E, Dubin A (2021) Microcirculation alterations in severe COVID-19 pneumonia. J Crit Care 61:73–75. https://doi.org/10.1016/j.jcrc.2020.10.002
Kaplan JE, Saba TM (1978) Platelet removal from the circulation by the liver and spleen. Am J Phys 235:H314–H320. https://doi.org/10.1152/ajpheart.1978.235.3.H314
Kazantzis D, Sergentanis TN, Machairoudia G, Dimitriou E, Kroupis C, Theodossiadis G, Theodossiadis P, Chatziralli I (2023) Correlation between imaging morphological findings and laboratory biomarkers in patients with retinal vein occlusion. Ophthalmol Ther 12:1239–1249. https://doi.org/10.1007/s40123-023-00677-1
Kerneur C, Cano CE, Olive D (2022) Major pathways involved in macrophage polarization in cancer. Front Immunol 17(13):1026954. https://doi.org/10.3389/fimmu.2022.1026954
Khan AO, Reyat JS, Hill H, Bourne JH, Colicchia M, Newby ML, Rayes J (2022) Preferential uptake of SARS-CoV-2 by pericytes potentiates vascular damage and permeability in an organoid model of the microvasculature. Cardiovasc Res 118:3085–3096. https://doi.org/10.1093/cvr/cvac097
Konradt C, Hunter CA (2018) Pathogen interactions with endothelial cells and the induction of innate and adaptive immunity. Eur J Immunol 48:1607–1620
Kossiva L, Thirios A, Panagouli E, Panos A, Lampidi S, Bacopoulou F, Tsolia M, Tsitsika A (2021) A case of COVID-19-related thrombocytopenia and leukopenia in an adolescent with mild symptoms. Children (Basel) 8:509. https://doi.org/10.3390/children8060509
Kuba K, Yamaguchi T, Penninger JM (2021) Angiotensin-converting enzyme 2 (ACE2) in the pathogenesis of ARDS in COVID-19. Front Immunol 12:732690. https://doi.org/10.3389/fimmu.2021.732690
Kuter DJ (1996) The physiology of platelet production. Stem Cells Journal 14:88–101. https://doi.org/10.1002/stem.5530140711
Lambadiari V, Mitrakou A, Kountouri A, Thymis J, Katogiannis K, Korakas E, Ikonomidis I (2021) Association of COVID-19 with impaired endothelial glycocalyx, vascular function and myocardial deformation 4 months after infection. Eur J Heart Fail 23:1916–1926. https://doi.org/10.1002/ejhf.2326
Lanza K, Perez LG, Costa LB, Cordeiro TM, Palmeira VA, Ribeiro VT et al (2020) Covid-19: the renin–angiotensin system imbalance hypothesis. Clin Sci (Lond) 134:1259–1264. https://doi.org/10.1042/CS20200492

Lapi D, Stornaiuolo M, Sabatino L, Sommella E, Tenore G, Daglia M, Novellino E (2020) The pomace extract Taurisolo protects rat brain from ischemia-reperfusion injury. Front Cell Neurosci 28(14):3. https://doi.org/10.3389/fncel.2020.00003

Lapi D, Cammalleri M, Dal Monte M, Di Maro M, Santillo M, Belfiore A, Nasti G, Damiano S, Trio R, Chiurazzi M, De Conno B (2021) The effects of angiotensin II or angiotensin 1-7 on rat pial microcirculation during hypoperfusion and reperfusion injury: role of redox stress. Biomolecules 11(12):1861

Lefrançais E, Ortiz-Muñoz G, Caudrillier A, Mallavia B, Liu F, Sayah DM, Thornton EE, Headley MB, David T, Coughlin SR, Krummel MF, Leavitt AD, Passegué E, Looney MR (2017) The lung is a site of platelet biogenesis and a reservoir for haematopoietic progenitors. Nature 544: 105–190. https://doi.org/10.1038/nature21706

Leppkes M, Knopf J, Naschberger E, Lindemann A, Singh J, Herrmann I, Herrmann M (2020) Vascular occlusion by neutrophil extracellular traps in COVID-19. EBioMedicine 58:102925. https://doi.org/10.1016/j.ebiom.2020.102925

Levi M, Thachil J, Iba T, Levye JH (2020) Coagulation abnormalities and thrombosis in patients with COVID-19. Lancet Haematol 7:e438–e440. https://doi.org/10.1016/S2352-3026(20) 30145-9

Li X, Ma X (2020) Acute respiratory failure in COVID-19: is it "typical" ARDS? Crit Care 24:198. https://doi.org/10.1186/s13054-020-02911-9

Lin J, Yan H, Chen H, He C, Lin C, He H, Zhang S, Shi S, Lincorresponding K (2021) COVID-19 and coagulation dysfunction in adults: a systematic review and meta-analysis. J Med Virol 93: 934–944. https://doi.org/10.1002/jmv.26346

Lodigiani C, Iapichino G, Carenzo L, Cecconi M, Ferrazzi P, Sebastian T, Kucher N, Studt JD, Sacco C, Bertuzzi A, Sandri MT, Barco S (2020) Venous and arterial thromboembolic complications in COVID-19 patients admitted to an academic hospital in Milan, Italy. Thromb Res 191:9–14. https://doi.org/10.1016/j.thromres.2020.04.024

Long B, Carius BM, Chavez S, Liang SY, Brady WJ, Koyfman A, Gottlieb M (2022) Clinical update on COVID-19 for the emergency clinician: presentation and evaluation. Am J Emerg Med 54:46–57. https://doi.org/10.1016/j.ajem.2022.01.028

Lv J, Wang Z, Qu Y, Zhu H, Zhu Q, Tong W, Huang B (2021) Distinct uptake, amplification, and release of SARS-CoV-2 by M1 and M2 alveolar macrophages. Cell Discovery 7:24. https://doi.org/10.1038/s41421-021-00258-1

Mahoney M, Damalanka VC, Tartell MA, Chung DE, Lourenco AL, Pwee D, 2 Janetka JW (2021) A novel class of TMPRSS2 inhibitors potently block SARS-CoV-2 and MERS-CoV viral entry and protect human epithelial lung cells. *bioRxiv*. 6;2021.05.06.442935. https://doi.org/10.1101/2021.05.06.442935

Man SM, Kanneganti TD (2016) Converging roles of caspases in inflammasome activation, cell death and innate immunity. Nat Rev Immunol 16:7–21. https://doi.org/10.1038/nri.2015.7

Martínez-Salazar B, Holwerda M, Stüdle C, Piragyte I, Mercader N, Engelhardt B, Rieben R, Döring Y (2022) COVID-19 and the vasculature: current aspects and Long-term consequences. Front Cell Dev Biol 10:824851

Masaru I, Horiuchi M (2009) Devil and angel in the renin-angiotensin system: ACE-angiotensin II-AT1 receptor axis vs. ACE2-angiotensin-(1-7)-mas receptor axis. Hypertens Res 32:533–536. https://doi.org/10.1038/hr.2009.74

Mason DR, Beck PL, Muruve DA (2012) Nucleotide-binding oligomerization domain-like receptors and inflammasomes in the pathogenesis of non-microbial inflammation and diseases. J Innate Immun 4:16–30. https://doi.org/10.1159/000334247

Mayadas TN, Cullere X, Lowell CA (2014) The multifaceted functions of neutrophils. Annu Rev Pathol 9:181–218. https://doi.org/10.1146/annurev-pathol-020712-164023

McQuaid C, Montagne A (2022) SARS-CoV-2 and vascular dysfunction: a growing role for pericytes. Cardiovasc Res cvac143. https://doi.org/10.1093/cvr/cvac143

Middleton EA, He XY, Denorme F, Campbell RA, Ng D, Salvatore SP, Yost CC (2020) Neutrophil extracellular traps contribute to immunothrombosis in COVID-19 acute respiratory distress syndrome. Blood 3(136):1169–1179. https://doi.org/10.1182/blood.2020007008

Mizurini DM, Hottz ED, Bozza PT, Monteiro RQ (2021) Fundamentals in COVID-19-associated thrombosis: molecular and cellular aspects. Front Cardiovasc Med 8:785738

Mohammadhosayni M, Mohammadi FS, Ezzatifar F, Gorabi AM, Khosrojerdi A, Aslani S, Mohammadi H (2021) Matrix metalloproteinases are involved in the development of neurological complications in patients with coronavirus disease 2019. Int Immunupharmacol 100: 108076

Murphy PM, Baggiolini M, Charo IF, Hébert R, Horuk CA, Matsushima K, Miller LH, Oppenheim J, Power CA (2000) International union of pharmacology. XXII. Nomenclature for chemokine receptors. Pharmacol Rev 52:145–176

Nägele MP, Haubner B, Tanner FC, Ruschitzka F, Flammer AJ (2020) Endothelial dysfunction in COVID-19: current findings and therapeutic implications. Atherosclerosis 314:58–62. https://doi.org/10.1016/j.atherosclerosis.2020.10.014

Natalello G, De Luca G, Gigante L, Campochiaro C, De Lorenzis E, Verardi L, Dagna L (2021) Nailfold capillaroscopy findings in patients with coronavirus disease 2019: broadening the spectrum of COVID-19 microvascular involvement. Microvasc Res 133:104071. https://doi.org/10.1016/j.mvr.2020.104071

Nejat R, Torshizi MF, Najafi DJ (2023) S protein, ACE2 and host cell proteases in SARS-CoV-2 cell entry and infectivity; is soluble ACE2 a two blade sword? A narrative review. Vaccine 11(2):204. https://doi.org/10.3390/vaccines11020204

Ni X, Yang X, Yang D, Bao J, Li R, Xiao Y, Hou C, Wang H, Liu J, Yang D, Xu Y, Cao Z, Gao Z (2020) Role of angiotensin-converting enzyme 2 (ACE2) in COVID-19. Crit Care 24:422. https://doi.org/10.1186/s13054-020-03120-0

Nicosia FR, Ligresti G, Caporarello N, Akilesh S, Ribatti D (2021) COVID-19 vasculopathy: mounting evidence for an indirect mechanism of endothelial injury. Am J Pathol 191:1374–1384. https://doi.org/10.1016/j.ajpath.2021.05.007

Noris M, Benigni A, Remuzzi G (2020) The case of complement activation in COVID-19 multiorgan impact. Kidney Int 98:314–322. https://doi.org/10.1016/j.kint.2020.05.013

Oki T, Kitaura J, Eto K et al (2006) Integrin αIIbβ3 induces the adhesion and activation of mast cells through interaction with fibrinogen. J Immunol 176:52–60. https://doi.org/10.4049/jimmunol.176.1.52

Osiaevi I, Schulze A, Evers G, Harmening K, Vink H, Kümpers P, Mohr M, Rovas A (2023) Persistent capillary rarefication in long COVID syndrome. Angiogenesis 26:53–61. https://doi.org/10.1007/s10456-022-09850-9

Ospina-Tascón GA, Calvache AJN, Quiñones E, Madriñan HJ, Valencia JD, Bermúdez WF, de Backer D (2017) Microcirculatory blood flow derangements during severe preeclampsia and HELLP syndrome. Pregnancy Hypertens 10:124–130. https://doi.org/10.1016/j.preghy.2017.07.140

Otifi HM, Adiga BK (2022) Endothelial dysfunction in Covid-19 infection. Am J Med Sci 363: 281–287. https://doi.org/10.1016/j.amjms.2021.12.010

Oudit Z, Kassiri C, Jiang PP, Liu SM, Poutanen JM, Penninger J, Butany GY (2009) SARS-coronavirus modulation of myocardial ACE2 expression and inflammation in patients with SARS. Eur J Clin Investig 39:618–625. https://doi.org/10.1111/j.1365-2362.2009.02153.x

Pine AB, Meizlish ML, Goshua G, Chang CH, Zhang H, Bishai J, Chun HJ (2020) Circulating markers of angiogenesis and endotheliopathy inCOVID-19. Pulm Circul 10:1–4. https://doi.org/10.1177/2045894020966547

Prasad KSS, Andre P, He M, Bao M, Manganello J, Phillips DR (2003) Soluble CD40 ligand induces β3 integrin tyrosine phosphorylation and triggers platelet activation by outside-in signaling. PNAS 100:12367–12371. https://doi.org/10.1073/pnas.2032886100

Prota AE, Campbell JA, Schelling P, Forrest JC, Watson MJ, Peters TR, Aurrand-Lions M, Imhof BA, Dermody TS, Stehle T (2003) Crystal structure of human junctional adhesion molecule 1:

implications for reovirus binding. PNAS 100:5366–5371. https://doi.org/10.1073/pnas.0937718100
Quintero-Fabián S, Arreola R, Becerril-Villanueva E, Torres-Romero JC, Arana-Argáez V, Lara-Riegos J, Ramírez-Camacho MA, Alvarez-Sánchez MA (2019) Role of matrix Metalloproteinases in angiogenesis and cancer. Front Oncol 9:1370. https://doi.org/10.3389/fonc.2019.01370
Raadsen M, Du Toit J, Langerak T, van Bussel B, van Gorp E, Goeijenbier M (2021) Thrombocytopenia in virus infections. J Clin Med 10:877. https://doi.org/10.3390/jcm10040877
Raeeszadeh-Sarmazdeh M, Do LD, Hritz BG (2020) Metalloproteinases and their inhibitors: potential for the development of new therapeutics. Cells 9:1313. https://doi.org/10.3390/cells9051313
Ranieri VM, Rubenfeld GD, Thompson BT et al (2012) Acute respiratory distress syndrome: the Berlin definition. JAMA 307:2526–2533. https://doi.org/10.1001/jama.2012.5669
Ranucci M, Ballotta A, Di Dedda U, Bayshnikova E, Dei Poli M, Resta M et al (2020) The procoagulant pattern of patients with COVID-19 acute respiratory distress syndrome. J Thromb Haemost 18:1747–1751
Roh JD, Kitchen RR, Guseh JS, McNeill JN, Aid M, Martinot AJ, Rosenzweig A (2022) Plasma proteomics of COVID-19–associated cardiovascular complications. JACC Basic Transl Sci 75: 425–441. https://doi.org/10.1016/j.jacbts.2022.01.013
Rohlfing A, Rath D, Geisler T, Gawaz M (2021) Platelets and COVID-19. Hamostaseologie 41: 379–385. https://doi.org/10.1055/a-1581-4355
Romagnoli S, Peris A, De Gaudio AR, Geppetti P (2020) SARS-CoV-2 and COVID-19: from the bench to the bedside. Physiol Rev 100:1415–1850. https://doi.org/10.1152/physrev.00020.2020
Romanowska-Kocejko M, Dudziak M, Hellmann M (2022) Nicotinamide adenine dinucleotide fluorescence to assess microvascular disturbances in post-COVID-19 patients. Cardiol J 29: 154–156. https://doi.org/10.5603/CJ.a2021.0132
Ross EA, Devitt A, Johnson JR (2021) Macrophages: the good, the bad, and the gluttony. Front Immunol 12:708186. https://doi.org/10.3389/fimmu.2021.708186
Rovas A, Osiaevi I, Buscher K, Sackarnd J, Tepasse PR, Fobker M, Kümpers P (2021) Microvascular dysfunction in COVID-19: the MYSTIC study. Angiogenesis 24:145–157. https://doi.org/10.1007/s10456-020-09753-7
Sabioni L, De Lorenzo A, Lamas C, Muccillo F, Caire H, Neto CF, Estato V, Tibirica E (2021) Systemic microvascular endothelial dysfunction and disease severity in COVID-19 patients: evaluation by laser Doppler perfusion monitoring and cytokine/chemokine analysis. Microvasc Res 134:104119. https://doi.org/10.1016/j.mvr.2020.104119
Sabioni L, De Lorenzo A, Castro-Faria-Neto HC, Estato V, Tibirica E (2023) Long-term assessment of systemic microcirculatory function and plasma cytokines after coronavirus disease 2019 (COVID-19). Braz J Infect Dis 27:102719. https://doi.org/10.1016/j.bjid.2022.102719
Sefik E, Qu R, Junqueira C, Kaffe E, Mirza H, Zhao J, Flavell RA (2022) Inflammasome activation in infected macrophages drives COVID-19 pathology. Nature 606:585–593. https://doi.org/10.1038/s41586-022-04802-1
Silva LS, Boldrini VO, Marques AM, Yasuda CL (2023a) Post-COVID myocarditis with antiheart antibodies persistence: clues for autoimmunity? Clin Cardiol 46:350–351. https://doi.org/10.1002/clc.23979
Silva MJA, Ribeiro LR, Gouveia MIM, dos Reis Marcelino B, Silva dos Santos C, Lima KVB, Lima LNGC (2023b) Hyperinflammatory response in COVID-19: a systematic review. Viruses 15:553. https://doi.org/10.3390/v15020553
Somers VK, Kara T, Xie J (2020) Progressive hypoxia: a pivotal pathophysiologic mechanism of COVID-19 pneumonia. Mayo Clin Proc 95:2339–2342. https://doi.org/10.1016/j.mayocp.2020.09.015
Sprenkeler EGG, Tool ATJ, Henriet SSV, van Bruggen R, Kuijpers TW (2022) Formation of neutrophil extracellular traps requires actin cytoskeleton rearrangements. Blood 26(139): 3166–3180. https://doi.org/10.1182/blood.2021013565

Suberviola B, Fito EC (2022) Management of hyperinflammation in COVID-19 patients. Rev Esp Quimioter 35:6–9. https://doi.org/10.37201/req/s03.02.2022

Sulli A, Gotelli E, Bica PF, Schiavetti I, Pizzorni C, Aloè T, Cutolo M (2022) Detailed videocapillaroscopic microvascular changes detectable in adult COVID-19 survivors. Microvasc Res 142:104361. https://doi.org/10.1016/j.mvr.2022.104361

Swanson KV, Deng M, Ting JPY (2019) The NLRP3 inflammasome: molecular activation and regulation to therapeutics. Nat Rev Immunol 19:477–489. https://doi.org/10.1038/s41577-019-0165-0

Tang Y, Liu J, Zhang D, Xu Z, Ji J, Wen C (2020) Cytokine storm in COVID-19: the current evidence and treatment strategies. Front Immunol 10(11):1708. https://doi.org/10.3389/fimmu.2020.01708

Tomaiuolo M, Brass LF, Stalker TJ (2017) Regulation of platelet activation and coagulation and its role in vascular injury and arterial thrombosis. Interv Cardiol Clin 6:1–12. https://doi.org/10.1016/j.iccl.2016.08.001

Troyano-Hernáez P, Reinosa R, Holguín A (2021) Evolution of SARS-CoV-2 envelope, membrane, nucleocapsid, and spike structural proteins from the beginning of the pandemic to September 2020: a global and regional approach by epidemiological week. Viruses 13(2):243. https://doi.org/10.3390/v13020243

Voiriot G, Dorgham K, Bachelot G, Fajac A, Morand-Joubert L, Parizot C, Lamazière A (2022) Identifcation of bronchoalveolar and blood immune-infammatory biomarker signature associated with poor 28-day outcome in critically ill COVID-19 patients. Sci Rep 12:9502. https://doi.org/10.1038/s41598-022-13179-0

von Hundelshausen P, Weber C (2007) Platelets as immune cells: bridging inflammation and cardiovascular disease. Circ Res 5(100):27–40. https://doi.org/10.1161/01.RES.0000252802.25497.b7

Vrints CJM, Krychtiuk KA, Van Craenenbroeck EM, Segers VF, Price S, Heidbuchela H (2020) Endothelialitis plays a central role in the pathophysiology of severe COVID-19 and its cardiovascular complications. Acta Cardiol 19:1–16. https://doi.org/10.1080/0001538.2020.1846921

Wang L, Sievert D, Clark AE, Lee S, Federman H, Gastfriend BD, Gleeson JG (2021) A human three-dimensional neural-perivascular 'assembloid' promotes astrocytic development and enables modeling of SARS-CoV-2 neuropathology. Nat Med 27:1600–1606. https://doi.org/10.1038/s41591-021-01443-1

Wang Y, Chuang CY, Hawkins CL, Davies MJ (2022) Activation and inhibition of human matrix Metalloproteinase-9 (MMP9) by HOCl, myeloperoxidase and chloramines. Antioxidants (Basel) 11:1616. https://doi.org/10.3390/antiox11081616

Weyrich AS, Zimmerman GA (2013) Platelets in lung biology. Annu Rev Physiol 75:569–591. https://doi.org/10.1146/annurev-physiol-030212-183752

Wool GD, Miller JL (2020) The impact of COVID-19 disease on platelets and coagulation. Pathobiology 13:1–13. https://doi.org/10.1159/000512007

Wu CT, Lidsky PV, Xiao Y, Cheng R, Lee IT, Nakayama T, Jackson PK (2022) SARS-CoV-2 replication in airway epithelia requires motile cilia and microvillar reprogramming. Cell 186:112–130. https://doi.org/10.1016/j.cell.2022.11.030

Xu S, Ilyas I, Weng J (2023) Endothelial dysfunction in COVID-19: an overview of evidence, biomarkers, mechanisms and potential therapies. Acta Pharmacol Sin 44:695–709

Yang SC, Tsai YF, Pan YL, Hwang TL (2021) Understanding the role of neutrophils in acute respiratory distress syndrome. Biom J 44:439–446

Zaid Y, Guessous F (2022) The ongoing enigma of SARS-CoV-2 and platelet interaction. Pract Thromb Haemost 6:e12642. https://doi.org/10.1002/rth2.12642

Zanza C, Romenskaya T, Manetti AC, Franceschi F, La Russa R, Bertozzi G, Longhitano Y (2022) Cytokine storm in COVID-19: Immunopathogenesis and therapy. Medicine 58:144. https://doi.org/10.3390/medicina58020144

Zarbock A, Singbart K, Ley K (2006) Complete reversal of acid-induced acute lung injury by blocking of platelet-neutrophil aggregation. J Clin Invest 116:3211–3219. https://doi.org/10.1172/JCI29499

Zhang L, Richards A, Barrasa MA, Jaenisch R (2021) Reverse-transcribed SARS-CoV-2 RNA can integrate into the genome of cultured human cells and can be expressed in patient-derived tissues. PNAS 118(21):e2105968118. https://doi.org/10.1073/pnas.2105968118

Zhang L, Sarangi V, Liu D, Ho M, Grassi AR, Wei L, Moon I, Vierkant RA, Larson NB, Lazaridis KN, Athreya AP, Wang L, Weinshilboum R (2022a) ACE2 and TMPRSS2 SARS-CoV-2 infectivity genes: deep mutational scanning and characterization of missense variants. Hum Mol Genet 16(31):4183–4192. https://doi.org/10.1093/hmg/ddac157

Zhang Y, Zhang H, Zhang W (2022b) SARS-CoV-2 variants, immune escape, and countermeasures. Front Med 16:196–207. https://doi.org/10.1007/s11684-021-0906-x

Zhang H, Lv P, Jiang J, Liu Y, Yan R, Shu S, Hu B, Xiao H, Cai K, Yuan S, Li Y (2023) Advances in developing ACE2 derivatives against SARS-CoV-2. Lancet 16. https://doi.org/10.1016/S2666-5247(23)00011-3

Zharkikh EV, Loktionova YI, Fedorovich AA, Gorshkov AY, Dunaev AV (2023) Assessment of blood microcirculation changes after COVID-19 using wearable laser Doppler Flowmetry. Diagnostics (Basel) 13:920. https://doi.org/10.3390/diagnostics13050920

Chapter 15
Mural Cells, Microvessels, and Alzheimer's Disease

Stephanie K. Bonney, Cara D. Nielson, and Andy Y. Shih

Abstract Alzheimer's disease is the leading cause of dementia and has been historically characterized by amyloid-beta plaques within the brain parenchyma and neurofibrillary tangles within neurons. However, it is now clear that cerebrovascular pathology and associated impairments in cerebral blood flow, blood–brain barrier function, and clearance of brain waste are key contributors to Alzheimer's disease progression. This chapter will discuss the microvascular changes that occur in Alzheimer's disease, related dementias, and aging with an emphasis on mural cells and their dysfunction during disease. In particular, pericytes will be reviewed in relation to the varying microvascular zones in the brain and the different cellular and acellular components of the neurovascular unit. Our discussion will synthesize data from preclinical and clinical studies, including examples that have used classical and novel technologies to shed light on vascular contributions to dementia.

Keywords Alzheimer's disease · Mural cells · Blood–brain barrier · Cerebrovasculature · Cerebral blood flow · Neurovascular coupling · Brain waste clearance · Vascular basement membrane · White matter degeneration

S. K. Bonney
Seattle Children's Research Institute, Seattle, WA, USA

C. D. Nielson
Seattle Children's Research Institute, Seattle, WA, USA

Graduate Program in Neuroscience, University of Washington, Seattle, WA, USA

A. Y. Shih (✉)
Seattle Children's Research Institute, Seattle, WA, USA

Graduate Program in Neuroscience, University of Washington, Seattle, WA, USA

Department of Pediatrics, University of Washington, Seattle, WA, USA

Department of Bioengineering, University of Washington, Seattle, WA, USA
e-mail: Andy.Shih@Seattlechildrens.org

R. F. Tuma et al. (eds.), *Handbook of Microcirculation*, Physiology in Health and Disease, https://doi.org/10.1007/978-3-031-99609-2_15

15.1 Zones of the Brain Microvascular Network and Associated Mural Cells

The microvascular network of the brain is separated into distinct zones with specialized structure and function. These zones have been well studied in the cerebral cortex and will be used as a framework for discussion (Fig. 15.1), though the zones will differ based on brain region, and regional differences remain incompletely characterized. Pial vessels on the surface of the brain are composed of 2-dimensional, interconnected webs of arteries and veins that supply and drain blood into and out of the brain (Blinder et al. 2010). The pial artery/arteriole network is derived from branches of the major cerebral arteries, the middle, anterior, and posterior cerebral arteries (Blinder et al. 2010). The convergence of these major artery territories form the leptomeningeal collaterals that allow blood to flow between arterial territories (Chalothorn and Faber 2010). The pial venule network is a similarly interconnected web and acts to drain blood from the brain into the large sinuses, including the superior sagittal sinus that runs along the midline of the brain (Nguyen et al. 2011).

From the pial arteriole and venule plexus, penetrating vessels branch from the network and dive into the cortical tissue to supply and drain blood (Nguyen et al. 2011; Taylor et al. 2016; Nishimura et al. 2010). Penetrating arterioles carrying oxygen-rich blood and are larger in diameter, allowing multiple red blood cells to travel in parallel at high velocity within the vessel lumen. Smaller vessels branch from the penetrating arteriole to form the arteriole-capillary transition (ACT) zone, where red blood cells remain fast-moving, but travel slower than those in arterioles (Hartmann et al. 2021a). In the cerebral cortex, the ACT zone is generally defined as the first four branching orders from the penetrating arteriole. From the ACT, microvessels then branch further to form the highly interconnected capillary bed where the smallest diameter vessels are found (Grant et al. 2017). This transition from arteriole to capillary is marked by a shift in the expression of the contractile protein, α-smooth muscle actin, which drops from high to low/undetectable levels. In the downstream capillaries, red blood cells are larger in diameter than the lumen, forcing them to deform in the lumen and travel single file (Secomb 1991). Capillaries occupy the largest surface area within the brain and are the site of gas exchange, delivering vital oxygen and nutrients into the brain parenchyma. The capillary bed then transitions into the postcapillary venules (Kucharz et al. 2021). The postcapillary venules are generally the few branch orders of microvessels just before convergence of flow into the ascending venules, which carry deoxygenated blood back to pial venules and sinuses on the cortical surface.

Mural cells exist in a continuum of morphologies and transcriptional states (Vanlandewijck et al. 2018). However, to best describe their distinctions in structure and function at each microvascular zone, we provide different descriptions and names of the mural cells along the zones. In all cases, the mural cells lie on the abluminal side of the endothelial cell layer and have varying contributions to blood flow regulation and vascular stability (Fig. 15.1). Smooth muscle cells (SMCs) are

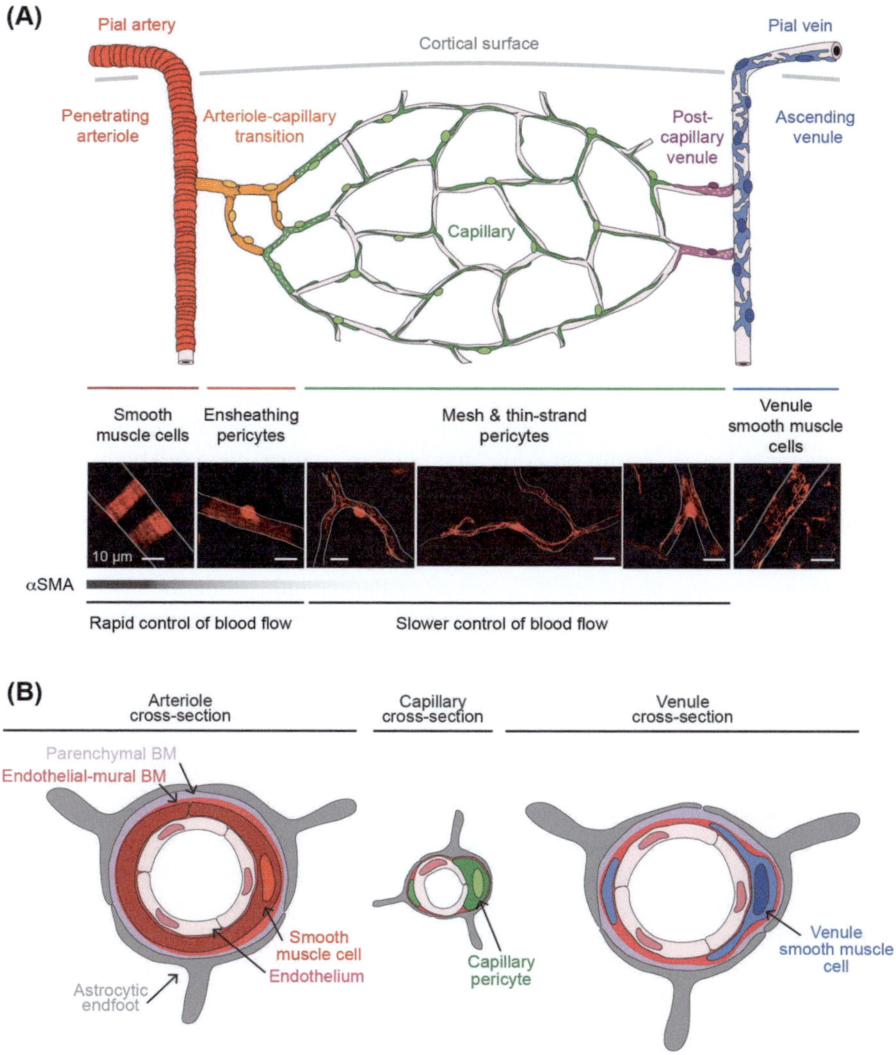

Fig. 15.1 Microvascular network of the healthy cerebral cortex and mural cell types. (**a**) Blood enters the brain via pial arteries and penetrating arterioles that dive into the brain parenchyma. Penetrating arterioles branch to form the arteriole-capillary transition zone which ramify into the capillary network. Blood flow through the capillaries is collected into postcapillary venules ultimately draining blood through ascending venules and pial veins. Penetrating arterioles are surrounded by smooth muscle cells (red). Ensheathing pericytes (orange) surround the arteriole-capillary transition zone. Mesh and thin-strand pericytes (green) cover capillaries and postcapillary venules. Venule smooth muscle cells (blue) surround ascending venules. High expression of α-smooth muscle actin (αSMA) in smooth muscle cells and ensheathing pericytes allow for rapid contractility and control of blood flow into the brain. Low to negligible expression of αSMA is found beyond the capillary zone, and thus, slower control of blood flow is observed. (**b**) Graphic depicting cross-sections of an arteriole, capillary and venule. Astrocytic endfeet (gray) attach to the outer parenchymal basement membrane (BM; light purple). Mural cells (red, green, or blue) and the endothelium (light pink) share the endothelial-mural BM (pink)

ring-shaped cells that enwrap pial arteries and penetrating arterioles (Hill et al. 2015). These cells have high expression levels of α-smooth muscle actin, making them dynamic regulators of arteriole diameter and blood flow. The ACT zone is surrounded by ensheathing pericytes, cells that have round, protruding cell bodies and complex processes that extend from the soma and completely enwrap the vessels. Like SMCs, ensheathing pericytes express α-smooth muscle actin (Grant et al. 2017; Hartmann et al. 2021b). They are also dynamic regulators of ACT zone vessel diameter, but have been shown to contract slower than SMCs (Hartmann et al. 2021b). Pial arteries, penetrating arterioles, and ACT zones dilate in response to neural activity through neurovascular coupling, increasing blood flow to active brain regions (Hill et al. 2015; Drew et al. 2011; Rungta et al. 2018; Hall et al. 2014; Cai et al. 2018).

The capillaries are surrounded by both mesh pericytes and thin strand pericytes, with mesh pericytes lying on the vessels closest to the ACT zone (Grant et al. 2017). Both cell types have a "bump-on-a-log" morphology, with a round cell body and thin processes that extend from the soma; however, mesh pericytes have more complex processes than thin-strand pericytes. These processes extend along the capillaries until they reach the tip of a neighboring pericyte process. In the normal brain, pericyte processes contact over 90% of the capillary length within the mouse cortex (Berthiaume et al. 2018a). Mesh and thin-strand pericytes are embedded in the basement membrane of capillaries, and form peg-and-socket interactions with endothelial cells (Ornelas et al. 2021). They also express low to undetectable levels of α-smooth muscle actin. Thus, they are not as contractile as SMCs or ensheathing pericytes. However, they do express contractile machinery that exists in other mural cell types (Vanlandewijck et al. 2018; Erdener et al. 2022), and have been shown to have contractile properties which allow them to regulate capillary tone and blood flow in response to vasoconstrictive signals and intramural blood pressure changes (Hartmann et al. 2021a; Peppiatt et al. 2006; Fernández-Klett et al. 2010; Berthiaume et al. 2018b; Glück et al. 2021; Klug et al. 2023). They may contribute to slower aspects of blood flow regulation, such as creation of basal capillary tone or slower phases of the hemodynamic response (Hartmann et al. 2021b; Rungta et al. 2018).

Postcapillary venules are covered by mesh pericytes similar in morphology to those seen in the capillary region closer to the ACT zone. The functional roles of postcapillary venule pericytes are not fully clear and it is unknown whether these cells are distinct from the mesh pericytes that are present on capillaries. Finally, ascending venules are surrounded by venular SMCs. These cells are stellate shaped and express low levels of α-smooth muscle actin, though are not particularly contractile. Their functions in vascular homeostasis remain elusive (Hill et al. 2015; Hartmann et al. 2021b).

The dysfunction of brain microvasculature can contribute to neurological disease by impairing blood flow, nutrient exchange and blood–brain barrier (BBB) function. These changes involve degeneration of the different mural cells types and vascular zones described above (Sweeney et al. 2016; Profaci et al. 2020). In this chapter, we will discuss how alterations to the microvascular network may contribute to dementia, with a focus on mural cell pathology in Alzheimer's disease (AD).

15.2 Alzheimer's Disease and Changes in the Brain Microvasculature

AD is the leading cause of dementia and the accumulation of amyloid-beta (Aβ) is a key disease-driving mechanism that can lead to degeneration of both neurons and the vasculature. A variety of mutations in the amyloid precursor protein (APP) and processing enzymes, like presenilin-1 (PSE1), have been identified to cause Aβ accumulation in human AD cases. The rest of the cases may be due to impaired Aβ degradation with age, in part due to reduction in neprilysin activity, an Aβ degrading enzyme (Iwata et al. 2001, 2002; Hellström-Lindahl et al. 2008). Abnormal processing of APP or degradation of Aβ lend to the "amyloid cascade hypothesis" (Barykin et al. 2017), where initial build-up of Aβ can promote Aβ aggregation into parenchymal plaques and the vessel wall, ultimately hindering the clearance of Aβ from the brain (Greenberg et al. 2020).

Cerebral amyloid angiopathy (CAA) is the build-up of Aβ along the vascular wall (Greenberg et al. 2020; Nicoll et al. 2004; Haglund et al. 2006; Eikelenboom et al. 2008; Thal et al. 2008). Diagnosis of CAA is most accurately made postmortem through pathological verification of Aβ accumulation along the brain vasculature. CAA weakens the vessel wall and leads to intracerebral hemorrhages, cerebral microbleeds, and cortical superficial siderosis, which can be detected through magnetic resonance imaging (MRI) or computed tomography (CT) scans.

Certain mutations in APP, such as the Swedish and Dutch mutations, have a higher propensity to cause CAA in humans (Greenberg et al. 2020). Mouse models that express these mutant forms of APP recapitulate aspects of human pathology and have been instrumental in understanding how vascular dysfunction and CAA contribute to cognitive decline. The APP-23 mouse model overexpresses APP with the Swedish mutations (KM670/671NL) and displays substantial vascular build-up of Aβ around large pial and penetrating arterioles at 16-months of age (Calhoun et al. 1999). These mice are prone to microbleeds in the brain (Reuter et al. 2016) and exhibit capillary occlusions and age-dependent blood flow abnormalities in the large arteries (Beckmann et al. 2003; Thal et al. 2009). Supported by this, human AD tissues from patients with the Swedish mutation have increased capillary occlusions associated with CAA (Thal et al. 2009). Mice that overexpress the Dutch mutation (E693Q) of human APP develop CAA and exhibit SMC degeneration and microbleeds by 29 months of age, as similarly observed in corresponding human AD tissue (Herzig et al. 2004).

Another commonly used mouse model of CAA is the Tg-SwDI mice which carry the Swedish, Dutch, and Iowa (D694N) mutations in human APP (Davis et al. 2004). These mice do not overexpress mutated human APP to the degree seen in APP-23 mice, but they exhibit earlier onset of CAA. CAA tends to occur more on the walls of capillaries, and less on pial and parenchymal arterioles compared to the APP-23 line, aligning more with what has been described as Type 1 CAA in sporadic CAA cases (Thal et al. 2002). Interestingly, Tg-SwDI mice were shown to have attenuated vascular responses to vasomodulators and neuronal activity at 3 months of age, prior to the presence of measurable Aβ build-up (Park et al. 2014). Although some studies have not observed neurovascular dysfunction in Tg-SwDI mice (Munting et al. 2021), Tg-SwDI mice display earlier cognitive impairment than mice with more prevalent parenchymal Aβ build-up like the Tg-5xFAD mice (APP mutations: Swedish, Florida (I716V), London (I717V) and PSE1 mutations: M146L and L286V) (Xu et al. 2014). These findings demonstrate that a decline in vascular function due to CAA may be an underlying factor contributing to dementia. However, there are numerous models of AD to explore (Szu and Obenaus 2021), and detailed studies of neurovascular pathology remain sparse and sometimes contradictory.

15.3 Vascular Transcriptomic Data from Human Alzheimer's Disease Tissue

Single-cell RNA sequencing is a powerful approach to study cell-type specific gene expression changes that occur in AD. However, vascular cells represent a small proportion of total brain cells, making enrichment for the vasculature necessary. Single-cell sequencing of mural cells is challenging since they are embedded in the basement membrane, often leading to low yields or damaged nuclei. An approach called VINE-seq (Vessel Isolation and Nuclei Extraction for Sequencing) was developed to overcome these challenges and enrich for vascular cell types in human tissue samples (Yang et al. 2022). This method enriches for the vasculature and relies on a gentle nuclei isolation protocol, followed by cleanup using sucrose and fluorescence activated cell sorting (FACS).

Use of VINE-seq demonstrated a collective reduction in the number of endothelial cells, SMCs, pericytes, and fibroblasts in human AD hippocampal and frontal cortex tissue compared to controls. Interestingly, two pericyte subpopulations were found to be intermingled along the capillary network in the human vasculature, as opposed to a single population in prior mouse studies (Vanlandewijck et al. 2018): Matrix-pericytes (M-pericytes) and Transport-pericytes (T-pericytes). M-pericytes were enriched with extracellular matrix organization genes and were selectively reduced in number with AD, whereas T-pericytes were enriched in small-molecule transmembrane transporters and were preserved in AD. VINE-seq also revealed that mural cells exhibited the largest gene expression changes in AD compared to other vascular cell types. Among the dysregulated signals were those involved in actin-mediated cell contraction and production/handling of reactive oxygen species, implicating deficiencies in pathways responsible for blood flow control and oxidative injury to vascular cells (Yang et al. 2022), perhaps involving a recent link between Aβ, reactive oxygen species and endothelin-1 production (Nortley et al. 2019).

A more recent single-cell transcriptomic study also examined vascular cell types but did not use VINE-seq vascular enrichment (Sun et al. 2023). This required sequencing of more cell nuclei to gain an adequate representation of vascular cell types in their data, and samples were collected across 6 brain regions. They found regional decreases in pericyte and endothelial cell counts in entorhinal cortex, hippocampus, and thalamus. However, in contrast to the prior study using VINE-seq, no overall difference in cell numbers were detected between controls and AD patients, except for an increase in fibroblast-like cells. Interestingly, gene expression analysis of pericytes revealed decreases in PDGFRβ expression in AD, which is consistent with a prior study showing marked decrease in vascular PDGFRβ using histology and RNA analysis (Smyth et al. 2022), suggesting a gradual decrease in signaling through PDGFB-PDGFRβ in AD.

15.4 Functions of Capillary Pericytes and Involvement in Cognitive Decline

15.4.1 Pericyte Loss in Aging and AD/ADRD

Pericyte loss occurs at an accelerated rate in AD and other dementias compared to normal aging. Postmortem neuropathological studies report a ~ 30–60% reduction in pericyte number/coverage in the frontal cortex, hippocampus, and white matter in AD patients relative to cognitively normal individuals, and this is worsened in carriers of the APOE4 allele which increases risk for AD (Sengillo et al. 2013; Halliday et al. 2016; Ding et al. 2020). However, it should be noted that pericyte loss in AD has not been uniformly reported and is likely brain region specific (Fernandez-Klett et al. 2020; Ding et al. 2021). Several studies have also reported elevated levels of soluble platelet derived growth factor receptor beta (sPDGFRβ) in the cerebral spinal fluid of patients with AD, suggesting degeneration of pericytes and release of pericyte-derived molecules (Miners et al. 2019; Montagne et al. 2015; Nation et al. 2019). Pericyte loss likely increases with disease progression, as PDGFRβ levels in the precuneus were shown to be significantly reduced in the final stages of disease compared to the early and mid-stages (Miners et al. 2017).

15.4.2 Animal Models to Study Pericyte Loss

Pericyte-deficient mouse models have been helpful in clarifying the physiological consequences of pericyte loss. Many groups have studied transgenic mouse models of congenital pericyte deficiency, including mice with null or hypomorphic alleles of PDGFRβ, as well as mice with reduced endothelial retention of platelet derived growth factor subunit B (PDGFB). The PDGFRβ/PDGFB signaling pathway is required for pericyte recruitment to the vessel wall during development and the vasculature does not properly develop in these mice (Hellström et al. 1999; Lindahl et al. 1997). Further, studies have revealed that pericyte loss to levels of ~50% or less leads to widespread BBB breakdown, neuroinflammation, neuron death, and impaired microvascular flow. Conditional, endothelial-specific ablation of PDGFB expression in young adult mice led to slow degradation of pericyte coverage and loss of BBB integrity over many months (Vazquez-Liebanas et al. 2021). Similarly, less severe congenital loss of one PDGFRB allele leads to neurovascular pathology in an age-related fashion (Bell et al. 2010). Together, these findings suggest that PDGFB-PDGFRB signaling is still needed to maintain vascular stability in the adult brain, and that supplementation of PDGF-BB may help to preserve pericyte coverage.

Some studies have examined the effects of pericyte loss using the more direct strategy of cell ablation. In one approach, human diphtheria toxin receptor (DTR) has been selectively expressed in CNS pericytes using cell-specific Cre drivers. Since mice are resilient to diphtheria toxin, ectopic expression of the human receptor engages selective pericyte death. These studies revealed that widespread pericyte loss in the adult brain over days causes marked impairment of blood flow, neurovascular coupling, and BBB integrity (Nikolakopoulou et al. 2019; Kisler et al. 2017). Interestingly, using a related strategy where diphtheria toxin (DTA) itself is expressed in PDGFRβ-positive cells, an approach that reduces pericyte coverage by ~75% in the retina, no overt loss of integrity was seen with the blood–retinal barrier despite changes to vascular structure (Park et al. 2017). Further, BBB dysfunction was also not observed. However, increased capillary flow stalls were reported in the brain and associated with local tissue hypoxia, suggesting impairment in blood flow can come before significant loss of BBB function (Choe et al. 2022). The discrepancy in findings between these two related approaches remains unresolved.

The complex outcomes of widespread brain pericyte loss can make it difficult to understand how gradual and regional pericyte loss is related to capillary dysfunction. Two approaches have been developed for targeted ablation of individual pericytes of the murine brain, both requiring cranial window implantation and two-photon microscopy. The first approach, 2phatal (two-photon chemical apoptotic targeted ablation), uses femtosecond pulsed laser light to excite Hoechst 33342, a cell-permeant DNA binding dye, inducing apoptosis of targeted cells (Hill et al. 2017). The second is an optical ablation technique in which two-photon irradiation is used to thermally ablate individual pericytes in the brain of transgenic mice with the endogenous expression of pericyte markers (Berthiaume et al. 2018a; Nielson et al. 2022). This approach revealed pericyte roles in capillary blood flow regulation and the phenomenon of pericyte remodeling, as described in more detail below (Hartmann et al. 2021b; Berthiaume et al. 2022).

While mouse models of pericyte loss have led to important key findings, it is also essential to study pericyte loss in the context of aging and AD. Analysis of the aged mouse cortex has revealed marked decreases in pericyte number and pericyte coverage of vessels, particularly in deeper cortical layers (Soto et al. 2015; Bennett et al. 2024). In some mouse models of AD involving expression of human mutant Aβ, degeneration of pericyte morphology and loss of coverage has been observed (Park et al. 2014), but this is not the case in all models (Skaaraas et al. 2021; Damisah et al. 2017). Combining these mouse models of AD with other comorbid factors, such as hypertension (Kruyer et al. 2015), accelerates pericyte loss and neurovascular dysfunction. Microvascular zones beyond the capillary bed are affected in AD mouse models, as evidenced by loss of arteriole and venule SMCs (Park et al. 2014; Lai et al. 2015; van Veluw et al. 2020). Disparate findings across models may be due to bonafide differences in how different Aβ mutations or expression patterns interact with the microvasculature. However, they may also result from a lack of detail of when and where pericyte loss occurs in different models, warranting deeper investigation in the most widely used models.

15.4.3 Pericyte Influence of Capillary Blood Flow

The influence of pericytes on capillary blood flow was long debated (Hartmann et al. 2021a). Capillaries have been shown to dilate prior to arterioles during neurovascular coupling (NVC) and constrict during cerebral ischemia (Hall et al. 2014). Further, a congenital model of pericyte deficiency and an adult-onset model of pericyte loss exhibit deficits in NVC and reduced cerebral blood flow (Nikolakopoulou et al. 2019; Kisler et al. 2017, 2020). However, other studies reported that NVC is solely influenced by arterioles (Hill et al. 2015), even though pericytes possess contractile abilities (Fernández-Klett et al. 2010). These contrasting findings were due to a lack of consensus on nomenclature within the vascular field. Past studies ha not distinguished the ensheathing pericytes on the ACT zone from the mesh and thin strand present on the capillaries, which have different gene expression patterns and contractile dynamics (Hartmann et al. 2021a). Further, congenital mouse models of pericyte deficiency and mouse models of adult-onset pericyte loss discussed above involve complex effects throughout the microvasculature, including BBB breakdown, neuron death, and edema. These outcomes make it difficult to disentangle physiological roles of pericytes in blood flow control from the pathophysiological effects of their widespread depletion.

To discern whether capillary pericytes (those with low/undetectable levels of α-smooth muscle actin) influence vessel tone and blood flow, recent studies have used an in vivo optogenetic approach (Hartmann et al. 2021b; O'Herron et al. 2022; Nelson et al. 2020). Pericytes were selectively stimulated with light in transgenic mice expressing different variants of channelrhodopsin, which are light-sensitive ion channels that promote cellular depolarization. The activation of pericytes led to a constriction of the underlying capillary coupled with a decrease in blood flow. However, this vessel constriction occurred on a slower time scale than the constriction that followed the activation of SMCs or ensheathing pericytes upstream. Using a separate optical manipulation, focal ablation of individual capillary pericytes using two-photon laser irradiation, revealed that pericytes regulate basal capillary tone (Hartmann et al. 2021b; Berthiaume et al. 2018a, 2022). The loss of pericytes led to a dilation and increase in blood flow in the uncovered vessel, providing evidence that pericytes influence basal capillary flow dynamics. However, whether pericytes with low/undetectable α-smooth muscle actin actively regulate blood flow during NVC is not completely understood. If so, they may contribute to slower aspects of the hemodynamic response.

One major point that is noted following pericyte loss is the local capillary dilations that can alter the distribution of blood flow within the capillary network (Fig. 15.2a). When vessels lose pericyte coverage and dilate, blood is diverted from neighboring capillary segments, increasing capillary flow heterogeneity (Berthiaume et al. 2022). The efficacy of oxygen extraction is dependent upon optimization of capillary flow patterns. During NVC, perfusion among capillaries in a network homogenizes, which leads to slowing of overall blood cell transit time to enhance oxygen extraction and promotes more even distribution of blood cells across the active tissue (Gutiérrez-Jiménez et al. 2016; Li et al. 2018, 2019). As capillary flow heterogeneity increases, the efficiency of oxygen extraction decreases (Jespersen and Østergaard 2012). Thus, the changes in capillary flow that occur after pericyte loss likely lead to inefficient oxygen extraction, creating areas of under oxygenated brain tissue. In fact, MRI studies have revealed disrupted capillary flow in patients with AD (Eskildsen et al. 2017). Faster speed of capillary flow has also been observed in congenital mouse models of pericyte deficiency, suggesting that the mean transit time of blood cells is abnormally increased (Watson et al. 2020). This demonstrates the importance of understanding how heterogeneity of capillary blood flow arises in AD and whether pericyte loss is a factor.

Pericyte loss also leads to an increase in capillary blood flow stalling. Stalls have been observed in both mouse models of optical pericyte ablation and adult-induced genetic pericyte depletion (Fig. 15.2a). These flow stalls have been linked to altered leukocyte-endothelial interactions, but may also involve the capillary flow steal effect related to abnormal flow distribution, as discussed above (Choe et al. 2022; Berthiaume et al. 2022). Stalled capillaries are more likely to regress, altering the structure of the capillary network and further increasing capillary flow heterogeneity and reducing oxygen extraction (Berthiaume et al. 2022). Increased cortical capillary stalls have been observed in both the Tg-APP/PS1 (APP Swedish mutation and exon 9 deletion in PSE1) and Tg-5xFAD mouse models of AD (Cruz Hernández et al. 2019). Interestingly, the administration of neutrophil antibodies reduced the number of capillary stalls and improved cerebral blood flow in these animals.

While pericyte loss is a common observation in AD, human brain slices exposed to Aβ demonstrated a strong constriction response in pericytes which may pathologically reduce blood flow within the capillary network prior to major pericyte loss (Nortley et al. 2019). Decreased blood flow is correlated with reduced cognitive performance and is likely a driver of the cognitive deficits symptomatic of AD. In patients with internal carotid artery occlusions, higher performance on a sustained attention task correlated with increased cerebral blood flow, and performance improved in all individuals when the occlusion was reversed (Marshall et al. 2001). When cerebral blood flow is reduced in mice, spatial memory decreases and Aβ deposition increases (Wang et al. 2016; Hattori et al. 2016). Further, cerebral blood flow deficits are a risk factor for dementia and are observed in AD patients and mouse models of AD (Dai et al. 2009; Wolters et al. 2017; Wiesmann et al. 2017; Li

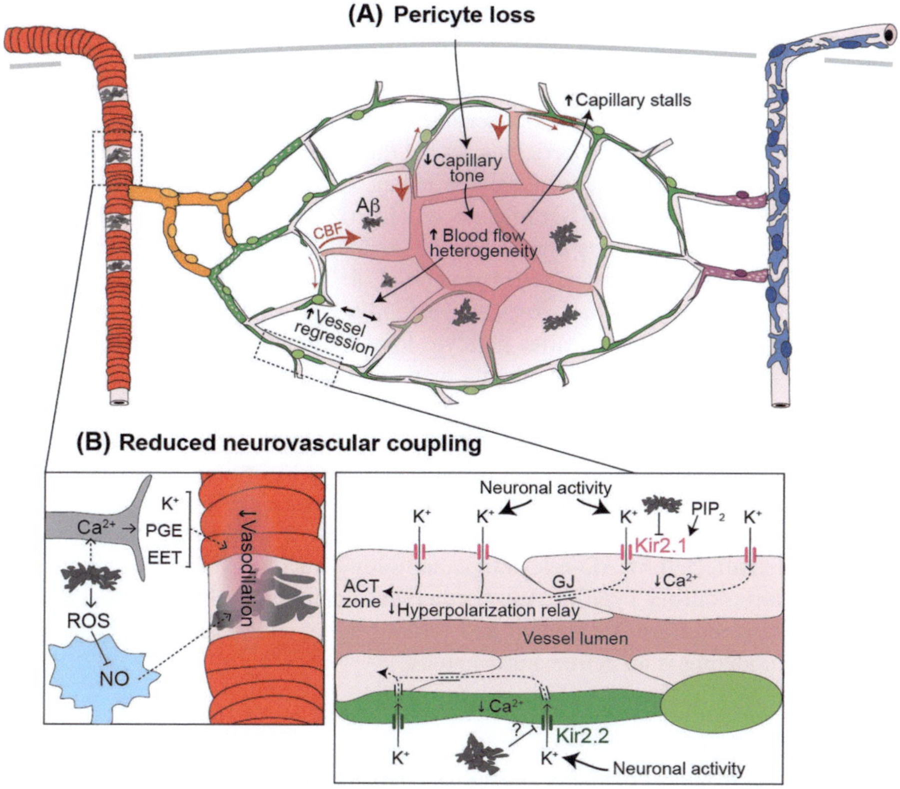

Fig. 15.2 Contributing factors to blood flow deficits in Alzheimer's disease. (**a**) Pericyte loss correlates with amyloid-beta (Aβ) build-up resulting in loss of capillary tone and increased blood flow heterogeneity. Cerebral blood flow (CBF) is then distributed unevenly through the capillary network resulting in capillary stalls and vessel regressions. (**b**) On the arteriole level (left), Aβ causes dysregulation in astrocytic-Ca2+ signaling which affects the release of the vasodilators K+, prostaglandins (PG), and epoxyeicosatrienoic acids (EET). Aβ also causes oxidative stress, affecting the production of nitric oxide (NO) from neurons. Together, these mechanisms attenuate vasodilation of arterioles during neurovascular coupling. Within the capillary zone (right), Aβ dampens Kir2.1 activity and sensing of neuronal activity via K+ on brain endothelial cells which normally initiates a hyperpolarization relay toward the upstream ACT zone through gap junctions. PIP2 can enhance Kir2.1 activity. Neuronal K+ is sensed via Kir2.2 on pericytes resulting in a hyperpolarization relay via gap junctions toward the upstream ACT zone; however, it is unknown if Kir2.2 is affected by Aβ presence

et al. 2014; Niwa et al. 2002). Thus, pericyte loss disrupts capillary blood flow under basal conditions, and together with compromise of the BBB, contributes to cerebral blood flow decrease and cognitive decline. However, it remains unclear when pericyte loss in AD occurs in relation to cerebral blood flow decreases, as pericyte changes are assessed postmortem in humans. Fluid biomarkers that report the health of pericytes (soluble platelet-derived growth factor) may be able to shed some light in future studies (Montagne et al. 2015).

15.4.4 Pericyte Structural Plasticity in Health and Aging

Pericytes are structurally plastic and can dynamically lengthen or retract their processes to negotiate their vascular territories (Berthiaume et al. 2018a). When a pericyte dies, the neighboring pericytes extend their processes into the regions of the endothelium that are left uncovered by pericytes. Growth stops when two pericyte processes meet end-to-end (Berthiaume et al. 2022). This phenomenon, called pericyte remodeling, restores capillary tone, reducing the capillary dilation seen after the loss of pericyte coverage (Hartmann et al. 2021b). Pericyte remodeling is an intrinsic repair mechanism to re-establish capillary flow patterns and homeostatic functions of the endothelium after pericyte loss (Berthiaume et al. 2018b). Aged mice exhibit slower pericyte remodeling than adult mice, leaving large stretches of endothelium uncovered for longer periods of time after pericyte loss (Berthiaume et al. 2022). Further, pericyte remodeling in aged mice only partially restores capillary tone to basal levels. These impairments lead to lasting changes in blood flow and vascular structure, likely exacerbating the negative effects of pericyte loss (Berthiaume et al. 2022). Future studies are needed to understand how pericyte remodeling is affected by AD-relevant pathobiology, such as with Aβ accumulation along capillaries during Type 1 CAA. Further, the signaling mechanisms that control pericyte remodeling are not known and finding therapeutic strategies to augment pericyte remodeling is an active area of study.

15.4.5 Pericyte Secretion of Neurotrophic Factors

Pericytes can support neuronal viability through mechanisms unrelated to blood supply and BBB function. A recent study demonstrated that loss of pericytes via diphtheria-mediated ablation in adult mice leads to a rapid decline in the number of neurons and loss of cognitive function (Nikolakopoulou et al. 2019). The behavioral deficits and neurodegeneration were attributed not only to blood flow deficits and BBB breakdown, but also loss of pleiotrophin levels, a neurotrophic factor highly expressed by brain pericytes. Interestingly, knock-down of pleiotrophin in pericytes in otherwise healthy mice only resulted in a decline in the neuronal populations in the presence of circulatory stress (Nikolakopoulou et al. 2019). Pericytes also support neural network function by secreting insulin-like growth factor 2 in response to increased neuronal activity, which promotes formation of long-term memories (Pandey et al. 2023). Studies in cultured human brain pericytes have revealed that pericytes can release a range of proregenerative growth factors, including brain-derived neurotrophic factor (BDNF), basic fibroblast growth factor (bFGF), beta-nerve growth factor (bNGF), vascular endothelial growth factor (VEGF), and placental growth factor (PLGF) when treated with the pericyte proliferation and migration signal, PDGF-BB (Gaceb et al. 2018). While these studies were performed in cultured cells, they raise the possibility that pericytes release factors to directly preserve neuronal viability and growth, in addition to supporting vascular growth. Interestingly, PLGF levels in the plasma were recently shown to be a highly sensitive biomarker for vascular contributions to cognitive impairment and dementia in humans (Hinman et al. 2023). PLGF is a member of the VEGF family and has been associated with pathological angiogenesis and inflammation in various tissues (Van Bergen et al. 2019). In general, little is known about growth factor release by pericytes and how this mechanism of neural support is affected in AD.

15.4.6 Disruption of Neurovascular Coupling Mechanisms in Alzheimer's Disease

To ensure healthy brain function, an adequate blood supply is needed to match the energy demands of neuronal activity. Findings from large cohort studies suggest that vascular dysregulation is an early driver in the progression of late-onset AD (Iturria-Medina et al. 2016). Mouse studies of the brain and retinal microvasculature have been instrumental in understanding the underlying mechanisms of NVC and how these processes are disrupted in dementia. In a healthy microvascular network, the capillary endothelium uses the inwardly rectifying potassium channel, Kir2.1, to "sense" the release of potassium (K^+) from neurons and astrocytes during neuronal activity (Fig. 15.2b) (Park et al. 2014). Sensation of K^+ by the endothelium leads to a rapid conduction of responses along the capillaries to upstream arterioles, leading to relaxation of arteriolar smooth muscle cells and increase of blood flow into the downstream capillaries. In Tg-5xFAD mice, Kir2.1 conductance is attenuated, and this is associated with weakened NVC responses. However, phosphatidylinositol 4,5 bisphosphate (PIP2), an endogenous phospholipid that activates Kir2.1 upon binding, can heighten Kir2.1 activity and restore NVC in Tg-5xFAD mice (Mughal et al. 2021) and other small vessel disease models (Dabertrand et al. 2021). Targeting PIP2 as a therapy for AD may improve NVC responses and preserve neuronal viability and function (Arancio 2008). The capillary endothelium is also known to propagate signals via the transient receptor potential ankyrin 1 (TRPA1) which propels retrograde Ca^{2+} responses via the Panx1 channel and ATP (Thakore et al. 2021). However, it is unknown if the endothelial-TRPA1-Panx1 pathway is disrupted in AD.

Pericytes may also play an important part in NVC, as diphtheria-mediated ablation of ~50% of pericytes in the cortex attenuates NVC (Kisler et al. 2020). Pericytes also sense neuronal activity, possibly via K^+ and Kir2.2-mediated mechanisms, which decrease pericyte Ca^{2+} concentrations and increase capillary diameter (Glück et al. 2021). This not only creates a local increase in blood flow, but also propagates a wave of hyperpolarization toward the upstream side of the network, leading to dilation of the ACT zone and arteriole. In retinal vascular networks, gap junctions between neighboring pericytes and between pericytes and the endothelium allow for the communication of these hyperpolarizing signals (Kovacs-Oller et al.

2020). Interestingly, the gap junction protein Cx43 forms string-like structures along the endothelium and were enriched closer to the arteriole side of the microvascular network, suggesting that structural features are in place to favor upstream conductance of hyperpolarizing signals (Ivanova et al. 2019). Cx43 is increased in regions of heavy Aβ burden; however, the effect of this has only been studied in the context of astroglial interactions in mouse models of AD (Yi et al. 2016). It is unknown if changes occur in endothelial-pericyte gap junctions in AD or aging, but such alterations could disrupt NVC. Further, it remains to be determined if pericyte-Kir2.2 activity and Ca^{2+} dynamics are altered in AD which could diminish the sensing of neuronal activity.

Worsened capillary regression is a key pathological feature of AD. Its occurrence can lead to pockets of under-perfused brain tissue (Moeini et al. 2018), and disruption of conductive capillary-to-arteriole signaling during neurovascular coupling (Steinman et al. 2021). Many factors may contribute to capillary regression, including vascular accumulation of Aβ, loss of proangiogenic signals, endothelial damage, pericyte loss, and capillary plugging by leukocytes or microthrombi (Berthiaume et al. 2022; Schager and Brown 2020; van Dinther et al. 2022).

Dilation of arterioles is necessary to supply oxygenated blood into the capillary network during neuronal activity, and this is facilitated by action of vasodilators. Vasodilators such as prostaglandins, epoxyeicosatrienoic acids, and K^+ are released from astrocytic endfeet that surround arterioles (Fig. 15.2b) (Attwell et al. 2010). This is in response to neurotransmission of glutamate which activates Ca^{2+} signaling in astrocytes via metabotropic glutamate receptors. This mechanism as a driving force for NVC has been debated (Nizar et al. 2013; Bonder and McCarthy 2014). However, recent work demonstrates that astrocytic-Ca^{2+} is at least responsible for amplifying NVC during sustained neuronal activity (Institoris et al. 2022). In addition to astrocytes, neuronal release of nitric oxide (NO) is capable of initiating dilation of nearby arterioles (Echagarruga et al. 2020).

Mouse models of AD, CAA, and aging mimic aspects of the impaired arteriole responses that contribute to reduced brain perfusion that is observed in AD patients (Viticchi et al. 2012; Hock et al. 1997). This is, in part, due to increased oxidative stress that can reduce the overall production of NO resulting in heightened vasoconstriction (Park et al. 2007) or promote the expression of the vasoconstrictive peptide, endothelin-1 (Nortley et al. 2019). Additionally, in aged mice, morphological changes to astrocytic endfeet have been noted which hinder astrocyte-SMC interactions, diminishing the effect of vasodilators (Duncombe et al. 2017). Furthermore, astrocytes exposed to Aβ have spontaneous Ca^{2+} fluctuations which may cause dysregulation in the expression and release of vasomodulators (Fig. 15.2b) (Lee et al. 2014). Physical damage to the endothelium and mural cell layer also reduces conductive responses and contractility of the vessel wall. Degeneration of SMCs along arterioles and pericytes along capillaries has been observed in postmortem AD tissue and mouse models of AD (Park et al. 2014; van Veluw et al. 2020; Vonsattel et al. 1991; Charidimou et al. 2017). Thus, it is likely that the defect in arteriole responses to NVC in AD is a multifaceted problem that includes loss of mural cells, alterations to astrocyte signaling and morphology, reduced capillary-to-arteriole signaling, and reduced NO production.

Regional differences in NVC have been noted, which may leave certain brain regions more vulnerable to pathology and cognitive decline (Sloan et al. 2010). For example, the hippocampus has reduced blood flow and oxygenation compared to the neocortex in healthy, young mice. The cerebral white matter is also particularly vulnerable, as will be discussed further below. This is attributed to lower vascular density and weaker NVC responses (Shaw et al. 2021). Additionally, the morphological diversity of astrocytes and the varying amount of contact they maintain with the vasculature in different brain regions are two factors that play a part in how rapid NVC responses occur following neuronal activity (Bernier et al. 2021). While these observations have been made, the underlying mechanism for these regional differences and if they contribute to dementia, remains poorly understood.

15.4.7 Loss of Blood–Brain Barrier Functionality in Aging and Dementia

The BBB tightly controls the movement of ions, molecules and cells between the blood and brain parenchyma (Profaci et al. 2020). The central component of the BBB is the brain endothelium and its tight-junctional complexes. However, signaling with neighboring pericytes, astrocytes, and the vascular basement membrane is necessary for optimal BBB structure and integrity. The influx and efflux of molecules into and out of the brain are tightly regulated through a variety of endothelial transporters. The brain endothelium also maintains lower levels of transcytosis and pores unlike the leakier vessels of some peripheral tissues.

BBB leakage is a common feature in AD and aging pathology, allowing harmful molecules and immune cells to enter the brain parenchyma from the blood. Studies have noted leakage of serum proteins such as albumin, immunoglobulin, and fibrinogen in postmortem histological examinations. Many leakage sites are associated with Aβ plaques and abnormal blood vessel morphology (Sengillo et al. 2013; Wisniewski et al. 1997; Ryu and McLarnon 2009). MRI studies have revealed that BBB leakage is a common occurrence early in AD progression (Montagne et al. 2015; Nation et al. 2019; Starr et al. 2009; van de Haar et al. 2016). High resolution-MR imaging in AD patients over time demonstrated that BBB leakage is concentrated in the hippocampus and worsening leakage correlated with greater cognitive decline (Montagne et al. 2015).

Breakdown of BBB properties in AD has been associated with loss of pericytes and disrupted endothelial tight junction integrity (Montagne et al. 2015). Loss of tight junction proteins, claudin-5 and occludin, is associated with increased Aβ build-up and reduced synaptic markers (Yamazaki et al. 2019). When treated with Aβ, cultured brain endothelial cells exhibit deficient barrier functionality (Veszelka et al. 2013), and cultured human pericytes rapidly degenerate (Verbeek et al. 1997), indicating that Aβ exposure can be toxic to vascular cells. Mice exhibiting pericyte deficiency in adulthood or with age have worsened BBB leakage and this is associated with heightened neuronal death (Vazquez-Liebanas et al. 2021; Bell et al. 2010; Nikolakopoulou et al. 2019. Pericytes are important for the suppression of transcytosis in the endothelium and polarization of astrocytic end feet (Daneman et al. 2010; Armulik et al. 2010; Ben-Zvi et al. 2014). This lends to the idea that endothelial tight junction instability and pericyte loss are driving features of BBB breakdown in AD. However, there are limited data on whether pericyte loss precedes the vascular insufficiencies seen in human AD, in part because pericyte loss and capillary flow defects are difficult to measure longitudinally in humans.

15.4.8 Removal of Amyloid-Beta from the Brain by Blood Vessels

The clearance of metabolic waste from the brain relies on several routes, including (1) the flow of cerebral spinal fluid (CSF) flow through perivascular routes and (2) extrusion of molecules through the BBB. Both of these vascular-dependent clearance routes appear to be compromised during AD and contribute to the buildup of Aβ and other metabolic waste products (Fig. 15.3a) (Bakker et al. 2016).

CSF is produced by the choroid plexus within the ventricles, which then perfuses into the subarachnoid space lining the surface of the brain. A portion of the CSF in the subarachnoid space enters in the perivascular spaces (often also referred to as Virchow-Robin spaces) of pial arteries and superficial aspects of the penetrating arterioles. From the perivascular space, the CSF then enters the brain and mixes with the interstitial fluid (ISF) of the neural parenchyma. Two pathways of CSF transit have been proposed, glial lymphatics (g-lymphatics) (Iliff et al. 2012) and intramural periarterial drainage (IPAD) (Carare et al. 2008), and these pathways differ primarily on their proposed route of fluid exit from the brain. In glymphatics, CSF flows into the brain from the perivascular space, supported by Aquaporin 4-mediated water flux through the endfeet of astrocytes that surround the arterioles and capillaries. The fluid then drains from the parenchyma into perivascular spaces surrounding venules (Iliff et al. 2012). The IPAD pathway also involves flow into the brain via arteriolar perivascular spaces and into the brain ISF. However, collection and drainage of waste occur along the basement membrane of arterioles (and potentially the ACT zone and capillaries), in a direction opposite to blood flow, back to the brain surface (Bakker et al. 2016). Venous drainage of fluid is not seen with studies supporting the IPAD model.

The anatomical basis and direction of flow in these CSF flow routes remains incompletely understood. For example, it is possible that the vascular basement membrane, which is porous and continuous with the perivascular space, serves as a conduit for CSF flow on parenchymal arterioles and throughout the capillary bed (Aldea et al. 2019). Alteration of vascular basement membrane thickness is

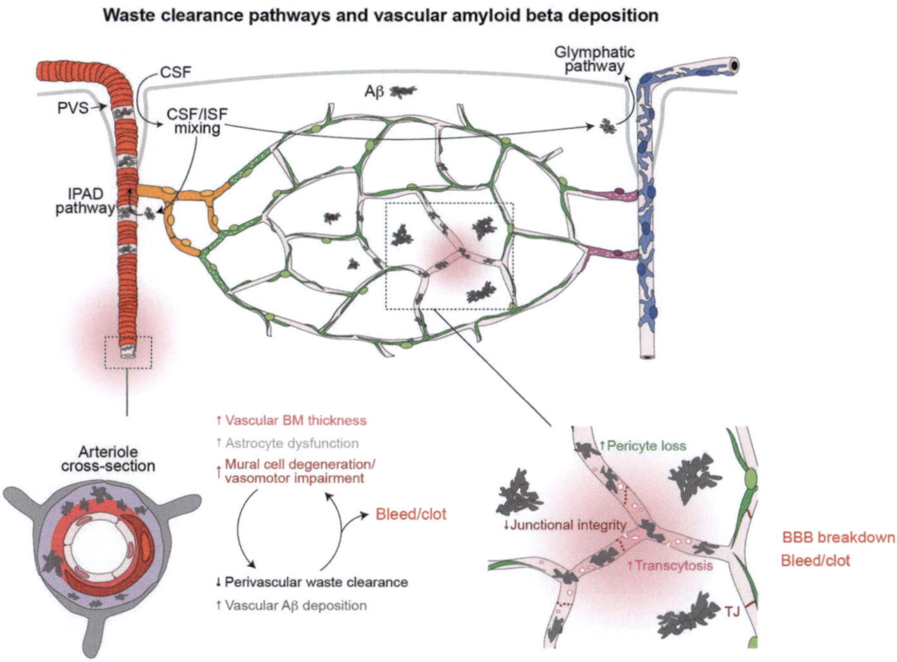

Fig. 15.3 Failure of waste clearance pathways and blood–brain barrier function in Alzheimer's disease. (**a**) Waste clearance relies on the entry of cerebral spinal fluid (CSF) through perivascular spaces (PVS) surrounding the penetrating arterioles and mixing with interstitial fluid (ISF). Two pathways of waste exiting the brain have been proposed where waste including amyloid-beta (Aβ) exits via the (1) glymphatic pathway through the perivascular spaces of venules or the (2) intramural periarterial drainage pathway (IPAD) through the vascular basement membrane of arterioles and capillaries. (**b**) Dysfunction of perivascular spaces demonstrated by cross-sections of an arteriole (left) and capillary (right) with the build-up of Aβ in the vascular basement membranes (BM). This results in thickening of the parenchymal (light purple) and endothelial-mural (pink) BMs, dysfunction of astrocytic Aquaporin-4 (Aqp4) water channels, mural cell degeneration, and impaired vasomotor activity. Together, these affect the removal of Aβ from the brain via the IPAD or glymphatic pathways. (**c**) Blood–brain barrier breakdown is associated with Aβ build-up, being characterized by pericyte loss, decreased junctional integrity, and likely increased transcytosis

commonly reported in aging and AD (Fig. 15.3b), as discussed further below, but its association with impaired brain clearance has not been studied thoroughly. Further, the relative contribution of these mechanisms to brain clearance and Aβ accumulation in CAA is not well understood, as approaches to dissect these pathways in vivo are lacking.

In both glymphatics and IPAD, the pulsatile activity of arteriole SMCs is thought to be a motive force (Aldea et al. 2019; Iliff et al. 2013). Oscillations in the diameter of penetrating arterioles occurs on slow timescales (vasomotion) and fast timescales (heart-beat), with both altering the size of the perivascular space creating a "pump" like driving force to push CSF–ISF exchange at glial endfeet and the parenchymal basement membrane (Fig. 15.3b). As such, weakened vasomotor activity due to loss of SMC coverage on arterioles may contribute to poor waste clearance and is a theory for vascular build-up of Aβ as observed in CAA (van Veluw et al. 2020). Further, in natural sleep and with some anesthetics, the size of the perivascular spaces increases by ~60% leading to an enhancement of CSF–ISF convective exchange. The dimensions of the perivascular spaces are important for how arteriole diameter changes affect CSF flow rate and resistance. The enlargement of the perivascular spaces is widely observed with aging and exacerbated in small vessel diseases (Wardlaw et al. 2020) and is likely to disrupt CSF–ISF exchange. The basis for perivascular space enlargement is not well understood and may involve increased CSF flow resistance or breakdown of arteriole wall components.

15.4.9 Removal of Amyloid-Beta via the Blood–Brain Barrier

The BBB facilitates the removal of waste across the endothelium through receptor-mediated transcytosis. Transfer of Aβ across the endothelium requires a two-step process involving uptake at the abluminal side by low-density lipoprotein receptor-1 (LRP-1) (Deane et al. 2004), which is a receptor for apolipoprotein E (ApoE) and ATP-dependent efflux at the luminal side by P-glycoprotein (Hartz et al. 2010). At the abluminal side, ApoE creates complexes with Aβ in the extracellular spaces, which binds to the LRP-1 receptor resulting in endocytosis and exocytosis of Aβ to the luminal side of the BBB for clearance into the blood. Deletion of endothelial-LRP-1 in Tg-5xFAD mice results in accelerated Aβ pathology and diminished spatial learning, demonstrating an important role for BBB removal of Aβ and the progression of AD (Storck et al. 2016).

Pericytes may participate in the removal of Aβ (Fig. 15.3C). Pericytes can internalize Aβ via LRP1-ApoE-mediated mechanisms (Ma et al. 2018). However, it is also thought that poor removal of Aβ due to high burden levels or presence of the ApoE4 isoform can contribute to pericyte death (Ma et al. 2018; Blanchard et al. 2020). The ApoE4 allele of the ApoE gene is a known genetic risk factor for development of AD in humans. ApoE4 still binds Aβ but has a higher affinity for the very-low-density lipoprotein receptor (VLDLR) which internalizes and transports across the BBB at a slower rate (Deane et al. 2008). ApoE4 can also contribute to microvascular damage and BBB leakage independent of Aβ. The ApoE4 allele augments a cyclophilinA-matrix metalloprotease 9 degradation pathway in pericytes (Bell et al. 2012; Montagne et al. 2021). Postmortem AD brain tissue from ApoE4 carriers has accelerated pericyte loss and BBB damage (Halliday et al. 2016). Thus, the BBB and pericytes contribute to the removal of Aβ but failure to keep pace with an increasing Aβ burden or impaired clearance mechanisms can hinder this process and further exacerbate Aβ pathology.

15.4.10 Changes to the Basement Membrane during Cognitive Decline

The vascular basement membrane (BM) is another important component in maintenance of vascular integrity, in addition to facilitating cellular crosstalk and waste clearance (Xu et al. 2019). The vascular BM consists of two separate layers, the parenchymal and endothelial-mural BMs. The endothelial-mural BM lies between the endothelial and mural cell layers and is found along all zones of the vascular network. The parenchymal BM is most prominent along arterioles and venules, sandwiched between the mural cell layer and the astrocytic endfeet where perivascular fibroblasts (PVFs) reside (Vanlandewijck et al. 2018; Bonney et al. 2021, 2022). Both BMs consist of glycoproteins from four major families: laminins, collagens, nidogens, and heparan sulfate proteoglycans. Endothelial cells, mural cells, PVFs, and astrocytes all anchor themselves to the BM through expression of specific integrins or dystroglycan receptors (Thomsen et al. 2017).

The vascular BM appears to thicken in AD and CAA. Collagen-IV, a major component of the endothelial-mural BM, is increased in several studies of postmortem AD tissue (Kalaria and Pax 1995; Farkas et al. 2000; Lepelletier et al. 2017; Magaki et al. 2018). Fibronectin and the heparan sulfate proteoglycans, agrin and perlecan, are also increased along capillaries in AD tissue (Lepelletier et al. 2017; Singh-Bains et al. 2019; Berzin et al. 2000). Proteomic studies on the brain microvasculature from AD patients with varying amounts of CAA reported an increase in collagen I, suggesting changes to the outer parenchymal BM of arterioles and venules (Ojo et al. 2021). Mouse models of AD and CAA have generally recapitulated these findings with increases in collagen-IV as well as laminin (Bourasset et al. 2009; Mehta et al. 2013; Merlini et al. 2011). However, some studies have reported a decrease in certain BM proteins like nidogen-2 (Keable et al. 2016). Decreases in collagen-IV are also reported in some AD mouse models that closely followed Aβ pathology throughout disease progression, suggesting that changes to the vascular BM may be a dynamic process (Hawkes et al. 2012, 2013; Kurata et al. 2011).

Early immunoelectron microscopy studies of AD postmortem tissue tracked Aβ deposition along the vasculature. On pial arteries and penetrating arterioles, Aβ fibrils were found within the parenchymal BM adjacent to astroglial endfeet (Fig. 15.3c, d). Larger Aβ deposits were found in the endothelial-mural BM and were associated with degeneration of SMCs. Along capillaries, Aβ accumulated with the endothelial-mural BM which becomes thickened, disorganized, and abnormally folded in regions with Aβ pathology (Lepelletier et al. 2017; Hawkes et al. 2013; Yamaguchi et al. 1992; Zarow et al. 1997). As mentioned above, BM thickening is likely due to an imbalance in the expression and degradation of BM proteins. However, BM thickening is observed prior to robust vascular Aβ accumulation and is associated with poor Aβ clearance (Merlini et al. 2011; Hawkes et al. 2013), suggesting that it may be an initiating factor that amplifies CAA. Laminin, nidogen, and collagen-IV can disrupt the formation of Aβ fibrils (Kiuchi et al. 2001,

2002), whereas perlecan can stabilize the formation of amyloid fibrils (Castillo et al. 1997). This raises the possibility that perivascular cells modulate their secretion of BM proteins and vascular Aβ accumulation occurs only when there is an alteration in the BM structure. Overall, changes in the vascular BM are likely to confound Aβ clearance and contribute to the vascular Aβ accumulation reported in at least 48% of AD patients (Jäkel et al. 2022).

15.4.11 Microvascular Instability and White Matter Degeneration

Damage to the cerebral white matter, composed of densely packed myelinated axon tracts, is a major component of AD pathology (Nasrabady et al. 2018) and can lead to accelerated cognitive decline by degrading the functional connectivity of brain networks (Zhao et al. 2020; Chen et al. 2019; Zhu et al. 2019). Under normal conditions, white matter regions are more susceptible to hypoperfusion due to lower capillary densities and greater distances to arterial perfusion sources, i.e., perfusion watershed (Moody et al. 1990; Brown and Thore 2011). With age, capillary density in the white matter is further reduced and associated with a reduction in proangiogenic signals that may be necessary to maintain capillary architecture throughout life (Schager and Brown 2020; Murugesan et al. 2012). There is also an increase in the number of string vessels in the white matter, which are thought to be regressed or collapsed capillaries (Challa et al. 2004).

Arteriole and venular changes also seem to contribute to white matter pathology in AD. Arteriole tortuosity is often seen with aging and worsened in AD and may contribute to increased resistance of blood flow. The occurrence of arteriole tortuosity appears to be more severe as the vessels pass into the white matter (Brown and Thore 2011). Venous collagenosis, which is thickening of the venule wall due to increased collagen deposition, was also recently emphasized as a potential vascular risk factor for white matter damage in AD (Lahna et al. 2022). These venous changes may impair drainage of blood, result in BBB leakage and vasogenic edema, and impair glymphatic perivascular drainage of CSF.

15.5 Conclusion

Dysfunction of the microvasculature is becoming increasingly recognized as a driver of cognitive impairment. Increased Aβ presence, mural cell loss, and gene expression changes in NVU cells likely all affect how the brain vasculature functions. These changes impair cerebral blood flow, neurovascular coupling, BBB function, and waste removal, and alter the composition of the vascular basement membrane. Together, these alterations can further increase Aβ burden, resulting in neurodegeneration and neuroinflammation. Strategies to improve cerebral blood flow, mural cell coverage, and waste clearance are actively being pursued as possible therapeutic avenues to reduce cognitive impairment during aging. The earliest, initiating events in the cascade leading to microvascular dysfunction remain unclear, but future longitudinal imaging studies may be able to chart the progression of vascular deficits at earlier stages. By investigating the sequence of events that occurs in pathology, we can better understand the key driving forces of disease and target these impairments earlier in disease development.

References

Aldea R et al (2019) Cerebrovascular smooth muscle cells as the drivers of intramural Periarterial drainage of the brain. Front Aging Neurosci 11:1

Arancio O (2008) PIP2: a new key player in Alzheimer's disease. Cell Sci 5(1):44–47

Armulik A et al (2010) Pericytes regulate the blood-brain barrier. Nature 468(7323):557–561

Attwell D et al (2010) Glial and neuronal control of brain blood flow. Nature 468(7321):232–243

Bakker EN et al (2016) Lymphatic clearance of the brain: perivascular, Paravascular and significance for neurodegenerative diseases. Cell Mol Neurobiol 36(2):181–194

Barykin EP et al (2017) Amyloid β modification: a key to the sporadic Alzheimer's disease? Front Genet 8:58

Beckmann N et al (2003) Age-dependent cerebrovascular abnormalities and blood flow disturbances in APP23 mice modeling Alzheimer's disease. J Neurosci 23(24):8453–8459

Bell RD et al (2010) Pericytes control key neurovascular functions and neuronal phenotype in the adult brain and during brain aging. Neuron 68:409–427

Bell RD et al (2012) Apolipoprotein E controls cerebrovascular integrity via cyclophilin a. Nature 485(7399):512–516

Bennett HC, Zhang Q, Wu Yt et al (2024) Aging drives cerebrovascular network remodeling and functional changes in the mouse brain. Nat Commun 15 6398 (2024). https://doi.org/10.1038/s41467-024-50559-8

Ben-Zvi A et al (2014) Mfsd2a is critical for the formation and function of the blood-brain barrier. Nature 509(7501):507–511

Bernier LP et al (2021) Location matters: navigating regional heterogeneity of the neurovascular unit. Front Cell Neurosci 15:696540

Berthiaume AA et al (2018a) Dynamic remodeling of pericytes in vivo maintains capillary coverage in the adult mouse brain. Cell Rep 22(1):8–16

Berthiaume AA et al (2018b) Pericyte structural remodeling in cerebrovascular health and homeostasis. Front Aging Neurosci 10:210

Berthiaume AA et al (2022) Pericyte remodeling is deficient in the aged brain and contributes to impaired capillary flow and structure. Nat Commun 13(1):5912

Berzin TM et al (2000) Agrin and microvascular damage in Alzheimer's disease. Neurobiol Aging 21(2):349–355

Blanchard JW et al (2020) Reconstruction of the human blood-brain barrier in vitro reveals a pathogenic mechanism of APOE4 in pericytes. Nat Med 26(6):952–963

Blinder P et al (2010) Topological basis for the robust distribution of blood to rodent neocortex. Proc Natl Acad Sci USA 107:12670–12675

Bonder DE, McCarthy KD (2014) Astrocytic Gq-GPCR-linked IP3R-dependent Ca2+ signaling does not mediate neurovascular coupling in mouse visual cortex in vivo. J Neurosci 34(39):13139–13150

Bonney SK et al (2021) Distinct features of brain perivascular fibroblasts and mural cells revealed by in vivo two-photon imaging. J Cereb Blood Flow Metab
Bonney SK et al (2022) Public volume electron microscopy data: an essential resource to study the brain microvasculature. Front Cell Dev Biol 10:849469
Bourasset F et al (2009) Reduction of the cerebrovascular volume in a transgenic mouse model of Alzheimer's disease. Neuropharmacology 56(4):808–813
Brown WR, Thore CR (2011) Review: cerebral microvascular pathology in ageing and neurodegeneration. Neuropathol Appl Neurobiol 37(1):56–74
Cai C et al (2018) Stimulation-induced increases in cerebral blood flow and local capillary vasoconstriction depend on conducted vascular responses. Proc Natl Acad Sci 15(25):E5796–E5804
Calhoun ME et al (1999) Neuronal overexpression of mutant amyloid precursor protein results in prominent deposition of cerebrovascular amyloid. Proc Natl Acad Sci 96(24):14088–14093
Carare RO et al (2008) Solutes, but not cells, drain from the brain parenchyma along basement membranes of capillaries and arteries: significance for cerebral amyloid angiopathy and neuroimmunology. Neuropathol Appl Neurobiol 34(2):131–144
Castillo GM et al (1997) Perlecan binds to the beta-amyloid proteins (a beta) of Alzheimer's disease, accelerates a beta fibril formation, and maintains a beta fibril stability. J Neurochem 69(6):2452–2465
Challa VR et al (2004) Increase of white matter string vessels in Alzheimer's disease. J Alzheimer's Dis 6(4):379–383
Chalothorn D, Faber JE (2010) Formation and maturation of the native cerebral collateral circulation. J Mol Cell Cardiol 49(2):251–259
Charidimou A et al (2017) Emerging concepts in sporadic cerebral amyloid angiopathy. Brain 140(7):1829–1850
Chen X et al (2019) Disrupted functional and structural connectivity within default mode network contribute to WMH-related cognitive impairment. Neuroimage: Clin 24:102088
Choe YG et al (2022) Pericyte loss leads to capillary stalling through increased leukocyte-endothelial cell interaction in the brain. Front Cell Neurosci 16:848764
Cruz Hernández JC et al (2019) Neutrophil adhesion in brain capillaries reduces cortical blood flow and impairs memory function in Alzheimer's disease mouse models. Nat Neurosci 22(3):413–420
Dabertrand F et al (2021) PIP2 corrects cerebral blood flow deficits in small vessel disease by rescuing capillary Kir2.1 activity. Proc Natl Acad Sci USA 118(17):e2025998118
Dai W et al (2009) Mild cognitive impairment and alzheimer disease: patterns of altered cerebral blood flow at MR imaging. Radiology 250(3):856–866
Damisah EC et al (2017) A fluoro-Nissl dye identifies pericytes as distinct vascular mural cells during in vivo brain imaging. Nat Neurosci 20(7):1023–1032
Daneman R et al (2010) Pericytes are required for blood-brain barrier integrity during embryogenesis. Nature 468(7323):562–566
Davis J et al (2004) Early-onset and robust cerebral microvascular accumulation of amyloid beta-protein in transgenic mice expressing low levels of a vasculotropic Dutch/Iowa mutant form of amyloid beta-protein precursor. J Biol Chem 279(19):20296–20306
Deane R et al (2004) LRP/amyloid beta-peptide interaction mediates differential brain efflux of Abeta isoforms. Neuron 43(3):333–344
Deane R et al (2008) ApoE isoform-specific disruption of amyloid beta peptide clearance from mouse brain. J Clin Invest 118(12):4002–4013
Ding R et al (2020) Loss of capillary pericytes and the blood-brain barrier in white matter in poststroke and vascular dementias and Alzheimer's disease. Brain Pathol 30(6):1087–1101
Ding R et al (2021) Loss with ageing but preservation of frontal cortical capillary pericytes in post-stroke dementia, vascular dementia and Alzheimer's disease. Acta Neuropathol Commun 9(1):130

Drew PJ, Shih AY, Kleinfeld D (2011) Fluctuating and sensory-induced vasodynamics in rodent cortex extends arteriole capacity. Proceedings of the National Academy of Sciences USA 108: 8473–8478

Duncombe J et al (2017) Ageing causes prominent neurovascular dysfunction associated with loss of astrocytic contacts and gliosis. Neuropathol Appl Neurobiol 43(6):477–491

Echagarruga CT et al (2020) nNOS-expressing interneurons control basal and behaviorally evoked arterial dilation in somatosensory cortex of mice. elife 9:e60533

Eikelenboom P et al (2008) Neuroinflammation in plaque and vascular beta-amyloid disorders: clinical and therapeutic implications. Neurodegener Dis 5(3–4):190–193

Erdener ŞE, Küreli G, Dalkara T (2022) Contractile apparatus in CNS capillary pericytes. Neurophotonics 9(2):021904

Eskildsen SF et al (2017) Increased cortical capillary transit time heterogeneity in Alzheimer's disease: a DSC-MRI perfusion study. Neurobiol Aging 50:107–118

Farkas E et al (2000) Pathological features of cerebral cortical capillaries are doubled in Alzheimer's disease and Parkinson's disease. Acta Neuropathol 100(4):395–402

Fernández-Klett F et al (2010) Pericytes in capillaries are contractile in vivo, but arterioles mediate functional hyperemia in the mouse brain. Proc Natl Acad Sci USA 107:22290–22295

Fernandez-Klett F et al (2020) Denser brain capillary network with preserved pericytes in Alzheimer's disease. Brain Pathol 30(6):1071–1086

Gaceb A et al (2018) Pericytes secrete pro-regenerative molecules in response to platelet-derived growth factor-BB. J Cereb Blood Flow Metab 38(1):45–57

Glück C et al (2021) Distinct signatures of calcium activity in brain mural cells. elife 10:e70591

Grant RI et al (2017) Organizational hierarchy and structural diversity of microvascular pericytes in adult mouse cortex. J Cereb Blood Flow Metab 39(3):411–425

Greenberg SM et al (2020) Cerebral amyloid angiopathy and Alzheimer disease – one peptide, two pathways. Nat Rev Neurol 16(1):30–42

Gutiérrez-Jiménez E et al (2016) Effect of electrical forepaw stimulation on capillary transit-time heterogeneity (CTH). J Cereb Blood Flow Metab 36(12):2072–2086

Haglund M et al (2006) Differential deposition of amyloid beta peptides in cerebral amyloid angiopathy associated with Alzheimer's disease and vascular dementia. Acta Neuropathol 111(5):430–435

Hall CN et al (2014) Capillary pericytes regulate cerebral blood flow in health and disease. Nature 508(7494):55–60

Halliday MR et al (2016) Accelerated pericyte degeneration and blood-brain barrier breakdown in apolipoprotein E4 carriers with Alzheimer's disease. J Cereb Blood Flow Metab 36(1):216–227

Hartmann DA, Coelho-Santos V, Shih AY (2021a) Pericyte control of blood flow across microvascular zones in the central nervous system. Annu Rev Physiol 84:331–354

Hartmann DA et al (2021b) Brain capillary pericytes exert a substantial but slow influence on blood flow. Nat Neurosci 24:633–645

Hartz AM, Miller DS, Bauer B (2010) Restoring blood-brain barrier P-glycoprotein reduces brain amyloid-beta in a mouse model of Alzheimer's disease. Mol Pharmacol 77(5):715–723

Hattori Y et al (2016) Gradual carotid artery stenosis in mice closely replicates Hypoperfusive vascular dementia in humans. J Am Heart Assoc 5(2):e002757

Hawkes CA et al (2012) Disruption of arterial perivascular drainage of amyloid-β from the brains of mice expressing the human APOE ε4 allele. PLoS One 7(7):e41636

Hawkes CA et al (2013) Regional differences in the morphological and functional effects of aging on cerebral basement membranes and perivascular drainage of amyloid-β from the mouse brain. Aging Cell 12(2):224–236

Hellström M et al (1999) Role of PDGF-B and PDGFR-beta in recruitment of vascular smooth muscle cells and pericytes during embryonic blood vessel formation in the mouse. Development 126(14):3047–3055

Hellström-Lindahl E, Ravid R, Nordberg A (2008) Age-dependent decline of neprilysin in Alzheimer's disease and normal brain: inverse correlation with a beta levels. Neurobiol Aging 29(2):210–221

Herzig MC et al (2004) Abeta is targeted to the vasculature in a mouse model of hereditary cerebral hemorrhage with amyloidosis. Nat Neurosci 7(9):954–960

Hill RA et al (2015) Regional blood flow in the Normal and ischemic brain is controlled by arteriolar smooth muscle cell contractility and not by capillary Pericytes. Neuron 87(1):95–110

Hill RA et al (2017) Targeted two-photon chemical apoptotic ablation of defined cell types in vivo. Nat Commun 8:15837

Hinman JD et al (2023) Placental growth factor as a sensitive biomarker for vascular cognitive impairment. Alzheimers Dement 19:3519

Hock C et al (1997) Decrease in parietal cerebral hemoglobin oxygenation during performance of a verbal fluency task in patients with Alzheimer's disease monitored by means of near-infrared spectroscopy (NIRS)—correlation with simultaneous rCBF-PET measurements. Brain Res 755(2):293–303

Iliff JJ et al (2012) A paravascular pathway facilitates CSF flow through the brain parenchyma and the clearance of interstitial solutes, including amyloid β. Sci Transl Med 4(147):147ra111

Iliff JJ et al (2013) Cerebral arterial pulsation drives paravascular CSF-interstitial fluid exchange in the murine brain. J Neurosci 33(46):18190–18199

Institoris A et al (2022) Astrocytes amplify neurovascular coupling to sustained activation of neocortex in awake mice. Nat Commun 13(1):7872

Iturria-Medina Y et al (2016) Early role of vascular dysregulation on late-onset Alzheimer's disease based on multifactorial data-driven analysis. Nat Commun 7:11934

Ivanova E, Kovacs-Oller T, Sagdullaev BT (2019) Domain-specific distribution of gap junctions defines cellular coupling to establish a vascular relay in the retina. J Comp Neurol 527(16):2675–2693

Iwata N et al (2001) Metabolic regulation of brain Abeta by neprilysin. Science 292(5521):15501552

Iwata N et al (2002) Region-specific reduction of a beta-degrading endopeptidase, neprilysin, in mouse hippocampus upon aging. J Neurosci Res 70(3):493–500

Jäkel L et al (2022) Prevalence of cerebral amyloid angiopathy: a systematic review and meta-analysis. Alzheimers Dement 18(1):10–28

Jespersen SN, Østergaard L (2012) The roles of cerebral blood flow, capillary transit time heterogeneity, and oxygen tension in brain oxygenation and metabolism. J Cereb Blood Flow Metab 32(2):264–277

Kalaria RN, Pax AB (1995) Increased collagen content of cerebral microvessels in Alzheimer's disease. Brain Res 705(1–2):349–352

Keable A et al (2016) Deposition of amyloid β in the walls of human leptomeningeal arteries in relation to perivascular drainage pathways in cerebral amyloid angiopathy. Biochim Biophys Acta 1862(5):1037–1046

Kisler K et al (2017) Pericyte degeneration leads to neurovascular uncoupling and limits oxygen supply to brain. Nat Neurosci 20(3):406–416

Kisler K et al (2020) Acute ablation of cortical Pericytes leads to rapid neurovascular uncoupling. Front Cell Neurosci 14:27

Kiuchi Y, Isobe Y, Fukushima K (2001) Entactin-induced inhibition of human amyloid beta-protein fibril formation in vitro. Neurosci Lett 305(2):119–122

Kiuchi Y et al (2002) Disassembly of amyloid beta-protein fibril by basement membrane components. Life Sci 70(20):2421–2431

Klug NR et al (2023) Intraluminal pressure elevates intracellular calcium and contracts CNS pericytes: role of voltage-dependent calcium channels. Proc Natl Acad Sci USA 120(9):e2216421120

Kovacs-Oller T et al (2020) The pericyte connectome: spatial precision of neurovascular coupling is driven by selective connectivity maps of pericytes and endothelial cells and is disrupted in diabetes. Cell Discovery 6:39

Kruyer A et al (2015) Chronic hypertension leads to neurodegeneration in the TgSwDI mouse model of Alzheimer's disease. Hypertension 66(1):175–182

Kucharz K et al (2021) Post-capillary venules are the key locus for transcytosis-mediated brain delivery of therapeutic nanoparticles. Nat Commun 12(1):4121

Kurata T et al (2011) Progressive neurovascular disturbances in the cerebral cortex of Alzheimer's disease-model mice: protection by atorvastatin and pitavastatin. Neuroscience 197:358–368

Lahna D et al (2022) Venous Collagenosis as pathogenesis of white matter Hyperintensity. Ann Neurol 92(6):992–1000

Lai AY et al (2015) Venular degeneration leads to vascular dysfunction in a transgenic model of Alzheimer's disease. Brain 138:1046–1058

Lee L, Kosuri P, Arancio O (2014) Picomolar amyloid-β peptides enhance spontaneous astrocyte calcium transients. J Alzheimer's Dis 38(1):49–62

Lepelletier FX et al (2017) Early changes in extracellular matrix in Alzheimer's disease. Neuropathol Appl Neurobiol 43(2):167–182

Li H et al (2014) Vascular and parenchymal amyloid pathology in an Alzheimer disease knock-in mouse model: interplay with cerebral blood flow. Mol Neurodegener 9:28

Li Y, Wei W, Wang RK (2018) Capillary flow homogenization during functional activation revealed by optical coherence tomography angiography based capillary velocimetry. Sci Rep 8(1):4107

Li B et al (2019) More homogeneous capillary flow and oxygenation in deeper cortical layers correlate with increased oxygen extraction. elife 8:e42299

Lindahl P et al (1997) Pericyte loss and microaneurysm formation in PDGF-B-deficient mice. Science 277:242–245

Ma Q et al (2018) Blood-brain barrier-associated pericytes internalize and clear aggregated amyloid-β42 by LRP1-dependent apolipoprotein E isoform-specific mechanism. Mol Neurodegener 13(1):57

Magaki S et al (2018) The effects of cerebral amyloid angiopathy on integrity of the blood-brain barrier. Neurobiol Aging 70:70–77

Marshall RS et al (2001) Recovery of brain function during induced cerebral hypoperfusion. Brain 124:1208–1217

Mehta DC, Short JL, Nicolazzo JA (2013) Altered brain uptake of therapeutics in a triple transgenic mouse model of Alzheimer's disease. Pharm Res 30(11):2868–2879

Merlini M et al (2011) Vascular β-amyloid and early astrocyte alterations impair cerebrovascular function and cerebral metabolism in transgenic arcAβ mice. Acta Neuropathol 122(3):293–311

Miners JS, Schulz I, Love S (2017) Differing associations between aβ accumulation, hypoperfusion, blood-brain barrier dysfunction and loss of PDGFRB pericyte marker in the precuneus and parietal white matter in Alzheimer's disease. J Cereb Blood Flow Metab 38(1):103–115

Miners JS et al (2019) CSF evidence of pericyte damage in Alzheimer's disease is associated with markers of blood-brain barrier dysfunction and disease pathology. Alzheimer's Res Ther 11(1):81

Moeini M et al (2018) Compromised microvascular oxygen delivery increases brain tissue vulnerability with age. Sci Rep 8(1):8219

Montagne A et al (2015) Blood-brain barrier breakdown in the aging human hippocampus. Neuron 85(2):296–302

Montagne A et al (2021) APOE4 accelerates advanced-stage vascular and neurodegenerative disorder in old Alzheimer's mice via cyclophilin a independently of amyloid-β. Nature Aging 1(6):506–520

Moody DM, Bell MA, Challa VA (1990) Features of the cerebral vascular pattern that predict vulnerability to perfusion or oxygenation deficiency: an anatomic study. Am J Neuroradiol 11: 431–439

Mughal A et al (2021) PIP2 improves cerebral blood flow in a mouse model of Alzheimer's disease. Function 2(2):zqab010
Munting LP et al (2021) Cerebral blood flow and cerebrovascular reactivity are preserved in a mouse model of cerebral microvascular amyloidosis. elife 10:e61279
Murugesan N et al (2012) Brain regional angiogenic potential at the neurovascular unit during normal aging. Neurobiol Aging 33(5):e1–e16
Nasrabady SE et al (2018) White matter changes in Alzheimer's disease: a focus on myelin and oligodendrocytes. Acta Neuropathol Commun 6(1):22
Nation DA et al (2019) Blood-brain barrier breakdown is an early biomarker of human cognitive dysfunction. Nat Med 25(2):270–276
Nelson AR et al (2020) Channelrhodopsin excitation contracts brain Pericytes and reduces blood flow in the aging mouse brain in vivo. Front Aging Neurosci 12:108
Nguyen J et al (2011) Occlusion of cortical ascending venules causes blood flow decreases, reversals in flow direction, and vessel dilation in upstream capillaries. J Cereb Blood Flow Metab 31:2243–2254
Nicoll JA et al (2004) Cerebral amyloid angiopathy plays a direct role in the pathogenesis of Alzheimer's disease. Pro-CAA position statement. Neurobiol Aging 25(5):589–597
Nielson CD et al (2022) In vivo single cell optical ablation of brain Pericytes. Front Neurosci 16: 974311
Nikolakopoulou AM et al (2019) Pericyte loss leads to circulatory failure and pleiotrophin depletion causing neuron loss. Nat Neurosci 22(7):1089–1098
Nishimura N et al (2010) Limitations of collateral flow after occlusion of a single cortical penetrating arteriole. J Cereb Blood Flow Metab 30:1914–1927
Niwa K et al (2002) Alterations in cerebral blood flow and glucose utilization in mice overexpressing the amyloid precursor protein. Neurobiol Dis 9(1):61–68
Nizar K et al (2013) Stimulus-induced vasodilation occurs without IP3 receptor activation and may precede astrocytic calcium increase. J Neurosci 33:8411–8422
Nortley R et al (2019) Amyloid β oligomers constrict human capillaries in Alzheimer's disease via signaling to pericytes. Science 365:6450
O'Herron PJ et al (2022) 3D optogenetic control of arteriole diameter in vivo. elife 11:e72802
Ojo JO et al (2021) Molecular pathobiology of the Cerebrovasculature in aging and in Alzheimers disease cases with cerebral amyloid Angiopathy. Front Aging Neurosci 13:658605
Ornelas S et al (2021) Three-dimensional ultrastructure of the brain pericyte-endothelial interface. J Cereb Blood Flow Metab 41:2185
Pandey K, Bessières B, Sheng SL, Taranda J, Osten P, Sandovici I, Constancia M, Alberini CM (2023) Neuronal activity drives IGF2 expression from pericytes to form long-term memory 111(23):3819–3836.e8. https://doi.org/10.1016/j.neuron.2023.08.030. Epub 2023 Oct 2. PMID: 37788670; PMCID: PMC10843759
Park L et al (2007) Nox2-derived reactive oxygen species mediate neurovascular dysregulation in the aging mouse brain. J Cereb Blood Flow Metab 27(12):1908–1918
Park L et al (2014) Age-dependent neurovascular dysfunction and damage in a mouse model of cerebral amyloid angiopathy. Stroke 45(6):1815–1821
Park DY et al (2017) Plastic roles of pericytes in the blood-retinal barrier. Nat Commun 8:15296
Peppiatt CM et al (2006) Bidirectional control of CNS capillary diameter by pericytes. Nature 443: 642–643
Profaci CP et al (2020) The blood-brain barrier in health and disease: important unanswered questions. J Exp Med 217(4):e20190062
Reuter B et al (2016) *Development of cerebral microbleeds in the APP23-transgenic mouse model of cerebral amyloid Angiopathy-a 9.4 tesla MRI study.* Frontiers in aging. Neuroscience 8:170
Rungta RL et al (2018) Vascular compartmentalization of functional hyperemia from the synapse to the pia. Neuron 99(2):362–375
Ryu JK, McLarnon JG (2009) A leaky blood-brain barrier, fibrinogen infiltration and microglial reactivity in inflamed Alzheimer's disease brain. J Cell Mol Med 13(9A):2911–2925

Schager B, Brown CE (2020) Susceptibility to capillary plugging can predict brain region specific vessel loss with aging. J Cereb Blood Flow Metab 40(12):2475–2490

Secomb TW (1991) Red blood cell mechanics and capillary blood rheology. Cell Biophys 18(3): 231–251

Sengillo JD et al (2013) Deficiency in mural vascular cells coincides with blood-brain barrier disruption in Alzheimer's disease. Brain Pathol 23(3):303–310

Shaw K et al (2021) Neurovascular coupling and oxygenation are decreased in hippocampus compared to neocortex because of microvascular differences. Nat Commun 12(1):3190

Singh-Bains MK et al (2019) Altered microglia and neurovasculature in the Alzheimer's disease cerebellum. Neurobiol Dis 132:104589

Skaaraas GHES et al (2021) Cerebral amyloid Angiopathy in a mouse model of Alzheimer's disease associates with upregulated angiopoietin and downregulated hypoxia-inducible factor. J Alzheimer's Dis 83(4):1651–1663

Sloan HL et al (2010) Regional differences in neurovascular coupling in rat brain as determined by fMRI and electrophysiology. NeuroImage 53(2):399–411

Smyth LCD et al (2022) Characterisation of PDGF-BB:PDGFRβ signalling pathways in human brain pericytes: evidence of disruption in Alzheimer's disease. Commun Biol 5(1):235

Soto I et al (2015) APOE stabilization by exercise prevents aging neurovascular dysfunction and complement induction. PLoS Biol 13(10):e1002279

Starr JM et al (2009) Blood-brain barrier permeability in Alzheimer's disease: a case-control MRI study. Psychiatry Res 171(3):232–241

Steinman J, Sun HS, Feng ZP (2021) Microvascular alterations in Alzheimer's disease. Front Cell Neurosci 14:618986

Storck SE et al (2016) Endothelial LRP1 transports amyloid-β(1-42) across the blood-brain barrier. J Clin Invest 126(1):123–136

Sun N et al (2023) Single-nucleus multiregion transcriptomic analysis of brain vasculature in Alzheimer's disease. Nat Neurosci 26(6):970–982

Sweeney MD, Ayyadurai S, Zlokovic BV (2016) Pericytes of the neurovascular unit: key functions and signaling pathways. Nat Neurosci 19(6):771–783

Szu JI, Obenaus A (2021) Cerebrovascular phenotypes in mouse models of Alzheimer's disease. J Cereb Blood Flow Metab 41(8):1821–1841

Taylor ZJ et al (2016) Microvascular basis for growth of small infarcts following occlusion of single penetrating arterioles in mouse cortex. J Cereb Blood Flow Metab 36(8):1357–1373

Thakore P et al (2021) Brain endothelial cell TRPA1 channels initiate neurovascular coupling. elife 10:e63040

Thal DR et al (2002) Two types of sporadic cerebral amyloid angiopathy. J Neuropathol Exp Neurol 61(3):282

Thal DR, Griffin WS, Braak H (2008) Parenchymal and vascular Abeta-deposition and its effects on the degeneration of neurons and cognition in Alzheimer's disease. J Cell Mol Med 12(5B):1848–1862

Thal DR et al (2009) Capillary cerebral amyloid angiopathy is associated with vessel occlusion and cerebral blood flow disturbances. Neurobiol Aging 30(12):1936–1948

Thomsen MS et al (2017) The vascular basement membrane in the healthy and pathological brain. J Cereb Blood Flow Metab 37(10):3300–3317

Van Bergen T et al (2019) The role of placental growth factor (PlGF) and its receptor system in retinal vascular diseases. Prog Retin Eye Res 69:116–136

van de Haar HJ et al (2016) Blood-brain barrier leakage in patients with early Alzheimer disease. Radiology 281(2):527–535

van Dinther M et al (2022) Assessment of microvascular rarefaction in human brain disorders using physiological magnetic resonance imaging. J Cereb Blood Flow Metab 42(5):718–737

van Veluw SJ et al (2020) Vasomotion as a driving force for Paravascular clearance in the awake mouse brain. Neuron 105(3):549–561

Vanlandewijck M et al (2018) A molecular atlas of cell types and zonation in the brain vasculature. Nature 554(7693):475–480
Vazquez-Liebanas E et al (2021) Adult-induced genetic ablation distinguishes PDGFB roles in blood-brain barrier maintenance and development. J Cereb Blood Flow Metab 2:264–279
Verbeek MM et al (1997) Rapid degeneration of cultured human brain pericytes by amyloid beta protein. J Neurochem 68(3):1135–1141
Veszelka S et al (2013) Docosahexaenoic acid reduces amyloid-β induced toxicity in cells of the neurovascular unit. J Alzheimer's Dis 36(3):487–501
Viticchi G et al (2012) Vascular predictors of cognitive decline in patients with mild cognitive impairment. Neurobiol Aging 33(6):1127
Vonsattel JP et al (1991) Cerebral amyloid angiopathy without and with cerebral hemorrhages: a comparative histological study. Ann Neurol 30(5):637–649
Wang L et al (2016) Chronic cerebral hypoperfusion induces memory deficits and facilitates aβ generation in C57BL/6J mice. Exp Neurol 283:353–364
Wardlaw JM et al (2020) Perivascular spaces in the brain: anatomy, physiology and pathology. Nat Rev Neurol 16(3):137–153
Watson AN et al (2020) Mild pericyte deficiency is associated with aberrant brain microvascular flow in aged PDGFRβ+/− mice. J Cereb Blood Flow Metab 40:2387
Wiesmann M et al (2017) Hypertension, cerebrovascular impairment, and cognitive decline in aged AβPP/PS1 mice. Theranostics 7(5):1277–1289
Wisniewski HM, Vorbrodt AW, Wegiel J (1997) Amyloid angiopathy and blood-brain barrier changes in Alzheimer's disease. Ann N Y Acad Sci 826:161–172
Wolters FJ et al (2017) Cerebral perfusion and the risk of dementia: a population-based study. Circulation 136(8):719–728
Xu W et al (2014) Cerebral microvascular rather than parenchymal amyloid-β protein pathology promotes early cognitive impairment in transgenic mice. J Alzheimer's Dis 38(3):621–632
Xu L, Nirwane A, Yao Y (2019) Basement membrane and blood-brain barrier. Stroke Vascul Neurol 4(2):78–82
Yamaguchi H et al (1992) Beta amyloid is focally deposited within the outer basement membrane in the amyloid angiopathy of Alzheimer's disease. An immunoelectron microscopic study. Am J Pathol 1:249–259
Yamazaki Y et al (2019) Selective loss of cortical endothelial tight junction proteins during Alzheimer's disease progression. Brain 42(4):1077–1092
Yang AC et al (2022) A human brain vascular atlas reveals diverse mediators of Alzheimer's risk. Nature 603(7903):885–892
Yi C et al (2016) Astroglial connexin43 contributes to neuronal suffering in a mouse model of Alzheimer's disease. Cell Death Differ 23(10):1691–1701
Zarow C et al (1997) Vascular basement membrane pathology and Alzheimer's disease. Ann N Y Acad Sci 826:147–160
Zhao J et al (2020) Alteration of functional connectivity in patients with Alzheimer's disease revealed by resting-state functional magnetic resonance imaging. Neural Regen Res 15(2):285–292
Zhu W et al (2019) Dysfunctional architecture underlies white matter Hyperintensities with and without cognitive impairment. J Alzheimer's Dis 71(2):461–476

Chapter 16
Imaging Microvascular Morphology and Hemodynamics

Yunke Ren, Janaka Senarathna, and Arvind P. Pathak

Abstract Recent advances in imaging technologies have unlocked significant insights into the role of the microcirculation in health and disease. The microcirculation comprises the body's tiniest blood vessels or microvessels and plays a crucial role delivering oxygen, nutrients, and metabolites throughout the body. This chapter introduces the reader to the arsenal of modern imaging tools that are at our disposal to interrogate, visualize, and characterize the microvascular morphology and hemodynamics that constitute the microcirculation. We describe the working principle, spatial resolution, underlying assumptions, and the advantages and limitations of each imaging technology. Our goal is to educate prospective users on how to select an imaging tool based on their specific microcirculatory research needs. Additionally, this chapter presents a host of examples to help readers familiarize themselves with how each imaging method can be utilized, ranging from applications that enhance our fundamental understanding of the microcirculation to those that catalyze the development of new drugs or tissue engineering constructs, and those that

Y. Ren
Depts. of Biomedical Engineering, Johns Hopkins University School of Medicine, Baltimore, MD, USA

J. Senarathna
Russell H. Morgan Department of Radiology and Radiological Sciences, Johns Hopkins University School of Medicine, Baltimore, MD, USA

A. P. Pathak (✉)
Depts. of Biomedical Engineering, Johns Hopkins University School of Medicine, Baltimore, MD, USA

Russell H. Morgan Department of Radiology and Radiological Sciences, Johns Hopkins University School of Medicine, Baltimore, MD, USA

Sidney Kimmel Comprehensive Cancer Center, Johns Hopkins University School of Medicine, Baltimore, MD, USA

Department of Electrical Engineering, Johns Hopkins University, Baltimore, MD, USA

Translational Tissue Engineering Center, Johns Hopkins University, Baltimore, MD, USA

Institute for Nanobiotechnology, Johns Hopkins University, Baltimore, MD, USA
e-mail: apathak2@jhmi.edu

R. F. Tuma et al. (eds.), *Handbook of Microcirculation*, Physiology in Health and Disease, https://doi.org/10.1007/978-3-031-99609-2_16

support image-guided surgery. Collectively, the breadth and scope of these applications underscore the essential role of imaging in advancing microcirculation research and accelerating diagnostic and therapeutic solutions for diseases involving the microcirculation. In summary, this chapter aims to serve as a comprehensive guide for those interested in leveraging imaging technologies to explore the microcirculation and characterize its structural and functional heterogeneity.

Keywords Blood vessel · Vasculature · Hemodynamic · Hypoxia · Imaging · Microcirculation · Microvascular · Perfusion · Computed tomography (CT) · Magnetic resonance imaging (MRI) · Near-infrared spectroscopy (NIRS) · Optical imaging · Positron emission tomography (PET) · Ultrasound (US) · Clinical · Translation

16.1 Introduction

Blood vessels form a complex network of interconnected channels that support the systemic circulation of oxygen, nutrients, and metabolites throughout the human body (Bozzone and Whittemore 2021). The "microcirculation" is the final stage of this hierarchical network, and consists of "microvessels" that transfer oxygen and nutrients to, and receive waste products from, each tissue or organ. The microcirculation has been the focus of preclinical and clinical research for several centuries. Different types of microvessels are involved in many, if not all, physiological functions and diseases. For example, the normal functioning of microvessels is essential to support homeostasis, while angiogenesis or the de novo formation of new microvessels is necessary to support wound healing (Johnson and Wilgus 2014) or the progression of many pathological conditions such as cancer (Folkman 1971) or inflammatory diseases (Szekanecz and Koch 2007). Additionally, the development of new therapeutics often targets the microcirculation, such as antiangiogenic or antivascular drugs for cancer therapy (Carmeliet and Jain 2011), and the integration of microvascular networks into tissue-engineered constructs to ensure cell viability (Ren et al. 2022). Therefore, imaging technologies and methods which permit spatial and temporal modulations of the microcirculation to be characterized, are essential for elucidating its role in health and disease, providing critical preclinical and clinical data to identify potential treatment targets, and validating the efficacy of novel therapeutic interventions (Rege et al. 2012).

However, imaging the microcirculation poses several challenges. In preclinical or small animal models that recapitulate human diseases, the diameter of microvessels can vary from approximately 250 μm for veins down to 4 μm for capillaries (Bert et al. 2008). Therefore, imaging these microvessels often requires high spatial resolution. In addition, since microvessels are embedded within biological tissue, imaging methods that provide high spatial resolution, e.g., optical imaging techniques, need to overcome tissue-induced light scattering to achieve sufficient depth for accessing the microvessels. Beyond imaging vascular morphology, it is equally important to assess hemodynamic parameters, including the blood flow within the

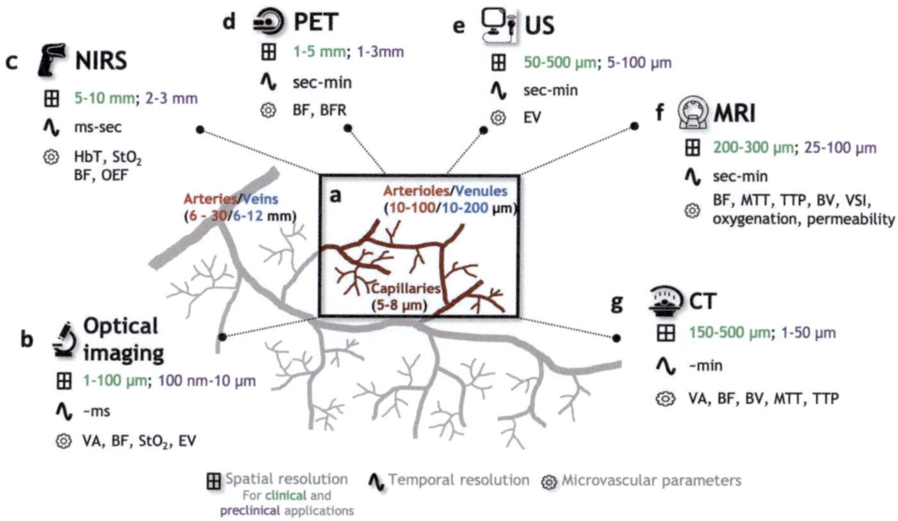

Fig. 16.1 A comparison of imaging technologies for the preclinical and clinical assessment of the microcirculation. (**a**) The circulatory system features large diameter vessels (~6–30 mm) alongside their microvascular counterparts (i.e., the "microcirculation") whose diameters span ~5–200 μm. Commercially available imaging modalities for direct visualization and interrogation of microvascular morphology and function include: (**b**) optical imaging; (**e**) ultrasound (US) imaging; and (**g**) computed tomography (CT). Other technologies that typically cannot image microvascular morphology due to their limited spatial resolution, but are sensitive to microvasculature-derived signals include: (**c**) near-infrared spectroscopy (NIRS); (**d**) positron emission tomography (PET); and (**f**) magnetic resonance imaging (MRI). Spatial and temporal resolutions shown apply to commercially available imaging systems. Abbreviations: *BF* blood flow, *BV* blood volume, *MTT* mean transit time, *TTP* time to peak, *VA* vessel architecture, *EV* erythrocyte velocity, *BFR* blood flow reserve, *StO2* oxygen saturation, *HbT* total hemoglobin, *OEF* oxygen extraction fraction, *VSI* vessel size index. (Adapted with permission from Guerraty et al. 2021)

microvessels and their interactions with surrounding cells and other microenvironmental factors (Ren et al. 2022). To address these challenges and push the boundaries of microcirculation research, scientists have made tremendous strides in imaging technologies. In the preclinical realm, optical imaging methods like confocal and two-photon microscopy are extensively utilized to explore microvascular biology (Benninger and Piston 2013; Foreman et al. 1996). These technologies have seen remarkable improvements, offering unprecedented resolution and functionality, with a wide range of specialized imaging methods for different applications (Rege et al. 2012). In the meanwhile, an increasing number of imaging methods previously used in the preclinical space are being translated into clinical applications (Guerraty et al. 2021). In this chapter, we will mainly focus on preclinical and translational imaging methods that are used to evaluate microvascular morphology and hemodynamics. A comparison of the microcirculation imaging methods described in this chapter is summarized in Fig. 16.1.

16.2 Preclinical Imaging of Microvascular Morphology

Imaging methods used in preclinical studies can be categorized broadly as those that are capable of imaging in live animals (i.e., in vivo) and those that image tissue samples obtained from euthanized animals (i.e., ex vivo). These technologies can also be classified as being two-dimensional (i.e., 2D or "widefield") or three-dimensional (i.e., 3D or "depth-resolved") imaging.

16.2.1 Imaging of Vascular Corrosion Casts and Intravascular Polymers

Vascular corrosion casting is a method that requires perfusing a wax-like resin or polymer into an animal's vascular system, which hardens to create a replica or "cast" of the vasculature after the surrounding tissue has been "corroded" away or optically cleared. (Aharinejad and Lametschwandtner 1992) Since the first time a solidifying material was used to create a vascular cast in 1663, (Aharinejad and Lametschwandtner 1992) this technique has been widely used in vascular research, and is often regarded as the "gold standard" for characterizing the morphology of the microvasculature. When conducted properly, a vascular corrosion cast can be imaged at extremely high resolutions, revealing the intricate microvascular morphology. Conventionally, imaging was primarily conducted using scanning electron microscopy (SEM) to obtain submicron details (Murakami et al. 1992; Ohtani et al. 1983). Later, researchers incorporated fluorescent dyes into the resin or polymer, which permitted the use of optical microscopy for imaging (Hillman et al. 2007). Recently, it was shown that multiple contrast agents can be combined with the polymer in a single labeling step, to enable multimodality and multiresolution imaging of the vasculature in diverse tissue types (Bhargava et al. 2022). Fig. 16.2 shows vascular corrosion casts imaged using different imaging modalities such as a stereoscope (Fig. 16.2a, b), SEM (Fig. 16.2c, d), and light-sheet microscopy (Fig. 16.2e, f) (Hillman et al. 2007; Verli et al. 2007). While the stereoscopic views provided an overview of the brain vasculature (Fig. 16.2a) and a magnified view of the cortical vessels (Fig. 16.2b), SEM provide high-resolution images of capillaries with diameters ranging from 4 to 6 μm (Fig. 16.2c), and enabled visualization of the endothelial imprints on the vessels (Fig. 16.2d).

The main drawback of vascular corrosion casting is that it involves a laborious process of euthanizing and perfusing animals. It begins with deep anesthesia of an experimental animal, followed by exsanguination using heparinized saline and perfusion with a fixative (e.g., buffered formalin). Subsequently, an x-ray opaque resin like Mercox© (Ladd Research, VT), Microfil® (Flow Tech Inc., MA), or BriteVu® (Scarlet Imaging, UT) is injected to fill the entire vascular network, including the microvessels, and then allowed to harden or polymerize. Historically, after the resin sets, the surrounding soft tissue would be dissolved using a strong

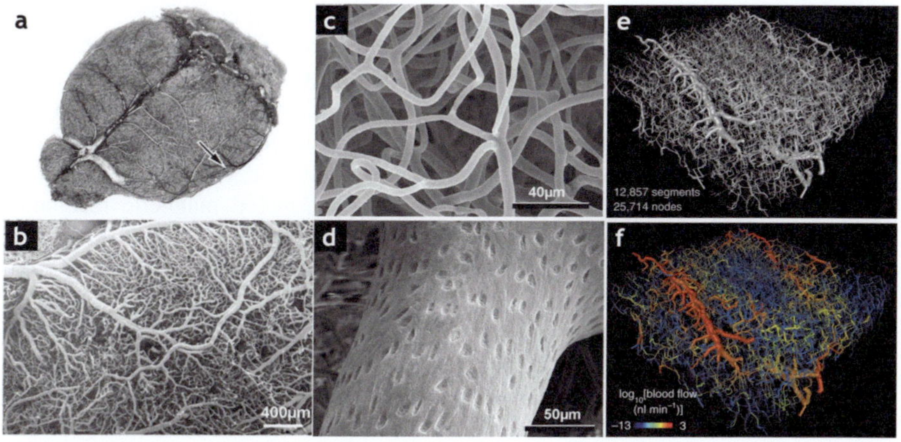

Fig. 16.2 Use of corrosion casting and intravascular polymers for interrogating microvascular morphology. (**a**–**d**) and (**e**–**f**) show applications of corrosion casting and intravascular polymers, respectively. (**a**) Stereoscopic overview of the murine brain vasculature, including the olfactory bulb and the prominent ascending middle cerebral artery (arrow) on the right hemisphere, with the cerebellum and pons visible in the upper right, demarcated by the transverse sinus. (**b**) Magnified view of the complex network of cortical microvessels. (**c**) Scanning electron microscopy (SEM) images depict the fine detail of cortical capillaries, with diameters of 4 to 6 μm and intercapillary spaces averaging around 30 μm. (**d**) SEM reveals the imprints left by endothelial cells on the surface of larger vessels, highlighting the elongated shapes of cells and nuclei along an artery. (**e**) A 0.33 μm resolution image of the thalamic microvasculature in a mouse brain acquired with 3D light sheet microscopy (LSM) covering a field of view of 300 × 300 × 100 μm (Folkman 1971). The tissue sample was perfused with an intravascular polymer mixture comprising of a CT contrast agent and fluorescently labeled MRI contrast agent, followed by optical clearing for imaging. (**f**) Simulated blood flow map for the thalamic microvasculature in (**e**), wherein each vessel is color coded by the scaled blood flow. (**a**–**d**) and (**e**–**f**) are adapted with permission from (Bhargava et al. 2022; Verli et al. 2007)

sodium or potassium hydroxide solution, completing the corrosion step. The resultant microvascular cast can then be cut into slices (ranging from 10 to 100 μm) for high-resolution imaging using SEM (Konerding 1991).

The development of microcomputed tomography (μCT) was a significant advance for imaging microvascular casts because it precluded the need for removing the soft tissue via corrosion or dissection. μCT exploits the contrast between the x-ray opaque resin and soft tissue for detailed visualization of the microvasculature (Zagorchev et al. 2010). With μCT, researchers can obtain high-resolution, 3D images of the vascular network, including the finest microvessels (1–30 μm) (Bhargava et al. 2022; Zagorchev et al. 2010). It not only simplifies the imaging workflow by eliminating the need for tissue dissolution, but also preserves the structural integrity of the sample, enabling comprehensive analysis, and re-imaging if required. In addition, researchers can employ tissue clearing agents to enable optical imaging of the microvascular cast, which is an alternative to μCT that also avoids the destructive corrosion step and permits even higher resolution 3D imaging (< 5 μm) of the entire vascular network (Bhargava et al. 2022; Zhu et al. 2023). In cases where fluorescent conjugates are added or mixed with the resin before injection, confocal or light-sheet microscopy can be utilized for imaging (Bhargava et al. 2022; Zhu et al. 2023; Bearden and Segal 2005; Wagner et al. 2006). For example, Bhargava et al. developed a multimodality imaging pipeline dubbed "VascuViz," where a water-soluble CT contrast agent (BriteVu, Scarlet Imaging LLC.) was combined with a fluorescently labeled magnetic resonance imaging (MRI) contrast agent (Galbumin-Rhodamine, BioPAL Inc.), allowing 3D imaging of the same microvasculature using μCT, optical microscopy, and MRI. Figure 16.2e shows the 3D image of the microvasculature in the thalamus of a mouse brain acquired with light-sheet microscopy at 0.33 μm resolution and labeled via VascuViz, which was then used to simulate blood flow in the thalamus (Fig. 16.2f).

Vascular corrosion casting has been applied to study the microcirculation in normal and abnormal tissues, such as tumors (Wagner et al. 2006; Hillman 2007). It allows for the computation of basic morphological parameters such as microvessel radius, length, tortuosity, segment density, fractional occupancy, etc. (Vasquez et al. 2023). However, since the technique involves animal exsanguination or blood removal, it can only provide a snapshot of the microvascular morphology at the time of euthanasia, and cannot provide information on the dynamic changes in microvascular morphology that are known to occur in live animals. Historically, vascular casts also had the added disadvantages of being brittle and hard to digitize. Although modern imaging methods have circumvented some of these hurdles, the issue of the cast fading or dissolving during later processing steps still poses a challenge.

16.2.2 Hemoglobin-Based Intrinsic Optical Signal (IOS) Imaging

Two-dimensional (2D) widefield optical imaging technologies are relatively straightforward, and are widely used for in vivo mapping of microvascular morphology. One such imaging technique uses the wavelength-dependent light absorption properties of erythrocyte-bound oxy- (HbO_2) and deoxyhemoglobin (HbR) as an endogenous or "intrinsic" optical imaging contrast (Hillman 2007) and is therefore known as intrinsic optical signal (IOS) imaging. To conduct IOS imaging, one needs a light source, typically used in conjunction with filters to generate specific wavelengths that are used to illuminate a biological tissue bed. Additional optics are needed to gather the light scattered back from the biological tissue and focus it onto a light detector or image sensor to create an image (Vasquez et al. 2023). Since the hemoglobin-rich blood within microvessels absorbs the incident light, they appear darker in these images and can be distinguished from the nonvascular background tissue. While image acquisition at multiple wavelengths can be combined to derive the different oxygenation states of blood corresponding to changes in the relative

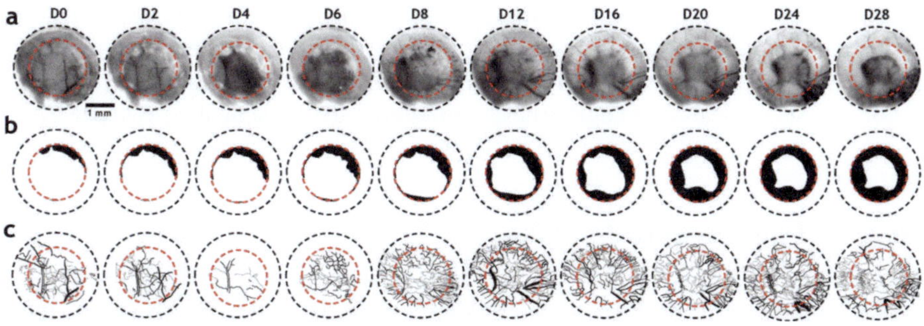

Fig. 16.3 Intrinsic optical signal (IOS) imaging of microvascular remodeling during calvarial defect healing. (**a**) IOS images enable the visualization of microvascular remodeling and bone formation within a calvarial defect. The dashed black circles indicate the 3 mm field of view (FoV) that is optically accessible via a cranial window, while the dashed red circles show the 2 mm defect. (**b**) Segmented images of bone highlight the substantial osteogenesis observed over the course of 4 weeks of bone healing. (**c**) Maps of the segmented microvasculature demonstrate robust angiogenesis occurs concurrently with bone regeneration (i.e., osteogenesis). (Adapted with permission from Ren et al. 2023a)

concentrations of HbO_2 and HbR (discussed in detail in the *Assessing Intravascular Oxygenation* section below), specific wavelengths known as isosbestic points are used to image the morphology of microvessels. These wavelengths, i.e., 500, 530, 570, and 797 nm, (Hillman 2007) are unique because their absorption by HbO_2 and HbR are identical. This feature makes it possible to measure changes in the total hemoglobin concentration (HbT, which is the sum of HbR and HbO_2 concentrations) independently from variations in blood oxygenation. HbT is often employed as a surrogate of blood volume or microvascular cross-sectional area by assuming a constant hematocrit, which enables IOS imaging to delineate microvascular structures. IOS imaging parameters such as magnification and field of view (FOV) can be varied to optimize microvessel visualization.

The main advantage of this method is that it utilizes a contrast mechanism that is intrinsic to the microcirculation and therefore does not require the injection of an exogenous probe or dye. This makes in vivo IOS imaging less invasive than other methods. However, due to light scattering, IOS imaging only permits the visualization of superficial blood vessels in tissue and requires direct optical access to the microvasculature of interest. Therefore, it is often coupled with an optical window to the organ or tissue being imaged. For example, a cranial window created via a craniotomy or by thinning a portion of the skull was used to evaluate the brain's microvasculature in vivo with IOS imaging (Ren et al. 2023a, b; Augustinaite and Kuhn 2021; Guevara et al. 2013). Although IOS has become a staple technique for modern neuroscience, (Vasquez et al. 2023; He et al. 2019; Johnston et al. 2013) recently, it has also been applied to evaluate angiogenesis and microvascular remodeling during bone healing and tissue engineering applications (Ren et al. 2023a, b; Mendez et al. 2018). Fig. 16.3 illustrates how IOS was used to track angiogenic evolution in vivo during a calvarial bone healing cascade over 4 weeks

(Ren et al. 2023a). Using the IOS images of a surgically created calvarial defect on mice (Fig. 16.3a), the bone (Fig. 16.3b), and microvascular structure (Fig. 16.3c) could be digitally segmented. This segmentation enabled the longitudinal quantification of osteogenesis and angiogenesis throughout the bone healing cascade.

16.2.3 Fluorescence Imaging

While utilizing intrinsic hemoglobin absorption to image the microvascular morphology is straightforward, fluorescence (FL) imaging offers an opportunity to enhance optical contrast and applicability. This imaging method relies on the phenomenon of Stokes shift, (Abian et al. 2020) in which some molecules (i.e., fluorophores) when illuminated or "excited" with certain wavelengths of light, undergo an energy transition that is accompanied by the emission of light at a longer wavelength (Rege et al. 2012; Ettinger and Wittmann 2014). Consequently, FL imaging requires an external light source (a lamp, laser, or LED) to optically excite the fluorophore. In the simplest optical imaging system for FL imaging, the emitted fluorescence can be separated from the excitation light via optical filters, and acquired using a lens system and a camera with a charge-coupled device (CCD) or complementary metal oxide (CMOS) sensor (Refaat et al. 2022). By incorporating these fluorescent molecules into the microvasculature and employing suitable optical filters in the path of emitted light, it is possible to enhance the contrast-to-noise ratio (CNR) of microvascular optical imaging. An example illustrating this CNR enhancement is shown in Figs. 16.4 and 16.7 where FL imaging via an intravenously injected dye (Fig. 16.4d) resulted in better delineation of the microvascular morphology than imaging under white (Fig. 16.4a) or green (Fig. 16.4b) illumination, and was comparable to a technique known as laser speckle contrast (LSC) which is covered later in this chapter (Fig. 16.4c).

Imaging microvascular morphology in vivo using fluorescence relies heavily on the intravenous administration of exogenous contrast agents or dyes. In contrast, ex vivo fluorescence imaging exploits the affinity of various molecules for the blood vessel endothelium. A popular approach is to use antibodies against receptors on endothelial cells. Numerous antibodies against endothelial epitopes, such as the von Willebrand factor, (Baluk and McDonald 2008) PECAM-1, E-selectin, P-selectin, CD31, CD34, CD105, and the vascular adhesion molecule (VCAM)-l, (Runnels et al. 2006; McDonald and Choyke 2003) have been used. Alternatively, endothelial lectins, a class of carbohydrate-binding proteins distinct from antibodies, have become a popular vascular label. (Rege et al. 2012) Lectins can be conjugated with fluorophores such as fluorescein isothiocyanate (FITC), and be used to label the microvasculature by intravenous injection or by applying such conjugates to fixed tissue sections, wherein the lectins bind to the carbohydrate components of the endothelial plasmalemma (Thurston et al. 1998; Robertson et al. 2015). Past research has established that lectin labeling is a sensitive and reliable method for visualizing vascular structures in a variety of tissues and organs, including brain, liver, spleen,

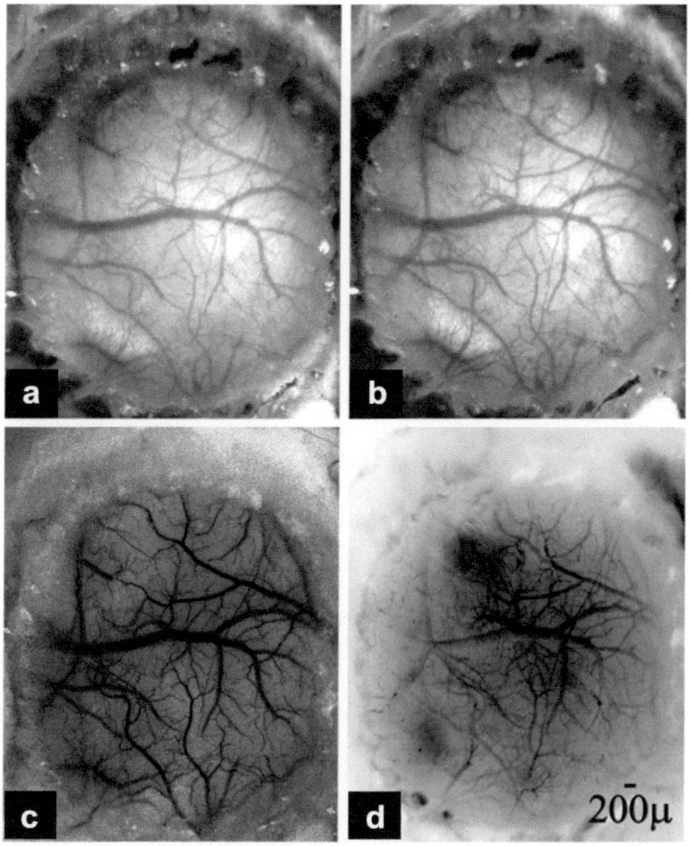

Fig. 16.4 Comparison of different in vivo optical imaging techniques for visualizing microvascular morphology. (**a**) Bright-field reflectance image using white light illumination of a 6 × 7.5 mm (Johnson and Wilgus 2014) region of rat cortex optically accessed via a thinned skull preparation. Images of the microvasculature obtained via: (**b**) bright-field green light reflectance; (**c**) laser speckle contrast; and (**d**) red fluorescence (with contrast inverted) of the same field of view (FoV). Fluorescence image was acquired after intravenous administration of rhodamine-dextran. (Adapted with permission from Murari et al. 2007)

kidney, small intestine, heart, and skeletal muscle (Robertson et al. 2015). They have also been used in cancer diagnosis applications. For example, Debbage et al. successfully differentiated efficiently and inefficiently perfused regions in a mouse mammary adenocarcinoma model via dual injections of HPA (helix pomatia agglutinin) lectin conjugated to different fluorophores (i.e., tetramethylrhodamine isothiocyanate or TRITC, and FITC), administered 15 minutes apart. (Debbage et al. 1998) Consequently, the tumor vasculature with intermittent perfusion was characterized by a mismatch in the lectin label between the red (TRITC) and green (FITC) FL imaging channels. Other fluorescent tracers used to image microvascular morphology include carbocyanine dyes DiI (1,1′-dioctadecyl-

3,3,3′,3′-tetramethylindocarbocyanine perchlorate) (Li et al. 2008; Stroh et al. 2005) and DiO (3,3′-dioctadecyloxacarbocyanine perchlorate), (Unthank et al. 1993) indocyanine green (ICG), (Nagahara et al. 2018) quantum dots, (Huang et al. 2015) and Evans blue (Honeycutt and O'Brien 2021).

16.2.4 Confocal Microscopy

In conventional fluorescence microscopy, light is uniformly directed through the sample across the field of view. This method works well in thin tissue sections, but for thicker tissue samples, light from planes above and below the focal plane will also be detected although they are not in focus. This out-of-focus light blurs the image, reducing the overall spatial resolution. Confocal microscopy circumvents this issue by including a pinhole in the light path to selectively exclude out-of-focus light from reaching the detector, preventing it from contributing to image blur during data collection (Elliott 2020). This selective focus provides effective optical sectioning, enhancing the clarity and specificity of the acquired images. During confocal imaging, both illumination and detection focus on the same diffraction-limited spot, and scanning this spot in two-dimension, e.g., the *X* and *Y* planes, creates a sharp image of a specific focal plane. While a single slice confocal image can be taken to investigate a specific location within a tissue sample, one can also move the region of interest along the third dimension (i.e., depth or the *Z* axis) to acquire images from progressively deeper tissue planes to construct a "z-stack" (Peterson 2010). This "optical sectioning" allows the reconstruction and quantitative analysis of the entire sample volume, providing a detailed and spatially accurate representation of the imaged structure in 3D (Jonkman et al. 2020).

Given its ability to achieve high-resolution 3D reconstructions of thick tissues, confocal imaging is frequently used to visualize microvascular structures ex vivo and in vivo, as well as their association and interaction with adjacent cells or microenvironmental factors (Foreman et al. 1996). Fig. 16.5 illustrates how Nakano et al. (2020) elegantly employed confocal microscopy to examine alterations in elements of the neurovascular unit in a rat model of retinopathy of prematurity (ROP). The study involved assessments of the morphology of microvessels and associated astrocytes. Using lectin to label the microvasculature and antiglial fibrillary acidic protein (GFAP) antibodies to label astrocytes, the study successfully identified changes in the distribution of astrocytes along different microvascular structures in control (Fig. 16.5a–c) and ROP animals (Fig. 16.5d–f). These images show ROP animals exhibited aggressive angiogenesis, tortuous arteries (Fig. 16.5e) and enlarged veins (Fig. 16.5f), along with a sparser and poor astrocytic coverage compared to the control group. Aside from this example from ophthalmology, confocal microscopy has played an important role in a wide variety of preclinical research areas related to the microvasculature, including investigating the dysregulation of the neurovascular unit in models of stroke, (Lochhead et al. 2023) imaging the abnormal microvasculature in tumor models, (Reitan et al.

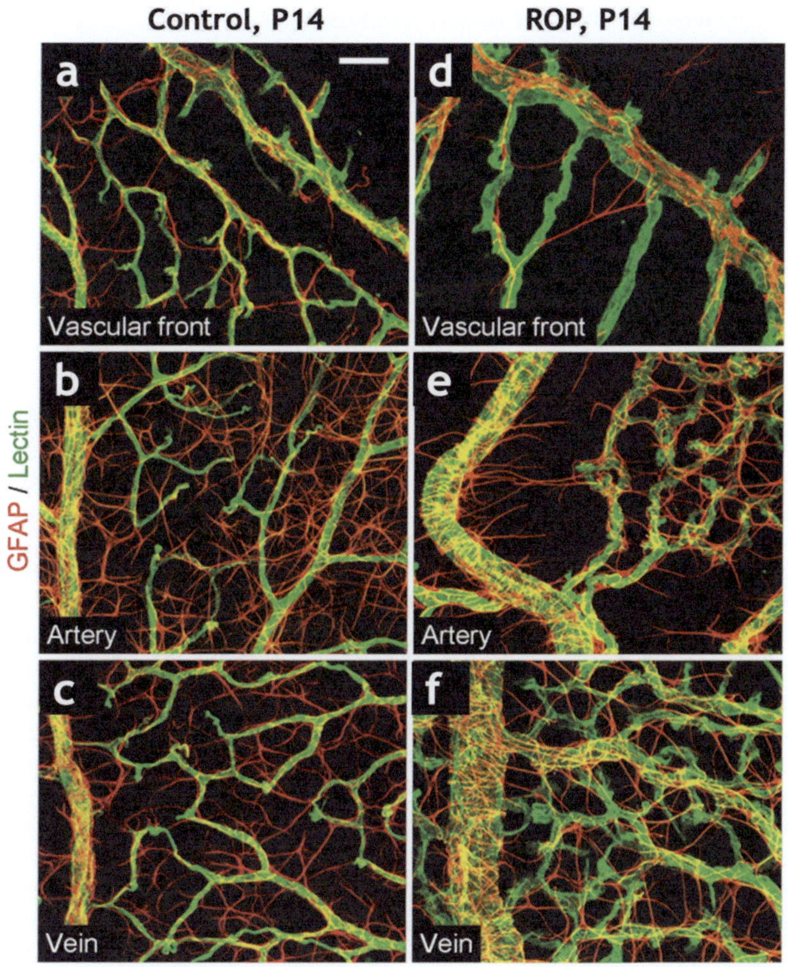

Fig. 16.5 Confocal microscopy of a "flat-mounted" retina stained for vascular endothelium and astrocytes. Confocal imaging of tomato lectin-FITC labeled vascular endothelial cells in the retinas of healthy rats reveals the distinct microvascular morphology of: (**a**) the vascular front; (**b**) an artery; and (**c**) a vein. The association of these vessels with neighboring astrocytes was visualized via fluorescence imaging of an astrocyte bound GFAP-Cy3 label (red channel). (**d**–**f**) Confocal images of similar microvascular locations in rats with "retinopathy of prematurity" (ROP), which show changes in microvascular morphology (green channel) and altered microvascular association of astrocytes (red channel). Scale bar = 25 μm. Abbreviations: *FITC* fluorescein isothiocyanate, *Cy3* Cyanine3. (Adapted with permission from Nakano et al. 2020)

2010) and exploring cardiovascular development, (Kolesova et al. 2021) among others.

16.2.5 Two- (or Multi)Photon Microscopy

Two- (or multi)photon microscopy is a superior alternative to confocal microscopy when acquiring high-resolution 3D images of the microvasculature in vivo or from ex vivo tissue sections. Similar to confocal microscopy, it provides better spatial resolution than conventional wide-field fluorescence imaging by limiting the scattering and blurring caused by out-of-focus light. The principles used by two-photon microscopy, though, diverge slightly from confocal microscopy. Instead of using a pinhole, two-photon microscopy focuses a beam of infrared laser illumination with twice the excitation wavelength of the dye, i.e., carrying roughly half the energy per photon than what is needed to excite the fluorophore, into a narrow spot or volume (Benninger and Piston 2013). Consequently, the simultaneous incidence of two photons is necessary to excite the fluorophore. Since the probability of two excitation photons occupying the same position at the exact same time outside the scanning focal spot is extremely low, this imaging method ensures there is no background signal generation and maintains a high CNR, even in the presence of substantial scattering at considerable imaging depths. These characteristics establish two-photon microscopy as superior to confocal microscopy, especially when imaging thicker (~1 mm) tissue samples (Miller et al. 2017). Researchers have exploited the high spatial resolution of this technique to complement the research domains addressed by confocal microscopy. For example, Lawson et al. employed multiphoton imaging to examine the evolving microvasculature during the developmental phase in a transgenic zebrafish model. (Lawson and Weinstein 2002)

Additionally, two- and multiphoton microscopy have been successfully used to characterize the relationship between the microvasculature and the lymphatics, (Wang et al. 2021) study neurovascular coupling, (Kumar et al. 2021) and monitor changes in tumor microvasculature associated with tumor progression (Tozer et al. 2005; Chen et al. 2019). Fig. 16.6b illustrates the use of two-photon microscopy to image the infiltration of the parenchyma around the microvessels by bone marrow-derived macrophages (BMDM) in a brain tumor model. (Chen et al. 2019) In contrast, Fig. 16.6a shows that very few BMDM were observed in the nontumor-bearing brain, because they could not traverse the blood–brain barrier (BBB) under normal physiological conditions. These observations generated by two-photon microscopy reveal important information about the relationship between the tumor vasculature and cell infiltration.

16.2.6 Lightsheet Microscopy

Lightsheet microscopy is a recently developed technique that enables high-resolution (~0.5 μm) 3D vascular imaging of large (>5 cm (Folkman 1971)) tissue samples (Schueth et al. 2023). Although the concept was first introduced by Richard Zsigmondy and Henry Siedentopf back in 1912, (Girkin and Carvalho 2018) it was

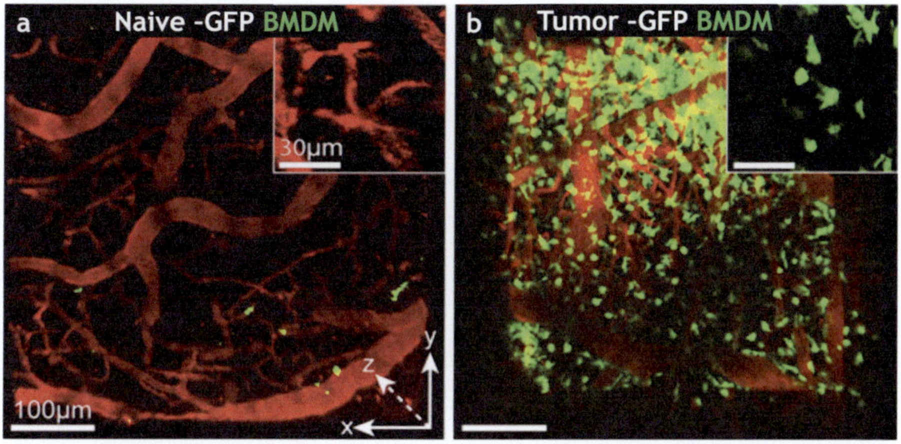

Fig. 16.6 Two-photon imaging of the microcirculation in a brain tumor microenvironment. (**a**) An image of the cerebral microvasculature in a nontumor-bearing mouse brain obtained via two-photon fluorescence imaging of an intravenously injected TRITC-Dextra dye (red channel). Few or no GFP-expressing bone marrow derived macrophages (BMDMs) were seen in the brain parenchyma (green channel). (**b**) In contrast, a similar image from a tumor-bearing mouse brain shows significant infiltration of GFP-expressing BMDMs (green channel). Scale bar = 100 μm; magnified inset scale = 30 μm. (Adapted with permission from (Chen et al. 2019)

only recent advancements in optical imaging tools, such as cameras, lasers, small model organisms, and tissue-clearing methods, that made it a reality and applicable to diverse applications. Lightsheet microscopy is able to achieve optical sectioning similar to confocal and two−/multiphoton microscopy, but its underlying principles are different. Conventional confocal and two-photon microscopy utilize laser point scanning techniques, in which a small point of focused light moves across different 3D positions to form a stack of images. In lightsheet microscopy, an entire imaging plane is illuminated with a thin "sheet" of light, and a camera is carefully focused onto this 2D plane akin to the acquisition of a widefield image (Girkin and Carvalho 2018; Stelzer et al. 2021). The resulting 3D image is generated by simultaneously moving these aligned excitation and detection planes to cover the entire 3D volume. With this working principle, lightsheet microscopy is more susceptible to scattering compared to confocal or two-photon microscopes and is therefore more suitable for imaging transparent or optically cleared samples. On the other hand, it offers advantages such as the ability to image larger samples (e.g., intact mouse brains), (Mano et al. 2018) low phototoxicity, and fast volumetric imaging speeds (Hillman et al. 2019).

Despite being a relatively recent development, lightsheet microscopy has been used to image the vasculature in a wide array of tissues, including the brain, (Mohamud Yusuf et al. 2020; Khouri et al. 2021) muscle, (Milgroom and Ralston 2016) and bone, (Rindone et al. 2021; Biswas et al. 2023) among others (Spangenberg et al. 2023; Glaser et al. 2022). Its capacity to map large sections of tissue has opened new avenues for developing vascular atlases and advancing

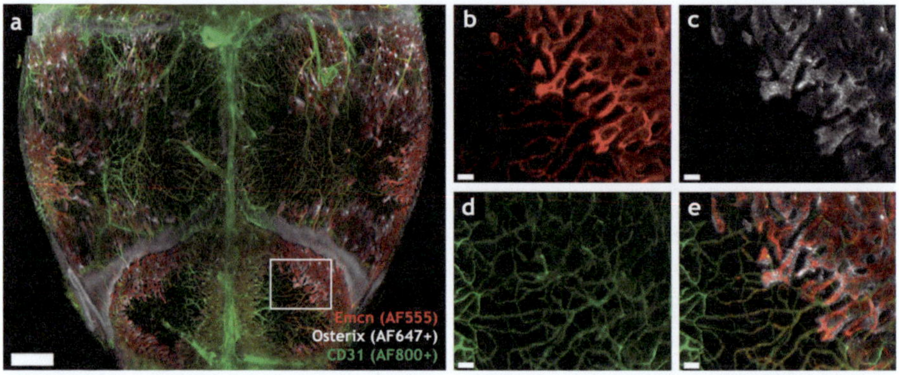

Fig. 16.7 3D imaging of the murine calvarium at single-cell resolution with light sheet microscopy (LSM). (**a**) A maximum intensity projection (MIP) showing microvessel morphology (red and green channels) and Osterix+ skeletal progenitor cells (white channel) in the parietal and posterior frontal bones of the murine calvarium. Both CD31 (green channel) and Endomucin (Emcn, red channel) are endothelial markers, and their differential expression was used to characterize distinct capillary subtypes. For example, $CD31^{hi}Emcn^{hi}$ (i.e., type H) vessels were known to be more associated with osteoprogenitor cells and functioned to support bone remodeling, while $CD31^{low}Emcn^{low}$ (i.e., type L) vessels showed less functional connections to bone. (**b–e**) Magnified views from the boxed region in (**a**) demonstrating LSM's capacity to image: (**b**) Endomucin-labeled blood vessels; (**c**) Osterix-labeled skeletal progenitor cells; (**d**) CD31-labeled blood vessels; (**e**) The composite image, shows the higher colocalization of skeletal progenitor cells (white channel) and type H vessels (i.e., vessels with both red and green label; adapted with permission from Rindone et al. 2021)

vascular systems biology applications (Bhargava et al. 2022). For example, Rindone et al. (2021) recently developed a quantitative light sheet microscopy (QLSM) pipeline to comprehensively map the microvascular environment in calvarial bone (Fig. 16.7). With this platform, they were able to create 3D maps at single-cell resolution (i.e., 1.3 μm) of microvascular subtypes and their spatial distributions in the calvarium during bone growth (Fig. 16.7a). By labeling the calvarial blood vessels with endomucin (Emcn, Fig. 16.7b) and CD31 (Fig. 16.7d) endothelial markers, they were able to identify vessels that highly expressed both markers (i.e., $CD31^{hi}Emcn^{hi}$, or type H vessels), which showed higher associations with osteoprogenitor cells labeled with osterix (Fig. 16.7c) and supported bone remodeling. These findings provide crucial insights into the role of the vascular phenotype in craniofacial bone defect healing that can aid the development of new tissue engineered constructs.

16.3 Preclinical Imaging of Microvascular Hemodynamics

16.3.1 Assessing Blood Flow

Evaluating blood flow is critical for understanding both normal physiological processes and disease states, as it directly impacts tissue perfusion and overall homeostasis. Since tracking blood flow necessitates live or in vivo imaging, it is usually more challenging to implement than ex vivo imaging of the microvascular morphology. This section highlights imaging methods that are suitable for evaluating blood flow in vivo.

16.3.2 Measuring Erythrocyte Velocity with Confocal or Two-Photon Microscopy

One way to assess blood flow is to measure the velocity of erythrocytes, or red blood cells (RBCs). The key approach involves using either a fluorescent label that makes flowing RBCs visible under the microscope, or employing a dye that makes the blood plasma fluorescent while leaving RBCs unlabeled, making them appear as dark spots against the fluorescing plasma (Rege et al. 2012).

RBCs can be tracked in 2D using "line scans" which are linear rasters of laser light. In this method, the vessel diameter and RBC flux are determined by imaging a region of the blood vessel with laser scans perpendicular to the plane of blood flow, using for example FITC-labeled plasma to visualize the microvessel lumen or fluorescently labeled RBCs for determining their flux. The RBC velocity is calculated from the slope of streaks in images acquired with lines scanning parallel or diagonal to the vessel, wherein the streak angle is inversely proportional to RBC velocity (Kornfield and Newman 2015). For example, Kornfield and Newman (2015) utilized ultrafast confocal line scans to track fluorescently labeled RBCs in the retinal microcirculation. This method enabled the accurate calculation of blood flow across vessels of varying diameters, facilitating direct measurements with minimal assumptions. Using 2D line scans to track RBCs offers a simple, direct method for quantifying absolute blood flow, providing valuable insights into hemodynamics.

Advanced imaging techniques like two-photon microscopy enable the direct observation of RBCs moving through microvessels, permitting measurements of their speed in 3D. For example, Schaffer et al. (2006) utilized a two-photon microscope to track the location of RBCs at various moments by capturing a time-series of images in a specific area within a microvessel. They determined flow velocity by calculating the distance RBCs traveled in a given period, as depicted in Fig. 16.8. As shown in Fig. 16.8a, clotting was induced in a targeted arteriole (green dot) using irradiation, and sequential images captured the evolution from partial (Fig. 16.8b) to complete (Fig. 16.8c) vessel occlusion via clot formation, as

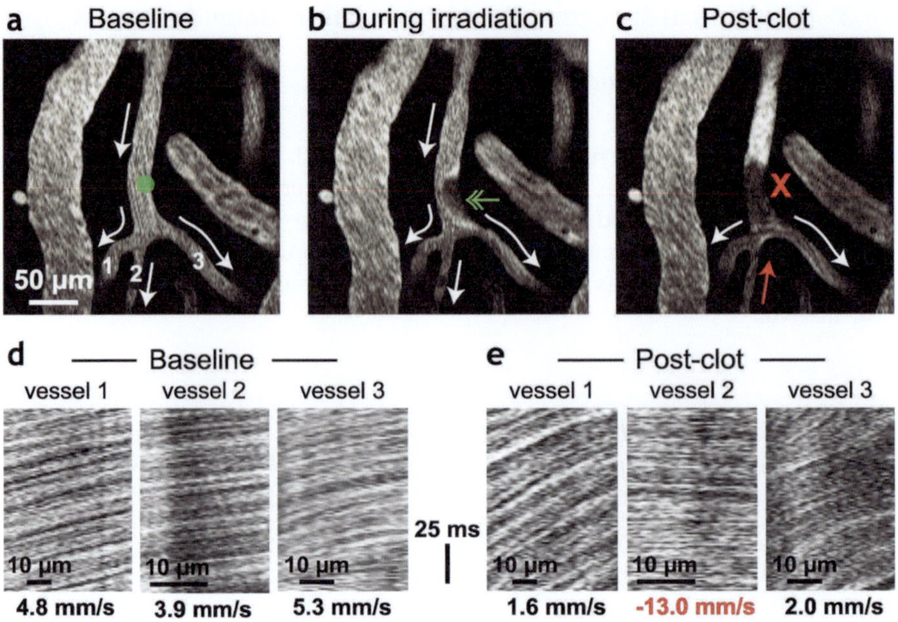

Fig. 16.8 Imaging targeted photothrombotic clotting in cortical blood vessels via two-photon laser scanning microscopy (TPLSM). Planar TPLSM images of photothrombotic clotting of a surface arteriole. (**a**) Baseline image in which the green circle marks target region for irradiation, and the white arrows indicate the blood flow direction. (**b**) Image showing partial microvascular occlusion (green double arrow) after the injection of Rose Bengal dye and 2-minutes of irradiation with the green laser. (**c**) Complete clotting of the target vessel (X) after an additional minute of irradiation. The red arrow indicates the reversal of blood flow in the center branch connected to the clotted vessel. (**d**) Baseline and (**e**) postclot line-scan data for the numbered vessels downstream from the target vessel in (a), with the corresponding changes in RBC velocity. Labeled in red, the center branch (vessel 2) of the clotted vessel exhibited a negative flow velocity, indicative of reversed blood flow. (Adapted with permission from (Schaffer et al. 2006)

well as changes in blood flow direction and speed. Baseline (Fig. 16.8d) and postclot (Fig. 16.8e) line-scan data helped further quantify the impact of clotting on RBC speed in downstream vessels. This method provided insights into the behavior of RBCs within individual microvessels, which is beneficial for studying conditions and diseases that alter microvascular blood flow. Tracking RBC velocity has also been done in physiological research across various organs, such as the kidneys, where Zimmerhackl et al. (1985) labeled RBCs to examine the impact of angiotensin on microvascular function; the cochlea, where Monfared et al. (2006) measured blood flow velocity via a microendoscope; and the brain, where Villringer et al. (1994) combined RBC tracking with confocal imaging to study capillary perfusion dynamics in rats.

16.3.3 Doppler Velocimetry

The main principle underlying Doppler velocimetry is the Doppler effect (Rege et al. 2012). When light is reflected from a moving scatterer, it experiences a shift in its frequency. The magnitude of this frequency shift is determined by the speed of the moving scatterer relative to the incoming light beam and is given by:

$$\Delta f = \frac{v \cdot \cos \theta}{c} f \tag{16.1}$$

Where v is the velocity of the scatterer, c is the velocity of light, θ is the angle between the direction of light and the motion of scatterers, f is the frequency of laser used, and Δf is the change in frequency.

When measuring blood flow in the microvasculature, the RBCs serve as the moving scatterers, and therefore, any change in frequency (Δf) is indicative of their velocity. Since RBC velocity is significantly lower than the speed of light, this frequency shift is minimal, and the detection of RBC speeds requires precise measurement techniques like interferometry. Laser Doppler Flowmetry (LDF) systems use wavelengths of light between 600 and 700 nm, depending on the light absorption properties of the targeted tissue (Metzger et al. 2011). Due to the complex structure of blood vessels and the multidirectional movement of RBCs, Δf is often a measure of the average blood flow within a given region-of-interest (ROI). Therefore, LDF is usually categorized under flowmetry rather than velocimetry.

Commercial LDF systems can achieve a spatial resolution of ~100 μm, while some research prototypes can achieve a resolution of ~25 μm (Pouratian and Toga 2002). LDF can monitor blood flow in real-time, making it suitable for continuous in vivo observation. However, LDF systems that employ line scanning may exhibit lower temporal resolution when covering a larger region-of-interest (ROI). The primary advantages of LDF include its ease of use and high temporal resolution (>100 Hz) (Mauritzon et al. 2022). While LDF does not offer absolute perfusion measurements (e.g., ml/min/g of tissue), its ability to assess relative perfusion changes suffices for a wide range of applications. For example, researchers utilized LDF to measure brain perfusion in rodents, (Sutherland et al. 2014) and investigate effects of neurological stimulation, (Peng et al. 2002, 2004; Rosengarten et al. 2003) stroke, subarachnoid hemorrhage, (Nakase et al. 1997; Muller et al. 1994; Bederson et al. 1995; Bogaert et al. 2000) and wound healing on blood flow (Raposio et al. 2017).

16.3.4 Ultrasound Imaging and Ultrasound Localization Microscopy (ULM)

Although conventional ultrasound imaging has been utilized to identify tumor angiogenesis in mouse breast cancer models, (Shelton et al. 2020) it typically offers spatial resolutions (>100 μm) that are too low for preclinical microvascular imaging, and has found wider application in clinical settings (please see the **Translational and Clinical Imaging of Microvascular Morphology and Hemodynamics** section for more details). Recent advancements in super-resolution ultrasound, however, have enhanced its utility for microvascular imaging (Song et al. 2023). Ultrasound localization microscopy (ULM) is an emerging super-resolution ultrasound technique that utilizes the principles of ultrasound imaging while providing a spatial resolution superior to that of conventional diffraction-limited ultrasound methods. ULM has been shown to achieve resolutions of <10 μm, creating detailed views of the microcirculation that were previously unattainable with conventional ultrasound technologies (Dencks and Schmitz 2023). ULM primarily relies on a specialized mode of acquisition that is distinct from conventional ultrasound acquisition modes such as the B-mode (i.e., Brightness-mode) or the Doppler mode, which are used for structural imaging and macroscopic flow visualization, (Carovac et al. 2011) respectively. The acquisition mode used in ULM is optimized for high-resolution, microvascular imaging because it leverages high frame rate (~kHz), contrast-enhanced imaging, and advanced computational algorithms to surpass the resolution of traditional ultrasound imaging (Dencks and Schmitz 2023).

The foundational principle of ULM is the precise tracking and localization of individual ultrasound scatterers. These scatterers are introduced into the microcirculation in the form of ultrasound contrast agents, which consist of an inert gas encapsulated by a shell that is called a microbubble (MB) and have radii of 1–3 μm (Ignee et al. 2016). Most ULM systems have a temporal resolution of thousands of frames per second, which allows them to track individual microbubbles within microvessels (Mano et al. 2023). After ultrasound image acquisition, various signal processing, localization, and reconstruction steps are performed to provide detailed information on blood flow at high spatial (<10 μm) and temporal resolution (>1000 Hz) (Dencks and Schmitz 2023). These processing steps usually require extensive computing power, because of the high frame rates, sophisticated signal processing, and complex data analyses that are needed to accurately track microbubbles and compute high-resolution images of blood flow (Errico et al. 2015).

ULM-derived perfusion measurements have revealed disease mechanisms, been used to evaluate treatment efficacy and diagnose conditions affecting the microcirculation. For example, Renaudin et al. (2022) adapted ULM to quantify functional hyperemia dynamics (i.e., the local increase in the brain's blood flow in response to neuronal activity) during brain activation, at a spatial resolution of 6.5 μm and a temporal resolution of 1 s in deep regions of the rat brain (Fig. 16.9). The spatial activation pattern of the whole brain during whisker stimulation (Fig. 16.9a), and the distinct increase in microbubble (MB) speed in a specific region of the brain,

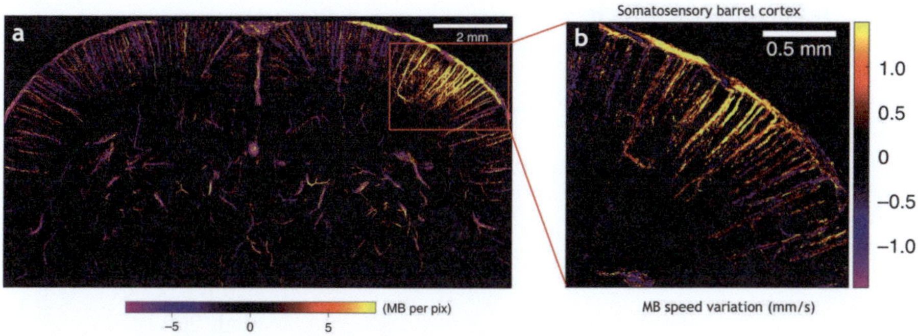

Fig. 16.9 Ultrasound localization microscopy (ULM) imaging of cortical blood flow variations during brain activation. (**a**) Map of the cortical vasculature acquired with ULM and color-coded by the microbubble (MB) density. Following a whisker barrel stimulation, one can clearly see the increase in MB density in the somatosensory barrel cortex. (**b**) Magnified view of the boxed region in (**a**), focusing on the somatosensory barrel cortex, in which higher MB speed variation was apparent, corresponding to the microvascular response to whisker stimulation. (Adapted with permission from Renaudin et al. 2022)

identified as the somatosensory or barrel cortex (Fig. 16.9b), can be clearly visualized and quantified using ULM. This approach not only demonstrated ULM's capability to capture dynamic microvascular changes with high precision, but also highlighted its potential for advancing our understanding of cerebral blood flow dynamics in response to neural activity. In addition, Denis et al. reported the use of ULM for visualizing blood flow in glomeruli, showing that it can also be used to enhance our understanding of organ function at the microvascular level. (Denis et al. 2023)

16.3.5 Laser Speckle Contrast (LSC) Imaging

Laser speckle contrast (LSC) imaging involves the use of a coherent (i.e., laser) light source to illuminate microvessels, resulting in back-scattered light from moving RBCs that could be collected through diffractive optics to create a spatiotemporal diffraction pattern on an image sensor called "laser speckles." This speckle pattern is quantified by a term called speckle contrast (K) and is calculated as (Johnson and Wilgus 2014):

$$K(x,y) = \frac{\sigma_N}{\mu_N} \tag{16.2}$$

Where σ_N and μ_N are the standard deviation and mean of the pixel intensities in a predefined neighborhood N around each pixel P(x,y). This neighborhood can be chosen in the temporal domain by collecting a stack of speckle images if spatial resolution needs to be prioritized or chosen in the spatial domain if temporal

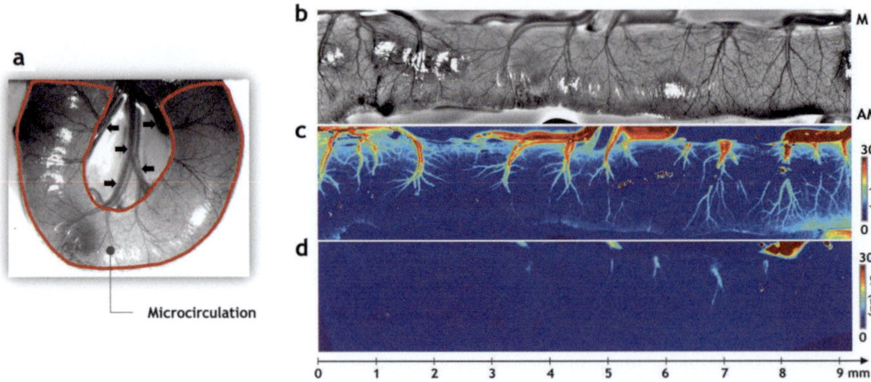

Fig. 16.10 Imaging intestinal blood flow in a murine model of necrotizing enterocolitis (NEC) with laser speckle contrast (LSC). (**a**) IOS image of a portion of the surgically exposed healthy gut in a mouse pup surgically exposed to provide optical access to. Black arrowheads indicate the mesenteric circulation, and the visible gut is outlined in red. (**b**) "Flat-map" of the intestinal microvasculature in (**a**) created via a rectilinear transformation. (**c**) Corresponding blood flow flat-map obtained with LSC after a similar rectilinear transformation on the healthy gut. (**d**) Blood flow map obtained with LSC after a similar rectilinear transformation on a NEC gut. Significantly less blood flow can be observed in the NEC gut. The "*x* axis" indicates the length along the flattened intestine. "M" and "AM" denote "mesenteric" and "antimesenteric" sides, respectively. "au" represents arbitrary blood flow units. (Adapted with permission from (Senarathna et al. 2022)

resolution is to be prioritized. Conversely, the neighborhood can also be selected in a spatiotemporal manner if a compromise between temporal and spatial resolutions is desired.

Next, since the local speckle contrast is inversely related to RBC velocity, it can be used as an approximation of blood flow (Briers et al. 2013). The primary benefit of using LSC is that in vivo perfusion can be measured by using the movement of RBCs as a "virtual contrast agent," (Murari et al. 2007) and therefore, this technique does not require the administration of any extrinsic contrast or dye. Moreover, it does not require many sophisticated optical components and is easy to implement. It is also minimally invasive and therefore ideal for longitudinal in vivo imaging. Figure 16.10c shows an example in which LSC was used to map the blood flow in intestinal microvessels in a murine model of necrotizing enterocolitis (NEC), which is a debilitating disease in infants. The curvilinear IOS image of the murine gut microcirculation (Fig. 16.10a) was transformed into a "flat map" via a rectilinear transformation to visualize the microvasculature (Fig. 16.10b), with its perfusion visualized via LSC (Fig. 16.10c) (Senarathna et al. 2022). Using this approach, Senarathna et al. were able to delineate the perfusion territories of gut microvessels or "intestinal vascular units" (IVUs) in healthy (Fig. 16.10b, c) and NEC guts (Fig. 16.10d) in vivo. By comparing the LSC images, they found a significant decrease of blood flow in NEC intestines compared to healthy ones, showcasing the potential of LSC in elucidating NEC pathogenesis. In spite of these advantages,

LSC imaging has limitations such as an inability to conduct optical sectioning, and assess only relative and not absolute perfusion. Consequently, LSC is primarily limited to mapping relative blood flow and its changes in superficial layers of an organ or tissue in most applications (Duncan and Kirkpatrick 2008).

16.3.6 Magnetic Resonance Imaging (MRI)

Magnetic resonance imaging (MRI) works by inducing the alignment of water protons in a patient's tissues when placed in a strong external magnetic field. Most clinical MRI scanners in use today are designed with a 1.5 Tesla (T) magnet as their core, and newer models can provide more powerful magnetic fields with magnitudes of 3.0 T and even 7.0 T (Wood et al. 2012). A radiofrequency (RF) pulse is used to perturb the bulk tissue magnetization, and its realignment (or relaxation) with the main applied magnetic field after the RF pulse is turned off, can be detected by an antenna. Since different tissues have different magnetic properties, they can be differentiated based on their differences in their relaxation characteristics (Grover et al. 2015). Using this principle, MRI can be exploited to quantify a wide range of microcirculatory properties, from intravascular oxygenation to perfusion through the microvasculature (Kennan et al. 1994; Pathak et al. 2008).

While MRI is widely used in the clinical space (please see the **Translational and Clinical Imaging of Microvascular Morphology and Hemodynamics** section), there have been numerous advancements in MRI technology that has expanded its utility for microvascular imaging in the preclinical space (Wei et al. 2020, 2023; Matsumoto et al. 2009; Lee et al. 2022). For example, Wei et al. (2020) investigated the longitudinal changes in cerebral physiological parameters across the adult lifespan in mice using MRI (Fig. 16.11). Specifically, they focused on measuring cerebral blood flow (CBF) in healthy C57BL/6 mice from 3 to 20 months of age, with imaging sessions at monthly intervals. To do this, they superimposed phase-contrast (PC) MRI images on time-of-flight (TOF) angiograms for four major feeding arterioles for visualization (Fig. 16.11a, b), and derived contrast-enhanced images (Fig. 16.11c) and phase velocity maps (Fig. 16.11d) depicting blood flow in these blood vessels. Their results indicated significant increases in CBF with age. This MRI study provided insights into age-related changes in brain physiology in mice, offering normative data on important physiological parameters. Understanding these alterations contributes to our comprehension of neurobiological aging and can inform research on neurodegenerative diseases and cognitive decline in people.

16.3.7 Assessing Intravascular Oxygenation

Another important hemodynamic parameter to monitor in vivo is intravascular oxygen saturation. Either the intrinsic signal from hemoglobin or that from extrinsic

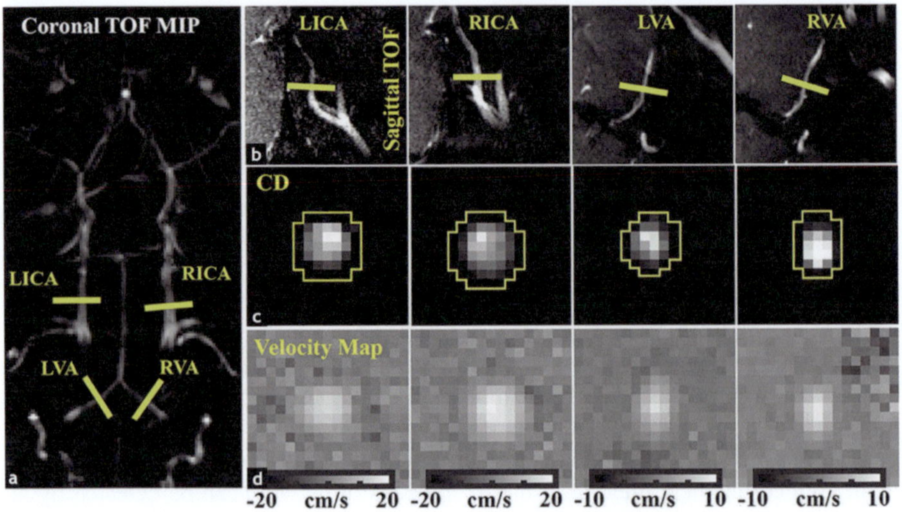

Fig. 16.11 Phase-contrast (PC) magnetic resonance imaging (MRI) of the murine brain microcirculation. (**a**) Coronal time-of-flight (TOF) angiogram aides the visualization of the four major feeding arteries, including left internal carotid artery (LICA), right internal carotid artery (RICA), left vertebral artery (LVA), and right vertebral artery (RVA). Yellow lines indicated locations of vessel cross-sections where PC MRI images were measured. (**b**) Sagittal view of the four vessels in (**a**) with vessel cross-sections superimposed. **(c)** Complex difference (CD) images corresponding to the cross-sections of the four arteries in (**a**). **(d)** Velocity maps corresponding to the cross-sections of the four arteries in (**a**) illustrating differences in their blood flow. (Adapted with permission from Wei et al. 2020)

probes can be used in combination with different imaging modalities to assess intravascular oxygen saturation (SO_2) or the partial pressure of oxygen (PO_2). Below we provide a detailed description for quantifying each of these parameters.

16.3.8 Multiwavelength IOS

IOS imaging at an isosbestic wavelength is effective for visualizing microvascular morphology as previously discussed. When conducted at different wavelengths, IOS can be used to derive intravascular SO_2. This method leverages the wavelength-dependent light absorption of oxyhemoglobin (HbO_2) and deoxyhemoglobin (HbR), from which SO_2 is calculated according to:

$$SO_2 = \frac{C_{HbO_2}}{C_{HbO_2} + C_{HbR}} \tag{16.3}$$

Where C_{HbO_2} is the concentration of oxyhemoglobin and C_{HbR} is the concentration of deoxyhemoglobin, which can be derived from their distinct absorption at visible and NIR wavelengths (Hillman 2007). The absorption coefficient μ_a can be calculated using the modified Beer-Lambert law:

$$\frac{I}{I_0} = \exp(-\mu_a \cdot PL + G) \tag{16.4}$$

Where I is the detected light intensity, I_0 the incident intensity, PL the path-length of light in the tissue, G a geometry-dependent factor, and μ_a the absorption coefficient of the tissue given by:

$$\mu_a = \xi_{HbO_2} C_{HbO_2} + \xi_{HbR} C_{HbR} \tag{16.5}$$

Where ξ is the wavelength-dependent species-specific absorption coefficient, which can be derived from the tabulated molar extinction coefficients for hemoglobin in water. Using these equations, the concentration of the HbO_2 and HbR can be calculated and used in Eq. (16.3) to derive the SO_2 for each pixel in the IOS image.

This method has been used to map the change in SO_2 in many preclinical models of neuronal stimulation. For example, Bouchard et al. (2009) successfully detected changes in the concentration of oxy- and deoxy-hemoglobin during neural stimulation, which were then used to calculate the corresponding changes in SO_2. Most neuroscience applications that employ IOS to map SO_2 make assumptions about the optical path-length (PL) in tissue. Although a number of models have been suggested for estimating PL, each is associated with its own degree of inaccuracy (Wolf et al. 2007). The most widely used approximations are the differential path-length factor (DPF) method and the diffusion approximation approach (Delpy et al. 1989; Arridge et al. 1992). A recent study by Ren et al. (2023b) introduced a new vascular-centric method to derive SO_2, which utilizes the underlying microvascular morphology to calculate the PL. With this approach, the authors circumvented the inaccuracies of conventional PL approximations, and successfully demonstrated that the SO_2 obtained through our technique showed strong spatiotemporal correlations with angiogenesis and osteogenesis during a four-week calvarial bone healing study, which was consistent with current literature reports. In contrast, results from the conventional PL method were poorly correlated with angiogenesis and osteogenesis. In addition, this method has proven invaluable in interrogating the brain's microcirculation during hypoxia and hyperoxia challenges (Fig. 16.12). Using IOS Imaging at 530 nm, the authors first captured the brain's microvascular structure (Fig. 16.12a). A time-course analysis (Fig. 16.12b) using a pulse oximeter tracked the physiological or systemic oxygen saturation (SpO_2) through phases of normoxia, hypoxia (indicated by reduced air flow for 3 min), and recovery, providing a dynamic view of oxygenation changes. SO_2 maps generated during these phases via the vascular-centric approach illustrated that the SO_2 significantly decreased during hypoxia (Fig. 16.12d) compared to normoxia (Fig. 16.12c and e), which also showed excellent agreement with changes in the systemic oxygenation. These results demonstrated that the vascular-centric method combined with multiwavelength IOS

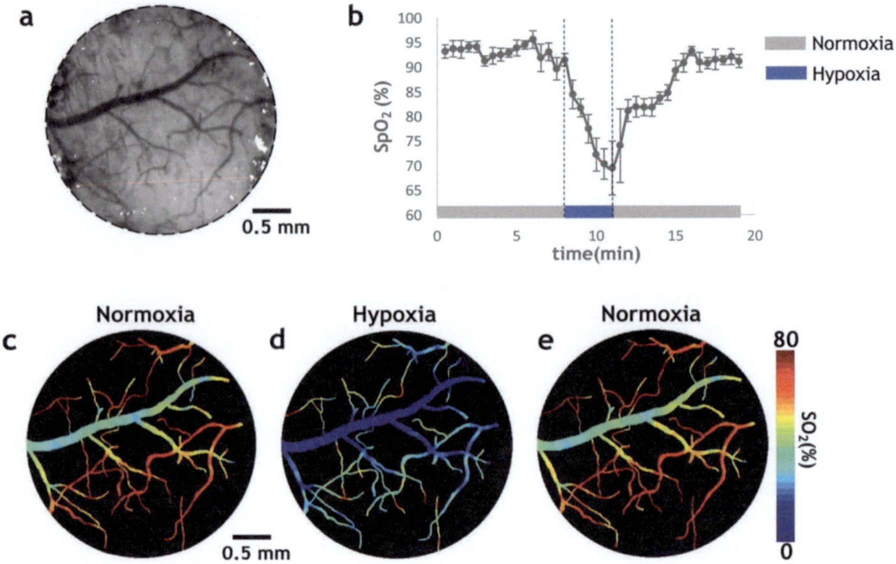

Fig. 16.12 Interrogating changes in cerebrovascular oxygenation during a hypoxic challenge via multiwavelength intrinsic optical signal (IOS) imaging. **(a)** An IOS image showing microvascular morphology of a portion of the murine cortex acquired under 530 nm (isosbestic) illumination. Microvascular morphology extracted from this image was used as input to a vascular-centric intravascular oxygen saturation (SO_2) computation paradigm. (**b**) The changes in systemic intravascular oxygen saturation measured via a pulse oximetry (SpO_2) during the hypoxia paradigm that was induced via reduced air flow as follows: normoxia for 8 min (gray bar); hypoxia, i.e., reduced air flow for 3 min (blue bar); and recovery for 8 min (gray bar). Maps showing the average intravascular SO_2 during: (**c**) normoxia; (**d**) hypoxia; (**e**) and recovery phases. Data in (**b**) are mean ± SD. (Adapted with permission from Ren et al. 2023b)

imaging allowed the characterization of the effect of hypoxia on cerebrovascular oxygenation.

The main advantage of IOS imaging for mapping intravascular SO_2 in vivo is that it is faster and easier to implement compared to the methods described in the ensuing sections (Ma et al. 2016). However, this technique is limited to generating 2D maps of intravascular SO_2 within superficial microvessels (within ~500 μm from the tissue surface (Hillman 2007)). In addition, this method can be susceptible to vascular remodeling that can accompany processes such as angiogenesis during tumor growth or wound healing (Ren et al. 2023b).

16.3.9 Near-Infrared Spectroscopy (NIRS)

Near-infrared spectroscopy (NIRS) belongs to a class of spectroscopy techniques first reported in 1977 (Jobsis 1977) and is also referred to as pulse oximetry. Similar to multiwavelength IOS imaging, NIRS leverages the differences in light absorption between HbO_2 and HbR to assess intravascular oxygenation. Unlike the visible spectrum utilized in IOS, which limits tissue penetration due to scattering and absorption, the NIR spectrum ranges from 700 to 1100 nm, allowing tissue penetration of up to several centimeters (Scheeren et al. 2012). NIRS has even been shown to penetrate bone when conducting trans-cranial cerebral oximetry, which is an important tool for noninvasive imaging of the brain's microcirculation (Barud et al. 2021).

Due to these advantages, NIRS has become one of the most prevalent methods for estimating blood oxygenation in vivo and has been utilized to validate other noninvasive techniques sensitive to intravascular oxygenation, such as blood-oxygen-level-dependent functional magnetic resonance imaging (BOLD-fMRI) (Sakatani et al. 2007). NIRS is most widely used for measuring brain and muscle oxygenation in vivo (Tisdall et al. 2007). It offers high temporal resolution, by acquiring data within seconds, and recent advancements in acquisition rates and real-time image reconstruction have further reduced this to ~160 milliseconds (Franceschini et al. 2000). Although the spatial resolution of NIRS is limited (~cm), complementing NIRS data with images acquired using higher-resolution techniques can provide a more detailed and informative assessments of vascular oxygenation (Pogue and Paulsen 1998). For example, a study by Cheung et al. combined NIRS and diffuse correlation spectroscopy to measure microvascular oxygenation, cerebral blood flow and hemoglobin concentration in the rat brain (Cheung et al. 2001). They were able to comprehensively monitor the regional microvascular dynamics during a hypercapnia challenge, revealing a 16% increase in oxygenation in the cortical vessels. These findings demonstrated the capability of NIRS to monitor changes in microvascular oxygenation in response to physiological stimuli, and showed the utility of integrating NIRS with other imaging modalities for studying microvascular physiology.

16.3.10 Photoacoustic Imaging (PAI)

Photoacoustic imaging (PAI) is a hybrid method that combines the advantages of optical and ultrasound imaging methods. It combines the high CNR and spatial resolution of optical imaging with the sensitivity and depth penetration of ultrasound imaging (Qiu et al. 2021). PAI utilizes a short laser pulse to illuminate its target which absorbs the light. The absorbed optical energy is then converted into heat and induces a transient rise in the local temperature. This increase in temperature results in a thermo-elastic expansion, leading to a local pressure elevation and the emission of an acoustic wave. (Li et al. 2018) Using this principle, PAI can detect the acoustic signal generated by optical absorption from endogenous tissue chromophores, including hemoglobin, melanin, and water; or exogenous contrast agents (e.g., nanoparticles and organic dyes) (Xia et al. 2014; Wu et al. 2014). PAI can be used to measure intravascular oxygenation based on different biophysical models (Li et al.

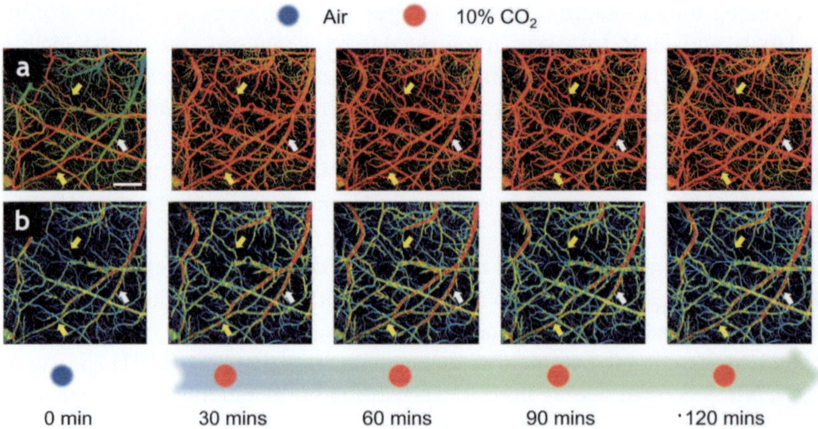

Fig. 16.13 Imaging changes in oxygen saturation (SO_2) and blood flow during a 120-minute hypercapnic challenge with multiparametric photoacoustic microscopy (PAM). Frames from continuous multiparametric PAM over 2 h of severe hypercapnia induced by the inhalation of 10% CO_2, resulted in an increase in (**a**) venous sO_2 (white arrows), while the arterial sO_2 (yellow arrows) remained unchanged. (**b**) The blood flow increased in both venous (white arrows) and arterial (yellow arrows) vessels. These parameters remained significantly elevated compared to baseline throughout the 2 h of hypercapnia. Arrows in yellow and white highlight alterations in arterioles and venules, respectively. Scale bar = 500 μm. (Adapted with permission from (Cao et al. 2021)

2018). These PAI methods either employ a linear model (Cox et al. 2012) that uses multiwavelength light absorption by HbO_2 and HbR just as in the previous two optical methods; or an optical transport model based method, (Kim et al. 2011) which estimates the optical fluence distribution in tissue; or an acoustic spectrum based model (Guo et al. 2012) that uses the acoustic frequency spectrum of the received PA signals at different wavelengths to quantify the absolute absorption coefficient based on Beer's law. PAI is also capable of fluorescence lifetime based pO_2 measurements, (Shao et al. 2013) where the excitation-state lifetime of an oxygen-sensitive fluorescent dye is used to derive the local pO_2.

PAI has been applied to assess the blood volume and SO_2 of the microvasculature in a wide variety of tissues, including the brain, (Rich and Seshadri 2015; Zhu et al. 2022) liver, (Karmacharya et al. 2020) bone, (Menger et al. 2022) and in tissue engineering applications (Shrestha et al. 2020). It is also a promising approach for tracking in vivo changes in tumor oxygenation (Nasri et al. 2023). An elegant example of this is shown in Fig. 16.13, in which cortex-wide, multiparametric photoacoustic microscopy (PAM) was used to characterize the impact of hypercapnia on cerebral hemodynamics and oxygen metabolism in the mouse brain (Cao et al. 2021). Through continuous multiparametric imaging spanning a 2-hour period of severe hypoxia, notable enhancements in venous SO_2 (Fig. 16.13a) and blood flow (Fig. 16.13b) were observed. These metrics remained elevated compared to baseline values throughout the 2-hour exposure to 10% CO_2. With its ability to map physiological changes in SO_2 and blood flow over large fields of view, PAI can be used to monitor complex dynamics within biological systems, offering invaluable insights for both basic research and clinical applications.

16.3.11 Two-Photon Phosphorescence Lifetime Microscopy (2PLM)

Two-photon phosphorescence lifetime microscopy (2PLM) is an advanced optical imaging technique that enables the measurement of the oxygen partial pressure (pO_2) within the microvasculature, offering high spatial (~0.5 μm) and temporal (~μs-ms) resolution. It combines the principles of two-photon excitation microscopy and phosphorescence lifetime measurements. In 2PLM, the excitation of phosphorescent probes by two-photon absorption results in light emission, whose lifetime is affected by the presence of oxygen (Abbasizadeh and Spencer 2021). This is due to a phenomenon known as quenching, where oxygen molecules interact with the excited phosphorescent probe, transferring energy away and thus shortening their emission lifetime. By quantifying the phosphorescence emission lifetime, 2PLM can provide a direct measure of oxygen concentration around the phosphorescent probe (Sakadzic et al. 2016). Different probes have been designed for 2PLM oxygen imaging, including PtP-C343 (Lecoq et al. 2011) and Oxyphor 2P (Esipova et al. 2019). These probes can be administered either into the bloodstream or directly into the tissue, making 2PLM versatile for imaging both, intravascular and tissue oxygenation.

For example, Fig. 16.14 shows the in vivo two-photon high-resolution pO_2 measurements in mouse cortical microvasculature via the integration of an optimized

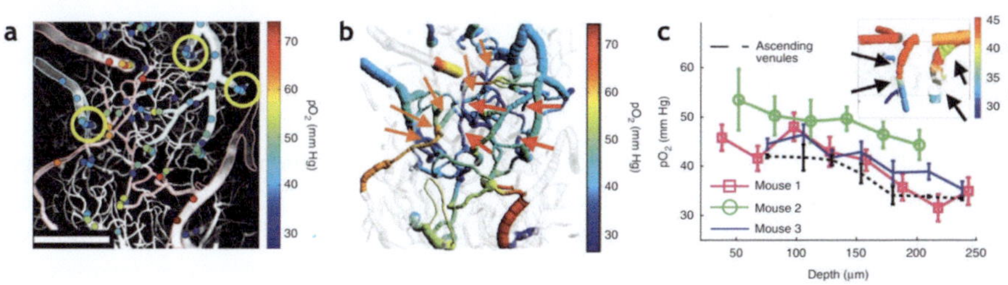

Fig. 16.14 Assessing partial oxygen pressure (pO_2) in cortical microvessels with two-photon phosphorescence lifetime microscopy (2PLM). (**a**) pO_2 measured in the cortical microvasculature in different vascular branches (colored dots indicate different pO_2) overlaid on a maximum intensity projection image of the microvascular structure (gray scale). Major pial arterioles and venules are outlined in red and blue, respectively. Scale bar = 200 μm. (b) Composite image showing pO_2 measurements in capillaries at a depth of 240 μm from the brain's surface (red arrows), and consecutive branches of the microvascular tree from a pial arteriole, to capillaries, to an ascending venule (orange arrows). (**c**) The dependence of pO_2 on cortical depth in three mice, excluding pial arterioles. Error bars represent the standard error. Inset shows a composite image of microvascular morphology and pO_2, with arrows indicating two ascending venules. (Adapted with permission from Sakadzic et al. 2010)

2PLM system with a two-photon-enhanced phosphorescent nanoprobe (Sakadzic et al. 2010). This method enabled measurements of pO_2 in individual capillary segments at depths of up to 250 μm with subsecond temporal resolution (Fig. 16.14a, b). The study demonstrated a consistent decrease in vascular pO_2 with increasing cortical depth (Fig. 16.14c), and an increase in pO_2 toward the cortical surface in ascending venules, suggesting potential oxygen efflux from tissue into the venous compartment. In addition to numerous studies conducted in the cerebral microvessels, (Lecoq et al. 2011) 2PLM has also been applied in other types of tissue including the retinal (Sencan et al. 2018) and intestinal capillaries (Sencan et al. 2021). The continued advancements in 2PLM technology hold great promise to facilitate our understanding of microvascular oxygenation in various tissue types.

16.4 Translational and Clinical Imaging of Microvascular Morphology and Hemodynamics

In contrast to preclinical imaging, clinical or translational imaging of the human microcirculation prioritizes noninvasiveness, patient safety, repeatability, and specificity. Moreover, the use of contrast agents or dyes in clinical settings is subject to stringent regulatory requirements (e.g., by the Food and Drug Administration of the United States, i.e., the FDA) and safety profiles, to minimize potential risks to patients and to ensure widespread utility. Despite these challenges, many preclinical imaging modalities have been successfully translated into the clinic. In this section, we will briefly introduce key clinical imaging modalities that are currently employed to assess microvascular morphology and hemodynamics in patients.

16.4.1 Optical Imaging

Many optical imaging methods described in the preclinical section have been translated and utilized into the clinical space. For example, LSC imaging has been used as an intraoperative imaging method to inform surgical decisions and improve surgical outcomes. It has provided real-time monitoring of blood flow in organs such as the liver, (Eriksson et al. 2014) stomach (Milstein et al. 2016), as well as during tumor resections (Richards et al. 2014) and revascularization surgeries (Hecht et al. 2009). In stroke patients undergoing decompressive hemicraniectomy, LSC has provided continuous observation of cortical hemodynamics to gain insights into cortical spreading depression (CSD) patterns (Woitzik et al. 2013) that may arise in the brain. In addition, FL imaging has played a crucial role in oncological surgeries, such as partial nephrectomies, where it aids in the identification and clamping of arteries feeding tumors to reduce blood loss (Tobis et al. 2012). Figure 16.15 illustrates the use of indocyanine green video angiography (ICG-VA) to assess

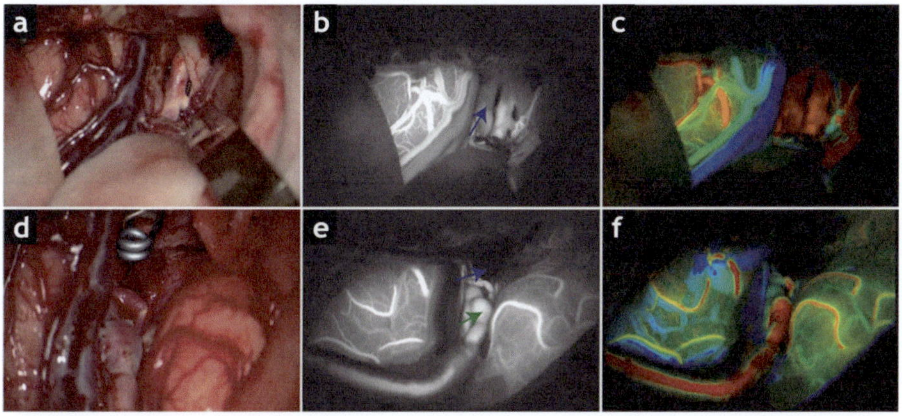

Fig. 16.15 Indocyanine green video angiography (ICG-VA) in cerebrovascular surgery. (**a**) Surgical exposure of an aneurysm, followed by (**b**) ICG-VA revealing the aneurysm, i.e., an abnormal bulge in the blood vessel wall (blue arrow), parent artery, and peripheral blood vessels. (**c**) A color-coded map generated using FLOW 800 software in which brain vessels are labeled with distinct colors post-ICG injection: arteries in red; cortical capillaries in yellow/green; and veins in blue. This color map illustrates the blood flow and degree of patency of the blood vessels within the FoV. (**d**) Following high-flow extracranial-to-intracranial (EC-IC) bypass (middle cerebral artery M3 segment-radial artery-ICA), the aneurysm was isolated at the initial part of the ICA and before the posterior communicating artery (PComA) branching. (**e**) Subsequent ICG-VA, and (**f**) FLOW 800 analysis showed resolution of the aneurysm (blue arrow) and fluent bypass vessel (green arrow). (Adapted with permission from Xue et al. 2021)

changes in cerebral blood flow during cerebrovascular surgery (Xue et al. 2021). Through the use of ICG-VA during cerebrovascular surgery, surgeons can enhance the visualization and management of complex vascular conditions. For example, an aneurysm (an abnormal bulge in the blood vessel wall) (Fig. 16.15a–c) and its resolution after surgery (Fig. 16.15d, e) were clearly visualized using ICG-VA, and subsequent flow analysis of the ICG tracer with specialized software confirmed treatment of the aneurysm.

In addition, by assuming intact neurovascular coupling and using microvascular-scale blood volume changes as a surrogate of neuronal activity, IOS imaging has been instrumental for functional brain mapping during neurosurgical procedures (Sato et al. 2017). Moreover, in a study involving 30 anesthetized patients, IOS imaging was able to successfully map brain function in 26 cases, offering new insights into the hierarchy of sensory response patterns within the human primary somatosensory cortex and demonstrating the technique's utility in accurately guiding neurosurgical decisions (Sato et al. 2017). Additionally, the concept of IOS has been applied to develop a clinical technique known as video-capillaroscopy, which is utilized to examine microvasculature in areas such as the nailfold (Lim et al. 2023) and oral mucosa (Scardina et al. 2009) of patients. For instance, in a study conducted by Abdelmaksoud et al., (Abdelmaksoud et al. 2022) Nailfold video-capillaroscopy (NVC) was used to noninvasively identify microvascular changes in patients with adolescent type I diabetes (T1D). It revealed significantly higher alterations in the microvasculature of adolescents with T1D compared to controls and therefore holds great promise for early assessment of T1D.

Another optical method that has been used extensively in the clinic is near-infrared spectroscopy (NIRS), which stands out for its noninvasive nature and portable device platforms, making it an invaluable tool for patient monitoring at the bedside. NIRS has proven especially useful in studying prematurely born infants afflicted with brain lesions, apnea, or undeveloped cerebrovascular autoregulation (Gagnon and Macnab 2005).

With their high spatial and temporal resolutions, these optical methods provide invaluable insights into the microvascular architecture and dynamics, capturing structural and functional details that are not visible with other methods such as MRI, US imaging, and CT. Despite these advantages, the application of optical imaging in clinical setting is constrained by its limited tissue penetration. Nonetheless, optical methods are usually simpler to use, cost-effective, and hold tremendous potential for miniaturization (Taruttis and Ntziachristos 2012; Rege et al. 2018). The evolution of clinical optical imaging technologies continues, which enhances the possibilities for novel noninvasive diagnostics and targeted surgical interventions in the future.

16.4.2 Positron Emission Tomography (PET) and Computed Tomography (CT)

As indicated by its name, Positron Emission Tomography (PET) involves imaging based on the detection of gamma radiation from the emission of positrons. It is a nuclear imaging technique that is mainly used for evaluating physiological processes rather than anatomy and involves the administration of small and safe amounts of radioactive material, called a radiotracer to the patient. It can be used to quantify blood flow as a measure of an organ's microcirculation and detect disease at the metabolic level. (Anand et al. 2009) To facilitate a more comprehensive diagnosis, PET images are often combined with those from other imaging modalities, such as CT, to enhances the visualization of areas of interest by correlating metabolic activity with anatomical location (van der Vaart et al. 2008). PET has been utilized for evaluating microvascular perfusion and detecting microvascular diseases. For example, perfusion cardiac PET is the most clinically advanced and widespread method for measuring myocardial blood flow (MBF) and myocardial blood flow reserve (MBFR) (Ziadi 2017). It has also been used to detect vascular plaque, imaging abdominal aortic aneurysms (AAA), vascular graft infection, and imaging vasculitis and aortitis (Liddy et al. 2020). For example, Fig. 16.16 shows a study that investigates the characterization of coronary artery plaques using PET-CT and radiofrequency intravascular ultrasound imaging, highlighting the difference between (Konerding 1991)F-sodium fluoride ((Konerding 1991)F-NaF, a

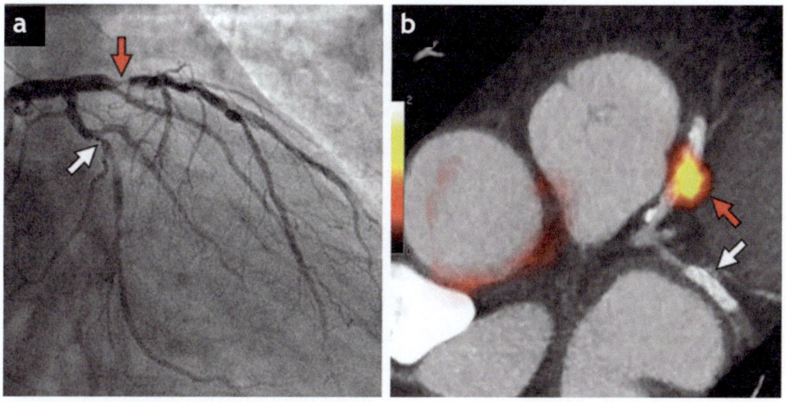

Fig. 16.16 Characterization of coronary artery plaques via combined positron emission tomography and computed tomography (PET-CT). (**a**) CT coronary angiography of a patient with myocardial infarction identified the location of the infarction (i.e., obstruction in blood flow, red arrow) and a nearby normal vessel (white arrow). (**b**) Electrocardiograph-gated PET was used to concurrently capture the uptake of a radiotracer, 18F-sodium fluoride (18F-NaF), after which images were reconstructed and fused with the CT coronary angiogram to form a PET-CT image. Only the constricted vessel (red arrow) exhibited increased 18F-NaF uptake on the PET-CT image, which was indicative of the presence of high-risk coronary plaque. This dual-modality approach demonstrates the potential of using PET-CT to identify high-risk plaques in blood vessels and inform treatment. (Adapted with permission from Joshi et al. 2014)

positron-emitting radiopharmaceutical used for imaging) positive and negative plaques in terms of their composition and associated risk factors for coronary artsery disease (Joshi et al. 2014). Fig. 16.16a illustrates the utility of CT-based coronary angiography in identifying a diseased blood vessel in a patient with coronary artery disease. Moreover, the concurrently acquired PET image clearly shows increased 18F-sodium fluoride (18F-NaF) uptake (Fig. 16.16b) indicative of the presence of high-risk coronary plaques, compared to a nearby normal vessel.

16.4.3 Ultrasound (US)

US imaging for studying the microcirculation utilizes the Doppler shift principle as introduced in the previous section. It employs high-frequency sound waves (~2–18 MHz in conventional methods (Guerraty et al. 2021) and 20–50 MHz for high-frequency devices (Goertz et al. 2003)) that are backscattered by RBCs, to visualize blood flow and evaluate microvascular functionality. US imaging often involves the use of microbubbles as an ultrasound contrast agent (UCA), which can act as echo-enhancers or therapeutic agents (Lee et al. 2017). Recent clinical advancements in cancer diagnostics have highlighted the efficacy of microbubbles, but their relatively large hydrodynamic radius (1–8 μm) has limited their applications in microvascular research (Perera et al. 2015). Currently, the field has developed bubbles with smaller radii termed "nanobubbles," which are nanoscale (~200–300 nm) ultrasound contrast agents (Exner and Kolios 2021). These smaller bubbles can move beyond the microvasculature into interstitial spaces, making this a promising technique for imaging the tumor microenvironment (Pellow et al. 2018). In addition, nanobubbles can reach higher densities in microvessels and therefore provide more interaction sites for cell-specific targeting for better imaging and therapeutic outcomes (Helfield et al. 2021). Despite the promising attributes of nano-UCAs, several challenges have impeded their seamless integration into clinical practice. These challenges include maintaining sufficient echogenicity within the clinically relevant ultrasound frequency range of 1–10 MHz and developing cost and time-efficient production processes for these nanoparticle formulations (Exner and Kolios 2021). At present, despite their potential, the clinical applications of nanobubbles still face challenges related to echogenicity, production costs, and scalability within the clinically relevant ultrasound frequency range (Helfield et al. 2021). However, ongoing research is successfully addressing many of these challenges, and creating utility for nanobubbles as effective diagnostic and therapeutic agents, broadening the scope of ultrasound imaging in clinical microvascular research and pushing the boundary for ultrasound-guided therapies (Exner and Kolios 2021).

16.4.4 Magnetic Resonance Imaging (MRI)

The principles of MRI were covered in a prior section (refer to **Preclinical Imaging of Microvascular Hemodynamics**). Although clinical MRI has traditionally exhibited low spatial resolution (>200 μm), recent advances originating from preclinical research have been actively translated into clinical applications. For example, innovative techniques such as "MR fingerprinting" allow the simultaneous measurement of multiple tissue properties in a single acquisition, (Panda et al. 2017) and the introduction of new contrast agents broadens MRI's application beyond the traditional. (Minami et al. 2014) For example, a recent study utilized MRI to derive an ultra-high-resolution angiography at 140 μm isotropic resolution and enabled the in-depth characterization of pial vascular structure and function in the human brain (Fig. 16.17) (Bollmann et al. 2022). The study leveraged the noninvasive time-of-flight magnetic resonance angiography (TOF-MRA), which uses the high signal intensity caused by inflowing water protons in the blood to generate contrast without the need for exogenous contrast agents. The arterial and venous vessels were differentiated based on their T2* values, wherein a higher concentration of deoxygenated hemoglobin in nonarterial vessels, like large pial veins and sinuses, leads to shorter T2* values (Fig. 16.17a). This method enabled the identification and subsequent removal of venous structures from the angiogram, helping to isolate the arterial vasculature for clearer image segmentation (Fig. 16.17b). These emerging applications show the potential of MRI to contribute to our understanding of complex biological systems and diseases, reinforcing its status as a powerful and

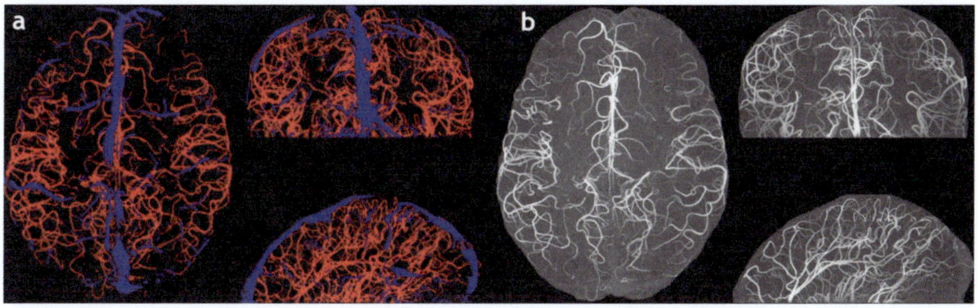

Fig. 16.17 Separation of microvascular subtypes in the brain via magnetic resonance imaging (MRI). The brain of a healthy human subject was imaged at 140 μm resolution on a 7 Tesla (T) MRI scanner. (**a**) Brain arteries (red) and veins (blue) were segmented on the basis of their T2∗ estimates. Since veins have higher deoxygenated hemoglobin concentrations, they have much shorter T2 ∗ s than the highly oxygenated arteries, and therefore, this difference in T2∗ relaxation times was used for segmentation of these two vessel types. (**b**) The segmented veins were then removed from the time-of-flight (TOF) MRI angiogram acquired over the same field of view. (Adapted with permission from Bollmann et al. 2022)

evolving tool in the diagnosis and study of conditions involving the microcirculation.

16.5 Summary

In summary, we have provided a comprehensive overview of techniques for imaging microvascular morphology and hemodynamics, ranging from the preclinical setting to translational and clinical applications. Each method's unique advantages, limitations, and practical applications were discussed, highlighting their suitability for distinct applications. We recommend that readers weigh multiple considerations before selecting the most appropriate imaging method for their application. These considerations include whether the primary focus is to assess microvascular morphology or function (i.e., hemodynamics); the spatial and temporal resolution needed; the accuracy and repeatability required; the need for absolute or relative measurements; alongside concerns regarding invasiveness, safety, and the field of view or coverage required. In addition, complex experiments that demand the assessment of multiple microvascular parameters may benefit from a multimodality approach that combines different imaging techniques to gain a more comprehensive understanding. We believe that continuous advancements in imaging technologies hold great promise to improve our understanding of the microcirculation, guide future research and enhance diagnostic and therapeutic strategies for patients with diseases that involve the microcirculation.

Acknowledgments This work was supported by NIH/NCI grant nos. 5R01CA196701-08 and 5R01CA237597-05.

References

Abbasizadeh N, Spencer JA (2021) Two-photon phosphorescence lifetime microscopy. Adv Exp Med Biol 3233:63–82

Abdelmaksoud AA, Daifallah SM, Salah NY, Saber AS (2022) Nail fold microangiopathy in adolescents with type 1 diabetes: relation to diabetic vascular complications. Microcirculation 29(6–7):e12771

Abian O, Vega S, Neira JL, Velazquez-Campoy A (2020) Chapter 17 – high-throughput screening for intrinsically disordered proteins by using biophysical methods. In: Pey AL (ed) Protein homeostasis diseases. Academic Press, pp 359–387

Aharinejad SH, Lametschwandtner A (1992) Corrosion casting – a historical review. In: Aharinejad SH, Lametschwandtner A (eds) Microvascular corrosion casting in scanning electron microscopy: techniques and applications. Springer Vienna, Vienna, pp 3–11

Anand SS, Singh H, Dash AK (2009) Clinical applications of PET and PET-CT. Med J Armed Forces India 65(4):353–358

Arridge SR, Cope M, Delpy DT (1992) The theoretical basis for the determination of optical pathlengths in tissue: temporal and frequency analysis. Phys Med Biol 37(7):1531–1560

Augustinaite S, Kuhn B (2021) Intrinsic optical signal imaging and targeted injections through a chronic cranial window of a head-fixed mouse. STAR Protocols 2(3):100779

Baluk P, McDonald DM (2008) Markers for microscopic imaging of lymphangiogenesis and angiogenesis. Ann N Y Acad Sci 1131:1–12

Barud M, Dabrowski W, Siwicka-Gieroba D et al (2021) Usefulness of cerebral oximetry in TBI by NIRS. J Clin Med 10(13)

Bearden SE, Segal SS (2005) Neurovascular alignment in adult mouse skeletal muscles. Microcirculation 12(2):161–167

Bederson JB, Germano IM, Guarino L (1995) Cortical blood flow and cerebral perfusion pressure in a new noncraniotomy model of subarachnoid hemorrhage in the rat. Stroke 26(6):1086–1091

Benninger RKP, Piston DW (2013) Two-photon excitation microscopy for the study of living cells and tissues. Curr Protoc Cell Biol 59

Bert M, Sabrina L, Marco D et al (2008) High-resolution tomographic imaging of microvessels. ProcSPIE:70780B

Bhargava A, Monteagudo B, Kushwaha P et al (2022) VascuViz: a multimodality and multiscale imaging and visualization pipeline for vascular systems biology. Nat Methods 19(2):242–254

Biswas L, Chen J, De Angelis J et al (2023) Lymphatic vessels in bone support regeneration after injury. Cell 186(2):382–397 e324

Bogaert L, Scheller D, Moonen J et al (2000) Neurochemical changes and laser Doppler flowmetry in the endothelin-1 rat model for focal cerebral ischemia. Brain Res 887(2):266–275

Bollmann S, Mattern H, Bernier M et al (2022) Imaging of the pial arterial vasculature of the human brain in vivo using high-resolution 7T time-of-flight angiography. elife 11:11

Bouchard MB, Chen BR, Burgess SA, Hillman EMC (2009) Ultra-fast multispectral optical imaging of cortical oxygenation, blood flow, and intracellular calcium dynamics. Opt Express 17(18):15670–15678

Bozzone DM, Whittemore S (2021) The circulatory system. Infobase Holdings, Inc., New York

Briers D, Duncan DD, Hirst E et al (2013) Laser speckle contrast imaging: theoretical and practical limitations. J Biomed Opt 18(6):066018

Cao R, Tran A, Li J et al (2021) Hemodynamic and oxygen-metabolic responses of the awake mouse brain to hypercapnia revealed by multi-parametric photoacoustic microscopy. J Cereb Blood Flow Metab 41(10):2628–2639

Carmeliet P, Jain RK (2011) Molecular mechanisms and clinical applications of angiogenesis. Nature 473(7347):298–307

Carovac A, Smajlovic F, Junuzovic D (2011) Application of ultrasound in medicine. Acta Inform Med 19(3):168–171

Chen Z, Ross JL, Hambardzumyan D (2019) Intravital 2-photon imaging reveals distinct morphology and infiltrative properties of glioblastoma-associated macrophages. Proc Natl Acad Sci USA 116(28):14254–14259

Cheung C, Culver JP, Takahashi K, Greenberg JH, Yodh AG (2001) In vivo cerebrovascular measurement combining diffuse near-infrared absorption and correlation spectroscopies. Phys Med Biol 46(8):2053–2065

Cox B, Laufer JG, Arridge SR, Beard PC (2012) Quantitative spectroscopic photoacoustic imaging: a review. J Biomed Opt 17(6):061202

Debbage PL, Griebel J, Ried M et al (1998) Lectin intravital perfusion studies in tumor-bearing-mice: micrometer-resolution, wide-area mapping of microvascular labeling, distinguishing efficiently and inefficiently perfused microregions in the tumor. J Histochem Cytochem 46(5): 627–639

Delpy DT, Arridge SR, Cope M et al (1989) Quantitation of pathlength in optical spectroscopy. Adv Exp Med Biol 248:41–46

Dencks S, Schmitz G (2023) Ultrasound localization microscopy. Z Med Phys 33(3):292–308

Denis L, Bodard S, Hingot V et al (2023) Sensing ultrasound localization microscopy for the visualization of glomeruli in living rats and humans. EBioMedicine 91:104578

Duncan DD, Kirkpatrick SJ (2008) Can laser speckle flowmetry be made a quantitative tool? J Opt Soc Am A Opt Image Sci Vis 25(8):2088–2094

Elliott AD (2020) Confocal microscopy: principles and modern practices. Curr Protoc Cytom 92(1): e68

Eriksson S, Nilsson J, Lindell G, Sturesson C (2014) Laser speckle contrast imaging for intraoperative assessment of liver microcirculation: a clinical pilot study. Med Devices (Auckl) 7:257–261

Errico C, Pierre J, Pezet S et al (2015) Ultrafast ultrasound localization microscopy for deep super-resolution vascular imaging. Nature 527(7579):499–502

Esipova TV, Barrett MJP, Erlebach E et al (2019) Oxyphor 2P: a high-performance probe for deep-tissue longitudinal oxygen imaging. Cell Metab 29(3):736–744

Ettinger A, Wittmann T (2014) Fluorescence live cell imaging. Methods Cell Biol 123:77–94

Exner AA, Kolios MC (2021) Bursting microbubbles: how Nanobubble contrast agents can enable the future of medical ultrasound molecular imaging and image-guided therapy. Curr Opin Colloid Interface Sci 54:54

Folkman J (1971) Tumor angiogenesis: therapeutic implications. N Engl J Med 285(21):1182–1186

Foreman DM, Bagley S, Moore J et al (1996) Three dimensional analysis of the retinal vasculature using immunofluorescent staining and confocal laser scanning microscopy. Br J Ophthalmol 80(3):246–251

Franceschini MA, Toronov V, Filiaci M, Gratton E, Fantini S (2000) On-line optical imaging of the human brain with 160-ms temporal resolution. Opt Express 6(3):49–57

Gagnon RE, Macnab AJ (2005) Near infrared spectroscopy (NIRS) in the clinical setting—an adjunct to monitoring during diagnosis and treatment. Spectroscopy 19:951895

Girkin JM, Carvalho MT (2018) The light-sheet microscopy revolution. J Opt 20(5):053002

Glaser AK, Bishop KW, Barner LA et al (2022) A hybrid open-top light-sheet microscope for versatile multi-scale imaging of cleared tissues. Nat Methods 19(5):613–619

Goertz DE, Yu JL, Kerbel RS, Burns PN, Foster FS (2003) High-frequency 3-D color-flow imaging of the microcirculation. Ultrasound Med Biol 29(1):39–51

Grover VP, Tognarelli JM, Crossey MM et al (2015) Magnetic resonance imaging: principles and techniques: lessons for clinicians. J Clin Exp Hepatol 5(3):246–255

Guerraty M, Bhargava A, Senarathna J, Mendelson AA, Pathak AP (2021) Advances in translational imaging of the microcirculation. Microcirculation 28(3):e12683

Guevara E, Sadekova N, Girouard H, Lesage F (2013) Optical imaging of resting-state functional connectivity in a novel arterial stiffness model. Biomed Opt Express 4(11):2332–2346

Guo Z, Favazza C, Garcia-Uribe A, Wang LV (2012) Quantitative photoacoustic microscopy of optical absorption coefficients from acoustic spectra in the optical diffusive regime. J Biomed Opt 17(6):066011

He Q, Arroyo ED, Smukowski SN et al (2019) Critical period inhibition of NKCC1 rectifies synapse plasticity in the somatosensory cortex and restores adult tactile response maps in fragile X mice. Mol Psychiatry 24(11):1732–1747

Hecht N, Woitzik J, Dreier JP, Vajkoczy P (2009) Intraoperative monitoring of cerebral blood flow by laser speckle contrast analysis. Neurosurg Focus 27(4):E11

Helfield B, Zou Y, Matsuura N (2021) Acoustically-stimulated Nanobubbles: opportunities in medical ultrasound imaging and therapy. Front Phys 9:9

Hillman EMC (2007) Optical brain imaging in vivo: techniques and applications from animal to man. J Biomed Opt 12(5):051402

Hillman EMC, Devor A, Bouchard MB et al (2007) Depth-resolved optical imaging and microscopy of vascular compartment dynamics during somatosensory stimulation. NeuroImage 35(1): 89–104

Hillman EMC, Voleti V, Li W, Yu H (2019) Light-sheet microscopy in neuroscience. Annu Rev Neurosci 42:295–313

Honeycutt SE, O'Brien LL (2021) Injection of Evans blue dye to fluorescently label and image intact vasculature. BioTechniques 70(3):181–185

Huang C, Ness VP, Yang X et al (2015) Spatiotemporal analyses of osteogenesis and angiogenesis via Intravital imaging in cranial bone defect repair. J Bone Miner Res 30(7):1217–1230

Ignee A, Atkinson NS, Schuessler G, Dietrich CF (2016) Ultrasound contrast agents. Endosc Ultrasound 5(6):355–362

Jobsis FF (1977) Noninvasive, infrared monitoring of cerebral and myocardial oxygen sufficiency and circulatory parameters. Science 198(4323):1264–1267

Johnson KE, Wilgus TA (2014) Vascular endothelial growth factor and angiogenesis in the regulation of cutaneous wound repair. Adv Wound Care (New Rochelle) 3(10):647–661

Johnston DG, Denizet M, Mostany R, Portera-Cailliau C (2013) Chronic in vivo imaging shows no evidence of dendritic plasticity or functional remapping in the Contralesional cortex after stroke. Cereb Cortex 23(4):751–762

Jonkman J, Brown CM, Wright GD, Anderson KI, North AJ (2020) Tutorial: guidance for quantitative confocal microscopy. Nat Protoc 15(5):1585–1611

Joshi NV, Vesey AT, Williams MC et al (2014) 18F-fluoride positron emission tomography for identification of ruptured and high-risk coronary atherosclerotic plaques: a prospective clinical trial. Lancet 383(9918):705–713

Karmacharya MB, Sultan LR, Kirkham BM et al (2020) Photoacoustic imaging for assessing tissue oxygenation changes in rat hepatic fibrosis. Diagnostics (Basel) 10(9)

Kennan RP, Zhong J, Gore JC (1994) Intravascular susceptibility contrast mechanisms in tissues. Magn Reson Med 31(1):9–21

Khouri K, Xie DF, Crouzet C et al (2021) Simple methodology to visualize whole-brain microvasculature in three dimensions. Neurophotonics. 8(2):025004

Kim S, Chen YS, Luke GP, Emelianov SY (2011) In vivo three-dimensional spectroscopic photoacoustic imaging for monitoring nanoparticle delivery. Biomed Opt Express 2(9): 2540–2550

Kolesova H, Olejnickova V, Kvasilova A, Gregorovicova M, Sedmera D (2021) Tissue clearing and imaging methods for cardiovascular development. iScience 24(4):102387

Konerding MA (1991) Scanning electron microscopy of corrosion casting in medicine. Scanning Microsc 5(3):851–865

Kornfield TE, Newman EA (2015) Measurement of retinal blood flow using fluorescently labeled red blood cells. eNeuro 2(2):ENEURO.0005

Kumar BS, Khot A, Chakravarthy VS, Pushpavanam S (2021) A network architecture for bidirectional neurovascular coupling in rat whisker barrel cortex. Front Comput Neurosci 15:638700

Lawson ND, Weinstein BM (2002) In vivo imaging of embryonic vascular development using transgenic zebrafish. Dev Biol 248(2):307–318

Lecoq J, Parpaleix A, Roussakis E et al (2011) Simultaneous two-photon imaging of oxygen and blood flow in deep cerebral vessels. Nat Med 17(7):893–898

Lee H, Kim H, Han H et al (2017) Microbubbles used for contrast enhanced ultrasound and theragnosis: a review of principles to applications. Biomed Eng Lett 7(2):59–69

Lee D, Le TT, Im GH, Kim SG (2022) Whole-brain perfusion mapping in mice by dynamic BOLD MRI with transient hypoxia. J Cereb Blood Flow Metab 42(12):2270–2286

Li Y, Song Y, Zhao L et al (2008) Direct labeling and visualization of blood vessels with lipophilic carbocyanine dye DiI. Nat Protoc 3(11):1703–1708

Li M, Tang Y, Yao J (2018) Photoacoustic tomography of blood oxygenation: a mini review. Photoacoustics. 10:65–73

Liddy S, Mallia A, Collins CD et al (2020) Vascular findings on FDG PET/CT. Br J Radiol 93(1113):20200103

Lim MWS, Setjiadi D, Dobbin SJH et al (2023) Nailfold video-capillaroscopy in the study of cardiovascular disease: a systematic review. Blood Press Monit 28(1):24–32

Lochhead JJ, Williams EI, Reddell ES et al (2023) High resolution multiplex confocal imaging of the neurovascular unit in health and experimental ischemic stroke. Cells 12(4)

Ma Y, Shaik MA, Kim SH et al (2016) Wide-field optical mapping of neural activity and brain haemodynamics: considerations and novel approaches. Philos Trans R Soc Lond Ser B Biol Sci 371(1705):20150360

Mano T, Albanese A, Dodt HU et al (2018) Whole-brain analysis of cells and circuits by tissue clearing and light-sheet microscopy. J Neurosci 38(44):9330–9337

Mano T, Grutman T, Ilovitsh T (2023) Versatile ultrasound-compatible microfluidic platform for in vitro microvasculature flow research and imaging optimization. ACS Omega 8(50): 47667–47677

Matsumoto S, Yasui H, Batra S et al (2009) Simultaneous imaging of tumor oxygenation and microvascular permeability using Overhauser enhanced MRI. Proc Natl Acad Sci USA 106(42): 17898–17903

Mauritzon S, Ginstman F, Hillman J, Wardell K (2022) Analysis of laser Doppler flowmetry long-term recordings for investigation of cerebral microcirculation during neurointensive care. Front Neurosci 16:1030805

McDonald DM, Choyke PL (2003) Imaging of angiogenesis: from microscope to clinic. Nat Med 9(6):713–725

Mendez A, Rindone AN, Batra N et al (2018) Phenotyping the microvasculature in critical-sized Calvarial defects via multimodal optical imaging. Tissue Eng Part C Methods 24(7):430–440

Menger MM, Körbel C, Bauer D et al (2022) Photoacoustic imaging for the study of oxygen saturation and total hemoglobin in bone healing and non-union formation. Photoacoustics 28: 100409

Metzger ZVI, Basrani B, Goodis HE (2011) Chapter 8 – instruments, materials, and devices. In: Hargreaves KM, Cohen S (eds) Cohen's pathways of the pulp, 10th edn. Mosby, St. Louis, pp 223–282

Milgroom A, Ralston E (2016) Clearing skeletal muscle with CLARITY for light microscopy imaging. Cell Biol Int 40(4):478–483

Miller DR, Jarrett JW, Hassan AM, Dunn AK (2017) Deep tissue imaging with multiphoton fluorescence microscopy. Curr Opin Biomed Eng 4:32–39

Milstein DMJ, Ince C, Gisbertz SS et al (2016) Laser speckle contrast imaging identifies ischemic areas on gastric tube reconstructions following esophagectomy. Medicine (Baltimore) 95(25): e3875

Minami Y, Yagyu Y, Murakami T, Kudo M (2014) Tracking navigation imaging of Transcatheter arterial chemoembolization for hepatocellular carcinoma using three-dimensional cone-beam CT angiography. Liver Cancer 3(1):53–61

Mohamud Yusuf A, Hagemann N, Schulten S et al (2020) Light sheet microscopy using FITC-albumin followed by immunohistochemistry of the same rehydrated brains reveals ischemic brain injury and early microvascular remodeling. Front Cell Neurosci 14:625513

Monfared A, Blevins NH, Cheung EL et al (2006) In vivo imaging of mammalian cochlear blood flow using fluorescence microendoscopy. Otol Neurotol 27(2):144–152

Muller TB, Haraldseth O, Unsgard G (1994) Characterization of the microcirculation during ischemia and reperfusion in the penumbra of a rat model of temporary middle cerebral artery occlusion: a laser Doppler flowmetry study. Int J Microcirc Clin Exp 14(5):289–295

Murakami T, Fujita T, Hinenoya H (1992) A modified method for vascular casting and scanning electron microscopy: its utility in the demonstration of rat pancreatic insulo-acinar and other blood vascular systems. Arch Histol Cytol 55(Suppl):233–238

Murari K, Li N, Rege A et al (2007) Contrast-enhanced imaging of cerebral vasculature with laser speckle. Appl Opt 46(22):5340–5346

Nagahara R, Onda N, Yamashita S et al (2018) Fluorescence tumor imaging by i.v. administered indocyanine green in a mouse model of colitis-associated colon cancer. Cancer Sci 109(5): 1638–1647

Nakano A, Kondo R, Kaneko Y et al (2020) Changes in components of the neurovascular unit in the retina in a rat model of retinopathy of prematurity. Cell Tissue Res 379(3):473–486

Nakase H, Kempski OS, Heimann A, Takeshima T, Tintera J (1997) Microcirculation after cerebral venous occlusions as assessed by laser Doppler scanning. J Neurosurg 87(2):307–314

Nasri D, Manwar R, Kaushik A, Er EE, Avanaki K (2023) Photoacoustic imaging for investigating tumor hypoxia: a strategic assessment. Theranostics 13(10):3346–3367

Ohtani O, Kikuta A, Ohtsuka A, Taguchi T, Murakami T (1983) Microvasculature as studied by the microvascular corrosion casting/scanning electron microscope method. I Endocrine and digestive system. Arch Histol Jpn 46(1):1–42

Panda A, Mehta BB, Coppo S et al (2017) Magnetic resonance fingerprinting-an overview. Curr Opin Biomed Eng 3:56–66

Pathak AP, Hochfeld WE, Goodman SL, Pepper MS (2008) Circulating and imaging markers for angiogenesis. Angiogenesis 11(4):321–335

Pellow C, Goertz DE, Zheng G (2018) Breaking free from vascular confinement: status and prospects for submicron ultrasound contrast agents. Wiley Interdiscip Rev Nanomed Nanobiotechnol 10(4):e1502

Peng X, Carhuapoma JR, Bhardwaj A et al (2002) Suppression of cortical functional hyperemia to vibrissal stimulation in the rat by epoxygenase inhibitors. Am J Physiol Heart Circ Physiol 283(5):H2029–H2037

Peng X, Zhang C, Alkayed NJ, Harder DR, Koehler RC (2004) Dependency of cortical functional hyperemia to forepaw stimulation on epoxygenase and nitric oxide synthase activities in rats. J Cereb Blood Flow Metab 24(5):509–517

Perera RH, Hernandez C, Zhou H et al (2015) Ultrasound imaging beyond the vasculature with new generation contrast agents. Wiley Interdiscip Rev Nanomed Nanobiotechnol 7(4):593–608

Peterson DA (2010) Confocal Microscopy. In: Kompoliti K, Metman LV (eds) Encyclopedia of movement disorders. Academic Press, Oxford, pp 250–252

Pogue BW, Paulsen KD (1998) High-resolution near-infrared tomographic imaging simulations of the rat cranium by use of a priori magnetic resonance imaging structural information. Opt Lett 23(21):1716–1718

Pouratian N, Toga AW (2002) Chapter 5 – optical imaging based on intrinsic signals. In: Toga AW, Mazziotta JC (eds) Brain mapping: the methods, 2nd edn. Academic Press, San Diego, pp 97–140

Qiu T, Lan Y, Gao W et al (2021) Photoacoustic imaging as a highly efficient and precise imaging strategy for the evaluation of brain diseases. Quant Imaging Med Surg 11(5):2169–2186

Raposio E, Bertozzi N, Moretti R, Grignaffini E, Grieco MP (2017) Laser Doppler Flowmetry and transcutaneous oximetry in chronic skin ulcers: a comparative evaluation. Wounds 29(7): 190–195

Refaat A, Yap ML, Pietersz G et al (2022) In vivo fluorescence imaging: success in preclinical imaging paves the way for clinical applications. J Nanobiotechnology 20(1):450

Rege A et al (2012) Optical imaging of microvascular morphology and perfusion. *Current.* Angiogenesis 1(3):243–260

Rege A, Cunningham SI, Liu Y et al (2018) Noninvasive assessment of retinal blood flow using a novel handheld laser speckle contrast imager. Transl Vis Sci Technol 7(6):7

Reitan NK, Thuen M, Goa PE, de Lange DC (2010) Characterization of tumor microvascular structure and permeability: comparison between magnetic resonance imaging and intravital confocal imaging. J Biomed Opt 15(3):036004

Ren Y, Senarathna J, Grayson WL, Pathak AP (2022) State-of-the-art techniques for imaging the vascular microenvironment in craniofacial bone tissue engineering applications. Am J Physiol Cell Physiol 323(5):C1524–C1538

Ren Y, Chu X, Senarathna J et al (2023a) Multimodality imaging reveals angiogenic evolution in vivo during calvarial bone defect healing. Angiogenesis 27:105

Ren Y, Senarathna J, Chu X, Grayson WL, Pathak AP (2023b) Vascular-centric mapping of in vivo blood oxygen saturation in preclinical models. Microvasc Res 148:104518

Renaudin N, Demené C, Dizeux A et al (2022) Functional ultrasound localization microscopy reveals brain-wide neurovascular activity on a microscopic scale. Nat Methods 19(8): 1004–1012

Rich LJ, Seshadri M (2015) Photoacoustic imaging of vascular hemodynamics: validation with blood oxygenation level-dependent MR imaging. Radiology 275(1):110–118

Richards LM, Towle EL, Fox DJ Jr, Dunn AK (2014) Intraoperative laser speckle contrast imaging with retrospective motion correction for quantitative assessment of cerebral blood flow. Neurophotonics 1(1):015006

Rindone AN, Liu X, Farhat S et al (2021) Quantitative 3D imaging of the cranial microvascular environment at single-cell resolution. Nat Commun 12(1):6219

Robertson RT, Levine ST, Haynes SM et al (2015) Use of labeled tomato lectin for imaging vasculature structures. Histochem Cell Biol 143(2):225–234

Rosengarten B, Lutz H, Hossmann KA (2003) A control system approach for evaluating somatosensory activation by laser-Doppler flowmetry in the rat cortex. J Neurosci Meth 130(1):75–81

Runnels JM, Zamiri P, Spencer JA et al (2006) Imaging molecular expression on vascular endothelial cells by in vivo immunofluorescence microscopy. Mol Imaging 5(1):31–40

Sakadzic S, Roussakis E, Yaseen MA et al (2010) Two-photon high-resolution measurement of partial pressure of oxygen in cerebral vasculature and tissue. Nat Methods 7(9):755–759

Sakadzic S, Yaseen MA, Jaswal R et al (2016) Two-photon microscopy measurement of cerebral metabolic rate of oxygen using periarteriolar oxygen concentration gradients. Neurophotonics. 3(4):045005

Sakatani K, Murata Y, Fujiwara N et al (2007) Comparison of blood-oxygen-level – dependent functional magnetic resonance imaging and near-infrared spectroscopy recording during functional brain activation in patients with stroke and brain tumors. J Biomed Opt 12(6): 062110–062118

Sato K, Nariai T, Momose-Sato Y, Kamino K (2017) Intraoperative intrinsic optical imaging of human somatosensory cortex during neurosurgical operations. Neurophotonics 4(3):031205

Scardina GA, Ruggieri A, Messina P (2009) Oral microcirculation observed in vivo by videocapillaroscopy: a review. J Oral Sci 51(1):1–10

Schaffer CB, Friedman B, Nishimura N et al (2006) Two-photon imaging of cortical surface microvessels reveals a robust redistribution in blood flow after vascular occlusion. PLoS Biol 4(2):e22

Scheeren TW, Schober P, Schwarte LA (2012) Monitoring tissue oxygenation by near infrared spectroscopy (NIRS): background and current applications. J Clin Monit Comput 26(4): 279–287

Schueth A, Hildebrand S, Samarska I et al (2023) Efficient 3D light-sheet imaging of very large-scale optically cleared human brain and prostate tissue samples. Commun Biol 6(1):170

Senarathna J, Kovler M, Prasad A et al (2022) In vivo phenotyping of the microvasculature in necrotizing enterocolitis with multicontrast optical imaging. Microcirculation 29(6–7):e12768

Sencan I, Esipova TV, Yaseen MA et al (2018) Two-photon phosphorescence lifetime microscopy of retinal capillary plexus oxygenation in mice. J Biomed Opt 23(12):1–9

Sencan I, Zong W, Alfadhel M et al (2021) Quantifying intestinal capillary oxygenation using two-photon phosphorescence lifetime microscopy. FASEB J 35(S1)

Shao Q, Morgounova E, Jiang C et al (2013) In vivo photoacoustic lifetime imaging of tumor hypoxia in small animals. J Biomed Opt 18(7):076019

Shelton SE, Stone J, Gao F, Zeng D, Dayton PA (2020) Microvascular ultrasonic imaging of angiogenesis identifies tumors in a murine spontaneous breast cancer model. Int J Biomed Imaging 2020:7862089

Shrestha B, DeLuna F, Anastasio MA, Yong Ye J, Brey EM (2020) Photoacoustic imaging in tissue engineering and regenerative medicine. Tissue Eng Part B Rev 26(1):79–102

Song P, Rubin JM, Lowerison MR (2023) Super-resolution ultrasound microvascular imaging: is it ready for clinical use? Z Med Phys 33(3):309–323

Spangenberg P, Hagemann N, Squire A et al (2023) Rapid and fully automated blood vasculature analysis in 3D light-sheet image volumes of different organs. Cell Rep Methods 3(3):100436

Stelzer EHK, Strobl F, Chang B-J et al (2021) Light sheet fluorescence microscopy. Nature Reviews Methods Primers 1(1):73

Stroh M, Zimmer JP, Duda DG et al (2005) Quantum dots spectrally distinguish multiple species within the tumor milieu in vivo. Nat Med 11(6):678–682

Sutherland BA, Rabie T, Buchan AM (2014) Laser Doppler flowmetry to measure changes in cerebral blood flow. Methods Mol Biol 1135:237–248

Szekanecz Z, Koch AE (2007) Mechanisms of disease: angiogenesis in inflammatory diseases. Nat Clin Pract Rheum 3(11):635–643

Taruttis A, Ntziachristos V (2012) Translational optical imaging. AJR Am J Roentgenol 199(2): 263–271

Thurston G, Murphy TJ, Baluk P, Lindsey JR, McDonald DM (1998) Angiogenesis in mice with chronic airway inflammation: strain-dependent differences. Am J Pathol 153(4):1099–1112

Tisdall MM, Tachtsidis I, Leung TS, Elwell CE, Smith M (2007) Near-infrared spectroscopic quantification of changes in the concentration of oxidized cytochrome c oxidase in the healthy human brain during hypoxemia. J Biomed Opt 12(2):024002

Tobis S, Knopf JK, Silvers CR et al (2012) Near infrared fluorescence imaging after intravenous indocyanine green: initial clinical experience with open partial nephrectomy for renal cortical tumors. Urology 79(4):958–964

Tozer GM, Ameer-Beg SM, Baker J et al (2005) Intravital imaging of tumour vascular networks using multi-photon fluorescence microscopy. Adv Drug Deliv Rev 57(1):135–152

Unthank JL, Lash JM, Nixon JC, Sidner RA, Bohlen HG (1993) Evaluation of carbocyanine-labeled erythrocytes for microvascular measurements. Microvasc Res 45(2):193–210

van der Vaart MG, Meerwaldt R, Slart RH et al (2008) Application of PET/SPECT imaging in vascular disease. Eur J Vasc Endovasc Surg 35(5):507–513

Vasquez B, Campos B, Cao A, Theint AT, Zeiger W (2023) High-sensitivity intrinsic optical signal imaging through flexible, low-cost adaptations of an upright microscope. eNeuro. 10(8): ENEURO.0046

Verli FD, Rossi-Schneider TR, Schneider FL, Yurgel LS, de Souza MA (2007) Vascular corrosion casting technique steps. Scanning 29(3):128–132

Villringer A, Them A, Lindauer U, Einhaupl K, Dirnagl U (1994) Capillary perfusion of the rat brain cortex. an in vivo confocal microscopy study. Circ Res 75(1):55–62

Wagner RC, Czymmek K, Hossler FE (2006) Confocal microscopy, computer modeling, and quantification of glomerular vascular corrosion casts. Microsc Microanal 12(3):262–268

Wang K, Pan Y, Tong S, Liang H, Qiu P (2021) Deep-skin multiphoton microscopy of lymphatic vessels excited at the 1700-nm window in vivo. Biomed Opt Express 12(10):6474–6484

Wei Z, Chen L, Hou X et al (2020) Age-related alterations in brain perfusion, venous oxygenation, and oxygen metabolic rate of mice: a 17-month longitudinal MRI study. Front Neurol 11:559
Wei Z, Liu H, Lin Z et al (2023) Non-contrast assessment of blood-brain barrier permeability to water in mice: an arterial spin labeling study at cerebral veins. NeuroImage 268:119870
Woitzik J, Hecht N, Pinczolits A et al (2013) Propagation of cortical spreading depolarization in the human cortex after malignant stroke. Neurology 80(12):1095–1102
Wolf M, Ferrari M, Quaresima V (2007) Progress of near-infrared spectroscopy and topography for brain and muscle clinical applications. J Biomed Opt 12(6):062104–062114
Wood R, Bassett K et al (2012) 1.5 tesla magnetic resonance imaging scanners compared with 3.0 tesla magnetic resonance imaging scanners: systematic review of clinical effectiveness. CADTH Technol Overv 2(2):e2201
Wu D, Huang L, Jiang MS, Jiang H (2014) Contrast agents for photoacoustic and thermoacoustic imaging: a review. Int J Mol Sci 15(12):23616–23639
Xia J, Yao J, Wang LV (2014) Photoacoustic tomography: principles and advances. Electromagn Waves (Camb) 147:1–22
Xue T, Deng R, Gao B et al (2021) Intraoperative indocyanine green video angiography (ICG-VA) with FLOW 800 software in complex intracranial aneurysm surgery. Chin Neurosurg J 7(1):28
Zagorchev L, Oses P, Zhuang ZW et al (2010) Micro computed tomography for vascular exploration. J Angiogenes Res 2:7
Zhu X, Huang Q, DiSpirito A et al (2022) Real-time whole-brain imaging of hemodynamics and oxygenation at micro-vessel resolution with ultrafast wide-field photoacoustic microscopy. Light Sci Appl 11(1):138
Zhu J, Liu X, Xu J et al (2023) Protocol for fine casting, imaging, and analysis of murine vascular networks with VALID. STAR Protoc 4(3):102441
Ziadi MC (2017) Myocardial flow reserve (MFR) with positron emission tomography (PET)/computed tomography (CT): clinical impact in diagnosis and prognosis. Cardiovasc Diagn Ther 7(2):206–218
Zimmerhackl B, Parekh N, Kucherer H, Steinhausen M (1985) Influence of systemically applied angiotensin II on the microcirculation of glomerular capillaries in the rat. Kidney Int 27(1): 17–24

Chapter 17
Biomimetic Models for Investigation of Microvascular Remodeling

Nien-Wen Hu, Arinola Lampejo, Ariana D. Suarez-Martinez, Nicholas Hodges, Mir Md Nasim Hossain, Peter Balogh, Bing Ren, Yong Huang, Jonathan J. Adorno, Shashwat S. Agarwal, Jacob C. Holter, Elton Y. Cao, Jonathan W. Song, Dorothy N. Beck, Shayn M. Peirce, and Walter L. Murfee

Abstract The challenges in investigating the cellular and system dynamics involved in microvascular remodeling motivates the development of models to recapitulate complex interactions. Both experimental and computational models offer biomimetic tools to incorporate, observe, and probe dynamics not always

---

The original version of the chapter has been revised. A correction to this chapter can be found at https://doi.org/10.1007/978-3-031-99609-2_18

N.-W. Hu · A. Lampejo · A. D. Suarez-Martinez · N. Hodges · W. L. Murfee (✉)
Department of Biomedical Engineering, University of Florida, Gainesville, FL, USA
e-mail: wmurfee@bme.ufl.edu

M. M. N. Hossain · P. Balogh
Department of Mechanical and Industrial Engineering, New Jersey Institute of Technology, Newark, NJ, USA

B. Ren · Y. Huang
Department of Mechanical and Aerospace Engineering, University of Florida, Gainesville, FL, USA

J. J. Adorno · J. C. Holter
Department of Biomedical Engineering, The Ohio State University, Columbus, OH, USA

S. S. Agarwal
Department of Chemical and Biomolecular Engineering, The Ohio State University, Columbus, OH, USA

E. Y. Cao
Department of Mechanical and Aerospace Engineering, The Ohio State University, Columbus, OH, USA

J. W. Song
Department of Mechanical and Aerospace Engineering, The Comprehensive Cancer Center, The Ohio State University, Columbus, OH, USA

D. N. Beck · S. M. Peirce
Department of Biomedical Engineering, University of Virginia, Charlottesville, VA, USA

R. F. Tuma et al. (eds.), *Handbook of Microcirculation*, Physiology in Health and Disease, https://doi.org/10.1007/978-3-031-99609-2_17

possible with in vivo methods. The objective of this chapter is to overview examples of experimental and computational modeling approaches that frame field leading research, applications, and opportunities. Major cellular players involved in microvascular growth will be introduced. Then, the need for integrative models will be emphasized by mapping challenges to the field of tissue engineering and providing specific research examples that define the range of approaches.

Keywords Microvascular remodeling · Angiogenesis · Lymphangiogenesis · Biomimetic model · Tissue engineering · Computational modeling

17.1 Introduction

The ability to mechanistically probe cell–cell interactions during physiological processes offers a powerful tool for scientific discovery and has motivated an area of research focused on biomimetic model development (Lampejo et al. 2022). In general, the goal of developing both experimental and computational biomimetic models is to mimic the complexity of an in vivo environment and allow researchers to study specific cellular, molecular, and environmental mechanisms in ways not possible with typical animal studies. In the context of microvascular network remodeling, which involves a variety of cell dynamics in a local tissue environment, one can appreciate the challenge (Lampejo et al. 2022). Angiogenesis, defined as the growth of new blood vessels from pre-existing vessels, is a multicellular process that involves the temporal and spatial coordination between endothelial cells, pericytes, nerves, growth factors, and macrophages (Stapor et al. 2013). This challenge is amplified when considering the coordination with other cell dynamics involved in system remodeling associated with lymphangiogenesis (Sweat et al. 2014) and neurogenesis (Hodges et al. 2020).

The objective of this chapter is to overview examples of experimental and computational modeling approaches that frame field leading research, applications, and opportunities. To establish the complexity of the model development challenge, major cellular players involved in angiogenesis and lymphangiogenesis will be introduced. We then aim to emphasize the need for integrative models by suggesting overlaps with the challenges that also motivate the field of tissue engineering, and provide specific research examples from leading microvascular research laboratories to define the scope of modeling approaches.

17.2 Players Involved in Microvascular Growth and Remodeling, the Needs for Biomimetic Models, and Overlaps with Tissue Engineering

Appreciation for the cellular players involved in microvascular growth and remodeling is realized when viewing an intact network (Fig. 17.1) (Stapor et al. 2013). In the blood vascular system, endothelial cells and perivascular cells are

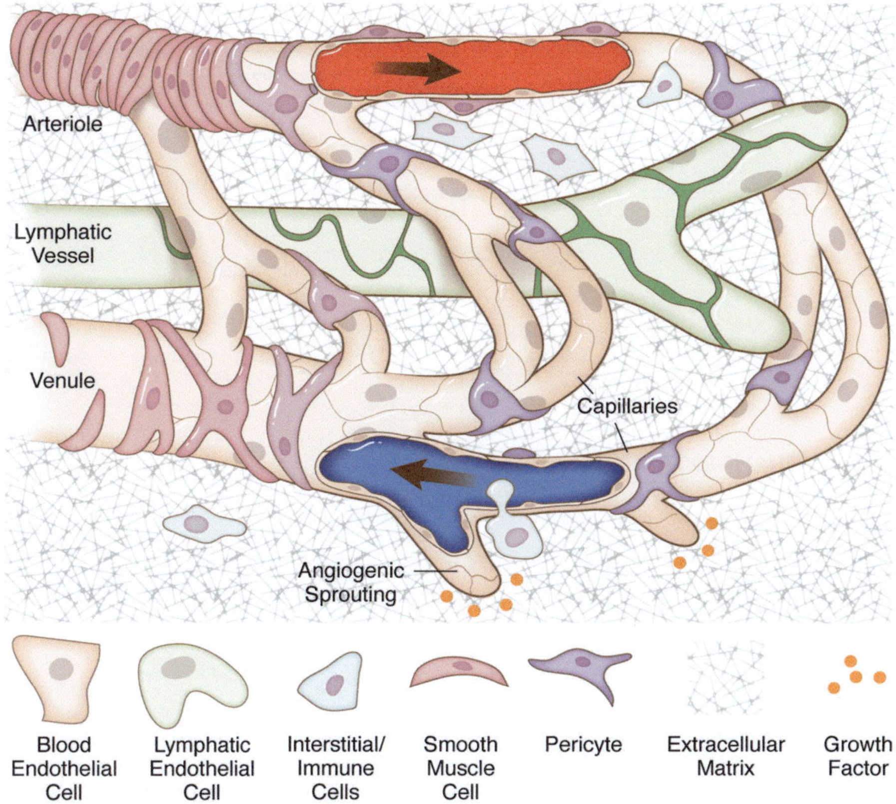

Fig. 17.1 Microvascular network cellular players. The microvascular system is comprised of distinct vessel types, including arterioles, capillaries, venules, and lymphatic vessels. Each vessel is constructed from specialized endothelial cells, accompanied by perivascular support cells such as smooth muscle cells and pericytes, which vary in morphology and function based on vessel type. Replicating the complexity of these networks during angiogenesis and lymphangiogenesis requires the integration of diverse cell types, organized into system-relevant patterns, alongside precise chemical cues, vessel perfusion, and a supportive matrix environment. Vascular remodeling and homeostasis depend on multicellular interactions and the coordination of cellular morphology and phenotypes, highlighting the importance of understanding cell–environment interactions for tissue engineering and the development of biomimetic microvascular models

present across the different vessel types: arterioles, capillaries, and venules (Peirce and Skalak 2003). The differences in cell morphologies and phenotypes correlate with vessel specific functions. For example, smooth muscle cells wrap tightly around arterioles, which is a characteristic morphology that aids in the ability of arterioles to change their diameter and preform vasoregulatory functions (Motherwell et al. 2017). Along blood capillaries, pericytes are present and tend to exhibit more elongated morphology (Stapor et al. 2014) For the lymphatic system, in contrast, initial lymphatic vessels, also referred to as the lymphatic capillaries, do not have any perivascular cell wrapping. In a microvascular network environment, there are also immune cells, and other interstitial cell populations such as fibroblasts, and stem

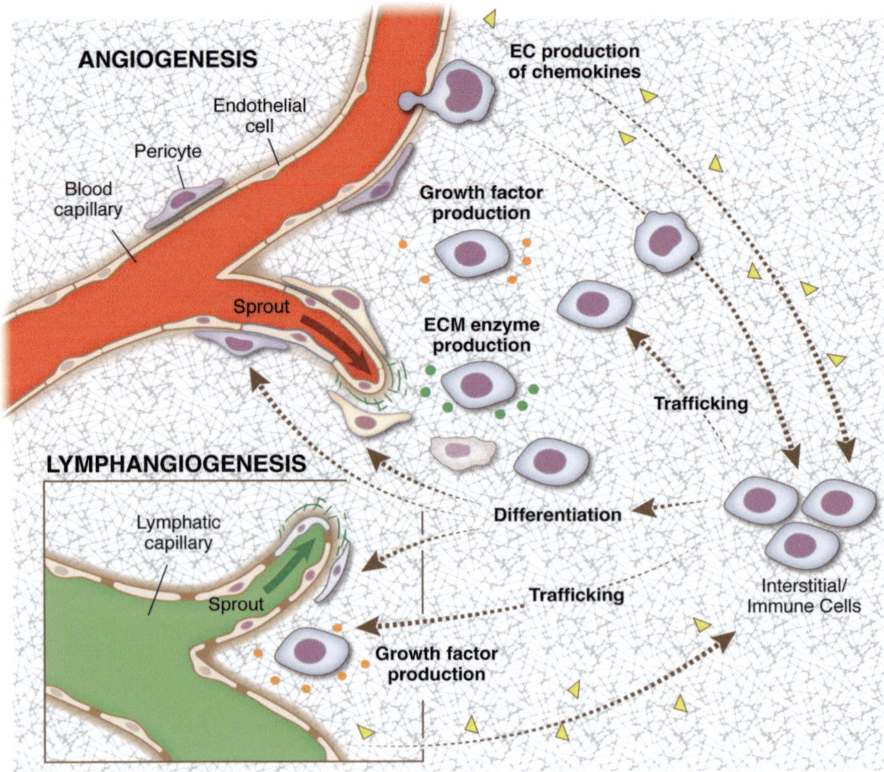

Fig. 17.2 Multicellular and system-level dynamics in angiogenesis and lymphangiogenesis. Angiogenesis and lymphangiogenesis involve complex interactions between multiple cell types and system-level processes. Capillary sprouting during vessel growth incorporates cell differentiation, migration, recruitment, proliferation, growth factor production, responses to fluid flow, and manipulation of the extracellular matrix. Interstitial and immune cells play a critical role in microvascular growth and neovascularization. These cells contribute through behaviors such as phagocytosis, trafficking, differentiation, proliferation, secretion of growth factors, enzymes, and chemokines, as well as extracellular matrix remodeling. These processes make interstitial and immune cells key targets for therapeutic modulation of vascular remodeling. These multiscale dynamics, overlap with processes like neurogenesis and underscore the need for physiologically relevant models to explore the interplay of cell and system-level interactions during microvascular and lymphatic development

cells, as well as extracellular matrix, chemical cues, and mechanics cues. The addition of processes like angiogenesis and lymphangiogenesis within this environment further highlights how the integration of coordinated interactions can have a profound impact on vascular development (Fig. 17.2). The fluid dynamics of blood flowing through microvascular networks and resulting hemodynamic transport characteristics and shear stresses add yet another complexity which influence such development. Other key dynamics include cell production of chemokines, cell trafficking, cell migration, cell recruitment, proliferation, and matrix degradation.

Given the various cell types and complex multicellular interactions involved in microvascular network growth and remodeling, understanding of the coordination between these processes requires a comprehensive view that allows for probing of one player or players and evaluating system-level responses. This complexity highlights the need for more physiologically relevant models. From a modeling and engineering perspective, critical questions are provoked. How much complexity is needed? How can we integrate the complexity? The answers depend on the scientific objective and the desired readouts. At a fundamental level, a model needs to include the different aspects of the real, physiologically relevant microenvironment including appropriate cell incorporation and patterning. The modeling challenges of recreating cellular and vessel spatiotemporal patterns is similar to the challenges faced when creating functional microvascular networks in tissue engineered constructs (Lampejo et al. 2022). Consequently, biomimetic development and tissue engineering research overlap and share a similar motivation with different applications.

17.3 Experimental Models

Angiogenesis and microvascular remodeling are dynamic and complex processes that span across length and time scales (Peirce and Skalak 2003). Experimental models can provide exquisitely detailed information about gene and protein expression (now at the single-cell level, thanks to techniques like single-cell RNA-seq), insights into cell–cell and cell–matrix interactions, and comprehensive descriptions of tissue-level function, including hemodynamics and oxygen delivery. However, experimental models also have significant limitations. For one, single-cell "-omics" data acquired from cultured cells and those extracted from living tissues are limited to snap shots in time. These data exclusively reflect the status of the cell at the instant the cell was removed from its environment. Hence, dynamic changes in gene expression, protein expression, metabolism, etc. at the level of the individual cell and within the context of its extracellular environment are impossible to deduce using current approaches, yet advances in the complexity incorporation are moving the field closer to realizing these readouts. Experimental approaches include cell culture models, microfluidic platforms, and tissue culture methods (Fig. 17.3). In vitro models typically consist of cells (monocultures or cocultures) in a controlled 2D or 3D microenvironment (Lampejo et al. 2022; Duval et al. 2017). 2D models keep the cultured cells parallel to the surface of the culture plate (Duval et al. 2017), while 3D models have the cells embedded in a matrix where the cells can move in all three directional planes (Lampejo et al. 2022). Microfluidic devices build on the complexity of basic 2D models by implementing microcirculatory network arrangement and flow (Motherwell and Murfee 2018). Platforms can integrate three matrix regions with cells to mimic the interfaces between microvascular and tissue resident spaces. Microfluidic based studies can be categorized as investigating dynamics associated with angiogenesis or microrheology (Fig. 17.4). Ex vivo tissue culture

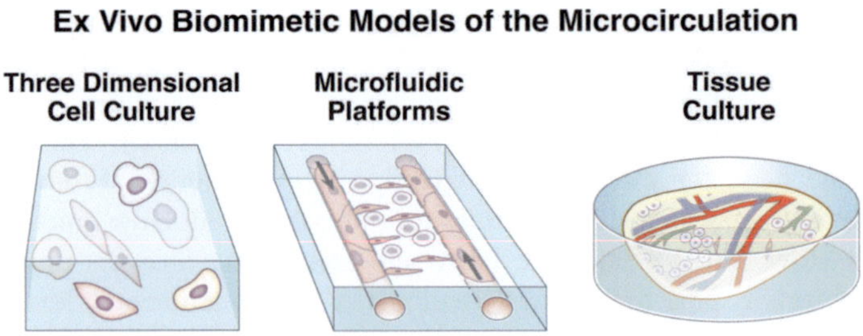

Fig. 17.3 Overview of experimental biomimetic models. 3D cell culture models use matrix scaffolds to host one to three cell types, providing a controlled environment for studying cell behavior and interactions. Microfluidic systems integrate fluidic channels lined with cells and matrix regions embedded with other cell types, enabling investigations into migration dynamics, responses to chemical gradients, and mechanical forces such as shear stress and interstitial flow. Tissue culture models attempt to bridge the gap between in vitro and in vivo systems, offering enhanced physiological relevance

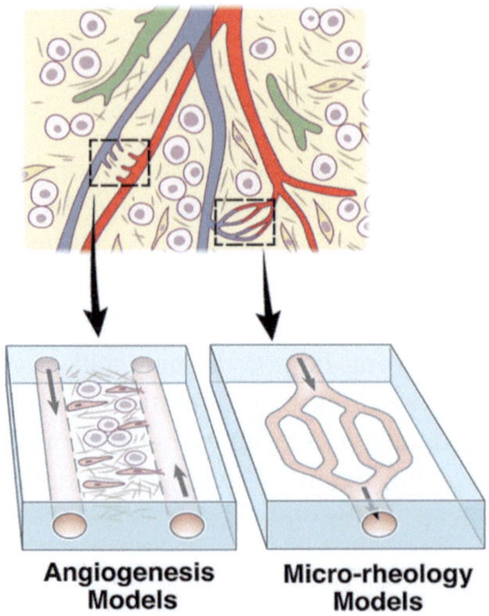

Fig. 17.4 Microfluidic platforms serve as powerful tools for exploring angiogenesis and microrheology. These systems allow for precise manipulation of the microenvironment within transparent setups, enabling real-time observation of vascular processes at the cellular level. In angiogenesis research, they are utilized to study the formation of new vessels, analyze the role of growth factors, assess extracellular matrix remodeling, and investigate the influence of mechanical forces like interstitial flow and shear stress. Additionally, microfluidic devices are used to examine how cell–cell interactions, such as those between endothelial cells and supporting cells, contribute to vascular development. In the context of blood disorders, these platforms support investigations into endothelial barrier dysfunction, including increased permeability, red blood cell damage, and blockage of small blood vessels

Experimental Models	Capabilities						
	Real Network	Angiogenesis	Lymph-angiogenesis	Perivascular Cells	Time-lapse	Flow	Native Micro-environment
2D Cell Culture	X	√	X	X	√	X	X
3D Cell Culture	X	√	X	√	√	X	X
Aortic Ring Assay	X	√	X	√	√	X	√
Microfluidics Device	X	√	√	X	√	√	X

Fig. 17.5 Comparison of experimental biomimetic models for studying microvascular remodeling. In vitro models, such as microfluidic systems, excel in providing precise control over the microenvironment, enabling time-lapse imaging, single-cell tracking, and, in some cases, perfusion. However, these models face limitations in replicating the complexity of intact microvascular networks and native multicellular environments. The rat mesentery culture model addresses some of these challenges by allowing time-lapse imaging of capillary sprouting and interactions between pericytes and endothelial cells within a native-like microvascular network. Despite their strengths, all in vitro models must contend with the broader challenge of replicating the diverse design criteria and physiological dynamics found in vivo systems, highlighting the ongoing need for improved biomimetic designs

models utilize an established physiological environment to monitor system-wide responses to changes in stimuli (Sweat et al. 2014). While each common model type simulates many of the biomimetic model needs that relate to physiological or pathological relevance (Fig. 17.5) and the incorporation of additional players are advancing the field, the gap between models and the real scenario continues to motivate the development of new approaches and validation studies in comparison to in vivo animal experiments. To highlight applications and opportunities, the subsequent section overviews examples of microfluidic and tissue culture modeling.

17.3.1 Example: Microfluidic Microvascular Modeling

Vessel-on-a-chip systems are recognized as important research tools as they have greatly contributed to advancing our understanding of the role of fluid forces (Song and Munn 2011; Helm et al. 2005), extracellular matrix (ECM) composition (Lugo-Cintrón et al. 2020), tissue stiffness (Bordeleau et al. 2017), and pathological conditions (Reid et al. 2017) in regulating vascular function. Numerous strategies have been employed for engineering microfabricated blood vessels in vitro. Examples include micropatterning based on PDMS (Akbari et al. 2017), stereolithography (Wang et al. 2020; Grigoryan et al. 2019), extrusion 3D printing (Hauser et al. 2021), xurography (Stallcop et al. 2018), and machine milling (Guckenberger et al. 2015), for reproducing vascular geometry and structures in vitro (Wong et al. 2020). These microfabrication techniques all serve the same goal of encapsulating endothelialized vessels within a platform that can: (1) recreate the complex structure and function of blood vessels of the microcirculation, (2) allow for the study of vascular biology-related mechanisms, and (3) enable facile experimental manipulation and analysis (Fleischer et al. 2020; Moses et al. 2021).

In addition to fabrication strategies, a key consideration for vessel-on-a-chip systems is the selection of ECM type. Several naturally derived ECM hydrogels are widely used for vessel-on-a-chip systems, including collagen, fibrin, and hyaluronic acid (HA). Often the selection of ECM is informed by the tissue type that is being modeled and its characteristic physiology and pathology. For instance, when designing a model for breast cancer, composite ECM hydrogels made from fibrillar collagen and HA are widely used to mimic the ECM conditions (e.g., stiffness) of breast tumors (Liu et al. 2018). Chang et al. utilized collagen/HA composite hydrogels to study the mechanisms that mediate blood vessel sprouting and permeability due to the interactions of ECM composition with isoforms of CXCL12, a family of chemokines that is a known driver of breast cancer and more than 20 other malignancies (Zhao et al. 2014). Using a vessel-on-a-chip model, it was shown that CXCL12-α promoted increased vascular permeability in collagen-only gels compared to collagen-HA gels, while, in contrast, CXCL12-γ led to increased vascular permeability in the collagen-HA phenotype compared to collagen-only hydrogels (Chang et al. 2020). Moreover, HA decreased vessel sprouting compared to collagen-only gels, which was reversed after blocking HA-receptor CD44.

With respect to brain microvasculature, Linville et al. used a removable rod to cast a 150 μm diameter perfusable lumen into a hydrogel comprised of fibrillar type I collagen (Linville et al. 2019). Although brain ECM is normally devoid of fibrillar collagen, the use of collagen I in this model was employed to provide the necessary mechanical integrity for lumen casting. The lumen was seeded with iPSC-derived human brain endothelial cells (dhBMECs) and served to mimic the blood–brain barrier (BBB), including physiologically low solute permeability and quiescent endothelial cells characteristic of brain microvasculature. The application of this model was further extended to study the phenomenon of blood–brain barrier opening (BBBO), or the reversible disruption of endothelial cell–cell junctions to enable transient entry into the brain (Stallcop et al. 2018). The authors observed that melittin, a membrane active peptide present in bee venom, transiently supported BBBO in their vessel-on-a-chip, with corroborating in vivo results in mice. Prospectively, this study demonstrates the potential of melittin-induced BBBO to enhance the delivery of therapeutics into the brain (Linville et al. 2021). Moreover, vessel-on-a chip models cast in collagen I have evolved to accommodate physiological environments of increasing complexity. Notably, Seo et al. fabricated a vessel-on-a-chip model that simulates the neurovascular unit by coculturing with six different cell types (pericytes, astrocytes, neurons, microglia, oligodendrocytes, and neural stem cells) embedded in the ECM surrounding a microvessel lined with brain endothelial cells (Seo et al. 2021). Zhao et al. also employed a model to investigate post inflammation leukocyte transmigration across the blood–brain barrier in a hierarchical artery, capillary and venule network (Zhao et al. 2022). They utilized a fiber pulling method to generate capillary lumen structures with diameters of about 7 μm.

In addition to natural ECM materials, numerous synthetic materials are of interest for vessel-on-a-chip applications and systems. Two examples include methacrylated gelatin, or GelMA and poly(ethylene glycol) diacrylate (PEGDA). GelMA is a modified version of gelatin reacted with methacrylic anhydride (MA), thereby making it a semisynthetic biomaterial (Van Den Bulcke et al. 2000). GelMA is highly biocompatible and has been shown to allow for proper cell viability, proliferation and motility (Piao et al. 2021). Additionally, GelMA can be easily manipulated by modulating the MA percentage to form gels with different swelling and mechanical properties (Seymour et al. 2021). Similarly, PEGDA is another synthetic material used for tissue engineering applications. Some of the advantages of PEGDA gels are their hydrophilic and porous properties: the porosity can be controlled by altering the molecular weight and concentration of PEG (Hillel et al. 2007). Furthermore, PEGDA can be paired with other biomaterials such as HA and gelatin to form hybrid hydrogels with a variety of favorable biological and physical properties. Nevertheless, one limitation shared by GelMA and PEGDA—unlike natural hydrogels—is that polymerization requires UV cross-linking, which can detrimentally affect the viability of stromal cells embedded with these gels. Thus, the careful consideration of ECM type in microfluidic fabrication design in a key component needed to optimize microfluidic device function and effectively simulate the physiological and environmental cues seen in a real tissue environment.

17.3.2 Example: Tissue Culture Modeling

Ex vivo tissue models offer an alternative approach that mimics in vivo complexity by making use of pre-existing tissue physiology while also providing increased experimental control and time-lapse capabilities associated with ex vivo culture. Examples include the retina explant model and the aortic ring which are common angiogenic models (Murakami et al. 2006; Iqbal et al. 2017; Katakia et al. 2020). Another tissue culture model that closely resembles native tissue and can be ex vivo is the rat mesentery culture model. The thinness of rat mesenteric tissue allows for easy viewing of an intact vascular network without additional embedding. The whole network view allows for the observation of pericytes, endothelial cells, and vascular smooth muscle cells in their normal physiological configuration. In addition to vasculature, this tissue also contains networks of nerves, lymphatics, as well as interstitial cells such as resident macrophages. Advantages of this model include incorporation of intact microvascular networks and lymphatics with supporting cells (Stapor et al. 2013), investigation of multicellular interactions (Stapor et al. 2013; Sweat et al. 2014) similar to in vivo, the ability to do time-lapse imaging during angiogenesis and lymphangiogenesis (Azimi et al. 2017), and the ability to do drug screenings (Azimi et al. 2015). Additionally, angiogenesis occurs mostly along venules (Stapor et al. 2013) and smooth muscle cells remain functional during culture (Motherwell et al. 2017). More recent applications have shown the use of this model for discovering blood and lymphatic vessel plasticity (Azimi et al. 2019) and for investigating stem cell fate and effects on vessel growth (Azimi et al. 2020; Hodges et al. 2022). The rat mesentery tissue culture also shows promise for the investigation of neurovascular interactions because it contains whole intact networks of both systems (Hodges et al. 2020). These advantages make the rat mesentery tissue culture model a useful biomimetic platform for studying spatial and systemic responses to stimuli, and a beneficial tool, along with cell and microfluidic models, to investigate microvascular remodeling.

17.4 Computational Models

Computational modeling shares a common premise with experimental modeling in that the goal is to integrate complexity allowing for quantitative characterizations of specific cell and system interactions in ways that can be achieved based on in vivo animal investigation. Generally, modeling approaches can be characterized as either trying to integrate empirical rule-based, cellular dynamics to learn dependent system-level effects or applying fundamental biotransport-based engineering relationships to characterize hemodynamic phenomena. This section provides examples of these categorical approaches and by doing so emphasizes the value of viewing microvascular complexity in silico.

17.4.1 *Example*: *Agent-Based Modeling (ABM) of Angiogenesis and Microvascular Patterning*

Currently, the only model system capable of dynamically representing and reporting single-cell genotypic and phenotypic characteristics for every cell within the context of a multicell tissue is a computational model. Agent-based models (ABMs) are becoming an increasingly popular computational modeling approach for representing individual, heterogeneous cells within a dynamically changing tissue environment (Metzcar et al. 2019; Vodovotz and An 2019). ABMs can simulate hundreds, thousands, or hundreds of thousands of autonomous, biological cells, or "cell agents" that interact with each other and with their local environment, which is typically discretized into a 2D or 3D lattice (Sivakumar et al. 2022; Thorne et al. 2007). Cell agents can communicate with neighboring cell agents in the same way that biological cells do, for example, by producing and secreting soluble ligands, such as growth factors and chemokines. These simulated biochemical ligands can diffuse to nearby cell agents and trigger intracellular signaling cascades via the binding of receptors expressed on their membranes, which modulate changes in gene and protein expression. In this manner, simulated biochemical signals transmitted across cell agents can evoke phenotypic changes in neighboring cell agents, such as cell differentiation, migration, proliferation, or apoptosis (Walpole et al. 2015). Cell agents can both modulate and sense their local, extracellular matrix environment. For example, cell agents can secrete extracellular matrix proteins (e.g., collagen, fibronectin, and elastin) that are deposited and cross-linked in the extracellular space (Longo et al. 2004). Likewise, cell agents can respond to local biochemical and biomechanical cues present in the extracellular matrix (Rikard et al. 2019). For example, a mechanically stiffened matrix can activate integrins and downstream signaling pathways in a cell agent to invoke changes in protein expression that lead to apoptosis (Virgilio et al. 2018, 2021).

How a simulated cell agent responds to a given input is dictated by hypothesized or known biological mechanisms substantiated by data reported in the literature. The set of mechanisms that dictate cell agent behaviors is called the ABM "rule set," and great care is taken by the model-builder to ensure that the rule set accurately and comprehensively simulates documented biological phenomena. Since ABM rules often include stochasticity, or randomness, even the most carefully curated rule set may yield a model that produces outputs that are nonphysiological. However, model outputs can be constrained to those that replicate biological observations by "parameter fitting," a process whereby parameters within the rule set are adjusted until the model outputs they generate accurately represent biological data within the range of reported statistical variation (Haase et al. 2024; Thorne et al. 2011). Furthermore, direct comparisons of model outputs to independent experimental data can confirm the "validity" of the ABM by demonstrating that ABM predictions fall within the realm of biological observations (Comlekoglu et al. 2023).

Over the past two decades, ABMs have been developed to study many different aspects of microvascular function and adaptation. For example, microvascular growth and remodeling has been simulated by representing endothelial cells, pericytes, and vascular smooth muscle cells as simulated cell agents (Peirce et al. 2004). These models have been used to predict rate-limiting mechanisms underpinning expansion or pruning of microvascular networks in different contexts, including both healthy and diseased (Walpole et al. 2017). In one of the first-ever published ABMs of microvascular growth and remodeling, Peirce et al. investigated how exogenously delivered proangiogenic growth factors, VEGF and angiopoietin-1, alter microvascular network patterning when delivered alone and in combination (Peirce et al. 2004). Model validation was performed by comparing ABM predictions to independent in vivo experiments conducted in a rat dorsal backpack window chamber. This study was one of the first demonstrations of how ABM can be used to prospectively design and engineer microvascular networks with specific network architectures using exogenous ligands. ABMs have also been used to study the recruitment of contractile perivascular cells, circulating mesenchymal progenitor cells, and monocytes following changes in hemodynamic-induced stimuli, including circumferential strain and shear stress (Peirce et al. 2004; Bailey et al. 2007, 2009). These models have revealed how heterogeneous cell dynamics contribute to arteriogenesis by going beyond what is possible to visualize and quantify using currently available experimental approaches. ABMs have also been used to compare the relative importance of endothelial cell juxtacrine signaling, mediated by NOTCH-1 receptor interactions with delta-like ligand 4 (DLL4), to paracrine signaling mediated by VEGF, in driving angiogenesis during normal development and in conditions where pericyte–endothelial interactions are disrupted, such as in the diabetic retina (Walpole et al. 2017; Ubezio et al. 2016; Bentley et al. 2008). This ABM simulated how different ratios of pericytes to endothelial cells altered spatial gradients of VEGF and DLL4 and suggested a new hypothesis for how pericyte dysfunction in diabetic retinopathy affects the dynamics of angiogenesis.

A new frontier for ABMs of microvascular growth and adaptation is in the design of biomaterials-based tissue engineered constructs, where efficient and effective vascularization is a requirement for successful tissue engraftment. Biomaterial design traditionally necessitates time-consuming and expensive experimentation to optimize various material compositions, porosities, degradation rates, and mechanical properties. As synthetic biomaterials become increasingly complex in their structure and biomedical applications, there is a pressing need for innovative computational modeling tools that enable efficient design and optimization of biomaterials. Given that ABM can predict how key biochemical and biomechanical cues in the extracellular environment activate intracellular signaling pathways to control cell behaviors, this predictive capability can be harnessed to enable the prospective design of novel hydrogels for inducing angiogenesis (Pruett et al. 2022).

For example, granular hydrogels are a newer class of biomaterials that have been shown to effectively induce and support neovessel growth, both when implanted in vivo (Pruett et al. 2021, 2023) and when seeded with endothelial and mesenchymal support cells in vitro *(*Grewal and Highley 2021*)*. However, there are multiple tunable parameters in the hydrogel fabrication process (e.g., hydrogel concentration, granular particle size, conjugated peptide coatings, etc.) that can be adjusted to achieve different hydrogel design features, such as different degrees of porosity and degradability. In current practice, the ideal set of parameters for achieving a desired outcome, in terms of vascular density or patterning metrics (e.g., number of branchpoints), are determined by time-consuming and potentially costly trial-and-error experimentation (Pfaff et al. 2021). However, ABMs can be used to simulate these different engineered environments and predict how different design features of the hydrogel will impact the extent of angiogenesis and/or vascular ingrowth (Fig. 17.6). Hence, a major benefit of ABMs is that they can serve as a "digital twin" for both in vivo and in vitro experiments, enabling time- and money-saving optimization of engineered tissue platforms for stimulating angiogenesis and desired microvascular network architectures.

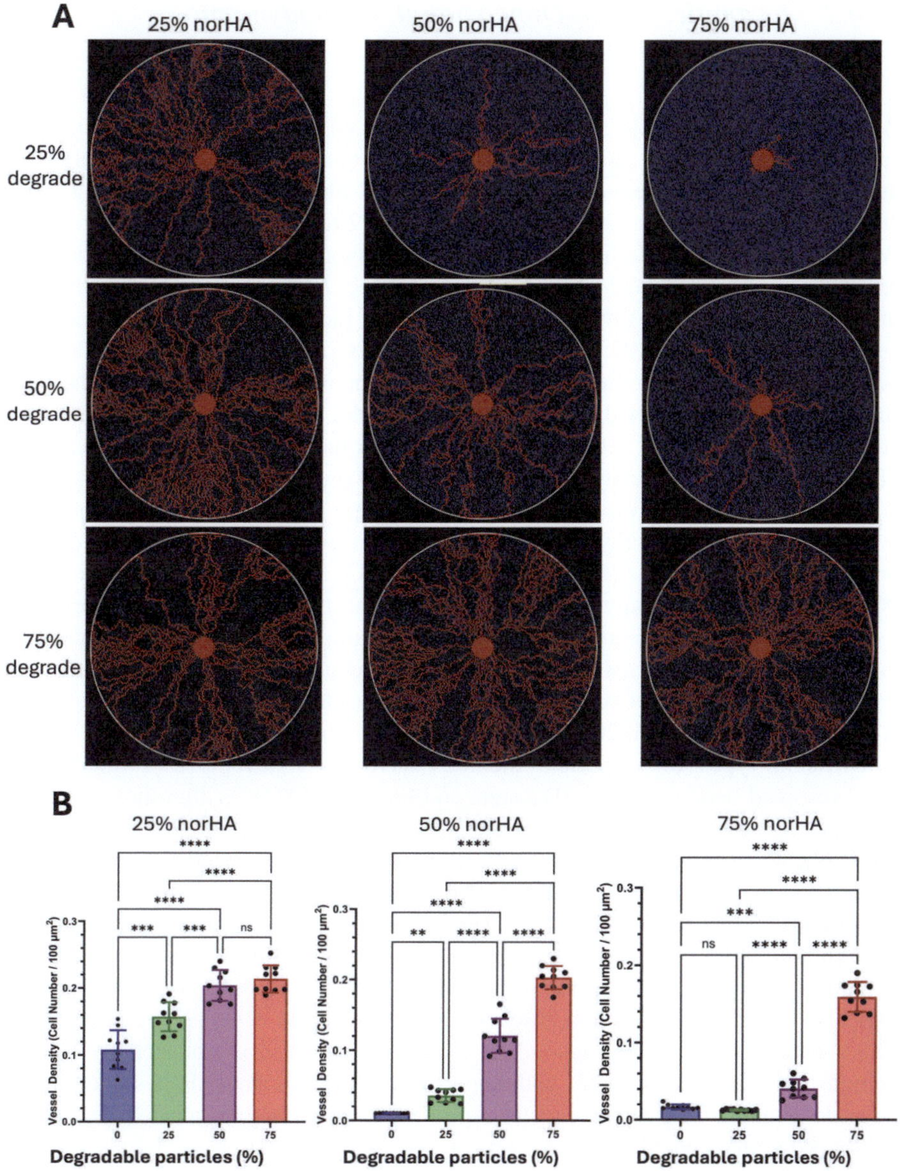

Fig. 17.6 Visual and quantified outputs of the ABM of angiogenic sprouting from an endothelial spheroid implanted into a simulated granular hydrogel. (**a**) Screenshots of representative ABM outputs of nine different combinations of norbornene-modified hyaluronic acid (NorHA) hydrogel particle percentages and degradability levels. Blue represents simulated norHA particles comprising the granular hydrogel, red represents simulated endothelial cells, and white circle shows ABM boundary, which represents the boundaries of a 6-well cell culture plate. (**b**) Quantification of ABM-predicted vessel densities (cells per area) for each simulated combination of norHA particle percentages and degradability levels. $N = 10$ simulations (seeded with same initial conditions) per combination. A one-way ANOVA was performed for all groups in each subplot followed by Tukey's HSD post hoc analysis (**$P < 0.01$, ***$P < 0.001$, ****$P < 0.0001$)

17.4.2 Example: Three-Dimensional, Hemodynamic Modeling

Within the past decade, in silico approaches have emerged for modeling microcirculatory hemodynamics in complex 3D vessel geometries which fully resolve the deformation and dynamics of each individual red blood cell (RBC) (Cheng et al. 2023; Balogh and Bagchi 2017; Huet and Wachs 2022; Ostalowski and Tan 2022; Wang et al. 2013; Rashidi et al. 2023). These modeling features are especially relevant in microvascular networks where vessel diameters are similar to the size of an RBC, which causes the particulate nature of blood flow to strongly influence hemodynamic characteristics including effective viscosity, wall shear stress, and the near-wall cell-free layer (Secomb 2017). Such characteristics, especially the wall shear stress, are established drivers of microvascular adaptation and growth (Galie et al. 2014; Anonymous. 2015), which emphasizes the significance and potential of 3D RBC-resolved computational fluid dynamics to provide new insights connecting hemodynamics to microvascular function. This type of high-fidelity 3D modeling, however, is computationally intensive and far from trivial to implement. Studies on physiological phenomena (Balogh and Bagchi 2017, 2018, 2019a, b; Hossain et al. 2023, 2024) have thus only begun to emerge in recent years due to i) the development of advanced algorithms to efficiently integrate 3D fluid and solid mechanics models, and ii) the increasing power of high-performance computing systems and the general availability of supercomputing resources to in silico researchers (Towns et al. 2014; Boerner et al. 2023).

Different types of numerical methods have been developed to model 3D RBC-resolved hemodynamics (Dasanna and Fedosov 2024). Within these methods a popular approach has been to integrate 3D solvers for the Navier–Stokes equations governing the fluid dynamics with finite element method-based deformable cell models via the immersed boundary method (IBM) (Cheng et al. 2023; Balogh and Bagchi 2017; Huet and Wachs 2022). The IBM facilitates fast and accurate solutions of highly complex time-dependent flows which fully resolve the deformation and dynamics of RBCs, and importantly, allows for the incorporation of phenomenological deformable cell models (Discher et al. 1998; Skalak et al. 1973; Evans 1973). The accuracy of such high-resolution modeling has been established by studies demonstrating good agreement with experiments for both single RBCs undergoing large deformation (e.g., micropipette aspiration (Balogh and Bagchi 2017), optical tweezers (Sinha and Graham 2015)), as well as suspensions of RBCs flowing through vessels (e.g., Fahraeus and Fahraeus–Lindqvist effects (Fedosov et al. 2010)) and complex microvasculatures (e.g., cell-free layer (Balogh and Bagchi 2019a), wall shear stress (Balogh and Bagchi 2019b)). Furthermore, 3D RBC-resolved models have enabled new insights into the cell-scale fluid dynamic origins of classic microcirculatory phenomena including, for example, the Zweifach–Fung effect which describes the disproportionate partitioning of RBCs flowing through bifurcations (Balogh and Bagchi 2018), as well as 3D spatial variations in the near-wall cell-free layer (Balogh and Bagchi 2019a; Hossain et al.

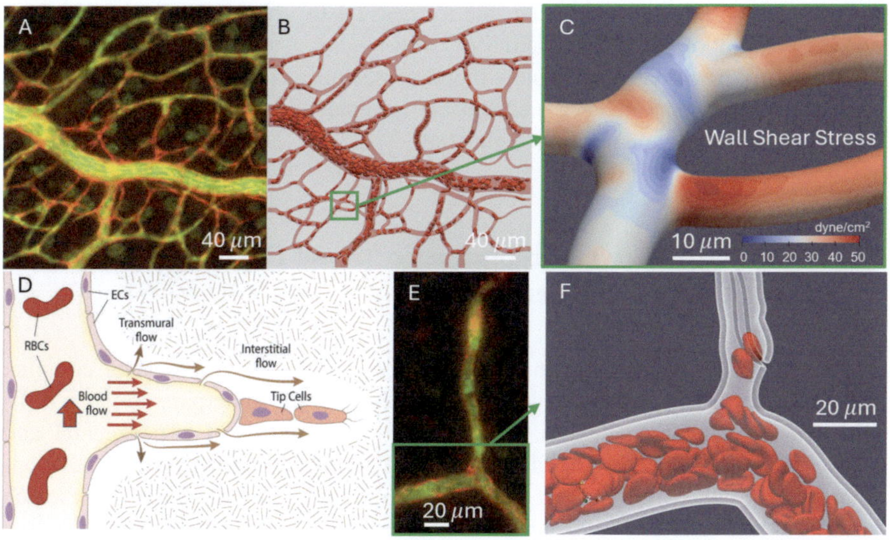

Fig. 17.7 Visualization and modeling of microvascular hemodynamics. (**a**) Representative image of an angiogenic network on rat mesentery tissue, highlighting the intricate vessel morphology. (**b**) Reconstructed 3D vessel network showcasing the simulation of red blood cell (RBC) flow, emphasizing spatial flow dynamics over time. (**c**) High-fidelity computational simulation demonstrating wall shear stress (WSS) distribution across the vessel network, with detailed spatial resolution at bifurcations and tortuous regions. (**d**) Schematic representation of blood flow dynamics near a sprout, illustrating RBC deformation, transmural flow, and interstitial fluid interactions in angiogenic microvascular regions. (**e**) Representative image of an angiogenic sprout on rat mesentery tissue, emphasizing the growth of new capillaries. (**f**) RBC-resolved simulation illustrating the detailed particulate nature of blood flow, highlighting RBC deformation and dynamics at a sprout branching off a vessel. These representations collectively demonstrate the integration of experimental imaging and 3D RBC-resolved simulations, emphasizing the ability of in silico models to capture localized WSS patterns and microcirculatory phenomena difficult to measure in vivo

2024) and wall shear stress (Hossain et al. 2023, 2024) due to in vivo microvessel complexity.

Integrating RBC-resolved simulations with in vivo experiments makes for a powerful combination in large-part because the high-fidelity modeling can provide access to information that is difficult to acquire through in vivo measurements. Localized 3D spatial patterns in relevant quantities such as wall shear stress and their time-dependent variations arise due to both the RBCs themselves and their interaction with complex 3D microvascular surfaces. For example, recent works which integrated RBC-resolved simulations and images of angiogenic microvascular networks (Fig. 17.7) demonstrated how rich 3D WSS patterns can occur at subendothelial cell length scales within bifurcations and mergers (Hossain et al. 2023) and due to vessel tortuosity (Hossain et al. 2024). Time-dependent WSS variations were further shown to follow the "footprints" of RBCs traversing a vessel (Hossain et al. 2023), adding to the complexity of such hemodynamic characteristics as they likely arise in vivo. This type of data is readily available from an in silico model but is difficult if not impossible to measure in vivo owing to the required degree of resolution in 3D space and time. New studies at this level of detail, such as that involving in vivo growth of new blood vessel sprouts (Fig. 17.7), for example, offer much promise to advance mechanistic understanding of how microcirculatory hemodynamics drive microvascular function.

17.5 Overlaps Between Biomimetic Model Development and Tissue Engineering

Despite the inspiring progress in tissue engineering, the development of complex and functional tissue constructs that mimic their natural counterparts remains difficult. The associated challenges motivate the development of experimental and computational biomimetic models. When considering biomimetic model development, recognition of the current state of biofabrication and biomaterials research fields provides a valuable perspective that helps push advancements in tissue engineering research. To date, a variety of techniques have been developed to construct perfusable channels within the bulk 3-dimensional constructs or scaffolds to mimic vessels. These techniques can be mainly implemented through casting and channel printing (Ren et al. 2023). During casting, a template is placed in a chamber, in which a hydrogel precursor which can later be cross-linked into a solid matrix is poured into. Perfusable channels can then be formed by removing the template from the cross-linked matrix. Such template materials include preformed objects (such as a needle or nylon strand) (Mannino et al. 2017; Ozkan et al. 2019; Mandrycky et al. 2020; Silvestri et al. 2020), molded templates (Golden and Tien 2007; Wang et al. 2014; Tocchio et al. 2015), and 3D-printed templates (Zhang et al. 2016; Lin et al. 2019; Kinstlinger et al. 2020; Maharjan et al. 2021). However, the resolution of the casting approach is limited due to the loss of template features from the hydrogel pouring process.

Fortunately, 3D bioprinting has emerged as a promising alternative for the fabrication of thick tissue constructs with perfusable channels. For example, a layer-by-layer printing method has been reported on (Zhao et al. 2012; Lee et al. 2014). During this process, a sacrificial template and a hydrogel precursor are sequentially printed on each layer, and the process repeats layer by layer to form a 3D construct, followed by the removal of the embedded sacrificial materials after the cross-linking of the printed hydrogel matrix. However, the alternate printing of the sacrificial and hydrogel matrix material makes this approach laborious. In addition, forming the complex layer-by-layer may introduce interface cracks between adjacent layers. Alternatively, light-assisted approaches have also been reported on to create perfusable channels within thick hydrogels. Examples include stereolithography (SLA) (Grigoryan et al. 2019) and digital light processing (DLP) (Zhu et al. 2017), in which the hydrogel matrix is selectively photo cross-linked layer by layer under photonic energy while the channel pattern is uncross-linked, and perfusable channels are formed after removing the uncross-linked hydrogel. However, ultraviolet (UV) light may injure living cells embedded in the hydrogel matrix. In this context, a sacrificial ink-assisted embedded printing in a hydrogel matrix has been developed to solve the aforementioned challenges (Skylar-Scott et al. 2019; Ren et al. 2022; Wu et al. 2022). During embedded 3D printing, a sacrificial ink is deposited in a yield-stress, self-healing supporting matrix and omnidirectionally supported by the surrounding matrix. Due to the yield-stress fluid property, the support matrix allows the translation of the printing nozzle and self-heals in the wake of the nozzle. The embedment of printed materials within the supporting matrix prevents the collapse of soft materials and ensures that print fidelity and structural integrity of printed structures could be achieved. As with other aforementioned methods, the sacrificial inks are removed from the matrix after the printing to form vessel-like hollows. In comparison with the conventional direct ink writing method, this embedded printing method enables the printing of soft cellular materials (e.g., low viscosity, low elastic modulus) and highly intricate structures (e.g., overhang and dangling structures). Currently, embedded sacrificial printing has been considered one of the most promising technologies for the creation of perfusable constructs. Incorporation of 3D bioprinting and biofabrication in model design may be an important strategy in mimicking bulk, perfusable tissue design and addressing the need for new biomimetic models that can advance microvascular remodeling research.

17.6 Summary

The experimental and computational examples detailed in this chapter frame the scope of contemporary modeling approaches that have been applied to understand the players and their spatiotemporal dynamics involved in microvascular growth and remodeling. In many ways, these examples showcase the impacts of recreating the complexity of multiple cell and system interactions in order to gain insights that cannot be made from observations in typical animal studies. While animal studies can provide a focus on specific molecular mechanistic effects, biomimetic modeling attempts to evaluate behaviors associated with the patterns of players across the scales of a microvascular network and can provide new views for scientific discovery. It is important to appreciate that challenges in developing such models overlap with similar challenges attributed to vascular tissue engineering, and the recognition of these overlaps serves to connect the different research areas.

References

Akbari E, Spychalski GB, Song JW (2017) Microfluidic approaches to the study of angiogenesis and the microcirculation. Microcirculation 24(5). https://doi.org/10.1111/MICC.12363

Anonymous. (2015) Implications of fluid shear stress in capillary sprouting during adult microvascular network remodeling. Mechanobiology of the. Endothelium:166–184. https://doi.org/10.1201/B18060-8

Azimi MS, Myers L, Lacey M et al (2015) An ex vivo model for anti-Angiogenic drug testing on intact microvascular networks. PLoS One 10(3):e0119227. https://doi.org/10.1371/JOURNAL.PONE.0119227

Azimi MS, Motherwell JM, Murfee WL (2017) An ex vivo method for time-lapse imaging of cultured rat mesenteric microvascular networks. J Vis Exp 2017(120):55183. https://doi.org/10.3791/55183

Azimi MS, Motherwell JM, Hodges NA et al (2019) Lymphatic-to-blood vessel transition in adult microvascular networks: a discovery made possible by a top-down approach to biomimetic model development. Microcirculation 27(2):e12595. https://doi.org/10.1111/MICC.12595

Azimi MS, Motherwell JM, Dutreil M et al (2020) A novel tissue culture model for evaluating the effect of aging on stem cell fate in adult microvascular networks. Geroscience 42(2):515. https://doi.org/10.1007/S11357-020-00178-0

Bailey AM, Thorne BC, Peirce SM (2007) Multi-cell agent-based simulation of the microvasculature to study the dynamics of circulating inflammatory cell trafficking. Ann Biomed Eng 35(6):916–936. https://doi.org/10.1007/S10439-007-9266-1

Bailey AM, Lawrence MB, Shang H et al (2009) Agent-based model of therapeutic adipose-derived stromal cell trafficking during ischemia predicts ability to roll on P-selectin. PLoS Comput Biol 5(2). https://doi.org/10.1371/JOURNAL.PCBI.1000294

Balogh P, Bagchi P (2017) Direct numerical simulation of cellular-scale blood flow in 3D microvascular networks. Biophys J 113(12):2815–2826. https://doi.org/10.1016/J.BPJ.2017.10.020

Balogh P, Bagchi P (2018) Analysis of red blood cell partitioning at bifurcations in simulated microvascular networks. Phys Fluids 30(5)

Balogh P, Bagchi P (2019a) The cell-free layer in simulated microvascular networks. J Fluid Mech 864:768–806. https://doi.org/10.1017/JFM.2019.45

Balogh P, Bagchi P (2019b) Three-dimensional distribution of wall shear stress and its gradient in red cell-resolved computational modeling of blood flow in in vivo-like microvascular networks. Physiol Rep 7(9). https://doi.org/10.14814/PHY2.14067

Bentley K, Gerhardt H, Bates PA (2008) Agent-based simulation of notch-mediated tip cell selection in angiogenic sprout initialisation. J Theor Biol 250(1):25–36. https://doi.org/10.1016/J.JTBI.2007.09.015

Boerner TJ, Deems S, Furlani TR et al (2023) ACCESS: advancing innovation: NSF'S advanced cyberinfrastructure coordination ecosystem: services & support. In: PEARC 2023 – computing for the common good: practice and experience in advanced research computing, pp 173–176. https://doi.org/10.1145/3569951.3597559

Bordeleau F, Mason BN, Lollis EM et al (2017) Matrix stiffening promotes a tumor vasculature phenotype. Proc Natl Acad Sci USA 114(3):492–497. https://doi.org/10.1073/PNAS.1613855114

Chang CW, Seibel AJ, Avendano A et al (2020) Distinguishing specific CXCL12 isoforms on their angiogenesis and vascular permeability promoting properties. Adv Healthc Mater 9(4). https://doi.org/10.1002/ADHM.201901399

Cheng X, Caruso C, Lam WA et al (2023) Marginated aberrant red blood cells induce pathologic vascular stress fluctuations in a computational model of hematologic disorders. Sci Adv 9(48). https://doi.org/10.1126/SCIADV.ADJ6423

Comlekoglu T, Dzamba BJ, Pacheco GG, et al (2023) Modeling the roles of cohesotaxis, cell-intercalation, and tissue geometry in collective cell migration of Xenopus mesendoderm. bioRxiv 2023; https://doi.org/10.1101/2023.10.16.562601

Dasanna AK, Fedosov DA (2024) Mesoscopic modeling of membranes at cellular scale. Eur Phys J Spec Top 233. https://doi.org/10.1140/EPJS/S11734-024-01177-4

Discher DE, Boal DH, Boey SK (1998) Simulations of the erythrocyte cytoskeleton at large deformation. II. Micropipette aspiration. Biophys J 75(3):1584. https://doi.org/10.1016/S0006-3495(98)74076-7

Duval K, Grover H, Han LH et al (2017) Modeling physiological events in 2D vs. 3D cell culture. Physiology (Bethesda) 32(4):266–277. https://doi.org/10.1152/PHYSIOL.00036.2016

Evans EA (1973) New membrane concept applied to the analysis of fluid shear- and micropipette-deformed red blood cells. Biophys J 13(9):941. https://doi.org/10.1016/S0006-3495(73)86036-9

Fedosov DA, Caswell B, Popel AS et al (2010) Blood flow and cell-free layer in microvessels. Microcirculation 17(8):615–628. https://doi.org/10.1111/J.1549-8719.2010.00056.X

Fleischer S, Tavakol DN, Vunjak-Novakovic G (2020) From arteries to capillaries: approaches to engineering human vasculature. Adv Funct Mater 30(37). https://doi.org/10.1002/ADFM.201910811

Galie PA, Nguyen DHT, Choi CK et al (2014) Fluid shear stress threshold regulates angiogenic sprouting. Proc Natl Acad Sci USA 111(22):7968–7973. https://doi.org/10.1073/PNAS.1310842111

Golden AP, Tien J (2007) Fabrication of microfluidic hydrogels using molded gelatin as a sacrificial element. Lab Chip 7(6):720–725. https://doi.org/10.1039/B618409J

Grewal MG, Highley CB (2021) Electrospun hydrogels for dynamic culture systems: advantages, progress, and opportunities. Biomater Sci 9(12):4228–4245. https://doi.org/10.1039/D0BM01588A

Grigoryan B, Paulsen SJ, Corbett DC et al (2019) Multivascular networks and functional intravascular topologies within biocompatible hydrogels. Science 364(6439):458–464. https://doi.org/10.1126/SCIENCE.AAV9750

Guckenberger DJ, De Groot TE, Wan AMD et al (2015) Micromilling: a method for ultra-rapid prototyping of plastic microfluidic devices. Lab Chip 15(11):2364–2378. https://doi.org/10.1039/C5LC00234F

Haase M, Comlekoglu T, Petrucciani A et al (2024) Agent-based model demonstrates the impact of nonlinear, complex interactions between cytokinces on muscle regeneration. elife 13:13. https://doi.org/10.7554/ELIFE.91924

Hauser PV, Chang HM, Nishikawa M et al (2021) Bioprinting scaffolds for vascular tissues and tissue vascularization. Bioengineering (Basel) 8(11). https://doi.org/10.3390/BIOENGINEERING8110178

Helm CLE, Fleury ME, Zisch AH et al (2005) Synergy between interstitial flow and VEGF directs capillary morphogenesis in vitro through a gradient amplification mechanism. Proc Natl Acad Sci USA 102(44):15779–15784. https://doi.org/10.1073/PNAS.0503681102

Hillel A, Shah P, Elisseeff J (2007) Hydrogels in cell encapsulation and tissue engineering. Biomed Polymers:57–82. https://doi.org/10.1533/9781845693640.57

Hodges NA, Barr RW, Murfee WL (2020) The maintenance of adult peripheral adult nerve and microvascular networks in the rat mesentery culture model. J Neurosci Methods 346:108923. https://doi.org/10.1016/J.JNEUMETH.2020.108923

Hodges NA, Lampejo AO, Shang H et al (2022) Viewing stromal vascular fraction de novo vessel formation and association with host microvasculature using the rat mesentery culture model. Microcirculation 29(6–7):e12758. https://doi.org/10.1111/MICC.12758

Hossain MMN, Hu NW, Abdelhamid M et al (2023) Angiogenic Microvascular Wall shear stress patterns revealed through three-dimensional red blood cell resolved modeling. Function (Oxf) 4(6). https://doi.org/10.1093/FUNCTION/ZQAD046

Hossain MMN, Hu NW, Kazempour A et al (2024) Hemodynamic characteristics of a tortuous microvessel using high-Fidelity red blood cell resolved simulations. Microcirculation 31(7). https://doi.org/10.1111/MICC.12875

Huet DP, Wachs A (2022) A Cartesian-octree adaptive front-tracking solver for immersed biological capsules in large complex domains. J Comput Phys 492:492. https://doi.org/10.1016/j.jcp.2023.112424

Iqbal F, Gratch YS, Szaraz P et al (2017) The aortic ring co-culture assay: a convenient tool to assess the Angiogenic potential of mesenchymal stromal cells in vitro. JoVE 2017(127):e56083. https://doi.org/10.3791/56083

Katakia YT, Duddu S et al (2020) Ex vivo model for studying endothelial tip cells: revisiting the classical aortic-ring assay. Microvasc Res 128:128. https://doi.org/10.1016/J.MVR.2019.103939

Kinstlinger IS, Saxton SH, Calderon GA et al (2020) Generation of model tissues with dendritic vascular networks via sacrificial laser-sintered carbohydrate templates. Nat Biomed Eng 4(9):916–932. https://doi.org/10.1038/S41551-020-0566-1

Lampejo AO, Hu NW, Lucas D et al (2022) A challenge for engineering biomimetic microvascular models: how do we incorporate the physiology? Front Bioeng Biotechnol 10. https://doi.org/10.3389/FBIOE.2022.912073

Lee VK, Kim DY, Ngo H et al (2014) Creating perfused functional vascular channels using 3D bio-printing technology. Biomaterials 35(28):8092–8102. https://doi.org/10.1016/J.BIOMATERIALS.2014.05.083

Lin NYC, Homan KA, Robinson SS et al (2019) Renal reabsorption in 3D vascularized proximal tubule models. Proc Natl Acad Sci USA 116(12):5399–5404. https://doi.org/10.1073/PNAS.1815208116/-/DCSUPPLEMENTAL

Linville RM, DeStefano JG, Sklar MB et al (2019) Human iPSC-derived blood-brain barrier microvessels: validation of barrier function and endothelial cell behavior. Biomaterials 190–191:24–37. https://doi.org/10.1016/J.BIOMATERIALS.2018.10.023

Linville RM, Komin A, Lan X et al (2021) Reversible blood-brain barrier opening utilizing the membrane active peptide melittin in vitro and in vivo. Biomaterials 275:275. https://doi.org/10.1016/J.BIOMATERIALS.2021.120942

Liu C, Lewin Mejia D, Chiang B et al (2018) Hybrid collagen alginate hydrogel as a platform for 3D tumor spheroid invasion. Acta Biomater 75:213–225. https://doi.org/10.1016/J.ACTBIO.2018.06.003

Longo D, Peirce SM, Skalak TC et al (2004) Multicellular computer simulation of morphogenesis: blastocoel roof thinning and matrix assembly in Xenopus laevis. Dev Biol 271(1):210–222. https://doi.org/10.1016/j.ydbio.2004.03.021

Lugo-Cintrón KM, Ayuso JM, White BR et al (2020) Matrix density drives 3D organotypic lymphatic vessel activation in a microfluidic model of the breast tumor microenvironment. Lab Chip 20(9):1586–1600. https://doi.org/10.1039/D0LC00099J

Maharjan S, Alva J, Cámara C et al (2021) Symbiotic photosynthetic oxygenation within 3D-bioprinted vascularized tissues. Matter 4(1):217–240. https://doi.org/10.1016/J.MATT.2020.10.022

Mandrycky C, Hadland B, Zheng Y (2020) 3D curvature-instructed endothelial flow response and tissue vascularization. Sci Adv 6(38). https://doi.org/10.1126/SCIADV.ABB3629

Mannino RG, Santiago-Miranda AN, Pradhan P et al (2017) 3D microvascular model recapitulates the diffuse large B-cell lymphoma tumor microenvironment in vitro. Lab Chip 17(3):407–414. https://doi.org/10.1039/C6LC01204C

Metzcar J, Wang Y, Heiland R et al (2019) A review of cell-based computational modeling in cancer biology. JCO Clin Cancer Inform 3(3):1–13. https://doi.org/10.1200/CCI.18.00069

Moses SR, Adorno JJ, Palmer AF et al (2021) Vessel-on-a-chip models for studying microvascular physiology, transport, and function in vitro. Am J Physiol Cell Physiol 320(1):C92–C105. https://doi.org/10.1152/AJPCELL.00355.2020

Motherwell J, Murfee WL (2018) Modelling microvascular pathology. Nat Biomed Eng 2(6): 349–350. https://doi.org/10.1038/s41551-018-0251-9

Motherwell JM, Azimi MS, Spicer K et al (2017) Evaluation of arteriolar smooth muscle cell function in an ex vivo microvascular network model. Sci Rep 7(1):1–12. https://doi.org/10.1038/s41598-017-02272-4

Murakami T, Suzuma K, Takagi H et al (2006) Time-lapse imaging of vitreoretinal angiogenesis originating from both quiescent and mature vessels in a novel ex vivo system. Invest Ophthalmol Vis Sci 47(12):5529–5536. https://doi.org/10.1167/IOVS.06-0373

Ostalowski K, Tan J (2022) Direct simulation of blood flow with heterogeneous cell suspensions in a patient-specific capillary network. PhFl 34(4):041912. https://doi.org/10.1063/5.0088342

Ozkan A, Ghousifam N, Hoopes PJ et al (2019) In vitro vascularized liver and tumor tissue microenvironments on a chip for dynamic determination of nanoparticle transport and toxicity. Biotechnol Bioeng 116(5):1201–1219. https://doi.org/10.1002/BIT.26919

Peirce SM, Skalak TC (2003) Microvascular remodeling: a complex continuum spanning angiogenesis to Arteriogenesis. Microcirculation 10(1):99–111. https://doi.org/10.1038/SJ.MN.7800172

Peirce SM, Van Gieson EJ, Skalak TC (2004) Multicellular simulation predicts microvascular patterning and in silico tissue assembly. FASEB J 18(6):731–733. https://doi.org/10.1096/FJ.03-0933FJE

Pfaff BN, Pruett LJ, Cornell NJ et al (2021) Selective and improved Photoannealing of microporous annealed particle (MAP) scaffolds. ACS Biomater Sci Eng 7(2):422–427. https://doi.org/10.1021/ACSBIOMATERIALS.0C01580

Piao Y, You H, Xu T et al (2021) Biomedical applications of gelatin methacryloyl hydrogels. Eng Regen 2:47–56. https://doi.org/10.1016/J.ENGREG.2021.03.002

Pruett LJ, Jenkins CH, Singh NS et al (2021) Heparin microislands in microporous annealed particle scaffolds for accelerated diabetic wound healing. Adv Funct Mater 31(35). https://doi.org/10.1002/ADFM.202104337

Pruett LJ, Taing AL, Singh NS et al (2022) In silico optimization of heparin microislands in microporous annealed particle hydrogel for endothelial cell migration. Acta Biomater 148: 171–180. https://doi.org/10.1016/J.ACTBIO.2022.05.049

Pruett LJ, Kenny HL, Swift WM et al (2023) De novo tissue formation using custom microporous annealed particle hydrogel provides long-term vocal fold augmentation. NPJ Regen Med 8(1). https://doi.org/10.1038/S41536-023-00281-8

Rashidi Y, Simionato G, Zhou Q et al (2023) Red blood cell lingering modulates hematocrit distribution in the microcirculation. Biophys J 122(8):1526–1537. https://doi.org/10.1016/J.BPJ.2023.03.020

Reid SE, Kay EJ, Neilson LJ et al (2017) Tumor matrix stiffness promotes metastatic cancer cell interaction with the endothelium. EMBO J 36(16):2373–2389. https://doi.org/10.15252/EMBJ.201694912

Ren B, Song K, Sanikommu AR et al (2022) Study of sacrificial ink-assisted embedded printing for 3D perfusable channel creation for biomedical applications. Appl. Phys Rev 9(1). https://doi.org/10.1063/5.0068329

Ren B, Jiang Z, Murfee WL et al (2023) Realizations of vascularized tissues: from in vitro platforms to in vivo grafts. Biophys Rev 4(1). https://doi.org/10.1063/5.0131972

Rikard SM, Athey TL, Nelson AR et al (2019) Multiscale coupling of an agent-based model of tissue fibrosis and a logic-based model of intracellular signaling. Front Physiol 10. https://doi.org/10.3389/FPHYS.2019.01481

Secomb TW (2017) Blood flow in the microcirculation. Annu Rev Fluid Mech 49:443–461

Seo S, Choi CH, Yi KS et al (2021) An engineered neurovascular unit for modeling neuroinflammation. Biofabrication 13(3). https://doi.org/10.1088/1758-5090/ABF741

Seymour AJ, Shin S, Heilshorn SC (2021) 3D printing of microgel scaffolds with tunable void fraction to promote cell infiltration. Adv Healthc Mater 10(18). https://doi.org/10.1002/ADHM.202100644

Silvestri VL, Henriet E, Linville RM et al (2020) A tissue-engineered 3D microvessel model reveals the dynamics of mosaic vessel formation in breast cancer. Cancer Res 80(19):4288–4301. https://doi.org/10.1158/0008-5472.CAN-19-1564

Sinha K, Graham MD (2015) Dynamics of a single red blood cell in simple shear flow. Phys Rev E Stat Nonlinear Soft Matter Phys 92(4):042710. https://doi.org/10.1103/PHYSREVE.92.042710/FIGURES/19/THUMBNAIL

Sivakumar N, Warner HV, Peirce SM et al (2022) A computational modeling approach for predicting multicell spheroid patterns based on signaling-induced differential adhesion. PLoS Comput Biol 18(11). https://doi.org/10.1371/JOURNAL.PCBI.1010701

Skalak R, Tozeren A, Zarda RP et al (1973) Strain energy function of red blood cell membranes. Biophys J 13(3):245–264. https://doi.org/10.1016/S0006-3495(73)85983-1

Skylar-Scott MA, Uzel SGM, Nam LL et al (2019) Biomanufacturing of organ-specific tissues with high cellular density and embedded vascular channels. Sci Adv 5(9). https://doi.org/10.1126/SCIADV.AAW2459

Song JW, Munn LL (2011) Fluid forces control endothelial sprouting. Proc Natl Acad Sci USA 108(37):15342–15347. https://doi.org/10.1073/PNAS.1105316108/-/DCSUPPLEMENTAL

Stallcop LE, Álvarez-García YR, Reyes-Ramos AM et al (2018) Razor-printed sticker microdevices for cell-based applications. Lab Chip 18(3):451–462. https://doi.org/10.1039/C7LC00724H

Stapor PC, Azimi MS, Ahsan T et al (2013) An angiogenesis model for investigating multicellular interactions across intact microvascular networks. Am J Physiol Heart Circ Physiol 304(2). https://doi.org/10.1152/AJPHEART.00552.2012

Stapor PC, Sweat RS, Dashti DC et al (2014) Pericyte dynamics during angiogenesis: new insights from new identities. J Vasc Res 51(3):163–174. https://doi.org/10.1159/000362276

Sweat RS, Sloas DC, Murfee WL (2014) VEGF-C induces Lymphangiogenesis and angiogenesis in the rat mesentery culture model. Microcirculation 21(6):532. https://doi.org/10.1111/MICC.12132

Thorne BC, Bailey AM, Peirce SM (2007) Combining experiments with multi-cell agent-based modeling to study biological tissue patterning. Brief Bioinform 8(4):245–257. https://doi.org/10.1093/BIB/BBM024

Thorne BC, Hayenga HN, Humphrey JD et al (2011) Toward a multi-scale computational model of arterial adaptation in hypertension: verification of a multi-cell agent based model. Front Physiol 2:2. https://doi.org/10.3389/FPHYS.2011.00020

Tocchio A, Tamplenizza M, Martello F et al (2015) Versatile fabrication of vascularizable scaffolds for large tissue engineering in bioreactor. Biomaterials 45:124–131. https://doi.org/10.1016/J.BIOMATERIALS.2014.12.031

Towns J, Cockerill T, Dahan M et al (2014) XSEDE: accelerating scientific discovery. Comput Sci Eng 16(5):62–74. https://doi.org/10.1109/MCSE.2014.80

Ubezio B, Blanco RA, Geudens I et al (2016) Synchronization of endothelial Dll4-notch dynamics switch blood vessels from branching to expansion. elife 5(APRIL2016). https://doi.org/10.7554/ELIFE.12167

Van Den Bulcke AI, Bogdanov B, De Rooze N et al (2000) Structural and rheological properties of methacrylamide modified gelatin hydrogels. Biomacromolecules 1(1):31–38. https://doi.org/10.1021/BM990017D

Virgilio KM, Martin KS, Peirce SM et al (2018) Agent-based model illustrates the role of the microenvironment in regeneration in healthy and mdx skeletal muscle. J Appl Physiol 125(5):1424–1439. https://doi.org/10.1152/JAPPLPHYSIOL.00379.2018

Virgilio KM, Jones BK, Miller EY et al (2021) Computational models provide insight into in vivo studies and reveal the complex role of fibrosis in mdx muscle regeneration. Ann Biomed Eng 49(2):536–547. https://doi.org/10.1007/S10439-020-02566-1

Vodovotz Y, An G (2019) Agent-based models of inflammation in translational systems biology: a decade later. Wiley Interdiscip Rev Syst Biol Med 11(6). https://doi.org/10.1002/WSBM.1460

Walpole J, Chappell JC, Cluceru JG et al (2015) Agent-based model of angiogenesis simulates capillary sprout initiation in multicellular networks. Integr Biol (Camb) 7(9):987–997. https://doi.org/10.1039/C5IB00024F

Walpole J, Mac Gabhann F, Peirce SM et al (2017) Agent-based computational model of retinal angiogenesis simulates microvascular network morphology as a function of pericyte coverage. Microcirculation 24(8). https://doi.org/10.1111/MICC.12393

Wang W, Diacovo TG, Chen J et al (2013) Simulation of platelet, thrombus and erythrocyte hydrodynamic interactions in a 3D arteriole with in vivo comparison. PLoS One 8(10). https://doi.org/10.1371/JOURNAL.PONE.0076949

Wang XY, Jin ZH, Gan BW et al (2014) Engineering interconnected 3D vascular networks in hydrogels using molded sodium alginate lattice as the sacrificial template. Lab Chip 14(15): 2709–2716. https://doi.org/10.1039/C4LC00069B

Wang WY, Lin D, Jarman EH et al (2020) Functional angiogenesis requires microenvironmental cues balancing endothelial cell migration and proliferation. Lab Chip 20(6):1153–1166. https://doi.org/10.1039/C9LC01170F

Wong JF, Mohan MD, Young EWK et al (2020) Integrated electrochemical measurement of endothelial permeability in a 3D hydrogel-based microfluidic vascular model. Biosens Bioelectron 147:147. https://doi.org/10.1016/J.BIOS.2019.111757

Wu Q, Song K, Zhang D et al (2022) Embedded extrusion printing in yield-stress-fluid baths. Matter 5(11):3775–3806. https://doi.org/10.1016/J.MATT.2022.09.003

Zhang YS, Davoudi F, Walch P et al (2016) Bioprinted thrombosis-on-a-chip. Lab Chip 16(21): 4097–4105. https://doi.org/10.1039/C6LC00380J

Zhao L, Lee VK, Yoo SS et al (2012) The integration of 3-D cell printing and mesoscopic fluorescence molecular tomography of vascular constructs within thick hydrogel scaffolds. Biomaterials 33(21):5325–5332. https://doi.org/10.1016/J.BIOMATERIALS.2012.04.004

Zhao S, Chang SL, Linderman JJ et al (2014) A comprehensive analysis of CXCL12 isoforms in breast cancer1,2. Transl Oncol 7(3):429–438. https://doi.org/10.1016/J.TRANON.2014.04.001

Zhao N, Guo Z, Kulkarni S et al (2022) Engineering the human blood-brain barrier at the capillary scale using a double-templating technique. Adv Funct Mater 32(30). https://doi.org/10.1002/ADFM.202110289

Zhu W, Qu X, Zhu J et al (2017) Direct 3D bioprinting of prevascularized tissue constructs with complex microarchitecture. Biomaterials 124:106–115. https://doi.org/10.1016/J.BIOMATERIALS.2017.01.042

Correction to: Biomimetic Models for Investigation of Microvascular Remodeling

Nien-Wen Hu, Arinola Lampejo, Ariana D. Suarez-Martinez, Nicholas Hodges, Mir Md Nasim Hossain, Peter Balogh, Bing Ren, Yong Huang, Jonathan J. Adorno, Shashwat S. Agarwal, Jacob C. Holter, Elton Y. Cao, Jonathan W. Song, Dorothy N. Beck, Shayn M. Peirce, and Walter L. Murfee

Correction to:
Chapter 17 in: R. F. Tuma et al. (eds.), *Handbook of Microcirculation*, Physiology in Health and Disease,
https://doi.org/10.1007/978-3-031-99609-2_17

The original version of Chapter 17 was inadvertently published with incorrect last name of the co-author. The name "Jonathan J. Adorn" has now been corrected to "Jonathan J. Adorno".

The updated version of this chapter can be found at
https://doi.org/10.1007/978-3-031-99609-2_17

R. F. Tuma et al. (eds.), *Handbook of Microcirculation*, Physiology in Health and Disease, https://doi.org/10.1007/978-3-031-99609-2_18

MIX
Papier aus verantwortungsvollen Quellen
Paper from responsible sources
FSC® C105338

If you have any concerns about our products,
you can contact us on
ProductSafety@springernature.com

In case Publisher is established outside the EU,
the EU authorized representative is:
Springer Nature Customer Service Center GmbH
Europaplatz 3, 69115 Heidelberg, Germany

Printed by Libri Plureos GmbH
in Hamburg, Germany

Operation Backfire

Britain's Cold War Guided Missile Programme

Graham M Simons

AIR WORLD

AIR WORLD

First published in Great Britain in 2026 by
Air World
An imprint of
Pen & Sword Books Ltd
Yorkshire – Philadelphia

ISBN: 9781036195229

A CIP catalogue record for this book is available from the British Library.

Typeset in 11pt Times by GMS Enterprises
Printed and bound by CPI Group (UK) Ltd.

The Publisher's authorised representative in the EU for product safety is Authorised Rep Compliance Ltd., Ground Floor, 71 Lower Baggot Street, Dublin D02 P593, Ireland.
www.arccompliance.com

For a complete list of Pen & Sword titles please contact
PEN & SWORD BOOKS LIMITED
47 Church Street, Barnsley, South Yorkshire, S70 2AS, England
E-mail: enquiries@pen-and-sword.co.uk
Website: www.pen-and-sword.co.uk
or
PEN AND SWORD BOOKS
1950 Lawrence Road, Havertown, PA 19083, USA
E-mail: uspen-and-sword@casematepublishers.com
Website: www.penandswordbooks.com

MIX
Paper | Supporting responsible forestry
FSC® C013604
FSC www.fsc.org